Absolute Value

If a is a real number, then

$$|a| = \begin{cases} a & \text{if } a \geq 0 \\ -a & \text{if } a < 0. \end{cases}$$

Lines

1. Slope: $m = \dfrac{y_2 - y_1}{x_2 - x_1}$
2. Slope-intercept equation: $y = mx + b$
3. Point-slope equation: $y - y_1 = m(x - x_1)$
4. Parallel lines: $m_1 = m_2$; Perpendicular lines: $m_1 m_2 = -1$

Distance Formula

Between (x_1, y_1) and (x_2, y_2): $d = \sqrt{(x_2 - x_1)^2 + (y_2 - y_1)^2}$

Circle

Center at (h, k), radius r: $(x - h)^2 + (y - k)^2 = r^2$

Arithmetic Sequence

nth term: $a_n = a_1 + (n - 1)d;$

Sum: $S_n = \frac{n}{2}(a_1 + a_n) = \frac{n}{2}[2a_1 + (n - 1)d]$

Geometric Sequence

nth term: $a_n = a_1 r^{n-1}$; Sum: $S_n = \dfrac{a_1(1 - r^n)}{1 - r}$

Infinite Geometric Series

Sum $= \dfrac{a_1}{1 - r}$ if $|r| < 1$

Vertical and Horizontal Shifts

Let c be a positive constant.

1. The graph of $y = f(x) + c$ is the graph of f raised c units.
2. The graph of $y = f(x) - c$ is the graph of f lowered c units.
3. The graph of $y = f(x + c)$ is the graph of f shifted c units to the left.
4. The graph of $y = f(x - c)$ is the graph of f shifted c units to the right.

Reflecting

The graph of $y = -f(x)$ is the graph of f reflected about the x-axis.

Stretching and Shrinking

If $c > 1$, the graph of $y = cf(x)$ is the graph of f stretched by a factor of c. If $0 < c < 1$, the graph of $y = cf(x)$ is the graph of f flattened out by a factor of c.

Vertical Line Test

Imagine a vertical line sweeping across a graph. If the vertical line at any position intersects the graph in more than one point, the graph is not the graph of a function.

Horizontal Line Test

Imagine a horizontal line sweeping down the graph of a function. If the horizontal line at any position intersects the graph in more than one point, the function is not one-to-one and its inverse is not a function.

ANNOTATED INSTRUCTOR'S EDITION

Intermediate Algebra

Dennis T. Christy
Robert Rosenfeld

Nassau Community College

WCB Wm. C. Brown Publishers
Dubuque, Iowa • Melbourne, Australia • Oxford, England

F. Hebert

Book Team

Editor *Paula-Christy Heighton*
Developmental Editor *Theresa Grutz*
Photo Editor *Rose Deluhery*
Publishing Services Coordinator/Design *Barbara J. Hodgson*

Wm. C. Brown Publishers
A Division of Wm. C. Brown Communications, Inc.

Vice President and General Manager *Beverly Kolz*
Vice President, Publisher *Earl McPeek*
Vice President, Director of Sales and Marketing *Virginia S. Moffat*
National Sales Manager *Douglas J. DiNardo*
Marketing Manager *Julie Joyce Keck*
Advertising Manager *Janelle Keeffer*
Director of Production *Colleen A. Yonda*
Publishing Services Manager *Karen J. Slaght*
Permissions/Records Manager *Connie Allendorf*

Wm. C. Brown Communications, Inc.

President and Chief Executive Officer *G. Franklin Lewis*
Corporate Senior Vice President, President of WCB Manufacturing *Roger Meyer*
Corporate Senior Vice President and Chief Financial Officer *Robert Chesterman*

Cover photo © Allan Power/Photo Researchers, Inc.

Cover/interior design by Schneck-DePippo Graphics

Illustrations by Schneck-DePippo Graphics

Copyedited by Carol I. Beal

A Times Mirror Company

ISBN 0-697-16401-2

Printed in the United States of America by Wm. C. Brown Communications, Inc., 2460 Kerper Boulevard, Dubuque, IA 52001

10 9 8 7 6 5 4 3 2 1

Chapter 1
p. 2: © Denis St. Seauvage/Photo Researchers, Inc.; p. 11: © Willie L. Hill, Jr./The Image Works; p. 28: © Bill Bachmann/The Image Works

Chapter 2
p. 42: © Bob Daemmrich/The Image Works; p. 49: © Kathy McLaughlin/The Image Works; p. 53: © Bob Daemmrich/The Image Works; p. 62: © Bob Daemmrich/The Image Works; p. 69: © Howard Dratch/The Image Works; p. 75: © Mark Antman/The Image Works

Chapter 3
p. 88: © Arni Katz/Unicorn Stock Photos; p. 98: © Bill Bachmann/The Image Works; p. 104: © James Pickerel/The Image Works; p. 111: © Judith Kramer/The Image Works; p. 119: © Aneal Vohra/Unicorn Stock Photos; p. 140: © Dick Young/Unicorn Stock Photos

Chapter 4
p. 160: © Ray Pfortner/Peter Arnold, Inc.; p. 169: © Steve Kaufman/Peter Arnold, Inc.; p. 174: © Bob Daemmrich/The Image Works; p. 183: © Matt Meadows/Peter Arnold, Inc.; p. 190: © Frank Pedrick/The Image Works

Chapter 5
p. 217: © Donald Dietz/Stock Boston; p. 228: © Gerard Vandystad/Photo Researchers, Inc.; p. 246: © Bob Daemmrich/The Image Works; p. 252: © Joseph Nettis/Stock Boston

Chapter 6
p. 267: © Holt Confer/The Image Works; p. 298: © Richard Pasley/Stock Boston; p. 305: © Michael Mathers/Peter Arnold, Inc.

Chapter 7
p. 338: © Ann Duncan/Tom Stack & Associates; p. 343: © Joe Sohm/The Image Works; p. 749: © Mark Antman/The Image Works

Chapter 8
p. 370: © Betts Anderson/Unicorn Stock Photos; p. 377: © Jeff Greenberg/MRP/Unicorn Stock Photos; p. 382: © Aneal Vohra/Unicorn Stock Photos; p. 389: © Aneal Vohra/Unicorn Stock Photos; p. 395: © Tierbild Okapia/Photo Researchers, Inc.; p. 400: © Scott Blackman/Tom Stack & Associates; p. 406: © FOVEA/Peter Arnold, Inc.; 8.211: © Hank Morgan/Photo Researchers, Inc.

Chapter 9
p. 429: © Charles Feil/Stock Boston; p. 438: © Richard Pasley/Stock Boston; p. 445: © Bob Daemmrich/Stock Boston; p. 454: © Tom McCarthy/Unicorn Stock Photos; p. 462: © Jim Olive/Peter Arnold, Inc.

Chapter 10
p. 477: © Jan Feingersh/Tom Stack & Associates; p. 486: © V. E. Horne/Unicorn Stock Photos; p. 495: © Frank Pedrick/The Image Works; p. 503: © Larry Mulvehill/The Image Works; p. 509: © Alan Carey/The Image Works; p. 515: © Burt Silverman

Chapter 11
p. 532: © James L. Shaffer; p. 540: © Keith Kent/Peter Arnold, Inc.; p. 548 and p. 552: © James L. Shaffer

To our parents

Contents

Chapter 5
Graphing Linear Equations and Inequalities in Two Variables

Chapter 6
Systems of Linear Equations and Inequalities

Chapter 7
Radicals and Complex Numbers

Chapter 8
Second-Degree Equations in One and Two Variables

Chapter 9
Functions

Chapter 10
Exponential and Logarithmic Functions

Chapter 11
Sequences, Series, and the Binomial Theorem

Preface

Audience

This book is intended for students who need a concrete approach to mathematics. It is written for college students who are proficient in beginning algebra. The thorough pedagogical features of the text and the associated ancillary package ensure that the student has a wealth of helpful material.

Approach

This book is written with the belief that current textbooks must provide a dynamic approach to problem solving that allows students to monitor their progress and that effectively integrates graphics and calculator use. Our approach in these four areas is explained next.

Problem-Solving Approach

Our experience is that students who take intermediate algebra learn best by "doing." Examples and exercises are crucial since it is usually in these areas that the students' main interactions with the material take place. The problem-solving approach contains brief, precisely formulated paragraphs, followed by many detailed examples. A relevant word problem introduces *every* section of the text, and word problems or other motivational problems are included in *every* section exercise set.

Graphics

A major component of a problem-solving approach with intuitive concept development is a strong emphasis on graphics. Students are given, and are encouraged to draw for themselves, visual representations of the concepts they are analyzing and the problems they are solving. Effective use of color enhances the many images in concept developments and in exercise sets.

Interactive Approach

Because students learn best by doing, progress check exercises are associated with each example problem in the text. By doing these exercises, students obtain immediate feedback on their understanding of the concept being discussed.

Calculator Use

The text encourages the use of calculators and discusses how they can be used effectively. It is assumed in the discussions that students have scientific calculators that use the algebraic operating system (AOS). Calculator illustrations show primarily the keystrokes required on a Texas Instruments TI-30-SLR+. In addition, at the end of *every* chapter graphing calculator sections are included that gradually introduce students to features of the Texas Instruments TI-81 calculator that are useful for intermediate algebra.

Features

- Problem-solving approach with intuitive concept developments
- Extensive and varied word problems and application problems
- "Think About It" exercises to develop critical-thinking skills
- Section introductions that include an interest-getting applied problem that is solved as an example in the text
- Discussions about basic concepts in geometry
- Effective use of color with an emphasis on graphics
- Over 6,000 exercises and 460 examples
- "Progress Check" exercises that allow for instant self-evaluation
- Problem sets of graduated difficulty that are closely matched to the example problems
- "Remember This" exercises that end each section and are crafted to provide smooth transition to the next section as well as spiral review of previous material

- Abundant chapter review exercises
- A chapter test and a cumulative review test for each chapter
- Boxes with labels for important definitions and rules
- "Note" and "Caution" remarks that provide helpful insights and point out potential student errors
- Unique chapter summaries that highlight specific objectives and key terms and concepts at the end of each chapter
- Instructions on calculator use
- Rigorous accuracy checking to avoid errors in the text
- Complete instructional package

Pedagogy

Section Introductions

In the spirit of problem solving, each section opens with a problem that should quickly involve students and teachers in a discussion of an important section concept. These problems are later worked out as an example in each section.

Keyed Section Objectives

Specific objectives of each section are listed at the beginning of the section, and the portion of the exposition that deals with each objective is signaled by numbered symbols such as [1].

Systematic Review

Students benefit greatly from a systematic review of previously learned concepts. At the end of each chapter there are a detailed chapter summary that includes a checklist of objectives illustrated by example problems, abundant chapter review exercises, a chapter test, and a cumulative review test. In addition, each section exercise set is concluded with a short set of "Remember This" exercises that review previous concepts, with particular emphasis on skills that will be needed in the next section.

"Think About It" Exercises

Each exercise set is followed immediately by a set of "Think About It" exercises. Although some of these problems are challenging, this section is not intended as a set of "mind bogglers." Instead, the goal is to help develop critical-thinking skills by asking students to create their own examples, express concepts in their own words, extend ideas covered in the section, and analyze topics slightly out of the mainstream. These exercises are an excellent source of nontemplate problems and problems that can be assigned for group work.

For the Instructor

The *Annotated Instructor's Edition* provides a convenient source for answers to exercise problems. Each answer is placed near the problem and appears in red. This feature eliminates the need to search through a separate answer key and helps instructors forecast effective problem assignments. Suggestions and comments based on our experiences in developmental mathematics are also provided to complement your teaching techniques, and relevant historical asides that can enliven the course material are often given. We mention three sources helpful for this historical material. *A History of Mathematics* by Victor J. Katz from HarperCollins, which emphasizes the influence of the most important textbooks of the past; *A History of Mathematical Notations* by Florian Cajori from Open Court, an older work that traces the origins and development of familiar mathematical symbols; and *The History of Mathematics: An Introduction,* 2e, by David M. Burton from Wm. C. Brown Publishers, a good discursive general introduction.

The *Instructor's Resource Manual* includes a guide to supplements that accompany *Intermediate Algebra,* solutions to "Think About It" exercises, reproducible tests, and transparency masters of key concepts and procedures.

The *Instructor's Solutions Manual* contains solutions to every section exercise in the text. These solutions are intended for the use of the instructor only.

WCB Computerized Testing Program provides you with an easy-to-use computerized testing and grade management program. No programming experience is required to generate tests randomly, by objective, by section, or by selecting specific test items. In addition, test items can be edited and new test items can be added. Also included with the *WCB Computerized Testing Program* is an on-line testing option which allows students to take tests on the computer. Tests can then be graded and the scores forwarded to the grade-keeping portion of the program.

The *Test Item File* is a printed version of the computerized testing program that allows you to examine all of the prepared test items and choose test items based on chapter, section, or objective. The objectives are taken directly from *Intermediate Algebra.*

For the Student

The *Student's Solutions Manual and Study Guide* provides a summary of the objectives, vocabulary, rules and formulas, and key concepts for each section. Detailed solutions are given for every-other odd-numbered section exercise.

Additional practice is available for each section, and each chapter ends with a sample practice test. The *Student's Solutions Manual and Study Guide* is available for student purchase through the bookstore.

Videotapes and Software that are text specific have been developed to reinforce the skills and concepts presented in *Intermediate Algebra*. Contact your Wm. C. Brown Publishers representative for detailed descriptions.

Acknowledgments

A project of this magnitude is a team effort that develops over many years with the input of many talented people. We are indebted to all who contributed. In particular, we wish to thank the reviewers of this text, who are listed separately; Carroll M. Schleppi, who checked for accuracy in our final manuscript; Carole D. Carney, who checked for accuracy in the typeset text; Bob Foley and Kathy Gutleber, who assisted with the exercise sets; Deborah Levine, who helped with the end-of-chapter material; Laura Hawkins, who did a prompt and accurate job of typing the manuscript; Carol Beal, who skillfully copyedited the manuscript; Dwala Canon, Eugenia M. Collins, Rose Deluhery, Theresa Grutz, Barbara Hodgson, Earl McPeek, and Linda Meehan, Wm. C. Brown Publishers; Deborah Schneck, Schneck-DePippo Graphics. To our parents, a special thank you. In each case they have always supported our efforts and taught us to persevere and overcome obstacles. Finally, but most important, we thank our wives, Margaret and Leda. They have given that special help and understanding only they could provide.

Dennis Christy
Robert Rosenfeld

Reviewers

Reba H. Davis
N. E. Mississippi Community College

William L. Grimes
Central Missouri State University

Patricia L. Hirschy
Asnuntuck Community Technical College

Sherri Hodge-Hardin
East Tennessee State University

Julia K. McDonald
University of Dubuque

J. Larry Martin
Missouri Southern State College

Susan Pfeifer
Butler County Community College

Sylvester Roebuck, Jr.
Olive-Harvey College

Jean C. Sanders
University of Wisconsin–Platteville

William S. Shirley
College of DuPage

Debbie Singleton
Lexington Community College

John Squires
Cleveland State Community College

Donna Szott
Community College of Allegheny County

Bob Verner
Capilano College

Jeannine Vigerust
New Mexico State University

George J. Witt
Glendale Community College

Intermediate Algebra

CHAPTER 1

Basic Concepts

The velocity of a projectile is a signed number that indicates both the speed and the direction of the projectile. Because up is usually denoted as a positive direction, a projectile traveling at a speed of 200 ft/second has a velocity of +200 ft/second when it is rising and −200 ft/second when it is falling. The relation between speed s and velocity v is given by the absolute value formula:

$$s = |v|.$$

Is a projectile moving faster when its velocity is 128 ft/second or when its velocity is −196 ft/second? (See Example 6 of Section 1.1.)

ALGEBRA IS a generalized version of arithmetic that requires a mastering of number concepts. This chapter first reviews set terminology and the basic properties and operations associated with real numbers. Then certain algebraic expressions are simplified, evaluated, and translated to complete the transition from arithmetic to algebra.

1.1 Real Numbers and Set Terminology

OBJECTIVES

1. **Graph real numbers.**
2. **Order real numbers and use inequality symbols.**
3. **Find the negative or opposite of a real number.**
4. **Find the absolute value of a real number.**
5. **Identify integers, rational numbers, irrational numbers, and real numbers.**
6. **Find the union or intersection of two sets.**

1 In algebra numerical relations are studied in a more general way by using symbols (such as x) that may be replaced by a number from some collection of numbers. Unless stated otherwise, it is assumed that a symbol like x may be replaced by any real number. Consequently, the rules that govern real numbers determine the procedures in algebra. A good way to describe real numbers is to interpret them geometrically by considering the number line in Figure 1.1. Every point on this line corresponds to a real number, and every real number corresponds to a point on this line. Therefore, this line is called the **real number line,** or simply the number line, and may be used to define a real number.

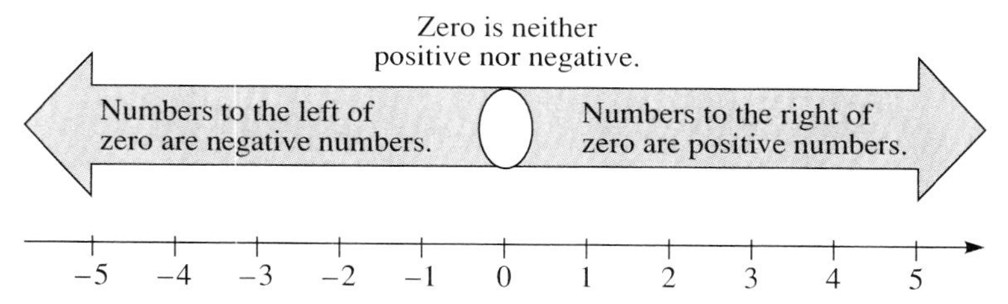
Zero is neither positive nor negative.
Numbers to the left of zero are negative numbers.
Numbers to the right of zero are positive numbers.
0
–5 –4 –3 –2 –1 0 1 2 3 4 5

Figure 1.1

Real Number

A real number is a number that can be represented as a point on the number line.

The word *real* for these numbers is due to Descartes (1637), who also coined *imaginary*.

The point on the number line corresponding to a number is called the **graph** of the number. A type of real number that is easy to graph is called an **integer.** The set* of integers is given by

$$\{\underbrace{\ldots, -4, -3, -2, -1}_{\text{negative integers}}, 0, \underbrace{1, 2, 3, 4, \ldots}_{\text{positive integers}}\},$$

Integer comes from Latin words meaning "untouched" or "whole"; *fraction* comes from a word meaning "broken."

where the three dots . . . called an **ellipsis** mean "and so on." As with all positive numbers, note that we customarily omit the sign on positive integers, so $+1 = 1$, $+2 = 2$, and so on.

*A **set** is simply a collection of objects, and we may describe a set by listing the objects or elements of the collection within braces.

EXAMPLE 1 Graph the integers 5, 0, and -3 on the number line.

Solution Place dots on the number line at 5, 0, and -3, as shown in Figure 1.2, to graph the given integers.

Figure 1.2

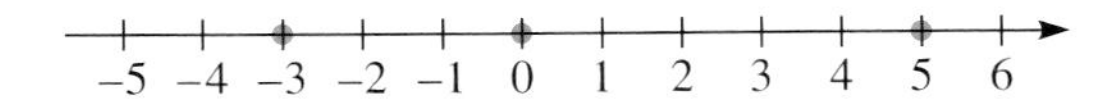

PROGRESS CHECK 1 Graph the integers -2, 7, and 0 on the number line.

Rational comes from *ratio*, but stress that the denominator cannot be zero.

Rational numbers and irrational numbers fill in the gaps on the number line between the integers. A number that can be written as a fraction with an integer in the numerator and a *nonzero* integer in the denominator is called a **rational number.** Using set-builder notation,* the set of rational numbers is given by

$$\left\{\frac{p}{q} : p \text{ and } q \text{ are integers, } q \text{ not equal to } 0\right\}.$$

Some examples of rational numbers are

$$\frac{4}{5}, \quad \frac{-2}{3}, \quad \frac{2}{-7}, \quad \frac{0}{6}, \quad 2.7\left(\text{or } \frac{27}{10}\right), \quad \text{and} \quad 5\left(\text{or } \frac{5}{1}\right).$$

Note that all integers are rational numbers, because each integer can be written with a denominator of 1. For instance,

$$5 = \frac{5}{1}, \qquad -3 = \frac{-3}{1}, \qquad \text{and} \qquad 0 = \frac{0}{1}.$$

Also, when a rational number is written in decimal form, it can be shown that the decimal must either terminate or be a repeating decimal. In the examples that follow, a bar is placed above the portion of any decimals that repeat.

Rational Number	Decimal Form	Classification
$\frac{1}{2}$	0.5	Terminating decimal
$\frac{9}{8}$	1.125	Terminating decimal
$\frac{1}{3}$	$0.3333\ldots = 0.\overline{3}$	Repeating decimal
$\frac{200}{99}$	$2.0202\ldots = 2.\overline{02}$	Repeating decimal

Real numbers that are not rational numbers are called **irrational numbers.** Such numbers in decimal form are neither terminating nor repeating decimals. Some examples of irrational numbers with their approximate decimal forms are

$$\sqrt{2} = 1.41421\ldots, \qquad \sqrt{3} = 1.73205\ldots, \qquad \text{and} \qquad \pi = 3.14159\ldots.$$

For many purposes, an irrational number like π (pi) may be approximated by rational numbers. Two well-known approximations for π are 3.14 and $\frac{22}{7}$. Example 2 shows how to graph rational numbers and irrational numbers.

Progress Check Answer

1.

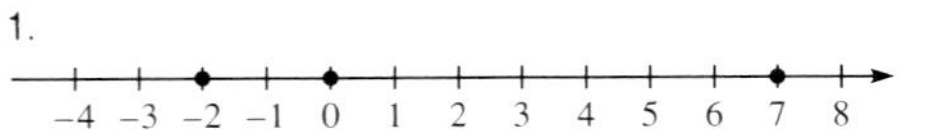

***Set-builder notation** writes sets in the form $\{x{:}x$ has property $P\}$, which is read, "the set of all elements x such that x has property P." The colon : is read "such that."

EXAMPLE 2 Graph the following numbers on the number line.

a. $-\frac{3}{4}$ **b.** $\sqrt{3}$

Solution See Figure 1.3.

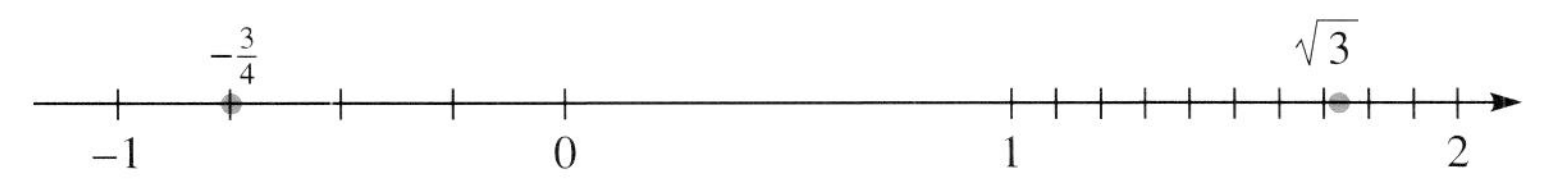

Figure 1.3

a. To graph the rational number $-\frac{3}{4}$, divide the line segment from 0 to -1 into four equal parts, and then place a dot on the third slash mark to the left of 0.

b. We approximate the graphs of irrational numbers like $\sqrt{3}$. By calculator, obtain an approximate value for $\sqrt{3}$.

3 [√‾] [1.7320508]

Because $\sqrt{3}$ is slightly more than 1.7, divide the segment from 1 to 2 into ten equal parts, and then place a dot slightly to the right of the seventh slash mark.

Note In practice it is usually sufficient to give a rough estimate of the graph of the types of numbers in this example.

PROGRESS CHECK 2 Graph the following numbers on the number line.

a. $-\frac{5}{3}$ **b.** $\sqrt{2}$

2 An important property of real numbers is that they can be put in numerical order. The **trichotomy property** states that if a and b are real numbers, then either a is less than b, a is greater than b, or a equals b. Statements involving inequality are symbolized as follows.

Statement	Read	Example
$a < b$	a is less than b	$3 < 7$
$a > b$	a is greater than b	$7 > 3$
$a \leq b$	a is less than or equal to b	$3 \leq 7$
$a \geq b$	a is greater than or equal to b	$3 \geq 3$
$a \neq b$	a is not equal to b	$7 \neq 3$

Note that statements like $a \geq b$ are true if either the "greater than" part is true or the "equal" part is true. It is not possible for both the "greater than" part and the "equal" part to be true simultaneously. Relations of "less than" and "greater than" can be seen easily on the number line, as shown in Figure 1.4. The graph of the larger number is to the right of the graph of the smaller number.

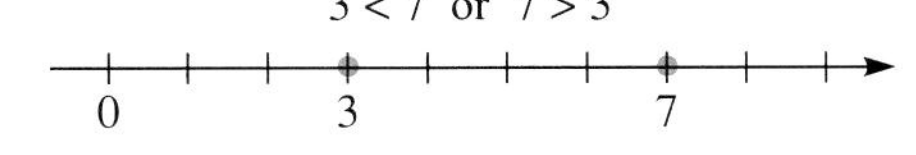

Figure 1.4

EXAMPLE 3 Classify each statement as true or false.

a. $-1 > -3$ **b.** $-1 \leq -3$ **c.** $-1 \geq -1$

Solution

a. $-1 > -3$ is read "-1 is greater than -3." Figure 1.5 shows that the graph of -1 is to the right of the graph of -3, so this statement is true.

b. $-1 \leq -3$ is read "-1 is less than or equal to -3." This statement is false because $-1 < -3$ is false and $-1 = -3$ is false.

c. $-1 \geq -1$ is read "-1 is greater than or equal to -1." Because "-1 is equal to -1" is true, $-1 \geq -1$ is true.

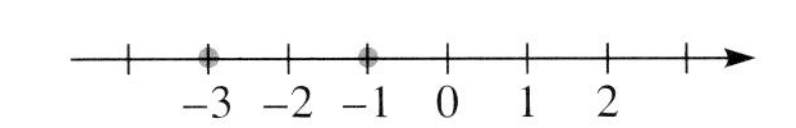

Figure 1.5

Progress Check Answer

2.

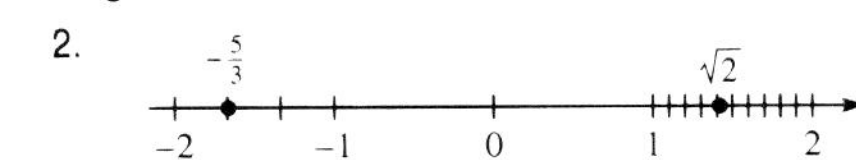

PROGRESS CHECK 3 Classify each statement as true or false.

a. $-4 > -2$ **b.** $-4 \geq -4$ **c.** $-4 \leq -2$

3 Two numbers on the number line that are the same distance from 0, but on opposite sides of 0, are called **negatives** or **opposites** of each other. To symbolize the negative of a number, place the symbol $-$ in front of the number. For instance, Figure 1.6 shows that the negative (or opposite) of 3 is -3 and the negative or opposite of -3, which may be written $-(-3)$, is 3. The latter example illustrates the double-negative rule.

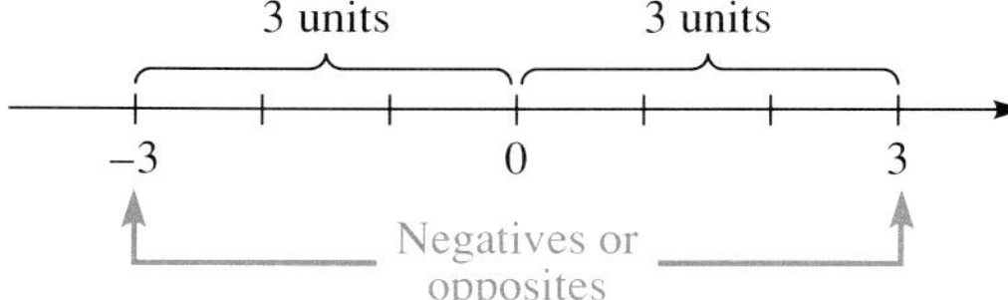

Figure 1.6

Double-Negative Rule

If a is a real number, then

$$-(-a) = a.$$

EXAMPLE 4 Find the negative, or opposite, $-a$, of each number.

a. $a = \frac{1}{2}$ **b.** $a = 0$ **c.** $a = -\sqrt{2}$

Solution

a. If $a = \frac{1}{2}$, then $-a = -(\frac{1}{2}) = -\frac{1}{2}$.
b. If $a = 0$, then $-a = -(0) = 0$. Zero is the only number that is its own negative.
c. By the double-negative rule, if $a = -\sqrt{2}$, then $-a = -(-\sqrt{2}) = \sqrt{2}$.

PROGRESS CHECK 4 Find the negative, or opposite, $-a$, of each number.

a. $a = -\frac{3}{4}$ **b.** $a = \pi$ **c.** $a = 0$

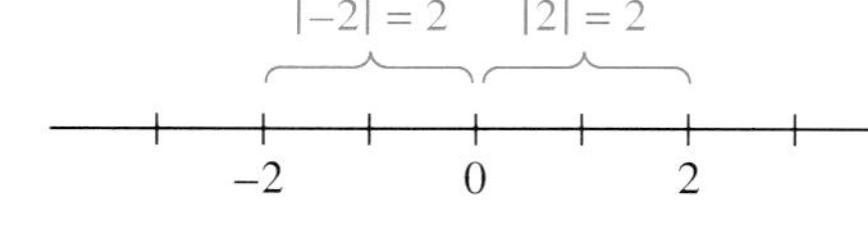

Figure 1.7

4 The **absolute value** of a real number a, denoted $|a|$, is the distance between a and 0 on the number line. For instance, $|2| = 2$, and $|-2| = 2$, as shown in Figure 1.7. The geometric interpretation of absolute value may be translated to the following algebraic rule.

Absolute Value

For any real number a,

$$|a| = a, \quad \text{if } a \geq 0,$$
$$|a| = -a, \quad \text{if } a < 0.$$

To illustrate this definition, note from Figure 1.7 that $|-2| = 2$. This result is obtained using the algebraic rule as follows: Because -2 is less than 0, use the formula $|a| = -a$; replacing a by -2 gives $|-2| = -(-2) = 2$. From both a geometric and an algebraic perspective, note that the absolute value of a number is never negative.

Progress Check Answers

3. (a) F (b) T (c) T
4. (a) $\frac{3}{4}$ (b) $-\pi$ (c) 0

EXAMPLE 5 Evaluate each expression.

a. $|2.4|$ **b.** $|-\frac{1}{2}|$ **c.** $-|-\frac{1}{2}|$

Solution

a. Figure 1.8 shows $|2.4| = 2.4$. Algebraically, 2.4 is greater than 0, so $|a| = a$ yields $|2.4| = 2.4$.

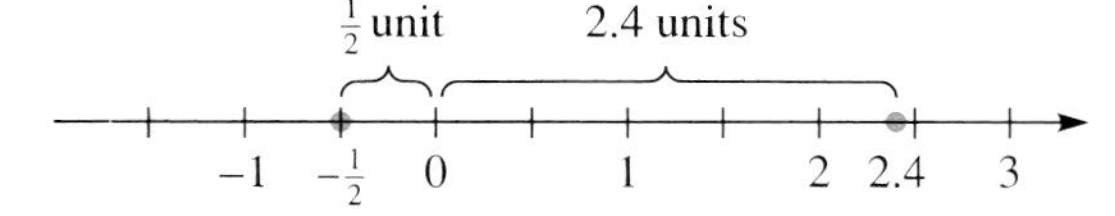

Figure 1.8

b. Figure 1.8 shows $|-\frac{1}{2}| = \frac{1}{2}$. Algebraically, $-\frac{1}{2}$ is less than 0, so $|a| = -a$ yields $|-\frac{1}{2}| = -(-\frac{1}{2}) = \frac{1}{2}$.

c. Because $|-\frac{1}{2}| = \frac{1}{2}$,

$$-|-\tfrac{1}{2}| = -(\tfrac{1}{2}) = -\tfrac{1}{2}.$$

PROGRESS CHECK 5 Evaluate each expression.

a. $|-\sqrt{2}|$ **b.** $-|-5|$ **c.** $|0|$

The chapter-opening problem illustrates a case where the numerical size of a number is more important than its sign, so the concept of absolute value is useful.

EXAMPLE 6 Solve the problem in the chapter introduction on page 2.

Solution Compare the speed at each velocity to determine when the projectile is moving faster.

$$s = |v| = |128 \text{ ft/second}| = 128 \text{ ft/second}$$
$$s = |v| = |-196 \text{ ft/second}| = 196 \text{ ft/second}$$

Since $196 > 128$, the projectile is moving faster when its velocity is -196 ft/second.

PROGRESS CHECK 6 Is a projectile moving faster when its velocity is -55 m/second or when its velocity is -85 m/second?

5 Real numbers, integers, rational numbers, and irrational numbers have been defined in this section. Depending on the use of the number, the distinction between these types of numbers may be important. Figure 1.9 summarizes these definitions and shows the relationships among the various sets of numbers. In set theory, when every element in set A is an element in set B, then A is called a **subset** of B. Note that all of the sets that have been discussed are subsets of the real numbers.

Real numbers (all numbers that can be represented by a point on the number line)

Rational numbers (quotients of two integers with a nonzero denominator; terminating or repeating decimals)	**Irrational numbers** (real numbers that are not rational numbers; nonterminating and nonrepeating decimals)
$\frac{8}{9}$, -2.5, $0.\overline{3}$	$\sqrt{2}$, $\sqrt{3}$, π
Integers . . ., −2, −1, 0, 1, 2, . . .	

Figure 1.9

Greek mathematicians showed before 300 B.C. that all real numbers were either rational or irrational.

Progress Check Answers

5. (a) $\sqrt{2}$ (b) -5 (c) 0
6. -85 m/second

EXAMPLE 7 From the set $\{2.5, \sqrt{7}, \sqrt{4}, 0, 0.\overline{6}, -\frac{9}{11}, -3, -\pi\}$, list all numbers that are in the following sets.

a. Integers
b. Rational numbers
c. Irrational numbers
d. Real numbers

Solution

a. $\sqrt{4}$ (which equals 2), 0, and -3 are integers.
b. 2.5, $\sqrt{4}$, 0, $0.\overline{6}$, $-\frac{9}{11}$, and -3 are rational numbers.
c. $\sqrt{7}$ and $-\pi$ are irrational numbers.
d. All the numbers in the set are real numbers. In Section 7.6 numbers like $\sqrt{-1}$, which are not real numbers, will be discussed.

PROGRESS CHECK 7 From $\left\{\frac{1}{2}, \frac{\pi}{2}, -\sqrt{2}, \sqrt{0}, 2, 0.2, 0.\overline{2}, -2\right\}$, list all numbers that are in the following sets.

a. Integers
b. Rational numbers
c. Irrational numbers
d. Real numbers

6 In the process of discussing the real numbers, set terminology has been helpful. We now add some other basic definitions involving sets that are useful in the study of numbers and algebra.

Set Definitions

1. **Equality of Sets** Two sets A and B are equal, written $A = B$, if and only if A and B have exactly the same elements.
2. **Union of Sets** The union of sets A and B, written $A \cup B$, is the set of elements that belong to A or to B or to both.
3. **Intersection of Sets** The intersection of sets A and B, written $A \cap B$, is the set of elements that belong to both A and B.
4. **Empty Set** The empty set, denoted by $\emptyset$, is the set containing no elements.

The empty set is sometimes called the "null" set.

Note that the definition of set equality makes use of the phrase "if and only if." This phrase is commonly used in mathematics definitions and theorems. The statement "*p* **if and only if** *q*" means "if *p*, then *q*"; and conversely, "if *q*, then *p*."

EXAMPLE 8 If $A = \{1,2,3,4\}$, $B = \{0,2,4\}$, and $C = \{1,3,5\}$, find each set.

a. $A \cup B$ **b.** $A \cap B$ **c.** $B \cap C$

Solution

a. The elements 0, 1, 2, 3, and 4 belong to A or B or both, so

$$\begin{aligned} A \cup B &= \{1,2,3,4\} \cup \{0,2,4\} \\ &= \{0,1,2,3,4\}. \end{aligned}$$

b. The elements 2 and 4 belong to both A and B, so

$$\begin{aligned} A \cap B &= \{1,2,3,4\} \cap \{0,2,4\} \\ &= \{2,4\}. \end{aligned}$$

Progress Check Answers

7. (a) $\sqrt{0}$, 2, -2 (b) $\frac{1}{2}$, $\sqrt{0}$, 2, 0.2, $0.\overline{2}$, -2 (c) $\pi/2$, $-\sqrt{2}$ (d) All

c. No element is common to both B and C, so the intersection of B and C is the empty set.

$$B \cap C = \{0,2,4\} \cap \{1,3,5\} = \emptyset$$

PROGRESS CHECK 8 If $A = \{0,1,2\}$, $B = \{3,4,5\}$, and $C = \{-1,0,1\}$, find each set.

a. $A \cup B$ **b.** $A \cap B$ **c.** $A \cap C$

EXAMPLE 9 Classify each statement as true or false.

a. If $A = \{1,2,3\}$ and $B = \{3,1,2\}$, then $A = B$.
b. The union of the set of rational numbers with the set of irrational numbers is the empty set.

Solution

a. The statement is true because A and B have exactly the same elements. Note that the order of the elements in a set is unimportant.
b. The statement is false because the union of two nonempty sets is never the empty set. In fact, the union of the set of rational numbers with the set of irrational numbers is the set of real numbers.

PROGRESS CHECK 9 Classify each statement as true or false.

a. The intersection of the set of rational numbers with the set of irrational numbers is the empty set.
b. $\emptyset = \{0\}$.

Progress Check Answers

8. (a) $\{0,1,2,3,4,5\}$ (b) $\emptyset$ (c) $\{0,1\}$
9. (a) T (b) F

EXERCISES 1.1

In Exercises 1–12, graph the given numbers on the real number line. To indicate scale, label slash marks at 0 and 1 in each example, and draw other slashes as needed.

1. 4, 0, -2

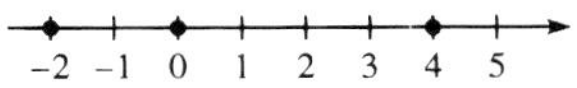

-2 -1 0 1 2 3 4 5

2. -5, 0, 1

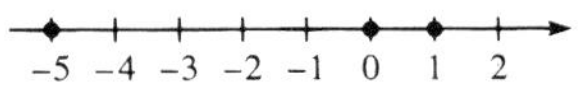

-5 -4 -3 -2 -1 0 1 2

3. 1, -2, 3

-2 -1 0 1 2 3 4

4. -2, 3, -4

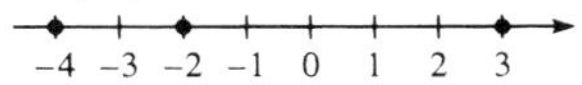

-4 -3 -2 -1 0 1 2 3

5. $\frac{1}{2}$, $\frac{1}{4}$, 0

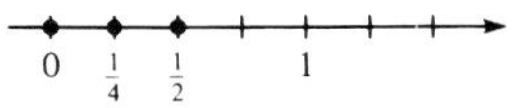

0 $\frac{1}{4}$ $\frac{1}{2}$ 1

6. $-\frac{1}{3}$, $-\frac{2}{3}$, 0

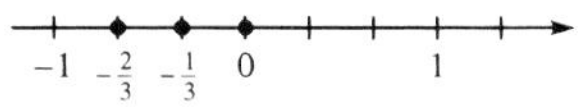

-1 $-\frac{2}{3}$ $-\frac{1}{3}$ 0 1

7. $\frac{8}{9}$, $\frac{9}{8}$, 0

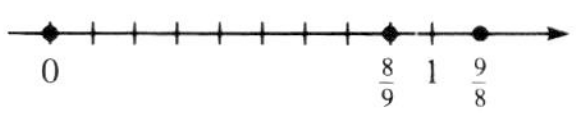

0 $\frac{8}{9}$ 1 $\frac{9}{8}$

8. $-\frac{3}{4}$, $-\frac{4}{3}$, 0

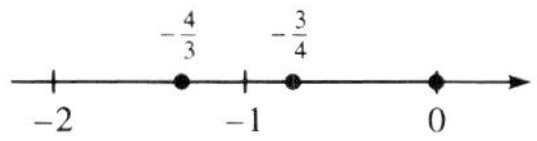

$-\frac{4}{3}$ $-\frac{3}{4}$
-2 -1 0

9. $\sqrt{2}$, $\sqrt{4}$, $\sqrt{8}$

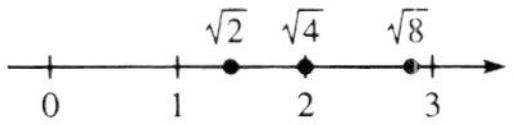

$\sqrt{2}$ $\sqrt{4}$ $\sqrt{8}$
0 1 2 3

10. $-\sqrt{5}$, $-\sqrt{10}$, $-\sqrt{15}$

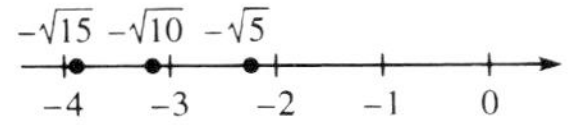

$-\sqrt{15}$ $-\sqrt{10}$ $-\sqrt{5}$
-4 -3 -2 -1 0

11. π, 2π, $\frac{\pi}{2}$

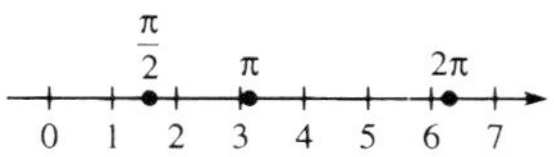

$\frac{\pi}{2}$ π 2π
0 1 2 3 4 5 6 7

12. $\sqrt{3}$, $\frac{\sqrt{3}}{2}$, $\frac{\sqrt{3}}{3}$

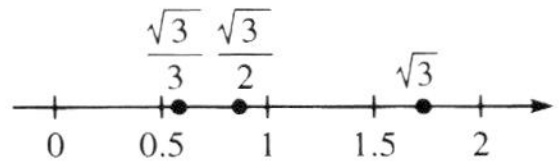

$\frac{\sqrt{3}}{3}$ $\frac{\sqrt{3}}{2}$ $\sqrt{3}$
0 0.5 1 1.5 2

In Exercises 13–18, classify each statement as true or false.

13. $-1 \geq -3$ True
14. $-2 < -4$ False
15. $2 \geq 2$ True
16. $-3 \leq -3$ True
17. $5 + 7 \leq 5 + 8$ True
18. $6 - 3 \leq 6 - 4$ False
19. What is the smallest number that can replace x so that $5 \leq x$ will be true? 5
20. What is the largest number that can replace y so that $y \leq 9$ will be true? 9
21. What is the largest number that can replace x so that $x \leq -4$ will be true? -4
22. What is the smallest number that can replace y so that $-1 \leq y$ will be true? -1

In Exercises 23–28, find the negative or opposite of a. That is, find $-a$. Draw the graph of both a and $-a$ in each case, and indicate which point is which.

23. $a = \frac{2}{3}$

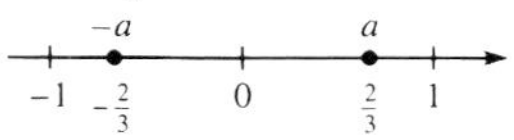

$-a$ a
-1 $-\frac{2}{3}$ 0 $\frac{2}{3}$ 1

24. $a = \frac{4}{5}$

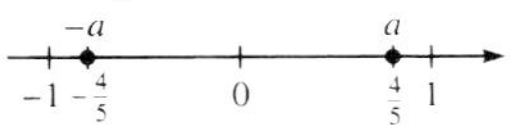

$-a$ a
-1 $-\frac{4}{5}$ 0 $\frac{4}{5}$ 1

25. $a = -1.2$

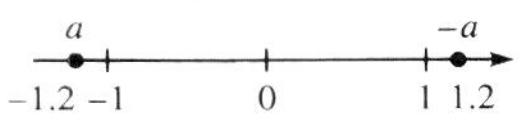

a $-a$
-1.2 -1 0 1 1.2

26. $a = -0.6$

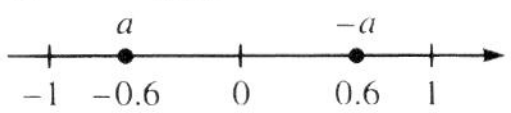

a $-a$
-1 -0.6 0 0.6 1

27. $a = 1\frac{2}{3}$

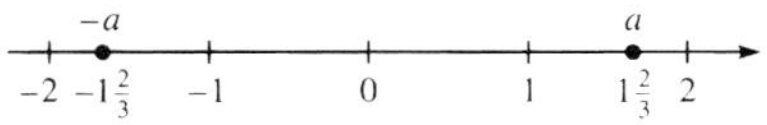

$-a$ a
-2 $-1\frac{2}{3}$ -1 0 1 $1\frac{2}{3}$ 2

28. $a = -\frac{5}{2}$

29. If the graph of a is on the right side of zero, on which side of zero is the graph of the following?

a. $-a$ Left **b.** $\frac{a}{2}$ Right

30. If the graph of b is on the left side of zero, on which side of zero is the graph of the following?

a. $-b$ Right **b.** $\frac{b}{2}$ Left

31. Yes or no? Are the absolute value of a and the absolute value of $-a$ always equal? Explain. Yes; a and $-a$ are equidistant from 0.

32. If a is greater than b, can you conclude that the absolute value of a is greater than the absolute value of b? Illustrate. No; $a = 1$, $b = -2$

33. Arrange these values in increasing order: $|-2|$, $|-1|$, -1, -2. $-2, -1, |-1|, |-2|$

34. Arrange these values in increasing order: $|-3|$, $-|3|$, -2, $|-2|$. $-|3|, -2, |-2|, |-3|$

In Exercises 35–42, evaluate the given expression.

35. $|1.7|$ 1.7
36. $|0.4|$ 0.4
37. $|-\frac{1}{5}|$ $\frac{1}{5}$
38. $|-2|$ 2
39. $-|4|$ -4
40. $-|-4|$ -4
41. $|3| + |-3|$ 6
42. $|\frac{1}{2}| + |-\frac{1}{2}|$ 1

43. If $m = 3$, then which of these is true?

a. $|m| = m$ **b.** $|m| = -m$ a

44. If $n = -4$, then which of these is true?

a. $|n| = n$ **b.** $|n| = -n$ b

45. Find a real number a for which $|a| > a$. a can be any negative number.

46. Find a real number a for which $|a| < a$. Not possible

47. Which statement is true?

a. The speed of a projectile equals the absolute value of its velocity.

b. The velocity of a projectile equals the absolute value of its speed. a

48. Which statement is true?

a. Velocity is never greater than speed.

b. Speed is never greater than velocity. a

49. Is a projectile moving faster when its velocity is -100 ft/second or $+30$ ft/second? -100 ft/second

50. Is a projectile moving faster when its velocity is -0.5 ft/second or $+0.4$ ft/second? -0.5 ft/second

51. For which change is the absolute value greater: a rise in a stock market index, from 3000 to 3030, or a drop from 3000 to 2910? Drop from 3000 to 2910

52. For which change in temperature is the absolute value greater: a rise from $-4°$ to $-1°$ or a drop from $-4°$ to $-5°$? Rise from $-4°$ to $-1°$

53. True or false? It is possible for a number to be both an integer and a rational number. True

54. True or false? It is possible for a number to be both an integer and an irrational number. False

55. True or false? It is possible for a number to be both rational and real. True

56. True or false? It is possible for a number to be both irrational and real. True

57. True or false? It is possible for a number to be both rational and irrational. False

58. True or false? It is possible for a number to be both integer and real. True

In Exercises 59–64, list all numbers in the given set that are in the following sets.

a. Integers **b.** Rational numbers
c. Irrational numbers **d.** Real numbers

59. $\{\frac{1}{2}, 3\frac{3}{4}, -0.1, \sqrt{9}\}$
a. $\sqrt{9}$ b. All c. None d. All

60. $\{\frac{10}{2}, \frac{2}{10}, -\frac{4}{5}, -\frac{5}{4}, \frac{6}{6}\}$
a. $\frac{10}{2}$, $\frac{6}{6}$ b. All c. None d. All

61. $\{3.4, 1, -5.3, \sqrt{5}, 0\}$
a. 1, 0 b. 3.4, 1, -5.3, 0 c. $\sqrt{5}$ d. All

62. $\{-1.2, \sqrt{3}, \pi, 999\}$
a. 999 b. -1.2, 999 c. $\sqrt{3}$, π d. All

63. $\{\sqrt{1}, \sqrt{2}, \sqrt{3}, \sqrt{4}, \sqrt{5}\}$
a. $\sqrt{1}$, $\sqrt{4}$ b. $\sqrt{1}$, $\sqrt{4}$ c. $\sqrt{2}$, $\sqrt{3}$, $\sqrt{5}$ d. All

64. $\left\{-\frac{1}{2}, -\frac{2}{1}, \frac{\pi}{1}, 0\right\}$
a. $-\frac{2}{1}$, 0 b. $-\frac{1}{2}$, $-\frac{2}{1}$, 0 c. $\frac{\pi}{1}$ d. All

65. Explain why 1.01001000100001 . . . is an irrational number but 1.010101 . . . is a rational number. 1.01001000100001 . . . never repeats, so it is irrational; 1.010101 . . . does repeat and so is rational.

66. Explain why 1.23456789101112 . . . is an irrational number but 1.234234234 . . . is a rational number. 1.23456789101112 . . . never repeats, so it is irrational; 1.234234234 . . . does repeat and so is rational.

For Exercises 67–74, use $A = \{-1, -2, 1, 2\}$, $B = \{-1, 1\}$, $C = \{-2, 2\}$, and $D = \{1, -1, 2, -2\}$.

67. True or false? $A = D$. True

68. True or false? $B = C$. False

69. Find $B \cup C$. $\{-1, 1, -2, 2\}$

70. Find $B \cap C$. $\emptyset$

71. Find $A \cap B$. $\{-1, 1\}$

72. Find $A \cup B$. $\{-1, -2, 1, 2\}$

73. True or false? $B \cup C = A$. True

74. True or false? $A \cap B = B$. True

For Exercises 75–84, use $A = \{1, \frac{1}{2}, \frac{1}{4}, \frac{1}{8}\}$, $B = \{0, 2, 4, 8\}$, $C = \{0, 1, 3, 9\}$, and $D = \{1, \frac{1}{3}, \frac{1}{9}\}$.

75. Find $B \cap C$. $\{0\}$

76. Find $A \cap B$. $\emptyset$

77. Find $C \cap D$. $\{1\}$

78. Find $A \cap D$. $\{1\}$

79. True or false? $B \cap C = A \cap B$. False

80. True or false? $C \cap D = A \cap D$. True

81. Find $B \cup C$. $\{0, 1, 2, 3, 4, 8, 9\}$

82. Find $A \cup D$. $\{1, \frac{1}{2}, \frac{1}{3}, \frac{1}{4}, \frac{1}{8}, \frac{1}{9}\}$

83. Find $B \cap B$. $\{0, 2, 4, 8\}$

84. Find $B \cup B$. $\{0, 2, 4, 8\}$

85. True or false? The intersection of the set of integers with the set of irrational numbers is empty. True

86. True or false? The intersection of the set of integers with the set of rational numbers is the set of integers. True

87. If two sets contain no elements in common, then their intersection is the empty set. True

88. If two nonempty sets contain no elements in common, then their union cannot be the empty set. True

89. What is the intersection of the set of odd integers with the set of even integers? $\emptyset$

90. What is the intersection of the set of positive real numbers with the set of negative real numbers? $\emptyset$

91. What is the intersection of the set of nonnegative real numbers with the set of nonpositive real numbers? $\{0\}$

92. What is the union of the set of odd integers with the set of even integers? All integers

THINK ABOUT IT

1. a. A common student error is to assume that $-a$ must represent a negative number. Explain why this assumption is incorrect.

b. Is it always true that $a > -a$?

2. a. What is the smallest number that can replace x so that $5 \leq x$ is true?

b. What is the smallest integer that can replace x so that $5 < x$ is true?

c. What is the smallest rational number that can replace x so that $5 < x$ is true?

3. The expression $-(-8) - 3$ illustrates three uses of the $-$ symbol. Describe the three uses.

4. Find the set of all replacements for a that make the statement true.

a. $|a| = 2$ **b.** $|a| = -3$ **c.** $|a| = a$

d. $|a| = -a$ **e.** $|a| = |-a|$

5. a. Given the dot for the graph of n, place dots for $n/2$ and $2n$.

0 n

b. Given the dot for the graph of m, place dots for $m/2$ and $2m$.

m 0

c. If a is some nonzero real number, is the graph of $a/2$ always closer to 0 than the graph of a?

d. Is the absolute value of $a/2$ always less than or equal to the absolute value of a? That is, is $|a/2| < |a|$?

REMEMBER THIS

1. Match the operation to the word for the answer.

a. Addition	**i.** Product	a. iv
b. Subtraction	**ii.** Difference	b. ii
c. Multiplication	**iii.** Quotient	c. i
d. Division	**iv.** Sum	d. iii

2. True or false? Zero is neither positive nor negative. True

3. True or false? Zero is an integer. True

4. True or false? 2 is greater than -15. True

5. $A = \{x: x > 0\}$, and $B = \{x: x < 1\}$. List five numbers which are simultaneously in both sets A and B. $\frac{1}{2}, \frac{1}{3}, \frac{1}{4}, \frac{1}{5}, \frac{1}{6}$; other answers are possible.

6. What number is the negative of -8? 8

7. If $|a| = 5$, what is $|-a|$? 5

8. $A = \{0,2,4\}$, $B = \{0,-2,-4\}$, and $C = \{0,-1,1\}$. True or false? $A \cap B$ equals $A \cap C$. True

9. Calculate $\frac{1}{2} + \frac{2}{3}$. $\frac{7}{6}$

10. Calculate $\frac{1}{2} \cdot \frac{2}{3}$. $\frac{1}{3}$

1.2 Operations on Real Numbers

The Nielsen rating system ranks TV shows on a special rating scale in which each rating point represents 921,000 homes. Over the period of a week the rating for a popular TV show changes from 15.3 to 13.8. Use

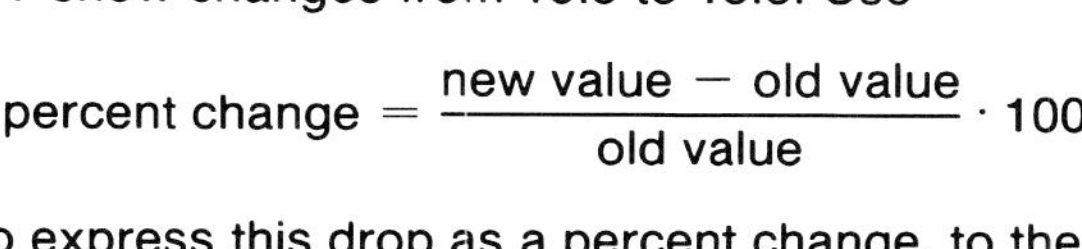

$$\text{percent change} = \frac{\text{new value} - \text{old value}}{\text{old value}} \cdot 100$$

to express this drop as a percent change, to the nearest hundredth of a percent. How many viewing homes did this show lose for this week, to the nearest ten thousand homes? (See Example 10.)

OBJECTIVES

1. Add, subtract, multiply, and divide real numbers.
2. Apply the definition of a positive integer exponent.
3. Evaluate expressions following the order of operations.
4. Solve applied problems involving positive and negative numbers.

1 Addition of real numbers may be visualized using number line diagrams, as explained in Example 1. Recall that the result of an addition is called a **sum.**

EXAMPLE 1 Find each sum by using the number line.

a. $-1 + 3$ **b.** $-1 + (-3)$

Solution

a. The addition $-1 + 3$ is shown in Figure 1.10. Begin by drawing an arrow with a length of 1 unit starting at 0 and pointing to the left. From the tip of the arrow at -1, we then draw an arrow to the right with a length of 3 units. Because the tip of this arrow is at 2, we conclude that $-1 + 3 = 2$.

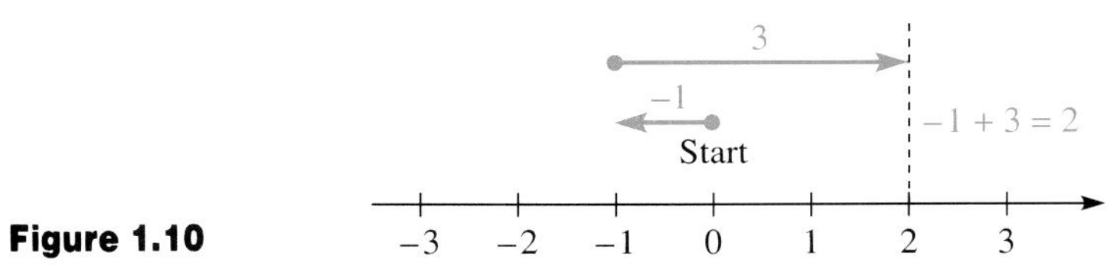

Figure 1.10

b. $-1 + (-3) = -4$, as shown in Figure 1.11.

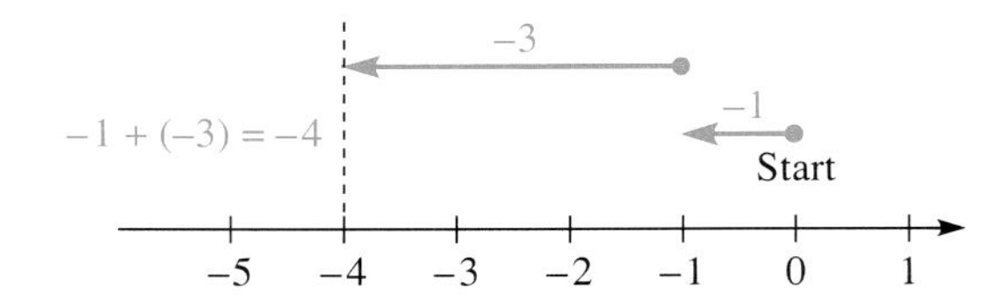

Figure 1.11

PROGRESS CHECK 1 Find each sum by using the number line.

a. $-4 + (-2)$ **b.** $-4 + 2$

In the number line method of addition, the lengths of the arrows are given by the absolute values of the numbers. Therefore, the concept of absolute value may be used to add real numbers efficiently with the help of the following procedures. Check to see that these methods are in agreement with the number line procedure.

Adding Real Numbers

Like Signs To add two real numbers with the same sign, add their absolute values and attach the common sign.

Unlike Signs To add two real numbers with different signs, subtract the smaller absolute value from the larger and attach the sign of the number with the larger absolute value.

Progress Check Answers

1. (a) (b)

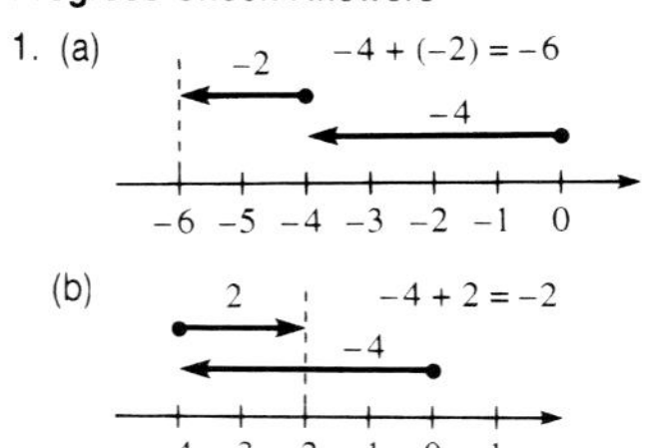

EXAMPLE 2 Find each sum.

a. $-17 + (-51)$ **b.** $23 + (-12)$ **c.** $-\frac{5}{9} + \frac{4}{9}$

Solution

a. $-17 + (-51) = -68$ *Think:* $|-17| = 17$, $|-51| = 51$, $17 + 51 = 68$, and the common sign is negative.

b. $23 + (-12) = 11$ *Think:* $|23| = 23$, $|-12| = 12$, $23 - 12 = 11$, and the positive number 23 has the larger absolute value.

c. $-\frac{5}{9} + \frac{4}{9} = -\frac{1}{9}$ *Think:* $|-\frac{5}{9}| = \frac{5}{9}$, $|\frac{4}{9}| = \frac{4}{9}$, $\frac{5}{9} - \frac{4}{9} = \frac{1}{9}$, and the negative number $-\frac{5}{9}$ has the larger absolute value.

PROGRESS CHECK 2 Find each sum.

a. $-8 + 15$ **b.** $64 + (-81)$ **c.** $-\frac{4}{7} + (-\frac{2}{7})$

In algebra subtraction is defined as the addition of the opposite. For instance, to subtract 7 from 3, add the opposite of 7 to 3.

$$3 - 7 = 3 + \overbrace{(-7)}^{\text{opposite of } 7} = -4$$

change $-$ to $+$

The result of a subtraction is called a **difference,** so the difference $3 - 7$ equals -4. In general, we will often switch between subtraction and addition by using the following definition.

Definition of Subtraction

If a and b are real numbers, then

$$a - b = a + (-b).$$

EXAMPLE 3 Find each difference.

a. $-16 - (-6)$ **b.** $-6 - (-16)$ **c.** $-\frac{1}{4} - \frac{2}{3}$

Solution

a. By the double-negative rule the opposite of -6 is 6, so

$$-16 - (-6) = -16 + 6 = -10.$$

b. Add the opposite of -16, which is 16, to -6, to get

$$-6 - (-16) = -6 + 16 = 10.$$

c. Use $a - b = a + (-b)$ and add fractions, using 12 as the least common denominator.

$$-\frac{1}{4} - \frac{2}{3} = -\frac{1}{4} + (-\frac{2}{3}) = -(\frac{1}{4} + \frac{2}{3})$$
$$= -(\frac{3}{12} + \frac{8}{12}) = -\frac{11}{12}$$

Caution In this example parts **a** and **b** illustrate that $a - b$ and $b - a$ give opposite answers, so be careful to write numbers in the correct order in an expression involving subtraction. With respect to the minus sign, the number we subtract "from" goes on the left, and the number to subtract goes on the right.

Progress Check Answers

2. (a) 7 (b) -17 (c) $-\frac{6}{7}$

PROGRESS CHECK 3 Find each difference.

a. $-8 - (-15)$ **b.** $-15 - (-8)$ **c.** $-\frac{1}{2} - \frac{3}{5}$

Multiplying numbers a and b may be symbolized in algebra by

$$ab, \quad a \cdot b, \quad a(b), \quad (a)b, \quad \text{or} \quad (a)(b).$$

The division of a by b may be written as

$$a \div b, \quad a/b, \quad \text{or} \quad \frac{a}{b}.$$

Multiplication and division of real numbers are related in that

$$\frac{a}{b} = c \quad \text{if and only if} \quad a = bc.$$

For instance, $8/2 = 4$ is equivalent to $8 = 2 \cdot 4$. The result of a multiplication is called a **product,** and the result of a division is called a **quotient.** Products and quotients of nonzero numbers are computed as follows.

Products and Quotients of Nonzero Real Numbers

Same Sign To multiply (or divide) two real numbers with the same sign, multiply (or divide) their absolute values, and make the sign of the product (or quotient) positive.

Different Signs To multiply (or divide) two real numbers with different signs, multiply (or divide) their absolute values, and make the sign of the product (or quotient) negative.

EXAMPLE 4 Find each product or quotient.

a. $(-4)(-7)$ **b.** $\left(-\frac{2}{3}\right)\left(\frac{4}{7}\right)$ **c.** $\frac{75}{-15}$

Solution

a. $(-4)(-7) = 28$ *Think:* $4 \cdot 7 = 28$, and the sign of the product is positive.

b. $\left(-\frac{2}{3}\right)\left(\frac{4}{7}\right) = -\frac{8}{21}$ *Think:* $\frac{2}{3} \cdot \frac{4}{7} = \frac{8}{21}$, and the sign of the product is negative.

c. $\frac{75}{-15} = -5$ *Think:* $\frac{75}{15} = 5$, and the sign of the quotient is negative.

PROGRESS CHECK 4 Find each product or quotient.

a. $4(-1.2)$ **b.** $\left(-\frac{1}{2}\right)\left(-\frac{6}{7}\right)$ **c.** $\frac{-48}{8}$

Multiplication and division involving the number 0 merit special consideration. First, the product of any number and zero is zero, so

$$a \cdot 0 = 0 \cdot a = 0$$

for any real number a. To illustrate the possible outcomes for a quotient involving zero, consider the cases of $0/2$, $2/0$, and $0/0$.

Progress Check Answers

3. (a) 7 (b) -7 (c) $-\frac{11}{10}$

4. (a) -4.8 (b) $\frac{3}{7}$ (c) -6

$\frac{0}{2} = 0$ because $0 = 2 \cdot 0$.

$\frac{2}{0}$ is undefined because no number times 0 is 2.

$\frac{0}{0}$ is undefined because any number times 0 is 0.

These examples illustrate the two rules we will use for a division involving 0.

Division Involving 0

1. 0 divided by any nonzero number is 0.
2. Division by 0 is undefined.

EXAMPLE 5 Find each product or quotient.

a. $0(-2)$ **b.** $\frac{0}{-2}$ **c.** $\frac{-2}{0}$ **d.** $\frac{-2(0)}{-2 + 2}$

Solution

a. $0(-2) = 0$ *Think:* The product of 0 and any nonzero number is 0.

b. $\frac{0}{-2} = 0$ *Think:* 0 divided by any nonzero number is 0.

c. $\frac{-2}{0}$ is undefined. *Think:* Division by 0 is undefined.

d. $\frac{-2(0)}{-2 + 2} = \frac{0}{0}$, which is undefined because division by 0 is undefined.

PROGRESS CHECK 5 Find each product or quotient.

a. $0 \div 0$ **b.** $0(0)$ **c.** $5/0$ **d.** $\frac{5 + (-5)}{-5}$

Two numbers are called **reciprocals** of each other if the product of the numbers is 1. For example, 7 and $\frac{1}{7}$ are reciprocals, and $-\frac{3}{5}$ and $-\frac{5}{3}$ are reciprocals, because

$$7 \cdot \tfrac{1}{7} = 1 \quad \text{and} \quad (-\tfrac{3}{5})(-\tfrac{5}{3}) = 1.$$

Because no number multiplied by 0 is 1, the number 0 has no reciprocal. These examples illustrate that the reciprocal of a nonzero number a is $1/a$, and the reciprocal of a nonzero fraction a/b is b/a. Using the concept of a reciprocal, division may be defined in terms of multiplication as follows.

Definition of Division

If a and b are real numbers with $b \neq 0$, then

$$a \div b = a \cdot \frac{1}{b}.$$

This definition says that to divide a by b, multiply a by the reciprocal of b.

Progress Check Answers

5. (a) Undefined (b) 0 (c) Undefined (d) 0

EXAMPLE 6 Divide using the reciprocal definition of division.

a. $48 \div (-3)$ **b.** $(-\frac{17}{25}) \div (-\frac{3}{5})$

Solution

a. $48 \div (-3) = 48(-\frac{1}{3})$ Product of 48 and the reciprocal of -3
$= -16$

b. $(-\frac{17}{25}) \div (-\frac{3}{5}) = (-\frac{17}{25}) \cdot (-\frac{5}{3})$ Product of $-\frac{17}{25}$ and the reciprocal of $-\frac{3}{5}$
$= \frac{17}{15}$

PROGRESS CHECK 6 Divide using the reciprocal definition of division.

a. $(-56) \div (-7)$ **b.** $(-\frac{7}{9}) \div (\frac{5}{12})$

2 In a product each number that is multiplied is called a **factor** of the product. Products often contain repeated factors, and an alternative way of writing an expression like $2 \cdot 2 \cdot 2$ is 2^3. We call 2^3 an **exponential expression** with **base** 2 and **exponent** 3. The number 2^3, or 8, is called the **third power** of 2. In general, the expression a^n, where n is a positive integer, means to use a as a factor n times.

Positive Integer Exponent

If a is a real number and n is a positive integer, then

$$a^n = \underbrace{a \cdot a \cdot a \cdots a}_{n \text{ factors}}.$$

When a product contains many factors, the following sign rules are helpful.

1. A product of nonzero factors is positive if the number of negative factors is even.
2. A product of nonzero factors is negative if the number of negative factors is odd.
3. A product is 0 if one or more factors is 0.

EXAMPLE 7 Multiply.

a. $(-2)^5$ **b.** $(-3)^4$ **c.** -3^4

Solution

a. $(-2)^5 = -32$ *Think:* $2^5 = 32$, and the sign of the product is negative because there are an odd number of negative factors.

b. $(-3)^4 = 81$ *Think:* $3^4 = 81$, and the sign of the product is positive because there are an even number of negative factors.

c. $-3^4 = -81$ *Think:* $3^4 = 81$, and the opposite of 3^4 is -81.

Caution In this example, note the difference in meaning of $(-3)^4$ and -3^4.

$$(-3)^4 = (-3)(-3)(-3)(-3)$$
$$-3^4 = -(3 \cdot 3 \cdot 3 \cdot 3)$$

In general, $(-a)^n$ denotes the nth power of $-a$, while $-a^n$ denotes the opposite of a^n.

PROGRESS CHECK 7 Multiply.

a. $(-2)^6$ **b.** -2^6 **c.** $(-3)^5$

Progress Check Answers

6. (a) 8 (b) $-\frac{28}{15}$

7. (a) 64 (b) -64 (c) -243

3 Without a priority for performing operations, different values may be possible for an expression involving more than one operation. To avoid this uncertainty, use this agreed-upon order of operations.

Order of Operations

1. Perform all operations within grouping symbols, like parentheses, first. If there is more than one symbol of grouping, simplify the innermost symbol of grouping first, and simplify the numerator and denominator of a fraction separately.
2. Evaluate powers of a number.
3. Multiply or divide working from left to right.
4. Add or subtract working from left to right.

A popular mnemonic for this order is Please Excuse My Dear Aunt Sally. See "Think About It" Exercise 1.

EXAMPLE 8 Evaluate $7 + 5(6 - 10)^2$.

Solution Follow the order of operations given above.

$$\begin{aligned} 7 + 5(6 - 10)^2 &= 7 + 5(-4)^2 && \text{Operate within parentheses.} \\ &= 7 + 5(16) && \text{Evaluate powers.} \\ &= 7 + 80 && \text{Multiply.} \\ &= 87 && \text{Add.} \end{aligned}$$

Stress that the addition is done last. A common student error is to add first.

PROGRESS CHECK 8 Evaluate $8 - 4(2 - 5)^3$.

When you simplify fractions, it is important to note that if a and b are any real numbers, with $b \neq 0$, then

$$-\frac{a}{b} = \frac{-a}{b} = \frac{a}{-b}.$$

For instance $-\frac{8}{2}$, $\frac{-8}{2}$, and $\frac{8}{-2}$ are all equal to -4. Simplified fractions are rarely written in the form $\frac{a}{-b}$.

EXAMPLE 9 Evaluate $\frac{-3(1 - 8)}{2 - 5^2}$.

Solution Simplify the numerator and the denominator independently using the order of operations.

$$\begin{aligned} \frac{-3(1 - 8)}{2 - 5^2} &= \frac{-3(-7)}{2 - 25} && \text{Operate within parentheses in the numerator; evaluate a power in the denominator.} \\ &= \frac{21}{-23} && \text{Multiply in the numerator; subtract in the denominator.} \\ &= -\frac{21}{23} && \frac{a}{-b} = -\frac{a}{b} \end{aligned}$$

PROGRESS CHECK 9 Evaluate $\frac{4 + 2^3}{5(4 - 15)}$.

Progress Check Answers

8. 116
9. $-\frac{12}{55}$

4 In applied problems involving positive and negative numbers, the following interpretations are usually associated with the signs of the numbers.

Positive direction	To the right, up, rise, above, forward, gain
Negative direction	To the left, down, fall, below, backward, loss

Like phrases are interpreted in a similar way. For instance, in the section-opening problem the ratings *drop* for the TV show is associated with a negative number.

EXAMPLE 10 Solve the problem in the section introduction on page 11.

Solution The new rating for the show is 13.8 and the old rating is 15.3. Substitute these values in the expression for percent change and then simplify.

$$\begin{aligned}\text{Percent change} &= \frac{13.8 - 15.3}{15.3} \cdot 100 \\ &= \frac{-1.5}{15.3} \cdot 100 \\ &= (-0.0980392) \cdot 100 \quad \text{Divide by calculator.} \\ &= -9.80392\end{aligned}$$

To the nearest hundredth of a percent, the percent change in ratings is -9.80 percent. This means the show lost 9.80 percent of viewing homes with respect to the previous week.

To find how many viewing homes were lost, the drop in ratings points is given by

$$13.8 - 15.3 = -1.5.$$

Each ratings point represents 921,000 homes, so the product of -1.5 and 921,000 gives the drop in viewing homes.

$$-1.5(921{,}000) = -1{,}381{,}500$$

To the nearest ten thousand homes, the show lost 1,380,000 viewing homes.

PROGRESS CHECK 10 Over the following week the rating for the TV show changed from 13.8 to 12.9. Redo the questions in Example 10 using these ratings values.

Progress Check Answer

10. -6.52 percent, $-830{,}000$

EXERCISES 1.2

In Exercises 1–4, select the correct word to describe the given expression (sum, difference, product, quotient).

1. $-1 + 5$ Sum **2.** $1-2(-3)$ Product
3. $7 - (-1)$ Difference **4.** $-8 \div 3$ Quotient

In Exercises 5–8, select the sketch that corresponds to the given sum.

5. $-2 + 1$ b

a.

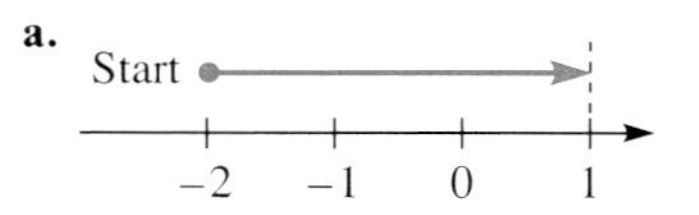

b.

6. $1 + (-2)$ b

a.

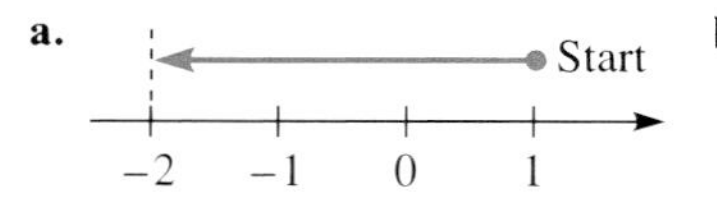

b.

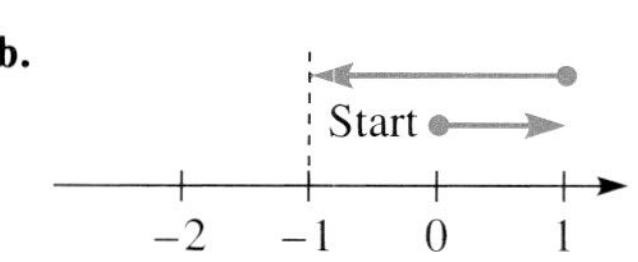

7. $-1 + (-2)$ a

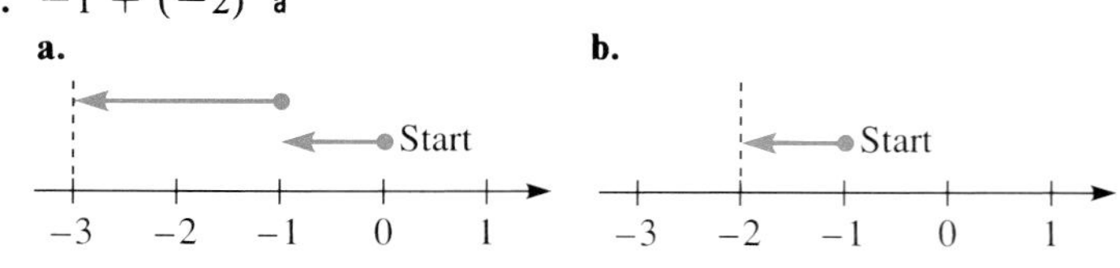

8. $-3 + 4$ b

a.

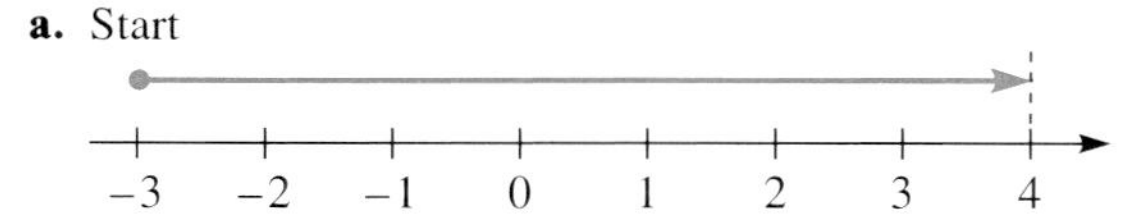

b.

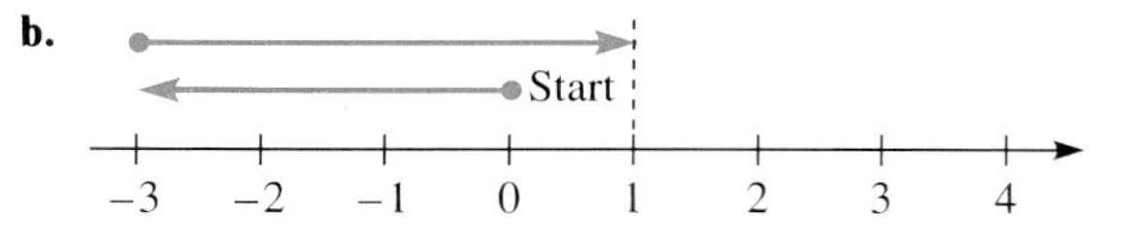

In Exercises 9–14, find the given sum by using the number line.

9. $-2 + 1$ -1 **10.** $-3 + 1$ -2 **11.** $2 + (-5)$ -3

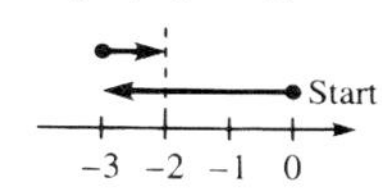

12. $3 + (-4)$ -1 **13.** $-3 + (-1)$ -4

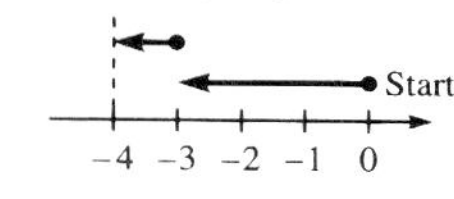

14. $-2 + (-4)$ -6

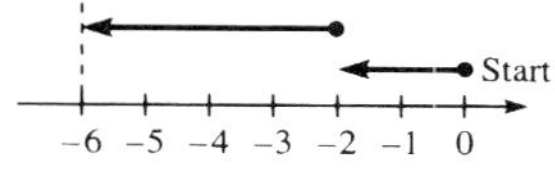

In Exercises 15–20, translate the expressions into English words. Use *minus* to indicate subtraction. Use *negative* to indicate sign. Use *opposite* for clarity when appropriate.

15. $1 - 5$ 1 minus 5 **16.** $-3 - 1$ Negative 3 minus 1
17. $3 - (-5)$ 3 minus negative 5
18. $-1 - (-3)$ Negative 1 minus negative 3
19. $-a - a$ Negative *a* minus *a*
20. $-(-4) - 2$ The opposite of negative 4 minus 2

In Exercises 21–24, translate the phrases into symbols.

21. 7 minus negative 3 $7 - (-3)$
22. Negative 3 minus 1 $-3 - 1$
23. Negative x minus negative 4 $-x - (-4)$
24. The opposite of negative x minus y $-(-x) - y$

In Exercises 25–44, find the given sum.

25. $-12 + 20$ 8 **26.** $-12 + 28$ 16
27. $-60 + 14$ -46 **28.** $-81 + 10$ -71
29. $-4 + (-97)$ -101 **30.** $-11 + (-12)$ -23
31. $-39 + (-13)$ -52 **32.** $-21 + (-8)$ -29
33. $1 + (-101)$ -100 **34.** $30 + (-47)$ -17
35. $57 + (-1)$ 56 **36.** $27 + (-18)$ 9
37. $-1.4 + 1.2$ -0.2 **38.** $-6.2 + (-3.6)$ -9.8
39. $\frac{2}{3} + (-\frac{5}{3})$ -1 **40.** $\frac{1}{9} + (-\frac{10}{9})$ -1
41. $-\frac{1}{2} + (-\frac{1}{2})$ -1 **42.** $-\frac{1}{3} + (-\frac{1}{3})$ $-\frac{2}{3}$
43. $\frac{1}{3} + (-\frac{1}{2})$ $-\frac{1}{6}$ **44.** $\frac{3}{4} + (-\frac{2}{3})$ $\frac{1}{12}$

In Exercises 45–50, rewrite each subtraction expression as an equivalent addition expression. Do not compute the answer.

45. $5 - (-2)$ $5 + 2$ **46.** $x - (-y)$ $x + y$
47. $-1 - (-2)$ $-1 + 2$ **48.** $-3 - (-4)$ $-3 + 4$
49. $x - y$ $x + (-y)$ **50.** $6 - 9$ $6 + (-9)$

In Exercises 51–72, find each difference.

51. $-1 - 6$ -7 **52.** $-6 - 1$ -7
53. $-2 - (-4)$ 2 **54.** $-4 - (-2)$ -2
55. $3 - (-7)$ 10 **56.** $4 - (-9)$ 13
57. $5 - 8$ -3 **58.** $1 - 10$ -9
59. $-21 - 10$ -31 **60.** $-11 - 17$ -28
61. $-42 - (-8)$ -34 **62.** $-12 - (-19)$ 7
63. $33 - (-47)$ 80 **64.** $71 - (-49)$ 120
65. $1.6 - (-1.6)$ 3.2 **66.** $-4.2 - (-1.7)$ -2.5
67. $-\frac{1}{4} - \frac{1}{3}$ $-\frac{7}{12}$ **68.** $-\frac{2}{3} - \frac{4}{5}$ $-\frac{22}{15}$
69. $-\frac{3}{4} - (-\frac{2}{3})$ $-\frac{1}{12}$ **70.** $-\frac{1}{5} - (-\frac{2}{3})$ $\frac{7}{15}$
71. $\frac{1}{6} - \frac{1}{3}$ $-\frac{1}{6}$ **72.** $\frac{3}{5} - \frac{9}{10}$ $-\frac{3}{10}$

73. Which of these statements is *not* always true?
a. The product of two positive numbers is positive.
b. The sum of two positive numbers is positive.
c. The quotient of two positive numbers is positive.
d. The difference of two positive numbers is positive. d

74. Which of these statements is *always* true?
a. The product of two negative numbers is negative.
b. The sum of two negative numbers is negative.
c. The quotient of two negative numbers is negative.
d. The difference of two negative numbers is negative. b

75. Which one of these is a product?
a. $(\frac{3}{4}) - (\frac{4}{3})$ **b.** $(\frac{3}{4})(-\frac{4}{3})$ b

76. Which one of these is a product?
a. $(-\frac{1}{2})(-\frac{3}{5})$ **b.** $(-\frac{1}{2}) - (\frac{3}{5})$ a

In Exercises 77–88, evaluate each expression.

77. $(-2)(-3)$ 6 **78.** $(-4)(1.8)$ -7.2
79. $\left(\frac{3}{4}\right)\left(-\frac{4}{3}\right)$ -1 **80.** $\frac{-16}{2}$ -8
81. $\frac{-12}{-2}$ 6 **82.** $11(-11)$ -121
83. $-1(14)$ -14 **84.** $-1(-1)$ 1
85. $0(-8)$ 0 **86.** $\frac{0}{-4}$ 0
87. $\frac{-3}{0}$ Undefined **88.** $\frac{3(0)}{-3 + 3}$ Undefined

In Exercises 89–92, write the reciprocal of each number.

89. -2 $-1/2$ **90.** $1/8$ 8 **91.** -1 -1 **92.** $\frac{0}{3}$ No reciprocal

In Exercises 93–96, rewrite each quotient as a product. Do not compute the answer.

93. $30 \div \frac{1}{3}$ $30(3)$ **94.** $\frac{3}{8} \div \left(-\frac{3}{8}\right)$ $\left(\frac{3}{8}\right)\left(-\frac{8}{3}\right)$
95. $\frac{3}{5} \div 2$ $\left(\frac{3}{5}\right)\left(\frac{1}{2}\right)$ **96.** $\frac{1}{2} \div (-2)$ $\left(\frac{1}{2}\right)\left(-\frac{1}{2}\right)$

In Exercises 97–104, divide using the reciprocal definition of division.

97. $27 \div \left(-\frac{1}{2}\right)$ -54 **98.** $-12 \div \frac{1}{3}$ -36
99. $\frac{1}{3} \div 2$ $\frac{1}{6}$ **100.** $\frac{1}{5} \div 4$ $\frac{1}{20}$
101. $21 \div \left(-\frac{3}{2}\right)$ -14 **102.** $-36 \div \frac{4}{5}$ -45
103. $-\frac{6}{10} \div \left(-\frac{3}{5}\right)$ 1 **104.** $-\frac{4}{3} \div \left(-\frac{2}{6}\right)$ 4

In Exercises 105–108, indicate whether the expression represents a positive or a negative number. Do not compute the number.

105. -20^6 Negative **106.** $(-20)^6$ Positive
107. $-\left(\frac{3}{5}\right)^2$ Negative **108.** $\left(-\frac{3}{5}\right)^2$ Positive

In Exercises 109–118, find the value of the given expression by multiplication.

109. $(-3)^2$ 9
110. -4^2 -16
111. -3^2 -9
112. $(-4)^2$ 16
113. $(-1)^7$ -1
114. $(-1)^6$ 1
115. -1^6 -1
116. -1^7 -1
117. $(-\frac{1}{2})^4$ $\frac{1}{16}$
118. $-(\frac{1}{2})^4$ $-\frac{1}{16}$

In Exercises 119–126, evaluate the given expression. Pay careful attention to the order of operations.

119. $5 - 2(5 - 2)$ -1
120. $6 - 3(6 - 3)$ -3
121. $3 + 2(5 - 7)^2$ 11
122. $1 + 3(2 - 5)^2$ 28
123. $-3 + 2(1 - 5)^3$ -131
124. $-4 + 3(5 - 8)^3$ -85
125. $\dfrac{-3(7 - 5)}{-5 - 2^2}$ $\frac{2}{3}$
126. $\dfrac{3 + 3^2}{2 - 2^2}$ -6

In Exercises 127–130, use the formula for percent change given in Example 10.

127. The population of a city decreased from 497,000 to 460,000 over a 10-year period. Express this as a percent change, to the nearest hundredth of a percent. -7.44 percent

128. A stock market index dropped from 3002 to 2843 in one day. Express this as a percent change, to the nearest hundredth of a percent. -5.30 percent

129. During a recession a worker took a pay cut from \$290 to \$270 per week. Express this as a percent change, to the nearest hundredth of a percent. -6.90 percent

130. During a drought the contents of a town's reservoir dropped from 80 percent full to 35 percent full. Express this drop as a percent change, to the nearest hundredth of a percent. -56.25 percent

THINK ABOUT IT

1. **a.** Explain how the acronym PEMDAS may be used to remember the agreed-upon order of operations.
   b. Show that $2 + 3x$ does not equal $5x$ by substituting 4 for x in both expressions. Explain in terms of the agreed-upon order of operations why some students mistakenly write $2 + 3x = 5x$.
2. Show the difference in meaning between $(-a)^n$ and $-a^n$ by writing each expression in factored form. In words, what does each expression represent?
3. **a.** Find values for s and t so that $|s + t| \neq |s| + |t|$.
   b. Find values for s and t so that $|s - t| \neq |s| - |t|$.
   c. Is it possible to find values for s and t so that $|st| \neq |s| \cdot |t|$?
4. When you divide two numbers, does it matter which one is inverted before multiplying? Which statements in this list are correct?
   a. $4 \div 2 = 4 \times \frac{1}{2}$
   b. $4 \div 2 = \frac{1}{4} \times 2$
   c. $\frac{1}{2} \div \frac{1}{4} = \frac{1}{2} \times 4$
   d. $\frac{1}{2} \div \frac{1}{4} = 2 \times \frac{1}{4}$
   e. $s \div t = s \times \dfrac{1}{t}$
   f. $s \div t = \dfrac{1}{s} \times t$
5. **a.** If $a = 3/5$, what is $1/a$?
   b. Rewrite each fraction using $\div$ (and parentheses, where necessary).
   $$\frac{2 - 5}{8 + 4}, \quad \frac{a}{b - 1}, \quad \frac{|a - 2|}{5}$$
   c. Rewrite each quotient in part **b** as a product.

REMEMBER THIS

1. To find the product of 3, 6, and 5, does it matter which two numbers are multiplied together first? No
2. If $x = -3$ and $y = 8$, find $x + y$. 5
3. If $a = -2$, $b = 3$, and $c = -4$, find $a + (b + c)$. -3
4. Which is greater, $(2 \cdot 3)(-4)$ or $2 \cdot 3 - 4$? $2 \cdot 3 - 4$
5. True or false? $17(29 + 31) = 17(29) + 17(31)$. True
6. Write an expression which represents the product of a and $7 + y$. $a(7 + y)$
7. Graph all the even integers from -4 to 2.

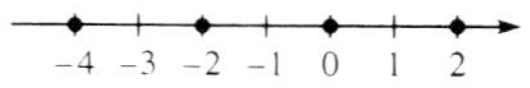

8. Arrange these real numbers in increasing order: $\{|-3|, \sqrt{5}, -1, 0, \frac{5}{3}\}$. $-1, 0, \frac{5}{3}, \sqrt{5}, |-3|$
9. True or false? All these numbers are rational: $\{-1, \frac{2}{3}, 0, \sqrt{4}, 1.6\}$. True
10. True or false? If $A = \{1,2,3\}$ and $B = \{3,2,1\}$, then $A \cap B = A \cup B$. True

1.3 Properties of Real Numbers

If subtractions for doors and windows are disregarded, then a paperhanger can choose between the formulas

$$A = 8a + 8b + 8c + 8d$$

or

$$A = 8(a + b + c + d)$$

to find the total wall area in the illustrated room. The two right-hand expressions in the formulas are equal according to which property of real numbers? (See Example 8.)

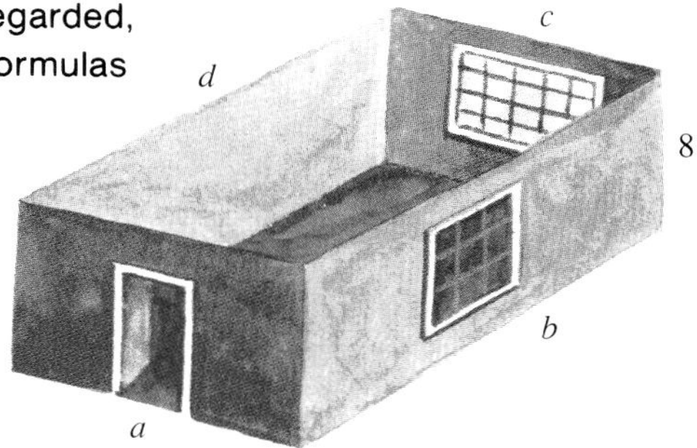

OBJECTIVES

1. **Identify and use commutative and associative properties.**
2. **Identify and use identity and inverse properties.**
3. **Identify and use the distributive property.**
4. **Use $-a = -1 \cdot a$ and the distributive property to remove parentheses.**

Unless it is stated otherwise, it is assumed in algebra that a symbol like x may be replaced by any real number. Consequently, the rules that govern real numbers determine the methods of computation in algebra. In this section we discuss the basic properties of real numbers with respect to addition and multiplication. As we develop the rules of algebra, keep in mind that they must always be in agreement with the properties of real numbers.

1 When two numbers are added or multiplied, the order in which they are written does not affect the result. These properties are called **commutative properties.**

Commutative Properties

For any real numbers a and b,

$$a + b = b + a \quad \text{Commutative property of addition}$$
$$ab = ba. \quad \text{Commutative property of multiplication}$$

"Think About It" Exercise 1 is a dictionary assignment for the meaning of *commute, associate,* and *distribute.*

Both commutative properties are illustrated in Example 1.

EXAMPLE 1 Use $x = -3$ and $y = 8$ to illustrate the following.

a. $x + y = y + x$ **b.** $xy = yx$

Solution

a. Replace x by -3 and y by 8, and find each sum separately.

$$x + y = -3 + 8 = 5$$
$$y + x = 8 + (-3) = 5$$

Thus, $x + y = y + x$ for the given values.

b. Replace x by -3 and y by 8, and find each product separately.

$$xy = -3(8) = -24$$
$$yx = 8(-3) = -24$$

Thus, $xy = yx$ if $x = -3$ and $y = 8$.

Note In this example replacing x and y by given values utilizes the **substitution property,** which states that if a and b are real numbers and $a = b$, then either may replace the other without affecting the truth or falsity of the statement.

PROGRESS CHECK 1 Use $x = -2$ and $y = -3$ to illustrate the following.

a. $x + y = y + x$ **b.** $xy = yx$

Another important pair of properties say that we obtain the same result if we change the grouping of numbers in an addition problem or in a multiplication problem. These properties are called **associative properties.**

Associative Properties

For any real numbers a, b, and c,

$$(a + b) + c = a + (b + c) \quad \text{Associative property of addition}$$
$$(ab)c = a(bc). \quad \text{Associative property of multiplication}$$

Example 2 illustrates these two properties.

EXAMPLE 2 Use $a = -3$, $b = 2$, and $c = 7$ to illustrate the following.

a. $(a + b) + c = a + (b + c)$ **b.** $(ab)c = a(bc)$

Solution According to the order of operations, perform all operations within parentheses first.

a. Find each sum separately, using the given values.

$$(a + b) + c = (-3 + 2) + 7 = -1 + 7 = 6$$
$$a + (b + c) = -3 + (2 + 7) = -3 + 9 = 6$$

Thus, for the specified numbers, $(a + b) + c = a + (b + c)$.

b. Find each product separately, using the given values.

$$(ab)c = (-3 \cdot 2) \cdot 7 = (-6) \cdot 7 = -42$$
$$a(bc) = -3 \cdot (2 \cdot 7) = -3 \cdot (14) = -42$$

Thus, $(ab)c = a(bc)$ if $a = -3$, $b = 2$, and $c = 7$.

PROGRESS CHECK 2 Use $a = 4$, $b = -8$, and $c = 6$ to illustrate the following.

a. $(a + b) + c = a + (b + c)$ **b.** $(ab)c = a(bc)$

The next example emphasizes the distinction between commutative properties and associative properties. Remember that in sums and products, commutative properties permit a change in written order, while associative properties permit a change in grouping.

EXAMPLE 3 Label each statement as an example of one of the commutative properties or one of the associative properties.

a. $8(xy) = 8(yx)$ **b.** $(x + 5) + 1 = x + (5 + 1)$

Progress Check Answers

1. (a) Both equal -5. (b) Both equal 6.
2. (a) Both equal 2. (b) Both equal -192.

Solution

a. This statement illustrates the commutative property of multiplication because the factors x and y are written in different orders on the two sides of the equal sign.

b. This statement illustrates the associative property of addition. On the left side of the equal sign parentheses group x and 5, but on the right side they group 5 and 1.

PROGRESS CHECK 3 Label each statement as an example of one of the commutative properties or one of the associative properties.

a. $2(ax) = (2a)x$ **b.** $3 + (x + 7) = 3 + (7 + x)$

2 Identity and inverse properties are also basic properties of the real numbers with respect to addition and multiplication. These properties are formally stated next.

Identity and Inverse Properties

	Addition	**Multiplication**
Identity Properties	There is a unique real number 0 such that for every real number a $a + 0 = 0 + a = a$.	There is a unique real number 1 such that for every real number a $a \cdot 1 = 1 \cdot a = a$.
Inverse Properties	For every real number a, there is a unique real number, denoted by $-a$, such that $a + (-a) = (-a) + a = 0$.	For every real number a except zero, there is a unique real number, denoted by $1/a$, such that $a \cdot \frac{1}{a} = \frac{1}{a} \cdot a = 1$.

Because of these properties, 0 is called the additive identity, 1 is called the multiplicative identity, $-a$ is called the additive inverse (or opposite) of a, and $1/a$ is called the multiplicative inverse (or reciprocal) of a.

EXAMPLE 4 Use one of the identity or inverse properties to simplify each statement, and indicate the property that was used.

a. $0 + (-8)$ **b.** $-9 \cdot -\frac{1}{9}$ **c.** $-2 \cdot 1$ **d.** $18 + (-18)$

Solution

a. By the addition identity property, $0 + (-8) = -8$.
b. By the multiplication inverse property, $-9 \cdot -\frac{1}{9} = 1$.
c. By the multiplication identity property, $-2 \cdot 1 = -2$.
d. By the addition inverse property, $18 + (-18) = 0$.

PROGRESS CHECK 4 Use one of the identity or inverse properties to simplify each statement, and indicate the property that was used.

a. $1 \cdot (-6)$ **b.** $-6 + 6$ **c.** $-6 + 0$ **d.** $\frac{1}{6} \cdot 6$

3 Up to this point each property has involved *only* addition or *only* multiplication. The crucial property that involves *both* addition and multiplication is the **distributive property.** To illustrate this property, consider Figure 1.12 (page 24) and note that the area of the large rectangle may be expressed either as a product or as a sum.

Progress Check Answers

3. (a) Associative, multiplication (b) Commutative, addition

4. (a) -6, identity, multiplication (b) 0, addition, inverse (c) -6, identity, addition (d) 1, multiplication inverse

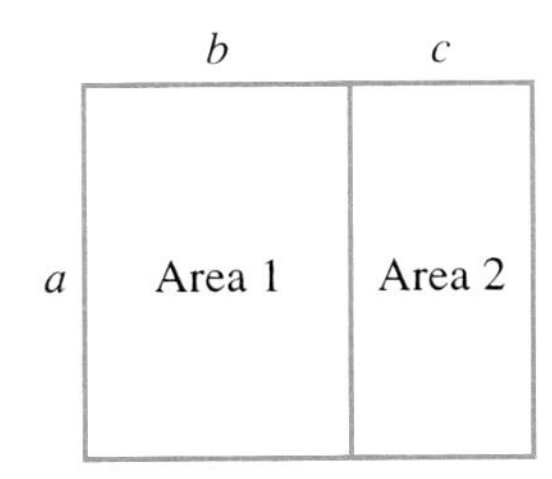

Figure 1.12

$$\text{Area as a product} = \text{length} \cdot \text{width} = a(b + c)$$
$$\text{Area as a sum} = \text{area 1} + \text{area 2} = ab + ac$$

Because both expressions represent the same number, we have

$$a(b + c) = ab + ac.$$

This equation, which relates a product and a sum, expresses the distributive property.

More formally, this is the distributive property of "multiplication over addition." See "Think About It" Exercise 2 for another distributive property.

Distributive Property

For any real numbers a, b, and c,

$$a(b + c) = ab + ac.$$

Example 5 illustrates the distributive property in arithmetic terms.

EXAMPLE 5 If $a = 3$, $b = 4$, and $c = 5$, show that $a(b + c) = ab + ac$.

Solution Evaluate the expression first on the left of the equal sign and then on the right.

$$a(b + c) = 3(4 + 5) = 3(9) = 27$$
$$ab + ac = 3(4) + 3(5) = 12 + 15 = 27$$

Thus, $a(b + c) = ab + ac$ for the given values of a, b, and c.

PROGRESS CHECK 5 If $a = 5$, $b = 6$, and $c = 7$, show that $a(b + c) = ab + ac$.

Using properties and definitions from this chapter, the distributive property may take alternative forms. For instance, by the commutative property of multiplication,

$$a(b + c) = ab + ac \quad \text{may be expressed as} \quad (b + c)a = ba + ca,$$

and because subtraction is defined in terms of addition, we have

$$a(b - c) = ab - ac \quad \text{and} \quad (b - c)a = ba - ca.$$

The distributive property also can be extended to forms like

$$a(b + c + d + \cdots + n) = ab + ac + ad + \cdots + an.$$

An important application of the distributive property is to multiply and remove parentheses, as shown in the next example.

EXAMPLE 6 Multiply using the distributive property.

a. $5(x + 4)$ **b.** $(y - 3)4$ **c.** $-3(5x + 2)$ **d.** $-1(x - y + 5)$

Solution

a. $5(x + 4) = 5 \cdot x + 5 \cdot 4 = 5x + 20$
b. $(y - 3)4 = y \cdot 4 - 3 \cdot 4 = 4y - 12$
c. $-3(5x + 2) = -3(5x) + (-3)(2) = -15x - 6$
d. $-1(x - y + 5) = -1 \cdot x + (-1)(-y) + (-1)(5) = -x + y - 5$

PROGRESS CHECK 6 Multiply using the distributive property.

a. $8(x - 9)$ **b.** $(x + 10)5$ **c.** $-4(3x + 1)$ **d.** $-2(y + x - 1)$

Progress Check Answers

5. Both equal 65.
6. (a) $8x - 72$ (b) $5x + 50$ (c) $-12x - 4$ (d) $-2y - 2x + 2$

The previous example had the effect of changing products into sums. But because all steps in that example are reversible, the distributive property can also be used to rewrite certain sums (or differences) as products. For this purpose we use the distributive property in a form like

$$ab + ac = a(b + c).$$

EXAMPLE 7 Use the distributive property to rewrite each expression as a product.

a. $7 \cdot 3 + 7 \cdot 4$ **b.** $dx + dy$ **c.** $11x - 2x$

Solution

a. $7 \cdot 3 + 7 \cdot 4 = 7(3 + 4)$
b. $dx + dy = d(x + y)$
c. $11x - 2x = (11 - 2)x$

PROGRESS CHECK 7 Use the distributive property to rewrite each expression as a product.

a. $4 \cdot 15 + 4 \cdot 7$ **b.** $19x - 19y$ **c.** $2y - 7y$

EXAMPLE 8 Solve the problem in the section introduction on page 21.

Solution According to the extended form of the distributive property,

$$8a + 8b + 8c + 8d \quad \text{and} \quad 8(a + b + c + d)$$

are equal. Both area formulas will always give the same result, but one may be more convenient than the other in a specific application.

PROGRESS CHECK 8 Alternative formulas for the sum of the integers from 1 to n are $S = \frac{1}{2}n(1 + n)$ and $S = \frac{1}{2}n(n + 1)$. The equality of $\frac{1}{2}n(1 + n)$ and $\frac{1}{2}n(n + 1)$ illustrates which property of real numbers?

4 There are two simple ways to find the opposite of a number. The first way is to simply change its sign, and the second is to multiply it by -1. In symbols, the second method states that for any real number a,

$$-a = -1 \cdot a.$$

This property, when used with the distributive property, determines the procedure for removing parentheses that are preceded by a negative sign.

EXAMPLE 9 Rewrite each statement without parentheses.

a. $-(3x + 1)$ **b.** $-(2 - y)$

Solution

a. $-(3x + 1) = -1 \cdot (3x + 1)$ $-a = -1 \cdot a.$
$= -1(3x) + (-1)(1)$ Distributive property.
$= -3x - 1$ Simplify.

b. $-(2 - y) = -1 \cdot (2 - y)$ $-a = -1 \cdot a.$
$= -1(2) + (-1)(-y)$ Distributive property.
$= -2 + y$ Simplify.

PROGRESS CHECK 9 Rewrite each statement without parentheses.

a. $-(10x - 17)$ **b.** $-(-13 + y)$

Progress Check Answers

7. (a) $4(15 + 7)$ (b) $19(x - y)$ (c) $(2 - 7)y$
8. Commutative property of addition
9. (a) $-10x + 17$ (b) $13 - y$

EXERCISES 1.3

For Exercises 1–4, use the given values to illustrate the following.

a. $x + y = y + x$ **b.** $xy = yx$
c. $x - y \neq y - x$ **d.** $x \div y \neq y \div x$

1. $x = -2, y = 6$
a. $x + y = y + x = 4$
b. $xy = yx = -12$
c. $x - y = -8, y - x = 8$
d. $x \div y = -\frac{1}{3}, y \div x = -3$

2. $x = 6, y = -3$
a. $x + y = y + x = 3$
b. $xy = yx = -18$
c. $x - y = 9, y - x = -9$
d. $x \div y = -2, y \div x = -\frac{1}{2}$

3. $x = -\frac{1}{2}, y = -\frac{1}{3}$
a. $x + y = y + x = -\frac{5}{6}$
b. $xy = yx = \frac{1}{6}$
c. $x - y = -\frac{1}{6}, y - x = \frac{1}{6}$
d. $x \div y = \frac{3}{2}, y \div x = \frac{2}{3}$

4. $x = -\frac{2}{3}, y = -\frac{3}{4}$
a. $x + y = y + x = -\frac{17}{12}$
b. $xy = yx = \frac{1}{2}$
c. $x - y = \frac{1}{12}, y - x = -\frac{1}{12}$
d. $x \div y = \frac{8}{9}, y \div x = \frac{9}{8}$

In Exercises 5–8, use the given values to illustrate the following.

a. $(a + b) + c = a + (b + c)$
b. $(ab)c = a(bc)$
c. $(a - b) - c \neq a - (b - c)$
d. $(a \div b) \div c \neq a \div (b \div c)$

5. $a = 12, b = -4, c = 3$
a. $(a + b) + c = a + (b + c) = 11$
b. $(ab)c = a(bc) = -144$
c. $(a - b) - c = 13, a - (b - c) = 19$
d. $(a \div b) \div c = -1, a \div (b \div c) = -9$

6. $a = -6, b = 3, c = 2$
a. $(a + b) + c = a + (b + c) = -1$
b. $(ab)c = a(bc) = -36$
c. $(a - b) - c = -11, a - (b - c) = -7$
d. $(a \div b) \div c = -1, a \div (b \div c) = -4$

7. $a = \frac{1}{2}, b = \frac{1}{3}, c = -\frac{1}{4}$
a. $(a + b) + c = a + (b + c) = \frac{7}{12}$
b. $(ab)c = a(bc) = -\frac{1}{24}$
c. $(a - b) - c = \frac{5}{12}, a - (b - c) = -\frac{1}{12}$
d. $(a \div b) \div c = -6, a \div (b \div c) = -\frac{3}{8}$

8. $a = \frac{2}{3}, b = -\frac{3}{4}, c = \frac{1}{2}$
a. $(a + b) + c = a + (b + c) = \frac{5}{12}$
b. $(ab)c = a(bc) = -\frac{1}{4}$
c. $(a - b) - c = \frac{11}{12}, a - (b - c) = \frac{23}{12}$
d. $(a \div b) \div c = -\frac{16}{9}, a \div (b \div c) = -\frac{4}{9}$

In Exercises 9–16, label each statement as an example of one of the commutative or one of the associative properties.

9. $(3a)x = 3(ax)$ Associative, multiplication
10. $3(ax) = 3(xa)$ Commutative, multiplication
11. $(a + x) + y = (x + a) + y$ Commutative, addition
12. $(a + x) + y = a + (x + y)$ Associative, addition
13. $(3a + 4) + x = x + (3a + 4)$ Commutative, addition
14. $(xa + b) + y = (ax + b) + y$ Commutative, multiplication
15. $(ax + y) + z = ax + (y + z)$ Associative, addition
16. $a[(xy)m] = [a(xy)]m$ Associative, multiplication
17. What is the sum of the number a and its additive inverse? 0
18. What is the product of the number b and its multiplicative inverse? Assume $b \neq 0$. 1
19. What is another name for the additive inverse of x?
Negative or opposite of x
20. What is another name for the multiplicative inverse of y?
Reciprocal of y
21. What is the sum of the number n and the additive identity? n
22. What is the product of the number m and the multiplicative identity? m
23. If $a - b = 7$, what is the value of $b - a$? -7
24. If $a \div b = 5$, what is the value of $b \div a$? $\frac{1}{5}$

In Exercises 25–30, use one of the additive or inverse properties to simplify the given statement. Indicate the property used.

25. $-4 + 0$ -4; addition identity
26. $-7 \cdot 1$ -7; multiplication identity
27. $(-\frac{4}{5}) \cdot (-\frac{5}{4})$ 1; multiplication inverse
28. $(-\frac{3}{5}) + (\frac{3}{5})$ 0; addition inverse
29. $2 + (-2)$ 0; addition inverse
30. $-2 \cdot (-\frac{1}{2})$ 1; multiplication inverse

In Exercises 31–34, use the given values to illustrate the following.

a. $x(y + z) = xy + xz$
b. $x(y - z) = xy - xz$

31. $x = -1, y = 2, z = -3$ a. $1 = 1$ b. $-5 = -5$
32. $x = 3, y = -2, z = -3$ a. $-15 = -15$ b. $3 = 3$
33. $x = -2, y = -4, z = 4$ a. $0 = 0$ b. $16 = 16$
34. $x = \frac{1}{2}, y = 3, z = 3$ a. $3 = 3$ b. $0 = 0$

In Exercises 35–42, multiply using the distributive property.

35. $4(x + 5)$ $4x + 20$
36. $(x - 8)2$ $2x - 16$
37. $-2(3x + 1)$ $-6x - 2$
38. $-3(2x + 1)$ $-6x - 3$
39. $-1(x - y + 2)$ $-x + y - 2$
40. $-1(-x + y - 5)$ $x - y + 5$
41. $\frac{1}{3}(3x - y)$ $x - \frac{1}{3}y$
42. $-\frac{2}{3}(6x + 3)$ $-4x - 2$

In Exercises 43–52, use the distributive property to rewrite each expression as a product.

43. $3 \cdot 8 + 3 \cdot 9$ $3(8 + 9)$
44. $2 \cdot 7 - 2 \cdot 5$ $2(7 - 5)$
45. $8x + 11x$ $(8 + 11)x$
46. $-3y + 2y$ $(-3 + 2)y$
47. $aw + a \cdot 1$ $a(w + 1)$
48. $bn + b$ $b(n + 1)$
49. $3a + 3b + 3c$ $3(a + b + c)$
50. $7a + 2a + 3a$ $(7 + 2 + 3)a$
51. $2x - x$ $(2 - 1)x$
52. $-5y - 2y$ $(-5 - 2)y$

In Exercises 53–56 two versions of a formula are given. Name the property of real numbers according to which the two versions are equivalent.

53. $A = \frac{1}{2}(bh)$ and $A = (\frac{1}{2}b)h$ Associative, multiplication
54. $P = 2\ell + 2w$ and $P = 2(\ell + w)$ Distributive
55. $C = \frac{5}{9}(F - 32)$ and $C = \frac{5}{9}F - \frac{5}{9}(32)$ Distributive
56. $A = \frac{1}{2}[(a + b)h]$ and $A = \frac{1}{2}[h(a + b)]$ Commutative, multiplication

57. To find the area of a triangle, you can multiply half the base times the height, or you can multiply half the height times the base. That is, $A = (\frac{1}{2}b)h$ and $A = (\frac{1}{2}h)b$ are both correct formulas. To show that the formulas are equivalent, a series of steps is needed. Name the property used at each step.
$A = (\frac{1}{2}b)h$ is equivalent to $A = \frac{1}{2}(bh)$. Associative, multiplication
$A = \frac{1}{2}(bh)$ is equivalent to $A = \frac{1}{2}(hb)$. Commutative, multiplication
$A = \frac{1}{2}(hb)$ is equivalent to $A = (\frac{1}{2}h)b$. Associative, multiplication

58. To find the sum of the integers from 1 to n, you can multiply the average of the first and last integers by n, or you can multiply the sum of the first and last by half of n. That is, $S = [\frac{1}{2}(1 + n)]n$ and $S = (1 + n)(\frac{1}{2}n)$ are both correct formulas. Show that the formulas are equivalent by naming the property used at each step shown.
$S = [\frac{1}{2}(1 + n)]\ n$ is equivalent to $S = \frac{1}{2}[(1 + n)n]$.
Associative, multiplication
$S = \frac{1}{2}[(1 + n)n]$ is equivalent to $S = \frac{1}{2}[n(1 + n)]$.
Commutative, multiplication
$S = \frac{1}{2}[n(1 + n)]$ is equivalent to $S = (\frac{1}{2}n)(1 + n)$.
Associative, multiplication
$S = (\frac{1}{2}n)(1 + n)$ is equivalent to $S = (1 + n)(\frac{1}{2}n)$
Commutative, multiplication

In Exercises 59–68, rewrite each statement without parentheses.

59. $-(a + b)$ $-a - b$
60. $-(3x + 4)$ $-3x - 4$
61. $-(1 - x)$ $-1 + x$
62. $-(3 - 2y)$ $-3 + 2y$
63. $-(-7 + 3n)$ $7 - 3n$
64. $-(-1 + w)$ $1 - w$
65. $-(2x - 3y + 4z)$ $-2x + 3y - 4z$
66. $-(x - y - z)$ $-x + y + z$
67. $-(x - 2y - 3z)$ $-x + 2y + 3z$
68. $-(-1 + x - 5n)$ $1 - x + 5n$

THINK ABOUT IT

1. Use a dictionary to find meanings for the words *commute*, *associate*, and *distribute* that are in agreement with the concepts expressed by the commutative, associative, and distributive properties stated in this section.
2. **a.** If $A = \{1,2,3\}$, $B = \{3,4,5\}$ and $C = \{0,2,4\}$, show that
$$A \cap (B \cup C) = (A \cap B) \cup (A \cap C)$$
and
$$A \cap (B \cap C) = (A \cap B) \cap C.$$
b. Name the properties illustrated in part **a.**
3. The operation * on the set {T,E,R,M} is defined in the following table.

*	T	E	R	M
T	R	M	T	E
E	M	T	E	R
R	T	E	R	M
M	E	R	M	T

a. Which element is the identity element?
b. Find the inverse of element M.

4. A common mental shortcut for computing a 15 percent tip at a restaurant is to first find 10 percent and then 5 percent, as shown in this illustration.

Bill = \$24
10 percent of bill = 2.40
5 percent of bill = 1.20
Tip = 3.60

This works because $0.10x + 0.05x = 0.15x$.
a. Which property of real numbers is used for this technique?
b. Use this technique to find 15 percent tips on these meals. (Try to do it mentally.) Meal cost: \$50, \$8, \$36, \$90
5. There are 6 different orders in which you can write the product of the real numbers a, b, and c. Each order has 2 possible groupings. For instance $(ab)c$, $a(bc)$, $(ac)b$, and $a(cb)$ are 4 of the 12 possibilities.
a. Write out all 12 possibilities.
b. Let $a = 3$, $b = 4$, $c = 5$ and show that all 12 arrangements have the same value.
c. What properties of real numbers are illustrated in these examples?

REMEMBER THIS

1. Do ab and $a + b$ represent the same calculation with a and b? No
2. Do a/b and $a \div b$ represent the same calculation with a and b? Yes
3. Evaluate $6x + 4x$ when $x = 7$. Evaluate $10x$ when $x = 7$. Are both answers the same? Yes; both are 70.
4. Evaluate $5 - (4 + 3)$ and $5 - 4 - 3$. Are they equal? Yes; both are -2.
5. The amount of space inside a square is called its (area; volume)? Area
6. Calculate $-1 + 87$. 86
7. Calculate $-1(87)$. -87
8. If the product of m and n is 1, then what name is given to m and n? Reciprocals
9. Multiply -11^2. -121
10. Evaluate $1 - 2(4 - 3)^2$. -1

1.4 Algebraic Expressions and Calculator Computation

If the annual inflation rate for the price of a new boat remains fixed at 6 percent, then a new boat that costs \$16,000 today will cost $16{,}000\,(1 + 0.06)^{15}$ dollars in 15 years. Use a scientific calculator to evaluate this expression and predict the price in 15 years to the nearest dollar. (See Example 6.)

OBJECTIVES

1. **Simplify algebraic expressions by combining like terms.**
2. **Evaluate algebraic expressions.**
3. **Evaluate an expression on a scientific calculator.**
4. **Apply geometry formulas.**
5. **Translate between verbal expressions and algebraic expressions.**

1 The symbols which represent numbers in algebra are classified into two types: variables and constants. A **variable** is a symbol that may be replaced by different numbers in a particular problem, while a **constant** is a symbol that represents the same number throughout a particular problem. An expression that combines variables and constants using the operations of arithmetic is called an **algebraic expression.** For instance,

$$2\ell + 2w, \qquad x^2 - 4x + 5, \qquad \text{and} \qquad \tfrac{1}{2}bh$$

are algebraic expressions. Those parts of an algebraic expression separated by plus (+) signs are called **terms** of the expression, and the terms in the above expressions are identified in Example 1.

EXAMPLE 1 Identify the term(s) in the following expressions.

a. $2\ell + 2w$ **b.** $x^2 - 4x + 5$ **c.** $\frac{1}{2}bh$

Solution

a. $2\ell + 2w$ is an algebraic expression with two terms, 2ℓ and $2w$.
b. $x^2 - 4x + 5$ may be written as $x^2 + (-4x) + 4$ and is an algebraic expression with three terms, x^2, $-4x$, and 5.
c. Because $\frac{1}{2}bh$ is a product (instead of a sum or difference), it is an algebraic expression with one term, $\frac{1}{2}bh$.

PROGRESS CHECK 1 Identify the term(s) in the following expressions.

a. $1 - x$ **b.** πr^2 **c.** $-16t^2 + 96t + 144$

If a term is a product of constants and variables, then the constant factor is called the (**numerical**) **coefficient** of the term. For example, the coefficient of $2w$ is 2, the coefficient of x^2 is 1 (since $x^2 = 1 \cdot x^2$), and the coefficient of $-x$ is -1 (since

Progress Check Answers

1. (a) $1, -x$ (b) πr^2 (c) $-16t^2, 96t, 144$

$-x = -1 \cdot x$). If two terms have exactly the same variables with the same exponents (such as $5xy^2$ and $-8xy^2$), they are called **like terms.** The distributive property specifies that like terms may be combined by adding their numerical coefficients and that only like terms may be combined.

EXAMPLE 2 Simplify by combining like terms.

a. $2x + 5x$ **b.** $-y - 8y$ **c.** $6x - y + x + 5y$

Solution

a. $2x + 5x = (2 + 5)x = 7x$
b. $-y - 8y = (-1 - 8)y = -9y$
c. $6x - y + x + 5y = (6 + 1)x + (-1 + 5)y = 7x + 4y$

Note For the simplification in part **c,** remember that we may reorder and regroup terms in any useful way because of the commutative and associative properties of addition, and that this is usually done *mentally*.

PROGRESS CHECK 2 Simplify by combining like terms.

a. $6y + 4y$ **b.** $-3b - b$ **c.** $8x - 3y + 6x - 5y$

Sometimes, it is necessary to remove the parentheses which group together certain terms in order to simplify algebraic expressions. Parentheses are removed by applying the distributive property, as discussed in Section 1.3. When there is more than one symbol of grouping, it is usually better to remove the innermost symbol of grouping first.

EXAMPLE 3 Remove the symbols of grouping and combine like terms.

a. $-(5 - x) + 3(2x + 1)$ **b.** $10x - 4[3x - (1 - x)]$

Solution

a. $-(5 - x) + 3(2x + 1)$
$= -1(5 - x) + 3(2x + 1)$ Use $-(5 - x) = -1(5 - x)$.
$= -5 + x + 6x + 3$ Distributive property.
$= 7x - 2$ Combine like terms.

b. $10x - 4[3x - (1 - x)]$
$= 10x - 4[3x - 1 + x]$ Distributive property (remove parentheses).
$= 10x - 4[4x - 1]$ Combine like terms inside the brackets.
$= 10x - 16x + 4$ Distributive property (remove brackets).
$= -6x + 4$ Combine like terms.

PROGRESS CHECK 3 Remove the symbols of grouping and combine like terms.

a. $-(3 - x) + 2(4x - 3)$ **b.** $6x - 2[4x - (1 - 3x)]$

2 When the numerical values of the symbols in an algebraic expression are known, then the expression can be evaluated by substituting the known values and performing the indicated operations.

EXAMPLE 4 Evaluate each expression given that $x = 3$ and $y = -8$.

a. $2x - y$ **b.** $x^2 + 5y^2$ **c.** $\dfrac{y + 6}{x - 1}$

Solution Substitute 3 for x and -8 for y, and then simplify.

a. $2x - y = 2 \cdot 3 - (-8)$ Replace x by 3 and y by -8.
$= 6 + 8$ Multiply and remove parentheses.
$= 14$ Add.

Progress Check Answers

2. (a) $10y$ (b) $-4b$ (c) $14x - 8y$
3. (a) $9x - 9$ (b) $-8x + 2$

b. $x^2 + 5y^2 = 3^2 + 5 \cdot (-8)^2$ Replace x by 3 and y by -8.
$= 9 + 5 \cdot 64$ Evaluate powers.
$= 9 + 320$ Multiply.
$= 329$ Add.

c. $\dfrac{y + 6}{x - 1} = \dfrac{-8 + 6}{3 - 1}$ Replace x by 3 and y by -8.
$= \dfrac{-2}{2}$ Simplify the numerator and denominator independently.
$= -1$ Divide.

PROGRESS CHECK 4 Evaluate each expression given that $x = 2$ and $y = -5$.

a. $y - 2x$ **b.** $3x^2 + y^2$ **c.** $\dfrac{6x}{8 - y}$

Note that keystroke sequences are not the same for scientific calculators and graphing calculators.

3 Evaluating expressions using a scientific calculator can be fast and accurate, but the algebraic rules that have been programmed into the calculator must be understood. A calculator using the algebraic operating system (AOS) calculates in the following order.

1. Keys that operate on the single number in the display are done immediately. Such keys are called function keys, and common function keys used in this course include square [x^2], square root [$\sqrt{\ }$], reciprocal [$1/x$], sign change [$+/-$], logarithmic [log], and exponential [10^x].
2. Powers [y^x] or [x^y] and roots [$\sqrt[x]{y}$] or [$y^{1/x}$] are calculated after the function keys.
3. Multiplication [×] and division [÷] have the next priority.
4. Addition [+] and subtraction [−] come last.

The equals key [=] completes all operations. Parentheses keys [(], [)] can be used to change these built-in priorities when required. Pressing close parentheses [)] automatically completes any operations done after the previous open parentheses [(]. Consider carefully the logic involved in the following examples, and keep in mind that in many cases other keystroke sequences are possible.

EXAMPLE 5 Evaluate on a scientific calculator.

a. $3(4 + 5)$ **b.** $\dfrac{5^2 - 4}{3^2 - 6}$

Solution

a. $3(4 + 5) = 27$. There are two common approaches for evaluating this expression on a calculator.

Method 1 Input the problem following the rules for order of operations. Begin inside the parentheses and add 4 and 5; then multiply by 3.

4 [+] 5 [=] [×] 3 [=] [27]

Method 2 Input the problem as it appears, including the parentheses, and let the calculator compute according to the agreed-upon order of operations. Note that although the multiplication symbol is omitted in the problem statement, it must be keyed in for the calculator sequence.

3 [×] [(] 4 [+] 5 [)] [=] [27]

Progress Check Answers

4. (a) -9 (b) 37 (c) $\frac{12}{13}$

b. To evaluate a fraction, divide the numerator by the denominator. Use the square key $\boxed{x^2}$ to evaluate the second power of a number. Two methods are given.

Method 1 Group the numerator and denominator with parentheses, because the given problem may be expressed as

$$(5^2 - 4) \div (3^2 - 6).$$

Now key in the problem as it appears.

$\boxed{(}$ 5 $\boxed{x^2}$ $\boxed{-}$ 4 $\boxed{)}$ $\boxed{\div}$ $\boxed{(}$ 3 $\boxed{x^2}$ $\boxed{-}$ 6 $\boxed{)}$ $\boxed{=}$ $\boxed{\quad 7}$

Method 2 Compute the denominator and store it. Then compute the numerator and divide by the stored number.

3 $\boxed{x^2}$ $\boxed{-}$ 6 $\boxed{=}$ $\boxed{\text{STO}}$ 5 $\boxed{x^2}$ $\boxed{-}$ 4 $\boxed{=}$ $\boxed{\div}$ $\boxed{\text{RCL}}$ $\boxed{=}$ $\boxed{\quad 7}$

On some calculators the store key looks like $\boxed{\text{M}_{\text{in}}}$ or $\boxed{\text{X} \to \text{M}}$, while the recall key is labeled $\boxed{\text{MR}}$ or $\boxed{\text{RM}}$.

PROGRESS CHECK 5 Evaluate on a scientific calculator.

a. $9(5 - 1)$ **b.** $\dfrac{5^2 + 3}{4^2 - 9}$

EXAMPLE 6 Solve the problem in the section introduction on page 28.

Solution To evaluate $16{,}000(1 + 0.06)^{15}$, use the power key $\boxed{y^x}$. Once again, we may input the problem by following the order of operations (method 1), or we may key in the problem as it appears (method 2) and let the priorities built into the calculator produce the desired result.

Method 1 1 $\boxed{+}$ .06 $\boxed{=}$ $\boxed{y^x}$ 15 $\boxed{\times}$ 16,000 $\boxed{=}$ $\boxed{38{,}344.931}$

Method 2 16,000 $\boxed{\times}$ $\boxed{(}$ 1 $\boxed{+}$ .06 $\boxed{)}$ $\boxed{y^x}$ 15 $\boxed{=}$ $\boxed{38{,}344.931}$

To the nearest dollar, the predicted price in 15 years is $38,345.

PROGRESS CHECK 6 Evaluate $12{,}000(1 + 0.06)^{15}$ on a scientific calculator. This number predicts the price in 15 years of an item that costs $12,000 today and increases in price by 6 percent each year. Round to the nearest integer.

4 Combining or evaluating algebraic expressions occurs often in geometry problems. For reference purposes, Figures 1.13 and 1.14 provide some formulas from geometry that will be required for this course.

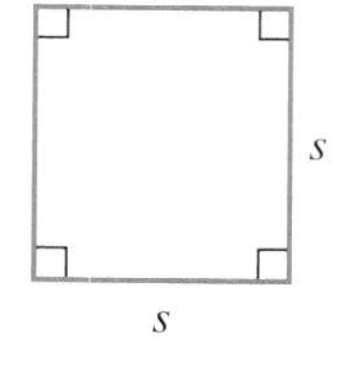

Square
$P = 4s$
$A = s^2$

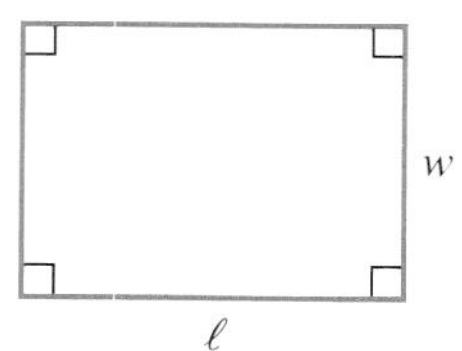

Rectangle
$P = 2\ell + 2w$
$A = \ell w$

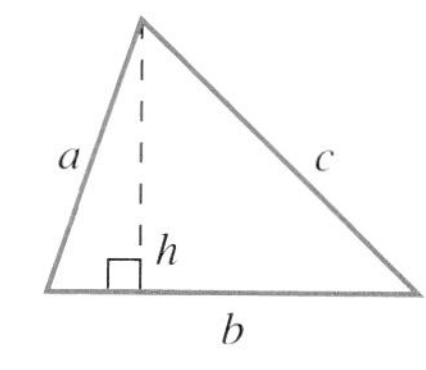

Triangle
$P = a + b + c$
$A = \frac{1}{2}bh$

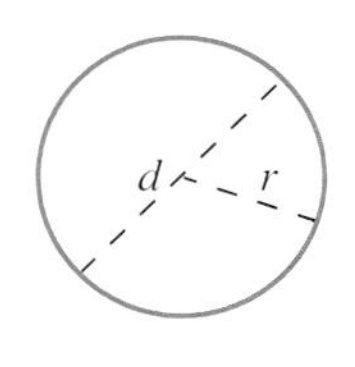

Circle
$C = \pi d$ or $C = 2\pi r$
$A = \pi r^2$

Figure 1.13 Two-dimensional figures: Formulas for perimeter P, circumference C, and area A.

Progress Check Answers
5. (a) 36 (b) 4
6. 28,759

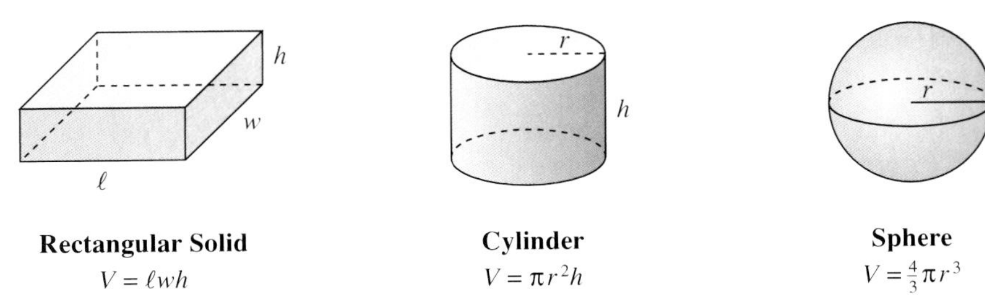

Figure 1.14 Three-dimensional figures: Formulas for volume V.

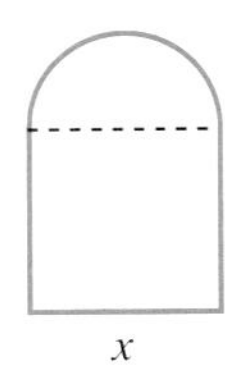

Figure 1.15

EXAMPLE 7 A window is shaped in the form of a semicircle mounted on a square, as shown in Figure 1.15. If the side length of the square is given by x, find the distance around the window in terms of x. Use 3.14 for π.

Solution Since the circumference of a circle is given by $C = \pi d$, the distance around the curved semicircular portion of the window is given by $C = \pi d/2$. The side length of the square is x, and the side length is also the diameter of the semicircle, so

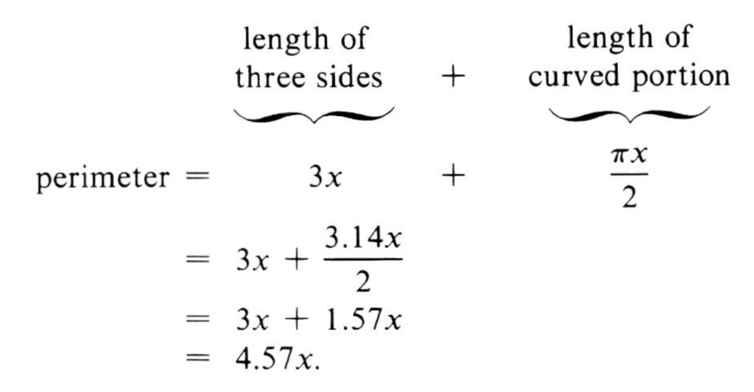

length of three sides + length of curved portion
perimeter = 3x + πx/2
= 3x + 3.14x/2
= 3x + 1.57x
= 4.57x.

In terms of x, the distance around the window is $4.57x$.

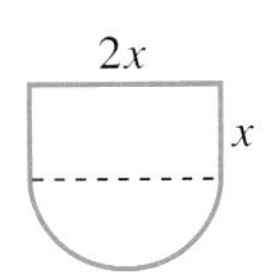

Figure 1.16

PROGRESS CHECK 7 Find the perimeter in terms of x of the tract of land in Figure 1.16. Use 3.14 for π.

EXAMPLE 8 A circular backyard aboveground pool has a diameter of 20 ft. See Figure 1.17.

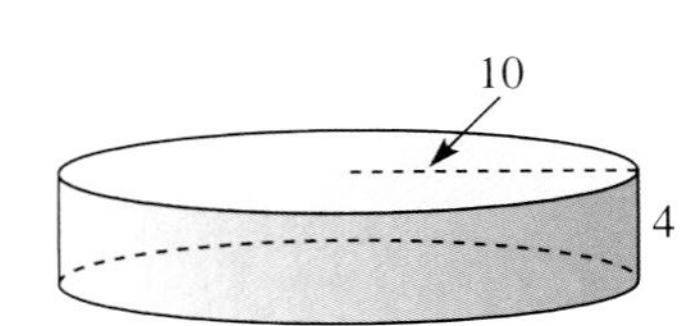

Figure 1.17

a. What volume of water (to the nearest cubic foot) will it take to fill the pool to a depth of 4 ft?

b. Use the fact that one cubic foot of water holds about 7.5 gallons (gal) to express the volume in gallons of water.

c. How long will it take to fill this pool if the water enters at 6 gal/minute? (Give the answer to the nearest hour.)

Solution

a. Since the shape of the pool is a cylinder, the formula to use is $V = \pi r^2 h$, where $r = 10$ ft and $h = 4$ ft.

$$\begin{aligned} V &= \pi r^2 h \\ &= 3.14(10^2)4 \quad \text{Replace } \pi \text{ by 3.14, } r \text{ by 10, and } h \text{ by 4.} \\ &= 1{,}256 \end{aligned}$$

To the nearest foot, the volume is 1,256 ft^3.

b. Each cubic foot holds 7.5 gal of water, so the filled pool holds $7.5 \times 1{,}256$, or 9,420, gal.

c. At 6 gal/minute, filling this pool will take 9,420/6, or 1,570, minutes, which equals about 26 hours.

Use of the π key on the calculator may give more accurate answers. Usually, the difference will be small. In this example, use of the π key yields 1,257 ft^3.

Progress Check Answer

7. $7.14x$

PROGRESS CHECK 8 Redo Example 8, but assume that the pool will be filled to a depth of 3 ft.

5 To develop proficiency in the language of algebra, you must be able to translate between verbal expressions and algebraic expressions. The following chart shows some algebraic expressions that may be used to translate typical verbal expressions that involve arithmetic operations.

Operation	Verbal Expression	Algebraic Expression
Addition	The sum of a number x and 1	$x + 1$, or $1 + x$
	A number y plus 2	$y + 2$, or $2 + y$
	A number w increased by 3	$w + 3$, or $3 + w$
	4 more than a number b	$b + 4$, or $4 + b$
	Add 5 and a number d	$d + 5$, or $5 + d$
Subtraction	The difference of a number x and 6	$x - 6$
	A number y minus 5	$y - 5$
	A number w decreased by 4	$w - 4$
	3 less than a number b	$b - 3$
	Subtract 11 from a number d	$d - 11$
	Subtract a number d from 11	$11 - d$
Multiplication	The product of a number x and 5	$5x$
	4 times a number t	$4t$
	A number y multiplied by 7	$7y$
	Twice a number w	$2w$
	$\frac{1}{3}$ of a number n	$\frac{1}{3}n$
	25 percent of a number P	$0.25P$
Division	The quotient of a number x and 8	$x \div 8$, or $x/8$
	A number y divided by 2	$y \div 2$, or $y/2$
	The ratio of a number a to a number b	$a \div b$, or a/b, or $a:b$

EXAMPLE 9 Translate each statement to an algebraic expression.

a. The price, p dollars, of a CD player increased by 15 dollars.
b. 9 less than the average a.
c. 6 percent of the amount of sales x.
d. Divide the sum of a and b by 2.

Solution Translate according to the chart above.

a. p dollars increased by 15 dollars is expressed as $p + 15$ or $15 + p$ dollars.
b. 9 less than a is expressed as $a - 9$. Note that $9 - a$ is not correct here.
c. In decimal form 6 percent is written as 0.06. Thus,

$$\underbrace{\text{6 percent}}_{0.06}\ \underset{\cdot}{\text{of}}\ \underbrace{\text{the amount of sales}}_{x.}$$

This expression may be written more simply as $0.06x$.

d. The sum of a and b is symbolized as $a + b$. Thus, the division of this sum by 2 can be written as

$$\frac{a + b}{2} \quad \text{or} \quad (a + b) \div 2.$$

Note Since division by 2 is defined as multiplication by $\frac{1}{2}$ (reciprocal), this expression could also have been written as $\frac{1}{2}(a + b)$.

Progress Check Answers

8. (a) 942 ft^3 (b) 7,065 gal (c) 20 hours

PROGRESS CHECK 9 Translate each statement to an algebraic expression.

a. The price, p dollars, of a TV set decreased by 50 dollars.
b. 1 more than the average a.
c. 15 percent of the taxable income i.
d. Divide 4 less than the number x by 2.

In Example 10 the translation process is reversed.

EXAMPLE 10 Translate each algebraic expression to a verbal expression.

a. $2x + 5y$ **b.** $a - 3b$ **c.** $4(x + 1)$

Solution There may be more than one correct translation, but care must be taken to keep the correct order of operations.

a. The sum $2x + 5y$ may be read as "the sum of 2 times x and 5 times y."
b. The difference $a - 3b$ means "subtract 3 times b from a."
c. The product $4(x + 1)$ may be stated as "4 times the sum of x and 1."

With experience you will often find the algebraic expression easier to comprehend than the corresponding verbal expression.

PROGRESS CHECK 10 Translate each algebraic expression to a verbal expression.

a. $9a - b$ **b.** $x + 3y$ **c.** $4(\ell + w + d)$

Progress Check Answers

9. (a) $p - 50$ (b) $a + 1$ (c) $0.15i$ (d) $(x - 4) \div 2$

10. (a) Subtract b from 9 times a. (b) The sum of x and 3 times y (c) 4 times the sum of ℓ and w and d

EXERCISES 1.4

In Exercises 1–12, identify the term(s) in the given expressions.

1. $3n + 5w$ $3n, 5w$
2. $-2\ell + 7x$ $-2\ell, 7x$
3. $2x - 4y$ $2x, -4y$
4. $\frac{1}{2}x - \frac{3}{2}y$ $\frac{1}{2}x, -\frac{3}{2}y$
5. $x^2 - 2x - 4$ $x^2, -2x, -4$
6. $y^2 + 3y - 8$ $y^2, 3y, -8$
7. $\frac{1}{2}mn$ $\frac{1}{2}mn$
8. $-2xy$ $-2xy$
9. $1 - a$ $1, -a$
10. $7 + x$ $7, x$
11. $3abc$ $3abc$
12. $(3a)(4b)$ $(3a)(4b)$

In Exercises 13–24, if possible, simplify by combining like terms. Indicate if no simplification is possible.

13. $5x + 6x$ $11x$
14. $3y + 2y$ $5y$
15. $3a + 3a$ $6a$
16. $3a - 3a$ 0
17. $2w + w$ $3w$
18. $-u + 2u$ u
19. $3v - 2$ Cannot be simplified
20. $-3v + 4w$ Cannot be simplified
21. $3x - y + x - 5y$ $4x - 6y$
22. $x - 3y - 5x + 2y$ $-4x - y$
23. $3x^2 - 1 + 2x^2 + 1$ $5x^2$
24. $2x^2 + x - x^2 + x$ $x^2 + 2x$

In Exercises 25–36, remove the symbols of grouping and combine like terms.

25. $-(1 - x) + 2(x - 1)$ $3x - 3$
26. $-(y - 2) + 3(2 - y)$ $-4y + 8$
27. $3 - (x - 2)$ $-x + 5$
28. $2 - (3 - x)$ $x - 1$
29. $(4x^2 - x) - (3x^2 - x)$ x^2
30. $(x^2 - x + 1) - (x^2 - x + 2)$ -1
31. $-2(x - 3) + 3(x - 4)$ $x - 6$
32. $-3(x - 2) + 2(x - 3)$ $-x$
33. $6x - 4[x - (1 + x)]$ $6x + 4$
34. $x - 2[3x - (x + 2)]$ $-3x + 4$
35. $3(x - 1) - 4[x - (2 - x)]$ $-5x + 5$
36. $-2(3x - 1) - 3[2x - (1 + x)]$ $-9x + 5$

In Exercises 37–50, evaluate the expression using the values given for the symbols.

37. $3x - y;\ x = 2, y = -2$ 8
38. $x - 3y;\ x = 2, y = -2$ 8
39. $3(x - y) - 3(x + y);\ x = -1, y = 2$ -12
40. $-2(2x + y) + 4(x + y);\ x = -1, y = 2$ 4
41. $x^2 + 3y^2 + 1;\ x = -3, y = -2$ 22
42. $2x^2 - 3y^2 - 1;\ x = 2, y = -5$ -68
43. $-x^2 + y^2;\ x = 2, y = -5$ 21
44. $-3x^2 - 2y^2;\ x = 2, y = -5$ -62
45. $\dfrac{y + 3}{x - 4};\ x = -4, y = -3$ 0
46. $\dfrac{3 - 2y}{6x + 12};\ x = -2, y = 1$ Undefined
47. $\dfrac{7x}{4 - y};\ x = -1, y = -3$ -1
48. $\dfrac{1 - (x + 2)}{3 - (y - 1)};\ x = -1, y = 3$ 0
49. $\dfrac{2 - (x - 1)^2}{2(x - 1)^2};\ x = -1$ $-\frac{1}{4}$
50. $\dfrac{4 - (2 - x)^2}{3(2 - x)^2};\ x = -2$ $-\frac{1}{4}$

In Exercises 51–64, use a scientific calculator to evaluate the given expression. Round decimal answers to the nearest hundredth.

51. $14(15 + 6)$ 294
52. $17(13 + 9)$ 374
53. $-28(54 - 61)$ 196
54. $-35(27 - 44)$ 595
55. $\dfrac{4^2 - 3}{4^2 - 2}$ $\frac{13}{14} = 0.93$
56. $\dfrac{5^2 + 3}{3^2 - 2}$ 4

57. $\dfrac{12 - 78^2}{16 + 21^2}$ -13.29

58. $\dfrac{19 - 62^2}{109 - 17^2}$ 21.25

59. $32 - 15^2(32 - 15)^2$ $-64{,}993$

60. $\dfrac{27 - 51^2}{(27 - 51)^2}$ -4.47

61. $5^4 - 4^5$ -399

62. $6^5 - 5^6$ $-7{,}849$

63. $12^4 - 12^3 + 12^2 - 12$ $19{,}140$

64. $11^4 - 11^3 + 11^2 - 11$ $13{,}420$

65. With annual inflation at 3.1 percent, the cost of a present-day \$15,000 automobile will equal $\$15{,}000(1 + 0.031)^{10}$ in 10 years. Evaluate this cost to the nearest hundred dollars. \$20,400

66. If real estate values in a neighborhood continue to inflate at 4.6 percent annually, the price of a \$110,000 house in today's market will be $\$110{,}000(1 + 0.046)^7$ in seven years. Evaluate this cost to the nearest thousand dollars. \$151,000

67. Rent for an apartment today in Centerville is \$400 per month. With annual inflation of 6 percent, what will the rent be (to the nearest dollar) in 5 years? Calculate $400(1 + 0.06)^5$.
\$535 per month

68. In one industry the average wage today is \$7.48 per hour. With annual inflation of 5.5 percent, what will this be in 10 years? Calculate $7.48(1 + 0.055)^{10}$. \$12.78 per hour

69. Find and simplify an expression for the perimeter of the figure shown. Use 3.14 for π. $9.14x$

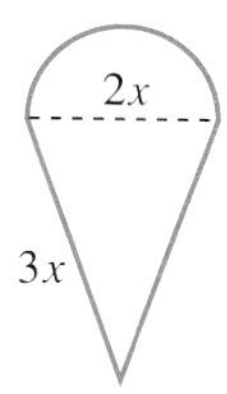

70. Find and simplify an expression for the perimeter of the figure shown. Use 3.14 for π. The curves are semicircles. $9.42x$

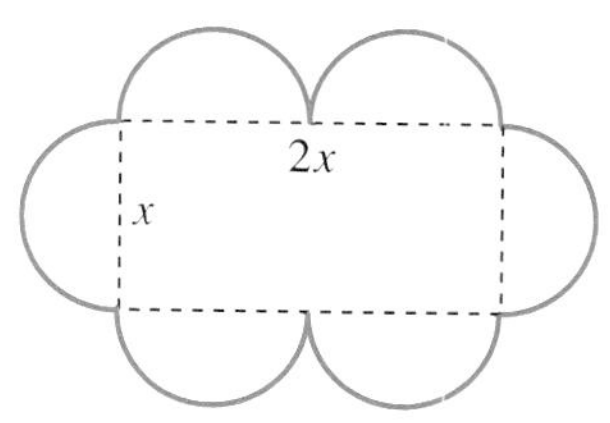

71. Find and simplify an expression for the area of the figure shown. The curves are semicircles. Use 3.14 for π.

$3.14 + 8x + 6.28x^2$

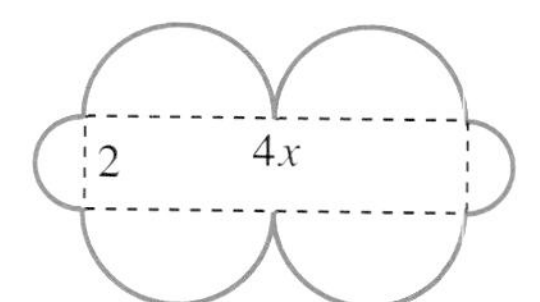

72. Find and simplify an expression for the area of the figure shown. a^2

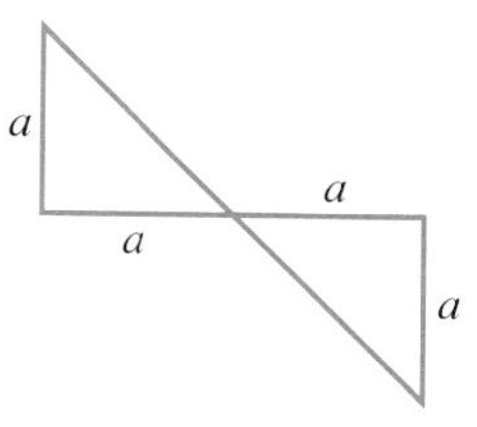

73. **a.** A cylindrical vat for storing brine in a pickle factory has a diameter of 10 ft and a depth of 10 ft. If it is filled to within 1 ft of the top, how much liquid will it hold? Give the answer to the nearest cubic foot. $707\ \text{ft}^3$

b. Express the volume in gallons if 1 ft^3 of liquid holds about 7.5 gal. Give the answer to the nearest 10 gal.
5,300 gal

74. Redo Exercise 73, but now assume that the vat is filled to within 2 ft of the top. a. $628\ \text{ft}^3$
b. 4,710 gal

75. As part of a bank vault lock assembly, a metal cube of side 6 in. has a cylindrical hole of radius 1 in. drilled through it.

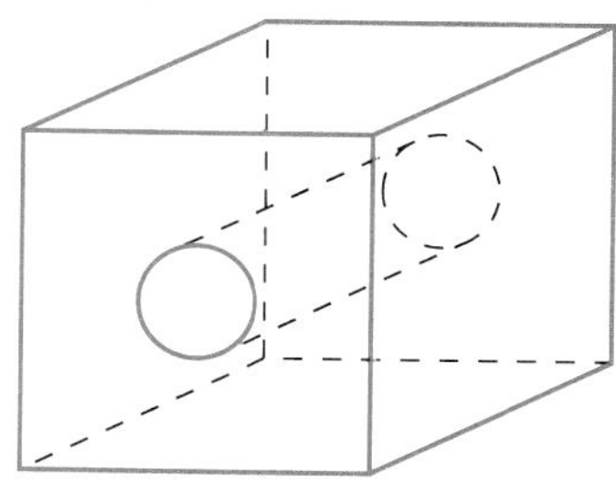

a. Find the volume of the remaining piece of metal. Round to the nearest cubic inch. $197\ \text{in.}^3$

b. If the metal weighs 0.3 lb/in.³, find the weight of the part to the nearest pound. 59 lb

76. A bowling ball has an 8-in. diameter.

a. Find its volume to the nearest cubic inch. $V = 268\ \text{in.}^3$

b. If the material weighs 1 oz/in.³, find the weight of the ball to the nearest pound. (There are 16 oz in 1 lb.)
$W = 17$ lb

In Exercises 77–88, translate the given statement to an algebraic expression.

77. 5 percent of the amount of sales t $0.05t$

78. 12 percent of the rental fee f $0.12f$

79. 0.5 percent of the value of the mutual fund v $0.005v$

80. 0.85 percent of the insurance premium p $0.0085p$

81. \$5 more than the price of the videotape v $v + 5$

82. \$6 less than the list price l $l - 6$

83. The sum of the price p and the sales tax, which is 6 percent of the price $p + 0.06p$

84. The list price l minus the 20 percent discount on the list price $l - 0.20l$

85. The sum of a and the product of b and c $a + bc$

86. The product of a and the sum of b and c $a(b + c)$

87. The quotient of x and two more than x $\dfrac{x}{x + 2}$

88. The sum of the quotient of x and 2 and x $\dfrac{x}{2} + x$

89. $\dfrac{x + 2}{a}$ and $x + \dfrac{2}{a}$ are two unequal algebraic expressions. Translate each into a verbal expression. The quotient of 2 more than x and a; the sum of x and the quotient of 2 and a

90. $(x - 2) + a$ and $x - (2 + a)$ are two unequal algebraic expressions. Translate each into a verbal expression. The sum of 2 less than x and a; the difference of x and 2 more than a

In Exercises 91–96, translate each algebraic expression into a verbal expression.

91. $3x + y$ The sum of 3 times x and y

92. $3(x + y)$ 3 times the sum of x and y

93. $\frac{x}{yz}$ x divided by the product of y and z

94. $\frac{x}{y} \cdot z$ The product of x divided by y and z

95. $0.05(x + y)$ 5 percent of the sum of x and y

96. $0.06(x - 10)$ 6 percent of the difference of x and 10

THINK ABOUT IT

1. a. Make up two terms which are like terms with respect to $3xy^2$. Why are they like terms?

b. Give two terms which are unlike terms with respect to $3xy^2$. Explain why they are unlike terms.

2. a. Circle A has radius 1 ft and circle B has radius 4 ft. How much larger is circumference B than circumference A?

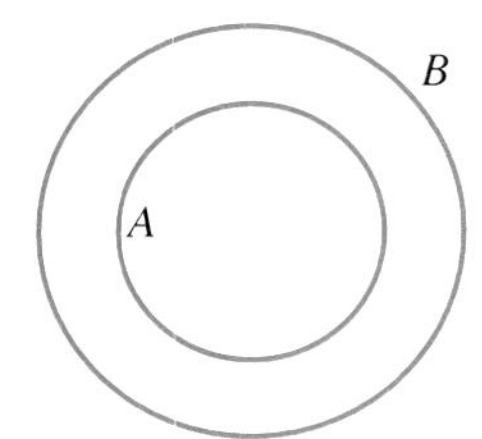

b. Circle C has radius 1,001 ft and circle D has radius 1,004 ft. How much larger is circumference D than circumference C?

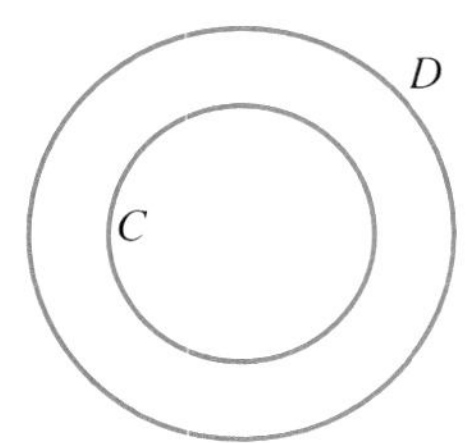

c. Imagine a child 3 ft tall who walks around the earth's equator. The radius of the equator is about 4,000 miles (mi). In one trip around the equator, how much further does a point on the top of the child's head go than a point on the bottom of the child's feet?

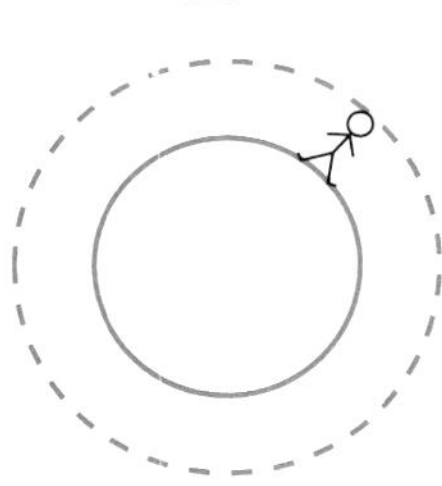

3. Simplify $\frac{2}{3}\{\frac{1}{2} - \frac{1}{8}[4x - 2(8x + 3)]\}$. Test your result by letting $x = 2$.

4. a. 632 may be expressed as $100 \cdot 6 + 10 \cdot 3 + 2$. Using this approach, how would 236 be expressed? Note that 236 is the number you get by reversing the digits of 632.

b. Is the difference $632 - 236$ (evenly) divisible by 99?

c. Is the difference $154 - 451$ divisible by 99?

d. Any three-digit number, abc, may be expressed as $100a + 10b + c$. Express cba in this way

e. Using this expanded notation, represent the difference abc − cba.

f. Explain why it follows that this difference must be evenly divisible by 99.

g. True or false? If you take any three-digit number, reverse the digits, and subtract the smaller from the larger, the answer will always be divisible by 99.

h. Make up a similar generalization about two-digit numbers.

5. Write an algebraic expression that corresponds to each sequence on a scientific calculator.

a. b [+] c [×] a [=]

b. b [+] c [=] [×] a [=]

c. b [y^x] n [×] a [=]

d. a [×] b [=] [y^x] n [=]

REMEMBER THIS

1. If x is replaced by 3 in $5x - 1 = 14$, does a true statement result? Yes

2. Is it true that $(x - 2)(x + 3)$ equals 0 when x equals -3? Yes

3. In the expression $3x + 5$, x is called a (variable, constant, coefficient). Variable

4. Translate into English: $A \neq -7$. A is not equal to negative 7.

5. To what power does y appear in the expression $-3x^4y^5$? 5

6. What is the result of adding 1 to $5x - 1$? $5x$

7. Rewrite $3(4 + 5)$ using the commutative property of multiplication. $(4 + 5)3$

8. Find the product of 12 and $\frac{1}{4}x + \frac{1}{3}$. $3x + 4$

9. True or false? The reciprocal of a negative number is also a negative number. True

10. Evaluate $\frac{-3 + 2^3}{(-3 + 2)^3}$. -5

Numbers and Operations

At the end of each chapter some pertinent features of the TI-81 graphing calculator are presented. We assume that you also have a copy of the manual which comes with the calculator to use for further reference. If you have some other graphing calculator, it should not be too difficult to adapt these discussions to your machine.

The TI-81 has three distinct capabilities useful in this course: It can do numeric calculations, it is programmable, and it can display graphs. In this section we point out the use of some of the more frequently used keys for calculation, especially those that are unlike the keys on a non-graphing calculator.

Observe that the numbers are all on grey keys, and the four arithmetic operations are on dark blue keys. Especially note that there are two different keys with minus signs: a grey key [(−)] for making negative signs, and a blue key [−] for doing subtractions. Be careful to use these keys correctly; beginning students often confuse them, and errors result.

There is a special key for raising numbers to powers. It is the black key [∧]. For example, 3^5 is entered as 3 [∧] 5. Also, there is a special key for the second power, a black key [x^2]. Thus, you have two ways to evaluate 13^2:

a. 13 [∧] 2 [ENTER] **b.** 13 [x^2] [ENTER]

Both sequences give the correct answer, 169.

In some expressions multiplication is implied, which means it is not always necessary to enter a multiplication symbol. For instance, the product of 3 and 4 can be entered either as 3 [(] 4 [)] or as 3 [×] 4.

As you work, the screen fills with symbols. This shows your original entry as well as the answer. Pressing the black key [CLEAR] will erase the screen.

EXAMPLE 1 Evaluate these expressions using the graphing calculator.

a. $\sqrt{2}$, to 4 decimal places **b.** 1.35^5, to 2 decimal places
c. $12{,}000(1 + 0.06)^{15}$ **d.** $-1.4 - 3.8$
e. $(-3)^2$ **f.** $|-2| + 3$

Solution

a. The square root symbol is in blue above the [x^2] key. To activate it, you press the blue key marked [2nd] followed by the [x^2] key. This writes a $\sqrt{\ }$ symbol on the screen. (We will use the format where 2nd functions are shown in brackets preceded by the [2nd] key symbol.) Therefore, to evaluate $\sqrt{2}$, use this sequence of keystrokes.

[2nd] [$\sqrt{\ }$] 2 [ENTER]

You will see 1.414213562 on the right-hand side of the screen. Thus, to 4 decimal places, the answer is 1.4142.

b. Use the following keystrokes.

1.35 [∧] 5 [ENTER]

To 2 decimal places, the answer is 4.48.

c. We can use implied multiplication as follows.

12,000 [(] 1 [+] .06 [)] [∧] 15 [ENTER]

The answer to 2 decimal places is 28,758.70.

d. This example uses both keys with minus signs.

[(−)] 1.4 [−] 3.8 [ENTER]

The answer is −5.2.

e. Recall that $(-3)^2$ is not the same as -3^2. It will be necessary to key in the parentheses to prevent the calculator from computing -3^2 as the negative of 3^2.

[(] [(−)] 3 [)] [x^2] [ENTER]

The correct answer is 9. If you see −9, you have not entered the values correctly.

f. The absolute value key is a black key with [x^{-1}] printed on it and the letters ABS printed in blue above it. When you press [2nd] followed by [x^{-1}], the letters abs appear on the screen. Use these keys to evaluate $|-2| + 3$.

[2nd] [ABS] [(−)] 2 [+] 3 [ENTER]

The answer 5 appears on the right side of the screen.

EXERCISES

Evaluate these expressions using the graphing calculator.

1. $\sqrt{11.5}$, to 4 decimal places — Ans. 3.3912
2. 1.01^4, to 8 decimal places — Ans. 1.04060401
3. $1{,}450(1 - 0.03)^{10}$, to 2 decimal places — Ans. 1069.26
4. $-3.87 - (-2.04)$, to 2 decimal places — Ans. -1.83
5. $\dfrac{(|-23| + 5)}{-4}$ — Ans. -7

Use the graphing calculator for these exercises from the chapter.

Section	*Exercises*
1.1	41
1.2	39, 101, 123
1.3	27
1.4	43, 59, 67

Chapter 1 SUMMARY

OBJECTIVES CHECKLIST Specific chapter objectives are summarized below along with numbered example problems from the text that should clarify the objectives. If you do not understand any objectives or do not know how to do the selected problems, then restudy the material.

1.1 Can you:

1. **Graph real numbers?**
   Graph the number $-\frac{3}{4}$ on the number line. [Example 2a]
2. **Order real numbers and use inequality symbols?**
   Classify the statement $-1 \leq -3$ as true or false. [Example 3b]
3. **Find the negative or opposite of a real number?**
   Find the negative, or opposite, $-a$, if $a = -\sqrt{2}$. [Example 4c]
4. **Find the absolute value of a real number?**
   Evaluate the expression $|2.4|$. [Example 5a]
5. **Identify integers, rational numbers, irrational numbers, and real numbers?**
   From the set $\{2.5, \sqrt{7}, \sqrt{4}, 0, 0.\overline{6}, -\frac{9}{11}, -3, -\pi\}$, list all numbers that are in the set of rational numbers. [Example 7b]
6. **Find the union or intersection of two sets?**
   If $A = \{1,2,3,4\}$ and $B = \{0,2,4\}$, find $A \cup B$. [Example 8a]

1.2 Can you:

1. **Add, subtract, multiply, and divide real numbers?**
   Find the difference $-6 - (-16)$. [Example 3b]
2. **Apply the definition of a positive integer exponent?**
   Multiply $(-2)^5$. [Example 7a]
3. **Evaluate expressions following the order of operations?**
   Evaluate $7 + 5(6 - 10)^2$. [Example 8]
4. **Solve applied problems involving positive and negative numbers?**
   Solve the problem in the section introduction on page 11. [Example 10]

1.3 Can you:

1. **Identify and use commutative and associative properties?**
   Use $a = -3$, $b = 2$, and $c = 7$ to illustrate that $(a + b) + c = a + (b + c)$. [Example 2a]
2. **Identify and use identity and inverse properties?**
   Use one of the identity or inverse properties to simplify the statement $0 + (-8)$, and indicate the property that was used. [Example 4a]

3. **Identify and use the distributive property?**
   Multiply $5(x + 4)$ using the distributive property. [Example 6a]
4. **Use $-a = -1 \cdot a$ and the distributive property to remove parentheses?**
   Rewrite the statement $-(2 - y)$ without parentheses. [Example 9b]

1.4 Can you:

1. **Simplify algebraic expressions by combining like terms?**
   Simplify $6x - y + x + 5y$ by combining like terms. [Example 2c]
2. **Evaluate algebraic expressions?**
   Evaluate $x^2 + 5y^2$ given that $x = 3$ and $y = -8$. [Example 4b]
3. **Evaluate an expression on a scientific calculator?**
   To the nearest integer evaluate $16{,}000(1 + 0.06)^{15}$ on a scientific calculator. [Example 6]
4. **Apply geometry formulas?**
   A circular backyard aboveground pool has a diameter of 20 ft. (See Figure 1.17.) What volume of water (to the nearest cubic foot) will it take to fill the pool to a depth of 4 ft? [Example 8a]
5. **Translate between verbal expressions and algebraic expressions?**
   Translate to an algebraic expression : 6 percent of the amount of sales x. [Example 9c]

KEY TERMS

Absolute value (1.1)
Algebraic expression (1.4)
Associative properties (1.3)
Base (1.2)
Coefficient (1.4)
Commutative properties (1.3)
Constant (1.4)
Difference (1.2)
Distributive property (1.3)
Ellipsis (1.1)
Empty set (1.1)
Equality of sets (1.1)
Exponent (1.2)
Exponential expression (1.2)
Factor (1.2)
Graph (of a number) (1.1)
Identity properties (1.3)
If and only if (1.1)
Integer (1.1)
Intersection of sets (1.1)
Inverse properties (1.3)
Irrational number (1.1)
Like terms (1.4)
Negative direction (1.2)
Negatives (or opposites) (1.1)
Positive direction (1.2)
Product (1.2)
Quotient (1.2)
Rational number (1.1)
Real number (1.1)
Real number line (1.1)
Reciprocals (1.2)
Set (1.1)
Set-builder notation (1.1)
Subset (1.1)
Substitution property (1.3)
Sum (1.2)
Terms (1.4)
Third power (1.2)
Trichotomy property (1.1)
Union of sets (1.1)
Variable (1.4)

KEY CONCEPTS AND PROCEDURES

Section	Key Concepts or Procedures to Review
1.1	■ Definitions of integers, rational numbers, irrational numbers, real numbers, absolute value, equality of sets, union of sets, intersection of sets, and empty set ■ Methods to graph real numbers on the number line ■ Relationships among the various types of numbers ■ Statements involving inequalities: $a < b,\quad a > b,\quad a \le b,\quad a \ge b,\quad a \ne b$ ■ Double-negative rule: $-(-a) = a$

Section	Key Concepts or Procedures to Review
1.2	■ Methods to add, subtract, multiply, and divide real numbers ■ Definitions of subtraction, division, and positive integer exponents ■ Division involving zero ■ Order of operations ■ Interpretation of the signs of real numbers
1.3	■ Statements of the commutative, associative, identity, inverse, and distributive properties of real numbers ■ Methods to simplify expressions using the basic properties of real numbers
1.4	■ Methods to simplify algebraic expressions by removing symbols of grouping and by combining like terms ■ Methods to evaluate algebraic expressions ■ Procedures for using a scientific calculator to evaluate expressions ■ Basic formulas from geometry ■ Guidelines for translating verbal expressions to algebraic expressions and vice versa

CHAPTER 1 REVIEW EXERCISES

1.1

1. Graph the number $-\frac{3}{2}$ on the number line.

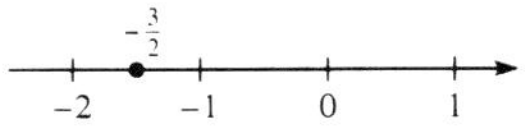

2. Classify the following statement as true or false: $-3 \leq 1$. True
3. Find $-a$ given that $a = -6$. 6
4. Evaluate $-\left|\frac{3}{4}\right|$. $-\frac{3}{4}$
5. From the set $\{3.1, \sqrt{9}, -\sqrt{4}, \sqrt{2}, 0, 0.3\overline{1}, \frac{4}{5}, -7, \pi\}$, list all numbers that are integers. $\sqrt{9}, -\sqrt{4}, 0, -7$
6. If $A = \{-2,0,2\}$ and $B = \{-3,-2,-1,0\}$, find $A \cap B$. $\{-2,0\}$

1.2

7. Find the sum: $-\frac{2}{5} + (-\frac{3}{5})$. -1
8. Find the product or quotient: $\dfrac{4(0)}{-3 + 3}$. Undefined
9. Multiply -2^4. -16
10. Evaluate $-2 + 8(1 - 4)^2$. 70
11. Over the course of a day, the price of one share of stock in a particular company fell from \$12.00 to \$10.80. Express this drop in share price as a percent change. Use the formula

$$\text{percent change} = \frac{\text{new value} - \text{old value}}{\text{old value}} \cdot 100.$$

-10 percent

1.3

12. Label the statement $(a + 3) + 2 = a + (3 + 2)$ as an example of one of the commutative properties or one of the associative properties. Associative property of addition
13. Use one of the identity or inverse properties to simplify the statement $-5 \cdot (-\frac{1}{5})$, and indicate the property that was used. 1; multiplication inverse property
14. Use the distributive property to write the expression $5a - 5b$ as a product. $5(a - b)$
15. Rewrite $-(-7 + x)$ without parentheses. $7 - x$
16. Multiply $-1(-2 - a + b)$ using the distributive property. $2 + a - b$

1.4

17. Remove the symbols of grouping and combine like terms: $2(3y - 1) - (3 - 2y)$. $8y - 5$
18. Evaluate the expression $\dfrac{4x}{6 - y}$ given that $x = -1$ and $y = -6$. $-\frac{1}{3}$
19. Evaluate $4{,}000(1 + 0.06)^{15}$ on a scientific calculator. Round to the nearest integer. 9,586
20. A circular tablecloth has a radius of 30 in. Find the distance around the edge of the tablecloth to the nearest inch. 188 in.
21. Translate to an algebraic expression: Divide 10 less than the number x by 5. $\dfrac{x - 10}{5}$

ADDITIONAL REVIEW EXERCISES

22. From the set $\{8, 0, \sqrt{5}, \sqrt{100}, -\frac{3}{5}, 0.\overline{1}, -2, \pi, 3.2\}$, list all numbers that are irrational. $\sqrt{5}, \pi$
23. If $A = \{0,3,6\}$ and $B = \{-1,1,4\}$, find $A \cap B$. $\emptyset$
24. True or false? The union of the set of rational numbers with the set of irrational numbers is the set of real numbers. True
25. Label the statement $3(ab) = 3(ba)$ as an example of one of the commutative properties or one of the associative properties. Commutative property of multiplication
26. If $a = -2$, $b = 3$, and $c = 5$, show that $a(bc) = (ab)c$. Both equal -30.

27. Use one of the identity or inverse properties to simplify the statement $-4 + 0$, and indicate the property that was used. -4; identity, addition
28. Find the difference $2 - (-3)$. 5
29. Find the sum $9 + (-15)$. -6
30. Find the product $(-\frac{1}{2})(-\frac{4}{5})$. $\frac{2}{5}$
31. Find the quotient $\frac{0}{-3}$. 0
32. Divide $-27 \div 3$. -9
33. Identify the terms in the expression $3s^2 - 4s + 12$. $3s^2$, $-4s$, 12
34. Simplify $-4a + 2b - 7 + 4a - 2$ by combining like terms. $2b - 9$
35. Evaluate $\frac{2x^2}{3y}$ given that $x = -3$ and $y = -6$. -1
36. A window is shaped in the form of a semicircle mounted on a square, as shown in Figure 1.15. If the side length of the square is 4 ft, find the area of the window. Use 3.14 for π, and round to the nearest hundredth. 22.28 ft²
37. A gift box is shaped like a rectangular solid with the length, width, and height all measuring the same. If the height of the box is given by x, find the volume of the box in terms of x. x^3
38. Graph the numbers $-\sqrt{2}$, 0, and $\frac{1}{2}$ on the number line.

    $-\sqrt{2}$ $\frac{1}{2}$
    -2 -1.5 -1 0 1

39. Given that $a = -\frac{5}{6}$, find the negative, or opposite, $-a$. $\frac{5}{6}$
40. Evaluate $-|-\sqrt{3}|$. $-\sqrt{3}$
41. Multiply $(-2)^6$. 64
42. Evaluate $-2 - (3 - 4)^4$. -3
43. Evaluate $\frac{-4(1 - 5)}{5 - 4^2}$. $-\frac{16}{11}$
44. If $a = 2$, $b = 3$, and $c = 4$, show that $a(b + c) = ab + ac$. Both equal 14.
45. Multiply $-5(-3x + 2)$ using the distributive property. $15x - 10$
46. Rewrite $-(6y + 2)$ without parentheses. $-6y - 2$
47. Remove the symbols of grouping and combine like terms: $8y - 2[4y - (1 - y)]$. $-2y + 2$
48. Translate to an algebraic expression: Four more than the product of a number x and 2. $2x + 4$
49. Translate the expression $a + 4b$ to a verbal expression. The sum of a and 4 times b
50. Use the distributive property to rewrite $ax + ay$ as a product. $a(x + y)$
51. Classify as true or false: $0 > -2$. True
52. Is a projectile moving faster when its velocity is -96 ft/second or when its velocity is 72 ft/second? -96 ft/second
53. An investor bought 400 shares of stock in a fast-food company for $41.50 per share and sold them all at $38.30 per share. How much money did the investor lose on this investment all together? $1,280
54. Evaluate $8{,}000(1 + 0.05)^{12}$ on a scientific calculator. Round to the nearest integer. 14,367

CHAPTER 1 TEST

1. What property is illustrated by this statement: $(a + b)(-3) = -3(a + b)$? Commutative property of multiplication
2. Graph the numbers $\frac{5}{2}$, 3, and -1 on the number line.

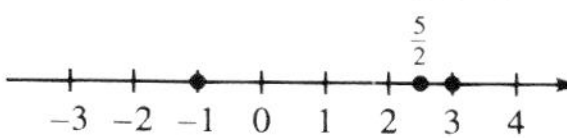

3. Translate to an algebraic expression: Half of the sum of x and y. $\frac{1}{2}(x + y)$
4. Which of the following is an irrational number?
   a. -2 **b.** 0 **c.** $\sqrt{9}$ **d.** $\sqrt{11}$ **e.** 0.5 d
5. Simplify $6x + 2y - 1 - 4x - 3y + 2$ by combining like terms. $2x - y + 1$
6. Evaluate $2(x - 1) + y^2$ given that $x = -1$ and $y = -2$. 0
7. Remove the symbols of grouping and combine like terms: $-6x - 2(x - 4) + 1$. $-8x + 9$
8. Use the distributive property to write the expression $2x + 2y$ as a product. $2(x + y)$
9. Translate the expression $a - 2b$ to a verbal expression. The difference between a and two times b
10. Which of the following statements is false?
    a. $0 > -2$ **b.** $-3 \geq -1$ **c.** $-4 \leq 5$
    d. $6 > 0$ **e.** $-4 < 0$ b
11. If $A = \{-3, -2, -1\}$ and $B = \{-2, -1, 0\}$, find $A \cup B$. $\{-3, -2, -1, 0\}$
12. A gardener wishes to plant flowers in a planter box shaped like a rectangular solid of length 36 in., height 11 in., and width 12 in. How much soil (by volume) will be needed to fill the planter to a height of 10 in.? 4,320 in.³
13. Evaluate $3{,}000(1 + 0.06)^{11}$ on a scientific calculator. Round to the nearest integer. 5,695
14. Rewrite $-(a - 2b - c)$ without parentheses. $-a + 2b + c$

Evaluate.

15. $-(5)^3$ -125
16. $|-7|$ 7
17. $-8 - 4$ -12
18. $(-5) + (-2) + 4$ -3
19. $\frac{4(-3)}{(-2)(-1)}$ -6
20. $\frac{-4 - (-4)}{-2 + 5}$ 0

CHAPTER

2 Linear Equations and Inequalities

José is offered two jobs selling video equipment. In one store he would be paid \$200 per week plus a 10 percent commission on each sale. In the other the pay is a straight 15 percent commission. At a certain point the second job begins to pay more than the first. This point can be found by solving the equation $200 + 0.10x = 0.15x$, where x represents the amount of sales at which both jobs pay the same. Find x, and also determine José's income at this point. (See Example 5 of Section 2.1.)

AN ESSENTIAL problem-solving skill in algebra is the ability to set up and solve equations and inequalities. This chapter focuses on topics associated with first-degree (or linear) equations and inequalities and considers many word problems and literal equations (formulas) that illustrate the usefulness of these topics.

2.1 Linear Equations in One Variable

OBJECTIVES

1 **Classify an equation as a conditional equation, an identity, or a false equation.**

2 **Identify linear equations in one variable.**

3 **Solve linear equations using properties of equality.**

1 Many problems which are analyzed by mathematics begin with writing equations to express some key relationship. Then the solution to the equation is used to solve the problem. A first step in studying algebra, then, is to be able to read, write, and work with equations. By definition, an **equation** is a statement that two expressions are *equal*. Therefore, every equation must contain an equality sign, =. For example, $5x - 1 = 14$ and $y^2 = 25$ are equations, but the expression $3x - 5$ is *not* an equation.

Stress that an equation must have an equal sign. Students are very careless about their use of the equal sign.

Note that if x is replaced by 3 in $5x - 1 = 14$, a true statement results. Therefore, we call 3 a **solution** of this equation. In the equation $y^2 = 25$ there are two values, 5 and -5, which are solutions. The set of all the solutions to an equation is called its **solution set,** and to *solve an equation* means to find its solution set.

Some equations are special because their solution set consists of *all* real numbers for which both sides are defined. For example, $3 + x = x + 3$ is true when x is replaced by *any* real number. Such equations are called **identities.** In contrast, some equations, like $x = x + 2$, have *no* solutions. They are called **false equations,** and their solution set is the empty set, $\emptyset$. The most familiar type of equation, like $5x - 1 = 14$, is true for *some* real numbers and false for others. Equations of this type are called **conditional equations.** All three types of equations are useful in solving applied problems.

EXAMPLE 1 Classify each equation as an identity, a false equation, or a conditional equation.

a. $x + 7 = 7 + x$ **b.** $x = x - 2$ **c.** $2x = 14$

Solution

a. The equation $x + 7 = 7 + x$ is an illustration of the commutative property of addition and so is true for all real numbers. Thus, it is an identity.

b. The equation $x = x - 2$ says that a number is equal to the number which is two less than itself. Because this is not possible, this equation is a false equation. Substitution of trial values will show that the two sides can never be equal.

c. $2x = 14$ is a conditional equation since the solution set consists only of the number 7, which can be seen by inspection.

Note Example 6 of this section shows other ways to recognize identities and false equations.

PROGRESS CHECK 1 Classify each equation as an identity, a false equation, or a conditional equation.

a. $3x = 12$ **b.** $x + 4 = x + 3$ **c.** $2x = x + x$

2 This chapter focuses on linear equations in one variable, because they have widespread application and can be solved by comparatively simple methods. By definition, a **linear equation in one variable** is an equation that can be written in the form

$$ax + b = 0$$

Progress Check Answers

1. (a) Conditional equation (b) False equation (c) Identity

Stress that linear equations are associated with graphs of *lines*. Whenever possible, connect new definitions to words students already know.

where a and b are real numbers, with $a \neq 0$. In this type of equation there is only one variable, and it appears to the first power but not to any higher power. For this reason it may also be called a **first-degree equation.** It is called linear because (as shown in Section 5.1) it is associated with a graph that is a straight line.

EXAMPLE 2 For each of the following, state whether or not it is a linear equation in one variable.

a. $x + 17 = 14$ **b.** $y^2 = 25$ **c.** $x + y = 10$ **d.** $4x + 8$

Solution

a. Yes; $x + 17 = 14$ is a linear equation in one variable. Note that this equation can be written as $1x + 3 = 0$, which has the standard form $ax + b = 0$.
b. No; this equation is not a linear equation because the variable y appears to the second power.
c. No; this equation is not a linear equation in one variable because there are two variables.
d. No; $4x + 8$ is not any type of equation (it is an expression) because there is no equals sign.

PROGRESS CHECK 2 For each of the following, state whether or not it is a linear equation in one variable.

a. $x^3 + x = 3$ **b.** $3x - 4 = 6$ **c.** $3x - 4$ **d.** $x + y = 12$

3 The general method for solving a linear equation in one variable is to start with the given equation and replace it by simpler and simpler equations that must have the same solution set until the solution set is clear. Two equations which have the same solution set are called **equivalent equations.** For instance, $5x - 1 = 14$ and $5x = 15$ are equivalent equations because they both have the same solution set, $\{3\}$. The equation $x = 3$ is the simplest one which is equivalent to $5x - 1 = 14$.

The standard approach to finding equivalent equations is to perform the same arithmetic operation on both sides of the current equation, such as adding the same number to both sides. This principle is based on the following properties of equality.

Properties of Equality

Let A, B, and C be algebraic expressions.

	Comment
Addition Property If $A = B$, then $A + C = B + C$.	Adding the same number to both sides of an equation produces an equivalent equation.
Subtraction Property If $A = B$, then $A - C = B - C$.	Subtracting the same number from both sides of an equation produces an equivalent equation.
Multiplication Property If $A = B$ and $C \neq 0$, then $AC = BC$.	Multiplying both sides of an equation by the same nonzero number produces an equivalent equation.
Division Property If $A = B$ and $C \neq 0$, then $\frac{A}{C} = \frac{B}{C}$.	Dividing both sides of an equation by the same nonzero number produces an equivalent equation.

Progress Check Answer

2. Only **b**

Consider carefully how these properties are used to solve the equations in the next example.

EXAMPLE 3 Solve $5x - 1 = 14$.

Solution The goal is to isolate x. Because 1 is subtracted from $5x$, we begin by adding 1 to both sides of the equation.

$$\begin{aligned} 5x - 1 &= 14 && \text{Given equation.} \\ 5x - 1 + 1 &= 14 + 1 && \text{Add 1 to both sides.} \\ 5x &= 15 \end{aligned}$$

Since at this point x is multiplied by 5, we continue by dividing both sides by 5.

$$\begin{aligned} \frac{5x}{5} &= \frac{15}{5} && \text{Divide both sides by 5.} \\ x &= 3 \end{aligned}$$

Thus, 3 is the solution to the equation. To check this, we replace x by 3 in the original equation.

Check

$$\begin{aligned} 5x - 1 &= 14 && \text{Original equation.} \\ 5(3) - 1 &\stackrel{?}{=} 14 && \text{Replace } x \text{ by 3.} \\ 15 - 1 &\stackrel{?}{=} 14 \\ 14 &\stackrel{\checkmark}{=} 14 \end{aligned}$$

Since $14 = 14$ is a true statement, the solution checks, and the solution set is $\{3\}$.

PROGRESS CHECK 3 Solve $3x - 5 = 7$.

Encourage clear writing. Suggest that students write one complete equation per line and that they describe how that line follows from the previous one.

Point out that it makes just as much sense to multiply both sides by $\frac{1}{5}$; some students may prefer a consistent approach through multiplication.

Some linear equations contain fractions. For these equations it is usually easier to first clear the equation of fractions by multiplying both sides by the least common denominator (LCD). If a side of the equation contains more than one term, the multiplication on that side is carried out according to the distributive property.

Some students may prefer to refer to the least common multiple (LCM) of the denominators.

EXAMPLE 4 Solve $\frac{3}{4}x + \frac{1}{3} = \frac{1}{6}x$.

Solution The LCD is 12, so multiplication by 12 will clear fractions and leave a simpler equivalent equation.

$$\begin{aligned} \frac{3}{4}x + \frac{1}{3} &= \frac{1}{6}x && \text{Given equation.} \\ 12\left(\frac{3}{4}x + \frac{1}{3}\right) &= 12\left(\frac{1}{6}x\right) && \text{Multiply both sides by 12.} \\ 9x + 4 &= 2x \end{aligned}$$

Because terms involving x appear on both sides of the equation, we subtract $9x$ from both sides, which isolates the variable term on one side of the equation.

$$\begin{aligned} 9x + 4 - 9x &= 2x - 9x && \text{Subtract } 9x \text{ from both sides.} \\ 4 &= -7x \\ \frac{4}{-7} &= \frac{-7x}{-7} && \text{Divide both sides by } -7. \\ -\frac{4}{7} &= x \end{aligned}$$

The solution set is therefore $\{-\frac{4}{7}\}$. Check this result by substituting $-\frac{4}{7}$ for x in the original equation.

Remind students that $\frac{-a}{b}$, $\frac{a}{-b}$, and $-\frac{a}{b}$ are all equivalent.

Progress Check Answer

3. $\{4\}$

PROGRESS CHECK 4 Solve $\frac{1}{2}x + \frac{3}{4} = \frac{5}{8}$.

The examples shown so far illustrate the general approach to solving linear equations. The method is summarized below.

Solution of Linear Equations

1. Simplify each side of the equation if necessary. This step involves mainly combining like terms, sometimes preceded by using the distributive property to remove parentheses. If necessary, clear the equation of fractions.
2. Use the addition or subtraction properties of equality, if necessary, to write an equivalent equation of the form $cx = n$. To accomplish this, write equivalent equations with all terms involving the unknown on one side of the equation and all constant terms on the other side.
3. Use the multiplication or division properties of equality to solve $cx = n$ if $c \neq 1$. The result of this step will read $x =$ number.
4. Check the solution by substituting it in the original equation.

Using this procedure, we can now solve the problem which opens the chapter.

EXAMPLE 5 Solve the problem which opens the chapter on page 42.

Solution We are asked to solve $200 + 0.10x = 0.15x$, where x represents the amount of sales at which both jobs pay the same. This equation is derived as follows.

pay at store 1		*pay at store 2*
200 plus 10 percent of sales	equals	15 percent of sales
$200 + 0.10x$	$=$	$0.15x$

It is not necessary to do any simplification, so we begin with step 2 of the general procedure.

This is an opportunity to stress the distinction between *factors* and *terms*, a major point but one which sinks in slowly.

$$200 + 0.10x - 0.10x = 0.15x - 0.10x \quad \text{Subtract } 0.10x \text{ from both sides.}$$
$$200 = 0.05x$$
$$\frac{200}{0.05} = \frac{0.05x}{0.05} \quad \text{Divide both sides by 0.05.}$$
$$4{,}000 = x$$

Check

$$200 + 0.10x = 0.15x \quad \text{Original equation.}$$
$$200 + 0.10(4{,}000) \stackrel{?}{=} 0.15(4{,}000) \quad \text{Replace } x \text{ by 4,000.}$$
$$200 + 400 \stackrel{?}{=} 600$$
$$600 \stackrel{\checkmark}{=} 600$$

The check shows that at \$4,000 in sales both jobs pay \$600. For sales above \$4,000 the straight 15 percent commission pays more. For sales below \$4,000 the first job pays more.

Note As a first step, we could have multiplied both sides by 100 to clear the decimals. This would have yielded the equation $20{,}000 + 10x = 15x$, which also has 4,000 as a solution.

Progress Check Answer

4. $\{-\frac{1}{4}\}$

PROGRESS CHECK 5 Redo Example 5, but assume that one store pays $200 per week plus a 12 percent commission, while the other pays $150 plus a 16 percent commission.

The next example shows what happens when the steps for solving a linear equation are applied to a false linear equation.

EXAMPLE 6 Solve the equation $4(x - 3) + 1 = 3(x - 2) + x$.

Solution The first step in solving this equation is to simplify both sides by using the distributive property and combining like terms.

$$\begin{aligned} 4(x - 3) + 1 &= 3(x - 2) + x && \text{Given equation.} \\ 4x - 12 + 1 &= 3x - 6 + x && \text{Distributive property.} \\ 4x - 11 &= 4x - 6 && \text{Combine like terms.} \end{aligned}$$

At this point we can recognize that this is a false equation, because it says that 11 less than some number is the same as 6 less than that same number, which is impossible. But if we do not spot this contradiction, we would continue. A reasonable next step is to subtract $4x$ from both sides.

$$\begin{aligned} 4x - 11 - 4x &= 4x - 6 - 4x && \text{Subtract } 4x \text{ from both sides.} \\ -11 &= -6 \end{aligned}$$

This result is certainly a false statement, which means that the original equation was also false, and so there are no numbers in the solution set. Thus, the solution set is $\emptyset$.

Note If a linear equation is an *identity*, then applying the usual steps will eventually lead to an equation where both sides are exactly the same, such as $2x + 3 = 2x + 3$. When you reach such an equation, you can conclude that the original equation is an identity.

PROGRESS CHECK 6 Solve $2(x + 1) = 3(x + 2) - x$.

Progress Check Answers

5. Both jobs pay $350 on $1,250 in sales.

6. $\emptyset$

EXERCISES 2.1

In Exercises 1–14, classify each equation as an identity, a false equation, or a conditional equation.

1. $x - x = 0$ Identity
2. $x - 2 = 10$ Conditional equation
3. $x - 4 = 4 - x$ Conditional equation
4. $3x = x(3)$ Identity
5. $4(x + 1) = 4 + 4x$ Identity
6. $x = x + 1$ False equation
7. $x = x - 1$ False equation
8. $2x + x = 3x$ Identity
9. $x + 2 = x - 1$ False equation
10. $4x = 3 + x$ Conditional equation
11. $x + x = 2$ Conditional equation
12. $x - x = 1$ False equation
13. $x + (2x - 5) = 4x - (5 + x)$ Identity
14. $(x + 2)(x - 2) = x^2 - 4$ Identity

In Exercises 15–28, state whether the given equation is or is not a linear equation in one variable.

15. $x = 4y$ No
16. $2x - 3 = 7$ Yes
17. $x = 4$ Yes
18. $x + y = 3$ No
19. $2x + 7$ No
20. $x + 2x = 42$ Yes
21. $2x - 3 = y$ No
22. $y^3 + 8 = 0$ No
23. $3x + 12 = 0$ Yes
24. $y - 3$ No
25. $xy = 5$ No
26. $x^2 + 2x + 3 = 0$ No
27. $x^2 = 49$ No
28. $3x + 4 = 5$ Yes

In Exercises 29–82, solve the given equation.

29. $5x + 8 = -2$ $\{-2\}$
30. $12x + 48 = 12$ $\{-3\}$
31. $-2 = 10 - 4x$ $\{3\}$
32. $20 - 6x = 2$ $\{3\}$
33. $2(x + 6) = 14$ $\{1\}$
34. $4 - 3x = -2x$ $\{4\}$
35. $3x + 24 = 6x + 6$ $\{6\}$
36. $19 - 4x = x + 4$ $\{3\}$
37. $3x - 4 = 4x - 11$ $\{7\}$
38. $5x - 3 = x + 5$ $\{2\}$
39. $7x + 31 = 20 - 4x$ $\{-1\}$
40. $3x + 7 = 4x - 5$ $\{12\}$
41. $5x - 2x = 36 - x$ $\{9\}$
42. $6x - 6 = 7x - 17$ $\{11\}$
43. $12x + 84 = 100 - 4x$ $\{1\}$
44. $200 + 12x = 4x + 8$ $\{-24\}$
45. $8x + 6 - 2x = 7 + 2x - 13$ $\{-3\}$
46. $2x + 5 + 3x - 4 = x + 6 + 5x$ $\{-5\}$
47. $7(2x + 5) - 6(x + 8) = 7$ $\{\frac{5}{2}\}$
48. $5(3 + 4x) - 6(7x - 5) = 5x - 36$ $\{3\}$
49. $3x - 18 = 3(x + 7)$ False equation; $\emptyset$
50. $5x + 8 = 5(x + 2) - 2$ Identity; all real numbers
51. $3(x - 4) + x = 4x - 12$ Identity; all real numbers
52. $2x + 5 + 3x - 4 = 2 + 5x$ False equation; $\emptyset$
53. $2 - (7x + 5) = 13 - 3x$ $\{-4\}$
54. $19 = 4 - 3(4 - 3x)$ $\{3\}$

55. $x + (3x - 5) = 5x - (2x + 2)$ {3}
56. $2 - 5(x - 3) = 1 + (x - 8)$ {4}
57. $16 = (3x - 3) - (x - 7)$ {6}
58. $3(x - 3) - 2(x - 2) = 7$ {12}
59. $0.2x + 4.5 = 1.2x$ {4.5}
60. $0.30x = 0.28 - 0.05x$ {0.8}
61. $0.27x - 0.39 = 0.17 + 0.19x$ {7}
62. $0.24x - 0.56 = 0.19x - 0.26$ {6}
63. $6x + 4 = 7x + 3 - x$ False equation; ∅
64. $7 - 4x + 3 = 9x + 10 - 13x$ Identity; all real numbers
65. $5x + 1 = 3 + 6x - 2 - x$ Identity; all real numbers
66. $2x - 3 = 4 - x + 3(x + 1)$ False equation; ∅
67. $\frac{3x}{2} = 21$ {14}
68. $\frac{4x - 3}{5} = 5$ {7}
69. $\frac{x - 1}{2} = \frac{x + 1}{3}$ {5}
70. $\frac{3 - 5x}{6} = 3$ {−3}
71. $\frac{x}{2} - \frac{x}{3} = 4$ {24}
72. $\frac{x}{4} - \frac{x}{6} = \frac{1}{2}$ {6}
73. $\frac{2x}{3} + \frac{3x}{4} = 102$ {72}
74. $\frac{x}{4} + \frac{x}{2} = \frac{1}{4}$ $\frac{1}{3}$
75. $\frac{3}{5}x - \frac{21}{5} + 3 = \frac{3}{10}x - \frac{3}{5}$ {2}
76. $\frac{5}{8}x - \frac{3}{8} + 1 = \frac{1}{2}x - \frac{1}{2}$ {−9}
77. $\frac{3}{2}x = \frac{2}{3}x + \frac{5}{2}$ {3}
78. $\frac{3}{2}x - \frac{2}{3} = \frac{4}{3}x + \frac{5}{2}$ {19}
79. $\frac{2}{3}x - \frac{3}{5} = -\frac{1}{3} - \frac{2}{5}x$ {$\frac{1}{4}$}
80. $\frac{2}{5} - \frac{1}{7}x = \frac{3}{7} - \frac{2}{5}x$ {$\frac{1}{9}$}
81. $\frac{x}{2} + \frac{x}{3} + \frac{x}{6} = 24$ {24}
82. $\frac{x}{6} + \frac{x}{16} + \frac{x}{8} = 17$ {48}

In Exercises 83–88 a linear equation is given whose solution can be used to solve the problem. Solve the equation and answer the question.

83. For selling a house, the real estate agent gets 6 percent of the sales price as a commission. A couple wants to sell their house and end up with \$150,000 after they pay the commission. Find the required selling price of the house by letting x represent the unknown and solving the equation $x - 0.06x = 150{,}000$. Round the answer to the nearest dollar. \$159,574
84. In a vacation community the real estate salespeople get a 12 percent commission on rentals. An owner wants to rent out a vacation house for one month and end up with \$2,000 income. Find the required rental price for the house by letting x represent the unknown and solving the equation $x - 0.12x = 2{,}000$. Round to the nearest dollar. \$2,273
85. The agent for a professional athlete gets 10 percent of the pro's annual income. How much does the athlete have to earn so that his own share will be \$1 million? Use the equation $x - 0.10x = 1{,}000{,}000$, where x represents the pro's income. Round to the nearest dollar. \$1,111,111
86. The total amount (including a 6 percent sales tax) for the sale of an automobile was \$18,550. How much did the auto cost before the tax was added? Solve $x + 0.06x = 18{,}550$. \$17,500
87. One sales job pays \$160 per week plus a 10 percent commission on sales. A second pays \$130 per week plus a 12 percent commission. At what point of sales do the jobs pay the same? Solve $160 + 0.10x = 130 + 0.12x$. \$1,500
88. A waiter can be paid \$300 per week or \$100 per week plus 15 percent of the cost of all the meals the waiter serves. How many dollars worth of meals would have to be served so that the second scheme paid more? Find out (to the nearest dollar) by solving $300 = 100 + 0.15x$, where x stands for the cost of the meals. \$1,333
89. You have \$20 to spend in a restaurant, which must cover the menu price, a 15 percent tip, and a 6 percent sales tax. What is the most expensive meal (x) you can order? Write an equation and solve it. $x + 0.15x + 0.06x = 20$; \$16.52
90. Refer to Exercise 89. Suppose you decide to leave a 20 percent tip. Now what is the most expensive meal you can order? $x + 0.20x + 0.06x = 20$; \$15.87
91. You can rent a stall in a marketplace for one day for \$50. You will sell crafts which cost you \$5 apiece for \$6.50 each. How many items (x) must you sell to make \$100 profit? Write an equation and solve it. $6.50x - 5.00x - 50 = 100$; 100 items
92. Refer to Exercise 91. Suppose the rental fee for the stall is increased to \$60 for the day. Now how many items (x) must you sell to make \$100 profit? $6.50x - 5.00x - 60 = 100$; 107 items

THINK ABOUT IT

1. a. Write in words what this equation says, and then explain why it must be an identity: $x + x = 2x$.
   b. Write in words what this equation says. Is it an identity or a false equation? $x - 1 = x + 1$.
2. What are equivalent equations? Explain and give an example.
3. What is the difference between a conditional equation and an identity?
4. If an equation has the form $ax + b = c$, what is the first step in solving for x? What is the second step? (Assume $a \neq 0$.)
5. When an equation contains fractions, why is it often a good idea to multiply both sides by the LCD of the fractions?

REMEMBER THIS

1. What is the formula for the volume of a circular cylinder with radius r and height h? $V = \pi r^2 h$
2. True or false? π is exactly equal to 3.14. False
3. Evaluate $A = \frac{1}{3}(2f + m)$ when $f = 89$ and $m = 80$. 86
4. What is the result of subtracting $5x$ from $5x - y$? $-y$
5. What is the result of subtracting b from 0? $-b$
6. Simplify $x - (3x - 1)$. $-2x + 1$
7. Use a calculator to compute $3(1 + 0.04)^{10}$ to the nearest hundredth. 4.44
8. Simplify $3x - 2(x - 1)$. $x + 2$
9. Rewrite $3(x + y)$ using the distributive property. $3x + 3y$
10. True or false? $-3 \leq -3$. True

2.2 Formulas and Literal Equations

For a report on weather in Canada a student must convert many temperature measurements from degrees Celsius to degrees Fahrenheit. In a reference book the student finds the formula $C = \frac{5}{9}(F - 32)$ that can be used to make the conversions, but inefficiently. Solve this formula for F in terms of C to obtain an efficient formula for the student's needs. (See Example 3.)

OBJECTIVES

1 **Find the value of a variable in a formula when given values for the other variables.**

2 **Solve a given formula or literal equation for a specified variable.**

1 To analyze relationships, we often use **literal equations,** which are equations that contain two or more letters. The letters may represent any mix of variables and constants. Common examples of literal equations are formulas, such as the simple interest formula $I = Prt$, and general forms of equation such as $ax + b = 0$. Because literal equations and formulas are types of equations, we may use the equation-solving techniques developed in this chapter to analyze them.

EXAMPLE 1 The formula $V = \frac{1}{3}\pi r^2 h$ gives the volume of a cone, shown in Figure 2.1, where the radius is r and the height is h. Find h when $r = 3$ in. and $V = 15$ in.^{3}. Use 3.14 to approximate π. Round the answer to the nearest hundredth.

Solution Substitute the given values into the formula and then solve for h.

$$V = \tfrac{1}{3}\pi r^2 h \quad \text{Given formula.}$$
$$15 = \tfrac{1}{3}(3.14)(3)^2 h \quad \text{Substitute given values.}$$
$$15 = 9.42h \quad \text{Simplify.}$$
$$\frac{15}{9.42} = \frac{9.42h}{9.42} \quad \text{Divide both sides by 9.42.}$$
$$1.59 = h \quad \text{Rounding to the nearest hundredth.}$$

So the height of the cone is approximately 1.59 in.

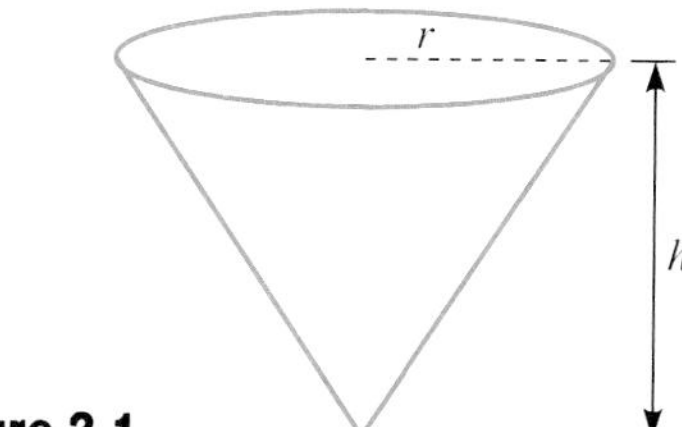

Figure 2.1

You could point out that the volume of any right solid (like a circular cylinder) is the product of the base area and the height. But if the figure comes to a point, the volume is $\frac{1}{3}$ the previous value.

Note When you check the solution by replacing each letter by its value, you will get $15 = 14.9778$, which indicates that 1.59 is approximately right. For more accuracy you would need to include more decimal places in the approximation for π. This problem occurs whenever irrational numbers (like π) are replaced by rational approximations (like 3.14).

PROGRESS CHECK 1 Use the formula in Example 1 to find h when $r = 6$ in. and $V = 44$ in.^{3}. Use 3.14 for π, and round the answer to the nearest hundredth.

EXAMPLE 2 If a professor counts a final exam (f) twice as much as a midterm exam (m), then a formula for averaging them is $A = \frac{1}{3}(2f + m)$. A is called a *weighted average* of f and m. If a student gets 80 on the midterm, what grade is needed on the final so that A equals 86?

Solution Replace the variables by their given values and solve for f. In the process, multiply both sides by 3 to clear fractions.

$$\begin{aligned} A &= \tfrac{1}{3}(2f + m) && \text{Given formula.} \\ 86 &= \tfrac{1}{3}(2f + 80) && \text{Substitute given values.} \\ 258 &= 2f + 80 && \text{Multiply both sides by 3.} \\ 178 &= 2f && \text{Subtract 80 from both sides.} \\ 89 &= f && \text{Divide both sides by 2.} \end{aligned}$$

Thus, a grade of 89 is required on the final. Verify that this solution checks exactly.

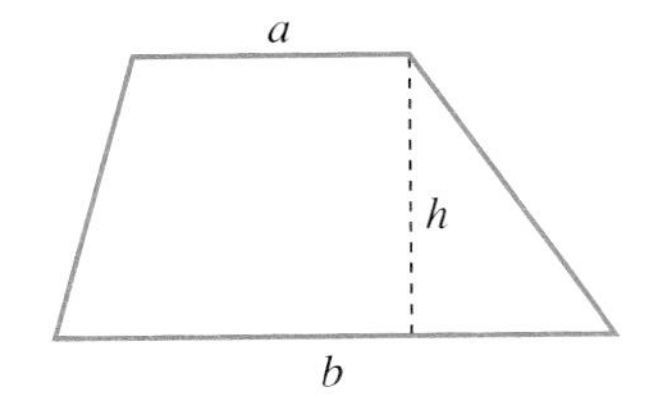

Figure 2.2

PROGRESS CHECK 2 The formula $A = \frac{1}{2}(a + b)h$ gives the area of a trapezoid with height h and bases a and b; see Figure 2.2. Find b if $h = 9$ centimeters (cm), $a = 11$ cm, and $A = 108$ cm^2.

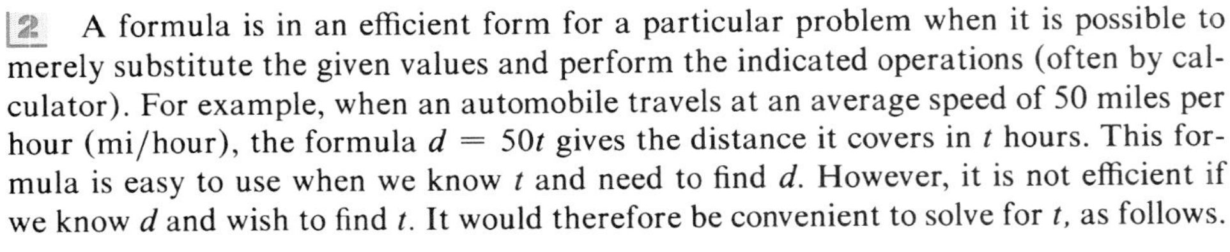
2 A formula is in an efficient form for a particular problem when it is possible to merely substitute the given values and perform the indicated operations (often by calculator). For example, when an automobile travels at an average speed of 50 miles per hour (mi/hour), the formula $d = 50t$ gives the distance it covers in t hours. This formula is easy to use when we know t and need to find d. However, it is not efficient if we know d and wish to find t. It would therefore be convenient to solve for t, as follows.

$$\begin{aligned} d &= 50t && \text{Given equation.} \\ \frac{d}{50} &= \frac{50t}{50} && \text{Divide both sides by 50.} \\ \frac{d}{50} &= t \end{aligned}$$

Thus $d = 50t$ is equivalent to $t = d/50$, and we may select the more useful version of the formula depending on the given information.

EXAMPLE 3 Solve the problem in the section introduction on page 49.

This may be a new procedure for some students. Remind students that the product of a number and its reciprocal is 1.

Solution We will solve for F so that the form of the result is the common formula for converting from degrees Celsius to degrees Fahrenheit. We begin by multiplying both sides of the equation by $\frac{9}{5}$ to eliminate the fraction $\frac{5}{9}$ from the right-hand side.

$$\begin{aligned} C &= \tfrac{5}{9}(F - 32) && \text{Given formula.} \\ \tfrac{9}{5}C &= \tfrac{9}{5} \cdot \tfrac{5}{9}(F - 32) && \text{Multiply both sides by } \tfrac{9}{5}. \\ \tfrac{9}{5}C &= F - 32 \\ \tfrac{9}{5}C + 32 &= F && \text{Add 32 to both sides.} \end{aligned}$$

Thus, the student should use the formula $F = \frac{9}{5}C + 32$ to make the temperature conversions.

Progress Check Answers

1. 1.17 in.
2. $b = 13$ cm

Note In this example it is also logical to clear fractions as a first step by multiplying both sides by 9. This approach leads to the result

$$F = \frac{9C + 160}{5}.$$

When rearranging formulas, there may be more than one useful version.

PROGRESS CHECK 3 Solve $A = P + Prt$ for t.

In Chapters 5 and 6 in questions dealing with graphs of lines, we will need to solve equations of the form $ax + by = c$ for y. This type of problem is discussed in Example 4.

EXAMPLE 4 Solve $5x - y = 2$ for y.

Solution The goal is to isolate y.

$$\begin{aligned} 5x - y &= 2 && \text{Given equation.} \\ 5x - 5x - y &= 2 - 5x && \text{Subtract } 5x \text{ from both sides.} \\ -y &= 2 - 5x \\ (-1)(-y) &= -1(2 - 5x) && \text{Multiply both sides by } -1. \\ y &= -2 + 5x && \text{Simplify using the distributive property on the right.} \end{aligned}$$

It is instructive to solve this equation two ways and point out that both approaches are efficient. A second approach starts by adding y to both sides.

The answer may be written as $y = -2 + 5x$ or $y = 5x - 2$.

PROGRESS CHECK 4 Solve $5x - y = -7$ for y.

The final example of this section derives the solution of the general form of a linear equation in one variable. This example proves that each linear equation has exactly one solution, $-b/a$. The quest for the general solution of equations of higher degrees was one of the great problems in the history of mathematics.

EXAMPLE 5 Solve $ax + b = 0$ with $a \neq 0$ for x.

Solution Isolate x in the usual way.

$$\begin{aligned} ax + b &= 0 && \text{Given equation.} \\ ax + b - b &= 0 - b && \text{Subtract } b \text{ from both sides.} \\ ax &= -b \\ \frac{ax}{a} &= \frac{-b}{a} && \text{Divide both sides by } a. \\ x &= -\frac{b}{a} \end{aligned}$$

The restriction $a \neq 0$ is needed, because $-b/0$ is undefined.

PROGRESS CHECK 5 Solve $rx - s = t$ with $r \neq 0$ for x.

Progress Check Answers

3. $t = \frac{A - P}{Pr}$
4. $y = 5x + 7$ or $y = 7 + 5x$
5. $x = \frac{s + t}{r}$

EXERCISES 2.2

In Exercises 1–32, find the value of the indicated variable in each formula. Use $\pi = 3.14$. Round your answer to the nearest hundredth when necessary.

1. $A = \frac{1}{2}bh$; find h when $A = 54$ and $b = 9$. 12
2. $I = Prt$; find P when $I = 150$, $r = 0.06$, and $t = 1$. 2500
3. $F = ma$; find a when $m = 10$ and $F = 0$. 0
4. $F = ma$; find a when $m = 20$ and $F = 0$. 0
5. $V = s^3$; find V when $s = 3.20$. 32.77
6. $V = \frac{4}{3}\pi r^3$; find V when $r = 6$. 904.32
7. $A = \frac{1}{2}h(a + b)$; find b when $A = 20$, $h = 2$, and $a = 6$. 14
8. $A = \frac{1}{2}h(a + b)$; find a when $A = 120$, $h = 12$, and $b = 3$. 17
9. $2x + 5y = -26$; find y if $x = -3$. −4
10. $7x + 3y = 7$; find y if $x = 4$. −7
11. $F = \frac{9}{5}C + 32$; if $F = 212$, find C. 100

12. $F = \frac{9}{5}C + 32$; if $F = 32$, find C. 0
13. $R = \alpha + x\beta$; find x if $\alpha = 4.35$, $\beta = 5.55$, and $R = 7.29$. 0.53
14. $R = \alpha + x\beta$; find x if $\alpha = 2.9$, $\beta = 7.62$, and $R = 10.83$. 1.04
15. $7x + 5y = 11$ and $x = 3$; find y. −2
16. $4y - 3x = 29$ and $y = 2$; find x. −7
17. $\frac{1}{2}x = \frac{2}{3}y + 4$ and $x = 24$; find y. 12
18. $\frac{2}{5}y = \frac{4}{3}x + 2$ and $x = 3$; find y. 15
19. $V = \pi r^2 h$, $V = 200$, and $r = 2.5$; find h. 10.19
20. $V = \ell wh$, $V = 63$, $\ell = 7$, and $h = 1.5$; find w. 6
21. $S = t - \frac{1}{2}gt^2$, $S = -15$, and $t = 1$; find g. 32
22. $T = 2\pi rh + \pi r^2$, $T = 31.4$, and $r = 2$; find h. 1.5
23. $\ell = a + (n - 1)d$, $\ell = 59$, $a = 2$, and $d = 3$; find n. 20
24. $T = mg - mf$, $T = 56$, $g = 37$, and $f = 9$; find m. 2
25. $A = \dfrac{(n-2)180}{n}$ and $A = 60$; find n. 3
26. $T = \dfrac{n+1}{2n}$ and $T = 4$; find n. 0.14
27. $A = \dfrac{\pi r^2 n}{90}$, $r = 4$, and $A = 50$; find n. 89.57
28. $V = 2\pi r(r + h)$, $V = 300$, and $r = 5$; find h. 4.55
29. $S = \dfrac{a}{1-r}$, $S = 12$, and $a = 7$; find r. 0.42
30. $A = \pi r^2 x + \pi r^3$, $A = 50$, and $r = 2$; find x. 1.98
31. $\ell = ar^{n-1}$, $\ell = 175$, $n = 3$, and $r = 5$; find a. 7
32. $C = \dfrac{nE}{1 + nr}$, $C = 1$, $n = 3$, and $r = 5$; find E. $\frac{16}{3}$

In Exercises 33–64, solve each formula for the variable indicated.

33. $A = \frac{1}{2}bh$ for b $\quad b = \dfrac{2A}{h}$
34. $V = \ell wh$ for w $\quad w = \dfrac{V}{\ell h}$
35. $V = \frac{4}{3}\pi r^3$ for r^3 $\quad r^3 = \dfrac{3V}{4\pi}$
36. $V = \pi r^2 h$ for h $\quad h = \dfrac{V}{\pi r^2}$
37. $\ell = a + (n - 1)d$ for a $\quad a = \ell - (n-1)d$
38. $\ell = a + (n - 1)d$ for d $\quad d = \dfrac{\ell - a}{n - 1}$
39. $\ell = a + (n - 1)d$ for n $\quad n = \dfrac{\ell + d - a}{d}$
40. $A = \dfrac{p + q + r}{3}$ for q $\quad q = 3A - p - r$
41. $S = \frac{1}{2}n(a + \ell)$ for a $\quad a = \dfrac{2S - n\ell}{n}$
42. $S = \dfrac{a}{1 - r}$ for r $\quad r = \dfrac{a - S}{-S}$ or $r = \dfrac{S - a}{S}$
43. $y = mx + b$, $m \neq 0$, for x $\quad x = \dfrac{y - b}{m}$
44. $w = \frac{1}{2}g^2 t$ for t $\quad t = \dfrac{2w}{g^2}$
45. $A = \dfrac{\pi r^2 n}{90}$ for n $\quad n = \dfrac{90A}{\pi r^2}$
46. $A = \frac{1}{2}h(a + b)$ for b $\quad b = \dfrac{2A - ah}{h}$
47. $3x - y = 0$ for y $\quad y = 3x$
48. $4x - 2y = 0$ for y $\quad y = 2x$
49. $3x - y = 7$ for y $\quad y = 3x - 7$
50. $4x - y = 12$ for y $\quad y = 4x - 12$
51. $x - y - 3 = 0$ for y $\quad y = x - 3$
52. $2x - y + 2 = 0$ for y $\quad y = 2x + 2$
53. $-0.6x - 0.5y = 2$ for y $\quad y = -\frac{6}{5}x - 4$
54. $0.5x - 0.2y = 0$ for y $\quad y = \frac{5}{2}x$
55. $\dfrac{x - y}{4} = 3$ for y $\quad y = x - 12$
56. $\dfrac{x + y}{2} = 8$ for y $\quad y = -x + 16$
57. $\frac{2}{3}x + \frac{1}{4}y = 4$ for y $\quad y = -\frac{8}{3}x + 16$
58. $\frac{4}{5}x - \frac{1}{2}y = 2$ for y $\quad y = \frac{8}{5}x - 4$
59. $x - y = -3$ for y $\quad y = x + 3$
60. $2x - y = -2$ for y $\quad y = 2x + 2$
61. $rx - s = 0$, $r \neq 0$, for x $\quad x = \dfrac{s}{r}$
62. $py + q = 0$, $p \neq 0$, for y $\quad y = \dfrac{-q}{p}$
63. $ax - b = c$, $a \neq 0$, for x $\quad x = \dfrac{c + b}{a}$
64. $mx + b = y$, $x \neq 0$, for m $\quad m = \dfrac{y - b}{x}$

65. The formula $h = \dfrac{w + 220}{5.5}$ shows the approximate relationship of "normal" weight to height for adult males, where h is height in inches and w is weight in pounds.
  a. Solve this formula for w. $w = 5.5h - 220$
  b. How many pounds should a 5 ft 6 in. man "normally" weigh? 143

66. The formula $h = \dfrac{w + 200}{5}$ shows the approximate relationship of "normal" weight to height for adult females, where h is height in inches and w is weight in pounds.
  a. Solve this formula for w. $w = 5h - 200$
  b. How many pounds should a 5-ft 6-in. woman "normally" weigh? 130

67. The formula for converting kilometers (k) to miles (m) is $m = 0.6214k$.
  a. Solve the formula for k. $k = 1.609m$
  b. The circumference of the earth is about 25,000 mi. Find this distance to the nearest thousand kilometers (km). 40,000 km

68. The formula for converting centimeters (c) to inches (i) is $i = 0.3937c$.
  a. Solve the formula for c. $c = 2.54i$
  b. A yardstick is 36 in. long. Find this length to the nearest centimeter. 91 cm

69. The speed of sound in air is about 1,100 ft/second. The formula is given as $d = 1{,}100t$, where d is the distance in feet and t is the time in seconds.
  a. Solve this formula for t. Use fractional form. $t = \dfrac{d}{1{,}100}$
  b. To the nearest tenth of a second, how long does it take a yodel to echo back from a surface which is 750 ft away? (*Hint:* The sound must travel 1,500 ft.) 1.4 seconds

70. The area of an ellipse is given by $A = \pi ab$, where a and b are as shown.

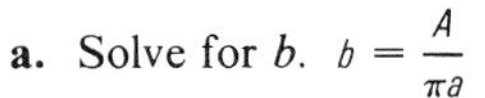

  a. Solve for b. $b = \dfrac{A}{\pi a}$
  b. If the area is 226 cm² and a is 6 cm, find b to the nearest hundredth of a centimeter. 12.00 cm

THINK ABOUT IT

1. The area of an ellipse is given by $A = \pi ab$, where a and b are as shown in the figure. What happens to the formula for A if a and b are equal? Explain, by referring to the diagram, why your answer is sensible.

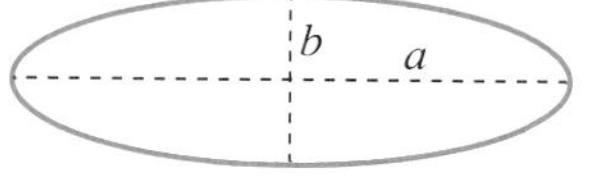

2. The area of a trapezoid (see Figure 2.2) is given by $A = \frac{1}{2}h(a + b)$.
   a. What happens to the formula if a and b are equal? For instance, suppose they are both equal to b. Why is this sensible?
   b. What happens to the formula if a shrinks to zero? Why is this sensible?
3. Explain why the formulas $A = \frac{1}{2}bh$ and $A = bh/2$ are equivalent.
4. The formula $S = \frac{1}{2}n(n + 1)$ gives the sum of the integers from 1 to n.
   a. To see that the formula is reasonable, let $n = 3$ and use the formula to show that $1 + 2 + 3$ is equal to 6.
   b. Find the sum of the integers from 1 to 999.
   c. Explain why these three versions of the formula are equivalent. Then use each version to find the sum of the numbers from 1 to 100.
      i. $S = \dfrac{(n + 1)}{2} \cdot n$ ii. $S = (n + 1) \cdot \dfrac{n}{2}$
      iii. $S = [(n + 1)(n)] \div 2$
5. A formula of the form $y = kx^p$ is called a **power function.**
   a. Solve the formula for k. (*Hint:* Divide both sides by the expression that is multiplying k.)
   b. Show that the formula for the volume of a sphere, $V = \frac{4}{3}\pi r^3$, is a power function by stating the values of k and p in this formula.

REMEMBER THIS

1. A's income is $\frac{1}{4}$ of B's income. If x represents B's income, what expression represents A's income? $\frac{1}{4}x$
2. If x represents some integer, what expression represents the next consecutive integer after x? $x + 1$
3. What is the sum of the angles of a triangle? 180°
4. What is the sum of x and $10{,}000 - x$? 10,000
5. A container holds x oz of a liquid which is 30 percent alcohol. What expression represents the amount of alcohol in the container? $0.30x$
6. Solve $-3x + 9 = 1$. $\{\frac{8}{3}\}$
7. Evaluate $\dfrac{a + b^2}{c - d^2}$ if $a = -1$, $b = -2$, $c = -3$, and $d = -4$. $-\frac{3}{19}$
8. Which property of real numbers is used to state that $(x - 4)(-1)$ equals $-x + 4$? Distributive
9. Is -5^4 a positive or negative quantity? Negative
10. Which of these numbers are irrational? $\{-3, 0, \sqrt{2}, \sqrt{121}, \sqrt{200}\}$. $\sqrt{2}$, $\sqrt{200}$

2.3 Applications

When A and B filled out their joint income tax return, A's income was $\frac{1}{4}$ of B's income, and their total income was $47,500. Find each person's individual income. (See Example 1.)

OBJECTIVES

1. **Solve word problems by translating phrases and setting up and solving equations.**
2. **Prove certain statements about integers.**
3. **Solve problems involving geometric figures, percentages, uniform motion, and liquid mixtures.**

1 In this section we combine the previous study of linear equations with the translation of verbal phrases into mathematical statements in order to discuss how to solve

several basic types of word problems. The following steps are recommended by both mathematics and reading specialists as a general approach to solving word problems.

To Solve a Word Problem

1. Read the problem several times. The first reading is a preview and is done quickly to obtain a general idea of the problem. The objective of the second reading is to determine exactly what you are asked to find. Write this down. Finally, read the problem carefully and note what information is given. If possible, display the given information in a sketch or chart.
2. Let a variable represent an unknown quantity (which is usually the quantity you are asked to find). Write down precisely what the variable represents. If there is more than one unknown, represent these unknowns in terms of the original variable.
3. Set up an equation that expresses the relationship between the quantities in the problem.
4. Solve the equation.
5. Answer the question.
6. Check the answer by interpreting the solution in the context of the word problem.

As a first example, we solve the problem which opens this section.

EXAMPLE 1 Solve the section-opening problem on page 53.

Solution We start by using the given condition that one income is $\frac{1}{4}$ of the other. Thus, if we let

$$x = \text{B's income,}$$

$$\text{then} \quad \tfrac{1}{4}x = \text{A's income.}$$

Next, we incorporate the given information that the sum of the incomes must be $47,500.

$$\underbrace{\text{A's income}}_{\frac{1}{4}x} \;\underset{+}{\overset{\text{plus}}{\downarrow}}\; \underbrace{\text{B's income}}_{x} \;\underset{=}{\overset{\text{equals}}{\downarrow}}\; \underbrace{\text{joint income.}}_{47{,}500}$$

Solve the Equation

$$\begin{aligned} 4(\tfrac{1}{4}x + x) &= 4(47{,}500) && \text{Multiply both sides by 4.} \\ x + 4x &= 190{,}000 \\ 5x &= 190{,}000 \\ \frac{5x}{5} &= \frac{190{,}000}{5} && \text{Divide both sides by 5.} \\ x &= \$38{,}000 \end{aligned}$$

Answer the Question B's income is $38,000. Thus A's income is one-fourth of that, or $9,500.

Check the Answer Because $9,500 is $\frac{1}{4}$ of $38,000, and $9,500 plus $38,000 equals $47,500, the solution checks.

PROGRESS CHECK 1 A classic old puzzle which many people answer incorrectly without algebra is this: Together, a bottle and a cork cost $1.10, and the bottle costs a dollar more than the cork. What is the price of each? Try guessing the answer before you use algebra.

Progress Check Answer

1. Bottle $1.05, cork 5¢

2 Historically, the study of algebra has always been connected with problems about the properties of numbers. The next example shows the power of algebra to prove statements that must be true for *infinite* sets of numbers, even though it is not possible to write them all down.

EXAMPLE 2 Show that the sum of *any* 5 consecutive integers is equal to 5 times the middle one.

It is helpful to illustrate this example by a few particular cases.

Solution We begin by letting x represent the smallest of the 5 integers. Then

$x + 1$ = second consecutive integer
$x + 2$ = third consecutive integer (the middle one)
$x + 3$ = fourth consecutive integer
$x + 4$ = fifth consecutive integer.

Set Up an Equation

The sum of five consecutive integers equals 5 times the middle integer.

$$\underbrace{x + (x + 1) + (x + 2) + (x + 3) + (x + 4)}_{\text{The sum of five consecutive integers}} \overset{\text{equals}}{=} \underbrace{5(x + 2)}_{\text{5 times the middle integer.}}$$

To prove the desired statement, we *show that the statement is an identity.*

$$x + (x + 1) + (x + 2) + (x + 3) + (x + 4) = 5(x + 2) \quad \text{Equation which represents the statement to be proved.}$$
$$5x + 10 = 5(x + 2) \quad \text{Combine like terms.}$$
$$5x + 10 = 5x + 10 \quad \text{Distributive property.}$$

The resulting identity, $5x + 10 = 5x + 10$, is true for all replacements of x, and so *any* 5 consecutive integers have the desired property.

PROGRESS CHECK 2 Show that the sum of any 3 consecutive integers is equal to 3 times the middle integer.

3 The rest of this section is devoted to illustrations of certain typical problems which lead to linear equations in one variable and can therefore be solved by the methods studied so far. These problems can serve as models for others which you may have to solve.

Geometric Problems

EXAMPLE 3 One of the two acute angles in a right triangle is 20° greater than the other. What are the measures of all the angles?

Solution By definition, a right triangle contains a right angle, so one of the angles measures 90°. To find the measures of the acute angles, let

x = measure of the smaller acute angle;
then $x + 20$ = measure of the larger acute angle.

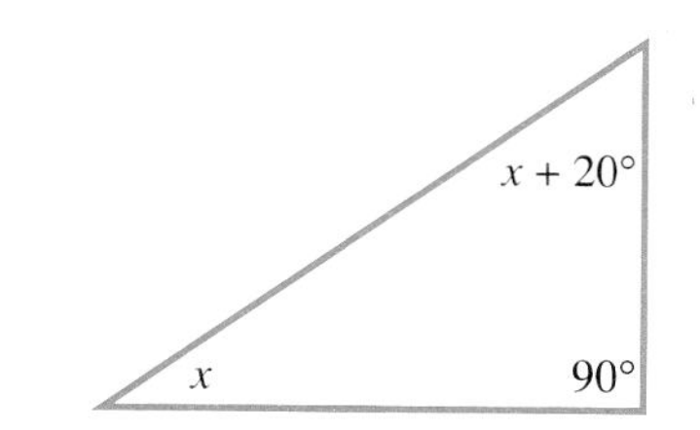

Figure 2.3

Draw a sketch of the problem as in Figure 2.3. Because the sum of the angle measures in a triangle is 180°, we can *set up an equation.*

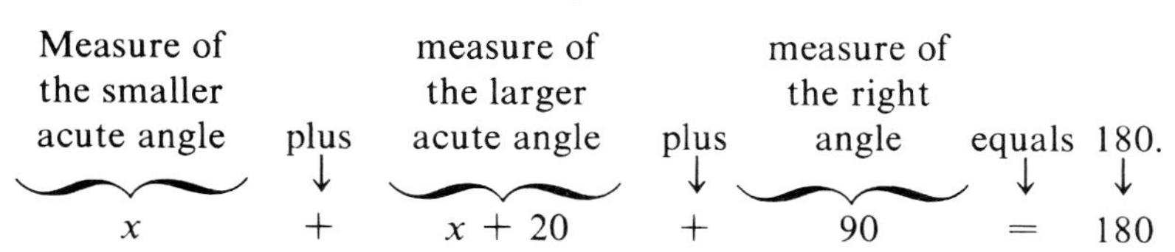

Measure of the smaller acute angle
plus
measure of the larger acute angle
plus
measure of the right angle
equals
180.

$$x + (x + 20) + 90 = 180$$

Progress Check Answer

2. $3x + 3 = 3(x + 1)$

Solve the Equation

$$2x + 110 = 180 \quad \text{Combine like terms.}$$
$$2x = 70 \quad \text{Subtract 110 from each side.}$$
$$x = 35 \quad \text{Divide both sides by 2.}$$

Answer the Question The angle measures are 35°, 55° (from 35 + 20), and 90°.

Check the Answer Because 35 + 55 + 90 = 180, and 55 is 20 greater than 35, the solution checks.

Note In any right triangle the sum of the measures of the two acute angles is 90°. Therefore, in this example we can find x by solving

$$x + (x + 20) = 90.$$

From geometry, any two angles whose measures add to 90° are called **complementary,** and any two angles whose measures add to 180° are called **supplementary.**

PROGRESS CHECK 3 If one acute angle of a right triangle is 2° less than the other, what are the measures of all the angles in the triangle?

Percentage Problems

You may want to point out that this is a special case of $I = Prt$ where $t = 1$.

For problems involving annual interest the basic relationship is $I = Pr$, where P represents principal, r is the annual interest rate, and I is the amount of interest earned in one year.

EXAMPLE 4 Last year, a student made two loans, one to pay for a car purchase (at a 9 percent annual interest rate) and one to pay for college tuition (at 5 percent). The total loan amount was $10,000, and the total interest paid for the year was $620. What was the amount of each loan?

Solution To find the amount borrowed at each rate, let

$$x = \text{amount borrowed at 9 percent;}$$
then $$10{,}000 - x = \text{amount borrowed at 5 percent.}$$

It is helpful to analyze each loan in a chart format.

Loan	Principal ·	Interest Rate =	Interest
Car	x	0.09	$0.09x$
Tuition	$10{,}000 - x$	0.05	$0.05(10{,}000 - x)$

Set up an equation using 620 as the total amount of interest.

$$\underbrace{0.09x}_{\text{Interest on car loan}} \underset{\text{plus}}{+} \underbrace{0.05(10{,}000 - x)}_{\text{interest on tuition}} \underset{\text{equals}}{=} \underbrace{620}_{\text{total interest.}}$$

Solve the Equation

$$0.09x + 500 - 0.05x = 620 \quad \text{Distributive property.}$$
$$0.04x + 500 = 620 \quad \text{Combine like terms.}$$
$$0.04x = 120 \quad \text{Subtract 500 from each side.}$$
$$x = \frac{120}{0.04} \quad \text{Divide both sides by 0.04.}$$
$$x = 3{,}000$$

Progress Check Answer

3. 44°, 46°, 90°

Answer the Question Since \$3,000 was borrowed for the car loan, then \$7,000 was borrowed for tuition.

Check the Answer The interest on the car loan was 0.09(\$3,000), which is \$270; and the interest on the tuition loan was 0.05(\$7,000), which is \$350—for a total interest of \$620.

PROGRESS CHECK 4 You win \$30,000 in a state lottery. You are advised to put some of the money in a risky investment that earns 15 percent annual interest and the rest in a safe investment that earns 7 percent annual interest. If you want the total interest to be \$3,000 annually, how much should be invested at each rate?

Uniform Motion Problems

Uniform motion is the name given to problems which can be analyzed using the formula $d = rt$, where d represents the distance traveled in time t by an object moving at a constant rate r. This formula applies to objects moving at a *constant* (or uniform) speed and to objects whose *average* speed is involved. Analyzing the problem with both a chart and a sketch is recommended here.

It is helpful to first ask students a question like this: If it takes you 2 hours to go 100 mi, how fast were you going? Then discuss the distinction between uniform speed and average speed over an interval.

EXAMPLE 5 On a video display an air traffic controller notices two planes 120 mi apart and flying toward each other on a collision course. One plane is flying at 500 mi/hour; the other is flying 300 mi/hour. How much time is there for the controller to prevent a crash?

Solution We need to find how long it takes for the planes to reach each other. Let

$$t = \text{time until the planes meet.}$$

Note that both planes travel for a time t. Now analyze the problem in a chart format, using $d = rt$.

Plane	Rate (mi/hour) ·	Time (hours) =	Distance (mi)
Slower plane	300	t	$300t$
Faster plane	500	t	$500t$

To *set up an equation,* sketch the situation as in Figure 2.4. From the sketch we can see that

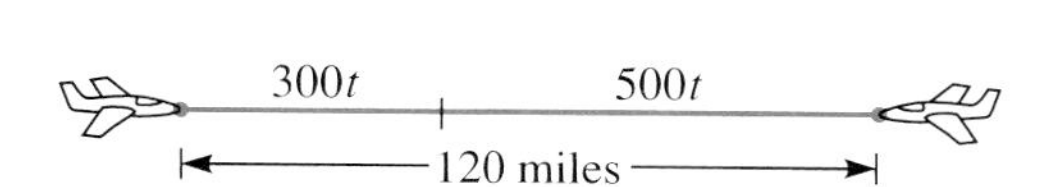

Figure 2.4

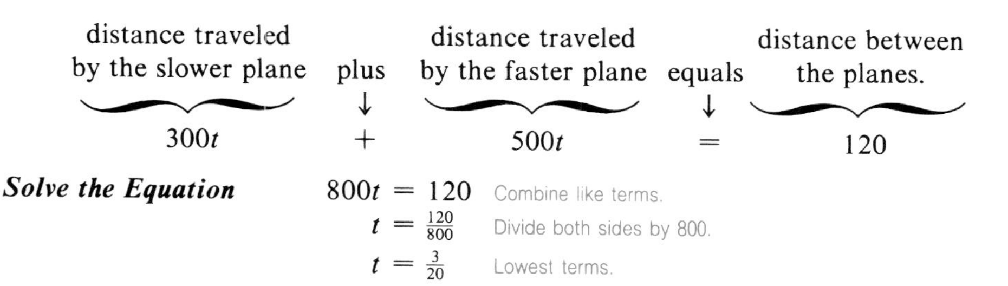

distance traveled by the slower plane
plus
distance traveled by the faster plane
equals
distance between the planes.
300t
+
500t
=
120

Solve the Equation

$$800t = 120 \quad \text{Combine like terms.}$$
$$t = \frac{120}{800} \quad \text{Divide both sides by 800.}$$
$$t = \frac{3}{20} \quad \text{Lowest terms.}$$

Answer the Question The planes reach each other in $\frac{3}{20}$ hour. Thus, the controller has $\frac{3}{20}$ hour, or 9 minutes, to prevent a collision.

Check the Answer In $\frac{3}{20}$ hour the slower plane travels $300(\frac{3}{20})$, or 45, mi, while the faster plane travels $500(\frac{3}{20})$, or 75, mi. The combined distance is 120 mi, so the answer checks.

Progress Check Answer

4. \$18,750 at 7 percent; \$11,250 at 15 percent

PROGRESS CHECK 5 Two snails are crawling toward each other in a straight line at constant speed. One crawls 1 ft/hour. The other crawls 1.5 ft/hour. If they start 5 ft apart, how long will it take them to meet?

Liquid Mixture Problems

To solve problems about liquid mixtures, you need to apply the concept of percentage in the following context:

$$\begin{pmatrix}\text{percent of}\\ \text{an ingredient}\end{pmatrix} \cdot \begin{pmatrix}\text{amount of}\\ \text{solution}\end{pmatrix} = \begin{pmatrix}\text{amount of}\\ \text{the ingredient}\end{pmatrix}.$$

For example, the *amount* of oil in 10 liters of a solution that is 30 percent oil is 0.30(10), or 3, liters. Once again, a chart is recommended for this type of problem.

EXAMPLE 6 A machine shop has two large containers that are each filled with a mixture of oil and gasoline. Container A contains 2 percent oil (and 98 percent gasoline). Container B contains 6 percent oil (and 94 percent gasoline). How much of each should be used to obtain 18 quarts (qt) of a new mixture that contains 5 percent oil?

Solution To find the correct mixture, let

Point out that, in general, if the sum of two numbers is n, then one can be called x and the other $n - x$.

$$x = \text{amount used from container A,}$$
$$\text{so} \quad 18 - x = \text{amount used from container B.}$$

As recommended, we analyze the problem with a chart.

Solution	Percent Oil ·	Amount of Solution (qt) =	Amount of Oil (qt)
Container A	2	x	$0.02x$
Container B	6	$18 - x$	$0.06(18 - x)$
New solution	5	18	$0.05(18)$

To *set up an equation,* we reason that the amount of oil in the new solution is the sum of the amounts contributed by the solutions from containers A and B.

$$\underbrace{0.02x}_{\text{Amount of oil from container A}} \overset{\text{plus}}{+} \underbrace{0.06(18 - x)}_{\text{amount of oil from container B}} \overset{\text{equals}}{=} \underbrace{0.05(18)}_{\text{amount of oil in new solution.}}$$

Solve the Equation

$$\begin{aligned} 0.02x + 1.08 - 0.06x &= 0.90 && \text{Distributive property.}\\ -0.04x + 1.08 &= 0.90 && \text{Combine like terms.}\\ -0.04x &= -0.18 && \text{Subtract 1.08 from both sides.}\\ x &= 4.5 && \text{Divide both sides by } -0.04.\end{aligned}$$

Answer the Question Mix 4.5 qt from container A with 18 − 4.5, or 13.5, qt from container B.

Check the Answer The new solution contains 4.5 + 13.5, or 18, qt. Also, 4.5 qt from container A contains 0.02(4.5) = 0.09 qt of oil, while 13.5 qt from container B contains 0.06(13.5) = 0.81 qt of oil. Thus, the new mixture contains 0.90 qt of oil in 18 qt of mixture. Because 0.90/18 = 0.05 = 5 percent, the new mixture does contain 5 percent oil, and the solution checks.

Progress Check Answer

5. 2 hours

PROGRESS CHECK 6 A chemist has two acid solutions, one 30 percent acid by volume and the other 70 percent acid. How much of each must be used to obtain 10 liters of a solution that is 42 percent acid?

Progress Check Answer

6. 7 liters of 30 percent solution, 3 liters of 70 percent solution

EXERCISES 2.3

1. Two families go apple picking. How can they divide the 300 apples they pick so that one family gets four times as many as the other family? 240 for one family, 60 for the other
2. You separate your collection of 681 marbles into two boxes so that one box has twice as many as the other. How many marbles are in each box? 227, 454
3. A boutique sells an ensemble consisting of a dress and jacket for \$170. If the jacket costs \$40 more than the dress, how much is the jacket? \$105
4. A shopper purchases two belts for \$37. If one belt costs \$12 more than the other, how much is the *more expensive* belt? \$24.50
5. When a 20-ft plank is cut into two pieces, one of the pieces is 11 ft shorter than the other. How long is each piece? 15.5, 4.5 ft
6. If a total of 88 people are working on a freighter, and there are 72 more crew members than officers, how many crew members are on the ship? 80
7. A 48-ft length of cable is cut so that one piece is 7 times the length of the other. How long is each piece? 42, 6 ft
8. A 20-ft-long strip of molding is cut into three different-size lengths so that the middle piece is 3 times the length of the shortest one, and the longest piece is 2 times the length of the middle one. How long is each piece of molding? 2, 6, and 12 ft
9. Three grassy areas in a park total 36,000 ft². The two larger ones have the same area, and the smaller one is $\frac{1}{4}$ the size of either large area.
   a. Find the size of each of the grassy areas. 16,000 ft², 4,000 ft²
   b. Will a concrete patio which takes up 4,000 ft² fit into the smaller grassy area? Yes, exactly
10. Three storage bins in a silo hold a total of 567 bushels of corn. The two larger bins each hold the same volume, which is 10 times the volume of the smallest bin. How many bushels will the smallest bin hold? 27
11. Two partners agree to share the \$60,000 profit from their business. If one partner receives $\frac{2}{3}$ of what the other partner receives, how much profit does each receive? \$36,000, \$24,000
12. Your grades so far are 80, 85, and 93. There is one more test to take. What is the lowest grade you can get on the fourth test and still average 84.5? (*Hint:* The sum of the grades when divided by 4 must equal 84.5.) 80
13. If your grades on the first two unit tests are each 85, is it still possible to bring your average up to 89.5 if only one more test is to be taken? What is the lowest grade, to the nearest whole number, that will do it? Yes; 99
14. The production staff manufactured 55,000 widgets in April and 48,500 widgets in May. The company will show a second-quarter profit if the average production for the quarter (April, May, and June) is 54,000 or more widgets. What is the least number of widgets that must be manufactured in June to show a profit for the second quarter? 58,500
15. Assembly line robots work 24-hour days and never need a vacation; but when repairs are needed, production must be stopped. On a particular robot assembly line, repairs are deemed necessary whenever the line produces an average of less than 1,500 finished pieces per hour. During the first 3 hours of production, the line has turned out 1,501, 1,504, and 1,495 pieces, respectively. What's the least number of pieces that must be manufactured during the next hour so that the assembly line is not taken out of service for repairs? 1,500
16. Four less than 3 times a number is 8 more than twice the number. Find the number. 12
17. Six more than 4 times a number is 4 less than 5 times the number. What is the number? 10
18. One-fourth of a number is equal to 12 less than the number. What is the number? 16
19. One-fourth of a number is equal to 12 more than the number. What is the number? -16
20. The sum of three consecutive integers is 132. What are the integers? 43, 44, 45
21. Two numbers add to 883. One-third of the larger number is 51 more than 3 times the smaller one. Find the two numbers. 73, 810
22. The sum of two numbers is 519. If $\frac{1}{3}$ the larger is 7 less than $\frac{1}{2}$ the smaller, what are the numbers? 216, 303

In Exercises 23–30, write and solve equations to prove statements about integers.

23. Show that the sum of any three consecutive integers is 3 less than 3 times the largest. $x + (x + 1) + (x + 2) = 3(x + 2) - 3$
24. Show that the sum of any five consecutive integers is 10 less than 5 times the largest. $x + (x + 1) + (x + 2) + (x + 3) + (x + 4) = 5(x + 4) - 10$
25. Show that the sum of any four consecutive *odd* integers is 4 more than 4 times the second *odd* integer. $x + (x + 2) + (x + 4) + (x + 6) = 4(x + 2) + 4$
26. Show that the sum of any three consecutive *even* integers is 3 times the middle integer. $x + (x + 2) + (x + 4) = 3(x + 2)$
27. Show that the sum of three consecutive integers can*not* be 47. $x + (x + 1) + (x + 2) = 47$; no integer solution
28. Show that the sum of three consecutive odd integers can*not* be 158. $x + (x + 2) + (x + 4) = 158$; no integer solution
29. Given three consecutive integers, show that 3 times the middle integer is equal to 3 less than 3 times the largest integer. $3(x + 1) = 3(x + 2) - 3$
30. Show that the sum of three consecutive even integers is 6 more than 3 times the smallest. $x + (x + 2) + (x + 4) = 3x + 6$

In Exercises 31–46, use known geometric relationships to build equations to solve the problem.

31. In a right triangle the measure of one of the acute angles is 3 times the measure of the other. What is the measure of the larger acute angle? 67.5°

32. The measure of one of the acute angles of a right triangle is $1\frac{1}{2}$ times the measure of the other. What are the measures of the acute angles? 54°, 36°

33. In a right triangle one of the acute angles is $\frac{2}{3}$ the other. What are the measures of all the angles in the triangle? 54°, 36°, 90°

34. In a right triangle one of the acute angles is $\frac{1}{5}$ the other. What is the measure of the smaller acute angle? 15°

35. In a right triangle one acute angle is 22° more than the other. Find the measure of the larger acute angle. 56°

36. In a triangle the measure of the middle angle is $\frac{1}{3}$ the measure of the largest and 2 times the measure of the smallest. Find the measures of all the angles. 20°, 40°, 120°

37. The angles of a triangle are represented by x, $x - 35$, and $x + 50$. What is the measure of each? 20°, 55°, 105°

38. Find the measures of the angles of a triangle if they are represented by $2x - 20$, $x + 30$, and $x - 10$. 70°, 75°, 35°

39. The measures of the angles of a triangle are three consecutive even integers. Find the measures of all the angles. 58°, 60°, 62°

40. The sum of the angles of a quadrilateral is 360°. If the measures of the angles of a quadrilateral are four consecutive odd integers, find them. 87°, 89°, 91°, 93°

41. Two angles are supplementary. If the measure of one of them is 5 more than 4 times the other, what are the measures of the angles? 35°, 145°

42. If the smaller of two supplementary angles has a measure 15° less than $\frac{1}{2}$ the measure of the larger angles, what is the measure of the smaller angle? 50°

43. If the measure of two complementary angles are represented by $2x + 3$ and $3x - 8$, what are the measures of these angles? 41°, 49°

44. A square patio always has more square feet of area than any other rectangular shape with the same perimeter. If a landscape designer has 128 linear feet of a special border brick, what are the dimensions of the largest concrete patio that could use these bricks as a border? 32 ft × 32 ft

45. A scalene triangle has sides represented by x, $2x$, and $3x - 2$. If the perimeter is 52 cm, how long are each of the sides? 9, 18, 25 cm

46. In an isosceles triangle the base is 43 ft. If the perimeter is 215 ft, how long are the two equal sides? 86 ft each

Exercises 47–58 involve annual interest. Use the relationship $I = Pr$, where P is principal, r is the annual interest rate, and I is the amount of interest earned in one year.

47. Fifteen thousand dollars is to be invested in two different accounts, one which earns 8 percent a year and another which earns 6 percent a year. How much should be put in each account so that the total annual interest earned is $925? $1,250 at 8 percent, $13,750 at 6 percent

48. A certain amount of money is invested at 12 percent interest per year and twice that amount is invested at 9 percent annual interest. Together, the two accounts earned annual interest of $5,400. How much is invested in each account? $18,000 at 12 percent, $36,000 at 9 percent

49. Money in one account is earning 7 percent annually and another account with $2,000 more earns 11 percent annually. Altogether, the two accounts earn $2,700 a year. How much money is invested in each account? $13,777.78 at 7 percent, $15,777.78 at 11 percent

50. A stereo system costs $750, and with sales tax the total bill comes to $810. What is the tax rate? 8 percent

51. If the tax on a radio which retails for $27 is $1.89, what is the tax rate? 7 percent

52. The population of a city increased 2 percent in one year to reach about 3 million people. To the nearest thousand people, about what was the population at the beginning of the year? 2,941,000

53. Residential real estate values in a popular vacation spot increased 5 percent during the last year. If a home is now valued at $200,000, what was it worth at the beginning of the year (to the nearest $500)? $190,500

54. The final bill for the sofa you purchased totaled $1,349. If this included 8.25 percent sales tax plus a $50 delivery fee, what was the original price of the sofa? The delivery fee was not taxed. $1,200

55. The trustees of a $375,000 fund that provides scholarship monies need to generate $26,000 worth of income each year. Their advisors recommend investing in two different financial instruments, one very safe but yielding only 5 percent annually and another yielding 12 percent annually but more risky.

  a. To the nearest thousand dollars, how much should be allocated to each investment? $271,000 at 5 percent, $104,000 at 12 percent

  b. What should be the split if they also want to generate an additional $10,000 for investment? $129,000 at 5 percent, $246,000 at 12 percent

56. An accountant recommends two investments to a client: one earning 10 percent annually and one earning 15 percent annually. These investments should cover the cost of the client's rent ($900 per month) for a year. If there is $86,000 to invest, how should the money be split between the two investments? $42,000 at 10 percent, $44,000 at 15 percent

57. You are considering the purchase of a testing device capable of diagnosing over 750 different computer chip failures occurring in cars manufactured after 1988. Without yearly upgrades to handle the chips in the latest cars, the value of this machine depreciates 15 percent each year. If this machine is already one year old, no upgrades have been made, and it is currently valued at $21,000, to the nearest dollar, how much was it worth new? $24,706

58. Refer to Exercise 57. Suppose the current value is \$24,000.
   a. How much was the machine worth new? \$28,235
   b. How much will it be worth after one more year has passed? \$20,400

Exercises 59–64 involve uniform motion and employ the relationship $d = rt$ in their solution. A sketch and a chart will make solving them easier.

59. Cities A and B are joined by a railroad line 351 mi long. One train leaves city A and another train leaves city B at exactly the same time, heading toward each other. If the A train is moving at 55 mi/hour and the B train is averaging 62 mi/hour, when will the two trains pass each other? In 3 hours
60. Two fishing boats leave the harbor at the same time and using the same float plan. One averages 17 knots (nautical miles) per hour and the other averages 14.5 knots per hour. How long will it take for the boats to be 10 nautical miles apart? 4 hours
61. A moving van makes the trip along the total length of Highway 4 in about 6 hours. Another, newer moving van makes the same trip in 5.5 hours by traveling 4 mi/hour faster. How fast does each van travel? How long is Highway 4? 44 mi/hour and 48 mi/hour; 264 mi
62. Two bicyclists make the same trip. The first takes about 3.5 hours, and the second, traveling about 1 mi/hour faster, only takes 3 hours. How many miles is the trip? 21 mi
63. One plane takes 2 hours and 45 minutes for a particular flight. With a head wind, the same plane goes 20 mi/hour slower and requires 3 hours to reach its destination. How far apart are the airports? 660 mi
64. A spacecraft moves through intergalactic space at a rate of 15,000 mi/hour. Unknown to the crew, a huge asteroid traveling at 7,000 mi/hour is hurtling directly toward the spacecraft. The craft and the asteroid are presently 165,000 mi apart. How long will it take for them to be within scanning range (22,000 mi) of each other, at which time the crew can make a course correction to avoid impact? 6.5 hours

In Exercises 65–72, use a chart and the concept of percentage, as follows: (percent of ingredient) · (amount of solution) = (amount of ingredient).

65. You want to winterize the plumbing in a trailer home you have in the mountains. To do this, you need an antifreeze and water mixture with at least 75 percent antifreeze. You already have 3 gal of a 50 percent antifreeze solution. How much pure antifreeze must you add to produce a 75 percent solution? How many gallons will you end up with? 3, 6 gal
66. You have 2 gal of 75 percent antifreeze and want to dilute it to a 25 percent solution. How much water should you add? How many gallons will you end up with? 4, 6 gal
67. Two acid mixtures, one 6 percent acid and the other 8 percent acid, are to be combined to make 8 gal of a 6.5 percent acid solution. How much of each should be used? 6 gal of 6 percent, 2 gal of 8 percent
68. How much 25 percent acid solution and how much 45 percent acid solution should you mix to get 7 liters of a solution that is 40 percent acid? 1.75 liters of 25 percent, 5.25 liters of 45 percent
69. Jewelers often use alloys to add strength, color, or other properties to the metals in jewelry. If a jeweler has two alloys, one 35 percent gold and one 25 percent gold, to be combined by melting them down, how much of each should be used to produce 0.25 oz of 30 percent gold alloy? 0.125 oz of each
70. If a copper alloy has 8 percent copper and another alloy has 15 percent copper, how much of each should be used to produce 20 lb of 11 percent copper alloy? About 8.57 lb of 15 percent alloy, 11.43 lb of 8 percent alloy
71. One cup of vinegar is mixed with 3 cups of a solution which is 15 percent vinegar and 85 percent oil. What is the percentage of vinegar in the final mixture? 36.25 percent
72. Two cups of liquid laundry detergent are added to 8 gal of wash water. What is the percentage of detergent in the laundry water? (*Note:* 1 gal = 16 cups.) About 1.5 percent

THINK ABOUT IT

1. In a word problem x represents one unknown quantity and $15 - x$ represents another. Which one of these can you then determine numerically?
   a. Their sum   b. Their difference
2. Make up a word problem involving a piece of string cut into two pieces which could be solved by the equation $15 - x = 2x$.
3. A chemist has two mixtures, 30 percent acid and 20 percent acid. Can they be mixed to form 6 liters of a 40 percent solution? Set up the equation for this and see what the solution tells you. What is the difficulty?
4. A car goes 120 mi at 40 mi/hour and then 120 mi at 60 mi/hour.
   a. What is its average speed for the whole trip? Recall that average speed equals total distance divided by total time.
   b. For routine solution of problems like this one, the **harmonic mean** may be used. The harmonic mean of two numbers a and b is defined as $m = 2ab/(a + b)$. Find the harmonic mean of 40 and 60.
5. You have 60 in. of string to shape into a rectangle with perimeter equal to 60 in. Make a chart which shows various possibilities for the length and width. Do they all have the same area? What dimensions appear to give the maximum area? The minimum area?

REMEMBER THIS

1. If x is replaced by -3, does $x + 2 \geq -4$ become a true statement? Yes
2. True or false? $a \leq a$, where a is any real number. True
3. Are 1 and 2 both solutions of the equation $x^2 - 3x + 2 = 0$? Yes
4. Which of these is a linear equation in one variable?
   a. $3x + 2 = 0$ **b.** $x^2 + 3x + 2 = 0$ a
5. What is the result of multiplying $5x - 2$ by -1? $-5x + 2$
6. Solve $y = 3x - 4$ for x. $x = \frac{y + 4}{3}$
7. Is the equation $3x + 4 = 3x + 5$ an identity or a false equation? False equation
8. Do these equations have the same solution set?
   a. $3x = 0$ **b.** $3x = 3x + 1$ No, $0 \neq \emptyset$
9. Simplify $x - [2 - (x - 1)]$. $2x - 3$
10. Rewrite this expression using the associative property of addition: $(3ab + 2cd) + 5ef$. $3ab + (2cd + 5ef)$

2.4 Linear Inequalities in One Variable

A recording company plans to produce and sell x compact discs of a particular album. The production cost includes a one-time, "up-front" recording cost of \$150,000 followed by a duplicating cost of \$2 per CD. The company receives \$8 on the sale of each CD. For what number of CDs sold is the revenue greater than the cost? (See Example 6.)

OBJECTIVES

1. Specify solution sets of linear inequalities by using graphs and interval notation.
2. Solve linear inequalities by applying properties of inequalities.

In applications it is often an inequality which is the key relationship to be described. For instance, the recording company in the opening problem wants revenue to be *greater than* cost so that the CD is profitable. Writing such inequalities demands the use of the symbols reviewed in the following table.

Symbol	Meaning
$<$	Less than
$\leq$	Less than or equal to
$>$	Greater than
$\geq$	Greater than or equal to

1 An **inequality** is a statement that relates expressions by using the inequality symbols above. For example,

$$x + 2 \geq -4, \qquad x > 2, \qquad \text{and} \qquad 0 \leq x < 8$$

are inequalities. As with equations, a **solution** of an inequality is a value for the variable that makes the inequality true, while the **solution set** is the set made up of *all* the solutions of the inequality. Thus, the solution set of the inequality $x > 2$ is the set of all real numbers greater than 2. One way to describe this infinite set of numbers is to graph it on the number line, as shown in Figure 2.5(a). The parenthesis at 2 in this figure means that 2 is not included in the solution set, and the arrow specifies all real numbers greater than 2. To graph a set like $\{x : x \geq 2\}$, we show that 2 is included in the solution set by putting a bracket at this point, as shown in Figure 2.5(b).

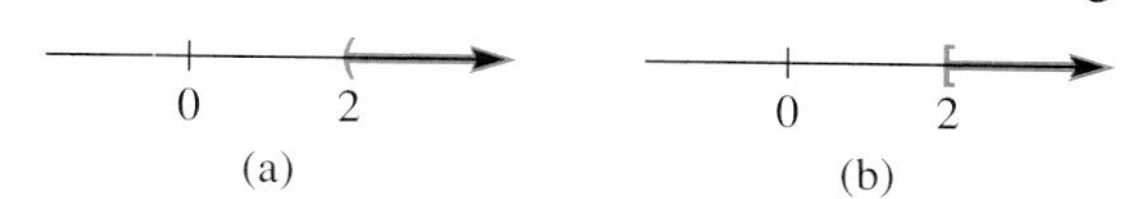

Figure 2.5

Sets of real numbers that may be represented graphically as half lines or the entire number line are examples of **intervals** that may be expressed conveniently by using **interval notation,** as outlined in the following chart.

Set Notation	Graph	Interval Notation
$\{x : x > a\}$	a	(a, ∞)
$\{x : x \geq a\}$	a	$[a, \infty)$
$\{x : x < a\}$	a	$(-\infty, a)$
$\{x : x \leq a\}$	a	$(-\infty, a]$
$\{x : x$ is a real number$\}$	0	$(-\infty, \infty)$

Note that the symbols ∞, read "infinity," and $-\infty$ are not real numbers but are convenient symbols that help us designate intervals that are unbounded in a positive or negative direction.

The symbol ∞ for infinity was first used by John Wallis (1616–1703), a professor at Oxford University in England.

EXAMPLE 1 Write the solution set to each inequality in interval notation, and graph the interval.

a. $x \geq -1$ **b.** $x < \frac{3}{2}$

Solution

a. Draw a number line. Place a bracket at -1, and then draw an arrow from the bracket to the right, as in Figure 2.6(a). The bracket means that the number -1 is a member of the solution set, and the arrow specifies all real numbers greater than -1. In interval notation $[-1,\infty)$ represents this set of numbers.

b. Figure 2.6(b) shows the graph of $\{x : x < \frac{3}{2}\}$, which is written as $(-\infty,\frac{3}{2})$ in interval notation. Note that in both the graph and the interval notation, the parenthesis indicates that $\frac{3}{2}$ is not a member of the solution set.

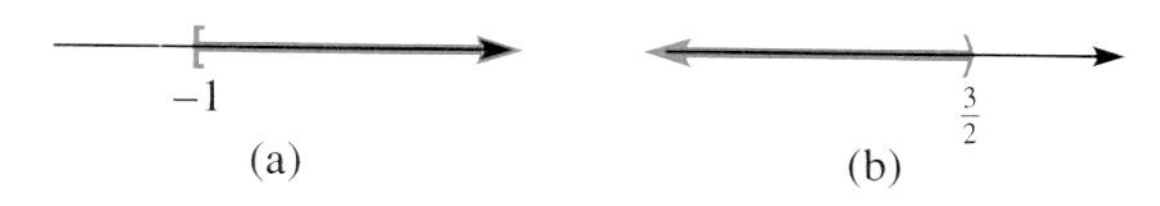

Figure 2.6

Note As shown in Figure 2.7, it is also common notation to use an open circle in a graph instead of a parenthesis to indicate an endpoint that is not included in the solution set, and to use a solid dot instead of a bracket when the endpoint is included. In this text we choose to use parentheses and brackets because this method reinforces writing intervals in interval notation and is also more common in higher-level mathematics.

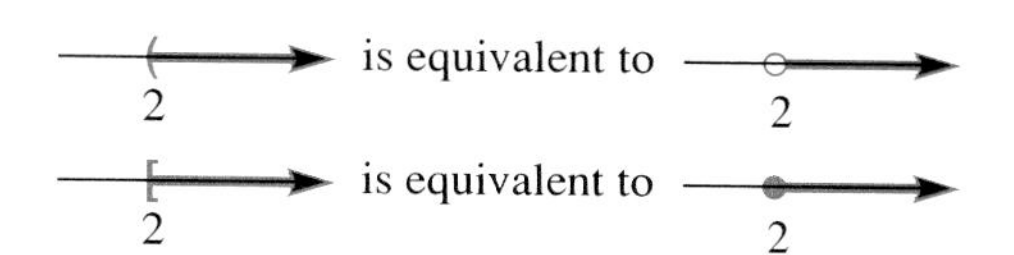

Figure 2.7

PROGRESS CHECK 1 Write the solution set of each inequality in interval notation, and graph the interval.

a. $x \le -\frac{1}{2}$ **b.** $x > 0$

2 By analogy to equations, a **first-degree** or **linear inequality** results if the equal sign in a linear equation is replaced by one of the inequality symbols. For example,

$$-5x + 1 < 16, \qquad x \ge 2, \qquad \text{and} \qquad 5(x - 1) > 2x + 1$$

are linear inequalities. The procedures for solving linear equalities are similar to the procedures for solving linear equations. The key idea is to create equivalent but simpler inequalities at each step until the solution set is clear. However, to create equivalent inequalities, we apply the following properties of inequalities. Although these properties are given only for <, similar properties may be stated for the other inequality symbols.

Properties of Inequalities

Let A, B, and C be algebraic expressions.

	Comment
1. If $A < B$, then $A + C < B + C$ and $A - C < B - C$.	The direction of the inequality is preserved when the same expression is added to (or subtracted from) both sides of an inequality.
2. If $A < B$ and $C > 0$, then $AC < BC$ and $\frac{A}{C} < \frac{B}{C}$.	The direction of the inequality is preserved when both sides of an inequality are multiplied (or divided) by the same positive expression.
3. If $A < B$ and $C < 0$, then $AC > BC$ and $\frac{A}{C} > \frac{B}{C}$.	The direction of the inequality is reversed when both sides of an inequality are multiplied (or divided) by the same negative expression.

Note that multiplying or dividing on both sides of an inequality demands special care because there are two cases: one for positive multipliers and one for negative multipliers. To see why, notice that the true inequality $-5 < 2$ leads to a true inequality if both sides are multiplied by *positive* 3.

$$-5(3) < 2(3) \quad \text{Multiply both sides by 3.}$$
$$-15 < 6 \quad \text{True inequality.}$$

But $-5 < 2$ leads to a false inequality when both sides are multiplied by *negative* 3.

$$-5(-3) < 2(-3) \quad \text{Multiply both sides by } -3.$$
$$15 < -6 \quad \text{False inequality.}$$

To obtain a true statement when multiplying both sides by -3, the direction of the inequality must be reversed.

$$-5 < 2 \quad \text{Original inequality.}$$
$$-5(-3) > 2(-3) \quad \text{Multiply both sides by } -3 \text{ and } \textit{reverse} \text{ the inequality symbol.}$$
$$15 > -6 \quad \text{True inequality.}$$

Progress Check Answers

1. (a) $(-\infty, -\frac{1}{2}]$ (b) $(0, \infty)$

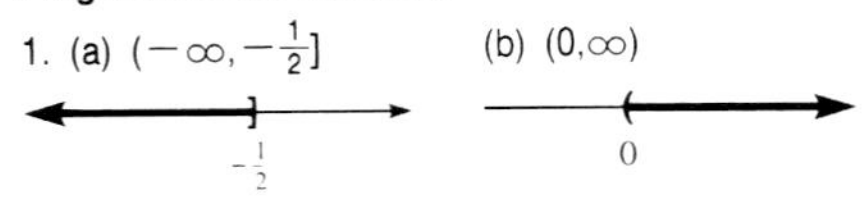

Always remember to reverse the direction of the inequality when multiplying or dividing by a negative number on both sides of an inequality. This step is also called reversing the *sense* of the inequality.

Warn students that it is a bad idea to divide both sides of an inequality by a *variable* unless they know whether it stands for a positive or a negative number, because they will not know if the inequality sign should be left alone or reversed.

EXAMPLE 2 Solve $-5x + 1 < 16$. Express the solution set graphically and in interval notation.

Solution As with equations, the goal is to isolate x on one side of the inequality.

$$-5x + 1 < 16$$

$$-5x + 1 - 1 < 16 - 1 \quad \text{Subtract 1 from both sides.}$$

$$-5x < 15$$

$$\frac{-5x}{-5} > \frac{15}{-5} \quad \text{Divide both sides by } -5 \text{ and reverse the inequality sign.}$$

$$x > -3$$

Thus, all real numbers greater than -3 make the inequality a true statement. The solution set is written as $(-3,\infty)$ and is graphed as shown in Figure 2.8.

Figure 2.8 −3

PROGRESS CHECK 2 Solve $-3x - 7 > 5$. Express the solution set graphically and in interval notation.

The next example contrasts the steps for solving linear inequalities with the four-step procedure for solving linear equations given in Section 2.1.

EXAMPLE 3 Solve $-5(x - 1) > 4x - 13$, and graph the solution set.

Solution We follow similar steps to the general procedure for solving linear equations.

1. Simplify the left side of the inequality by removing parentheses.

$$-5(x - 1) > 4x - 13$$

$$-5x + 5 > 4x - 13 \quad \text{Distributive property.}$$

2. Write an inequality of the form $cx > n$ by adding or subtracting on both sides of the inequality.

$$-9x + 5 > -13 \quad \text{Subtract } 4x \text{ from both sides.}$$

$$-9x > -18 \quad \text{Subtract 5 from both sides.}$$

3. Isolate x by multiplying or dividing on both sides of the inequality.

$$\frac{-9x}{-9} < \frac{-18}{-9} \quad \text{Divide both sides by } -9 \text{ and reverse the inequality.}$$

$$x < 2$$

4. Although we cannot check the entire solution set because it is an infinite set of numbers, we can at least select one number less than 2 and check that our result is reasonable. Picking 1 yields

$$-5(x - 1) > 4x - 13 \quad \text{Original inequality.}$$

$$-5(1 - 1) > 4(1) - 13 \quad \text{Replace } x \text{ with 1.}$$

$$0 > -9. \quad \text{True inequality.}$$

The check confirms that 1 is a solution, so $x < 2$ is reasonable. The solution set is $(-\infty,2)$, as shown in Figure 2.9.

Figure 2.9

PROGRESS CHECK 3 Solve $5(x - 1) > 2x + 1$, and graph the solution set.

Progress Check Answers

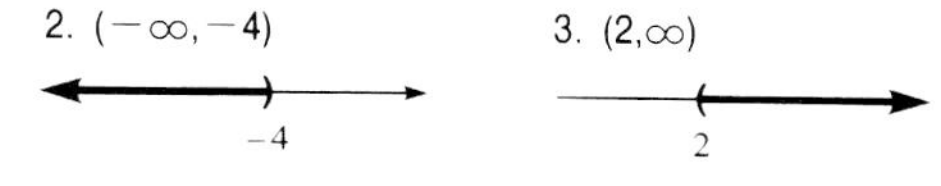

The two inequalities $2 < 5$ and $5 > 2$ have the same meaning, and in general, the inequality

$$a > b \quad \text{is equivalent to} \quad b < a.$$

This equivalence means that the left and right sides of an inequality are interchangeable provided that the inequality symbol is reversed. As illustrated in Example 4, sometimes one version is preferable because it is easier to visualize.

EXAMPLE 4 Solve $3 - x \leq 3x$, and graph the solution set.

Solution For this inequality an efficient method is to isolate x on the right side of the inequality.

$$3 - x \leq 3x$$
$$3 \leq 4x \quad \text{Add } x \text{ to both sides.}$$
$$\tfrac{3}{4} \leq x \quad \text{Divide both sides by 4.}$$

Then, the answer is easier to visualize if we interchange sides and reverse the inequality symbol, to get

$$x \geq \tfrac{3}{4}.$$

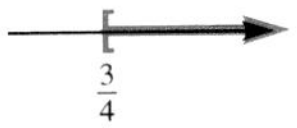

Substituting any number greater than $\frac{3}{4}$ in the original inequality will show that our answer is reasonable. The solution set is $[\frac{3}{4},\infty)$, as graphed in Figure 2.10.

Figure 2.10 $\frac{3}{4}$

PROGRESS CHECK 4 Solve $6 - 2x \geq x$, and graph the solution set.

Some inequalities are true for all real numbers, while others can never be true. The next example illustrates one of these cases.

EXAMPLE 5 Solve $x > x - 3$.

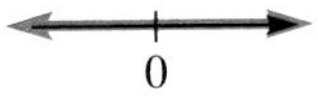

Solution Because any number is greater than the number that is 3 smaller than it, this inequality is always true. Thus, the solution set is the set of all real numbers, which is graphed as in Figure 2.11 and is written as $(-\infty,\infty)$ in interval notation. If the solution set is not clear when $x > x - 3$ is examined, then a useful next step is to subtract x from both sides.

Figure 2.11 0

$$x > x - 3$$
$$x - x > x - x - 3 \quad \text{Subtract } x \text{ from both sides.}$$
$$0 > -3$$

At this point we know that the original inequality is *equivalent* to $0 > -3$, which is always true. So $x > x - 3$ is true for all replacements of x.

PROGRESS CHECK 5 Solve $x < x - 3$.

The section-opening problem illustrates a business application of linear inequalities.

EXAMPLE 6 Solve the problem that opens the section on page 62.

Solution If we let x represent the number of compact discs produced and sold, then $150{,}000 + 2x$ represents the production cost and $8x$ represents the revenue. We need to find values of x for which

$$\underbrace{\text{revenue}}_{8x} \quad \underbrace{\text{is greater than}}_{>} \quad \underbrace{\text{cost.}}_{150{,}000 + 2x}$$

Progress Check Answers

4. $(-\infty,2]$ 2

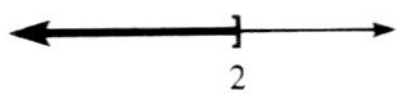

5. $\emptyset$

Solve the Inequality $6x > 150{,}000$ Subtract $2x$ from both sides.
$x > 25{,}000$ Divide both sides by 6.

Answer the Question Revenue is greater than cost when the company produces and sells more than 25,000 CDs.

Check the Answer The revenue from the sale of 25,000 CDs is \$8(25,000) = \$200,000, while the cost of producing 25,000 CDs is \$150,000 + \$2(25,000) = \$200,000. Therefore, 25,000 CDs is the break-even point, and revenue is greater than cost after this point.

PROGRESS CHECK 6 Redo Example 6, but assume that the producer's cost is given by $168{,}000 + 3x$ and the revenue is given by $10x$.

Progress Check Answer

6. Revenue is greater than cost when more than 24,000 CDs are sold.

EXERCISES 2.4

In Exercises 1–8, write the solution set to each inequality in interval notation and then graph the interval.

1. $x \leq 4$ $(-\infty, 4]$

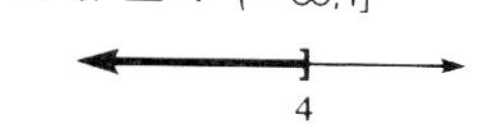

4

2. $x > 2$ $(2, \infty)$

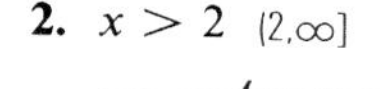

2

3. $x \geq -1$ $[-1, \infty)$

−1

4. $x < -4$ $(-\infty, -4)$

−4

5. $x < 0$ $(-\infty, 0)$

0

6. $x \geq 0$ $[0, \infty)$

0

7. $x \geq \frac{1}{2}$ $[\frac{1}{2}, \infty)$

$\frac{1}{2}$

8. $x \leq -\frac{2}{3}$ $(-\infty, -\frac{2}{3}]$

$-\frac{2}{3}$

For Exercises 9–50, solve the given inequality. Express the solution set graphically and in interval notation.

9. $2x + 1 < 7$ $(-\infty, 3)$

3

10. $4x - 3 < 13$ $(-\infty, 4)$

4

11. $4 - 4x \leq -3x$ $[4, \infty)$

4

12. $5 + 3y \leq 9y + 5$ $[0, \infty)$

0

13. $x + \frac{3}{8} \geq \frac{7}{8}$ $[\frac{1}{2}, \infty)$

$\frac{1}{2}$

14. $y - \frac{1}{2} \geq \frac{3}{4}$ $[\frac{5}{4}, \infty)$

$\frac{5}{4}$

15. $84 > -6y$ $(-14, \infty)$

−14

16. $-6x - 4 < 26$ $(-5, \infty)$

−5

17. $3x + 4 \leq -29 - \frac{2}{3}x$ $(-\infty, -9]$

−9

18. $8x + 6 - 2x \leq 7 + 2x - 10$ $(-\infty, -\frac{9}{4}]$

$-\frac{9}{4}$

19. $5x + 5 < 6 + x - 4$ $(-\infty, -\frac{3}{4})$

$-\frac{3}{4}$

20. $7 + 6x < 14x + 35 - 6x$ $(-14, \infty)$

−14

21. $-2x + 1 < 7$ $(-3, \infty)$

−3

22. $-4x - 3 < 13$ $(-4, \infty)$

−4

23. $-5x + 7 > -8$ $(-\infty, 3)$

3

24. $-6x - 3 > 15$ $(-\infty, -3)$

−3

25. $7 - x \leq 3$ $[4, \infty)$

4

26. $4 - x \leq -3$ $[7, \infty)$

7

27. $12 - x \geq 16$ $(-\infty, -4]$

−4

28. $20 - x \geq 18$ $(-\infty, 2]$

2

29. $10 - 4x < -6$ $(4, \infty)$

4

30. $-3x + 8 > -19$ $(-\infty, 9)$

9

31. $2(x + 1) > x - 3$ $(-5, \infty)$

−5

32. $7(x + 2) > 4x + 2$ $(-4, \infty)$

−4

33. $4(x - 2) < 3x + 1$ $(-\infty, 9)$

9

34. $5(x - 3) \leq 3x + 1$ $(-\infty, 8]$

8

35. $-7(x - 1) \geq 11x - 29$ $(-\infty, 2]$

2

36. $-2(x - 3) < 2x + 18$ $(-3, \infty)$

−3

37. $-10(x + 7) < 50 + 2x$ $(-10, \infty)$

−10

38. $-3(x + 3) > 5x + 7$ $(-\infty,-2)$

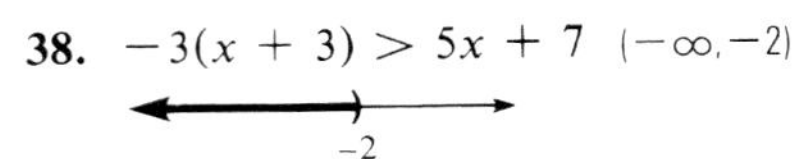

-2

39. $-7(x - 3) \le 33 - x$ $[-2,\infty)$

-2

40. $-4(x - 4) \ge 7x - 6$ $(-\infty,2]$

2

41. $5 - 3x \ge 2x$ $(-\infty,1]$

1

42. $2(3 - x) \ge x$ $(-\infty,2]$

2

43. $10 - 7x \le 3x$ $[1,\infty)$

1

44. $-5 - x \le 4x$ $[-1,\infty)$

-1

45. $x > x - 1$ $(-\infty,\infty)$

0

46. $x > x + 1$ $\emptyset$

47. $2(x + 1) \ge 3 + 2x$ $\emptyset$

48. $2(x + 1) \ge -3 + 2x$ $(-\infty,\infty)$

0

49. $2x + x < 4x + 4 - x$ $(-\infty,\infty)$

0

50. $-5x - 3 - x < 4x - 5 - 10x$ $\emptyset$

51. Sally calculates her business overhead (rent, phone, electricity, etc.) to be $2,600 per month. The materials for each floral piece she creates cost $20. If she can sell each arrangement for $150, how many arrangements must she sell each month in order to show a profit? More than 20

52. Sally's overhead increases to $3,000 per month, and the raw materials cost rises to $25 per arrangement. If she is still only getting $150 per arrangement sold, how many such arrangements must she sell to show a profit now? More than 24

53. To qualify for a sports program, a child must weigh more than 60 lb. This is expressed as $p > 60$. Use the formula $p = 2.205k$ to express this requirement in kilograms. Round to the nearest hundredth of a kilogram. $k > 27.21$

54. For a chemistry experiment to work, the temperature of a solid must be kept below 200 in the Kelvin scale. This is expressed as $K < 200$. Use the formula $K = 273.15 + \frac{5}{9}(F - 32)$ to express this requirement in Fahrenheit. Round to the nearest hundredth of a degree. $F < -99.67$

THINK ABOUT IT

1. **a.** A student multiplied both sides of the inequality $-\frac{1}{2}x < 5$ by negative 2 and forgot to reverse the inequality. What is the relation between the solution set he got and the correct solution set? Is there any number that is in both solution sets?
   b. If you multiply both sides of $3x - 4 \le 8$ by negative 3 and forget to reverse the direction of the inequality, you get $-9x + 12 \le -24$. What is the relation between the solution sets of the two inequalities? Is there any number that is in both solution sets?
2. Why do you reverse the inequality when you multiply both sides of an inequality by a negative number?
3. If you purchase a discount card for $20 from a video store that entitles you to 15 percent off on all purchases, how much must you purchase for the card to save you over $25?
4. Give examples that show that these statements can be *false*.
   a. If $a > b$, then $ac > bc$.
   b. If $a > b$, then $a/c > b/c$.
   c. If $a > b$, then $a^2 > b^2$.
5. By definition, $a < b$ if and only if $b - a$ is positive. Use the definition to prove that if $a < b$ and $c < 0$, then $ac > bc$.

REMEMBER THIS

1. True or false? 86 is in the solution sets of both of these inequalities: $84.5 \le x$ and $x < 89.5$. True
2. Find the intersection of $A = \{5,6,7,8,9,10\}$ and $B = \{8,9,10,11,12\}$. $\{8,9,10\}$
3. Find the union of $A = \{0,1\}$ and $B = \{2,3\}$. $\{0,1,2,3\}$
4. Show that it is not possible for the sum of two consecutive integers to equal 0 by solving $x + (x + 1) = 0$.
   $x = -\frac{1}{2}$ is not an integer.
5. The perimeter of a rectangle is 9 in., and the length is twice the width. Write and solve an equation to determine the length and width. $2x + 2(2x) = 9$; width $= \frac{3}{2}$ in., length $= 3$ in.
6. Solve $y = mx + b$ for x. $x = \frac{y - b}{m}$
7. Solve $\frac{1}{3}x + \frac{1}{4} = \frac{1}{2}$. $\{\frac{3}{4}\}$
8. Write an expression for the sum of the reciprocal and the opposite of n. $\frac{1}{n} + (-n)$
9. Simplify $(|-3| - |3|)(|-4| + |4|)$. 0
10. Evaluate $\frac{(4 - x)^2}{4 - x^2}$ when $x = 3$. $-\frac{1}{5}$

2.5 Compound Inequalities

In a philosophy class an average from 84.5 up to but not including 89.5 results in a grade of B⁺. A student has grades of 93, 76, and 90 for the first three tests. With one test left, find all possible grades on the last test that result in a grade of B⁺. (See Example 4.)

OBJECTIVES

1 **Solve compound inequalities involving *and* statements.**

2 **Solve compound inequalities involving *or* statements.**

In the section-opening problem the average (a) of the student's test results leads to a grade of B^+ when a satisfies the pair of inequalities

$$84.5 \leq a \quad \text{and} \quad a < 89.5.$$

When two inequalities are joined by the word *and* or *or,* the result is called a **compound inequality.** For an *and* statement to be true, both inequalities must be true *simultaneously,* whereas an *or* statement is true when *at least one* of the inequalities is true. We focus first on solving inequalities with *and.*

Point out that in contrast to the meaning in ordinary conversation, the meaning of *or* in mathematics is assumed to be *inclusive.*

1 To see which numbers satisfy

$$84.5 \leq a \quad \text{and} \quad a < 89.5$$

we first graph, in Figure 2.12, each inequality separately. Then, we find the set of numbers common to these two graphs, which are the real numbers between 84.5 and 89.5, including 84.5 and excluding 89.5.

The statement "$84.5 \leq a$ and $a < 89.5$" is usually expressed in compact form and written as $84.5 \leq a < 89.5$. As suggested by Figure 2.12, this set is written as [84.5,89.5) in interval notation. The following chart shows the different methods for expressing intervals that may be represented graphically as line segments.

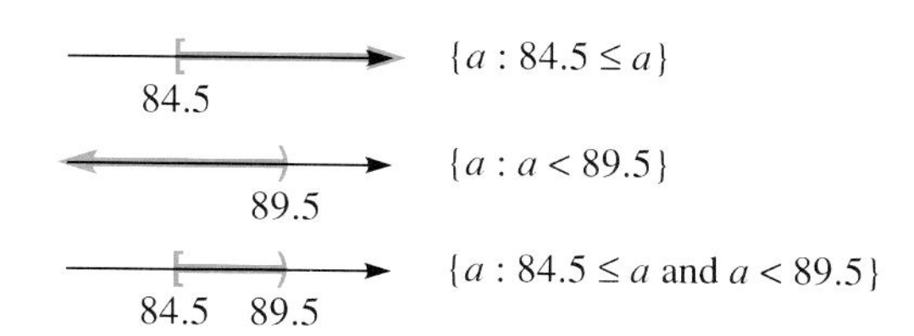

Figure 2.12

Type of Interval	Set Notation	Graph	Interval Notation
Open interval	$\{x : a < x < b\}$	(a, b) open endpoints	(a, b)
Closed interval	$\{x : a \leq x \leq b\}$	[a, b] closed endpoints	$[a, b]$
Half-open interval	$\{x : a \leq x < b\}$	[a, b)	$[a, b)$
	$\{x : a < x \leq b\}$	(a, b]	$(a, b]$

EXAMPLE 1 Find the solution set to each inequality in interval notation, and graph the interval.

a. $x < 2$ and $x > 0$ **b.** $x > 2$ and $x \geq 0$
c. $x > 2$ and $x < 0$ **d.** $0 < x < 2$

Solution

a. The set of numbers that satisfies both $x < 2$ and $x > 0$ simultaneously is the set of numbers between 0 and 2, as shown in Figure 2.13(a). In interval notation this set of numbers is written as $(0,2)$.

b. Figure 2.13(b) shows that the numbers common to the solution set of $x > 2$ and the solution set of $x \geq 0$ are the numbers that are greater than 2. Therefore, $(2,\infty)$ is the solution set.

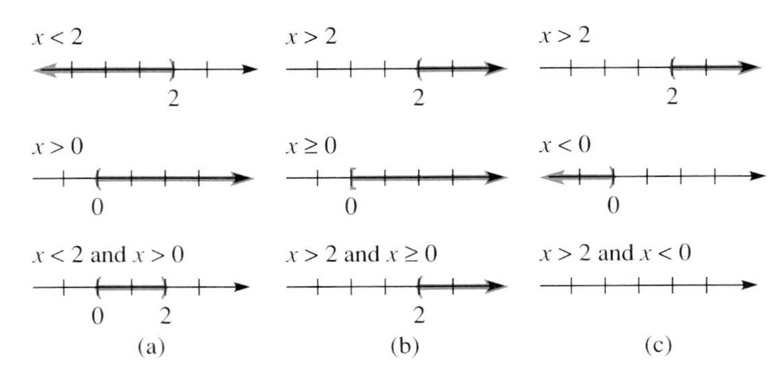

Figure 2.13

c. No number is both greater than 2 and less than 0, as can be seen in Figure 2.13(c). So the solution set is $\emptyset$.

d. $0 < x < 2$ may be read "x is between 0 and 2," and this compound inequality is compact form for "$0 < x$ and $x < 2$." This interval is graphed as in Figure 2.14 and is written as $(0,2)$ in interval notation. Note that the inequalities in parts **a** and **d** in this example are equivalent.

$(0, 2) = \{x : 0 < x < 2\}$

0 2

Figure 2.14

PROGRESS CHECK 1 Find the solution set to each inequality in interval notation, and graph the interval.

a. $x > 1$ and $x \geq 3$ **b.** $x < 1$ and $x \geq 3$
c. $x \geq 1$ and $x \leq 3$ **d.** $1 \leq x \leq 3$

The methods used in Example 1 suggest the following general procedure for solving a compound inequality involving *and*. For this procedure, recall from Section 1.1 that the *intersection* of two sets is the set of elements common to the two sets.

Solution of *And* Inequalities

To solve a compound inequality involving *and:*

1. Solve separately each inequality in the compound inequality.
2. Find the intersection of the solution sets of the separate inequalities.

EXAMPLE 2 Solve $x - 4 > 0$ and $2x \leq 3x - 12$. Express the solution set graphically and in interval notation.

Progress Check Answers

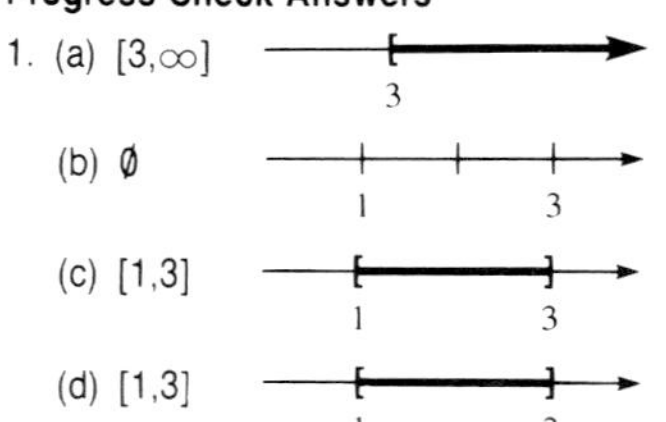

Solution First, solve each inequality separately.

$$\begin{aligned} x - 4 &> 0 \\ x &> 4 \end{aligned} \qquad \text{and} \qquad \begin{aligned} 2x &\leq 3x - 12 \\ -x &\leq -12 \\ x &\geq 12 \end{aligned}$$

The intersection of the solution sets of these two inequalities is then $\{x : x \geq 12\}$, as illustrated in Figure 2.15. In interval notation, $[12,\infty)$ is the solution set.

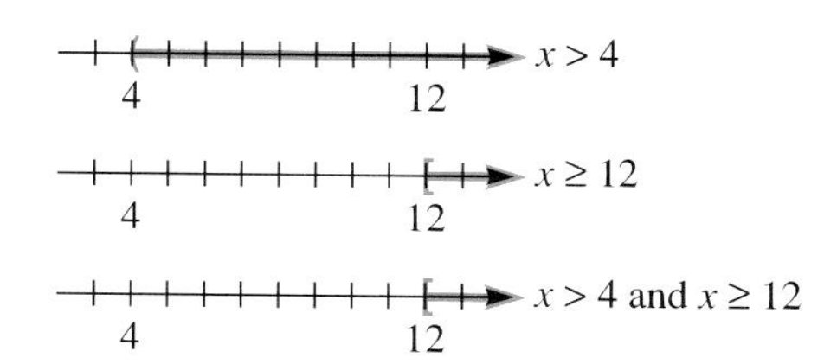

Figure 2.15

PROGRESS CHECK 2 Solve $x - 4 < 0$ and $2x \geq 3x - 12$. Express the solution set graphically and in interval notation.

When a compound inequality involving *and* is written in compact form, a solution can often be obtained without rewriting in an expanded form, as shown in the next two examples.

EXAMPLE 3 Solve $-64 < 192 - 32t < 64$. Express the solution set graphically and in interval notation.

Solution Although the given inequality is actually an abbreviated form of the pair of inequalities $-64 < 192 - 32t$ and $192 - 32t < 64$, we can use the properties of inequalities carefully and stay in the compact form. The goal is to isolate t in the middle member of the compound inequality.

$$-64 < 192 - 32t < 64$$

$$-64 - 192 < 192 - 192 - 32t < 64 - 192 \qquad \text{Subtract 192 from each member.}$$

$$-256 < -32t < -128$$

$$\frac{-256}{-32} > t > \frac{-128}{-32} \qquad \text{Divide each member by } -32 \text{ and reverse the inequality signs.}$$

$$8 > t > 4$$

$$4 < t < 8 \qquad a > b \text{ is equivalent to } b < a.$$

Thus, t is between 4 and 8, so the solution set is the interval $(4,8)$, which is graphed in Figure 2.16.

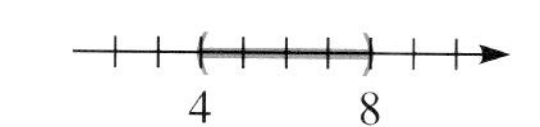

Figure 2.16

Note In obtaining the solution set in this example, observe that an inequality like

$$8 > t > 4 \qquad \text{is equivalent to} \qquad 4 < t < 8,$$

but that we do *not* write $(8,4)$ for the solution set, because in interval notation a must be less than b when writing (a,b).

PROGRESS CHECK 3 Solve $-128 \leq 160 - 32t \leq 128$. Express the solution set graphically and in interval notation.

EXAMPLE 4 Solve the problem in the section introduction on page 69.

Solution A grade of B^+ results if the average (a) satisfies $84.5 \leq a < 89.5$. Since the average is the sum of the four test grades divided by 4, letting x represent the grade on the last test yields

$$84.5 \leq \frac{93 + 76 + 90 + x}{4} < 89.5$$

$$84.5 \leq \frac{259 + x}{4} < 89.5$$

$$338 \leq 259 + x < 358 \qquad \text{Multiply each member by 4.}$$

$$79 \leq x < 99. \qquad \text{Subtract 259 from each member.}$$

Progress Check Answers

2. $(-\infty,4)$

3. $[1,9]$

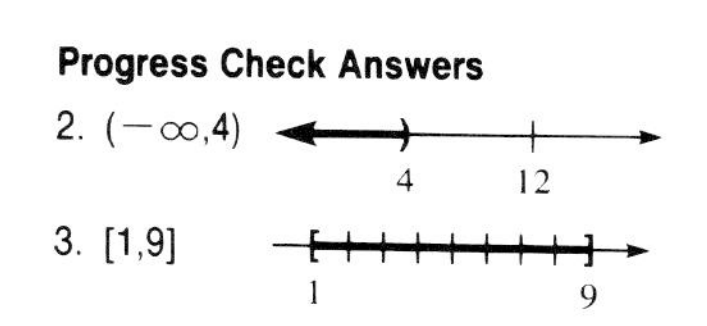

Many students do not realize that sometimes they cannot improve their average significantly when one test remains. Ask them to try this: A student gets 60s on the first three tests. What grade is needed on the fourth test to give an average over 75?

Answer the Question Any grade in the last test from 79 up to but not including 99 results in a grade of B^+. In interval notation, the answer is $[79,99)$.

Check the Answer If the last grade is 79, then the average of the test grades is $(93 + 76 + 90 + 79)/4 = 84.5$, the lowest average that produces B^+. If the last test grade is 99, the average is $(93 + 76 + 90 + 99)/4 = 89.5$, the excluded endpoint of the B^+ interval. The solution checks.

PROGRESS CHECK 4 An average from 74.5 up to but not including 79.5 earns a C^+. If your first three grades are 68, 83, and 75, find all possible grades on the fourth exam that result in a grade of C^+.

2 Recall from Section 1.1 that the *union* of two sets A and B is denoted by $A \cup B$ and is the set of all elements belonging to A or B or both. Because an *or* statement is true when at least one of the statements is true, we find the solution set to a compound inequality involving *or* by finding the union of the solution sets of its component inequalities.

Solution of *Or* Inequalities

To solve a compound inequality involving *or:*

1. Solve separately each inequality in the compound inequality.
2. Find the union of the solution sets of the separate inequalities.

EXAMPLE 5 Solve $x - 3 < -2$ or $x - 3 > 2$. Express the solution set graphically and in interval notation.

Solution First, solve each inequality separately.

$$x - 3 < -2 \quad \text{or} \quad x - 3 > 2$$
$$x < 1 \qquad\qquad x > 5$$

The union of the solution sets of these two inequalities is graphed in Figure 2.17. In interval notation this solution set is written as $(-\infty,1) \cup (5,\infty)$.

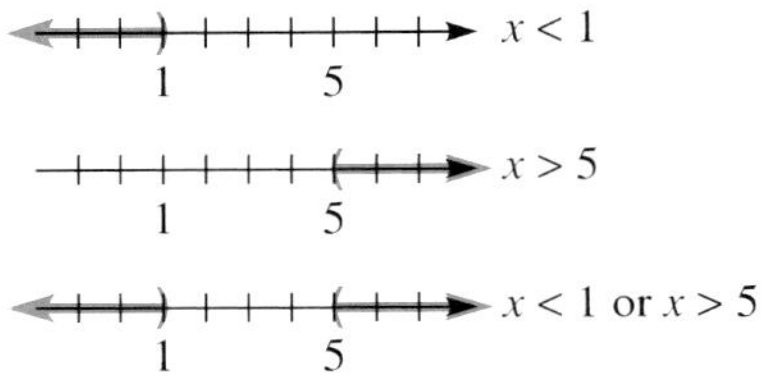

$x < 1$ (1, 5)

$x > 5$ (1, 5)

$x < 1$ or $x > 5$ (1, 5)

Figure 2.17

Caution Only an *and* compound inequality can be written in compact form. Common student errors occur when statements like "$x < 1$ or $x > 5$" are shortened to

Wrong		Wrong
$1 > x > 5$	or	$5 < x < 1$.

PROGRESS CHECK 5 Solve $2x - 1 < -3$ or $2x - 1 > 3$. Express the solution set graphically and in interval notation.

The next example considers the types of solutions that result when we find the union of two half lines. Solving compound inequalities involving linear inequalities and the word *or* requires this type of analysis in the second step of the general solution procedure.

EXAMPLE 6 Find the solution set to each inequality in interval notation, and graph the interval.

a. $x < 0$ or $x > 2$ **b.** $x > 0$ or $x < 2$ **c.** $x \geq 0$ or $x > 2$

Solution

a. The solution set is the union of the numbers to the left of 0 with the numbers to the right of 2, as shown in Figure 2.18(a). In interval notation, we write

Progress Check Answers

4. [72,92]

5. $(-\infty,-1) \cup (2,\infty)$ (−1, 2)

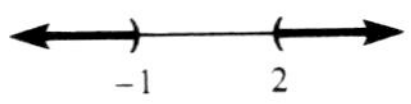

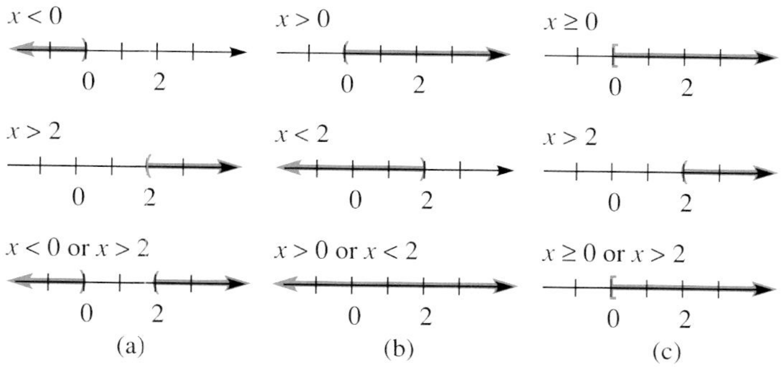

$x < 0$ | $x > 0$ | $x \geq 0$

$x > 2$ | $x < 2$ | $x > 2$

$x < 0$ or $x > 2$ | $x > 0$ or $x < 2$ | $x \geq 0$ or $x > 2$

(a) (b) (c)

Figure 2.18

$(-\infty,0) \cup (2,\infty)$. Linear inequalities joined by *or* are usually graphed as two distinct half lines on a number line, as in this example.

b. Every number is either greater than 0 or less than 2, as can be seen in Figure 2.18(b). Thus, the solution set graphs as the entire number line, which is written as $(-\infty,\infty)$ in interval notation.

c. Figure 2.18(c) shows that the numbers in the union of $\{x : x \geq 0\}$ with $\{x : x > 2\}$ are the numbers that are greater than or equal to 0. Write $[0,\infty)$ to specify this solution set in interval notation.

PROGRESS CHECK 6 Find the solution set to each inequality in interval notation, and graph the interval.

a. $x > -1$ or $x < 1$ **b.** $x \leq -1$ or $x \geq 1$ **c.** $x \leq 0$ or $x < 1$

Progress Check Answers

6. (a) $(-\infty,\infty)$ 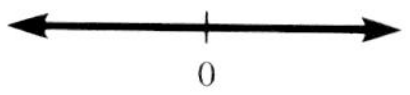
(b) $(-\infty,-1] \cup [1,\infty)$
(c) $(-\infty,1)$

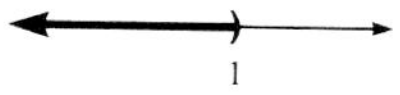

EXERCISES 2.5

In Exercises 1–34, find the solution set for each inequality. Express your answer in interval notation, and graph the interval.

1. $x < 3$ and $x > 1$ $(1,3)$

2. $x > 1$ and $x \geq 3$ $[3,\infty)$

3. $x > 3$ and $x < 1$ $\emptyset$

4. $1 < x < 3$ $(1,3)$

5. $-2 \leq x \leq 5$ $[-2,5]$

6. $x > 0$ and $x \geq 1$ $[1,\infty)$

7. $x > -1$ and $x \leq 4$ $(-1,4]$

8. $x \leq 7$ and $x \geq 3$ $[3,7]$

9. $x \leq -1$ and $x \leq -5$ $(-\infty,-5]$

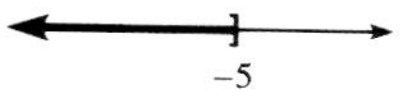

10. $x < -2$ and $x > 3$ $\emptyset$

11. $x - 3 < 0$ and $3x \geq 2x - 5$ $[-5,3)$

12. $5x - 2 > 8$ and $x - 3 > 0$ $(3,\infty)$

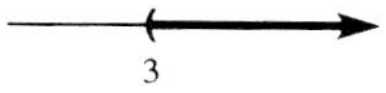

13. $x - 4 > 1$ and $x > 2x - 3$ $\emptyset$

14. $x + 2 > -1$ and $-2x \geq 6$ $\emptyset$

15. $-2x + 1 < 7$ and $4x + 5 \leq 2$ $(-3,-\frac{3}{4}]$

16. $-x < 2$ and $-6x - 4 > 26$ $\emptyset$

17. $3x - 6 \geq 2x - 14$ and $-6x \geq 12$ $[-8,-2]$

18. $-2x + 1 < 7$ and $-x < 0$ $(0,\infty)$

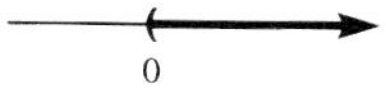

19. $8 - 3x > -19$ and $3x - 1 \leq 4x$ $[-1,9)$

20. $2(x + 1) \geq x - 3$ and $5(x - 3) < 3x + 1$ $[-5,8)$

21. $x + 5 < 0$ and $x < -5$ $(-\infty,-5)$

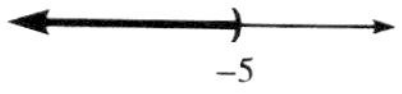

22. $5x - 2 < 28$ and $4x + 2 > 5x$ $(-\infty,2)$

23. $-12 \leq 4 - 8x < 12$ $(-1,2]$

24. $-36 \le 4 - 8x \le 36$ $[-4,5]$

-4 5

25. $-6 \le 4x + 10 < 6$ $[-4,-1)$

-4 -1

26. $-5 < -25 + 10t < 5$ $(2,3)$

2 3

27. $-6 < 6 - 4t < -2$ $(2,3)$

2 3

28. $-16 < 48 - 64x \le 16$ $[\frac{1}{2},1)$

$\frac{1}{2}$ 1

29. $-240 < 256 - 16x < 176$ $(5,31)$

5 31

30. $-112 < 168 - 56x \le 112$ $[1,5)$

1 5

31. $-24 \le 32 - 4t < 24$ $(2,14]$

2 14

32. $-154 \le -63 - 21t < 154$ $(-\frac{31}{3},\frac{13}{3}]$

$-\frac{31}{3}$ $\frac{13}{3}$

33. $-15 \le -7 - 2t \le 21$ $[-14,4]$

-14 4

34. $-400 < 500 + 20t \le 100$ $[-45,-20]$

-45 -20

35. In a typical school, an average from 80 up to but not including 90 earns a B. If your first two grades are 87 and 73, find all possible grades in the third exam which result in a grade of B. Assume that 100 is the highest possible score. $[80,100]$

36. At the Royal Rover Obedience School, dogs are rated as "well behaved" if they achieve an average score of 93 up to but not including 107 in a series of obedience tests. If your dog has already received grades of 91 and 72, what is the range of possible scores on the last exam for a "well-behaved" rating? Assume that the possible scores are 1 to 200. $[116,158)$

37. For a laboratory experiment to work, the temperature of a solution must be kept between 37° and 39° F. Thus $37 < F < 39$. Use the formula $F = \frac{9}{5}C + 32$ to find an equivalent range of values in the Celsius scale by solving $37 < \frac{9}{5}C + 32 < 39$. Round values to the nearest hundredth. $2.78 < C < 3.89$

38. The element mercury is a liquid between $-38.87°$ and $356.58°$ C. This range of temperature values is therefore given by $-38.87 < C < 356.58$. Express this in degrees Fahrenheit by using the formula $C = \frac{5}{9}(F - 32)$ and solving the resulting compound inequality. Round values to the nearest hundredth. $-37.97 < F < 673.84$

In Exercises 39–62, express the solution set graphically and in interval notation.

39. $x > 7$ or $x < 5$ $(-\infty,5) \cup (7,\infty)$

5 7

40. $x \ge 5$ or $x > 7$ $[5,\infty)$

5

41. $x < 0$ or $x \ge -2$ $(-\infty,\infty)$

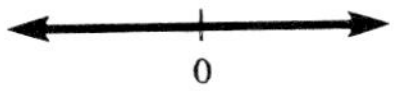

0

42. $x > 0$ or $x < -3$ $(-\infty,-3) \cup (0,\infty)$

-3 0

43. $x \le 5$ or $x < 7$ $(-\infty,7)$

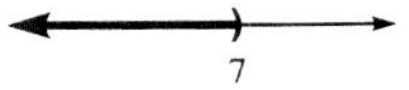

7

44. $x < -2$ or $x \ge -5$ $(-\infty,\infty)$

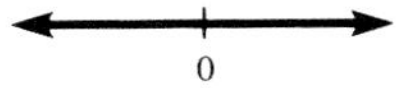

0

45. $x + 2 < -3$ or $x + 2 > 3$ $(-\infty,-5) \cup (1,\infty)$

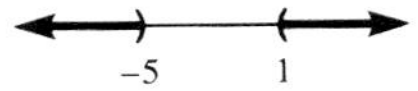

-5 1

46. $x - 4 > 5$ or $x - 4 < -5$ $(-\infty,-1) \cup (9,\infty)$

-1 9

47. $x + 2 \ge -3$ or $x + 2 \le 3$ $(-\infty,\infty)$

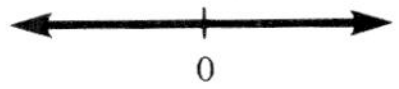

0

48. $x - 4 < 5$ or $x - 4 \ge -5$ $(-\infty,\infty)$

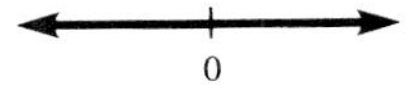

0

49. $x + 1 < 7$ or $x - 2 \le 6$ $(-\infty,8]$

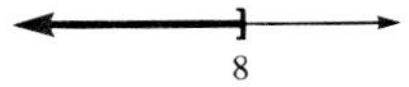

8

50. $x - 5 > -4$ or $x - 8 \ge -1$ $(1,\infty)$

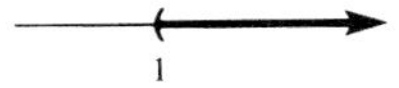

1

51. $x - 3 < -2$ or $x + 4 \le -3$ $(-\infty,1)$

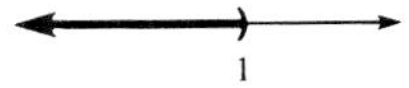

1

52. $x + 5 \ge 3$ or $x + 2 > 8$ $[-2,\infty)$

-2

53. $x + 1 < 0$ or $x - 2 \ge 0$ $(-\infty,-1) \cup [2,\infty)$

-1 2

54. $x - 3 \le 4$ or $x - 1 \ge -1$ $(-\infty,\infty)$

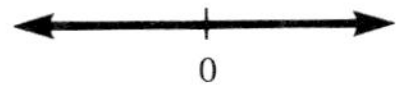

0

55. $x + 1 \ge 0$ or $x - 2 \le 0$ $(-\infty,\infty)$

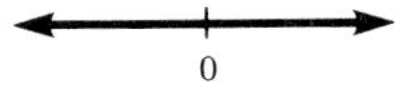

0

56. $x - 3 \le -4$ or $x - 1 \ge 1$ $(-\infty,-1] \cup [2,\infty)$

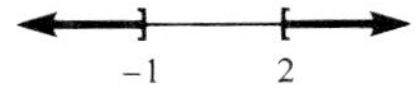

-1 2

57. $x - 3 \le -4$ and $x - 1 \ge 1$ $\emptyset$

58. $x - 3 \le 4$ and $x - 1 \ge -1$ $[0,7]$

0 7

59. $x + 1 \ge 0$ and $x - 2 \le 0$ $[-1,2]$

-1 2

60. $x + 1 < 0$ and $x - 2 \ge 0$ $\emptyset$

61. $x \le 7$ or $x + 5 \ge 17$ $(-\infty,7] \cup [12,\infty)$

7 12

62. $x - 3 \ge -4$ or $x \le 7$ $(-\infty,\infty)$

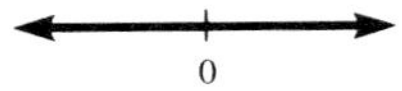

0

THINK ABOUT IT

1. Why is $-3 < x < 5$ called a "compound" inequality?
2. Solve $x < 3x - 12 < 2x$.
3. Solve $m - 2ts < x < m + 2ts$ for m. Assume $s > 0$. This is an inequality used in statistics to estimate the mean of a large group of people based on data from a survey of part of that group.
4. Explain why $5 < x < 1$ is a statement which contradicts itself.
5. In solving $a < bx < c$ for x, explain why you should not divide by b unless some other information is given. What is the needed information?

REMEMBER THIS

1. Is -2 in the solution set of $|2x - 4| < 3$? No
2. If $x = -3$, is it true that $|x| = -x$? Yes
3. Why is there no solution to $|x| = -8$? Absolute value of x can never be negative.
4. If the speed of an object moving in a straight line is 192 ft/second, what distance will it cover in 2 seconds? 384 ft
5. Graph the solution set for $-5x + 2 < 6$.

6. In a machine shop one can contains a mixture of 2 percent oil and 98 percent gasoline. Another can holds pure oil. How much of each should be used to get 8 qt of a new mixture that contains 50 percent oil? Round to the nearest tenth of a quart. 4.1 qt of 2 percent oil, 3.9 qt of 100 percent oil
7. The formula $A = \frac{1}{2}h(b_1 + b_2)$ gives the area of a trapezoid of height h with bases b_1 and b_2. If $A = 18 \text{ in.}^2$, $h = 6$ in., and one base is 1.5 in., find the other base. 4.5 in.
8. Solve $\frac{1}{2}x = \frac{2}{3}x - \frac{4}{5}$. $\left\{\frac{24}{5}\right\}$
9. If x represents the before-tax price of a car, what expression represents the total cost including a 5 percent sales tax? $x + 0.05x = 1.05x$
10. What number is the identity for multiplication? 1

2.6 Absolute Value Equations and Inequalities

A projectile fired vertically up from the ground with an initial velocity of 192 ft/second will hit the ground 12 seconds later, and the speed y of the projectile in terms of the elapsed time t is given by

$$y = |192 - 32t|.$$

For what values of t is the speed of the projectile 48 ft/second? (See Example 3.)

OBJECTIVES

1. **Solve equations involving absolute value.**
2. **Solve inequalities of the form $|ax + b| < c$.**
3. **Solve inequalities of the form $|ax + b| > c$.**

1 To solve equations and inequalities that involve absolute value, it is useful to consider both algebraic and geometric perspectives of the absolute value concept. Recall from Section 1.2 that the absolute value of a real number a, denoted $|a|$, is the distance between a and 0 on the number line. For instance, $|2| = 2$ and $|-2| = 2$, as shown in Figure 2.19.

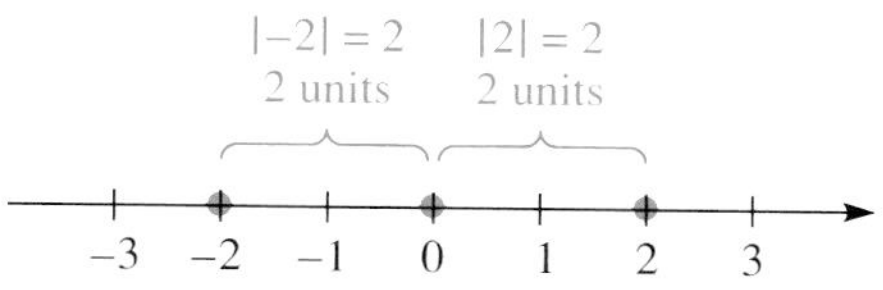

Figure 2.19

We see that if a is a positive number or 0, then $|a| = a$; while if a is a negative number then $|a| = -a$. Thus, the algebraic definition of $|a|$ is

$$|a| = \begin{cases} a, & \text{if } a \geq 0, \\ -a, & \text{if } a < 0. \end{cases}$$

Example 1 shows how these definitions can be used to solve equations like $|x| = 3$.

EXAMPLE 1 Solve $|x| = 3$. Use both an algebraic and a geometric interpretation.

Solution By definition, $|x|$ equals either x or $-x$. Setting these two expressions equal to 3 gives

$$x = 3 \quad \text{or} \quad -x = 3,$$
$$\text{so} \quad x = 3 \quad \text{or} \quad x = -3.$$

Both solutions check, and the solution set is $\{3, -3\}$.

To solve $|x| = 3$ from a geometric perspective, look for the two numbers that are 3 units from 0 on the number line. The required numbers are 3 and -3, as shown in Figure 2.20.

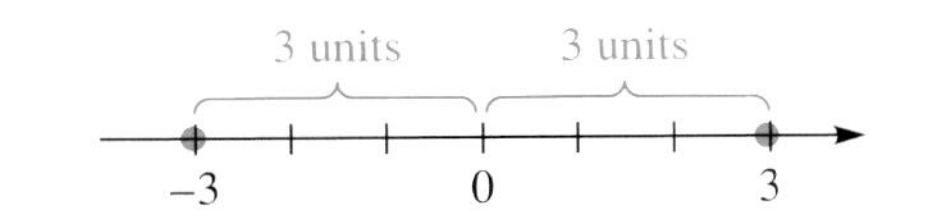

Figure 2.20

PROGRESS CHECK 1 Solve $|x| = 8$. Use both an algebraic and a geometric interpretation.

Example 1 illustrates that for $c > 0$,

$$|x| = c \quad \text{implies} \quad x = c \quad \text{or} \quad x = -c.$$

Other first-degree expressions may replace x in this result to provide a general procedure for solving absolute value equations of the form $|ax + b| = c$.

> **Solution of $|ax + b| = c$**
>
> If $|ax + b| = c$ and $c > 0$, then
>
> $$ax + b = c \quad \text{or} \quad ax + b = -c.$$

To solve $|ax + b| = c$ when $c = 0$, note that it is only necessary to solve $ax + b = 0$, because 0 is the only number whose absolute value is 0. Also, equations like $|x| = -3$ have no solution, since the absolute value of a number is never negative; so in general, $|ax + b| = c$ has no solution if $c < 0$.

EXAMPLE 2 Solve $|2x + 3| = 5$.

Solution $|2x + 3| = 5$ is solved as follows.

$$\begin{aligned} 2x + 3 &= 5 & \text{or} \quad 2x + 3 &= -5 \\ 2x &= 2 & \text{or} \quad 2x &= -8 \\ x &= 1 & \text{or} \quad x &= -4 \end{aligned}$$

Both solutions check (verify this), and the solution set is $\{1, -4\}$.

PROGRESS CHECK 2 Solve $|4x - 6| = 18$.

EXAMPLE 3 Solve the problem in the section introduction on page 75.

Solution Replacing y by 48 in the formula $y = |192 - 32t|$ gives $48 = |192 - 32t|$, which implies

$$\begin{aligned} 192 - 32t &= 48 & \text{or} \quad 192 - 32t &= -48 \\ -32t &= -144 & \text{or} \quad -32t &= -240 \\ t &= 4.5 & \text{or} \quad t &= 7.5. \end{aligned}$$

Progress Check Answers

1. $\{8, -8\}$

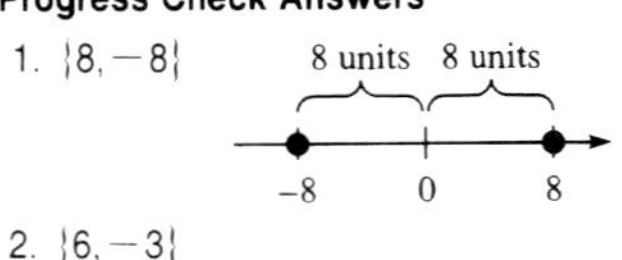

2. $\{6, -3\}$

Both solutions check since both values for t are meaningful in the context of the problem, while

$$|192 - 32(4.5)| = |48| = 48 \quad \text{and} \quad |192 - 32(7.5)| = |-48| = 48.$$

Thus, the projectile attains a speed of 48 ft/second after 4.5 seconds (on its way up) and again (on its way down) when 7.5 seconds have elapsed.

PROGRESS CHECK 3 For what values of t is the projectile considered in Example 3 moving at a speed of 144 ft/second?

The next example shows how to solve absolute value equations of the form $|ax + b| = |cx + d|$.

EXAMPLE 4 Solve $|3x + 2| = |5 - 2x|$.

Solution By the definition of absolute value, $|3x + 2| = \pm(3x + 2)$, and $|5 - 2x| = \pm(5 - 2x)$. Setting these two expressions equal to each other produces two distinct cases.

$$3x + 2 = 5 - 2x \quad \text{or} \quad 3x + 2 = -(5 - 2x)$$

Then solving each of these equations separately gives

$$\begin{aligned} 3x + 2 &= 5 - 2x &\quad \text{or} \quad 3x + 2 &= -(5 - 2x) \\ 5x &= 3 &\quad \text{or} \quad 3x + 2 &= -5 + 2x \\ x &= \tfrac{3}{5} &\quad \text{or} \quad x &= -7. \end{aligned}$$

Both $\frac{3}{5}$ and -7 check in the original equation, and the solution set is $\{\frac{3}{5}, -7\}$.

Note The given equation is also satisfied when

$$-(3x + 2) = -(5 - 2x) \quad \text{or} \quad -(3x + 2) = 5 - 2x.$$

However, multiplication of both sides of these equations by -1 shows that these two equations are equivalent to the two we solved and so need not be considered.

PROGRESS CHECK 4 Solve $|3 - x| = |2x + 1|$.

The idea that absolute value can define a distance on the number line may be extended and used to interpret expressions of the form $|x - a|$, which occur often in higher mathematics. Given any two points a and b, the distance between them on the number line is $|a - b|$, as illustrated in Figure 2.21. Note that $|a - b| = |b - a|$, so the order in the subtraction is not significant. This geometric perspective may be used to analyze equations of the form $|x - \text{constant}| = \text{constant}$, as discussed in Example 5.

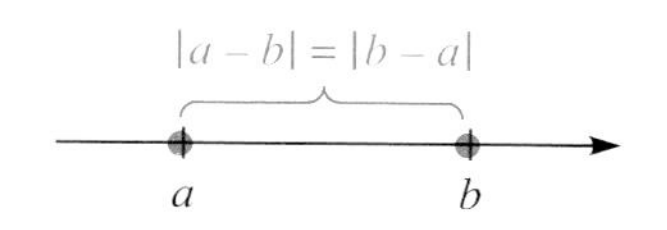

Figure 2.21

EXAMPLE 5 Solve $|x - 1| = 4$. Use both an algebraic and a geometric interpretation.

Solution By algebraic methods, $|x - 1| = 4$ implies

$$\begin{aligned} x - 1 &= 4 &\quad \text{or} \quad x - 1 &= -4 \\ x &= 5 &\quad \text{or} \quad x &= -3. \end{aligned}$$

Thus, the solution set is $\{5, -3\}$. From a geometric perspective, $|x - 1|$ represents the distance between 1 and some number x on the number line. So to solve $|x - 1| = 4$, we find the two numbers that are 4 units from 1 on the number line. Figure 2.22 confirms that the required numbers are 5 and -3.

Figure 2.22

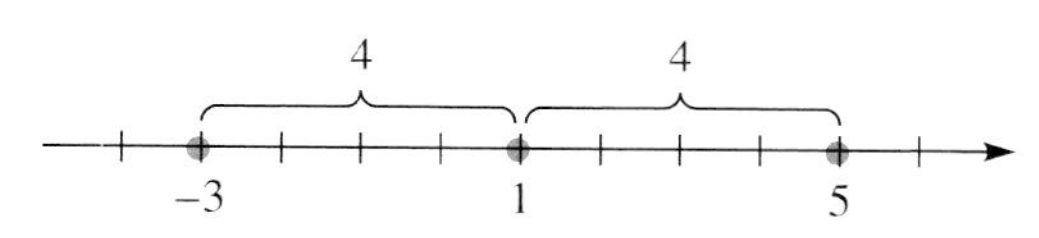

Progress Check Answers

3. 1.5 seconds, 10.5 seconds
4. $\{-4, \frac{2}{3}\}$

PROGRESS CHECK 5 Solve $|x - 2| = 7$. Use both an algebraic and a geometric interpretation.

2 In many applications involving absolute value the inequality signs ($<, \leq, >, \geq$) express the required relation in a problem. For instance, we can determine when the speed of the projectile described in the section-opening problem is *less than* 48 ft/second by solving

$$|192 - 32t| < 48.$$

To solve such inequalities, first consider that

$$|x| < 3$$

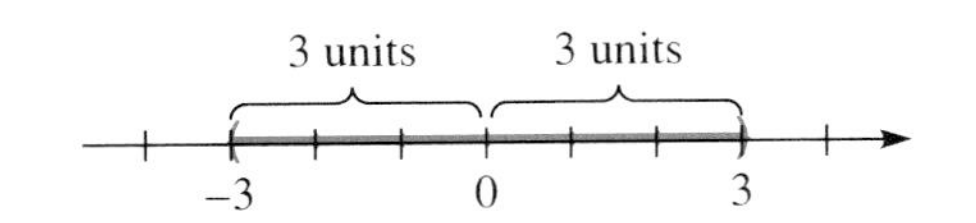

Figure 2.23

can be solved from a geometric viewpoint by finding all numbers that are less than 3 units from 0 on the number line. This set of numbers is graphed in Figure 2.23 and is expressed in set-builder notation by $\{x : -3 < x < 3\}$ and in interval notation by $(-3,3)$. This example illustrates that for $c > 0$

$$|x| < c \qquad \text{implies} \qquad -c < x < c,$$

which gives a general procedure for solving inequalities of the form $|ax + b| < c$.

Solution of $|ax + b| < c$

If $|ax + b| < c$ and $c > 0$, then

$$-c < ax + b < c.$$

Note that the inequality symbol $\leq$ may replace $<$ in our discussion to this point, so by similar reasoning, an inequality like $|x| \leq 3$ implies $-3 \leq x \leq 3$.

EXAMPLE 6 Solve $|x - 3| \leq 2$. Use both an algebraic and a geometric interpretation.

Solution $|x - 3| \leq 2$ implies

$$-2 \leq x - 3 \leq 2.$$

Now solve this compound inequality by the methods of the preceding section.

$$-2 \leq x - 3 \leq 2$$
$$1 \leq x \leq 5 \qquad \text{Add 3 to each member.}$$

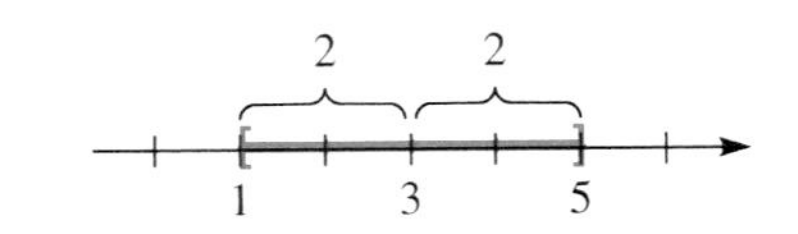

Figure 2.24

Thus, $|x - 3| \leq 2$ is true when x is any number between 1 and 5, inclusive, and the solution set is $[1,5]$. From a geometric viewpoint, $|x - 3| \leq 2$ requires that the distance on the number line between 3 and some number x be less than or equal to 2 units. Figure 2.24 shows that such an interval starts at 1, ends at 5, and includes both endpoints.

PROGRESS CHECK 6 Solve $|x - 1| \leq 6$. Use both an algebraic and a geometric interpretation.

EXAMPLE 7 Determine when the speed of the projectile described in the section-opening problem is less than 48 ft/second by solving $|192 - 32t| < 48$.

Solution $|192 - 32t| < 48$ implies $-48 < 192 - 32t < 48$.

$$-48 < 192 - 32t < 48$$
$$-240 < -32t < -144 \qquad \text{Subtract 192 from each member.}$$
$$7.5 > t > 4.5 \qquad \text{Divide each member by } -32 \text{ and reverse the sense of the inequalities.}$$

Progress Check Answers

5. $\{9, -5\}$

6. $[-5,7]$

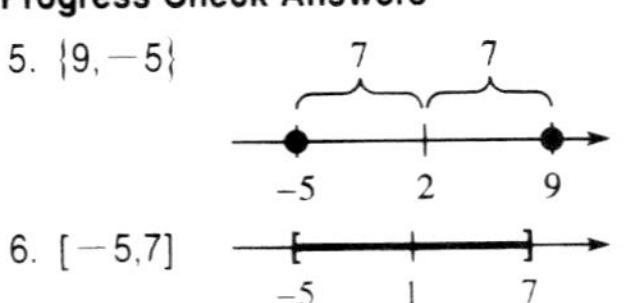

Thus, the speed of the projectile is less than 48 ft/second when the elapsed time is between 4.5 seconds and 7.5 seconds. This answer is written in interval notation as (4.5 seconds, 7.5 seconds).

PROGRESS CHECK 7 Find when the speed of the projectile described in the section-opening problem is less than 16 ft/second by solving $|192 - 32t| < 16$.

3 Finally, to solve absolute value inequalities of the form $|ax + b| > c$, consider that an inequality like

$$|x| > 3$$

can be solved by finding all numbers greater than 3 units from 0 on the number line. Figure 2.25 specifies this set of numbers graphically, in set-builder notation, and in interval notation. As suggested by this example, if $c > 0$,

$$|x| > c \quad \text{implies} \quad x < -c \quad \text{or} \quad x > c.$$

So inequalities of the form $|ax + b| > c$ may be solved as follows.

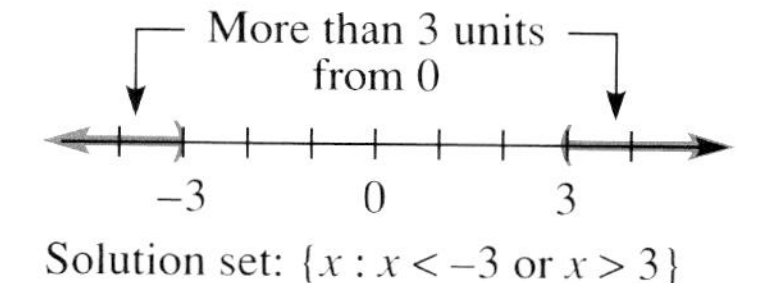

Figure 2.25

Solution of $|ax + b| > c$

If $|ax + b| > c$ and $c > 0$, then

$$ax + b < -c \quad \text{or} \quad ax + b > c.$$

EXAMPLE 8 Solve $|x - 1| > \frac{1}{2}$. Use both an algebraic and a geometric interpretation.

Solution Algebraically, the given inequality translates to the compound statement

$$x - 1 < -\tfrac{1}{2} \quad \text{or} \quad x - 1 > \tfrac{1}{2}.$$

We then add 1 to both sides in both inequalities to get

$$x < \tfrac{1}{2} \quad \text{or} \quad x > \tfrac{3}{2}.$$

So the solution set in interval notation is $(-\infty,\frac{1}{2}) \cup (\frac{3}{2},\infty)$. Figure 2.26 shows that this answer is in agreement from a geometric viewpoint, because the numbers located more than $\frac{1}{2}$ unit from 1 on the number line are either to the left of $\frac{1}{2}$ or to the right of $\frac{3}{2}$.

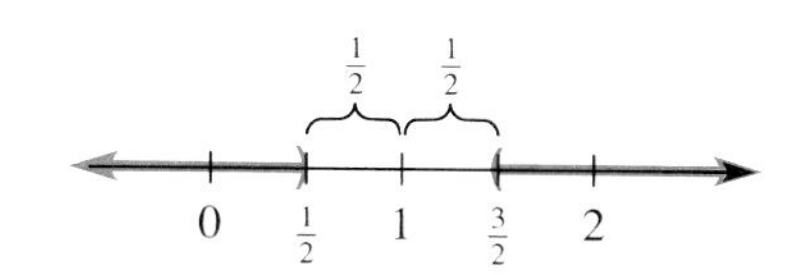

Figure 2.26

PROGRESS CHECK 8 Solve $|x - 15| > 5$. Use both an algebraic and a geometric interpretation.

EXAMPLE 9 Solve $|5 - 3x| \geq 4$.

Solution $|5 - 3x| \geq 4$ implies

$$5 - 3x \leq -4 \quad \text{or} \quad 5 - 3x \geq 4.$$

Then solving each of these inequalities separately gives

$$\begin{aligned} 5 - 3x &\leq -4 \quad \text{or} \quad 5 - 3x \geq 4 \\ -3x &\leq -9 \quad \text{or} \quad -3x \geq -1 \qquad \text{Subtract 5 from both sides of both inequalities.} \\ x &\geq 3 \quad \text{or} \quad x \leq \tfrac{1}{3}. \qquad \text{Divide by } -3 \text{ and reverse the inequality symbols.} \end{aligned}$$

The solution set determined by this compound statement is graphed as in Figure 2.27 and is expressed in interval notation as $(-\infty,\frac{1}{3}] \cup [3,\infty)$.

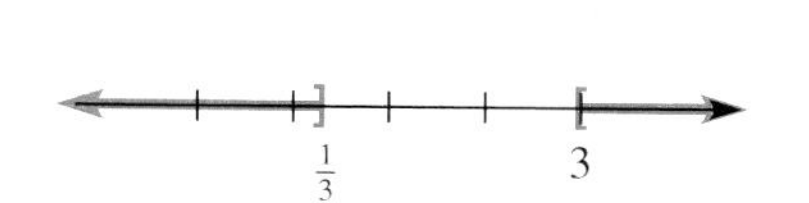

Figure 2.27

Progress Check Answers

7. (5.5 seconds, 6.5 seconds)
8. $(-\infty,10) \cup (20,\infty)$

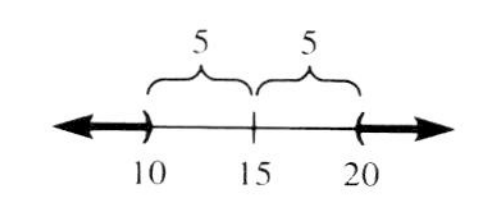

Progress Check Answers

9. $(-\infty, -4] \cup [5, \infty)$

PROGRESS CHECK 9 Solve $|1 - 2x| \geq 9$. Express the solution set graphically and in interval notation.

EXERCISES 2.6

In Exercises 1–8, solve the given equation using both an algebraic and a geometric interpretation.

1. $|x| = 1$ $\{1, -1\}$

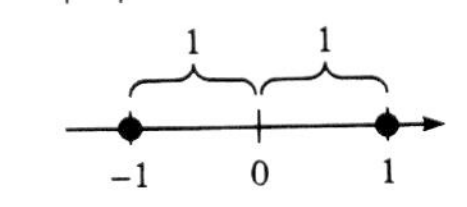

2. $|x| = 6$ $\{-6, 6\}$

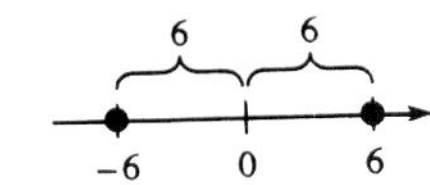

3. $|x| = 7$ $\{-7, 7\}$

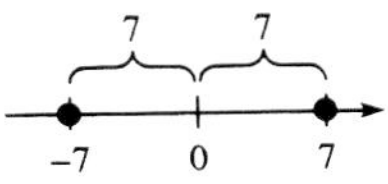

4. $|x| = -4$ $\emptyset$

5. $|x| = -3$ $\emptyset$

6. $|x| = 5$ $\{-5, 5\}$

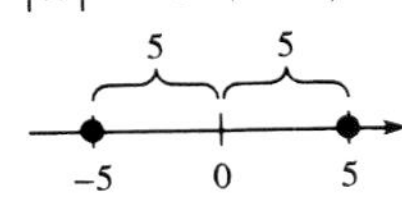

7. $|x| = 0$ $\{0\}$

8. $|x| + 4 = 4$ $\{0\}$

In Exercises 9–20, solve the given equation.

9. $|3x - 4| = 5$ $\{-\frac{1}{3}, 3\}$

10. $|3x + 9| = 15$ $\{2, -8\}$

11. $|7x - 63| = 0$ $\{9\}$

12. $|2x - 5| = 57$ $\{31, -26\}$

13. $|4x + 5| = 35$ $\{7.5, -10\}$

14. $|3x - 2| = 0$ $\{\frac{2}{3}\}$

15. $|4x - 12| = 36$ $\{-6, 12\}$

16. $|14x - 21| = 63$ $\{6, -3\}$

17. $|10x + 5| = 45$ $\{4, -5\}$

18. $|3x - 4| = -2$ $\emptyset$

19. $|2x - 3| = -4$ $\emptyset$

20. $|8x - 64| = 56$ $\{1, 15\}$

21. The speed of a projectile fired vertically from the ground in terms of time elapsed (t) is given by $s = |240 - 32t|$. For what values of t is the projectile's speed 80 ft/second? 5, 10 seconds

22. The speed of a projectile fired vertically from the ground is given by $s = |49 - 32t|$, where t = time elapsed. For what value of t is the projectile's speed 25 ft/second? 0.75, 2.3125 seconds

23. Find all values for which the absolute value of $8 - 3x$ is 100. $\{-\frac{92}{3}, 36\}$

24. Find all values for which the absolute value of $8 + 3x$ is 100. $\{\frac{92}{3}, -36\}$

In Exercises 25–32, solve the given equation.

25. $|3x + 2| = |5x - 2|$ $\{0, 2\}$

26. $|7x - 3| = |9x + 5|$ $\{-4, -\frac{1}{8}\}$

27. $|5 - x| = |3x + 2|$ $\{0.75, -3.5\}$

28. $|4 - x| = |5x + 6|$ $\{-\frac{1}{3}, -\frac{5}{2}\}$

29. $|7x - 8| = |9x + 12|$ $\{-10, -\frac{1}{4}\}$

30. $|10 - 13x| = |9x - 12|$ $\{1, -\frac{1}{2}\}$

31. $|12x - 11| = |20x - 19|$ $\{1, \frac{15}{16}\}$

32. $|18x - 16| = |26 - 10x|$ $\{1.5, -1.25\}$

In Exercises 33–52, solve using both an algebraic and a geometric interpretation.

33. $|x - 3| = 5$ $\{8, -2\}$

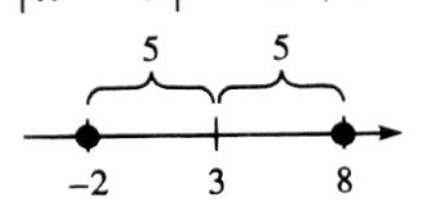

34. $|x - 4| = 3$ $\{1, 7\}$

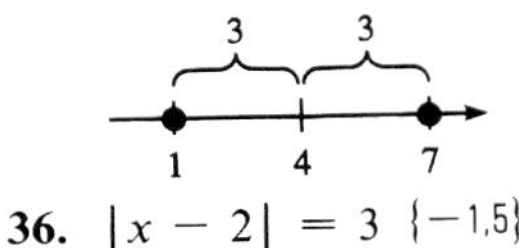

35. $|x - 1| = 6$ $\{-5, 7\}$

36. $|x - 2| = 3$ $\{-1, 5\}$

37. $|x + 1| = 4$ $\{-5, 3\}$

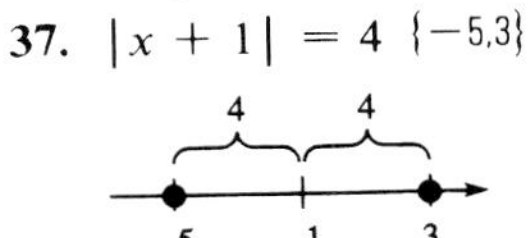

38. $|x + 2| = 7$ $\{-9, 5\}$

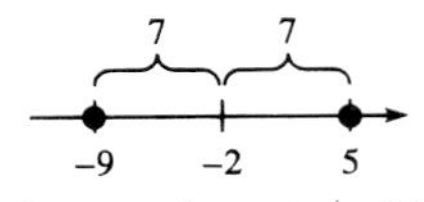

39. $|x + 10| = 8$ $\{-18, -2\}$

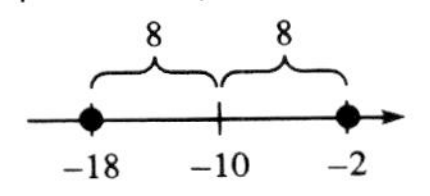

40. $|x + 3| = 4$ $\{-7, 1\}$

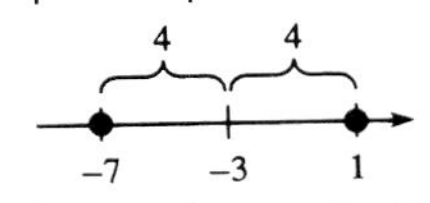

41. $|x - 6| \leq 6$ $[0, 12]$

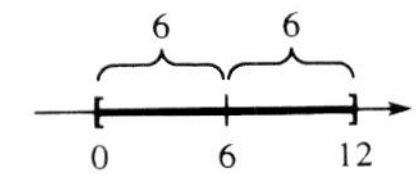

42. $|x - 2| \leq 8$ $[-6, 10]$

43. $|x - 5| \leq 7$ $[-2, 12]$

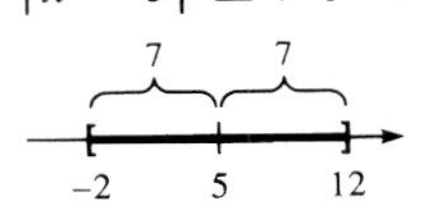

44. $|x - 6| \leq 5$ $[1, 11]$

45. $|x + 10| < 8$ $(-18, -2)$

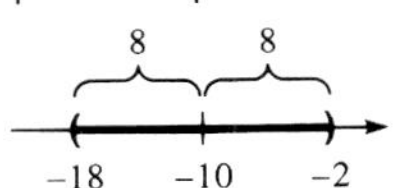

46. $|x + 3| < 4$ $(-7, 1)$

47. $|x - 1| < 7$ $(-6, 8)$

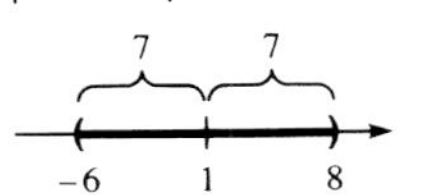

48. $|x + 3| \leq 2$ $[-5, -1]$

49. $|x + 8| \leq 2$ $[-10, -6]$

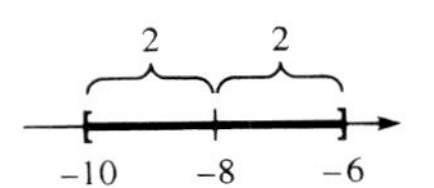

50. $|x - 7| \leq 9$ $[-2, 16]$

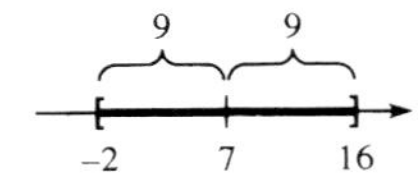

51. $|5 - x| < 14$ $(-9, 19)$

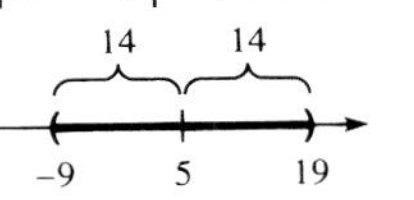

52. $|2 - x| < 5$ $(-3, 7)$

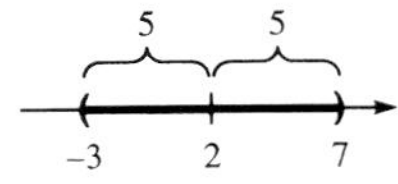

53. A projectile is fired vertically from the ground. Find when the speed of the projectile is less than 80 ft/second if speed s is given by $|240 - 32t|$. Express your answer graphically and in interval notation. $(5,10)$

5 10

54. A projectile is fired vertically from the ground. If speed $s = |49 - 32t|$, when is the speed of the projectile less than 25 ft/second? Express your answer graphically and in interval notation. $(0.75, 2.3125)$

0.75 2.3125

55. Find all values for which the absolute value of $25 - 4x$ is greater than 100. Express your answer graphically and in interval notation. $(-\infty,-18.75) \cup (31.25,\infty)$

−18.75 31.25

56. Find all values for which the absolute value of $25 + 5x$ is less than 100. Express your answer graphically and in interval notation. $(-25,15)$

−25 15

In Exercises 57–62, solve by using the geometric interpretation of absolute value. Check by an algebraic approach.

57. $|x - 1| > 3$ $(-\infty,-2) \cup (4,\infty)$

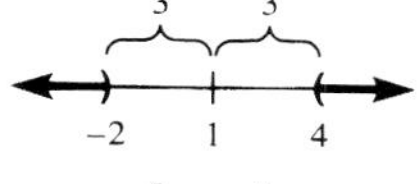

58. $|x - 3| \geq 5$ $(-\infty,-2] \cup [8,\infty)$

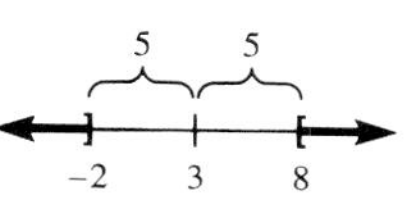

59. $|x - 12| \geq 7$ $(-\infty,5] \cup [19,\infty)$

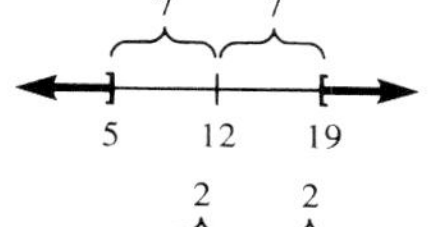

60. $|x + 3| > 2$ $(-\infty,-5) \cup (-1,\infty)$

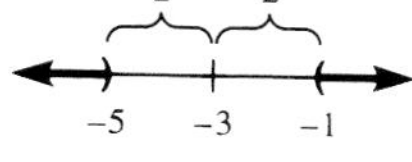

61. $|x + 2| > 3$ $(-\infty,-5) \cup (1,\infty)$

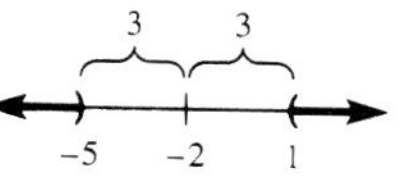

62. $|x + 5| \geq 7$ $(-\infty,-12] \cup [2,\infty)$

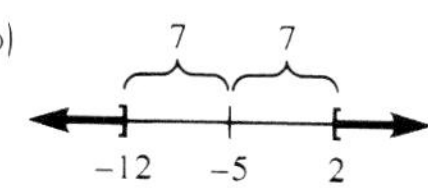

In Exercises 63–74, give the solution set graphically and in interval notation.

63. $|1 - 3x| \geq 5$ $(-\infty,-\frac{4}{3}] \cup [2,\infty)$

64. $|4 - 5x| > 11$ $(-\infty,-\frac{7}{5}) \cup (3,\infty)$

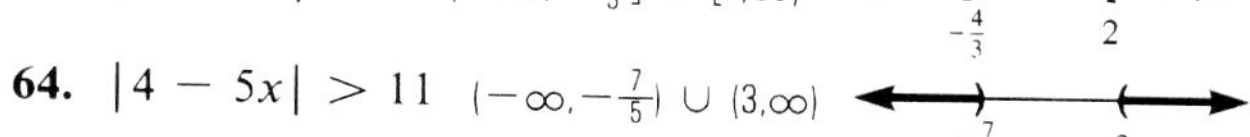

65. $|3 - 4x| \geq 13$ $(-\infty,-2.5] \cup [4,\infty)$

66. $|7 - 3x| > 13$ $(-\infty,-2) \cup (\frac{20}{3},\infty)$

67. $|3x - 8| \geq 1$ $(-\infty,\frac{7}{3}] \cup [3,\infty)$

68. $|5x + 2| \geq 3$ $(-\infty,-1] \cup [\frac{1}{5},\infty)$

69. $|3 - 7x| > 11$ $(-\infty,-\frac{8}{7}) \cup (2,\infty)$

70. $|1 - 4x| > 19$ $(-\infty,-4.5) \cup (5,\infty)$

71. $|1 - 4x| \leq 19$ $[-4.5,5]$

72. $|2x - 3| \leq 7$ $[-2,5]$

73. $|4x - 5| > 3$ $(-\infty,\frac{1}{2}) \cup (2,\infty)$

74. $|4x - 5| < 3$ $(\frac{1}{2},2)$

THINK ABOUT IT

1. Explain how the inequality $|x + 2| < 1$ may be solved geometrically. (*Hint:* Rewrite $x + 2$ in the form $a - b$.)
2. Write the given interval using inequalities and absolute value.
   a. $[-5,5]$ **b.** $(-\infty,-5) \cup (5,\infty)$
   c. $(6,8)$ **d.** $(-\infty,6] \cup [8,\infty)$
3. A property of absolute value is $|a/b| = |a|/|b|$. Use this property to solve each inequality. Assume $x \neq 0$.
   a. $\left|\frac{1}{x}\right| < 3$ **b.** $\left|\frac{2}{x}\right| \geq 1$
4. A famous inequality with wide application in advanced mathematics is called the absolute value inequality. It states that for any real values of a and b, $|a + b|$ can never be greater than $|a| + |b|$.
   a. Find values for a and b so that the two expressions are equal. There are many correct choices.
   b. Find values for a and b so that $|a + b|$ is less than $|a| + |b|$. There are many correct choices.
5. On a math test there was an equation that looked much too difficult to solve, so a student guessed that there was some kind of trick involved. Can you find the solution set of this equation just by looking at it?
$$\tfrac{4}{3}\pi\left|\sqrt{2.3x} - \sqrt[3]{x^2}\right| + 5 = 0$$

REMEMBER THIS

1. What is the meaning of $(-5)^4$? $(-5)(-5)(-5)(-5)$
2. In the expression a^n the letter a is called the (base, exponent, power, exponent). Base
3. Only one of these statements is correct. Which is it?
   a. $3^2 \cdot 3^2 = 9^2$ **b.** $3^2 \cdot 3^2 = 9^4$ a
4. Evaluate both of these expressions when $x = 2$.
   a. $-4x^3 \cdot 7x^3$ **b.** $-28x^6$ Both equal $-1{,}792$
5. True or false? $(\frac{3}{4})^2 < \frac{3}{4}$. True
6. Solve the compound inequality $x - 5 > 0$ and $2x \geq 3x - 10$. Express the solution graphically and in interval notation. $(5, 10]$

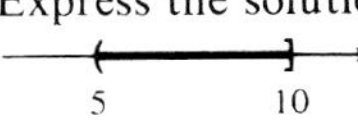

7. A meal plus a 15 percent tip plus a 5 percent tax came to \$17.40. Both the tip and the tax were applied only to the price of the meal. What was the price of the meal before tax and tip? \$14.50
8. Solve $A = \frac{1}{2}(a + b)$ for a. $a = 2A - b$
9. Solve $5x - 3(x - 2) = 5x - 3(x + 2)$. $\emptyset$
10. Which of these expressions are undefined?
    a. $\frac{3}{0}$ **b.** $\frac{0}{3}$ **c.** $\frac{-1}{3}$ **d.** $\frac{3}{-1}$ a

GRAPHING CALCULATOR

Evaluating Expressions Using Last Entry and [ANS]

In these examples we show one way to evaluate an algebraic expression for several different values of the variable.

EXAMPLE 1 Evaluate the expression $(x - 4)^3$ for $x = 1.2$, -4.7, and 3.01.

Solution This method uses a feature called Last Entry, which recalls the last expression entered so that you can alter it. Observe that the word ENTRY is printed in blue above the [ENTER] key.

Step 1 Evaluate $(x - 4)^3$ for $x = 1.2$ as usual, like this:

[(] 1.2 [−] 4 [)] [∧] 3

Press [ENTER] to see the answer, -21.952.

Step 2 Using Last Entry allows us to start again, as follows:
Press [2nd] [ENTRY]. This causes (1.2 − 4)∧3 to reappear for editing.
Use the left arrow key to put the cursor on the 1.
Press [INS] (which stands for "insert").
Key in -4.7.
Press [DEL] three times to delete the 1.2.
Press [ENTER].
The answer is -658.503.

If you repeat this procedure, replacing -4.7 by 3.01, you will see the answer $-.970299$. ⌟

EXAMPLE 2 **a.** First, evaluate $1 + \sqrt{3}$; then use that answer to evaluate $1 + \sqrt{3} - \sqrt{2}$.
b. First, evaluate $1 + \sqrt{3}$; then use that answer to evaluate $10 \div (1 + \sqrt{3})$.

Solution

a. This can be done easily by just continuing the problem.

Step 1 Key in 1 [+] [2nd] [√] 3.
Press [ENTER] to see the answer, 2.732050808.

Step 2 Continue with the subtraction by pressing [−] [2nd] [√] 2. Press [ENTER]; the answer is 1.317837245. Notice (Figure 2.28) that as soon as you press the subtraction key, the letters Ans appear on the screen. Ans stands for "answer," and it is always equal to the previous answer. You can check this by pressing the sequence [2nd] [ANS] [ENTER].

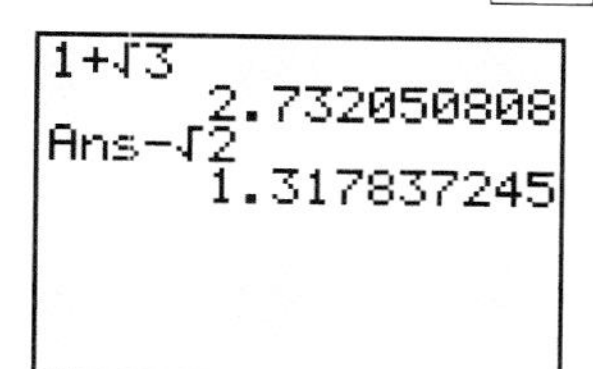

Figure 2.28

b. Step 1 Evaluate $1 + \sqrt{3}$ as before to get 2.732050808.

Step 2 Key in 10 [÷] [2nd] [ANS]. Press [ENTER]. As shown in Figure 2.29, the answer is 3.660254038.

```
1+√3
      2.732050808
10/Ans
      3.660254038
```

Figure 2.29

EXERCISES

1. Use the Last Entry feature to evaluate the expression $(x - 1)^3$ for $x = 1.1$, 2.2, and 3.3. Ans. .001; 1.728; 12.167
2. **a.** Evaluate $\sqrt{2} + \sqrt{3}$. Ans. 3.1463
   b. Use the answer from part **a** to evaluate $\dfrac{5}{\sqrt{2} + \sqrt{3}}$. Ans. 1.5892
   (Report both answers to four decimal places.)

Use the graphing calculator to check the answers for these exercises.

Section	*Exercises*
2.1	29, 59, 71
2.2	1, 7
2.3	5, 11, 65
2.6	13, 17, 25

Chapter 2 SUMMARY

OBJECTIVES CHECKLIST Specific chapter objectives are summarized along with numbered example problems from the text that should clarify the objectives. If you do not understand any objectives or do not know how to do the selected problems, then restudy the material.

2.1 Can you:

1. **Classify an equation as a conditional equation, an identity, or a false equation?**
   Classify the equation $x + 7 = 7 + x$ as an identity, a false equation, or a conditional equation. [Example 1a]
2. **Identify linear equations in one variable?**
   State whether or not $x + 17 = 14$ is a linear equation in one variable. [Example 2a]
3. **Solve linear equations using properties of equality?**
   Solve $\frac{1}{4}x + \frac{1}{3} = \frac{5}{6}x$. [Example 4]

2.2 Can you:

1. **Find the value of a variable in a formula when given values for the other variables?**
   The formula $V = \frac{1}{3}\pi r^2 h$ gives the volume of a cone with radius r and height h. Find h when $r = 3$ in. and $V = 15$ in.^{3}. Use 3.14 to approximate π. Round the answer to the nearest hundredth. [Example 1]

2. Solve a given formula or literal equation for a specified variable?
Solve $C = \frac{5}{9}(F - 32)$ for F. [Example 3]

2.3 Can you:

1. Solve word problems by translating phrases and setting up and solving equations?
When A and B filled out their joint income tax return, A's income was $\frac{1}{4}$ of B's income, and their total income was $47,500. Find each person's individual income. [Example 1]

2. Prove certain statements about integers?
Show that the sum of *any* 5 consecutive integers is equal to 5 times the middle one. [Example 2]

3. Solve problems involving geometric figures, percentages, uniform motion, and liquid mixtures?
One of the two acute angles in a right triangle is 20° greater than the other. What are the measures of all the angles? [Example 3]

2.4 Can you:

1. Specify solution sets of linear inequalities by using graphs and interval notation?
Write the solution set to $x < \frac{3}{2}$ in interval notation, and graph the interval. [Example 1b]

2. Solve linear inequalities by applying properties of inequalities?
Solve $-5(x - 1) > 4x - 13$, and graph the solution set. [Example 3]

2.5 Can you:

1. Solve compound inequalities involving *and* statements?
Solve $-64 < 192 - 32t < 64$. Express the solution set graphically and in interval notation. [Example 3]

2. Solve compound inequalities involving *or* statements?
Solve $x - 3 < -2$ or $x - 3 > 2$. Express the solution set graphically and in interval notation. [Example 5]

2.6 Can you:

1. Solve equations involving absolute value?
Solve $|2x + 3| = 5$. [Example 2]

2. Solve inequalities of the form $|ax + b| < c$?
Solve $|x - 3| \le 2$. Use both an algebraic and a geometric interpretation. [Example 6]

3. Solve inequalities of the form $|ax + b| > c$?
Solve $|x - 1| > \frac{1}{2}$. Use both an algebraic and a geometric interpretation. [Example 8]

KEY TERMS

Complementary angles (2.3)
Compound inequality (2.5)
Conditional equation (2.1)
Equation (2.1)
Equivalent equations (2.1)
False equation (2.1)
First-degree equation (2.1)
First-degree (or linear) inequality (2.4)
Identity (2.1)
Inequality (2.4)
Interval (2.4)
Interval notation (2.4)
Linear equation in one variable (2.1)
Literal equation (2.2)
Solution of an equation (2.1)
Solution of inequality (2.4)
Solution set (2.1)
Supplementary angles (2.3)

KEY CONCEPTS AND PROCEDURES

Section	Key Concepts or Procedures to Review
2.1	■ Procedure for solving a linear equation in one variable ■ Properties of equality
2.2	■ Formulas and literal equations are solved for a given variable by using the equation-solving principles given in Section 2.1.
2.3	■ Guidelines for setting up and solving word problems ■ Methods for proving certain statements about integers
2.4	■ Methods for representing sets of real numbers graphically or using interval notation ■ Properties of inequalities ■ Methods to solve linear inequalities
2.5	■ Procedures for solving compound inequalities involving *and* statements or *or* statements
2.6	■ Definition of absolute value ■ Methods to solve absolute value equations and inequalities and to interpret them geometrically ■ $\lvert a - b\rvert$ is the distance between points a and b on the number line. ■ If $\lvert ax + b\rvert = c$ and $c > 0$, then $ax + b = c$ or $ax + b = -c$. ■ If $\lvert ax + b\rvert < c$ and $c > 0$, then $-c < ax + b < c$. ■ If $\lvert ax + b\rvert > c$ and $c > 0$, then $ax + b < -c$ or $ax + b > c$.

CHAPTER 2 REVIEW EXERCISES

2.1

1. Classify the equation $x + 7 = x - 6$ as an identity, a false equation, or a conditional equation. False equation
2. State whether the equation $x^2 + 2 = 4$ is a linear equation in one variable. Not a linear equation
3. Solve $6x - 5 = -8$. $\{-\frac{1}{2}\}$
4. Solve $\frac{1}{2}x - \frac{1}{3} = -\frac{10}{3}$. $\{-6\}$
5. Solve $2(x + 1) + 4 = 5(x - 2) + 7$. $\{3\}$

2.2

6. Use the formula $A = \frac{1}{2}(a + b)h$ to find A when $a = 2$ in., $b = 10$ in., and $h = 3$ in. 18 in.
7. Use the formula $Z = \frac{1}{3}(af + m)$ to find m when $f = 76$, $a = 2$, and $Z = 78$. 82
8. Solve $c = \dfrac{a + b}{2}$ for a. $a = 2c - b$
9. Solve $3x - y = 8$ for y. $y = 3x - 8$
10. Solve $ax - c = bd$ with $a \neq 0$ for x. $x = \dfrac{bd + c}{a}$

2.3

11. When A and B filled out their joint income tax return, A's income was $\frac{1}{3}$ of B's income, and their total income was \$56,800. Find each person's individual income. \$14,200, \$42,600
12. Show that the sum of any 7 consecutive integers is equal to 7 times the middle number. $7x + 21 = 7(x + 3)$
13. One of the two acute angles in a right triangle is 40° greater than the other. What are the measures of all the angles? 25°, 65°, 90°
14. A total of \$20,000 was invested. Part of it was invested at 8 percent and the remainder was invested at 4 percent. The total interest for the year was \$1,000. How much money was invested at each rate? \$5,000 at 8 percent, \$15,000 at 4 percent
15. Two bicyclists are moving toward each other in a straight line. One is moving at 10 mi/hour; the other is moving at 15 mi/hour. If they start 30 mi apart, how long will it take until they meet? 1 hour and 12 minutes
16. A chemist has two acid solutions, one 20 percent acid and the other 8 percent acid. How much of each must be used to obtain 100 liters of a solution that is 17 percent acid? 75 liters of 20 percent, 25 liters of 8 percent

2.4

17. Write the solution set to $x \geq 2$ in interval notation, and graph the interval. $[2, \infty)$

18. Solve $-x + 3 < 15$. Express the solution set graphically and in interval notation. $(-12, \infty)$

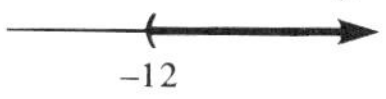

19. Solve $-2(x + 1) > 3x - 7$, and graph the solution set. $(-\infty, 1)$

20. Solve $4x - 1 \geq x - 7$, and graph the solution set. $[-2, \infty)$

21. Solve $x < x + 2$. $(-\infty, \infty)$

2.5

22. Find the solution set to $2 < x < 4$ in interval notation, and graph the interval. $(2,4)$

Solve each inequality. Express the solution set graphically and in interval notation.

23. $x - 3 < 0$ and $2x \geq x + 1$ $[1,3)$

24. $48 < 80 + 16t < 96$ $(-2,1)$

25. $x + 7 < 9$ or $x - 6 > -2$ $(-\infty, 2) \cup (4, \infty)$

26. $x > 1$ or $x < 3$ $(-\infty, \infty)$

2.6

27. Solve $|3x - 4| = 11$. $\{-\frac{7}{3}, 5\}$

28. Solve $|5 - x| = |4x - 5|$. $\{0,2\}$

29. Solve $|x - 3| = 4$. Use both an algebraic and a geometric interpretation. $\{-1,7\}$

30. Solve $|x - 4| \leq 1$. Use both an algebraic and a geometric interpretation. $[3,5]$

31. Solve $|2 + 3x| \geq 1$. Express the solution set graphically and in interval notation. $(-\infty, -1] \cup [-\frac{1}{3}, \infty)$

ADDITIONAL REVIEW EXERCISES

32. Which of these is a linear equation in one variable?
a. $x^2 - 1 = -8$ **b.** $2x + y = 8$
c. $-3x + 1 = x + 4$ c

33. Which of these equations is an identity?
a. $-2x + 6 = -2x$ **b.** $-2(x - 3) = -2x + 6$
c. $-2x + 6 = 12$ b

Solve.

34. $-4x - 19 = -9$ $\{-\frac{5}{2}\}$

35. $-3(x + 6) - 4 = 2(x - 1)$ $\{-4\}$

36. $\frac{3}{4}x + \frac{5}{12} = -\frac{1}{3}$ $\{-1\}$

37. $|-4x + 6| = 78$ $\{-18, 21\}$

38. $|-x + 1| = |6x - 13|$ $\{2, \frac{12}{5}\}$

39. One of the two acute angles in a right triangle is 5 times the other. What are the measures of all the angles? 15°, 75°, 90°

40. At 9 A.M. you started from your home, traveling by car at 40 mi/hour. At 11 A.M. your sister started after you on the same road, traveling at 55 mi/hour. At what time did your sister overtake you? 4:20 P.M.

41. One solution of sulfuric acid is 35 percent acid and another is 10 percent acid. How many liters of each must be used to make 95 liters of a new solution that is 20 percent acid? 38 liters of 35 percent, 57 liters of 10 percent

42. A man weighs 80 lb more than twice the weight of his son. How much do they each weigh if their combined weight is 218 lb? Man, 172 lb; son, 46 lb

43. An investment of $4,600 is made for one year, yielding $255. Part of the money is invested at 6 percent simple interest and the rest at 5 percent simple interest. How much is invested at each rate? $2,100 at 5 percent, $2,500 at 6 percent

Solve for the indicated variable.

44. $-6x - 3y = 12$; for y $y = -2x - 4$

45. $A = \dfrac{a + b + c}{3}$; for b $b = 3A - a - c$

46. $c(a + b) = d$; for a $a = \dfrac{d}{c} - b$

47. $S = \frac{1}{2}(a + L)$; for L, if $a = 3$ and $S = 52$ 101

Solve. Express the solution set graphically and in interval notation.

48. $|x + 7| \leq 5$ $[-12, -2]$

49. $x \leq -2$ $(\infty, -2]$

50. $-x > -x + 2$ $\emptyset$

51. $3x + 5 > -3x - 7$ $(-2, \infty)$

52. $0 \leq x + 1 \leq 4$ $[-1, 3]$

53. $x + 3 > 4$ or $x + 1 < -3$ $(-\infty, -4) \cup (1, \infty)$

54. $-(x - 8) > 4x + 12$ $(-\infty, -\frac{4}{5})$

55. $x + 2 \leq 0$ and $-2x > 3x + 5$ $(-\infty, -2]$

56. $x > 7$ or $x < 5$ $(-\infty, 5) \cup (7, \infty)$

57. $|4 - 2x| \geq 10$ $(-\infty, -3] \cup [7, \infty)$

58. $-3x + 9 \leq -12$ $[7, \infty)$

59. $-33 < 12 + 9t < 84$ $(-5, 8)$

60. $|x + 1| \geq 0$ $(-\infty, \infty)$

CHAPTER 2 TEST

1. Which of these is a conditional equation?
   a. $-3x + 7 = 46$ **b.** $2x + 4 > 9$ **c.** $3x = 3x$ a
2. One of the two acute angles of a right triangle is 6° less than 2 times the other. What are the measures of the acute angles?
   32°, 58°
3. Two cars traveling in the same direction on a straight road began their trip at noon. One car was traveling at 50 mi/hour; the other was traveling at 54 mi/hour. At what time will the cars be 14 mi apart? 3:30 P.M.

In Questions 4–7, solve for x.

4. $4 + 2(x + 3) = -8(x + 5)$ $\{-5\}$
5. $\frac{1}{2}x - \frac{1}{3} = \frac{1}{12}$ $\{\frac{5}{6}\}$
6. $|4x - 5| = |-2x + 13|$ $\{-4, 3\}$
7. $|-2x - 8| = 12$ $\{-10, 2\}$

In Questions 8–10, solve for the indicated variable.

8. $P = \dfrac{x - y - z}{10}$; for y $y = x - z - 10P$
9. $5x + 3y = 6$; for y $y = -\frac{5}{3}x + 2$
10. $ax + b = n$; for x, if $a = 2$, $b = 3$, $n = 12$ $x = \frac{9}{2}$

In Questions 11–15, solve the given inequality. Express the solution set in interval notation.

11. $-x \geq 3$ $(-\infty, -3]$
12. $x + 7 > x - 1$ $(-\infty, \infty)$
13. $-1 \leq x - 2 \leq 5$ $[1, 7]$
14. $|2x - 1| < 9$ $(-4, 5)$
15. $|-3x| < 0$ $\emptyset$

In Questions 16–20, solve the given inequality. Express the solution set graphically.

16. $|6 - x| \geq 9$

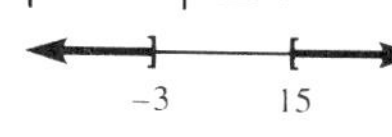

17. $-2(x - 5) < x + 8$

$\frac{2}{3}$

18. $x + 3 \geq 0$ and $5 - 2x \leq 5$

0

19. $2x + 1 < 5$ or $x + 1 > 4$

2 3

20. $-7 \leq -3 + 4x \leq 9$

−1 3

CUMULATIVE TEST 2

1. Solve the formula $P = 2\ell + 2w$ for w. $w = \dfrac{P - 2\ell}{2}$
2. List the numbers in the set of positive integers. $\{1, 2, 3, \ldots\}$
3. Evaluate $\dfrac{2a^2 + b}{2(a^2 + b)}$ when $a = -1$ and $b = \frac{1}{2}$. $\frac{5}{6}$
4. A 20 percent discount resulted in a TV selling for \$352. What was the original price? \$440
5. Simplify $-4 - [-3 - (-2 - 1)]$. -4
6. What property is illustrated by the statement $2(6n) = (2 \cdot 6)n$?
   Associative property of multiplication
7. Rewrite the expression $-(2x - 3y + 5)$ without parentheses.
   $-2x + 3y - 5$
8. What is the circumference of a circle with diameter 8 in.? (Use 3.14 for π.) 25.12 in.
9. Translate to an algebraic expression: Two times the sum of x and y. $2(x + y)$
10. Evaluate $-|-4|$. -4
11. Remove the symbols of grouping and combine like terms: $-2(a - 3) + 4(3a + 1)$. $10a + 10$
12. Evaluate $3{,}000(1 + 0.07)^9$ using a scientific calculator. Round to the nearest hundredth. 5,515.38
13. One of the two acute angles in a right triangle is 5° less than the other. What are the measures of the acute angles?
    47.5°, 42.5°

In Questions 14–17, solve the given equation.

14. $3x + 12 = -9x + 20$ $\{\frac{2}{3}\}$
15. $|-3x - 11| = 2$ $\{-\frac{13}{3}, -3\}$
16. $\frac{1}{3}x - \frac{1}{6} = \frac{1}{4}$ $\{\frac{5}{4}\}$
17. $|-3x - 4| = |2x - 1|$ $\{-\frac{3}{5}, -5\}$

In Questions 18–20, solve the given inequality. Express the solution set graphically and in interval notation.

18. $7x - 4 \geq 8x - 6$ $(-\infty, 2]$

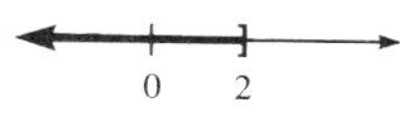

19. $-3 < -7 + 4x < 13$ $(1, 5)$

20. $|x + 6| < 9$ $(-15, 3)$

CHAPTER 3

Exponents and Polynomials

You would love to buy a fancy sports car which costs \$35,000. You can come up with a \$5,000 deposit, but you must finance the rest. The bank offers you a 4-year loan with an annual interest rate of 12 percent. Compute your monthly payment E by using the formula

$$E = \frac{Pr}{1 - (1 + r)^{-n}}.$$

P is the amount of the loan, r is the monthly interest rate, and n is the number of months the loan runs. (See Example 10 in Section 3.1.)

USING THE definitions and properties of exponents is a crucial skill in algebra. In this chapter we show how to work with exponents that are integers and especially how to use them in connection with polynomials. As you will see, polynomials are algebraic expressions with terms of the form

$$(\text{real number})x^{\text{nonnegative integer}}.$$

Typical polynomial terms look like $5x^3$ or $5x^2$ or $-2x$. Polynomial expressions are widely used in solving applied problems.

3.1 Integer Exponents

OBJECTIVES

1. **Interpret and write positive integer exponents.**
2. **Apply the product property of exponents.**
3. **Use zero as an exponent.**
4. **Use negative integer exponents.**
5. **Apply the quotient property of exponents.**
6. **Apply the power properties of exponents.**
7. **Evaluate integer exponents on a calculator.**

1 Recall from Chapter 1 that a positive integer exponent is a shortcut way of expressing a repeated factor. That is,

6^3 is shorthand for $6 \cdot 6 \cdot 6$,
$(-5)^2$ is shorthand for $(-5) \cdot (-5)$,
a^n, where n is a positive integer, is shorthand for

$$\underbrace{a \cdot a \cdot a \cdot a \cdots a.}_{n \text{ factors}}$$

Thus by a^n, where n is a positive integer, we mean to use a as a factor n times. The expression a^n is read "**the *n*th power of *a***" and a^n is called an **exponential expression** with **base** a and **exponent** n. For the first power of a, remember that $a^1 = a$.

EXAMPLE 1 Write each expression using exponents. In each case, identify the base and the exponent.

a. $8 \cdot 8 \cdot 8 \cdot 8$ **b.** $(p/q) \cdot (p/q)$ **c.** n
d. $(-1) \cdot (-1) \cdot (-1)$ **e.** $-2 \cdot 2 \cdot 2$

Solution Apply the definition of a positive integer exponent.

a. $8 \cdot 8 \cdot 8 \cdot 8 = 8^4$. The base is 8 and the exponent is 4.
b. $(p/q) \cdot (p/q) = (p/q)^2$. The base is p/q and the exponent is 2. Note that parentheses are needed to show that the base is p/q.
c. $n = n^1$. The base is n and the exponent is 1.
d. $(-1) \cdot (-1) \cdot (-1) = (-1)^3$. The base is (-1) and the exponent is 3.
e. $-2 \cdot 2 \cdot 2 = (-1)2 \cdot 2 \cdot 2 = (-1)2^3$, which is the opposite or negative of 2^3. Thus $-2 \cdot 2 \cdot 2 = -2^3$. The base is 2 and the exponent is 3.

Caution Be careful to distinguish between the form $(-a)^n$, as in part **d,** and the form $-a^n$, as in part **e.** In the expression $(-a)^n$ the parentheses indicate that the base is $-a$. The absence of parentheses in $-a^n$ means that the expression symbolizes the negative of a^n, or $(-1)a^n$.

Ask students to evaluate both $(-2)^4$ and -2^4 on a scientific calculator.

PROGRESS CHECK 1 Write each expression using exponents. In each case identify the base and the exponent.

a. $3 \cdot 3 \cdot 3 \cdot 3 \cdot 3$ **b.** $-4 \cdot 4$
c. $(-1) \cdot (-1)$ **d.** $(1/7) \cdot (1/7) \cdot (1/7)$

Progress Check Answers

1. (a) 3^5; 3, 5 (b) -4^2; 4, 2 (c) $(-1)^2$; -1, 2 (d) $(1/7)^3$; 1/7, 3

2 We now develop a rule for the product of two exponential expressions which have the same base. For instance, consider the product $3^2 \cdot 3^4$. We can reason that

$$3^2 \cdot 3^4 = \underbrace{(3 \cdot 3)}_{\text{2 factors}} \cdot \underbrace{(3 \cdot 3 \cdot 3 \cdot 3)}_{\text{4 factors}} = \underbrace{3 \cdot 3 \cdot 3 \cdot 3 \cdot 3 \cdot 3}_{\text{6 factors}} = 3^6.$$

Note that the exponent 6 on the resulting product is the sum of the original exponents (2 and 4). This example illustrates the product property of exponents.

Product Property of Exponents

If m and n are positive integers and a is a real number, then

$$a^m \cdot a^n = a^{m+n}.$$

Remember that this property applies only to exponential expressions with the *same base*. Computing a product this way is really just counting the number of times a appears as a factor.

EXAMPLE 2 Use the product property of exponents to simplify each expression.

a. $3^4 \cdot 3^3$ **b.** $y^3 \cdot y^5$ **c.** $(-5) \cdot (-5)^7$ **d.** b^5b^4b

Solution

a. $3^4 \cdot 3^3 = 3^{4+3} = 3^7 = 2{,}187$
b. $y^3 \cdot y^5 = y^{3+5} = y^8$
c. Because $(-5) = (-5)^1$, we have

$$(-5) \cdot (-5)^7 = (-5)^1 \cdot (-5)^7 = (-5)^{1+7} = (-5)^8 = 390{,}625.$$

d. The three factors have a common base, so add all three exponents.

$$b^5b^4b = b^{5+4+1} = b^{10}$$

PROGRESS CHECK 2 Simplify each expression.

a. $2^2 \cdot 2^3$ **b.** $(-y)^6(-y)^4$ **c.** $(-1)^5 \cdot (-1)$ **d.** c^2c^5c

EXAMPLE 3 Find (a) the product and (b) the sum of $-4x^3$ and $7x^3$ in simplest form.

Solution

a. To find the product, we *multiply* the expressions.

$$\begin{aligned} -4x^3 \cdot 7x^3 &= (-4 \cdot 7)(x^3 \cdot x^3) && \text{Reorder and group.} \\ &= (-4 \cdot 7)x^6 && \text{Product property of exponents.} \\ &= -28x^6 && \text{Simplify.} \end{aligned}$$

b. To find the sum, we *add* the expressions. Simplify by combining like terms.

$$\begin{aligned} -4x^3 + 7x^3 &= (-4 + 7)x^3 && \text{Distributive property.} \\ &= 3x^3 && \text{Simplify.} \end{aligned}$$

PROGRESS CHECK 3 Find (a) the product and (b) the sum of $-3x^4$ and $2x^4$ in simplest form.

Progress Check Answers

2. (a) 32 (b) $(-y)^{10}$ (c) 1 (d) c^8
3. (a) $-6x^8$ (b) $-x^4$

3 In the product property of exponents we carefully stated that the exponents were *positive* integers. Because it is a goal of algebra to be able to use *any* real number as an exponent, we now introduce the idea of using zero and negative integer exponents. The guiding principle is that all rules which work for positive integer exponents should continue to work with zero and negative integer exponents.

The idea that any new rule should include the important features of old, more restricted ones was called the *principle of the permanence of equivalent forms* by George Peacock (1791–1858), one of the founders of modern abstract algebra.

We start by applying the product property of exponents to the product of 3^2 and 3^0. Using the product property, we have

$$3^2 \cdot 3^0 = 3^{2+0} = 3^2.$$

When we multiply 3^2 by 3^0, the result is 3^2. Thus by the identity property of multiplication 3^0 must equal 1, and we can make the following definition.

Zero Exponent

If a is a nonzero real number, then $a^0 = 1$.

EXAMPLE 4 Evaluate each expression. Assume $x \neq 0$.

a. $3^0 + 2^0$ **b.** $(-8 + 3)^0$ **c.** $(5x)^0$ **d.** $5x^0$

Solution Apply the zero exponent definition.

a. $3^0 = 1$ and $2^0 = 1$, so $3^0 + 2^0 = 1 + 1 = 2$.
b. $(-8 + 3)^0 = (-5)^0 = 1$. Note that $(-8 + 3)^0$ is not equal to $(-8)^0 + 3^0$.
c. The parentheses indicate that a nonzero number is raised to the zero power, so $(5x)^0 = 1$. The assumption $x \neq 0$ is needed because 0^0 is undefined.
d. Because $5x^0 = 5^1 \cdot x^0$, we have $5x^0 = 5(1) = 5$.

PROGRESS CHECK 4 Evaluate each expression. Assume $x \neq 0$.

a. $3^0 + (-3)^0$ **b.** $(8 - 2)^0$ **c.** $(-4x)^0$ **d.** $-4x^0$

4 Next we want to have a definition for exponents that are negative integers. Consider the result of applying the product property of exponents to $3^2 \cdot 3^{-2}$.

$$3^2 \cdot 3^{-2} = 3^{2+(-2)} = 3^0 = 1$$

When we multiply 3^2 by 3^{-2}, the result is 1. Thus, by the inverse property of multiplication 3^{-2} is the reciprocal of 3^2, or $3^{-2} = \frac{1}{3^2}$. It follows that we can extend our previous laws of exponents by making the following definition.

Negative Exponent

If a is a nonzero real number and n is an integer, then

$$a^{-n} = \frac{1}{a^n}.$$

This definition means that a^{-n} is the reciprocal of a^n.

Progress Check Answers

4. (a) 2 (b) 1 (c) 1 (d) -4

EXAMPLE 5 Write each expression with positive exponents, and evaluate, where possible. Assume $x \neq 0$.

a. 4^{-3} **b.** $(-5)^{-3}$ **c.** $3^{-1} - 2^{-1}$ **d.** $(\frac{2}{5})^{-4}$ **e.** $3x^{-2}$

Solution Use the negative exponent definition and the principles for operations on real numbers from Section 1.2.

a. $4^{-3} = \dfrac{1}{4^3} = \dfrac{1}{64}$

b. $(-5)^{-3} = \dfrac{1}{(-5)^3} = \dfrac{1}{-125} = -\dfrac{1}{125}$

c. $3^{-1} - 2^{-1} = \dfrac{1}{3^1} - \dfrac{1}{2^1} = \dfrac{2}{6} - \dfrac{3}{6} = -\dfrac{1}{6}$

d. $\left(\dfrac{2}{5}\right)^{-4} = \dfrac{1}{(\frac{2}{5})^4} = \dfrac{1}{\frac{16}{625}} = 1 \cdot \dfrac{625}{16} = \dfrac{625}{16}$

e. $3x^{-2} = 3 \cdot \dfrac{1}{x^2} = \dfrac{3}{x^2}$

It is worth stressing this caution, since students find negative exponents nonintuitive.

Caution The sign of the exponent does not give you the sign of the resulting expression. For instance,

$$4^{-3} \text{ is positive; } (-5)^{-3} \text{ is negative.}$$

Negative exponents represent reciprocals which may or may not be negative numbers.

PROGRESS CHECK 5 Write each expression with positive exponents, and evaluate, where possible. Assume $x \neq 0$.

a. 100^{-1} **b.** $(-100)^{-1}$ **c.** $4^{-2} - 3^{-2}$ **d.** $(\frac{3}{4})^{-3}$ **e.** $4x^{-3}$

5 When we multiply exponential expressions that have the same base, we *add* their exponents. So it is natural to expect that when we divide exponential expressions with the same base, we should *subtract* their exponents. To see that $a^m/a^n = a^{m-n}$, consider the following three cases.

Case	Example Using Positive Integer Exponent Definition	Example Using $a^m/a^n = a^{m-n}$
$m > n$	$\dfrac{7^6}{7^2} = \dfrac{7 \cdot 7 \cdot 7 \cdot 7 \cdot 7 \cdot 7}{7 \cdot 7} = 7^4$	$\dfrac{7^6}{7^2} = 7^{6-2} = 7^4$
$m = n$	$\dfrac{7^2}{7^2} = \dfrac{7 \cdot 7}{7 \cdot 7} = 1$	$\dfrac{7^2}{7^2} = 7^{2-2} = 7^0 = 1$
$m < n$	$\dfrac{7^2}{7^6} = \dfrac{7 \cdot 7}{7 \cdot 7 \cdot 7 \cdot 7 \cdot 7 \cdot 7} = \dfrac{1}{7^4}$	$\dfrac{7^2}{7^6} = 7^{2-6} = 7^{-4} = \dfrac{1}{7^4}$

We could give similar examples where m and n are zero or negative integers. Thus we may state the quotient property of exponents.

Quotient Property of Exponents

If m and n are integers and a is a nonzero real number, then

$$\frac{a^m}{a^n} = a^{m-n}.$$

Progress Check Answers

5. (a) $\frac{1}{100}$ (b) $-\frac{1}{100}$ (c) $-\frac{7}{144}$ (d) $\frac{64}{27}$ (e) $4/x^3$

EXAMPLE 6 Simplify and write the result using only positive exponents. Assume $a \neq 0$ and $x \neq 0$.

a. $\dfrac{a^7}{a^4}$ **b.** $\dfrac{3^{-2}}{3^2}$ **c.** $\dfrac{4^2x^3}{4^5x^3}$ **d.** $\dfrac{10^{-5}x^2}{10^{-6}x^3}$

Solution Apply the quotient property of exponents.

a. $\dfrac{a^7}{a^4} = a^{7-4} = a^3$

b. $\dfrac{3^{-2}}{3^2} = 3^{-2-2} = 3^{-4} = \dfrac{1}{3^4} = \dfrac{1}{81}$

c. $\dfrac{4^2x^3}{4^5x^3} = 4^{2-5}x^{3-3} = 4^{-3}x^0 = \dfrac{1}{4^3} \cdot 1 = \dfrac{1}{64}$

d. $\dfrac{10^{-5}x^2}{10^{-6}x^3} = 10^{-5-(-6)}x^{2-3} = 10^1x^{-1} = 10 \cdot \dfrac{1}{x} = \dfrac{10}{x}$

PROGRESS CHECK 6 Simplify and write the result using only positive integers. Assume $a \neq 0$ and $y \neq 0$.

a. $\dfrac{4^5}{4^3}$ **b.** $\dfrac{a^{-4}}{a^4}$ **c.** $\dfrac{5^8y^5}{5^8y^6}$ **d.** $\dfrac{3^{-1}y}{3y^7}$

6 Another very important property of exponents concerns raising to a power an expression that contains exponents. For example, what is the result when x^2 is raised to the fourth power?

$$(x^2)^4 \quad \text{means} \quad x^2 \cdot x^2 \cdot x^2 \cdot x^2 = x^{2+2+2+2} = x^8.$$

Note that the exponent 8 in the result is the product of the original exponents (2 and 4). This example illustrates the power-to-a-power property of exponents, which also holds for zero and negative exponents.

Power-to-a-Power Property of Exponents

If m and n are integers and a is a nonzero real number, then

$$(a^m)^n = a^{mn}.$$

This property says that to raise an exponential expression to a power, use the same base and multiply the exponents.

EXAMPLE 7 Simplify each expression. Assume a and x are not zero.

a. $(10^3)^2$ **b.** $(x^{-3})^4$ **c.** $[(-a)^0]^2$ **d.** $(2^5)^{-1}$

Solution Apply the power-to-a-power property of exponents.

a. $(10^3)^2 = 10^{3 \cdot 2} = 10^6$, or 1,000,000 (1 million)

b. $(x^{-3})^4 = x^{-3 \cdot 4} = x^{-12} = \dfrac{1}{x^{12}}$

c. $[(-a)^0]^2 = (-a)^{0 \cdot 2} = (-a)^0 = 1$

d. $(2^5)^{-1} = 2^{5(-1)} = 2^{-5} = \dfrac{1}{2^5} = \dfrac{1}{32}$

PROGRESS CHECK 7 Simplify each expression.

a. $(5^3)^4$ **b.** $(y^{-6})^5$ **c.** $[(-4)^2]^0$ **d.** $(3^{-4})^{-1}$

Progress Check Answers

6. (a) 16 (b) $1/a^8$ (c) $1/y$ (d) $1/(9y^6)$
7. (a) 5^{12} (b) $1/y^{30}$ (c) 1 (d) 81

Because we will often need to raise products and quotients to powers, it is a good idea to establish clear procedures to do so. First, consider the result when $3x$ is raised to the third power.

$$\begin{aligned} (3x)^3 &= (3x)(3x)(3x) && \text{Definition of positive integer exponent.} \\ &= (3 \cdot 3 \cdot 3)(xxx) && \text{Reorder and regroup.} \\ &= 3^3x^3 && \text{Definition of positive integer exponent.} \\ &= 27x^3 && \text{Simplify.} \end{aligned}$$

You can see that in raising the product $3x$ to the third power, we just raised each individual factor to that power. A similar result can be seen when a fraction is raised to a power. For example,

$$\left(\frac{3}{4}\right)^3 = \frac{3}{4} \cdot \frac{3}{4} \cdot \frac{3}{4} = \frac{3 \cdot 3 \cdot 3}{4 \cdot 4 \cdot 4} = \frac{3^3}{4^3}$$

These examples illustrate the following power properties, which hold also for zero and negative exponents.

Power Properties of Products and Quotients

If n is an integer and a and b are real numbers, then

$$(ab)^n = a^nb^n \quad \text{and} \quad \left(\frac{a}{b}\right)^n = \frac{a^n}{b^n}, \quad b \neq 0.$$

EXAMPLE 8 Simplify each expression, and write the result using only positive exponents. Assume $x \neq 0$.

a. $(-3x^4y)^2$ **b.** $\left(\frac{-a}{7x}\right)^2$ **c.** $(5x^{-3})^{-2}$

Solution Apply the necessary power properties.

a. $$\begin{aligned} (-3x^4y)^2 &= (-3)^2(x^4)^2y^2 && \text{Product-to-a-power property.} \\ &= 9x^8y^2 && \text{Power-to-a-power property.} \end{aligned}$$

b. $$\begin{aligned} \left(\frac{-a}{7x}\right)^2 &= \frac{(-1)^2a^2}{7^2x^2} && \text{Quotient- and product-to-a-power properties.} \\ &= \frac{a^2}{49x^2} && \text{Simplify.} \end{aligned}$$

c. $$\begin{aligned} (5x^{-3})^{-2} &= 5^{-2}(x^{-3})^{-2} && \text{Product-to-a-power property.} \\ &= 5^{-2}x^6 && \text{Power-to-a-power property.} \\ &= \frac{1}{5^2}x^6 && \text{Negative exponent definition.} \\ &= \frac{x^6}{25} && \text{Simplify.} \end{aligned}$$

PROGRESS CHECK 8 Simplify each expression, and write the result using only positive exponents. Assume $y \neq 0$ and $b \neq 0$.

a. $(-2ax^4)^3$ **b.** $\left(\frac{-x}{4y}\right)^4$ **c.** $(2b^{-1})^{-3}$

Progress Check Answers

8. (a) $-8a^3x^{12}$ (b) $x^4/(256y^4)$ (c) $b^3/8$

An important special case of a fraction raised to a power is a fraction raised to a negative power. In Example 5d we showed one way to do this. But often such problems may be done more easily by using

$$\left(\frac{a}{b}\right)^{-n} = \left(\frac{b}{a}\right)^{n}, \qquad a, b \neq 0.$$

We can verify this property from our own previous methods.

$$\left(\frac{a}{b}\right)^{-n} = \frac{1}{(a/b)^n} = \frac{1}{a^n/b^n} = 1 \cdot \frac{b^n}{a^n} = \frac{b^n}{a^n} = \left(\frac{b}{a}\right)^{n}$$

We illustrate this principle in the next example.

EXAMPLE 9

a. Evaluate $\left(\frac{4}{3}\right)^{-2}$. **b.** Simplify $\left(\frac{4x^2}{5}\right)^{-2}$.

Solution

a. $\left(\frac{4}{3}\right)^{-2} = \left(\frac{3}{4}\right)^{2}$ — $\left(\frac{a}{b}\right)^{-n} = \left(\frac{b}{a}\right)^{n}$.

$= \frac{3^2}{4^2}$ — Quotient-to-a-power property.

$= \frac{9}{16}$ — Simplify.

b. $\left(\frac{4x^2}{5}\right)^{-2} = \left(\frac{5}{4x^2}\right)^{2}$ — $\left(\frac{a}{b}\right)^{-n} = \left(\frac{b}{a}\right)^{n}$.

$= \frac{5^2}{4^2(x^2)^2}$ — Quotient- and product-to-a-power properties.

$= \frac{25}{16x^4}$ — Simplify.

PROGRESS CHECK 9

a. Evaluate $\left(\frac{3}{5}\right)^{-3}$. **b.** Simplify $\left(\frac{2x^3}{3}\right)^{-3}$.

7 Negative integer exponents can be evaluated on a calculator with the power function key [y^x]. For example, 5^{-3} is computed as

5 [y^x] 3 [+/−] [=] [0.008].

An exponent of -1 is best computed with the reciprocal key, so 4^{-1} is simply

4 [$1/x$] [0.25].

If your calculator does not permit usage of the power key [y^x] with a negative base, you will need to enter a positive base and then decide the correct sign yourself. Thus you would compute $(-5)^{-3}$ by evaluating 5^{-3} as above, and then you would decide that the correct answer is *negative* 0.008, because a negative base raised to an odd power is negative.

Progress Check Answers

9. (a) $\frac{125}{27}$ (b) $27/(8x^9)$

EXAMPLE 10 Solve the chapter-opening problem on page 88.

Solution The car costs \$35,000. The loan will be \$35,000 − \$5,000 = \$30,000. The annual interest rate is 12 percent, so the monthly interest rate is (12 percent)/12, or 1 percent. The time is 4 years, which means $n = 48$ months. We substitute these values in the given formula.

$$E = \frac{Pr}{1-(1+r)^{-n}} \quad \text{Formula for equal monthly payments.}$$
$$= \frac{(30{,}000)(0.01)}{1-(1+0.01)^{-48}} \quad \text{Substitution.}$$
$$= \frac{300}{1-1.01^{-48}} \quad \text{Simplify.}$$
$$= \frac{300}{1-0.6202604} \quad \text{Calculator evaluation of negative exponent.}$$
$$= \frac{300}{0.3797395} \quad \text{Simplify.}$$
$$= 790.02 \quad \text{Simplify to the nearest cent.}$$

The monthly payment will be \$790.02 for 4 years. Note that this schedule gives a total of \$37,920.96, of which \$7,920.96 is interest.

Note The calculator evaluation in the example may be performed efficiently by first computing the denominator and storing it.

1 [−] 1.01 [y^x] 48 [+/−] [=] [STO] [0.3797396]

Then compute the numerator and divide by the stored number.

30,000 [×] 0.01 [÷] [RCL] [=] [790.01506]

Progress Check Answer

10. \$996.43

PROGRESS CHECK 10 Redo Example 10, but assume that the loan is for 3 years.

EXERCISES 3.1

In Exercises 1–8, write each expression using exponents. In each case, identify the base and the exponent.

1. $6 \cdot 6 \cdot 6$ 6^3; 6, 3
2. $(1/3)(1/3)(1/3)$ $(1/3)^3$; 1/3, 3
3. $(-a)(-a)(-a)(-a)$ $(-a)^4$; $-a$, 4
4. $(-b)(-b)$ $(-b)^2$; $-b$, 2
5. $-3 \cdot 3 \cdot 3 \cdot 3$ -3^4; 3, 4
6. $-x \cdot x \cdot x$ $-x^3$; x, 3
7. m m^1; m, 1
8. 7 7^1; 7, 1

In Exercises 9–16, use the product property of exponents to simplify the given expression.

9. $4^5 \cdot 4^3$ 4^8
10. $2^5 \cdot 2^5$ 2^{10}
11. $x^2 \cdot x^3$ x^5
12. $a^5 \cdot a^4$ a^9
13. $y \cdot y^3$ y^4
14. $(-a)(-a)^2$ $(-a)^3$
15. $m \cdot m^2 \cdot m^3$ m^6
16. $x^6 \cdot x \cdot x^5$ x^{12}

In Exercises 17–22, find (a) the product and (b) the sum of the given expressions.

17. $-2x^2$ and $3x^2$ $-6x^4$, x^2
18. $6a^4$ and $-5a^4$ $-30a^8$, a^4
19. x^5 and x^5 x^{10}, $2x^5$
20. y^3 and y^3 y^6, $2y^3$
21. $-x$ and x $-x^2$, 0
22. $-n^3$ and n^3 $-n^6$, 0

In Exercises 23–32, evaluate the given expression. Assume x and y are not equal to zero.

23. $4^0 + 2^0 + 1^0$ 3
24. $1^0 + 3^0 + 9^0$ 3
25. $(4 + 2 + 1)^0$ 1
26. $(1 + 3 + 9)^0$ 1
27. $3x^0$ 3
28. $-5y^0$ -5
29. $(3x)^0$ 1
30. $(-5y)^0$ 1
31. $(-5)^0$ 1
32. $-x^0$ -1

In Exercises 33–44, write each expression with positive exponents, and evaluate, where possible. Assume x does not equal zero.

33. 2^{-1} $\frac{1}{2}$
34. 3^{-1} $\frac{1}{3}$
35. 4^{-2} $\frac{1}{16}$
36. 5^{-2} $\frac{1}{25}$
37. $(\frac{3}{5})^{-1}$ $\frac{5}{3}$
38. $(\frac{1}{4})^{-1}$ 4
39. $2^{-3} - 3^{-2}$ $\frac{1}{72}$
40. $3^{-2} - 4^{-1}$ $-\frac{5}{36}$
41. $2x^{-3}$ $\frac{2}{x^3}$
42. $4x^{-2}$ $\frac{4}{x^2}$
43. $(2x)^{-3}$ $\frac{1}{8x^3}$
44. $(4x)^{-2}$ $\frac{1}{16x^2}$

In Exercises 45–56, simplify the given expression, and write the result using only positive exponents. Assume a, c, and x do not equal zero.

45. $\frac{a^5}{a^2}$ a^3
46. $\frac{c^7}{c}$ c^6
47. $\frac{2^{-3}}{2^3}$ $\frac{1}{2^6}$
48. $\frac{4^{-1}}{4}$ $\frac{1}{4^2}$
49. $\frac{4^{-2}x^2}{4^{-3}x^2}$ 4
50. $\frac{3^{-4}x^5}{3^{-3}x^5}$ $\frac{1}{3}$
51. $\frac{5x^6}{5^{-2}x^3}$ $125x^3$
52. $\frac{3x^4}{3^{-1}x^2}$ $9x^2$
53. $\frac{3^4x^{-1}}{3x^2}$ $\frac{27}{x^3}$
54. $\frac{2^3a^{-3}}{2a^3}$ $\frac{4}{a^6}$
55. $\frac{a^2c^4x^{-1}}{a^3c^{-1}x^{-1}}$ $\frac{c^5}{a}$
56. $\frac{a^{-1}c^2x^{-1}}{a^{-4}c^2x}$ $\frac{a^3}{x^2}$

In Exercises 57–66, simplify the given expression, and write the result using only positive exponents. Assume x and a are not zero.

57. $(2^3)^4$ 2^{12}
58. $(4^3)^2$ 4^6
59. $(x^{-1})^2$ $\frac{1}{x^2}$
60. $(x^2)^{-3}$ $\frac{1}{x^6}$
61. $[(-a)^2]^{-2}$ $\frac{1}{a^4}$
62. $[(-a)^3]^{-3}$ $-\frac{1}{a^9}$
63. $(2^3)^{-4}$ $\frac{1}{2^{12}}$
64. $[(-4)^3]^{-2}$ $\frac{1}{4^6}$
65. $[(x^{-3})^0]^2$ 1
66. $[(x^4)^0]^{-3}$ 1

In Exercises 67–78, simplify each expression, and express the result with positive exponents.

67. $(2x^2)^3$ $8x^6$
68. $(3x^2)^4$ $81x^8$
69. $(-2x^3y^2)^3$ $-8x^9y^6$
70. $(-3xy^3)^2$ $9x^2y^6$
71. $\left(\frac{-a}{5x}\right)^2$ $\frac{a^2}{25x^2}$
72. $\left(\frac{-5b^2}{3x}\right)^2$ $\frac{25b^4}{9x^2}$
73. $(3x^{-2})^{-3}$ $\frac{x^6}{27}$
74. $(2a^{-1})^{-2}$ $\frac{a^2}{4}$
75. $(\frac{2}{3})^{-2}$ $\frac{9}{4}$
76. $(\frac{1}{4})^{-3}$ 64
77. $\left(\frac{2x^2}{3}\right)^{-3}$ $\frac{27}{8x^6}$
78. $\left(\frac{3}{4x}\right)^{-2}$ $\frac{16x^2}{9}$

In Exercises 79–82, use a calculator to evaluate the given expressions. Give the answer to three decimal places.

79. 1.08^{20} 4.661
80. 1.09^{20} 5.604
81. 0.54^{-10} 474.310
82. 0.08^{-4} 24,414.063

In Exercises 83 and 84, use the formula for equal monthly payments, given in the problem which opens the chapter, to find the monthly payment for the given loan.

83. A $100,000 mortgage with a 9.6 percent annual interest rate for 30 years $848.16
84. A $100,000 mortgage with a 9.6 percent annual interest rate for 20 years $938.67

THINK ABOUT IT

1. a. Find a replacement for x so that x^4 is larger than x^3.
   b. Find a replacement for x so that x^3 is larger than x^4.
   c. Can x^3 ever be equal to x^4?
   d. Why can you not say which is larger, x^3 or x^4?
2. a. Is it true that $(2^3)^4$ equals $(4^3)^2$?
   b. Is it true that $(3^4)^5$ equals $(5^4)^3$?
   c. In general, is it true that $(a^b)^c$ equals $(c^b)^a$?
3. Arrange these in increasing order: 3^{-2}, -2^3, -3^2, 2^{-3}.
4. The statement "Any number raised to the zero power equals 1" is not quite correct. Why not?
5. A student reasoned that $2^3 \cdot 5^2$ must equal 10^5, because "when you multiply, you add the exponents." What crucial idea did the student forget?

REMEMBER THIS

1. Simplify: $10^3 \cdot 10^4$. 10^7
2. Simplify: $10^{-2} \cdot 10^2$. 1
3. True or false? If you multiply 41.378 by 10^2, the decimal place is moved two positions to the right; and if you multiply it by 10^3, the decimal place is moved three positions to the right. True
4. Simplify: $5^n \cdot 5^3$. 5^{n+3}
5. $(3y^2)^3$ equals $(3y^2)(3y^2)(3y^2)$, which equals $3 \cdot 3 \cdot 3 \cdot y^2 \cdot y^2 \cdot y^2$, which simplifies to ______. $27y^6$
6. Solve: $|x - 3| = 1$. $\{4,2\}$
7. What property is used when $3(x + 5)$ is rewritten as $3x + 15$? Distributive
8. Solve the formula $P = 2\ell + 2w$ for ℓ. $\ell = \frac{P - 2w}{2}$
9. After receiving a 10 percent raise, a worker's gross weekly pay was $231. What was it before the raise? $210
10. Graph the solution set of $-1 \le 2x + 1 \le 3$.

3.2 More on Exponents and Scientific Notation

The energy, in electron volts, of any form of radiation is given by $E = h\tau$, where h is Planck's constant, which equals 4.146×10^{-15}, and τ is the frequency of the radiation in hertz units. If the energy is more than 4.5 electron volts, it can break up water molecules, like those in the cells of your body. Use the formula to show that the ultraviolet light in sunshine, which has a frequency of 1.5×10^{15} hertz, is dangerous to your skin. (See Example 6.)

OBJECTIVES

1. **Simplify more complex exponential expressions.**
2. **Simplify expressions with literal exponents.**
3. **Convert numbers from standard notation to scientific notation, and vice versa.**
4. **Perform computations using scientific notation.**

1 We continue the application of exponents to more complex expressions where we must use various combinations of the exponent properties.

EXAMPLE 1 Simplify each expression, and write the result using only positive exponents. Assume $x \neq 0$ and $y \neq 0$.

a. $(2y^2)^4(y^5)^3$ **b.** $\left(\frac{y^2}{y^{-1}}\right)^3$ **c.** $\left(\frac{-3x^2y}{5xy^{-1}}\right)^{-2}$

Solution

a.
$$\begin{aligned}(2y^2)^4(y^5)^3 &= 2^4y^8y^{15} && \text{Power properties.}\\ &= 2^4y^{8+15} && \text{Product property of exponents.}\\ &= 16y^{23} && \text{Simplify.}\end{aligned}$$

b.
$$\begin{aligned}\left(\frac{y^2}{y^{-1}}\right)^3 &= (y^{2-(-1)})^3 && \text{Quotient property of exponents}\\ &= (y^3)^3 && \text{Simplify.}\\ &= y^9 && \text{Power-to-a-power property.}\end{aligned}$$

c.
$$\begin{aligned}\left(\frac{-3x^2y}{5xy^{-1}}\right)^{-2} &= \left(\frac{5xy^{-1}}{-3x^2y}\right)^2 && \left(\frac{a}{b}\right)^{-n} = \left(\frac{b}{a}\right)^n.\\ &= \left(\frac{5}{-3}x^{1-2}y^{-1-1}\right)^2 && \text{Quotient property of exponents.}\\ &= \left(\frac{5}{-3}x^{-1}y^{-2}\right)^2 && \text{Simplify.}\\ &= \frac{25}{9}x^{-2}y^{-4} && \text{Product- and quotient-to-a-power properties.}\\ &= \frac{25}{9}\frac{1}{x^2}\frac{1}{y^4} && \text{Negative exponent definition.}\\ &= \frac{25}{9x^2y^4} && \text{Simplify.}\end{aligned}$$

Note Alternative methods are often possible. For instance, we could have simplified the expression in part **b** as follows.

$$\left(\frac{y^2}{y^{-1}}\right)^3 = \frac{(y^2)^3}{(y^{-1})^3} = \frac{y^6}{y^{-3}} = y^{6-(-3)} = y^9$$

PROGRESS CHECK 1 Simplify the given expressions. Assume $x \neq 0$ and $y \neq 0$.

a. $(2b^3)^3(b^2)^5$ **b.** $\left(\frac{x^3}{x^{-2}}\right)^4$ **c.** $\left(\frac{2x^{-2}y^5}{3x^{-3}y}\right)^{-2}$

2 In many algebraic expressions, variables or arbitrary constants are used to represent exponents. We can apply the exponent properties of the previous section if we assume that the variables in the exponents represent integers. Letters used as exponents are called **literal exponents.**

EXAMPLE 2 Simplify the given expression. Assume $x \neq 0$ and n is an integer.

a. $2^{n+3} \cdot 2^{n+1}$ **b.** $\frac{x^{3n+2}}{x^3}$ **c.** $\frac{3}{3^{-n}}$ **d.** $(x^{n-1})^{-3}$

Solution

a. $2^{n+3} \cdot 2^{n+1} = 2^{(n+3)+(n+1)}$ Product property of exponents.
$= 2^{2n+4}$ Combine like terms in the exponent.

b. Apply the quotient property of exponents.

$$\frac{x^{3n+2}}{x^3} = x^{3n+2-3} = x^{3n-1}$$

c. Note that $3 = 3^1$, and apply the quotient property of exponents.

$$\frac{3}{3^{-n}} = 3^{1-(-n)} = 3^{1+n}$$

d. Apply the power-to-a-power property of exponents.

$$(x^{n-1})^{-3} = x^{(n-1)(-3)} = x^{-3n+3}$$

PROGRESS CHECK 2 Simplify the expression. Assume $x \neq 0$ and n is an integer.

a. $5^n \cdot 5^n$ **b.** $\frac{x^{3n}}{x^{2n}}$ **c.** $\frac{7^n}{7}$ **d.** $(x^{3-n})^{-1}$

3 Numbers that appear in scientific and technical work are often either very large or very small. For example, the official definition of 1 second of time involves the number of oscillations of a cesium atom, and this atom oscillates about 9,190,000,000 times per second. An example of a very small number is the length of time it takes light to travel 1 m. This number is about 0.00000000334 second. To work effectively with such numbers, we often write them in scientific notation form. A positive number N is expressed in scientific notation when it is written in the form

$$N = m \times 10^k,$$

where $1 \leq m < 10$ and k is an integer. For example,

$$9{,}190{,}000{,}000 = (9.19)(1{,}000{,}000{,}000) = 9.19 \times 10^9$$
$$0.00000000334 = (3.34)(1/1{,}000{,}000{,}000) = 3.34 \times 10^{-9}.$$

Note that the effect of multiplying a number by 10^9 is to move the decimal point 9 places to the right, while the effect of multiplying a number by 10^{-9} is to move the decimal point 9 places to the left. This observation leads to the following procedure for converting a number from standard notation to scientific notation.

Progress Check Answers

1. (a) $8b^{19}$ (b) x^{20} (c) $9/(4x^2y^8)$
2. (a) 5^{2n} (b) x^n (c) 7^{n-1} (d) x^{n-3}

To Convert to Scientific Notation

1. Immediately after the first nonzero digit of the number, place an apostrophe (').
2. Starting at the apostrophe, count the number of places to the decimal point. If you move to the right, your count is expressed as a positive number. If you move to the left, the count is negative.
3. The apostrophe indicates the position of the decimal in the factor between 1 and 10; the count represents the exponent to be used in the factor which is a power of 10.

The following chart illustrates how this procedure is used. The direction of the counting is indicated by the arrow.

Number	=	Number from 1 to 10	×	Power of 10
7′86,440,000.	=	7.8644	×	10^8
0.0005′9	=	5.9	×	10^{-4}
6′.3	=	6.3	×	10^0

To express a negative number in scientific notation, you simply use the procedure above and write a negative sign before it.

EXAMPLE 3 A radio signal travels at the speed of light, about 186,000 mi/second, and takes about 0.0000054 second to travel 1 mi. Write these numbers in scientific notation.

Solution Follow the procedure outlined above.

$$1'86{,}000. = 1.86 \times 10^5$$

$$0.000005'4 = 5.4 \times 10^{-6}$$

PROGRESS CHECK 3 Write each number in scientific notation.

a. 2,734,000,000 **b.** 0.85

To convert a number from scientific notation to standard notation, just carry out the indicated multiplication. Move the decimal point to the right when the power of 10 has a positive exponent, and move it to the left when the power of 10 has a negative exponent. Insert zeros, as needed, to indicate the position of the decimal point.

EXAMPLE 4 Express each number in standard notation.

a. 1.69×10^4 **b.** 1.69×10^{-4}

Solution

a. To multiply by 10^4, we move the decimal point 4 places to the right.

$$1.69 \times 10^4 = 1.6900. = 16{,}900$$

b. To multiply by 10^{-4}, we move the decimal point 4 places to the left.

$$1.69 \times 10^{-4} = 0.0001.69 = 0.000169$$

PROGRESS CHECK 4 Express each number in standard notation.

a. 8.6×10^{-3} **b.** 8.6×10^3

Progress Check Answers

3. (a) 2.734×10^{-9} (b) 8.5×10^{-1}
4. (a) 0.0086 (b) 8,600

4 Scientific notation makes it easy to multiply and divide because we can make effective use of the product and quotient properties of exponents. We demonstrate this in the next example.

EXAMPLE 5 Perform the indicated operations, and express the result in standard notation.

a. $(4 \times 10^{-4})(8 \times 10^{7})$ **b.** $0.095/5{,}000$

Solution

a.
$$\begin{aligned}(4 \times 10^{-4})(8 \times 10^{7}) &= (4 \cdot 8)(10^{-4}10^{7}) && \text{Reorder and regroup.}\\ &= (4 \cdot 8) \times 10^{-4+7} && \text{Product property of exponents.}\\ &= 32 \times 10^{3} && \text{Simplify.}\\ &= 32{,}000 && \text{Standard notation.}\end{aligned}$$

b.
$$\begin{aligned}\frac{0.095}{5{,}000} &= \frac{9.5 \times 10^{-2}}{5 \times 10^{3}} && \text{Scientific notation.}\\ &= \frac{9.5}{5} \times \frac{10^{-2}}{10^{3}} && \frac{ac}{bd} = \frac{a}{b} \cdot \frac{c}{d}\\ &= \frac{9.5}{5} \times 10^{-2-3} && \text{Quotient property of exponents.}\\ &= 1.9 \times 10^{-5} && \text{Simplify.}\\ &= 0.000019 && \text{Standard notation.}\end{aligned}$$

PROGRESS CHECK 5 Perform the indicated operations, and express the result in standard notation.

a. $(7 \times 10^{5})(3 \times 10^{-1})$ **b.** $0.00081/600$

Scientific calculators are programmed to work with scientific notation. To enter a number in scientific notation, first enter the significant digits of the number from 1 to 10, press [EE] or [EXP], and finally, enter the exponent of the power of 10. For example, to enter 9.4×10^{-2}, press

9.4 [EE] 2 [+/−].

The display looks like

[9.4 − 02]

with the exponent appearing on the right in the display. Read the owner's manual to your calculator to learn its scientific notation capabilities. Here are two common features you need to keep in mind.

1. If a calculation results in an answer with too many digits for display, the calculator automatically displays the answer in scientific notation.

Example: 38 [y^x] 6 [=] [3.0109 09]

The display shown here is common for a calculator with an eight-digit display. Note that the result shows only five significant digits followed by the correct exponent. However, the exact value of 38^6 is 3,010,936,384, and the calculator carries this answer internally. Check your owner's manual to determine the accuracy you can expect in displayed and internal values.

2. Entries in standard and scientific notation may be mixed in the same problem.

Example: $(6.7 \times 10^{3}) + 125$
6.7 [EE] 3 [+] 125 [=] [6.825 03]

Some calculators would display the result in this example in standard notation.

Progress Check Answers

5. (a) 210,000 (b) 0.00000135

Max Planck (1858–1947), a German physicist, was the originator of the quantum theory of energy.

EXAMPLE 6 Solve the problem in the section introduction on page 98.

Solution We substitute the given values into the formula for energy and then simplify.

$$\begin{aligned} E &= h\tau \\ &= (4.146 \times 10^{-15})(1.5 \times 10^{15}) \\ &= (4.146)(1.5)(10^{-15+15}) \\ &= 6.219 \times 10^{0} \\ &= 6.219 \end{aligned}$$

This computation may be verified by calculator as follows.

4.146 [EE] 15 [+/−] [×] 1.5 [EE] 15 [=] [6.219]

The energy of the ultraviolet light of sunshine is 6.219 electron volts, which is more than the 4.5 electron volts needed to break up water molecules.

PROGRESS CHECK 6 A radio station broadcasting at 550 on the AM band sends out radiation with a frequency of 550 kilohertz, or 550,000 hertz. Use the formula in Example 6 to compute the energy of these radio waves. Are they strong enough to break up water molecules?

Progress Check Answer

6. 2.28×10^{-9}; no

EXERCISES 3.2

In Exercises 1–18, simplify each expression, and write the result using only positive exponents. Assume all variables are not equal to zero.

1. $(3x^3)^2(x^2)^3$ $9x^{12}$
2. $(x^2)^4(2x)^3$ $8x^{11}$
3. $(-2x)^2(3x^2)$ $12x^4$
4. $(-3x)^3(2x)^2$ $-108x^5$
5. $(3x^{-1})^2(2x^2)^{-1}$ $\frac{9}{2x^4}$
6. $(5x^{-2})^{-1}(3x^{-1})^2$ $\frac{9}{5}$
7. $\left(\frac{y}{y^{-1}}\right)^3$ y^6
8. $\left(\frac{y^2}{y^{-2}}\right)^3$ y^{12}
9. $\left(\frac{2x^{-1}}{x}\right)^2$ $\frac{4}{x^4}$
10. $\left(\frac{3x^{-2}}{x}\right)^3$ $\frac{27}{x^9}$
11. $\left(\frac{-2xy}{3xy^{-2}}\right)^{-1}$ $-\frac{3}{2y^3}$
12. $\left(\frac{-3x^2y^{-1}}{4xy^{-1}}\right)^{-2}$ $\frac{16}{9x^2}$
13. $\frac{(ab)^{-2}}{a^2b^{-3}}$ $\frac{b}{a^4}$
14. $\frac{(a^2b)^{-3}}{a^{-2}b^{-3}}$ $\frac{1}{a^4}$
15. $\frac{(2x)^0(3y)^{-1}}{2x^03y^{-1}}$ $\frac{1}{18}$
16. $\frac{3x^0(2y)^{-3}}{(3x)^02y^{-3}}$ $\frac{3}{16}$
17. $\frac{(2y^3)^{-3}(4x^{-2})^2}{(3x^2)^{-2}(3y^{-1})^3}$ $\frac{2}{3y^6}$
18. $\frac{(5x^2)^{-3}(3y^{-1})^4}{(5y^2)^{-2}(9x^{-3})^2}$ $\frac{1}{5}$

In Exercises 19–36, simplify the given expression. Assume x does not equal zero and n is an integer.

19. $2^n \cdot 2^{n+1}$ 2^{2n+1}
20. $3^{m-1} \cdot 3^{m+1}$ 3^{2m}
21. $3^2 \cdot 3^n$ 3^{n+2}
22. $4^2 \cdot 4^{n-2}$ 4^n
23. $5^{m-1} \cdot 5^{1-m}$ 1
24. $x^{2-n} \cdot x^{n-2}$ 1
25. $\frac{x^{2n+1}}{x}$ x^{2n}
26. $\frac{x^{2n+1}}{x^n}$ x^{n+1}
27. $\frac{2^n}{2^{n-1}}$ 2
28. $\frac{4^{2n}}{4^{2n+1}}$ $\frac{1}{4}$
29. $(x^{n+1})^2$ x^{2n+2}
30. $(x^{n+2})^3$ x^{3n+6}
31. $(x^{n-1})^{-1}$ x^{-n+1}
32. $(x^{-n+4})^{-2}$ x^{2n-8}
33. $(2^{2n+1})^3$ 2^{6n+3}
34. $(3^{3n-2})^{-2}$ 3^{-6n+4}
35. $\frac{n2^{n+1}}{n^32^{n-1}}$ $\frac{4}{n^2}$
36. $\frac{n^{n-1}}{n^{1-n}}$ n^{2n-2}

In Exercises 37–48, write the given numbers in scientific notation.

37. The speed of sound in air: 1,100 ft/second 1.1×10^3
38. The speed of sound in water: 4,220 ft/second 4.22×10^3
39. The distance from the earth to the sun: 93 million mi 9.3×10^7
40. The distance from the earth to the moon: 384 million m 3.84×10^8
41. The age of moon rocks: 3.9 billion years 3.9×10^9
42. The radius of the sun: 696 million m 6.96×10^8
43. The radius of a hydrogen atom: 0.00000000005 m 5×10^{-11}
44. The average speed of an atom of oxygen at 0° C: 461.4 m/second 4.614×10^2
45. $-579{,}000{,}000$ -5.79×10^8
46. $-123{,}456{,}000$ -1.23456×10^8
47. -0.00032 -3.2×10^{-4}
48. -0.0123 -1.23×10^{-2}

In Exercises 49–60, write the given number in standard notation.

49. A human body which weighs about 175 lb contains about 3.4×10^{27} atoms. 3,400,000,000,000,000,000,000,000,000
50. The mass of an electron is about 9.1×10^{-28} g. 0.000 000 000 000 000 000 000 000 000 91
51. The wavelength of X-rays is about 3×10^{-9} cm. 0.000 000 003
52. A molecule of oxygen in the air at sea level experiences about 3.55×10^9 collisions per second. 3,550,000,000
53. The electrical charge on a single electron is 1.6×10^{-19} coulomb. 0.000 000 000 000 000 000 16
54. The speed of light in space is 3×10^5 km/second. 300,000
55. The rest mass of an electron is 5.11×10^5 electron volts. 511,000
56. The length of the classical electron radius is 2.82×10^{-15} m. 0.000 000 000 000 002 82
57. -10^{-2} -0.01
58. -10^{-4} -0.0001
59. -3.5×10^{-1} -0.35
60. -4.601×10^3 $-4{,}601$

In Exercises 61–66, use scientific notation to perform the given operations.

61. $(2.4 \times 10^2) \times (3.0 \times 10^{-2})$ 7.2

62. $(-1.8 \times 10^{-1}) \times (5.0 \times 10^3)$ $-9 \times 10^2 = -900$

63. $\dfrac{-2.8 \times 10^2}{4.0 \times 10^{-2}}$ $-7 \times 10^3 = -7{,}000$

64. $\dfrac{6.0 \times 10^{-3}}{1.2 \times 10^{-4}}$ $5 \times 10^1 = 50$

65. $(3 \times 10^8)^2$ 9×10^{16}

66. $(-2 \times 10^{-3})^3$ -8×10^{-9}

67. The mass of an electron is about 9.1×10^{-31} kg. The mass of a proton is 1,836 times the mass of an electron. Express the mass of a proton in scientific notation. 1.671×10^{-27} kg

68. The speed of light in space is 3×10^5 km/second. The distance from the earth to the sun is about 150 million km. Use the formula $d = rt$ to find the time it takes sunlight to reach the earth. 500 seconds; 8.3 minutes

69. If a person whose heart beats about 70 times per minute lives 70 years, how many times will the heart beat in a lifetime? Express the answer in scientific notation. 2.58×10^9

70. Use the formula $E = h\tau$ from Example 6 to calculate the energy in electron volts of a microwave oven which has $\tau = 2.45 \times 10^9$ hertz. Recall that $h = 4.146 \times 10^{-15}$. Express the result in scientific notation. 1.016×10^{-5} electron volts

71. The U.S. federal deficit in 1989 was about \$3 trillion. If you spent \$1,000 per second, how long would it take to spend \$3 trillion? Express the answer in seconds in scientific notation. Then express the answer in years. 3×10^9 seconds; about 95 years

72. In 1975 the U.S. gross national product was valued at \$2,695.0 billion, while in 1985 it was valued at \$3,618.7 billion. Both figures are expressed in "constant 1982 dollars" to account for inflation. Express the amount of the increase in scientific notation. \$923.7 billion = $\$9.237 \times 10^{11}$

73. In 1989 the U.S. federal government collected about \$975.5 billion in revenue, of which \$425.2 billion came from individual income taxes. Express each number in scientific notation, and compute the percentage of the total revenue accounted for by individual income taxes. $\dfrac{4.252 \times 10^{11}}{9.755 \times 10^{11}} \times 100 = 43.6$ percent

74. In 1989 the U.S. government spent about \$298 billion for national defense, out of a total budget of about \$1.14 trillion. Express both numbers in scientific notation, and then find the percentage of the total that was spent on national defense. $\dfrac{2.98 \times 10^{11}}{1.14 \times 10^{12}} \times 100 = 26.1$ percent

75. Use the formula $E = mc^2$ to find out how much energy is contained in 2 kg of mass. E is the energy in joules, m is the mass in kilograms, and $c = 3 \times 10^8$ m/second is the speed of light. One joule of energy is the equivalent of 2.4×10^{-10} tons of TNT. 1.8×10^{17} joules, 43.2 million tons of TNT

76. Use the formula $E = mc^2$ to find m (in kilograms) if c, the speed of light, is 3×10^8 m/second and E is 2.4×10^{17} joules. This will give you the amount of matter used to create the explosive force of the largest hydrogen bomb ever exploded. 2.7 kg

THINK ABOUT IT

1. a. What number is 10 times larger than 4.56×10^{18}?
   b. What number is 100 times smaller than 2.08×10^{12}?
2. a. The correct measurement in an experiment is 3.2×10^3. Which is a more serious error, to record it as 4.2×10^3 or as 3.2×10^4?
   b. When using scientific notation, why might a scientist be more concerned about the power of 10 than the other part of the number?
3. An angstrom is the name given to the length of 10^{-10} m. It is a unit often used to measure wavelengths of light. It is named after the Swedish physicist Anders Angstrom. One fermi is the name given to the length 10^{-15} m, and it is often the unit used for expressing sizes of nuclear particles. It was named after the Italian physicist Enrico Fermi. How many fermis does it take to equal 1 angstrom?
4. Our galaxy, the Milky Way, is like a flat spiral in space with a diameter of about 10^{21} m. The solar system, consisting of our sun and the planets orbiting it, is like a flat oval, fitting into a space about 10^{13} m across. About how many times bigger in diameter is the Milky Way than the solar system?
5. Find the product of $m \times 10^n$ and $n \times 10^m$.

REMEMBER THIS

1. The expression $4x^2 + 3x + 1$ has how many terms? 3
2. True or false? $3x^2 - 5x$ equals $3x^2 + (-5x)$. True
3. True or false? $\dfrac{2}{x^4}$ equals $2x^{-4}$. True
4. Which of the terms in this expression are "like" terms? $5x^3 - 4x^2 + 4x^3 - 5$. $5x^3$ and $4x^3$
5. Does $3x^4 + 6x^4$ simplify to $9x^4$ or to $9x^8$? $9x^4$
6. Use a calculator to evaluate $\dfrac{3{,}000}{1 - 1.08^{-24}}$ to two decimal places. 3,561.67
7. In the expression $(-4)^3$, -4 is called the ______ and 3 is called the ______. Base, exponent
8. What is the name of this set of numbers? $\{-1, -2, -3, \ldots\}$. Negative integers
9. What is the reciprocal of 5? $\frac{1}{5}$
10. Solve: $3(4x + 1) - 2x = 3$. $\{0\}$

3.3 Addition and Subtraction of Polynomials

If there are n people in a room and everyone shakes hands with everyone else, then the total number of handshakes $H(n)$ is given by $H(n) = \frac{n^2}{2} - \frac{n}{2}$. Find $H(8)$ to determine the number of handshakes when 8 people are present. (See Example 5.)

OBJECTIVES

1. **Identify polynomials.**
2. **State the degree of a polynomial, and identify monomials, binomials, and trinomials.**
3. **Evaluate polynomials, and use $P(x)$ notation.**
4. **Add polynomials.**
5. **Subtract polynomials.**

1 *Polynomial* is the name given to a particular type of algebraic expression. Some examples of polynomials are

$$3x^2 + 12x + 16, \qquad \tfrac{1}{3}\pi r^3, \qquad \text{and} \qquad x^2 + y^2.$$

A **polynomial** is defined as an algebraic expression which may be written as a finite sum of terms that contain only nonnegative integer exponents when each term is expressed without variables in the denominators. For a polynomial which has only one variable, say x, each term must be expressible in the form

$$ax^n,$$

where n is a nonnegative integer and a is a real number constant. Some expressions which will help you to identify polynomials are given in Example 1.

EXAMPLE 1 Which of the following algebraic expressions are polynomials?

a. $x^2 - 5x + 3$ **b.** $\frac{y}{3}$ **c.** $\frac{2}{x}$

Solution

a. Recall that the *terms* of an algebraic expression are the parts which are separated by plus (+) signs. Therefore, $x^2 - 5x + 3$, which may be written as $x^2 + (-5x) + 3$, is an algebraic expression with three terms: x^2, $-5x$, and 3. Furthermore, you can see that each term contains only *nonnegative integer exponents* on the variable x, because $-5x = -5x^1$ and $3 = 3x^0$. Therefore, $x^2 - 5x + 3$ is a polynomial.

b. $\frac{y}{3}$ is an algebraic expression with one term which may be written as $\frac{1}{3}y$. The exponent on y is the nonnegative integer 1. Thus, $\frac{y}{3}$ is a polynomial.

c. Rewriting $\frac{2}{x}$ in the form ax^n results in $2x^{-1}$. Since the exponent on the variable is a *negative* integer, $\frac{2}{x}$ is *not* a polynomial.

PROGRESS CHECK 1 Which of the following algebraic expressions are polynomials?

a. $-\frac{8}{x^3}$ **b.** $y^3 - 8$ **c.** $\frac{-z^4}{8}$

2 Polynomials that contain one, two, or three terms have special names, as shown in the following table.

Number of Terms in Polynomial	Name
1	Monomial
2	Binomial
3	Trinomial

Also, a polynomial is characterized by its **degree,** with the **degree of a monomial** being the number of variable factors in the term. This number is given by the sum of the exponents on its variables. The **degree of a polynomial** is defined to be the same as the degree of its highest-degree term. Thus, the degree of $2x^4y^3$ is 7, the degree of $24xy$ is 2, and the degree of $2x^4y^3 + 24xy$ is 7 (since $2x^4y^3$ is the highest-degree term).

EXAMPLE 2 Give the degree of each polynomial, and identify polynomials that are monomials, binomials, or trinomials.

a. $3x^4 - 5x^3$ **b.** $12x^2y + x^3y - 50xy$ **c.** -1

Solution

a. Since $3x^4 - 5x^3$ has two terms, it is a binomial. The degree of $3x^4$ is 4, because the variable x is a factor 4 times; similarly, the degree of $-5x^3$ is 3. Since $3x^4$ is the highest-degree term, $3x^4 - 5x^3$ is a polynomial of degree 4. Note that when a term contains only one variable, then the degree of the term is simply its exponent.

b. Since $12x^2y + x^3y - 50xy$ is a polynomial with three terms, it is a trinomial. The degree of the polynomial is 4, because the degree of x^3y is 4 and the degree of each of the remaining terms is less than 4.

c. Because the polynomial -1 has only one term, it is a monomial. Moreover, since -1 may be written as $-1x^0$, it is a polynomial of degree 0. In the same way, any nonzero constant is a polynomial of degree 0. The special polynomial which consists just of the number 0 does not have a degree, since it is not possible to distinguish 0 from $0x$ or $0x^2$ or $0x^n$ or any other power of x.

Note It is often convenient to arrange the terms of a polynomial so that the degrees of the terms are in descending order. In this notation we write the polynomial in part **b** as

$$x^3y + 12x^2y - 50xy$$

This arrangement makes it easy to identify the degree of the polynomial and to spot the absence of certain lower-degree terms, and it simplifies computations which involve more than one polynomial.

PROGRESS CHECK 2 Give the degree of each polynomial, and identify polynomials that are monomials, binomials, or trinomials.

a. $7x^3 - 6x^2 + 5x$ **b.** 6 **c.** $xy^3 - 8xy^2$

3 In applications we often have to evaluate a polynomial for some known value of a variable. To do this, replace all occurrences of a variable by its given value, and then perform the indicated operations.

Progress Check Answers

1. **b** and **c**
2. (a) 3; trinomial (b) 0; monomial (c) 4; binomial

EXAMPLE 3 Evaluate $4x^3 - 5x^2 + 6x - 11$ for $x = 4$. Check the result on a calculator.

Solution Substitute 4 for x and simplify.

$$\begin{aligned} 4x^3 - 5x^2 + 6x - 11 &= 4(4)^3 - 5(4)^2 + 6(4) - 11 \\ &= 4 \cdot 64 - 5 \cdot 16 + 24 - 11 \\ &= 189 \end{aligned}$$

CALCULATOR CHECK Use the power key $\boxed{y^x}$ to compute 4^3 and the square key $\boxed{x^2}$ to compute 4^2.

$$4 \boxed{\times} 4 \boxed{y^x} 3 \boxed{-} 5 \boxed{\times} 4 \boxed{x^2} \boxed{+} 6 \boxed{\times} 4 \boxed{-} 11 \boxed{=} \boxed{\quad 189}$$

The calculation may also be done using the memory keys. This allows you to enter x just once.

$$4 \boxed{\text{STO}} 4 \boxed{\times} \boxed{\text{RCL}} \boxed{y^x} 3 \boxed{-} 5 \boxed{\times} \boxed{\text{RCL}} \boxed{x^2} \boxed{+} 6$$
$$\boxed{\times} \boxed{\text{RCL}} \boxed{-} 11 \boxed{=} \boxed{\quad 189}$$

PROGRESS CHECK 3 Evaluate $2x^5 - 9x^3 - x^2 + 3$ for $x = 5$. Check the result on a calculator.

When you evaluate polynomials, it is useful to associate a letter, customarily P, with a polynomial name. We may then denote a polynomial in x, like $5x - 2$, by

$$P(x) = 5x - 2$$

and a polynomial in t, like $t^2 - 2t + 1$, by

$$P(t) = t^2 - 2t + 1.$$

Note that the variable in the polynomial appears in the parentheses and that the notation $P(x)$ is read "P of x" or "P at x." Using this notation, we may denote the value of the polynomial $P(x)$ when $x = 1$ by writing $P(1)$. If $P(x) = 5x - 2$, we find $P(1)$ by substituting 1 for x in this equation.

$$\begin{aligned} P(x) &= 5x - 2 && \text{Given equation.} \\ P(1) &= 5(1) - 2 && \text{Replace } x \text{ by 1.} \\ &= 3 && \text{Simplify.} \end{aligned}$$

The result $P(1) = 3$ indicates that the value of the given polynomial is 3 when $x = 1$. [Remember that the notation $P(x)$ does not mean P times x.]

EXAMPLE 4 If $P(x) = 2x^2 - 5x + 3$, find $P(4)$ and $P(-1)$.

Solution Replace x by the number inside the parentheses and simplify.

$$\begin{aligned} P(x) &= 2x^2 - 5x + 3 \\ P(4) &= 2(4)^2 - 5(4) + 3 \\ &= 32 - 20 + 3 \\ &= 15 \end{aligned} \qquad \begin{aligned} P(x) &= 2x^2 - 5x + 3 \\ P(-1) &= 2(-1)^2 - 5(-1) + 3 \\ &= 2 + 5 + 3 \\ &= 10 \end{aligned}$$

Thus, $P(4) = 15$ and $P(-1) = 10$.

PROGRESS CHECK 4 If $P(x) = 3x^2 - x + 4$, find $P(3)$ and $P(-2)$.

Sometimes, it is more natural to denote a polynomial by using symbols other than P and x, as illustrated by the section-opening problem.

EXAMPLE 5 Solve the problem in the section introduction on page 104.

Progress Check Answers

3. 5,103
4. 28; 18

Solution Substitute 8 for n in the given formula and simplify.

$$H(n) = \frac{n^2}{2} - \frac{n}{2} \quad \text{Given formula.}$$

$$H(8) = \frac{8^2}{2} - \frac{8}{2} \quad \text{Replace } n \text{ by 8.}$$

$$= 28 \quad \text{Simplify.}$$

Thus, there will be 28 handshakes when 8 people are present.

PROGRESS CHECK 5 Use the formula in Example 5 to compute the number of handshakes when 10 people are present. That is, find $H(10)$.

4 The procedure for adding polynomials follows from the distributive property, which implies first that we may combine like terms by combining their numerical coefficients and second that only like terms may be combined. For example,

$$7x^3 + 2x^3 = (7 + 2)x^3 = 9x^3,$$
$$5ab - 2ab = (5 - 2)ab = 3ab,$$
$$2x^2 + 3x \text{ does not simplify.}$$

In polynomial addition, it is common to display the problem by aligning like terms in columns, as shown in the following example.

Encourage students to write the word *add* or *subtract* when they use the column method.

EXAMPLE 6 Add $6x^2 - 5x + 1$ and $-5x^2 + 8x - 7$.

Solution The given problem may be written as follows.

$$\begin{array}{rr} & 6x^2 - 5x + 1 \\ \text{Add:} & -5x^2 + 8x - 7 \\ \hline \end{array}$$

Now add $6x^2$ and $-5x^2$; then add $-5x$ and $8x$; and finally, add 1 and -7. The result is $x^2 + 3x - 6$, as displayed below.

$$\begin{array}{rr} & 6x^2 - 5x + 1 \\ \text{Add:} & -5x^2 + 8x - 7 \\ \hline & x^2 + 3x - 6 \end{array}$$

Therefore, $x^2 + 3x - 6$ is the sum of the two given polynomials.

Caution A common *mistake* at this point is to add exponents when adding polynomials. But the addition of exponents is associated with the *multiplication* of polynomials, as shown in our work from Section 3.1. Compare these two examples as a reminder.

$$\text{Addition:} \quad 8x^3 + 2x^3 = (8 + 2)x^3 = 10x^3$$
$$\text{Multiplication:} \quad (8x^3)(2x^3) = (8 \cdot 2)(x^3 \cdot x^3) = 16x^6$$

PROGRESS CHECK 6 Add $-8x^2 + x - 3$ and $2x^2 - 7x + 6$.

Sometimes in simplifying expressions, you may find it easier to add polynomials without writing them in columns, but this involves removing parentheses. We demonstrate this idea in Example 7, which is equivalent to Example 6.

EXAMPLE 7 Simplify $(6x^2 - 5x + 1) + (-5x^2 + 8x - 7)$.

Solution Removing parentheses depends on the distributive property, as you can see by thinking of the given problem as

$$1(6x^2 - 5x + 1) + 1(-5x^2 + 8x - 7).$$

Progress Check Answers

5. 45
6. $-6x^2 - 6x + 3$

Therefore, when parentheses are removed, we obtain

$$6x^2 - 5x + 1 - 5x^2 + 8x - 7.$$

Now proceed as follows.

$$\begin{aligned} &6x^2 - 5x + 1 - 5x^2 + 8x - 7 \\ &= (6x^2 - 5x^2) + (-5x + 8x) + (1 - 7) && \text{Reorder and regroup.} \\ &= x^2 + 3x - 6 && \text{Combine like terms.} \end{aligned}$$

Thus, the given expression simplifies to $x^2 + 3x - 6$.

PROGRESS CHECK 7 Simplify $(-3x^2 - x + 2) + (-x^2 + 3x - 5)$.

5 The procedure for subtracting polynomials is based on the method we use for subtracting real numbers. Recall that in arithmetic we may subtract b from a by adding to a the negative of b. In symbols, this is written as $a - b = a + (-b)$. In terms of polynomials the negative of a polynomial is obtained by changing the signs on all of the terms. For example, the negative of $x^2 - 3x - 4$ is written as $-(x^2 - 3x - 4)$ and simplifies as shown next.

$$\begin{aligned} -(x^2 - 3x - 4) &= -1(x^2 - 3x - 4) \\ &= -x^2 + 3x + 4 \end{aligned}$$

Thus, $-x^2 + 3x + 4$ is the negative of $x^2 - 3x - 4$, and we can obtain the result by just changing the signs on all of the terms. This leads to the following method for polynomial subtraction.

To Subtract Polynomials

To subtract two polynomials, change the signs on all of the terms in the polynomial that is being subtracted, and then add the polynomials.

EXAMPLE 8 Subtract $x^2 - 3x - 4$ from $2x^2 - 4x + 3$.

Solution The given problem may be written as

$$(2x^2 - 4x + 3) - (x^2 - 3x - 4).$$

According to the subtraction procedure above, we change the signs on all of the terms in $x^2 - 3x - 4$ and then add the polynomials.

$$\begin{aligned} &(2x^2 - 4x + 3) - (x^2 - 3x - 4) \\ &= (2x^2 - 4x + 3) + (-x^2 + 3x + 4) \\ &= (2x^2 - x^2) + (-4x + 3x) + (3 + 4) \\ &= x^2 - x + 7 \end{aligned}$$

PROGRESS CHECK 8 Subtract $4x^2 - x + 2$ from $x^2 - 6x - 2$.

For some problems you may prefer to arrange a polynomial subtraction with like terms aligned in columns.

EXAMPLE 9 Subtract $3x^4 - x^3 + 6x$ from $2x^4 + x^2 - 5x$.

Solution Align like terms vertically and display the problem as follows.

$$\begin{array}{rl} & 2x^4 \qquad\quad + x^2 - 5x \\ \text{Subtract:} & \underline{3x^4 - x^3 \qquad\quad + 6x} \end{array}$$

Progress Check Answers

7. $-4x^2 + 2x - 3$
8. $-3x^2 - 5x - 4$

Now change the signs on all of the terms in $3x^4 - x^3 + 6x$ and then add.

$$\begin{array}{lrrrr} \text{Add:} & 2x^4 & & +x^2 & -5x \\ & -3x^4 & +x^3 & & -6x \\ \hline & -x^4 & +x^3 & +x^2 & -11x \end{array}$$

Thus, the answer is $-x^4 + x^3 + x^2 - 11x$.

PROGRESS CHECK 9 Subtract $8x^3 + 2x^2 - 3$ from $9x^3 + 4x - 8$.

Progress Check Answer

9. $x^3 - 2x^2 + 4x - 5$

EXERCISES 3.3

In Exercises 1–12, state whether the given algebraic expression is a polynomial.

1. $x + 3$ Y **2.** $y - 2$ Y **3.** $3x^2 - 2x + 1$ Y **4.** $4x^3 - x$ Y

5. $\frac{x}{2} - 1$ Y **6.** $\frac{3x^2}{5} + x$ Y **7.** $\frac{1}{x} + 2x$ N **8.** $x^2 - \frac{3}{x}$ N

9. $3x^{-2} + 5x^{-1} + 4$ N **10.** $2x^{-3} + x^3$ N

11. $\frac{1}{3}$ Y **12.** -3.5 Y

In Exercises 13–22, give the degree of the polynomial, and identify any polynomials that are monomials, binomials, or trinomials.

13. $4x^5 + 2x - 1$ 5, trinomial **14.** $-3x^2 + x + 1$ 2, trinomial

15. $-4x^2$ 2, monomial **16.** $\frac{1}{3}x$ 1, monomial

17. $2 + x^2 - 5x$ 2, trinomial **18.** $5 - x^2$ 2, binomial

19. 3 0, monomial **20.** 0 No degree, monomial

21. $x^2 + y$ 2, binomial **22.** $x + 2xy + y$ 2, trinomial

In Exercises 23–32, evaluate the polynomial for the given value of x; then check the result on the calculator.

23. $x^2 + x + 1$; $x = 1$ 3 **24.** $2x^2 + 3x + 4$; $x = 1$ 9

25. $2 + 3x$; $x = 4$ 14 **26.** $3 - 4x$; $x = -1$ 7

27. $1.7x^8$; $x = -1$ 1.7 **28.** $3.2x^9$; $x = -1$ −3.2

29. $x^2 + x$; $x = \frac{1}{2}$ $\frac{3}{4}$ **30.** $x^2 - x$; $x = -\frac{1}{2}$ $\frac{3}{4}$

31. $5x^{15} - 3x^{12} + 7x^2 - 2$; $x = 0$ −2

32. $1 + 2x - 3x^2 + 4x^3 - 5x^4$; $x = 0$ 1

In Exercises 33–38, find the value of the given expressions.

33. If $P(x) = 3x - 3$, find $P(4)$ and $P(0)$. 9, −3

34. If $P(x) = 3 - 3x$, find $P(-4)$ and $P(0)$. 15, 3

35. If $Q(x) = 3x^2 - 2x + 1$, find $Q(-1)$ and $Q(1)$. 6, 2

36. If $Q(x) = x^3 - 8$, find $Q(2)$ and $Q(-1)$. 0, −9

37. If $P(x) = 2x + 1$ and $Q(x) = x^2$, find $P(1) + Q(2)$. $3 + 4 = 7$

38. If $P(x) = 1 - x^2$ and $Q(x) = x^2 - 1$, find $P(5) + Q(5)$. 0

39. If $P(x) = 5x^2 + 1$, is it true that $P(3) = P(-3)$? True

40. If $P(x) = 5x + 1$, is it true that $P(3) = P(-3)$? False

In Exercises 41–46, answer the question by evaluating the given polynomial.

41. If there are n teams in a league and each plays the other K times, then the total number of games is given by $\frac{K}{2}(n^2 - n)$. Find the total number of games if there are 5 teams and each plays the other 3 times. 30

42. Use the formula from Exercise 41 to find the total number of games if there are 14 teams and each plays the other 14 times. 1,274

43. A curious formula was presented in the eighteenth century by the Swiss mathematician Euler. It is $P(n) = n^2 + n + 41$. For integer values of n from 0 to 39, it produces prime numbers. Confirm that $P(0)$, $P(1)$, and $P(2)$ are prime numbers. Show that $P(40)$ is not prime by noting that $P(40)$ is a perfect square. 41; 43; 47; $1{,}681 = 41^2$

44. Another formula for generating certain prime numbers is $P(n) = n^2 - 79n + 1{,}601$. It yields primes for integer values of n from 0 to 79. Confirm that $P(39)$ and $P(40)$ yield the same prime number. Show that $P(80)$ is not prime by noting that $P(80)$ is a perfect square. 41; $1{,}681 = 41^2$

45. The "triangular" numbers are those that can be arranged in the following pattern (like bowling pins). This way of describing numbers goes back to the Pythagorean society of ancient Greece.

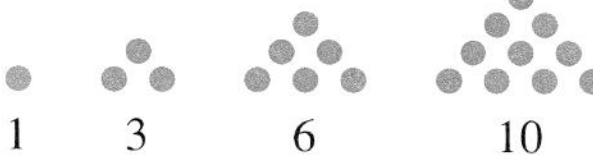

A formula for the nth triangular number is $T(n) = \frac{1}{2}(n^2 + n)$.

a. Confirm that the 4th and 5th triangular numbers are 10 sand 15 by evaluating $T(4)$ and $T(5)$. $\frac{1}{2}(16 + 4) = 10$, $\frac{1}{2}(25 + 5) = 15$

b. Find the 100th triangular number. 5,050

46. The formula $P(n) = \frac{n^2}{9}$ can be used to find numbers which satisfy $X^3 + Y^3 = Z^2$. Pick any rational numbers for n. Then take $X = P(n)$, $Y = 2P(n)$, and $Z = nP(n)$. Show that it works for $n = 3$ and $n = 6$. This problem dates back to the work of Arab mathematicians in the eleventh century. $1^3 + 2^3 = 3^2$; $4^3 + 8^3 = 24^2$

In Exercises 47–58, find the sum of the given polynomials.

47. $x^2 + 2x + 3$; $3x^2 - 4x + 5$ $4x^2 - 2x + 8$

48. $3x^2 - x + 2$; $2x^2 + 3x - 1$ $5x^2 + 2x + 1$

49. $-x^2 + 7x - 1$; $x^2 - 6x + 3$ $x + 2$

50. $3x^2 - 4x + 6$; $-3x^2 + 4x - 1$ 5

51. $x^3 + 1$; $x^2 - 2$ $x^3 + x^2 - 1$

52. $x^4 - 16$; $x^2 - 4$ $x^4 + x^2 - 20$

53. $x + 1$; $x - 2$; $x + 3$ $3x + 2$

54. $x - 3$; $x - 4$; $x - 5$ $3x - 12$

55. $\frac{x^2}{2} + \frac{x}{3} + \frac{1}{4}$; $\frac{3x^2}{2} + \frac{4x}{3} - \frac{1}{8}$ $2x^2 + \frac{5x}{3} + \frac{1}{8}$

56. $\frac{x^2}{3} - \frac{x}{2}$; $\frac{x}{3} - \frac{x^2}{2}$ $-\frac{x^2}{6} - \frac{x}{6}$

57. $x^3 - x$; $x^2 - 1$; x^4 $x^4 + x^3 + x^2 - x - 1$

58. $1 - x^2$; $x^2 - x^3$; $x^3 - x^4$ $1 - x^4$

In Exercises 59–70, find the difference by subtracting the second polynomial from the first one.

59. $2x - 9$; $x - 4$ $x - 5$
60. $-3x + 2$; $-x + 3$ $-2x - 1$
61. $x^2 - x + 1$; $x^2 + x - 1$ $-2x + 2$
62. $3x^2 + x - 4$; $x^2 - 2x - 4$ $2x^2 + 3x$
63. $3x^2 + 2x + 1$; $2x^2 + x + 2$ $x^2 + x - 1$
64. $3x^2 - x - 5$; $x^2 - 2x + 1$ $2x^2 + x - 6$
65. $x^3 - 1$; $x^2 - 3x$ $x^3 - x^2 + 3x - 1$
66. $x^3 + x$; $x^2 + 2$ $x^3 - x^2 + x - 2$
67. $2x^2 - 4$; $2x^2 - 4$ 0
68. $2x^2 - 5$; $5 - 2x^2$ $4x^2 - 10$
69. $\frac{x^2}{2} + \frac{x}{3} - \frac{1}{4}$; $\frac{x^2}{4} + \frac{x}{6} + \frac{1}{8}$ $\frac{x^2}{4} + \frac{x}{6} - \frac{3}{8}$
70. $-\frac{x^2}{2} + x$; $\frac{x^2}{2} + 2x - 3$ $-x^2 - x + 3$

In Exercises 71–82, simplify the given expression.

71. $(7x^2 - 6x + 2) + (-4x^2 + 7x - 8)$ $3x^2 + x - 6$
72. $(-5x^2 + 4x - 8) + (2x^2 - 9x - 3)$ $-3x^2 - 5x - 11$
73. $(7x^2 - 6x + 2) - (-4x^2 + 7x - 8)$ $11x^2 - 13x + 10$
74. $(-5x^2 + 4x - 8) - (2x^2 - 9x - 3)$ $-7x^2 + 13x - 5$
75. $(2x^3 - 3x^2 + 4x) + (3x^2 - 4x^3 + 2)$ $-2x^3 + 4x + 2$
76. $(-x^3 + x^2 - x + 1) + (2x^3 - 2x^2 + 2x - 2)$ $x^3 - x^2 + x - 1$
77. $(3x^2 + 2) - (3x^2 - 2x + 2)$ $2x$
78. $(1 + x^2 - x^3) - (x^3 + x^2 - 1)$ $-2x^3 + 2$
79. $\left(\frac{x^2}{2} + \frac{x}{5} - \frac{1}{3}\right) - \left(\frac{x^2}{4} + \frac{x}{7} - \frac{2}{3}\right)$ $\frac{x^2}{4} + \frac{2x}{35} + \frac{1}{3}$
80. $\left(\frac{x^4}{16} - \frac{x^2}{4} + \frac{x}{2} - 1\right) - \left(x^4 - \frac{x^2}{2} + \frac{x}{4} - \frac{1}{16}\right)$ $-\frac{15x^4}{16} + \frac{x^2}{4} + \frac{x}{4} - \frac{15}{16}$
81. $(x^2 + 3x - 2) - (x^2 - 3x + 2) - (2x^2 + x + 1)$ $-2x^2 + 5x - 5$
82. $(3x^2 - 1) - (x - 1) - (3 - x + x^2)$ $2x^2 - 3$

In Exercises 83 and 84, find a polynomial expression for the perimeter.

83. $4x^2 + 20x + 6$

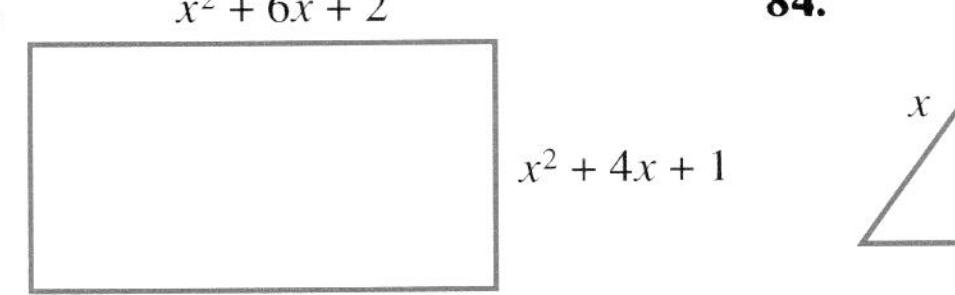

84. $3x^2 + 2x - 5$

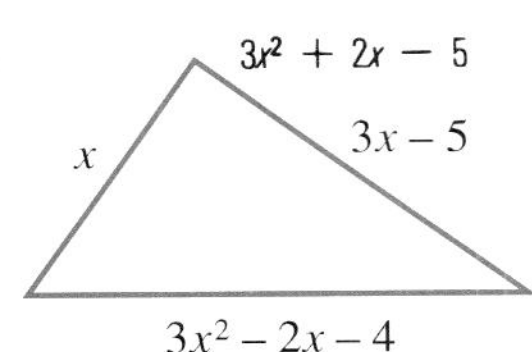

In Exercises 85–88, solve the equation for x.

85. $x^2 + 3x - 5 = x^2 + 2x + 1$ $\{6\}$
86. $2x^2 + 5x - 1 = 2x^2 - x - 7$ $\{-1\}$
87. $(3x^2 + 4x - 2) - (3x^2 - x + 1) = (x - 4) + (2x + 7)$ $\{3\}$
88. $(x^2 - 1) + (x - x^2) = 1 - x$ $\{1\}$

THINK ABOUT IT

1. A continuing addition like the one shown is called a *telescoping series,* because one term of each entry "cancels out" one term of the previous entry.
   Series: $(\frac{1}{2} - \frac{1}{3}) + (\frac{1}{3} - \frac{1}{4}) + (\frac{1}{4} - \frac{1}{5}) + (\frac{1}{5} - \frac{1}{6})$
   This makes it easy to simplify and evaluate the expression.
   a. The sum of the given series is $\frac{1}{2} - \frac{1}{6}$, or $\frac{1}{3}$. Why?
   b. Find a simpler expression for the polynomial below; then find $P(-2)$.
   $$P(n) = (n - n^2) + (n^2 - n^3) + (n^3 - n^4) + (n^4 - n^5)$$
   c. Find a simpler expression for this series. Is it a polynomial?
   $$\left(\frac{1}{n} - \frac{1}{n+1}\right) + \left(\frac{1}{n+1} - \frac{1}{n+2}\right) + \left(\frac{1}{n+2} - \frac{1}{n+3}\right) + \left(\frac{1}{n+3} - \frac{1}{n+4}\right) + \left(\frac{1}{n+4} - \frac{1}{n+5}\right)$$
2. a. When you evaluate a polynomial for $x = 1$, explain why the result is just the sum of the coefficients.
   b. If $P(x) = 3x^5 - x^4 + 2x^3 - 4x^2 + 6x - 3$, find $P(1)$.
3. The term -3 in $P(x)$ of part **b** of Exercise 2 is called the "constant term," because it contains no variables. Explain why, when $P(x)$ is any polynomial, $P(0)$ is just the constant term.
4. True or false? A polynomial in x can never be undefined. Explain your answer.
5. Given $P(x) = 3x^2 + 5x + 7$, find the following.
   a. $P(y)$, by replacing x by y
   b. $P(2t)$, by replacing x by $2t$
   c. Does $P(2t)$ equal $2P(t)$?

REMEMBER THIS

1. Multiply: $4x^3 \cdot 7x^3$. $28x^6$
2. According to the distributive property, $x(n + m)$ is equal to ______. $xn + xm$
3. True or false? "Seven minus x" is equivalent to "seven plus negative x." True
4. The *subtraction* expression $3x - 4y$ is equivalent to what *addition* expression? $3x + (-4y)$
5. Multiply: $(-x)(-2x^2)(-3x^3)$. $-6x^6$
6. Express 586 trillion in scientific notation. 5.86×10^{14}
7. True or false? If a^n is positive, then a^{-n} is negative. False
8. Solve: $|x - 3| = 10$. $\{-7, 13\}$
9. Which of these numbers are not rational? $\sqrt{2}, \sqrt{3}, \sqrt{4}, \sqrt{\frac{4}{9}}$. $\sqrt{2}, \sqrt{3}$
10. A 10 percent price increase resulted in a TV selling for $638. What was the original price? $580

3.4 Multiplication of Polynomials

A manufacturer has been selling about 2,000 lawn tractors per year for \$2,000 each. The sales department believes that for each \$100 the price is raised, sales will fall by 100 units, so the formula $(2{,}000 + 100x)(2{,}000 - 100x)$ gives the estimated revenue, where x represents the number of \$100 price increases. Multiply the two factors, and write a polynomial expression for the total estimated revenue. Explain, according to this expression, why raising the price is not a good idea. (See Example 7.)

OBJECTIVES

1. **Multiply a monomial by a monomial.**
2. **Multiply two polynomials.**
3. **Use the FOIL method to multiply binomials.**
4. **Multiply binomials that are the sum and difference of two terms.**
5. **Use the formulas for the square of a binomial.**
6. **Multiply binomials, and then solve linear equations.**

1 The simplest polynomial multiplication is the product of two monomials. To compute this type of product, follow the methods of Section 3.1. As a reminder, we repeat the solution to Example 3a from that section.

$$\begin{aligned} -4x^3 \cdot 7x^3 &= (-4 \cdot 7)(x^3 \cdot x^3) && \text{Reorder and regroup.} \\ &= (-4 \cdot 7)x^{3+3} && \text{Product property of exponents.} \\ &= -28x^6 && \text{Simplify.} \end{aligned}$$

In summary, to multiply two monomials, you first multiply the numerical coefficients and then multiply the variable factors, using the appropriate properties of exponents.

EXAMPLE 1 Find each product.

a. $(-3y^5)(5y^3)$ **b.** $(-2x^3)(\frac{3}{2}y^3)$ **c.** $(3x^2y)(5x^4y^3)$

Solution Multiply the numerical coefficients and multiply the variable factors.

a. $(-3y^5)(5y^3) = (-3 \cdot 5)(y^5 \cdot y^3) = -15y^{5+3} = -15y^8$
b. $(-2x^3)(\frac{3}{2}y^3) = (-2 \cdot \frac{3}{2})(x^3 \cdot y^3) = -3x^3y^3$
c. $(3x^2y)(5x^4y^3) = (3 \cdot 5)(x^2 \cdot x^4)(y \cdot y^3) = 15x^{2+4}y^{1+3} = 15x^6y^4$

PROGRESS CHECK 1 Find each product.

a. $(-2x^4)(-3x^3)$ **b.** $(-\frac{4}{3}x)(9y^2)$ **c.** $(6a^4b^2)(7a^4b)$

2 The general procedure for multiplying polynomials uses the distributive property in the forms

$$a(b + c) = ab + ac \quad \text{and} \quad (a + b)c = ac + bc,$$

or in an extended form such as

$$a(b + c + \cdots + n) = ab + ac + \cdots + an.$$

Progress Check Answers
1. (a) $6x^7$ (b) $-12xy^2$ (c) $42a^8b^3$

EXAMPLE 2 Find each product.

a. $-3x(5x^2 - 2x + 1)$ **b.** $(3x + 4y)2x$

Solution In applying the distributive property, note that we frequently switch between addition and subtraction by using the subtraction definitions: $a - b = a + (-b)$.

a. $-3x(5x^2 - 2x + 1) = (-3x)(5x^2) + (-3x)(-2x) + (-3x)(1)$
$= -15x^3 + 6x^2 - 3x$

b. $(3x + 4y)2x = (3x)(2x) + (4y)(2x)$
$= 6x^2 + 8xy$

PROGRESS CHECK 2 Find each product.

a. $3x^2(x^2 + 2x - 5)$ **b.** $(8y - x)4y$

When *both* polynomials contain more than one term, the distributive property is used more than once. Note that according to the distributive property,

$$(x + 2)(\quad) \quad \text{equals} \quad x(\quad) + 2(\quad),$$

no matter what expression is inside the parentheses. For instance,

$$(x + 2)(x + 4) \quad \text{equals} \quad x(x + 4) + 2(x + 4)$$

Then by using the distributive property again, we get

$$x(x + 4) + 2(x + 4) = x^2 + 4x + 2x + 8 = x^2 + 6x + 8.$$

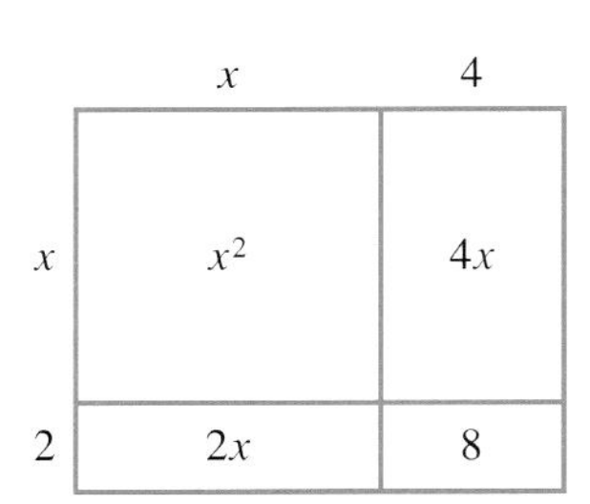

Figure 3.1

Therefore, $(x + 2)(x + 4) = x^2 + 6x + 8$, and the rectangle shown in Figure 3.1 gives a geometric interpretation of this result in terms of area. Close inspection shows that in this example we multiplied each term of the first polynomial by each term of the other. This leads to the following general procedure.

To Multiply Polynomials

To multiply two polynomials, multiply each term of one polynomial by each term of the other polynomial, and then combine like terms.

In Example 3 we find products by stressing the distributive property. Then in Example 4 we repeat the problem of Example 3b, using the procedure stated above with a vertical arrangement that makes it easy to combine like terms in the product.

EXAMPLE 3 Find each product.

a. $(2x - 3)(2x + 5)$ **b.** $(3x^2 - 4x + 1)(x - 2)$

Point out that there is more than one order that will lead to the correct answer. Parts **a** and **b** are done in different orders as an illustration.

Solution Use the distributive property twice and simplify.

a. $(2x - 3)(2x + 5) = 2x(2x + 5) - 3(2x + 5)$
$= 4x^2 + 10x - 6x - 15$
$= 4x^2 + 4x - 15$

b. $(3x^2 - 4x + 1)(x - 2) = (3x^2 - 4x + 1)x + (3x^2 - 4x + 1)(-2)$
$= 3x^3 - 4x^2 + x - 6x^2 + 8x - 2$
$= 3x^3 - 10x^2 + 9x - 2$

Progress Check Answers

2. (a) $3x^4 + 6x^3 - 15x^2$ (b) $32y^2 - 4xy$

PROGRESS CHECK 3 Find each product.

a. $(3x - 2)(4x + 1)$ **b.** $(5x^2 - 3x + 1)(x - 3)$

EXAMPLE 4 Find the product of $3x^2 - 4x + 1$ and $x - 2$.

Solution Multiply each term of the first polynomial by each term of the second. Place like terms in the products under each other; then add.

$$\begin{array}{rrrrl} & 3x^2 & - 4x & + 1 & \\ & & x & - 2 & \\ \hline & 3x^3 & - 4x^2 & + x & \quad \text{This line equals } x(3x^2 - 4x + 1). \\ \text{Add:} & & - 6x^2 & + 8x - 2 & \quad \text{This line equals } -2(3x^2 - 4x + 1). \\ \hline & 3x^3 & - 10x^2 & + 9x - 2 & \end{array}$$

The product is $3x^3 - 10x^2 + 9x - 2$, as before.

PROGRESS CHECK 4 Find the product of $5x^2 - 3x + 1$ and $x - 3$ using a vertical arrangement.

3 Because so many problems involve the product of two *binomials,* mental shortcuts are often used. According to our general procedure, to multiply two binomials, you multiply each term of one binomial by each term of the other binomial and then combine like terms. A simple method for doing this is shown below. This method is called the FOIL method and the letters F, O, I, and L denote the products of the first, outer, inner, and last terms respectively.

"Think About It" Exercise 2 asks for other mnemonics to show that FOIL is just a memory aid.

$$\overset{\text{F}\quad\text{L}}{(a + b)(c + d)} = \overset{\text{F}}{ac} + \overset{\text{O}}{ad} + \overset{\text{I}}{bc} + \overset{\text{L}}{bd}$$
$$\underset{\text{O}}{\overset{\text{I}}{}}$$

To multiply using the FOIL method, add the following products.

1. Multiply the **F**irst terms: *ac*.
2. Multiply the **O**uter terms: *ad*.
3. Multiply the **I**nner terms: *bc*.
4. Multiply the **L**ast terms: *bd*.

In the following example, note that the outer product and the inner product are often like terms and are therefore combined.

EXAMPLE 5 Multiply using the FOIL method.

a. $(x + 5)(x + 2)$ **b.** $(2x - 5)(3x - 1)$ **c.** $(5x - y)(5x + y)$

Solution

a. This first example is easiest because all terms are positive.

$$\begin{aligned} (x + 5)(x + 2) &= \overset{\text{F}}{x(x)} + \overset{\text{O}}{x(2)} + \overset{\text{I}}{5(x)} + \overset{\text{L}}{5(2)} \\ &= x^2 + \underbrace{2x + 5x}_{} + 10 \\ &= x^2 + \quad 7x \quad + 10 \end{aligned}$$

Thus $(x + 5)(x + 2) = x^2 + 7x + 10$.

Progress Check Answers

3. (a) $12x^2 - 5x - 2$ (b) $5x^3 - 18x^2 + 10x - 3$
4. $5x^3 - 18x^2 + 10x - 3$

b. Be careful to keep track of the signs.

$$(2x - 5)(3x - 1) = \overset{F}{2x(3x)} + \overset{O}{2x(-1)} + \overset{I}{(-5)(3x)} + \overset{L}{(-5)(-1)}$$
$$= 6x^2 - 2x - 15x + 5$$
$$= 6x^2 - 17x + 5$$

c. Note that the outer and inner pairs are like terms.

$$(5x - y)(5x + y) = \overset{F}{5x(5x)} + \overset{O}{5x(y)} + \overset{I}{(-y)(5x)} + \overset{L}{(-y)(y)}$$
$$= 25x^2 + 5xy - 5xy - y^2$$
$$= 25x^2 - y^2$$

Note It should be a goal to become so proficient with the FOIL method that the work displayed above is done *mentally*, so that you can often just write down the answer.

PROGRESS CHECK 5 Multiply by the FOIL method.

a. $(x + 3)(x + 7)$ **b.** $(4x - 3)(5x - 2)$ **c.** $(7x + 2y)(7x - 2y)$

4 A product which comes up often and has a particularly simple result involves the sum and difference of the same two terms. Note in Example 5c that the factors in $(5x - y)(5x + y)$ differ only in the operation between $5x$ and y, while the result of the multiplication is $(5x)^2$ and y^2 with a minus sign between them. The outer and inner products were like terms which added to 0 and so do not appear in the final answer. This result is summarized in the following formula.

Product of the Sum and Difference of Two Expressions

The product of the sum and difference of two terms is the square of the first term minus the square of the second term.

$$(a + b)(a - b) = a^2 - b^2$$

EXAMPLE 6 Find each product. Use $(a + b)(a - b) = a^2 - b^2$.

a. $(x + 5)(x - 5)$ **b.** $(1 + 3x)(1 - 3x)$ **c.** $(3x - 4y)(3x + 4y)$

Solution

a. Substitute x for a and 5 for b in the given formula.

$$(a + b)(a - b) = a^2 - b^2$$
$$\downarrow \quad \downarrow \quad \downarrow \quad \downarrow \qquad \downarrow \quad \downarrow$$
$$(x + 5)(x - 5) = x^2 - 5^2$$
$$= x^2 - 25$$

b. Here $a = 1$ and $b = 3x$, so

$$(1 + 3x)(1 - 3x) = 1^2 - (3x)^2$$
$$= 1 - 9x^2.$$

Progress Check Answers

5. (a) $x^2 + 10x + 21$ (b) $20x^2 - 23x + 6$ (c) $49x^2 - 4y^2$

c. Replace a by $3x$ and b by $4y$. The product is the square of the first term minus the square of the second term.

$$\begin{aligned}(3x - 4y)(3x + 4y) &= (3x)^2 - (4y)^2 \\ &= 9x^2 - 16y^2\end{aligned}$$

PROGRESS CHECK 6 Find each product. Use $(a + b)(a - b) = a^2 - b^2$.

a. $(x + 3)(x - 3)$ **b.** $(3 + 2x)(3 - 2x)$ **c.** $(4x - 7y)(4x + 7y)$

EXAMPLE 7 Solve the problem that opens the section on page 111.

Solution We can use the formula for $(a + b)(a - b)$.

$$\begin{aligned}(2{,}000 + 100x)(2{,}000 - 100x) &= (2{,}000)^2 - (100x)^2 \\ &= 4{,}000{,}000 - 10{,}000x^2\end{aligned}$$

Therefore, the estimated revenue is $4{,}000{,}000 - 10{,}000x^2$ dollars, where x is the number of times the price is raised by \$100. Because x^2 can never be negative, this shows that each price increase of \$100 will cause the estimated revenue to decrease. For instance, if $x = 1$ (the price goes to \$2,100), then the estimated revenue in dollars will be $4{,}000{,}000 - 10{,}000(1)^2 = 3{,}990{,}000$. If $x = 2$ (the price goes to \$2,200), the revenue will be $4{,}000{,}000 - 10{,}000(2)^2 = 3{,}960{,}000$. The higher the price goes, the lower the estimated revenue becomes.

PROGRESS CHECK 7 For the company described in Example 7, if each \$50 increase in price will lose 50 sales, the expression for estimated revenue becomes $(2{,}000 + 50x)(2{,}000 - 50x)$. Find the product, and then analyze whether raising the price is a good idea in this case.

5 Other special products which come up frequently involve raising a binomial to a power, most commonly squaring or cubing. In the next few examples we illustrate these products.

EXAMPLE 8 Expand.

a. $(a + b)^2$ **b.** $(a - b)^2$

Solution

a. $(a + b)^2$ means $(a + b)(a + b)$, and to *expand* $(a + b)^2$ means to multiply this expression out.

$$\begin{aligned}(a + b)(a + b) &= a(a) + a(b) + b(a) + b(b) \\ &= a^2 + ab + ba + b^2 \\ &= a^2 + 2ab + b^2\end{aligned}$$

Therefore, the expanded form of $(a + b)^2$ is $a^2 + 2ab + b^2$. A geometric interpretation of this result in terms of area is shown in Figure 3.2.

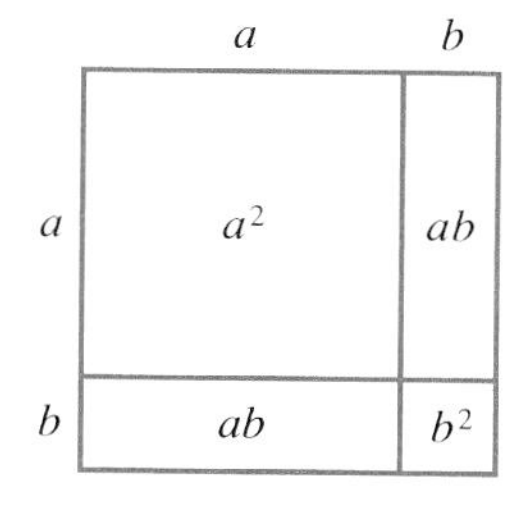

Figure 3.2

b. Write $(a - b)^2$ as $(a - b)(a - b)$. Then

$$\begin{aligned}(a - b)(a - b) &= a^2 - ab - ba + b^2 \\ &= a^2 - 2ab + b^2\end{aligned}$$

Thus, $(a - b)^2 = a^2 - 2ab + b^2$.

PROGRESS CHECK 8 Expand.

a. $(m + n)^2$ **b.** $(m - n)^2$

The problem of squaring a binomial occurs often enough that you should memorize the following formulas, derived in Example 8.

Progress Check Answers

6. (a) $x^2 - 9$ (b) $9 - 4x^2$ (c) $16x^2 - 49y^2$
7. $4{,}000{,}000 - 2{,}500x^2$; no
8. (a) $m^2 + 2mn + n^2$ (b) $m^2 - 2mn + n^2$

Square-of-a-Binomial Formulas

To square a binomial, use

$$(a + b)^2 = a^2 + 2ab + b^2$$
or
$$(a - b)^2 = a^2 - 2ab + b^2.$$

The square of a binomial is the square of the first term, plus or minus twice the product of the two terms, plus the square of the second term.

This reminder cannot be mentioned too often. Point out that, in general, $(a + b)^n$ will not equal $a^n + b^n$. Perhaps you should ask students to see what happens for $n = 1, 2, 3, 4$.

Caution Remember to include the middle term, the ab term. In particular, notice that $(a + b)^2$ does not equal $a^2 + b^2$, because this answer leaves out the middle term.

EXAMPLE 9 Use the square-of-a-binomial formulas to find these products.

a. $(x + 3)^2$ **b.** $(2x - 5)^2$ **c.** $(2x + 3y)(2x + 3y)$

Solution

a. Use the formula for $(a + b)^2$ with $a = x$ and $b = 3$.

$$\begin{aligned}(a + b)^2 &= a^2 + 2ab + b^2\\ (x + 3)^2 &= (x)^2 + 2(x)(3) + (3)^2\\ &= x^2 + 6x + 9\end{aligned}$$

b. Use the formula for $(a - b)^2$ with $a = 2x$ and $b = 5$.

$$\begin{aligned}(a - b)^2 &= a^2 - 2ab + b^2\\ (2x - 5)^2 &= (2x)^2 - 2(2x)(5) + (5)^2\\ &= 4x^2 - 20x + 25\end{aligned}$$

c. Because $(2x + 3y)(2x + 3y) = (2x + 3y)^2$, the product is the square of the first term, plus twice the product of the two terms, plus the square of the second term.

$$\begin{aligned}(a + b)^2 &= a^2 + 2ab + b^2\\ (2x + 3y)^2 &= (2x)^2 + 2(2x)(3y) + (3y)^2\\ &= 4x^2 + 12xy + 9y^2\end{aligned}$$

PROGRESS CHECK 9 Find each product. Use the formulas for the square of a binomial.

a. $(x + 1)^2$ **b.** $(2x - 7)^2$ **c.** $(y - 3x)(y - 3x)$

EXAMPLE 10 Expand $(a + b)^3$.

Solution $(a + b)^3$ means $(a + b)(a + b)(a + b)$. To multiply this expression, first multiply $(a + b)(a + b)$, as follows.

$$(a + b)(a + b) = a^2 + 2ab + b^2$$

Now find the product of this result and the third factor of $a + b$.

$$\begin{array}{lr} & a^2 + 2ab + b^2\\ & a + b\\ \hline & a^3 + 2a^2b + ab^2\\ \text{Add:} & a^2b + 2ab^2 + b^3\\ \hline & a^3 + 3a^2b + 3ab^2 + b^3\end{array}$$

This line equals $a(a^2 + 2ab + b^2)$.
This line equals $b(a^2 + 2ab + b^2)$.

Progress Check Answers

9. (a) $x^2 + 2x + 1$ (b) $4x^2 - 28x + 49$ (c) $y^2 - 6xy + 9x^2$

In expanded form $(a + b)^3$ equals $a^3 + 3a^2b + 3ab^2 + b^3$.

PROGRESS CHECK 10 Expand $(x - h)^3$.

6 The procedures for multiplying binomials, together with the methods for solving a linear equation, enable us to solve the problem which follows.

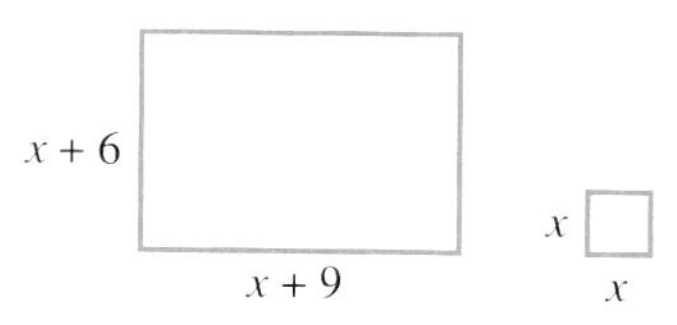

Figure 3.3

EXAMPLE 11 The area of the rectangle exceeds the area of the square in Figure 3.3 by 99 feet². Find the dimensions of the rectangle.

Solution *Set up an equation* using $A = \ell w$, $A = s^2$, and the given information.

Area of rectangle	minus	area of square	equals	99.
$(x + 6)(x + 9)$	$-$	x^2	$=$	99

Solve the Equation

$$x^2 + 9x + 6x + 54 - x^2 = 99 \quad \text{FOIL multiplication.}$$
$$15x + 54 = 99 \quad \text{Combine like terms.}$$
$$15x = 45 \quad \text{Subtract 54 from both sides.}$$
$$x = 3 \quad \text{Divide both sides by 15.}$$

Answer the Question The dimensions of the rectangle are 9 ft by 12 ft.

Check the Answer The area of the rectangle is $9 \cdot 12 = 108$ ft². The area of the square is $3 \cdot 3 = 9$ ft². The difference in area is 99 ft², so the solution is correct.

PROGRESS CHECK 11 The length and width of a rectangle are given by $x + 1$ and $x - 2$, and the side length of a square is given by x. Find the length of the rectangle if the area of the square exceeds the area of the rectangle by 7 in.².

Progress Check Answers

10. $x^3 - 3x^2h + 3xh^2 - h^3$
11. 6 in.

EXERCISES 3.4

In Exercises 1–26, find each product.

1. $(-2x^3)(7x^5)$ $-14x^8$
2. $(8y^2)(-3y^3)$ $-24y^5$
3. $\left(\frac{4x^2}{5}\right)\left(\frac{5x^3}{4}\right)$ x^5
4. $(\frac{3}{2}x)(\frac{3}{2}x^3)$ $\frac{9}{4}x^4$
5. $\left(\frac{x^5}{3}\right)(12x^{-2})$ $4x^3$
6. $(24x^8)\left(\frac{x^{-4}}{8}\right)$ $3x^4$
7. $(4x^2y)(5xy^3)$ $20x^3y^4$
8. $(3xy)(-7x)$ $-21x^2y$
9. $3x(4x^2 - 5x + 6)$ $12x^3 - 15x^2 + 18x$
10. $5x(2x^2 - 3x - 1)$ $10x^3 - 15x^2 - 5x$
11. $-2y(y^2 - 3y + 5)$ $-2y^3 + 6y^2 - 10y$
12. $-4y(-3y^2 + y - 2)$ $12y^3 - 4y^2 + 8y$
13. $(x^2 - 2xy + y^2)xy$ $x^3y - 2x^2y^2 + xy^3$
14. $(x - 2y - 3)(-3x^2y^2)$ $-3x^3y^2 + 6x^2y^3 + 9x^2y^2$
15. $x^{-3}(3x^9 + x^6 + 2x^3)$ $3x^6 + x^3 + 2$
16. $x^{-2}(2x^6 - 3x^4 - 4x^2)$ $2x^4 - 3x^2 - 4$
17. $(3x + 2)(4x + 3)$ $12x^2 + 17x + 6$
18. $(2x + 5)(3x + 7)$ $6x^2 + 29x + 35$
19. $(4y^2 + 3y - 2)(y + 3)$ $4y^3 + 15y^2 + 7y - 6$
20. $(x^2 - 2x + 3)(x - 1)$ $x^3 - 3x^2 + 5x - 3$
21. $(x - 1)(x^2 + x + 1)$ $x^3 - 1$
22. $(y + 2)(y^2 - 2y + 4)$ $y^3 + 8$
23. $(x^2 + 2x + 1)(x^2 + 2x + 1)$ $x^4 + 4x^3 + 6x^2 + 4x + 1$
24. $(x^2 - 6x + 9)(x^2 - 6x + 9)$ $x^4 - 12x^3 + 54x^2 - 108x + 81$
25. $(x^2 + 2xy + y^2)(x + y)$ $x^3 + 3x^2y + 3xy^2 + y^3$
26. $(x^2 - 2xy + y^2)(x - y)$ $x^3 - 3x^2y + 3xy^2 - y^3$

In Exercises 27–38, multiply using the FOIL method.

27. $(x + 3)(x + 5)$ $x^2 + 8x + 15$
28. $(x + 5)(x + 6)$ $x^2 + 11x + 30$
29. $(3x - 1)(2x - 4)$ $6x^2 - 14x + 4$
30. $(4x - 3)(5x - 4)$ $20x^2 - 31x + 12$
31. $(2x + \frac{1}{2})(x - 2)$ $2x^2 - \frac{7}{2}x - 1$
32. $(x + 7)(3x - 4)$ $3x^2 + 17x - 28$
33. $(a + 2b)(a - 3b)$ $a^2 - ab - 6b^2$
34. $(5a - 4b)(4a + 5b)$ $20a^2 + 9ab - 20b^2$
35. $(2a^2 - 4)(3a + 2)$ $6a^3 + 4a^2 - 12a - 8$
36. $(a^2 - 1)(a - 1)$ $a^3 - a^2 - a + 1$
37. $(m^2 + n^2)(m + n)$ $m^3 + m^2n + mn^2 + n^3$
38. $\left(\frac{m^2}{4} + 2\right)\left(\frac{m}{2} - 2\right)$ $\frac{m^3}{8} - \frac{m^2}{2} + m - 4$

Exercises 39–52 can be done using special product formulas. Give the name of the pattern. Then give the result.

39. $(x + 7)(x - 7)$ Sum and difference of two expressions; $x^2 - 49$
40. $(y - 3)(y + 3)$ Sum and difference of two expressions; $y^2 - 9$
41. $(2y - 5)(2y + 5)$ Sum and difference of two expressions; $4y^2 - 25$
42. $(4x + 1)(4x - 1)$ Sum and difference of two expressions; $16x^2 - 1$
43. $(5x + 3y)(5x - 3y)$ Sum and difference of two expressions; $25x^2 - 9y^2$
44. $(2x - y)(2x + y)$ Sum and difference of two expressions; $4x^2 - y^2$
45. $(y + 9)^2$ Square of binomial; $y^2 + 18y + 81$
46. $(x + 10)^2$ Square of binomial; $x^2 + 20x + 100$
47. $(2x - 7)^2$ Square of binomial; $4x^2 - 28x + 49$
48. $(6 - 5y)^2$ Square of binomial; $36 - 60y + 25y^2$
49. $(x + 3y)^2$ Square of binomial; $x^2 + 6xy + 9y^2$

50. $(2x - 5y)^2$ Square of binomial; $4x^2 - 20xy + 25y^2$
51. $(x^2 + x)^2$ Square of binomial; $x^4 + 2x^3 + x^2$
52. $(y^2 - 3y)^2$ Square of binomial; $y^4 - 6y^3 + 9y^2$
53. Expand $(m - x)^3$. $m^3 - 3m^2x + 3mx^2 - x^3$
54. Expand $(x + d)^3$. $x^3 + 3x^2d + 3xd^2 + d^3$
55. Expand $(x + 2)^3$. $x^3 + 6x^2 + 12x + 8$
56. Expand $(y - 5)^3$. $y^3 - 15y^2 + 75y - 125$
57. Expand $(2x - y)^3$. $8x^3 - 12x^2y + 6xy^2 - y^3$
58. Expand $(x + 2y)^3$. $x^3 + 6x^2y + 12xy^2 + 8y^3$
59. Use multiplication to explain why the expression $(10 - x)(10 + x)$ can never be greater than 100.
Because $x^2 \geq 0$, $100 - x^2$ cannot be more than 100.
60. Use multiplication to explain why $(8 + 3x)(8 - 3x)$ can never be greater than 64. Because $9x^2 \geq 0$, $64 - 9x^2$ cannot be greater than 64.
61. **a.** Write a polynomial expression for the area of the rectangle. $25 - x^2$
b. For what value of x will the area be maximum? $x = 0$

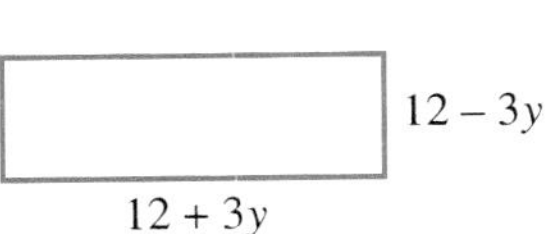

62. **a.** Write a polynomial expression for the area of the rectangle. $144 - 9y^2$
b. For what value of y will the area be a maximum? $y = 0$

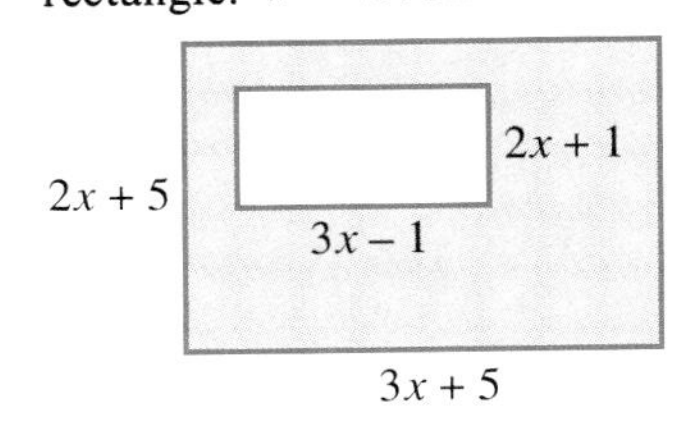

63. The shaded area equals 55 in.^{2}. Find the area of the larger rectangle. $A = 63$ in.^{2}

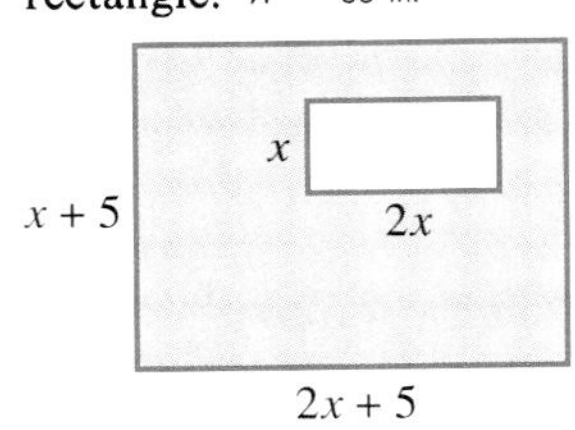

64. The shaded area is 146 cm^2. Find the area of the smaller rectangle. $A = 154$ cm.^{2}

THINK ABOUT IT

1. **a.** Show that $(a - b)^2$ equals $(b - a)^2$ by expanding both expressions.
b. What is the relationship between $a - b$ and $b - a$?
c. Without doing the work, would you expect $(a - b)^3$ and $(b - a)^3$ to be equal? Explain.
2. **a.** What do the letters FOIL stand for?
b. Would the FILO method give the same answer?
c. How many different ways can you arrange the letters {F,O,I,L}?
3. **a.** The figure shown demonstrates geometrically that $(a + b)^2 = a^2 + 2ab + b^2$. Verify this.
b. Draw and label a similar picture for a cube to find an expansion for $(a + b)^3$. The volume of the cube will be made up of the sum of eight smaller pieces.

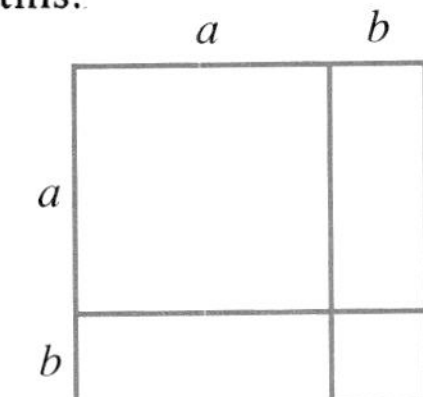

4. **a.** If you *multiply* a polynomial of degree 3 times a polynomial of degree 4, what degree will the product polynomial have? Show this by an example.
b. If you *add* a polynomial of degree 3 to a polynomial of degree 4, what degree will the sum polynomial have? Give an example.
c. Show that if you add two polynomials of degree 3, you cannot say, *in general*, what the degree of the sum will be. Give examples where the degree of the sum is 3, 2, 1, 0.
5. Find these products.
a. $(x - 1)(x + 1)$ **b.** $(x - 1)(x^2 + x + 1)$
c. $(x - 1)(x^3 + x^2 + x + 1)$
d. Assume that the pattern above holds and find $(x - 1)(x^{20} + x^{19} + x^{18} + \cdots + x + 1)$.
e. This product can be found the hard way or the easy way. Do the calculation both ways. Find $(d^4 + e^4)(d^2 + e^2)(d + e)(d - e)$.

REMEMBER THIS

1. The word *product* refers to which operation? Multiplication
2. What is the degree of this polynomial: $2x + 3y$? 1
3. Express 60 as a product of prime factors. $2^2 \cdot 3 \cdot 5$
4. Divide $\dfrac{30x^4y}{6x^2y}$. $5x^2$
5. True or false? $x(m + n) + y(m + n)$ is equal to $m(x + y) + n(x + y)$. True
6. Find $P(-1)$ if $P(x) = 3x^2 - 4x + 2$. 9
7. Simplify $(6x^2 - 5x + 1) + (-6x^2 + 5x + 1)$. 2
8. Simplify $\left(\dfrac{y^3}{y^{-1}}\right)^{-2}$, and write the result using only positive exponents. $\dfrac{1}{y^8}$
9. In the expression $5x^2 + 7x - 2$, 5 is called the ______ of x^2. Coefficient
10. From home Pat can bike to town in half an hour or walk to town in 1 hour and 40 minutes. Pat's biking speed is 7 mi/hour faster than her walking speed. What is the distance from Pat's home to town? 5 mi

3.5 Factoring Polynomials with Common Factors

If a storekeeper raises a price 20 percent and then the following week reduces this new price by 20 percent, the final price can be represented as $(x + 0.20x) - 0.20(x + 0.20x)$, where x is the original price. Factor out the greatest common factor, $x + 0.20x$; then simplify and interpret the result. Is the final price more or less than the original price? (See Example 5.)

OBJECTIVES

1. **Find the greatest common factor (GCF) of a set of terms.**
2. **Factor out the greatest common factor from a polynomial.**
3. **Factor by grouping.**

We will often need to transform a given expression into a more useful form. One technique for doing this is called factoring, which is the reverse of multiplication. Using our earlier work on multiplying polynomials as a basis, we can now discuss several procedures for factoring polynomials. Notice that when you multiply polynomials, you essentially are changing a product into a sum. For instance,

$$\underset{\text{product}}{5(x + y)} = \underset{\text{sum}}{5x + 5y}.$$

The reverse process, changing a given sum into a product, is called **factoring** (because the components of a product are called factors). We show these relationships below. To change a product into a sum, we multiply.

$$5(x + y) \xrightarrow{\text{multiplying}} 5x + 5y$$

To change a sum into a product, we factor.

$$5(x + y) \xleftarrow{\text{factoring}} 5x + 5y$$

1 The first step in factoring a sum is to find the greatest factor that is common to each of the terms. For example:

Polynomial	Greatest Common Factor
$5x + 5y$	5
$16x + 24$	8
$2x^2 - 4x$	$2x$

In each case we identify the *greatest* common factor, or **GCF.** For instance, although 2 and 4 are common factors of $16x$ and 24, 8 is the *largest* number that divides both terms. It is not always obvious what the GCF is, so it will be useful to review a procedure for finding the GCF of a set of numbers.

To Find the Greatest Common Factor (GCF)

1. Write each number as a product of prime factors.
2. The GCF is the product of all the common prime factors, with each prime number appearing the fewest number of times it appears in any one of the factorizations. If there are no common prime factors, the GCF is 1.

EXAMPLE 1 Find the greatest common factor of 60 and 72.

Solution Express 60 and 72 in prime factored form.

$$60 = 2 \cdot 2 \cdot 3 \cdot 5 = 2^2 \cdot 3 \cdot 5$$
$$72 = 2 \cdot 2 \cdot 2 \cdot 3 \cdot 3 = 2^3 \cdot 3^2$$

The prime number 2 appears in both factorizations, and the fewest number of times it appears is twice. Similarly, the prime number 3 appears in both factorizations, and the fewest number of times it appears is once. The prime number 5 is not a common factor. Thus,

$$\text{GCF} = 2^2 \cdot 3 = 12.$$

The largest number that is a factor of both 60 and 72 is 12.

PROGRESS CHECK 1 Find the GCF of 90 and 72.

When variables are involved, the method is similar. Just as the GCF of $2^2 \cdot 3 \cdot 5$ and $2^3 \cdot 3^2$ is $2^2 \cdot 3$, the GCF of x^2yz and x^3y^2 is x^2y. With respect to variables, the GCF is the product of all common variable factors, with each variable appearing the fewest number of times it appears in any one of the terms.

EXAMPLE 2 Find the GCF of $10x^3$, $5x^4$, $15x^2$.

Solution The GCF of 10, 5, and 15 is 5. The GCF of x^3, x^4, and x^2 is x^2. Thus, the GCF of the three terms is $5x^2$.

PROGRESS CHECK 2 Find the GCF of $8y^5$, $20y^4$, and $12y^3$.

2 The first step in factoring polynomials is to identify the GCF of all the terms. Then, use the distributive property to factor it out. For instance, the GCF of the terms in $10x^2 + 8x$ is $2x$, and we factor it out as follows.

$$10x^2 + 8x = 2x \cdot 5x + 2x \cdot 4 \qquad \text{Think: } \frac{10x^2}{2x} = 5x; \frac{8x}{2x} = 4.$$
$$= 2x(5x + 4) \qquad \text{Distributive property.}$$

Thus, $2x(5x + 4)$ is the factored form of $10x^2 + 8x$. Because factoring reverses multiplying, you should check factoring answers by multiplication. For this example,

$$2x(5x + 4) = 2x \cdot 5x + 2x \cdot 4 = 10x^2 + 8x,$$

which checks.

Progress Check Answers

1. 18
2. $4y^3$

Here is a summary of the above procedure.

To Factor Out the GCF

1. Find the GCF of the terms in the polynomial.
2. Express each term in the polynomial as a product with the GCF as one factor.
3. Factor out the GCF using the distributive property.
4. Check the answer through multiplication.

EXAMPLE 3 Factor out the greatest common factor.

a. $8x + 16$ **b.** $25x^4 - 15x^3 + 10x^2$ **c.** $30x^4y + 12x^2y^3$

Solution

a. The GCF of $8x$ and 16 is 8.

$$8x + 16 = 8 \cdot x + 8 \cdot 2 \quad \text{Think: } \frac{8x}{8} = x;\ \frac{16}{8} = 2.$$
$$= 8(x + 2) \quad \text{Distributive property.}$$

The answer $8(x + 2)$ checks, because multiplying 8 and $x + 2$ gives $8x + 16$.

b. The GCF of $25x^4$, $-15x^3$, and $10x^2$ is $5x^2$.

$$25x^4 - 15x^3 + 10x^2$$
$$= 5x^2(5x^2) - 5x^2(3x) + 5x^2(2) \quad \frac{25x^4}{5x^2} = 5x^2;\ \frac{-15x^3}{5x} = -3x;\ \frac{10x^2}{5x^2} = 2.$$
$$= 5x^2(5x^2 - 3x + 2) \quad \text{Distributive property.}$$

The answer checks through multiplication.

c. The GCF of $30x^4y$ and $12x^2y^3$ is $6x^2y$.

$$30x^4y + 12x^2y^3$$
$$= 6x^2y(5x^2) + 6x^2y(2y^2) \quad \frac{30x^4y}{6x^2y} = 5x^2;\ \frac{12x^2y^3}{6x^2y} = 2y^2.$$
$$= 6x^2y(5x^2 + 2y^2) \quad \text{Distributive property.}$$

The answer checks through multiplication.

PROGRESS CHECK 3 Factor out the greatest common factor.

a. $16x - 10$ **b.** $16y^4 + 4y^3 - 12y$ **c.** $18x^5y^3 + 32x^3y^4$

Two special situations are shown in Example 4. Examine these cases carefully.

EXAMPLE 4 Factor out the greatest common factor.

a. $x^2 + x$ **b.** $x(x - 3) + 2(x - 3)$

Solution

a. The GCF of x^2 and x is x.

$$x^2 + x = x \cdot x + x \cdot 1 \quad \text{Note that we write } x \text{ as } x \cdot 1.$$
$$= x(x + 1) \quad \text{Distributive property.}$$

This factorization is special because one of the terms is the GCF itself. When this happens, it is helpful to express that term as GCF · 1 before applying the distributive property because the 1 is needed in the answer. It is especially important in problems like this to check that multiplying out your answer produces the original polynomial.

Part **a** illustrates a type of problem that many students find strange. It is worth discussing carefully.

Progress Check Answers

3. (a) $2(8x - 5)$ (b) $4y(4y^3 + y^2 - 3)$ (c) $2x^3y^3(9x^2 + 16y)$

b. The greatest common factor is the *binomial* $x - 3$. Factoring out the GCF using the distributive property gives

$$x(x-3) + 2(x-3) = \overset{\text{GCF}}{(x-3)}(x+2).$$

The answer $(x - 3)(x + 2)$ can be converted back to $x(x - 3) + 2(x - 3)$ by using the distributive property.

PROGRESS CHECK 4 Factor out the greatest common factor.

a. $18x^2 - 6x$ **b.** $x(x + 1) + 4(x + 1)$

The problem which opens this section shows an application of the two special situations discussed in Example 4.

EXAMPLE 5 Solve the problem in the section introduction on page 119.

Solution We will factor out the GCF, which is $(x + 0.20x)$, and then simplify.

$$\begin{aligned} &(x + 0.20x) - 0.20(x + 0.20x) \\ &= 1(x + 0.20x) - 0.20(x + 0.20x) && \text{GCF} = 1 \cdot \text{GCF.} \\ &= (x + 0.20x)(1 - 0.20) && \text{Factor out the GCF.} \\ &= (1.20x)(0.80) && \text{Combine like terms.} \\ &= 0.96x && \text{Simplify.} \end{aligned}$$

The final price is 96 percent of the original price.

PROGRESS CHECK 5 If a storekeeper reduces a price by 50 percent and then increases this new price the following month by 50 percent, the final price can be represented as $(x - 0.50x) + 0.50(x - 0.50x)$, where x is the original price. Factor out the GCF, $x - 0.50x$; then simplify and interpret the result. Is the final price more or less than the original price?

3 Some polynomials where the GCF of all the terms is 1 can still be factored by a method called **factoring by grouping.** For example, to factor

$$mx + my + nx + ny,$$

we first arrange the terms into smaller groups which have a useful GCF, and we rewrite the expression as

$$(mx + my) + (nx + ny).$$

When we factor out the common factor in each group, we have

$$m(x + y) + n(x + y).$$

Now we can factor out the common binomial factor $x + y$, giving

$$(x + y)(m + n)$$

as the factored version of the original polynomial. You might check also that if we had grouped the original expression as

$$(mx + nx) + (my + ny),$$

we would have obtained the same answer.

EXAMPLE 6 Factor by grouping.

a. $6x^2 + 12x + 7x + 14$ **b.** $x^2 - 3x - 3x + 9$

Progress Check Answers

4. (a) $6x(3x - 1)$ (b) $(x + 1)(x + 4)$
5. $0.75x$; less

Solution

a. $6x^2 + 12x + 7x + 14$

$$= \overbrace{(6x^2 + 12x)}^{\text{common factor } 6x} + \overbrace{(7x + 14)}^{\text{common factor } 7}$$ Group terms with common factors.

$= 6x(x + 2) + 7(x + 2)$ Factor in each group.

$= (x + 2)(6x + 7)$ Factor out the common binomial factor.

b. In the following solution we choose to factor out -3 instead of 3 from the group on the right so that we get the same binomial factor in both groups.

$x^2 - 3x - 3x + 9$

$= (x^2 - 3x) + (-3x + 9)$ Group terms with a common factor.

$= x(x - 3) - 3(x - 3)$ Factor in each group.

$= (x - 3)(x - 3)$ Factor out the common binomial factor.

$= (x - 3)^2$ Simplify the notation.

PROGRESS CHECK 6 Factor by grouping.

a. $2x^2 + 10x + 3x + 15$ **b.** $x^2 - 4x - 4x + 16$

Progress Check Answers

6. (a) $(2x + 3)(x + 5)$ (b) $(x - 4)^2$

EXERCISES 3.5

In Exercises 1–8, find the greatest common factor (GCF) of the given expressions.

1. 54, 66 6

2. 54, 72 18

3. $2^2 \cdot 3 \cdot 5$, $2 \cdot 3^2 \cdot 7$ 6

4. $2^3 \cdot 3^2 \cdot 5$, $2^2 \cdot 3^3 \cdot 5^2$ 180

5. $12x^3$, $4x^2$, $6x$ $2x$

6. $18y$, $24y^2$, $32y^3$ $2y$

7. $14xy$, $21x^2$ $7x$

8. $36x^2y^2$, $18xy$ $18xy$

In Exercises 9–38, factor out the GCF.

9. $6x + 9$ $3(2x + 3)$

10. $8y + 4$ $4(2y + 1)$

11. $12x^2 + 8x$ $4x(3x + 2)$

12. $9y^2 - 12y$ $3y(3y - 4)$

13. $3x^2y + 5xy^2$ $xy(3x + 5y)$

14. $2s^3t + 4s^2t^2$ $2s^2t(s + 2t)$

15. $-4x^2y + 6x^3y$ $2x^2y(-2 + 3x)$

16. $-9xy^2 + 6xy^3$ $3xy^2(-3 + 2y)$

17. $m^2 - m$ $m(m - 1)$

18. $n + n^2$ $n(1 + n)$

19. $3x^2 + 6x^3 + 9x^4$ $3x^2(1 + 2x + 3x^2)$

20. $12y^4 + 8y^3 - 4y^2$ $4y^2(3y^2 + 2y - 1)$

21. $12x - 16x^2 - 8x^3$ $4x(3 - 4x - 2x^2)$

22. $15x^3 - 20x^2 - 25x$ $5x(3x^2 - 4x - 5)$

23. $-n^6 + n^4 - n^2$ $n^2(-n^4 + n^2 - 1)$

24. $-2a^4 + 3a^5 + 4a^6$ $a^4(-2 + 3a + 4a^2)$

25. $8xy^3 + 64x^2y^2 + 8x^3y$ $8xy(y^2 + 8xy + x^2)$

26. $3m^3n - 9m^2n^2 + 3mn^3$ $3mn(m^2 - 3mn + n^2)$

27. $9ab^2c^5 - 36a^2b^3c^4 - 81a^3b^4c^3$ $9ab^2c^3(c^2 - 4abc - 9a^2b^2)$

28. $8p^4q^7r^4 - 4p^5q^8r^3 + 2p^6q^9r^2$ $2p^4q^7r^2(4r^2 - 2pqr + p^2q^2)$

29. $x(x + 3) + 7(x + 3)$ $(x + 3)(x + 7)$

30. $5(x - 8) - x(x - 8)$ $(x - 8)(5 - x)$

31. $3y^2(x - 1) + 2(x - 1)$ $(x - 1)(3y^2 + 2)$

32. $t(2s + 3) - r(2s + 3)$ $(2s + 3)(t - r)$

33. $(x + 1)(x - 2) + (x + 3)(x - 2)$ $(x - 2)(2x + 4)$

34. $(x + 4)(y + 3) + (x + 4)(y - 3)$ $(x + 4)(2y)$

35. $(x + 4)(y + 3) - (x + 4)(y - 3)$ $6(x + 4)$

36. $(x + 10)(y - 9) - 9(x + 10)$ $(x + 10)(y - 18)$

37. $(x + 1)^3 - 2(x + 1)^2 + 3(x + 1)$
$(x + 1)[(x + 1)^2 - 2(x + 1) + 3]$

38. $(a + b)^4 - 3(a + b)^3 - 5(a + b)^2$
$(a + b)^2[(a + b)^2 - 3(a + b) - 5]$

In Exercises 39–50, factor by grouping.

39. $2x^2 + 4x + 3x + 6$ $(x + 2)(2x + 3)$

40. $3x^2 + 9x + 2x + 6$ $(x + 3)(3x + 2)$

41. $15x^2 + 5x - 12x - 4$ $(3x + 1)(5x - 4)$

42. $12x^2 - 18x - 10x + 15$ $(2x - 3)(6x - 5)$

43. $y^2 - 5y - 5y + 25$ $(y - 5)^2$

44. $y^2 + 6y + 6y + 36$ $(y + 6)^2$

45. $9a^2 + 12ab + 12ab + 16b^2$ $(3a + 4b)^2$

46. $25c^2 - 10cd - 10cd + 4d^2$ $(5c - 2d)^2$

47. $6m^2 + 2mn - 3mn - n^2$ $(2m - n)(3m + n)$

48. $6p^2 - 8pq - 9pq + 12q^2$ $(2p - 3q)(3p - 4q)$

49. $1 + x + a + ax$ $(1 + a)(1 + x)$

50. $4 - 5a - 4b + 5ab$ $(4 - 5a)(1 - b)$

51. If a storekeeper raises a price 30 percent and one week later cuts this new price by 30 percent, the final price can be represented as $(x + 0.30x) - 0.30(x + 0.30x)$, where x is the original price. Factor out $x + 0.30x$; then simplify and interpret the result. Is the final price more or less than the original price? $0.91x$; less

52. If a storekeeper raises a price 70 percent and one week later cuts this new price by 70 percent, the final price can be represented as $(x + 0.70x) - 0.70(x + 0.70x)$, where x is the original price. Factor out $x + 0.70x$; then simplify and interpret the result. Is the final price more or less than the original price? $0.51x$; less

53. What is the overall result of raising a price by 10 percent and then raising that price by 20 percent? Factor out $x + 0.10x$ in $(x + 0.10x) + 0.20(x + 0.10x)$ and simplify to find out. $1.32x$; a 32 percent increase of the original price

54. What is the overall result of raising a price by 20 percent and then raising that price by 10 percent? Factor out $x + 0.20x$ in $(x + 0.20x) + 0.10(x + 0.20x)$ and simplify to find out. $1.32x$; a 32 percent increase of the original price

THINK ABOUT IT

1. What does the word *common* mean in *greatest common factor?*
2. True or false? The *lowest* common integer factor of any two positive integers is 1. Explain.
3. **a.** A pair of positive integers whose GCF is 1 is called *relatively prime.* Which of these pairs are relatively prime? (6,35), (2,21), (7,35).
   b. Is a pair of two different prime numbers also relatively prime?
4. Look up *factor* in the dictionary. What does it have to do with *factory?* English use of this word in algebra dates back to the seventeenth century.
5. Find two expressions whose greatest common factor is the following.
   a. $5x^2$
   b. $3x - 1$

REMEMBER THIS

1. True or false? The product of the two binomials $x + 4$ and $x + 5$ is a trinomial. True
2. Find a pair of integers whose product is 14 and whose sum is -9. $-7, -2$
3. Find a pair of integers whose product is -30 and whose sum is 13. $-2, 15$
4. True or false? If $a = 1$, then $ax^2 + bx + c$ is the same as $x^2 + bx + c$. True
5. Which of these numbers are not integers? $-2, 0, \sqrt{4}, \frac{3}{5}, \sqrt{5}$. $\frac{3}{5}, \sqrt{5}$
6. Expand $(3x - 4)^2$. $9x^2 - 24x + 16$
7. For the figures shown, the difference between the areas is 7. Find the larger area. 15

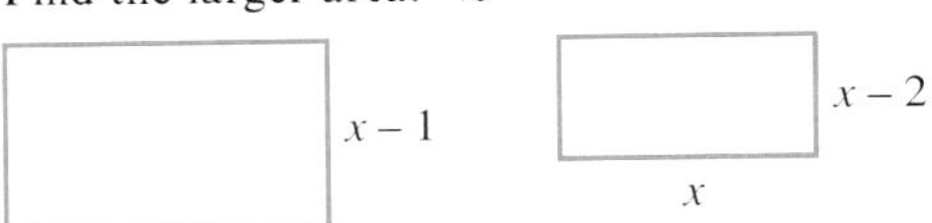

8. Simplify $\left(\frac{-2x^3y}{3xy^{-2}}\right)^{-2}$. Write the result using only positive exponents. $\frac{9}{4x^4y^6}$
9. Evaluate $E = \frac{Pr}{1 - (1 + r)^{-n}}$ if $P = 20{,}000$, $r = 0.01$, and $n = 60$. This gives the monthly payment for repaying a five-year loan at 12 percent annual interest. \$444.89
10. Graph the solution set of $-2 \le 3x + 4 \le 13$.

3.6 Factoring Trinomials

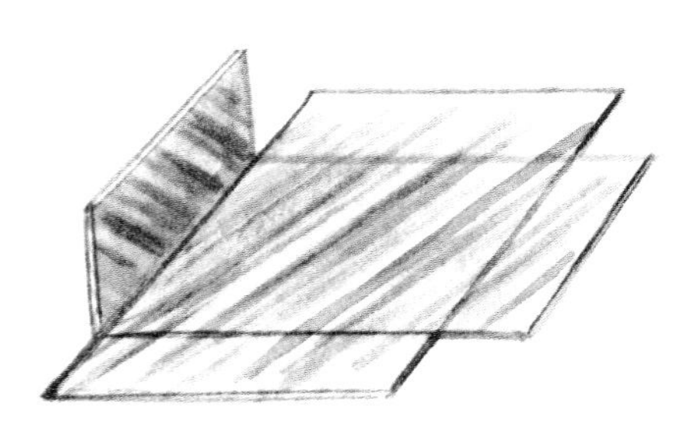

A thin sheet of metal forms a square with sides equal to 12 inches. The corners will be cut out as shown, and the sides will be folded up to make an open box. To find out what size corners we need to cut out to get the box with maximum volume, we can solve $12x^2 - 96x + 144 = 0$. Factor the left side of this equation. (See Example 6.)

OBJECTIVES

1. **Factor $ax^2 + bx + c$ with $a = 1$.**
2. **Factor $ax^2 + bx + c$ with $a \neq 1$ by reversing FOIL.**
3. **Factor $ax^2 + bx + c$ with $a \neq 1$ by the *ac* method.**
4. **Factor trinomials by substitution.**

1 Since factoring is the reverse of multiplying, we shall see that factoring trinomials with integer coefficients can often be related to the multiplication of two binomials with integer coefficients. Notice what happens when we multiply binomials of the form $x + m$ and $x + n$.

$$(x + m)(x + n) = x^2 + nx + mx + mn$$

At this point there are four terms, but the two middle terms have x as a common factor and can be combined to give

$$(x + m)(x + n) = x^2 + (m + n)x + mn,$$

which is a trinomial. This equation provides a model for factoring trinomials of the form $ax^2 + bx + c$ when $a = 1$. In Examples 1–4 we demonstrate how to apply the model, but note that we shall be interested only in binomial factors with *integer* coefficients.

"Think About It" Exercise 5 discusses factoring over the rationals.

EXAMPLE 1 Factor $x^2 + 9x + 14$.

Solution We match the given trinomial to the factoring model.

$$\begin{aligned} x^2 + (m + n)x + mn &= (x + m)(x + n) \\ x^2 + \quad 9x \quad + 14 &= (x + ?)(x + ?) \end{aligned}$$

The question marks stand for two integers m and n such that $mn = 14$ and $m + n = 9$. We must therefore find two integers whose product is 14 and whose sum is 9. Because both 14 and 9 are positive, we need consider only positive possibilities for m and n. It is easier to work first with the product, because there are fewer ways to make a product than a sum using integers.

Possible Factor Pairs of 14	Associated Sums
$1 \cdot 14 = 14$	$1 + 14 = 15$
$2 \cdot 7 = 14$	$2 + 7 = 9$

The combination of 2 and 7 is the required pair. We replace the question marks by 2 and 7 and check the solution further by a FOIL multiplication. We get

$$\begin{aligned} (x + 2)(x + 7) &= x^2 + 2x + 7x + 14 \\ &= x^2 + 9x + 14, \end{aligned}$$

which is the original trinomial. Thus, $x^2 + 9x + 14$ factors as $(x + 2)(x + 7)$. According to the commutative property of multiplication, it is also correct to write $(x + 7)(x + 2)$.

PROGRESS CHECK 1 Factor $x^2 + 7x + 10$.

EXAMPLE 2 Factor $x^2 + 10 - 11x$.

Solution As a preliminary step, we write the terms in order of decreasing powers of x to make it easier to use the factoring model.

$$x^2 + 10 - 11x = x^2 - 11x + 10$$

Referring to the model, we therefore want to find two integers whose product is 10 and whose sum is -11. Because the product is positive and the sum is negative, we need consider only negative values for m and n.

Progress Check Answer

1. $(x + 2)(x + 5)$

Possible Factor Pairs of 10	Associated Sums
$(-1)(-10) = 10$	$-1 + (-10) = -11$
$(-2)(-5) = 10$	$-2 + (-5) = -7$

The required integers are -1 and -10, so

$$x^2 - 11x + 10 = (x - 1)(x - 10).$$

You should check this factorization through multiplication.

PROGRESS CHECK 2 Factor $x^2 - 13x + 12$.

From these examples observe that when you factor $ax^2 + bx + c$ with $a = 1$ into $(x + m)(x + n)$, *m and n will have the same sign when c is positive.* Furthermore, their sign will match the sign of b. In Example 3 we consider the case when the last term, c, in the trinomial is negative.

EXAMPLE 3 Factor $t^2 + 13t - 30$.

Solution The variable is t instead of x, but the form of the problem still matches the factoring model. We therefore seek two integers whose product is -30 and whose sum is 13. Because the product is negative, m and n must have *opposite* signs.

Possible Factor Pairs of -30	Associated Sums
$1(-30) = -30$	$1 + (-30) = -29$
$-1(30) = -30$	$-1 + 30 = 29$
$2(-15) = -30$	$2 + (-15) = -13$
$-2(15) = -30$	$-2 + 15 = 13$
$3(-10) = -30$	$3 + (-10) = -7$
$-3(10) = -30$	$-3 + 10 = 7$
$5(-6) = -30$	$5 + (-6) = -1$
$-5(6) = -30$	$-5 + 6 = 1$

The integers -2 and 15 are the required pair, so

$$t^2 + 13t - 30 = (t - 2)(t + 15).$$

A good classroom exercise here is to have one student secretly multiply two binomials of his or her own choice and present the product to the class for factoring.

PROGRESS CHECK 3 Factor $s^2 - 4s - 12$.

In the next example we extend this method to factor trinomials with two variables.

EXAMPLE 4 Factor $x^2 - 3ax - 10a^2$.

Solution The factorization fits the form $(x + m)(x + n)$, where m and n are two *polynomials* whose product is $-10a^2$ and whose sum is $-3a$. Because of the negative sign in the product, m and n must have opposite signs.

Possible Factor Pairs of $-10a^2$	Associated Sums
$a(-10a) = -10a^2$	$a + (-10a) = -9a$
$-a(10a) = -10a^2$	$-a + 10a = 9a$
$2a(-5a) = -10a^2$	$2a + (-5a) = -3a$
$-2a(5a) = -10a^2$	$-2a + 5a = 3a$

Progress Check Answers

2. $(x - 12)(x - 1)$
3. $(s - 6)(s + 2)$

The required polynomials are $2a$ and $-5a$, which we check through multiplication.

$$\begin{aligned}(x + 2a)(x - 5a) &= x^2 - 5ax + 2ax - 10a^2\\ &= x^2 - 3ax - 10a^2\end{aligned}$$

Thus, $x^2 - 3ax - 10a^2$ factors into $(x + 2a)(x - 5a)$.

PROGRESS CHECK 4 Factor $x^2 - 3bx - 28b^2$.

Even though all the preceding examples were factorable, many trinomials cannot be factored. For these trinomials it is impossible to find a pair of integers which give the desired sum and product. For example, let us try to factor $x^2 + 2x + 5$. Because the constant term is positive and the sign of the middle term is positive, we look for two positive integers whose product is 5 and whose sum is 2.

Possible Factor Pairs of 5	Associated Sums
$1 \cdot 5 = 5$	$1 + 5 = 6$

The only pair of integers whose product is 5 is $(1)(5)$, but this does not have the desired sum of 2. This means that $x^2 + 2x + 5$ cannot be factored using integer coefficients. A polynomial like this which cannot be factored into two polynomial factors of positive degree with integer coefficients is called a **prime polynomial** over the set of integers.

EXAMPLE 5 Show that $y^2 + 2y - 6$ is a prime polynomial.

Solution We look for two integers whose product is -6 and whose sum is 2.

Possible Factor Pairs of -6	Associated Sums
$1(-6) = -6$	$1 + (-6) = -5$
$-1(6) = -6$	$-1 + 6 = 5$
$2(-3) = -6$	$2 + (-3) = -1$
$-2(3) = -6$	$-2 + 3 = 1$

Upon inspection, we find that no pair of integers satisfies the requirements. Thus, $y^2 + 2y - 6$ is a prime polynomial.

PROGRESS CHECK 5 Show that $y^2 - 12y - 20$ is a prime polynomial.

In some problems, factoring procedures should be applied more than once. The phrase *factor completely* may be used as a reminder to continue factoring until none of the factors can be factored again. We illustrate this by solving the section-opening problem.

EXAMPLE 6 Solve the section-opening problem on page 124.

Solution As a general strategy for the first step, any common factors should be factored out.

$$\begin{aligned}12x^2 - 96x + 144 &= 12(x^2 - 8x + 12) && \text{Factor GCF.}\\ &= 12(x - 6)(x - 2) && -6 \text{ and } -2 \text{ have product 12 and sum } -8.\end{aligned}$$

PROGRESS CHECK 6 Factor completely $4y^4 - 28y^3 + 48y^2$.

Examples 1–6 all match a model where the coefficient of x^2 is one. But when this coefficient does not equal one, the trinomial $ax^2 + bx + c$ is harder to factor because it is more difficult to see how to get the middle term. To see why this is so, look at the following product.

$$(px + m)(qx + n) = (pq)x^2 + (pn + qm)x + mn$$

Progress Check Answers

4. $(x - 7b)(x + 4b)$
5. There are no integers whose product is -20 and whose sum is -12.
6. $4y^2(y - 4)(y - 3)$

In this model the factors for the last term, m and n, do not now necessarily sum to the middle coefficient, which instead of being just $m + n$ is now $pn + qm$. We will show two methods for factoring trinomials where a does not equal one. In different situations you may prefer one method to the other.

2 Method 1 for factoring trinomials: reversing FOIL

This method of factoring is based on the fact that the middle term in the trinomial to be factored is the sum of the outer product and the inner product from the FOIL multiplication. For simplicity, we will factor terms like $3x^2$ into $3x$ and x, rather than $-3x$ and $-x$, because this will allow us to carry over the sign restrictions for m and n discussed earlier.

EXAMPLE 7 Factor $2x^2 + 7x + 6$.

Solution By reversing FOIL, the first term $2x^2$ is the result of multiplying the first terms in the FOIL method. Thus, a possible factorization begins

$$2x^2 + 7x + 6 = (2x + m)(x + n).$$

The constant term, 6, is the result of multiplying the last terms in the FOIL method, so m and n must have a product equal to 6. To get this, we might use 6 and 1, or 3 and 2. We don't need to try negative integers for m and n because the middle term has a positive coefficient. Thus,

$$2x^2 + 7x + 6 \stackrel{?}{=} \begin{array}{c}(2x + 6)(x + 1)\\ \text{or}\\ (2x + 1)(x + 6)\\ \text{or}\\ (2x + 3)(x + 2)\\ \text{or}\\ (2x + 2)(x + 3)\end{array}$$

The middle term is the sum of the inner and outer products.

$$(2x + 6)(x + 1): \quad 2x + 6x = 8x \qquad (2x + 1)(x + 6): \quad 12x + x = 13x$$

$$(2x + 3)(x + 2): \quad 4x + 3x = 7x \qquad (2x + 2)(x + 3): \quad 6x + 2x = 8x$$

Thus, the correct factorization is $2x^2 + 7x + 6 = (2x + 3)(x + 2)$.

PROGRESS CHECK 7 Factor $3x^2 + 17x + 10$.

In Example 7 we did not really have to check two of the possible factorizations. It was not necessary to test $(2x + 6)(x + 1)$ and $(2x + 2)(x + 3)$. The reason is that 2 is a common factor of $2x + 6$ in the first case and of $2x + 2$ in the second, so

$$(2x + 6)(x + 1) = 2(x + 3)(x + 1)$$
$$\text{and} \quad (2x + 2)(x + 3) = 2(x + 1)(x + 3).$$

These factorizations imply that 2 is a common factor of the original trinomial, $2x^2 + 7x + 6$, which is not the case. It is only necessary to test factorizations that are possible according to the following guideline.

Progress Check Answer

7. $(3x + 2)(x + 5)$

Requirement for Binomial Factors

If the terms of a trinomial have no common integer factors (except 1 and -1), then the same is true for both of its binomial factors.

EXAMPLE 8 Factor $10x^2 - 9x + 2$.

Solution To start, we notice that the terms of the trinomial have no common factor besides 1 and -1. The term $10x^2$ is the result of multiplying the first terms in the FOIL method. Thus,

$$10x^2 - 9x + 2 \stackrel{?}{=} \begin{matrix} (10x + ?)(x + ?) \\ \text{or} \\ (5x + ?)(2x + ?). \end{matrix}$$

The constant term, 2, is the result of multiplying the two last terms in the FOIL method, and the only way to get the product 2 is with the factors $(-1)(-2)$. We can't use $(1)(2)$ because the middle coefficient term is negative.

$$10x^2 - 9x + 2 \stackrel{?}{=} \begin{matrix} (10x - 1)(x - 2) \\ \text{or} \\ (10x - 2)(x - 1) \\ \text{or} \\ (5x - 1)(2x - 2) \\ \text{or} \\ (5x - 2)(2x - 1) \end{matrix}$$

Using the requirement for binomial factors, we can eliminate $(10x - 2)(x - 1)$ and $(5x - 1)(2x - 2)$, because they have binomial factors with a common factor of 2 (and we know that 2 is not a common factor of the original trinomial). So we need to test only the two remaining possibilities to see if either gives a middle term equal to $-9x$.

$$(10x - 1)(x - 2) \qquad\qquad (5x - 2)(2x - 1)$$
$$-20x - x = -21x \qquad\qquad -5x - 4x = -9x$$

Thus, $10x^2 - 9x + 2 = (5x - 2)(2x - 1)$.

PROGRESS CHECK 8 Factor $6x^2 - 7x + 2$.

EXAMPLE 9 Factor completely $12x^3 + 10x^2 - 8x$.

Solution First, we can factor out the GCF, which is $2x$.

$$12x^3 + 10x^2 - 8x = 2x(6x^2 + 5x - 4)$$

Now suppose we try to factor $6x^2 + 5x - 4$ as $(2x + 1)(3x - 4)$.

$$(2x + 1)(3x - 4)$$
$$-8x + 3x = -5x$$

This result is the *opposite* of the middle term we want, so we switch the signs on the factors of -4, from $(1)(-4)$ to $(-1)(4)$.

$$(2x - 1)(3x + 4)$$
$$8x - 3x = 5x$$

Progress Check Answer

8. $(3x - 2)(2x - 1)$

Thus, $6x^2 + 5x - 4 = (2x - 1)(3x + 4)$, and so the complete factorization is

$$12x^3 + 10x^2 - 8x = 2x(2x - 1)(3x + 4).$$

PROGRESS CHECK 9 Factor completely $18x^4 - 33x^3 - 6x^2$.

When trinomials have a negative leading term, it simplifies matters to first factor out -1 and then proceed as in our previous examples.

EXAMPLE 10 Factor $-5x^2 + 7x + 6$.

Solution First, we factor out -1 to make the sign of the squared term positive.

$$-5x^2 + 7x + 6 = -1(5x^2 - 7x - 6)$$

You can check that $5x^2 - 7x - 6 = (5x + 3)(x - 2)$. Therefore, the complete factorization is

$$-5x^2 + 7x + 6 = -1(5x + 3)(x - 2) \qquad \text{or} \qquad -(5x + 3)(x - 2).$$

PROGRESS CHECK 10 Factor $-3x^2 + 11x + 20$.

3 Method 2 for factoring trinomials: *ac* method

Because this is a generalization of the previous method, it can be used even when $a = 1$. It is particularly valuable for showing the importance of the distributive property.

The *ac* method for factoring $ax^2 + bx + c$ often involves less trial and error than the previous approach, particularly if you cannot mentally eliminate many of the possibilities. It is called the *ac* method because it refers to the coefficients in the trinomial model that are called a and c. Note also that the *ac* method relies on *factoring by grouping,* which we discussed in Section 3.5.

EXAMPLE 11 Factor $5x^2 + 9x + 4$ by the *ac* method.

Solution Follow the steps below.

Step 1 **Find two integers whose product is *ac* and whose sum is *b*.**
If this is not possible, then the trinomial cannot be factored.

For $5x^2 + 9x + 4$, we have $a = 5$, $b = 9$, and $c = 4$. Thus, $ac = 5 \cdot 4 = 20$. We look for two integers whose product is 20 and whose sum is 9. Because the product and sum are both positive, we need only consider positive integers.

Possible Factor Pairs of 20	Associated Sums
$1 \cdot 20 = 20$	$1 + 20 = 21$
$2 \cdot 10 = 20$	$2 + 10 = 12$
$4 \cdot 5 = 20$	$4 + 5 = 9$

The required integers are 4 and 5.

Step 2 **Replace *b* by the sum of the two integers from step 1 and then distribute *x*.**

$$\begin{aligned} &5x^2 + 9x + 4 \\ &= 5x^2 + (4 + 5)x + 4 && \text{Replace 9 by 4 + 5.} \\ &= 5x^2 + 4x + 5x + 4 && \text{Distributive property.} \end{aligned}$$

Step 3 **Factor by grouping.**

$$\begin{aligned} &5x^2 + 4x + 5x + 4 \\ &= (5x^2 + 4x) + (5x + 4) && \text{Group terms with common factors.} \\ &= x(5x + 4) + 1(5x + 4) && \text{Factor in each group.} \\ &= (5x + 4)(x + 1) && \text{Factor out the common binomial factor.} \end{aligned}$$

Progress Check Answers

9. $3x^2(6x + 1)(x - 2)$
10. $-(3x + 4)(x - 5)$

Thus, $5x^2 + 9x + 4 = (5x + 4)(x + 1)$. You can check that this is correct through multiplication.

Note You could also have grouped as follows.

$$\begin{aligned} 5x^2 + 9x + 4 &= (5x^2 + 5x) + (4x + 4) \\ &= 5x(x + 1) + 4(x + 1) \\ &= (x + 1)(5x + 4) \end{aligned}$$

Both answers are correct.

PROGRESS CHECK 11 Factor $2x^2 + 11x + 9$ by the *ac* method.

EXAMPLE 12 Factor $12x^2 + 31x - 30$.

Solution Because there are so many factors of 12 and 30, trial and error may take a long time, and so we use the *ac* method.

Step 1 For this trinomial, $a = 12$, $b = 31$, and $c = -30$. Thus, $ac = 12(-30) = -360$. We look for two integers whose product is -360 and whose sum is 31. With a little trial and error, we find that they are -9 and 40.

Step 2 Replace 31 by $-9 + 40$ and then distribute x.

$$\begin{aligned} &12x^2 + 31x - 30 \\ &= 12x^2 + (-9 + 40)x - 30 && \text{Replace 31 by } -9 + 40. \\ &= 12x^2 - 9x + 40x - 30 && \text{Distributive property.} \end{aligned}$$

Step 3 Factor by grouping.

$$\begin{aligned} &(12x^2 - 9x) + (40x - 30) \\ &= 3x(4x - 3) + 10(4x - 3) && \text{Factor in each group.} \\ &= (4x - 3)(3x + 10) && \text{Factor out the common binomial factor.} \end{aligned}$$

Thus, $12x^2 + 31x - 30 = (4x - 3)(3x + 10)$. Check this result.

PROGRESS CHECK 12 Factor $12x^2 + 19x - 18$.

4 As will be discussed more fully in Chapter 8, a polynomial expressed in the form $ax^2 + bx + c$ $(a \neq 0)$, where the highest power of x is 2, is called a **quadratic form** in x. The variable x can be replaced by other expressions without changing this form. For example, $(t - 5)^2 + 8(t - 5) + 7$ is a quadratic form in $t - 5$. Similarly, $2x^4 + 3x^2 - 9$, which is the same as $2(x^2)^2 + 3(x^2) - 9$, is a quadratic form in x^2. Such quadratic forms can often be factored by the methods of this section.

EXAMPLE 13 Factor.

a. $(t - 5)^2 + 8(t - 5) + 7$ **b.** $2x^4 + 3x^2 - 9$

Solution

a. $(t - 5)^2 + 8(t - 5) + 7$ fits the form $x^2 + 8x + 7$ with $x = t - 5$. Because $x^2 + 8x + 7$ factors as $(x + 7)(x + 1)$, we substitute $t - 5$ for x to get

$$\begin{aligned} (t - 5)^2 + 8(t - 5) + 7 &= [(t - 5) + 7][(t - 5) + 1] \\ &= (t + 2)(t - 4). \end{aligned}$$

b. $2x^4 + 3x^2 - 9$ fits the form $2y^2 + 3y - 9$ with $y = x^2$. Then, factoring $2y^2 + 3y - 9$ gives $(2y - 3)(y + 3)$; so by replacing y with x^2, we have

$$2x^4 + 3x^2 - 9 = (2x^2 - 3)(x^2 + 3).$$

Note By "seeing" the quadratic form, you may be able to factor directly without writing the substitution steps.

Progress Check Answers

11. $(2x + 9)(x + 1)$
12. $(4x + 9)(3x - 2)$

Progress Check Answers

13. (a) $(t - 6)(t - 5)$ (b) $(3x^2 + 1)(x^2 - 2)$

PROGRESS CHECK 13 Factor.

a. $(t - 3)^2 - 5(t - 3) + 6$ **b.** $3x^4 - 5x^2 - 2$

EXERCISES 3.6

In Exercises 1–30, factor completely the given trinomial.

1. $x^2 + 8x + 15$ $(x + 3)(x + 5)$
2. $x^2 + 11x + 28$ $(x + 4)(x + 7)$
3. $y^2 + 10y + 9$ $(y + 1)(y + 9)$
4. $y^2 + 13y + 12$ $(y + 1)(y + 12)$
5. $t^2 + 8t + 12$ $(t + 6)(t + 2)$
6. $t^2 + 11t + 18$ $(t + 9)(t + 2)$
7. $x^2 - 8x + 15$ $(x - 3)(x - 5)$
8. $x^2 - 16x + 15$ $(x - 15)(x - 1)$
9. $y^2 - 25y + 24$ $(y - 24)(y - 1)$
10. $y^2 - 14y + 24$ $(y - 12)(y - 2)$
11. $c^2 - 9c + 18$ $(c - 3)(c - 6)$
12. $c^2 - 7c + 6$ $(c - 1)(c - 6)$
13. $x^2 + 5x - 24$ $(x - 3)(x + 8)$
14. $x^2 - x - 20$ $(x - 5)(x + 4)$
15. $y^2 - 2y - 8$ $(y - 4)(y + 2)$
16. $y^2 + 2y - 24$ $(y - 4)(y + 6)$
17. $s^2 - s - 6$ $(s - 3)(s + 2)$
18. $s^2 + s - 30$ $(s - 5)(s + 6)$
19. $x^2 + 3ax + 2a^2$ $(x + a)(x + 2a)$
20. $x^2 + 6ax + 8a^2$ $(x + 2a)(x + 4a)$
21. $y^2 - 7by + 12b^2$ $(y - 4b)(y - 3b)$
22. $y^2 - 10by + 16b^2$ $(y - 2b)(y - 8b)$
23. $d^2 + bd - 2b^2$ $(d - b)(d + 2b)$
24. $d^2 + bd - 6b^2$ $(d - 2b)(d + 3b)$
25. $2x^2 + 16x + 30$ $2(x + 3)(x + 5)$
26. $3x^2 + 18x + 24$ $3(x + 2)(x + 4)$
27. $4y^2 - 12y + 8$ $4(y - 1)(y - 2)$
28. $5y^2 - 20y + 15$ $5(y - 1)(y - 3)$
29. $6z^2 - 12z - 48$ $6(z + 2)(z - 4)$
30. $4z^2 + 8z - 60$ $4(z + 5)(z - 3)$

In Exercises 31–34, find any trinomials which are prime polynomials.

31. $x^2 + 2x + 1, x^2 + 2x + 2, x^2 + 3x + 2$ $x^2 + 2x + 2$
32. $x^2 + 6x + 8, x^2 + 9x + 8, x^2 + 8x + 9$ $x^2 + 8x + 9$
33. $y^2 - 3y + 3, y^2 - 3y + 2, y^2 - 4y + 3$ $y^2 - 3y + 3$
34. $y^2 - 4y + 4, y^2 - 5y + 4, y^2 - 4y + 5$ $y^2 - 4y + 5$

In Exercises 35–46, factor completely.

35. $15x^2 + 31x + 14$ $(3x + 2)(5x + 7)$
36. $6x^2 + 19x + 15$ $(3x + 5)(2x + 3)$
37. $6t^2 - 7t + 2$ $(2t - 1)(3t - 2)$
38. $9c^2 - 21c + 10$ $(3c - 5)(3c - 2)$
39. $4y^2 - 4y - 15$ $(2y + 3)(2y - 5)$
40. $4m^2 + 13m - 12$ $(4m - 3)(m + 4)$
41. $16x^3 - 4x^2 - 6x$ $2x(2x + 1)(4x - 3)$
42. $6x^3 + 14x^2 + 4x$ $2x(x + 2)(3x + 1)$
43. $3y^4 - 9y^3 + 6y^2$ $3y^2(y - 1)(y - 2)$
44. $6y^4 - 15y^3 - 9y^2$ $3y^2(y - 3)(2y + 1)$
45. $12a^3b + 20a^2b^2 + 8ab^3$ $4ab(3a + 2b)(a + b)$
46. $24a^3b - 4a^2b^2 - 20ab^3$ $4ab(a - b)(6a + 5b)$

In Exercises 47–60 the given polynomial is quadratic in form. Factor each one.

47. $(t + 3)^2 + 6(t + 3) + 8$ $(t + 5)(t + 7)$
48. $(t - 1)^2 + 5(t - 1) + 6$ $(t + 2)(t + 1)$
49. $6(r + 2)^2 - 11(r + 2) - 10$ $(2r - 1)(3r + 8)$
50. $12(r - 5)^2 + 17(r - 5) - 5$ $(4r - 21)(3r - 10)$
51. $4(x - 2)^2 - 8(x - 2) + 3$ $(2x - 5)(2x - 7)$
52. $9(x - 3)^2 - 18(x - 3) + 8$ $(3x - 11)(3x - 13)$
53. $y^4 + 8y^2 + 15$ $(y^2 + 3)(y^2 + 5)$
54. $y^4 + 11y^2 + 30$ $(y^2 + 5)(y^2 + 6)$
55. $6t^4 - 29t^2 + 9$ $(3t^2 - 1)(2t^2 - 9)$
56. $6t^4 - 19t^2 + 10$ $(3t^2 - 2)(2t^2 - 5)$
57. $x^2y^2 + 3xy + 2$ $(xy + 1)(xy + 2)$
58. $x^2y^2 + 5xy + 6$ $(xy + 2)(xy + 3)$
59. There are two numbers for which the number plus twice its square equals 55. They can be found by solving $x + 2x^2 - 55 = 0$. Factor the left side of this equation. $(2x + 11)(x - 5) = 0$
60. A ball is thrown down from a roof 160 feet high with an initial velocity of 48 ft/sec. To find the number of seconds it takes to hit the ground you can solve $16t^2 + 48t - 160 = 0$. Factor the left side of this equation. $16(t + 5)(t - 2) = 0$

THINK ABOUT IT

1. **a.** Why is $2x^2 - 4x + 5$ called a trinomial?
   b. The expression abc has three factors. Is it a trinomial?
2. A trinomial of the form $ax^2 + bx + c$ is factorable over the integers if and only if $b^2 - 4ac$ is a perfect square. Use this rule to show that $2x^2 + 3x + 4$ cannot be factored but that $2x^2 + 3x - 2$ can be factored.
3. **a.** If $x^2 + bx + 11$ is factorable, what are the only two possible values of b?
   b. If $x^2 + bx + c$ is factorable and c is a prime number, what are the only possible values of b?
4. If you know one factor of a trinomial, you can find the other by division. Find the second factor of $6x^2 - 37x - 119$ if one factor is $3x + 7$. Then use the result to complete the equation $6x^2 - 37x - 119 = (3x + 7)(\quad)$.
5. In the text we said that $5x + 2y$ cannot be factored because we restricted factoring to the set of *integers*. But if you allow factoring over the set of *rational* numbers, then this expression can be factored. For example, you can factor out 5 to get $5x + 2y = 5(x + \frac{2}{5}y)$.
   a. Check this result by multiplication.
   b. Factor out 2 from $5x + 2y$.
   c. Factor out 4 from $5x + 2y$.
   d. If you allow factoring over the rationals, is there any limit to the number of ways the expression $5x + 2y$ can be factored?
   e. If you restrict factoring to the set of *integers*, is there any limit to the number of ways the expression $5x + 2y$ can be factored?

REMEMBER THIS

1. Expand $(x + a)^2$. $x^2 + 2ax + a^2$
2. Which of these equations is an identity?
   a. $(x + 3)^3 = x^3 + 3^3$ b. $(x + 3)^2 = x^2 + 3^2$
   c. $(x + 3)^1 = x^1 + 3^1$ d. $(x + 3)^0 = x^0 + 3^0$ c
3. Solve $(x + 3)^3 = x^3 + 3^3$. $\{-3, 0\}$
4. Multiply $(y + c)(y - c)$. $y^2 - c^2$
5. Multiply $(a - b)(a^2 + ab + b^2)$. $a^3 - b^3$
6. Factor out the GCF: $30x^4y + 24x^2y^3$. $6x^2y(5x^2 + 4y^2)$
7. Factor by grouping: $3x^2 + 15x + x + 5$. $(3x + 1)(x + 5)$
8. What property states that $3(5n)$ is equal to $(3 \cdot 5)n$? Associative property of multiplication
9. Simplify $1 - [2 - (3 - 4)]$. -2
10. \$50,000 is invested in two parts, one at 5 percent and one at 9 percent. The total return is \$4,100. How much is invested at 9 percent? \$40,000

3.7 Special Factoring Models

A golf ball is made with a smaller spherical core inside, as shown. If the radius of the golf ball is R and the radius of the core is r, then the part of the ball which is made up of the outer material is given by the difference of the volumes of the two spheres, $V = \frac{4}{3}\pi R^3 - \frac{4}{3}\pi r^3$. Express this formula differently by factoring it completely. (See Example 5.)

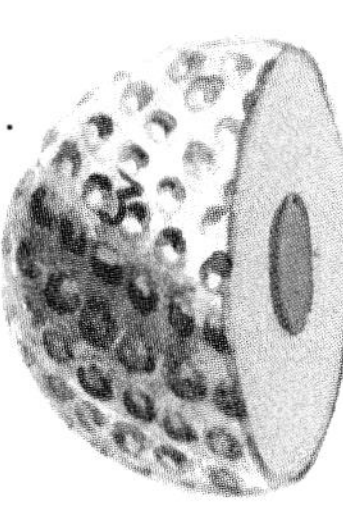

OBJECTIVES

1. **Identify and factor perfect square trinomials.**
2. **Factor the difference of two squares.**
3. **Factor the sum or difference of two cubes.**

Certain patterns appear so often in factoring problems that models for factoring them *should be memorized.* In this section we illustrate several such factoring formulas.

1 Perfect square trinomials

The factored form of $x^2 + 10x + 25$ is $(x + 5)(x + 5)$, which can also be written as $(x + 5)^2$. This means that the trinomial $x^2 + 10x + 25$ is the square of the binomial $x + 5$. A trinomial that is the square of a binomial is called a **perfect square trinomial.** You can factor a perfect square trinomial efficiently by using formulas we worked out earlier for the square of a binomial.

Factoring Models for Perfect Square Trinomials

$$a^2 + 2ab + b^2 = (a + b)^2$$
$$a^2 - 2ab + b^2 = (a - b)^2$$

These models may be useful if two terms in the trinomial to be factored are perfect squares. If they are, you should test the possibility that the trinomial is a perfect square trinomial, as shown in the next example.

EXAMPLE 1 Factor each trinomial.

a. $x^2 + 6x + 9$ **b.** $4x^2 - 12xy + 9y^2$ **c.** $9x^2 + 50x + 25$

Solution

a. Because x^2 is the square of x and $9 = 3^2$, we suspect that $x^2 + 6x + 9$ might be a perfect square trinomial that factors into $(x + 3)^2$. To test this, we must see if twice the product of x and 3 is the same as the middle term, $6x$, in the original trinomial.

$$2 \cdot x \cdot 3 \stackrel{?}{=} 6x$$
$$6x \stackrel{\checkmark}{=} 6x \quad \text{True}$$

Thus, the trinomial is a perfect square, and the factored form of $x^2 + 6x + 9$ is $(x + 3)^2$. We can check this result as shown below.

$$\begin{array}{ccccccccc} (a + b)^2 & = & a^2 & + & 2\ a\ \ b & + & b^2 & & \\ & & \downarrow & & \downarrow\ \ \downarrow & & \downarrow & & \\ (x + 3)^2 & = & x^2 & + & 2(x)(3) & + & 3^2 & = & x^2 + 6x + 9 \end{array}$$

b. Because $4x^2 = (2x)^2$ and $9y^2 = (3y)^2$, we suspect that $4x^2 - 12xy + 9y^2$ is a perfect square trinomial of the form $(a - b)^2$, taking $a = 2x$ and $b = 3y$. We just have to see if the middle term, $-12xy$, matches $-2ab$ in the model.

$$-2ab \stackrel{?}{=} -12xy$$
$$-2(2x)(3y) \stackrel{?}{=} -12xy \quad a = 2x \text{ and } b = 3y.$$
$$-12xy \stackrel{\checkmark}{=} -12xy \quad \text{True.}$$

Thus, the trinomial is a perfect square and $4x^2 - 12xy + 9y^2 = (2x - 3y)^2$.

c. Because $9x^2 = (3x)^2$ and $25 = 5^2$, we suspect that $9x^2 + 50x + 25$ might be a perfect square trinomial that factors into $(3x + 5)^2$. To check, we test to see if twice the product of $3x$ and 5 is the same as the middle term in the trinomial, $50x$.

$$2(3x)(5) \stackrel{?}{=} 50x$$
$$30x = 50x \quad \text{False}$$

This means that the trinomial is not a perfect square and that $(2x + 5)^2$ is not a correct factorization. By going back to reverse FOIL or the *ac* method, we can show, however, that

$$9x^2 + 50x + 25 = (9x + 5)(x + 5).$$

PROGRESS CHECK 1 Factor each trinomial.

a. $x^2 - 14x + 49$ **b.** $16x^2 - 40xy + 25y^2$ **c.** $x^2 + 20x + 36$

2 Difference of squares

Another useful pattern for factoring is based on the product of the sum and difference of the same two terms.

$$(a + b)(a - b) = a^2 - b^2$$

This pattern implies that the difference of two squares is always factorable.

Progress Check Answers

1. (a) $(x - 7)^2$ (b) $(4x - 5y)^2$ (c) $(x + 18)(x + 2)$

Factoring Model for a Difference of Squares

$$a^2 - b^2 = (a + b)(a - b)$$

Keep in mind that even powers of variables, like x^2, y^4, t^6, and so on, are perfect squares.

EXAMPLE 2 Factor each difference of squares.

a. $y^2 - 16$ **b.** $25y^2 - 64x^2$ **c.** $16x^6 - t^4$ **d.** $(x + y)^2 - 25$

Solution

a. Using the factoring model for a difference of squares, we have $a = y$ and $b = 4$.

$$\begin{array}{rcccl} & a^2 - b^2 &=& (a + b)(a - b) \\ & \downarrow \quad \downarrow & & \downarrow \quad \downarrow \quad \downarrow \quad \downarrow \\ y^2 - 16 = & y^2 - 4^2 &=& (y + 4)(y - 4) \end{array}$$

b. Because $25y^2 = (5y)^2$ and $64x^2 = (8x)^2$, we replace a by $5y$ and b by $8x$ in the factoring model.

$$\begin{array}{rcccl} & a^2 - b^2 &=& (a + b)\ (a - b) \\ & \swarrow \quad \downarrow & & \downarrow \quad \downarrow \quad \downarrow \quad \downarrow \\ 25y^2 - 64x^2 = & (5y)^2 - (8x)^2 &=& (5y + 8x)(5y - 8x) \end{array}$$

c. Both $16x^6$ and t^4 are squares, since $16x^6$ is the square of $4x^3$ and t^4 is the square of t^2.

$$\begin{array}{rcccl} & a^2 - b^2 &=& (a + b)\ (a - b) \\ & \downarrow \quad \downarrow & & \downarrow \quad \downarrow \quad \downarrow \quad \downarrow \\ 16x^6 - t^4 = & (4x^3)^2 - (t^2)^2 &=& (4x^3 + t^2)(4x^3 - t^2) \end{array}$$

d. This expression matches the model with $a = x + y$ and $b = 5$.

$$\begin{array}{rcl} & a^2 \quad - b^2 = & (a \quad + b)\ (a \quad - b) \\ & \downarrow \qquad \downarrow & \downarrow \qquad \downarrow \quad \downarrow \qquad \downarrow \\ (x + y)^2 - 25 = & (x + y)^2 - 5^2 = & [(x + y) + 5][(x + y) - 5] \\ & = & (x + y + 5)(x - y - 5) \end{array}$$

PROGRESS CHECK 2 Factor each difference of squares.

a. $x^2 - 121$ **b.** $9t^2 - 36c^2$ **c.** $25x^8 - 9y^{10}$ **d.** $(s - t)^2 - 81$

In the next example more than one factoring procedure is required in order to factor completely.

EXAMPLE 3 Factor completely.

a. $4x^3 - 4x$ **b.** $x^4 - 16$

Solution

a. We first factor out the GCF, which is $4x$.

$$4x^3 - 4x = 4x(x^2 - 1)$$

Then we factor $x^2 - 1$ as the difference of squares.

$$4x^3 - 4x = 4x(x + 1)(x - 1)$$

Progress Check Answers

2. (a) $(x + 11)(x - 11)$ (b) $(3t + 6c)(3t - 6c)$ (c) $(5x^4 + 3y^5)(5x^4 - 3y^5)$ (d) $(s - t + 9)(s - t - 9)$

b. This expression can be factored as the difference of squares.

$$x^4 - 16 = (x^2 + 4)(x^2 - 4)$$

Notice now that $x^2 - 4$ is again a difference of squares, which factors into $(x + 2)(x - 2)$. Thus,

$$x^4 - 16 = (x^2 + 4)(x + 2)(x - 2).$$

In general, $a^2 + b^2 = (a + b + \sqrt{2ab})(a + b - \sqrt{2ab})$.

Caution Note that in Example 3b the expression $x^2 + 4$ could not be factored. In general, *the sum of squares $a^2 + b^2$ cannot be factored over the integers* but is always a prime polynomial.

PROGRESS CHECK 3 Factor completely.

a. $3x - 3xy^2$ **b.** $81y^4 - 1$

3 Sum or difference of two cubes

Expressions of the form $a^3 + b^3$ and $a^3 - b^3$ can always be factored, as can be seen by considering the two special products shown next.

$$\begin{aligned}(a + b)(a^2 - ab + b^2) &= a^3 - a^2b + ab^2 + a^2b - ab^2 + b^3 \\ &= a^3 + b^3\end{aligned}$$

Similarly,

$$\begin{aligned}(a - b)(a^2 + ab + b^2) &= a^3 + a^2b + ab^2 - a^2b - ab^2 - b^3 \\ &= a^3 - b^3.\end{aligned}$$

By reversing these products, we obtain factoring models for the sum and the difference of two cubes.

Factoring Models for Sums and Differences of Cubes

$$a^3 + b^3 = (a + b)(a^2 - ab + b^2)$$
$$a^3 - b^3 = (a - b)(a^2 + ab + b^2)$$

To use these models, you must identify appropriate replacements for a and b in the expression to be factored and then substitute in these formulas.

EXAMPLE 4 Factor.

a. $x^3 + 125$ **b.** $8y^3 - 1$

Solution

a. Use the formula for the sum of two cubes. Replace a with x and b with 5 (because $125 = 5^3$).

$$\begin{aligned} a^3 + b^3 &= (a + b)(a^2 - a\,b + b^2) \\ &\downarrow \\ x^3 + 125 = x^3 + 5^3 &= (x + 5)[x^2 - x(5) + 5^2] \\ &= (x + 5)(x^2 - 5x + 25)\end{aligned}$$

b. To factor $8y^3 - 1$, use the formula for the difference of two cubes. Because $8y^3 = (2y)^3$ and $1 = 1^3$, replace a with $2y$ and b with 1.

$$\begin{aligned} a^3 - b^3 &= (a - b)(a^2 + a\,b + b^2) \\ &\downarrow \\ 8y^3 - 1 = (2y)^3 - 1^3 &= (2y - 1)[(2y)^2 + (2y)(1) + 1^2] \\ &= (2y - 1)(4y^2 + 2y + 1)\end{aligned}$$

Progress Check Answers

3. (a) $3x(1 + y)(1 - y)$
(b) $(3y + 1)(3y - 1)(9y^2 + 1)$

Caution The trinomial factors $a^2 - ab + b^2$ and $a^2 + ab + b^2$ in these factoring models are prime polynomials. There is no point in trying to factor them further.

PROGRESS CHECK 4 Factor.

a. $y^3 + 1$ **b.** $8x^3 - 27$

EXAMPLE 5 Solve the problem in the section introduction on page 133.

Solution We are asked to factor $\frac{4}{3}\pi R^3 - \frac{4}{3}\pi r^3$ completely.

$$\begin{aligned} \tfrac{4}{3}\pi R^3 - \tfrac{4}{3}\pi r^3 &= \tfrac{4}{3}\pi(R^3 - r^3) && \text{Factor out } \tfrac{4}{3}\pi. \\ &= \tfrac{4}{3}\pi(R - r)(R^2 + Rr + r^2) && \text{Difference of cubes.} \end{aligned}$$

Thus, $V = \frac{4}{3}\pi(R - r)(R^2 + Rr + r^2)$ is a factored version of the formula.

PROGRESS CHECK 5 A metal piece of a certain lock consists of a smaller cube (side $= s$) mounted on top of a larger cube (side $= b$). The total weight of this piece is given by $W = 8s^3 + 27b^3$, where 8 and 27 are the respective densities of the small and large cubes. Express this formula differently by factoring it completely.

Example 6 serves as a reminder that you should always first check for common factors. It also points out that powers of variables that are multiples of 3, like x^9 and y^6, are perfect cubes.

EXAMPLE 6 Factor completely.

a. $5x^3 - 40$ **b.** $2x^9y + 54y^7$

Solution

a. First, factor out the common factor, 5, and then apply the factoring model for the difference of two cubes.

$$\begin{aligned} 5x^3 - 40 &= 5(x^3 - 8) && \text{Factor out 5.} \\ &= 5(x - 2)(x^2 + 2x + 4) && \text{Difference of cubes.} \end{aligned}$$

b. First, factor out the common factor, $2y$.

$$2x^9y + 54y^7 = 2y(x^9 + 27y^6)$$

Because $x^9 = (x^3)^3$ and $27y^6 = (3y^2)^3$, we use x^3 for a and $3y^2$ for b in the model for the sum of two cubes.

$$\begin{aligned} 2x^9y + 54y^7 &= 2y(x^3 + 3y^2)[(x^3)^2 - (x^3)(3y^2) + (3y^2)^2] \\ &= 2y(x^3 + 3y^2)(x^6 - 3x^3y^2 + 9y^4) \end{aligned}$$

PROGRESS CHECK 6 Factor completely.

a. $2x^3 + 128$ **b.** $3x^{15}y^2 - 24y^{11}$

Progress Check Answers

4. (a) $(y + 1)(y^2 - y + 1)$
(b) $(2x - 3)(4x^2 + 6x + 9)$
5. $(2s + 3b)(4s^2 - 6sb + 9b^2)$
6. (a) $2(x + 4)(x^2 - 4x + 16)$
(b) $3y^2(x^5 - 2y^3)(x^{10} + 2x^5y^3 + 4y^6)$

EXERCISES 3.7

In Exercises 1–12, factor each trinomial.

1. $x^2 + 10x + 25$ $(x + 5)^2$
2. $y^2 + 12y + 36$ $(y + 6)^2$
3. $9x^2 + 6x + 1$ $(3x + 1)^2$
4. $9x^2 + 12x + 4$ $(3x + 2)^2$
5. $4x^2 - 12x + 9$ $(2x - 3)^2$
6. $9x^2 - 30x + 25$ $(3x - 5)^2$
7. $16a^2 - 24ab + 9b^2$ $(4a - 3b)^2$
8. $4c^2 - 20cd + 25d^2$ $(2c - 5d)^2$
9. $16x^2 + 30x + 9$ $(8x + 3)(2x + 3)$
10. $9x^2 + 15x + 4$ $(3x + 1)(3x + 4)$
11. $y^2 - 13y + 36$ $(y - 9)(y - 4)$
12. $4y^2 - 25y + 25$ $(4y - 5)(y - 5)$

In Exercises 13–28, factor each expression.

13. $m^2 - 25$ $(m + 5)(m - 5)$

14. $n^2 - 16$ $(n + 4)(n - 4)$

15. $9 - x^2$ $(3 + x)(3 - x)$

16. $1 - x^2$ $(1 + x)(1 - x)$

17. $4x^2 - 9y^2$ $(2x + 3y)(2x - 3y)$

18. $16y^2 - 9x^2$ $(4y + 3x)(4y - 3x)$

19. $25a^2c^2 - 36b^2d^2$ $(5ac + 6bd)(5ac - 6bd)$

20. $100x^2y^2 - 49z^2$ $(10xy + 7z)(10xy - 7z)$

21. $16t^6 - 1$ $(4t^3 + 1)(4t^3 - 1)$

22. $9 - 25s^6$ $(3 + 5s^3)(3 - 5s^3)$

23. $9x^4 - 25y^8$ $(3x^2 + 5y^4)(3x^2 - 5y^4)$

24. $X^{10} - Y^{12}$ $(X^5 + Y^6)(X^5 - Y^6)$

25. $(x + y)^2 - 9$ $(x + y + 3)(x + y - 3)$

26. $(x - y)^2 - 4$ $(x - y + 2)(x - y - 2)$

27. $(2x + 3y)^2 - 16z^2$ $(2x + 3y + 4z)(2x + 3y - 4z)$

28. $(x + 2y)^2 - 9$ $(x + 2y + 3)(x + 2y - 3)$

In Exercises 29–36, factor completely.

29. $3x^3 - 3x$ $3x(x + 1)(x - 1)$

30. $5y^3 - 20y$ $5y(y + 2)(y - 2)$

31. $2x^3y - 18xy^3$ $2xy(x + 3y)(x - 3y)$

32. $36x^3y - 4xy^3$ $4xy(3x + y)(3x - y)$

33. $x^4 - 81$ $(x^2 + 9)(x + 3)(x - 3)$

34. $1 - y^4$ $(1 + y^2)(1 + y)(1 - y)$

35. $16x^4 - 81y^8$ $(4x^2 + 9y^4)(2x + 3y^2)(2x - 3y^2)$

36. $x^8 - 256$ $(x^4 + 16)(x^2 + 4)(x + 2)(x - 2)$

In Exercises 37–48, factor the given expression. Check by multiplication.

37. $x^3 + 1$ $(x + 1)(x^2 - x + 1)$

38. $x^3 - 1$ $(x - 1)(x^2 + x + 1)$

39. $8m^3 + 27n^3$ $(2m + 3n)(4m^2 - 6mn + 9n^2)$

40. $64p^3 - 27r^3$ $(4p - 3r)(16p^2 + 12pr + 9r^2)$

41. $x^3y^3 + 8$ $(xy + 2)(x^2y^2 - 2xy + 4)$

42. $27 - a^3b^3$ $(3 - ab)(9 + 3ab + a^2b^2)$

43. $x^6 + 1$ $(x^2 + 1)(x^4 - x^2 + 1)$

44. $x^6 - 8$ $(x^2 - 2)(x^4 + 2x^2 + 4)$

45. $2m^3 - 16$ $2(m - 2)(m^2 + 2m + 4)$

46. $3n^3 - 81p^3$ $3(n - 3p)(n^2 + 3pn + 9p^2)$

47. $2xy^9 + 16x^7$ $2x(y^3 + 2x^2)(y^6 - 2x^2y^3 + 4x^4)$

48. $24x^{10}y^3 + 3x$ $3x(2x^3y + 1)(4x^6y^2 - 2x^3y + 1)$

49. A hole is cut from a rectangular solid as shown in the figure. The volume of the remaining solid is given by $4a^3 - 4ab^2$. Factor this expression for the volume.

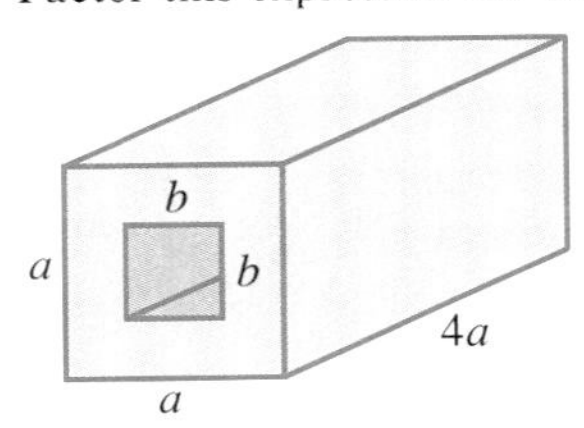

$4a(a + b)(a - b)$

50. A rectangular solid with height h has a square cross section of side s. If a corner piece with two sides equal to x is removed, the volume of the solid that remains is $s^2h - x^2h$. Factor this to find another expression for the volume. $h(s + x)(s - x)$

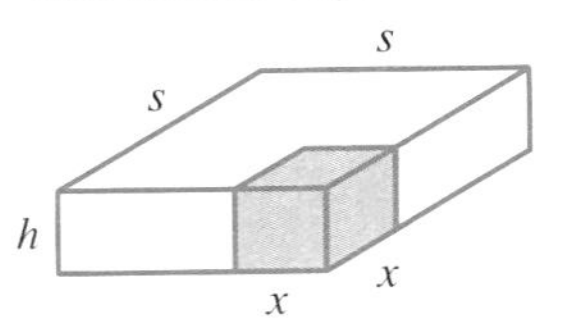

51. The sketch shown consists of three areas whose sum is $c^2 + 2cd + d^2$.

a. Factor this polynomial to find another expression for this area. $(c + d)^2$

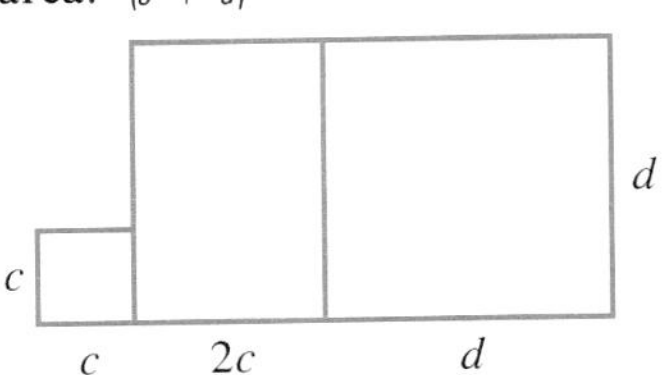

b. Show that the three pieces can be cut up and rearranged to make a square whose dimensions correspond to the two factors from part **a.**

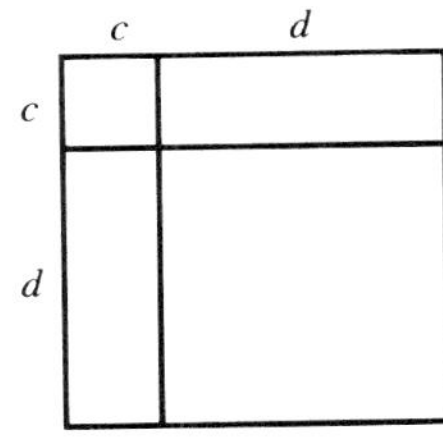

52. The sketch shown consists of three areas whose sum is $r^2 + 2r(2r + 1) + (2r + 1)^2$.

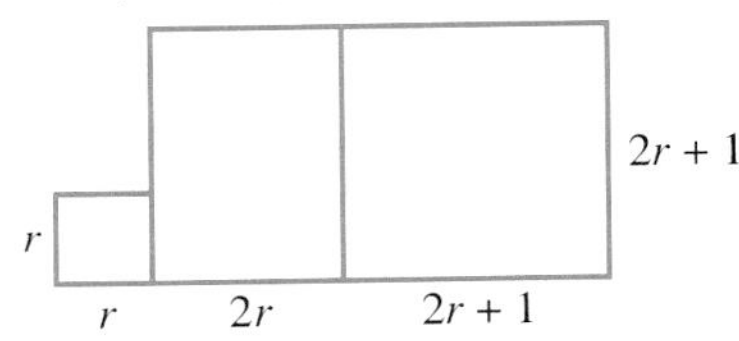

a. Simplify and factor this polynomial to find another expression for this area. $(3r + 1)^2$

b. Show that the three pieces can be cut up and rearranged to make a square whose dimensions correspond to the two factors from part **a.**

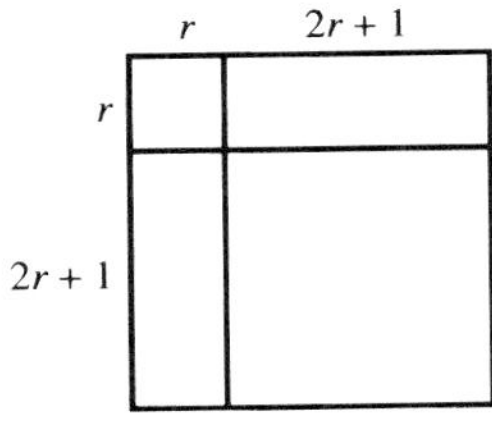

53. The area of a washer can be expressed as $\pi R^2 - \pi r^2$, where R is the radius of the larger circle and r is the radius of the smaller circle. (See the figure.) A ring-shaped region of this type is called an **annulus.** Factor this polynomial to find another expression for the area. $\pi(R + r)(R - r)$

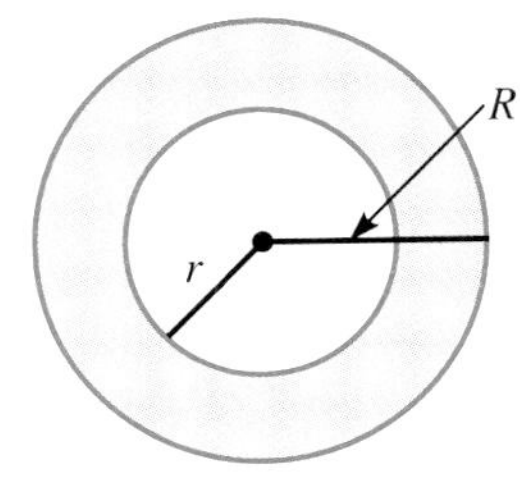

54. The total shaded area shown in the figure is given by

$\pi(a+2)^2 - \pi(a+1)^2 + \pi \cdot 1^2$.

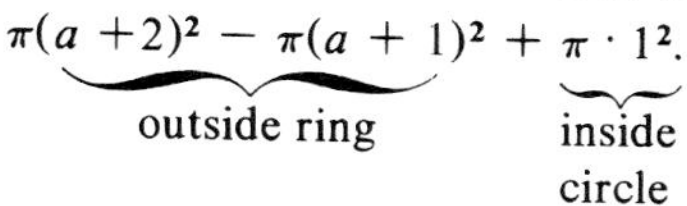

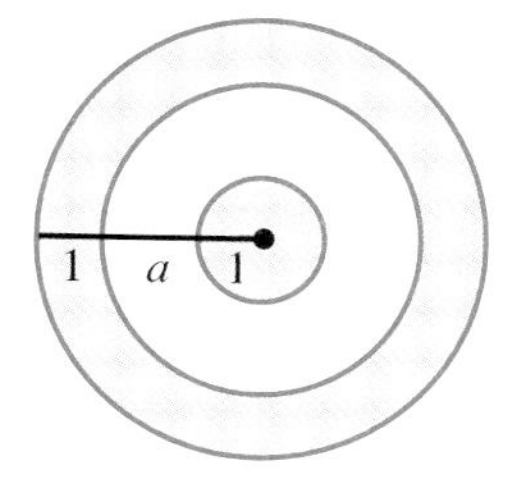

a. Simplify this expression and factor it. $2\pi(a+2)$

b. Find an expression for the circumference of the largest circle. $2\pi(a+2)$

THINK ABOUT IT

1. Later in the text we shall see that in solving quadratic equations, it can be useful to recognize perfect square expressions. Find the value of b that would make each of these perfect squares.

a. $x^2 + bx + 9$

b. $x^2 + bx + 25$

c. $x^2 + bx + 36$

d. Explain how you determined the middle coefficient in parts **a**, **b**, and **c** of this exercise; then find an expression for b that would make $x^2 + bx + s^2$ a perfect square.

2. The expression $x^2 + 6x$ can be represented as the area of the accompanying figure.

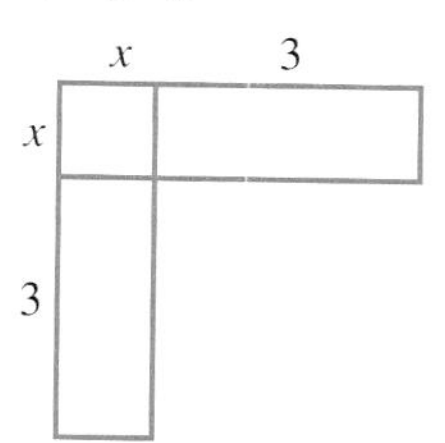

a. Show that this is correct by finding each of the three areas and their sum.

b. If you add a piece at the missing corner, you can complete the picture of a square, as shown by the dotted lines in the figure. What is the area of the piece that is needed to complete the square?

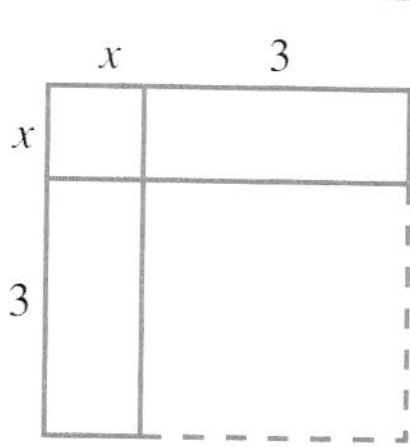

c. When you add 9 to $x^2 + 6x$, you can say that you *completed the square*. Explain why.

3. a. Following the work in Exercise 2, draw a figure to represent $x^2 + 8x$.

b. What is the area of the piece that is needed to complete the square?

4. Find the value of c that will make each of these expressions a perfect square. That is, find the number that will complete the square.

a. $x^2 + 10x + c$ **b.** $x^2 + 12x + c$ **c.** $x^2 + 14x + c$

d. Explain how, in solving parts **a**, **b**, and **c** of this exercise, you can find the value of the constant.

5. The factoring model for the factorization of a "perfect cube" polynomial can be found by expanding $(a + b)^3$. Find this model.

REMEMBER THIS

1. Divide $a^3 + b^3$ by $a + b$. $a^2 - ab + b^2$

2. Factor by grouping: $x^2 + mx + nx + mn$. $(x + n)(x + m)$

3. Factor out the GCF: $5x^3y - 15x^2y + 10xy$. $5xy(x^2 - 3x + 2)$

4. Factor $x^2 + 10x + 16$. $(x + 2)(x + 8)$

5. Factor $x^2 + x - 12$. $(x + 4)(x - 3)$

6. Factor $6x^2 - 35x - 6$. $(6x + 1)(x - 6)$

7. Find $P(-2)$ if $P(y) = y^3 - y + 6$. 0

8. Simplify $\frac{2}{2^{-m}}$. 2^{m+1}

9. Express 0.000421 in scientific notation. 4.21×10^{-4}

10. The formula for the surface area S of a cylinder is given by $S = 2\pi rh + 2\pi r^2$, where h is the height of the cylinder and r is the radius. Solve the formula for h; then find h when $r = 1$ and $S = 4\pi$. $h = \frac{S - 2\pi r^2}{2\pi r}$; 1

3.8 General Factoring Strategy

A diver jumps from a cliff that is 64 ft above the water. The diver's height above the water (y) after t seconds is given by the formula

$$y = 64 - 16t^2.$$

Factor completely the right side of the formula. (See Example 3.)

OBJECTIVES

1 **Apply a general strategy to factor polynomials systematically.**

1 There is such a variety of factoring problems that you may find it helpful to have a general strategy for solving such problems in a systematic way. For convenience, we collect in one table all of the factoring models we have used.

Summary of Factoring Models

Model	Name
1. $ab + ac = a(b + c)$	Common factor
2. $x^2 + (m + n)x + mn = (x + m)(x + n)$	Trinomial ($a = 1$)
3. $(pq)x^2 + (pn + qm)x + mn = (px + m)(qx + n)$	General trinomial
4. $a^2 + 2ab + b^2 = (a + b)^2$	Perfect square trinomial
5. $a^2 - 2ab + b^2 = (a - b)^2$	Perfect square trinomial
6. $a^2 - b^2 = (a + b)(a - b)$	Difference of squares
7. $a^3 + b^3 = (a + b)(a^2 - ab + b^2)$	Sum of cubes
8. $a^3 - b^3 = (a - b)(a^2 + ab + b^2)$	Difference of cubes

Referring to these models, you may use the following steps as guidelines for factoring polynomials.

Guidelines for Factoring a Polynomial

1. Factor out any common factors (if present) as the first factoring procedure.
2. Check for factorizations according to the number of terms in the polynomial.

   Two terms Look for a difference of squares or cubes or for a sum of cubes. Then apply models 6, 7, or 8, if applicable. Remember that $a^2 + b^2$ is a prime polynomial.

   Three terms Is the trinomial a perfect square trinomial? If yes, use models 4 or 5. If no, and the coefficient of the squared term is 1, try to use model 2. If no, and the coefficient of the squared term is not 1, use FOIL reversal or the *ac* method.

   Four terms Try factoring by grouping.
3. Make sure that no factors of two or more terms can be factored again.

Also, remember that as a general rule, you can always check your factoring by multiplying your answer.

EXAMPLE 1 Factor completely $6x^2 - 48x + 72$.

Solution Following step 1 in the guidelines above, we first factor out the GCF, which is 6.

$$6x^2 - 48x + 72 = 6(x^2 - 8x + 12)$$

Next, the polynomial $x^2 - 8x + 12$ has three terms, and we can match it to factoring model 2. It factors into $(x - 2)(x - 6)$, which cannot be factored further. Thus, the final factorization is

$$6x^2 - 48x + 72 = 6(x - 2)(x - 6).$$

PROGRESS CHECK 1 Factor completely $4x^2 - 28x + 48$.

EXAMPLE 2 Factor completely $3x^2 - 2x - 16$.

Solution There are no common factors to factor out, and because the coefficient of the squared term is not 1, this matches model 3. We can use FOIL reversal or the *ac* method to obtain

$$3x^2 - 2x - 16 = (3x - 8)(x + 2).$$

Since both factors are prime, this is the complete factorization.

PROGRESS CHECK 2 Factor completely $5x^2 + 38x - 16$.

EXAMPLE 3 Solve the problem that opens the section on page 140.

Solution We must factor $64 - 16t^2$. We first factor out 16.

$$64 - 16t^2 = 16(4 - t^2)$$

The factor $4 - t^2$ is a difference of two terms. Careful inspection shows that it matches model 6, the difference of squares, where $a^2 = 4$ and $b^2 = t^2$. Therefore, it factors as

$$4 - t^2 = (2 + t)(2 - t).$$

Thus, the complete factorization is

$$64 - 16t^2 = 16(2 + t)(2 - t).$$

PROGRESS CHECK 3 If the cliff is 144 ft above the water, the formula in Example 3 becomes $y = 144 - 16t^2$. Factor completely the right side of this formula.

EXAMPLE 4 Factor completely $16s^4 - t^4$.

Solution There are no common factors to factor out, but the polynomial matches model 6, the difference of squares, where $a = 4s^2$ and $b = t^2$. Therefore, it factors as

$$16s^4 - t^4 = (4s^2 + t^2)(4s^2 - t^2).$$

Because $4s^2 - t^2$ is itself a difference of squares, where $a = 2s$ and $b = t$, we can factor further to get

$$16s^4 - t^4 = (4s^2 + t^2)(2s + t)(2s - t),$$

which is the complete factorization.

PROGRESS CHECK 4 Factor completely $81 - w^4$.

Progress Check Answers

1. $4(x - 4)(x - 3)$
2. $(5x - 2)(x + 8)$
3. $16(3 + t)(3 - t)$
4. $(9 + w^2)(3 + w)(3 - w)$

EXAMPLE 5 Factor completely $y^2 + 3y + by + 3b$.

Solution Here we try factoring by grouping, because the polynomial contains four terms with no factors common to all four.

$$\begin{aligned} y^2 + 3y + by + 3b &= y(y + 3) + b(y + 3) \\ &= (y + 3)(y + b) \end{aligned}$$

PROGRESS CHECK 5 Factor completely $x^2 + ax + bx + ab$.

EXAMPLE 6 Factor completely $x^6 - 64$.

"Think About It" Exercise 2 asks the student to do the factorization first as a difference of cubes.

Solution First, we note that there are no common factors. This expression is unusual because it matches the models for both a difference of squares and a difference of cubes. Such expressions factor much more easily if you apply the difference-of-squares model first.

$$x^6 - 64 = (x^3)^2 - 8^2 = (x^3 + 8)(x^3 - 8)$$

Now factor $x^3 + 8$ by the sum-of-cubes model and $x^3 - 8$ by the difference-of-cubes model to get the complete factorization.

$$x^6 - 64 = (x + 2)(x^2 - 2x + 4)(x - 2)(x^2 + 2x + 4)$$

PROGRESS CHECK 6 Factor completely $y^6 - 1$.

Progress Check Answers

5. $(x + b)(x + a)$
6. $(y + 1)(y^2 - y + 1)(y - 1)(y^2 + y + 1)$

EXERCISES 3.8

In Exercises 1–48, factor completely the given expressions. If the polynomial cannot be factored, write "Prime." In Exercises 45–48, clear parentheses first.

1. $x^2 + x$ $x(x + 1)$
2. $4y^2 - 8$ $4(y^2 - 2)$
3. $18x^2 + 3x - 3$ $3(3x - 1)(2x + 1)$
4. $12x^2 + 10x - 12$ $2(3x - 2)(2x + 3)$
5. $2y^3 + 4y^2 - 70y$ $2y(y - 5)(y + 7)$
6. $5x^3 + 30x^2 + 25x$ $5x(x + 1)(x + 5)$
7. $10a^2 + 29a + 10$ $(2a + 5)(5a + 2)$
8. $14a^2 - 53a + 14$ $(2a - 7)(7a - 2)$
9. $b^2 + b + 1$ Prime
10. $c^2 + c + 2$ Prime
11. $16d^4 - e^4$ $(4d^2 + e^2)(2d + e)(2d - e)$
12. $d^8 - e^8$ $(d^4 + e^4)(d^2 + e^2)(d + e)(d - e)$
13. $x^2 + 2x + cx + 2c$ $(x + c)(x + 2)$
14. $x^2 + 5x + nx + 5n$ $(x + n)(x + 5)$
15. $abc + 2ac + bc + 2c$ $(a + 1)(b + 2)c$
16. $abc - ac - ab + a$ $(c - 1)(b - 1)a$
17. $64x^6 - 1$ $(2x + 1)(4x^2 - 2x + 1)(2x - 1)(4x^2 + 2x + 1)$
18. $x^6y^6 - 64$ $(xy + 2)(x^2y^2 - 2xy + 4)(xy - 2)(x^2y^2 + 2xy + 4)$
19. $(a + b)^2 + 2(a + b) + 1$ $(a + b + 1)^2$
20. $(m - n)^2 - 4(m - n) + 4$ $(m - n - 2)^2$
21. $4s^4t^4 - s^2t^2$ $s^2t^2(2st + 1)(2st - 1)$
22. $9s^2t^2 - s^4t^4$ $s^2t^2(3 + st)(3 - st)$
23. $x^4 - 3x^2 - 4$ $(x + 2)(x - 2)(x^2 + 1)$
24. $y^4 - 11y^2 + 18$ $(y + 3)(y - 3)(y^2 - 2)$
25. $(x + y)^2 - 9$ $(x + y + 3)(x + y - 3)$
26. $(2x - y)^2 - 16$ $(2x - y + 4)(2x - y - 4)$
27. $n^2 + 16$ Prime
28. $25m^2 + 36$ Prime
29. $xy^3 + 6x^2y^2 + 9x^3y$ $xy(y + 3x)^2$
30. $m^3n - 2m^2n^2 + mn^3$ $mn(m - n)^2$
31. $24w^2 - 86w - 15$ $(6w + 1)(4w - 15)$
32. $60z^2 + 67z - 6$ $(5z + 6)(12z - 1)$
33. $a^2 - 6a + 9 - b^2$ $(a - 3 + b)(a - 3 - b)$
34. $b^2 + 8b + 16 - a^2$ $(b + 4 + a)(b + 4 - a)$
35. $n^3 + 1{,}000$ $(n + 10)(n^2 - 10n + 100)$
36. $125 - m^3$ $(5 - m)(25 + 5m + m^2)$
37. $(a + b)^3 + c^3$ $(a + b + c)[(a + b)^2 - (a + b)c + c^2]$
38. $(a + 1)^3 - 8$ $(a - 1)(a^2 + 4a + 7)$
39. $2x^2 - 3x - 4$ Prime
40. $3x^2 - 4x - 5$ Prime
41. $x^2y^2 + xy - 12$ $(xy + 4)(xy - 3)$
42. $x^2y^2 - 2xy - 8$ $(xy - 4)(xy + 2)$
43. $x^6 - 4a^2$ $(x^3 - 2a)(x^3 + 2a)$
44. $4y^6 - b^4$ $(2y^3 - b^2)(2y^3 + b^2)$
45. $(x + 1)^2 + c(x + 2) - 1$ $(x + c)(x + 2)$
46. $(x + 2)^2 + n(x + 2) + 2(n - 2)$ $(x + 4)(x + n)$
47. $(x - 3)^2 + 3(2c - 3) - cx$ $(x - 6)(x - c)$
48. $x^2 - a(x - 1) - a(b + 1) + bx$ $(x - a)(x + b)$
49. Because the moon is less massive than the earth, the pull of gravity at its surface is weaker, and so objects fall more slowly there. The formula for the height of an object dropped from d ft above the moon's surface is roughly $y = d - 4t^2$, where t is the elapsed time in seconds.
    a. What is the formula for an object dropped from 225 ft? $y = 225 - 4t^2$
    b. Express the result of part **a** in factored form. $y = (15 + 2t)(15 - 2t)$
    c. Show that the object hits the surface in 7.5 seconds. When $t = 7.5$, $y = 0$

50. On earth, the formula for the height of an object dropped from d ft above the ground is $y = d - 16t^2$, where t is elapsed time in seconds.

a. What is the formula for an object dropped from 225 ft?
$y = 225 - 16t^2$

b. Express the result of part **a** in factored form.
$y = (15 + 4t)(15 - 4t)$

c. Show that the object hits the ground in 3.75 seconds.
When $t = 3.75$, $y = 0$

51. Find an expression for the perimeter of this rectangle, and then write it in factored form.

$P = 2(a + b)^2$

$b^2 + ab$

$a^2 + ab$

52. Find an expression for the surface area of the sides of this box, and then write it in factored form. Do not include the top and bottom. $A = 2(a + b)^2$

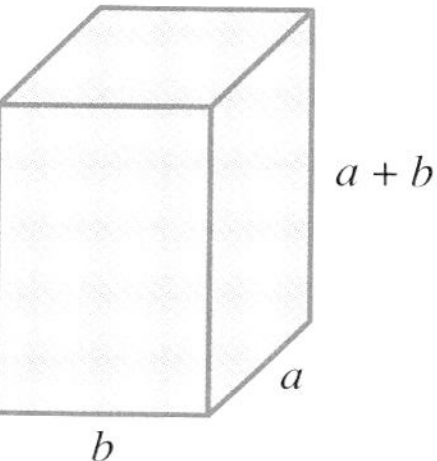

53. Find an expression for the shaded area, and write it in factored form. $A = (a + b)(a - b)$

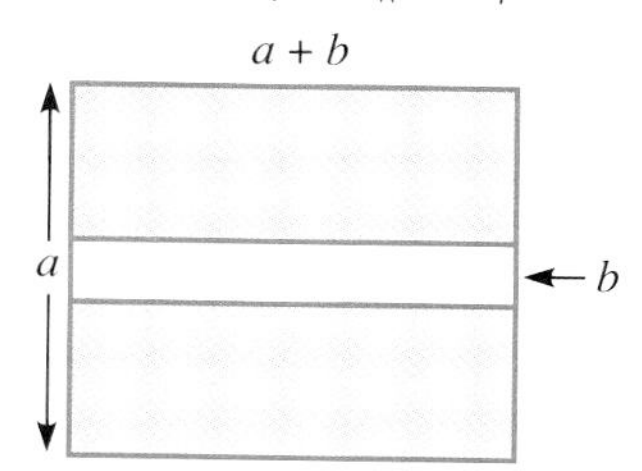

54. Find an expression for the shaded area, and then write it in factored form. $A = (a - b)^2$

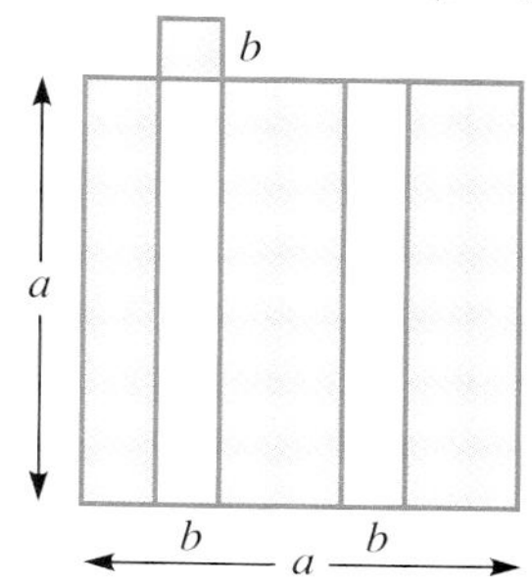

THINK ABOUT IT

1. $x - 1$ can be written as $(x - 1)(1)$.
$x^2 - 1$ can be written as $(x - 1)(x + 1)$.
$x^3 - 1$ can be written as $(x - 1)(x^2 + x + 1)$.

a. Can $x^4 - 1$ be written as $(x - 1)(x^3 + x^2 + x + 1)$?

b. How can you factor $x^5 - 1$?

2. In Example 6 of this section we factored $x^6 - 64$ completely as $(x + 2)(x - 2)(x^2 - 2x + 4)(x^2 + 2x + 4)$ by first treating $x^6 - 64$ as a difference of squares. Try the factorization again, starting by thinking of $x^6 - 64$ as a difference of *cubes*. Show that the two solutions are equivalent.

3. In many applications it is convenient to factor out the variable which appears to the lowest power. Try this approach when the variable has a negative or fractional exponent. Rational number exponents will be considered in detail in Section 7.1.

a. Factor out $x^{1/2}$ from $5x^{5/2} + 4x^{3/2} + 3x^{1/2}$.

b. Factor out x^{-3} from $3x^{-1} + 4x^{-2} + 5x^{-3}$.

c. Factor out $x^{-3/2}$ from $3x^{1/2} + 4x^{-1/2} + 5x^{-3/2}$.

4. Factor $x - y$ as a difference of *squares* by using fractional exponents. (*Hint:* x is the square of what?)

5. Factor $x - y$ as a difference of *cubes* by using fractional exponents. (*Hint:* x is the cube of what?)

REMEMBER THIS

1. What is the degree of this polynomial? $3x^2 + 5x + 1$. 2

2. True or false? If the product of two numbers is zero, then at least one of the numbers must be zero. If false, show an example. True

3. True or false? If the product of two numbers is one, then at least one of the numbers must be one. If false, show an example.
False; $\frac{1}{2} \cdot 2 = 1$

4. Show by substitution that both 0 and 5 are solutions to $4x^2 = 20x$. $4 \cdot 0^2 \stackrel{?}{=} 20 \cdot 0, 0 \stackrel{\checkmark}{=} 0$; $4 \cdot 5^2 \stackrel{?}{=} 20 \cdot 5, 100 \stackrel{\checkmark}{=} 100$.

5. Which of these are true?

a. $0^2 + 1^2 = 2^2$ **b.** $1^2 + 2^2 = 3^2$ **c.** $2^2 + 3^2 = 4^2$
d. $3^2 + 4^2 = 5^2$ **e.** $4^2 + 5^2 = 6^2$ d

6. Factor $x^2 - 16y^2$. $(x + 4y)(x - 4y)$

7. Factor $4y^2 - 10y - 6$. $2(2y + 1)(y - 3)$

8. Multiply 3×10^5 times 2×10^9. 6×10^{14}

9. Solve $\frac{x - 3}{2} > 2x + \frac{3}{2}$. Graph the solution set. $\{x: x > -2\}$

-2 x

10. How much water must be added to 5 gal of a solution which is 75 percent antifreeze to dilute it to 50 percent antifreeze? 2.5 gal

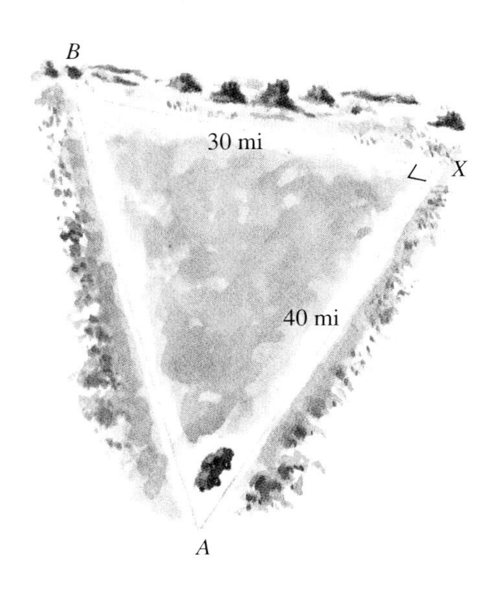

See "Think About It" Exercise 1 for the meaning of *quadratic*.

3.9 Solving Equations by Factoring

A person can drive from A to X to B at 60 mi/hour or directly from A to B at 40 mi/hour. Which way takes less time? (See Example 9.)

OBJECTIVES

1. **Solve certain quadratic equations using factoring.**
2. **Solve higher-degree equations using factoring.**
3. **Solve applied problems that lead to quadratic equations.**

1 When we attempt to solve equations which contain polynomials that are higher than the first degree, we shall find factoring to be an extremely useful tool. We start by looking at polynomial equations of degree 2, which are also called quadratic equations.

Definition of Quadratic Equation

A **second-degree** or **quadratic equation** is an equation that can be written in the form

$$ax^2 + bx + c = 0,$$

where a, b, and c are real numbers with $a \neq 0$.

Using factoring to solve quadratic equations depends on an idea called the zero product principle, which says in words that the product of two expressions is equal to zero if and only if one or both of those expressions is equal to zero. We state this principle formally in a box, and then we show how to use it.

Zero Product Principle

For any numbers a and b,

$$ab = 0 \quad \text{if and only if} \quad a = 0 \quad \text{or} \quad b = 0.$$

EXAMPLE 1 Solve $4x^2 = 20x$.

Solution Rewrite the given equation so that one side is equal to zero. Then, factor and apply the zero product principle.

$4x^2 = 20x$	Original equation.
$4x^2 - 20x = 0$	Rewrite the equation so one side is 0 by subtracting $20x$ from both sides.
$4x(x - 5) = 0$	Factor the nonzero side.
$4x = 0 \quad \text{or} \quad x - 5 = 0$	Set each factor equal to 0 according to the zero product principle.
$x = \frac{0}{4} \qquad x = 5$	Solve each linear equation.
$x = 0$	

Check each possible solution by substituting it in the original equation.

$$4x^2 = 20x \qquad 4x^2 = 20x$$
$$4(0)^2 \stackrel{?}{=} 20(0) \qquad 4(5)^2 \stackrel{?}{=} 20(5)$$
$$0 \stackrel{\checkmark}{=} 0 \qquad 100 \stackrel{\checkmark}{=} 100$$

Thus, the solution set is $\{0,5\}$.

Note Each "possible solution" from the above procedure will always check if there are no errors in the work. The check is needed only to catch any mistakes.

PROGRESS CHECK 1 Solve $3x^2 = 18x$.

Example 1 illustrates a general method for solving quadratic equations provided that we can factor the nonzero side. The steps are outlined below. In Chapter 8 we will show other approaches to use when factoring does not work.

Factoring Method for Solving Quadratic Equations

1. If necessary, change the form of the equation so that one side is zero.
2. Factor the nonzero side of the equation.
3. Set each factor equal to zero, and obtain the solution(s) by solving the resulting equations.
4. Check each solution by substituting it in the original equation.

Because a second-degree polynomial factors into *two* factors, quadratic equations usually have two solutions. But Example 2 shows that it is possible for a quadratic equation to have only one solution.

EXAMPLE 2 Solve $y^2 + 9 = 6y$.

Solution Follow the steps given.

$y^2 + 9 = 6y$	Original equation.
$y^2 - 6y + 9 = 0$	Rewrite the equation so one side is 0.
$(y - 3)(y - 3) = 0$	Factor the nonzero side.
$y - 3 = 0$ or $y - 3 = 0$	Set each factor equal to 0.
$y = 3$ $\quad$ $y = 3$	Solve each linear equation.

The solution, 3, checks, because $3^2 + 9 = 6(3)$ is a true statement, and the solution set is $\{3\}$.

PROGRESS CHECK 2 Solve $y^2 + 4 = -4y$.

Some equations are already in factored form, with zero on one side, when they are given; so the first couple of steps in the procedure can be skipped.

EXAMPLE 3 Solve $(5x + 3)(x - 2) = 0$.

Solution We can start immediately by setting each factor equal to zero.

$(5x + 3)(x - 2) = 0$	Original equation.
$5x + 3 = 0$ or $x - 2 = 0$	Set each factor equal to 0.
$5x = -3$ $\quad$ $x = 2$	Solve each linear equation.
$x = -\frac{3}{5}$	

Progress Check Answers

1. $\{0,6\}$
2. $\{-2\}$

Check

$$\begin{aligned} (5x+3)(x-2) &= 0 \\ [5(-\tfrac{3}{5})+3](-\tfrac{3}{5}-2) &\stackrel{?}{=} 0 \\ (0)(-\tfrac{13}{5}) &\stackrel{?}{=} 0 \\ 0 &\stackrel{\checkmark}{=} 0 \end{aligned} \qquad \begin{aligned} (5x+3)(x-2) &= 0 \\ [5(2)+3](2-2) &\stackrel{?}{=} 0 \\ (13)(0) &\stackrel{?}{=} 0 \\ 0 &\stackrel{\checkmark}{=} 0 \end{aligned}$$

Thus, the solution set is $\{-\frac{3}{5},2\}$.

Caution The equation $(5x+3)(x-2)=0$ is written in a form that allows us to immediately apply the zero product principle. We can do so not only because one side is in factored form but also because *one side is equal to zero.* Be careful about an equation like

$$(x+3)(x+2)=6,$$

because the fact that the product of two expressions equals 6 (or any other nonzero value) does not tell you anything crucial about either factor. In cases like this it is necessary to multiply the two factors on the left as a first step and then rewrite the equation with zero on one side.

PROGRESS CHECK 3 Solve $(3x-7)(x+3)=0$.

2 The factoring method associated with the zero product principle may be applied to solve higher-degree equations provided we can factor the polynomial on the nonzero side of the equation.

EXAMPLE 4 Solve each equation.

a. $(3x-1)(x+5)(x-8)=0$ **b.** $y^3=16y$

Solution

a. Because one side is already zero and the nonzero side is in factored form, we just set each factor equal to zero and solve each resulting equation.

$(3x-1)(x+5)(x-8)=0$		Original equation.
$3x-1=0$ or $x+5=0$ or $x-8=0$		Set each factor equal to 0.
$x=\frac{1}{3}$ $\quad x=-5$ $\quad x=8$		Solve each equation.

You can verify that all three solutions check, and so the solution set is $\{\frac{1}{3},-5,8\}$.

b. First, rewrite the equation so that one side is zero.

$$\begin{aligned} y^3 &= 16y \\ y^3-16y &= 0 \end{aligned}$$

Now factor completely.

$$\begin{aligned} y(y^2-16) &= 0 \\ y(y+4)(y-4) &= 0 \end{aligned}$$

Setting each factor equal to zero and solving gives

$$y=0 \quad \text{or} \quad y+4=0 \quad \text{or} \quad y-4=0$$
$$y=-4 \qquad\qquad y=4.$$

All three solutions check, because $0^3=16(0)$, $(-4)^3=16(-4)$, and $4^3=16(4)$ are all true statements. The solution set is $\{0,-4,4\}$.

PROGRESS CHECK 4 Solve each equation.

a. $(3x+2)(2x-3)(x-1)=0$ **b.** $x^3=4x$

Progress Check Answers

3. $\{\frac{7}{3},-3\}$

4. (a) $\{-\frac{2}{3},\frac{3}{2},1\}$ (b) $\{0,2,-2\}$

3 Quadratic equations appear in many word problems. As a sample of this variety, we will discuss four situations which commonly involve quadratic equations:

1. Formulas which contain quadratic expressions
2. Number relations
3. Geometric figures
4. The Pythagorean theorem

The next example is a word problem in which a formula is given. Using this formula to solve the problem leads directly to a quadratic equation. Many applications from the physical sciences include such formulas.

EXAMPLE 5 The height (y) of a projectile that is shot directly up from the ground with an initial velocity of 64 ft/second is given by the formula

$$y = 64t - 16t^2,$$

where t is the elapsed time in seconds. For what value(s) of t is the projectile 48 ft off the ground?

Solution We begin by replacing y by 48 in the given formula.

$$y = 64t - 16t^2$$
$$48 = 64t - 16t^2$$

This is the quadratic equation we need to solve. We rewrite it so that one side is zero. We choose to do this in a way that makes the coefficient of the squared term positive. Then we follow the usual procedure.

$$16t^2 - 64t + 48 = 0$$
$$16(t^2 - 4t + 3) = 0$$
$$16(t - 1)(t - 3) = 0$$

Because the constant factor, 16, cannot be zero, we need only set the two factors which contain a variable equal to zero.

$$t - 1 = 0 \quad \text{or} \quad t - 3 = 0$$
$$t = 1 \qquad\qquad t = 3$$

Both solutions check, and the projectile attains a height of 48 ft after 1 second (on the way up) and again when 3 seconds have elapsed (on its way down).

Caution In this example both solutions of the quadratic equation lead to sensible answers in the application, but this is not always the case. In fact, it is common for one of the solutions to not make sense. Thus, it is especially important in applications to check that all solutions lead to reasonable answers to the original question.

PROGRESS CHECK 5 In Example 5, for what value(s) of t is the projectile 64 ft off the ground?

Problems about number relations often lead to quadratic equations, as shown in the next example. Such problems therefore often have two distinct solutions.

EXAMPLE 6 The sum of the squares of two consecutive integers is 181. What are the integers?

Solution Using x to represent the smaller integer and $x + 1$ to represent the next consecutive integer, we can *set up an equation.*

Progress Check Answer
5. 2 seconds

The sum of the squares of two consecutive integers is 181.

$$x^2 + (x + 1)^2 = 181$$

Solve the Equation

$$\begin{aligned} x^2 + x^2 + 2x + 1 &= 181 && \text{Expand } (x+1)^2. \\ 2x^2 + 2x + 1 &= 181 && \text{Combine like terms.} \\ 2x^2 + 2x - 180 &= 0 && \text{Subtract 181 from both sides.} \\ 2(x^2 + x - 90) &= 0 && \text{Factor out GCF.} \\ 2(x - 9)(x + 10) &= 0 && \text{Factor the trinomial.} \\ x - 9 = 0 \quad \text{or} \quad x + 10 &= 0 && \text{Set each variable factor equal to 0.} \\ x = 9 \qquad\qquad x &= -10 && \text{Solve each linear equation.} \end{aligned}$$

Answer the Question There are two pairs of integers that answer the question. If x is 9, then $x + 1$ is 10; and if x is -10, then $x + 1$ is -9. One solution is 9 and 10, and the second solution is -10 and -9.

Check the Answer Both proposed solutions check, because both $9^2 + 10^2 = 181$ and $(-10)^2 + (-9)^2 = 181$ are true, and each answer is a pair of consecutive integers.

PROGRESS CHECK 6 The product of two consecutive integers is 132. Find the integers.

Word problems that involve areas of geometric figures often lead to quadratic equations. To solve them quickly, you should be familiar with the basic formulas for the area and perimeter of circles, triangles, and rectangles. As pointed out in the caution in Example 5, it is important to check that all solutions lead to sensible answers.

EXAMPLE 7 The area of a square is 5 more than the perimeter. Find the side length for the square measured in meters.

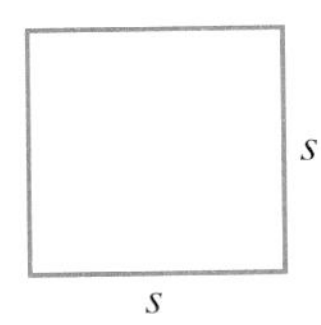

Figure 3.4

Solution We represent the length of the side by s, as shown in Figure 3.4. It follows then that the area is s^2 and the perimeter is $4s$. Now *set up an equation.*

The area is 5 more than the perimeter.

$$s^2 = 4s + 5$$

Solve the Equation

$$\begin{aligned} s^2 - 4s - 5 &= 0 && \text{Rewrite so one side is 0.} \\ (s - 5)(s + 1) &= 0 && \text{Factor the nonzero side.} \\ s - 5 = 0 \quad \text{or} \quad s + 1 &= 0 && \text{Set each factor equal to 0.} \\ s = 5 \qquad\qquad s &= -1 && \text{Solve each linear equation.} \end{aligned}$$

Answer the Question The solution $s = -1$ does not make sense in this problem because a square cannot have a negative side length, so the only sensible solution is that the side length is 5 m.

Check the Answer If $s = 5$, then the area is 25 and the perimeter is 20. Because 25 is 5 more than 20, the solution checks.

PROGRESS CHECK 7 The area of a square is 12 more than the perimeter. Find the side length of the square if the length is measured in feet.

Progress Check Answers

6. 11, 12 and -12, -11 7. 6 ft

Many word problems are solved with the help of geometric sketches that involve **right triangles,** which are triangles that contain a 90° angle. The longest side of a right triangle, which is always the side opposite the 90° angle, is called the **hypotenuse,** while the other two sides are called the **legs** of the triangle. Since ancient times people whose work involves right angles, like surveyors and builders, have known a crucial fact about right triangles that distinguishes them from all other triangles. This fact is stated in the Pythagorean theorem, which is named for a mathematician who lived in Greece more than 2,000 years ago and who offered one proof of this fact.

Hypotenuse comes from Greek words meaning "stretched under," as the base of a triangle whose vertex is a right angle lies "under" the right angle, and the vertex stands on the legs of the triangle.

Pythagorean Theorem

In a right triangle with legs of length a and b and hypotenuse of length c,

$$c^2 = a^2 + b^2.$$

In words, the square of the length of the hypotenuse is equal to the sum of the squares of the lengths of the two legs.

Examples 8 and 9 show two applications of the Pythagorean theorem.

EXAMPLE 8 The length of a rectangle is 3 more than the width, and a diagonal measures 15 m. Find the dimensions of the rectangle.

Solution To find the dimensions, let

$$x = \text{width},$$
$$\text{so} \quad x + 3 = \text{length}.$$

Now sketch the situation as in Figure 3.5 and note that the diagonal is the hypotenuse of a right triangle. We may then use the Pythagorean theorem to *set up an equation.*

$$c^2 = a^2 + b^2$$
$$15^2 = (x + 3)^2 + x^2 \quad \text{Replace } c \text{ by 15, } a \text{ by } x + 3\text{, and } b \text{ by } x.$$

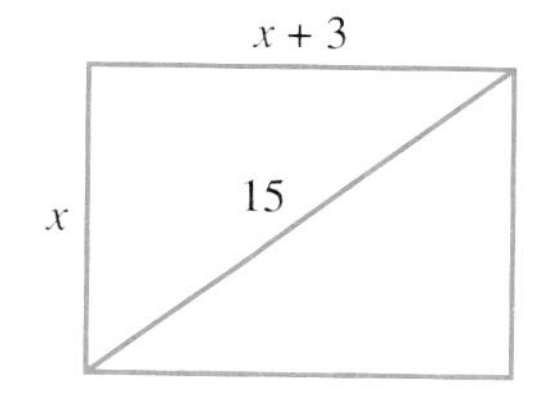

Figure 3.5

Solve the Equation

$$225 = x^2 + 6x + 9 + x^2 \quad \text{Expand and evaluate powers.}$$
$$0 = 2x^2 + 6x - 216 \quad \text{Rewrite so one side is 0.}$$
$$0 = 2(x^2 + 3x - 108) \quad \text{Factor out 2.}$$
$$0 = 2(x - 9)(x + 12) \quad \text{Factor } x^2 + 3x - 108.$$
$$x - 9 = 0 \quad \text{or} \quad x + 12 = 0 \quad \text{Set each variable factor equal to 0.}$$
$$x = 9 \qquad x = -12 \quad \text{Solve each linear equation.}$$

Answer the Question Reject the negative solution. The width is 9 m and the length is $9 + 3 = 12$ m.

Check the Answer Because 12 is 3 more than 9 and $15^2 = 9^2 + 12^2$ is a true statement, a rectangle that measures 9 m by 12 m satisfies the given conditions.

PROGRESS CHECK 8 The length of a rectangle is 2 more than the width, and a diagonal measures 10 ft. Find the dimensions of the rectangle.

EXAMPLE 9 Solve the problem on page 144 that opens this section.

Solution We will use the relationship that time = distance/speed, but first we will use the Pythagorean theorem to find the distance from A to B.

Step 1 Find the distance from A to B. From the simplified sketch of the problem in Figure 3.6, we see that this distance is the hypotenuse of a right triangle. We *set up an equation* based on the Pythagorean theorem.

$$c^2 = a^2 + b^2$$
$$x^2 = 30^2 + 40^2 \quad \text{Replace } c \text{ by } x\text{, } a \text{ by 30, and } b \text{ by 40.}$$

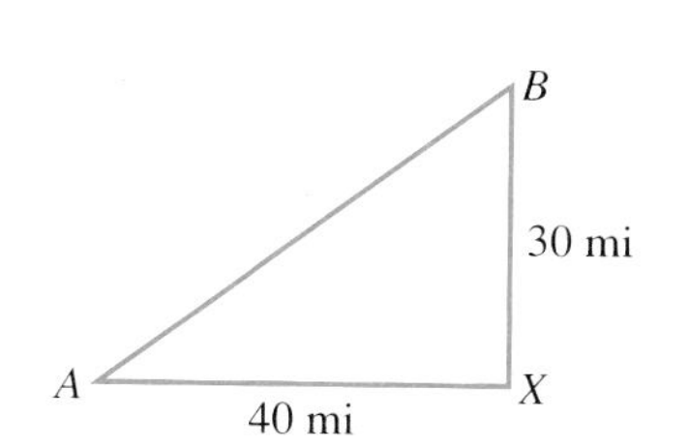

Figure 3.6

Solve the Equation

$$x^2 = 2{,}500 \quad \text{Evaluate powers and add.}$$
$$x^2 - 2{,}500 = 0 \quad \text{Subtract 2,500 from each side.}$$
$$(x + 50)(x - 50) = 0 \quad \text{Factor the nonzero side.}$$
$$x + 50 = 0 \quad \text{or} \quad x - 50 = 0 \quad \text{Set each factor equal to 0.}$$
$$x = -50 \qquad x = 50 \quad \text{Solve each linear equation.}$$

Progress Check Answer

8. Length = 8 ft, width = 6 ft

Answer the Question Only the positive solution is meaningful here. The distance from A to B is 50 mi.

Check the Answer Because $50^2 = 30^2 + 40^2$ is a true statement, the solution is correct.

Step 2 Find the time it takes to travel each route. Recall that the speed along the legs of the triangle is 60 mi/hour, while the speed along the hypotenuse is 50 mi/hour.

$$\text{From } A \text{ to } X \text{ the time is } \frac{\text{distance}}{\text{speed}} = \frac{40 \text{ mi}}{60 \text{ mi/hour}} = \frac{2}{3} \text{ hour} = 40 \text{ minutes.}$$

$$\text{From } X \text{ to } B \text{ the time is } \frac{\text{distance}}{\text{speed}} = \frac{30 \text{ mi}}{60 \text{ mi/hour}} = \frac{1}{2} \text{ hour} = 30 \text{ minutes.}$$

Therefore, to go from A to X to B takes 70 minutes.

$$\text{From } A \text{ to } B \text{ the time is } \frac{\text{distance}}{\text{speed}} = \frac{50 \text{ mi}}{40 \text{ mi/hour}} = \frac{5}{4} \text{ hour} = 75 \text{ minutes.}$$

Conclusion It is quicker to go the long way from A to X to B.

PROGRESS CHECK 9 A person can drive from A to X to B as diagrammed in the figure at 55 mi/hour or from A to B at 40 mi/hour. Which way takes less time?

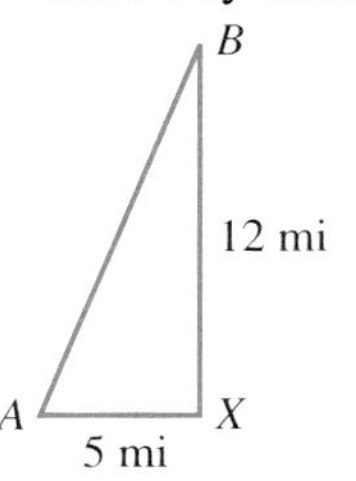

Progress Check Answer

9. From A to X to B

EXERCISES 3.9

In Exercises 1–42, solve each equation.

1. $5x^2 = 30x$ $\{0,6\}$
2. $7x^2 = 21x$ $\{0,3\}$
3. $4y^2 = -12y$ $\{0,-3\}$
4. $3y^2 = -21y$ $\{0,-7\}$
5. $5x^2 = 18x$ $\{0,\frac{18}{5}\}$
6. $2y^2 = -7y$ $\{0,-\frac{7}{2}\}$
7. $x^2 - 49 = 0$ $\{7,-7\}$
8. $x^2 = 1$ $\{1,-1\}$
9. $3z^2 - 3 = 0$ $\{1,-1\}$
10. $5z^2 - 45 = 0$ $\{3,-3\}$
11. $y^2 + 10 = 7y$ $\{2,5\}$
12. $x^2 + 12 = 7x$ $\{3,4\}$
13. $x^2 + 7x = -6$ $\{-1,-6\}$
14. $y^2 + 2y = -1$ $\{-1\}$
15. $z^2 = 2 - z$ $\{1,-2\}$
16. $z^2 = 6 - z$ $\{2,-3\}$
17. $2x^2 - 3x - 2 = 0$ $\{-\frac{1}{2},2\}$
18. $6x^2 - 13x + 6 = 0$ $\{\frac{2}{3},\frac{3}{2}\}$
19. $12y^2 - 7y + 1 = 0$ $\{\frac{1}{3},\frac{1}{4}\}$
20. $20y^2 + 9y + 1 = 0$ $\{-\frac{1}{4},-\frac{1}{5}\}$
21. $2x^2 - 2x + 4 = x^2 + 2x$ $\{2\}$
22. $3x^2 - 4x + 4 = 2x^2 + 2x - 5$ $\{3\}$
23. $(4x)(3x - 1) = 0$ $\{0,\frac{1}{3}\}$
24. $2y(2y - 5) = 0$ $\{0,\frac{5}{2}\}$
25. $(y - 1)(y + 2) = 0$ $\{1,-2\}$
26. $(x + 2)(x - 3) = 0$ $\{-2,3\}$
27. $(x - 3)(x + 5) = 9$ $\{4,-6\}$
28. $(y + 3)(y + 1) = 15$ $\{2,-6\}$
29. $(3z + 1)(2z - 5) = -5$ $\{0,\frac{13}{6}\}$
30. $(2x - 1)(3x - 1) = 1$ $\{0,\frac{5}{6}\}$
31. $(x - 8)(x + 8) + 15 = 0$ $\{7,-7\}$
32. $(x - 5)(x + 5) + 9 = 0$ $\{4,-4\}$
33. $(x - 412)(x + 6.02)(x + \pi) = 0$ $\{412,-6.02,-\pi\}$
34. $(y - 87)(y + 64)(y - 1.6) = 0$ $\{87,-64,1.6\}$
35. $(2y - 3)(3y - 4)(4y - 5) = 0$ $\{\frac{3}{2},\frac{4}{3},\frac{5}{4}\}$
36. $(3y - 2)(4y - 3)(5y - 4) = 0$ $\{\frac{2}{3},\frac{3}{4},\frac{4}{5}\}$
37. $x^3 = x$ $\{0,1,-1\}$
38. $z^3 = 25z$ $\{0,5,-5\}$
39. $x^3 - 3x^2 + 2x = 0$ $\{0,1,2\}$
40. $x^3 + 3x^2 + 2x = 0$ $\{0,-1,-2\}$
41. $x - \frac{40}{x} = 3$ $(x \neq 0)$ $\{-5,8\}$
42. $x - 1 = \frac{30}{x}$ $(x \neq 0)$ $\{6,-5\}$

In Exercises 43–48, use the zero product principle to make up an equation whose solution set is given.

43. $\{3,5\}$ $x^2 - 8x + 15 = 0$
44. $\{-5,-7\}$ $x^2 + 12x + 35 = 0$
45. $\{6,-6\}$ $x^2 - 36 = 0$
46. $\{10,-10\}$ $x^2 - 100 = 0$
47. $\{2,4,6\}$ $x^3 - 12x^2 + 44x - 48 = 0$

48. $\{-1, -3, -5\}$ $x^3 + 9x^2 + 23x + 15 = 0$

49. The height (y) in feet of a baseball that is thrown straight up from the ground with an initial velocity of 33 mi/hour is given approximately by $y = 48t - 16t^2$, where t is elapsed time in seconds. (*Note:* 33 mi/hour is about 48 ft/second.) For what value(s) of t is the ball 36 ft off the ground? $t = \frac{3}{2}$ seconds

50. For what value(s) of t is the ball described in Exercise 49 32 ft off the ground? $t = 1, 2$ seconds

51. The height (y) of a hammer dropped from the roof of the tallest building in Atlanta, the C & S Plaza, is given by the formula $y = 1{,}024 - 16t^2$. How long will it take the hammer to hit the ground? 8 seconds

52. The height (y) of a diver who steps off a cliff 64 ft above the water is given by $y = 64 - 16t^2$. To find how long it takes the diver to hit the water, set the height equal to zero and solve the resulting equation. 2 seconds

53. The sum of the squares of two consecutive integers is 113. Find the integers. 7, 8 and -7, -8

54. The product of two consecutive integers is zero. Find the integers. -1, 0 and 0, 1

55. The sum of two numbers is 3 and their product is -40. Find the numbers. (*Hint:* Let x represent one number and $-40/x$ represent the other.) 8 and -5

56. The sum of two numbers is one and their product is -30. Find the number. (*Hint:* Let x represent one number and $-30/x$ represent the other.) -5 and 6

57. The volume and the surface area of a cube are equal. Find the dimensions of the cube, and find the volume and area. Side = 6; area = volume = 216

58. The volume of a cube is half the surface area. Find the dimensions of the cube, and find the volume and area. Side = 3; volume = 27; surface area = 54

59. The area of a square is 21 more than the perimeter. Find the area. The unit of length is feet. 49 ft²

60. The area of a square is equal to its perimeter. Find the area. The unit of length is centimeters. 16 cm²

61. The area of a rectangle is 24 in.². The perimeter is 28 in. Find its dimensions. (*Hint:* Let x represent the length and $24/x$ the width.) Length = 12 in.; width = 2 in.

62. The area of a rectangle is 60 m². The perimeter is 32 m. Find its dimensions. (*Hint:* Let x represent the length and $60/x$ the width.) Length = 10 m; width = 6 m

63. The length of a rectangle is 4 more than the width and the diagonal measures 20 yards (yd). Find the area of the rectangle. 192 yd²

64. The width of a rectangle is 9 ft. The diagonal is 9 less than twice the length. Find the area. 108 ft²

65. A person can drive from A to X to B at 60 mi/hour or directly from A to B at 50 mi/hour. Which way takes less time?

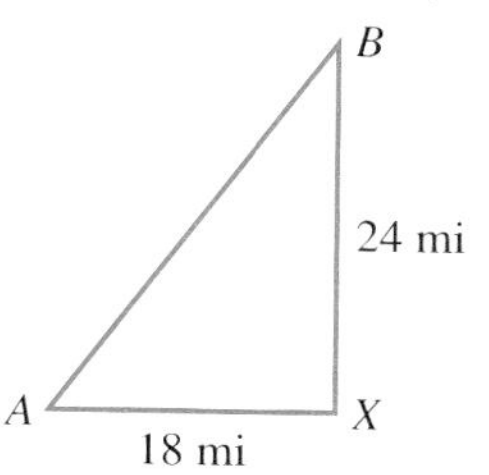

A to B takes 36 minutes; A to X to B takes 42 minutes.

66. A person can drive from A to X to B at 45 mi/hour or directly from A to B at 30 mi/hour. Which way takes less time?

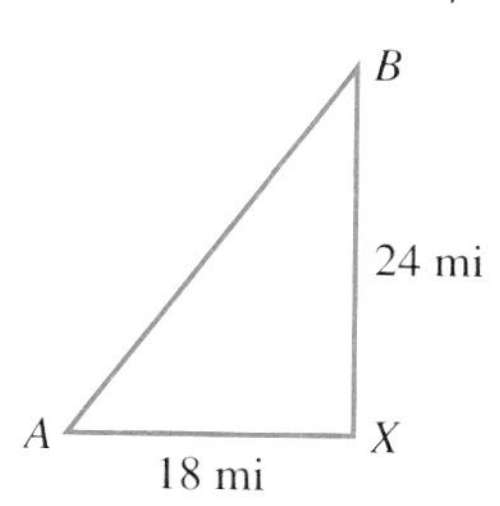

A to B takes 1 hour; A to X to B takes 0.93 hour.

THINK ABOUT IT

1. Second-degree equations are called quadratic because *to quadrate* means "to make square," and any quadratic equation can be rewritten into the form $(x + a)^2 = b$, which is the square of a binomial equal to a constant.
   a. Show that the equation $x^2 + 6x - 7 = 0$ is equivalent to $(x + 3)^2 = 16$ by solving the first equation and then showing that both of its solutions are also solutions of the second equation.
   b. Show by expanding the left side of the second equation in part **a** and then simplifying that you get the first equation.

2. Why are both words *zero* and *product* crucial in the zero product principle?

3. Why is it not correct to solve $(x - 2)(x + 5) = 10$ by setting each factor equal to 10? Show that the results you get this way are wrong.

4. Use the zero product principle to make up a quadratic equation with solution set $\{7\}$.

5. Why can't a quadratic equation have three distinct solutions?

REMEMBER THIS

1. Which of these expressions are undefined? 0/3, 3/0, 0/0. 3/0 and 0/0
2. For what value of x will the denominator of the fraction $\frac{7}{3x-2}$ be equal to 0? $\frac{2}{3}$
3. True or false? No matter what real number replaces x, the expression $x^2 + 2$ can never be equal to zero. True
4. Express 18 as a product of prime factors. $18 = 2 \cdot 3^2$
5. What is the sum of $(a - b)$ and $(b - a)$? 0
6. True or false? $x - y$ is always equal to $(-1)(y - x)$. True
7. Solve $x^2 - 7x + 10 = 0$ by factoring. {2,5}
8. Simplify $(2x^2 - x + 1) - (3x^2 - 2x + 1)$. $-x^2 + x$
9. Graph the solution set of $3x + 2 < x + 6$.

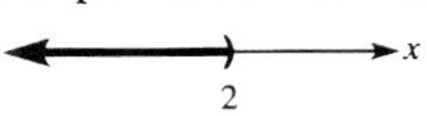

10. Together, two items cost $2.40. One of the items costs $2 more than the other. What is the price of the more expensive item? $2.20

GRAPHING CALCULATOR

Assigning Numerical Values and Algebraic Expressions to Variables

Notice that above the grey [8] key on the right side there is a letter P. The other letters are found similarly in alphabetical order above other keys. After Z are five other characters, including a comma and quotation marks. To produce one of these letters or characters on the screen, you must press the key marked [ALPHA] before you press the desired key. When you press [ALPHA], the cursor on the screen becomes a blinking A, which signals that the next key pressed will show up as an alphabetical character.

EXAMPLE 1 Write the letter P on the screen.

Solution Press [ALPHA]; then press the key with the grey P above it. You should see a P on the screen. If you then press [ENTER], you will see whatever numerical value is currently assigned to the letter P. ┘

Single letters may be used to name memory cells in which you can save numbers. The key for doing this is the black key marked [STO▸], which is an abbreviation for "store in." Pressing [STO▸] causes the cursor to become an A, because the calculator expects you to press a letter next.

EXAMPLE 2 a. Save the value 2.8 in Z and −0.5 in W.
b. Use the saved values to compute Z(W + 3).

Solution a. These sequences will save the values.

2.8 [STO▸] [Z] [ENTER]
−0.5 [STO▸] [W] [ENTER]

b. This sequence leads to the correct answer, 7.

[ALPHA] [Z] [(] [ALPHA] [W] [+] 3 [)] [ENTER] ┘

The black key [Y=] at the top of the calculator produces a list of variables labeled Y_1 to Y_4. These variables are used to store expressions involving the letter X. Example 3 shows one application of this key.

EXAMPLE 3 a. Assign the expression $X^2 - X$ to the variable Y_1.
b. Evaluate the expression Y_1 for X = 8 and −9.

Solution

a. Step 1 Press [Y=].

Step 2 Use the arrow keys to put the cursor just after the equals sign on the line for Y_1. Key in the expression $X^2 - X$ using the key marked [X|T] to write the letter X. (This is quicker than [ALPHA] [X].) The screen should appear as shown in Figure 3.7. At this point, the calculator knows that Y_1 represents $X^2 - X$.

```
:Y1■X²-X
:Y2=
:Y3=
:Y4=
```

Figure 3.7

Step 3 Press [2nd] [QUIT] to get back to the home screen.

b. Step 1 Press 8 [STO▸] [X] to assign the value 8 to X.

Step 2 Press [2nd] [Y-VARS] and then 1 so that Y_1 appears on the screen. Then press [ENTER] to see the value of $X^2 - X$ when $X = 8$. The answer 56 should appear. To get the answer when $X = -9$, just assign -9 to X and repeat step 2. The correct answer is 90.

EXERCISES

1. Save the value 3.7 in Z and -0.6 in W; then use the saved values to compute $W(Z + 3)$. Ans. -4.02
2. Assign the expression $X^2 - X + 1$ to the variable Y_1; then evaluate it for $X = -2.1$ and 3.5. Ans. 7.51, 9.75

Use the graphing calculator for these exercises from the chapter.

Note On the TI-81 the black key [EE] is used for scientific notation. Thus 2.4×10^6 is keyed in as 2.4 [EE] 6.

Section	*Exercises*
3.1	57, 81
3.2	63
3.3	33, 37

Chapter 3 SUMMARY

OBJECTIVES CHECKLIST Specific chapter objectives are summarized below along with numbered example problems from the text that should clarify the objectives. If you do not understand any objectives or do not know how to do the selected problems, then restudy the material.

3.1 **Can you:**

1. **Interpret and write positive integer exponents?**
   Write the expression $(p/q) \cdot (p/q)$ using exponents. Identify the base and the exponent. [Example 1b]
2. **Apply the product property of exponents?**
   Use the product property of exponents to simplify $y^3 \cdot y^5$. [Example 2b]
3. **Use zero as an exponent?**
   Evaluate the expression $5x^0$. Assume $x \neq 0$. [Example 4d]
4. **Use negative integer exponents?**
   Write $3^{-1} - 2^{-1}$ with positive exponents, and evaluate it. [Example 5c]
5. **Apply the quotient property of exponents?**
   Simplify $\frac{3^{-2}}{3^2}$ and write the result using only positive exponents. [Example 6b]
6. **Apply the power properties of exponents?**
   Simplify $(-3x^4y)^2$, and write the result using only positive exponents. [Example 8a]
7. **Evaluate integer exponents on a calculator?**
   Evaluate $E = \dfrac{Pr}{1 - (1 + r)^{-n}}$ when $P = 30{,}000$, $r = 0.01$, and $n = 48$. [Example 10]

3.2 Can you:

1. **Simplify more complex exponential expressions?**
   Simplify $\left(\dfrac{-3x^2y}{5xy^{-1}}\right)^{-2}$, and write the result using only positive exponents. Assume $x \neq 0$ and $y \neq 0$. [Example 1c]
2. **Simplify expressions with literal exponents?**
   Simplify $2^{n+3} \cdot 2^{n+1}$. [Example 2a]
3. **Convert numbers from standard notation to scientific notation, and vice versa?**
   Express 186,000 and 0.0000054 in scientific notation. [Example 3]
4. **Perform computations using scientific notation?**
   Multiply $(4 \times 10^{-4})(8 \times 10^7)$ using scientific notation, and express the result in standard notation. [Example 5a]

3.3 Can you:

1. **Identify polynomials?**
   Which of the following algebraic expressions are polynomials? $x^2 - 5x + 3$; $\dfrac{y}{3}$; $\dfrac{2}{x}$. [Example 1]
2. **State the degree of a polynomial, and identify monomials, binomials, and trinomials?**
   Give the degree of $3x^4 - 5x^3$, and identify it as a monomial, binomial, or trinomial. [Example 2a]
3. **Evaluate polynomials, and use $P(x)$ notation?**
   If $P(x) = 2x^2 - 5x + 3$, find $P(4)$ and $P(-1)$. [Example 4]
4. **Add polynomials?**
   Add $6x^2 - 5x + 1$ and $-5x^2 + 8x - 7$. [Example 6]
5. **Subtract polynomials?**
   Subtract $x^2 - 3x - 4$ from $2x^2 - 4x + 3$. [Example 8]

3.4 Can you:

1. **Multiply a monomial by a monomial?**
   Find $(-3y^5)(5y^3)$. [Example 1a]
2. **Multiply two polynomials?**
   Find the product $(3x^2 - 4x + 1)(x - 2)$. [Example 3b]
3. **Use the FOIL method to multiply binomials?**
   Multiply $(2x - 5)(3x - 1)$ using the FOIL method. [Example 5b]
4. **Multiply binomials that are the sum and difference of two terms?**
   Find $(3x - 4y)(3x + 4y)$. [Example 6c]
5. **Use the formulas for the square of a binomial?**
   Use the square-of-a-binomial formula to find $(2x - 5)^2$. [Example 9b]
6. **Multiply binomials, and then solve linear equations?**
   Solve $(x + 6)(x + 9) - x^2 = 99$. [Example 11]

3.5 Can you:

1. **Find the greatest common factor (GCF) of a set of terms?**
   Find the GCF of $10x^3$, $5x^4$, $15x^2$. [Example 2]
2. **Factor out the greatest common factor from a polynomial?**
   Factor out the GCF from $25x^4 - 15x^2 + 10x^2$. [Example 3b]
3. **Factor by grouping?**
   Factor $6x^2 + 12x + 7x + 14$ by grouping. [Example 6a]

3.6 Can you:

1. **Factor $ax^2 + bx + c$ with $a = 1$?**
   Factor $x^2 + 10 - 11x$. [Example 2]
2. **Factor $ax^2 + bx + c$ with $a \neq 1$ by reversing FOIL?**
   Factor $2x^2 + 7x + 6$. [Example 7]
3. **Factor $ax^2 + bx + c$ with $a \neq 1$ by the *ac* method?**
   Factor $12x^2 + 31x - 30$ by the *ac* method. [Example 12]
4. **Factor trinomials by substitution?**
   Factor $2x^4 + 3x^2 - 9$. [Example 13b]

3.7 Can you:

1. **Identify and factor perfect square trinomials?**
   Factor $4x^2 - 12xy + 9y^2$. [Example 1b]
2. **Factor the difference of two squares?**
   Factor $25y^2 - 64x^2$. [Example 2b]
3. **Factor the sum or difference of two cubes?**
   Factor $x^3 + 125$. [Example 4a]

3.8 Can you:

1. **Apply a general strategy to factor polynomials systematically?**
   Factor completely $16s^4 - t^4$. [Example 4]

3.9 Can you:

1. **Solve certain quadratic equations using factoring?**
   Solve $y^2 + 9 = 6y$. [Example 2]
2. **Solve higher-degree equations using factoring?**
   Solve $y^3 = 16y$. [Example 4b]
3. **Solve applied problems that lead to quadratic equations?**
   The height (y) of a projectile that is shot directly up from the ground with an initial velocity of 64 ft/second is given by the formula

$$y = 64t - 16t^2,$$

   where t is the elapsed time in seconds. For what value(s) of t is the projectile 48 ft off the ground? [Example 5]

KEY TERMS

ac method (3.6)
Base (3.1)
Binomial (3.3)
Degree of a monomial (3.3)
Degree of a polynomial (3.3)
Difference of cubes (3.7)
Difference of squares (3.7)
Exponent (3.1)
Exponential expression (3.1)
Factoring (3.5)
Factoring by grouping (3.5)
FOIL method (3.4)
GCF, greatest common factor (3.5)
Hypotenuse (3.9)
Legs of a right triangle (3.9)
Literal exponents (3.2)
Monomial (3.3)
Negative exponent (3.1)
*n*th power of *a* (3.1)
Perfect square trinomial (3.7)
Polynomial (3.3)
Power (3.1)
Prime polynomial (3.6)
Pythagorean theorem (3.9)
Quadratic equation (3.9)
Quadratic form (3.6)
Right triangles (3.9)
Scientific notation (3.2)
Sum of cubes (3.7)
Trinomial (3.3)
Zero exponent (3.1)
Zero product principle (3.9)

KEY CONCEPTS AND PROCEDURES

Section	Key Concepts or Procedures to Review
3.1	■ Laws of exponents (m and n denote integers) **1.** $a^m a^n = a^{m+n}$ **2.** $a^0 = 1$ $(a \neq 0)$ **3.** $a^{-n} = \frac{1}{a^n}$ $(a \neq 0)$ **4.** $\frac{a^m}{a^n} = a^{m-n}$ $(a \neq 0)$ **5.** $(a^m)^n = a^{mn}$ **6.** $(ab)^n = a^n b^n$ **7.** $\left(\frac{a}{b}\right)^n = \frac{a^n}{b^n}$ $(b \neq 0)$
3.2	■ Methods to convert a number from standard notation to scientific notation, and vice versa
3.3	■ Methods to evaluate polynomials ■ Methods to add and subtract polynomials
3.4	■ Methods to multiply polynomials ■ FOIL multiplication method ■ Product models **1.** $(a + b)(a - b) = a^2 - b^2$ **2.** $(a + b)^2 = a^2 + 2ab + b^2$ **3.** $(a - b)^2 = a^2 - 2ab + b^2$
3.5	■ Methods to find and factor out the greatest common factor (GCF) of the terms in a polynomial ■ Methods to factor by grouping
3.6	■ Methods to factor a trinomial of the form $ax^2 + bx + c$
3.7	■ Factoring models for an expression that is a perfect square trinomial, that is a difference of squares, or that is the sum or difference of cubes
3.8	■ Guidelines for factoring a polynomial
3.9	■ Definition of quadratic equation ■ Zero product principle: $a \cdot b = 0$ if and only if $a = 0$ or $b = 0$. ■ Factoring method for solving quadratic equations ■ Pythagorean theorem

CHAPTER 3 REVIEW EXERCISES

3.1

1. Write $(-b)(-b)(-b)(-b)(-b)(-b)$ using exponents. Identify the base and the exponent. $(-b)^6$; base $= -b$; exponent $= 6$
2. Use the product property to simplify $x \cdot x^5$. x^6
3. Evaluate $4y^0$ if $y = 12$. 4
4. Simplify $\frac{3^4b^{-4}}{3^3b^4}$, and write the result using only positive exponents. Assume $b \neq 0$. $\frac{3}{b^8}$
5. Evaluate $E = \frac{Pr}{1-(1+r)^{-n}}$ if $r = \frac{0.108}{12} = 0.009$, $P = 100{,}000$ and $n = 20 \cdot 12 = 240$. This gives the monthly payment for a \$100,000 loan at 10.8 percent annual interest over 20 years. \$1,018.61

3.2

6. Simplify $\left(\frac{-3ab}{4ab^{-3}}\right)^{-1}$, and write the result using only positive exponents. Assume a and b are not zero. $\frac{-4}{3b^4}$
7. Simplify $\frac{3^n}{3^{n-1}}$, where n is an integer. 3
8. A certain radio station broadcasts at a frequency of about 1,260,000 hertz. Write this number in scientific notation. 1.26×10^6
9. Use scientific notation to compute $\frac{-2.4 \times 10^3}{3.0 \times 10^{-3}}$. -8×10^5; $-800{,}000$
10. If one red blood cell contains 270,000,000 hemoglobin molecules, about how many molecules of hemoglobin are there in 2 million red blood cells? 5.4×10^{14}

3.3

11. Is $5x^{-3} + 4x^{-2} + 3$ a polynomial? No, because the exponents are negative
12. For $-4 + x^3 - 3x^2 + x$, give the degree and identify it as a monomial, binomial, trinomial, or none of these. Degree $= 3$; none of these
13. If $P(x) = 2x^2 - 3x + 4$, find $P(-1)$ and $P(1)$. $P(-1) = 9$; $P(1) = 3$
14. Find the sum of $-2y^2 - 6y - 4$ and $3y^2 + 6y - 8$. $y^2 - 12$
15. Simplify $(x^2 - 4x + 1) - (-2x^2 - 4x - 3) - (x^2 + x + 1)$. $2x^2 - x + 3$

3.4

16. Multiply $\left(\frac{3a^4}{7}\right)\left(\frac{7a^2}{3}\right)$. a^6
17. Multiply $(x - 3)(x^2 + 2x - 4)$. $x^3 - x^2 - 10x + 12$
18. Use FOIL to find $(2x + 5)(4x - 3)$. $8x^2 + 14x - 15$
19. Expand $(2x + y)^2$. $4x^2 + 4xy + y^2$
20. The shaded area shown in the diagram equals 160 in.^{2}. Find the area of the larger rectangle. $x = 2$; $A = 168$ in.^{2}

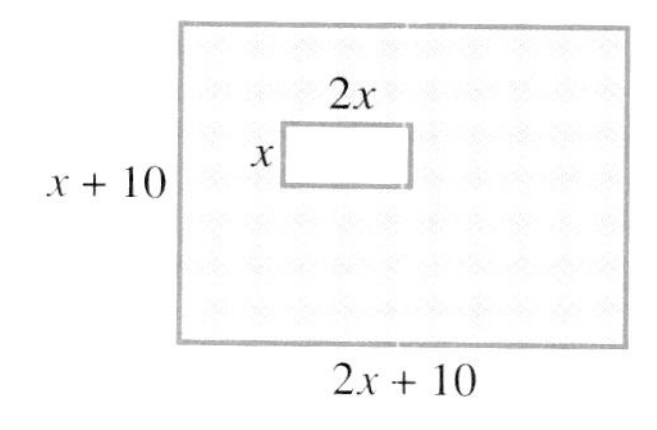

3.5

21. Find the GCF of $40x^3y^3$ and $32xy^2$. $8xy^2$
22. Factor out the GCF from $5x^3y + 4xy^3$. $xy(5x^2 + 4y^2)$
23. Factor out the GCF from $7(y - 3) - y(y - 3)$. $(y - 3)(7 - y)$
24. Factor by grouping: $6x^2 + 3x + 8x + 4$. $(2x + 1)(3x + 4)$
25. What is the overall result of raising a price by 10 percent and then raising that price by 10 percent? Factor out $x + 0.10x$ from $(x + 0.10x) + 0.10(x + 0.10x)$ and simplify to find out. $1.21x$; the result is a 21 percent increase of the original price.

3.6

Factor completely.

26. $x^2 - 7x + 10$ $(x - 2)(x - 5)$
27. $y^2 + 5y - 24$ $(y + 8)(y - 3)$
28. $x^2 - 3ax + 2a^2$ $(x - a)(x - 2a)$
29. $18x^4 + 3x^2 - 1$ $(6x^2 - 1)(3x^2 + 1)$
30. Which of these are prime polynomials?
    a. $x^2 + x + 1$ b. $x^2 + 2x + 1$ c. $x^2 + 4x + 3$ a

3.7

Factor completely.

31. $9x^2 - 24x + 16$ $(3x - 4)^2$
32. $9x^2 - 25y^2$ $(3x - 5y)(3x + 5y)$
33. $6y^3 - 6y$ $6y(y + 1)(y - 1)$
34. $x^3y^3 + 27$ $(xy + 3)(x^2y^2 - 3xy + 9)$
35. $c^2 - cd - 2d^2$ $(c + d)(c - 2d)$

3.8

36. Factor $2x^3 - 18x + 28x$. $2x(x - 2)(x - 7)$
37. Factor $abc + ac + 2bc + 2c$. $c(a + 2)(b + 1)$
38. Factor $b^6 - c^6$. $(b + c)(b - c)(b^2 - bc + c^2)(b^2 + bc + c^2)$
39. Factor $x^4 - h^4$. $(x^2 + h^2)(x + h)(x - h)$
40. Find an expression for the perimeter of this rectangle, and write it in factored form. $2(a - b)^2$

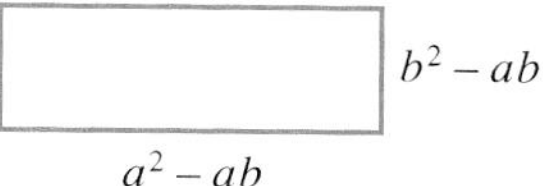

3.9

41. Solve $2x^2 - 9x + 4 = 0$. $\{\frac{1}{2}, 4\}$
42. Solve $(x - 3)(x + 3) + 5 = 0$. $\{-2, 2\}$
43. Solve $(2y - 1)(3y + 4)(4y + 1) = 0$. $\{\frac{1}{2}, -\frac{4}{3}, -\frac{1}{4}\}$
44. Solve $x + 1 = \frac{42}{x}$, $x \neq 0$. $\{-7, 6\}$
45. If the area of a square is equal to twice its perimeter, find the area. 64 square units

ADDITIONAL REVIEW EXERCISES

46. Express 0.000108 in scientific notation. 1.08×10^{-4}
47. Which of the following is a trinomial of degree 2?
    a. $4x^3 + 6x^2$ b. $7x^2 + 3x - 1$ c. $2x^3 + 4x + 6$ b
48. Evaluate $-2x^{-1}$ if $x = \frac{1}{2}$. -4

49. Write $(x + y)(x + y)(x + y)$ using exponents. Identify the base and the exponent. $(x + y)^3$; base: $x + y$; exponent: 3

50. Express 2.3×10^5 in standard notation. 230,000

51. Expand $(x - 2y)^2$. $x^2 - 4xy + 4y^2$

52. The height (y) of a projectile that is shot directly up from the ground with an initial velocity of 80 ft/second is given by the formula

$$y = 80t - 16t^2,$$

where t is the elapsed time in seconds. For what value(s) of t is the projectile 64 ft off the ground? $t = 1, 4$

53. The product of two consecutive integers is 240. Find the integers. -15 and -16; 15 and 16

54. The length of a rectangle is 5 more than the width and the diagonal measures 25 ft. Find the area of the rectangle. 300 ft²

55. Factor out the GCF: $8x^3y^2 + 24x^2y^3 + 12xy^2$. $4xy^2(2x^2 + 6xy + 3)$

56. Factor out the GCF: $m(m + 2) + 3(m + 2)$. $(m + 2)(m + 3)$

57. Factor by grouping: $2x^2 - 6x + 5x - 15$. $(x - 3)(2x + 5)$

58. Multiply and express in scientific notation: $(4.2 \times 10^{12})(2.5 \times 10^{-18})$. 1.05×10^{-5}

59. If $P(x) = -2x^3 - 3x^2 + x - 1$, find $P(-2)$. 1

Solve.

60. $x^2 + 12 = 7x$ $\{3,4\}$

61. $8x^3 = 72x$ $\{-3,0,3\}$

62. $9y^2 = 36y$ $\{0,4\}$

63. $(2y + 5)(y - 3) = 0$ $\{-\frac{5}{2}, 3\}$

Simplify and write the result using only positive exponents.

64. $(2a^3)^4(b^4)^2$ $16a^{12}b^8$

65. $(3xy^{-2})^3$ $\frac{27x^3}{y^6}$

66. $2^{x+2} \cdot 2^x$ 2^{2x+2}

67. $(-x)^4(-x)^3(-x)$ $(-x)^8$ or x^8

68. $\frac{3^{-2}x^5}{3^{-4}x^{-2}}$ $9x^7$

69. $\left(\frac{3a^2}{7}\right)^{-2}$ $\frac{49}{9a^4}$

Perform the indicated operation and simplify.

70. $(2 + 5x)(2 - 5x)$ $4 - 25x^2$

71. $(32y^{10})\frac{y^{-6}}{4}$ $8y^4$

72. $(a - 3b - 1)(-4a^2b^2)$ $-4a^3b^2 + 12a^2b^3 + 4a^2b^2$

73. $(4x^2 - 2x + 3) + (-x^2 - 3x + 1)$ $3x^2 - 5x + 4$

74. $(3x + 5)(2x - 6)$ $6x^2 - 8x - 30$

75. $(3y^2 - 2y + 4)(y - 3)$ $3y^3 - 11y^2 + 10y - 12$

76. $\left(\frac{x^2}{2} - \frac{2x}{5} - \frac{4}{9}\right) - \left(\frac{3x^2}{2} - \frac{3x}{5} + \frac{5}{9}\right)$ $-x^2 + \frac{x}{5} - 1$

Factor completely.

77. $x^2 + 12x + 32$ $(x + 8)(x + 4)$

78. $a^3 + 64$ $(a + 4)(a^2 - 4a + 16)$

79. $12x^2 - 31x - 30$ $(4x + 3)(3x - 10)$

80. $8x^2 - 15x - 2$ $(8x + 1)(x - 2)$

81. $s^2 - 8s - 20$ $(s - 10)(s + 2)$

82. $y^2 - 3by - 10b^2$ $(y - 5b)(y + 2b)$

83. $2m^3 + 26m^2 + 60m$ $2m(m + 10)(m + 3)$

84. $x^2 + 7 - 8x$ $(x - 7)(x - 1)$

85. $4x^2 + 12xy + 9y^2$ $(2x + 3y)^2$

86. $2y^2 - 13y + 18$ $(2y - 9)(y - 2)$

87. $16y^4 - 1$ $(2y - 1)(2y + 1)(4y^2 + 1)$

88. $t^4 + 2t^2 - 3$ $(t - 1)(t + 1)(t^2 + 3)$

89. $25s^8 - 9t^4$ $(5s^4 - 3t^2)(5s^4 + 3t^2)$

90. $24x^4 + 52x^3 + 24x^2$ $4x^2(3x + 2)(2x + 3)$

91. $s^2t^3 - 27s^8$ $s^2(t - 3s^2)(t^2 + 3s^2t + 9s^4)$

CHAPTER 3 TEST

1. Simplify $3y^4 \cdot 5y^5$ $15y^9$

2. Evaluate $\frac{16^{-1}}{4^{-3}}$. 4

3. Simplify $3^m \cdot 3$. 3^{m+1}

4. Express 1,400 in scientific notation. 1.4×10^3

5. Which of these are polynomials?

a. $3 - x^2 + 5x$ **b.** $2x + \frac{1}{3x^2}$ **c.** 7 **d.** $\frac{x}{5}$ All except b

6. What is the degree of $4x^5 - 3x^2 + 2$? 5

7. Find the product: $(2x^2 - 5x + 3)(x - 1)$. $2x^3 - 7x^2 + 8x - 3$

For Questions 8–15, factor the given expression completely. If it cannot be factored, write "Prime polynomial."

8. $5x^2 + 10x$ $5x(x + 2)$

9. $x^2 + 10x + 9$ $(x + 1)(x + 9)$

10. $4a^2 - 4a + 1$ $(2a - 1)^2$

11. $16y^2 - 25$ $(4y + 5)(4y - 5)$

12. $n^3 + 1$ $(n + 1)(n^2 - n + 1)$

13. $x^2 + 4$ Prime polynomial

14. $6x^2 - 5x - 6$ $(2x - 3)(3x + 2)$

15. $3a^3b^2 + 3a^2b - 6a$ $3a(ab - 1)(ab + 2)$

16. Factor by grouping: $3x^2 - 6x + 2x - 4$. $(3x + 2)(x - 2)$

For problems 17–19, solve the given equation, show the solution set, and show the check for each solution.

17. $(x - 3)^2 = x^2 + 18$ $\{-\frac{3}{2}\}$; check: $\frac{81}{4} = \frac{81}{4}$

18. $(2x + 1)(3x - 5)x = 0$ $\{-\frac{1}{2}, \frac{5}{3}, 0\}$; all checks give $0 = 0$.

19. $x^2 + 4x = -4$ $\{-2\}$; check: $(-2)^2 + 4(-2) = -4$

20. The area of a triangle is 30 in.², and the height is 4 in. less than the base. Find the height of the triangle. 6 in.

CUMULATIVE TEST 3

1. What is the name of this set of numbers? $\{\ldots,-3,-2,-1,0,1,2,3,\ldots\}$. The integers
2. Which property of the real numbers justifies replacing bc by cb? Commutative property of multiplication
3. a. Simplify $\sqrt{n^2-(n+2)(n-2)}$. 2
   b. Evaluate $\sqrt{30{,}421^2-(30{,}423)(30{,}419)}$. 2
4. Evaluate $\dfrac{3n-4m^2}{\frac{10}{3}m+n}$ when $m=-1$ and $n=\frac{1}{3}$. 1
5. Why is $\frac{1}{2}$ called a rational number? It is the ratio of two integers.
6. Simplify $-3x-2(-5x+4)$. $7x-8$
7. Solve $12x-5=19$. $\{2\}$
8. Solve $A=2\pi r$ for r. $r=A/(2\pi)$
9. A 10-ft metal rod is cut into three pieces so that the longest piece is twice as long as the shortest and 1 ft longer than the middle-sized one. Find the lengths of all three pieces. 2.2, 3.4, and 4.4 ft
10. Graph the solution set of $3x+4<2x-1$.

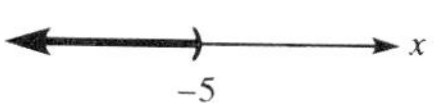

11. Graph the solution set of $|5-x|\le 2$.

12. Solve $|-2x+5|<7$. $\{x: -1<x<6\}$ or $(-1,6)$
13. Write $2^{-1}+4^{-1}$ with positive exponents and evaluate it. $\frac{1}{2}+\frac{1}{4}=\frac{3}{4}$
14. Arrange these values in *increasing* order; no calculations are necessary: $\dfrac{143}{1.04^{-12}}$, 143×1.04^{-12}, 143^{-12}. 143^{-12}, 143×1.04^{-12}, $\dfrac{143}{1.04^{-12}}$
15. Simplify $\left(\dfrac{-4x^3y^2}{8xy^{-1}}\right)^{-3}$, and write the result using only positive exponents. $\dfrac{-8}{x^6y^9}$
16. Divide $\dfrac{3.33\times 10^{-2}}{1.11\times 10^{-3}}$. 30
17. Make up an example of a binomial of degree 5. Many correct answers; an example is x^5+2.
18. If $P(x)=3x+1$ and $Q(x)=2x-1$, find $\dfrac{P(3)}{Q(3)}$. 2
19. Factor $2x^2+3x-2$. $(2x-1)(x+2)$
20. Solve $2x^2+7x=4$. $\{\frac{1}{2},-4\}$

CHAPTER 4

Rational Expressions

If the probability that a driver will be in an automobile accident next year is represented by p, then the **odds** that this event will not happen is given by

$$\frac{1-p}{p}.$$

a. Find the odds that a driver will not be in an automobile accident next year if $p = 0.06$.

b. For what value of p is the odds undefined? (See Example 2 of Section 4.1.)

IN THIS chapter we examine expressions which are quotients of two polynomials. They are called rational expressions by analogy to the definition of rational *numbers,* and we will see that the principles for working with these expressions are direct extensions of the methods that govern fractions in ordinary arithmetic.

4.1 Simplification of Rational Expressions

OBJECTIVES

1. **Find all values of a variable which make a rational expression undefined.**
2. **Determine whether a pair of fractions are equivalent.**
3. **Express rational expressions in lowest terms.**
4. **Build up a rational expression to an equivalent expression with a specified denominator.**

1 In the chapter-opening problem the expression for the odds

$$\frac{1-p}{p}$$

is an example of a rational expression. The numerator is the polynomial $1 - p$, and the denominator is the polynomial p.

Rational Expression

A **rational expression** is an expression of the form

$$\frac{A}{B},$$

where A and B are polynomials with $B \neq 0$.

Some other examples of rational expressions are

$$\frac{1}{3}, \quad \frac{y^2+4}{4y-7}, \quad \text{and} \quad \frac{xy}{y+x}.$$

In the definition above we state that B, the polynomial in the denominator, cannot be zero because division by zero is undefined (as discussed in Section 1.2). It will therefore be important when using rational expressions to note any values for a variable which lead to a zero denominator, because they will cause the expression to be meaningless.

EXAMPLE 1 Find all values of the variable which make each rational expression undefined.

a. $\dfrac{1}{2x-5}$ **b.** $\dfrac{y}{y^2-16}$ **c.** $\dfrac{x-1}{x^2+1}$

Solution We must find any values which make the *denominators* equal to zero.

a. Set the denominator $2x - 5$ equal to zero and solve.

$$\begin{aligned} 2x - 5 &= 0 \\ 2x &= 5 \\ x &= \tfrac{5}{2} \end{aligned}$$

The only number that makes $1/(2x - 5)$ undefined is $\frac{5}{2}$.

b. Set the denominator $y^2 - 16$ equal to zero and solve.

$$y^2 - 16 = 0$$
$$(y + 4)(y - 4) = 0$$
$$y + 4 = 0 \quad \text{or} \quad y - 4 = 0$$
$$y = -4 \qquad\qquad y = 4$$

Thus, $y/(y^2 - 16)$ is undefined when y equals -4 and when y equals 4.

c. Because x^2 is always greater than or equal to zero, no matter what the value of x is, the denominator $x^2 + 1$ is always greater than or equal to one. Thus, no replacements for x leads to a denominator of zero. This means that there are no real values for which $(x - 1)/(x^2 + 1)$ is undefined.

PROGRESS CHECK 1 Find all values of the variable which make each rational expression undefined.

a. $\dfrac{x - 5}{6x + 2}$ **b.** $\dfrac{1}{x^2 - 2x - 15}$ **c.** $\dfrac{y^2 - 25}{3}$

EXAMPLE 2 Solve the problem in the chapter introduction on page 160.

Solution

a. Replace p by 0.06 in the expression given for the odds. If $p = 0.06$, then

$$\frac{1 - p}{p} = \frac{1 - 0.06}{0.06} = \frac{0.94}{0.06} = \frac{94}{6} = \frac{47}{3}.$$

The value of the odds against being in an automobile accident next year is $\frac{47}{3}$. This would commonly be expressed as "the odds against being in an automobile accident next year are 47 to 3."

b. By setting the denominator equal to zero, we see that the value of p for which the odds against being in an accident is undefined is $p = 0$. (*Note:* Also, since probabilities are never outside the range of 0 to 1, the only sensible values for p in the odds expression are in the interval $0 < p \leq 1$.)

PROGRESS CHECK 2 In Example 2, if we ask for the odds that a driver *will* be in an automobile accident during the next year, then the expression for the odds is

$$\frac{p}{1 - p}.$$

a. Find the odds that a race car driver will be in a racing accident next year if $p = 0.6$.

b. For what value of p is this expression for odds undefined?

2 When different fractions have the same value (such as $\frac{1}{2}$ and $\frac{5}{10}$), they are called equivalent, and it is useful to have a simple method for determining if two fractions are equivalent.

Equivalence of Fractions

Let a, b, c, and d be real numbers, with b and $d \neq 0$.

$$\frac{a}{b} = \frac{c}{d}, \quad \text{if and only if} \quad ad = bc.$$

The products ad and bc in the box above are often called **cross products**.

Progress Check Answers

1. (a) $-\frac{1}{3}$ (b) $5, -3$ (c) None
2. (a) 3 to 2 (b) 1

EXAMPLE 3 Determine whether each pair of fractions are equivalent.

a. $\frac{1}{3}, \frac{2}{6}$ **b.** $\frac{-5}{3}, \frac{5}{-3}$ **c.** $\frac{27}{41}, \frac{2}{3}$ **d.** $\frac{2 \cdot 3}{5 \cdot 3}, \frac{2}{5}$

Solution Use the equivalence-of-fractions principle and compare cross products.

a. Because $1(6) = 3(2)$, $\frac{1}{3}$ and $\frac{2}{6}$ are equivalent fractions.

b. Because $(-5)(-3) = 3(5)$, $\frac{-5}{3}$ and $\frac{5}{-3}$ are equivalent fractions.

c. Because $27(3) \neq 41(2)$, $\frac{27}{41}$ and $\frac{2}{3}$ are not equivalent fractions.

d. Because $(2 \cdot 3)5 = (5 \cdot 3)2$, $\frac{2 \cdot 3}{5 \cdot 3}$ and $\frac{2}{5}$ are equivalent fractions.

PROGRESS CHECK 3 Determine whether each pair of fractions are equivalent.

a. $\frac{3}{9}, \frac{1}{3}$ **b.** $\frac{2}{7}, \frac{4}{13}$ **c.** $\frac{-3}{-5}, \frac{3}{5}$ **d.** $\frac{3 \cdot 6}{3 \cdot 7}, \frac{6}{7}$

3 Although there are an unlimited variety of equivalent forms for a fraction, it is often necessary to express a fraction in lowest terms. To express a fraction in lowest terms, we apply the fundamental principle of fractions, which allows us to divide out all nonzero *factors* which appear in both the numerator and the denominator.

Fundamental Principle of Fractions

If a, b, and k are real numbers, with $b \neq 0$ and $k \neq 0$, then

$$\frac{ak}{bk} = \frac{a}{b}.$$

Note that this principle follows from the criteria for the equivalence of two fractions, since $(ak)b$ is equal to $(bk)a$. This implies that reducing a fraction to lowest terms does not alter its value.

EXAMPLE 4 Express $\frac{18}{54}$ in lowest terms.

Solution We write each number as a product of prime factors to see which factors are common. We can express 18 as $2 \cdot 3 \cdot 3$ and 54 as $2 \cdot 3 \cdot 3 \cdot 3$.

$$\frac{18}{54} = \frac{2 \cdot 3 \cdot 3}{2 \cdot 3 \cdot 3 \cdot 3}$$ Write 18 and 54 in prime factored form.

$$= \frac{\overset{1}{\cancel{2}} \cdot \overset{1}{\cancel{3}} \cdot \overset{1}{\cancel{3}}}{\underset{1}{\cancel{2}} \cdot \underset{1}{\cancel{3}} \cdot \underset{1}{\cancel{3}} \cdot 3}$$ Divide out the common factor $2 \cdot 3 \cdot 3$ according to the fundamental principle.

$$= \frac{1}{3}$$

In lowest terms, $\frac{18}{54}$ is expressed as $\frac{1}{3}$.

PROGRESS CHECK 4 Express $\frac{24}{22}$ in lowest terms.

The method for reducing rational expressions to lowest terms is a direct extension of the one for ordinary fractions, and we state it in the following box.

Progress Check Answers

3. (a) Yes (b) No (c) Yes (d) Yes
4. $\frac{12}{11}$

Fundamental Principle of Rational Expressions

If A, B, and K are polynomials, with $B \neq 0$ and $K \neq 0$, then

$$\frac{AK}{BK} = \frac{A}{B}.$$

We illustrate the principle in Example 5.

EXAMPLE 5 Express $\dfrac{2x + 8}{5x + 20}$ in lowest terms.

Solution First, factor completely the numerator and the denominator to see if there are any common factors.

$$\frac{2x + 8}{5x + 20} = \frac{2(x + 4)}{5(x + 4)}$$

Then we divide out this common factor according to the fundamental principle, provided $x \neq -4$ (which would make the denominator equal to zero).

Dividing out common *factors* cannot be stressed too much. Students make lots of mistakes canceling incorrectly. Illustrations like $\dfrac{2+3}{7+3} \neq \dfrac{2}{7}$ are helpful.

$$\frac{2\overset{1}{\cancel{(x + 4)}}}{5\underset{1}{\cancel{(x + 4)}}} = \frac{2}{5}$$

In lowest terms, $\dfrac{2x + 8}{5x + 20} = \dfrac{2}{5}$, provided that $x \neq -4$.

PROGRESS CHECK 5 Express $\dfrac{3x - 18}{4x - 24}$ in lowest terms.

On the basis of Example 5, we state a two-step procedure for reducing a rational expression to lowest terms.

To Reduce Rational Expressions to Lowest Terms

1. Factor completely the numerator and the denominator of the fraction.
2. Divide out nonzero factors that are common to the numerator and the denominator according to the fundamental principle.

EXAMPLE 6 Express each rational expression in lowest terms.

a. $\dfrac{4x + 12}{x^2 + 2x - 3}$ **b.** $\dfrac{y^2 - 4}{(y + 2)^2}$

Solution Factor and use the fundamental principle.

a. $\dfrac{4x + 12}{x^2 + 2x - 3} = \dfrac{4(x + 3)}{(x - 1)(x + 3)} = \dfrac{4}{x - 1}$, for $x \neq -3$ and $x \neq 1$

b. $\dfrac{y^2 - 4}{(y + 2)^2} = \dfrac{(y + 2)(y - 2)}{(y + 2)(y + 2)} = \dfrac{y - 2}{y + 2}$, for $y \neq -2$

$\dfrac{4x + 12}{x^2 + 2x - 3}$ reduces to $\dfrac{4}{x - 1}$ because both fractions have the same value for any admissible value of x. Have students try a few values for x, like 0, 1, 2, -3.

PROGRESS CHECK 6 Express each rational expression in lowest terms.

a. $\dfrac{x^2 - 9}{5x - 15}$ **b.** $\dfrac{x^2 - 5x + 6}{x^2 - x - 6}$

Progress Check Answers

5. $\frac{3}{4}$, $x \neq 6$

6. (a) $\dfrac{x + 3}{5}$, $x \neq 3$ (b) $\dfrac{x - 2}{x + 2}$, $x \neq 3, -2$

In Example 6 we were careful to point out the values for which the given expressions were undefined, but from this point on we will not usually list such values. It should be understood that any necessary restrictions always apply.

In the next example we deal with a special case, which is discussed further in the note which follows the solution.

EXAMPLE 7 Express $\frac{x}{3x^2 + x}$ in lowest terms.

Solution We factor the denominator and apply the fundamental principle.

$$\frac{x}{3x^2 + x} = \frac{x}{x(3x + 1)} = \frac{1}{3x + 1}$$

Note Because the common factor x makes up the entire numerator, it must be replaced by 1 when we divide it out.

$$\frac{x}{x(3x + 1)} = \frac{\overset{1}{\cancel{x}}}{\underset{1}{\cancel{x}}(3x + 1)} = \frac{1}{3x + 1}$$

Also, remember that the fundamental principle applies *only to factors,* not to terms. Factors are the components in a *product,* and terms are the components in a *sum.* Dividing out *terms* leads to the following kind of mistake.

$$\frac{x}{3x^2 + x} = \frac{\overset{1}{\cancel{x}}}{3x^2 + \underset{1}{\cancel{x}}} = \frac{1}{3x^2 + 1} \qquad \textbf{Wrong}$$

Point out that the mistake can be noticed by checking the cross products.

PROGRESS CHECK 7 Express $\frac{y}{y^2 - y}$ in lowest terms.

It is important when reducing rational expressions to understand what happens when we reverse the order in which numbers are written in addition and subtraction. In addition, $a + b$ is equal to $b + a$; but in subtraction, $a - b$ is the *opposite* of $b - a$. In symbols, $a - b$ is equal to $-1(b - a)$. Thus,

$$\frac{a + b}{b + a} = 1 \quad \text{but} \quad \frac{a - b}{b - a} = \frac{-1(b - a)}{b - a} = -1.$$

Stress *in words* that $b - a$ can always be replaced by $-1(a - b)$ and that an expression divided by its negative yields -1.

Both of these concepts are used in the next example.

EXAMPLE 8 Express each rational expression in lowest terms.

a. $\frac{y^2 + by}{b^2 + by}$ **b.** $\frac{2 - r}{r - 2}$ **c.** $\frac{y - x}{x^2 - y^2}$

Solution

a. We will use the fact that $y + b = b + y$.

$$\frac{y^2 + by}{b^2 + by} = \frac{y(y + b)}{b(b + y)} = \frac{y}{b}$$

b. This quotient must equal -1, because the quotient of two (nonzero) opposites is always -1.

$$\frac{2 - r}{r - 2} = \frac{-1(r - 2)}{r - 2} = -1$$

Progress Check Answer

7. $\frac{1}{y - 1}$

c. $x^2 - y^2$ factors into $(x + y)(x - y)$, and $y - x$ is the opposite of $x - y$, so

$$\frac{y - x}{x^2 - y^2} = \frac{-1(x - y)}{(x + y)(x - y)} = \frac{-1}{x + y}.$$

Note In Section 1.2 we showed that the placement of a negative sign in a fraction is arbitrary. Thus, the answer in part **c** could be expressed in any one of three ways:

$$\frac{-1}{x + y} = -\frac{1}{x + y} = \frac{1}{-(x + y)}.$$

In general, for polynomials A and B, $B \neq 0$,

$$-\frac{A}{B} = \frac{-A}{B} = \frac{A}{-B}.$$

The form $A/-B$, however, is rarely used.

PROGRESS CHECK 8 Express each rational expression in lowest terms.

a. $\frac{x - 1}{1 - x}$ **b.** $\frac{4 - y^2}{y - 2}$ **c.** $\frac{y - x}{x^2 - 2xy + y^2}$

4 Up until this point the fundamental principle of fractions has been used to express a fraction in lowest terms. A second application of this principle occurs when adding (or subtracting) fractions, as shown below.

fundamental principle

Adding fractions: $\frac{1}{9} + \frac{2}{3} = \frac{1}{9} + \frac{2 \cdot 3}{3 \cdot 3} = \frac{1}{9} + \frac{6}{9} = \frac{7}{9}$

The next example shows how to build up a rational expression to an equivalent expression with a specified denominator (in preparation for adding and subtracting rational expressions in Section 4.3).

EXAMPLE 9 Build up each rational expression to an equivalent expression with the indicated denominator.

a. $\frac{5}{12x^2y} = \frac{?}{36x^2y^3}$ **b.** $\frac{x}{x + 3} = \frac{?}{x^2 - 9}$

Solution Use the fundamental principle to build up each fraction.

a. Because $(36x^2y^3)/(12x^2y) = 3y^2$, build up to the indicated form by multiplying in the numerator and denominator by $3y^2$.

$$\frac{5}{12x^2y} = \frac{5 \cdot 3y^2}{12x^2y \cdot 3y^2} = \frac{15y^2}{36x^2y^3}$$

b. Because $x^2 - 9$ factors as $(x + 3)(x - 3)$, we multiply by $x - 3$ in the numerator and denominator.

$$\frac{x}{x + 3} = \frac{x(x - 3)}{(x + 3)(x - 3)} = \frac{x^2 - 3x}{x^2 - 9}$$

PROGRESS CHECK 9 Rewrite each expression with the indicated denominator.

a. $\frac{4}{9xy^3} = \frac{?}{36x^2y^3}$ **b.** $\frac{3x}{x - 2} = \frac{?}{x^2 - 4}$

Progress Check Answers

8. (a) -1 (b) $-1(2 + y)$ (c) $\frac{-1}{x - y}$

9. (a) $\frac{16x}{36x^2y^3}$ (b) $\frac{3x^2 + 6x}{x^2 - 4}$

EXERCISES 4.1

In Exercises 1–14, find all values of the variable which make each rational expression undefined.

1. $\frac{1}{3x-4}$ $\frac{4}{3}$
2. $\frac{2}{4x-5}$ $\frac{5}{4}$
3. $\frac{3}{y+2}$ -2
4. $\frac{3}{2y+1}$ $-\frac{1}{2}$
5. $\frac{x}{7}$ No value
6. $\frac{3y}{5}$ No value
7. $\frac{8}{x}$ 0
8. $\frac{-7}{y}$ 0
9. $\frac{x}{x^2-9}$ 3, -3
10. $\frac{2x}{x^2-1}$ 1, -1
11. $\frac{x^2-16}{x-5}$ 5
12. $\frac{y^2-4}{y+8}$ -8
13. $\frac{y}{y^2+2}$ No value
14. $\frac{y+3}{y^2+9}$ No value

In Exercises 15–20, determine whether the given pair of fractions are equivalent.

15. $\frac{11}{55}, \frac{111}{555}$ Yes
16. $\frac{22}{77}, \frac{222}{777}$ Yes
17. $\frac{-8}{3}, \frac{8}{-3}$ Yes
18. $\frac{6}{-5}, \frac{-6}{5}$ Yes
19. $\frac{3}{5}, \frac{9}{25}$ No
20. $\frac{4}{7}, \frac{16}{49}$ No

In Exercises 21–54, write each rational expression in lowest terms. Indicate if the expression is already in lowest terms.

21. $\frac{24}{120}$ $\frac{1}{5}$
22. $\frac{105}{42}$ $\frac{5}{2}$
23. $\frac{30}{24}$ $\frac{5}{4}$
24. $\frac{121}{132}$ $\frac{11}{12}$
25. $\frac{3(x+7)}{4(x+7)}$ $\frac{3}{4}$
26. $\frac{5(y-2)}{6(y-2)}$ $\frac{5}{6}$
27. $\frac{5y+1}{5y+2}$ Already in lowest terms
28. $\frac{6x-5}{2x-5}$ Already in lowest terms
29. $\frac{x^2-1}{x+1}$ $x-1$
30. $\frac{t+2}{t^2-4}$ $\frac{1}{t-2}$
31. $\frac{w^2+3w+2}{w^2+5w+6}$ $\frac{w+1}{w+3}$
32. $\frac{z^2-4z+3}{z^2-5z+6}$ $\frac{z-1}{z-2}$
33. $\frac{3s-6}{4s-8}$ $\frac{3}{4}$
34. $\frac{5s+15}{6s+18}$ $\frac{5}{6}$
35. $\frac{2(x+y)^2}{5(x+y)^5}$ $\frac{2}{5(x+y)^3}$
36. $\frac{3z(x-y)^2}{4z^2(x-y)}$ $\frac{3(x-y)}{4z}$
37. $\frac{4y^2-4y-3}{2y^2-7y+6}$ $\frac{2y+1}{y-2}$
38. $\frac{6y^2-5y-6}{6y^2-13y+6}$ $\frac{3y+2}{3y-2}$
39. $\frac{3m}{3m^2+6m}$ $\frac{1}{m+2}$
40. $\frac{4n}{8n^2-4n}$ $\frac{1}{2n-1}$
41. $\frac{a+2}{3a+6}$ $\frac{1}{3}$
42. $\frac{a-1}{2a-2}$ $\frac{1}{2}$
43. $\frac{m+1}{1+m}$ 1
44. $\frac{3x+2}{2+3x}$ 1
45. $\frac{n-5}{5-n}$ -1
46. $\frac{2n-3}{3-2n}$ -1
47. $\frac{x^2-7x}{49-7x}$ $\frac{-x}{7}$
48. $\frac{x^2-5x}{25-5x}$ $\frac{-x}{5}$
49. $\frac{3-x}{x^2-9}$ $\frac{-1}{x+3}$
50. $\frac{2-3x}{9x^2-4}$ $\frac{-1}{3x+2}$
51. $\frac{x^2-1}{x^3-1}$ $\frac{x+1}{x^2+x+1}$
52. $\frac{y^3+8}{y^2-4}$ $\frac{y^2-2y+4}{y-2}$
53. $\frac{a^3+a^2}{a^3+1}$ $\frac{a^2}{a^2-a+1}$
54. $\frac{2b^3-4b^2}{b^3-8}$ $\frac{2b^2}{b^2+2b+4}$

In Exercises 55–66, build up the rational expression to an equivalent expression with the indicated denominator.

55. $\frac{3}{5xy^2} = \frac{?}{10x^2y^2}$ $\frac{6x}{10x^2y^2}$
56. $\frac{7}{3xy^2} = \frac{?}{21x^3y^4}$ $\frac{49x^2y^2}{21x^3y^4}$
57. $\frac{6}{m} = \frac{?}{8m(n+2)}$ $\frac{48(n+2)}{8m(n+2)}$
58. $\frac{1}{n} = \frac{?}{2n(m+3)}$ $\frac{2(m+3)}{2n(m+3)}$
59. $\frac{x-2}{5x} = \frac{?}{20x^2}$ $\frac{4x(x-2)}{20x^2}$
60. $\frac{y+1}{2y} = \frac{?}{10y^3}$ $\frac{5y^2(y+1)}{10y^3}$
61. $\frac{3}{x-5} = \frac{?}{x^2-25}$ $\frac{3(x+5)}{x^2-25}$
62. $\frac{2}{x+1} = \frac{?}{x^2-1}$ $\frac{2(x-1)}{x^2-1}$
63. $\frac{y}{y+1} = \frac{?}{y^2-4y-5}$ $\frac{y(y-5)}{y^2-4y-5}$
64. $\frac{3y}{y-2} = \frac{?}{y^2-3y+2}$ $\frac{3y(y-1)}{y^2-3y+2}$
65. $\frac{x-1}{x-2} = \frac{?}{x^2-4}$ $\frac{(x-1)(x+2)}{x^2-4}$
66. $\frac{x+2}{x+3} = \frac{?}{x^2-9}$ $\frac{(x+2)(x-3)}{x^2-9}$

67. The time T in seconds that it takes a 4-ft simple pendulum to go back and forth once is called its period and is given by the formula
$$T = \frac{4\pi}{\sqrt{g}},$$
where g is the force due to gravity (expressed in appropriate units). This means, for example, that a simple pendulum will swing more slowly on the moon (where the force of gravity is smaller) than on the earth.
   a. Find the period for this pendulum if $g = 25$. About 2.5 seconds
   b. For what value of g is the period undefined? What would happen to this pendulum in a spaceship where there was no gravity? $g = 0$; it would not swing back and forth when released.

68. An object attached to a spring oscillates when pulled and released. In an ideal spring with no friction, the formula for the period of this oscillation is $T = 2\pi\sqrt{m/k}$, where k is a constant which describes the stiffness of the spring and m is the mass of the object.
   a. Find T when $m = 4$ and $k = 16$. $T = \pi$
   b. For what value of k is the period undefined? $k = 0$

69. The percent change in the value of an investment in one year is given by $\frac{V_n - V_o}{V_o} \cdot 100$, where V_o is the original value and V_n is the value now.
 a. Find the percent change in an investment that grows from \$1,200 to \$1,500 in one year. 25 percent
 b. For what value of V_o is the percent change undefined? 0

70. The percent error of a measurement is given by $\frac{x_m - x_t}{x_t} \cdot 100$, where x_t is the true value and x_m is the measured value.
 a. Find the percent error of a measurement of 0.97 in. if the true length is 1.00 in. -3 percent
 b. For what value of x_t is the percent error undefined? 0

71. The percentage of profit based on the cost to the store for a sale is given by $\frac{s - c}{c} \cdot 100$, where s is the selling price and c is the original cost to the store.
 a. Find the percentage of profit on an item which costs the store \$5 and is sold for \$10. 100 percent
 b. For what value of c is the percentage of profit undefined? 0

72. The percentage of depreciation of a piece of manufacturing equipment in one year is given by $\frac{V_f - V_o}{V_o} \cdot 100$, where V_f is the value at the end of the year and V_o is the value at the beginning of the year.
 a. For what value of V_o is this percentage undefined? 0
 b. What is the percentage of depreciation for a machine which costs \$100,000 but is only worth \$80,000 after one year? 20 percent

73. The width of a rectangle is 3 less than the length. The ratio of the area to the perimeter is given by $\frac{x(x - 3)}{2x + 2(x - 3)}$.

x

$x - 3$

 a. For what value of x is the ratio undefined? $\frac{3}{2}$
 Why is this not a sensible value for x in the first place?
 x must be greater than 3.
 b. For what value of x is this ratio equal to 1? (Then the area and perimeter are equal.) 6

74. The ratio of the area of a square to its perimeter is given by $s^2/4s$.
 a. For what value of s is this ratio undefined? 0
 b. For what value of s is this ratio equal to 2? (Then the area of the square is twice the perimeter.) 8

THINK ABOUT IT

1. What does the concept of "rational expression" have to do with the word *ratio?*
2. Explain why $\frac{x - 2}{x^2 + 1}$ can never be undefined.
3. Make up a rational expression that is undefined when $x = 2$ and equals 10 when $x = 3$.
4. Explain why this statement is not quite right: When the top and bottom of a fraction are multiplied by the same number, the resulting fraction and the original fraction are equivalent.
5. Some people say informally that you reduce $\frac{5x}{6x}$ to $\frac{5}{6}$ by "canceling" the x's. They also say that $5 + 2x - 2x$ equals 5 because $2x$ and $-2x$ "cancel out." Explain *in terms of arithmetic operations* why these two uses of the word *cancel* are not the same. (Consequently, many people prefer to avoid the word altogether.)

REMEMBER THIS

1. Find the product: $-3 \cdot \frac{2}{9}$. $-\frac{2}{3}$
2. Divide: $\frac{4}{7} \div -2$. $-\frac{2}{7}$
3. Factor $3x + 6$. $3(x + 2)$
4. Divide $\frac{(x + 4)^6}{(x + 4)^2}$. $(x + 4)^4$
5. True or false? $(x + 4)^2 = x^2 + 16$. False
6. Solve $(x + 1)(x - 2)(x + 3) = 0$. $\{-1, 2, -3\}$
7. Factor $9y^2 - 16x^4$. $(3y + 4x^2)(3y - 4x^2)$
8. Evaluate $3 - 2^2(1 - 2^2)$. 15
9. Solve $a = b/c$ for c. $c = b/a$
10. Find the dimensions of the rectangle given that the area is 24 cm². $x = 5$; length $= 8$ cm, width $= 3$ cm

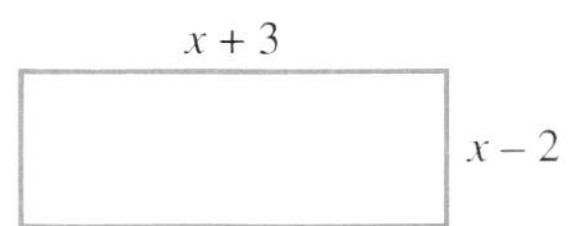

4.2 Multiplication and Division of Rational Expressions

A medical study is set up to compare the risk of illness for people who are exposed to a pollutant to the risk of illness for people who are not exposed. The quotient of these two risks is an important statistic called the **relative risk.** Suppose there are n people in each group. If s people get sick in the nonexposed group, and $s + x$ people get sick in the exposed group, then the relative risk is given by

$$\frac{s + x}{n} \div \frac{s}{n}.$$

Express this division in lowest terms. (See Example 4.)

OBJECTIVES

1. **Multiply rational expressions.**
2. **Divide rational expressions.**

1 The methods for multiplying algebraic expressions are based directly on the methods for multiplying fractions in arithmetic. Recall that the product of two or more fractions is the product of their numerators divided by the product of their denominators. For example,

$$\frac{3}{8} \cdot \frac{5}{7} = \frac{3 \cdot 5}{8 \cdot 7} = \frac{15}{56},$$

$$\frac{1}{3} \cdot \frac{3}{4} \cdot \frac{4}{5} = \frac{1 \cdot \overset{1}{\cancel{3}} \cdot \overset{1}{\cancel{4}}}{\underset{1}{\cancel{3}} \cdot \underset{1}{\cancel{4}} \cdot 5} = \frac{1}{5}.$$

The second example shows that when the same factor appears in both the numerator and the denominator, it is usually easier to divide out this factor before multiplying. We follow these **same procedures** to multiply rational expressions.

Multiplication of Rational Expressions

If $\frac{A}{B}$ and $\frac{C}{D}$ are rational expressions, with $B \neq 0$ and $D \neq 0$, then

$$\frac{A}{B} \cdot \frac{C}{D} = \frac{AC}{BD}.$$

In Example 1, note that factoring and the fundamental principle are used to express the products in lowest terms.

EXAMPLE 1 Multiply, and express each product in lowest terms.

a. $6x^2 \cdot \dfrac{5y}{3x^3} \cdot \dfrac{2}{y}$ **b.** $\dfrac{2x + 6}{x^2} \cdot \dfrac{x}{x^2 - 9}$ **c.** $\dfrac{t^2 + 4t - 5}{t^2 - 2t - 3} \cdot \dfrac{t^2 - 4t + 3}{t^2 + 6t + 5}$

Solution

a.
$$6x^2 \cdot \frac{5y}{3x^3} \cdot \frac{2}{y} = \frac{6x^2}{1} \cdot \frac{5y}{3x^3} \cdot \frac{2}{y} \qquad \text{Write } 6x^2 \text{ as } \frac{6x^2}{1}.$$
$$= \frac{6x^2 \cdot 5y \cdot 2}{1 \cdot 3x^3 \cdot y} \qquad \text{Multiply fractions.}$$
$$= \frac{2 \cdot 3 \cdot x \cdot x \cdot 5 \cdot y \cdot 2}{3 \cdot x \cdot x \cdot x \cdot y} \qquad \text{Write in prime factored form.}$$
$$= \frac{20}{x} \qquad \text{Write in lowest terms using the fundamental principle.}$$

b. In general, factor completely all numerators and denominators as a first step.

$$\frac{2x + 6}{x^2} \cdot \frac{x}{x^2 - 9} = \frac{2(x + 3)}{x \cdot x} \cdot \frac{x}{(x + 3)(x - 3)} \qquad \text{Factor completely.}$$
$$= \frac{2 \cdot (x + 3) \cdot x}{x \cdot x \cdot (x + 3) \cdot (x - 3)} \qquad \text{Multiply fractions.}$$
$$= \frac{2}{x(x - 3)} \qquad \text{Write in lowest terms using the fundamental principle.}$$

Unless there is a clear reason to do otherwise, it is sensible to leave answers in factored form.

It is not necessary to multiply out expressions like $x(x - 3)$ for the final answer to these problems.

c.
$$\frac{t^2 + 4t - 5}{t^2 - 2t - 3} \cdot \frac{t^2 - 4t + 3}{t^2 + 6t + 5}$$
$$= \frac{(t + 5)(t - 1)}{(t - 3)(t + 1)} \cdot \frac{(t - 3)(t - 1)}{(t + 5)(t + 1)} \qquad \text{Factor completely.}$$
$$= \frac{(t + 5)(t - 1)(t - 3)(t - 1)}{(t - 3)(t + 1)(t + 5)(t + 1)} \qquad \text{Multiply fractions.}$$
$$= \frac{(t - 1)^2}{(t + 1)^2} \qquad \text{Write in lowest terms using the fundamental principle.}$$

PROGRESS CHECK 1 Find (in lowest terms) the given products.

a. $\dfrac{7y}{5x^5} \cdot 20x^2 \cdot \dfrac{1}{y}$ **b.** $\dfrac{x^2 - 8x + 16}{x^3} \cdot \dfrac{x^2}{x - 4}$ **c.** $\dfrac{y^2 - 5y + 6}{y^2 + 3y + 2} \cdot \dfrac{y^2 - 2y - 3}{y^2 - 4}$

Example 2 shows how exponent properties and definitions may be used to express a product in lowest terms.

EXAMPLE 2 Multiply $\dfrac{(x + 4)^6}{a^8b^5} \cdot \dfrac{a^2b^5}{(x + 4)^3}$. Express the answer in lowest terms.

Solution Recall that $a^m/a^n = a^{m-n}$, $a^{-n} = 1/a^n$, and $b^0 = 1$.

$$\frac{(x + 4)^6}{a^8b^5} \cdot \frac{a^2b^5}{(x + 4)^3} = \frac{(x + 4)^6a^2b^5}{a^8b^5(x + 4)^3} \qquad \text{Multiply fractions.}$$
$$= a^{2-8}b^{5-5}(x + 4)^{6-3} \qquad \text{Quotient property of exponents.}$$
$$= a^{-6}b^0(x + 4)^3 \qquad \text{Simplify exponents.}$$
$$= \frac{(x + 4)^3}{a^6} \qquad \text{Negative and zero exponent definitions.}$$

Progress Check Answers

1. (a) $\dfrac{28}{x^3}$ (b) $\dfrac{x - 4}{x}$ (c) $\dfrac{(y - 3)^2}{(y + 2)^2}$

PROGRESS CHECK 2 Multiply $\frac{(x-5)^2}{a^3y^3} \cdot \frac{a^3y^9}{(x-5)^6}$. Express the answer in lowest terms.

2 The procedure for division of rational expressions is the same as for ordinary fractions. To divide two fractions, invert the divisor to find its reciprocal, and then multiply. For example,

$$\frac{3}{4} \div \frac{5}{8} = \frac{3}{4} \cdot \frac{8}{5} = \frac{6}{5}.$$

Division of Rational Expressions

If $\frac{A}{B}$ and $\frac{C}{D}$ are rational expressions, with $B \neq 0$, $C \neq 0$, and $D \neq 0$, then

$$\frac{A}{B} \div \frac{C}{D} = \frac{A}{B} \cdot \frac{D}{C} = \frac{AD}{BC}.$$

EXAMPLE 3 Perform each division, and express the quotient in lowest terms.

a. $\frac{33}{x^2} \div \frac{3}{x}$ **b.** $\frac{(y-2)^2}{5y} \div \frac{y^2-4}{15y}$ **c.** $\frac{3-x}{y^2-y} \div \frac{x^2-9}{y^2-2y+1}$

Solution In each case convert the division problem to multiplication.

a.
$$\frac{33}{x^2} \div \frac{3}{x} = \frac{33}{x^2} \cdot \frac{x}{3} \quad \text{Multiply } \frac{33}{x^2} \text{ by the reciprocal of } \frac{3}{x}.$$
$$= \frac{3 \cdot 11 \cdot x}{x \cdot x \cdot 3} \quad \text{Multiply and factor.}$$
$$= \frac{11}{x} \quad \text{Lowest terms.}$$

b.
$$\frac{(y-2)^2}{5y} \div \frac{y^2-4}{15y} = \frac{(y-2)^2}{5y} \cdot \frac{15y}{y^2-4} \quad \text{Multiply by the reciprocal of the divisor.}$$
$$= \frac{(y-2)(y-2)}{5y} \cdot \frac{3 \cdot 5 \cdot y}{(y+2)(y-2)} \quad \text{Factor completely.}$$
$$= \frac{(y-2)(y-2) \cdot 3 \cdot 5 \cdot y}{5 \cdot y(y+2)(y-2)} \quad \text{Multiply fractions.}$$
$$= \frac{3(y-2)}{y+2} \quad \text{Lowest terms.}$$

c. To obtain lowest terms in this example, recognize that $3 - x$ and $x - 3$ are opposites.

$$\frac{3-x}{y^2-y} \div \frac{x^2-9}{y^2-2y+1} = \frac{3-x}{y^2-y} \cdot \frac{y^2-2y+1}{x^2-9} \quad \text{Multiply by the reciprocal.}$$
$$= \frac{3-x}{y(y-1)} \cdot \frac{(y-1)(y-1)}{(x+3)(x-3)} \quad \text{Factor completely.}$$
$$= \frac{(3-x)(y-1)(y-1)}{y(y-1)(x+3)(x-3)} \quad \text{Multiply fractions.}$$
$$= \frac{-1(x-3)(y-1)(y-1)}{y(y-1)(x+3)(x-3)} \quad \text{Replace } 3-x \text{ by } -1(x-3).$$
$$= \frac{-1(y-1)}{y(x+3)} \text{ or } \frac{1-y}{y(x+3)} \quad \text{Lowest terms.}$$

Progress Check Answer

2. $\frac{y^6}{(x-5)^4}$

PROGRESS CHECK 3 Do each division and express the answer in lowest terms.

a. $\dfrac{16}{x^4} \div \dfrac{12}{x^2}$ **b.** $\dfrac{(t+5)^2}{4t} \div \dfrac{t^2-25}{12t^2}$ **c.** $\dfrac{y-1}{x^2+x} \div \dfrac{y-y^2}{x^2+4x+3}$

EXAMPLE 4 Solve the problem in the section introduction on page 169.

Solution By division we get a simpler expression for the relative risk.

$$\frac{s+x}{n} \div \frac{s}{n} = \frac{s+x}{n} \cdot \frac{n}{s} \qquad \text{Division definition.}$$
$$= \frac{(s+x) \cdot n}{ns} \qquad \text{Multiply fractions.}$$
$$= \frac{s+x}{s} \qquad \text{Lowest terms.}$$

The relative risk is $\dfrac{s+x}{s}$.

PROGRESS CHECK 4 If r people in the exposed group get sick, and $r - x$ people in the nonexposed group get sick, the relative risk is $\dfrac{r}{n} \div \dfrac{r-x}{n}$. Do the division and get a simpler expression for the relative risk.

Progress Check Answers

3. (a) $\dfrac{4}{3x^2}$ (b) $\dfrac{3(t+5)}{t-5)}$ (c) $\dfrac{-1(x+3)}{xy}$

4. $\dfrac{r}{r-x}$

EXERCISES 4.2

In Exercises 1–20, multiply, and express each product in lowest terms.

1. $3x^2 \cdot \dfrac{4}{x^2}$ 12

2. $5y^3 \cdot \dfrac{3}{y^3}$ 15

3. $x^2 \cdot \dfrac{3}{x} \cdot \dfrac{y}{x}$ $3y$

4. $y^3 \cdot \dfrac{4x}{y} \cdot \dfrac{2}{y^2}$ $8x$

5. $\dfrac{2x+6}{x^2} \cdot \dfrac{5x}{x+3}$ $\dfrac{10}{x}$

6. $\dfrac{5y-15}{y^2} \cdot \dfrac{2y}{y-3}$ $\dfrac{10}{y}$

7. $\dfrac{5x-5}{2x+4} \cdot \dfrac{2x+3}{6x-6}$ $\dfrac{5(2x+3)}{12(x+2)}$

8. $\dfrac{2x-8}{y+2} \cdot \dfrac{5y+10}{3x-12}$ $\dfrac{10}{3}$

9. $\dfrac{t^2-1}{t^2-4} \cdot \dfrac{t+2}{t+1}$ $\dfrac{t-1}{t-2}$

10. $\dfrac{s^2-9}{4s^2-4} \cdot \dfrac{s-1}{s-3}$ $\dfrac{s+3}{4(s+1)}$

11. $\dfrac{x^2+4x+4}{x^2+6x+9} \cdot \dfrac{x+3}{x+2}$ $\dfrac{x+2}{x+3}$

12. $\dfrac{9x^2+6x+1}{4x^2+4x+1} \cdot \dfrac{2x+1}{3x+1}$ $\dfrac{3x+1}{2x+1}$

13. $\dfrac{y^2-2y-8}{y^2-2y-15} \cdot \dfrac{y^2+5y+6}{y^2-4}$ $\dfrac{(y-4)(y+2)}{(y-5)(y-2)}$

14. $\dfrac{x^2+x-6}{x^2-9} \cdot \dfrac{x^2-x-6}{x^2+5x+6}$ $\dfrac{x-2}{x+3}$

15. $\dfrac{t^2-t-2}{t^2-t-12} \cdot \dfrac{t^2-3t-4}{t^2+t-6}$ $\dfrac{(t+1)^2}{(t+3)^2}$

16. $\dfrac{s^2+6s+8}{s^2-6s+8} \cdot \dfrac{s^2-2s-8}{s^2+2s-8}$ $\dfrac{(s+2)^2}{(s-2)^2}$

17. $\dfrac{6x^2+7x-3}{15x^2-17x+4} \cdot \dfrac{10x^2-3x-4}{4x^2+8x+3}$ 1

18. $\dfrac{12x^2-x-6}{6x^2+x-2} \cdot \dfrac{10x^2+3x-4}{20x^2+x-12}$ 1

19. $\dfrac{x^2-3x-4}{x+1} \cdot \dfrac{x+2}{8+2x-x^2}$ -1

20. $\dfrac{x^2-9}{x+3} \cdot \dfrac{x+2}{6+x-x^2}$ -1

In Exercises 21–26, multiply the given expressions, and express the answer in lowest terms.

21. $\dfrac{(x+2)^3}{a^3b^3} \cdot \dfrac{ab}{(x+2)^5}$ $\dfrac{1}{a^2b^2(x+2)^2}$

22. $\dfrac{(y-5)^3}{ab^3} \cdot \dfrac{a^2b^4}{(y-5)^2}$ $ab(y-5)$

23. $\dfrac{(x+1)^5(x-2)^4}{6a} \cdot \dfrac{3a}{(x+1)^6(x-2)^3}$ $\dfrac{x-2}{2(x+1)}$

24. $\dfrac{3a(x+3)^4(x+4)^3}{a^3(x+4)(x+5)} \cdot \dfrac{a(x+3)^2(x+5)}{(x+3)^6(x+4)^2}$ $\dfrac{3}{a}$

25. $\dfrac{x^2+2x+1}{x^2+4x+4} \cdot \dfrac{(x+2)^3}{(x+1)^3}$ $\dfrac{x+2}{x+1}$

26. $\dfrac{y^2-6y+9}{(x+4)^5} \cdot \dfrac{x^2+8x+16}{(y-3)^4}$ $\dfrac{1}{(x+4)^3(y-3)^2}$

In Exercises 27–40, divide the given expressions.

27. $\dfrac{24}{x^5} \div \dfrac{12}{x^2}$ $\dfrac{2}{x^3}$

28. $\dfrac{18}{x} \div \dfrac{6}{x^5}$ $3x^4$

29. $\dfrac{x+1}{y} \div \dfrac{(x+1)^2}{y}$ $\dfrac{1}{x+1}$

30. $\dfrac{(y-4)^3}{x^3} \div \dfrac{y-4}{x}$ $\dfrac{(y-4)^2}{x^2}$

31. $\dfrac{(y-3)^2}{x^2} \div \dfrac{(y-3)^4}{x^5}$ $\dfrac{x^3}{(y-3)^2}$

32. $\dfrac{5t}{(t+5)^3} \div \dfrac{t^5}{(t+5)^6}$ $\dfrac{5(t+5)^3}{t^4}$

33. $\dfrac{w^2 - 9}{25w^2 - 4} \div \dfrac{w - 3}{5w - 2}$ $\quad \dfrac{w + 3}{5w + 2}$

34. $\dfrac{w^2 - 16}{w^2 - 4} \div \dfrac{(w - 4)^2}{(w - 2)^2}$ $\quad \dfrac{(w + 4)(w - 2)}{(w - 4)(w + 2)}$

35. $\dfrac{4 - x}{x^2 - x} \div \dfrac{x^2 - 4x}{(x - 1)^2}$ $\quad \dfrac{1 - x}{x^2}$

36. $\dfrac{x - 5}{10x} \div \dfrac{25 - x^2}{6x^2}$ $\quad \dfrac{-3x}{5x + 25}$

37. $\dfrac{t + 2}{t^2 + 7t + 12} \div \dfrac{(t + 2)^4}{t^2 + 5t + 6}$ $\quad \dfrac{1}{(t + 2)^2(t + 4)}$

38. $\dfrac{2t^2 + 8t + 6}{t^2 + 6t + 9} \div \dfrac{t^2 + 4t + 3}{(t + 3)^5}$ $\quad 2(t + 3)^3$

39. $(x - y)^2 \div (x^2 - y^2)$ $\quad \dfrac{x - y}{x + y}$

40. $(3x + 2y)^2 \div (9x^2 - 4y^2)$ $\quad \dfrac{3x + 2y}{3x - 2y}$

41. In the calculation of probabilities by a tree diagram, the probability that any particular sequence of events will occur is found by multiplying probabilities along the appropriate path in the tree. In the tree diagram shown, the fractions represent the probabilities. Use the given tree diagram to find the probability of (a) success at both stages and (b) failure at both stages.

a. $\dfrac{y}{n}$ $\quad$ b. $\dfrac{n - x - a}{n}$

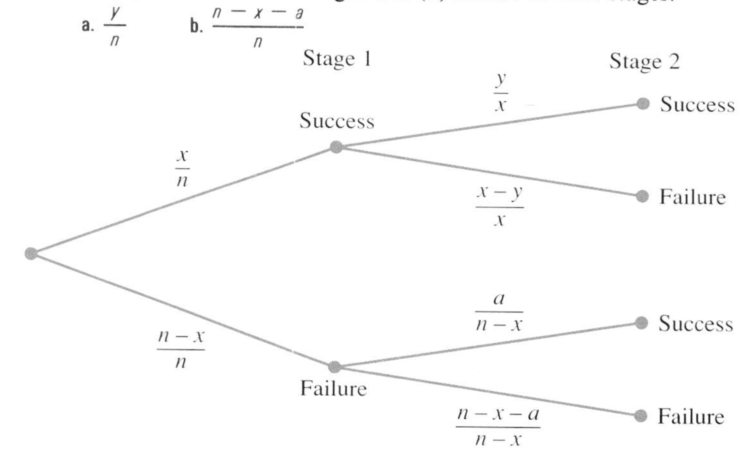

42. Use the tree diagram in Exercise 41 to find the probability of (a) success at stage 1 followed by failure at stage 2 and (b) failure at stage 1 followed by success at stage 2.

a. $\dfrac{x - y}{n}$ $\quad$ b. $\dfrac{a}{n}$

43. If the dimensions of a trapezoid are as shown, find the area.

$\dfrac{x + w}{2}$

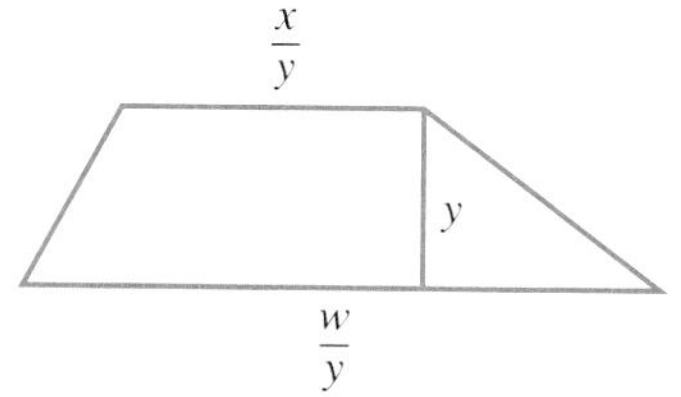

44. If the area and base of a triangle are as shown, find the height.

$\dfrac{2(x - 5)}{x + 5}$

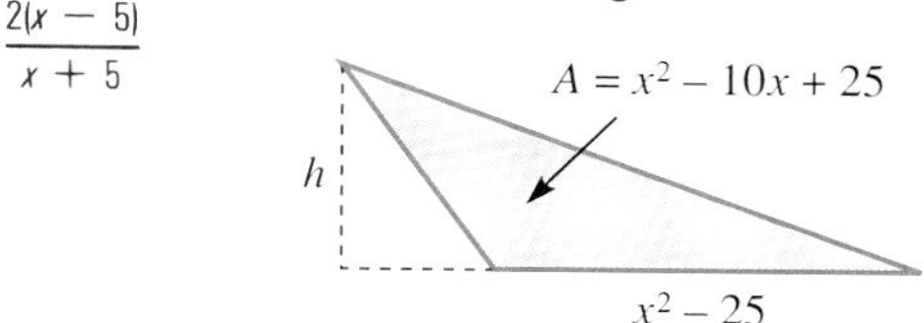

45. The odds *against* an event occurring equals the probability that it does *not* occur divided by the probability that it *does* occur. Find the odds against event A if the probability that A occurs is y/m and the probability that A does not occur is $1 - (y/m)$.

$\dfrac{m - y}{y}$

46. The odds *in favor* of an event equals the probability that it occurs divided by the probability that it does *not* occur. Find the odds in favor of event A if the probability that A occurs is a/n and the probability that A does *not* occur is $1 - (a/n)$. $\quad \dfrac{a}{n - a}$

THINK ABOUT IT

1. Show that it is wrong to reduce $\dfrac{x + a}{y + a}$ to $\dfrac{x}{y}$ by using cross multiplication to show that the two fractions are not equivalent.

2. Simplify these expressions.

a. $\dfrac{n}{n + 1} \cdot \dfrac{n + 1}{n + 2}$

b. $\dfrac{n}{n + 1} \cdot \dfrac{n + 1}{n + 2} \cdot \dfrac{n + 2}{n + 3}$

c. $\dfrac{n}{n + 1} \cdot \dfrac{n + 1}{n + 2} \cdot \dfrac{n + 2}{n + 3} \cdot \dfrac{n + 3}{n + 4}$

3. A professor said, "There is no reason to ever do division. Every division problem can be expressed as a multiplication problem."

a. What was the professor getting at?

b. How would you express $x \div y$ as a multiplication problem?

c. Show that $8 \cdot 0.25$ gives the same answer as $8 \div 4$. Why should this be so?

4. Divide $\dfrac{(x - a)^n}{x^{-5}}$ by $\dfrac{(x - a)^{n - 5}}{x^2}$.

5. Archimedes (287–212 B.C.) was one of the greatest mathematicians of all times. Among other things, he made a remarkable finding about a cylinder and its inscribed sphere (see the sketch). He proved that the ratio of the two volumes is the same as the ratio of the two surface areas. Use the given formulas to find out what he discovered. (*Hint:* Express h in terms of r, so that r is the only variable.)

Volume of a cylinder $= \pi r^2 h$

Volume of a sphere $= \frac{4}{3}\pi r^3$

Surface area of cylinder $= 2\pi r h + 2\pi r^2$

Surface area of sphere $= 4\pi r^2$

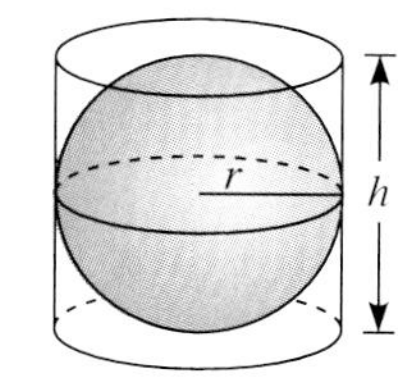

REMEMBER THIS

1. Write as a single fraction: $\frac{3}{7} - \frac{1}{3}$. $\frac{2}{21}$
2. Write as a single fraction: $\frac{3}{5} - \frac{3}{-5}$. $\frac{6}{5}$
3. What is the LCD for $\frac{5}{12}$ and $\frac{7}{100}$? 300
4. Simplify $(x + h)^2 - x^2$, and express in factored form. $h(2x + h)$
5. For what value(s) of p is this expression undefined? $\frac{5}{p^2 - 9}$. $-3, 3$
6. Are 33/7 and 3,333/707 equivalent fractions? Yes
7. Express $\frac{4m + 12}{m^2 + 2m - 3}$ in lowest terms. $\frac{4}{m - 1}$
8. Solve $x^2 + 3x = 28$. $\{-7, 4\}$
9. Solve $-2x + 3 < 5$. $\{x: x > -1\}; (-1, \infty)$
10. The difference of the squares of two consecutive positive integers is 31. What are the two consecutive integers? 15 and 16

4.3 Addition and Subtraction of Rational Expressions

In a certain game the player either wins d dollars or loses d dollars. If the probability of winning is s/n, then the average amount of money players can expect to win in this game is given by

$$\frac{sd}{n} - \frac{(n - s)d}{n},$$

which is called the "expected value" of the game. Express this difference as a single fraction. (See Example 3.)

OBJECTIVES

1. **Add and subtract rational expressions with the same denominator.**
2. **Add and subtract rational expressions with opposite denominators.**
3. **Find the least common denominator (LCD).**
4. **Add and subtract rational expressions with unlike denominators.**

1 The method for adding (or subtracting) rational expressions *which have the same denominator* is the same as the method for adding or subtracting such fractions in arithmetic. The sum (or difference) is equal to the sum (or difference) of the numerators divided by the common denominator. For example,

$$\frac{2}{7} + \frac{1}{7} = \frac{2 + 1}{7} = \frac{3}{7} \quad \text{and} \quad \frac{6}{5} - \frac{3}{5} = \frac{6 - 3}{5} = \frac{3}{5}.$$

The equivalent procedure for rational expressions is given in the next box.

Addition or Subtraction of Rational Expressions

If $\frac{A}{B}$ and $\frac{C}{B}$ are rational expressions, with $B \neq 0$, then

$$\frac{A}{B} + \frac{C}{B} = \frac{A + C}{B} \quad \text{and} \quad \frac{A}{B} - \frac{C}{B} = \frac{A - C}{B}.$$

EXAMPLE 1 Write as a single fraction in lowest terms.

a. $\frac{3x}{9} + \frac{x}{9}$ **b.** $\frac{3}{y + 5} - \frac{5}{y + 5}$

Solution

a. $\frac{3x}{9} + \frac{x}{9} = \frac{3x + x}{9}$ Sum of the numerators divided by the common denominator.

$= \frac{4x}{9}$ Add in the numerator.

b. $\frac{3}{y + 5} - \frac{5}{y + 5} = \frac{3 - 5}{y + 5}$ Difference of the numerators divided by the common denominator.

$= \frac{-2}{y + 5}$ Subtract in the numerator.

PROGRESS CHECK 1 Write as a single fraction in lowest terms.

a. $\frac{5}{4t} - \frac{2}{4t}$ **b.** $\frac{x}{x - 3} + \frac{2x}{x - 3}$

The next example, particularly the subtraction case, shows types that students commonly get wrong if they do not use parentheses in the numerators.

EXAMPLE 2 Write as a single fraction in lowest terms.

a. $\frac{x^2 + 2x}{x^2 + 1} + \frac{3x^2 - 4x}{x^2 + 1}$ **b.** $\frac{2x + 1}{3x - 5} - \frac{x + 4}{3x - 5}$

c. $\frac{4x + 1}{x^2 - 4} - \frac{3x - 1}{x^2 - 4}$.

Solution Because the numerators contain more than one term, we enclose them in parentheses.

a. $\frac{x^2 + 2x}{x^2 + 1} + \frac{3x^2 - 4x}{x^2 + 1} = \frac{(x^2 + 2x) + (3x^2 - 4x)}{x^2 + 1}$ Add the fractions.

$= \frac{4x^2 - 2x}{x^2 + 1}$ Simplify in the numerator.

b. $\frac{2x + 1}{3x - 5} - \frac{x + 4}{3x - 5} = \frac{(2x + 1) - (x + 4)}{3x - 5}$ Subtract the fractions.

$= \frac{2x + 1 - x - 4}{3x - 5}$ Remove parentheses.

$= \frac{x - 3}{3x - 5}$ Combine like terms.

Stress the importance of writing the parentheses after the minus sign. Omitting them is a major source of error.

Progress Check Answers

1. (a) $\frac{3}{4t}$ (b) $\frac{3x}{x - 3}$

Caution In this example, note that we insert and then remove parentheses to ensure that we subtract the entire numerator of the fraction on the right. A common error, when students do not use parentheses, is to subtract only the first term, as shown below.

$$\frac{2x+1}{3x-5} - \frac{x+4}{3x-5} = \frac{2x+1-x+4}{3x-5} \qquad \textbf{Wrong}$$

c. $\frac{4x+1}{x^2-4} - \frac{3x-1}{x^2-4} = \frac{(4x+1)-(3x-1)}{x^2-4}$ Subtract the fractions.

$= \frac{4x+1-3x+1}{x^2-4}$ Remove parentheses.

$= \frac{x+2}{x^2-4}$ Simplify in the numerator.

$= \frac{x+2}{(x+2)(x-2)}$ Factor completely.

$= \frac{1}{x-2}$ Lowest terms.

PROGRESS CHECK 2 Write as a single fraction in lowest terms.

a. $\frac{3x^2-6}{2x^2+x} + \frac{3x^2+1}{2x^2+x}$ **b.** $\frac{6x-5}{x-1} - \frac{6x-1}{x-1}$

c. $\frac{3x+4}{x^2-1} + \frac{-2x-5}{x^2-1}$

EXAMPLE 3 Solve the problem in the section introduction on page 174.

Solution Simplify the expression given for the expected value.

$\frac{sd}{n} - \frac{(n-s)d}{n} = \frac{sd-[(n-s)d]}{n}$ Subtract the fractions.

$= \frac{sd-[nd-sd]}{n}$ Distributive property.

$= \frac{sd-nd+sd}{n}$ Remove parentheses.

$= \frac{2sd-nd}{n}$ Simplify the numerator.

The expected value is given in a single fraction as $\frac{2sd-nd}{n}$.

PROGRESS CHECK 3 In a certain game the player will either lose d dollars or win $2d$ dollars. If the probability of winning is s/n, then the expected value is given by

$$\frac{2sd}{n} - \frac{(n-s)d}{n}.$$

Express this difference as a single fraction.

2 To add or subtract rational expressions when one denominator is the *opposite* of the other, choose either fraction and multiply both its numerator and denominator by -1. This will make both denominators the same.

Progress Check Answers

2. (a) $\frac{6x^2-5}{2x^2+x}$ (b) $\frac{-4}{x-1}$ (c) $\frac{1}{x+1}$

3. $\frac{3sd-nd}{n}$

EXAMPLE 4 Write as a single fraction in lowest terms.

a. $\frac{2}{y} - \frac{7}{-y}$ **b.** $\frac{3x+4}{4-3x} + \frac{1-x}{3x-4}$ **c.** $\frac{t^2}{t-3} + \frac{9}{3-t}$

Solution For the common denominator we usually choose to make the coefficient of the highest power of the variable a positive number.

a.
$$\frac{2}{y} - \frac{7}{-y} = \frac{2}{y} - \frac{7(-1)}{-y(-1)} \quad \text{Fundamental principle.}$$
$$= \frac{2}{y} - \frac{-7}{y} \quad \text{Perform the } -1 \text{ multiplications.}$$
$$= \frac{2-(-7)}{y} \quad \text{Subtract the fractions.}$$
$$= \frac{9}{y} \quad \text{Simplify.}$$

b.
$$\frac{3x+4}{4-3x} + \frac{1-x}{3x-4} = \frac{(3x+4)(-1)}{(4-3x)(-1)} + \frac{1-x}{3x-4} \quad \text{Fundamental principle.}$$
$$= \frac{-3x-4}{3x-4} + \frac{1-x}{3x-4} \quad \text{Perform the } -1 \text{ multiplications.}$$
$$= \frac{(-3x-4)+(1-x)}{3x-4} \quad \text{Add the fractions.}$$
$$= \frac{-4x-3}{3x-4} \quad \text{Simplify.}$$

Note This problem could also have been done by multiplying numerator and denominator in the right-hand fraction by -1. The answer would be $\frac{4x+3}{4-3x}$, which is equivalent to the answer above.

c.
$$\frac{t^2}{t-3} + \frac{9}{3-t} = \frac{t^2}{t-3} + \frac{9(-1)}{(3-t)(-1)} \quad \text{Fundamental principle.}$$
$$= \frac{t^2}{t-3} + \frac{-9}{t-3} \quad \text{Perform the } -1 \text{ multiplications.}$$
$$= \frac{t^2-9}{t-3} \quad \text{Add the fractions.}$$
$$= \frac{(t+3)(t-3)}{t-3} \quad \text{Factor completely.}$$
$$= t+3 \quad \text{Lowest terms.}$$

PROGRESS CHECK 4 Write as a single fraction in lowest terms.

a. $\frac{2}{-x} - \frac{2}{x}$ **b.** $\frac{t+3}{t-3} + \frac{2-3t}{3-t}$ **c.** $\frac{y^2}{y-2} + \frac{y+2}{2-y}$

3 Working with rational expressions which have *different denominators* is often easier if they are first rewritten as equivalent expressions with the same denominator, usually the least common denominator, or LCD. Example 5 reviews the procedure for finding the LCD of fractions in arithmetic.

Progress Check Answers

4. (a) $-\frac{4}{x}$ (b) $\frac{4t+1}{t-3}$ (c) $y+1$

EXAMPLE 5 Find the least common denominator for $\frac{5}{12}$ and $\frac{7}{100}$.

Solution To find the LCD, first express 12 and 100 as products of prime factors.

$$12 = 2 \cdot 2 \cdot 3$$
$$100 = 2 \cdot 2 \cdot 5 \cdot 5$$

The LCD is the product of all the different prime factors, with each prime number appearing the most number of times it appears in any one factorization. Thus,

$$\text{LCD} = 2 \cdot 2 \cdot 3 \cdot 5 \cdot 5 = 300.$$

PROGRESS CHECK 5 Find the least common denominator for $\frac{1}{12}$ and $\frac{5}{30}$.

To find the least common denominator for rational expressions, use a similar procedure.

To Find the LCD

1. Factor each denominator completely.
2. The LCD is the product of all the different factors, with each factor appearing the most number of times it appears in any one factorization.

This procedure is illustrated in Example 6.

EXAMPLE 6 Find the least common denominator for the following.

a. $\dfrac{5}{12x^2y}$ and $\dfrac{4}{9xy^3}$ **b.** $\dfrac{x}{2x - 2}$ and $\dfrac{3}{x^2 - 1}$

Solution

a. First, factor each denominator completely.

$$12x^2y = 2 \cdot 2 \cdot 3 \cdot x \cdot x \cdot y$$
$$9xy^3 = 3 \cdot 3 \cdot x \cdot y \cdot y \cdot y$$

The LCD will contain the factors 2, 3, x, and y. In any factorization the greatest number of times that 2, 3, and x appear is twice, and that y appears is three times. Thus,

$$\text{LCD} = 2 \cdot 2 \cdot 3 \cdot 3 \cdot x \cdot x \cdot y \cdot y \cdot y = 36x^2y^3.$$

b. Factor each denominator completely.

$$2x - 2 = 2(x - 1)$$
$$x^2 - 1 = (x + 1)(x - 1)$$

The least common denominator will contain the factors 2, $x - 1$, and $x + 1$. The most number of times each factor appears in any one factorization is once. Thus,

$$\text{LCD} = 2(x + 1)(x - 1).$$

PROGRESS CHECK 6 Find the LCD for the following.

a. $\dfrac{1}{8xy^2}$ and $\dfrac{7}{6x^2y}$ **b.** $\dfrac{1}{x^2 + 5x}$ and $\dfrac{x}{3x + 15}$

Progress Check Answer

5. 60 6. (a) $24x^2y^2$ (b) $3x(x + 5)$

4 Example 7 reviews the arithmetic for adding fractions with different denominators. Then we present the equivalent procedure for rational expressions.

EXAMPLE 7 Add $\frac{5}{12} + \frac{7}{100}$. Express the result in lowest terms.

Solution As shown in Example 5, the LCD is 300. We divide 300 by each denominator to find the factor to use in the fundamental principle. Because $\frac{300}{12} = 25$ and $\frac{300}{100} = 3$, we may add the fractions as shown below.

$$\frac{5}{12} + \frac{7}{100} = \frac{5 \cdot 25}{12 \cdot 25} + \frac{7 \cdot 3}{100 \cdot 3} \quad \text{Fundamental principle.}$$

$$= \frac{125}{300} + \frac{21}{300} \quad \text{Perform the multiplications.}$$

$$= \frac{146}{300} \quad \text{Add the fractions.}$$

$$= \frac{73}{150} \quad \text{Lowest terms.}$$

PROGRESS CHECK 7 Add $\frac{7}{18} + \frac{17}{30}$. Express the result in lowest terms.

To Add or Subtract Rational Expressions

1. Completely factor each denominator, and find the LCD.
2. For each fraction, obtain an equivalent fraction by applying the fundamental principle and multiplying the numerator and the denominator of the fraction by the factors of the LCD that are not contained in the denominator of that fraction.
3. Add or subtract the numerators, and divide this result by the common denominator.
4. Express the answer in lowest terms.

EXAMPLE 8 Write as a single fraction in lowest terms.

a. $\frac{1}{6x} + \frac{5}{8}$ **b.** $\frac{5}{12x^2y} - \frac{4}{9xy^3}$

Solution

a. Since $6x$ factors as $2 \cdot 3 \cdot x$ and 8 factors as $2 \cdot 2 \cdot 2$, the most number of times that 2 appears is three, and that 3 and x appear is once. Thus, the LCD is $2 \cdot 2 \cdot 2 \cdot 3 \cdot x$, or $24x$. We divide the LCD by each denominator to get the factor to use in the fundamental principle step. Because $\frac{24x}{6x} = 4$ and $\frac{24x}{8} = 3x$, we have

$$\frac{1}{6x} + \frac{5}{8} = \frac{1 \cdot 4}{6x \cdot 4} + \frac{5 \cdot 3x}{8 \cdot 3x} \quad \text{Fundamental principle.}$$

$$= \frac{1 \cdot 4 + 5 \cdot 3x}{24x} \quad \text{Add the fractions.}$$

$$= \frac{4 + 15x}{24x} \text{ or } \frac{15x + 4}{24x} \quad \text{Simplify in the numerator.}$$

Progress Check Answer

7. $\frac{43}{45}$

b. The LCD is $36x^2y^3$ (as explained in Example 6). Then,

$$\frac{5}{12x^2y} - \frac{4}{9xy^3} = \frac{5(3y^2)}{12x^2y(3y^2)} - \frac{4(4x)}{9xy^3(4x)} \quad \text{Fundamental principle.}$$
$$= \frac{5(3y^2) - 4(4x)}{36x^2y^3} \quad \text{Subtract the fractions.}$$
$$= \frac{15y^2 - 16x}{36x^2y^3}. \quad \text{Simplify in the numerator.}$$

PROGRESS CHECK 8 Write as a single fraction in lowest terms.

a. $\dfrac{5}{6x} + \dfrac{2}{9}$ **b.** $\dfrac{3}{8xy^2} - \dfrac{7}{6x^2y}$

EXAMPLE 9 Write as a single fraction in lowest terms.

a. $\dfrac{y}{y-3} + \dfrac{2}{y}$ **b.** $\dfrac{x}{x+3} - \dfrac{18}{x^2-9}$ **c.** $\dfrac{3x+1}{x^2+x-2} + \dfrac{x}{x^2-2x+1}$

Solution

a. Because neither denominator factors, the LCD is $y(y-3)$, which is the product of the denominators.

$$\frac{y}{y-3} + \frac{2}{y} = \frac{y(y)}{(y-3)(y)} + \frac{2(y-3)}{y(y-3)} \quad \text{Fundamental principle.}$$
$$= \frac{y(y) + 2(y-3)}{y(y-3)} \quad \text{Add the fractions.}$$
$$= \frac{y^2 + 2y - 6}{y(y-3)} \quad \text{Simplify in the numerator.}$$

The numerator does not factor, so we stop here.

b. Because $x^2 - 9$ factors as $(x+3)(x-3)$, the LCD is $(x+3)(x-3)$. Note that in this example an extra step is needed to reduce the answer to lowest terms.

$$\frac{x}{x+3} - \frac{18}{x^2-9}$$
$$= \frac{x}{x+3} - \frac{18}{(x+3)(x-3)} \quad \text{Factor completely.}$$
$$= \frac{x(x-3)}{(x+3)(x-3)} - \frac{18}{(x+3)(x-3)} \quad \text{Fundamental principle.}$$
$$= \frac{x(x-3) - 18}{(x+3)(x-3)} \quad \text{Subtract the fractions.}$$
$$= \frac{x^2 - 3x - 18}{(x+3)(x-3)} \quad \text{Simplify in the numerator.}$$
$$= \frac{(x+3)(x-6)}{(x+3)(x-3)} \quad \text{Factor in the numerator.}$$
$$= \frac{x-6}{x-3} \quad \text{Lowest terms.}$$

Progress Check Answers

8. (a) $\dfrac{15 + 4x}{18x}$ (b) $\dfrac{9x - 28y}{24x^2y^2}$

c. To find the LCD, we determine that $x^2 + x - 2$ factors as $(x + 2)(x - 1)$, while $x^2 - 2x + 1$ factors as $(x - 1)(x - 1)$. The most number of times that $x - 1$ appears is twice and that $x + 2$ appears is once. Thus, the LCD is $(x + 2)(x - 1)^2$.

$$\frac{3x+1}{x^2+x-2} + \frac{x}{x^2-2x+1}$$

$$= \frac{3x+1}{(x+2)(x-1)} + \frac{x}{(x-1)(x-1)}$$ Factor completely.

$$= \frac{(3x+1)(x-1)}{(x+2)(x-1)(x-1)} + \frac{x(x+2)}{(x-1)(x-1)(x+2)}$$ Fundamental principle.

$$= \frac{(3x+1)(x-1) + x(x+2)}{(x+2)(x-1)^2}$$ Add the fractions.

$$= \frac{3x^2 - 3x + x - 1 + x^2 + 2x}{(x+2)(x-1)^2}$$ Remove parentheses in the numerator.

$$= \frac{4x^2 - 1}{(x+2)(x-1)^2}$$ Simplify.

Although the numerator factors as $(2x + 1)(2x + 1)$, neither of these factors appear in the denominator, so the above result represents lowest terms.

PROGRESS CHECK 9 Write as a single fraction in lowest terms.

a. $\frac{t}{t-2} + \frac{3}{t}$ **b.** $\frac{2t}{t-1} - \frac{4}{t^2-1}$ **c.** $\frac{x-1}{x^2-x-6} + \frac{3x}{x^2-6x+9}$

Progress Check Answers

9. (a) $\frac{t^2+3t-6}{t(t-2)}$ (b) $\frac{2(t+2)}{t+1}$ (c) $\frac{4x^2+2x+3}{(x-3)^2(x+2)}$

EXERCISES 4.3

In Exercises 1–12, rewrite the given expression as a single fraction; then express it in lowest terms.

1. $\frac{5x}{4} + \frac{3x}{4}$ $2x$
2. $\frac{12y}{5} - \frac{7y}{5}$ y
3. $\frac{3}{x} - 1$ $\frac{3-x}{x}$
4. $\frac{2}{t} + 1$ $\frac{2+t}{t}$
5. $\frac{3y}{y+5} - \frac{2y-5}{y+5}$ 1
6. $\frac{x-7}{2x-3} + \frac{x+4}{2x-3}$ 1
7. $\frac{x^2+2x}{x+1} + \frac{x+2}{x+1}$ $x+2$
8. $\frac{y^2-y}{y+2} - \frac{y+8}{y+2}$ $y-4$
9. $\frac{2w+5}{w^2-9} - \frac{w+2}{w^2-9}$ $\frac{1}{w-3}$
10. $\frac{3z-2}{z^2-1} + \frac{1-2z}{z^2-1}$ $\frac{1}{z+1}$
11. $\frac{x^2+3x}{x^2+2x+1} + \frac{2}{x^2+2x+1}$ $\frac{x+2}{x+1}$
12. $\frac{x^2-x-8}{x^2-6x+9} - \frac{7-3x}{x^2-6x+9}$ $\frac{x+5}{x-3}$

In Exercises 13–20, write each expression as a single fraction; then express it in lowest terms.

13. $\frac{1}{x} + \frac{1}{-x}$ 0
14. $\frac{2}{y} - \frac{3}{-y}$ $\frac{5}{y}$
15. $\frac{2}{x-4} - \frac{2}{4-x}$ $\frac{4}{x-4}$
16. $\frac{u}{2u-v} + \frac{v}{v-2u}$ $\frac{u-v}{2u-v}$
17. $\frac{y+1}{2y-3} - \frac{y-1}{3-2y}$ $\frac{2y}{2y-3}$
18. $\frac{2x-4}{3x-1} + \frac{x-5}{1-3x}$ $\frac{x+1}{3x-1}$
19. $\frac{t^2}{t-5} + \frac{25}{5-t}$ $t+5$
20. $\frac{t^2-7t}{t-5} - \frac{10}{5-t}$ $t-2$

In Exercises 21–32, find the lowest common denominator for the given fractions.

21. $\frac{3}{5xy^2}, \frac{4}{7x^2y}$ $35x^2y^2$
22. $\frac{1}{2x^3y^2}, \frac{1}{3xy}$ $6x^3y^2$
23. $\frac{5}{4x^3y^2z}, \frac{7}{6xy^2z^3}$ $12x^3y^2z^3$
24. $\frac{3}{8x^4y^2}, \frac{5}{12y^2z}$ $24x^4y^2z$
25. $\frac{3}{x^2+2x}, \frac{x}{x+2}$ $x(x+2)$
26. $\frac{a}{x^2+x}, \frac{b}{x^2-x}$ $x(x+1)(x-1)$
27. $\frac{y+3}{y^2+2y+1}, \frac{y+2}{y^2-1}$ $(y+1)^2(y-1)$
28. $\frac{y-3}{4y^2+12y+9}, \frac{y-1}{4y^2-9}$ $(2y+3)^2(2y-3)$
29. $\frac{n}{2n+2}, \frac{n}{3n-3}$ $6(n+1)(n-1)$
30. $\frac{n+12}{6n+12}, \frac{n-16}{8n-16}$ $24(n+2)(n-2)$
31. $\frac{m+1}{3m-1}, \frac{3m-1}{6m}$ $6m(3m-1)$
32. $\frac{m-3}{2m}, \frac{m}{2m+3}$ $2m(2m+3)$

In Exercises 33–64, write the given expression as a single fraction in lowest terms.

33. $\dfrac{1}{2x} + \dfrac{1}{3}$ $\quad \dfrac{3 + 2x}{6x}$

34. $\dfrac{5}{4x} + \dfrac{3}{8y}$ $\quad \dfrac{10y + 3x}{8xy}$

35. $\dfrac{3}{5y} - \dfrac{3}{10y^2}$ $\quad \dfrac{6y - 3}{10y^2}$

36. $\dfrac{y}{6x} - \dfrac{x}{7y}$ $\quad \dfrac{7y^2 - 6x^2}{42xy}$

37. $\dfrac{a}{xy^2} + \dfrac{b}{x^2y}$ $\quad \dfrac{ax + by}{x^2y^2}$

38. $\dfrac{1}{xyz} - \dfrac{1}{xy}$ $\quad \dfrac{1 - z}{xyz}$

39. $\dfrac{1}{x} - x$ $\quad \dfrac{1 - x^2}{x}$

40. $yz + \dfrac{y}{z}$ $\quad \dfrac{yz^2 + y}{z}$

41. $\dfrac{x}{x - 1} + \dfrac{1}{x}$ $\quad \dfrac{x^2 + x - 1}{x(x - 1)}$

42. $\dfrac{2y}{y - 5} + \dfrac{3}{y}$ $\quad \dfrac{2y^2 + 3y - 15}{y(y - 5)}$

43. $\dfrac{2}{x} - \dfrac{x}{x + 3}$ $\quad \dfrac{-x^2 + 2x + 6}{x(x + 3)}$

44. $\dfrac{1}{u} - \dfrac{u}{2u - 1}$ $\quad \dfrac{-u^2 + 2u - 1}{u(2u - 1)}$

45. $\dfrac{1}{v + 2} - \dfrac{1}{v^2 - 4}$ $\quad \dfrac{v - 3}{v^2 - 4}$

46. $\dfrac{1}{v + 4} + \dfrac{v}{v^2 - 16}$ $\quad \dfrac{2v - 4}{v^2 - 16}$

47. $\dfrac{-3}{2x - 3} + \dfrac{6x}{4x^2 - 9}$ $\quad \dfrac{-9}{4x^2 - 9}$

48. $\dfrac{2}{3x + 1} - \dfrac{6x}{9x^2 - 1}$ $\quad \dfrac{-2}{9x^2 - 1}$

49. $\dfrac{x}{x - 2} - \dfrac{3x + 2}{x^2 - 4}$ $\quad \dfrac{x + 1}{x + 2}$

50. $\dfrac{x}{x + 3} + \dfrac{5x - 3}{x^2 - 9}$ $\quad \dfrac{x - 1}{x - 3}$

51. $\dfrac{y}{y^2 + 3x + 2} + \dfrac{1}{y^2 + 4x + 3}$ $\quad \dfrac{y^2 + 4y + 2}{(y + 1)(y + 2)(y + 3)}$

52. $\dfrac{y + 1}{y^2 + 5y + 6} - \dfrac{y + 3}{y^2 + 3y + 2}$ $\quad \dfrac{-4}{(y + 1)(y + 3)}$

53. $\dfrac{x}{x^2 + 3x + 2} - \dfrac{1}{x^2 + 2x + 1}$ $\quad \dfrac{x^2 - 2}{(x + 1)^2(x + 2)}$

54. $\dfrac{-1}{x^2 - 2x - 8} + \dfrac{x - 2}{x^2 - 8x + 16}$ $\quad \dfrac{x^2 - x}{(x - 4)^2(x + 2)}$

55. $\dfrac{4}{x + 3} - \dfrac{16}{(x + 3)^2}$ $\quad \dfrac{4x - 4}{(x + 3)^2}$

56. $\dfrac{x}{x - 5} + \dfrac{4x^2}{(x - 5)^2}$ $\quad \dfrac{5x^2 - 5x}{(x - 5)^2}$

57. $1 + \dfrac{2}{x + 1} - \dfrac{8}{(x + 1)^2}$ $\quad \dfrac{x^2 + 4x - 5}{(x + 1)^2}$

58. $1 - \dfrac{2}{x - 2} + \dfrac{1}{(x - 2)^2}$ $\quad \dfrac{x^2 - 6x + 9}{(x - 2)^2}$

59. $\dfrac{3}{t^2 - t} + \dfrac{2}{t^2 + t} + \dfrac{1}{t^2}$ $\quad \dfrac{t + 5}{t^2 - 1}$

60. $\dfrac{t + 2}{t^2 - 2t} + \dfrac{t - 2}{t^2 + 2t} + \dfrac{2}{t^2}$ $\quad \dfrac{4t}{t - 4}$

61. $\dfrac{2b}{a^2 - b^2} + \dfrac{1}{a - b} + \dfrac{1}{a + b}$ $\quad \dfrac{2}{a - b}$

62. $\dfrac{6a + 3b}{4a^2 - 9b^2} - \dfrac{1}{2a + 3b} - \dfrac{1}{2a - 3b}$ $\quad \dfrac{1}{2a - 3b}$

63. $\dfrac{x - y}{2x^2 + 3xy + y^2} - \dfrac{2x - y}{x^2 + 2xy + y^2}$ $\quad \dfrac{-3x^2}{(x + y)^2(2x + y)}$

64. $\dfrac{x + 2y}{3x^2 - 8xy + 4y^2} - \dfrac{3x + 2y}{x^2 - 4xy + 4y^2}$ $\quad \dfrac{-8x^2}{(x - 2y)^2(3x - 2y)}$

65. Find the perimeter of this rectangle. 2

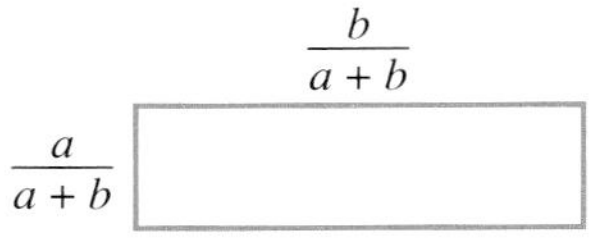

66. Find the perimeter of this rectangle. 4

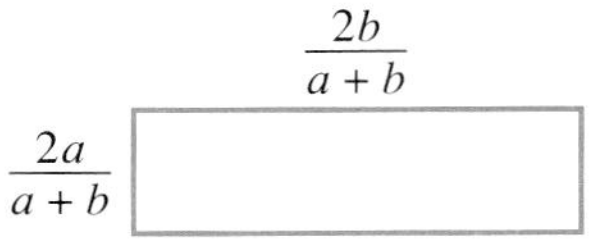

67. a. Show that $\dfrac{1}{n + 1} + \dfrac{1}{n(n + 1)} = \dfrac{1}{n}$.

This equation provides a way to find two fractions with 1 in the numerator which sum to another fraction with 1 in the numerator.

b. Find two fractions whose sum is $\frac{1}{3}$ by letting $n = 3$ in the formula of part **a.**

a. $\dfrac{n}{n(n + 1)} + \dfrac{1}{n(n + 1)} = \dfrac{n + 1}{n(n + 1)} = \dfrac{1}{n}$ $\quad$ b. $\frac{1}{4} + \frac{1}{12} = \frac{1}{3}$

68. a. Show that the reciprocal of $\dfrac{1}{2}\left(\dfrac{1}{a} + \dfrac{1}{b}\right)$ is $\dfrac{2ab}{a + b}$.

b. The expression $\dfrac{2ab}{a + b}$ is called the harmonic mean of a and b. Find the harmonic mean of 6 and 12.

a. $\dfrac{1}{\frac{1}{2}\left(\frac{1}{a} + \frac{1}{b}\right)} = \dfrac{2ab}{a + b}$ $\quad$ b. 8

69. The weighted average of a and b is given by $\left(\dfrac{x}{n}\right)a + \left(\dfrac{n - x}{n}\right)b$. Express this as a single fraction.

$\dfrac{ax - bx + bn}{n}$

70. If the chance of success in a business venture is s/n, then the chance of not obtaining success is $(n - s)/n$. Find the sum of these probabilities. 1

THINK ABOUT IT

1. Two expressions are negatives of each other if their sum is zero.

a. What is the negative of $-3x$?

b. What is the negative of $\dfrac{4}{x}$?

c. What is the negative of $2x^2 - 3x + 1$?

d. What is the negative of $\dfrac{x - 1}{x + 2}$?

2. Which of these sums are equal to zero?

a. $\dfrac{x - 2}{x - 1} + \dfrac{2 - x}{x - 1}$ $\quad$ **b.** $\dfrac{x - 2}{x - 1} + \dfrac{x - 2}{1 - x}$

c. $\dfrac{x - 2}{x - 1} + \dfrac{1 - x}{2 - x}$

3. Express each sum as a single fraction.

a. $1 + \dfrac{1}{x}$ $\quad$ **b.** $1 + \dfrac{1}{x} + \dfrac{1}{x^2}$ $\quad$ **c.** $1 + \dfrac{1}{x} + \dfrac{1}{x^2} + \cdots + \dfrac{1}{x^{10}}$

4. a. Multiply $\left(1 - \frac{1}{n}\right)\left(1 - \frac{1}{n+1}\right)$ by FOIL.
   b. Multiply $\left(1 - \frac{1}{n}\right)\left(1 - \frac{1}{n+1}\right)$ by first writing each factor as a single fraction. Compare this to the answer from part **a.**
   c. Multiply $\left(1 - \frac{1}{n}\right)\left(1 - \frac{1}{n+1}\right)\left(1 - \frac{1}{n+2}\right)$.
   d. Multiply $\left(1 - \frac{1}{n}\right)\left(1 - \frac{1}{n+1}\right)\cdots\left(1 - \frac{1}{n+9}\right)$.

5. The Egyptians in ancient times expressed most fractional quantities as sums of unit fractions, which are fractions with numerators equal to 1. For example, the answer given in the Rhind papyrus (1650 B.C.) to the question of how to divide 6 loaves among 10 men is $\frac{1}{2} + \frac{1}{10}$ (but we would write $\frac{6}{10}$ or $\frac{3}{5}$).
   a. Show that $\frac{x+y}{xy} = \frac{1}{x} + \frac{1}{y}$.
   b. For the equation in part **a,** let $x = 3$ and $y = 5$ to express $\frac{8}{15}$ as a sum of two unit fractions.
   c. Use the equality given in part **a** to find two unit fractions whose sum is $\frac{2}{7}$.

REMEMBER THIS

1. Express as a single fraction: $1 + \frac{1}{t}$. $\frac{t+1}{t}$
2. Multiply $2x^2 \cdot \frac{6y}{x^3} \cdot \frac{5}{3y}$. $\frac{20}{x}$
3. Apply the distributive property to $10(3 - \frac{9}{10})$, and simplify. 21
4. Apply the distributive property to $y\left(1 - \frac{1}{y}\right)$ and simplify. $y - 1$
5. Multiply $x^2(x^{-1} + 3x^{-2})$. $x + 3$
6. Are these two expressions equivalent? $\frac{x}{ab}$ and $\frac{x^2}{a^2b^2}$. No
7. Solve $3x^2 + 13x - 10 = 0$. $\{\frac{2}{3}, -5\}$
8. True or false? $3^{-2} > 2^{-3}$. False
9. Solve $\frac{3}{y} + y = \frac{3}{y}$. No solution
10. A one-year loan of $2,000 has simple interest equal to $250. What annual interest rate is this? Use $I = prt$. 12.5 percent

4.4 Complex Fractions

In the determination of the reliability of medical screening tests a statistic called the **false positive rate,** which is a quotient of fractional expressions, is calculated. In one study the formula for the false positive rate was

$$F_+ = \frac{1 - \frac{1}{y}}{1 + \frac{10a - n}{ny}}.$$

Rewrite the formula by simplifying the right-hand side of the equation. (See Example 4.)

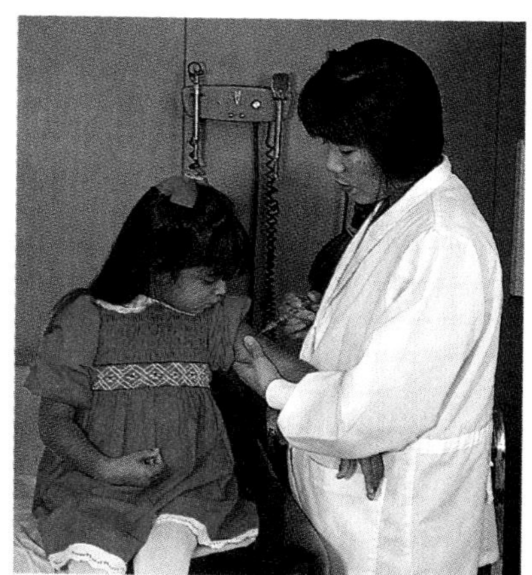

OBJECTIVES

1 **Simplify complex fractions.**

2 **Simplify complex fractions arising from negative exponents.**

1 A **complex fraction** is a fraction in which the numerator or the denominator or both involve fractions. In the diagram below, the complex fraction given in the section-opening problem is used to show the components of a complex fraction.

$$\frac{1 - \dfrac{1}{y}}{1 + \dfrac{10a - n}{ny}}$$

- $1 - \frac{1}{y}$ ← Numerator of complex fraction
- ← Primary fraction bar
- $1 + \frac{10a - n}{ny}$ ← Denominator of complex fraction

Every complex fraction can be rewritten as an equivalent rational expression in which the numerator and denominator do *not* include fractions. This is called simplifying the complex fraction. Two methods for simplifying are illustrated in Example 1.

EXAMPLE 1 Simplify the complex fraction $\dfrac{3 - \dfrac{9}{10}}{\dfrac{2}{5} + 1}$.

Solution

Both methods show that the fraction simplifies to $\frac{3}{2}$.

Method 1 First, obtain a single fraction in the numerator and in the denominator of the complex fraction.

$$\frac{3 - \dfrac{9}{10}}{\dfrac{2}{5} + 1} = \frac{\dfrac{3 \cdot 10}{1 \cdot 10} - \dfrac{9}{10}}{\dfrac{2}{5} + \dfrac{1 \cdot 5}{1 \cdot 5}} = \frac{\dfrac{30}{10} - \dfrac{9}{10}}{\dfrac{2}{5} + \dfrac{5}{5}} = \frac{\dfrac{21}{10}}{\dfrac{7}{5}}$$

Then because a/b is equivalent to $a \div b$, we get

$$\frac{\dfrac{21}{10}}{\dfrac{7}{5}} = \frac{21}{10} \div \frac{7}{5} = \frac{21}{10} \cdot \frac{5}{7} = \frac{21 \cdot 5}{10 \cdot 7} = \frac{3}{2}.$$

Method 2 First, find the LCD of all the fractions in the numerator and the denominator of the complex fraction.

$$\text{LCD of } \frac{3}{1}, \frac{9}{10}, \frac{2}{5}, \text{ and } \frac{1}{1} \text{ is 10.}$$

Then multiply the numerator and the denominator of the complex fraction by this LCD and simplify the result.

$$\frac{3 - \dfrac{9}{10}}{\dfrac{2}{5} + 1} = \frac{10\left(3 - \dfrac{9}{10}\right)}{10\left(\dfrac{2}{5} + 1\right)} = \frac{10 \cdot 3 - 10 \cdot \dfrac{9}{10}}{10 \cdot \dfrac{2}{5} + 10 \cdot 1} = \frac{30 - 9}{4 + 10} = \frac{21}{14} = \frac{3}{2}$$

PROGRESS CHECK 1 Simplify $\dfrac{5 - \dfrac{1}{3}}{\dfrac{1}{6} + 3}$. Use both methods of Example 1.

Progress Check Answer

1. $\frac{28}{19}$

The two methods illustrated in Example 1 can also be used for simplifying complex fractions which contain rational expressions.

Methods to Simplify Complex Fractions

Method 1 (Obtain single fractions and divide): Obtain single fractions in both the numerator and the denominator of the complex fraction. Then divide by multiplying by the reciprocal of the denominator.

Method 2 (Multiply using the LCD): Find the LCD of all the fractions that appear in the numerator and the denominator of the complex fraction. Then multiply the numerator and the denominator of the complex fraction by the LCD, and simplify the results.

EXAMPLE 2 Simplify the complex fraction $\dfrac{\dfrac{2}{x} - 5}{1 + \dfrac{1}{x}}$.

Solution Consider carefully how both methods may be used to simplify the complex fraction to $\dfrac{2 - 5x}{x + 1}$.

Method 1 First, obtain single fractions in the numerator and denominator.

$$\frac{\dfrac{2}{x} - 5}{1 + \dfrac{1}{x}} = \frac{\dfrac{2}{x} - \dfrac{5x}{x}}{\dfrac{1x}{x} + \dfrac{1}{x}} = \frac{\dfrac{2 - 5x}{x}}{\dfrac{x + 1}{x}}$$

Then divide by multiplying by the reciprocal of the denominator.

$$\frac{2 - 5x}{x} \div \frac{x + 1}{x} = \frac{2 - 5x}{x} \cdot \frac{x}{x + 1} = \frac{2 - 5x}{x + 1}$$

Method 2 The LCD of $2/x$, $5/1$, $1/1$, and $1/x$ is x. Therefore, we multiply the numerator and the denominator of the complex fraction by x and then simplify.

$$\frac{\dfrac{2}{x} - 5}{1 + \dfrac{1}{x}} = \frac{x\left(\dfrac{2}{x} - 5\right)}{x\left(1 + \dfrac{1}{x}\right)} = \frac{x \cdot \dfrac{2}{x} - x \cdot 5}{x \cdot 1 + x \cdot \dfrac{1}{x}} = \frac{2 - 5x}{x + 1}$$

PROGRESS CHECK 2 Simplify the complex fraction $\dfrac{\dfrac{4}{y} + 1}{2 - \dfrac{3}{y}}$.

Whether to use method 1 or method 2 for simplifying complex fractions is a personal choice, but the next two examples include some guidelines that you should consider.

Progress Check Answer

2. $\dfrac{4 + y}{2y - 3}$

EXAMPLE 3 Simplify the complex fraction $\dfrac{\dfrac{1}{a}+\dfrac{1}{b}}{\dfrac{2}{a+b}}$.

Solution The LCD, which is $ab(a+b)$, is relatively complex, and the denominator is already a single fraction. Therefore, we will choose method 1 and begin by adding in the numerator.

$$\frac{\dfrac{1}{a}+\dfrac{1}{b}}{\dfrac{2}{a+b}}=\frac{\dfrac{1b}{ab}+\dfrac{1a}{ba}}{\dfrac{2}{a+b}}=\frac{\dfrac{b+a}{ab}}{\dfrac{2}{a+b}}$$

Then divide by multiplying by the reciprocal of the denominator.

$$\frac{b+a}{ab}\div\frac{2}{a+b}=\frac{b+a}{ab}\cdot\frac{a+b}{2}=\frac{(a+b)^2}{2ab}$$

PROGRESS CHECK 3 Simplify the complex fraction $\dfrac{\dfrac{1}{y}+\dfrac{2}{x}}{\dfrac{3}{2y+x}}$.

EXAMPLE 4 Solve the problem in the section introduction on page 183.

Solution The LCD of $1/1$, $1/y$, and $(10a-n)/ny$ is ny. The LCD is relatively simple, and method 1 would require us to obtain single fractions in both the numerator and the denominator. Therefore, we select method 2.

$$F_+=\frac{1-\dfrac{1}{y}}{1+\dfrac{10a-n}{ny}}$$ Given formula.

$$=\frac{ny\left(1-\dfrac{1}{y}\right)}{ny\left(1+\dfrac{10a-n}{ny}\right)}$$ Multiply numerator and denominator by ny.

$$=\frac{ny\cdot 1-ny\cdot\dfrac{1}{y}}{ny\cdot 1+ny\,\dfrac{10a-n}{ny}}$$ Distributive property.

$$=\frac{ny-n}{ny+10a-n}$$ Simplify.

PROGRESS CHECK 4 Medical screening tests also have false negative rates. A formula for one such rate is given next. Simplify the expression on the right side of the formula.

$$F_-=\frac{\dfrac{10-an}{ny}}{9-\dfrac{10a-9n}{ny}}$$

Example 5 points out that it is necessary to check to see that final results are expressed in lowest terms.

Progress Check Answers

3. $\dfrac{(x+2y)^2}{3xy}$ 4. $\dfrac{10-an}{9ny-10a+9n}$

EXAMPLE 5 Simplify the complex fraction $\dfrac{\dfrac{y}{x} - 2}{\dfrac{y^2}{x^2} - 4}$.

Solution The LCD is x^2, and method 2 is a good choice.

$$\frac{\dfrac{y}{x} - 2}{\dfrac{y^2}{x^2} - 4} = \frac{x^2\left(\dfrac{y}{x} - 2\right)}{x^2\left(\dfrac{y^2}{x^2} - 4\right)}$$ Multiply in the numerator and denominator by x^2.

$$= \frac{xy - 2x^2}{y^2 - 4x^2}$$ Distributive property.

$$= \frac{x(y - 2x)}{(y + 2x)(y - 2x)}$$ Factor.

$$= \frac{x}{y + 2x}$$ Lowest terms.

PROGRESS CHECK 5 Simplify the complex fraction $\dfrac{\dfrac{1}{16} - \dfrac{1}{n^2}}{\dfrac{1}{4} + \dfrac{1}{n}}$.

2 Problems involving negative exponents often lead to complex fractions, as shown in Example 6.

EXAMPLE 6 Simplify $\dfrac{x^{-1} + 1}{1 - x^{-2}}$.

Solution Begin by using the definition of a negative exponent to rewrite the expression with no negative exponents.

$$\frac{x^{-1} + 1}{1 - x^{-2}} = \frac{\dfrac{1}{x} + 1}{1 - \dfrac{1}{x^2}}$$ Definition of negative exponent.

$$= \frac{x^2\left(\dfrac{1}{x} + 1\right)}{x^2\left(1 - \dfrac{1}{x^2}\right)}$$ Multiply numerator and denominator by the LCD, x^2.

$$= \frac{x + x^2}{x^2 - 1}$$ Distributive property.

$$= \frac{x(x + 1)}{(x + 1)(x - 1)}$$ Factor.

$$= \frac{x}{x - 1}$$ Lowest terms.

PROGRESS CHECK 6 Simplify $\dfrac{2x^{-1} - 1}{4x^{-2} - 1}$.

Progress Check Answers

5. $\dfrac{n - 4}{4n}$ 6. $\dfrac{x}{2 + x}$

EXERCISES 4.4

In Exercises 1–30, simplify the complex fraction.

1. $\dfrac{1-\frac{1}{3}}{\frac{5}{6}+4}$ Answer: $\frac{4}{29}$

2. $\dfrac{2+\frac{5}{8}}{\frac{1}{3}-\frac{1}{4}}$ Answer: $\frac{63}{2}$

3. $\dfrac{\frac{1}{x}}{\frac{2}{x^2}}$ Answer: $\frac{x}{2}$

4. $\dfrac{\frac{3}{x^2}}{\frac{4}{x}}$ Answer: $\frac{3}{4x}$

5. $\dfrac{\frac{1}{n}+m}{\frac{1}{n}-m}$ Answer: $\frac{1+mn}{1-mn}$

6. $\dfrac{n+\frac{1}{m}}{n-\frac{1}{m}}$ Answer: $\frac{mn+1}{mn-1}$

7. $\dfrac{1+\frac{1}{x}}{2-\frac{1}{x}}$ Answer: $\frac{x+1}{2x-1}$

8. $\dfrac{3-\frac{2}{x^2}}{4-\frac{3}{x^2}}$ Answer: $\frac{3x^2-2}{4x^2-3}$

9. $\dfrac{\frac{1}{a}}{\frac{2}{a}-\frac{3}{b}}$ Answer: $\frac{b}{2b-3a}$

10. $\dfrac{\frac{3}{a}-\frac{4}{b}}{\frac{5}{ab}}$ Answer: $\frac{3b-4a}{5}$

11. $\dfrac{\frac{1}{x+y}}{\frac{1}{x}-\frac{1}{y}}$ Answer: $\frac{xy}{y^2-x^2}$

12. $\dfrac{\frac{x-y}{x+y}}{\frac{1}{y}-\frac{1}{x}}$ Answer: $\frac{xy}{x+y}$

13. $\dfrac{\frac{3}{w}+w}{\frac{9}{w^2}-w^2}$ Answer: $\frac{w}{3-w^2}$

14. $\dfrac{\frac{16}{v^2}-\frac{9}{w^2}}{\frac{4}{v}-\frac{3}{w}}$ Answer: $\frac{4w+3v}{vw}$

15. $\dfrac{a+\frac{1}{b}+\frac{1}{c}}{c+\frac{1}{b}+\frac{1}{a}}$ Answer: $\frac{a^2bc+ac+ab}{abc^2+ac+bc}$

16. $\dfrac{1-\frac{1}{a}-\frac{1}{b}}{b+\frac{1}{a}+1}$ Answer: $\frac{ab-b-a}{ab^2+b+ab}$

17. $\dfrac{\frac{x}{x+y}}{\frac{1}{x+y}+\frac{y}{x+y}}$ Answer: $\frac{x}{1+y}$

18. $\dfrac{\frac{m}{m-n}}{\frac{m}{m-n}+\frac{n}{m-n}}$ Answer: $\frac{m}{m+n}$

19. $\dfrac{\frac{4}{m-n}}{\frac{2}{n-m}}$ Answer: -2

20. $\dfrac{\frac{x^2}{x-y}}{\frac{x}{y-x}}$ Answer: $-x$

21. $\dfrac{3+\frac{4}{h}}{2-\frac{h+6}{hk}}$ Answer: $\frac{3hk+4k}{2hk-h-6}$

22. $\dfrac{\frac{1}{k}-\frac{h+k}{hk}}{\frac{2}{h}-\frac{h-k}{hk}}$ Answer: $\frac{-k}{3k-h}$

23. $\dfrac{\frac{x}{x+y}-\frac{y}{x-y}}{\frac{y}{x+y}-\frac{x}{x-y}}$ Answer: $-\frac{x^2-2xy-y^2}{x^2+y^2}$

24. $\dfrac{\frac{a}{a+b}-\frac{b}{a-b}}{\frac{b}{a+b}+\frac{a}{a-b}}$ Answer: $\frac{a^2-2ab-b^2}{a^2+2ab-b^2}$

25. $\dfrac{\frac{1}{x}+\frac{x}{x+2}}{\frac{3}{x^2-2x-8}}$ Answer: $\frac{x^3-3x^2-2x-8}{3x}$

26. $\dfrac{\frac{2x}{x^2-2x-3}}{\frac{1}{x}+\frac{2}{x+1}}$ Answer: $\frac{2x^2}{3x^2-8x-3}$

27. $\dfrac{\frac{1}{x}+\frac{2}{x^2}+\frac{3}{x-1}}{\frac{4}{(x-1)^2}}$ Answer: $\frac{(x-1)(4x^2+x-2)}{4x^2}$

28. $\dfrac{\frac{2}{x^2}+\frac{3}{x}}{\frac{1}{(x-1)^2}+\frac{3}{x-1}}$ Answer: $\frac{(3x+2)(x-1)^2}{x^2(3x-2)}$

29. $\dfrac{\frac{x}{y}-\frac{y}{x}}{\frac{x}{y^2}-\frac{y}{x^2}}$ Answer: $\frac{(x+y)xy}{x^2+xy+y^2}$

30. $\dfrac{\frac{1}{j}+\frac{1}{k}}{\frac{k}{j^2}+\frac{j}{k^2}}$ Answer: $\frac{jk}{k^2-jk+j^2}$

In Exercises 31–36, simplify the complex fractions.

31. $\dfrac{x^{-2}-1}{x^{-1}-1}$ Answer: $\frac{1+x}{x}$

32. $\dfrac{x^{-2}-2}{x^{-4}-4}$ Answer: $\frac{x^2}{1+2x^2}$

33. $\dfrac{n^{-1}+n^{-2}}{1+n}$ Answer: $\frac{1}{n^2}$

34. $\dfrac{2n^{-2}-3n^{-3}}{2n-3}$ Answer: $\frac{1}{n^3}$

35. $\dfrac{y+5+6y^{-1}}{y+1-6y^{-1}}$ Answer: $\frac{y+2}{y-2}$

36. $\dfrac{2y-5-3y^{-1}}{2y-7-4y^{-1}}$ Answer: $\frac{y-3}{y-4}$

37. In photography the focal length of a lens is given by $f=\dfrac{1}{\frac{1}{d}+\frac{1}{a}}$, where d is the distance from some object to the lens and a is the distance of its image from the lens. Simplify the complex fraction in this equation. Answer: $f=\frac{ad}{a+d}$

38. Simplify the following formula for the harmonic mean of three numbers a, b, and c.

$$M=\frac{3}{\frac{1}{a}+\frac{1}{b}+\frac{1}{c}}$$

Answer: $\frac{3abc}{bc+ac+ab}$

39. **a.** Simplify the following formula for the weighted mean of a and b, where w_1 and w_2 are the weights.

$$m=\frac{a+\left(\frac{w_2}{w_1}\right)b}{1+\frac{w_2}{w_1}}$$

Answer: $m=\frac{w_1a+w_2b}{w_1+w_2}$

b. Show that if w_1 and w_2 are equal, then this is just the ordinary mean (average) of a and b. Answer: $m=\frac{a+b}{2}$

40. Simplify this complex fraction, which approximates $\sqrt{28}$:

$$5+\frac{1}{3+\frac{1}{2}}.$$

Answer: $\frac{37}{7}$

THINK ABOUT IT

1. A complex fraction which approximates π is $3 + \cfrac{1}{7 + \cfrac{1}{16}}$.
   a. Simplify this to one improper fraction.
   b. Express the result of part **a** as a decimal, and compare it with π (which equals 3.14159265 to eight decimal places).
2. Solve for d: $f = \cfrac{1}{\cfrac{1}{d} + \cfrac{1}{a}}$.
3. Simplify.
   a. $\cfrac{x^{-1}}{1 + \cfrac{1}{x^{-1}}}$
   b. $\cfrac{x^{-2}}{1 + \cfrac{1}{x^{-2}}}$
   c. $\cfrac{x^{-n}}{1 + \cfrac{1}{x^{-n}}}$
4. Consider these expressions.
   $\cfrac{1}{2 + \cfrac{1}{2}}$, $\cfrac{1}{2 + \cfrac{1}{2 + \cfrac{1}{2}}}$, and

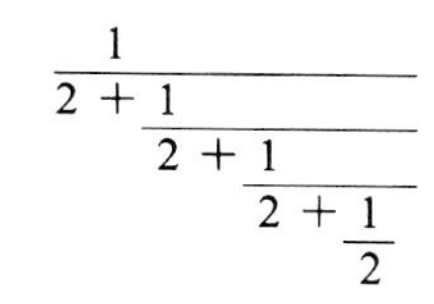

   a. Show that they simplify to $\frac{2}{5}$, $\frac{5}{12}$, and $\frac{12}{29}$, respectively.
   b. If you keep going like this, the expression is called a *continued fraction*. There is a pattern to the fractions in part **a**. The next fraction is $\frac{29}{70}$. What comes after that?
   c. If you express each fraction as a decimal and add 1 to each, you may see that the answers are getting closer and closer to $\sqrt{2}$. Find an approximation to $\sqrt{2}$ on a calculator, and see how these answers compare.
5. Consider these expressions.

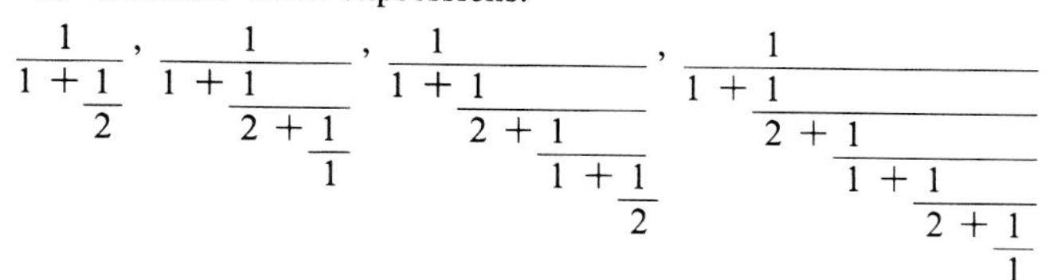

   a. Show that they simplify to $\frac{2}{3}$, $\frac{3}{4}$, $\frac{8}{11}$, and $\frac{11}{15}$, respectively.
   b. There is a pattern to the fractions in part **a**. The next fraction is $\frac{30}{41}$. What are the next three fractions in the sequence?
   c. If you express each fraction in the sequence as a decimal and add 1 to each, you may see that the answers are getting closer and closer to $\sqrt{3}$. Find an approximation to $\sqrt{3}$ on a calculator, and see how these answers compare.

REMEMBER THIS

1. Simplify $\frac{an^3}{n}$ by applying the quotient property of exponents. Assume $n \neq 0$. an^2
2. Rewrite the expression $(a + b) \cdot \frac{1}{c}$ according to the distributive property. $\frac{a}{c} + \frac{b}{c}$
3. Which of these expressions are monomials?
   a. $\frac{1}{x}$ b. $\frac{x}{1}$ c. $\frac{2x^2}{3}$ d. $x - 4$ b and c
4. Which of these are true?
   a. $\frac{15 + 2}{2} = 15$
   b. $\frac{15 + 2}{2} = 15 + 1$
   c. $\frac{15 + 2}{2} = 7.5 + 1$ c
5. Match the correct value to each name.

$$3\overline{)14}\quad \text{quotient } 4,\ 12,\ \text{remainder } 2$$

   a. Quotient = 4
   b. Remainder = 2
   c. Divisor = 3
   d. Dividend = 14
6. Write as a single fraction: $\frac{1}{x + n} - \frac{1}{n}$. $\frac{-x}{n(x + n)}$
7. Simplify $\frac{2n + 2}{n^2} \cdot \frac{n}{n^2 - 1}$. $\frac{2}{n(n - 1)}$
8. Find a fraction with denominator x^2y^2 which is equivalent to $3y/x$. $\frac{3xy^3}{x^2y^2}$
9. Solve $9x^2 = 6x - 1$. $\{\frac{1}{3}\}$
10. A student got an 86 and an 87 on the first two tests. What grades on the third test will make the average more than 89.5? Assume that the highest possible grade is 100.
   $\{x: 95.5 < x \leq 100\}$, or $(95.5, 100]$

4.5 Division of Polynomials

When solving the formula $S = \frac{1}{2}n(a + \ell)$ for ℓ, a student obtains the answer $\ell = \frac{2S}{n} - a$. However, the answer given in the text is $\ell = \frac{2S - na}{n}$. Is the student's answer also correct? (See Example 3.)

OBJECTIVES

1 **Divide a polynomial by a monomial.**

2 **Divide a polynomial by a polynomial with at least two terms.**

1 The simplest case of polynomial division is division of one monomial by another, by applying the quotient property of exponents, as shown in Section 3.1. This type of problem is reviewed briefly in Example 1. Because division by zero is undefined, assume throughout this section that we exclude any value for a variable that results in a zero denominator.

EXAMPLE 1 Find each quotient. Write the result using only positive exponents.

a. $\frac{16x^6}{8x}$ **b.** $\frac{6y^2z^4}{-9y^3z^4}$

Solution Divide the numerical coefficients and divide the variable factors.

a. $\frac{16x^6}{8x} = \frac{16}{8}x^{6-1} = 2x^5$

b. $\frac{6y^2z^4}{-9y^3z^4} = \frac{6}{-9}y^{2-3}z^{4-4} = -\frac{2}{3}y^{-1}z^0 = -\frac{2}{3} \cdot \frac{1}{y} \cdot 1 = -\frac{2}{3y}$

One also says that the ring of polynomials is not "closed" under division.

Note The result of the addition, subtraction, or multiplication of two polynomials is always another polynomial. But as shown in Example 1b, the quotient of two polynomials is not necessarily a polynomial.

PROGRESS CHECK 1 Find each quotient. Write the result using only positive exponents.

a. $\frac{20y^5}{5y^3}$ **b.** $\frac{32xz^2}{-4x^4z^2}$

Progress Check Answers

1. (a) $4y^2$ (b) $-\frac{8}{x^3}$

The procedure for dividing a polynomial which has more than one term by a monomial follows from the distributive property, because every division problem can also be expressed as a multiplication problem, using the reciprocal of the divisor. Thus, we may divide $a + c$ by b as follows.

$$(a + c) \div b = (a + c) \cdot \frac{1}{b} \qquad \text{Rewrite as multiplication.}$$
$$= a \cdot \frac{1}{b} + c \cdot \frac{1}{b} \qquad \text{Distributive property.}$$
$$= \frac{a}{b} + \frac{c}{b} \qquad \text{Simplify.}$$

This result leads to the following efficient way to perform this type of division.

To Divide a Polynomial by a Monomial

To divide a polynomial by a monomial, divide each term of the polynomial by the monomial. In symbols,

$$\frac{a + c}{b} = \frac{a}{b} + \frac{c}{b}, \qquad b \neq 0.$$

From a different viewpoint, this procedure can be thought of as the reverse of adding fractions with a common denominator, where we write

$$\frac{a}{b} + \frac{c}{b} = \frac{a + c}{b}, \qquad b \neq 0.$$

In addition we combine two or more fractions into one; in division by a monomial we split one fraction into two (or more) fractions.

EXAMPLE 2

a. Divide $8x^3 - 4x^2 + 3x$ by $4x^2$ **b.** Divide $\dfrac{-15x^2y^2 - 2x^2y}{-5xy}$.

Solution

a. Divide each term of $8x^3 - 4x^2 + 3x$ by $4x^2$, and simplify.

$$\frac{8x^3 - 4x^2 + 3x}{4x^2} = \frac{8x^3}{4x^2} - \frac{4x^2}{4x^2} + \frac{3x}{4x^2}$$
$$= 2x - 1 + \frac{3}{4x}$$

Note The answer to a division problem can be checked by multiplying the answer by the divisor. In this case $4x^2\left(2x - 1 + \dfrac{3}{4x}\right) = 8x^3 - 4x^2 + 3x$, so the answer checks.

b. Because of the negative sign in the denominator, we choose to express the subtraction in the numerator as an addition expression, using the relationship $a - b = a + (-b)$.

$$\frac{-15x^2y^2 - 2x^2y}{-5xy} = \frac{-15x^2y^2}{-5xy} + \frac{-2x^2y}{-5xy}$$
$$= 3xy + \frac{2}{5}x$$

PROGRESS CHECK 2

a. Divide $10x^4 + 5x^2 - 4x$ by $5x^2$. **b.** Divide $\dfrac{5x^2y - 8xy^2}{-2xy}$.

Progress Check Answers

2. (a) $2x^2 + 1 - \dfrac{4}{5x}$ (b) $-\frac{5}{2}x + 4y$

EXAMPLE 3 Solve the problem in the section introduction on page 190.

Solution We may start from the answer given in the text and divide each term of $2S - na$ by n.

$$\ell = \frac{2S - na}{n}$$
$$= \frac{2S}{n} - \frac{na}{n} = \frac{2S}{n} - a$$

Thus, the student's answer is also correct.

Caution Be sure to divide *all* terms of the numerator by the divisor. A common student error in Example 3 is to divide n only into na and incorrectly write $\ell = 2S - a$, forgetting that the $2S$ term must also be divided by n.

PROGRESS CHECK 3 When solving the formula $C = \frac{5}{9}(F - 32)$ for F, a student obtains the answer $F = \frac{9}{5}C + 32$. If the answer in the text key is $F = \dfrac{9C + 160}{5}$, is the student's answer also correct?

2 Division of a polynomial by a polynomial with two or more terms is similar to long division in arithmetic. To illustrate, consider the division of 146 by 13 shown below, and note the names of the key components.

$$\begin{array}{rrl} & 12 & \leftarrow \text{Quotient} \\ \text{Divisor} \rightarrow 13 & \overline{)158} & \leftarrow \text{Dividend} \\ & \underline{13} & \\ & 28 & \\ & \underline{26} & \\ & 2 & \leftarrow \text{Remainder} \end{array}$$

The result may be expressed as

$$\frac{158}{13} = 12 + \frac{2}{13},$$

and in general,

$$\frac{\text{dividend}}{\text{divisor}} = \text{quotient} + \frac{\text{remainder}}{\text{divisor}}.$$

Dividing 13 into 158 is like dividing $x + 3$ into $x^2 + 5x + 8$, where x stands for 10. Note how the steps compare as we go through the problem.

The first step is to write both polynomials with descending powers.

$$x + 3\overline{)x^2 + 5x + 8}$$

Divide the first term of the dividend by the first term of the divisor to obtain the first term of the quotient.

$$\begin{array}{rl} & x \\ x + 3 & \overline{)x^2 + 5x + 8} \end{array} \qquad \text{Think: } \frac{x^2}{x} = x.$$

Next, multiply the entire divisor by the first term of the quotient and subtract this product from the dividend.

$$\begin{array}{rrl} & & x \\ & x + 3 & \overline{)x^2 + 5x + 8} \\ \text{Subtract:} & & \underline{x^2 + 3x\phantom{{}+8}} \\ & & \phantom{x^2+{}}2x + 8 \end{array} \qquad \text{Think: } x(x + 3) = x^2 + 3x.$$

Use the remainder as the new dividend and repeat the above procedure until the remainder is of lower degree than the divisor.

Progress Check Answer

3. Yes

$$\begin{array}{rl} & \underline{\;x + 2\;\;\;\;\;\;\;\;\;} \\ x + 3 & \big)\, x^2 + 5x + 8 \\ \text{Subtract:} & \underline{x^2 + 3x} \\ & \;\;\;\;\;\;\; 2x + 8 \\ \text{Subtract:} & \;\;\;\;\;\;\; \underline{2x + 6} \\ & \;\;\;\;\;\;\;\;\;\;\;\;\;\; 2 \end{array}$$

Thus, the quotient is $x + 2$, the remainder is 2, and the answer may be written as

$$\frac{x^2 + 5x + 8}{x + 3} = x + 2 + \frac{2}{x + 3}.$$

The procedure is summarized in the box below.

Long Division of Polynomials

1. Arrange the terms of the dividend and the divisor with descending powers. *If a lower power is absent in the dividend, write 0 as its coefficient.*
2. Divide the first term of the dividend by the first term of the divisor to obtain the first term of the quotient.
3. Multiply the entire divisor by the first term of the quotient, and subtract this result from the dividend.
4. Use the remainder as the new dividend and repeat the above procedure until the remainder is of lower degree than the divisor.

The next two examples illustrate the procedure.

EXAMPLE 4 Divide $3x^2 - 7x - 5$ by $x - 3$.

Solution As shown in color, the key divisions are that $3x^2$ divided by x is $3x$, and $2x$ divided by x is 2.

$$\frac{3x^2}{x} = 3x \qquad \frac{2x}{x} = 2$$

$$\begin{array}{rll} & \underline{\;3x + 2\;\;\;\;\;\;\;\;\;\;\;} & \\ x - 3 & \big)\, 3x^2 - 7x - 5 & \\ \text{Subtract:} & \underline{3x^2 - 9x} & \text{This line is } 3x(x - 3). \\ & \;\;\;\;\;\;\;\;\; 2x - 5 & \\ \text{Subtract:} & \;\;\;\;\;\;\;\;\; \underline{2x - 6} & \text{This line is } 2(x - 3). \\ & \;\;\;\;\;\;\;\;\;\;\;\;\;\;\;\;\; 1 & \end{array}$$

The quotient is $3x + 2$, the remainder is 1, and we may write

$$\frac{3x^2 - 7x - 5}{x - 3} = 3x + 2 + \frac{1}{x - 3}.$$

Note Check that dividend = (divisor)(quotient) + remainder. In this example,

$$\begin{aligned} (x - 3)(3x + 2) + 1 &= 3x^2 + 2x - 9x - 6 + 1 \\ &= 3x^2 - 7x - 5, \end{aligned}$$

which is the correct dividend, so the answer checks.

PROGRESS CHECK 4 Divide $3x^2 - 2x + 6$ by $x + 2$, and check your answer.

Progress Check Answer

4. $3x - 8 + \dfrac{22}{x + 2}$

EXAMPLE 5 Divide $\dfrac{y^3 + 1}{y + 1}$.

Solution The dividend is a polynomial of degree 3 with no second-degree or first-degree terms. Following step 1 of the procedure given in the box, $0x^2$ and $0x$ are inserted so that like terms align vertically.

$$
\begin{array}{rrrrrl}
 & & y^2 & - y & + 1 & \\
y + 1 \,) & y^3 & + 0y^2 & + 0y & + 1 & \quad \frac{y^3}{y} = y^2,\ -\frac{y^2}{y} = -y, \text{ and } \frac{y}{y} = 1. \\
\text{Subtract:} & y^3 & + y^2 & & & \quad \text{This line is } y^2(y + 1). \\
 & & -y^2 & + 0y & + 1 & \\
\text{Subtract:} & & -y^2 & - y & & \quad \text{This line is } -y(y + 1). \\
 & & & y & + 1 & \\
\text{Subtract:} & & & y & + 1 & \quad \text{This line is } 1(y + 1). \\
 & & & & 0 &
\end{array}
$$

A zero remainder is not usually written in the answer, so

$$\frac{y^3 + 1}{y - 1} = y^2 - y + 1.$$

PROGRESS CHECK 5 Divide $\dfrac{x^3 - 1}{x - 1}$.

In the final example the divisor has degree 2. This focuses attention on step 4 of the procedure, which says to stop as soon as the remainder is of lower degree than the divisor.

EXAMPLE 6 Divide $\dfrac{3x^4 - 9x^3 - x^2 + 7x - 10}{3x^2 + 2}$.

Solution Be careful to align like terms, as shown in the division below.

$$
\begin{array}{rrrrrrl}
 & x^2 & - 3x & - 1 & & & \\
3x^2 + 2 \,) & 3x^4 & - 9x^3 & - x^2 & + 7x & - 10 & \quad \frac{3x^4}{3x^2} = x^2,\ \frac{-9x^3}{3x^2} = -3x, \text{ and } \frac{-3x^2}{3x^2} = -1. \\
\text{Subtract:} & 3x^4 & & + 2x^2 & & & \quad \text{This line is } x^2(3x^2 + 2). \\
 & & - 9x^3 & - 3x^2 & + 7x & - 10 & \\
\text{Subtract:} & & - 9x^3 & & - 6x & & \quad \text{This line is } -3x(3x^2 + 2). \\
 & & & - 3x^2 & + 13x & - 10 & \\
\text{Subtract:} & & & - 3x^2 & & - 2 & \quad \text{This line is } -1(3x^2 + 2). \\
 & & & & 13x & - 8 &
\end{array}
$$

Stop here, because the remainder is of lower degree than the divisor. The answer is

$$\frac{3x^4 - 9x^3 - x^2 + 7x - 10}{3x^2 + 2} = x^2 - 3x - 1 + \frac{13x - 8}{3x^2 + 2}.$$

PROGRESS CHECK 6 Divide $\dfrac{2x^4 - 6x^3 + 5x^2 + 5x - 3}{2x^2 + 3}$.

Progress Check Answers

5. $x^2 + x + 1$
6. $x^2 - 3x + 1 + \dfrac{14x - 6}{2x^2 + 3}$

EXERCISES 4.5

In Exercises 1–40, find each quotient.

1. $\dfrac{24x^7}{6x}$ $4x^6$
2. $\dfrac{30x^5}{5x^3}$ $6x^2$
3. $\dfrac{4x^2y^2}{8xy^3}$ $\dfrac{x}{2y}$
4. $\dfrac{2xy^4}{18x^3y^3}$ $\dfrac{y}{9x^2}$
5. $\dfrac{-30x^{10}y^{10}z^{10}}{10x^{10}y^{12}z^8}$ $-\dfrac{3z^2}{y^2}$
6. $\dfrac{14xy^2z}{-4x^2yz}$ $-\dfrac{7y}{2x}$
7. Divide $12n^3 - 4n^2 + n$ by $2n^2$. $6n - 2 + \dfrac{1}{2n}$
8. Divide $15m^5 - 10m^3 + 5m$ by $5m^2$. $3m^3 - 2m + \dfrac{1}{m}$
9. Divide $24xy - 18x^2y^2$ by $-6xy^2$. $-\dfrac{4}{y} + 3x$

10. Divide $-10x^2y + 12xy^2 - 14xy$ by $-2xy$. $5x - 6y + 7$

11. Divide $-9rs^2t + 14rs$ by $12st^2$. $-\frac{3rs}{4t} + \frac{7r}{6t^2}$

12. Divide $-6rs^2t^3 + 12r^3s^2t$ by $9r^2s^2t^2$. $-\frac{2t}{3r} + \frac{4r}{3t}$

13. $\frac{6x - 9y + 12}{-3xy}$ $-\frac{2}{y} + \frac{3}{x} - \frac{4}{xy}$

14. $\frac{8y^2 - 12x^2 + 16x^2y^2}{-4xy}$ $-\frac{2y}{x} + \frac{3x}{y} - 4xy$

15. $\frac{a^2 + 2ab + b^2}{ab}$ $\frac{a}{b} + 2 + \frac{b}{a}$

16. $\frac{a^3 + 3a^2b + 3ab^2 + b^3}{a^2b^2}$ $\frac{a}{b^2} + \frac{3}{b} + \frac{3}{a} + \frac{b}{a^2}$

17. Divide $3x^2 + 2x + 1$ by $x - 1$. $3x + 5 + \frac{6}{x - 1}$

18. Divide $x^2 + 2x + 3$ by $x + 3$. $x - 1 + \frac{6}{x + 3}$

19. Divide $4y^2 - 2y - 1$ by $y + 2$. $4y - 10 + \frac{19}{y + 2}$

20. Divide $-3y^2 + 13y + 12$ by $y - 3$. $-3y + 4 + \frac{24}{y - 3}$

21. Divide $3y^2 - 13y + 12$ by $y - 3$. $3y - 4$

22. Divide $2y^2 + 5y - 12$ by $y + 4$. $2y - 3$

23. Divide $\frac{6x^2 + 7x - 2}{2x + 3}$. $3x - 1 + \frac{1}{2x + 3}$

24. Divide $\frac{4x^2 - 4x - 17}{2x - 5}$. $2x + 3 - \frac{2}{2x - 5}$

25. Divide $\frac{x^3 + 8}{x + 2}$. $x^2 - 2x + 4$

26. Divide $\frac{x^3 - 27}{x - 3}$. $x^2 + 3x + 9$

27. Divide $\frac{n^3 - 1}{n + 1}$. $n^2 - n + 1 - \frac{2}{n + 1}$

28. Divide $\frac{m^3 + 8}{m - 2}$. $m^2 + 2m + 4 + \frac{16}{m - 2}$

29. Divide $4x^2 + 9$ by $2x + 3$. $2x - 3 + \frac{18}{2x + 3}$

30. Divide $x^2 + 16$ by $x + 4$. $x - 4 + \frac{32}{x + 4}$

31. Divide $9x^4 + 5x^2 + x + 3$ by $3x - 1$.
$3x^3 + x^2 + 2x + 1 + \frac{4}{3x - 1}$

32. Divide $4x^4 + x^2 + x - 21$ by $2x + 3$. $2x^3 - 3x^2 + 5x - 7$

33. Divide $\frac{2x^4 + 2x^3 + 3x^2 + 2x - 4}{2x^2 + 1}$. $x^2 + x + 1 + \frac{x - 5}{2x^2 + 1}$

34. Divide $\frac{6a^4 - 3a^3 - 11a^2 + 2a + 5}{3a^2 - 1}$.
$2a^2 - a - 3 + \frac{a + 2}{3a^2 - 1}$

35. Divide $2c^4 + c^3 - 5c^2 + 5c - 4$ by $c^2 - c + 1$.
$2c^2 + 3c - 4 - \frac{2c}{c^2 - c + 1}$

36. Divide $3d^4 + 5d^3 + 2d^2 + 1$ by $3d^2 - d + 1$.
$d^2 + 2d + 1 - \frac{d}{3d^2 - d + 1}$

37. If $P(x) = x^4 + x^3 - x^2 + 2x + 3$ and $Q(x) = x + 1$, find $P(x) \div Q(x)$. $x^3 - x + 3$

38. If $P(x) = x^5 - 2x^3 + 5x^2 - 10$ and $Q(x) = x^2 - 2$, find $P(x) \div Q(x)$. $x^3 + 5$

39. Find $m^4 + 2m^3 + 3m^2 + 2m + 1$ divided by $m^2 + m + 1$.
$m^2 + m + 1$

40. Find $m^4 - 2m^3 + 3m^2 - 2m + 1$ divided by $m^2 - m + 1$.
$m^2 - m + 1$

In Exercises 41–44 we have given two solutions which resulted from two different ways of solving the given equation. Show that the two solutions are equivalent.

41. Solve $A = \frac{1}{2}(a + b)h$ for a: $a = \frac{2A - bh}{h}$; $a = \frac{2A}{h} - b$.
$\frac{2A - bh}{h} = \frac{2A}{h} - \frac{bh}{h} = \frac{2A}{h} - b$

42. Solve $x = \mu + z\sigma$ for σ: $\sigma = \frac{x - \mu}{z}$; $\sigma = \frac{x}{z} - \frac{\mu}{z}$.
$\frac{x - \mu}{z} = \frac{x}{z} - \frac{\mu}{z}$

43. Solve $y = \frac{a(x + b)}{c}$ for x: $x = \frac{cy - ab}{a}$; $x = \frac{cy}{a} - b$.
$\frac{cy - ab}{a} = \frac{cy}{a} - \frac{ab}{a} = \frac{cy}{a} - b$

44. Solve $\frac{1 + w}{-1} = \frac{p}{1 - p}$ for p: $p = \frac{1 + w}{w}$; $p = \frac{1}{w} + 1$.
$\frac{1 + w}{w} = \frac{1}{w} + \frac{w}{w} = \frac{1}{w} + 1$

THINK ABOUT IT

1. If the remainder is 6, the quotient is 4, and the divisor is 7, what is the dividend? What division problem is this?
2. If the dividend is $x^2 + 3x - 2$, the divisor is $x - 5$, and the remainder is 38, what is the quotient? What division problem is this?
3. In this section of the text all the problems were chosen to work out "nicely" with integers. Here's one that involves fractions. Divide $3x^2 - x + 4$ by $5x + 1$.
4. What happens if you use the long division procedure to divide a polynomial by one with a higher degree? We start such a problem for you. Divide 3 by $x + 2$.

$$\begin{array}{r} 3x^{-1} - 6x^{-2} \\ x + 2\overline{)3 \qquad\qquad} \\ \underline{3 + 6x^{-1}} \\ -6x^{-1} \\ \underline{-6x^{-1} - 12x^{-2}} \\ 12x^{-2} \end{array}$$

   a. What is the next term in the quotient?
   b. Will this procedure ever terminate?
   c. Why do we stop long division when the degree of the remainder is smaller than the degree of the divisor?
5. Divide $3x^n + 2x^{n-1} + 5$ by x^{-n}.

REMEMBER THIS

1. True or false? When a positive integer is divided by 3, the only possible remainders are 0, 1, and 2. True
2. Let $P(x)$ stand for $x^2 + 2x + 3$. Find the remainder when $P(x)$ is divided by $x - 10$; then find $P(10)$. Both answers are 123.
3. Let $P(x) = x^2 - 8x + 15$. Is 3 a solution of $P(x) = 0$? Yes
4. Multiply $(x - 2)(x - 4)(x + 1)$. $x^3 - 5x^2 + 2x + 8$
5. Simplify $\frac{a + n/b}{b + n/a}$. $\frac{a}{b}$
6. Write as a single fraction: $\frac{3}{x - 4} + \frac{5}{4 - x}$. $\frac{-2}{x - 4}$
7. Factor $y^3 + 64$. $(y + 4)(y^2 - 4y + 16)$
8. Simplify $\frac{2.8 \times 10^{-8}}{1.4 \times 10^{-9}}$. $2 \times 10^1 = 20$
9. Evaluate $3 \cdot 2^{-3} + 2 \cdot 3^{-2}$. $\frac{43}{72}$
10. Solve $3(x^2 + \frac{1}{3}) = 3x^2 + 1$. This is an identity; solution set is all real numbers.

4.6 Synthetic Division

A rectangular piece of cardboard 12 in. by 18 in. is to be used to make an open-top box by cutting a small square from each corner and bending up the sides. If the volume of the box is to be 216 in.³, then the side length x for each square cutout may be found by solving $(18 - 2x)(12 - 2x)x = 216$, which is equivalent to

$$x^3 - 15x^2 + 54x - 54 = 0.$$

Use the remainder theorem to check that 3 is a solution of this equation. (See Example 4.)

x

OBJECTIVES

1. **Divide a polynomial by a polynomial of the form $x - b$ using synthetic division.**
2. **Evaluate a polynomial using the remainder theorem.**
3. **Apply the remainder theorem to solving equations involving polynomials.**

1 A special case of polynomial division occurs with the division of any polynomial by a polynomial of the form $x - b$. This type of division problem is important from both a practical and a theoretical viewpoint and may be performed by a shortcut method called **synthetic division.** To understand the basis for this shortcut, consider the division of $4x^3 + 3x - 5$ by $x - 2$ using the long division procedure.

$$\begin{array}{rl}
 & 4x^2 + 8x + 19 \\
x - 2 \,\big)\!\!\!\! & 4x^3 + 0x^2 + 3x - 5 \\
\text{Subtract:} & \underline{4x^3 - 8x^2} \\
 & 8x^2 + 3x - 5 \\
\text{Subtract:} & \underline{8x^2 - 16x} \\
 & 19x - 5 \\
\text{Subtract:} & \underline{19x - 38} \\
 & 33
\end{array}$$

To shorten this procedure, we first note that there is no need to write all the powers of x. When the polynomials are written with terms in descending powers of x, then only the coefficients are needed. Also, we note that all the coefficients shown in color may be omitted because they are predictable and unnecessary repetitions. Therefore, an abbreviation of this division that uses only the necessary coefficients is as shown below.

$$\begin{array}{rrrrr} & 4 & 8 & 19 & \\ -2\,) & 4 & 0 & 3 & -5 \\ \text{Subtract:} & & -8 & -16 & -38 \\ & & 8 & 19 & 33 \end{array}$$

If we bring down 4 as the first entry in the bottom row, all the coefficients of the quotient appear. The arrangement may then be shortened as follows.

$$\begin{array}{rrrrrll} -2 \,| & \boxed{4} & 0 & 3 & -5 & \leftarrow \text{Coefficients of dividend} & \text{(row 1)} \\ \text{Subtract:} & & -8 & -16 & -38 & & \text{(row 2)} \\ \text{Coefficients of quotient} \rightarrow & 4 & 8 & 19 & 33 & \leftarrow \text{Remainder} & \text{(row 3)} \end{array}$$

Finally, by replacing -2 with 2, which is the value of b, we may change the sign of each number in row 2 and add at each step instead of subtracting. The final arrangement for synthetic division is then as follows.

$$\begin{array}{rrrrrll} 2 \,| & 4 & 0 & 3 & -5 & & \text{(row 1)} \\ \text{Add:} & & 8 & 16 & 38 & & \text{(row 2)} \\ & 4 & 8 & 19 & 33 & \rightarrow \text{Remainder} & \text{(row 3)} \end{array}$$

Quotient: $4x^2 + 8x + 19$

Use this example as a basis for understanding the synthetic division procedure outlined next.

Synthetic Division

To divide a polynomial $P(x)$ by $x - b$:

1. Form row 1 by writing the coefficients of the terms in the dividend $P(x)$. The dividend must be written in descending powers and 0 entered as the coefficient of any missing term. Write the value of b to the left of these coefficients.
2. Bring down the first dividend entry as the first coefficient in the quotient.
3. Multiply this quotient coefficient by b. Place the result under the next number in row 1, and then add.
4. Repeat the procedure in step 3 until all entries in row 1 have been used.
5. The last number in the bottom row is the remainder. The other numbers in the bottom row are, from left to right, the coefficients of descending powers of the quotient.

It is important to remember that synthetic division applies only to division by a polynomial of the form $x - b$. Because this divisor is a first-degree polynomial, the degree of the polynomial in the quotient is always one less than the degree of the polynomial in the dividend.

See the "Think About It" exercises for discussion of other cases.

EXAMPLE 1 Use synthetic division to divide $5x^3 - x^2 + 4x + 3$ by $x + 1$.

Solution Follow the steps outlined above. For this division the value of b is -1, because $x + 1$ is written $x - (-1)$ when the divisor is expressed in the form $x - b$.

$$\begin{array}{r|rrrr} -1 & 5 & -1 & 4 & 3 \\ & & -5 & 6 & -10 \\ \hline & 5 & -6 & 10 & -7 \end{array}$$

Quotient: $5x^2 - 6x + 10$; -7: Remainder

Computation steps
1. Bring down 5.
2. $5(-1) = -5$; $-1 + (-5) = -6$.
3. $-6(-1) = 6$; $4 + 6 = 10$.
4. $10(-1) = -10$; $3 + (-10) = -7$.

Using the quotient and remainder shown above, the answer is

$$\frac{5x^3 - x^2 + 4x - 3}{x + 1} = 5x^2 - 6x + 10 + \frac{-7}{x + 1}.$$

PROGRESS CHECK 1 Use synthetic division to divide $4x^3 + 3x^2 - x + 5$ by $x + 2$.

EXAMPLE 2 Use synthetic division to divide $5x^4 + x - 7$ by $x - 2$. Express the result in the form

dividend = (divisor)(quotient) + remainder.

Solution Use 0's as the coefficients of the missing x^3 and x^2 terms. The divisor is $x - 2$, so $b = 2$. Now divide.

$$\begin{array}{r|rrrrr} 2 & 5 & 0 & 0 & 1 & -7 \\ & & 10 & 20 & 40 & 82 \\ \hline & 5 & 10 & 20 & 41 & 75 \end{array} \to \text{Remainder}$$

Quotient: $5x^3 + 10x^2 + 20x + 41$

The answer in the form requested is

$$5x^4 + x - 7 = (x - 2)(5x^3 + 10x^2 + 20x + 41) + 75.$$

PROGRESS CHECK 2 Use synthetic division to divide $3x^4 - x^2 + 8$ by $x - 3$. Express the result in the form requested in Example 2.

2 The result in Example 2 shows that

$$P(x) = 5x^4 + x - 7$$

may be written as

$$P(x) = (x - 2)(5x^3 + 10x^2 + 20x + 41) + 75.$$

When $x = 2$, the factor $x - 2 = 0$. Thus,

$$P(2) = 0 + 75 = 75.$$

The value of the polynomial when $x = 2$ is the same as the remainder obtained when $P(x)$ is divided by $x - 2$. This illustrates the following theorem.

Remainder Theorem

If a polynomial $P(x)$ is divided by $x - b$, the remainder is $P(b)$.

Progress Check Answers

1. $4x^2 - 5x + 9 - \frac{13}{x + 2}$
2. $3x^4 - x^2 + 8 = (x - 3)(3x^3 + 9x^2 + 26x + 78) + 242$

In Section 3.3 we showed that $P(b)$ may be found by substituting b for x in the equation given for $P(x)$. The remainder theorem provides an alternative method. That is, we find $P(b)$ by determining the remainder when $P(x)$ is divided by $x - b$. This approach is often simpler, because the remainder may be found by synthetic division. Besides simplicity, the remainder theorem method has other uses, and we will consider one of these applications in Example 5.

EXAMPLE 3 If $P(x) = 3x^4 + 2x^3 + x^2 - 7x - 9$, find $P(-2)$ by (a) direct substitution and (b) the remainder theorem.

Solution

a. To use the direct substitution method, replace x by -2 in the given equation and simplify.

$$\begin{aligned} P(-2) &= 3(-2)^4 + 2(-2)^3 + (-2)^2 - 7(-2) - 9 \\ &= 48 - 16 + 4 + 14 - 9 \\ &= 41 \end{aligned}$$

b. To use the remainder theorem method, divide $P(x)$ by $x - (-2)$ using synthetic division.

$$\begin{array}{r|rrrrr} -2 & 3 & 2 & 1 & -7 & -9 \\ & & -6 & 8 & -18 & 50 \\ \hline & 3 & -4 & 9 & -25 & 41 \end{array}$$

Since the remainder is 41, $P(-2) = 41$.

PROGRESS CHECK 3 If $P(x) = 2x^4 + 5x^3 - 4x^2 + x - 8$, find $P(-3)$ by (a) direct substitution and (b) the remainder theorem.

3 The remainder theorem is often applied when solving equations involving polynomials. In this context, the initial application of the remainder theorem is usually to check to see that a number is a solution to such an equation.

EXAMPLE 4 Solve the problem in the section introduction on page 196.

Solution To determine whether 3 is a solution of

$$x^3 - 15x^2 + 54x - 54 = 0$$

using the remainder theorem, let $P(x)$ equal the polynomial on the left side of the equal sign. Then 3 is a solution if and only if $P(3) = 0$. Synthetic division gives

$$\begin{array}{r|rrrr} 3 & 1 & -15 & 54 & -54 \\ & & 3 & -36 & 54 \\ \hline & 1 & -12 & 18 & 0. \end{array}$$

Since the remainder is 0, $P(3) = 0$, so 3 is a solution. Thus, one way to form the desired box involves cutting a 3-in. square from each corner.

PROGRESS CHECK 4 Use the remainder theorem to determine whether 4 is a solution of the equation given in Example 4.

When one solution to an equation involving polynomials is found, it may be possible to use the zero product principle (discussed in Section 3.9) in conjunction with the remainder theorem and factoring to find all the solutions.

Progress Check Answers

3. $P(-3) = -20$

4. No; $P(4) = -14 \neq 0$

EXAMPLE 5 Solve $x^3 - 5x^2 + 2x + 8 = 0$, given that 2 is one of the solutions.

Solution First, let $P(x)$ equal the polynomial on the left side of the equal sign and verify that 2 is a solution by the remainder theorem method.

$$\begin{array}{r|rrrr} 2 & 1 & -5 & 2 & 8 \\ & & 2 & -6 & -8 \\ \hline & 1 & -3 & -4 & 0 \end{array}$$

As expected, the remainder is 0, so $P(2) = 0$ and 2 is a solution. Because the remainder is 0 the division above indicates that $P(x)$ factors as

$$x^3 - 5x^2 + 2x + 8 = (x - 2)(x^2 - 3x - 4),$$

so by factoring completely we get

$$\begin{aligned} x^3 - 5x^2 + 2x + 8 &= 0 \\ (x - 2)(x^2 - 3x - 4) &= 0 \\ (x - 2)(x - 4)(x + 1) &= 0. \end{aligned}$$

Applying the zero product principle to the last equation gives solutions of 2, 4, and -1, so the solution set is $\{2,4,-1\}$.

PROGRESS CHECK 5 Solve $x^3 - 2x^2 - 5x + 6 = 0$, given that 1 is one of the solutions.

Progress Check Answer

5. $\{1,-2,3\}$

EXERCISES 4.6

In Exercises 1–12, use synthetic division.

1. $4x^3 - 2x^2 + 3x - 1$ divided by $x - 2$ $\quad 4x^2 + 6x + 15 + \dfrac{29}{x - 2}$
2. $x^3 + 2x^2 - 3x + 4$ divided by $x - 3$ $\quad x^2 + 5x + 12 + \dfrac{40}{x - 3}$
3. $2y^3 - 4y^2 + 6y - 8$ divided by $y + 1$ $\quad 2y^2 - 6y + 12 - \dfrac{20}{x + 1}$
4. $y^3 + 3y^2 - 5y + 7$ divided by $y + 2$ $\quad y^2 + y - 7 + \dfrac{21}{y + 2}$
5. $-5t^4 + 4t^3 - t^2 + 6t - 4$ divided by $t - 1$ $\quad -5t^3 - t^2 - 2t + 4$
6. $-t^4 - 4t^3 + 2t^2 + 8t - 1$ divided by $t + 4$ $\quad -t^3 + 2t - \dfrac{1}{t + 4}$
7. $(x^4 - x^2 + 1) \div (x + 1)$ $\quad x^3 - x^2 + \dfrac{1}{x + 1}$
8. $(x^5 - x^3 + x) \div (x - 1)$ $\quad x^4 + x^3 + 1 + \dfrac{1}{x - 1}$
9. $(n^3 + 8) \div (n + 2)$ $\quad n^2 - 2n + 4$
10. $(n^5 - 32) \div (n - 2)$ $\quad n^4 + 2n^3 + 4n^2 + 8n + 16$
11. $\dfrac{1 - 2y + 4y^2 + 8y^3}{y + 1}$ $\quad 8y^2 - 4y + 2 - \dfrac{1}{y + 1}$
12. $\dfrac{2 - 3y^2 + 4y - y^3}{3 + y}$ $\quad -y^2 + 4 - \dfrac{10}{y + 3}$

In Exercises 13–18, use synthetic division; then express the result in the form dividend = (divisor)(quotient) + remainder.

13. Divide $6x^4 - x + 8$ by $x - 2$.
    $6x^4 - x + 8 = (x - 2)(6x^3 + 12x^2 + 24x + 47) + 102$
14. Divide $5x^4 - 2x^2 + 1$ by $x + 1$.
    $5x^4 - 2x^2 + 1 = (x + 1)(5x^3 - 5x^2 + 3x - 3) + 4$
15. Divide $2t^3 - 5t^2 + 7t - 30$ by $t - 3$.
    $2t^3 - 5t^2 + 7t - 30 = (t - 3)(2t^2 + t + 10) + 0$
16. Divide $3t^3 + 10t^2 - 6t + 8$ by $t + 4$.
    $3t^3 + 10t^2 - 6t + 8 = (t + 4)(3t^2 - 2t + 2) + 0$
17. $(x^3 + 1) \div (x - 1)$ $\quad x^3 + 1 = (x - 1)(x^2 + x + 1) + 2$
18. $(x^5 - 3x^3 - 4x + 6) \div (x - 2)$
    $x^5 - 3x^3 - 4x + 6 = (x - 2)(x^4 + 2x^3 + x^2 + 2x) + 6$

In Exercises 19–24, find the desired value by (a) direct substitution and (b) the remainder theorem.

19. $P(x) = 4x^3 - 2x^2 + 3x - 1$. Find $P(2)$. 29
20. $P(x) = x^3 + 2x^2 - 3x + 4$. Find $P(3)$. 40
21. $Q(y) = 2y^3 - 4y^2 + 6y - 8$. Find $Q(-1)$. -20
22. $Q(y) = y^3 + 3y^2 - 5y + 7$. Find $Q(-2)$. 21
23. $P(n) = -5n^4 + 4n^3 - n^2 + 6n - 4$. Find $P(1)$. 0
24. $P(n) = 2n^3 - 5n^2 + 7n - 30$. Find $P(3)$. 0

In Exercises 25–30, use the remainder theorem to determine if the given value is a solution of the equation.

25. Is 3 a solution of $x^4 - 5x^3 - 4x^2 + 20x + 30 = 0$? Yes
26. Is -4 a solution of $2x^4 + 9x^3 + 9x^2 + 18x - 8 = 0$? Yes
27. Is -2 a solution of $-4y^3 - 8y^2 + y + 2 = 0$? Yes
28. Is -5 a solution of $y^3 + 4y^2 - 4y + 5 = 0$? Yes
29. Is -3 a solution of $x^3 - 3x^2 + 3x - 9 = 0$? No
30. Is 2 a solution of $x^4 - 2x^3 + 4x^2 - 8x + 16 = 0$? No

In Exercises 31–36, use the given information to solve the equation.

31. Solve $x^3 - 6x^2 + 11x - 6 = 0$, given that 2 is one of the solutions. $\{1,2,3\}$
32. Solve $y^3 + 3y^2 - 10y - 24 = 0$, given that -2 is a solution. $\{-2,3,-4\}$

33. Solve $6y^3 - 11y^2 + 6y - 1 = 0$, given that 1 is a solution.
$\{1, \frac{1}{2}, \frac{1}{3}\}$

34. Solve $6x^3 - 25x^2 + 18x + 9 = 0$, given that 3 is a solution.
$\{3, -\frac{1}{3}, \frac{3}{2}\}$

35. Solve $x^3 + 3x^2 + 3x + 1 = 0$, given that -1 is a solution.
$\{-1\}$

36. Solve $x^3 - 9x^2 + 24x - 20 = 0$, given that 2 is one solution.
$\{2, 5\}$

37. The dimensions of a rectangular solid are three consecutive integers, as shown in the figure, and the surface area is 28 more than the volume. Find the dimensions of the solid. First, check that the surface area equals $6x^2 - 2$ and the volume equals $x^3 - x$. Then solve $(6x^2 - 2) - (x^3 - x) = 28$. Use the remainder theorem to show that 3 is one solution of this equation; then solve the problem. How many such solids are there?
There are 2; 2, 3, 4 and 4, 5, 6.

$x - 1$ $x + 1$ x

38. The dimensions of a rectangular solid are three consecutive integers, as shown in the figure, and the surface area is 34 more than the volume. Find the dimensions of the solid. Note that the area equals $6x^2 - 2$, and the volume equals $x^3 - x$; then solve $(6x^2 - 2) - (x^3 - x) = 34$. Use the remainder theorem to show that 4 is the *only integer* solution; then solve the problem.
3, 4, 5; the quadratic polynomial cannot be factored.

$x - 1$ $x + 1$ x

39. In the sixteenth century, the French mathematician Francois Viète worked out a method for solving any equation of the form $x^3 + 3ax = 2b$, where a and b can be any real numbers.

a. Let $a = -1$ and $b = 1$ to get an equation of this form.
$x^3 - 3x = 2$

b. Given that one of the solutions is 2, use the remainder theorem to find any other solutions. $\{-1, 2\}$

40. Refer to Exercise 39, and let $a = -9$ and $b = -27$ to get an equation of the desired form. Given that one of the solutions is 3, use the remainder theorem to find any other solutions.
$x^3 - 27x = -54$; $\{-6, 3\}$

THINK ABOUT IT

1. In the text all the synthetic division exercises avoided fractions. This is not a necessary feature of synthetic division. Use synthetic division to divide $4x^3 - 11x^2 + 2x + 5$ by $x - \frac{3}{4}$. Find the quotient and the remainder.

2. In the text we said that synthetic division works when the divisor has the form $x - b$, where the coefficient of x is 1.

a. Why would synthetic division not work in a problem like $2x^3 + 9x^2 + x - 12$ divided by $2x + 3$?

b. Divide $2x^3 + 9x^2 + x - 12$ by $2x + 3$ using *long division.*

c. Instead of dividing by $2x + 3$ where synthetic division does not work, you could factor out the 2 to get $2(x + \frac{3}{2})$ and then use $x + \frac{3}{2}$ as the divisor in synthetic division. At the end you would therefore have to divide the quotient by 2 to get the correct answer. Try it this way.

3. How would you modify the problem so synthetic division may be used to divide $x^3 - 2x^2 + 3x - 4$ by $2 - x$? Figure out a way and then try it.

4. a. True or false? If the remainder in synthetic division is 0, then the divisor is a factor of the dividend.

b. Use synthetic division to show that $x - 2$ is a factor of $3x^4 - 6x^3 - 5x^2 + 11x - 2$. Then express $3x^4 - 6x^3 - 5x^2 + 11x - 2$ in factored form.

5. a. Check that $x + 1$ is a factor of $x^3 + x^2 + x + 1$ by using synthetic division.

b. Show that $x + 1$ is not a factor of $x^4 + x^3 + x^2 + x + 1$.

c. Without doing any calculations, but using your answers in parts **a** and **b,** explain why $x + 1$ is or is not a factor of $x^{99} + x^{98} + x^{97} + \cdots + x + 1$.

REMEMBER THIS

1. Simplify $6\left(\frac{5x}{6} + \frac{1}{3}\right)$ by applying the distributive property.
$5x + 2$

2. Simplify $(y^2 + y - 6)\left(\frac{3}{y - 2} + \frac{2}{y + 3}\right)$ by applying the distributive property. $5(y + 1)$

3. Solve $60m + 180 = 96m$. $\{5\}$

4. For what value(s) of x is the expression $\frac{5}{x - 3} - \frac{30}{x^2 - 9}$ undefined? $3, -3$

5. If a job takes 20 minutes to complete, then what fraction is completed in 1 minute? $\frac{1}{20}$

6. When $x^3 - 2x^2 + 3x - 4$ is divided by $x - 1$, what is the remainder? $P(1) = -2$

7. Factor by grouping: $3m^2 + 21m + 6m + 42$. $(m + 7)(3m + 6)$

8. Multiply $x^3(5x^{-3} + 3x^{-1})$. $5 + 3x^2$

9. An inheritance of \$20,000 was split and invested in two accounts, one of which earns an annual interest rate of 6 percent and another which earns 10 percent. After one year the total interest earned was \$1,760. How much was invested at 10 percent? \$14,000

10. Solve $x^2 + 3x = 1 + 3x$. $\{1, -1\}$

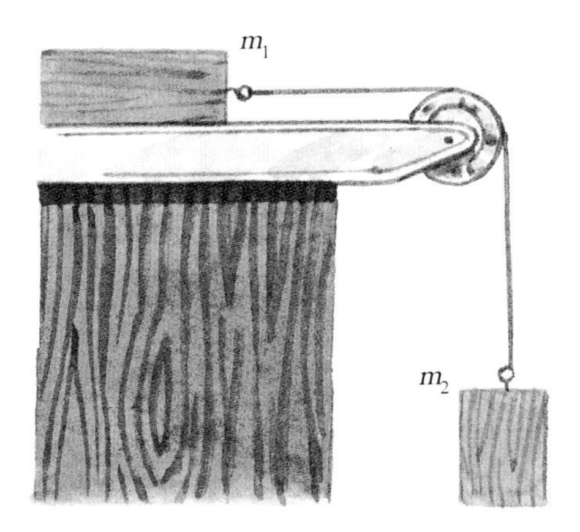

4.7 Solving Equations Containing Rational Expressions

In a physics experiment a block of mass m_1 on a smooth horizontal surface is pulled by a string which is attached to a block of mass m_2 hanging over a pulley. Ignoring friction and the mass of the pulley, the formula for the tension (T) in the string is

$$T = \frac{32m_1m_2}{m_1 + m_2}.$$

Find m_1 if $m_2 = 3$ units and $T = 60$ lb. The unit for mass in this notation is called a slug. (See Example 4.)

OBJECTIVES

1. **Solve equations that contain rational expressions.**
2. **Solve formulas that contain rational expressions for a specified variable.**
3. **Set up and solve work problems.**
4. **Set up and solve uniform motion problems.**

Stress the contrast between solving an *equation,* where you multiply both *sides* by the same number, and rewriting a *single fraction,* where you multiply the *top and bottom* by the same number. Remind students to always write complete equations, showing both sides and the equals sign.

1 As illustrated by the problem that opens this section, many applications of algebra involve equations that contain rational expressions. The usual first step for solving such an equation is to multiply both sides of the equation by the least common denominator of all fractions that appear in it. This results in an equation that does not contain fractions and that may often be solved by methods we have already discussed.

EXAMPLE 1 Solve $\frac{5x}{6} + \frac{1}{3} = \frac{x}{2}$.

Solution Multiply both sides of the equation by the LCD, 6.

$$6\left(\frac{5x}{6} + \frac{1}{3}\right) = 6\left(\frac{x}{2}\right) \quad \text{Multiply both sides by 6.}$$
$$6\left(\frac{5x}{6}\right) + 6\left(\frac{1}{3}\right) = 6\left(\frac{x}{2}\right) \quad \text{Distributive property.}$$
$$5x + 2 = 3x \quad \text{Simplify.}$$
$$2 = -2x \quad \text{Subtract } 5x \text{ from both sides.}$$
$$-1 = x \quad \text{Divide both sides by } -2.$$

To check this solution, replace x by -1 in the original equation.

$$\frac{5(-1)}{6} + \frac{1}{3} \stackrel{?}{=} \frac{-1}{2}$$
$$\frac{-5}{6} + \frac{1}{3} \stackrel{?}{=} \frac{-1}{2} \quad \text{Simplify.}$$
$$\frac{-1}{2} \stackrel{\checkmark}{=} \frac{-1}{2} \quad \text{Add fractions and simplify.}$$

Thus, the solution set is $\{-1\}$.

PROGRESS CHECK 1 Solve $\frac{x}{5} + 1 = \frac{x+2}{10}$.

Progress Check Answer
1. $\{-8\}$

EXAMPLE 2 Solve $\frac{4}{x} - 3 = \frac{5}{2x}$.

Solution To remove fractions, multiply both sides of the equation by the LCD, $2x$. Note that this step requires the restriction that $x \neq 0$ to ensure that we are multiplying both sides of the equation by a nonzero number.

$$2x\left(\frac{4}{x} - 3\right) = 2x\left(\frac{5}{2x}\right) \quad \text{Multiply both sides by } 2x.$$

$$2x\left(\frac{4}{x}\right) - 2x(3) = 2x\left(\frac{5}{2x}\right) \quad \text{Distributive property.}$$

$$8 - 6x = 5 \quad \text{Simplify.}$$

$$-6x = -3 \quad \text{Subtract 8 from both sides.}$$

$$x = \tfrac{1}{2} \quad \text{Divide both sides by } -6 \text{ and simplify.}$$

Both sides of the original equation equal 5 when $x = \frac{1}{2}$, so $\frac{1}{2}$ checks and the solution set is $\{\frac{1}{2}\}$.

PROGRESS CHECK 2 Solve $\frac{5}{x} - 4 = \frac{10}{3x}$.

EXAMPLE 3 Solve $\frac{5}{y^2 + y - 6} = \frac{3}{y - 2} + \frac{2}{y + 3}$.

Solution Because $y^2 + y - 6$ equals $(y - 2)(y + 3)$, the LCD is $(y - 2)(y + 3)$. Assuming that $y \neq 2$ and $y \neq -3$, start by multiplying both sides of the original equation by the LCD.

$$(y - 2)(y + 3)\frac{5}{y^2 + y - 6} = (y - 2)(y + 3)\left(\frac{3}{y - 2} + \frac{2}{y + 3}\right) \quad \text{Multiply both sides by } (y - 2)(y + 3).$$

$$\frac{(y - 2)(y + 3) \cdot 5}{y^2 + y - 6} = \frac{(y - 2)(y + 3) \cdot 3}{y - 2} + \frac{(y - 2)(y + 3) \cdot 2}{y + 3} \quad \text{Distributive property.}$$

$$5 = (y + 3)3 + (y - 2)2 \quad \text{Fundamental principle of rational expressions.}$$

$$5 = 3y + 9 + 2y - 4 \quad \text{Distributive property.}$$

$$5 = 5y + 5 \quad \text{Simplify.}$$

$$0 = 5y \quad \text{Subtract 5 from both sides.}$$

$$0 = y \quad \text{Divide both sides by 5.}$$

The solution checks as shown next.

Check

$$\frac{5}{0^2 + 0 - 6} \stackrel{?}{=} \frac{3}{0 - 2} + \frac{2}{0 + 3} \quad \text{Replace } y \text{ by 0 in the original equation.}$$

$$\frac{5}{-6} \stackrel{?}{=} \frac{3}{-2} + \frac{2}{3}$$

$$\frac{-5}{6} \stackrel{\checkmark}{=} \frac{-5}{6}$$

The solution set is $\{0\}$.

PROGRESS CHECK 3 Solve $\frac{6}{x^2 - 2x - 8} = \frac{5}{x - 4} + \frac{1}{x + 2}$.

Progress Check Answers

2. $\{\frac{5}{12}\}$
3. $\{0\}$

EXAMPLE 4 Solve the problem in the section introduction on page 202.

Solution After replacing T and m_2 by their given values, multiply both sides of the equation by the LCD, which is $m_1 + 3$. (The denominator of the left-hand side is equal to 1.)

$$T = \frac{32m_1m_2}{m_1 + m_2} \quad \text{Given formula.}$$
$$60 = \frac{32m_1(3)}{m_1 + 3} \quad \text{Replace } T \text{ by 60 and } m_2 \text{ by 3.}$$
$$60 = \frac{96m_1}{m_1 + 3} \quad \text{Simplify.}$$
$$60(m_1 + 3) = 96m_1 \quad \text{Multiply both sides by } m_1 + 3.$$
$$60m_1 + 180 = 96m_1 \quad \text{Distributive property.}$$
$$180 = 36m_1 \quad \text{Subtract } 60m_1 \text{ from both sides.}$$
$$5 = m_1 \quad \text{Divide both sides by 36.}$$

Because $60 = \frac{32(5)(3)}{5 + 3}$ is a true statement, 5 checks and m_1 is a mass of 5 slugs.

PROGRESS CHECK 4 Use the formula in Example 4 to find m_2 if $T = 48$ lb and $m_1 = 2$ slugs.

Clearing fractions in the next two examples leads to quadratic equations that can be solved by the methods of Section 3.9.

EXAMPLE 5 Solve $x - \frac{10}{x} = 3$.

Solution Multiply both sides of the equation by the LCD, x.

$$x\left(x - \frac{10}{x}\right) = x \cdot 3 \quad \text{Multiply both sides by } x.$$
$$x \cdot x - x\left(\frac{10}{x}\right) = x \cdot 3 \quad \text{Distributive property.}$$
$$x^2 - 10 = 3x \quad \text{Simplify.}$$
$$x^2 - 3x - 10 = 0 \quad \text{Subtract } 3x \text{ from both sides.}$$
$$(x - 5)(x + 2) = 0 \quad \text{Factor the nonzero side.}$$
$$x - 5 = 0 \quad \text{or} \quad x + 2 = 0 \quad \text{Set each factor equal to 0.}$$
$$x = 5 \qquad\qquad x = -2 \quad \text{Solve each linear equation.}$$

Because $5 - \frac{10}{5}$ and $-2 - \frac{10}{-2}$ both simplify to 3, both solutions check. The solution set is $\{5, -2\}$.

PROGRESS CHECK 5 Solve $y + \frac{6}{y} = 5$.

The next example shows that the methods of this section may lead to **extraneous solutions,** which are apparent solutions that do not check in the original equation and are therefore not part of the solution set. When solving equations containing rational expressions, check for extraneous solutions at values for the variable that make the LCD zero.

EXAMPLE 6 Solve $1 = \frac{5}{x - 3} - \frac{30}{x^2 - 9}$.

Progress Check Answers

4. 6 slugs 5. $\{2,3\}$

Solution Because $x^2 - 9$ factors as $(x + 3)(x - 3)$, the LCD is $(x + 3)(x - 3)$. We therefore begin by multiplying both sides of the equation by $(x + 3)(x - 3)$, assuming $x \neq -3$ and $x \neq 3$.

$$(x+3)(x-3) \cdot 1 = (x+3)(x-3)\left(\frac{5}{x-3} - \frac{30}{x^2-9}\right)$$ Multiply both sides by the LCD.

$$(x+3)(x-3) \cdot 1 = (x+3)(x-3)\frac{5}{x-3} - (x+3)(x-3)\frac{30}{x^2-9}$$ Distributive property.

$$(x+3)(x-3) = 5(x+3) - 30$$ Fundamental principle.

$$x^2 - 9 = 5x + 15 - 30$$ Multiply.

$$x^2 - 5x + 6 = 0$$ Rewrite so one side is 0.

$$(x-2)(x-3) = 0$$ Factor.

$$x - 2 = 0 \quad \text{or} \quad x - 3 = 0$$ Set each factor equal to 0.

$$x = 2 \qquad\qquad x = 3$$ Solve each equation.

The solution $x = 2$ does check as shown next, but the restriction $x \neq 3$ means that 3 is an extraneous solution and should not be included in the solution set. Substitution of 3 for x shows that this extraneous solution leads to division by zero.

Check

$$1 \stackrel{?}{=} \frac{5}{2-3} - \frac{30}{2^2-9} \qquad\qquad 1 \stackrel{?}{=} \frac{5}{3-3} - \frac{30}{3^2-9}$$

$$1 \stackrel{?}{=} -5 - (-6) \qquad\qquad 1 \neq \frac{5}{0} - \frac{30}{0}$$

$$1 \stackrel{\checkmark}{=} 1 \qquad\qquad \text{3 is an extraneous solution.}$$

The solution set is $\{2\}$.

PROGRESS CHECK 6 Solve $1 = \dfrac{12}{x^2 - 4} - \dfrac{3}{x - 2}$.

2 As mentioned in Section 2.2, it is sometimes necessary to convert a formula to a form that is more efficient for a particular problem. The methods of this section may be used to rearrange formulas that contain fractions.

EXAMPLE 7 Solve the formula $t = \dfrac{v - v_0}{v}$ for v.

Solution The general approach is to first get rid of fractions and then isolate all terms containing the desired variable on one side of the equation, where it can be factored out.

$$t = \frac{v - v_0}{v}$$ Given formula.

$$tv = v - v_0$$ Multiply both sides by v.

$$tv - v = -v_0$$ Subtract v from both sides.

$$v(t - 1) = -v_0$$ Factor out v.

$$v = \frac{-v_0}{t - 1}$$ Divide both sides by $t - 1$.

PROGRESS CHECK 7 Solve the formula in the section-opening problem for m_1.

Progress Check Answers

6. $\{-5\}$

7. $m_1 = \dfrac{Tm_2}{T - 32m_2}$

3 One type of application which involves equations with rational expressions is called a **work problem.** The goal in such problems is to find the time needed to complete a job when two or more people or machines work together. The basis for analyzing work problems is to assume a *constant work rate:* **If a job requires t units of time to complete, then $1/t$ of the job is completed in one unit of time.** For example, if a pump can empty a storage tank in 4 hours, then (assuming a constant pumping rate) it empties $\frac{1}{4}$ of the tank for each hour of pumping.

EXAMPLE 8 One robot assembly line (A) can fill an order in 20 minutes. Another (B) can fill the order in 30 minutes. How long would it take the two lines working together to fill the order? (Assume that the two machines do not interfere with one another.)

Solution Let x represent the number of minutes required to process the order with *both* lines operating. Then apply the basic idea outlined above.

Line A: This line fills the order in 20 minutes, so it processes $\frac{1}{20}$ of the order in 1 minute.

Line B: This line fills the order in 30 minutes, so it processes $\frac{1}{30}$ of the order in 1 minute.

Together: The assembly lines together fill the order in x minutes, so they process $1/x$ of the order in 1 minute.

Now *set up an equation* as follows.

$$\underbrace{\text{Part done by line A in 1 minute}}_{\frac{1}{20}} + \underbrace{\text{part done by line B in 1 minute}}_{\frac{1}{30}} = \underbrace{\text{part done by both lines in 1 minute.}}_{\frac{1}{x}}$$

Solve the Equation

$$60x\left(\frac{1}{20} + \frac{1}{30}\right) = 60x\left(\frac{1}{x}\right) \quad \text{Multiply both sides by the LCD, } 60x.$$

$$60x\left(\frac{1}{20}\right) + 60x\left(\frac{1}{30}\right) = 60x\left(\frac{1}{x}\right) \quad \text{Distributive property.}$$

$$3x + 2x = 60 \quad \text{Simplify.}$$

$$5x = 60 \quad \text{Combine like terms.}$$

$$x = 12 \quad \text{Divide both sides by 5.}$$

The general solution to $1/a + 1/b = 1/x$ is $x = ab/(a + b)$. See Exercise 44.

Answer the Question Together, the two assembly lines complete the order in 12 minutes.

Check the Answer In 12 minutes line A does $12 \cdot \frac{1}{20}$ of the job while line B does $12 \cdot \frac{1}{30}$ of the job. Because $\frac{12}{20} + \frac{12}{30} = 1$, the whole order is completed, and the solution checks.

PROGRESS CHECK 8 A lawn can be mowed in 20 minutes with a ride-on mower and in 80 minutes with a self-propelled mower. How long will it take to mow the lawn using the two mowers together?

Progress Check Answer

8. 16 minutes

4 In Section 2.3 we solved certain uniform motion problems using the formula

$$\text{distance} = \text{rate} \cdot \text{time} \qquad \text{or} \qquad d = rt.$$

But for uniform motion problems in which two events take the same time, it is more useful to rewrite the formula as

$$\text{time} = \frac{\text{distance}}{\text{rate}} \quad \text{or} \quad t = \frac{d}{r}.$$

This formula will lead to equations containing fractions.

EXAMPLE 9 An experimental light plane, powered by the pilot's pedaling, flew a distance of $\frac{1}{2}$ mi against a 2-mi/hour wind. With that same wind behind it, it flew 1 mi in the same time. What is the speed of the plane in still air?

Solution

Let x represent the plane's rate in still air. Then,

$$x - 2 = \text{rate against the wind (subtract wind speed)}$$
and
$$x + 2 = \text{rate with the wind (add wind speed).}$$

Now analyze the problem in a chart format using $t = d/r$.

Direction	Distance ÷	Rate =	Time
Against wind	0.5	$x - 2$	$\frac{0.5}{x-2}$
With the wind	1	$x + 2$	$\frac{1}{x+2}$

To *set up an equation* use the given condition that both trips require the same amount of time.

$$\underbrace{\frac{0.5}{x-2}}_{\text{Time against wind}} \quad \overset{\text{equals}}{=} \quad \underbrace{\frac{1}{x+2}}_{\text{time with wind.}}$$

Solve the Equation

$$0.5(x + 2) = 1(x - 2) \quad \text{Multiply both sides by } (x - 2)(x + 2).$$
$$0.5x + 1 = x - 2 \quad \text{Distributive property.}$$
$$1 = 0.5x - 2 \quad \text{Subtract } 0.5x \text{ from both sides.}$$
$$3 = 0.5x \quad \text{Add 2 to both sides.}$$
$$6 = x \quad \text{Divide both sides by 0.5.}$$

Answer the Question The plane's rate in still air is 6 mi/hour.

Check the Answer If $x = 6$, then the flight against the wind takes $\frac{0.5}{6-2} = \frac{0.5}{4}$ $= 0.125$ hour, and the trip with the wind takes $\frac{1}{6+2} = \frac{1}{8} = 0.125$ hour. The times are the same, so the solution checks.

PROGRESS CHECK 9 An improved version of the plane described in Example 9 flew 1 mi with a 3-mi/hour wind pushing it in the same amount of time it took to fly 0.6 mi against that wind. What is the speed of the plane in still air?

Progress Check Answer

9. 12 mi/hour

EXERCISES 4.7

In Exercises 1–42, solve the given equation.

1. $\dfrac{3x}{4} - \dfrac{1}{2} = \dfrac{x}{8}$ $\{\frac{4}{5}\}$

2. $\dfrac{1}{3} + \dfrac{x}{6} = \dfrac{x}{9}$ $\{-6\}$

3. $\dfrac{y}{3} - 4 = \dfrac{y}{4}$ $\{48\}$

4. $\dfrac{2y}{3} - \dfrac{3y}{4} = 1$ $\{-12\}$

5. $5 - \dfrac{u}{2} = u$ $\{\frac{10}{3}\}$

6. $1 + \dfrac{2v}{3} = \dfrac{3v}{2}$ $\{\frac{6}{5}\}$

7. $\dfrac{2x+1}{2} = \dfrac{3x+2}{3}$ No solution; false equation

8. $\dfrac{3x+1}{2} = \dfrac{6x+2}{4}$ All real numbers; identity

9. $\dfrac{1}{x} + \dfrac{2}{x} = \dfrac{3}{x}$ Identity; all real numbers except 0

10. $\dfrac{1}{x} - \dfrac{2}{x} = \dfrac{3}{x}$ No solution; false equation

11. $\dfrac{1}{n} + 2 = \dfrac{3}{2n}$ $\{\frac{1}{4}\}$

12. $\dfrac{3}{m} - 2 = \dfrac{1}{3m}$ $\{\frac{4}{3}\}$

13. $\dfrac{4}{a} + \dfrac{3}{2} + \dfrac{2}{3a} = 0$ $\{-\frac{28}{9}\}$

14. $\dfrac{5}{2} - \dfrac{4}{3a} - \dfrac{3}{4a} = 0$ $\{\frac{5}{6}\}$

15. $\dfrac{x-1}{x+1} = \dfrac{x}{2x+2}$ $\{2\}$

16. $\dfrac{x}{x-1} = \dfrac{x+1}{3x-3}$ $\{\frac{1}{2}\}$

17. $\dfrac{4}{x} - \dfrac{1}{2x^2} = 0$ $\{\frac{1}{8}\}$

18. $\dfrac{5}{x^2} + \dfrac{1}{5x} = 0$ $\{-25\}$

19. $\dfrac{6}{x+1} - \dfrac{4}{x-1} = \dfrac{10}{x^2-1}$ $\{10\}$

20. $\dfrac{1}{y-2} - \dfrac{2}{y^2-4} = \dfrac{3}{y+2}$ $\{3\}$

21. $\dfrac{5}{2m-3} - \dfrac{3m-1}{4m^2-9} = \dfrac{2}{2m+3}$ $\{-\frac{22}{3}\}$

22. $\dfrac{2}{3m+1} + \dfrac{4}{3m-1} = \dfrac{6m+8}{9m^2-1}$ $\{\frac{1}{2}\}$

23. $\dfrac{6}{x+2} + \dfrac{4}{x-4} = \dfrac{5}{x^2-2x-8}$ $\{\frac{21}{10}\}$

24. $\dfrac{2}{x-3} - \dfrac{1}{x+5} = \dfrac{1}{x^2+2x-15}$ $\{-12\}$

25. $\dfrac{1}{3x-1} - \dfrac{2}{x-2} = \dfrac{5x}{3x^2-7x+2}$ $\{0\}$

26. $\dfrac{5}{3x+2} + \dfrac{x+13}{6x^2+x-2} = \dfrac{4}{2x-1}$ $\{0\}$

27. $\dfrac{1}{s^2+3s} = \dfrac{2}{s^2+4s+3}$ $\{1\}$

28. $\dfrac{6}{s^2-s} = \dfrac{3}{s^2-3s+2}$ $\{4\}$

29. $x + \dfrac{1}{x} = \dfrac{5}{2}$ $\{\frac{1}{2}, 2\}$

30. $y - \dfrac{1}{y} = \dfrac{8}{3}$ $\{-\frac{1}{3}, 3\}$

31. $3y - \dfrac{5}{2} - \dfrac{1}{2y} = 0$ $\{-\frac{1}{6}, 1\}$

32. $\dfrac{6}{5y} + \dfrac{4}{5} - 2y = 0$ $\{-\frac{3}{5}, 1\}$

33. $2 = \dfrac{x}{x^2-1} - \dfrac{2}{x-1}$ $\{-\frac{1}{2}, 0\}$

34. $3 = \dfrac{2x}{x^2-9} + \dfrac{x+9}{x+3}$ $\{0, 4\}$

35. $1 = \dfrac{6}{x-1} - \dfrac{12}{x^2-1}$ $\{5\}$

36. $1 = \dfrac{7}{x-4} - \dfrac{56}{x^2-16}$ $\{3\}$

37. $\dfrac{2}{y^2-1} = 1 - \dfrac{1}{y+1}$ $\{2\}$

38. $\dfrac{5}{y+2} + \dfrac{20}{y^2-4} + 1 = 0$ $\{-3\}$

39. $\dfrac{e+4}{e+2} - \dfrac{e-2}{e-4} = \dfrac{e^2-13}{e^2-2e-8}$ $\{1, -1\}$

40. $\dfrac{e-3}{e-1} + \dfrac{e-1}{e+3} = \dfrac{22}{e^2+2e-3}$ $\{4, -4\}$

41. $\dfrac{3}{x} + \dfrac{2}{x-1} = \dfrac{4x-12}{x^2-2x+1}$ $\{-1, -3\}$

42. $\dfrac{3}{2y} + \dfrac{3y+4}{y^2-8y+16} = \dfrac{4}{y-4}$ $\{-4, -12\}$

43. The harmonic mean H of two numbers a and b is given by $H = \dfrac{2ab}{a+b}$. If $H = 8$ and $a = 6$, find b. $b = 12$

44. Solve $\dfrac{1}{a} + \dfrac{1}{b} = \dfrac{1}{x}$ for x. Show that x is half the harmonic mean defined in Exercise 43. $x = \dfrac{ab}{a+b} = \frac{1}{2}H$

45. If a resistor of R_1 ohms is connected in parallel to a resistor of R_2 ohms, then the total resistance R is given by the formula $\dfrac{1}{R} = \dfrac{1}{R_1} + \dfrac{1}{R_2}$. Find R_1 when $R = 6$ ohms and $R_2 = 10$ ohms. $R_1 = 15$ ohms

46. Use the formula in Exercise 45 to find R_2 if $R = 4$ ohms and $R_1 = 12$ ohms. $R_2 = 6$ ohms

47. If the total cost of producing n units of a product consists of \$1,000 in fixed costs and \$10 per unit, then the average cost per unit A is given by $A = \dfrac{10n + 1{,}000}{n}$. How many units should be produced for the average cost to be \$20 per unit? 100 units

48. In Exercise 47, how many units should be produced for the average cost to be \$15 per unit? 200 units

In Exercises 49–54, solve the given equation for the specified variable.

49. Solve $w = \dfrac{1-p}{p}$ for p. $p = \dfrac{1}{1+w}$

50. Solve $p = \dfrac{1}{1+w}$ for w. $w = \dfrac{1-p}{p}$

51. Solve $\dfrac{1}{f} = \dfrac{1}{d_1} + \dfrac{1}{d_2}$ for d_1. $d_1 = \dfrac{fd_2}{d_2 - f}$

52. Solve $\dfrac{1}{a} = \dfrac{1}{b} - \dfrac{1}{c}$ for c. $c = \dfrac{ab}{a-b}$

53. Solve $P = \dfrac{S}{1+ni}$ for i. $i = \dfrac{S-P}{Pn}$

54. Solve $S = \dfrac{P}{1-nd}$ for d. $d = \dfrac{S-P}{nS}$

55. One person in an office can enter some data into a computer in 4 hours. Another person can do the job in 6 hours. If they work together from separate keyboards, how long will it take to enter the data? 2.4 hours

56. One carpenter can install the plasterboard for a job in 20 hours, while another can install it in 16 hours. If they work together, will the job be done in less than 10 hours? Yes, in about 8 hours 54 minutes

57. Coin sorter A can process a sack of coins in 20 minutes; sorter B can process the sack in 30 minutes. How long would it take the two machines working together to process the sack of coins? 12 minutes

58. A mail-processing machine can sort 50,000 pieces of mail in 1 hour. A newer one takes 40 minutes to do the same job.
- **a.** If they operate together, how long will it take them to sort 50,000 pieces of mail? 24 minutes
- **b.** If they operate together, how long will it take them to sort 80,000 pieces of mail? 38.4 minutes

59. A pump can empty a tank in 100 minutes. Show algebraically that two such pumps working together will cut the time in half. $\frac{1}{100} + \frac{1}{100} = \frac{1}{x}$; $x = 50$ minutes

60. One machine can complete a job in 2 hours. At what rate would a second machine need to operate so that together they could complete the job in $\frac{1}{2}$ hour? $\frac{1}{2} + \frac{1}{x} = \frac{1}{0.5}$; $x = \frac{2}{3}$; rate $= 1$ job in $\frac{2}{3}$ hours

61. A ferry can travel 20 mi downstream in the same time it can travel 10 mi upstream.
- **a.** If the current of the river is 2 mi/hour, what is the boat's rate in still water? 6 mi/hour
- **b.** How long would it take the boat to travel 16 mi upstream? 4 hours
- **c.** About how long would it take to make a round-trip of 8 mi each way if it took 15 minutes to turn around? $3\frac{1}{4}$ hours

62. A ferry can travel 12 mi downstream in the same time it can travel 8 mi upstream.
- **a.** If the current of the river is 1 mi/hour, what is the boat's rate in still water? 5 mi/hour
- **b.** How long would it take the boat to travel 15 mi downstream? 2.5 hours
- **c.** About how long would it take to make a round-trip of 6 mi each way if it took half an hour to turn around? 3 hours

63. A bicycle racer goes for a fixed time 10 mi along a level desert road moving the pedals at a constant rate against a 10 mi/hour head wind. Then the cyclist turns around and goes for that same time with the wind pushing. On the return trip the cyclist covers 30 mi. Cycling at this rate, how long would it take to cover 50 mi with a 5-mi/hour wind pushing from behind? 2 hours

64. Two small children who run at the same speed are playing on a moving sidewalk in an airport. The sidewalk is $\frac{1}{4}$ mi long and moves at 1 mi/hour. One child starts at each end, and they run toward each other. They meet in 3 minutes. What is their running speed on regular ground? 2.5 mi/hour

65. A small group of hikers climbed a 1.5-mi trail up Mount Pisgah in 2 hours. How much time should they spend coming down the same trail so that their average rate of hiking for the entire hike is 1 mi/hour? 1 hour

66. Two friends canoed downstream with a 2-mi/hour current and then canoed back paddling at the same rate. Their effective speed upstream was $\frac{1}{5}$ their downstream speed.
- **a.** What is their rate of speed in still water? 3 mi/hour
- **b.** If they paddled downstream for 1 hour, how long did the whole trip take? 6 hours
- **c.** What is their average effective speed for the whole trip? $1\frac{2}{3}$ mi/hour

67. A plane flies 360 mi from A to B with a 50-mi/hour tail wind. It flies back against this wind in twice the time.
- **a.** What is the speed of the plane in still air? 150 mi/hour
- **b.** How long did it take to fly each way? 1.8 hours from A to B; 3.6 hours from B to A
- **c.** What is the average speed for the round trip? 133.33 mi/hour

THINK ABOUT IT

1. People who do a lot of probability calculations often use shortcuts for solving some equations. For example, to solve the equation $\frac{1}{x-1} = \frac{2}{3}$ for x, they first notice that the top plus the bottom on the left equals x. To get a new simpler proportion, they say, "Replace each fraction by its (top plus its bottom) over its top." This would yield $x = \frac{5}{2}$ immediately.
- **a.** Check that $\frac{5}{2}$ is the solution to the original equation.
- **b.** Solve $\frac{3}{3-x} = \frac{4}{3}$ by first replacing each fraction by its "top minus bottom over top."
- **c.** Solve $\frac{p}{1+p} = \frac{1}{4}$ using "top over (bottom minus top)."

2. Explain why no real number is a solution to $\frac{5x^2}{6} + \frac{1}{3} = \frac{x^2}{2}$.

3. Here are several versions of a classic puzzle.
- **a.** The weight of a brick is half a pound plus half the weight of a brick. How many pounds does the brick weigh?
- **b.** The weight of a brick is $\frac{1}{4}$ of a pound plus $\frac{1}{4}$ of a brick. How many pounds does the brick weigh?
- **c.** The brick weighs $\frac{4}{5}$ of a pound plus $\frac{4}{5}$ of a brick. How many pounds does the brick weigh?
- **d.** This is the problem stated in general. The weight w of brick is a certain fraction f of a pound plus that fraction of a brick. How many pounds does the brick weigh? (Solve for w in terms of f.)
- **e.** For which version of this problem is the answer that the brick weighs 10 lb?

4. A car travels at x mi/hour from A to B and returns on the same route at y mi/hour. Show that the average speed for the round-trip is the harmonic mean of x and y. (See Exercise 43.)

5. Show that in Exercise 59 the value 100 was not necessary to the problem. Show that two similar machines working together will cut the time in half regardless of how long they take working separately.

REMEMBER THIS

1. Is the equation $3x + 4y = 23$ satisfied when $x = 1$ and $y = 5$? Yes
2. Is the equation $3x + 4y = 23$ satisfied when $x = 5$ and $y = 1$? No
3. True or false? If two lines in the same plane are perpendicular, then at their intersection there are four right angles. True
4. A mail order company charges \$13 per compact disc plus \$3 shipping for any order. Write an equation for the total cost (C) of an order in terms of the number (n) of discs ordered. $C = 13n + 3$
5. Use synthetic division to divide $x^3 - 2x^2 + x - 2$ by $x - 3$. $x^2 + x + 4 + \frac{10}{x-3}$
6. Simplify $\frac{a - b/c}{c}$. $\frac{ac-b}{c^2}$
7. Subtract $\frac{1}{x+1} - \frac{1}{x-1}$. $\frac{-2}{x^2-1}$
8. Factor $24x^2y - 2xy - 40y$. $2y(3x - 4)(4x + 5)$
9. Solve $(x - 1)(x - 2) = 2$. $\{0,3\}$
10. In a state with a 6 percent sales tax on meals the total bill for a meal was \$13.25. A patron wanted to leave a tip equal to 20 percent of the pretax cost of the meal. How much was the tip? \$2.50

GRAPHING CALCULATOR

Programming

A program is just a sequence of steps carried out by the calculator automatically. For a program to work, two stages are necessary: You must enter the steps of the program into the calculator (which is done in Edit mode), and then you must start the program going (which is called *executing* the program). To illustrate, we construct a program for substituting values into different expressions.

EXAMPLE 1

a. Write a program that can be used to evaluate an expression involving x.
b. Use the program to evaluate $x^2 - 3x + 4$ for $x = 1, 0, -1$.
c. Use the same program to evaluate $(7 - 3x)^3$ for $x = 1, 3$.

Solution

a. The calculator can hold many programs. This one will be saved as Program 1 with the title EVAL (to signify that it is an evaluating program). The steps of the program contain some special words which are copied from various lists that can be shown on the screen. When you need one of these words, refer to the table given next.

Table of special programming words

Word	Keys to Press
Lbl	[PRGM], then [1]
GoTo	[PRGM], then [2]
Disp	[PRGM], then arrow to [I/O], then [1]
Input	[PRGM], then arrow to [I/O], then [2]

Also, remember that to write Y_1 on the screen, you press [2nd] [Y-VARS], then [1].

Step 1 Press the key PRGM. The screen appears as shown in Figure 4.1.

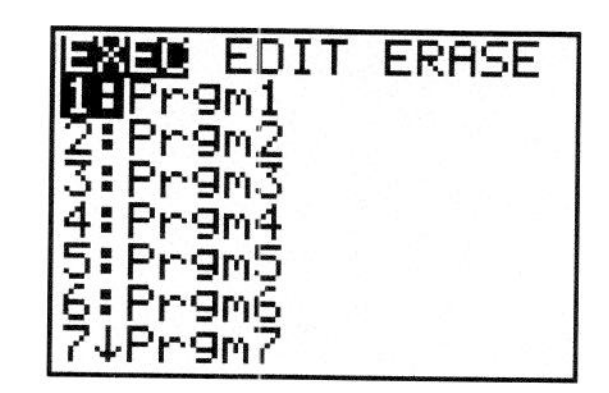
EXEC EDIT ERASE
1:Prgm1
2:Prgm2
3:Prgm3
4:Prgm4
5:Prgm5
6:Prgm6
7↓Prgm7

Figure 4.1

Step 2 Use the arrow key to put the cursor on EDIT and press 1. The calculator is now in Edit mode, and the screen shows "Prgm1:" followed by the blinking A. To name the program EVAL, type EVAL and then press ENTER.

Step 3 Enter the steps as shown in Figure 4.2, remembering to press ENTER at the end of each line. To produce quotation marks, use ALPHA and the plus key. For a blank space, use ALPHA and the grey zero key. After all steps have been entered, press 2nd [QUIT] to get back to the home screen. At this point, the program is saved and is ready to be used.

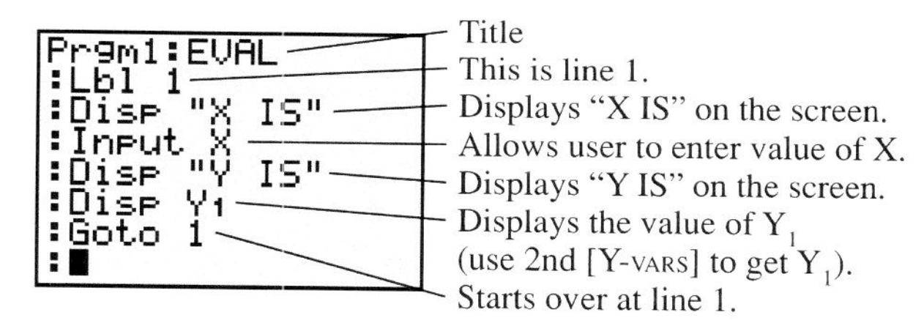
Prgm1:EVAL
:Lbl 1
:Disp "X IS"
:Input X
:Disp "Y IS"
:Disp Y1
:Goto 1
:
Title
This is line 1.
Displays "X IS" on the screen.
Allows user to enter value of X.
Displays "Y IS" on the screen.
Displays the value of Y1
(use 2nd [Y-VARS] to get Y1).
Starts over at line 1.

Figure 4.2

b. First, before you start the program, the expression to be evaluated must be assigned the name Y_1, because there is a reference to Y_1 in the program. This is done as follows.

Step 1 Press Y=; then enter $x^2 - 3x + 4$ for Y_1.

Step 2 Press PRGM and observe that the cursor is on EXEC. Press 1 to start program 1. "Prgm1" should appear on screen. Press ENTER. The program is now working. It asks you for an x value. Since the first required value is $x = 1$, key in 1 and press ENTER. The answer Y = 2 should appear.
Continue keying in the rest of the x values. The correct answers are Y = 4 when X = 0, and Y = 8 when X = −1.
To stop the program, press 2nd [QUIT].

c. All that is necessary is to redefine Y_1. So press Y=. Press CLEAR to erase the old expression, and key in (7 − 3X)∧3. Restart the program as in part **b,** step 2. The correct answers are Y = 64 when X = 1, and Y = −8 when X = 3.

A crucial feature of any program is that it only has to be written once. It can be saved indefinitely in the calculator (until you erase it) and used whenever needed.

Exercises

Use the program EVAL to evaluate each expression for the given values. Round answers to the nearest thousandth.

1. $-x^2 + 2x - 4$ for $x = -1, 3.2$ Ans. −7, −7.84
2. $\frac{1}{5}x^2 - \frac{1}{4}x$ for $x = \frac{1}{2}, \frac{1}{4}, \frac{3}{4}$ Ans. −0.075, −0.05, −0.075
3. $\dfrac{1}{2 + x^2}$ for $x = 2.1, 2.2, 2.3$ Ans. 0.156, 0.146, 0.137

Chapter 4 SUMMARY

OBJECTIVES CHECKLIST Specific chapter objectives are summarized below along with numbered example problems from the text that should clarify the objectives. If you do not understand any objectives or do not know how to do the selected problems, then restudy the material.

4.1 Can you:

1. **Find all values of a variable which make a rational expression undefined?**
   Find all values of y which make the expression $\dfrac{y}{y^2 - 16}$ undefined. [Example 1b]

2. **Determine whether a pair of fractions are equivalent?**
   Determine whether the fractions $\frac{27}{41}$ and $\frac{2}{3}$ are equivalent. [Example 3c]

3. Express rational expressions in lowest terms?

Express $\dfrac{y^2 + by}{b^2 + by}$ in lowest terms. [Example 8a]

4. Build up a rational expression to an equivalent expression with a specified denominator?

Build up $\dfrac{x}{x + 3} = \dfrac{?}{x^2 - 9}$. [Example 9b]

4.2 Can you:

1. Multiply rational expressions?

Multiply $\dfrac{2x + 6}{x^2} \cdot \dfrac{x}{x^2 - 9}$, and express the result in lowest terms. [Example 1b]

2. Divide rational expressions?

Divide $\dfrac{3 - x}{y^2 - y} \div \dfrac{x^2 - 9}{y^2 - 2y + 1}$, and express the result in lowest terms. [Example 3c]

4.3 Can you:

1. Add and subtract rational expressions with the same denominator?

Write $\dfrac{2x + 1}{3x - 5} - \dfrac{x + 4}{3x - 5}$ as a single fraction in lowest terms. [Example 2b]

2. Add and subtract rational expressions with opposite denominators?

Write $\dfrac{3x + 4}{4 - 3x} + \dfrac{1 - x}{3x - 4}$ as a single fraction in lowest terms. [Example 4b]

3. Find the least common denominator (LCD)?

Find the least common denominator for $\dfrac{x}{2x - 2}$ and $\dfrac{3}{x^2 - 1}$. [Example 6b]

4. Add and subtract rational expressions with unlike denominators?

Write $\dfrac{3x + 1}{x^2 + x - 2} + \dfrac{x}{x^2 - 2x + 1}$ as a single fraction in lowest terms. [Example 9c]

4.4 Can you:

1. Simplify complex fractions?

Simplify $\dfrac{\dfrac{1}{a} + \dfrac{1}{b}}{\dfrac{2}{a + b}}$. [Example 3]

2. Simplify complex fractions arising from negative exponents?

Simplify $\dfrac{x^{-1} + 1}{1 - x^{-2}}$. [Example 6]

4.5 Can you:

1. Divide a polynomial by a monomial?

Divide $\dfrac{-15x^2y^2 - 2x^2y}{-5xy}$. [Example 2b]

2. Divide a polynomial by a polynomial with at least two terms?

Divide $3x^2 - 7x - 5$ by $x - 3$. [Example 4]

4.6 Can you:

1. Divide a polynomial by a polynomial of the form $x - b$ using synthetic division?

Use synthetic division to divide $5x^4 + x - 7$ by $x - 2$. [Example 2]

2. **Evaluate a polynomial using the remainder theorem?**
If $P(x) = 3x^4 + 2x^3 + x^2 - 7x - 9$, find $P(-2)$ by using the remainder theorem. [Example 3b]

3. **Apply the remainder theorem to solve equations involving polynomials?**
Solve $x^3 - 5x^2 + 2x + 8 = 0$, given that 2 is one of the solutions. [Example 5]

4.7 Can you:

1. **Solve equations that contain rational expressions?**
Solve $\frac{5}{y^2 + y - 6} = \frac{3}{y - 2} + \frac{2}{y + 3}$. [Example 3]

2. **Solve formulas that contain rational expressions for a specified variable?**
Solve the formula $t = \frac{v - v_0}{v}$ for v. [Example 7]

3. **Set up and solve work problems?**
One robot assembly line (A) can fill an order in 20 minutes. Another (B) can fill the order in 30 minutes. How long would it take the two lines working together to fill the order? (Assume that the two machines do not interfere with one another.) [Example 8]

4. **Set up and solve uniform motion problems?**
An experimental light plane, powered by the pilot's pedaling, flew a distance of $\frac{1}{2}$ mi against a 2-mi/hour wind. With that same wind behind it, it flew 1 mi in the same time. What is the speed of the plane in still air? [Example 9]

KEY TERMS

Complex fraction (4.4)
Cross products (4.1)
Equivalent fractions (4.1)
Extraneous solution (4.7)
Least common denominator (4.3)
Rational expression (4.1)
Remainder theorem (4.6)
Synthetic division (4.6)
Work problem (4.7)

KEY CONCEPTS AND PROCEDURES

Section	Key Concepts or Procedures to Review
4.1	■ Definition of rational expression ■ Equivalence of fractions: $\frac{a}{b} = \frac{c}{d}$ if and only if $ad = bc$ $(b, d \neq 0)$ ■ Fundamental principle of fractions: $\frac{ak}{bk} = \frac{a}{b}$ $(b, k \neq 0)$ ■ Fundamental principle of rational expressions: $\frac{AK}{BK} = \frac{A}{B}$ $(B, K \neq 0)$ ■ Methods to reduce rational expressions to lowest terms ■ Methods to build up a rational expression to an equivalent expression with an indicated denominator
4.2	■ Methods to multiply and divide rational expressions
4.3	■ Methods to add and subtract rational expressions and to find the least common denominator

Section	Key Concepts or Procedures to Review
4.4	■ Definition of a complex fraction ■ Methods to simplify a complex fraction
4.5	■ Procedure for dividing a polynomial by a monomial: $\frac{a+c}{b} = \frac{a}{b} + \frac{c}{b} \quad (b \neq 0)$ ■ Long division of polynomials
4.6	■ Synthetic division procedure for the division of a polynomial by $x - b$ ■ Remainder theorem ■ Method for solving certain equations involving polynomials when one solution is known
4.7	■ Methods to solve an equation containing rational expressions ■ When solving equations containing rational expressions, check solutions in the original equation and reject extraneous solutions. ■ Procedures for setting up and solving work problems and uniform motion problems

CHAPTER 4 REVIEW EXERCISES

4.1

1. Find all values of x which make the expression $\frac{x}{x^2 - 4}$ undefined. $-2, 2$
2. Determine whether the fractions $\frac{4}{5}$ and $\frac{16}{25}$ are equivalent. No
3. Express $\frac{-3x + 15}{x^3 - 5x^2}$ in lowest terms. $\frac{-3}{x^2}$
4. Rewrite the expression with the indicated denominator: $\frac{y - 3}{2y} = \frac{?}{8y^2}$. $4y^2 - 12y$
5. Evaluate $T = \frac{8\pi}{\sqrt{g}}$ if $g = 16$. This gives the time T in seconds for a 16-ft pendulum to go back and forth, where g is the force due to gravity. For what nonnegative value of g is T undefined? 2π; $g = 0$

4.2

Multiply, and express each product in lowest terms.

6. $\frac{3x - 12}{5y + 15} \cdot \frac{y + 3}{4x - 16}$ $\frac{3}{20}$
7. $x^4 \cdot \frac{5}{x^2} \cdot \frac{2y}{x^2}$ $10y$
8. $\frac{(x - 2)^3}{3a^2b} \cdot \frac{3a^3b^4}{(x - 2)^2}$ $ab^3(x - 2)$

Divide, and express the answer in lowest terms.

9. $\frac{50}{y} \div \frac{25}{y^6}$ $2y^5$
10. $\frac{(t + 2)^3}{t^3} \div \frac{(t + 2)}{t}$ $\frac{(t + 2)^2}{t^2}$

4.3

11. Find the lowest common denominator for $\frac{7}{6x^3y^2z}$ and $\frac{5}{9x^2yz}$. $18x^3y^2z$

Write each expression as a single fraction in lowest terms.

12. $\frac{3w + 7}{w^2 - 4} - \frac{2w + 5}{w^2 - 4}$ $\frac{1}{w - 2}$
13. $\frac{a}{3a - b} + \frac{b}{b - 3a}$ $\frac{a - b}{3a - b}$
14. $\frac{2}{x} - \frac{x}{x + 5}$ $\frac{-x^2 + 2x + 10}{x(x + 5)}$
15. $\frac{-1}{x^2 - x - 6} + \frac{x - 2}{x^2 - 6x + 9}$ $\frac{x^2 - x - 1}{(x - 3)^2(x + 2)}$

4.4

Simplify the complex fraction.

16. $\frac{2 + \frac{1}{x}}{1 - \frac{1}{x}}$ $\frac{2x + 1}{x - 1}$
17. $\frac{\frac{25}{a^2} - \frac{4}{b^2}}{\frac{5}{a} - \frac{2}{b}}$ $\frac{5b + 2a}{ab}$
18. $\frac{\frac{1}{a} + \frac{a}{a + 3}}{\frac{2}{a^2 + a - 6}}$ $\frac{a^3 - a^2 + a - 6}{2a}$
19. $\frac{3n^{-3} + 2n^{-2}}{2n + 3}$ $\frac{1}{n^3}$

4.5

Find each quotient.

20. $\frac{-9a^4b^2 - 2a^3b^2}{-3a^2b^2}$ $3a^2 + \frac{2}{3}a$
21. $\frac{x^4 + 2x^3y^2 + 2x^2y^3 + y^4}{x^3y^3}$ $\frac{x}{y^3} + \frac{2}{y} + \frac{2}{x} + \frac{y^2}{x^3}$
22. $(x^2 - 2x + 3) \div (x - 3)$ $x + 1 + \frac{6}{x^2 - 2x + 3}$

23. $\dfrac{n^3+1}{n-1}$ $n^2+n+1+\dfrac{2}{n-1}$

24. $(x^5-2x^3+5x^2-10)\div(x^3+5)$ x^2-2

4.6

25. Use synthetic division to divide $4x^3-2x^2+3x-1$ by $x+2$. $4x^2-10x+23+\dfrac{-47}{x+2}$

26. Use synthetic division to divide $2x^4-x^2+6$ by $x-2$. Use the form dividend = (divisor)(quotient) + remainder to express the result. $2x^4-x^2+6=(x-2)(2x^3+4x^2+7x+14)+34$

27. If $P(x)=3x^4+x^3-x^2-4x-1$, use the remainder theorem to find $P(-2)$. 43

28. Use the remainder theorem to determine if 4 is a solution of $x^4-4x^3+2x^2-15x+28=0$. Yes

29. Solve $x^3-2x^2-x+2=0$, given that $x=-1$ is one of the solutions. $\{-1,1,2\}$

4.7

30. Solve $\dfrac{3}{x}-1=\dfrac{18}{4x}$. $-\frac{3}{2}$

31. Solve $\dfrac{9}{x^2-x-2}=\dfrac{4}{(x+1)}+\dfrac{3}{(x-2)}$.
No solution; $x=2$ is extraneous.

32. Solve $1=\dfrac{6}{x^2-4}+\dfrac{3}{x+2}$. $\{-1,4\}$

33. One man working alone can paint a room in 3 hours; another can paint it in 5 hours. How long will it take the two men working together to paint the room? $1\frac{7}{8}$ hours

34. A bicyclist pedaled 2 mi against a 5-mi/hour wind. With the same wind at her back, she pedaled 4 mi in the same amount of time. At what speed would the bicyclist have been pedaling if there were no wind? 15 mi/hour

ADDITIONAL REVIEW EXERCISES

Perform the indicated operations and/or simplify.

35. $\dfrac{m}{m+5}+\dfrac{4m}{m+5}$ $\dfrac{5m}{m+5}$

36. $\dfrac{2-a}{b^2+b}\div\dfrac{a^2-4}{b^2-3b-4}$ $\dfrac{4-b}{b(a+2)}$

37. $\dfrac{5}{6a^3b^2}-\dfrac{1}{8ab^3}$ $\dfrac{20b-3a^2}{24a^3b^3}$

38. $\dfrac{16ab^4}{-2a^3b^4}$ (use only positive exponents) $\dfrac{-8}{a^2}$

39. $\dfrac{3x-9}{x^4}\cdot\dfrac{x^2}{4x-12}$ $\dfrac{3}{4x^2}$

40. $\dfrac{7a^3b-4ab^3}{-2ab}$ $-\frac{7}{2}a^2+2b^2$

41. $\dfrac{5x+4}{x^2-9}-\dfrac{3x-2}{x^2-9}$ $\dfrac{2}{x-3}$

42. $(3x^2+7x+4)\div(x+2)$ $3x+1+\dfrac{2}{x+2}$

43. $\dfrac{x^2}{x-1}+\dfrac{x^2+4}{1-x}$ $\dfrac{-4}{x-1}$

44. $(2y^4-3y^3-2y^2+3y+5)\div(y-2)$ $2y^3+y^2+3+\dfrac{11}{y-2}$

45. $\dfrac{(y-2)^7}{x^9w^3}\cdot\dfrac{x^4w^3}{(y-2)^5}$ $\dfrac{(y-2)^2}{x^5}$

46. $\dfrac{3sd}{n}-\dfrac{(n-s)d}{n}$ $\dfrac{4sd-nd}{n}$

47. $\frac{5}{12}+\frac{7}{20}$ $\frac{23}{30}$

48. $(y^3+y+1)\div(y-1)$ $y^2+y+2+\dfrac{3}{y-1}$

49. $\dfrac{x-1}{x^2-x-12}+\dfrac{2x}{x^2-8x+16}$ $\dfrac{3x^2+x+4}{(x-3)(x-4)^2}$

50. $(4x^3+2x^2-x+3)\div(x-2)$ [Express the result in the form dividend = (divisor)(quotient) + (remainder).]
$4x^3+2x^2-x+3=(x-2)(4x^2+10x+19)+41$

Solve.

51. $\dfrac{3}{2y}-1=\dfrac{4}{y}$ $\{-\frac{5}{2}\}$

52. $\dfrac{4}{x^2+3x+2}=\dfrac{1}{x+2}+\dfrac{3}{x+1}$ $\{-\frac{3}{4}\}$

53. $\dfrac{-x}{4}+\dfrac{3}{12}=\dfrac{3x}{8}$ $\{\frac{2}{5}\}$

54. $\dfrac{1}{x-3}-\dfrac{6}{x^2-9}=1$ $\{-2\}$

55. $y+\dfrac{5}{y}=-6$ $\{-1,-5\}$

56. $a=\dfrac{b+c}{b}$, for b $b=\dfrac{c}{a-1}$

57. $x^3-3x^2-13x+15=0$, given that 1 is one of the solutions $\{-3,1,5\}$

58. A lawn can be mowed in 30 minutes with a ride-on mower and in 1 hour with a self-propelled mower. How long will it take to mow the lawn using the two mowers together? 20 minutes

59. A boat can travel 3 mi downstream in the same time it can travel 1 mi upstream. If the current of the river is 3 mi/hour, what is the boat's rate in still water? 6 mi/hour

60. Find all values of x which make the expression $\dfrac{2}{x^2+2x-3}$ undefined. $\{-3,1\}$

61. Rewrite the expression with the indicated denominator: $\dfrac{5}{7x^2y}=\dfrac{?}{56x^4y^3}$. $40x^2y^2$

62. Find the least common denominator for $\dfrac{3}{2x+4}$ and $\dfrac{x}{x^2+5x+6}$. $2(x+2)(x+3)$

63. If $P(x)=-x^4-3x^3+2x^2+4x-8$, find $P(-3)$. -2

64. Determine whether 4 is a solution of $x^3-4x^2-7x+28$. Yes

Simplify.

65. $\frac{36}{42}$ $\frac{6}{7}$

66. $\dfrac{-3x+6}{5x-10}$ $-\frac{3}{5}$

67. $\dfrac{x^2-5x+4}{x^2-6x+8}$ $\dfrac{x-1}{x-2}$

68. $\dfrac{3m+3a}{a^2+am}$ $\dfrac{3}{a}$

69. $\dfrac{6-\frac{1}{2}}{\frac{1}{4}+5}$ $\dfrac{22}{21}$

70. $\dfrac{\frac{3}{y}-2}{3+\frac{4}{y}}$ $\dfrac{3-2y}{3y+4}$

71. $\dfrac{4-x}{x^2-16}$ $\dfrac{-1}{x+4}$

72. $\dfrac{\frac{2}{a}+\frac{1}{b}}{\frac{3}{a+2b}}$ $\dfrac{(a+2b)^2}{3ab}$

73. $\dfrac{\frac{1}{25}-\frac{1}{x^2}}{\frac{1}{x}-\frac{1}{5}}$ $\dfrac{-(x+5)}{5x}$

74. $\dfrac{x^{-1}+x^{-2}}{1+x}$ $\dfrac{1}{x^2}$

CHAPTER 4 TEST

1. Find all values of x for which $\frac{x+1}{x^2-4}$ is undefined. 2, −2
2. Express in lowest terms: $\frac{2a-4}{2a^2-a-6}$. $\frac{2}{2a+3}$
3. Find the lowest common denominator for $\frac{a+3}{a^3-a}$ and $\frac{a-2}{a^2-2a-3}$. $a(a-1)(a+1)(a-3)$
4. Find a fraction with denominator a^3b^3 which is equivalent to $\frac{2b}{a^2}$. $\frac{2ab^4}{a^3b^3}$

For Questions 5–10, perform the indicated operation and express the answer in lowest terms.

5. $\frac{x^2-1}{4x-12}\cdot\frac{x^2-6x+9}{x^2-2x-3}$ $\frac{x-1}{4}$
6. $\frac{(y+2)^6(y-1)^5}{4b}\cdot\frac{2b}{(y+2)^7(y-1)^4}$ $\frac{y-1}{2(y+2)}$
7. $\frac{x^3-6x^2}{(x+2)^2}\div\frac{6-x}{x^2+2x}$ $\frac{-x^3}{x+2}$
8. $\frac{x^2+2x-3}{x^2-8x+16}-\frac{1+5x}{x^2-8x+16}$ $\frac{x+1}{x-4}$
9. $\frac{3t+2}{2t-1}+\frac{t+1}{1-2t}$ $\frac{2t+1}{2t-1}$
10. $\frac{1}{2ab^2}-\frac{1}{3a^2b}$ $\frac{3a-2b}{6a^2b^2}$
11. Simplify $\frac{1+\frac{1}{xy}}{\frac{1}{x}+\frac{1}{y}}$. $\frac{xy+1}{y+x}$
12. Simplify $\frac{3n^{-2}-4n^{-3}}{3n-4}$. $\frac{1}{n^3}$
13. Simplify $\frac{-25x^5y^3z}{10x^2y^3z^5}$, and write the results using only positive exponents. $\frac{-5x^3}{2z^4}$
14. Divide $\frac{-6x^2+8y^2-2x^2y^2}{-2xy}$. $\frac{3x}{y}-\frac{4y}{x}+xy$
15. Divide $4x^2-4x+3$ by $2x+1$. $2x-3+\frac{6}{2x+1}$
16. Use synthetic division to divide $x^3+6x^2+5x-10$ by $x+4$. Use the form dividend = (divisor)(quotient) + remainder to express the result. $x^3+6x^2+5x-10=(x^2+2x-3)(x+4)+2$
17. Given $P(x)=4x^3+x^2-3x+1$, use the remainder theorem to find $P(-2)$. -21
18. Solve $2y^3-3y^2-3y+2$, given that 2 is a solution. $\{2,-1,\frac{1}{2}\}$
19. Solve $\frac{8}{x+3}-\frac{6}{x-3}=\frac{14}{x^2-9}$. $\{28\}$
20. A mail-processing machine can sort 50,000 pieces of mail in 1 hour. An older machine takes $1\frac{1}{2}$ hours to do the same job. If they work together, how long will it take them to sort 50,000 pieces of mail? 36 minutes

CUMULATIVE TEST 4

1. What property of the real numbers justifies replacing $3(x+y)$ by $3x+3y$? Distributive property
2. Evaluate $\frac{4a^2-b^2}{b-a}$ when $a=\frac{1}{2}$, $b=2$. -2
3. Which of these is *not* an integer?
   (a) 0 **(b)** $\frac{1}{2}$ **(c)** $\sqrt{9}$ **(d)** -3 b
4. Simplify $-2(5x+1)-(-x+8)$. $-9x-10$
5. Solve $-8x+11=-53$. $\{-8\}$
6. Solve $v=\frac{1}{2}gt^2$ for g. $g=\frac{2v}{t^2}$
7. The length of a rectangle is 2 in. less than 3 times the width. The perimeter is 116 in. Find the dimensions of the rectangle. Width 15 in., length 43 in.
8. Solve $-5x-2>-3x+8$, and write the answer in interval notation. $(-\infty,-5)$
9. Graph the solution set of $|6+x|\le 3$.

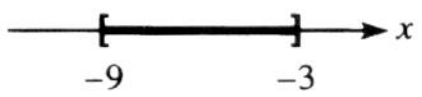

10. Solve $|-3x+1|>1$, and write the answer in interval notation. $(-\infty,0)\cup(\frac{2}{3},\infty)$
11. Evaluate $8^{-1}+2^{-2}+5^0$. $\frac{11}{8}$
12. Simplify $\frac{(6xy^2)^2}{(x^3y)^{-1}}$, and write the result using only positive exponents. $36x^5y^5$
13. If $P(x)=4x-2$ and $Q(x)=-2x+3$, find $\frac{P(3)}{Q(1)}$. 10
14. Factor $3x^2+5x-12$. $(3x-4)(x+3)$
15. Solve $2x^2-7x+6=0$. $\{\frac{3}{2},2\}$
16. Divide $\frac{x+2}{3x^2-3}\div\frac{x^2+4x+4}{x+1}$, and express the result in lowest terms. $\frac{1}{3(x-1)(x+2)}$
17. Add $\frac{x-1}{(x+1)^2}+\frac{2}{x^2+4x+3}$, and express the result in lowest terms. $\frac{x^2+4x-1}{(x+3)(x+1)^2}$
18. Simplify $\frac{\frac{x}{y^2}+\frac{y}{x^2}}{xy}$. $\frac{x^3+y^3}{x^3y^3}$
19. Use synthetic division to divide $2x^3-3x^2+4$ by $x-3$. $2x^2+3x+9+\frac{31}{x-3}$
20. Solve $\frac{7}{x^2-5x+6}=\frac{4}{x-3}+\frac{1}{2-x}$. $\{4\}$

CHAPTER 5 Graphing Linear Equations and Inequalities in Two Variables

A "shareware" computer program is sold very inexpensively (or given free) so that potential users can try it at no risk. People who like the program and continue to use it are on their honor to then send the author the official, more expensive price and become a registered user. At one software distributor, an order of computer shareware disks costs \$3 per disk plus \$4 shipping per order. An equation for the total cost (C) of an order in terms of the number (n) of disks ordered is $C = 3n + 4$. Graph this equation. (See Example 6 of Section 5.1.)

USING GRAPHS to analyze the relationship between two variables is one of the most useful tools of mathematics, and a good way to begin this topic is to consider so-called *linear* relationships. They are called linear because their graphs are straight lines.

5.1 Graphing Linear Equations

OBJECTIVES

1 **Determine if an ordered pair is a solution of an equation.**

2 **Graph ordered pairs.**

3 **Graph linear equations by plotting ordered-pair solutions.**

4 **Graph linear equations by using intercepts.**

5 **Graph $x = a$ or $y = b$.**

1 Equations like $\ell + w = 50$ and $y = 3x - 2$ express a relationship between *two* variables. Solutions of such equations are, therefore, *pairs* of numbers that show corresponding values of the variables. In the following chart we illustrate this by listing some solutions to the equation $y = 3x - 2$.

If x Equals	Then $y = 3x - 2$	Thus, a Solution Is
2	$3(2) - 2 = 4$	$x = 2, y = 4$
1	$3(1) - 2 = 1$	$x = 1, y = 1$
0	$3(0) - 2 = -2$	$x = 0, y = -2$
−1	$3(-1) - 2 = -5$	$x = -1, y = -5$
−2	$3(-2) - 2 = -8$	$x = -2, y = -8$

Solutions of this type are abbreviated using notation called **ordered pairs.** For example, the solution "$x = 2, y = 4$" is written

(2,4).
value of x ↑ ↑ corresponding value of y

By general agreement, the x value is always listed first, so the order of the numbers in the pair is significant. For instance, in the equation $y = 3x - 2$, (2,4) is a solution but (4,2) is not. The ordered pair (4,2) means "$x = 4, y = 2$", which is not a solution of $y = 3x - 2$ because $2 = 2(4) - 1$ is false.

EXAMPLE 1 Determine if the ordered pair is a solution of the equation.

a. $2x + y = 5$; (1,3) **b.** $y = 2x - 3$; $(-1,1)$

Solution Replace x and y by the appropriate components of the given ordered pair.

a. (1,3) means $x = 1$, $y = 3$. Then,

$$2x + y = 5 \quad \text{Given equation.}$$
$$2(1) + (3) \stackrel{?}{=} 5 \quad \text{Replace } x \text{ by 1 and } y \text{ by 3.}$$
$$5 \stackrel{\checkmark}{=} 5. \quad \text{Simplify.}$$

Since $5 = 5$ is a true statement, (1,3) *is* a solution of $2x + y = 5$.

b. $(-1,1)$ means $x = -1$, $y = 1$.

$$y = 2x - 3$$
$$1 \stackrel{?}{=} 2(-1) - 3$$
$$1 \stackrel{?}{=} -5$$

Since $1 = -5$ is a false statement, $(-1,1)$ is *not* a solution of $y = 2x - 3$.

PROGRESS CHECK 1 Determine if the ordered pair is a solution of the equation.

a. $x - 5y = 5$; $(0,-1)$ **b.** $y = -2x + 5$; $(1,2)$

2 Ordered pairs can be represented graphically by using the **Cartesian (or rectangular) coordinate system.** This system is named in honor of the French mathematician and philosopher René Descartes and is formed from the intersection of two real number lines at right angles. The values for x (or the first component in the ordered pairs) are represented on a horizontal number line, and the values for y (or the second component in the ordered pairs) are represented on a vertical number line. These two lines are called the ***x*-axis** and the ***y*-axis,** and they intersect at their common zero point, which is called the **origin** (see Figure 5.1).

Descartes (1596–1650) published *La Geometrie* in 1637. This was the first modern joining of algebra and geometry by the assignment of equations to curves. In his work it was not necessary that the axes intersect at right angles.

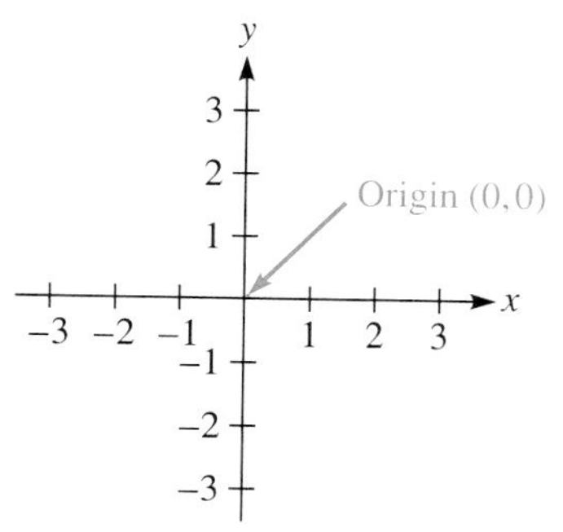

Figure 5.1

Any ordered pair can be represented as a point in this coordinate system. The first component indicates the distance of the point to the right or left of the vertical axis. The second component indicates the distance of the point above or below the horizontal axis. These components are called the **coordinates** of the point, and the point is called the **graph** of the ordered pair.

The name *coordinate* is due to Leibniz (1692).

EXAMPLE 2 Graph the following ordered pairs.

a. $(4,3)$ **b.** $(-2,-1)$

Solution

a. To graph $(4,3)$, start at the origin and go 4 units to the right and 3 units up, as shown in Figure 5.2(a).

b. To graph $(-2,-1)$, start at the origin and go 2 units to the left and 1 unit down, as shown in Figure 5.2(b).

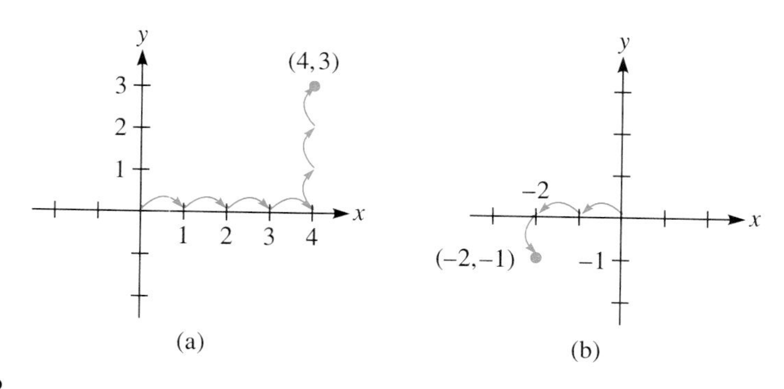

Figure 5.2

PROGRESS CHECK 2 Graph the following ordered pairs.

a. $(3,4)$ **b.** $(-1,-2)$

Progress Check Answers

1. (a) Yes (b) No 2.

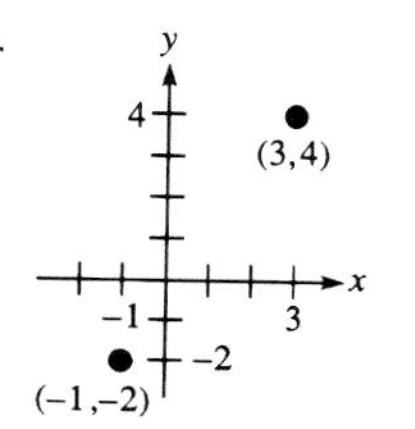

Quadrant means one-fourth, and it was originally associated with one-fourth of a circle. Astronomical measuring instruments called quadrants were made in the shape of a quarter circle and were used hundreds of years before telescopes.

The Cartesian coordinate system divides the plane into four regions called **quadrants.** The quadrant in which both x and y are positive is designated the first quadrant. The remaining quadrants are labeled in a counterclockwise direction. Figure 5.3 shows the name of each quadrant as well as the sign of x and y in that quadrant.

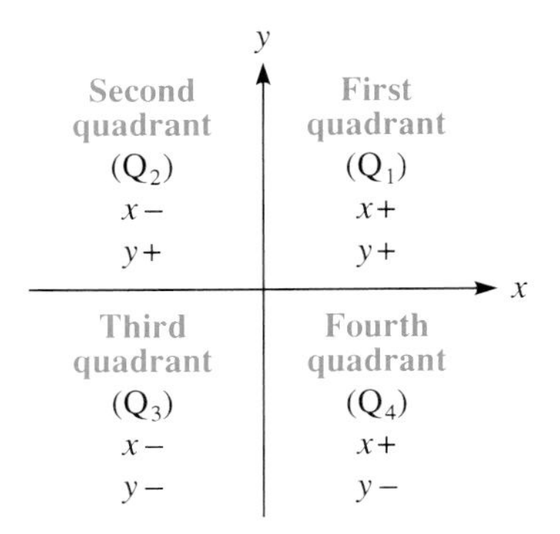

Figure 5.3

EXAMPLE 3 Graph the following ordered pairs, and indicate the quadrant location of each point.

a. $(-3,2)$ **b.** $(3,-4)$ **c.** $(0,2)$

Solution See Figure 5.4.

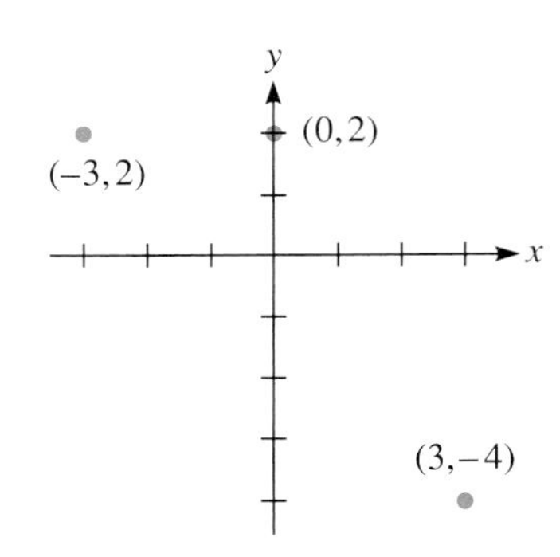

Figure 5.4

a. To graph $(-3,2)$, start at the origin and go 3 units to the left and 2 units up. The graph $(-3,2)$ lies in quadrant 2 (or Q_2).
b. To graph $(3,-4)$, start at the origin and go 3 units to the right and 4 units down. The graph $(3,-4)$ lies in quadrant 4 (or Q_4).
c. To graph $(0,2)$, start at the origin and go 0 units to the right and 2 units up, which means $(0,2)$ lies on the y-axis. Any point that lies on an axis is not located in a quadrant.

PROGRESS CHECK 3 Graph the following ordered pairs, and indicate the quadrant location of each point.

a. $(3,-2)$ **b.** $(-3,0)$ **c.** $(-1,3)$

3 As shown above, any given ordered pair of numbers is represented by a specific point in the Cartesian coordinate system. This correspondence enables us to draw a geometric picture called the graph of an equation in two variables.

Progress Check Answers

3.

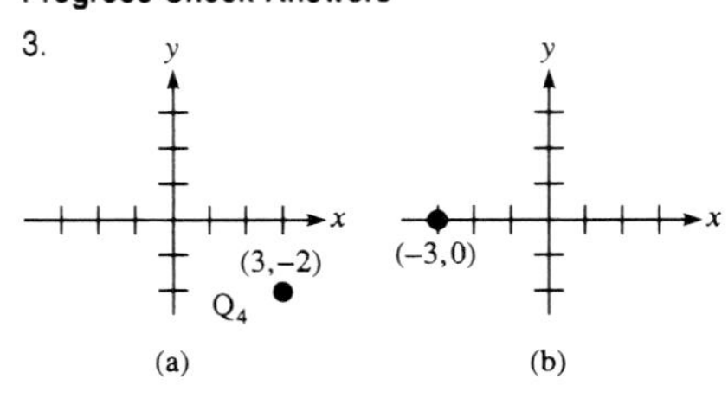

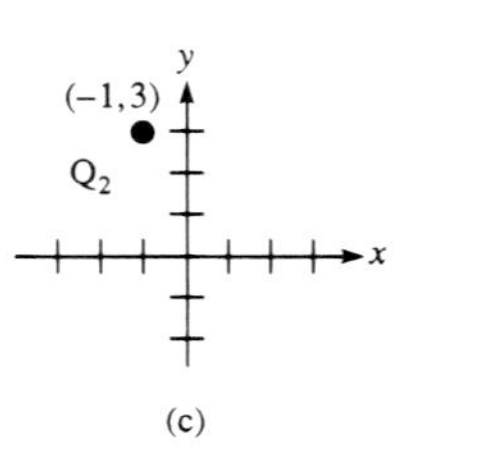

> **Graph**
>
> The **graph** of an equation in two variables is the set of all points in a coordinate system that correspond to ordered pair solutions of the equation.

The use of the word *graph* for the picture of a mathematical curve is due to Sylvester (1878).

One method for determining the graph of an equation in two variables is to simply choose several convenient values for one variable and then use them to determine a list of ordered-pair solutions. If you plot enough of these solutions, a trend may emerge, and the graph can be completed by following the established pattern. This method is shown in Example 4.

EXAMPLE 4 Graph the equation $y = 3x + 1$.

For the use of the graphing calculator to graph lines, see the end of the chapter.

Solution A convenient set of values to use for x is the set of integers from 2 to -2, which we use to make a list of ordered-pair solutions.

If $x =$	Then $y = 3x + 1$	Thus, the Ordered-Pair Solutions Are
2	$3(2) + 1 = 7$	$(2,7)$
1	$3(1) + 1 = 4$	$(1,4)$
0	$3(0) + 1 = 1$	$(0,1)$
-1	$3(-1) + 1 = -2$	$(-1,-2)$
-2	$3(-2) + 1 = -5$	$(-2,-5)$

These ordered pairs are graphed in Figure 5.5, where they appear to all lie in a straight line. In fact, the graph of $y = 3x + 1$ is the straight line shown in Figure 5.6. Every ordered pair that satisfies $y = 3x + 1$ corresponds to a point on this line, and every point on the line corresponds to an ordered pair that satisfies $y = 3x + 1$.

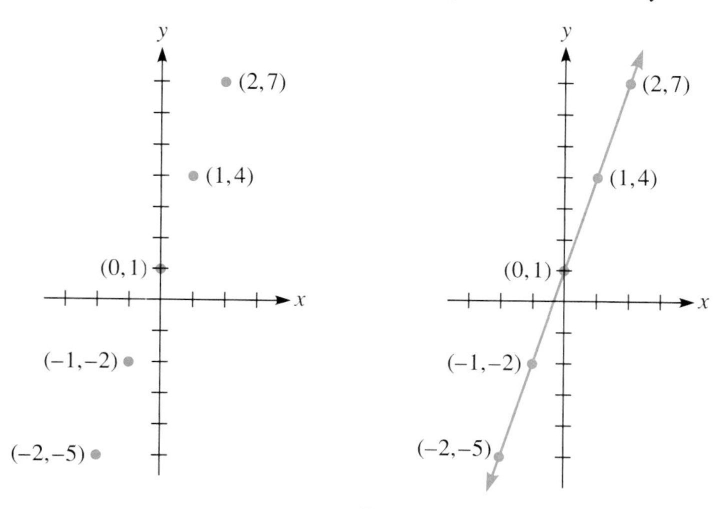

Figure 5.5 **Figure 5.6**

PROGRESS CHECK 4 Graph the equation $y = 2x - 1$.

Progress Check Answer

4.

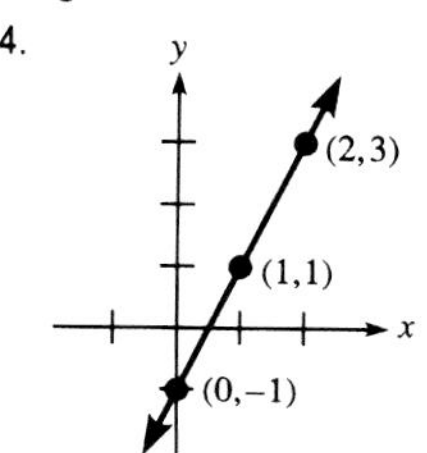

The list of *all* the solutions of most equations in two variables is an infinite set of ordered pairs. Rather than try to decide which of these points to plot, we often try to determine the basic shape of the graph *from the form of the equation*. In this chapter the focus is on *linear* equations, which are defined in the next box.

The word *line* comes from the Greek word for thread and is related to the word *linen*. Carpenters still snap a stretched chalked string to draw a straight line on a flat surface.

Linear Equations in Two Variables

A **linear equation in two variables** is an equation that can be written in the general form

$$Ax + By = C,$$

where A, B, and C are real numbers with A and B not both zero. The graph of a linear equation in two variables is a straight line.

In a linear equation any variable which has a nonzero coefficient must appear only to the *first power*. Thus, we see that the equation in Example 4, $y = 3x + 1$, is linear, and therefore its graph is a straight line. Rewritten in general form, this equation becomes $-3x + y = 1$. Other examples of linear equations are

$$4x - 3y = 0, \qquad y = 2, \qquad \text{and} \qquad x = -3.$$

In order to draw the graph of a *linear* equation, it is sufficient to find any two distinct points in the graph and draw a line through them. But in practice, it is recommended that a third point be determined as a check. If you plot three points for a linear equation and they are not in a straight line, then there is an error in your work.

EXAMPLE 5 Graph $y = -x - 2$.

Solution The equation $y = -x - 2$ is a linear equation (because the variables x and y appear to the first power), and therefore its graph is a straight line. We draw the line by first finding three distinct points in its graph. We will find these points by letting x equal 2, 0, and -2, but we could have chosen any three values.

If $x =$	Then $y = -x - 2$	Thus, the Ordered-Pair Solutions Are
2	$-(2) - 2 = -4$	$(2,-4)$
0	$-(0) - 2 = -2$	$(0,-2)$
-2	$-(-2) - 2 = 0$	$(-2,0)$

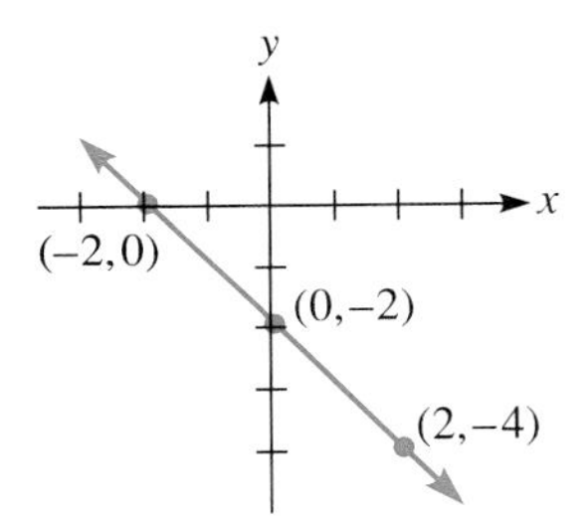

Figure 5.7

These three points are plotted in Figure 5.7, and the line is drawn through them.

PROGRESS CHECK 5 Graph $y = -x - 1$.

In applied problems some adjustments may be necessary so that the graph is restricted to sensible values for the variables.

EXAMPLE 6 Solve the problem in the chapter introduction on page 217.

Solution The variables are n and C, so the axes are labeled accordingly. Also, from the context of the problem, only nonnegative values of n are sensible. To determine three points on the line, we let n equal 0, 1, and 2, as shown next. The graph is pictured in Figure 5.8.

If $n =$	Then $C = 3n + 4$	Thus, the Ordered-Pair Solutions Are
0	$3(0) + 4 = 4$	$(0,4)$
1	$3(1) + 4 = 7$	$(1,7)$
2	$3(2) + 4 = 10$	$(2,10)$

Progress Check Answer

5.

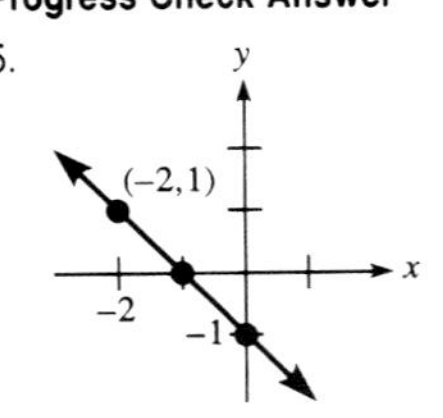

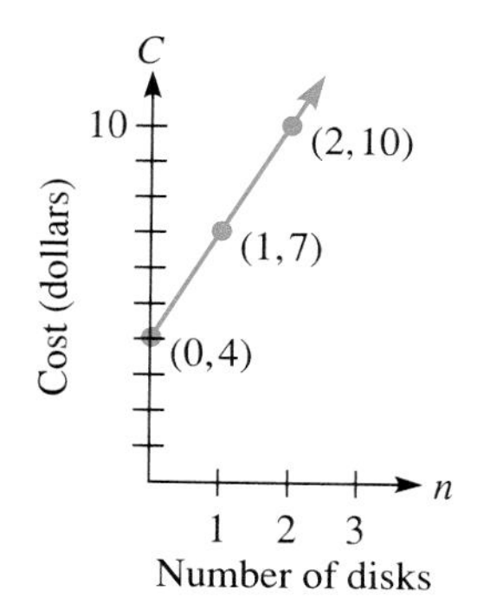

Figure 5.8

Note Technically, the graph should consist only of distinct unconnected points that correspond to nonnegative *integer* values of n, since the purchaser can order only a whole number of disks. But in common practice, the dots are connected to obtain a better visual image of the relationship.

PROGRESS CHECK 6 It costs a company \$10 to manufacture one unit of a particular product. Therefore, the total cost C of manufacturing n units of this product is given by $C = 10n$. Graph this equation.

4 Graphing linear equations using intercepts

The point where a graph crosses the x-axis (if it does cross the x-axis) is called the **x-intercept.** Because this point is *on* the x-axis, its *second component must be zero.* Similarly, the point where a graph crosses the y-axis is called the **y-intercept,** and its *first component must be zero.* This leads to a simple way to find the intercepts.

Point out the spelling of *intercept;* students may confuse it with *intersect.* Originally, *intercept* referred to the *distance* between the origin and the point of intersection. *Intercept* means to capture between.

To Find Intercepts

To find the x-intercept $(a,0)$, let $y = 0$ and solve for x.
To find the y-intercept $(0,b)$, let $x = 0$ and solve for y.

Graphing linear equations by drawing a line through the intercepts is especially useful when the linear equation is given in general form, not solved for y or x.

EXAMPLE 7 Graph $2x + 3y = -6$ by using the x- and y-intercepts.

Solution To find the x-intercept, let $y = 0$ and solve for x.

$$2x + 3y = -6 \quad \text{Given equation.}$$
$$2x + 3(0) = -6 \quad \text{Replace } y \text{ by 0.}$$
$$2x = -6 \quad \text{Simplify.}$$
$$x = -3 \quad \text{Divide both sides by 2.}$$

Thus, the x-intercept is $(-3,0)$.

To find the y-intercept, let $x = 0$ and solve for y.

$$2x + 3y = -6 \quad \text{Given equation.}$$
$$2(0) + 3y = -6 \quad \text{Replace } x \text{ by 0.}$$
$$3y = -6 \quad \text{Simplify.}$$
$$y = -2 \quad \text{Divide both sides by 3.}$$

The y-intercept is $(0,-2)$.

Progress Check Answer

6.

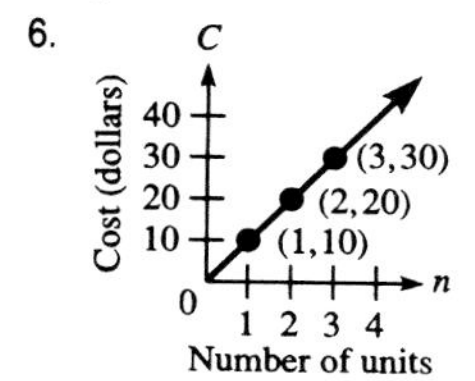

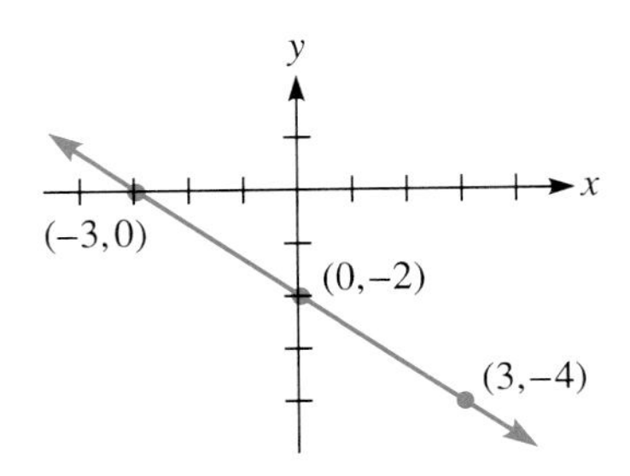

Figure 5.9

For a checking point we arbitrarily let $x = 3$, which leads to the solution $(3,-4)$. The two intercepts and the checking point are plotted and a line is drawn through them to get the graph in Figure 5.9.

Note Because the x-coordinate of the y-intercept is always zero, it is also conventional to define the y-intercept as just the value of y where the graph intersects the y-axis. With this definition, the y-intercept of the graph in Figure 5.9 is simply -2, rather than $(0,-2)$. In this text we choose to always write both coordinates as a reminder that an intercept is a point on the graph. Similar remarks hold for the x-intercept.

PROGRESS CHECK 7 Graph $x - 2y = 4$ by using the x- and y-intercepts.

5 Graphing horizontal and vertical lines

Linear equations which correspond to vertical and horizontal lines have particularly simple forms; they may be written in the form $x = a$ or $y = b$, as shown next.

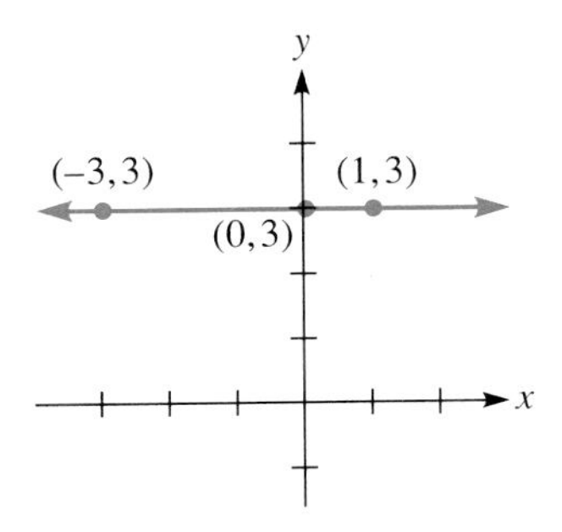

Figure 5.10

EXAMPLE 8 Graph the line $y = 3$.

Solution The linear equation $y = 3$ is equivalent to $0x + y = 3$, which implies that any ordered pair of the form $(a,3)$ is a solution. Thus, a few of the solutions are $(1,3)$, $(0,3)$, and $(-3,3)$. Since all these points have the *same second component,* they are all at the same height; and therefore the graph is a *horizontal* line, as shown in Figure 5.10.

PROGRESS CHECK 8 Graph $y = -1$.

EXAMPLE 9 Graph $x = -1$.

Solution The linear equation $x = -1$ is equivalent to $x + 0y = -1$, which means that x equals -1 for all values of y. Thus, all points of the graph have the *same x-coordinate,* implying that the graph will be a *vertical* line. Using $(-1,1)$, $(-1,2)$, and $(-1,3)$ as three arbitrary points in the graph leads to the line in Figure 5.11.

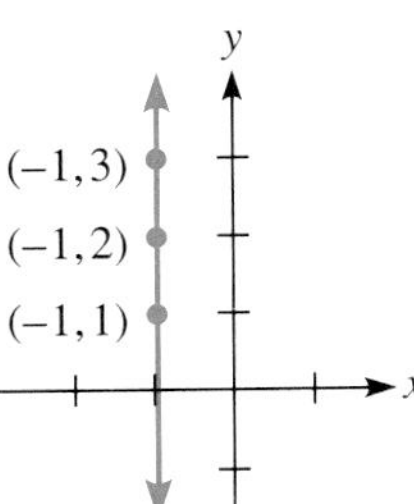

Figure 5.11

PROGRESS CHECK 9 Graph $x = 3$.

Examples 8 and 9 lead to the general principles described in the box below.

> **To Graph $x = a$ and $y = b$**
>
> 1. The graph of the linear equation $x = a$ is a vertical line that contains the point $(a,0)$. If $a = 0$, then the equation is $x = 0$, and the line is the y-axis.
> 2. The graph of the linear equation $y = b$ is a horziontal line that contains the point $(0,b)$. If $b = 0$, then the equation is $y = 0$, and the line is the x-axis.

Progress Check Answers

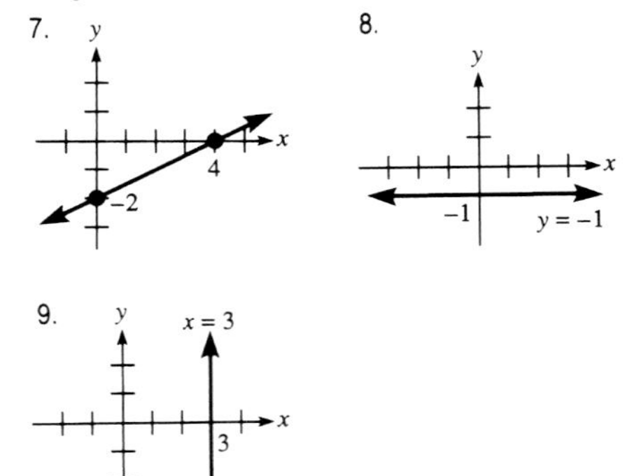

EXERCISES 5.1

In Exercises 1–6, determine if the ordered pair is a solution of the equation.

1. $3x - y = 1$; $(1,2)$ Yes

2. $3y - x = 1$; $(1,2)$ No

3. $\dfrac{1 + x}{3x} = y$; $(0,1)$ No

4. $y = \dfrac{x - 1}{2x}$; $(1,0)$ Yes

5. $4x + 3y + 1 = 0$; $(-\frac{1}{2}, \frac{1}{3})$ Yes

6. $x^2 + y^2 = 25$; $(3,-4)$ Yes

In Exercises 7–18, graph the ordered pair, and indicate its quadrant.

7. $(2,4)$

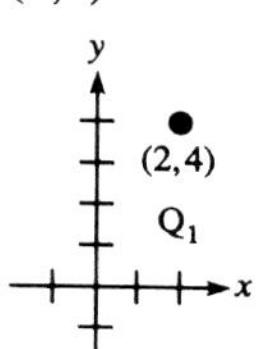

8. $(3,1)$

9. $(4,-2)$

10. $(1,-3)$

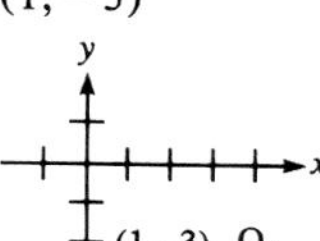

11. $(-5,2)$

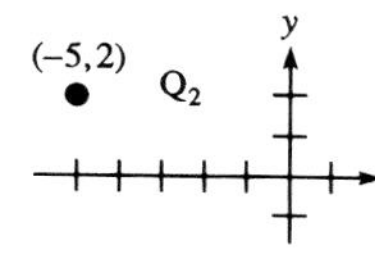

12. $(-1,4)$

13. $(-3,-3)$

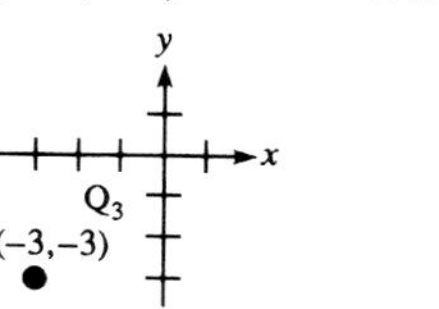

14. $(-1,-4)$

15. $(2,0)$ No quadrant

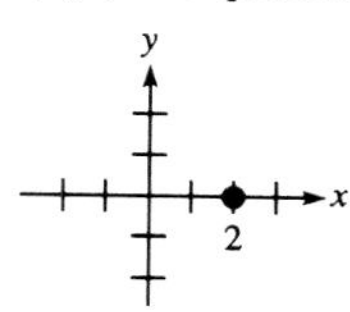

16. $(0,0)$ No quadrant

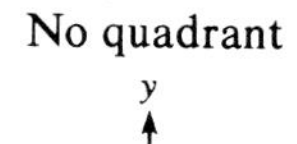

17. $(-3,0)$ No quadrant

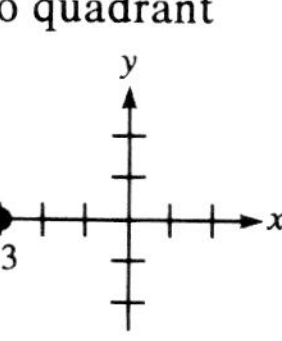

18. $(0,4)$ No quadrant

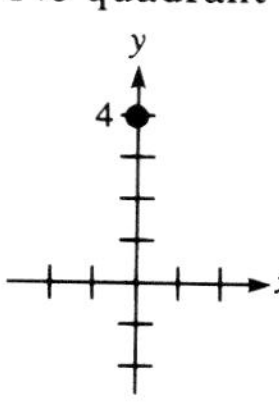

In Exercises 19–22, complete each ordered pair so that it is a solution to the given equation.

19. $y = x + 1$; $(0,\ \ )$, $(1,\ \ )$, $(2,\ \ )$ $(0,1)$, $(1,2)$, $(2,3)$

20. $y = 2x - 3$; $(0,\ \ )$, $(1,\ \ )$, $(2,\ \ )$ $(0,-3)$, $(1,-1)$, $(2,1)$

21. $y = 4$; $(0,\ \ )$, $(1,\ \ )$, $(2,\ \ )$ $(0,4)$, $(1,4)$, $(2,4)$

22. $y = -3$; $(0,\ \ )$, $(1,\ \ )$, $(2,\ \ )$ $(0,-3)$, $(1,-3)$, $(2,-3)$

In Exercises 23–42, graph the given equation.

23. $y = x + 3$

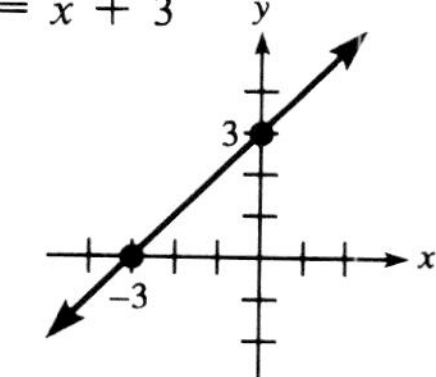

24. $y = x - 2$

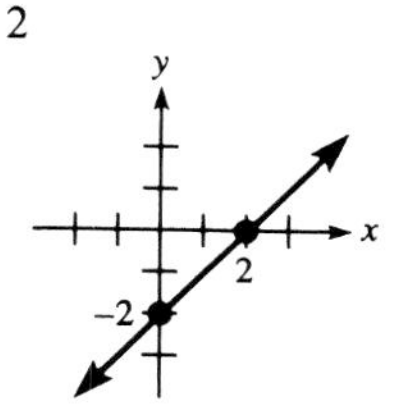

25. $y = -x - 3$

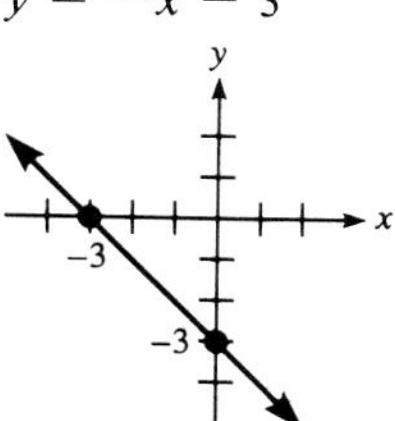

26. $y = -x + 1$

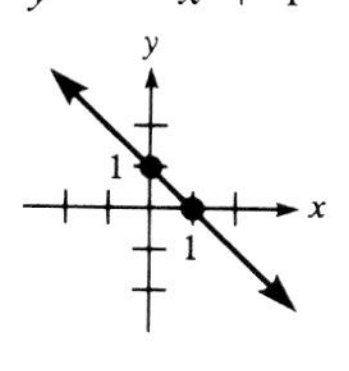

27. $2x + 5y = -10$

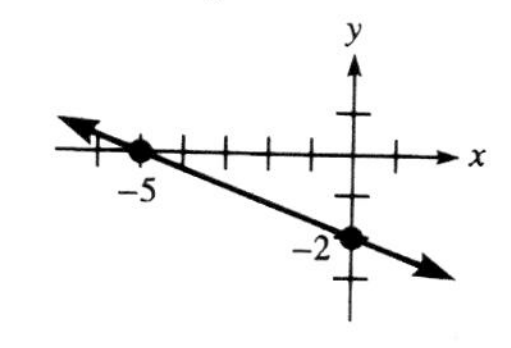

28. $3x - 4y = -12$

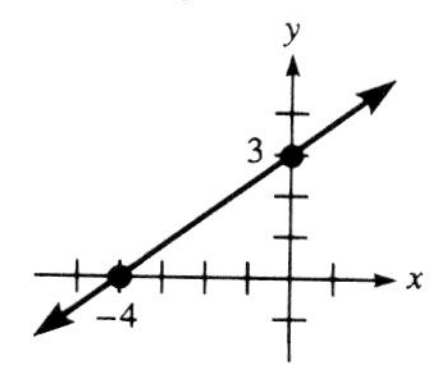

29. $-5x + y = 5$

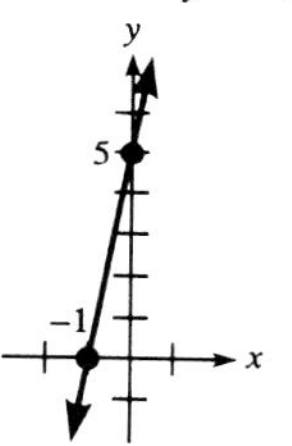

30. $-4x - 2y = 8$

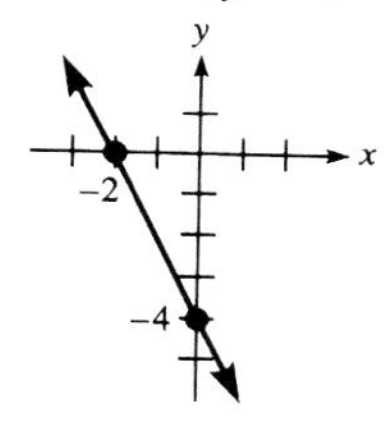

31. $x - y = 0$

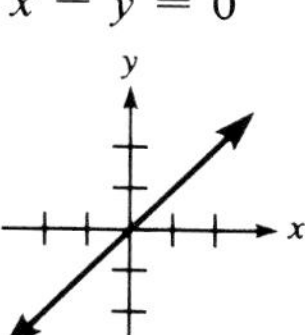

32. $x + y = 0$

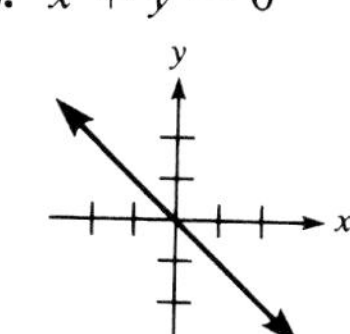
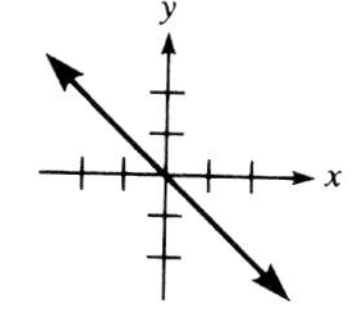
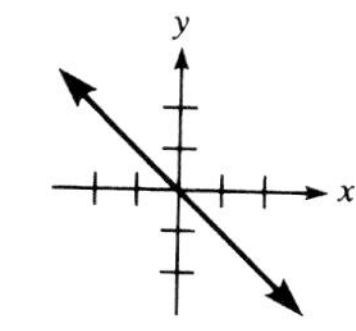
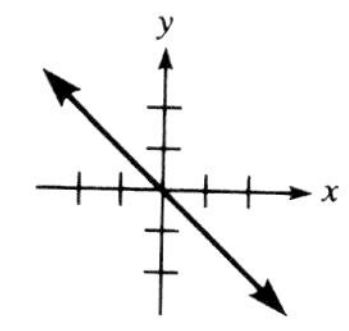
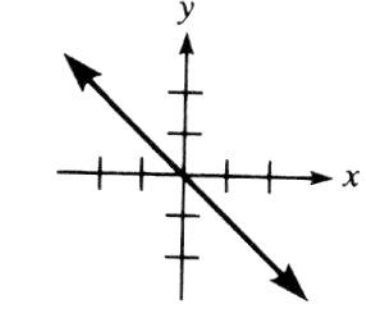

33. $2x - 4y = 9$

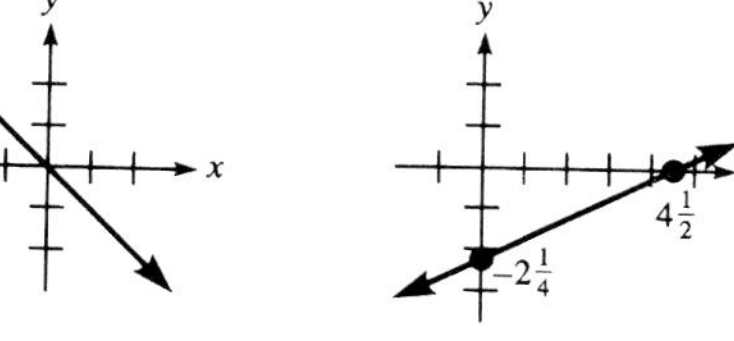

34. $3x + y = 10$

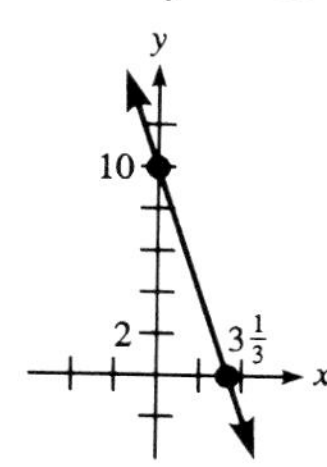

35. $y = 3$

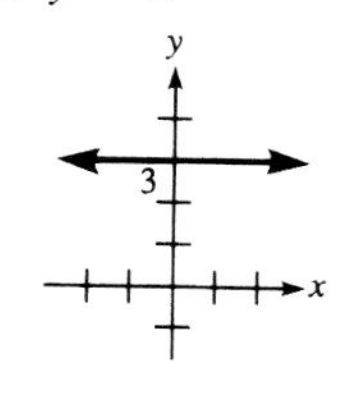

36. $y = -2$

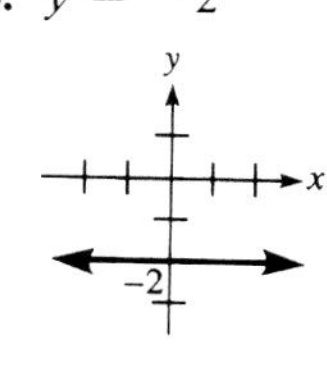

37. $x = 2$

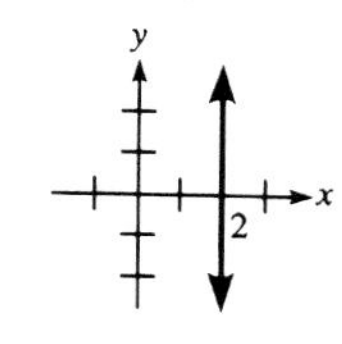

38. $x = -1$

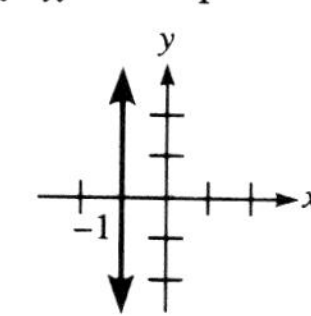

39. $y = 0$

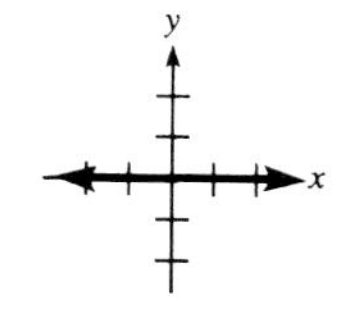

40. $x = 0$

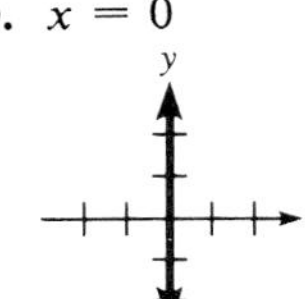

41. $3x - 2 = 0$

42. $3y + 6 = 0$

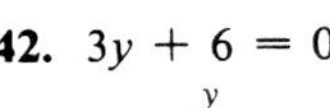

43. The sales tax (T) in a state is 4 percent of the purchase price (P), which is given by $T = 0.04P$. Graph this equation with P as the horizontal axis.

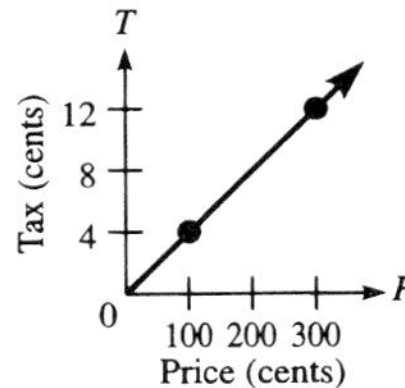

44. In one state the tax (T) in cents on gasoline is 10¢ per gallon (g). Therefore, the tax on a purchase is given by $T = 10g$. Graph this equation with g as the horizontal axis.

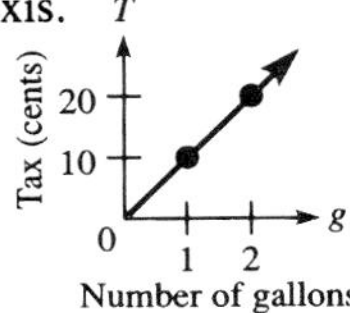

45. A car rents for $60 a week plus 20¢ per mile.

a. Write an equation for the weekly cost C. Let m represent the weekly mileage. $C = 60 + 0.20m$

b. Graph the equation with m as the horizontal axis.

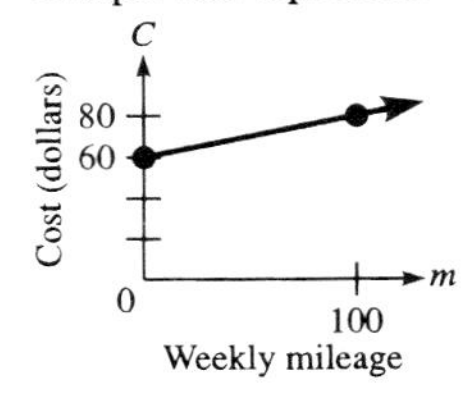

46. The daily pay (P) for a carpenter is $20 per hour for labor plus $10 for lunch and travel.

a. Write an equation for daily pay, where h represents the number of hours worked. $P = 10 + 20h$

b. Graph the equation with h as the horizontal axis.

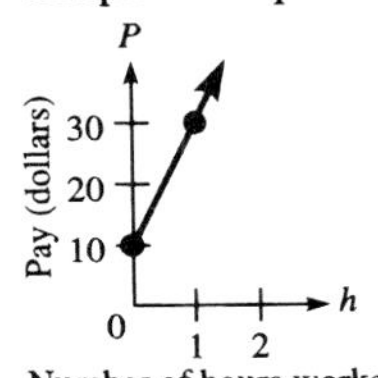

47. The charge C for photo developing is $3.00 per roll plus 30¢ per print.

a. Write an equation for this relationship using n for the number of prints. $C = 3 + 0.30n$

b. Graph the equation with n as the horizontal axis.

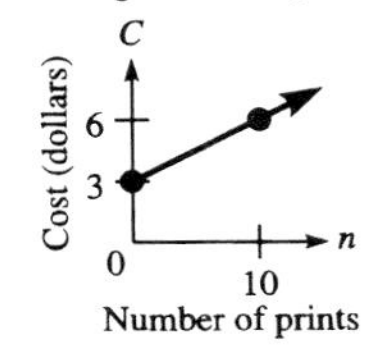

48. The weight w of a truckload of automobiles is 10,000 lb plus 3,000 lb per automobile. Assume the truck can hold no more than 6 autos.

a. Write an equation for the relationship using n for the number of automobiles. $w = 10{,}000 + 3{,}000n$

b. Graph the equation with n as the horizontal axis.

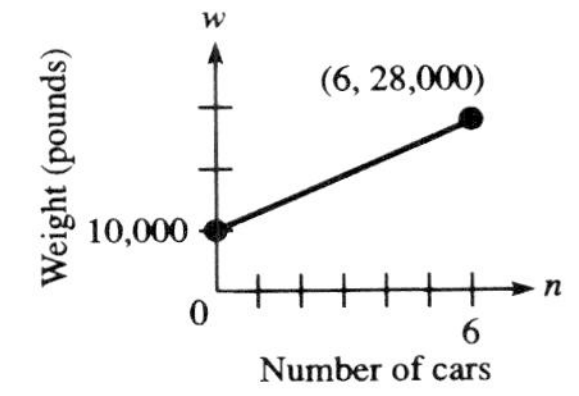

THINK ABOUT IT

1. a. Because $y = \dfrac{(2x + 1)(x - 2)}{(x - 2)}$ is not defined for $x = 2$, its graph has a hole in it when $x = 2$. Draw the graph.

b. Graph $y = \dfrac{(2x + 1)(x - 2)(x - 3)}{(x - 2)(x - 3)}$.

2. The second equation below has the same form as the first, but x and y have been switched.

$$y = 3x + 2$$
$$x = 3y + 2$$

a. Are they both linear equations?

b. Solve the second equation for y.

c. Complete these charts, and describe the relationships between the two sets of ordered pairs.

For $y = 3x + 2$

x	y
0	
1	
2	

For $y = \dfrac{x - 2}{3}$

x	y
2	
5	
8	

d. Find any ordered pair which is a solution in *both* equations by solving $3x + 2 = \dfrac{x - 2}{3}$. Show that it satisfies both equations.

3. **a.** Graph the line $y = 3$.
   b. How many points are on the line $y = 3$ between (0,3) and (1,3)?
   c. Why is the following question difficult? Are there twice as many points on the line $y = 3$ between (0,3) and (2,3) as there are between (0,3) and (1,3)? A mathematician who made a major contribution to answering this question is Georg Cantor (1845–1918). A lively account of his life and work is given in *Men of Mathematics* by E. T. Bell, Simon and Schuster, 1937.

4. The graphs of ordered *pairs* of variables are conveniently drawn on a *two*-dimensional coordinate system. By analogy, the graphs of ordered *triples* of variables may be drawn on a *three*-dimensional coordinate system like the one shown. The graph of the point (1,2,3) is drawn.
   Graph these points.
   a. (2,3,4) **b.** (3,5,0)

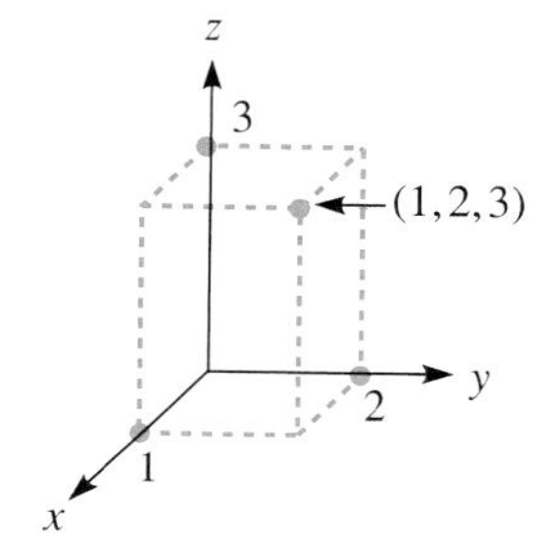

5. **a.** Graph this set of four points and connect them.

x	0	1	2	3
y	0	3	5	6

   b. On the same graph from part **a,** add these points, which keep y the same but replace x by its opposite.

x	0	-1	-2	-3
y	0	3	5	6

   c. Explain why the procedure illustrated in parts **a** and **b** will always produce a picture in which the right side is a reflection of the left side. Such a graph is called *symmetric about the y-axis.*
   d. By looking at the chart, decide which of these sets of ordered pairs has a graph which is symmetric about the y-axis. Then draw each graph to check your answer.

I.

x	y
3	1
-3	1
4	2
-4	2
5	3
-5	3
0	1

II.

x	y
1	1
1	-1
2	-2
2	2
3	3
3	-3
1	0

REMEMBER THIS

1. A printing shop charges \$20 for 500 copies. Which expression represents the cost per copy: 500/20 or 20/500? 20/500
2. Two lines in a plane that never intersect are called ______ . Parallel
3. A quadrilateral has ______ sides. Four
4. Solve $d = rt$ for r. $r = d/t$
5. Solve $\frac{1}{x} - x = \frac{3}{2}$. $\{\frac{1}{2}, -2\}$
6. Simplify: $\frac{3}{x+1} + \frac{2}{x - \frac{1}{2}}$. $\frac{10x + 1}{(x+1)(2x-1)}$
7. Factor $100a^2 - 49b^2$. $(10a + 7b)(10a - 7b)$
8. Subtract $3x^2 + 4x - 2$ from $2x^2 - 4x - 2$. $-x^2 - 8x$
9. Solve $|3x| < 6$. $(-2,2)$ or $\{x: -2 < x < 2\}$
10. The area of a rectangle is 5.25 in.² and the length is 2 in. more than the width. Find the dimensions of the rectangle. (*Hint:* Use multiplication to rid the equation of decimals.)
   Width $= \frac{3}{2}$ in.; length $= \frac{7}{2}$ in.

5.2 The Slope of a Line and the Distance Formula

A certain olympic ski jump drops 60 m over a distance of 86 m. Measure the steepness of this jump by finding its slope. (See Example 3.)

OBJECTIVES

1. **Find and interpret the slope of a line.**
2. **Determine if lines are parallel, perpendicular, or neither by using slope.**
3. **Interpret line graphs.**
4. **Find the distance between two points.**

1 An important characteristic of a line is its steepness, or the degree to which it slants. Mathematically, the concept of **slope** is used to measure the inclination of the line with respect to the horizontal axis. Two familiar concrete illustrations of the slope of a line are the grade of a roadway and the pitch of a roof. To introduce the definition of slope, we consider the line in Figure 5.12 that contains the arbitrary points (x_1,y_1) and (x_2,y_2). Notice how subscripts are used as a reminder that x_1 and y_1 are the coordinates of point 1, and that x_2 and y_2 are the coordinates of point 2. To find the slope, calculate the vertical change, $y_2 - y_1$, which is called the **rise,** and divide this number by the horizontal change, $x_2 - x_1$, which is called the **run.**

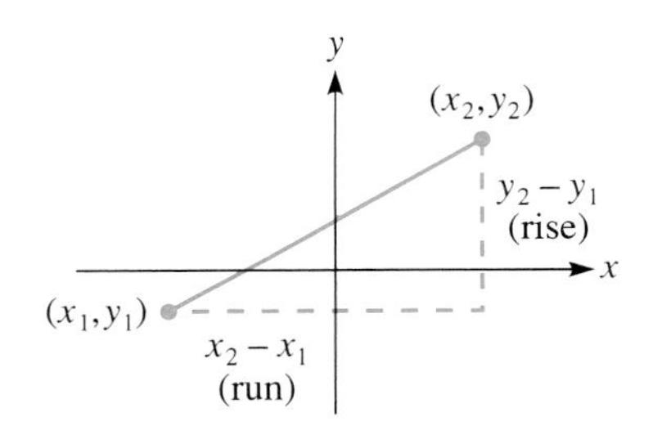

Figure 5.12

Slope of a Line

If (x_1,y_1) and (x_2,y_2) are any two distinct points on a line, with $x_1 \neq x_2$, then the slope m of the line is

$$m = \frac{\text{rise}}{\text{run}} = \frac{\text{change in } y}{\text{change in } x} = \frac{y_2 - y_1}{x_2 - x_1}.$$

EXAMPLE 1 Find the slope of the line through the given points.

a. $(-1,2)$ and $(3,4)$ **b.** $(-3,3)$ and $(1,0)$

Solution

a. If we label $(-1,2)$ as point 1, then $x_1 = -1$, $y_1 = 2$, $x_2 = 3$, and $y_2 = 4$. The slope formula now gives

$$m = \frac{\text{rise}}{\text{run}} = \frac{y_2 - y_1}{x_2 - x_1} = \frac{4 - 2}{3 - (-1)} = \frac{2}{4} = \frac{1}{2}.$$

The slope is $\frac{1}{2}$, which means that y increases 1 unit for each 2-unit increase in x, as shown in Figure 5.13.

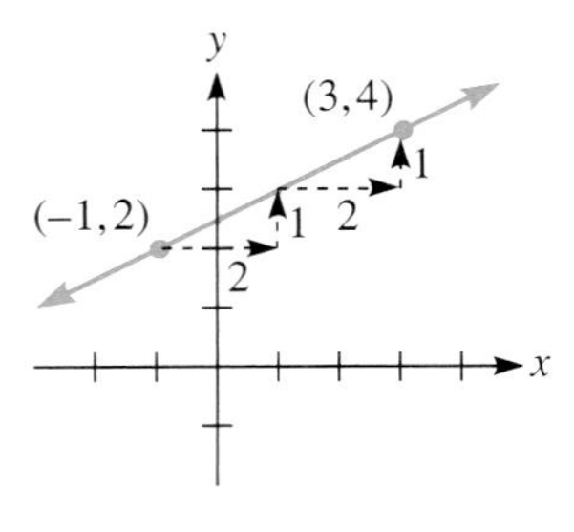

Figure 5.13

Note The slope is not affected by the way in which the points are labeled. In this example, if we label $(3,4)$ as point 1, then $x_1 = 3$, $y_1 = 4$, $x_2 = -1$, and $y_2 = 2$. So

$$m = \frac{y_2 - y_1}{x_2 - x_1} = \frac{2 - 4}{-1 - 3} = \frac{-2}{-4} = \frac{1}{2}.$$

b. We let $x_1 = -3$, $y_1 = 3$, $x_2 = 1$, and $y_2 = 0$ and substitute in the slope formula.

$$m = \frac{\text{rise}}{\text{run}} = \frac{y_2 - y_1}{x_2 - x_1} = \frac{0 - 3}{1 - (-3)} = \frac{-3}{4}$$

A slope of $-3/4$ indicates that as x increases 4 units, y *decreases* 3 units, as shown in Figure 5.14.

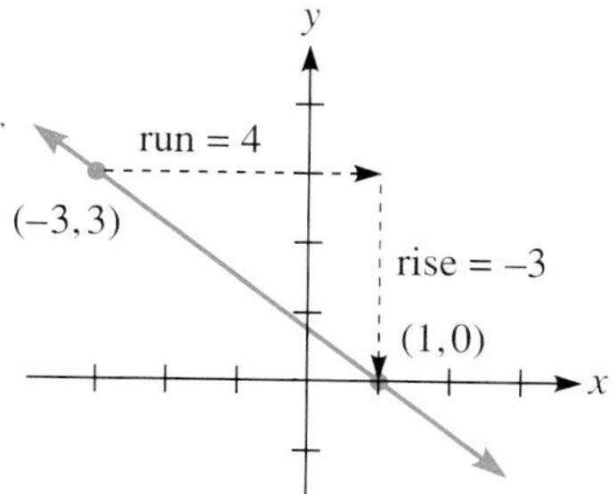

Figure 5.14

PROGRESS CHECK 1 Find the slope of the line through the given points.

a. $(1,-1)$ and $(5,3)$ **b.** $(1,-1)$ and $(-2,3)$

When the slope formula is applied to lines that are horizontal or vertical, then special cases occur, as shown next.

EXAMPLE 2 Find the slope of the line through the given points.

a. $(-1,4)$ and $(2,4)$ **b.** $(3,5)$ and $(3,1)$

Solution

a. Because both points have the same y-coordinate, the line through them is horizontal. (See Figure 5.15.) Applying the slope formula gives

$$m = \frac{\text{rise}}{\text{run}} = \frac{y_2 - y_1}{x_2 - x_1} = \frac{4 - 4}{2 - (-1)} = \frac{0}{3} = 0.$$

The slope of every horizontal line is zero because the rise is always zero.

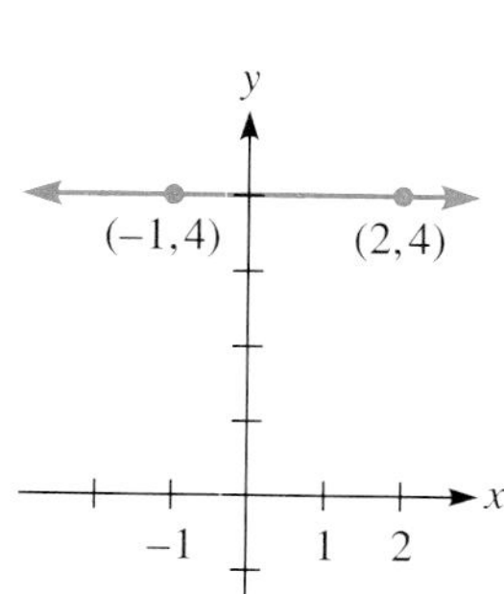

Figure 5.15

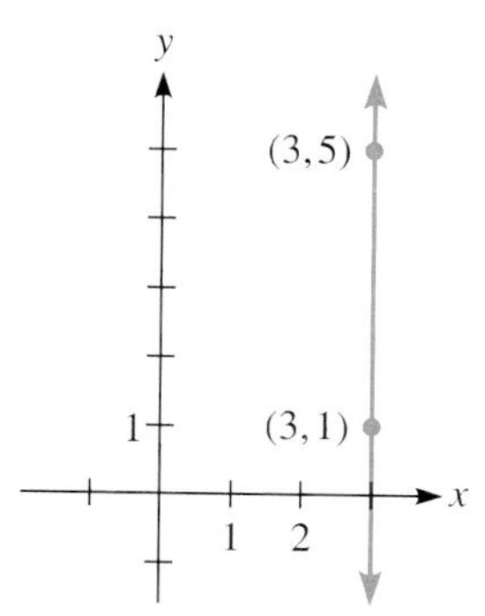

Figure 5.16

b. Because the two points have the same x-coordinate, the line through them is vertical. (See Figure 5.16.) As a consequence, the run will be zero, and the slope is not defined.

$$m = \frac{\text{rise}}{\text{run}} = \frac{y_2 - y_1}{x_2 - x_1} = \frac{1 - 5}{3 - 3} = \frac{-4}{0}$$

It follows that **the slope of every vertical line is undefined.**

PROGRESS CHECK 2 Find the slope of the line through the given points.

a. $(6,1)$ and $(0,1)$ **b.** $(-2,-3)$ and $(-2,1)$

Progress Check Answers

1. (a) 1 (b) $-\frac{4}{3}$
2. (a) 0 (b) Undefined

Examples 1 and 2 have illustrated cases in which the slope of a line is positive, negative, zero, or undefined. These cases are summarized in Figure 5.17.

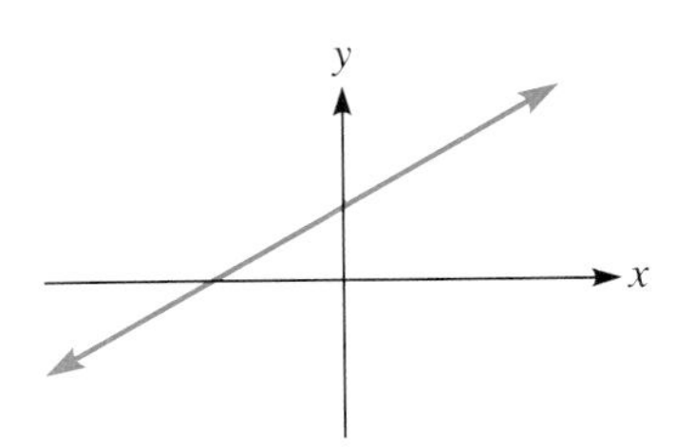

Slope is positive.
Line is higher on the right.
y increases as x increases.

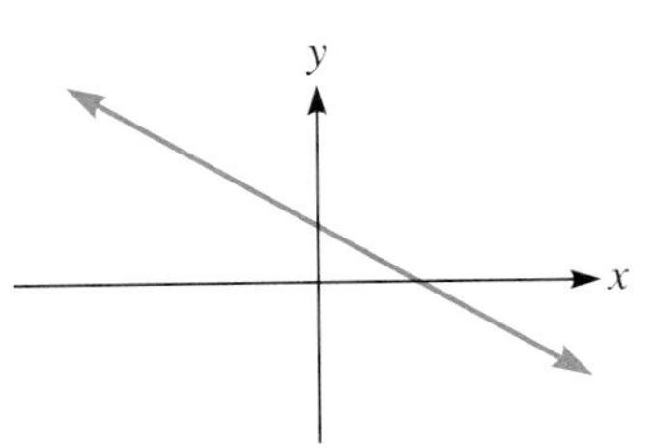

Slope is negative.
Line is lower on the right.
y decreases as x increases.

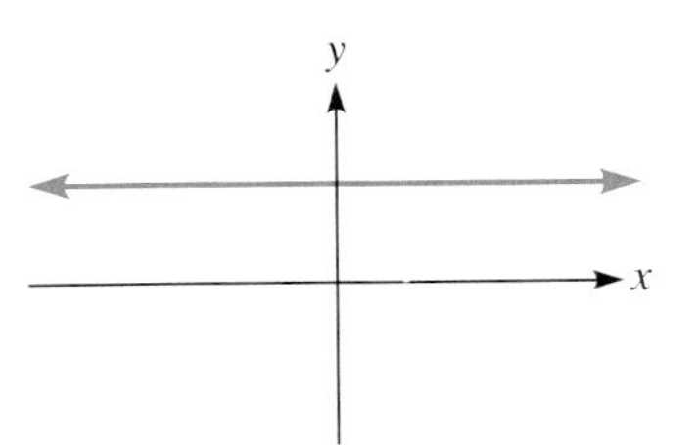

Slope is zero.
Line is horizontal.
y remains constant as x increases.

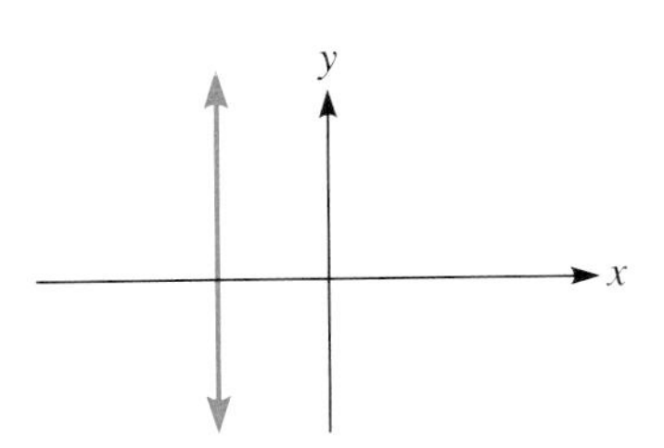

Slope is undefined.
Line is vertical.
x remains constant as y increases.

Figure 5.17

The next two examples call for an interpretation of the slope in applied problems.

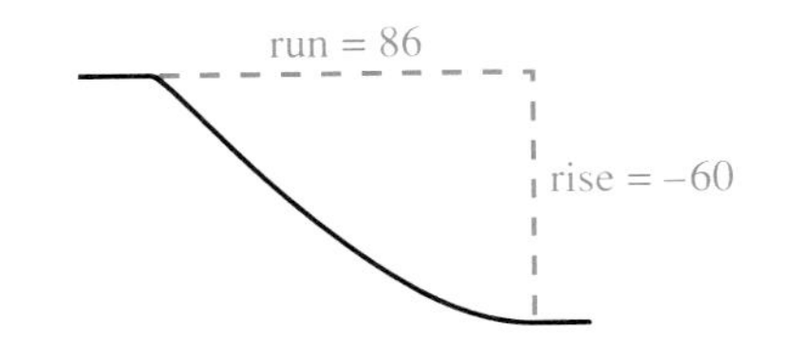

Figure 5.18

EXAMPLE 3 Solve the problem in the section introduction on page 228.

Solution It is reasonable to measure the steepness of this ski jump by slope because its shape is almost a straight line. As shown in Figure 5.18, the rise is represented by −60, because the height of the ski jump *falls* as the skier goes forward.

$$m = \frac{\text{rise}}{\text{run}} = \frac{-60}{86} = -0.70$$

The slope shows that the ski jump falls about 0.7 m per horizontal meter. It can be shown that a line with this slope makes about a 35° angle with the horizontal.

PROGRESS CHECK 3 Find the slope (pitch) of the wall of an A-frame ski lodge that rises 20 ft vertically through a horizontal distance of 15 ft.

In many applications, the slope of a line describes a *rate* relationship between two variables. This interpretation is illustrated in the next example.

EXAMPLE 4 A printing shop charges \$20 for 500 copies of a flyer and \$25 for 750 copies. If the relation between the number of copies (x) and cost (y) graphs as a line, calculate and interpret the slope.

Solution Figure 5.19 shows the graph of the relation between x and y. We let $x_1 = 500$, $y_1 = 20$, $x_2 = 750$, and $y_2 = 25$ and substitute in the slope formula.

$$m = \frac{\text{change in } y}{\text{change in } x} = \frac{y_2 - y_1}{x_2 - x_1} = \frac{25 - 20}{750 - 500} = \frac{5}{250} = 0.02$$

Progress Check Answer
3. $\frac{4}{3}$

In this problem, because the units for the numerator are dollars and the units for the denominator are numbers of copies, the slope describes the change in cost in *dollars per copy.* Therefore, a slope of 0.02 means that it costs 0.02 dollar (or 2 cents) for each additional copy. If the units for x and y had been reversed, so that x represented dollars and y represented copies, then the slope would have been 50 *copies per dollar.* The answers are equivalent, because 50 copies per dollar is the same as 2 cents per copy. In Figure 5.19 we see that the minimum charge for just setting up the job is $10, as given by the y-intercept.

In applied problems it is a good idea to ask students to give the units for the slope. Whenever possible, ask students to interpret the slope in the context of the problem.

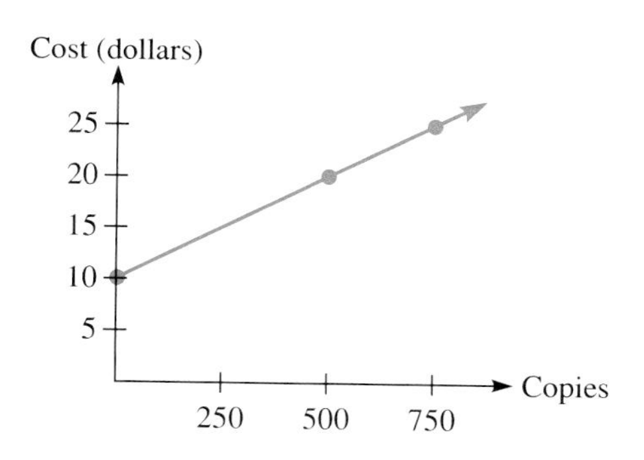

Figure 5.19

PROGRESS CHECK 4 Redo the problem in Example 4, but assume that the cost is $50 for 500 copies and $58 for 700 copies.

2 Slope is a convenient device for analyzing parallel and perpendicular lines. Two lines in a plane that never intersect are called **parallel.** Because they "go in the same direction," it is intuitively clear to say that lines are parallel if they have the *same* slope. Two lines in a plane are **perpendicular** if they meet at right angles. It can be shown mathematically that this happens when the product of their slopes is -1. These properties are summarized in the following box; but note that vertical lines are excluded, because the slope of vertical lines is undefined.

Parallel and Perpendicular Lines

1. Two nonvertical lines are parallel if and only if their slopes are equal.
2. Two nonvertical lines are perpendicular if and only if the product of their slopes is -1. The slope of one is the negative of the reciprocal of the slope of the other.

For the next example, recall that a **parallelogram** is a quadrilateral in which both pairs of opposite sides are parallel, while a **rectangle** is a parallelogram in which all four angles are right angles.

EXAMPLE 5 A quadrilateral has vertices at $A(0,1)$, $B(6,3)$, $C(7,0)$, and $D(1,-2)$.

a. Show that quadrilateral $ABCD$ is a parallelogram.
b. Show that parallelogram $ABCD$ is a rectangle.

Solution Quadrilateral $ABCD$ is shown in Figure 5.20.

a. First, compute the slope of each of the four sides.

$$m_{AB} = \frac{3-1}{6-0} = \frac{2}{6} = \frac{1}{3}$$
$$m_{BC} = \frac{0-3}{7-6} = \frac{-3}{1} = -3$$
$$m_{CD} = \frac{0-(-2)}{7-1} = \frac{2}{6} = \frac{1}{3}$$
$$m_{AD} = \frac{-2-1}{1-0} = \frac{-3}{1} = -3$$

Each pair of opposite sides is parallel because $m_{AB} = m_{CD} = \frac{1}{3}$, and $m_{BC} = m_{AD} = -3$. Thus, $ABCD$ is a parallelogram.

b. All four angles are right angles, because

$$m_{AB} \cdot m_{BC} = m_{AB} \cdot m_{AD} = m_{CD} \cdot m_{BC} = m_{CD} \cdot m_{AD} = \tfrac{1}{3}(-3) = -1.$$

Thus, parallelogram $ABCD$ is a rectangle.

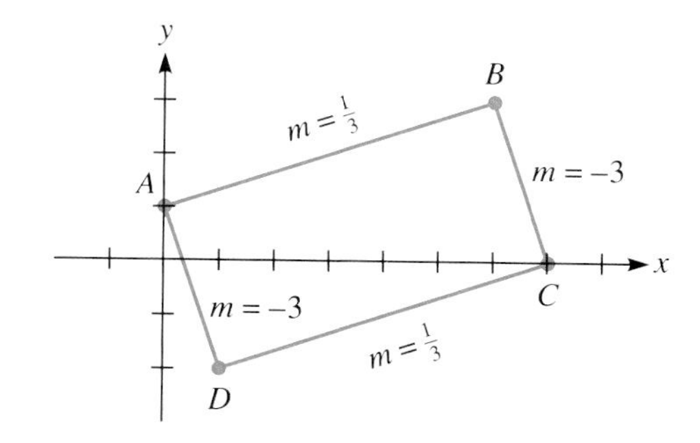

Figure 5.20

Progress Check Answer

4. Slope = 0.04; cost is 4¢ per copy.

PROGRESS CHECK 5 A quadrilateral has vertices at $A(0,1)$, $B(6,4)$, $C(7,2)$, and $D(1,-1)$.

a. Show that quadrilateral $ABCD$ is a parallelogram.
b. Show that quadrilateral $ABCD$ is a rectangle.

3 In some applications relationships between variables are revealed by observations rather than described by equations. Such relationships are often pictured by a type of graph called a **line graph.** In such graphs points are plotted that correspond to observations, and the points are then connected with line segments. The resulting image often captures significant changes and trends in the relationship. In Example 6, note how the ideas of this chapter help in the interpretation of line graphs.

EXAMPLE 6 A pot of water was heated on a kitchen stove and the temperature recorded every minute for 7 minutes. The line graph in Figure 5.21 shows the results.

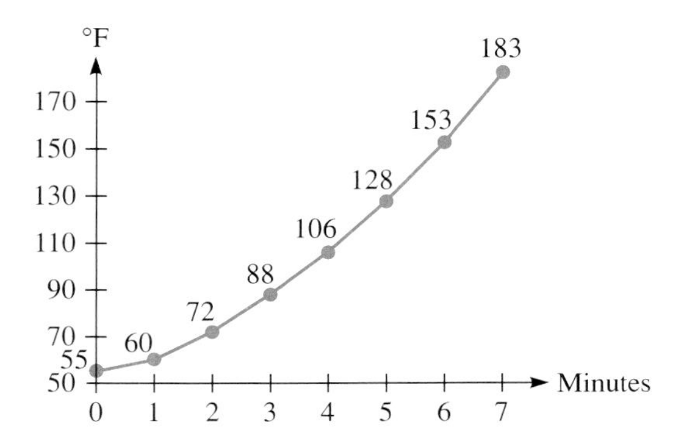

Figure 5.21

a. What was the original temperature of the water? What was the temperature after 1 minute?
b. How long did it take the temperature to rise to 88°?
c. Over which interval did the temperature rise most?
d. Find the slope of a line connecting the first and last points. Interpret the slope.

Solution

a. The starting temperature is given by the y-intercept; it was 55°. To find the temperature after 1 minute, locate 1 on the horizontal axis, and draw a vertical line through it that intersects the graph. See Figure 5.22. From that point draw a horizontal line to the temperature axis to see that the temperature after 1 minute was 60°. Note that to make the reading easier, the temperatures are written right on the graph.

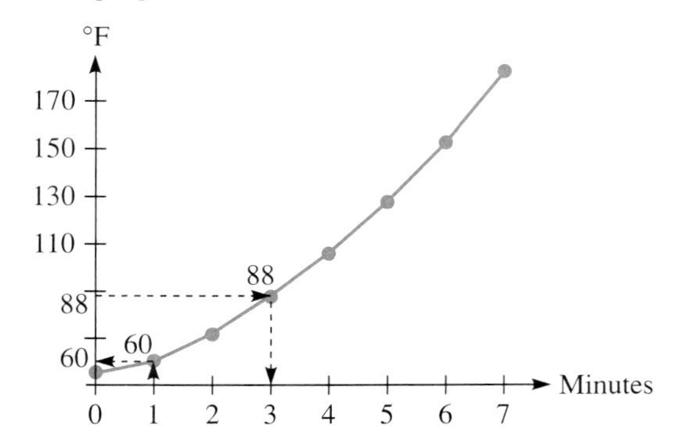

Figure 5.22

Progress Check Answers

5. (a) $m_{AB} = m_{CD} = \frac{1}{2}$; $m_{BC} = m_{AD} = -2$
(b) $\frac{1}{2}(-2) = -1$

b. Locate 88° on the temperature axis, draw a horizontal line to intersect the graph, and then draw a vertical line down to the time axis. The time is 3 minutes. See Figure 5.22.

c. The temperature rose most over the interval for which the line has the largest slope. This can be seen to be the interval from 6 to 7 minutes, where the rise is 30° and the run is 1 minute, for a slope of 30.

d. The slope of the line connecting the first and last points in the graph is $m = \frac{183 - 55}{7 - 0} = \frac{128}{7} = 18.29$. This slope means that, on average, the temperature increased about 18° per minute.

Caution Note that the water is heating up faster and faster. In each succeeding minute there is a greater rise in temperature. But also note the danger in assuming that a graph would keep going the same way forever if the experiment continued. Because the water would boil at 212°, the whole nature of the results would change at that point. Guessing what will happen to a graph beyond the range of values actually observed is called **extrapolation,** and this must always be done with great care.

PROGRESS CHECK 6 Answer these questions based on Figure 5.21.

a. What was the water temperature after 5 minutes?
b. At what time was the temperature equal to 153°?
c. Over which interval did the temperature increase the least?
d. For the first 5 minutes, what was the average temperature increase per minute?

4 Besides the slope formula, another useful formula for analyzing line segments is the distance formula. If $P_1(x_1,y_1)$ and $P_2(x_2,y_2)$ are two points on a line, then the length of line segment P_1P_2 is defined as the distance between P_1 and P_2. It is easiest to find the distance between two points on a horizontal or vertical line. If the points lie on the same horizontal line, the distance between the points is given by the absolute value of the difference between their x-coordinates. If two points lie on the same vertical line, the distance between them is given by the absolute value of the difference between their y-coordinates.

EXAMPLE 7 Find the distance between the given points.

a. $(-4,-1)$ and $(-4,-3)$ **b.** $(-2,2)$ and $(4,2)$

Solution See Figure 5.23.

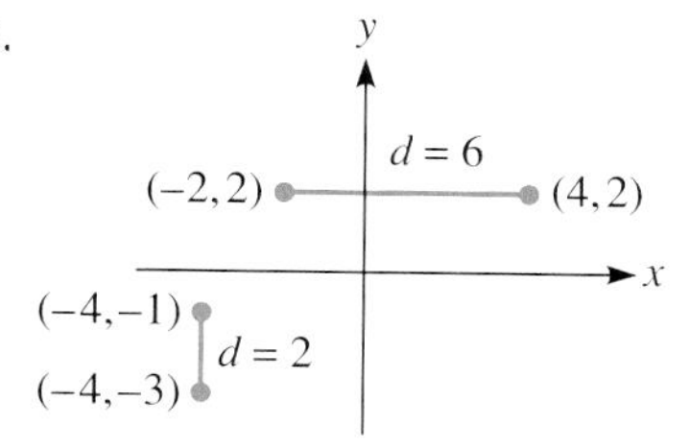

Figure 5.23

a. The points $(-4,-1)$ and $(-4,-3)$ have the same x-coordinate, so

$$d = |y_2 - y_1| = |-3 - (-1)| = |-2| = 2.$$

b. The points $(-2,2)$ and $(4,2)$ have the same y-coordinate, so

$$d = |x_2 - x_1| = |4 - (-2)| = |6| = 6.$$

PROGRESS CHECK 7 Find the distance between the given points.

a. $(3,5)$ and $(3,-3)$ **b.** $(-7,-1)$ and $(0,-1)$

Progress Check Answers

6. (a) 128° (b) 6 minutes (c) 0 to 1 (d) 14.6° per minute
7. (a) 8 (b) 7

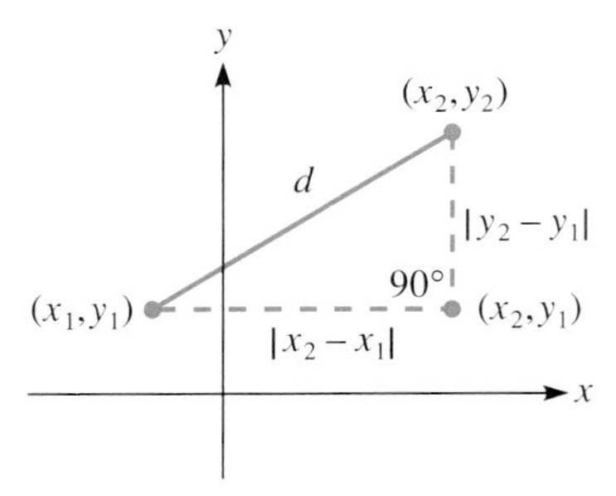

Figure 5.24

To find the distance between two points that do not lie on the same horizontal or the same vertical line, we may construct a right triangle, as shown in Figure 5.24, and apply the Pythagorean theorem (which was discussed earlier in Section 3.9). Because the distance d is the length of the hypotenuse in the right triangle, we get

$$d^2 = |x_2 - x_1|^2 + |y_2 - y_1|^2$$
$$d = \sqrt{(x_2 - x_1)^2 + (y_2 - y_1)^2}.$$

This result gives the distance formula. Note that absolute value symbols are not needed in this formula because the square of any real number is never negative.

Distance Formula

The distance d between (x_1,y_1) and (x_2,y_2) is

$$d = \sqrt{(x_2 - x_1)^2 + (y_2 - y_1)^2}.$$

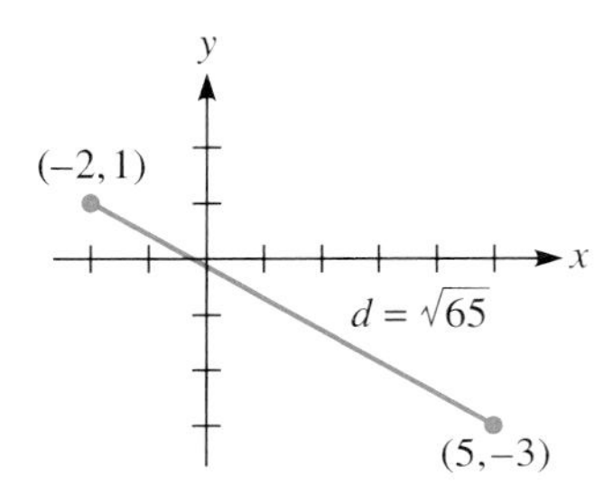

Figure 5.25

EXAMPLE 8 Find the distance between $(-2,1)$ and $(5,-3)$.

Solution Let $x_1 = -2$, $y_1 = 1$, $x_2 = 5$, and $y_2 = -3$, and use the distance formula.

$$\begin{aligned} d &= \sqrt{(x_2 - x_1)^2 + (y_2 - y_1)^2} \\ &= \sqrt{[5 - (-2)]^2 + (-3 - 1)^2} \\ &= \sqrt{7^2 + (-4)^2} \\ &= \sqrt{49 + 16} \\ &= \sqrt{65} \end{aligned}$$

See Figure 5.25.

PROGRESS CHECK 8 Find the distance between $(-3,2)$ and $(-4,-2)$.

EXAMPLE 9 Show that $(-5,-3)$, $(-4,1)$ and $(0,0)$ are vertices of a right triangle using (a) the distance formula and (b) the slope formula.

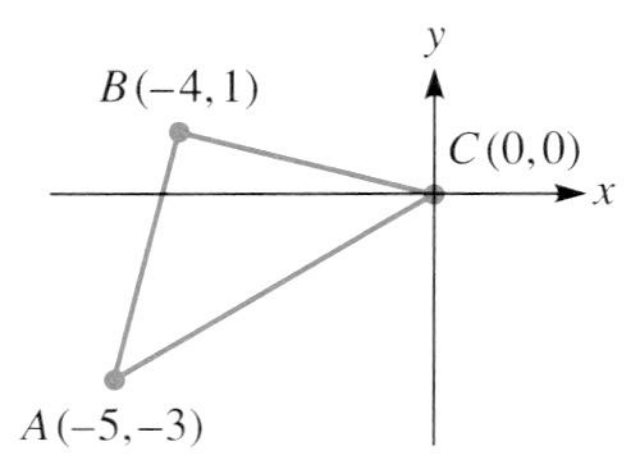

Figure 5.26

Solution Label $(-5,-3)$ as point A, $(-4,1)$ as point B, and $(0,0)$ as point C, as shown in Figure 5.26.

a. By the distance formula,

$$\begin{aligned} d_{AB} &= \sqrt{[-5 - (-4)]^2 + (-3 - 1)^2} = \sqrt{17}, \\ d_{BC} &= \sqrt{(-4 - 0)^2 + (1 - 0)^2} = \sqrt{17}, \\ d_{AC} &= \sqrt{(-5 - 0)^2 + (-3 - 0)^2} = \sqrt{34}. \end{aligned}$$

Because $(\sqrt{34})^2 = (\sqrt{17})^2 + (\sqrt{17})^2$, the Pythagorean theorem holds; so ABC is a right triangle with angle B as the right angle.

b. By the slope formula,

$$\begin{aligned} m_{AB} &= \frac{-3 - 1}{-5 - (-4)} = \frac{-4}{-1} = 4, \\ m_{BC} &= \frac{1 - 0}{-4 - 0} = \frac{1}{-4} = -\frac{1}{4}, \\ m_{AC} &= \frac{-3 - 0}{-5 - 0} = \frac{-3}{-5} = \frac{3}{5}. \end{aligned}$$

Because $m_{AB} \cdot m_{BC} = 4(-\frac{1}{4}) = -1$, angle B is a right angle, and so ABC is a right triangle.

PROGRESS CHECK 9 Show that $A(-2,0)$, $B(3,5)$, and $C(4,2)$ are vertices of a right triangle using (a) the distance formula and (b) the slope formula.

Progress Check Answers

8. $\sqrt{17}$
9. (a) $d_{AB} = \sqrt{50}$, $d_{BC} = \sqrt{10}$, $d_{AC} = \sqrt{40}$; $(\sqrt{10})^2 + (\sqrt{40})^2 = (\sqrt{50})^2$ (b) $m_{AB} = 1$, $m_{BC} = -3$, $m_{AC} = \frac{1}{3}$; angle C is a right angle.

EXERCISES 5.2

In Exercises 1–12, find the slope of the line through the given points, and find the distance between them.

1. (1,1) and (4,5) $\frac{4}{3}$; 5
2. (2,1) and (3,4) 3; $\sqrt{10}$
3. (1,4) and (4,1) −1; $\sqrt{18}$
4. (3,3) and (6,1) $-\frac{2}{3}$; $\sqrt{13}$
5. (−4,−3) and (1,−1) $\frac{2}{5}$; $\sqrt{29}$
6. (−3,−1) and (4,−3) $-\frac{2}{7}$; $\sqrt{53}$
7. (−2,1) and (3,−1) $-\frac{2}{5}$; $\sqrt{29}$
8. (−3,2) and (2,5) $\frac{3}{5}$; $\sqrt{34}$
9. (0,1) and (3,1) 0; 3
10. (−3,−3) and (1,−3) 0; 4
11. (0,0) and (0,−8) Undefined; 8
12. (−2,−4) and (−2,−3) Undefined; 1
13. Find the slope of a mountain trail that rises 80 ft over a horizontal distance of 100 ft. 0.8
14. Find the slope of a segment of highway that falls 88 yards (yd) over a horizontal distance of 500 yd. −0.176
15. The world's steepest roller coaster drops a vertical distance of 192 ft over a horizontal distance of 110 ft. Find the slope of this portion of the coaster. −1.75
16. The Great Pyramid of Khufu, completed about 2700 B.C. at Giza, Egypt, is the tallest pyramid ever built. It has a square base 756 ft per side and rises to a height of about 480 ft. Thus, its sides rise 480 ft over a run of 378 ft. Find the slope. 1.27
17. The weekly cost y of renting a car is \$100 if the weekly mileage x is 100 mi and \$120 if the mileage is 200 mi. If the relationship between x and y graphs as a line, calculate and interpret the slope. $m = 0.20$; mileage charge is 20¢ per mile.
18. Redo the problem in Exercise 17, but assume that the weekly charge is \$150 for 100 mi and \$185 for 200 mi. $m = 0.35$; mileage charge is 35¢ per mile.
19. Shipping costs (c) within zone 1 charged by a delivery service have a linear relationship with the weight (w) of the parcel. The charge is \$2.38 for a 10-oz parcel and \$2.98 for a 20-oz parcel. Graph the line, and calculate and interpret the slope. $m = 0.06$; rate is 6¢ per ounce.

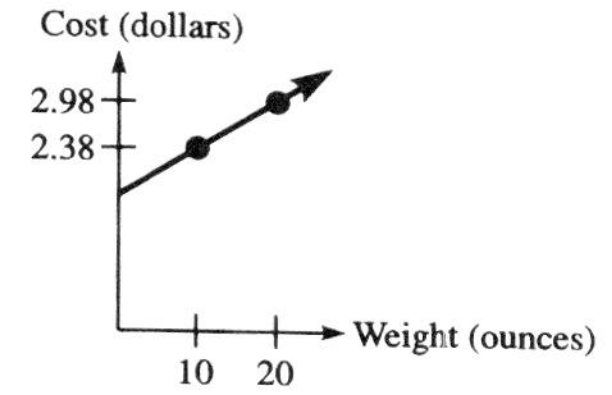

20. The company described in Exercise 19 has the following rates for zone 7: For 10 oz the charge is \$7.79, and for 20 oz the charge is \$11.49. Graph the line; then calculate and interpret the slope. $m = 0.37$; rate is 37¢ per ounce.

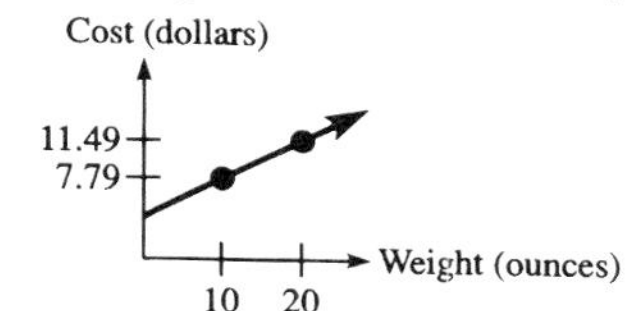

21. The well to a rural home can fill a 60-gal tank in 10 minutes. Assume the relationship between gallons and time can be represented by a line through the origin. Graph the line; then calculate and interpret the slope. $m = 6$; well provides 6 gal/minute.

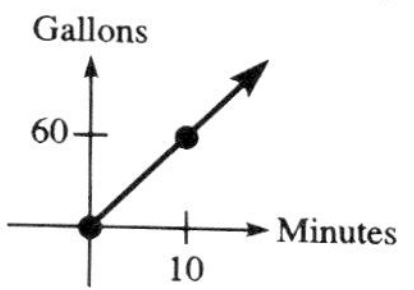

22. A plane flies from New York City to Burlington, Vermont, a distance of 272 mi, in 40 minutes. Assume the relationship between the distance this plane covers and time can be represented by a line through the origin. Graph the line; then calculate and interpret the slope. $m = 6.8$; plane flies 6.8 mi/minute, or 408 mi/hour.

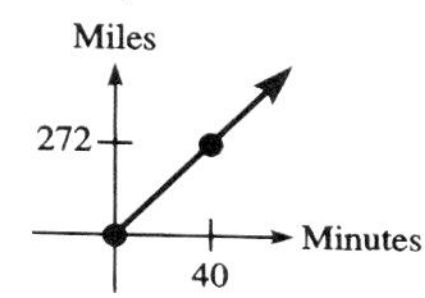

23. The consumer price index (CPI) in 1980 for all items in the market basket was 82, while in 1990 it was 130. If you graph these data as two points on a line showing the relationship between the CPI and the year, then the slope represents the *average* annual rate of inflation. Find the average rate of inflation for the years 1980 to 1990. 4.8 percent per year
24. The world population in 1975 was 4 billion, and in 1990 it was 5.3 billion. If you graph these data as two points on a line, then the slope represents the average annual rate of population growth. Find the average annual growth rate (in millions per year) for the years 1975 to 1990. 86.7 million people per year
25. A quadrilateral has vertices at $A(0,0)$, $B(4,1)$, $C(5,-3)$, and $D(1,-4)$.
    a. Show that $ABCD$ is a parallelogram. $m_{AB} = m_{CD} = \frac{1}{4}$; $m_{AD} = m_{BC} = -4$
    b. Show that $ABCD$ is a rectangle. $\frac{1}{4}(-4) = -1$
    c. Show that the lengths of both diagonals are equal. $\sqrt{34} = \sqrt{34}$
26. A quadrilateral has vertices at $A(4,-1)$, $B(2,-4)$, $C(-4,0)$, and $D(-2,3)$.
    a. Show that $ABCD$ is a parallelogram. $m_{AB} = m_{CD} = \frac{3}{2}$; $m_{AD} = m_{BC} = -\frac{2}{3}$
    b. Show that $ABCD$ is a rectangle. $\frac{3}{2}(-\frac{2}{3}) = -1$
    c. Show that the lengths of both diagonals are equal. $\sqrt{65} = \sqrt{65}$
27. Show that quadrilateral $ABCD$ is a parallelogram but *not* a rectangle. The vertices are $A(2,0)$, $B(3,3)$, $C(7,3)$, and $D(6,0)$. Confirm that the diagonals are unequal in length. $m_{AB} = m_{CD} = 3$; $m_{BC} = m_{AD} = 0$; $3 \cdot 0$ does not equal -1; $\sqrt{34} \neq \sqrt{18}$

28. Show that quadrilateral $ABCD$ is a parallelogram but *not* a rectangle. The vertices are $A(-2,0)$, $B(2,0)$, $C(5,-2)$, and $D(1,-2)$. Confirm that the diagonals are unequal in length. $m_{AB} = m_{CD} = 0$; $m_{AD} = m_{BC} = -\frac{2}{3}$; $0 \cdot (-\frac{2}{3})$ does not equal -1; $\sqrt{53} \neq \sqrt{5}$

29. Two lines are perpendicular.

a. If the slope of one is 5, then the slope of the other is ______. $-\frac{1}{5}$

b. If the slope of one is 0, then the slope of the other is ______. Undefined

c. If the slope of one is 0.7, then the slope of the other is ______. $-\frac{10}{7}$

30. Two lines are perpendicular.

a. If the slope of one is $-\frac{1}{2}$, then the slope of the other is ______. 2

b. If the slope of one is $\frac{2}{3}$, then the slope of the other is ______. $-\frac{3}{2}$

c. If the slope of one is -0.23, then the slope of the other is ______. $\frac{100}{23}$

For Exercises 31–34, use the fact that if A, B, and C are on the same straight line (collinear), then the slope of AB equals the slope of AC.

31. Determine whether $A(-5,-2)$, $B(1,-1)$, and $C(13,1)$ are collinear. Yes; $m_{AB} = m_{AC} = \frac{1}{6}$

32. Determine whether $A(-4,2)$, $B(-1,-1)$, and $C(2,-4)$ are collinear. Yes; $m_{AB} = m_{AC} = -1$

33. Determine whether $A(-3,-1)$, $B(1,0)$, and $C(9,1)$ are collinear. No; $m_{AB} = \frac{1}{4}$, $m_{AC} = \frac{1}{6}$

34. Determine whether $A(-3,2)$, $B(-2,-6)$, and $C(0,-23)$ are collinear. No; $m_{AB} = -8$, $m_{AC} = -\frac{25}{3}$

In Exercises 35–38, use the given line graph to answer the questions.

35. The line graph shows the progress of the Dow-Jones Industrial Average for one day.

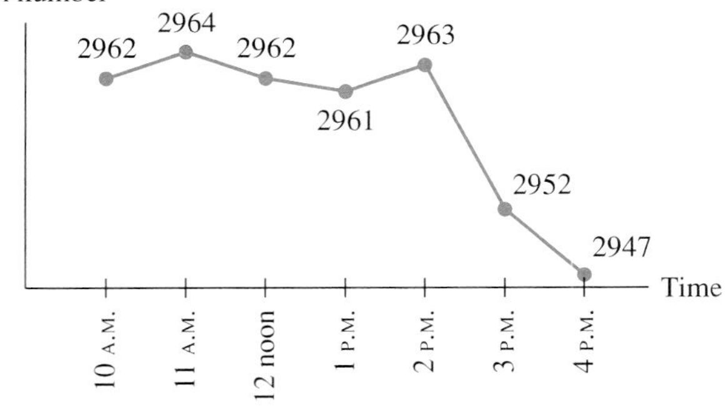

a. What was the Dow-Jones index at noon? 2962

b. At what time was the index at its lowest value of the day? 4 P.M.

c. Over which hourly interval did the index change most? 2–3 P.M.

d. On which hourly interval did the index change least? 12 noon–1 P.M.

e. During which hourly interval did the index drop below 2950? 3–4 P.M.

36. The line graph shows some hourly temperatures at a weather station for one day.

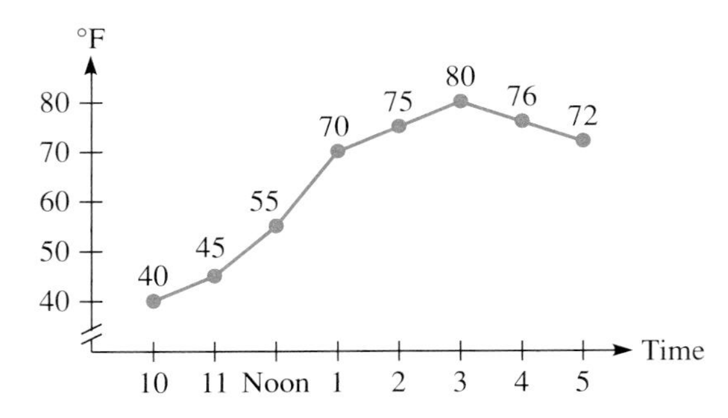

a. What was the temperature at 10 A.M.? 40°

b. In which hourly interval did the temperature first reach 60°? 12 noon–1 P.M.

c. Between which two hourly readings was the temperature change greatest? 12 noon–1 P.M.

d. What was the maximum temperature recorded over this period? 80°

e. How much did the temperature drop between 3 and 5 P.M.? 8°

37. In 1920, the number of icebergs sighted per month south of Newfoundland was recorded. The data are shown in a line graph.

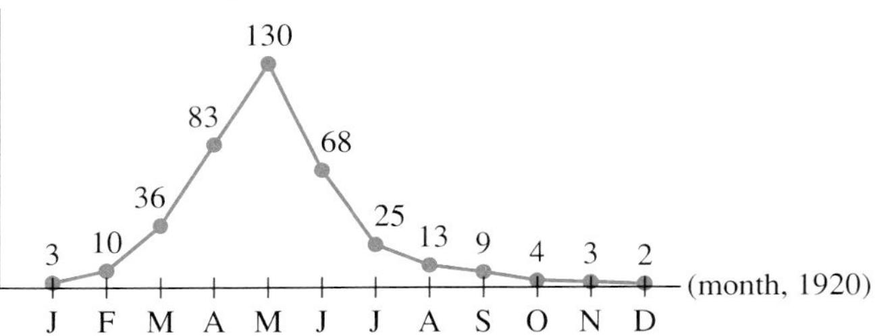

[These data appear in *Data Analysis and Regression* by J. W. Tukey and F. Mosteller (Addison-Wesley, 1977); they got it from Sir Napier Shaw's *Manual of Meteorology* (Cambridge University Press, 1942).]

a. What was the least number of icebergs sighted? In what month? 2; December

b. What was the greatest number of icebergs sighted? In what month? 130; May

c. Between what two months was there the greatest increase in the number of sightings? March and April; also April and May

d. Between what two months was there the greatest decrease in the number of sightings? May and June

38. This graph shows, in *millions*, the number of pieces of mail sent out by the U.S. House of Representatives over several years. [The graph is based on one in E. R. Tufte, *Visual Display of Quantitative Information* (Graphics Press, 1983).]

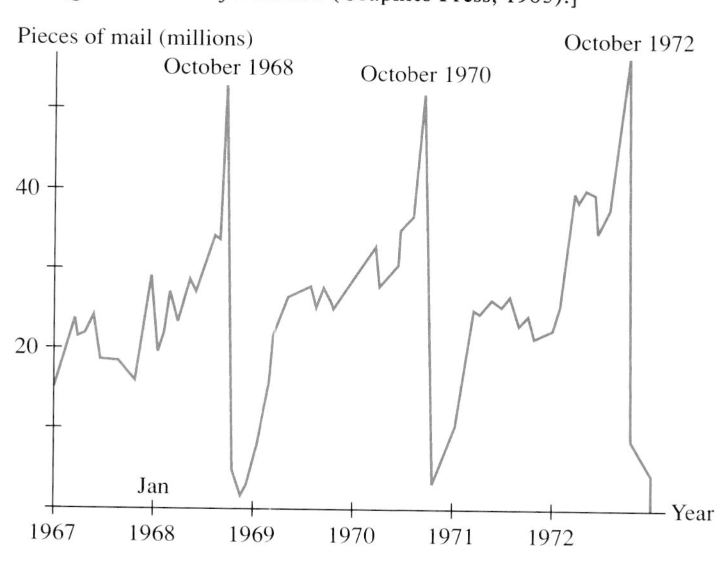

a. The graph exhibits periodic behavior. When does it peak? October of even years
b. Can you explain the behavior of the graph? Peaks occur before election.
c. In what months were more than 50 million pieces mailed? October 1968, 1970, 1972
d. Interpret the slope of the segment between two consecutive months. Change in number of pieces mailed per month
e. What is the interpretation of the steep negative slope after October 1970? Sharp reduction in the number of pieces mailed

THINK ABOUT IT

1. Railroad schedules in Europe and Japan often make intuitive use of the concept of slope. A simplified example is shown.

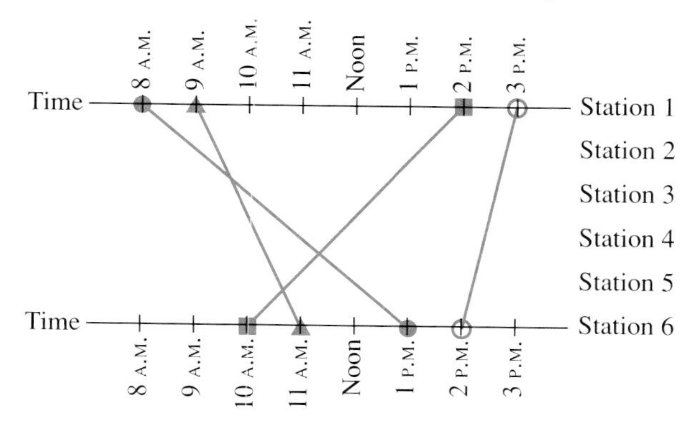

a. A train leaves station 1 at 8 A.M. What times does it arrive at station 6?
b. A train arrives at station 6 at 11 A.M.; when does it leave station 1?
c. What is the interpretation of the line with squares at its endpoints?
d. What is the interpretation of the *intersection* of two lines?
e. What does it mean if one line is *steeper* than another? [For a more complex real example, see page 24 of *Envisioning Information* by E. R. Tufte (Graphics Press, 1990), where a "wondrously complex" graph shows the timetable from 1937 for a Java railroad line.]

2. Since $\frac{-2}{3}$ is equal to $\frac{2}{-3}$, this means that a rise of -2 with a run of 3 describes the same slope as a rise of 2 and a run of -3. Show that this is correct by starting at the origin and drawing these rises and runs to find a second point.
3. Find the slopes of the lines which connect the origin to each of these points.
a. (1,0)
b. (1,1)
c. (1,5)
d. (1,20)
e. (1,1 million)
f. Why is it said informally that a vertical line has "infinite" slope?
4. A graph starts at (0,1). The next point is $(1,\frac{1}{2})$. Then $(2,\frac{1}{4})$. Then $(3,\frac{1}{8})$.
a. What are the next two points?
b. What is the 100th point in this sequence?
c. Is each point lower than the previous one?
d. Will the points ever reach the x-axis?
e. Which of these shapes best describes the graph which connects the points?

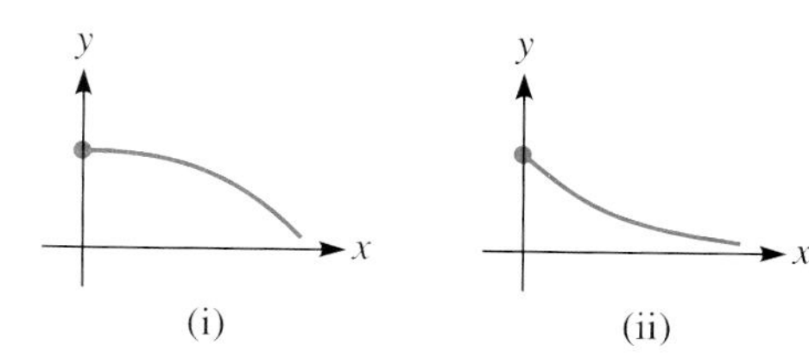

5. A simple proof of the Pythagorean theorem follows from expressing the area of this figure two ways. The figure is built up from four copies of a right triangle, with legs a and b and hypotenuse c.
   a. The area of the large square is $(a + b)^2$. Why?
   b. Find another expression for the area of the large square by adding the areas of the pieces which make it up.
   c. Equate the expressions from parts **a** and **b**; then simplify to get the Pythagorean theorem.

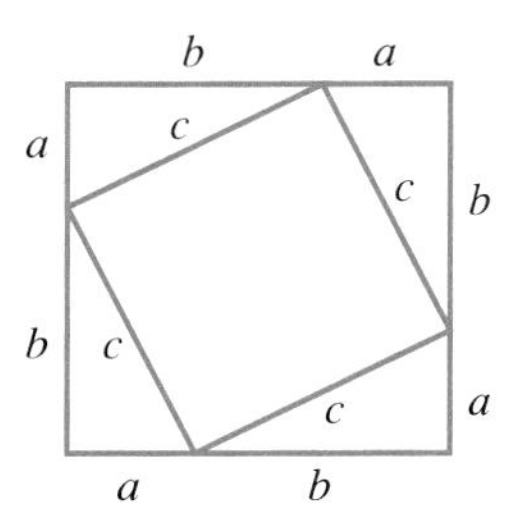

REMEMBER THIS

1. Solve $ax + by = c$ for y. $y = \dfrac{c - ax}{b}$
2. In the equation $-3x + 5y = 715$, what is the value of y when $x = 220$? 275
3. The point $(0,b)$ is a solution to $3x + 5y = 3$. Find b. $\frac{3}{5}$
4. Solve $y = mx + b$ for m. $m = \dfrac{y - b}{x}$
5. Graph $y = -3$. Is this a horizontal or a vertical line? Horizontal

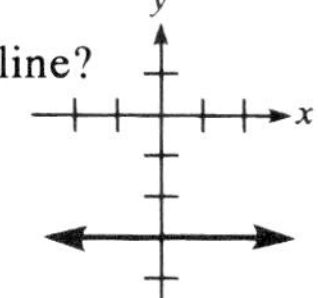

6. Simplify $\left(\dfrac{4x^{-2}y}{12x^{-1}y^{-2}}\right)^{-2}$. $\dfrac{9x^2}{y^6}$
7. Solve $2/x = 3 - x$. $\{1,2\}$
8. Expand $(n + 1)^3$. $n^3 + 3n^2 + 3n + 1$
9. For what value of b will $x^2 + bx + 9$ be a perfect square trinomial? 6
10. The difference of the squares of two consecutive odd positive integers is 48. What are the integers? 11 and 13

5.3 Equations of a Line

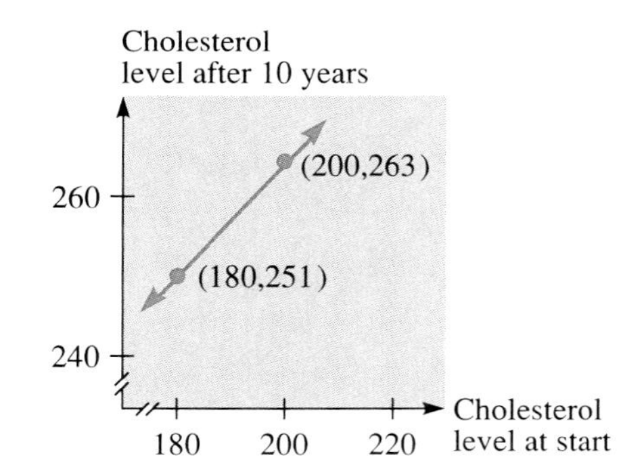

As part of a 10-year medical research project, a statistician noted that the cholesterol level of subjects tended to go up over the 10-year period. Those who started with a level at about 180 went up to 251. Those who started at about 200 went up to 263.

a. The relationship between the beginning level and the level 10 years later was graphed as a line, as shown in the figure. Write the equation of this line in the form $Ax + By = C$.

b. What level does this equation predict for someone whose original level is 220? (See Example 3.)

OBJECTIVES

1 **Write an equation for a line given its slope and a point on the line.**

2 **Write an equation for a line given two points on the line.**

3 **Find the slope and *y*-intercept given an equation for the line.**

4 **Graph and write an equation for a line given the slope and *y*-intercept.**

5 **Solve problems involving the equations of parallel and perpendicular lines.**

The equation for a line can be written in several equivalent forms. As mentioned in Section 5.1, the form $Ax + By = C$ is called the *general form* or the *standard form*. In this section we will consider two other useful forms: the *point-slope form* and the *slope-intercept form*.

Stress that the choice of form depends on the application. They are all equivalent. It is not that some are right and some are wrong.

1 The position of a line in a plane is determined when one point on the line and the slope are given. Therefore, with just this information, the equation of the line can be determined. Consider any nonvertical line with slope m that passes through the point (x_1,y_1), as shown in Figure 5.27. If (x,y) represents any other point on this line, then the slope definition gives

$$\frac{y - y_1}{x - x_1} = m.$$

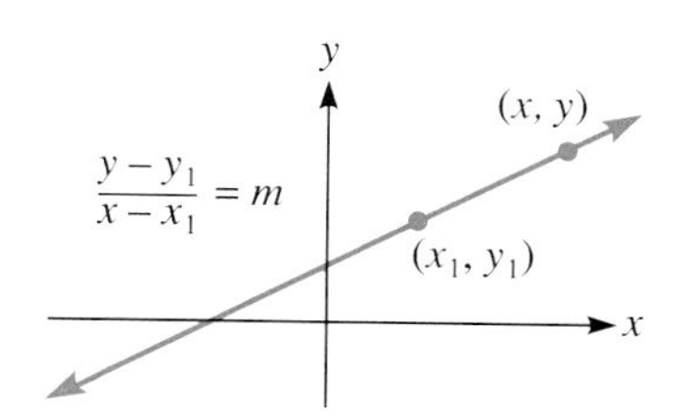

Figure 5.27

If we multiply both sides by $x - x_1$, this equation becomes

$$y - y_1 = m(x - x_1),$$

and the result is called the *point-slope form* of the equation of a line.

> **Point-Slope Equation**
>
> An equation of the line with slope m passing through (x_1,y_1) is
>
> $$y - y_1 = m(x - x_1).$$
>
> This equation is called the **point-slope form** of the equation of a line.

EXAMPLE 1 Find an equation of the line through $(3,-2)$ with slope -5. Express the answer in the general form $Ax + By = C$.

Solution We are given $x_1 = 3$, $y_1 = -2$, and $m = -5$. Substitute these numbers in the point-slope equation and simplify to the form requested.

$$y - y_1 = m(x - x_1) \quad \text{Point-slope equation.}$$
$$y - (-2) = -5(x - 3) \quad \text{Replace } y_1 \text{ by } -2,\ m \text{ by } -5, \text{ and } x_1 \text{ by } 3.$$
$$y + 2 = -5x + 15 \quad \text{Remove parentheses.}$$
$$5x + y + 2 = 15 \quad \text{Add } 5x \text{ to both sides.}$$
$$5x + y = 13 \quad \text{Subtract 2 from both sides.}$$

In general form, the equation of the line is $5x + y = 13$.

PROGRESS CHECK 1 Express in general form an equation for the line through $(4,-3)$ with slope -1.

2 Because the position of a line is determined when two points on it are known, this is also sufficient information to find an equation for the line. The equation can be found by first finding the slope and then using the point-slope form of the equation.

EXAMPLE 2 Find an equation for the line through the points $(-2,5)$ and $(1,6)$. Express the answer in the general form.

Solution First, find the slope of the line. (See Figure 5.28.)

$$m = \frac{y_2 - y_1}{x_2 - x_1} = \frac{6 - 5}{1 - (-2)} = \frac{1}{3}$$

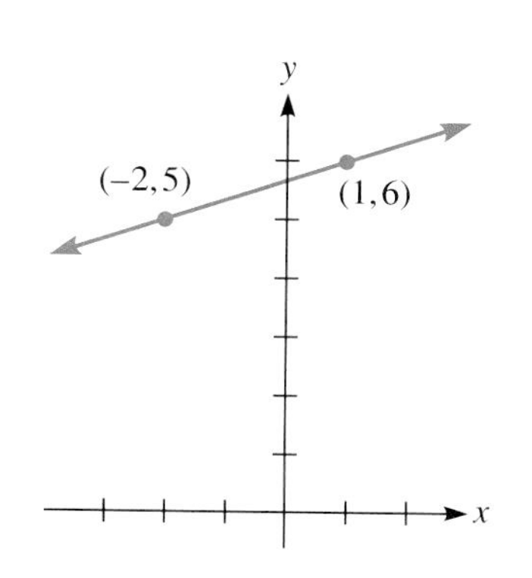

Figure 5.28

Progress Check Answer

1. $x + y = 1$

Now use the point-slope equation with $m = \frac{1}{3}$ and *either* $(-2,5)$ or $(1,6)$ for (x_1,y_1). We use $x_1 = 1$ and $y_1 = 6$, as follows.

$$y - y_1 = m(x - x_1) \quad \text{Point-slope equation.}$$
$$y - 6 = \tfrac{1}{3}(x - 1) \quad \text{Replace } y_1 \text{ by 6, } m \text{ by } \tfrac{1}{3}, x_1 \text{ by 1.}$$
$$3(y - 6) = 1(x - 1) \quad \text{Multiply both sides by 3.}$$
$$3y - 18 = x - 1 \quad \text{Remove parentheses.}$$
$$-x + 3y = 17 \quad \text{On both sides, subtract } x \text{ and add 18.}$$

In general form, the equation of the line is $-x + 3y = 17$.

Note To confirm that either point can be used, we show the results of using $(-2,5)$ as (x_1,y_1) in the point-slope equation.

$$y - 5 = \tfrac{1}{3}[x - (-2)] \quad \text{Replace } y_1 \text{ by 5, } m \text{ by } \tfrac{1}{3}, x_1 \text{ by } -2.$$
$$3(y - 5) = x - (-2) \quad \text{Multiply both sides by 3.}$$
$$3y - 15 = x + 2 \quad \text{Remove parentheses.}$$
$$-x + 3y = 17 \quad \text{On both sides, subtract } x \text{ and add 15.}$$

This equation is the same result as before.

PROGRESS CHECK 2 Find an equation for the line through the points $(-5,3)$ and $(4,1)$. Express the answer in general form.

EXAMPLE 3 Solve the problem in the section introduction on page 238.

Solution

a. We are asked to find in general form the equation of the line through $(180,251)$ and $(200,263)$. As in Example 2, first find the slope.

$$m = \frac{y_2 - y_1}{x_2 - x_1} = \frac{263 - 251}{200 - 180} = \frac{12}{20} = \frac{3}{5}$$

Now use the point-slope equation with one of the points, say $(180,251)$, as follows.

$$y - y_1 = m(x - x_1) \quad \text{Point-slope equation.}$$
$$y - 251 = \tfrac{3}{5}(x - 180) \quad \text{Replace } y_1 \text{ by 251, } m \text{ by } \tfrac{3}{5}, x_1 \text{ by 180.}$$
$$5(y - 251) = 3(x - 180) \quad \text{Multiply both sides by 5.}$$
$$5y - 1{,}255 = 3x - 540 \quad \text{Remove parentheses.}$$
$$-3x + 5y = 715 \quad \text{On both sides, subtract } 3x \text{ and add 1,255.}$$

In general form, the equation of the line is $-3x + 5y = 715$.

b. To find the predicted cholesterol level for someone who started with a level of 220, replace x by 220 in the equation that defines the relation, and solve for y.

$$-3x + 5y = 715 \quad \text{Equation of the line.}$$
$$-3(220) + 5y = 715 \quad \text{Replace } x \text{ by 220.}$$
$$5y = 1{,}375 \quad \text{Add 3(220), or 660, to both sides.}$$
$$y = 275 \quad \text{Divide both sides by 5.}$$

The prediction is that 10 years later the cholesterol level will be 275.

PROGRESS CHECK 3 In a medical study a drug was found to lower patients' diastolic blood pressure in a way that could be graphed as a line. On average, the drug reduced a pressure of 96 to 87, and a pressure of 111 to 97, as shown in the figure.

a. Write an equation in general form for the line shown.

b. Use the equation to predict, to the nearest unit, the diastolic blood pressure for someone whose pressure is 100 before using the drug.

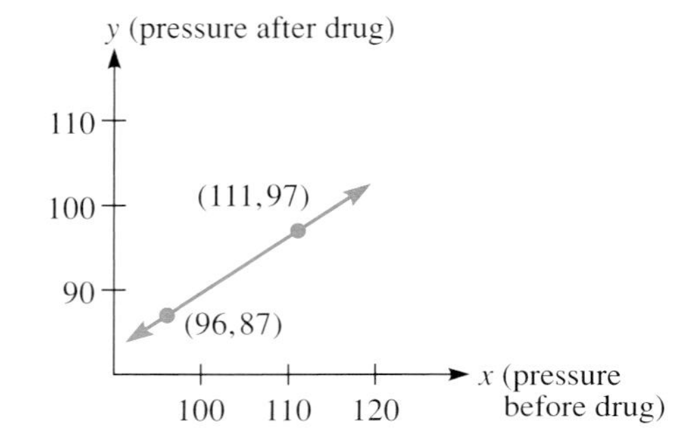

Progress Check Answers

2. $2x + 9y = 17$ 3. (a) $2x - 3y = -69$ (b) 90

3 The point-slope equation is used extensively to find the equation of a line, but it is not very helpful for graphing lines or for interpreting linear relationships in applications. For these purposes it will often be more convenient to use the *slope-intercept form,* which we develop next. This form follows from the point-slope form by using the y-intercept as the point. Note that in Figure 5.29 we use the letter b to denote the y-coordinate of the point where the graph crosses the y-axis, so the point $(0,b)$ is the y-intercept. Then we proceed using $(0,b)$ to replace (x_1,y_1).

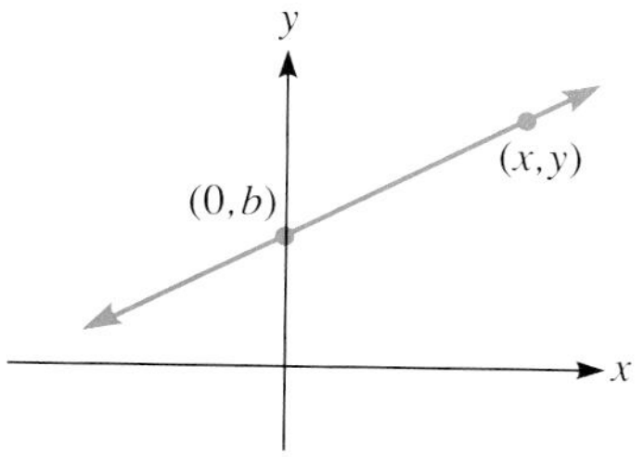

Figure 5.29

$$y - y_1 = m(x - x_1) \quad \text{Point-slope equation.}$$
$$y - b = m(x - 0) \quad \text{Replace } y_1 \text{ by } b \text{ and } x_1 \text{ by } 0.$$
$$y - b = mx \quad \text{Simplify.}$$
$$y = mx + b \quad \text{Add } b \text{ to both sides.}$$

This result is called the *slope-intercept form* of the equation of a line. When an equation is written in this form, it is easy to find the slope and y-intercept.

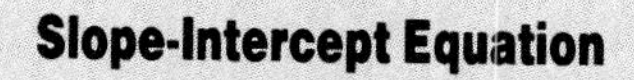

The graph of the equation

$$y = mx + b$$

is a line with slope m and y-intercept $(0,b)$.

EXAMPLE 4 Find the slope and the y-intercept of the line defined by the following equations.

a. $y = 5x + 4$ **b.** $3x + 4y = -5$

Solution

a. The equation is given in the form $y = mx + b$, with $m = 5$ and $b = 4$. Thus,

the slope of the line is 5, → $y = 5x + 4$ ← and the y-intercept is (0,4).

b. First, express the equation in the form $y = mx + b$.

$$3x + 4y = -5 \quad \text{Given equation.}$$
$$4y = -3x - 5 \quad \text{Subtract } 3x \text{ from both sides.}$$
$$y = -\tfrac{3}{4}x - \tfrac{5}{4} \quad \text{Divide both sides by 4.}$$

Matching the equation to the form $y = mx + b$, we see that

$$m = -\tfrac{3}{4} \quad \text{and} \quad b = -\tfrac{5}{4}.$$

Thus, the slope is $-\frac{3}{4}$ and the y-intercept is $(0, -\frac{5}{4})$.

PROGRESS CHECK 4 Find the slope and the y-intercept of the line defined by the following equations.

a. $y = -2x + 3$ **b.** $4x - 3y = 12$

4 When the slope and the y-intercept of a line are known, its equation can be written directly and it can be graphed, as shown in Example 5.

Progress Check Answers

4. (a) -2, (0,3) (b) $\frac{4}{3}$, $(0,-4)$

EXAMPLE 5 The slope of a line is $-\frac{1}{3}$ and the y-intercept is (0,2).

a. Find the equation of the line in slope-intercept form.
b. Graph the line.

Solution

a. By substituting $m = -\frac{1}{3}$ and $b = 2$ in the slope-intercept equation $y = mx + b$, we determine that the equation is

$$y = -\tfrac{1}{3}x + 2.$$

b. The y-intercept (0,2) is one point on the line. To find another point, interpret a slope of $-\frac{1}{3}$ to mean that when x increases 3 units, y decreases 1 unit. By starting at (0,2) and going 3 units to the right and 1 unit down, we obtain (3,1) as a second point on the line. Drawing a line through these two points produces the graph in Figure 5.30.

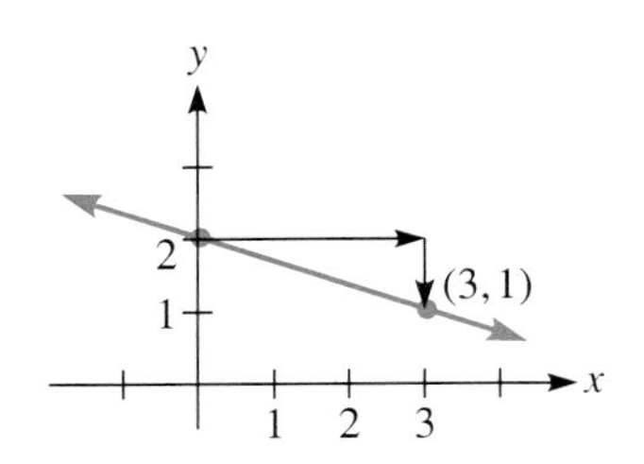

Figure 5.30

PROGRESS CHECK 5 The slope of a line is $-\frac{3}{4}$ and the y-intercept is 5.

a. Find the equation of the line in slope-intercept form.
b. Graph the line.

5 The slope-intercept form of the equation of a line provides a technique for solving problems that deal with parallel and perpendicular lines. Several such problems are illustrated next.

EXAMPLE 6 Determine if the graphs of $x + 2y = 3$ and $2x + 4y = 5$ are distinct parallel lines.

Solution Two distinct lines are parallel if they have the same slope but different y-intercepts. Therefore, we begin by expressing the equations in slope-intercept form in order to determine the slope and the intercept.

$$\begin{aligned} x + 2y &= 1 \\ 2y &= -x + 1 \\ y &= -\tfrac{1}{2}x + \tfrac{1}{2} \end{aligned} \qquad \begin{aligned} 2x + 4y &= 5 \\ 4y &= -2x + 5 \\ y &= -\tfrac{2}{4}x + \tfrac{5}{4} \\ y &= -\tfrac{1}{2}x + \tfrac{5}{4} \end{aligned}$$

From these equations we get the following results.

Slope: $-\frac{1}{2}$ | Slope: $-\frac{1}{2}$
y-intercept: $(0,\frac{1}{2})$ | y-intercept: $(0,\frac{5}{4})$

The lines have the same slope and different y-intercepts, so they are distinct parallel lines. They are sketched in Figure 5.31.

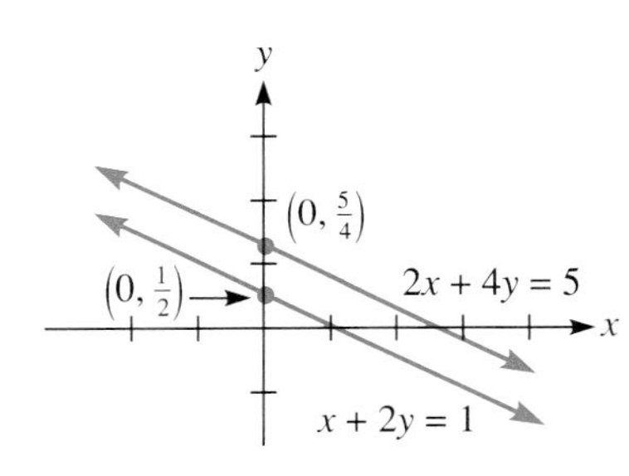

Figure 5.31

PROGRESS CHECK 6 Determine if the graphs of $2x + y = 1$ and $2x + y = 3$ are distinct parallel lines.

EXAMPLE 7 Find an equation for the line which passes through the point (3,4) and is parallel to the line $x - 2y = 1$.

Solution Since a point on the line is given, the main problem is to find the slope. We find it from the given line.

$$\begin{aligned} x - 2y &= 1 && \text{Given line.} \\ -2y &= -x + 1 && \text{Subtract } x \text{ from both sides.} \\ y &= \tfrac{1}{2}x - \tfrac{1}{2} && \text{Divide both sides by } -2 \text{ and simplify.} \end{aligned}$$

Progress Check Answers

5.

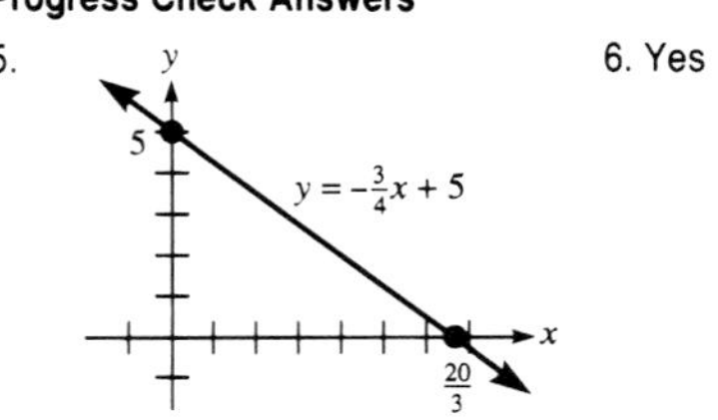

6. Yes

This equation is slope-intercept form, so the slope of the given line is $\frac{1}{2}$. Because the slopes of parallel lines are equal, we next find an equation for the line through (3,4) with slope $\frac{1}{2}$.

$$\begin{aligned} y - y_1 &= m(x - x_1) && \text{Point-slope form.} \\ y - 4 &= \tfrac{1}{2}(x - 3) && \text{Replace } y_1 \text{ by 4, } m \text{ by } \tfrac{1}{2}\text{, } x_1 \text{ by 3.} \\ 2(y - 4) &= x - 3 && \text{Multiply both sides by 2.} \\ 2y - 8 &= x - 3 && \text{Remove parentheses.} \\ x - 2y &= -5 && \text{On both sides, subtract } 2y \text{ and add 3.} \end{aligned}$$

Therefore, an equation for the line which passes through (3,4) and which is parallel to $x - 2y = 1$ is $x - 2y = -5$.

PROGRESS CHECK 7 Find an equation for the line which passes through (0,0) and is parallel to $2x + y = 3$.

EXAMPLE 8 Find an equation for the line through the origin which is perpendicular to $3x - y = 10$.

Solution Lines are perpendicular if the product of their slopes is -1. So we begin by finding the slope of the given line.

$$\begin{aligned} 3x - y &= -10 && \text{Given equation.} \\ 3x + 10 &= y && \text{Add 10 and } y \text{ to both sides.} \end{aligned}$$

This is the slope-intercept form, with $m = 3$ and $b = 10$. Therefore, the slope of the given line is 3. It follows that any line perpendicular to it must have slope equal to $-\frac{1}{3}$.

Now use the slope-intercept form to determine the equation of a line through the origin with slope $-\frac{1}{3}$. Since the line goes through the origin, its intercept is (0,0).

$$\begin{aligned} y &= mx + b && \text{Slope-intercept form.} \\ y &= -\tfrac{1}{3}x + 0 && \text{Replace } m \text{ by } -\tfrac{1}{3} \text{ and } b \text{ by 0.} \\ y + \tfrac{1}{3}x &= 0 && \text{Add } \tfrac{1}{3}x \text{ to both sides.} \end{aligned}$$

Thus, the desired equation is $y + \frac{1}{3}x = 0$, or $x + 3y = 0$.

Note Instead of using the slope-intercept form, we could have used the point-slope form, taking the point as (0,0). The result is the same.

PROGRESS CHECK 8 Find an equation for the line which is perpendicular to $2x - y = 5$ and passes through the origin.

Progress Check Answers

7. $y = -2x$

8. $y = -\frac{1}{2}x$

EXERCISES 5.3

In Exercises 1–6, find an equation for the line through the given point and having the given slope. Express the answer in the form $Ax + By = C$.

1. Point (4,8); slope $= -3$ $3x + y = 20$
2. Point $(-1,0)$; slope $= 2$ $2x - y = -2$
3. Point $(2,-5)$; slope $= \frac{3}{4}$ $3x - 4y = 26$
4. Point $(-2,-3)$; slope $-\frac{2}{3}$ $2x + 3y = -13$
5. Point (5,1); horizontal line $y = 1$
6. Point $(-2,0)$; horizontal line $y = 0$

In Exercises 7–12, find an equation for the line through the given points. Express the answer in general form.

7. (0,0) and (4,2) $x - 2y = 0$
8. $(-2,5)$ and (0,0) $5x + 2y = 0$
9. $(2,-1)$ and $(-1,-2)$ $x - 3y = 5$
10. $(-5,-1)$ and $(-1,3)$ $x - y = -4$
11. $(-2,2)$ and $(3,-2)$ $4x + 5y = 2$
12. $(-3,-2)$ and $(5,-5)$ $3x + 8y = -25$

In Exercises 13–20, find the slope and y-intercept of the given line.

13. $y = 2x + 1$ Slope $= 2$; y-intercept is (0,1).
14. $y = -3x - 2$ Slope $= -3$; y-intercept is $(0,-2)$.
15. $2x + 3y = 4$ Slope $= -\frac{2}{3}$; y-intercept is $(0,\frac{4}{3})$.
16. $3x - 4y = 5$ Slope $= \frac{3}{4}$; y-intercept is $(0,-\frac{5}{4})$.
17. $y = 6$ Slope $= 0$; y-intercept is (0,6).
18. $2y - 5 = 0$ Slope $= 0$; y-intercept is $(0,\frac{5}{2})$.
19. $x = 2$ Slope undefined; no y-intercept
20. $3x + 1 = 0$ Slope undefined; no y-intercept

In Exercises 21–26, (a) write the equation of the line in slope-intercept form, and (b) graph the line.

21. The slope is $-\frac{1}{2}$ and the y-intercept is $(0,1)$.

a. $y = -\frac{1}{2}x + 1$ b.

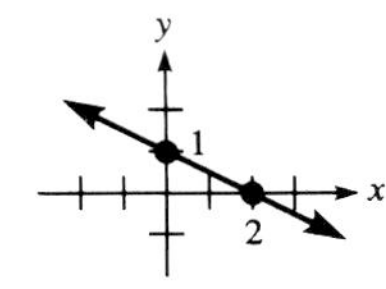

22. The slope is $\frac{3}{4}$ and the y-intercept is $(0,-3)$.

a. $y = \frac{3}{4}x - 3$ b.

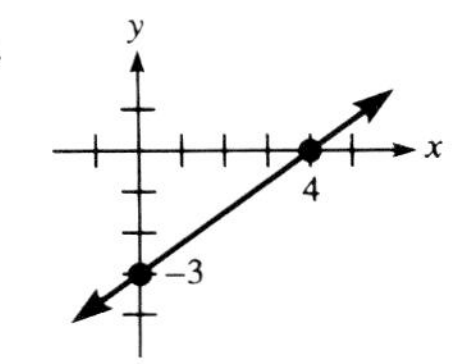

23. The slope is 6 and the y-intercept is $(0,\frac{1}{2})$.

a. $y = 6x + \frac{1}{2}$ b.

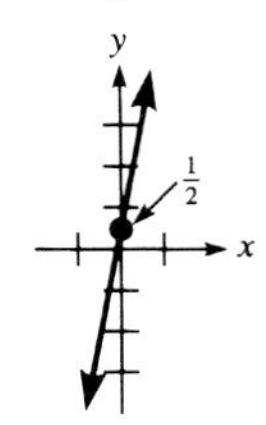

24. The slope is -2 and the y-intercept is $(0,-\frac{3}{5})$.

a. $y = -2x - \frac{3}{5}$ b.

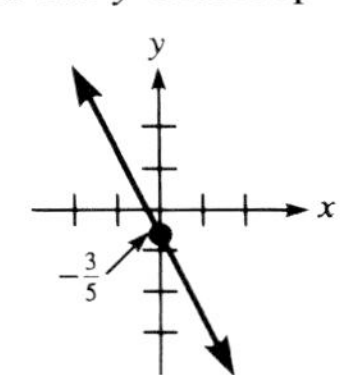

25. The slope is 0 and the y-intercept is $(0,-2)$.

a. $y = -2$ b.

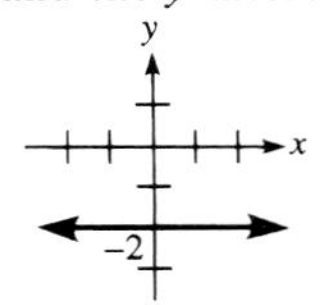

26. The slope is 0 and the y-intercept is $(0,\frac{1}{3})$.

a. $y = \frac{1}{3}$ b.

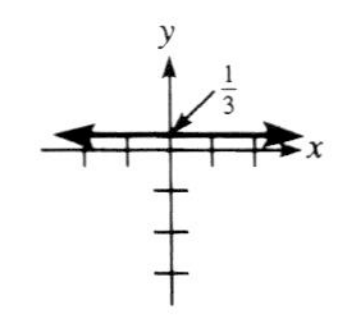

In Exercises 27–34, determine if the two lines are distinct parallel lines.

27. $3x + y = -2;\ 6x + 2y = 4$ Yes

28. $2x - y = 1;\ 6x - 3y = -3$ Yes

29. $x + y = 1;\ x + y = 2$ Yes

30. $x - y = 10;\ x - y = 0$ Yes

31. $2x + 3y = 4;\ 4x + 6y = 8$ Same line

32. $x - 2y = 1;\ 2x - 4y = 2$ Same line

33. $x = 1;\ x = 4$ Yes

34. $y = -3;\ y = 2$ Yes

In Exercises 35–46, find an equation for the line through the given point which satisfies the stated condition.

35. Through $(1,2)$ parallel to $x + 3y = 4$ $x + 3y = 7$

36. Through $(3,0)$ parallel to $2x - 3y = 1$ $2x - 3y = 6$

37. Through $(-4,-2)$ parallel to $y = \frac{1}{3}x - 2$ $y = \frac{1}{3}x - \frac{2}{3}$

38. Through $(0,-3)$ parallel to $y = -\frac{1}{2}x + 4$ $y = -\frac{1}{2}x - 3$

39. Through $(3,2)$ parallel to $x = 1$ $x = 3$

40. Through $(5,3)$ parallel to $y = 0$ $y = 3$

41. Through $(1,3)$ perpendicular to $3x + 2y = 5$ $2x - 3y = -7$

42. Through $(-3,-2)$ perpendicular to $4x - y = 1$ $x + 4y = -11$

43. Through the origin perpendicular to $x + y = 10$ $x - y = 0$

44. Through the origin perpendicular to $x - y = 2$ $x + y = 0$

45. Through $(1,3)$ perpendicular to $y = 2$ $x = 1$

46. Through $(-4,-2)$ perpendicular to $x = -3$ $y = -2$

47. The relationship between the number of heating degree-days in a month and the average number of kilowatt-hours of electricity billed by a utility company for that month is approximately linear and is well illustrated by the line shown in the graph.

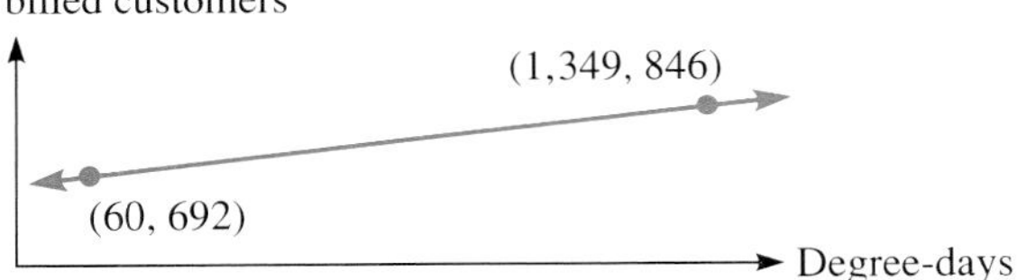

a. Write the equation of this line in the form $y = mx + b$. To maintain accuracy over such a large domain, compute the slope to five decimal places, and round the intercept to two decimal places. $y = 0.11947x + 684.83$

b. To the nearest integer, what kilowatt-hour billing corresponds to 800 degree-days? 780 kilowatt-hours

48. The relationship between shipping charge and weight for one company was given by a linear relationship, as shown in the graph.

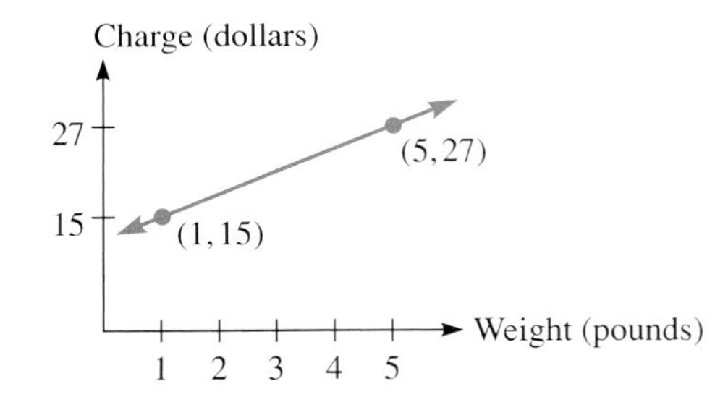

a. Write an equation for this line in the general form $Ax + By = C$. $-3x + y$

b. What charge corresponds to a weight of 3.5 lb? $22.50

49. The trend over the years 1971 to 1989 shows that the gradual increase in the percentage of women who are in the work force is approximately linear, with two of the points shown in the graph.

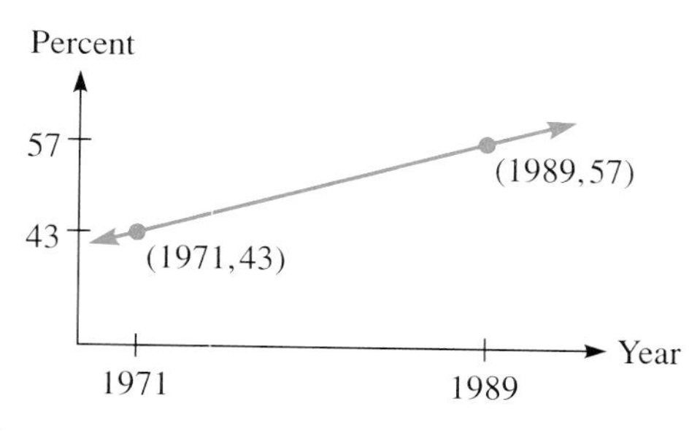

a. Find the equation of this line in the form $y = mx + b$. To maintain accuracy over such a large domain, find m to four decimal places and b to two decimal places.
$y = 0.7778x - 1{,}490.04$

b. If you assume that this pattern continues, what percentage is predicted for 1995? 61.67 percent

50. An "on-line" telephone information service charges a fixed connection charge plus an hourly rate. The relationship between time connected and total charge is linear. Two points are shown in the figure.

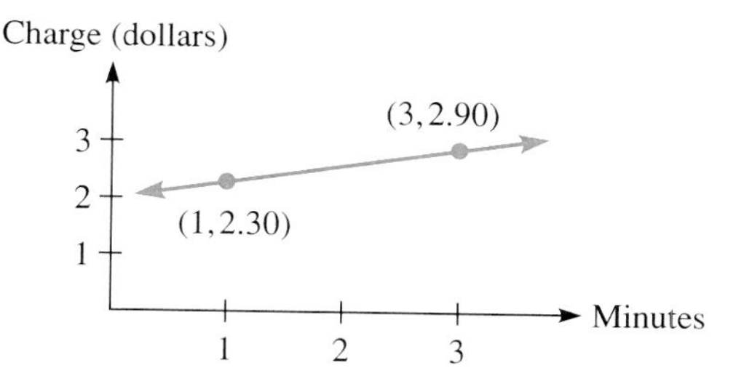

a. Find an equation for this relation in the general form $Ax + By = C$. $0.30x - y = -2$

b. What is the connection charge? \$2.00

c. What is the cost for a 1-minute 30-second call? \$2.45

d. What is the *hourly* rate? \$18 per hour

THINK ABOUT IT

1. a. Write the slope-intercept equation of a line with slope 3 and y-intercept $(0,2)$.

b. Alter the equation in part **a** so that the line has a hole in it when $x = 5$. (That is, make the value of y undefined when $x = 5$.)

2. a. Find an equation in general form of the line whose x-intercept is $(m,0)$ and whose y-intercept is $(0,m)$.

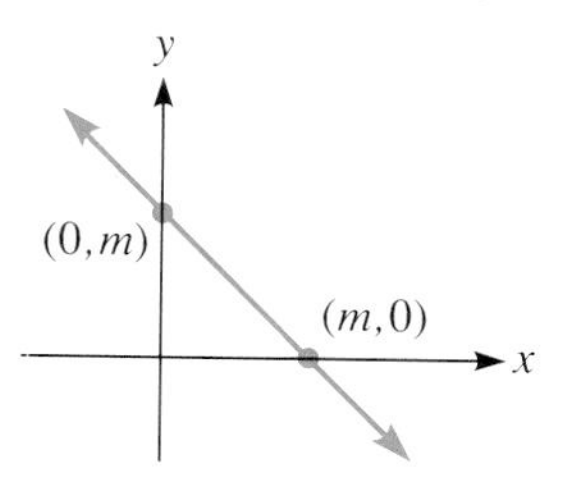

b. What is the slope of the line in part **a**?

3. Suppose that the slope of a line equals the value where it crosses the y-axis. Find the x-intercept.

4. a. Find the slope of $3x + y = 1$.

b. Find the slope of $3x + y = 2$.

c. Find the slope of $3x + y = c$, where c is some real constant.

d. Find the slope of $ax + by = c$.

e. An equation is given that fits the form $ax + by = c$. True or false? If you change just the constant c, you get the equation of a line parallel to the original one.

5. The shortest distance between two points is given by the length of the straight line between them. This fact may lead to a quick solution of certain problems which appear more complicated.

Here is such an example: Two places, A and B, near a river are shown in the figure. What point on the bank is such that the trip from A to the shore to B is a minimum? That would be a good docking place for a boat which carried supplies for both places.

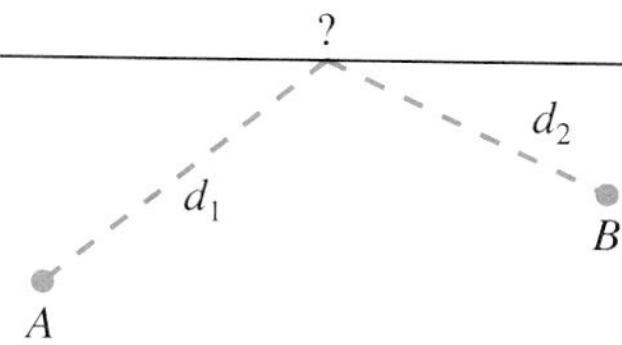

The simple solution is shown in the next sketch, where B' is a reflection of B on the other side of the river. The straight line from A to B' crosses the river at point P.

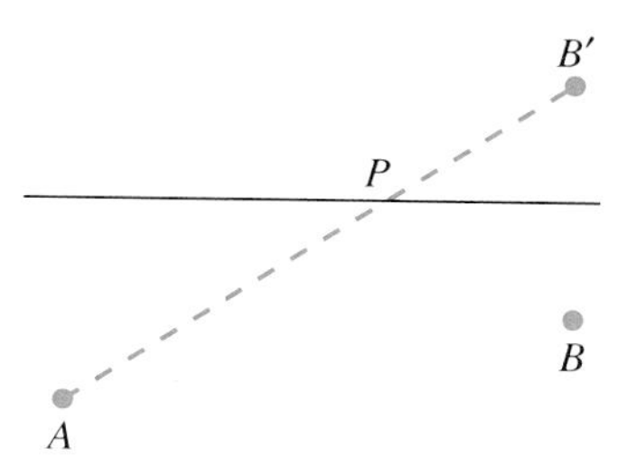

a. Explain why point P is the desired docking place.

b. If the distance from A to the river is 3 mi, the distance from B to the river is 1 mi, and the distance along the river as shown in the sketch is 10 mi, find the smallest possible total distance from A to the shore and then to B.

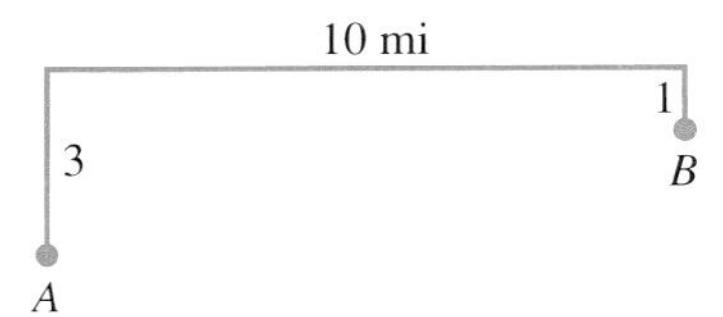

c. It follows from the next sketch that the two right triangles are similar, and that their sides are therefore proportional. Why?

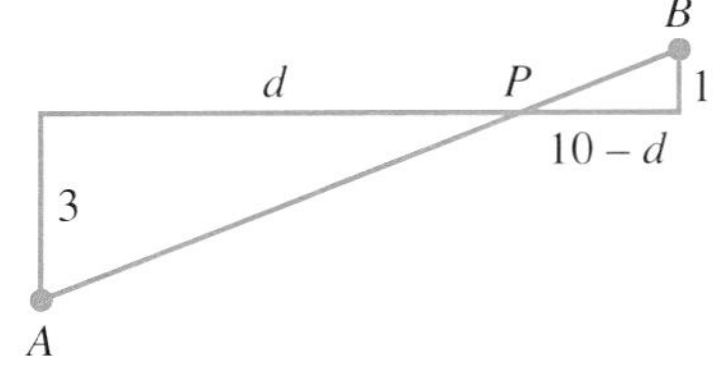

d. Set up and solve a proportion to determine how far along the shore point P is.

REMEMBER THIS

1. Which inequality represents the phrase "x is at least 40"?
   a. $x \le 40$ **b.** $x < 40$ **c.** $x > 40$ **d.** $x \ge 40$ d
2. What numerical inequality results when the ordered pair $(-1,0)$ is substituted into the algebraic inequality $y \le 2x + 4$? $0 \le 2$
3. True or false? The graph of $y = 2x + 4$ is a straight line. True
4. Is the statement $5 \ge 5$ true or false? True
5. Graph $x = -1$.

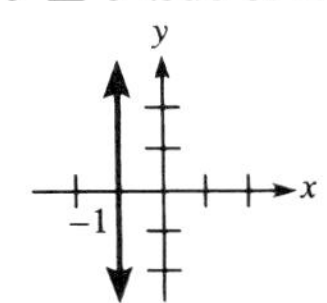

6. Find the distance between (0,0) and (1,1). $\sqrt{2}$
7. Use the remainder theorem to prove that $x - 1$ is a factor of $6x^5 - x^4 + 3x^3 - 5x^2 - x - 2$. $P(1) = 0$
8. Divide $3^{-1}x^{-2}y^{-3}$ by $3^{-3}x^{-1}y^{-2}$. $\frac{9}{xy}$
9. Solve $6x^3 - 14x^2 = 0$. $\{0, \frac{7}{3}\}$
10. The perimeter of a rectangle exceeds the area by 4, and the length is $\frac{4}{3}$ the width. Find the dimensions of the rectangle.
    Two solutions: width $= \frac{3}{2}$, length $= 2$; width $= 2$, length $= \frac{8}{3}$

5.4 Graphing Linear Inequalities

A car loan has a 10 percent annual interest rate. A loan for furniture has a 16 percent annual interest rate. A family can afford over the year at most $1,000 in interest payments for these two purchases. This restriction is given by $0.10x + 0.16y \le 1{,}000$ where x is the amount of the car loan and y is the amount of the furniture loan.

a. Graph this inequality for $x \ge 0$ and $y \ge 0$.
b. If the family makes a $4,000 car loan, what is the most they can borrow for furniture?
(See Example 5.)

OBJECTIVES

1. Determine if an ordered pair is a solution of an inequality.
2. Graph linear inequalities.
3. Graph linear inequalities restricted to $x \ge 0$, $y \ge 0$.

1 The problem that opens the section expresses a relationship between two variables as an algebraic *inequality,* to correspond to the phrase "at most." As with equations which involve two variables, a solution of an inequality in two variables is an ordered pair that makes the inequality a true statement.

EXAMPLE 1 Determine if $(-1,0)$ is a solution of $y \leq 2x + 4$.

Solution $(-1,0)$ means $x = -1, y = 0$. Then,

$y \leq 2x + 4$ Given inequality.
$0 \leq 2(-1) + 4$ Replace y by 0 and x by -1.
$0 \leq 2.$ Simplify.

Recall that $0 \leq 2$ is true if either $0 < 2$ or $0 = 2$ is true. Because $0 < 2$ is true, $(-1,0)$ is a solution of $y \leq 2x + 4$.

PROGRESS CHECK 1 Determine if $(0,-5)$ is a solution of $2x - y \leq 12$.

2 By analogy to equations, a **linear inequality** results if the equal sign in a linear equation is replaced by one of the inequality symbols. The graph of a linear inequality is related to the graph of a line. To see this, consider the inequality $y \leq 2x + 4$. The solution set of this inequality consists of all ordered pairs that satisfy

$$y < 2x + 4 \quad \text{or} \quad y = 2x + 4.$$

In Figure 5.32(a) we have drawn the line $y = 2x + 4$. Note that this line separates the plane into two regions. Each region consists of the set of points on one side of the line and is called a **half plane.** The solution set to $y < 2x + 4$ is given by the half plane *below* the line (because the relationship is *less* than), while the solution set to $y > 2x + 4$ is given by the half plane *above* the line. Thus, we graph $y \leq 2x + 4$ as shown in Figure 5.32(b), using shading to indicate that the half plane in the solution set is below the line.

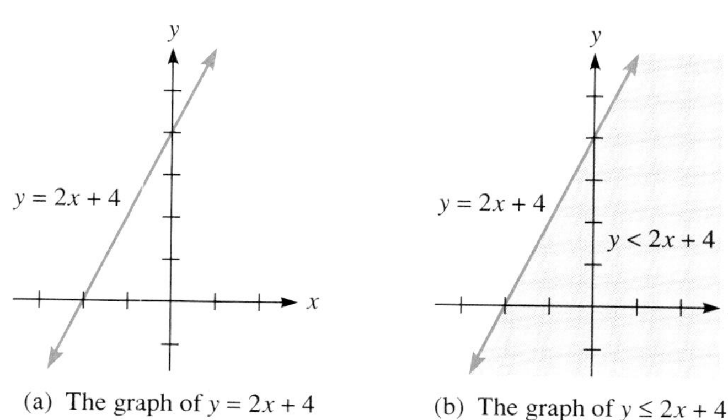

(a) The graph of $y = 2x + 4$ (b) The graph of $y \leq 2x + 4$

Figure 5.32

When the inequality symbol is either $<$ or $>$, then the points on the line are not part of the solution set, and this is indicated by drawing the line as a dashed line, as shown in Example 2.

EXAMPLE 2 Graph $y > 2x + 4$.

Solution From the previous discussion, the solution set is given by the half plane *above* $y = 2x + 4$. This graph is shown in Figure 5.33, where the line itself is drawn as a dashed line to show that the points on the line are not included in the solution set.

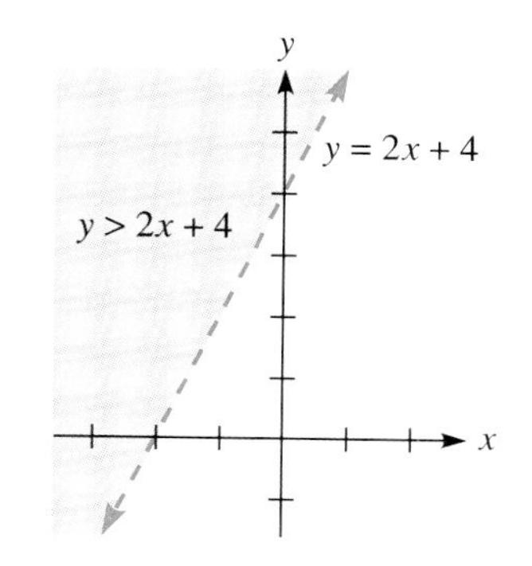

Figure 5.33

Progress Check Answer

1. Yes

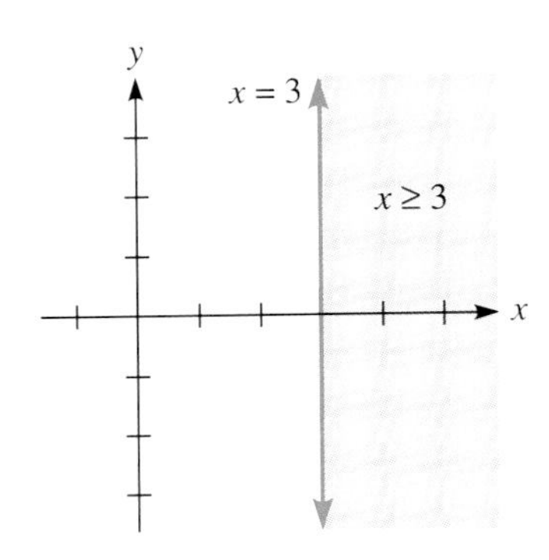

Figure 5.34

PROGRESS CHECK 2 Graph $y < -2x + 1$.

The next example deals with the case of a vertical line boundary for a solution set.

EXAMPLE 3 Graph $x \geq 3$.

Solution To see $x \geq 3$ as an inequality in two variables, note that $x \geq 3$ is equivalent to $x + 0y \geq 3$. Because the inequality symbol is $\geq$, the line itself is drawn solid. Then the half plane to the right of the line is shaded, because the x-coordinates of all points to the right of this line are greater than 3. The graph is shown in Figure 5.34.

PROGRESS CHECK 3 Graph $x < -1$.

When the given inequality is not written in the form where it is solved for y or x, it may not be obvious which half plane to use in the solution set. An easy way to decide in such cases is to pick some **test point** that is not on the line and substitute the coordinates of this point into the inequality. If the resulting statement is true, then shade the half plane containing the test point. If it is not true, shade the half plane on the other side of the line. The origin (0,0) is a convenient point to use in this test, provided that the origin is not on the line.

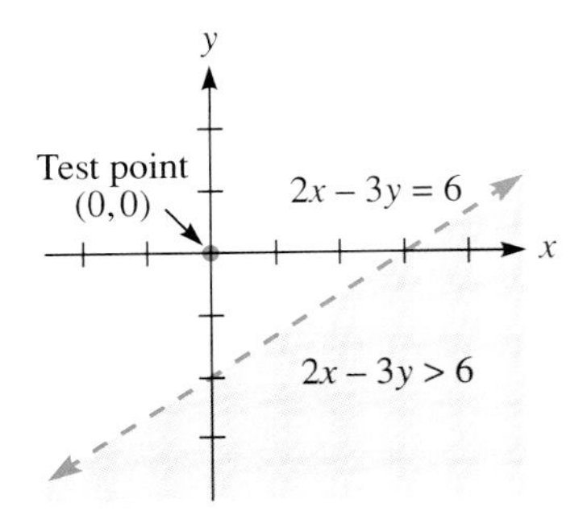

Figure 5.35

EXAMPLE 4 Graph $2x - 3y > 6$.

Solution First, graph $2x - 3y = 6$ as a dashed line, because of the $>$ sign. This line can be drawn quickly by plotting its intercepts, $(0,-2)$ and $(3,0)$. Then using the origin as a test point, substitute 0 for both x and y in the given inequality.

$$
\begin{aligned}
2x - 3y &> 6 && \text{Given inequality.}\\
2(0) - 3(0) &> 6 && \text{Replace } x \text{ by 0 and } y \text{ by 0.}\\
0 &> 6 && \text{Simplify.}
\end{aligned}
$$

Because $0 > 6$ is a false statement, (0,0) is not a solution, so shade the half plane not containing the origin. Figure 5.35 shows the graph of $2x - 3y > 6$.

PROGRESS CHECK 4 Graph $x - 2y < 4$.

3 It is quite common for applied problems which involve inequalities in x and y to have meaning only for nonnegative numbers. In such cases the graph should be restricted to quadrant 1 and its boundaries.

EXAMPLE 5 Solve the problem in the section introduction on page 246.

Solution

a. We need to graph $0.10x + 0.16y \leq 1{,}000$, where $x \geq 0$ and $y \geq 0$, because the amount of a loan cannot be negative. First, graph $0.10x + 0.16y = 1{,}000$ for $x \geq 0$ and $y \geq 0$ by determining that the intercepts are (10,000, 0) and (0, 6,250), and then drawing a solid line segment with these intercepts as endpoints. Then to see what region to shade, choose (0,0) as a test point.

$$
\begin{aligned}
0.10x + 0.16y &\leq 1{,}000 && \text{Given inequality.}\\
0.10(0) + 0.16(0) &\leq 1{,}000 && \text{Replace } x \text{ by 0 and } y \text{ by 0.}\\
0 &\leq 1{,}000 && \text{Simplify.}
\end{aligned}
$$

Because $0 \leq 1{,}000$ is true, we shade the points on the same side of the line segment as the origin and use the nonnegative x- and y-axes as boundary lines, as shown in Figure 5.36.

Progress Check Answers

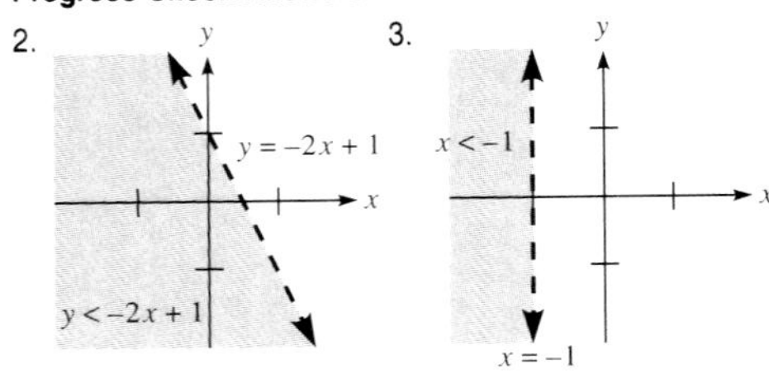

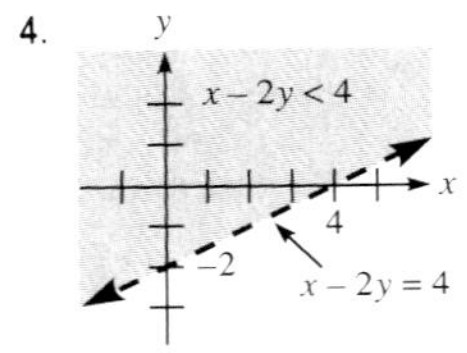

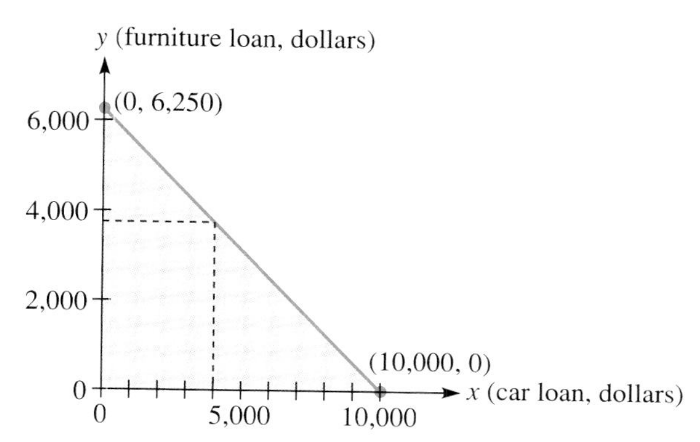

Figure 5.36

b. Since the car loan is given as $4,000, let $x = 4{,}000$ and solve for y.

$$0.10x + 0.16y \leq 1{,}000 \quad \text{Given inequality.}$$
$$0.10(4{,}000) + 0.16y \leq 1{,}000 \quad \text{Replace } x \text{ by 4,000.}$$
$$400 + 0.16y \leq 1{,}000 \quad \text{Simplify.}$$
$$0.16y \leq 600 \quad \text{Subtract 400 from both sides.}$$
$$y \leq 3{,}750 \quad \text{Divide both sides by 0.16}$$

The furniture loan can be no more than $3,750.

PROGRESS CHECK 5 A costume designer needs to buy some decorative ribbon. Gold ribbon costs $4 per foot; silver costs $3 per foot. The total amount spent on ribbon cannot be more than $240. If x and y represent the number of feet of gold and silver ribbon purchased, respectively, then $4x + 3y \leq 240$ expresses the expense restriction.

a. Graph this inequality for $x \geq 0$ and $y \geq 0$.

b. If 21 ft of gold ribbon are purchased, what is the longest piece of silver ribbon that can be afforded?

In summary, we have used the following procedures to graph linear inequalities.

To Graph a Linear Inequality

1. Graph the linear equation that results when the inequality symbol is replaced by =. Draw this boundary line as a solid line if the inequality symbol is $\leq$ or $\geq$, and draw a dashed line for $<$ or $>$.
2. Shade the half plane on one side of the boundary line as follows.

Case 1: The inequality is solved for y or x.

Inequality begins	Shade the half plane
$y <$	Below the line
$y >$	Above the line
$x <$	Left of the line
$x >$	Right of the line

Case 2: The inequality is not solved for y or x. Choose a test point that is not on the boundary line, and substitute the coordinates of this point into the inequality.

Resulting inequality is	Shade the half plane
True	Containing the test point
False	Not containing the test point

Progress Check Answers

5. (a)

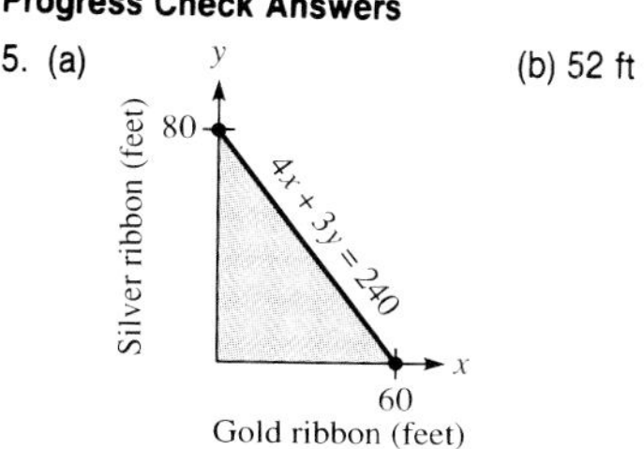

(b) 52 ft

EXERCISES 5.4

In Exercises 1–6, determine if the given ordered pair is a solution of

1. Is $(-2,-1)$ a solution of $y \leq 3x + 1$? No
2. Is $(-1,-2)$ a solution of $y \leq 3x + 1$? Yes
3. Is $(0,0)$ a solution of $x - y < 2$? Yes
4. Is $(0,0)$ a solution of $x - y \geq 2$? No
5. Is $(-2,-8)$ a solution of $y \geq 2x - 5$? Yes
6. Is $(-6,-9)$ a solution of $y \geq \frac{2}{3}x - 5$? Yes

In Exercises 7–24, graph the given inequality.

7. Graph $y > 3x - 2$.

8. Graph $y > -2x + 3$.

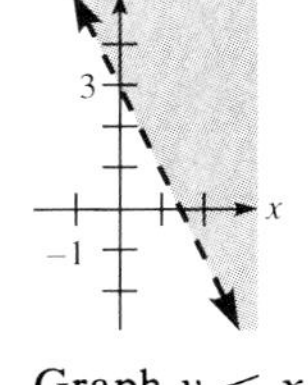

9. Graph $y < -x - 2$.

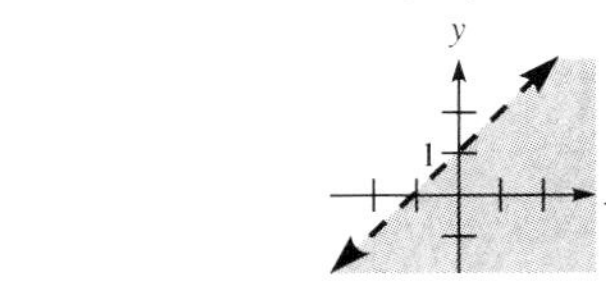

10. Graph $y < x + 1$.

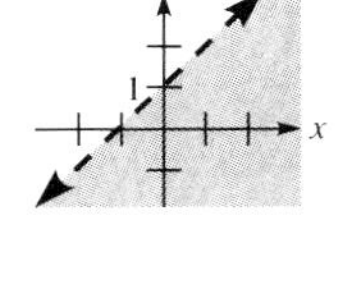

11. Graph $y \leq \frac{1}{2}x - 1$.

12. Graph $y \leq -\frac{2}{3}x + 3$.

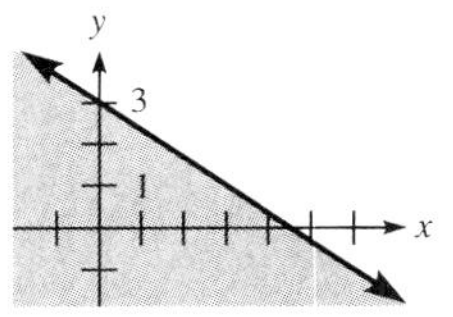

13. Graph $y \geq 5x - 5$.

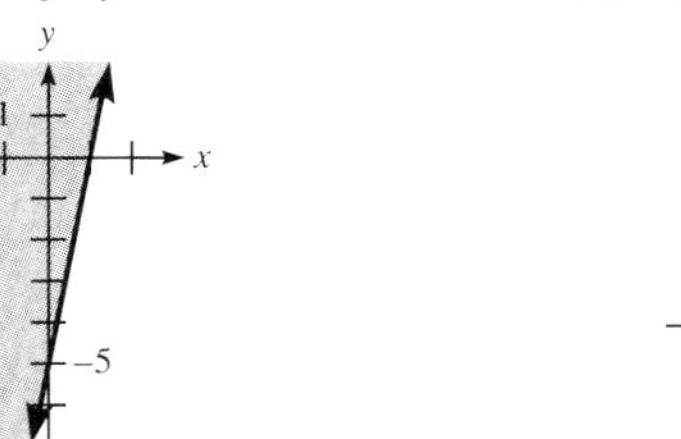

14. Graph $y \geq -6x + 3$.

15. Graph $y > 1$.

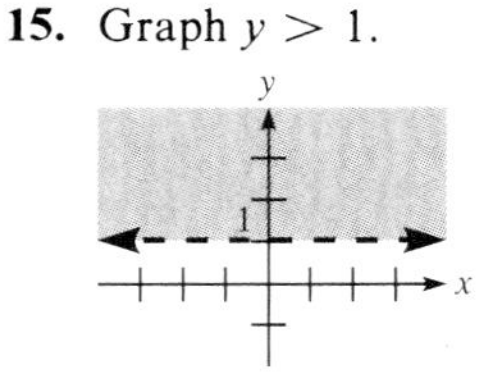

16. Graph $y < -\frac{1}{2}$.

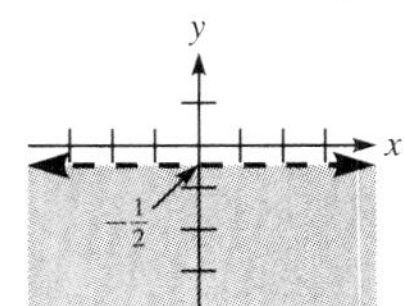

17. Graph $x \leq 2$.

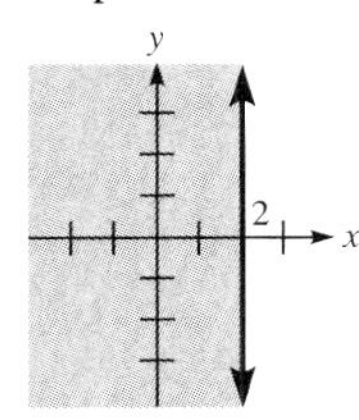

18. Graph $x \geq 0$.

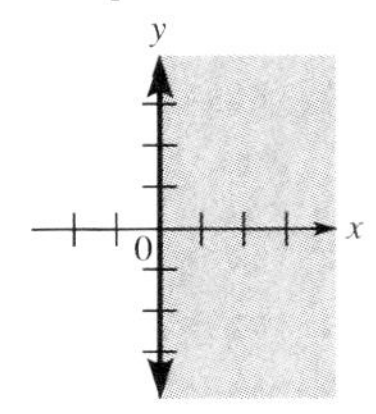

19. Graph $2x - 3y < 0$.

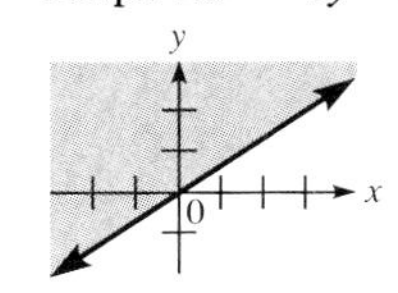

20. Graph $5x - 3y > 0$.

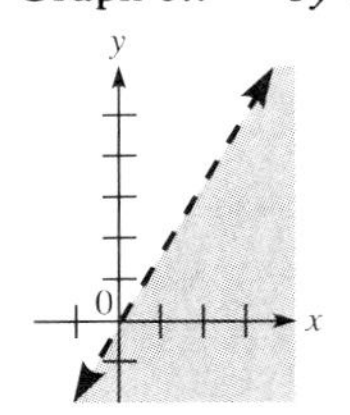

21. Graph $x + 4y \geq 8$.

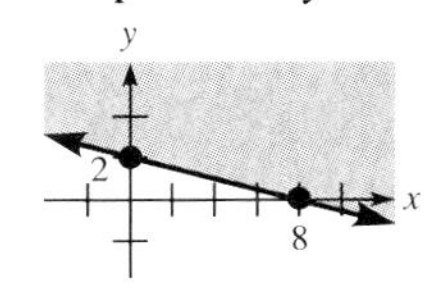

22. Graph $4x - 3y \leq 12$.

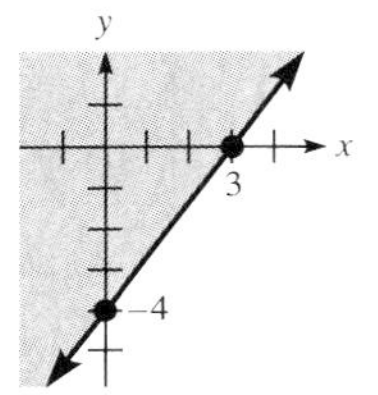

23. Graph $2x - 5y < 10$.

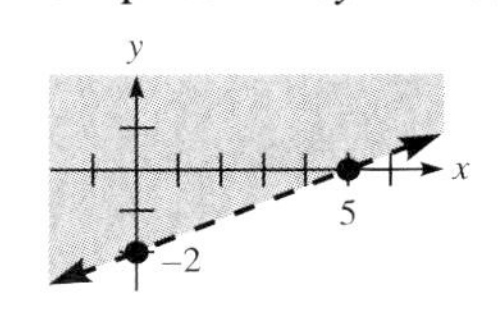

24. Graph $5x + y < 5$.

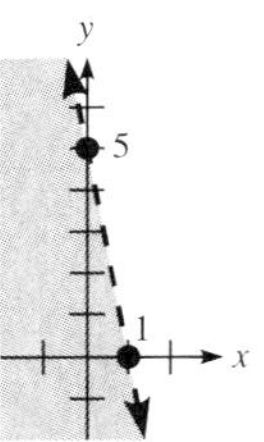

25. A fast-food outlet makes a 10 percent profit on food and a 50 percent profit on drink. For the total profit to be at least \$100 per day, the inequality $0.10x + 0.50y \geq 100$ must be true, where x is the daily sales on food and y is the daily sales on drink.
   a. Graph the inequality.

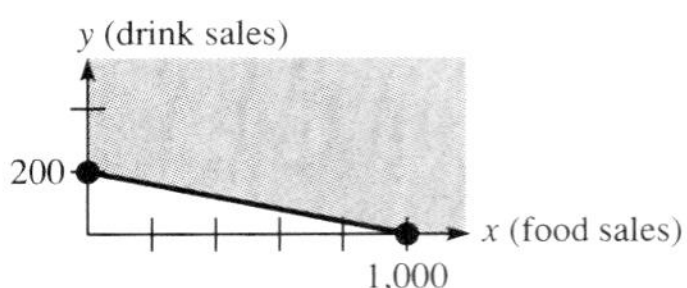

   b. Interpret the intercepts. y-intercept, (0,200): Selling \$200 worth of drinks and no food makes a profit of \$100. x-intercept, (1,000, 0): Selling \$1,000 worth of food and no drinks makes a profit of \$100.
   c. If food sales are \$400, what is the minimum amount they must make in drink sales to achieve the desired daily profit? \$120

26. Redo Exercise 25 for a profit on food of 20 percent and a profit on drink of 40 percent.

a.

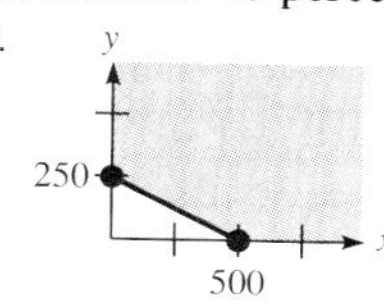

b. y-intercept, (0, 250): Selling \$250 worth of drinks and no food makes a profit of \$100.
x-intercept, (500, 0): Selling \$500 worth of food and no drinks makes a profit of \$100.

c. \$50

27. A couple has \$200,000 to spend on building a new house in the country. If land costs \$2,000 per acre and building costs are \$80 per square foot, then the inequality $2{,}000x + 80y \leq 200{,}000$ describes the restrictions on their purchase.

a. What do x and y each stand for in the inequality?
x is the number of acres of land; y is the number of square feet in the house.

b. Graph the inequality.

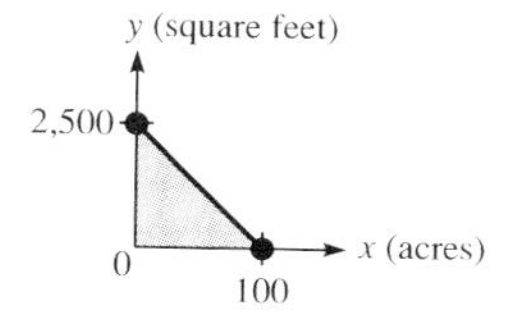

c. If they decide that their house must be at least 1,800 ft², what amount of land can they buy? 28 acres at most

d. If they decide that they want to own at least 10 acres of land, what size house can they afford? 2,250 ft² at most

28. Redo Exercise 27, but assume that land costs \$1,500 per acre and building costs are \$100 per square foot.

a. The equation is $1{,}500x + 100y \leq 200{,}000$. x is the number of acres; y is the number of square feet in the house.

b.

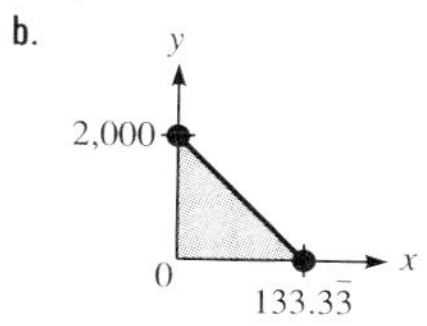

c. 13.333 acres at most

d. 1,850 ft² at most

THINK ABOUT IT

1. If two lines intersect, then the point of intersection is on both lines.

a. Show that the lines $y = 2x + 1$ and $y = 4x - 1$ intersect at (1,3) by showing that (1,3) satisfies both equations.

b. The x value of the point of intersection can be found algebraically by equating the two expressions for y from part **a.** Show that the solution to $2x + 1 = 4x - 1$ is $x = 1$.

c. The y value of the point of intersection can be found by substituting the solution for x in either equation. Show that replacing x by 1 in either equation yields $y = 3$.

d. By algebra, find the point of intersection of $y = 3x + 5$ and $y = -x + 1$. Sketch both lines to confirm your result.

2. The region where the solution sets of *two* linear inequalities *overlap* is called their **intersection.** It consists of all points which satisfy *both* inequalities. Graph the intersection of $2x - 4y \geq 8$ and $x \leq 2$.

3. Show that the intersection of $y \geq 0$, $x \geq 2y$, and $x + y \leq 5$ is made up of a triangle and its interior. What are the vertices of this triangle?

4. The idea of graphing a line in two dimensions extends to graphing a *plane* in three dimensions. Instead of two axes for x and y, there are *three* axes for x, y, and z and three intercepts.

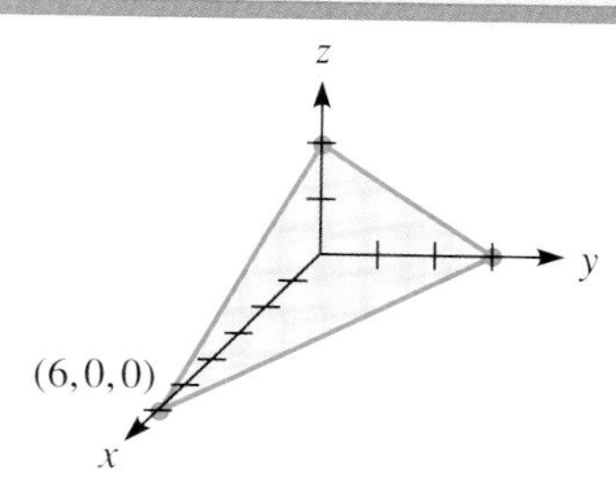

A part of the graph of $x + 2y + 3z = 6$ is shown in the figure. It is a tilted triangular region which contains the solutions of $x + 2y + 3z = 6$ for which x, y, and z are not negative.

a. The coordinates of the x-intercept are (6,0,0). What are the coordinates of the y-intercept? Of the z-intercept?

b. Make a sketch of the plane $2x + 3y + 4z = 12$, and label the three intercepts.

5. Referring to the graph in part **a** of the previous exercise, we have written a related *inequality:* $0 \leq x + 2y + 3z \leq 6$.

a. Which of these points are in the solution set? (0,0,0), (1,1,1), $(1,\frac{1}{2},\frac{1}{3})$, (3,2,1), $(-1-1,-1)$.

b. What shape in space is made by the complete solution set of the inequality in part **a?**

REMEMBER THIS

1. If $y = kx$, where k is some positive number, what happens to y as x increases? y increases also.
2. Solve $y = k/x$ for x. $x = k/y$
3. Solve $F = \dfrac{km_1m_2}{d^2}$ for k. $k = \dfrac{Fd^2}{m_1m_2}$
4. Find an equation for a line through the origin with slope 3. $y = 3x$
5. What is the slope of a line which is perpendicular to $y = 2x + 5$? Slope $= -\frac{1}{2}$
6. Graph the line $x + y = 1$.

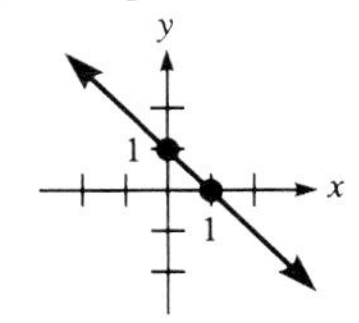

7. Solve $\dfrac{x+3}{x-1} + \dfrac{x+1}{x-2} = 4$. $\{\frac{3}{2}, 5\}$
8. Write 459 trillion in scientific notation. 4.59×10^{14}
9. Solve $a = \dfrac{b+c}{b}$ for b. $b = \dfrac{c}{a-1}$
10. For a special sale a store cuts its appliance prices by 15 percent. A washing machine goes on sale for \$399. To the nearest dollar, what is its original price? \$469

5.5 Variation

The time T it takes to cook a turkey varies directly with its weight w. If it takes 3 hours and 36 minutes to cook a 12-lb turkey, how long will it take to cook a 20-lb turkey? (See Example 3.)

OBJECTIVES

Solve problems involving the following.

1. **Direct variation**
2. **Inverse variation**
3. **Variation of powers of variables**
4. **Variation among more than two variables.**

1 In many applications of algebra the relationship between variables is stated in the language of variation. The simplest type of variation occurs when one variable is expressed as the product of a constant and another variable. This is called **direct variation.**

> **y Varies Directly as x**
>
> The statement "y varies directly as x" means that
>
> $$y = kx,$$
>
> where k is a constant called the **variation constant.**

In some applications, the relationship $y = kx$ is also described by saying that y is **proportional** to x, and that k is the constant of proportionality.

EXAMPLE 1 Write a variation equation for the given statement, and identify the variation constant if it is known.

a. The circumference C of a circle varies directly as its diameter d.
b. The property tax T on a home varies directly as the assessed value v of the home.

Solution

a. C varies directly as d means that $C = kd$. Because the formula for the circumference of a circle is $C = \pi d$, the variation constant k has the fixed value π.
b. T varies directly as v means that $T = kv$. The variation constant k depends on the property tax rate, which is fixed for any particular tax district.

PROGRESS CHECK 1 Write a variation equation for the given statement, and identify the variation constant if it is known.

a. If a car travels at a constant speed of 45 mi/hour, then the distance d traveled varies directly as the time t.
b. The sales tax T on a purchase varies directly as the price p of the item.

Direct variation between two variables can also be described graphically. The equation $y = kx$ which expresses direct variation between x and y graphs as a straight line through the origin with slope k. The graph is shown in Figure 5.37 for k positive.

The value of k can be determined if one pair of values [besides (0,0)] for the variables is known. Then that value of k may be used to find other corresponding values of the variables.

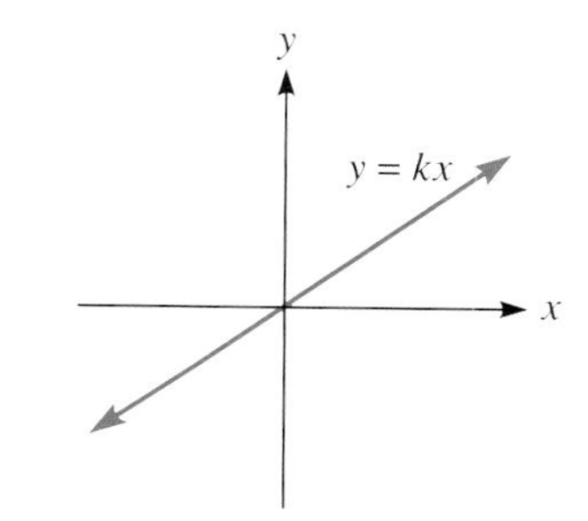

Figure 5.37

EXAMPLE 2 y varies directly as x, and $y = 4$ when $x = 5$. Find y when $x = 15$.

Solution Since y varies directly as x, we have

$$y = kx.$$

To find k, replace y by 4 and x by 5.

$$4 = k(5)$$
$$\tfrac{4}{5} = k$$

Therefore, the variation equation is $y = \frac{4}{5}x$. Then when $x = 15$,

$$y = \tfrac{4}{5}(15) = 12.$$

Thus, $y = 12$ when $x = 15$.

PROGRESS CHECK 2 y varies directly as x, and $y = 2$ when $x = 3$. Find y when $x = 27$.

EXAMPLE 3 Solve the problem in the section introduction on page 252.

Solution Since cooking time T varies directly as weight w,

$$T = kw.$$

To find k, replace T by 3 hours and 36 minutes (216 minutes) and w by 12 lb.

$$216 = k(12)$$
$$\tfrac{216}{12} = k, \quad \text{or} \quad k = 18.$$

Thus, $T = 18w$, which means the cooking time is 18 minutes per pound of turkey.

Next, replace w by 20 to find the cooking time for a 20-lb turkey.

$$T = 18(20) = 360$$

The cooking time is 360 minutes, or 6 hours.

Progress Check Answers
1. (a) $d = kt$; $k = 45$ (b) $T = kp$
2. 18

PROGRESS CHECK 3 The weight of an object on the moon varies directly as the weight of the object on earth. An object that weighs 102 lb on earth weighs 17 lb on the moon. How much will a person who weighs 132 lb on earth weigh on the moon?

2 In some relationships one variable decreases as another increases. If this happens in such a way that the product of the two variables is constant, then we say that the variables **vary inversely,** or that one is *inversely proportional* to the other.

y Varies Inversely as x

The statement "y varies inversely as x" means $xy = k$, or

$$y = \frac{k}{x},$$

where k is a constant called the **variation constant.**

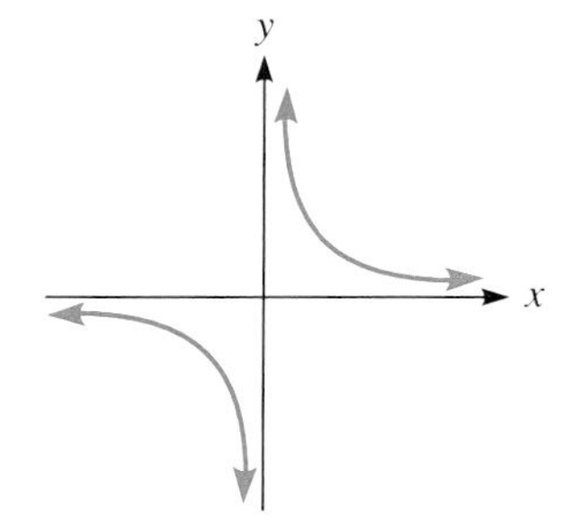

Figure 5.38

Figure 5.38 shows the graph of the variation equation $y = k/x$ for $k > 0$. (Note that the equation has no meaning for x or y equal to zero.) This curve, which is called a **hyperbola,** is discussed more fully in Section 8.7, but at this point we can use the graph to get a visual impression of the concept of inverse variation.

EXAMPLE 4 y varies inversely as x, and $y = 3$ when $x = 8$. Find y when $x = 10$.

Solution Since y varies inversely as x, we have

$$y = \frac{k}{x}.$$

To find k, replace y by 3 and x by 8, and solve for k.

$$3 = \frac{k}{8}$$

$$24 = k$$

Thus, $y = 24/x$. When $x = 10$, $y = 24/10 = 2.4$.

PROGRESS CHECK 4 y varies inversely as x, and $y = 2$ when $x = 9$. Find y when $x = 5$.

EXAMPLE 5 For anything which travels at a constant rate r, the time t required to travel a *fixed distance* d varies inversely as r. For instance, sound travels through water at a constant rate of about 4,800 ft/second, while it travels through air at about 1,100 ft/second. If it takes 2 seconds for sound to go from A to B in water, about how long will it take sound to go from A to B in air?

Solution Since t varies inversely as r,

$$t = \frac{k}{r}.$$

Perhaps point out that k just represents d. Some students may prefer to use d as the constant.

To find k, replace t by 2 and r by 4,800 (these are the given values for sound in water), and solve for k.

$$2 = \frac{k}{4{,}800}$$

Progress Check Answers

3. 22 lb 4. 3.6

This means that the distance from A to B is 9,600 ft. Thus, $t = \dfrac{9{,}600}{r}$.

For air, when $r = 1{,}100$,

$$t = \frac{9{,}600}{1{,}100} = 8.73.$$

It takes about 8.7 seconds for sound to travel from A to B.

PROGRESS CHECK 5 Sound travels through air at about 1,100 ft/second and through granite rock at about 20,000 ft/second. If it takes 0.1 second for sound to travel from A to B through granite, about how long will it take sound to travel from A to B through air?

3 The concept of variation extends to include direct and inverse variation of variables raised to specific powers.

EXAMPLE 6 y varies inversely as the cube of x, and $y = 64$ when $x = \frac{1}{2}$. Find y when $x = 1$.

Solution Since y varies inversely as x^3,

$$y = \frac{k}{x^3}.$$

Now, find k as follows.

$$64 = \frac{k}{(\frac{1}{2})^3} \qquad \text{Replace } y \text{ by 64 and } x \text{ by } \tfrac{1}{2}.$$

$$64 = \frac{k}{\frac{1}{8}} \qquad \text{Evaluate a power.}$$

$$8 = k \qquad \text{Multiply both sides by } \tfrac{1}{8}.$$

Thus, $y = 8/x^3$. When $x = 1$,

$$y = \frac{8}{1^3} = \frac{8}{1} = 8.$$

PROGRESS CHECK 6 y varies inversely as the square of x, and $y = 36$ when $x = \frac{1}{3}$. Find y when $x = 2$.

EXAMPLE 7 The weight (w) of a solid cube of uniform density varies directly as the cube of the side (s).

a. When the side is doubled, how many *times* heavier does the cube become?

b. A cube of ice that is 1 ft per side weighs about 62 lb. Can an average adult lift an ice cube that is 2 ft per side? How much does such a cube weigh?

Many students find this result surprising; the problem creates good classroom discussion. Ask them what happens to the area of a rectangle when the sides are doubled.

Solution

a. Since w varies directly as the cube of s,

$$w = ks^3.$$

Doubling the side means replacing s by $2s$.

$$w = k(2s)^3 = 8ks^3$$

Because $8ks^3$ is 8 times greater than ks^3, doubling the side increases the weight to 8 times the original weight.

b. Changing the side of the block of ice from 1 to 2 ft is doubling it. Therefore, from part **a,** the new weight is 8 times the old weight, so the new weight is $8 \cdot 62$, or 496 lb, which is far beyond the lifting capabilities of an average adult.

PROGRESS CHECK 7 **a.** What happens to the weight of a solid cube when the side is tripled? **b.** What is the weight of a cube of ice that is 3 ft per side?

Progress Check Answers

5. 1.8 seconds 6. $y = 1$

7. (a) Weight increases 27 times. (b) 1,674 lb

4 Another extension of direct and inverse variation between two variables is used to describe variation among *more than two* variables.

Note that both Example 8 and "Progress Check" Exercise 8 are illustrations of inverse square laws, which appear frequently in physics.

EXAMPLE 8 An important scientific law which is expressed in terms of variation is Newton's law of gravitation, which states that the gravitational force F between any two objects in the universe varies directly as the product of their masses (m_1 and m_2) and inversely as the square of the distance d between them. Write this relationship algebraically.

Solution Since F varies directly as the product m_1m_2, the equation must contain km_1m_2; and since F varies inversely as the square of the distance, the equation must contain k/d^2. These are combined to give

$$F = \frac{km_1m_2}{d^2}.$$

Note If y varies directly as the *product* of other variables, say x and w, we write $y = kxw$, and we say that y **varies jointly** as x and w. In this example an alternative way to say that F varies directly as the product m_1m_2 is to say that F varies jointly as m_1 and m_2.

PROGRESS CHECK 8 The resistance R of a wire to an electrical current varies directly as its length ℓ and inversely as the square of its diameter d. Write this relationship algebraically.

EXAMPLE 9 From the formula in Example 8, describe what happens to the gravitational attraction between two objects if both masses are cut in half and the distance between them is doubled.

Solution Replace m_1 by $\frac{1}{2}m_1$, m_2 by $\frac{1}{2}m_2$, and d by $2d$.

$$F = \frac{k(\frac{1}{2}m_1)(\frac{1}{2}m_2)}{(2d)^2} = \frac{\frac{1}{4}km_1m_2}{4d^2} = \frac{1}{16}\left(\frac{km_1m_2}{d^2}\right)$$

The resulting force is $\frac{1}{16}$ of the original.

PROGRESS CHECK 9 Using the formula from "Progress Check" Exercise 8, what happens to the resistance in a wire if the length is doubled and the diameter is cut in half?

Progress Check Answers

8. $R = \frac{k\ell}{d^2}$ 9. Resistance increases 8 times.

EXERCISES 5.5

In Exercises 1–18, write a variation equation for the given statement, and identify the variation constant if it is known.

1. The perimeter P of a square varies directly as the length of its side s. $P = ks$; $k = 4$
2. The perimeter P of a square varies directly as the length of the diagonal d. $P = kd$; $k = 4/\sqrt{2}$ or $2\sqrt{2}$
3. For a vehicle traveling 60 mi/hour, the distance d covered varies directly as the elapsed time t. $d = kt$; $k = 60$
4. The tax t on gasoline varies directly as the number of gallons g purchased. $t = kg$
5. The price p of carpeting varies directly as the area A to be covered. $p = kA$
6. The energy E required to heat a building varies directly as the volume V of the building. $E = kV$
7. In a manufacturing process the cost per unit c varies inversely with the number n of units produced. $c = k/n$
8. In a distribution of a fixed amount of money, the amount A per person varies inversely as the number n of people. $A = k/n$
9. The time t to complete a construction job varies inversely with the number n of people hired. $t = k/n$
10. The pressure p exerted by a weight on a floor varies inversely with the area A of contact. $p = k/A$
11. The delay d at a toll bridge varies inversely with the traffic volume v. $d = k/v$
12. For a fixed quantity of goods, the supply s varies inversely with the demand d. $s = k/d$
13. y varies directly as x^2 and inversely as z. $y = kx^2/z$
14. y varies directly as x and inversely as z^2. $y = kx/z^2$
15. f varies directly as the product of a and b and inversely as c^2. $f = kab/c^2$
16. f varies directly as c^3 and inversely as the product of a and b. $f = kc^3/(ab)$

17. g varies jointly as x and y and inversely as the square root of z. $g = kxy/\sqrt{z}$

18. g varies jointly as f_1 and f_2 and inversely as the square of d. $g = kf_1f_2/d^2$

In Exercises 19–34, solve for the specified variable.

19. y varies directly as x, and $y = 5$ when $x = 3$. Find y when $x = 4$. $\frac{20}{3}$

20. y varies directly as x, and $y = 2$ when $x = 7$. Find y when $x = 5$. $\frac{10}{7}$

21. u varies directly as v, and $u = -2$ when $v = 4$. Find u when $v = -2$. 1

22. u varies directly as v, and $u = 1$ when $v = -5$. Find u when $v = -2$. $\frac{2}{5}$

23. s is proportional to t, and $s = \frac{1}{2}$ when $t = 4$. Find s when $t = 2$. $\frac{1}{4}$

24. s is proportional to t, and $s = \frac{1}{3}$ when $t = \frac{1}{9}$. Find s when $t = -6$. -18

25. y varies inversely as x, and $y = 5$ when $x = 3$. Find y when $x = 4$. $\frac{15}{4}$

26. y varies inversely as x, and $y = 2$ when $x = 7$. Find y when $x = 5$. $\frac{14}{5}$

27. u varies inversely as v, and $u = -3$ when $v = 9$. Find u when $v = -9$. 3

28. u varies inversely as v, and $u = 1$ when $v = -3$. Find u when $v = 2$. $-\frac{3}{2}$

29. s varies inversely as t, and $s = \frac{1}{2}$ when $t = \frac{1}{4}$. Find s when $t = -8$. $-\frac{1}{64}$

30. s varies inversely as t, and $s = \frac{1}{3}$ when $t = 9$. Find s when $t = -6$. $-\frac{1}{2}$

31. y varies directly as x^2 and inversely as z; $y = 4$ when $x = 4$ and $z = 2$. Find y when $x = 2$ and $z = 4$. $\frac{1}{2}$

32. y varies directly as x and inversely as z^2; $y = \frac{1}{2}$ when $x = 8$ and $z = 2$. Find y when $x = 4$ and $z = -1$. 1

33. V varies jointly as ℓ, w, and h and inversely as n; $V = 5$ when $\ell = 2$, $w = 3$, $h = 4$, and $n = 6$. Find V when $\ell = 4$, $w = 3$, $h = 1$, and $n = 5$. 3

34. W varies jointly as A and h and inversely as d^2; $W = 1$ when $A = 9$, $h = 2$, and $d = 6$. Find W when $A = 9$, $h = 2$, and $d = 2$. 9

35. The cost of tiling the surface of a metal cube varies directly as the square of the length of the side.
    a. If it costs \$1,800 to tile a cube with 10-ft sides, what is the cost for tiling a cube with 5-ft sides? \$450
    b. What size cube can be tiled for \$162? Side = 3 ft

36. The cost of shipping goods by sea varies directly as the volume of the container. If the charge for 1,000 ft^3 is \$280, what is the charge for a container that is 3 ft by 4 ft by 5 ft? \$16.80

37. The interest earned on an investment at simple interest varies directly as the length of time the money is invested. If such an investment earns \$200 in 3 years, what will it earn in 10 years? \$666.67

38. The weekly earnings of a part-time employee vary directly as the number of hours worked. For an employee who makes \$111.60 for 18 hours of work, what are the earnings for 20 hours of work? \$124

39. Galileo, a sixteenth-century Italian scientist, did important experiments in motion. He rolled a ball down an inclined track and measured how far it traveled in equal units of time. From this he learned that the distance covered is directly proportional to the square of the elapsed time. If a ball rolling down a track covers 16 in. in 1 second, how far will it roll in 2 seconds? 64 in.

40. According to statistical theory, the margin of error in a survey to estimate the percentage of the population who support the president is inversely proportional to the square root of the number of people in the survey. Such a survey of 900 people has a 3 percent margin of error. How many people are needed to get a 2 percent margin of error? 2,025

41. The energy of motion in physics is called kinetic energy. The kinetic energy e of a moving object is directly proportional to the square of its velocity v. If the velocity of an automobile increases from 10 to 60 mi/hour, what effect does that have on the kinetic energy (which is an indicator of the destructive force of a crash)? (*Hint:* Replace v by $6v$, since 60 is 6 times 10.) Kinetic energy becomes 36 times as great.

42. In a storm the wind pressure p (which measures the destructive capacity of the storm) varies directly as the square of the wind velocity v. If the wind doubles in velocity, what happens to the destructive power of the storm? (*Hint:* The original value of v becomes $2v$.) Destructive power increases 4 times.

43. When a star collapses, the mass in its core is compressed to an extremely dense star. If it is dense enough, a black hole results. Both the size (volume) and the density of the resulting black hole are determined by the mass of the collapsing star. For instance, the density of the black hole is inversely proportional to the square of the mass of the star. A mass equal to 3 suns would be compressed to have a density of 2.1×10^{18} kilograms per cubic meter (kg/m^3). What density would result from a mass equal to 4 suns? 1.18×10^{18} kg/m^3

44. Kepler's third law about the orbits of the planets states that the square of the period p of the orbit is proportional to the cube of the larger axis a of the orbit. The earth's orbit takes 1 year. The larger axis of the orbit of Jupiter is about 5.2 times as long as the earth's. What is the approximate period for Jupiter's orbit? (*Hint:* After you find p^2, take the square root to find p.) About 11.86 years

THINK ABOUT IT

1. **a.** If y varies directly as x, does x vary directly as y?
   b. In an example, y varies directly as x, with variation constant equal to 10. Write this as an equation; then rewrite the equation to show that x varies directly as y. What is the new variation constant?

2. **a.** If y varies inversely as x, does x vary inversely as y?
   b. In an example, y varies inversely as x, with variation constant equal to 5. Write this as an equation; then rewrite the equation to show that x varies directly as y. What is the new variation constant?

3. a. What happens to the *circumference* of a circle when you triple the radius? Explain this in terms of direct variation.
   b. What happens to the *area* of a circle when you triple the radius? Explain this in terms of direct variation.
4. a. Confirm that for a right triangle as shown, the sum of the two smaller areas equals the larger area.

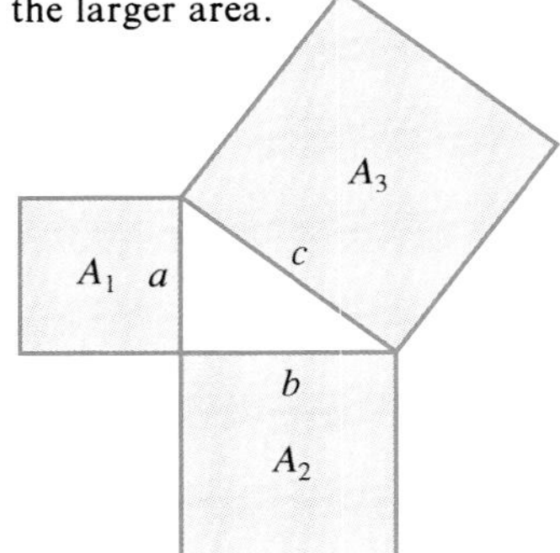

   b. Does the equality still hold if the squares are replaced by semicircles?

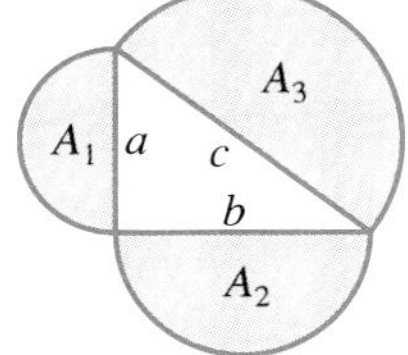

   c. An equilateral triangle is drawn on each side of a right triangle as shown. The area of an equilateral triangle is given by $A = (\sqrt{3}/4)s^2$, where s is the length of the side. Is it still true that the sum of the two smaller areas equals the larger area?

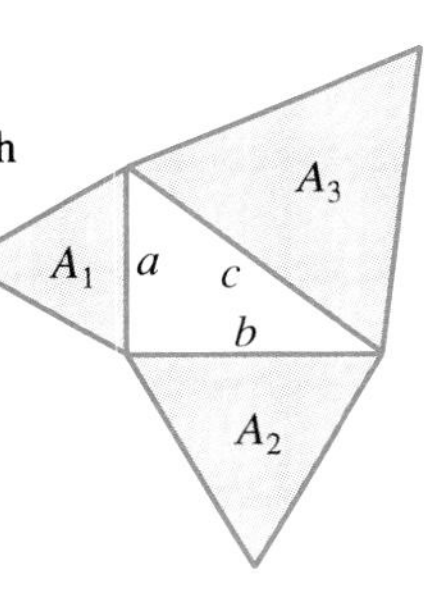

   d. Note that in parts **a**, **b**, and **c** the individual areas were directly proportional to the squares of the corresponding side of the original triangle. In part **a**, we had area $=$ $1 \cdot \text{side}^2$. In part **b**, we had area $= (\pi/4) \cdot \text{side}^2$. In part **c**, area $= (\sqrt{3}/4) \cdot \text{side}^2$. Show that generalizations of the Pythagorean theorem will hold for other shapes as long as their areas are directly proportional to the square of the side with the *same* constant of variation (k).
5. Consider a solid cube of uniform density. Its weight is directly proportional to its volume, which is directly proportional to the *cube* of its side. This weight rests on one face of the cube, whose area is directly proportional to the *square* of its side. The pressure which the cube exerts on its base is given in pounds per square foot (the weight of the cube divided by the area of the base).
   a. A 1-ft solid cube weighs 10 lb. What is the pressure on the base?
   b. If you double the sides of the cube in part **a**, what is the pressure on the base?
   c. If you triple the sides of the cube in part **a**, what is the pressure on the base?
   d. Every material will disintegrate and collapse if the pressure on it is large enough. Refer to parts **a**, **b**, and **c**, and explain why there is a limit to the possible size of a cube made of any given material.
   e. Explain informally why the discussion in parts **a–d** also limits the size of trees, elephants, and people. (It also shows why many giant monsters in science fiction movies are strictly imaginary.) [For a fascinating related discussion, see *On Being the Right Size* by J. B. S. Haldane (Oxford University Press, 1985).]

REMEMBER THIS

1. Show that the point (1,3) lies on both these lines: $3x + y = 6$ and $3x - y = 0$. (1,3) is a solution to both equations.
2. Add the polynomials $6x - 4y$ and $3x + 4y$. $9x$
3. Multiply both sides of this equation by the LCD and simplify: $\frac{3}{7}x - \frac{4}{5}y = 2$. $15x - 28y = 70$
4. True or false? If two distinct straight lines intersect, then there is some ordered pair that satisfies both equations. Explain. True; the point of intersection is on both lines.
5. True or false? If two lines in the same plane have different slopes, then they must intersect at some point. Explain. True; different slopes mean they are not parallel.
6. Graph the solution set of $x + y \leq 1$.

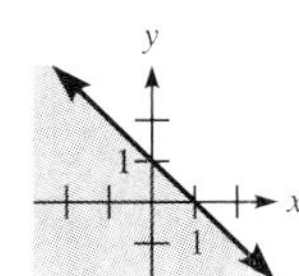

7. Find an equation for the line which includes $(-3,-3)$ and $(4,0)$. $3x - 7y = 12$
8. Simplify $\dfrac{1 - \frac{1}{x}}{1 - \frac{1}{x^2}}$. For what values of x is this complex fraction not defined? $\dfrac{x}{x+1}$; 0, 1, -1
9. Factor $x^3 - 27$. $(x - 3)(x^2 + 3x + 9)$
10. Determine if the points A, B, and C are on the same straight line: $A(0,0)$, $B(n,n)$, $C(n + 5, n + 5)$. Yes; $m_{AB} = 1$ and $m_{AC} = 1$; all three points lie on the line $y = x$.

GRAPHING CALCULATOR

Drawing Graphs of Lines

A graphing calculator can produce an unlimited variety of graphs from simple to complex. We introduce this feature of the calculator by focusing on graphs of linear equations. Before any linear equation can be graphed, it must be written in the form

$$y = \text{algebraic expression.}$$

EXAMPLE 1 **a.** Graph the equation $4x + y = 2$.
b. Use the graph from part **a** to estimate the intercepts.

Solution

a. Step 1 Solve for y to get $y = 2 - 4x$.

Step 2 Press the [Y=] key, and on the Y_1 line key in 2 − 4X. Note that the equal sign appears on a dark background. This means that the calculator is ready to draw the graph.

Step 3 Press [ZOOM] and then 6 to get the Standard graph setup. The screen should look like Figure 5.39. You can see that the graph is a line slanting down to the right.

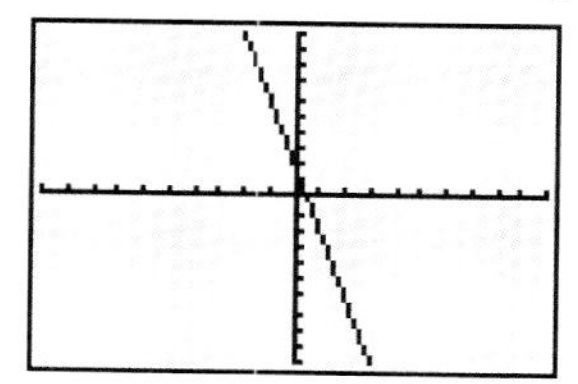

Figure 5.39

The Zoom Standard feature produces a coordinate system where x and y both extend from −10 to 10 with a scale of 1, which means that each dot on the axes represents 1 unit. If this setup is not suitable, it can be changed by using the [RANGE] key.

Note This is not the only sequence of keystrokes for drawing this graph. For instance, in step 3 you can press the [GRAPH] key instead of [ZOOM] Standard. The [GRAPH] key will plot the graph using whatever range assignments are current.

b. Continuing with the graph from part **a,** we use the [ZOOM] and [TRACE] keys to estimate the intercepts.

Step 1 We will zoom in on the part of the picture that shows both intercepts by using the Zoom Box feature. Press [ZOOM] and 1 for Zoom Box. Press the arrow keys until the cross is located near the point labeled P shown in Figure 5.40.

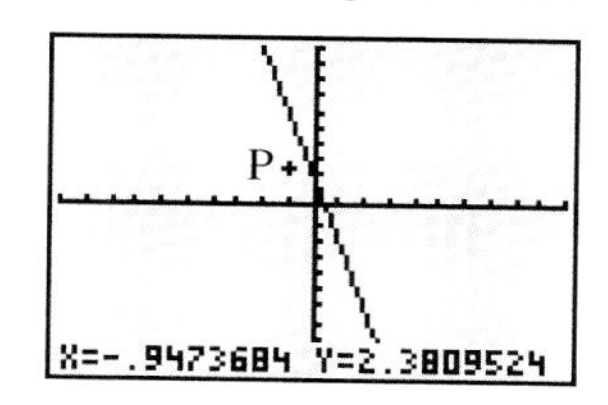

Figure 5.40

Press [ENTER]. A small blinking square appears. Press the arrow keys until the cross is located near the point labeled Q in Figure 5.41. This will draw a box, as shown in Figure 5.41.

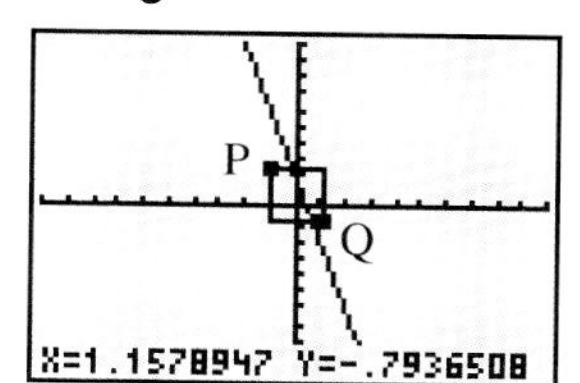

Figure 5.41

Press [ENTER]. This will now show only the part of the graph that was in the box. You can see that the x-intercept is between 0 and 1, and that the y-intercept is near 2.

Step 2 We will use the [TRACE] key to move along the graph.

Press [TRACE]. Note that a blinking square appears on a point of the graph and that its x- and y-coordinates appear at the bottom of the screen. Press the *left* and *right* arrow keys to move along the graph.

Try to stop on the x-intercept. That is, try to stop where $y = 0$. You will only be able to get close, because you are limited by the resolution of the graph, but it should be clear that the x-coordinate is near 0.5.

Step 3 To get a better estimate, we will zoom in using the Zoom In feature.
Press [ZOOM] and 2 to select Zoom In.
Press [ENTER]. This produces a new close-up view on which you can trace the graph more finely. You can repeat Zoom In and Trace as many times as you wish to improve the estimate. Note that when a coordinate gets very close to zero, the calculator uses scientific notation.

By a similar sequence of steps, you can determine that the y-intercept is near 2.

The next example illustrates the importance of the Range feature.

EXAMPLE 2 Graph $2x + 3y = -30$, and show both intercepts.

Solution

Step 1 Solve for y to get $y = (-30 - 2x)/3$. Press [Y=], and enter this expression for Y_1.

Step 2 Draw the graph using [ZOOM] Standard. The graph is shown in Figure 5.42. Note that the intercepts are not displayed because the axes do not extend far enough in the negative direction.

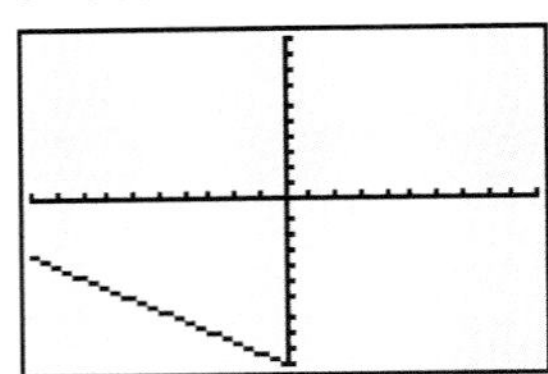

Figure 5.42

Step 3 Press [RANGE]. Replace X_{min} and Y_{min} by -20. This will extend both axes so that they reach from -20 to $+10$.

Step 4 Press [GRAPH]. Now both intercepts can be seen.

Caution If the display is *entirely blank* when you expect to see a graph, the most likely cause is that the range settings are not suitable. For instance, graphing $y = x + 25$ with [ZOOM] Standard will produce a blank display.

Exercises

Use the graphing calculator to display the graphs of these lines. Use the Range feature if necessary so that both intercepts appear on the screen at the same time.

1. $3x + y = 3$ Ans.

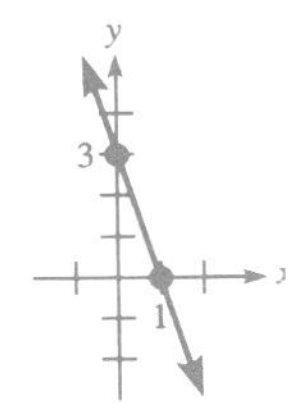

2. $3x + 2y = 30$ Ans.

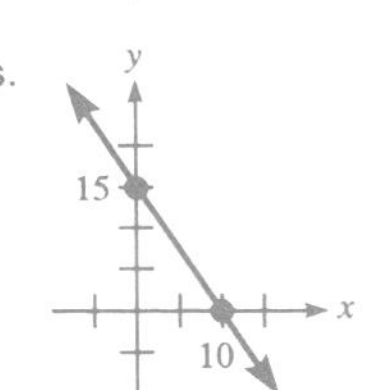

Use the graphing calculator to draw the lines given in these exercises from Section 5.1.
Exercises 5.1: 23, 27, 31, 35, 45

Chapter 5 SUMMARY

OBJECTIVES CHECKLIST Specific chapter objectives are summarized below along with numbered example problems from the text that should clarify the objectives. If you do not understand any objectives or do not know how to do the selected problems, then restudy the material.

5.1 Can you:

1. **Determine if an ordered pair is a solution of an equation?**
Determine if the ordered pair (1,3) is a solution of the equation $2x + y = 5$. [Example 1a]

2. **Graph ordered pairs?**
Graph the ordered pair $(3,-4)$, and indicate its quadrant location. [Example 3b]

3. **Graph linear equations by plotting ordered-pair solutions?**
   Graph $y = -x - 2$. [Example 5]

4. **Graph linear equations by using intercepts?**
   Graph $2x + 3y = -6$ by using the x- and y-intercepts. [Example 7]

5. **Graph $x = a$ or $y = b$?**
   Graph the line $y = 3$. [Example 8]

5.2 Can you:

1. **Find and interpret the slope of a line?**
   A printing shop charges $20 for 500 copies of a flyer and $25 for 750 copies. If the relation between the number of copies (x) and cost (y) graphs as a line, calculate and interpret the slope. [Example 4]

2. **Determine if lines are parallel, perpendicular, or neither by using slope?**
   A quadrilateral has vertices $A(0,1)$, $B(6,3)$, $C(7,0)$, and $D(1,-2)$. Show that quadrilateral $ABCD$ is a parallelogram. [Example 5a]

3. **Interpret line graphs?**
   A pot of water was heated on the stove and the temperature recorded every minute for 7 minutes. The line graph in Figure 5.21 shows the result (see page 232). Over which interval did the temperature rise most? [Example 6c]

4. **Find the distance between two points?**
   Find the distance between $(-2,1)$ and $(5,-3)$. [Example 8]

5.3 Can you:

1. **Write an equation for a line given its slope and a point on the line?**
   Find an equation of the line through $(3,-2)$ with slope -5. [Example 1]

2. **Write an equation for a line given two points on the line?**
   Find an equation for the line through the points $(-2,5)$ and $(1,6)$. [Example 2]

3. **Find the slope and y-intercept given an equation for the line?**
   Find the slope and the y-intercept of the line defined by the equation $3x + 4y = -5$. [Example 4b]

4. **Graph and write an equation for a line given the slope and y-intercept?**
   The slope of a line is $-\frac{1}{3}$ and the y-intercept is $(0,-2)$. Find the equation of the line in slope-intercept form, and graph the line. [Example 5]

5. **Solve problems involving the equations of parallel and perpendicular lines?**
   Find the equation of the line through the origin which is perpendicular to $3x - y = 10$. [Example 8]

5.4 Can you:

1. **Determine if an ordered pair is a solution of an inequality?**
   Determine if $(-1,0)$ is a solution of $y \leq 2x + 4$. [Example 1]

2. **Graph linear inequalities?**
   Graph $2x - 3y > 6$. [Example 4]

3. **Graph linear inequalities restricted to $x \geq 0$, $y \geq 0$?**
   Solve the problem in the section introduction on page 246. [Example 5]

5.5 Can you:

1. **Solve problems involving direct variation?**
   y varies directly as x, and $y = 4$ when $x = 5$. Find y when $x = 15$. [Example 2]

2. **Solve problems involving inverse variation?**
   y varies inversely as x, and $y = 3$ when $x = 8$. Find y when $x = 10$. [Example 4]

3. **Solve problems involving variations of powers of variables?**
The weight (w) of a solid cube varies directly as the cube of the side (s). When the side is doubled, how many times heavier does the cube become? [Example 7a]

4. **Solve problems involving variation among more than two variables?**
From the formula $F = \frac{km_1m_2}{d^2}$, describe what happens to the gravitational attraction (F) between two objects if both masses (m_1 and m_2) are cut in half and the distance (d) between them is doubled. [Example 9]

KEY TERMS

Cartesian coordinate system (5.1)
Coordinates (5.1)
Direct variation (5.5)
Extrapolation (5.2)
Graph (5.1)
Half plane (5.4)
Inverse variation (5.5)
Joint variation (5.5)
Linear equation (5.1)
Linear inequality (5.4)
Line graph (5.2)
Ordered pair (5.1)
Origin (5.1)
Parallel lines (5.2)
Parallelogram (5.2)
Perpendicular lines (5.2)
Point-slope equation (5.3)
Quadrant (5.1)
Rectangle (5.2)
Rectangular coordinate system (5.1)
Rise (5.2)
Run (5.2)
Slope-intercept equation (5.3)
Slope of a line (5.2)
Test point (5.4)
Variation constant (5.5)
x-axis (5.1)
x-intercept (5.1)
y-axis (5.1)
y-intercept (5.1)

KEY CONCEPTS AND PROCEDURES

Section	Key Concepts or Procedures to Review
5.1	■ Definition of the graph of an equation in two variables ■ General form of a linear equation in two variables: $Ax + By = C$ ■ Procedures for graphing a linear equation by plotting points and by using x- and y-intercepts ■ The graph of $x = a$ is a vertical line containing $(a,0)$; the graph of $y = b$ is a horizontal line containing $(0,b)$.
5.2	■ Slope formula: $m = \frac{\text{rise}}{\text{run}} = \frac{\text{change in } y}{\text{change in } x} = \frac{y_2 - y_1}{x_2 - x_1} \quad (x_2 \neq x_1)$ ■ For two lines with slopes m_1 and m_2: parallel lines: $m_1 = m_2$; perpendicular lines: $m_1m_2 = -1$ ■ Distance formula: $d = \sqrt{(x_2 - x_1)^2 + (y_2 - y_1)^2}$
5.3	■ Point-slope equation: $y - y_1 = m(x - x_1)$ ■ Slope-intercept equation: $y = mx + b$

Section	Key Concepts or Procedures to Review
5.4	■ Procedure for solving linear inequalities by graphing
5.5	■ The statement "y varies directly as x" means that $y = kx$ (k is the variation constant). ■ The statement "y varies inversely as x" means $xy = k$, or $y = k/x$. ■ The statement "y varies jointly as x and w" means $y = kxw$.

CHAPTER 5 REVIEW EXERCISES

5.1

1. Determine if the ordered pair $(2,-1)$ is a solution of the equation $-4x - y = -9$. No
2. Graph the ordered pair $(1,-3)$, and indicate its quadrant location.

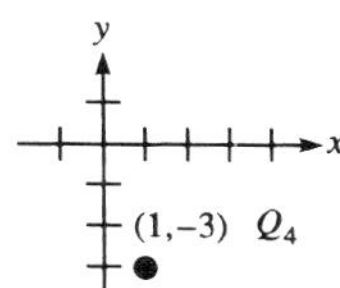

3. Graph the equation $y = 2x - 1$.

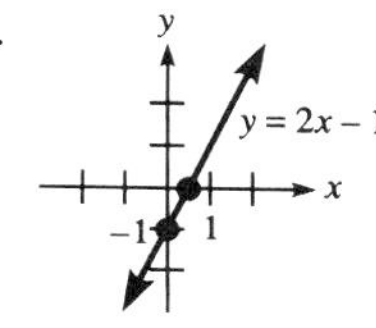

4. Graph $-x + y = -4$ by using the x- and y-intercepts.
x-intercept: (4,0); y-intercept: $(0,-4)$

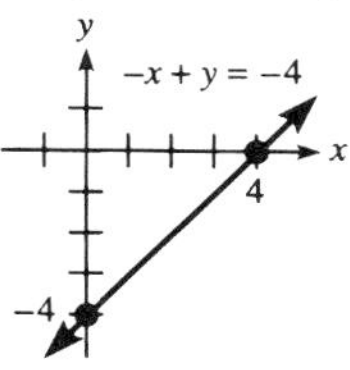

5. Graph $x = 2$.

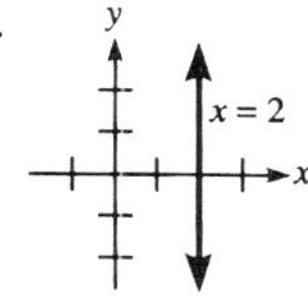

5.2

6. Find the slope of the line through the points $(-2,1)$ and $(4,5)$.
$\frac{2}{3}$
7. Shipping costs (c) charged by a delivery service have a linear relationship with the weight (w) of the parcel. The charge is \$1.98 for a 5-oz parcel and \$2.23 for a 10-oz parcel. Calculate and interpret the slope of the line. $m = 0.05$; rate is 5¢ per ounce.
8. A quadrilateral has vertices at $A(0,2)$, $B(1,0)$, $C(-1,-1)$, and $D(-2,1)$. Show that quadrilateral $ABCD$ is a rectangle.
$m_{AB} = m_{CD} = -2$, $m_{BC} = m_{AD} = \frac{1}{2}$; opposite sides are parallel and therefore it's a parallelogram. $m_{AB}m_{BC} = m_{AB}m_{AD} = m_{CD}m_{BC} = m_{AD}m_{CD} = -1$; angles are right angles and therefore it's a rectangle.

9. The line graph shows some hourly temperatures at a weather station for one day.

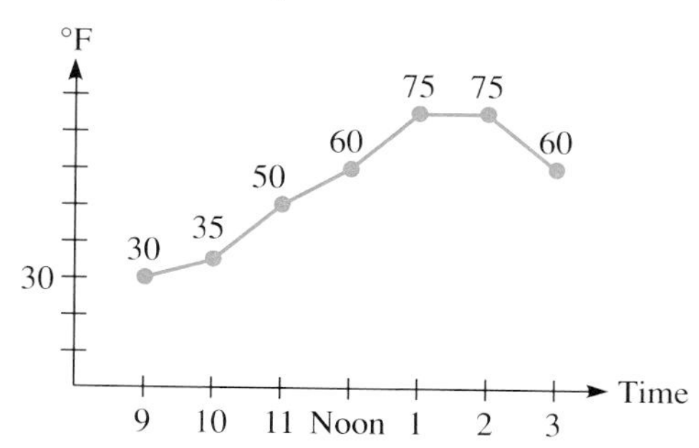

a. Between which two hourly readings was the temperature change the smallest? 1–2 P.M.
b. In which hourly interval did the temperature first reach 40°? 10–11 A.M.

10. Find the distance between the points $(4,7)$ and $(3,-2)$ $\sqrt{82}$

5.3

11. Find an equation in general form of the line through $(-1,3)$ with slope 2. $-2x + y = 5$
12. Find an equation for the line through the points $(3,2)$ and $(-1,-1)$. $3x + 4y = -1$
13. Find the slope and y-intercept of the line defined by the equation $-2x - y = 1$. $m = -2, b = -1$
14. Find the slope-intercept form of the equation of a line whose slope is -1 and whose y-intercept is $\frac{1}{2}$. Graph the line.
$y = -x + \frac{1}{2}$

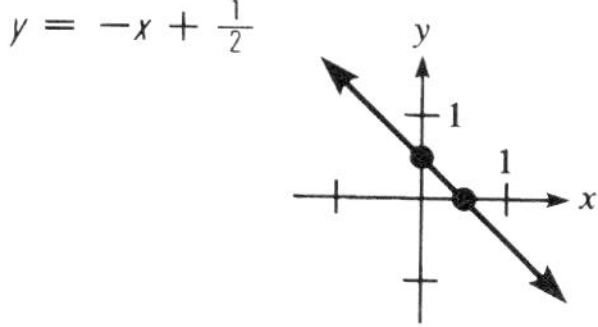

15. Find the equation of the line through $(0,5)$ that is parallel to $4x - 6y = 18$. $y = \frac{2}{3}x + 5$

5.4

16. Determine if $(1,-2)$ is a solution of $y \geq 2x + 3$. No
17. Graph $y \geq -\frac{2}{3}x + 3$.

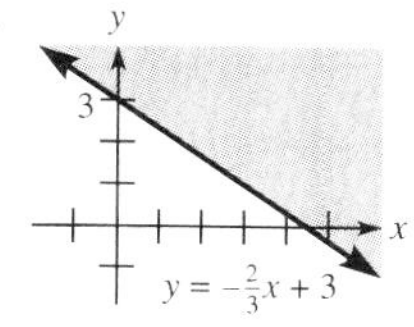

18. Graph $y < 1$.

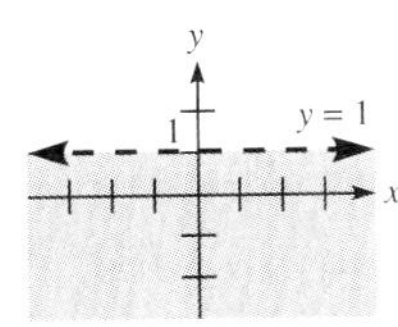

19. Graph $-2x + 3y \leq 6$.

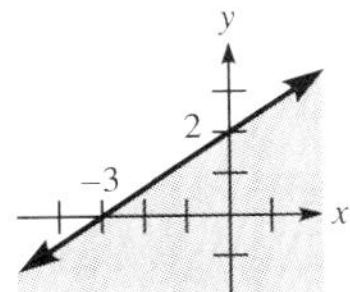

20. A craftsperson needs to buy some beads. Brass beads cost \$2 a bag, silver beads cost \$5 a bag, and the total amount spent on beads cannot be more than \$200. Let x and y represent the number of bags of brass and silver beads purchased, respectively.
   a. Write the inequality expressing the expense restriction.
   b. Graph the inequality for $x \geq 0$ and $y \geq 0$.
   c. Determine the greatest number of bags of silver beads that can be afforded if 36 bags of brass beads are purchased. a. $2x + 5y \leq 200$

b. c. 25 bags

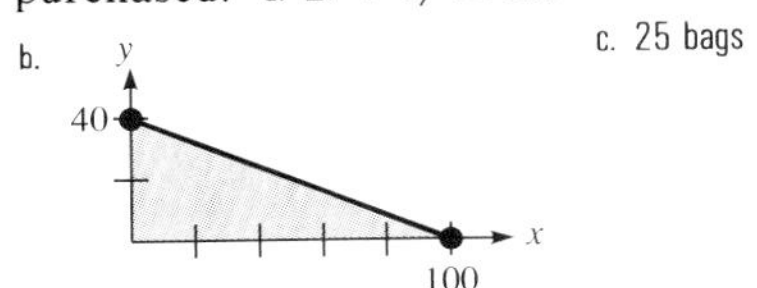

5.5

21. The cost (C) of tiling a floor varies directly as the area (A) of the floor. Write the relationship algebraically. $C = kA$
22. If y varies directly as x, and $y = 2$ when $x = 7$, find y when $x = -21$. -6
23. If y varies inversely as x, and $y = 4$ when $x = 50$, find y when $x = 5$. 40
24. Suppose that M varies jointly as a and b and inversely as c^2, and that $M = 20$ when $a = 5$, $b = 8$, and $c = 2$. Find M when $a = 4$, $b = 50$, and $c = 5$. 16
25. The weekly earnings of a part-time employee vary directly as the number of hours worked. For an employee who makes \$112.50 for 15 hours of work, what are the earnings for 19 hours of work? \$142.50

ADDITIONAL REVIEW EXERCISES

Graph.

26. $2x - y = 6$

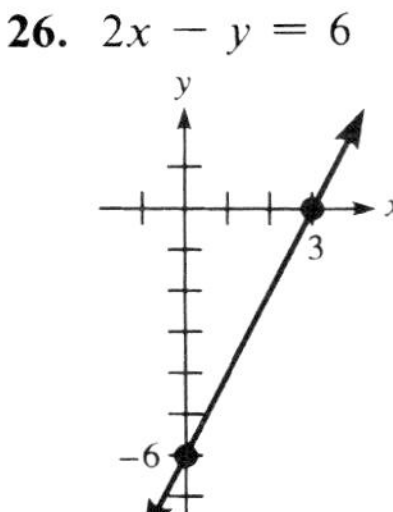

27. $y = \frac{1}{2}x + 1$

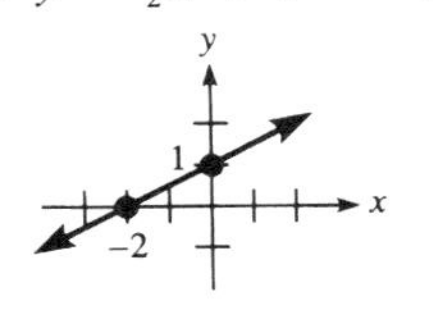

28. $y = -1$

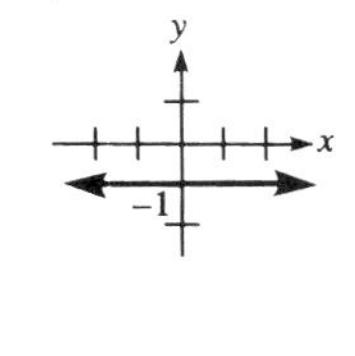

29. $x \geq 2$

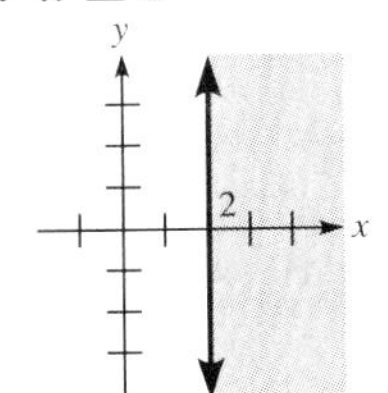

30. $5x + y > 5$

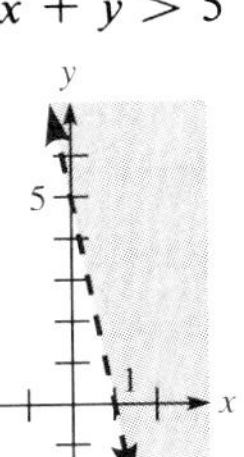

31. $x = -3$

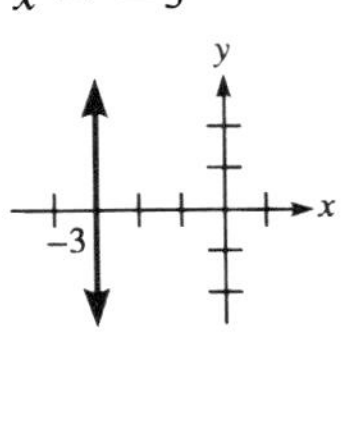

Find the equation of each line. Express it in general form.

32. With slope $\frac{1}{2}$ and y-intercept 1 $-x + 2y = 2$
33. Through (6,1) and (0,2) $x + 6y = 12$
34. Through (4,−3) with slope 1 $-x + y = -7$
35. Through (0,0), perpendicular to $4x - 4y = 7$ $x + y = 0$
36. Through (1,1), parallel to $4x - 4y = 7$ $-x + y = 0$
37. Complete each ordered pair so that it is a solution to the equation $y = -5x + 1$: (0,), (1,), (2,). (0,1), (1,−4), (2,−9)
38. Find the distance between the points (−5,0) and (−3,2). $2\sqrt{2}$
39. Find the slope of the line through the points (−5,0) and (−3,2). 1
40. Determine if (−4,−5) is a solution of $y = -\frac{1}{2}x - 7$. Yes
41. Are the lines $-2x - 3y = 4$ and $4x + 6y = -8$ parallel, perpendicular, or the same line? Same line
42. If y varies inversely as x, and $x = 12$ when $y = 4$, find y when $x = 16$. 3
43. If y varies directly as x, and $y = \frac{1}{2}$ when $x = 3$, find y when $x = 24$. 4
44. Suppose that A varies directly as B^2 and inversely as C, and that $A = 2$ when $B = 1$ and $C = 3$. Find A when $B = 3$ and $C = 6$. 9
45. Find the slope and y-intercept of the line $x + 4y = 0$. $-\frac{1}{4}$, (0,0)
46. Show that the points $A(0,0)$, $B(2,3)$, and $C(6,-4)$ are the vertices of a right triangle.
   $m_{AB} = \frac{3}{2}$, $m_{AC} = -\frac{2}{3}$; $m_{AB} \cdot m_{AC} = -1$, A is a right angle
47. A quadrilateral has vertices $A(0,2)$, $B(12,6)$, $C(14,0)$, and $D(2,-4)$. Show that quadrilateral $ABCD$ is a rectangle.
   $m_{AB} = m_{CD} = \frac{1}{3}$; $m_{BC} = m_{AD} = -3$; $\frac{1}{3}(-3) = -1$
48. A car rents for \$70 a week plus 15¢ per mile.
   a. Write an equation for the weekly cost C. Let m represent the weekly mileage. $C = 70 + 0.15m$
   b. Graph the equation.

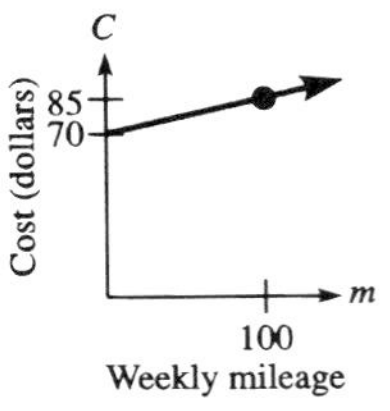

49. A gasoline pump can fill a 15-gal tank in 3 minutes. Assume the relationship between gallons and time can be represented by a line through the origin. Calculate and interpret the slope of the line. $m = 5$; pump provides 5 gal/minute.
50. The weight (w) of a solid cube varies directly as the cube of an edge (e). If a cube with edge 8 in. weighs 10 lb, what will a cube with edge 7 in. made of the same material weigh? Round to the nearest tenth of a pound. 6.7 lb

51. This line graph shows the progress of the Dow-Jones Industrial Average for one day.

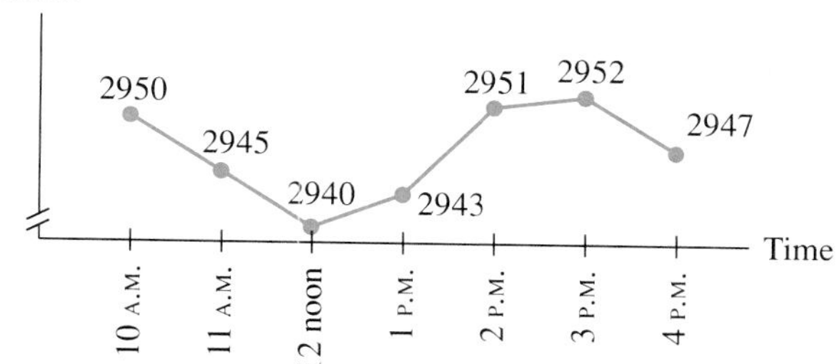

a. During which hourly interval did the index drop below 2945? 11 A.M.–12 noon

b. On which hourly interval did the index change the most? 1–2 P.M.

52. The owner of a used-book store buys both paperbacks and hardcovers at local sales. A paperback costs \$0.25 and a hardcover costs \$0.50. The total amount to be spent on books this month cannot be more than \$100. Let x and y represent the number of paperbacks and hardcovers purchased, respectively.

a. Write the inequality expressing the expense restriction. $0.25x + 0.50y \leq 100$

b. Graph the inequality for $x \geq 0$ and $y \geq 0$.

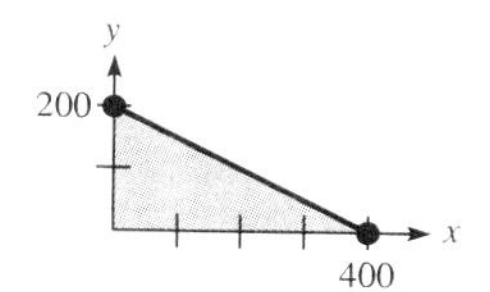

c. Determine the greatest number of hardcovers that can be afforded if 355 paperbacks are purchased. 22

CHAPTER 5 TEST

1. Find the distance between the points $(-5,1)$ and $(3,7)$. 10
2. Write the equation of the line through $(2,0)$ and $(0,-1)$ in slope-intercept form. $y = \frac{1}{2}x - 1$
3. Find the slope and y-intercept of the line $-5x + 10y = 15$. $m = \frac{1}{2}$, $b = \frac{3}{2}$
4. If y varies inversely as x, and $y = 12$ when $x = -2$, find y when $x = 6$. -4
5. A printing shop charges \$25 for 100 invitations and \$30 for 150. If the relation between the number of invitations (x) and the cost (y) graphs as a line, calculate and interpret the slope. $m = 0.10$; 10¢ per invitation
6. Determine whether the lines $4x + y = 1$ and $16x + 4y = -1$ are parallel, perpendicular, or the same line. Parallel
7. Write the relationship algebraically: w varies jointly as s and t and inversely as the square of r. $w = \frac{kst}{r^2}$
8. Complete each ordered pair so that it is a solution to $y = -x - 4$: (2,), (0,), (−2,). $(2,-6)$, $(0,-4)$, $(-2,-2)$
9. A quadrilateral has vertices at $A(-4,6)$, $B(-8,0)$, $C(4,-8)$, and $D(8,-2)$.

   a. Show that $ABCD$ is a parallelogram. $m_{AB} = m_{CD} = \frac{3}{2}$; $m_{AD} = m_{BC} = -\frac{2}{3}$

   b. Show that $ABCD$ is a rectangle. $\frac{3}{2}(-\frac{2}{3}) = -1$

Find the equation of each line. Express it in general form.

10. Through $(-5,1)$ and $(-8,4)$ $x + y = -4$
11. Through $(2,1)$ with slope 0 $y = 1$
12. With slope $\frac{2}{3}$ and y-intercept -1 $-2x + 3y = -3$
13. Through $(6,0)$ and perpendicular to $y = -3x$ $-x + 3y = -6$

Graph.

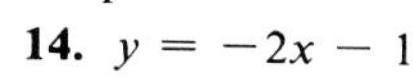

14. $y = -2x - 1$

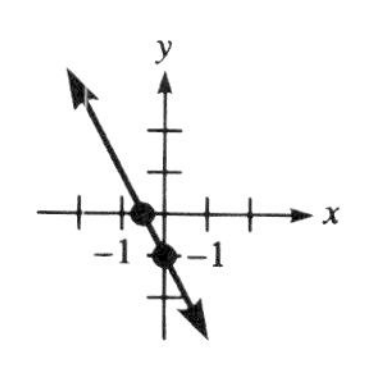

15. $x - 2y = 3$

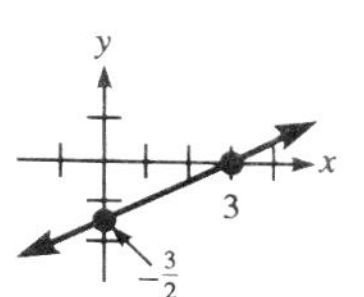

16. $y < 3$

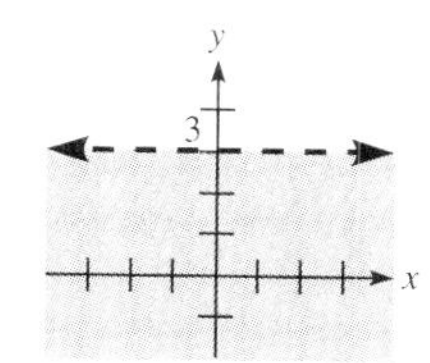

17. $-2x - 3y \geq 12$

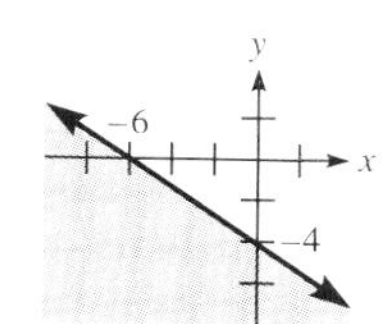

18. $x = 0$

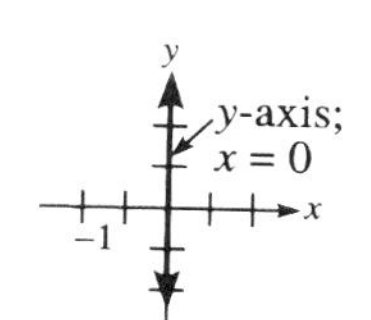

19. A taxi ride costs \$2 plus \$0.50 per mile traveled.

    a. Write an equation for the cost C. Let n represent the number of miles. $C = 0.50n + 2$

    b. Graph the equation with n as the horizontal axis.

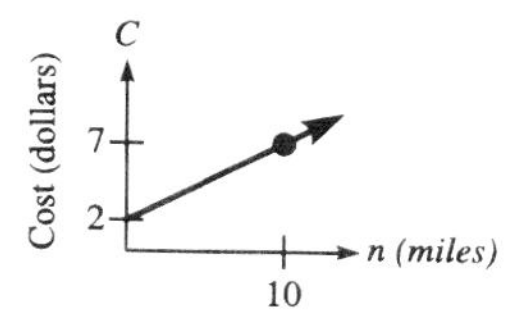

20. The distance d needed to stop a car after applying the brakes varies directly as the square of the speed r at which the car is traveling. If a car traveling 30 mi/hour needs 100 ft to stop, in how many feet will a car traveling 60 mi/hour be able to stop? 400 ft

CUMULATIVE TEST 5

1. Which of the following is *not* a rational number?
   a. $\sqrt{3}$ **b.** $\sqrt{9}$ **c.** $\frac{3}{4}$ **d.** 5 **e.** 0 a
2. Which of the following is an instance of the associative property of addition of real numbers?
   a. $a + b = b + a$ **b.** $(a + b) + c = a + (b + c)$
   c. $a + b = a + b$ **d.** $a + 0 = a$
   e. $a(b + c) = ab + ac$ b
3. Solve $-8x + 14 = -2x + 2$. $\{2\}$
4. Simplify $-5[3(1 - 6) - 2(-3)(-4)] - 7$. 188
5. Simplify $-(-2)^3 - 2^3 + 2^0$. 1
6. Simplify $(2x^3 - 4x^2 + 6x + 1) - (4x^3 + 4x^2 - 6x - 3)$.
   $-2x^3 - 8x^2 + 12x + 4$
7. Simplify $\dfrac{(3x + 6) - 2(3x)}{3}$. $-x + 2$
8. Solve $3x - 12 = 6y$ for x. $x = 2y + 4$
9. Evaluate $\dfrac{-b + \sqrt{b^2 - 4ac}}{2a}$ when $a = 1$, $b = -2$, and $c = 0$. 2
10. Solve $|4x - 1| \leq 7$, and write the results using interval notation. $[-\frac{3}{2}, 2]$
11. In his will, a man left half of his money to his wife, a third to his son, and the remainder, which was $2,500, to his daughter. How much money did the man have? $15,000
12. Solve $-5x - 3 > -2x + 9$, and graph the solution set.
    $(-\infty, -4)$ 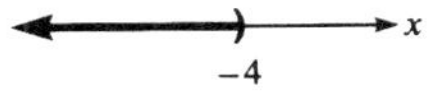

13. Factor $x^2 - 11x - 12$. $(x - 12)(x + 1)$
14. Express $\dfrac{2x^4 - 8x^3 + 4x^2}{3x^2 - 12x + 6}$ in lowest terms. $\dfrac{2x^2}{3}$
15. Find all values of y which make the expression $\dfrac{y + 2}{y^2 - 4}$ undefined. $-2, 2$
16. Add $\dfrac{5x + 2}{1 - 3x} + \dfrac{5x + 3}{3x - 1}$, and express the results in lowest terms. $\dfrac{1}{3x - 1}$
17. Solve $x^3 - 2x^2 - 5x + 6$, given that $x = 3$ is one of the solutions. $\{-2,1,3\}$
18. If y varies directly as x, and $y = 3$ when $x = 12$, find y when $x = 8$. 2
19. Find an equation for the line through the points $(6,-2)$ and $(3,7)$. $y = -3x + 16$
20. Find the distance between the points $(-1,-3)$ and $(4,9)$.
    13 units

CHAPTER 6

Systems of Linear Equations and Inequalities

A corporation operates in a state that levies a 7 percent tax on the income that remains after paying the federal tax. Meanwhile, the federal tax is 28 percent of the income that remains after paying the state tax. If, during the current year, a corporation has \$500,000 in taxable income, determine the state and federal income tax. (See Example 9 of Section 6.1.)

IN THE chapter-opening problem an accountant is faced with a dilemma: To find the state tax, the federal tax must be known; but to find the federal tax, the state tax must be known. This problem can be resolved mathematically by defining each tax using a linear equation in two variables, thereby forming a *pair* of equations that is called a **linear system of equations.** As discussed in Example 9 of Section 6.1, the linear system required for this accounting problem is

$$\begin{aligned} s &= 0.07(500{,}000 - f) \\ f &= 0.28(500{,}000 - s). \end{aligned}$$

Other examples of linear systems in two variables are

$$\begin{aligned} x + y &= 9 \\ 5x - y &= -3, \end{aligned} \qquad \begin{aligned} c &= 1{,}000 + 5n \\ c &= 4{,}000 + 3n, \end{aligned} \qquad \text{and} \qquad \begin{aligned} y &= -3x + 1 \\ x &= 2. \end{aligned}$$

Note that the pair of equations in the system must contain the same variables and that an equation like $x = 2$ may be written as $x + 0y = 2$ when expressed as a linear equation in two variables. This chapter considers several methods for solving linear systems in two and in three variables and applies these methods to analyze a wide variety of problems. Applied problems may also require the inequality relations, so this chapter concludes by considering systems of linear inequalities.

Chapter Outline

6.1 Systems of Linear Equations in Two Variables

OBJECTIVES

1. **Determine whether an ordered pair is a solution to a system of linear equations.**
2. **Solve a system of linear equations graphically.**
3. **Solve a system of linear equations by the addition-elimination method.**
4. **Solve a system of linear equations by the substitution method.**
5. **Use systems of linear equations to solve applied problems.**

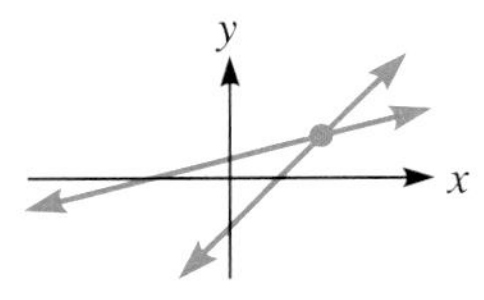

Consistent system

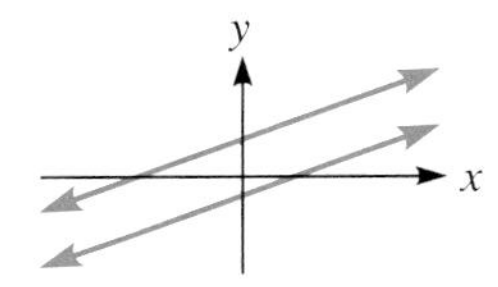

Inconsistent system

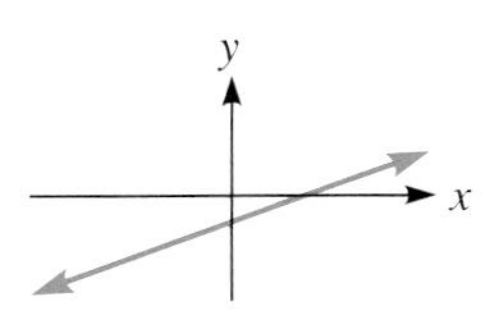

Dependent system

Figure 6.1

1 Recall that the graph of a linear equation in two variables is a straight line. Thus, the graph of a *system* of *two* such equations consists of *two* straight lines. There are three possible cases and they are represented in Figure 6.1.

Case 1 The equations represent two lines which intersect at one point and so have 1 point in common. This system is called **consistent.**

Case 2 The equations represent two distinct lines which are parallel and do not intersect at all and so have no points in common. This system is called **inconsistent.**

Case 3 Both equations represent the same line and so have all the points of that line in common. This system is called **dependent** (and consistent).

Solving a system of two linear equations is equivalent to finding all points the two lines have in common. Algebraically, the **solution set of a system of linear equations** consists of all ordered pairs that satisfy both equations at the same time (simultaneously). Example 1 shows how to determine whether a given ordered pair is a solution of a system.

EXAMPLE 1 Determine if the given ordered pairs are solutions of the system

$$\begin{aligned} x + y &= 12 \\ 5x - 2y &= 4. \end{aligned}$$

a. (4,8) **b.** (10,2)

Solution

a. (4,8) means $x = 4$, $y = 8$. Replace x by 4 and y by 8 in each equation.

$$\begin{aligned} x + y &= 12 \\ 4 + 8 &\stackrel{?}{=} 12 \\ 12 &= 12 \quad \text{True} \end{aligned} \qquad \begin{aligned} 5x - 2y &= 4 \\ 5(4) - 2(8) &\stackrel{?}{=} 4 \\ 4 &= 4 \quad \text{True} \end{aligned}$$

(4,8) is therefore a solution of the system because it satisfies both equations.

b. To check (10,2), replace x by 10 and y by 2 in each equation.

$$\begin{aligned} x + y &= 12 \\ 10 + 2 &\stackrel{?}{=} 12 \\ 12 &= 12 \quad \text{True} \end{aligned} \qquad \begin{aligned} 5x - 2y &= 4 \\ 5(10) - 2(2) &\stackrel{?}{=} 4 \\ 46 &= 4 \quad \text{False} \end{aligned}$$

Because (10,2) is not a solution of $5x - 2y = 4$, it is not a solution of the system.

PROGRESS CHECK 1 Determine if the given ordered pairs are solutions of the system

$$x + y = 25$$
$$6x - y = 3.$$

a. (4,21) **b.** (10,15)

2 Because the graph of a linear equation provides a picture of its solutions, we can find all common solutions to a pair of linear equations by drawing their graphs on the same coordinate system. The solution is given by all points where the lines intersect, and we specify the solution by giving the coordinates of such points.

EXAMPLE 2 Solve by graphing: $x + 2y = 4$
$y = x + 5.$

Solution $x + 2y = 4$ is a linear equation in general form that is easily graphed by finding intercepts. Letting $x = 0$ gives a y-intercept of (0,2), and letting $y = 0$ gives an x-intercept of (4,0). The equation $y = x + 5$ is in slope-intercept form and may be graphed by recognizing and using that the y-intercept is (0,5) and the slope is 1.

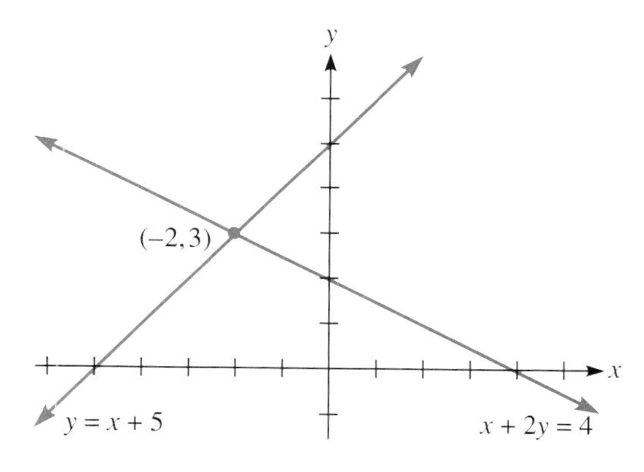

Figure 6.2

In Figure 6.2 we graph both of these equations on the same coordinate system, and it appears that the lines meet at (−2,3). We check this apparent solution.

$$x + 2y = 4 \qquad\qquad y = x + 5$$
$$-2 + 2(3) \stackrel{?}{=} 4 \qquad\qquad 3 \stackrel{?}{=} -2 + 5$$
$$4 = 4 \quad \text{True} \qquad\qquad 3 = 3 \quad \text{True}$$

Thus, the solution is (−2,3).

Note To obtain an accurate solution using the graphing method, draw the graphs carefully on graph paper. Even then, it may not be possible to read the exact coordinates of a solution like $(\frac{2}{7}, -\frac{3}{5})$. Algebraic methods for finding exact solutions will be considered next.

Remind students that even if they are using a graphing calculator, they may see only an approximate solution.

PROGRESS CHECK 2 Solve by graphing: $x + 2y = 6$
$y = x + 6.$

3 The graphing method for solving linear systems is good for illustrating the behavior of a system of equations in two variables and for estimating the solution. However, for finding exact solutions, algebraic methods are usually preferable. One such method, called the **method of addition-elimination,** is based on the property that

$$\text{if} \quad A = B$$
$$\text{and} \quad C = D,$$
$$\text{then} \quad A + C = B + D.$$

Progress Check Answers

1. (a) Yes (b) No
2.

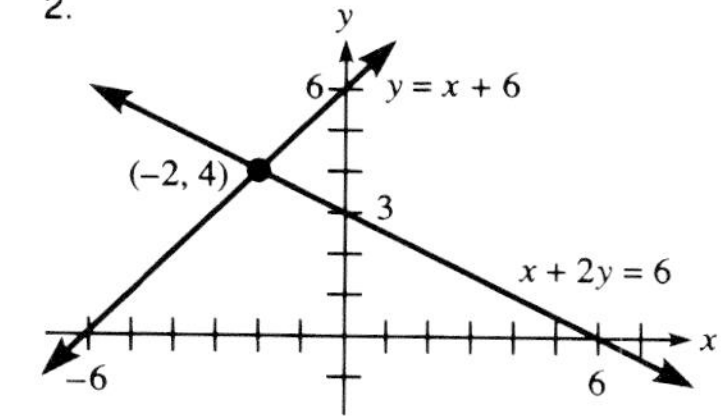

In words, adding equal quantities to equal quantities results in equal sums. For this method to result in the elimination of a variable, the coefficients of either x or y must be opposites, as in the system in Example 3.

EXAMPLE 3 Solve by addition-elimination:

$$\begin{aligned} 5x + 2y &= 4 \\ x - 2y &= 8. \end{aligned}$$

Solution Because the coefficients of y are opposites, add the equations to eliminate this variable, and solve the resulting equation for x.

$$\begin{aligned} 5x + 2y &= 4 \\ x - 2y &= 8 \\ \hline 6x &= 12 && \text{Add the equations.} \\ x &= 2 && \text{Divide both sides by 6.} \end{aligned}$$

The x-coordinate of the solution is 2. To find the y-coordinate, substitute 2 for x in either of the given equations.

$$\begin{aligned} 5x + 2y &= 4 \\ 5(2) + 2y &= 4 \\ 2y &= -6 \\ y &= -3 \end{aligned} \qquad \text{or} \qquad \begin{aligned} x - 2y &= 8 \\ 2 - 2y &= 8 \\ -2y &= 6 \\ y &= -3 \end{aligned}$$

Thus, the solution is $(2, -3)$. Check this solution in the usual way.

Caution When students solve a linear system of equations, a common error is to solve for only one variable and write answers like $x = 2$. Remember that it is important to find *both* coordinates of a solution.

PROGRESS CHECK 3 Solve by addition-elimination:

$$\begin{aligned} x - y &= 6 \\ 2x + y &= 9. \end{aligned}$$

The equations in a linear system often need to be transformed using properties of equality so that adding on both sides will eliminate a variable. In such a solution, labeling equations with numbers is useful, as illustrated in Example 4.

EXAMPLE 4 Solve by addition-elimination:

$$\begin{aligned} 6x + 10y &= 7 && (1) \\ 15x - 4y &= 3. && (2) \end{aligned}$$

Solution If equivalent equations are formed by multiplying both sides of equation (1) by 5 and both sides of equation (2) by -2, then the x variable can be eliminated.

$$\begin{aligned} 5(6x + 10y) = 5(7) \quad &\rightarrow \quad 30x + 50y = 35 && (3) \\ -2(15x - 4y) = -2(3) \quad &\rightarrow \quad -30x + 8y = -6 && (4) \\ \hline & \qquad 58y = 29 && \text{Add equations (3) and (4).} \\ & \qquad y = \tfrac{29}{58}, \text{ or } \tfrac{1}{2} && \text{Divide both sides by 58.} \end{aligned}$$

To find x, replace y by $\frac{1}{2}$ in equation (1) or equation (2).

$$\begin{aligned} 6x + 10y &= 7 \\ 6x + 10(\tfrac{1}{2}) &= 7 \\ 6x + 5 &= 7 \\ 6x &= 2 \\ x &= \tfrac{1}{3} \end{aligned} \qquad \text{or} \qquad \begin{aligned} 15x - 4y &= 3 \\ 15x - 4(\tfrac{1}{2}) &= 3 \\ 15x - 2 &= 3 \\ 15x &= 5 \\ x &= \tfrac{1}{3} \end{aligned}$$

The solution is $(\frac{1}{3}, \frac{1}{2})$. Check it in the original equations (1) and (2).

Progress Check Answer

3. $(5, -1)$

Note Sometimes the determined value for one of the solution coordinates is a fraction or a decimal that is awkward to use for finding the other coordinate. In such cases it may be easier to find the remaining coordinate value by returning to the original system of equations. For instance, in this example, after finding y, we may find x as follows.

$$\begin{aligned} 2(6x + 10y) = 2(7) \rightarrow 12x + 20y &= 14 \\ 5(15x - 4y) = 5(3) \rightarrow 75x - 20y &= 15 \\ \hline 87x \qquad &= 29 \\ x &= \tfrac{29}{87}, \text{ or } \tfrac{1}{3} \end{aligned}$$

PROGRESS CHECK 4 Solve by addition-elimination: $\begin{aligned} 6x + 6y &= 7 \\ 8x - 9y &= -2. \end{aligned}$

Example 5 points out what happens when the addition-elimination method is applied to inconsistent systems and dependent systems.

EXAMPLE 5 Solve each system.

a. $\begin{aligned} 2x + 4y &= 3 \\ 2x + 4y &= 9 \end{aligned}$ **b.** $\begin{aligned} \tfrac{1}{2}x - y &= \tfrac{5}{2} \\ -x + 2y &= -5 \end{aligned}$

Solution

a. If the bottom equation is multiplied by -1 and added to the top equation, then both x and y are eliminated.

$$\begin{aligned} 2x + 4y &= 3 \\ -1(2x + 4y) = 1(9) \rightarrow -2x - 4y &= -9 \\ \hline 0 &= -6 \quad \text{Add the equations.} \end{aligned}$$

The false equation $0 = -6$ indicates that there is no solution and that the system is inconsistent. The solution set for every inconsistent system is $\emptyset$. Graphing these two equations as in Figure 6.3 reveals that they represent distinct parallel lines, which never intersect.

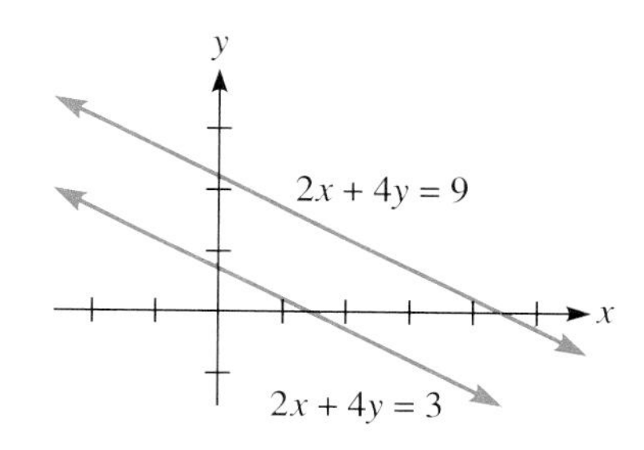

Figure 6.3

b. First, clear the top equation of fractions by multiplying both sides by 2. Then adding the equations in the resulting system eliminates both variables.

$$\begin{aligned} 2(\tfrac{1}{2}x - y) = 2(\tfrac{5}{2}) \rightarrow \quad x - 2y &= 5 \\ -x + 2y &= -5 \\ \hline 0 &= 0 \quad \text{Add the equations.} \end{aligned}$$

The equation $0 = 0$, which is always true, indicates that the system is dependent. Thus, the same line is the graph of both equations. Graphically, the solution set is the set of all points on that line, which may be specified by $\{(x,y)\colon -x + 2y = -5\}$.

Remind students that a solution is a point which the lines have in common; but since parallel lines have no point in common, there is no solution.

PROGRESS CHECK 5 Solve each system.

a. $\begin{aligned} -4x + y &= -1 \\ 8x - 2y &= 2 \end{aligned}$ **b.** $\begin{aligned} 2x - 3y &= 4 \\ x - \tfrac{3}{2}y &= 0 \end{aligned}$

Use Examples 3–5 as a basis for understanding the summary that follows of the addition-elimination method of this section.

Progress Check Answer

4. $(\tfrac{1}{2}, \tfrac{2}{3})$

5. (a) Dependent system; $\{(x,y)\colon -4x + y = -1\}$
(b) Inconsistent system; $\emptyset$

Addition-Elimination Method Summary

To solve a linear system by addition-elimination:

1. If necessary, write both equations in the form $ax + by = c$.
2. If necessary, multiply one or both equations by numbers that make the coefficients of either x or y opposites of each other. Add the two equations to eliminate a variable.
3. Solve the equation from step 2. A unique solution gives the value of one variable. If the equation is $0 = n$, where $n \neq 0$, the system is inconsistent. If the equation is $0 = 0$, the system is dependent.
4. Use the known coordinate value to find the other coordinate value through substitution in either of the original equations.
5. Check the solution in each of the original equations.

4 Another algebraic method for solving linear systems of equations is called the **method of substitution.** This method is most efficient when at least one equation in a linear system is solved for one of the variables, as in the system in Example 6.

EXAMPLE 6 Solve by substitution:
$$y = 1{,}000 + 5x$$
$$y = 4{,}000 + 3x.$$

Solution Both equations are solved for y. We choose to substitute $1{,}000 + 5x$ for y in the bottom equation and solve for x.

$$\begin{aligned} 1{,}000 + 5x &= 4{,}000 + 3x \\ 2x &= 3{,}000 \\ x &= 1{,}500 \end{aligned}$$

Next, replace x by 1,500 in either equation that is solved for y.

$$\begin{aligned} y &= 1{,}000 + 5x \\ y &= 1{,}000 + 5(1{,}500) \\ y &= 8{,}500 \end{aligned} \qquad \text{or} \qquad \begin{aligned} y &= 4{,}000 + 3x \\ y &= 4{,}000 + 3(1{,}500) \\ y &= 8{,}500 \end{aligned}$$

Thus, the solution is (1,500, 8,500).

PROGRESS CHECK 6 Solve by substitution:
$$y = 2{,}000 + 6x$$
$$y = 9{,}000 + 2x.$$

The substitution method also works well when at least one of the equations is easily solved for one of the variables. In this case, look to solve for a variable with a coefficient of 1 or -1.

EXAMPLE 7 Solve by substitution:
$$2x - 11y = 19 \qquad (1)$$
$$5y = x - 8. \qquad (2)$$

Solution We choose to solve for x in equation (2) because this choice leads to an equivalent equation that does not contain fractions.

$$5y = x - 8 \qquad (2)$$
$$5y + 8 = x \qquad (3)$$

Progress Check Answer

6. (1,750, 12,500)

Now substitute $5y + 8$ for x in equation (1) and solve for y.

$$\begin{aligned} 2(5y + 8) - 11y &= 19 \\ 10y + 16 - 11y &= 19 \\ -y &= 3 \\ y &= -3 \end{aligned}$$

From equation (3), $x = 5y + 8$, so

$$x = 5(-3) + 8 = -7.$$

Use equations (1) and (2) to confirm that the solution is $(-7,-3)$.

PROGRESS CHECK 7 Solve by substitution:

$$\begin{aligned} 4x &= y - 21 \\ 7x + 5y &= -3. \end{aligned}$$

The substitution method of this section is summarized next.

Substitution Method Summary

To solve a linear system by substitution:

1. If necessary, solve for a variable in one of the equations. Avoid fractions if possible by solving for a variable with a coefficient of 1 or -1.
2. Use the result in step 1 to make a substitution in the *other* equation. The result is an equation with one unknown.
3. Solve the equation from step 2. A unique solution gives the value of one variable. If the equation is never true, the system is inconsistent. If the equation is always true, the system is dependent.
4. Use the known coordinate value to find the other coordinate value through substitution in the equation in step 1.
5. Check the solution in each of the original equations.

5 A variety of word problems are conveniently analyzed by a linear system in two variables. In particular, many problems that were solved in Chapter 2 using one variable are also readily solved using two variables and a system of linear equations. Compare the next example with the methods shown in Example 4 of Section 2.3.

See "Think About It" Exercise 2 for an example of how ancient this approach is.

EXAMPLE 8 A $16,000 retirement fund was split into two investments, one portion at 11 percent annual interest and the rest at 6 percent. If the total annual interest is $1,400, how much was invested at each rate?

Solution To find the amount invested at each rate, let

x = amount invested at 11 percent

and y = amount invested at 6 percent.

Use $I = Pr$ and analyze the investments in a chart format.

Investment	Principal ·	Interest Rate =	Interest
1st account	x	0.11	$0.11x$
2nd account	y	0.06	$0.06y$

Progress Check Answer

7. $(-4,5)$

The two principals add to \$16,000, and the sum of the interests is \$1,400, so the required system is

$$\begin{aligned} x + y &= 16{,}000 \quad (1) \\ 0.11x + 0.06y &= 1{,}400 \quad (2) \end{aligned}$$

To solve this system, first multiply both sides of equation (2) by 100 to clear decimals.

$$\begin{aligned} x + y &= 16{,}000 \quad (1) \\ 11x + 6y &= 140{,}000 \quad (3) \end{aligned}$$

To eliminate y, we can multiply equation (1) by -6 on both sides and then add the result to equation (3).

$$\begin{aligned} -6x - 6y &= -96{,}000 \\ 11x + 6y &= 140{,}000 \\ \hline 5x \qquad &= 44{,}000 \quad \text{Add the equations.} \\ x &= 8{,}800 \end{aligned}$$

To find y, substitute 8,800 for x in equation (1).

$$\begin{aligned} x + y &= 16{,}000 \\ 8{,}800 + y &= 16{,}000 \quad \text{Replace } x \text{ by 8,800.} \\ y &= 7{,}200 \end{aligned}$$

Thus, \$8,800 was invested at 11 percent, and \$7,200 was invested at 6 percent.

Check \$8,800 + \$7,200 = \$16,000; and the first account earns 11 percent of \$8,800, or \$968, while the second earns 6 percent of \$7,200, or \$432, for a total of \$1,400 interest. The solution checks.

PROGRESS CHECK 8 How should a \$60,000 investment be split so that the total annual earnings are \$6,000 if one portion is invested at 8.5 percent annual interest and the rest at 10.5 percent?

Remind students that a system of equations is often referred to as a "set of simultaneous equations."

In the chapter-opening problem two variables are defined in terms of each other. Thus, they have interlocking solutions that lead naturally to simultaneous equations.

EXAMPLE 9 Solve the problem in the chapter introduction on page 267.

Solution Let s and f represent the state and federal income taxes, respectively, and write an equation for each in terms of the other.

State:	State tax	is ↓	7 percent	of ↓	income after federal deductions.	
	s	$=$	0.07	$\cdot$	$(500{,}000 - f)$	(1)
Federal:	Federal tax	is ↓	28 percent	of ↓	income after state deductions.	
	f	$=$	0.28	$\cdot$	$(500{,}000 - s)$	(2)

The resulting linear system with parentheses cleared is

$$\begin{aligned} s &= 35{,}000 - 0.07f \quad (3) \\ f &= 140{,}000 - 0.28s \quad (4) \end{aligned}$$

Since the equations are solved for a variable, we choose to use the substitution method and replace f by $140{,}000 - 0.28s$ in equation (3).

$$\begin{aligned} s &= 35{,}000 - 0.07(140{,}000 - 0.28s) \\ s &= 35{,}000 - 9{,}800 + 0.0196s \\ 0.9804s &= 25{,}200 \\ s &= \frac{25{,}200}{0.9804} \approx 25{,}704 \end{aligned}$$

Progress Check Answer

8. \$15,000 at 8.5 percent; \$45,000 at 10.5 percent

Then $f = 0.28(500{,}000 - s)$, so
$f = 0.28(500{,}000 - 25{,}704) \approx 132{,}803$.

Thus, to the nearest dollar, the state tax is \$25,704 and the federal tax is \$132,803.

Check $0.28(500{,}000 - 25{,}704) \approx 132{,}803$ and $0.07(500{,}000 - 132{,}803) \approx 25{,}704$, so the solution checks.

PROGRESS CHECK 9 Redo the problem in Example 9, but assume that the state and federal income tax rates are 5 percent and 32 percent, respectively, and the corporation has \$800,000 in taxable income.

Progress Check Answer

9. State tax: \$27,642; federal tax: \$247,155

EXERCISES 6.1

In Exercises 1–6, determine if the given ordered pairs are solutions of the system.

1. $x + y = 10$
   $3x - 2y = 20$
   a. $(8,2)$ Yes b. $(7,3)$ No

2. $2x + 3y = 6$
   $x - y = 3$
   a. $(0,72)$ No b. $(3,0)$ Yes

3. $\frac{x}{2} - \frac{y}{3} = 4$
   $\frac{x}{3} + \frac{y}{3} = 1$
   a. $(8,0)$ No b. $(6,-3)$ Yes

4. $\frac{x}{4} + \frac{y}{6} = 1$
   $3x = 2y$
   a. $(2,3)$ Yes b. $(4,0)$ No

5. $y = 3x + 1$
   $y = -3x + 1$
   a. $(1,4)$ No b. $(0,1)$ Yes

6. $y = -x + 2$
   $y = x - 2$
   a. $(2,0)$ Yes b. $(-2,4)$ No

In Exercises 7–20, solve by graphing. Indicate whether the system is consistent, inconsistent, or dependent.

7. $y = x + 2$
   $y = -x + 2$

(0,2); consistent

8. $y = x + 3$
   $y = -x + 3$

(0,3); consistent

9. $y = x - 2$
   $y = -x + 2$

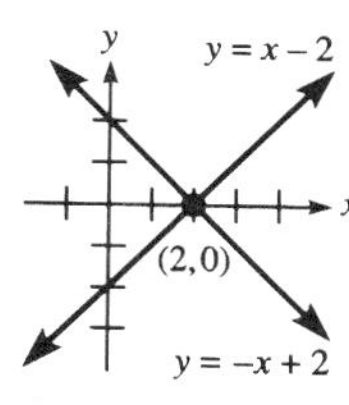

(2,0); consistent

10. $y = x - 3$
    $y = -x + 3$

(3,0); consistent

11. $5x = 4y$
    $2y = 5x - 10$

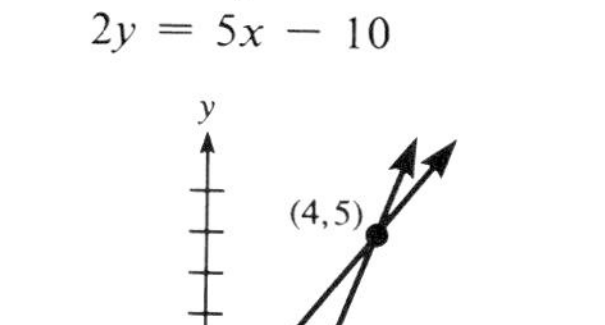

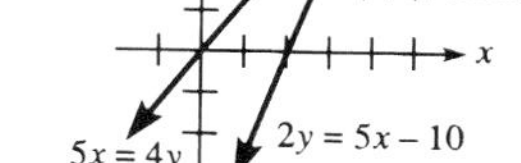

(4,5); consistent

12. $3y = x$
    $y = x - 2$

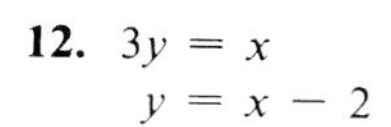
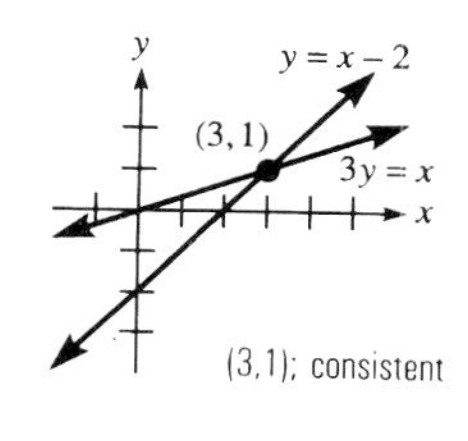

(3,1); consistent

13. $y = x + 2$
    $2x + 3y = 1$

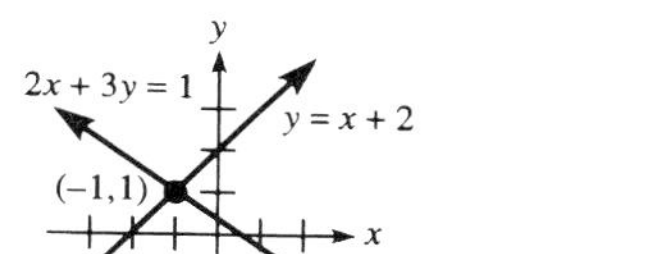

(−1,1); consistent

14. $x + 2y = 2$
    $2x + y = -2$

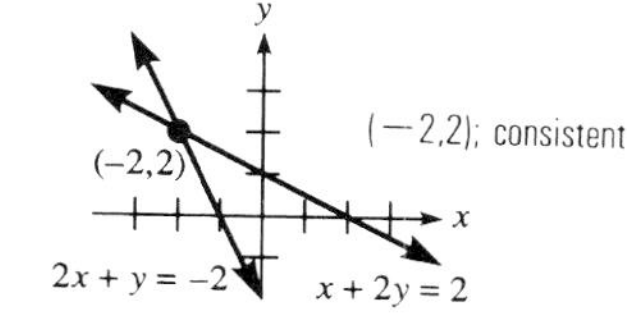

(−2,2); consistent

15. $x = 5$
    $y = -2$

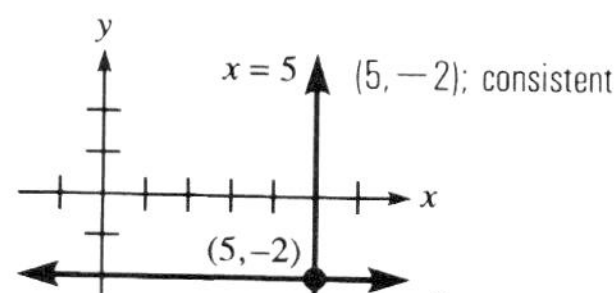

(5, −2); consistent

16. $y = 2$
    $x = -3$

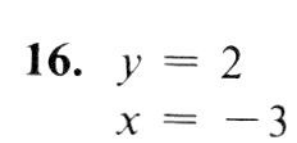
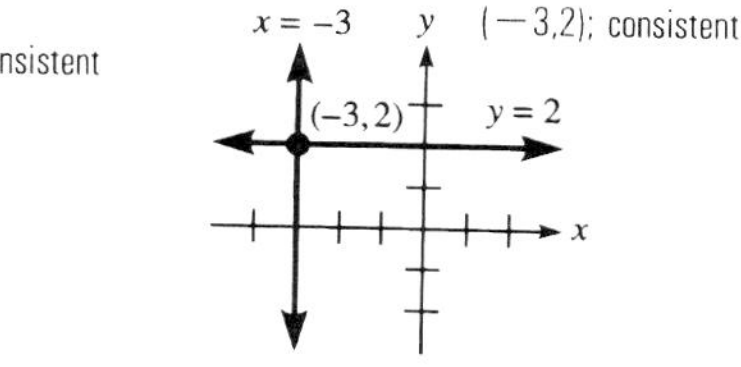

(−3,2); consistent

17. $x + y = 3$
    $x + y = 5$

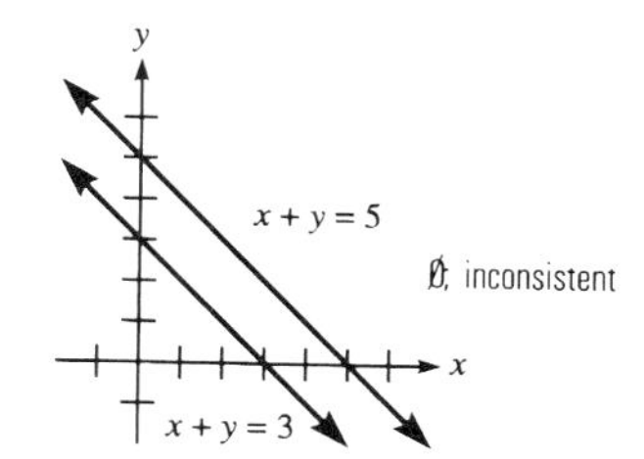

∅; inconsistent

18. $y = 3x + 2$
    $y = 3x - 2$

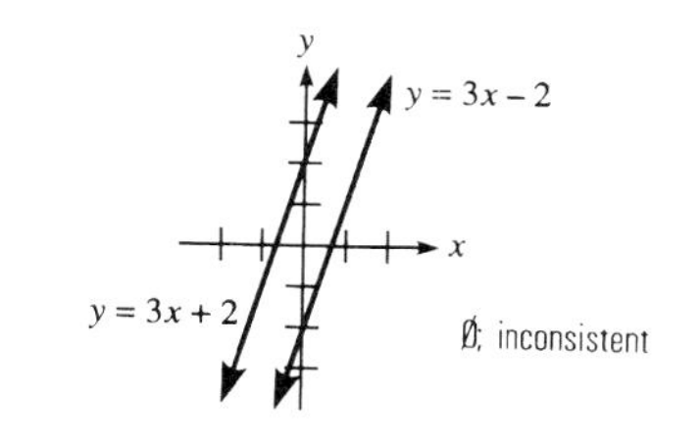

∅; inconsistent

19. 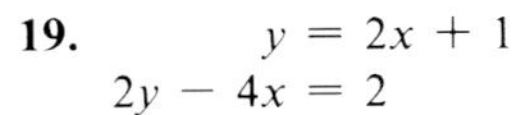

$$y = 2x + 1$$
$$2y - 4x = 2$$

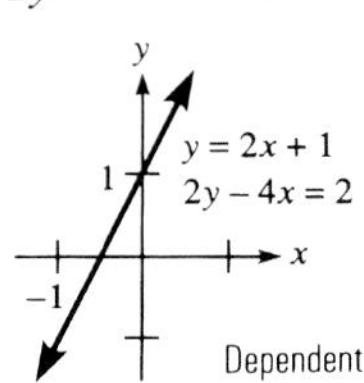

20. $2x + y = 5$
$2y = -4x + 10$

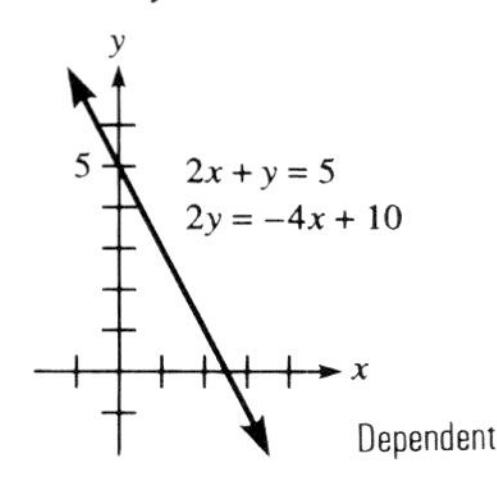

In Exercises 21–50, solve the system by using the addition-elimination method.

21. $4x + 3y = 7$
$x - 3y = 3$ $(2, -\frac{1}{3})$

22. $x - 2y = 1$
$x + 2y = 3$ $(2, \frac{1}{2})$

23. $3x - 2y = 1$
$-3x + y = 3$ $(-\frac{7}{3}, -4)$

24. $2x - 2y = 1$
$-2x + y = 3$ $(-\frac{7}{2}, -4)$

25. $y = 3x + 5$
$y = 5 - 3x$ $(0, 5)$

26. $y = 2x - 4$
$y = 4 - 2x$ $(2, 0)$

27. $3x + 2y = 10$
$3x - 4y = 1$ $(\frac{7}{3}, \frac{3}{2})$

28. $2x - 5y = 7$
$2x + 2y = 5$ $(\frac{39}{14}, -\frac{2}{7})$

29. $2x - 3y = 5$
$3x - 3y = 1$ $(-4, -\frac{13}{3})$

30. $3x - 4y = 7$
$2x - 4y = -2$ $(9, 5)$

31. $3x + 5y = 4$
$x - y = 1$ $(\frac{9}{8}, \frac{1}{8})$

32. $5x + 2y = 3$
$2x - y = 1$ $(\frac{5}{9}, \frac{1}{9})$

33. $x + 3y = 1$
$3x - 2y = 1$ $(\frac{5}{11}, \frac{2}{11})$

34. $x - 4y = 4$
$2x - 2y = 1$ $(-\frac{2}{3}, -\frac{7}{6})$

35. $2x + 3y = 5$
$3x + 2y = 5$ $(1, 1)$

36. $3x - 4y = 1$
$4x - 3y = 1$ $(\frac{1}{7}, -\frac{1}{7})$

37. $2x + 3y = 5$
$3x + 5y = 7$ $(4, -1)$

38. $3x + 4y = 10$
$4x + 4y = 9$ $(-1, \frac{13}{4})$

39. $2x + 5y = 3$
$5x + 7y = 11$ $(\frac{34}{11}, -\frac{7}{11})$

40. $2x - 3y = 4$
$5x + 2y = 1$ $(\frac{11}{19}, -\frac{18}{19})$

41. $3x + 2y = 1$
$-4x - 3y = 2$ $(7, -10)$

42. $-2x + 5y = 3$
$3x - 7y = 4$ $(41, 17)$

43. $2x + 5y = 7$
$4x + 10y = 6$ $\emptyset$

44. $-6x + 3y = 2$
$2x - y = 1$ $\emptyset$

45. $4y = x - 6$
$6y = \frac{3}{2}x - 2$ $\emptyset$

46. $2x = 3y + 1$
$3x = \frac{9}{2}y - 1$ $\emptyset$

47. $4x - 6y = 10$
$2x - 3y = 5$ Dependent system; $\{(x,y): 4x - 6y = 10\}$

48. $3x - 4y = 1$
$9x - 12y = 3$ Dependent system; $\{(x,y): 3x - 4y = 1\}$

49. $\frac{3}{2}x + \frac{5}{2}y = \frac{7}{2}$
$10y = -6x + 14$ Dependent system; $\{(x,y): 10y = -6x + 14\}$

50. $\frac{1}{3}x + \frac{2}{3}y = \frac{4}{3}$
$2y = 4 - x$ Dependent system; $\{(x,y): 2y = 4 - x\}$

In Exercises 51–64, solve each system by the method of substitution.

51. $y = 500 + 2x$
$y = 600 + x$ $(100, 700)$

52. $y = -1{,}000 + 10x$
$y = -2{,}000 + 15x$ $(200, 1{,}000)$

53. $y = 30 - 5x$
$y = 50 + 20x$ $(-\frac{4}{5}, 34)$

54. $y = 100 - 5x$
$y = 10 + 20x$ $(\frac{18}{5}, 82)$

55. $3x - 9y = 15$
$5y = x + 1$ $(14, 3)$

56. $3x + 7y = 21$
$3y = x - 15$ $(\frac{21}{2}, -\frac{3}{2})$

57. $2y - x = 7$
$y + 3x = 10$ $(\frac{13}{7}, \frac{31}{7})$

58. $3y + x = 4$
$y - 2x = 1$ $(\frac{1}{7}, \frac{9}{7})$

59. $2x - y = 5$
$2y - x = 5$ $(5, 5)$

60. $3x - y = 4$
$3y - x = 4$ $(2, 2)$

61. $y = 2x - 1$
$y = 2x - 2$ $\emptyset$

62. $2y + x = 6$
$2y + x = 0$ $\emptyset$

63. $x = 3y - 1$
$6y - 2x = 2$ Dependent system; $\{(x,y): x = 3y - 1\}$

64. $6x + y = 8$
$3x = 4 - \frac{1}{2}y$ Dependent system; $\{(x,y): 6x + y = 8\}$

In Exercises 65–100, solve the problem by setting up a system of two linear equations in two variables. Note carefully what each variable represents, and check that the solutions make sense.

65. A \$200,000 retirement fund was split into two investments, one portion (x) at 10 percent annual interest and the rest (y) at 6 percent. If the total annual interest is \$18,000, how much was invested at each rate? \$150,000 at 10 percent; \$50,000 at 6 percent

66. How should a \$50,000 investment be split so that the total annual earnings are \$5,000 if one portion is invested at 8.5 percent annual interest and the rest at 10.5 percent? \$12,500 at 8.5 percent; \$37,500 at 10.5 percent

67. One year a couple had total income equal to \$60,000 and their total tax bill came to \$14,850. If the income from their salaries was taxed at 28 percent and the income from their investments was taxed at 15 percent, how much income did they have in each category? \$45,000 salary; \$15,000 investments

68. A student borrowed a total of \$6,200 one year to pay for tuition and other expenses. The tuition loan had an annual interest rate of 8 percent, and the other loan had an annual interest rate of 10 percent. The combined interest payments for the year came to \$520. Find the amount of each loan. \$5,000 tuition; \$1,200 expenses

69. A company gives a bonus to a division manager based on this principle: The bonus rate is 15 percent of the division's after-tax income. The tax meanwhile is 28 percent of the division's income after the bonus is paid. If the division's income is \$400,000, to the nearest dollar what is the manager's bonus? \$45,094

70. Redo Exercise 69, but assume that the income is cut in half to \$200,000. Is the bonus half that found in Exercise 69? \$22,547; yes

71. A corporation operates in a state that levies a 4 percent tax on the income that remains after paying the federal tax. Meanwhile, the federal tax is 30 percent of the income that remains after paying the state tax. If, during the current year, a corporation has \$1 million in taxable income, determine (to the nearest dollar) the state and federal income tax. State tax: \$28,340; federal tax: \$291,498

72. Redo Exercise 71, but assume that the state tax rate is 8 percent instead of 4. If, during the current year, a corporation has \$1 million in taxable income, determine the state and federal income tax. Is the amount of the state tax double that found in Exercise 71? State tax: \$57,377; federal tax: \$282,787; no

73. On their joint income tax return, A's income was $\frac{1}{4}$ of B's income, and their total income was $47,500. Find each person's individual income. A's income: $9,500; B's income: $38,000

74. Together, a newspaper and a magazine cost $5, and the magazine costs $2 more than the newspaper. What is the price of each? (Try guessing the answer first. Most people guess wrong.) Cost of newspaper: $1.50; cost of magazine: $3.50

75. On a video display an air traffic controller notices two planes 90 mi apart and flying toward each other on a collision course. One plane is flying 500 mi/hour, and the other is flying 400 mi/hour. How much time is there for the controller to prevent a crash? 6 minutes

76. Two turtles are crawling toward each other in a straight line at constant speed. One goes at the rate of 2 ft in 10 minutes. The other moves 2.5 ft in 10 minutes. If they start 9 ft apart, how long will it take them to meet? 20 minutes

77. A machine shop has two large containers that are each filled with a mixture of oil and gasoline. Container A contains 2 percent oil (and 98 percent gasoline). Container B contains 5 percent oil (and 95 percent gasoline). How much of each should be used to obtain 6 gal of a new mixture that contains 3 percent oil? 4 gal of 2 percent solution; 2 gal of 5 percent solution

78. A chemist has two acid solutions, one 30 percent acid and the other 20 percent acid. How much of each must be used to obtain 10 liters of a solution that is 22 percent acid? 2 liters of 30 percent solution; 8 liters of 20 percent solution

79. A small company is trying to decide between two copy machines for the office. Machine A will cost $1,000 plus $0.04 per copy, while machine B will cost $4,000 plus $0.01 per copy. How many copies must be made for the cost to be the same? If they make 60,000 copies per year, about how long will it take before the costs are the same? 100,000 copies; 20 months

80. At mile marker 200 on a long interstate highway, a car going 60 mi/hour is 15 mi behind a car going 55 mi/hour. If they travel at constant speed, at what mile marker will the faster car catch up to the slower one? (Assume that the mile marker numbers are increasing.) Mile marker 380

81. Economists have made mathematical models which project production of various resources over time. For country A, the production index is given by $I = 1{,}000 + 60.5x$, where x is the number of years elapsed from the present. For country B, the index is given by $I = 1{,}120 + 58.1x$. According to these indexes, how many years will it take for country A to catch up to country B? 50 years

82. Refer to Exercise 81. Country C has a production index given by $I = 880 + 65.6x$. How long will it take country C to catch up to country B? 32 years

83. Find two numbers whose sum is 10 and whose difference is 15. 12.5 and -2.5

84. Find two numbers whose sum is $\frac{5}{8}$ and whose difference is $\frac{8}{5}$. $-\frac{39}{80}$ and $\frac{89}{80}$

For Exercises 85–88, recall that the sum of two complementary angles is 90° and the sum of two supplementary angles is 180°

85. Two angles are complementary. If the difference between the angle measures is 24°, find the angle measures. 57° and 33°

86. Two angles are supplementary. If the difference between the angle measures is 54°, find the angle measures. 117° and 63°

87. Two angles are complementary, and one is 4 times as large as the other. Find the smaller angle. 18°

88. Two angles are supplementary, and one is $\frac{3}{5}$ the size of the other. Find the larger angle. 112.5°

89. The length of a rectangle is 3 times the width, and the perimeter is 12 cm. Find the area. 6.75 cm²

90. The length of a rectangle is 4 times the width, and the perimeter is 12 cm. Find the area. 5.76 cm²

91. The perimeter of a rectangle is 22 in. and the length is 6 in. more than the width. Find the dimensions of the rectangle. 8.5 in. by 2.5 in.

92. The perimeter of a rectangle is 22 in. and the width is $\frac{2}{3}$ the length. How much longer is the length than the width? 2.2 in.

93. The total bill for a shopper who bought clothes and tools at a department store consisted of $180 in purchases and $9.20 in tax. The clothes were taxed at 4 percent and the tools at 6 percent. How much was spent on clothes (including tax)? $83.20

94. The total bill at a restaurant was $164.78, of which $7.78 was tax. The food was taxed at a 4 percent rate and the beverages at 10 percent. How much (including tax) was spent on food? $137.28

95. A shopper was offered a 20 percent discount on hardback books and a 10 percent discount on paperbacks. The regular price for the purchase was $254.70, on which the total discount was $40.44. How much did the shopper spend on each kind of book? $94.50 on paperbacks; $119.76 on hardbacks

96. A ski shop discounts red-tag items by 30 percent and yellow-tag items by 20 percent. A shopper who bought only red- and yellow-tagged items got a total discount of $120 on a purchase, bringing the bill down to $305. After the discount, how much did the shopper spend in each category? $60 on yellow-tag items; $245 on red-tag items

97. An airplane flying with the wind takes 48 minutes to fly 360 mi but 54 minutes to fly this distance against this wind. Find the wind speed and the speed of the plane with no wind. (*Hint:* 48 minutes = $\frac{48}{60}$ hour.) Plane speed = 425 mi/hour; wind speed = 25 mi/hour

98. Two friends paddle a canoe downstream a distance of 12 mi in 2 hours. Returning upstream, although they paddle at the same rate, takes them 3 hours. Find the speed of the current. 1 mi/hour

99. A child runs at her top speed for 0.1 mi the "wrong" way along a moving sidewalk in an airport. This takes 3 minutes. When she runs (also at top speed) the "right" way, it takes 1 minute. How fast is the sidewalk moving? $\frac{1}{30}$ mi/minute, or 2 mi/hour

100. Redo Exercise 99 but with the speed of the sidewalk changed. It now takes the child 1.2 minutes running the right way and 2 minutes the wrong way. What is the speed of the moving sidewalk? $\frac{1}{60}$ mi/minute, or 1 mi/hour

THINK ABOUT IT

1. a. When the equations in a system of two nonvertical lines are written in slope-intercept form, it is not difficult to classify the system as having exactly one solution, no solutions, or infinitely many solutions. How is this done?
   b. Find the number of solutions without solving the system.

$$\begin{aligned} 4x - 2y &= 5 \\ y &= 2x - 7 \end{aligned}$$

2. Clay tablets from more than 3,000 years ago show that the mathematicians of ancient Babylonia also solved systems of linear equations, but their approach was different from ours. One of their problems leads to this system.

$$\begin{aligned} \tfrac{2}{3}x - \tfrac{1}{2}y &= 500 \qquad (1) \\ x + y &= 1{,}800 \qquad (2) \end{aligned}$$

   The basic idea of their solution is to start with a wrong answer and fix it, so they start with the guess that x and y are equal. From equation 2, this gives $x = y = 900$ as a starting point.
   a. Letting x and y each equal 900 in the left-hand side of equation (1) gives 150, which is 350 short of the desired 500. Check this.
   b. What change in x and y will make up this shortage? Since (by equation 2) every increase of 1 in x must be accompanied by a decrease of 1 in y, they find that when x increases by 1, the left-hand side of equation (1) increases by $\frac{7}{6}$. Check that $\frac{2}{3}(1) - \frac{1}{2}(-1) = \frac{7}{6}$.
   c. To find out how many unit increases (u) will make up the shortage of 350, they solve $\frac{7}{6}u = 350$, which gives $u = 300$. Therefore, $x = 900 + 300 = 1{,}200$, and $y = 900 - 300 = 600$. Use the Babylonian method to solve

$$\begin{aligned} \tfrac{1}{2}x - \tfrac{1}{4}y &= 350 \\ x + y &= 1{,}000. \end{aligned}$$

   [This problem and others like it are described in *A History of Mathematics* by Victor J. Katz (HarperCollins, 1993).]

3. Which one of the figures would be used to graphically solve the system

$$\begin{aligned} 2x + y &= 4 \\ 3x - 5y &= 45? \end{aligned}$$

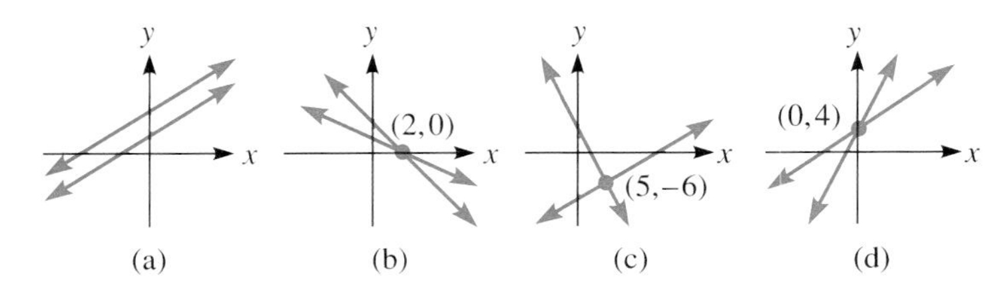

4. It is not too difficult to derive a formula for the solution of a system of two linear equations in two variables. The general system may be written as

$$\begin{aligned} a_1x + b_1y &= c_1 \\ a_2x + b_2y &= c_2. \end{aligned}$$

   a. Check that the solution is given by

$$x = \frac{c_1b_2 - b_1c_2}{a_1b_2 - a_2b_1} \quad \text{and} \quad y = \frac{a_1c_2 - a_2c_1}{a_1b_2 - a_2b_1}.$$

   To do this, substitute these expressions for x and y in the original system and show that an identity results.
   b. Derive the solution yourself by solving the original system by the addition-elimination method (*Hint:* To start, multiply both sides of the top equation by b_2 and both sides of the bottom equation by $-b_1$.)

5. The coordinate system we use in algebra is called rectangular because the x- and y-axes are perpendicular. Other systems are possible. For instance, let the y-axis be a vertical line, and let the x-axis make a 45° angle with it, as shown in the figure.

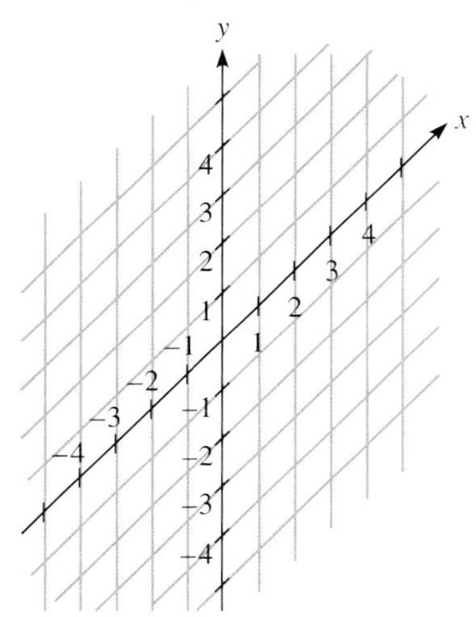

   a. Use the new system to graph this system: $x + y = 2$, $x + y = 3$.
   On a rectangular coordinate system the equations represent parallel lines. Are the lines still parallel in the new system?
   b. Use the new system to graph this system: $x + y = 7$, $y - x = 1$.
   On a rectangular coordinate system the lines intersect at (3,4). Do they still intersect at (3,4) in the new system?
   c. Explain why the choice of coordinate system cannot change the solution set of the system of linear equations.

REMEMBER THIS

1. Which of these are linear equations in one variable?
   a. $x^2 - 2x + 3 = 0$ b. $2x - 7 = 0$ c. $y = 4x + 6$ b
2. Is $2x - 3y + z = 3$ true when $x = -3$, $y = -1$, and $z = 6$?
   Yes
3. What equation results from adding the two equations?
$$\begin{aligned} 2x - y &= 7 \\ -x + 2y &= 6 \end{aligned} \quad x + y = 13$$
4. If a system of equations has no solution, then it is called
   a. inconsistent b. dependent c. nonlinear a
5. y varies directly as x^2 and $y = 6$ when $x = 2$. Find y when $x = 3$. $\frac{27}{2}$
6. Graph the inequality $y > 2x + 4$.

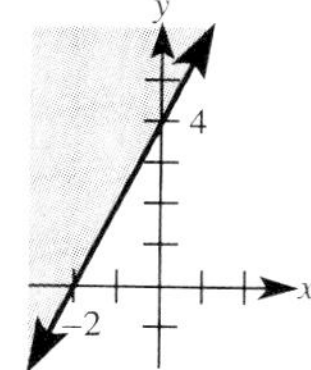

7. Find an equation for the line through $(3,-7)$ and $(3,-8)$. $x = 3$
8. Find the distance between $(-1,4)$ and $(3,-1)$. $\sqrt{41}$
9. Add, and express the result in lowest terms.
$$\frac{t-1}{2t-3} + \frac{t-2}{3-2t} \quad \frac{1}{2t-3}$$
10. The area of a triangle is 8 in.² and the height is 6 in. less than the base. Find the height of the triangle. 2 in.

6.2 Systems of Linear Equations in Three Variables

In electronics, applying Kirchhoff's laws to the circuit shown in the diagram yields the following system of equations.

$$\begin{aligned} I_1 + I_2 + I_3 &= 0 \\ R_1 I_1 - R_3 I_3 &= E_1 \\ R_2 I_2 - R_3 I_3 &= E_2 \end{aligned}$$

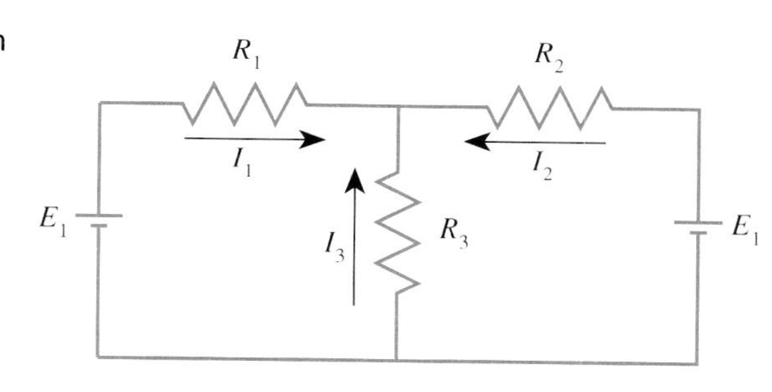

If $E_1 = 7$ volts, $E_2 = 2$ volts, $R_1 = 5$ ohms, $R_2 = 4$ ohms, and $R_3 = 9$ ohms, find values of currents I_1, I_2, and I_3, which are measured in amperes. (See Example 3.)

OBJECTIVES

1. **Determine whether an ordered triple is a solution of a system of linear equations.**
2. **Solve linear systems in three variables with exactly one solution.**
3. **Solve linear systems in three variables that are inconsistent or dependent.**

1 Often, problems involve many variables and many relationships among the variables, so it is useful to extend the methods of Section 6.1 beyond linear systems in two variables. In this section we deal with **linear (first-degree) equations in three variables,** which are equations of the form

$$ax + by + cz = d,$$

where a, b, c, and d are real numbers and x, y, and z are variables. A solution of such an equation is an **ordered triple** of numbers of the form (x,y,z) that satisfies the equation. Furthermore, the solution set of a system of three linear equations in three variables consists of all ordered triples that satisfy all the equations at the same time (simultaneously).

EXAMPLE 1 Determine if $(-3,-1,6)$ is a solution of the system.

$$\begin{aligned} 2x - 3y + z &= 3 \quad (1)\\ 5x - 4y + 2z &= 1 \quad (2)\\ 7x - 2y + 3z &= 2 \quad (3) \end{aligned}$$

Solution $(-3,-1,6)$ means $x = -3$, $y = -1$, and $z = 6$. Substitute and check to see if true statements result in all three equations.

$$\begin{aligned} 2x - 3y + z &= 3 & (1)\\ 2(-3) - 3(-1) + 6 &\stackrel{?}{=} 3 &\\ 3 &= 3 & \text{True} \end{aligned}$$

$$\begin{aligned} 5x - 4y + 2z &= 1 & (2)\\ 5(-3) - 4(-1) + 2(6) &\stackrel{?}{=} 1 &\\ 1 &= 1 & \text{True} \end{aligned}$$

$$\begin{aligned} 7x - 2y + 3z &= 2 & (3)\\ 7(-3) - 2(-1) + 3(6) &\stackrel{?}{=} 2 &\\ -1 &= 2 & \text{False} \end{aligned}$$

Thus, $(-3,-1,6)$ is a solution of equations (1) and (2) but not of equation (3). Because this ordered triple is not a solution of all three equations, it is not a solution of this system.

PROGRESS CHECK 1 Determine if $(\frac{1}{2},1,-2)$ is a solution of the system.

$$\begin{aligned} 8x - 3y - 3z &= 7\\ 6x + y + 2z &= 0\\ -2x + 4y + 3z &= -3 \end{aligned}$$

2 The next example discusses how the addition-elimination method of the previous section may be extended to solve a linear system in three variables.

EXAMPLE 2 Solve the system.

$$\begin{aligned} 2x - y + z &= 7 \quad (1)\\ -x + 2y - z &= 6 \quad (2)\\ 2x - 3y - 2z &= 9 \quad (3) \end{aligned}$$

Solution The initial goal is to obtain two equations in two variables which may be solved as in Section 6.1. To obtain a first equation, select any pair of equations in the system and use the addition method to eliminate one of the variables. For the given system, z is eliminated simply by adding equations (1) and (2).

$$\begin{aligned} 2x - y + z &= 7 \quad (1)\\ -x + 2y - z &= 6 \quad (2)\\ \hline x + y &= 13 \quad (4) \end{aligned}$$

To obtain a second equation, select a different pair of equations in this system and eliminate the *same* variable z. We choose equations (1) and (3) and eliminate z by multiplying both sides of equation (1) by 2 and adding the result to equation (3).

$$\begin{aligned} 2(2x - y + z) = 2(7) \rightarrow 4x - 2y + 2z &= 14\\ 2x - 3y - 2z &= 9\\ \hline 6x - 5y &= 23 \quad (5) \end{aligned}$$

Progress Check Answer

1. Solution

Equations (4) and (5) can now be used to obtain a linear system in two variables (our initial goal).

$$\begin{aligned} x + y &= 13 \quad (4) \\ 6x - 5y &= 23 \quad (5) \end{aligned}$$

To solve this system, we choose to eliminate y and solve for x, as shown next.

$$\begin{aligned} 5(x + y) = 5(13) \rightarrow \quad 5x + 5y &= 65 \\ 6x - 5y &= 23 \\ \hline 11x \qquad &= 88 \\ x &= 8 \end{aligned}$$

Then substituting 8 for x in equation (4) gives

$$\begin{aligned} 8 + y &= 13 \\ y &= 5. \end{aligned}$$

Finally, by replacing x by 8 and y by 5 in equation (1), we have

$$\begin{aligned} 2(8) - 5 + z &= 7 \\ z &= -4. \end{aligned}$$

Thus, the solution is $(8,5,-4)$. Check it in the original equations (1), (2), and (3).

PROGRESS CHECK 2 Solve the system.

$$\begin{aligned} 3x + 2y + z &= -3 \\ 2x + 3y + 2z &= 5 \\ -2x + y - z &= 3 \end{aligned}$$

The methods of Example 2 illustrate a general procedure for solving a linear system in three variables, which is summarized next.

To Solve a Linear System in Three Variables

1. Select any pair of equations in the system and use the addition method to eliminate one of the variables.
2. Choose a different pair of equations in the system and eliminate the *same* variable by using the addition method again.
3. Use the results of steps 1 and 2 to obtain a linear system in two variables, and solve this system by the methods of Section 6.1.
4. Substitute the values of the two variables obtained in step 3 into one of the original equations to find the value of the third variable.
5. Check the solution in all three of the original equations.

The next example illustrates that the general procedure just given may be shortened when at least one equation in a linear system in three variables has a missing term.

EXAMPLE 3 Solve the problem in the section introduction on page 279.

Solution Using the equations and substitutions given in the question leads to the system

$$\begin{aligned} I_1 + I_2 + I_3 &= 0 \quad (1) \\ 5I_1 \qquad - 9I_3 &= 7 \quad (2) \\ 4I_2 - 9I_3 &= 2. \quad (3) \end{aligned}$$

Progress Check Answer

2. $(-3,1,4)$

Equation (2) is missing an I_2 term, and equation (3) has no I_1 term. Select one of these variables, say I_1, for further elimination. Multiplying both sides of equation (1) by -5 and adding the result to equation (2) gives

$$\begin{array}{rl} -5(I_1 + I_2 + I_3) = -5(0) \rightarrow & -5I_1 - 5I_2 - 5I_3 = 0 \\ & \underline{5I_1 \qquad\quad - 9I_3 = 7} \\ & \qquad - 5I_2 - 14I_3 = 7. \qquad (4) \end{array}$$

Equations (3) and (4) now form a linear system in two variables, and I_3 may be found as follows.

$$\begin{array}{rl} 5(4I_2 - 9I_3) = 5(2) \rightarrow & 20I_2 - 45I_3 = 10 \\ 4(-5I_2 - 14I_3) = 4(7) \rightarrow & \underline{-20I_2 - 56I_3 = 28} \\ & -101I_3 = 38 \\ & I_3 = -\frac{38}{101} \end{array}$$

Because finding I_2 using I_3 involves a lot of work with fractions, we choose to find I_2 by again using equations (3) and (4).

$$\begin{array}{rl} 14(4I_2 - 9I_3) = 14(2) \rightarrow & 56I_2 - 126I_3 = 28 \\ -9(-5I_2 - 14I_3) = -9(7) \rightarrow & \underline{45I_2 + 126I_3 = -63} \\ & 101I_2 \qquad\quad = -35 \\ & I_2 = -\frac{35}{101} \end{array}$$

Finally, replacing I_2 by $-\frac{35}{101}$ and I_3 by $-\frac{38}{101}$ in equation (1) gives

$$\begin{aligned} I_1 - \tfrac{35}{101} - \tfrac{38}{101} &= 0 \\ I_1 - \tfrac{73}{101} &= 0 \\ I_1 &= \tfrac{73}{101}. \end{aligned}$$

Thus, $I_1 = \frac{73}{101}$ ampere, $I_2 = -\frac{35}{101}$ ampere, and $I_3 = -\frac{38}{101}$ ampere. Check this solution in all three of the original equations.

PROGRESS CHECK 3 Use the equations given in Example 3 and find I_1, I_2, and I_3 if $E_1 = 3$ volts, $E_2 = 10$ volts, $R_1 = 2$ ohms, $R_2 = 9$ ohms, and $R_3 = 5$ ohms.

3 When the application of our current methods for solving a linear system in three variables results in a false equation *at any step*, then the system is inconsistent and has no solution. Example 4 illustrates this case.

EXAMPLE 4 Solve the system.

$$\begin{aligned} 2x + y - z &= 2 \qquad (1) \\ x + 2y + z &= 5 \qquad (2) \\ x - y - 2z &= -2 \qquad (3) \end{aligned}$$

Solution We follow the steps of the general procedure and choose to eliminate z.

Step 1 Eliminate z using equations (1) and (2).

$$\begin{array}{rl} 2x + \ y - z = 2 & (1) \\ \underline{x + 2y + z = 5} & (2) \\ 3x + 3y \qquad = 7 & (4) \end{array}$$

Step 2 Eliminate z again using a different pair of equations, say (2) and (3).

$$\begin{array}{rl} 2(x + 2y + z) = 2(5) \rightarrow & 2x + 4y + 2z = 10 \\ & \underline{x - \ y - 2z = -2} \\ & 3x + 3y \qquad = 8 \qquad (5) \end{array}$$

Progress Check Answer

3. $I_1 = -\frac{8}{73}$ ampere, $I_2 = \frac{55}{73}$ ampere, $I_3 = -\frac{47}{73}$ ampere

Step 3 Solve the linear system in two variables resulting from steps 1 and 2.

$$\begin{array}{rr} & 3x + 3y = 7 \\ -1(3x + 3y) = -1(8) \rightarrow & \underline{-3x - 3y = -8} \\ & 0 = -1 \end{array}$$

The false equation $0 = -1$ indicates that the system is inconsistent and has no solution. The solution set for every inconsistent system is $\emptyset$.

Note At any step a false equation implies an inconsistent system. Therefore, if step 1 results in a false equation, then conclude without further work that the system is inconsistent.

PROGRESS CHECK 4 Solve the system.

$$\begin{aligned} x - y - z &= 5 \\ 4x + y + 3z &= 8 \\ -2x + 2y + 2z &= 7 \end{aligned}$$

A linear system in three variables may also be a dependent system, and the next example illustrates this case.

EXAMPLE 5 Solve the system.

$$\begin{aligned} x - y + 2z &= 0 \quad (1) \\ -x + 4y + z &= 0 \quad (2) \\ -x + 2y - z &= 0 \quad (3) \end{aligned}$$

Solution We follow steps of the general procedure and choose to eliminate x.

Step 1 Eliminate x using equations (1) and (2).

$$\begin{array}{rl} x - y + 2z = 0 & (1) \\ \underline{-x + 4y + z = 0} & (2) \\ 3y + 3z = 0 & (4) \end{array}$$

Step 2 Eliminate x again using a different pair of equations.

$$\begin{array}{rl} x - y + 2z = 0 & (1) \\ \underline{-x + 2y - z = 0} & (3) \\ y + z = 0 & (5) \end{array}$$

Step 3 Solve the system comprised of equations (4) and (5).

$$\begin{array}{rr} & 3y + 3z = 0 \\ -3(y + z) = -3(6) \rightarrow & \underline{-3y - 3z = 0} \\ & 0 = 0 \end{array}$$

The identity $0 = 0$ indicates that the system is dependent and that the number of solutions is infinite. In higher mathematics, specifying the solution set for the system in this example is considered.

Note It is also common to conclude that a system is dependent if *both* step 1 and step 2 in our current methods produce identities. In this case all the equations are equivalent, and the solution set is the set of all ordered triples satisfying any equation in the system. However, if one step produces an identity and the other step produces a false equation, then the system is inconsistent. Remember that a false equation *at any step* implies an inconsistent system.

A linear system with the constant term of zero for each equation is called a homogeneous system. A trivial solution is to let all variables equal zero. In this example all solutions fit the form $(3a, a, -a)$.

Progress Check Answer

4. Inconsistent system; $\emptyset$

PROGRESS CHECK 5 Solve the system.

$$\begin{aligned} 3x + 2y - z &= 3 \\ x - y + z &= 1 \\ 5x \qquad + z &= 5 \end{aligned}$$

As with linear equations in two variables, it is possible to interpret geometrically a linear system in three variables. However, an equation that may be written in the form $ax + by + cz = d$ graphs as a plane in a three-dimensional space. Consider carefully Figure 6.4, which summarizes the geometric solution: either exactly one point, no point, infinitely many points on a line, or infinitely many points in a plane.

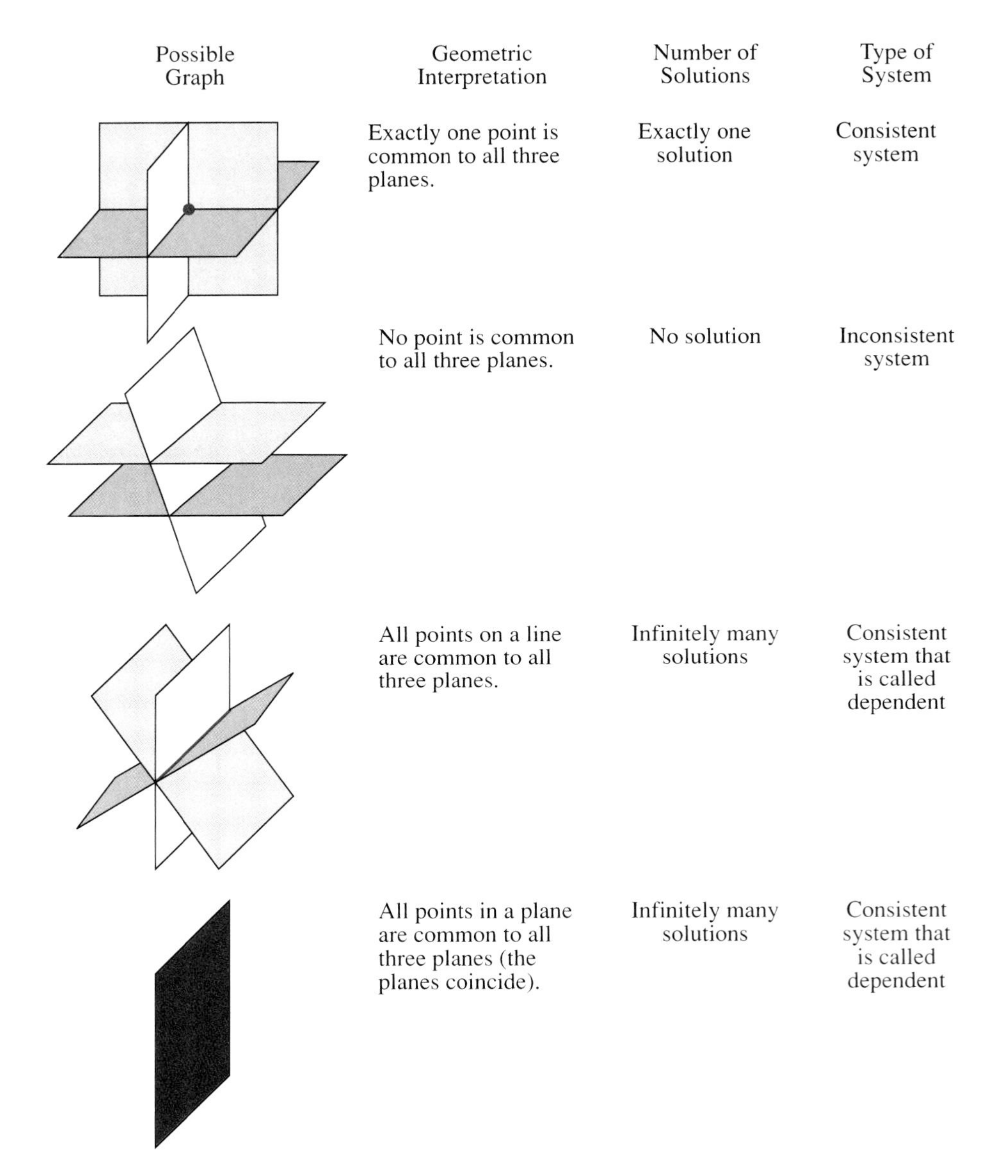

Possible Graph	Geometric Interpretation	Number of Solutions	Type of System
	Exactly one point is common to all three planes.	Exactly one solution	Consistent system
	No point is common to all three planes.	No solution	Inconsistent system
	All points on a line are common to all three planes.	Infinitely many solutions	Consistent system that is called dependent
	All points in a plane are common to all three planes (the planes coincide).	Infinitely many solutions	Consistent system that is called dependent

Figure 6.4

Progress Check Answer

5. Dependent system; infinitely many solutions

EXERCISES 6.2

In Exercises 1–6, determine if the given ordered triples are solutions of the system.

1. $$\begin{aligned} x + y + z &= 3 \\ 3x - 2y + z &= 2 \\ x + y - z &= 1 \end{aligned}$$

a. $(1,1,1)$ Yes
b. $(4,3,-4)$ No

2. $$\begin{aligned} 2x + 3y - z &= 1 \\ 2x - 3y - z &= -5 \\ 2x + y + z &= -1 \end{aligned}$$

a. $(-1,1,0)$ Yes
b. $(2,1,6)$ No

3. $$\begin{aligned} x - y + z &= 6 \\ 2x - y + z &= 7 \\ 2x - 2y + 2z &= 12 \end{aligned}$$

a. $(1,2,7)$ Yes
b. $(1,1,6)$ Yes

4. $$\begin{aligned} x + y + z &= 6 \\ x - y + z &= 4 \\ 2x + 2y + 2z &= 12 \end{aligned}$$

a. $(2,1,3)$ Yes
b. $(3,1,2)$ Yes

5. $$\begin{aligned} 4x - 3y + 5z &= 7 \\ 5x - 3y + 4z &= 5 \\ 3x - 4y + 5z &= 6 \end{aligned}$$

a. $(0,1,2)$ Yes
b. $(-\frac{1}{3},0,\frac{15}{9})$ No

6. $$\begin{aligned} 4x - 2y + z &= 1 \\ 2x + 4y + z &= 3 \\ x + y - z &= 1 \end{aligned}$$

a. $(\frac{1}{2},\frac{1}{2},0)$ Yes
b. $(2,1,-5)$ No

In Exercises 7–26, solve the given system.

7. $$\begin{aligned} x + y + z &= 6 \\ 2x + y - z &= 1 \\ 3x + 2y + z &= 10 \end{aligned}$$
$(1,2,3)$

8. $$\begin{aligned} 2x - y + z &= 2 \\ 3x - 2y - z &= -1 \\ x - 3y + 6z &= -1 \end{aligned}$$
$(2,3,1)$

9. $$\begin{aligned} 3x - 2y + z &= -3 \\ 3x + 2y + 3z &= 5 \\ x - y - 2z &= -6 \end{aligned}$$
$(-1,1,2)$

10. $$\begin{aligned} 2x - 3y + z &= 11 \\ x + 3y + 2z &= 4 \\ 3x - y - 3z &= 4 \end{aligned}$$
$(3,-1,2)$

11. $$\begin{aligned} x - y + 3z &= 4 \\ 2x + 3y - 3z &= -10 \\ 3x - 2y + z &= -4 \end{aligned}$$
$(-2,0,2)$

12. $$\begin{aligned} 4x - 2y - z &= -15 \\ -4x - y - 2z &= 6 \\ 2x - 3y + 3z &= 3 \end{aligned}$$
$(-3,0,3)$

13. $$\begin{aligned} x + 3y - 4z &= 0 \\ x - y + 2z &= 4 \\ x + 2y - z &= 4 \end{aligned}$$
$(2,2,2)$

14. $$\begin{aligned} 2x - y + 5z &= 18 \\ 2x - 3y - 4z &= -15 \\ x - 4y + 3z &= 0 \end{aligned}$$
$(3,3,3)$

15. $$\begin{aligned} x + 2y + 3z &= 4 \\ 2x + 2y + 4z &= 6 \\ 3x + 3y + 5z &= 6 \end{aligned}$$
$(-1,-2,3)$

16. $$\begin{aligned} 5x - 3y + z &= -16 \\ 4x - 3y + 2z &= -11 \\ 3x - 2y + 3z &= 5 \end{aligned}$$
$(2,11,7)$

17. $$\begin{aligned} x + y + z &= 9 \\ 2x + 3y + 4z &= 23 \\ 4x - 6y - 2z &= 0 \end{aligned}$$
$(5,3,1)$

18. $$\begin{aligned} x + y + z &= 12 \\ 3x - 2y + 5z &= 20 \\ 6x + 4y - 2z &= 48 \end{aligned}$$
$(6,4,2)$

19. $$\begin{aligned} 3x - 2y + 4z &= 9 \\ 2x - z &= 1 \\ y + 2z &= 1 \end{aligned}$$
$(1,-1,1)$

20. $$\begin{aligned} 2x - 4y + 3z &= 18 \\ 3x + y &= 4 \\ 2y + z &= -2 \end{aligned}$$
$(2,-2,2)$

21. $$\begin{aligned} x + 2y &= -1 \\ 2x + 3z &= 9 \\ 3y + 4z &= -2 \end{aligned}$$
$(3,-2,1)$

22. $$\begin{aligned} 2x - y &= 0 \\ 2y - 3z &= 13 \\ 3x - 4z &= 15 \end{aligned}$$
$(1,2,-3)$

23. $$\begin{aligned} 2x + 3y + 4z &= 3 \\ 4x + 3y - 4z &= 2 \\ 6x - 6y + 8z &= 3 \end{aligned}$$
$(\frac{1}{2},\frac{1}{3},\frac{1}{4})$

24. $$\begin{aligned} 3x - 4y + 3z &= 2 \\ 3x + 4y + 3z &= 4 \\ 4x + 4y + z &= 3 \end{aligned}$$
$(\frac{1}{3},\frac{1}{4},\frac{2}{3})$

25. $$\begin{aligned} x + y + z &= 0 \\ -x + y - z &= 0 \\ -x - y + z &= 0 \end{aligned}$$
$(0,0,0)$

26. $$\begin{aligned} 2x - y - z &= 0 \\ x - 2y - z &= 0 \\ x - y - 2z &= 0 \end{aligned}$$
$(0,0,0)$

In Exercises 27–34, show that the given system is inconsistent.

27. $$\begin{aligned} x + y + z &= 1 \\ x + y + z &= 2 \\ x + y + z &= 3 \end{aligned}$$

28. $$\begin{aligned} x + 2y + 3z &= 0 \\ 2x + 4y - z &= 2 \\ 3x + 6y - 4z &= 3 \end{aligned}$$

29. $$\begin{aligned} x - 2y - z &= 0 \\ 2x + y + 2z &= 1 \\ 3x - y + z &= 2 \end{aligned}$$

30. $$\begin{aligned} 3x + 2y - 4z &= 2 \\ x - y + 2z &= 1 \\ 2x + 3y - 6z &= 3 \end{aligned}$$

31. $$\begin{aligned} 2x - 3y + 4z &= 2 \\ -x - y + z &= 1 \\ 3x - 2y + 3z &= 4 \end{aligned}$$

32. $$\begin{aligned} 2x - y + 2z &= 3 \\ 2x - y - 3z &= 1 \\ 2x - y + 4z &= 3 \end{aligned}$$

33. $$\begin{aligned} x + y - z &= 3 \\ -x - y + z &= -3 \\ 2x + 2y - 2z &= 5 \end{aligned}$$

34. $$\begin{aligned} 3x - y - z &= 1 \\ 6x - 2y - 2z &= 2 \\ 9x - 3y - 3z &= 0 \end{aligned}$$

27–34. Answers will vary. Solutions lead to $0 = a$, with $a \neq 0$.

In Exercises 35–42, show that the given system is dependent.

35. $$\begin{aligned} x + 2y + 3z &= 1 \\ 2x + 4y - z &= 2 \\ 3x + 6y - 4z &= 3 \end{aligned}$$

36. $$\begin{aligned} 3x + y - z &= 1 \\ x - y + 2z &= -1 \\ 2x + 2y - 3z &= 2 \end{aligned}$$

37. $$\begin{aligned} 2x - 3y + 4z &= 4 \\ 4x + 2y - z &= -1 \\ 6x - y + 3z &= 3 \end{aligned}$$

38. $$\begin{aligned} x - y + z &= 1 \\ x + 2y - z &= 1 \\ x - 4y + 3z &= 1 \end{aligned}$$

39. $$\begin{aligned} 5x - 2y - 3z &= 3 \\ 2x - 5y + 3z &= -3 \\ x + 8y - 9z &= 9 \end{aligned}$$

40. $$\begin{aligned} 4x - 3y + 5z &= 1 \\ 2x - y - 3z &= 1 \\ 2x - 2y + 8z &= 0 \end{aligned}$$

41. $$\begin{aligned} x + y + z &= 0 \\ x - y - z &= 0 \\ 2x - 2y - 2z &= 0 \end{aligned}$$

42. $$\begin{aligned} -x + y + z &= 0 \\ x - y + z &= 0 \\ 3x - 3y - z &= 0 \end{aligned}$$

35–42. Answers will vary. Solutions lead to $0 = 0$.

In Exercises 43–48, use a system of three linear equations in three variables to solve the given problem.

43. Applying Kirchhoff's law to the electric circuit shown leads to the following system.

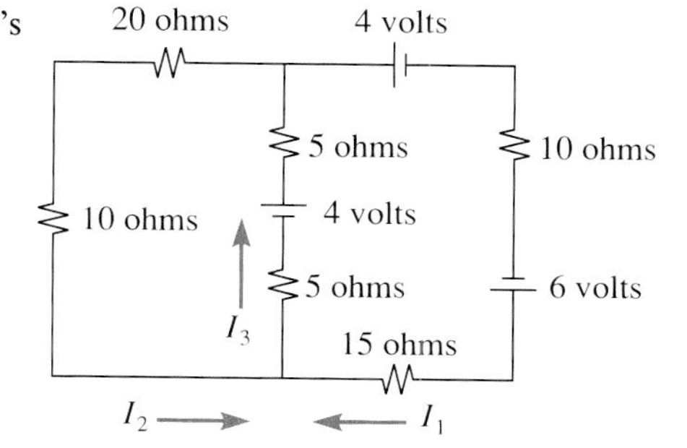

$$\begin{aligned} I_1 + I_2 - I_3 &= 0 \\ 30I_2 + 10I_3 &= 4 \\ 25I_1 + 10I_3 &= 6 \end{aligned}$$

Find the values of currents I_1, I_2, and I_3, in amperes.
$I_1 = \frac{2}{13}$ ampere, $I_2 = \frac{4}{65}$ ampere, $I_3 = \frac{14}{65}$ ampere

44. In the circuit of Exercise 43, if all the resistances are doubled, the resulting system is

$$\begin{aligned} I_1 + I_2 - \quad I_3 &= 0 \\ 60I_2 + 20I_3 &= 4 \\ 50I_1 \quad + 20I_3 &= 6. \end{aligned}$$

Solve the system for the three new currents. How do they compare with the answers in Exercise 43?
$I_1 = \frac{1}{13}$ ampere, $I_2 = \frac{2}{65}$ ampere, $I_3 = \frac{7}{65}$ ampere; they are half the previous values.

45. Given three points (not all in a straight line), it is possible to find a polynomial of degree 2 whose graph contains those points. Recall that the general form of a second-degree polynomial is $P(x) = ax^2 + bx + c$, where a, b, and c are real numbers. To find a second-degree polynomial whose graph contains (1,4), (2,9), and (−1,6), first replace x by 1 and $y = P(x)$ by 4 in the general form. This will give $4 = a + b + c$ as the first equation in the system. Repeat for the other two points to get the next two equations. Then solve for a, b, and c.
$4 = a + b + c$
$9 = 4a + 2b + c$
$6 = a - b + c$
Solution: $(2, -1, 3)$; polynomial: $P(x) = 2x^2 - x + 3$

46. Refer to Exercise 45, and find a second-degree polynomial whose graph contains (1,−2), (2,−3), and (3,−6).
$P(x) = -x^2 + 2x - 3$

47. The following problem is typical of those from algebra textbooks of 100 years ago. (It is from William J. Milne's *High School Algebra,* published by the American Book Company in 1892). Divide 125 into four parts such that, if the first be increased by 4, the second diminished by 4, the third multiplied by 4, and the fourth divided by 4, all these results will be equal. (*Hint:* Call the 4 parts a, b, c, and $125 - a - b - c$. The first equation is then $a + 4 = b - 4$.) 16, 24, 5, 80

48. Divide 180 into four parts such that if the first be increased by 5, the second diminished by 5, the third multiplied by 5, and the fourth divided by 5, all these results will be equal. 20, 30, 5, 125

THINK ABOUT IT

1. The following linear system in three variables is in **triangular form.**

$$\begin{aligned} 3x + 2y - \quad z &= 11 \qquad (1) \\ 5y + 3z &= 1 \qquad (2) \\ 4z &= 8 \qquad (3) \end{aligned}$$

Solve this system. Describe a general procedure for solving this type of system.

2. Extend the methods of this section to solve this system of four linear equations in four variables. (*Hint:* First eliminate one variable to get a new system of three linear equations in three variables.)

$$\begin{aligned} w - \quad x + \quad y + \quad z &= -4 \\ 2w - 3x + \quad y - \quad z &= -3 \\ w + 2x - 2y + 3z &= 1 \\ 3w + \quad x + 2y - 3z &= 9 \end{aligned}$$

3. Recall that a system of three linear equations in three variables is inconsistent if there is no point which is simultaneously in all three planes (as shown in Figure 6.4). Draw and describe two other possible graphs of inconsistent systems.

4. Solve for x, y, and z in terms of a, b, and c.

$$\begin{aligned} x + z &= a \\ x + y &= b \\ y + z &= c \end{aligned}$$

5. In Exercises 35–42 the systems are dependent and therefore have infinitely many solutions. For Exercises 35 and 42, find three of the solutions.

REMEMBER THIS

1. Solve $a_1b_2x - b_1a_2x = c$ for x. $x = \dfrac{c}{a_1b_2 - b_1a_2}$

2. Evaluate $-9(7) - (-2)(11)$. -41

3. Evaluate $2[0(-1) - 7(1)] - 4[-2(-1) - 3(1)] + 3[-2(7) - 3(0)]$. -52

4. Is $(-\frac{1}{2}, \frac{3}{5})$ a solution to $2x + 5y = 2$
$6x - 10y = -9$? Yes

5. Write a system of two linear equations for solving this problem. How many liters of a 5 percent solution of ammonia and how many liters of a 10 percent solution should be mixed to make n liters of a 6 percent solution? $0.05x + 0.10y = 0.06n$
$x + y = n$

6. Solve the system of equations: $y = 3x + 4$
$y - 3x = 2$. $\emptyset$

7. Find the equation of a line through the origin which is perpendicular to $y = 3x - 4$. $y = -\frac{1}{3}x$

8. Write this relationship algebraically: F varies jointly as m_1 and m_2 and inversely as the square of d. $F = \dfrac{km_1m_2}{d^2}$

9. Simplify: $\dfrac{1 + \dfrac{2}{a}}{b + \dfrac{3}{5}}$. $\dfrac{5a + 10}{5ab + 3a}$

10. Evaluate $3^{-1} + 3^0 + 3^1$. $\frac{13}{3}$

6.3 Determinants and Cramer's Rule

A window cleaner is made of water and ammonia. Use a linear system in two variables and Cramer's rule to show that it is not possible to combine one brand that is 5 percent ammonia with another that is 10 percent ammonia to form a mixture that is 15 percent ammonia. Assume there are ample supplies of the two brands, and try to obtain *n* liters that are 15 percent ammonia. (See Example 4.)

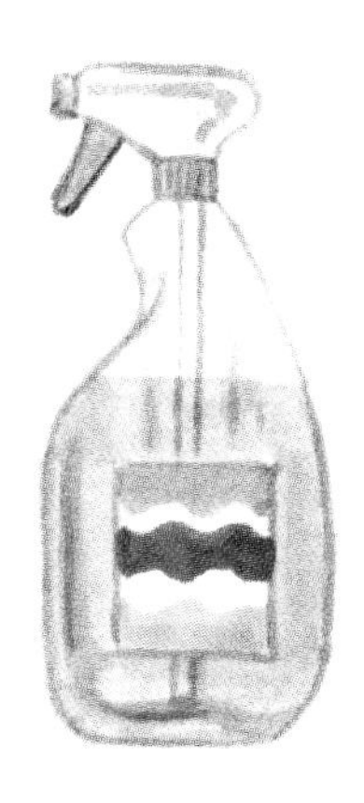

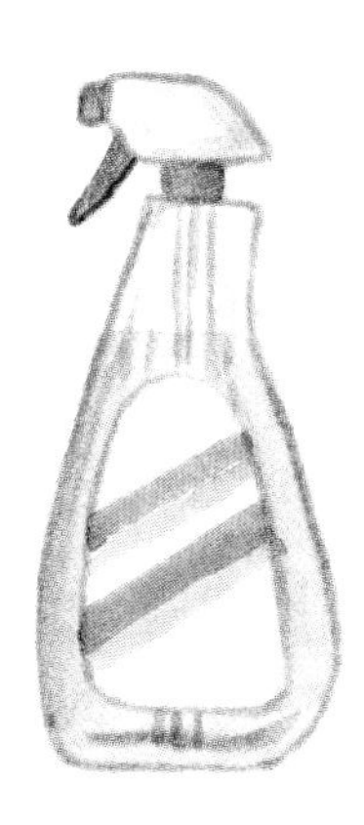

OBJECTIVES

1. **Evaluate a second-order determinant.**
2. **Solve a linear system in two variables using Cramer's rule.**
3. **Evaluate a third-order determinant.**
4. **Solve a linear system in three variables using Cramer's rule.**

1 A linear system in two variables with exactly one solution may be solved by formulas using a method called Cramer's rule. To see how these formulas originate, we now find the general solution of the system.

$$\begin{aligned} a_1x + b_1y &= c_1 \qquad (1)\\ a_2x + b_2y &= c_2 \qquad (2)\end{aligned}$$

If equivalent equations are formed by multiplying both sides of equation (1) by b_2 and both sides of equation (2) by $-b_1$, then x may be found as follows.

$$\begin{aligned} a_1b_2x + b_1b_2y &= c_1b_2 \qquad (3)\\ -b_1a_2x - b_1b_2y &= -b_1c_2 \qquad (4)\\ \hline a_1b_2x - b_1a_2x &= c_1b_2 - b_1c_2 \qquad \text{Add equations (3) and (4).}\\ x(a_1b_2 - b_1a_2) &= c_1b_2 - b_1c_2 \qquad \text{Factor out } x.\\ x &= \frac{c_1b_2 - b_1c_2}{a_1b_2 - b_1a_2} \qquad \text{Divide by } a_1b_2 - b_1a_2, \text{ with } a_1b_2 - b_1a_2 \neq 0.\end{aligned}$$

Similarly, multiplying both sides of equation (1) by $-a_2$ and both sides of equation (2) by a_1, and then adding, leads to

$$y = \frac{a_1c_2 - c_1a_2}{a_1b_2 - b_1a_2}.$$

Instead of memorizing these formulas in this form, we define what is called a determinant. Consider the expression $a_1b_2 - b_1a_2$, which is the denominator in the formulas for both x and y. This expression may be used to define the value of a second-order determinant as follows.

The general solution for two, three, and four equations was published by Colin Maclaurin (England) in 1748. Gabriel Cramer (Switzerland) also published a solution in 1750, in which he used a special notation with superscripts (as we use subscripts). Earlier versions which did not achieve popularity came from Seki Kowa (Japan, 1683) and G. W. Leibniz (Germany, 1693).

Our current vertical line notation for determinants is due to Arthur Cayley (1843).

Second-Order Determinant

A square array of numbers with two rows and two columns that is enclosed by vertical bars is called a **second-order determinant.** The value of a second-order determinant is given by

$$\begin{vmatrix} a_1 & b_1 \\ a_2 & b_2 \end{vmatrix} = a_1b_2 - b_1a_2.$$

Principal diagonal + Secondary diagonal −

$$\begin{vmatrix} a_1 & b_1 \\ a_2 & b_2 \end{vmatrix} = a_1b_2 - b_1a_2$$

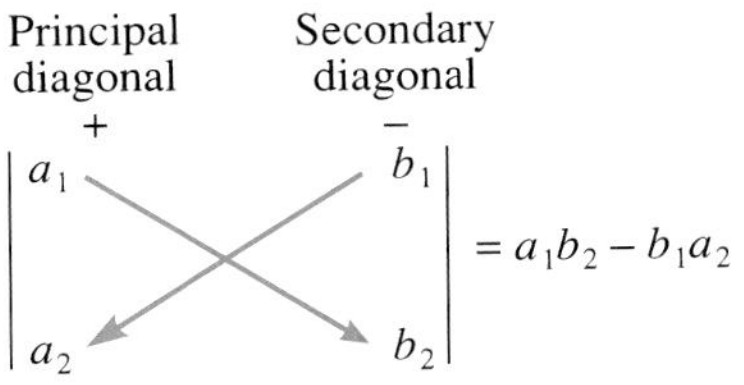

Figure 6.5

Figure 6.5 illustrates that the numbers a_1 and b_2 are elements of the principal diagonal and the numbers b_1 and a_2 are elements of the secondary diagonal. Note that the value of a second-order determinant is the product of the elements of the principal diagonal minus the product of the elements of the secondary diagonal.

EXAMPLE 1 Find the value of each determinant.

a. $\begin{vmatrix} 5 & 6 \\ 2 & 3 \end{vmatrix}$ **b.** $\begin{vmatrix} -9 & -2 \\ 11 & 7 \end{vmatrix}$ **c.** $\begin{vmatrix} a_1 & c_1 \\ a_2 & c_2 \end{vmatrix}$

Solution

a. $\begin{vmatrix} 5 & 6 \\ 2 & 3 \end{vmatrix} = 5(3) - 6(2) = 15 - 12 = 3$

b. $\begin{vmatrix} -9 & -2 \\ 11 & 7 \end{vmatrix} = -9(7) - (-2)(11) = -63 + 22 = -41$

c. $\begin{vmatrix} a_1 & c_1 \\ a_2 & c_2 \end{vmatrix} = a_1c_2 - c_1a_2$

PROGRESS CHECK 1 Find the value of each determinant.

a. $\begin{vmatrix} 6 & 8 \\ 5 & 7 \end{vmatrix}$ **b.** $\begin{vmatrix} 12 & -5 \\ 15 & -7 \end{vmatrix}$ **c.** $\begin{vmatrix} c_1 & b_1 \\ c_2 & b_2 \end{vmatrix}$

2 The formulas for the general solution of a linear system in two variables may now be stated efficiently in determinant form, as specified in Cramer's rule.

Cramer's Rule for 2 by 2 Systems

The solution to the system

$$a_1x + b_1y = c_1$$
$$a_2x + b_2y = c_2,$$

with $a_1b_2 - b_1a_2 \neq 0$, is

$$x = \frac{D_x}{D} = \frac{\begin{vmatrix} c_1 & b_1 \\ c_2 & b_2 \end{vmatrix}}{\begin{vmatrix} a_1 & b_1 \\ a_2 & b_2 \end{vmatrix}} = \frac{c_1b_2 - b_1c_2}{a_1b_2 - b_1a_2}$$

$$y = \frac{D_y}{D} = \frac{\begin{vmatrix} a_1 & c_1 \\ a_2 & c_2 \end{vmatrix}}{\begin{vmatrix} a_1 & b_1 \\ a_2 & b_2 \end{vmatrix}} = \frac{a_1c_2 - c_1a_2}{a_1b_2 - b_1a_2}.$$

Progress Check Answers

1. (a) 2 (b) −9 (c) $c_1b_2 - b_1c_2$

Note how the determinants in the formulas for x and y are formed. The determinant in both denominators is denoted by D and is formed from the coefficients of x and y. In the determinants in the numerators, the column containing the constants on the right side of the equations first replaces in D the coefficients of x to form D_x and then the coefficients of y to form D_y.

EXAMPLE 2 Use Cramer's rule to solve the system of equations.

$$\begin{aligned} 2x + 5y &= 3 \\ 6x - 10y &= -9 \end{aligned}$$

Solution First, evaluate D, which is the determinant formed from the coefficients of x and y.

$$D = \begin{vmatrix} 2 & 5 \\ 6 & -10 \end{vmatrix} = 2(-10) - 5(6) = -50$$

Second, evaluate D_x. To form this determinant, replace the column in D containing the *coefficients of* x by the column with the constants on the right side of the equations.

$$D_x = \begin{vmatrix} \mathbf{3} & 5 \\ \mathbf{-9} & -10 \end{vmatrix} = 3(-10) - 5(-9) = 15$$

Third, evaluate D_y. To form this determinant, replace the column in D containing the *coefficients of* y by the column with the constants on the right side of the equations.

$$D_y = \begin{vmatrix} 2 & \mathbf{3} \\ 6 & \mathbf{-9} \end{vmatrix} = 2(-9) - 3(6) = -36$$

Fourth, use Cramer's rule.

$$x = \frac{D_x}{D} = \frac{15}{-50} = -\frac{3}{10}$$

and

$$y = \frac{D_y}{D} = \frac{-36}{-50} = \frac{18}{25}$$

Thus, the solution is $(-\frac{3}{10}, \frac{18}{25})$. Check this solution in the given system.

PROGRESS CHECK 2 Solve by Cramer's rule:
$$\begin{aligned} 2x - 7y &= 4 \\ 6x + 14y &= -3. \end{aligned}$$

Cramer's rule may be used to find x and y whenever $D \neq 0$. If $D = 0$, then the system is either inconsistent or dependent and Cramer's rule does not apply.

In the 2×2 case if $D = D_x = D_y = 0$, then the system is dependent. If $D = 0$ and either D_x or D_y is not equal to zero, then the system is inconsistent.

EXAMPLE 3 Determine whether Cramer's rule applies to the following system. If it does, find the solution.

$$\begin{aligned} x - 1 &= 3y \\ 6y + 2 &= 2x \end{aligned}$$

Solution A condition in Cramer's rule is that both equations be written in the form $ax + by = c$, so we begin by transforming the given system to

$$\begin{aligned} x - 3y &= 1 \\ -2x + 6y &= -2. \end{aligned}$$

Now we find D.

$$D = \begin{vmatrix} 1 & -3 \\ -2 & 6 \end{vmatrix} = 1(6) - (-3)(-2) = 0$$

Because $D = 0$, Cramer's rule does not apply to the given system.

Progress Check Answer

2. $(\frac{1}{2}, -\frac{3}{7})$

Note When Cramer's rule does not apply, then elimination methods as considered in Sections 6.1, 6.2, and 6.4 are usually used to analyze the system. In the system in this example, multiplying both sides of $x - 3y = 1$ by -2 results in $-2x + 6y = -2$, so this system is dependent.

PROGRESS CHECK 3 Determine whether Cramer's rule applies to the following system. If it does, find the solution.

$$\begin{aligned} 3y &= 4x - 7 \\ 3x &= 4y \end{aligned}$$

To solve the section-opening problem, we recall from Section 2.3 that liquid mixture problems are analyzed using

$$\begin{pmatrix}\text{percent of}\\ \text{an ingredient}\end{pmatrix} \cdot \begin{pmatrix}\text{amount of}\\ \text{solution}\end{pmatrix} = \begin{pmatrix}\text{amount of}\\ \text{ingredient}\end{pmatrix}.$$

It is not intuitively obvious to many students that the percentage of ammonia in the mixture must be between the two original percentages.

EXAMPLE 4 Solve the problem in the section introduction on page 287.

Solution Let

$$\begin{aligned} x &= \text{amount used in liters of 5 percent solution} \\ y &= \text{amount used in liters of 10 percent solution,} \end{aligned}$$

and organize the key components in the problem in a chart format.

Solution	Percent ammonia	·	Amount of solution (liters)	=	Amount of ammonia (liters)
First brand	5		x		$0.05x$
Second brand	10		y		$0.10y$
New solution	15		n		$0.15n$

Because x liters of the first brand combine with y liters of the second brand to form n liters of the new solution,

$$x + y = n. \qquad (1)$$

Also, the amount of ammonia in the new solution is the sum of the amounts contributed by the two brands, so

$$0.05x + 0.10y = 0.15n. \qquad (2)$$

To solve the resulting system, we choose to first multiply both sides of equation (2) by 100 to clear decimals and then apply Cramer's rule to the system.

$$\begin{aligned} x + y &= n \\ 5x + 10y &= 15n \end{aligned}$$

Progress Check Answer

3. Cramer's rule applies; (4,3).

The three determinants defined in Cramer's rule are

$$D = \begin{vmatrix} 1 & 1 \\ 5 & 10 \end{vmatrix} = 1(10) - 1(5) = 5,$$

$$D_x = \begin{vmatrix} n & 1 \\ 15n & 10 \end{vmatrix} = n(10) - 1(15n) = -5n,$$

$$D_y = \begin{vmatrix} 1 & n \\ 5 & 15n \end{vmatrix} = 1(15n) - n(5) = 10n,$$

so

$$x = \frac{D_x}{D} = \frac{-5n}{5} = -n \quad \text{and} \quad y = \frac{D_y}{D} = \frac{10n}{5} = 2n.$$

The mathematical solution to the system is $x = -n$ and $y = 2n$. But in the context of the problem both x and y must be between 0 and n to make sense. Since the values in the solution are outside this range, the desired mixture is not possible.

PROGRESS CHECK 4 Show that it is not possible to mix a 30 percent acid solution with a 60 percent acid solution to get n liters of a 20 percent acid solution. What range of concentrations are possible when mixing a 30 percent solution with a 60 percent solution?

3 Cramer's rule may be extended to solve linear systems in three variables; but to apply it, we must learn to evaluate third-order determinants. We first define the **minor** of an element to be the determinant formed by deleting the row and column containing the given element. For example,

This basic approach and vocabulary for working with determinants is due to Augustin-Louis Cauchy (1815).

$$\text{minor of } a_1 = \begin{vmatrix} \not a_1 & \not b_1 & \not c_1 \\ \not a_2 & b_2 & c_2 \\ \not a_3 & b_3 & c_3 \end{vmatrix} = \begin{vmatrix} b_2 & c_2 \\ b_3 & c_3 \end{vmatrix}$$

$$\text{minor of } a_2 = \begin{vmatrix} \not a_1 & b_1 & c_1 \\ \not a_2 & \not b_2 & \not c_2 \\ \not a_3 & b_3 & c_3 \end{vmatrix} = \begin{vmatrix} b_1 & c_1 \\ b_3 & c_3 \end{vmatrix}$$

We can now show how to evaluate a third-order determinant using the first-column elements together with their respective minors.

Third-Order Determinant

A square array of numbers with three rows and three columns that is enclosed by vertical bars is called a **third-order determinant.** The value of a third-order determinant is given by

$$\begin{vmatrix} a_1 & b_1 & c_1 \\ a_2 & b_2 & c_2 \\ a_3 & b_3 & c_3 \end{vmatrix} = a_1 \begin{vmatrix} b_2 & c_2 \\ b_3 & c_3 \end{vmatrix} - a_2 \begin{vmatrix} b_1 & c_1 \\ b_3 & c_3 \end{vmatrix} + a_3 \begin{vmatrix} b_1 & c_1 \\ b_2 & c_2 \end{vmatrix}.$$

The method in this box is called the **expansion of the determinant by minors about the first column.**

Progress Check Answer

4. $\frac{4}{3}n$ and $-\frac{1}{3}n$ liters of 30 percent and 60 percent solutions, respectively, do not make sense; between 30 and 60 percent.

EXAMPLE 5 Evaluate the determinant by expansion by minors about the first column.

$$\begin{vmatrix} 2 & -2 & 3 \\ 4 & 0 & 7 \\ 3 & 1 & -1 \end{vmatrix}$$

Solution

$$\begin{aligned}
&\begin{vmatrix} 2 & -2 & 3 \\ 4 & 0 & 7 \\ 3 & 1 & -1 \end{vmatrix} \\
&= 2\begin{vmatrix} 2 & -2 & 3 \\ 4 & 0 & 7 \\ 3 & 1 & -1 \end{vmatrix} - 4\begin{vmatrix} 2 & -2 & 3 \\ 4 & 0 & 7 \\ 3 & 1 & -1 \end{vmatrix} + 3\begin{vmatrix} 2 & -2 & 3 \\ 4 & 0 & 7 \\ 3 & 1 & -1 \end{vmatrix} \\
&= 2\begin{vmatrix} 0 & 7 \\ 1 & -1 \end{vmatrix} - 4\begin{vmatrix} -2 & 3 \\ 1 & -1 \end{vmatrix} + 3\begin{vmatrix} -2 & 3 \\ 0 & 7 \end{vmatrix} \\
&= 2[0(-1) - 7(1)] - 4[-2(-1) - 3(1)] + 3[-2(7) - 3(0)] \\
&= 2(-7) - 4(-1) + 3(-14) \\
&= -14 + 4 - 42 \\
&= -52
\end{aligned}$$

PROGRESS CHECK 5 Evaluate the determinant by expansion by minors about the first column.

$$\begin{vmatrix} 1 & 1 & 2 \\ -5 & 2 & 2 \\ 2 & 0 & 3 \end{vmatrix}$$

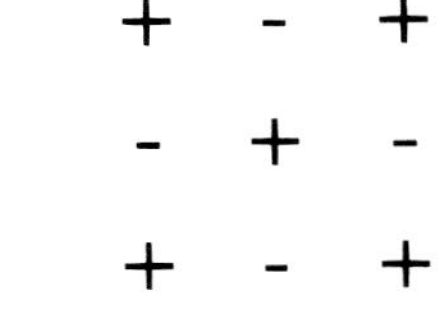

Figure 6.6

A third-order determinant may be evaluated by expanding the determinant by minors about *any* row or column. However, when an element is multiplied by its minor, the resulting product must be added or subtracted according to the sign pattern in Figure 6.6. That is, we add the products associated with elements in positions labeled +, and we subtract the products associated with elements in positions labeled −. To illustrate, the next example evaluates once again the determinant in Example 5. But this time we choose to expand by minors about the second column, because it is usually easier to expand about columns or rows that contain the most zero elements.

EXAMPLE 6 Evaluate the determinant by expansion by minors about the second column.

$$\begin{vmatrix} 2 & -2 & 3 \\ 4 & 0 & 7 \\ 3 & 1 & -1 \end{vmatrix}$$

Solution The sign pattern in Figure 6.6 for column 2 is −, +, −. Therefore, expansion by minors about the second column is as follows.

$$\begin{aligned}
\begin{vmatrix} 2 & -2 & 3 \\ 4 & 0 & 7 \\ 3 & 1 & -1 \end{vmatrix} &= -(-2)\begin{vmatrix} 4 & 7 \\ 3 & -1 \end{vmatrix} + 0\begin{vmatrix} 2 & 3 \\ 3 & -1 \end{vmatrix} - 1\begin{vmatrix} 2 & 3 \\ 4 & 7 \end{vmatrix} \\
&= 2(-25) + 0(\text{not needed}) - 1(2) \\
&= -50 - 2 \\
&= -52
\end{aligned}$$

Progress Check Answer

5. 17

PROGRESS CHECK 6 Evaluate the determinant by expansion by minors about the second column.

$$\begin{vmatrix} 1 & 1 & 2 \\ -5 & 2 & 2 \\ 2 & 0 & 3 \end{vmatrix}$$

4 In Cramer's rule for linear systems in three variables, consider how the formulas for x, y, and z follow an arrangement similar to that for 2 by 2 systems.

Cramer's Rule for 3 by 3 Systems

The solution to the system

$$\begin{aligned} a_1x + b_1y + c_1z &= d_1 \\ a_2x + b_2y + c_2z &= d_2 \\ a_3x + b_3y + c_3z &= d_3 \end{aligned} \qquad \textit{with} \qquad D = \begin{vmatrix} a_1 & b_1 & c_1 \\ a_2 & b_2 & c_2 \\ a_3 & b_3 & c_3 \end{vmatrix} \neq 0$$

is $x = D_x/D$, $y = D_y/D$, and $z = D_z/D$, where

$$D_x = \begin{vmatrix} \mathbf{d_1} & b_1 & c_1 \\ \mathbf{d_2} & b_2 & c_2 \\ \mathbf{d_3} & b_3 & c_3 \end{vmatrix}, \quad D_y = \begin{vmatrix} a_1 & \mathbf{d_1} & c_1 \\ a_2 & \mathbf{d_2} & c_2 \\ a_3 & \mathbf{d_3} & c_3 \end{vmatrix}, \quad \text{and} \quad D_z = \begin{vmatrix} a_1 & b_1 & \mathbf{d_1} \\ a_2 & b_2 & \mathbf{d_2} \\ a_3 & b_3 & \mathbf{d_3} \end{vmatrix}.$$

The determinant in all denominators is denoted by D and is formed from the coefficients of x, y, and z. When solving for x, the determinant D_x is obtained from D by using d_1, d_2, and d_3 as replacements for the corresponding a's (the coefficients of x). Similarly, the d's replace the coefficients of y to form D_y and replace the coefficients of z to form D_z.

EXAMPLE 7 Use Cramer's rule to solve this system of equations.

To save time on exams, consider asking students to find the value of only one variable.

$$\begin{aligned} -2x + y - z &= 3 \\ 2x + 3y + 2z &= 5 \\ 3x + 2y + z &= -3 \end{aligned}$$

Solution The four determinants defined in Cramer's rule are

$$D = \begin{vmatrix} -2 & 1 & -1 \\ 2 & 3 & 2 \\ 3 & 2 & 1 \end{vmatrix} = 11, \qquad D_x = \begin{vmatrix} \mathbf{3} & 1 & -1 \\ \mathbf{5} & 3 & 2 \\ \mathbf{-3} & 2 & 1 \end{vmatrix} = -33,$$

$$D_y = \begin{vmatrix} -2 & \mathbf{3} & -1 \\ 2 & \mathbf{5} & 2 \\ 3 & \mathbf{-3} & 1 \end{vmatrix} = 11, \qquad D_z = \begin{vmatrix} -2 & 1 & \mathbf{3} \\ 2 & 3 & \mathbf{5} \\ 3 & 2 & \mathbf{-3} \end{vmatrix} = 44.$$

Then

$$x = \frac{D_x}{D} = \frac{-33}{11} = -3, \qquad y = \frac{D_y}{D} = \frac{11}{11} = 1, \qquad \text{and} \qquad z = \frac{D_z}{D} = \frac{44}{11} = 4.$$

Thus, the solution is $(-3,1,4)$. Check this solution in all three equations in the given system.

Progress Check Answer

6. 17

Note For the purposes of Cramer's rule, zero elements are positioned in determinants to correspond to any missing terms, since a system like

$$\begin{aligned} 2x + 3y &= 10 \\ 3x + 2z &= 2 \\ 4y + z &= 6 \end{aligned} \quad \text{is equivalent to} \quad \begin{aligned} 2x + 3y + 0z &= 10 \\ 3x + 0y + 2z &= 2 \\ 0x + 4y + z &= 6. \end{aligned}$$

PROGRESS CHECK 7 Solve by Cramer's rule.

$$\begin{aligned} 3x + y + z &= 1 \\ x - y + z &= 3 \\ -2x - 2y + z &= 3 \end{aligned}$$

Progress Check Answer

7. $(\frac{1}{2}, -\frac{3}{2}, 1)$

EXERCISES 6.3

In Exercises 1–16, find the value of each determinant.

1. $\begin{vmatrix} 1 & 2 \\ 3 & 4 \end{vmatrix}$ -2

2. $\begin{vmatrix} 3 & 4 \\ 1 & 2 \end{vmatrix}$ 2

3. $\begin{vmatrix} 3 & -2 \\ 2 & -3 \end{vmatrix}$ -5

4. $\begin{vmatrix} 4 & -1 \\ 1 & -4 \end{vmatrix}$ -15

5. $\begin{vmatrix} 2 & 1 \\ 4 & 2 \end{vmatrix}$ 0

6. $\begin{vmatrix} 1 & -2 \\ 3 & -6 \end{vmatrix}$ 0

7. $\begin{vmatrix} a & b \\ c & d \end{vmatrix}$ $ad - bc$

8. $\begin{vmatrix} a & -b \\ c & -d \end{vmatrix}$ $-ad + bc$

9. $\begin{vmatrix} a & 2a \\ b & 2b \end{vmatrix}$ 0

10. $\begin{vmatrix} x & y \\ 3x & 3y \end{vmatrix}$ 0

11. $\begin{vmatrix} 1 & 0 \\ 0 & 1 \end{vmatrix}$ 1

12. $\begin{vmatrix} -1 & 0 \\ 0 & -1 \end{vmatrix}$ 1

13. $\begin{vmatrix} a & 0 \\ 0 & a \end{vmatrix}$ a^2

14. $\begin{vmatrix} 0 & b \\ b & 0 \end{vmatrix}$ $-b^2$

15. $\begin{vmatrix} \frac{3}{2} & \frac{1}{2} \\ \frac{1}{3} & 2 \end{vmatrix}$ $\frac{17}{6}$

16. $\begin{vmatrix} \frac{2}{3} & \frac{1}{4} \\ \frac{3}{4} & \frac{1}{3} \end{vmatrix}$ $\frac{5}{144}$

In Exercises 17–24 for the given system, write and evaluate each of these determinants: (a) D, (b) D_x, and (c) D_y.

17. $x + 2y = 3$
$2x + 3y = 4$

$D = \begin{vmatrix} 1 & 2 \\ 2 & 3 \end{vmatrix} = -1$

$D_x = \begin{vmatrix} 3 & 2 \\ 4 & 3 \end{vmatrix} = 1$

$D_y = \begin{vmatrix} 1 & 3 \\ 2 & 4 \end{vmatrix} = -2$

18. $x - 2y = 3$
$2x - 3y = 4$

$D = \begin{vmatrix} 1 & -2 \\ 2 & -3 \end{vmatrix} = 1$

$D_x = \begin{vmatrix} 3 & -2 \\ 4 & -3 \end{vmatrix} = -1$

$D_y = \begin{vmatrix} 1 & 3 \\ 2 & 4 \end{vmatrix} = -2$

19. $3x - 2y = 5$
$x + 3y = 1$

$D = \begin{vmatrix} 3 & -2 \\ 1 & 3 \end{vmatrix} = 11$

$D_x = \begin{vmatrix} 5 & -2 \\ 1 & 3 \end{vmatrix} = 17$

$D_y = \begin{vmatrix} 3 & 5 \\ 1 & 1 \end{vmatrix} = -2$

20. $5x + 3y = 2$
$3x - y = 0$

$D = \begin{vmatrix} 5 & 3 \\ 3 & -1 \end{vmatrix} = -14$

$D_x = \begin{vmatrix} 2 & 3 \\ 0 & -1 \end{vmatrix} = -2$

$D_y = \begin{vmatrix} 5 & 2 \\ 3 & 0 \end{vmatrix} = -6$

21. $x + 2y = 5$
$2x + 4y = 10$

$D = \begin{vmatrix} 1 & 2 \\ 2 & 4 \end{vmatrix} = 0$

$D_x = \begin{vmatrix} 5 & 2 \\ 10 & 4 \end{vmatrix} = 0$

$D_y = \begin{vmatrix} 1 & 5 \\ 2 & 10 \end{vmatrix} = 0$

22. $x + 3y = 1$
$x + 3y = 2$

$D = \begin{vmatrix} 1 & 3 \\ 1 & 3 \end{vmatrix} = 0$

$D_x = \begin{vmatrix} 1 & 3 \\ 2 & 3 \end{vmatrix} = -3$

$D_y = \begin{vmatrix} 1 & 1 \\ 1 & 2 \end{vmatrix} = 1$

23. $x + 5 = 3y$
$y = 7x - 1$

$D = \begin{vmatrix} 1 & -3 \\ -7 & 1 \end{vmatrix} = -20$

$D_x = \begin{vmatrix} -5 & -3 \\ -1 & 1 \end{vmatrix} = -8$

$D_y = \begin{vmatrix} 1 & -5 \\ -7 & -1 \end{vmatrix} = -36$

24. $y = 2x - 1$
$y = 3x + 2$

$D = \begin{vmatrix} -2 & 1 \\ -3 & 1 \end{vmatrix} = 1$

$D_x = \begin{vmatrix} -1 & 1 \\ 2 & 1 \end{vmatrix} = -3$

$D_y = \begin{vmatrix} -2 & -1 \\ -3 & 2 \end{vmatrix} = -7$

In Exercises 25–44, use Cramer's rule to solve the given system of equations. If Cramer's rule does not apply because $D = 0$, determine if the system is inconsistent or dependent.

25. $x + y = 1$
$x + 2y = 2$ (0,1)

26. $3x + y = 3$
$2x - y = 2$ (1,0)

27. $x + y = 0$
$x - y = 0$ (0,0)

28. $3x + y = 0$
$5x + 2y = 0$ (0,0)

29. $3x + 2y = 7$
$4x - y = 2$ (1,2)

30. $2x - 3y = -6$
$4x + y = 16$ (3,4)

31. $3x - 4y = 2$
$4y + 3x = -3$ $\left(-\frac{1}{6}, -\frac{5}{8}\right)$

32. $2x + 3y = 1$
$6y + 3x = 5$ $\left(-3, \frac{7}{3}\right)$

33. $y = 2x + 1$
$y = 3x + 2$ $(-1, -1)$

34. $y = 3x - 2$
$y = 2x + 5$ $(7, 19)$

35. $2x + 3y = 1$
$4x + 6y = 3$ $D = 0$; inconsistent system; $\emptyset$.

36. $x - 2y = 0$
$-3x + 6y = 1$ $D = 0$; inconsistent system; $\emptyset$.

37. $y = 2x + 1$
$y = 2x + 3$ $D = 0$; inconsistent system; $\emptyset$

38. $y = 2x - 1$
$y = 2x$ $D = 0$; inconsistent system; $\emptyset$

39. $x - y = 4$
$-2x + 2y = -8$ $D = 0$; dependent system; $\{(x, y): x - y = 4\}$

40. $-3x + 2y = 6$
$6x - 4y = -12$ $D = 0$; dependent system; $\{(x, y): -3x + 2y = 6\}$

41. $y = 3x + 4$
$3y - 9x = 12$ $D = 0$; dependent system; $\{(x, y): y = 3x + 4\}$

42. $y = 3x + 1$
$\frac{1}{2}y = \frac{3}{2}x + \frac{1}{2}$ $D = 0$; dependent system; $\{(x, y): y = 3x + 1\}$

43. $y = 5$
$x + 2y = 2$ $(-8, 5)$

44. $2x - 3y = 7$
$x = 1$ $\left(1, -\frac{5}{3}\right)$

In Exercises 45–54, evaluate the determinant.

45. $\begin{vmatrix} 1 & 1 & 1 \\ 2 & 2 & 2 \\ 3 & 3 & 3 \end{vmatrix}$ 0

46. $\begin{vmatrix} 1 & 2 & 3 \\ 2 & 4 & 6 \\ 3 & 6 & 9 \end{vmatrix}$ 0

47. $\begin{vmatrix} 1 & 0 & 0 \\ 0 & 1 & 0 \\ 0 & 0 & 1 \end{vmatrix}$ 1

48. $\begin{vmatrix} 2 & 0 & 0 \\ 0 & 2 & 0 \\ 0 & 0 & 2 \end{vmatrix}$ 8

49. $\begin{vmatrix} 1 & 2 & 4 \\ 2 & 4 & 1 \\ 4 & 1 & 2 \end{vmatrix}$ -49

50. $\begin{vmatrix} 1 & 3 & 9 \\ 3 & 9 & 1 \\ 9 & 1 & 3 \end{vmatrix}$ -676

51. $\begin{vmatrix} a & a & a \\ a & a & a \\ a & a & a \end{vmatrix}$ 0

52. $\begin{vmatrix} 1 & a & a^2 \\ a & a^2 & 1 \\ a^2 & 1 & a \end{vmatrix}$ $-(a^3 - 1)^2$

53. $\begin{vmatrix} 0 & 1 & 2 \\ 1 & 0 & 2 \\ 2 & 1 & 0 \end{vmatrix}$ 6

54. $\begin{vmatrix} 0 & 0 & 3 \\ 4 & 5 & 6 \\ 7 & 8 & 9 \end{vmatrix}$ -9

In Exercises 55–60, evaluate first by expansion about the first column and then by expansion about the first row. Both computations should give the same result.

55. $\begin{vmatrix} 1 & -1 & 1 \\ 2 & 0 & 2 \\ 3 & 1 & 2 \end{vmatrix}$ -2

56. $\begin{vmatrix} 2 & 5 & -1 \\ 0 & 1 & 4 \\ 3 & -2 & 6 \end{vmatrix}$ 91

57. $\begin{vmatrix} 1 & -2 & 1 \\ -2 & 1 & -2 \\ 1 & -1 & 1 \end{vmatrix}$ 0

58. $\begin{vmatrix} 3 & -2 & 1 \\ -2 & 0 & -2 \\ 1 & -2 & 1 \end{vmatrix}$ -8

59. $\begin{vmatrix} 1 & 2 & 3 \\ 4 & 5 & 6 \\ 7 & 8 & 9 \end{vmatrix}$ 0

60. $\begin{vmatrix} 2 & 3 & 4 \\ 5 & 6 & 7 \\ 8 & 9 & 10 \end{vmatrix}$ 0

In Exercises 61–64 for each system, write and evaluate D, D_x, D_y, and D_z.

61. $-2x + y - z = 2$
$x + 2y - 3z = 4$
$3x - y + 2z = -1$

$D = \begin{vmatrix} -2 & 1 & -1 \\ 1 & 2 & -3 \\ 3 & -1 & 2 \end{vmatrix} = -6$

$D_x = \begin{vmatrix} 2 & 1 & -1 \\ 4 & 2 & -3 \\ -1 & -1 & 2 \end{vmatrix} = -1$

$D_y = \begin{vmatrix} -2 & 2 & -1 \\ 1 & 4 & -3 \\ 3 & -1 & 2 \end{vmatrix} = -19$

$D_z = \begin{vmatrix} -2 & 1 & 2 \\ 1 & 2 & 4 \\ 3 & -1 & -1 \end{vmatrix} = -5$

62. $x + y + z = 0$
$2x - y - z = 0$
$3x + 2y + z = 0$

$D = \begin{vmatrix} 1 & 1 & 1 \\ 2 & -1 & -1 \\ 3 & 2 & 1 \end{vmatrix} = 3$

$D_x = \begin{vmatrix} 0 & 1 & 1 \\ 0 & -1 & -1 \\ 0 & 2 & 1 \end{vmatrix} = 0$

$D_y = \begin{vmatrix} 1 & 0 & 1 \\ 2 & 0 & -1 \\ 3 & 0 & 1 \end{vmatrix} = 0$

$D_z = \begin{vmatrix} 1 & 1 & 0 \\ 2 & -1 & 0 \\ 3 & 2 & 0 \end{vmatrix} = 0$

63. $3x + y + 4z = 3$
$x + 5y + 9z = 1$
$2x + 6y + 5z = 2$

$D = \begin{vmatrix} 3 & 1 & 4 \\ 1 & 5 & 9 \\ 2 & 6 & 5 \end{vmatrix} = -90 = D_x$

$D_y = \begin{vmatrix} 3 & 3 & 4 \\ 1 & 1 & 9 \\ 2 & 2 & 5 \end{vmatrix} = 0$

$D_z = \begin{vmatrix} 3 & 1 & 3 \\ 1 & 5 & 1 \\ 2 & 6 & 2 \end{vmatrix} = 0$

64. $2x + 7y + z = 2$
$8x + 2y + 8z = 2$
$x + 8y + 2z = 2$

$D = \begin{vmatrix} 2 & 7 & 1 \\ 8 & 2 & 8 \\ 1 & 8 & 2 \end{vmatrix} = -114$

$D_x = \begin{vmatrix} 2 & 7 & 1 \\ 2 & 2 & 8 \\ 2 & 8 & 2 \end{vmatrix} = -24$

$D_y = \begin{vmatrix} 2 & 2 & 1 \\ 8 & 2 & 8 \\ 1 & 2 & 2 \end{vmatrix} = -26$

$D_z = \begin{vmatrix} 2 & 7 & 2 \\ 8 & 2 & 2 \\ 1 & 8 & 2 \end{vmatrix} = 2$

In Exercises 65–76, solve the given system using Cramer's rule.

65. $5x - y + z = 2$
$-x + 2y + 5z = 3$
$2x - 3y - 4z = 1$ $\left(-\frac{1}{2}, -\frac{20}{7}, \frac{23}{14}\right)$

66. $3x + y - 2z = 2$
$4x - 2y + z = 0$
$-x + 3y - 5z = -2$ $(1, 3, 2)$

67. $2x + 3y - 4z = -1$
$3x - 3y + 6z = -1$
$6x + 9y - 4z = -1$ $\left(-\frac{1}{2}, \frac{1}{3}, \frac{1}{4}\right)$

68. $-x + 2y - 4z = -1$
$x + 2y + 4z = 3$
$3x + y + 2z = -1$
$(-1, \frac{1}{2}, \frac{3}{4})$

69. $x + y - 5z = 4$
$x - y + 5z = -4$
$2x + 2y - 5z = 9$
$(0, 5, \frac{1}{5})$

70. $x + y - 2z = 2$
$x - y + 2z = 0$
$-x + 3y + 4z = -3$
$(1, 0, -\frac{1}{2})$

71. $2x + 2y = 1$
$y + z = 1$
$x + z = 1$
$(\frac{1}{4}, \frac{1}{4}, \frac{3}{4})$

72. $2x + y = 1$
$y + z = 1$
$-x + 2z = 1$
$(\frac{1}{3}, \frac{1}{3}, \frac{2}{3})$

73. $2x - 3y + z = 0$
$4x - 6y + 2z = 1$
$x - 2y + z = 2$
Ø, inconsistent system

74. $x - 2y + 3z = 2$
$-2x + 3y + z = 1$
$6x - 9y - 3z = 3$
Ø, inconsistent system

75. $x + y + z = 1$
$2x + 2y + 2z = 2$
$3x + 3y + 3z = 3$
Dependent system

76. $x + y + z = 0$
$3x + 3y + 3z = 0$
$5x + 5y + 5z = 0$ Dependent system

In Exercises 77–90 each problem can be solved by a system of equations. Solve the appropriate system by using Cramer's rule.

77. A certain kind of cleaning solution consists of bleach and water. One batch is 25 percent bleach, while a second is 35 percent bleach. How much of each should be mixed to get 3 quarts (qt) of a solution which is one-third bleach? 80 oz of 35 percent solution; 16 oz of 25 percent solution

78. One perfume contains 1 percent essence of rose, while a second contains 1.8 percent essence of rose. How much of each should be combined to make 24 oz of perfume which contains 1.5 percent essence of rose? 9 oz of 1 percent solution; 15 oz of 1.8 percent solution

79. A $250,000 retirement fund was split into two investments, one portion (x) at 12.5 percent annual interest and the rest (y) at 8.5 percent. If the total annual interest is $26,050, how much was invested at each rate? $120,000 at 12.5 percent; $130,000 at 8.5 percent

80. How should a $50,000 investment be split so that the total annual earnings are $2,205 if one portion is invested at 6.25 percent annual interest and the rest at 3.75 percent? $13,200 at 6.25 percent; $36,800 at 3.75 percent

81. The sum of two numbers is -1, and the larger minus the smaller is 8. Find both numbers. -4.5, 3.5

82. The reciprocals of two numbers m and n have a sum of 2 and a difference of 1. Find the numbers. $\left(\textit{Hint:} \text{ let } x = \frac{1}{m} \text{ and let } y = \frac{1}{n}.\right)$ $\frac{2}{3}$, 2

83. A strand of barbed wire is 750 yards (yd) long. You wish to use it to make a rectangular enclosure that is twice as long as it is wide. What are the dimensions of the enclosure? Width = 125 yd; length = 250 yd

84. A strand of barbed wire is 750 yd long. You wish to use it to make a rectangular enclosure that is 4 times as long as it is wide. What are the dimensions of the enclosure? Width = 75 yd; length = 300 yd

85. A comparison is made between two liquids. In one container a liquid is evaporating at 1 liter per day. There are 300 liters of it to start with. The second is evaporating at 1.5 liters per day. There are 350 liters of it to start with.
  a. At what point do the two containers contain the same amount of liquid? 100 days
  b. At that time, what is the total amount of liquid left in both containers? 200 liters

86. Starting at 0° C, one piece of metal is being heated at the rate of 2.5° C per minute. At the same time another piece, started at 20° C, is being heated at the rate of 1° C per minute.
  a. After what amount of time will they be the same temperature? $13\frac{1}{3}$ minutes
  b. At what temperature will they be the same temperature? $33\frac{1}{3}$° C

87. A silversmith has three alloys, each containing some gold, some silver, and some copper. Alloy 1 contains 40 percent gold, 40 percent silver, and 20 percent copper. Alloy 2 contains 60 percent gold, 10 percent silver, and 30 percent copper. Alloy 3 contains 80 percent gold, 15 percent silver, and 5 percent copper. Can these alloys be melted and remixed to make 100 oz of an alloy with 58 percent gold, 20 percent silver, and 22 percent copper? Use the table to derive a system of three equations in three variables.

Alloy	Amount of alloy	Amount of gold	Amount of silver
1	x	$0.40x$	$0.40x$
2	y	$0.60y$	$0.10y$
3	z	$0.80z$	$0.15z$
New	100	0.58(100)	0.20(100)

Yes; 30 oz alloy 1, 50 oz alloy 2, 20 oz alloy 3

88. A silversmith has three alloys, each containing some gold, some silver, and some copper. Alloy 1 contains 45 percent gold, 45 percent silver, and 10 percent copper. Alloy 2 contains 60 percent gold, 10 percent silver, and 30 percent copper. Alloy 3 contains 80 percent gold, 15 percent silver, and 5 percent copper. Can these alloys be melted and remixed to make 100 oz of an alloy with 50 percent gold, 25 percent silver, and 25 percent copper? Make a table like the one in Exercise 87, and derive a system of three equations in three variables.
Cannot be done; solution of system calls for one amount to be negative

89. In triangle ABC angle A is 20° less than the sum of B and C; angle B is 40° less than the sum of A and C; and angle C is 120° less than the sum of A and B. Find the measure of each angle. $A = 80°$, $B = 70°$, $C = 30°$

90. In a triangle with sides a, b, and c, side a is one-third the sum of the other two sides; side b is one-half the sum of the other two sides; and side c is 2 less than the sum of the other two sides. Find the length of each side. $a = 3$, $b = 4$, $c = 5$

THINK ABOUT IT

1. **a.** Explain why the value of a determinant which has integer elements must be an integer.
   b. Why does it follow from part **a** that the solution to the system
$$\begin{aligned} ax + by &= c \\ dx + ey &= f, \end{aligned}$$
   where all coefficients are integers, must be a pair of *rational* numbers? (Assume $D \neq 0$.)
   c. Why does it follow from part **b** that the graphs of two lines given by $ax + by = c$ and $dx + ey = f$ can never intersect at $(\sqrt{2}, \sqrt{3})$ no matter what integers are chosen for the coefficients?
2. For what value(s) of a does the following system have exactly one solution? What is this solution in terms of a?
$$\begin{aligned} 2x + 3y &= 5 \\ ax + y &= 1 \end{aligned}$$
3. **a.** Show that $\begin{vmatrix} 3 & 5 \\ 6 & 10 \end{vmatrix} = 0$.
   b. Show that if the second row of a 2 by 2 determinant is a multiple of the first row, then the value of the determinant is zero. $\left(\textit{Hint:} \text{ Evaluate } \begin{vmatrix} a & b \\ ka & kb \end{vmatrix}.\right)$
   c. Find values for x and y which make the value of the determinant $\begin{vmatrix} 2 & 7 \\ x & y \end{vmatrix}$ equal to zero.
4. **a.** Fill in the blank with "rows" or "columns." In a determinant the _____ go vertically and the _____ go horizontally.
   b. Evaluate $\begin{vmatrix} 3 & 1 \\ 5 & 7 \end{vmatrix}$.
   c. Switch the columns in part **b** and evaluate again.
   d. Verify that, in general, $\begin{vmatrix} a & b \\ c & d \end{vmatrix}$ and $\begin{vmatrix} b & a \\ d & c \end{vmatrix}$ are opposites.
   e. Is it true that if you switch the *rows* in $\begin{vmatrix} a & b \\ c & d \end{vmatrix}$ you also get a sign change?
5. The evaluation of a determinant using expansion by minors can be extended to larger determinants. The pattern of signs shown in Figure 6.6 becomes
$$\begin{matrix} + & - & + & - \\ - & + & - & + \\ + & - & + & - \\ - & + & - & + \end{matrix}$$
   in the 4 by 4 case. Evaluate $\begin{vmatrix} 3 & 2 & 4 & 1 \\ 1 & 0 & 1 & 0 \\ 2 & 0 & 3 & 2 \\ 4 & 0 & 2 & 3 \end{vmatrix}$ using expansion by minors along the second column. Why is this the easiest column to use?

REMEMBER THIS

1. Solve the system. $7y + 2z = 1$
$$3z = 12 \quad (-1,4)$$
2. Add -2 times the first equation to the second equation.
$$\begin{aligned} x - y + z &= -1 \\ 2x - 3y - 2z &= 9 \end{aligned} \qquad -y - 4z = 11$$
3. In the determinant $\begin{vmatrix} a & b & c \\ d & e & f \\ g & h & i \end{vmatrix}$, what are the following entries?
   a. In *row* 3 g, h, i **b.** In *column* 1 a, d, g
4. Solve.
$$\begin{aligned} 2x + y - z &= 4 \\ x + 2y + z &= 5 \\ x - y - 2z &= -1 \end{aligned}$$
   Dependent system
5. True or false? The graphs of $x + y = 5$ and $x + y + 10$ are parallel lines. True
6. Evaluate $1 + 2(1 + .03)^{12}$. Round to the nearest hundredth. 3.85
7. Graph the solution of $5 - 2x < 3$. Also give the solution in interval notation. $(1, \infty)$

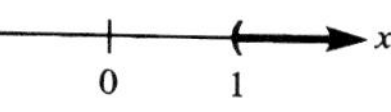

8. Factor $12x^2 - 16x - 3$. $(2x - 3)(6x + 1)$
9. Solve $\frac{1}{x} + 2 = -x$. $\{-1\}$
10. Find an equation for the line shown. $y = x + 1$

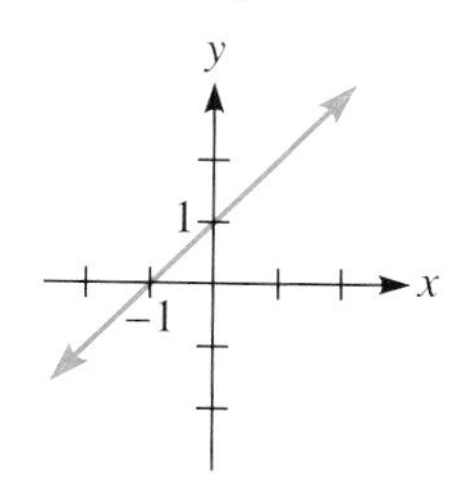

6.4 Triangular Form and Matrices

A 4,000-seat theater was sold out for a weekend concert. Total receipts on Friday night were $53,200 when seats in the orchestra, mezzanine, and balcony sold for $16, $12, and $8, respectively. On Saturday, prices were raised to $20, $15, and $8, and total receipts increased to $65,300. How many seats of each type are in the theater? (See Example 4.)

OBJECTIVES

1 **Solve a linear system by transforming the system to triangular form.**

2 **Solve a linear system using matrices.**

The method is named for Carl Friedrich Gauss (German, 1777–1855). The technique was developed for use in his derivation of the method of least squares. The idea was to reduce the amount of calculation required by Cramer's method. An identical approach was published in China during the Han dynasty almost 2,000 years earlier in a text called *Jiuzhang.*

1 **Gaussian elimination** is a systematic method that is commonly used by computers to analyze linear systems. To understand this method, first consider the following linear system in three variables.

$$\begin{aligned} 3x - 5y + z &= 3 \quad (1)\\ 7y + 2z &= 1 \quad (2)\\ 3z &= 12 \quad (3) \end{aligned}$$

This system is said to be in **triangular form** and is easy to solve. Equation (3), $3z = 12$, quickly gives $z = 4$. Then replacing z by 4 in equation (2) yields

$$\begin{aligned} 7y + 2(4) &= 1\\ 7y &= -7\\ y &= -1, \end{aligned}$$

and replacing z by 4 and y by -1 in equation (1) gives

$$\begin{aligned} 3x - 5(-1) + 4 &= 3\\ 3x &= -6\\ x &= -2. \end{aligned}$$

Thus, the solution is $(-2,-1,4)$.

The idea in Gaussian elimination is to transform any given system to triangular form by producing **equivalent systems** (ones with the same solution) using the operations that follow.

Operations that Produce Equivalent Systems

1. Interchange the order in which two equations are listed.
2. Multiply both sides of an equation by a nonzero number.
3. Add a multiple of one equation to another equation.

The three operations that produce equivalent systems are called the **elementary operations.** Note that the first operation clearly affects only the form of the system, and the other two operations were often used in the addition-elimination method of Section 6.1.

EXAMPLE 1 Solve by transforming to triangular form.

$$\begin{aligned} 2x - 3y - 2z &= 9 \\ x - y + z &= -1 \\ -x + 2y + z &= -2 \end{aligned}$$

Solution In triangular form x must be eliminated in all equations after the first one. This is usually easier to do when the coefficient of x in the first equation is 1, so begin by interchanging the order of the first two equations to obtain

$$\begin{aligned} x - y + z &= -1 \\ 2x - 3y - 2z &= 9 \\ -x + 2y + z &= -2. \end{aligned}$$

Now adding -2 times the first equation to the second equation gives

$$\begin{aligned} x - y + z &= -1 \\ - y - 4z &= 11 \\ -x + 2y + z &= -2. \end{aligned}$$

Remind students that there are two parts to this step but only the final result is written.

And adding the first equation to the third equation gives

$$\begin{aligned} x - y + z &= -1 \\ - y - 4z &= 11 \\ y + 2z &= -3. \end{aligned}$$

In triangular form y must be eliminated in all equations after the second one. For this system adding the second equation to the third equation eliminates y in the third equation.

$$\begin{aligned} x - y + z &= -1 \\ -y - 4z &= 11 \\ -2z &= 8 \end{aligned}$$

You might point out that for some systems it may be easier to work toward a triangular form like

$$\begin{aligned} a_1x + b_1y + c_1z &= d_1 \\ a_2x + b_2y &= d_2 \\ a_3x &= d_3. \end{aligned}$$

The system is now in triangular form. Using $-2z = 8$, we have $z = -4$, and back substitution gives

$$\begin{aligned} -y - 4(-4) &= 11 \\ -y &= -5 \\ y &= 5 \end{aligned} \qquad \text{and} \qquad \begin{aligned} x - 5 + (-4) &= -1 \\ x &= 8. \end{aligned}$$

Thus, the solution is $(8,5,-4)$. Check this solution in the original equations.

PROGRESS CHECK 1 Solve the system by transforming to triangular form.

$$\begin{aligned} 2x - 3y - z &= -9 \\ 3x - y + 6z &= 1 \\ x + 2y - 3z &= 0 \end{aligned}$$

2 Our current method may be shortened by writing down only the constants in the equations and leaving out all the x's, y's, and z's. The standard notation for such an

Progress Check Answer

1. $(-1,2,1)$

The word *matrix* for this array was coined by James Joseph Sylvester (English, 1814–1897) and put into common use by Arthur Cayley around 1855. The development of the basic ideas of matrix operations are due to these two men. In 1871 Sylvester became chair of the mathematics department at the newly opened Johns Hopkins University.

abbreviation utilizes matrices. A **matrix** is a rectangular array of numbers that is enclosed in brackets (or parentheses). Each number in the matrix is called an **entry** or **element** of the matrix. A system like

$$\begin{aligned} a_1x + b_1y + c_1z &= d_1 \\ a_2x + b_2y + c_2z &= d_2 \\ a_3x + b_3y + c_3z &= d_3 \end{aligned}$$

is abbreviated by the matrix

$$\left[\begin{array}{ccc:c} a_1 & b_1 & c_1 & d_1 \\ a_2 & b_2 & c_2 & d_2 \\ a_3 & b_3 & c_3 & d_3 \end{array}\right]$$

This matrix is called the **augmented matrix** of the system. Note that it consists of the coefficients of the variables and an additional column (separated by an optional dashed line) that contains the constants on the right side of the equals sign. When the elementary operations that produce equivalent systems of equations are restated in the language of matrices, we obtain operations, called **elementary row operations,** which are stated next.

Corresponding Operations for Solving a Linear System

Elementary Operations on Equations	Elementary Row Operations on Matrices
1. Interchange two equations.	**1.** Interchange two rows.
2. Multiply both sides of an equation by a nonzero number.	**2.** Multiply each entry in a row by a nonzero number.
3. Add a multiple of one equation to another.	**3.** Add a multiple of the entries in one row to another row.

A linear system may now be solved in matrix form by using elementary row operations to obtain matrices of equivalent systems until a system in triangular form is reached. In the next example both the matrix form and the equation form of the system will be displayed to reinforce the similarities in the methods.

EXAMPLE 2 Solve the system.

$$\begin{aligned} 2x + 2y - z &= -3 \\ 3x + y + z &= 1 \\ x - y + z &= 3 \end{aligned}$$

Show both the matrix form of the system and the corresponding equations.

Solution We use the operations above and proceed as follows.

Equation Form	Matrix Form
$\begin{aligned} 2x + 2y - z &= -3 \\ 3x + y + z &= 1 \\ x - y + z &= 3 \end{aligned}$	$\left[\begin{array}{ccc:c} 2 & 2 & -1 & -3 \\ 3 & 1 & 1 & 1 \\ 1 & -1 & 1 & 3 \end{array}\right]$
↓ Interchange equations 1 and 3.	↓ Interchange rows 1 and 3.
$\begin{aligned} x - y + z &= 3 \\ 3x + y + z &= 1 \\ 2x + 2y - z &= -3 \end{aligned}$	$\left[\begin{array}{ccc:c} 1 & -1 & 1 & 3 \\ 3 & 1 & 1 & 1 \\ 2 & 2 & -1 & -3 \end{array}\right]$

↓ Add -3 times the first equation to the second equation.

$$\begin{aligned} x - y + z &= 3 \\ 4y - 2z &= -8 \\ 2x + 2y - z &= -3 \end{aligned}$$

↓ Add -3 times each entry in row 1 to the corresponding entry in row 2.

$$\left[\begin{array}{rrr|r} 1 & -1 & 1 & 3 \\ 0 & 4 & -2 & -8 \\ 2 & 2 & -1 & -3 \end{array}\right]$$

↓ Add -2 times the first equation to the third equation.

$$\begin{aligned} x - y + z &= 3 \\ 4y - 2z &= -8 \\ 4y - 3z &= -9 \end{aligned}$$

↓ Add -2 times each entry in row 1 to the corresponding entry in row 3.

$$\left[\begin{array}{rrr|r} 1 & -1 & 1 & 3 \\ 0 & 4 & -2 & -8 \\ 0 & 4 & -3 & -9 \end{array}\right]$$

↓ Add -1 times the second equation to the third equation.

$$\begin{aligned} x - y + z &= 3 \\ 4y - 2z &= -8 \\ - z &= -1 \end{aligned}$$

↓ Add -1 times each entry in row 2 to the corresponding entry in row 3.

$$\left[\begin{array}{rrr|r} 1 & -1 & 1 & 3 \\ 0 & 4 & -2 & -8 \\ 0 & 0 & -1 & -1 \end{array}\right]$$

By further work with row operations this system can be transformed to the form $\left[\begin{array}{rrr|r} 1 & 0 & 0 & a \\ 0 & 1 & 0 & b \\ 0 & 0 & 1 & c \end{array}\right]$ which implies that the solution is $x = a$, $y = b$, and $z = c$. This method, which is described in "Think About It" Exercise 2, is called Gauss-Jordan elimination.

The last row or last equation tells us $-z = -1$, so $z = 1$. Then

$$\begin{aligned} 4y - 1(1) &= -8 \\ 4y &= -6 \\ y &= -\tfrac{3}{2} \end{aligned} \qquad \begin{aligned} x - (-\tfrac{3}{2}) + 1 &= 3 \\ x + \tfrac{5}{2} &= 3 \\ x &= \tfrac{1}{2}. \end{aligned}$$

Confirm in the original equations that the solution is $(\frac{1}{2}, -\frac{3}{2}, 1)$.

PROGRESS CHECK 2 Solve the system.

$$\begin{aligned} 5x + 4y + 3z &= 0 \\ x + y + z &= 0 \\ 6x + 3y + 2z &= 1 \end{aligned}$$

Show both the matrix form of the system and the corresponding equations.

Gaussian elimination may lead to considerable work with fractions even in simple linear systems, as shown in the next example.

EXAMPLE 3 Use matrix form to solve the system

$$\begin{aligned} 2x + 5y - 1 &= 0 \\ 3x - 2y - 11 &= 0. \end{aligned}$$

Solution Both equations must be written in the form $ax + by = c$, so begin by transforming the system to

$$\begin{aligned} 2x + 5y &= 1 \\ 3x - 2y &= 11. \end{aligned}$$

The augmented matrix for the system is

$$\left[\begin{array}{rr|r} 2 & 5 & 1 \\ 3 & -2 & 11 \end{array}\right].$$

For this system we first multiply each entry in row 1 by $\frac{1}{2}$. This choice results in a matrix with a 1 as the first entry in column 1.

$$\left[\begin{array}{rr|r} 1 & \frac{5}{2} & \frac{1}{2} \\ 3 & -2 & 11 \end{array}\right]$$

Progress Check Answer

2. $(\frac{1}{2}, -1, \frac{1}{2})$

Now add -3 times row 1 to row 2.

$$\left[\begin{array}{cc|c} 1 & \frac{5}{2} & \frac{1}{2} \\ 0 & -\frac{19}{2} & \frac{19}{2} \end{array}\right]$$

The last row corresponds to $-\frac{19}{2}y = \frac{19}{2}$, so $y = -1$. Then replacing y by -1 in the equation corresponding to row 1 gives

$$\begin{aligned} x + \tfrac{5}{2}(-1) &= \tfrac{1}{2} \\ x &= \tfrac{1}{2} + \tfrac{5}{2} \\ x &= 3 \end{aligned}$$

Thus, the solution is $(3, -1)$. Check this solution in the two original equations.

PROGRESS CHECK 3 Use matrix form to solve the system.

$$\begin{aligned} 2x - 3y - 13 &= 0 \\ 5x + 4y + 2 &= 0 \end{aligned}$$

EXAMPLE 4 Solve the problem in the section introduction on page 298.

Solution Let x, y, and z represent the number of seats in the orchestra, mezzanine, and balcony, respectively. Because the theater has 4,000 seats,

$$x + y + z = 4{,}000. \qquad (1)$$

On Friday all seats were sold, with orchestra, mezzanine, and balcony seats producing $16x$, $12y$, and $8z$ dollars in revenue, respectively. Total receipts were \$53,200, so

$$16x + 12y + 8z = 53{,}200. \qquad (2)$$

Similarly, using the higher ticket prices and total receipts from Saturday gives

$$20x + 15y + 8z = 65{,}300. \qquad (3)$$

The augmented matrix for the system of equations (1), (2), and (3) is

$$\left[\begin{array}{ccc|c} 1 & 1 & 1 & 4{,}000 \\ 16 & 12 & 8 & 53{,}200 \\ 20 & 15 & 8 & 65{,}300 \end{array}\right].$$

To obtain 0's in the first column after row 1, we add -16 times the first row to the second row, and -20 times the first row to the third row.

$$\left[\begin{array}{ccc|c} 1 & 1 & 1 & 4{,}000 \\ 0 & -4 & -8 & -10{,}800 \\ 0 & -5 & -12 & -14{,}700 \end{array}\right]$$

To obtain 0 in the second column after row 2, we first multiply each entry in row 2 by $-\frac{1}{4}$ to make the coefficient of y in the second equation a 1.

$$\left[\begin{array}{ccc|c} 1 & 1 & 1 & 4{,}000 \\ 0 & 1 & 2 & 2{,}700 \\ 0 & -5 & -12 & -14{,}700 \end{array}\right]$$

Now add 5 times the second row to the third row.

$$\left[\begin{array}{ccc|c} 1 & 1 & 1 & 4{,}000 \\ 0 & 1 & 2 & 2{,}700 \\ 0 & 0 & -2 & -1{,}200 \end{array}\right]$$

Progress Check Answer

3. $(2, -3)$

The third row tells us $-2z = -1{,}200$, so $z = 600$. Then, back substitution into the equations corresponding to row 2 and row 1 gives

$$\begin{aligned} y + 2z &= 2{,}700 \\ y + 2(600) &= 2{,}700 \\ y &= 1{,}500 \end{aligned} \quad \text{and} \quad \begin{aligned} x + y + z &= 4{,}000 \\ x + 1{,}500 + 600 &= 4{,}000 \\ x &= 1{,}900. \end{aligned}$$

Thus, the theater has 1,900 orchestra seats, 1,500 mezzanine seats, and 600 balcony seats. Confirm that this solution checks in the context of this application.

PROGRESS CHECK 4 A 2,000-seat theater was sold out for a weekend concert. Total receipts on Friday night were \$31,500 when seats in the orchestra, mezzanine, and balcony sold for \$20, \$15, and \$10, respectively. On Saturday prices were raised to \$25, \$20, and \$10, and total receipts increased to \$39,000. How many seats of each type are in the theater?

In the use of Gaussian elimination, if a row of zeros occurs in the augmented matrix to the left of the dashed line, then the system does not have exactly one solution. Such systems are either inconsistent or dependent. Consideration of the equation associated with the row in question should reveal which case applies.

Progress Check Answer

4. Orchestra: 800; mezzanine: 700; balcony: 500

EXERCISES 6.4

In Exercises 1–10, solve the system. If necessary, transform to triangular form.

1. $\begin{aligned} 3x - 4y - 5z &= 6 \\ 5y + 6z &= -4 \\ 7z &= 7 \end{aligned}$ (1, −2, 1)

2. $\begin{aligned} 2x + 3y - 4z &= -9 \\ 3y + z &= 5 \\ 3z &= 6 \end{aligned}$ (−2, 1, 2)

3. $\begin{aligned} x + 4y - z &= 1 \\ 3x + 2y &= 6 \\ 5x &= 10 \end{aligned}$ (2, 0, 1)

4. $\begin{aligned} x - 4y - 2z &= 7 \\ 2x - 7y &= 3 \\ 3x &= 15 \end{aligned}$ (5, 1, −3)

5. $\begin{aligned} x - 3y + 4z &= 11 \\ 2x + y + 3z &= 3 \\ 3x - 2y - 4z &= 3 \end{aligned}$ (1, −2, 1)

6. $\begin{aligned} x + y + z &= 1 \\ 3x - 4y + z &= 6 \\ 2x + y + 3z &= 5 \end{aligned}$ (0, −1, 2)

7. $\begin{aligned} 2x - 3y + 2z &= 5 \\ 4x - y - 2z &= -7 \\ x + y - z &= 1 \end{aligned}$ (3, 5, 7)

8. $\begin{aligned} 6x - 4y + z &= 2 \\ x + y - z &= 0 \\ 3x + y - z &= 4 \end{aligned}$ (2, 4, 6)

9. $\begin{aligned} x - 4y &= -7 \\ 2x + 3y &= -3 \end{aligned}$ (−3, 1)

10. $\begin{aligned} x + 3y &= 1 \\ -2x - 5y &= -4 \end{aligned}$ (7, −2)

In Exercises 11–16, solve the given system. Show both the matrix form of the system and the corresponding equations.

11. $\begin{aligned} 3x + 3y &= -9 \\ x + 2y &= -5 \end{aligned}$ (−1, −2)

12. $\begin{aligned} 2x + 3y &= 2 \\ x - y &= -1 \end{aligned}$ $(-\frac{1}{5}, \frac{4}{5})$

13. $\begin{aligned} x - 3y + 3z &= 1 \\ 2x + y + z &= 1 \\ 3x + 2y - z &= 0 \end{aligned}$ $(0, \frac{1}{3}, \frac{2}{3})$

14. $\begin{aligned} x + y + 2z &= 2 \\ 5x + y - 3z &= 1 \\ 3x - y + z &= -3 \end{aligned}$ $(-\frac{1}{4}, \frac{9}{4}, 0)$

15. $\begin{aligned} 2x + 3y - 4z &= -2 \\ 2x + y + 4z &= 4 \\ x - 2y + 2z &= 2 \end{aligned}$ $(\frac{1}{2}, 0, \frac{3}{4})$

16. $\begin{aligned} 3x + y + 2z &= 2 \\ x + 3y - z &= 3 \\ x - y - 3z &= -5 \end{aligned}$ $(-\frac{1}{2}, \frac{3}{2}, 1)$

In Exercises 17–24, solve the system using just the matrix form.

17. $\begin{aligned} 4x - 8y - 3 &= 0 \\ -2x + 2y + 1 &= 0 \end{aligned}$ $(\frac{1}{4}, -\frac{1}{4})$

18. $\begin{aligned} 6x + 15y &= 8 \\ -3x + 6y &= 5 \end{aligned}$ $(-\frac{1}{3}, \frac{2}{3})$

19. $\begin{aligned} 3x - 5y - 14 &= 0 \\ 2x + 3y - 3 &= 0 \end{aligned}$ (3, −1)

20. $\begin{aligned} 5x + 2y + 1 &= 0 \\ 3x - 4y + 11 &= 0 \end{aligned}$ (−1, 2)

21. $\begin{aligned} x - y + z &= 3 \\ 2x - y - z &= -1 \\ 3x + 2y - 3z &= 1 \end{aligned}$ (2, 2, 3)

22. $\begin{aligned} x - y + z &= 2 \\ 3x - y - z &= 4 \\ 2x + 3y - 2z &= -1 \end{aligned}$ (1, −1, 0)

23. $\begin{aligned} 2x - 3y - 2z &= 1 \\ 2x + 3y + 3z &= 5 \\ 3x - 2y - 4z &= -3 \end{aligned}$ (1, −1, 2)

24. $\begin{aligned} 2x + 5y - 3z &= -6 \\ -3x + 2y - 5z &= -10 \\ 5x - 3y + 2z &= -2 \end{aligned}$ (−1, 1, 3)

In Exercises 25–28 the problems can be solved by setting up appropriate systems. Solve the system by using matrix form.

25. A 1,000-seat theater was sold out for two concerts by a performer. Total receipts for the first concert were \$9,900 when seats in the orchestra, mezzanine, and balcony sold for \$15, \$10, and \$6, respectively. For the second concert, total receipts were \$12,200 when prices were \$18, \$12, and \$8. How many seats of each type are in the theater? 300 orchestra and mezzanine; 400 balcony

26. A ballpark has three prices of tickets for the bleachers and the upper and lower stands. For a series of three baseball games the sales were as follows: On day 1, they sold 10,000 of each price and took in \$900,000. On day 2, the number of middle-priced tickets increased by 5,000, while the others remained the same as on the previous day. Receipts were \$1,025,000. On day 3, the number of high-priced tickets increased by 5,000, while the others remained the same as on the previous day. Receipts were \$1,275,000. What prices were charged for each type of ticket? \$15, \$25, \$50

27. A used-records shop sells records, CDs, and cassettes. The CDs sell for \$8, while the records and cassettes sell for \$4. One week the sales totaled \$1,000. During the week they sold 10 more cassettes than records, and they sold twice as many CDs as records and cassettes combined. How many of each were sold? 30 cassettes, 20 records, 100 CDs

28. A gas station sells three grades of gasoline, regular, plus, and super, for \$1.10, \$1.15, and \$1.25 per gallon, respectively. One day sales were \$2,760. On this day it sold as much regular as the other two combined, and the amount of plus sold was the same as the amount of super. How many gallons of each were sold? If the station's profit is 5¢ per gallon, how much profit did it make for the day? 1,200 gal regular; 600 gal plus and super; profit: \$120

THINK ABOUT IT

1. Use Gaussian elimination to solve for x, y, and z in terms of a, b, and c.

$$\begin{aligned} x + z &= a \\ x + y &= b \\ y + z &= c \end{aligned}$$

2. In Example 4 of this section we stopped using matrix methods when we reached triangular form. We then completed the solution by back substitution. An alternative method, called **Gauss-Jordan elimination,** continues to obtain matrices of equivalent systems until reaching the form

$$\left[\begin{array}{ccc|c} 1 & 0 & 0 & a \\ 0 & 1 & 0 & b \\ 0 & 0 & 1 & c \end{array}\right].$$

From this form we can directly read that $x = a$, $y = b$, and $z = c$. Check the solution in Example 4 by Gauss-Jordan elimination.

3. Refer to Exercise 2, and use Gauss-Jordan elimination to solve

$$\begin{aligned} x + y + z &= 6 \\ x - y - z &= -4 \\ -x + y - z &= -2. \end{aligned}$$

4. a. Evaluate $\begin{vmatrix} 1 & 2 \\ 3 & 4 \end{vmatrix}$.

b. If -3 times row 1 is added to row 2, the resulting determinant is $\begin{vmatrix} 1 & 2 \\ 0 & -2 \end{vmatrix}$, which has a 0 as the first element of row 2. Evaluate this determinant, and compare the answer with the answer in part **a.**

c. Evaluate $\begin{vmatrix} 5 & 6 \\ 10 & 15 \end{vmatrix}$ by first adding -2 times row 1 to row 2.

d. In general, the value of a determinant is unchanged when a multiple of one row is added to another row. Find the value of the following determinant by using this property to obtain 0's as the first element in row 2 and row 3 before expanding by minors about the first column.

$$\begin{vmatrix} 1 & 2 & 1 \\ -2 & 1 & -1 \\ 3 & 3 & -1 \end{vmatrix}$$

5. Solve this system of four linear equations in four variables. Use Gaussian elimination.

$$\begin{aligned} u + v + w + x &= 0 \\ u + v \phantom{{}+w} + x &= -2 \\ v + w + x &= -3 \\ u + v + w \phantom{{}+x} &= 2 \end{aligned}$$

REMEMBER THIS

1. Find the intersection of $A = \{1,2,3,4,5,6\}$ and $B = \{4,5,6,7,8\}$. $\{4,5,6\}$

2. Graph the solution set of $y \leq 2x + 3$.

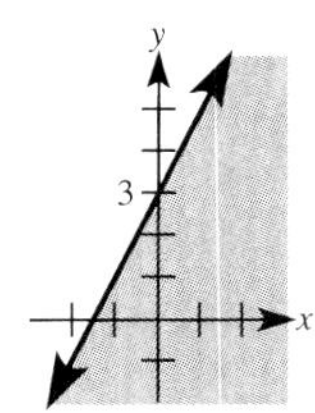

3. Is (5,5) a solution of *all* of these inequalities?

$$\begin{aligned} 9x + 12y &\leq 108 \\ x &\geq 3 \\ y &\geq 4 \end{aligned}$$ Yes

4. Some red cloth costs \$12 per yard, and some green cloth costs \$11 per yard. Write an expression for the total cost of x yd of red and y yd of green. $12x + 11y$

5. Solve by Cramer's rule: $2x + 4y = 1$, $x - 3y = -2$. $\left(-\frac{1}{2}, \frac{1}{2}\right)$

6. Evaluate $\begin{vmatrix} 3 & 4 \\ -1 & 2 \end{vmatrix}$. 10

7. Use synthetic division and the remainder theorem to see if -2 is a zero of $P(x) = 2x^3 - x^2 - 11x + 2$. No; $P(-2) = 4 \neq 0$

8. Solve $(2x - 2)(x - 3)(3x + 4) = 0$. $\{1, 3, -\frac{4}{3}\}$

9. Solve $ab - b = c$ for b. $b = \dfrac{c}{a - 1}$

10. Rewrite $5(a + b)$ using the commutative property of addition. $5(b + a)$

6.5 Systems of Linear Inequalities

For redecorating a room, you will buy some solid-color material at \$9 per yard and some with a print design at \$12 per yard. You can spend *no more than* \$108 but you need *at least* 3 yd of the solid-color material and at least 4 yd of the print.

a. Translate these requirements into a system of linear inequalities.
b. Solve the system in part **a** graphically.
c. Specify and interpret one solution that is in the solution set. (See Example 3.)

OBJECTIVES

1 **Solve systems of linear inequalities graphically.**
2 **Solve word problems that translate to systems of linear inequalities.**

1 Because the situation described in the opening problem involves the phrases "at least" and "no more than," translation into algebraic symbols will result in a system of *inequalities* which will have to be solved. An ordered pair is a **solution of a *system* of inequalities** if it is a solution to *every* individual inequality in the system. The set of *all* these ordered pairs makes up the solution *set* of the system. This set is called the **intersection** of the individual solution sets, and it is best specified by a graph found as follows: On the same coordinate system, graph the solutions to each inequality in the system, and then shade in the overlap (intersection) of these half planes. This procedure is illustrated in Example 1, which includes a brief review of how to graph a linear inequality.

EXAMPLE 1 Graph the solution set of the system

$$5x + y \leq 10$$
$$y > 2x + 4.$$

Solution First, graph $5x + y = 10$, as shown in Figure 6.7(a). Use a *solid* line, because the inequality symbol is $\leq$, which means that the points on the line are solutions. Then shade the half plane that contains the origin, because choosing (0,0) as a test point leads to a true inequality.

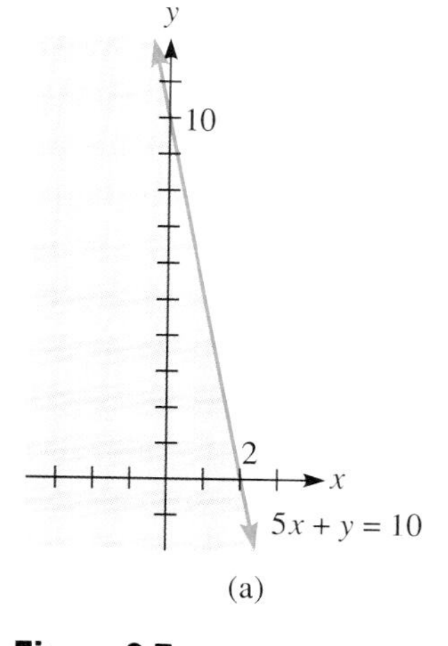

(a)

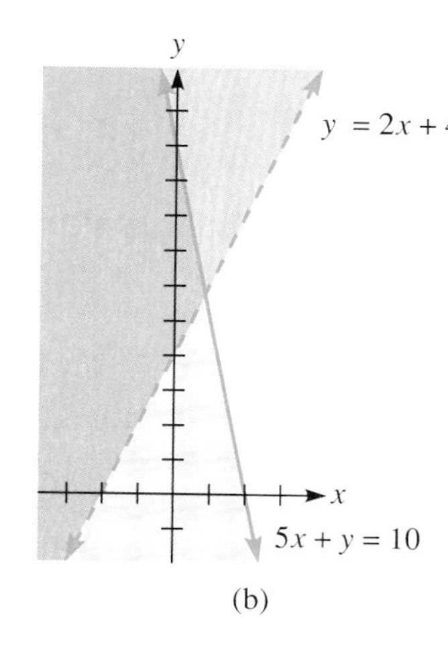

(b)

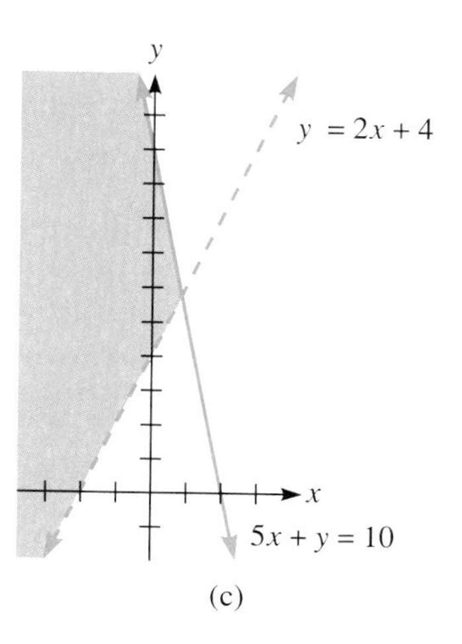

(c)

Figure 6.7

Second, on the same coordinate system, graph $y = 2x + 4$, as shown in Figure 6.7(b). Use a *dashed* line, because the inequality symbol is $>$, and so the points on the line are *not* solutions. Shade the half plane above the line, because the inequality begins $y >$.

Finally, the region that has been shaded twice, as in Figure 6.7(c), is the graph of the solution set of the system. It is the intersection of the two solution sets.

With respect to the boundary lines, note that the solution set of this system includes a portion of the line $5x + y = 10$ but excludes entirely the line $y = 2x + 4$.

PROGRESS CHECK 1 Graph the solution set of the system

$$\begin{aligned} x + 4y &< 8 \\ y &\leq 2x. \end{aligned}$$

EXAMPLE 2 Graph the solution set of the system

$$\begin{aligned} x + y &> 3 \\ 2x - y &> -4. \end{aligned}$$

Solution By choosing the origin as a test point in both cases, we determine that $x + y > 3$ graphs as the set of points above the line $x + y = 3$, while $2x - y > -4$ graphs as the set of points below the line $2x - y = -4$. The graph of the solution set of the system is the intersection of these two half planes that is shown in Figure 6.8.

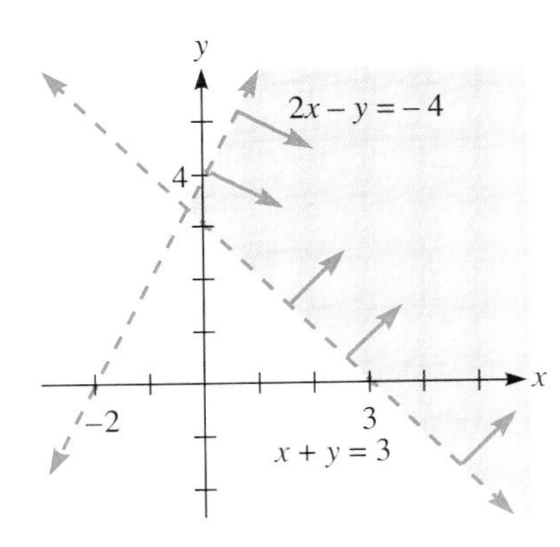

Figure 6.8

PROGRESS CHECK 2 Graph the solution set of the system

$$\begin{aligned} x + y &\geq 2 \\ x - 2y &\leq -4. \end{aligned}$$

2 To solve the section-opening problem, we must set up and then solve a system of linear inequalities.

EXAMPLE 3 Solve the problem in the section introduction on page 305.

Solution

a. First, let

$$\begin{aligned} x &= \text{amount of solid material in yards} \\ \text{and} \quad y &= \text{amount of print material in yards.} \end{aligned}$$

At \$9 per yard for solid and \$12 per yard for print, the cost limitation is given by

cost of solid plus cost of print is no more than \$108.

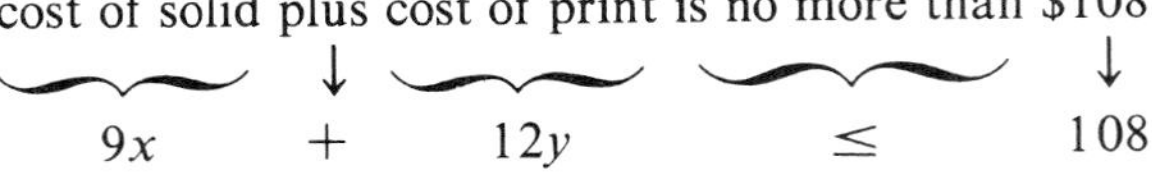

$$9x \quad + \quad 12y \quad \leq \quad 108$$

Progress Check Answers

1.

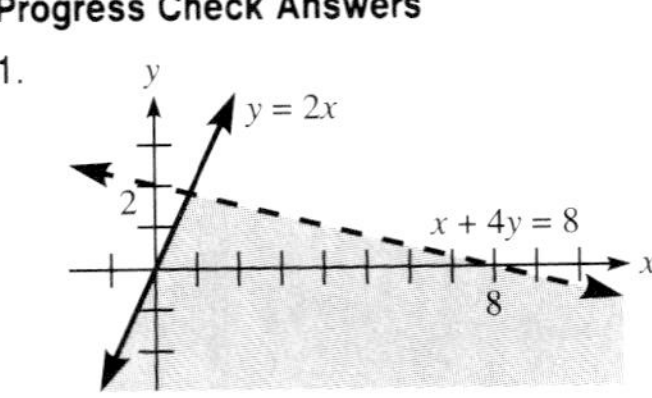

2.

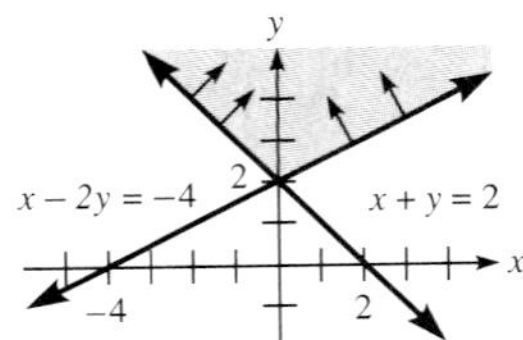

The requirement of at least 3 yd of the solid material translates to $x \geq 3$, while requiring at least 4 yd of the print gives $y \geq 4$.

Thus, the requested system is

$$\begin{aligned} 9x + 12y &\leq 108 \\ x &\geq 3 \\ y &\geq 4. \end{aligned}$$

b. To solve the system, we first graph the three individual solution sets. Using (0,0) as a test point, we find that the graph of $9x + 12y \leq 108$ is the half plane below the line $9x + 12y = 108$. The graph of $x \geq 3$ is the half plane to the right of the vertical line $x = 3$, and the graph of $y \geq 4$ is the half plane above the horizontal line $y = 4$. In all three cases the lines themselves are included in the solutions. Finally, the graphical solution to this system is the intersection of these three half planes shown in Figure 6.9.

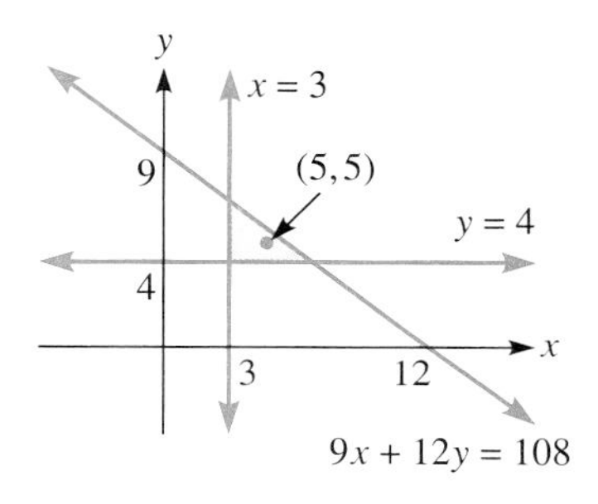

Figure 6.9

A natural question is, "What is the *maximum* number of yards of cloth you could buy given these constraints?" This is the type of question studied in linear programming. See "Think About It" Exercises 3 and 4.

c. One point which lies in the final shaded region is (5,5). This means that buying 5 yd of the solid and 5 yd of the printed material satisfies the requirements of spending no more than \$108 to purchase at least 3 yd of solid and at least 4 yd of print. Check that the total cost for this purchase is \$105.

PROGRESS CHECK 3 For a party you need to buy some ham at \$8 per pound and some cheese at \$4 per pound. You will need *at least* 1 lb of ham and *at least* 2 lb of cheese, and you wish to spend *no more than* \$32 on these two items. Answer the same three questions answered in Example 3.

Progress Check Answers

3. (a) $8x + 4y \leq 32$, $x \geq 1$, $y \geq 2$
(b)
(c) One point is (2,3).

EXERCISES 6.5

In Exercises 1–22, graph the solution sets of the system.

1. $y \leq x + 1$
$y \geq 2 - x$

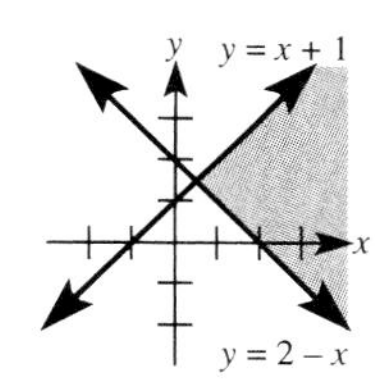

2. $y \geq -x$
$y \leq x - 3$

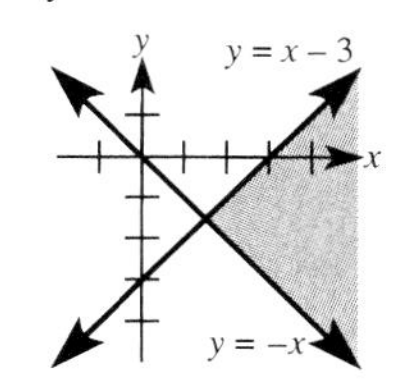

3. $2x - y < -1$
$2x + y < 4$

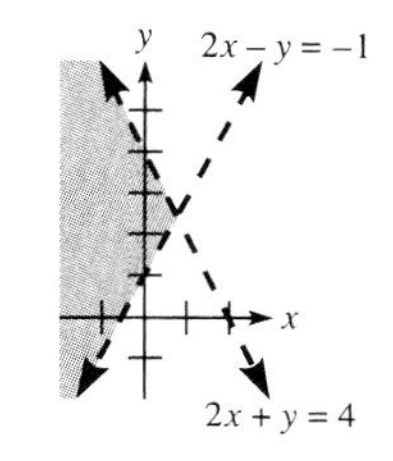

4. $y > x + 4$
$x - 2y > -6$

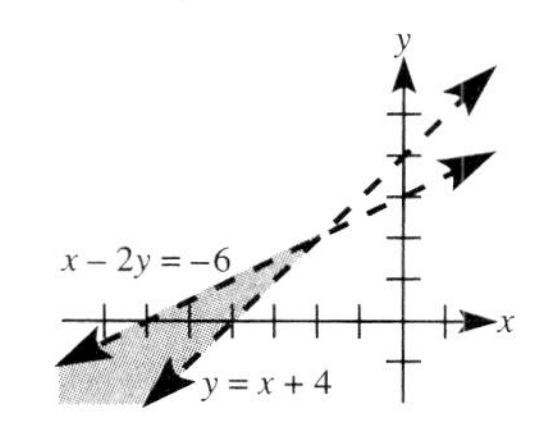

5. $x + 2y < 3$
$3x + y \geq 8$

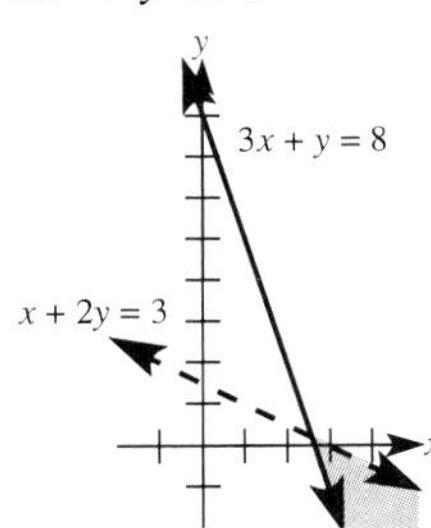

6. $2x + y > -2$
$x - 3y \geq -8$

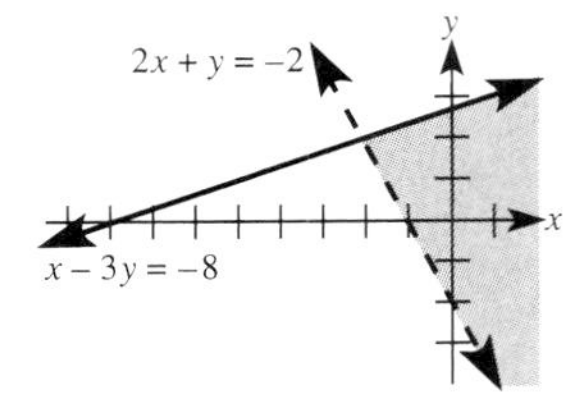

7. $y - \frac{1}{2}x < 2$
$y + \frac{1}{2}x < 3$

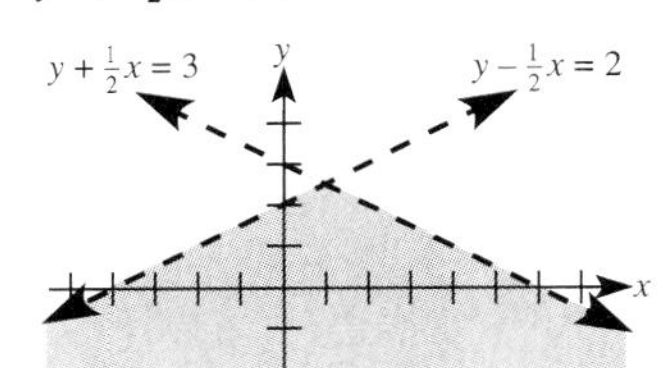

8. $y - \frac{1}{3}x \geq -3$
$y + \frac{1}{3}x > -1$

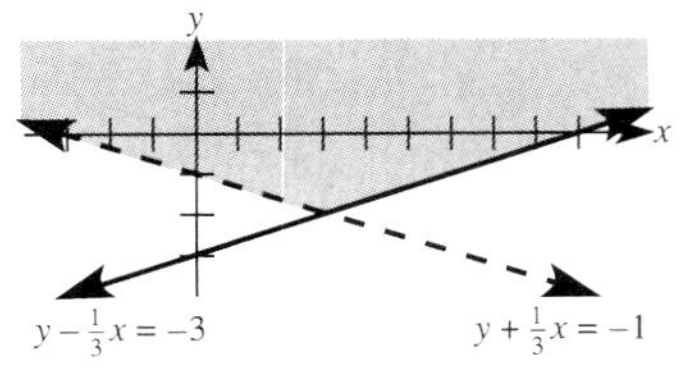

9. $4x + y < 4$
$x + 4y > 8$

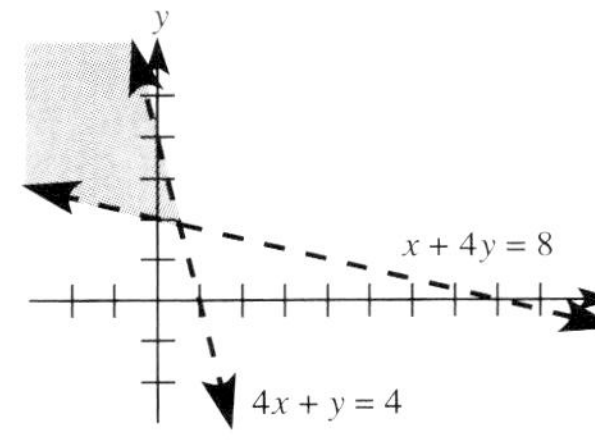

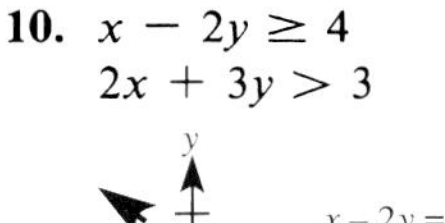

10. $x - 2y \geq 4$
$2x + 3y > 3$

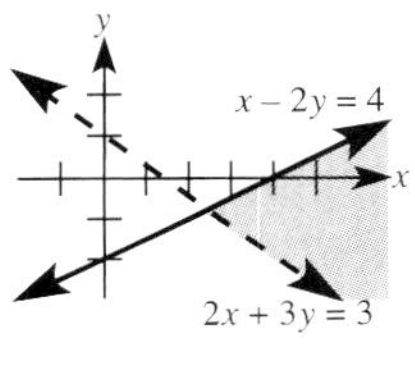

11. $y - 4 < 0$
$x + y \geq 2$

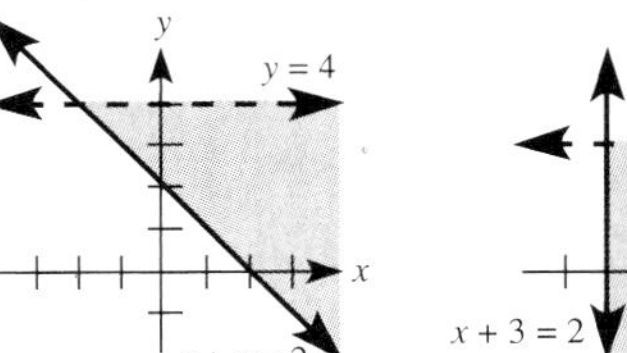

12. $y - 2 < 1$
$x + 3 \geq 2$

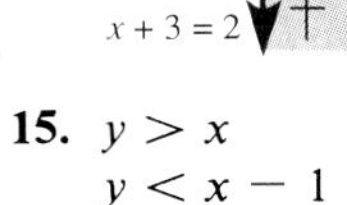

13. $x < 2y - 3$
$y < 0$

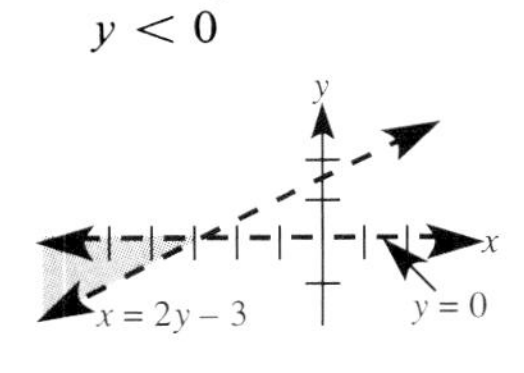

14. $y < -3x - 2$
$x > 0$

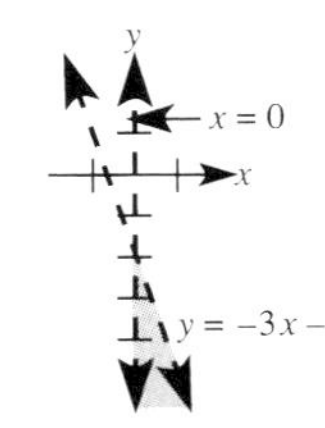

15. $y > x$
$y < x - 1$

No solution; the solution sets do not overlap.

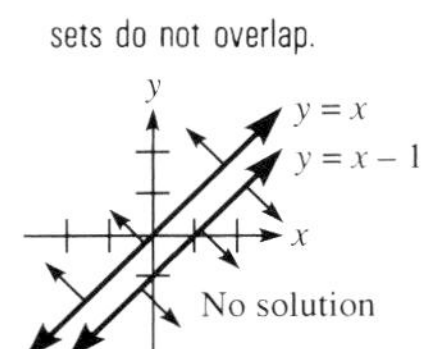

16. $y < x$
$y > x - 1$

17. $y > x + 2$
$y \geq x + 1$

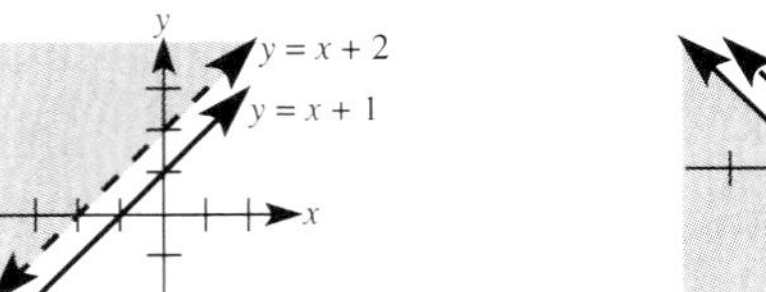
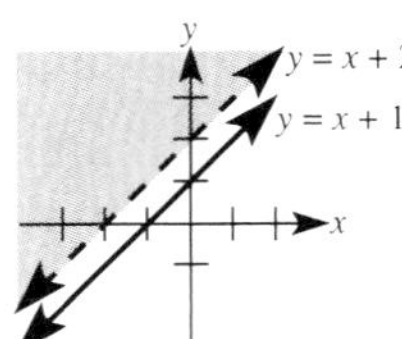

18. $x + y < 1$
$x + y \leq 0$

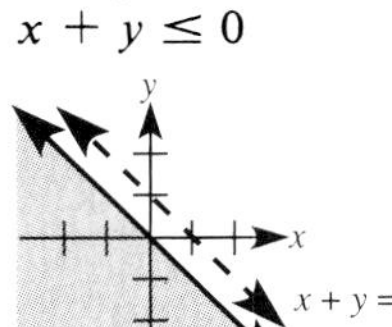

19. $2x + 3y \leq 6$
$x \geq 1$
$y \geq 1$

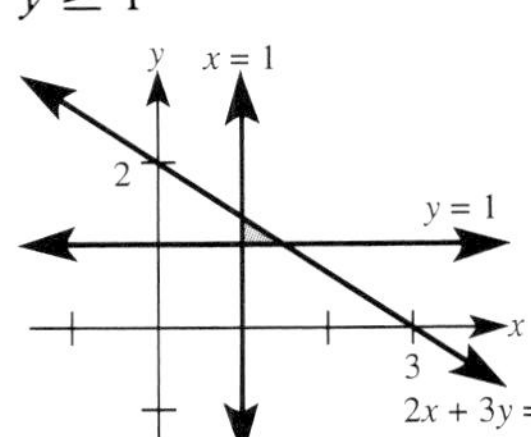

20. $x \geq 2$
$y \geq 2$
$x + y \leq 6$

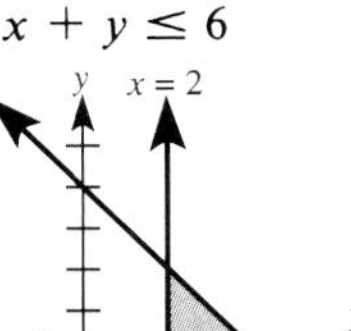

21. $2x + 3y \leq 12$
$y \leq 2x$
$y \geq \frac{1}{2}x$

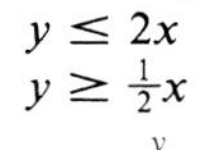
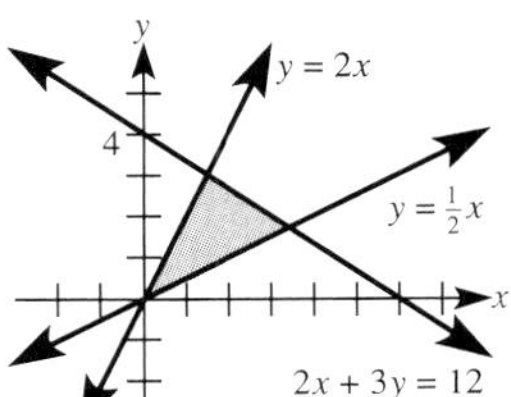

22. $x + 2y \leq 8$
$y \leq 3x$
$y \geq 2x$

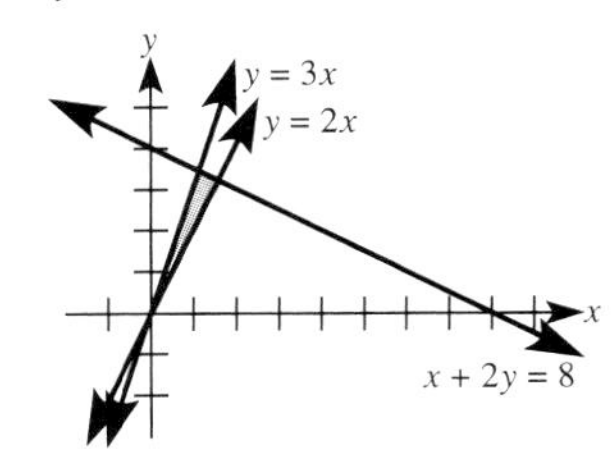

For 23–28, perform the following steps.

a. Translate the problem into a system of linear inequalities.
b. Solve the system graphically.
c. Specify and interpret one solution from the solution set.

23. For a costume design in a play you need to buy silver ribbon at \$3 per foot and gold ribbon at \$4 per foot. You need at least 20 ft of each, and you wish to spend no more than \$180 all together. How much of each can you buy?

a. $3x + 4y \leq 180$
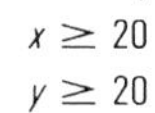
$x \geq 20$
$y \geq 20$

b.

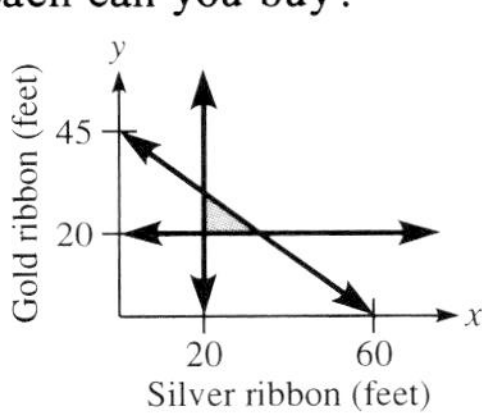

c. One solution is 25 ft of each; total cost: \$175.

24. For a party you will buy some cheese at \$6 per pound and some pasta salad at \$5 per pound. You can spend up to \$30 but you need at least 1 lb of each. How much of each can you buy?

a. $6x + 5y \leq 30$
$x \geq 1$
$y \geq 1$

b.

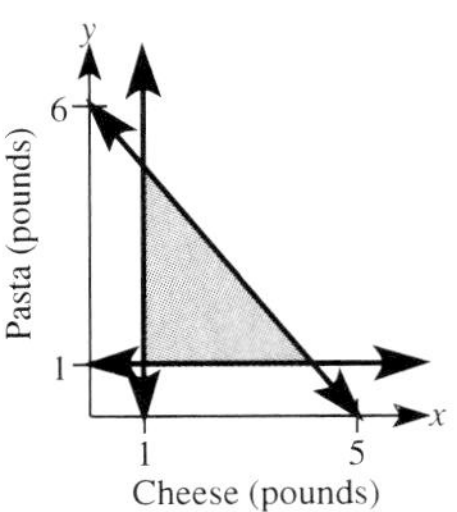

c. One solution is 2 lb of each, for a total cost of \$22.

25. A person on a special diet plan must consume less than 200 calories at lunch, which consists of low-fat yogurt and skim milk. The lunch must contain at least 50 calories from each of these foods. There are about 15 calories per ounce of yogurt and about 10 calories per ounce of skim milk. How many ounces of each can the meal contain?

a. $15x + 10y \leq 200$
$15x \geq 50$
$10y \geq 50$

b.

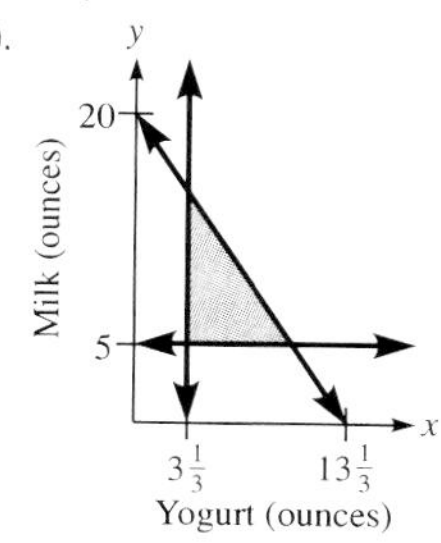

c. 6 oz of each provides about 150 calories.

26. A person on a diet plan must consume less than 500 calories at a meal, which consists of a serving of chicken and a serving of rice. A cookbook notes that there are about 200 calories in 8 oz of rice and about 180 calories in 3 oz of chicken. The meal must contain at least 150 calories from each source. How many ounces of each can the meal contain?

a. $25x + 60y \leq 500$
$25x \geq 150$
$60y \geq 150$

b.

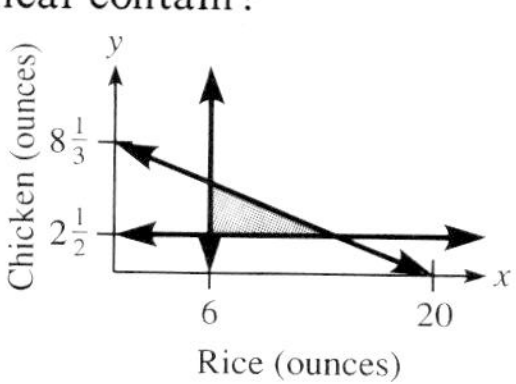

c. 7 oz of rice and 4 oz of chicken provides about 415 calories.

27. You need to buy some large and some small screws. The small ones each cost 3¢ and the large ones each cost 4¢. You have $5 to spend, which must cover the screws plus a 5 percent sales tax. You need at least 75 small and 50 large screws. How many of each can you buy?

a. $3x + 4y + 0.5(3x + 4y) \leq 500$
$3.15x + 4.20y \leq 500$
$x \geq 75$
$y \geq 50$

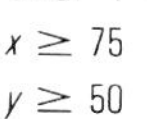

b.

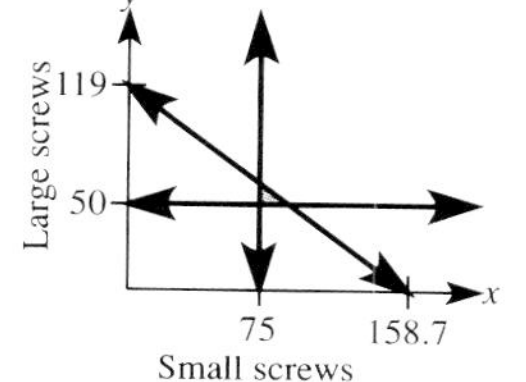

c. One solution is 80 small screws and 55 large screws, for a total cost of $4.83.

28. The maximum weight a certain shipper will handle for the contents of one package is 50 lb. One package is to contain some metal parts which each weigh 1 lb and some plastic parts which each weigh $\frac{1}{2}$ lb. The package must contain at least 20 plastic and 10 metal parts. How many of each can be shipped?

a. $1x + \frac{1}{2}y \leq 50$
$x \geq 10$
$y \geq 20$

b.

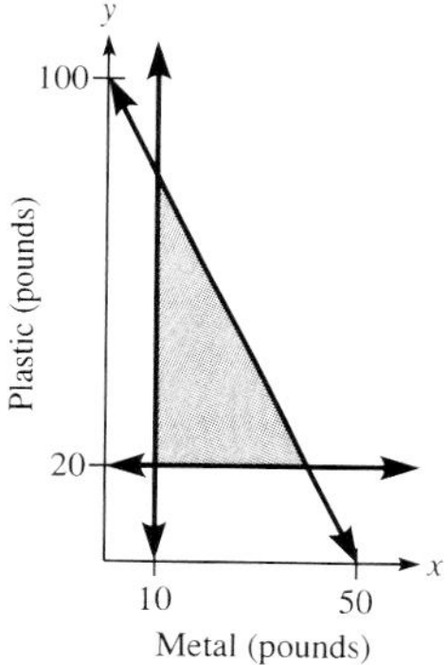

c. One solution is 20 metal pieces and 30 plastic pieces, for a total weight of 35 lb.

THINK ABOUT IT

1. Treat these systems of inequalities as relations in two variables, x and y. Graph their solution sets, and describe the solution set.
   a. $x \geq 2$, $x \leq 2$ **b.** $x > 2$, $x < 2$ **c.** $x \geq y$, $y \geq x$ **d.** $x \geq 0$, $x \leq 2$
2. **a.** Find a system of linear inequalities whose solution set is the entire first quadrant, including the axes.
   b. and **c.** Find a system of linear inequalities whose solution set is as shown.

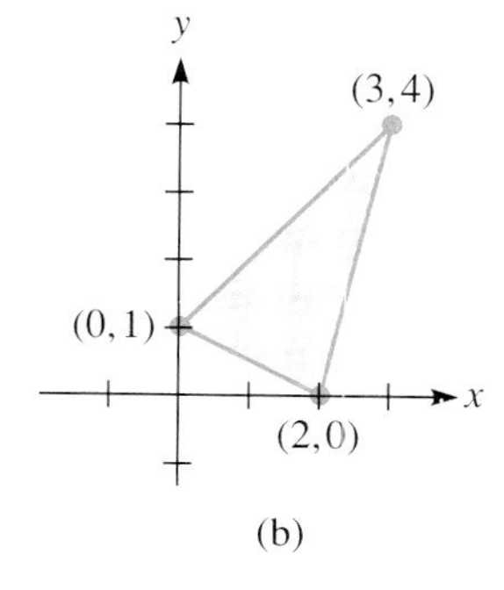

(b)

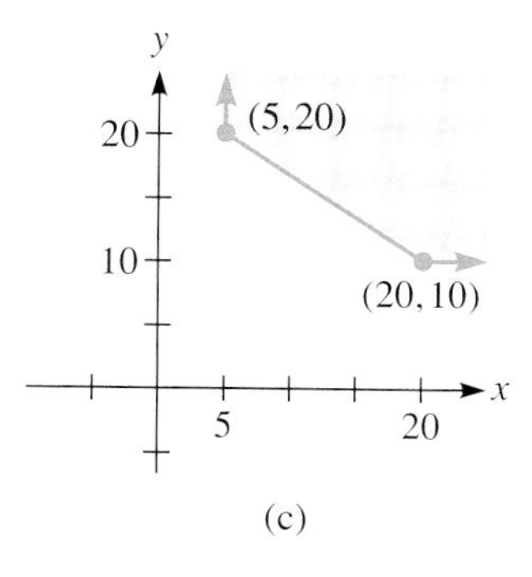

(c)

3. The area of mathematics called **linear programming** deals with the problem of optimization given certain constraints. An example based on the chapter-opening problem follows. Given the conditions of the problem, what is the *maximum number of yards* of material you can buy? That is, what values of x and y which satisfy the conditions of the problem will maximize $x + y$? You can find the answer by examining Figure 6.9. (*Hint:* As a general rule, the solution of this type of problem is a point where two boundary lines intersect.)
4. For Exercise 27, what is the maximum number of screws you can purchase? (See the hint for Exercise 3.)
5. Systems of inequalities in two variables are not necessarily linear. Use the fact that the equation $x^2 + y^2 = 9$ determines a circle with radius 3 and center at the origin to graph the solution set of this system.

$$x^2 + y^2 \leq 9$$
$$y \geq x$$

REMEMBER THIS

1. True or false? Both 4 and -4 are solutions to $x^2 = 16$. True
2. Which of these are rational numbers? $\{2, \sqrt{2}, 4, \sqrt{4}\}$. $2, 4, \sqrt{4}$
3. True or false? $-0.5 < (-0.5)^2$. True
4. Is there a real number whose square is -9? No
5. Solve using matrices: $3x - 4y = -2$, $12x + 12y = -1$. $\left(-\frac{1}{3}, \frac{1}{4}\right)$
6. A varies inversely as B. When A is 7, B is 8. Find B when A is 2. 28
7. Add and simplify: $\dfrac{3}{x-2} + \dfrac{4}{2-x}$. $\dfrac{-1}{x-2}$
8. The product of two numbers is 11, while their difference is 3.5. Find the numbers. -5.5 and -2; 2 and 5.5
9. Simplify $\dfrac{3x^{-2}y^3}{3x^2y^3}$. $\dfrac{1}{x^4}$
10. Graph all prime numbers between 10 and 20.

10 11 12 13 14 15 16 17 18 19 20 x

GRAPHING CALCULATOR

Displaying More Than One Graph at the Same Time

For some problems it will be valuable to see more than one graph on the same set of axes. This is demonstrated in Example 1.

EXAMPLE 1 Graph $y = 2x + 5$ and $y = 2x + 7$ on the same set of axes to see if they appear to be parallel.

Solution We will key in both equations before graphing.

Step 1 Press Y=.
On the first line put $2x + 5$; on the second line put $2x + 7$.

Step 2 Press ZOOM Standard. The screen will show both graphs, as in Figure 6.10. The lines do appear to be parallel.

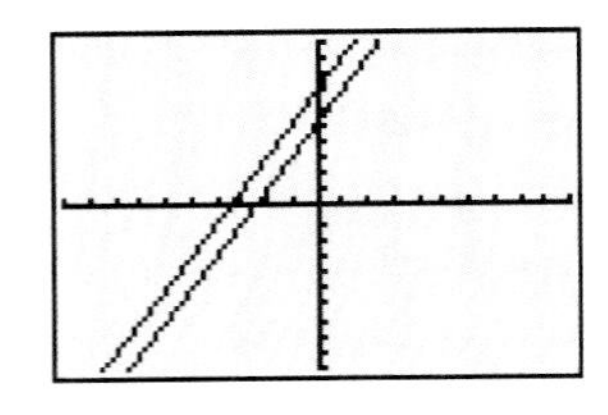

Figure 6.10

Note You can instruct the calculator to *not* draw a graph in its list by placing the cursor on the equals sign and pressing ENTER. This will change the appearance of the equals sign. Only the light equals sign on the dark background will result in a graph.

In the next example we wish to see if two lines look *perpendicular*. It will be necessary to use the Zoom Square feature to correct the built-in angular distortion present in the standard display. This distortion occurs because in the standard display the dots are spaced differently on the x- and y-axes. Zoom Square makes the spaces equal.

EXAMPLE 2 **a.** Graph $y = 2x + 5$ and $y = -\frac{1}{2}x + 4$ on the same set of axes to see if they look perpendicular.

b. Solve the system $y = 2x + 5$
$y = -\frac{1}{2}x + 4$.

Solution

a. Step 1 Press Y=; then key in both equations.

Caution Keying in [(−)] 1 [÷] 2 [X | T] [+] 4 will *not* give the right graph for the second equation. The calculator interprets this sequence as $\frac{-1}{2x} + 4$. A correct sequence is to use parentheses and key in [(] [(−)] 1 [÷] 2 [)] [X | T] [+] 4, or you could use decimals and key in [(−)] .5 [X | T] [+] 4.

Step 2 Press [ZOOM] [5] to activate [ZOOM] Square. The graphs look perpendicular.

b. The solution is given graphically by the point of intersection of the two lines. To estimate the coordinates of this point, use [TRACE] and [ZOOM] In until the coordinates can be read as accurately as you desire. You should see that the point of intersection is about (−0.4,4.2).

Note When [TRACE] is active, the up and down arrow keys cause the cursor to jump from one graph to the other, while the left and right keys cause the cursor to move along the graph.

Exercises

1. Graph $x + y = 5$ and $x + y = -5$ on the same set of axes to see that they look parallel. Ans.

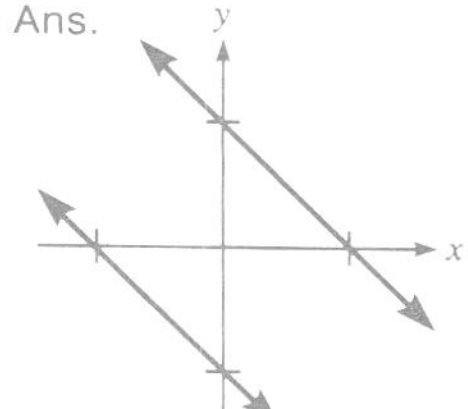

2. Graph $x + y = 5$ and $x - y = 5$ on the same set of axes to see that they look perpendicular. Ans.

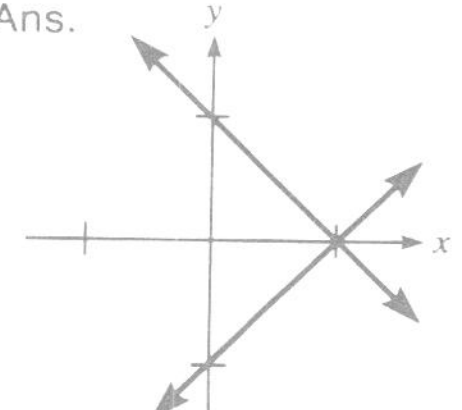

Use the graphing calculator to investigate these exercises from Section 6.1.

Exercises 6.1: 7, 11, 17, 19, 21, 51

Chapter 6 SUMMARY

OBJECTIVES CHECKLIST Specific chapter objectives are summarized below along with numbered example problems from the text that should clarify the objectives. If you do not understand any objectives or do not know how to do the selected problems, then restudy the material.

6.1 Can you:

1. **Determine whether an ordered pair is a solution to a system of linear equations?**
   Determine if (10,2) is a solution of the system $x + y = 12$
   $5x - 2y = 4.$ [Example 1b]

2. **Solve a system of linear equations graphically?**
   Solve by graphing: $x + 2y = 4$
   $y = x + 5.$ [Example 2]

3. **Solve a system of linear equations by the addition-elimination method?**
   Solve by addition-elimination: $6x + 10y = 7$
   $15x - 4y = 3.$ [Example 4]

4. **Solve a system of linear equations by the substitution method?**
   Solve by substitution: $y = 1{,}000 + 5x$
   $y = 4{,}000 + 3x.$ [Example 6]

5. **Use systems of linear equations to solve applied problems?**
   A \$16,000 retirement fund was split into two investments, one portion at 11 percent annual interest and the rest at 6 percent. If the total annual interest is \$1,400, how much was invested at each rate? [Example 8]

6.2 Can you:

1. **Determine whether an ordered triple is a solution of a system of linear equations?**
Determine if $(-3,-1,6)$ is a solution of the system.

$$\begin{aligned} 2x - 3y + z &= 3 \\ 5x - 4y + 2z &= 1 \\ 7x - 2y + 3z &= 2 \end{aligned}$$

[Example 1]

2. **Solve linear systems in three variables with exactly one solution?**
Solve the system.

$$\begin{aligned} 2x - y + z &= 7 \\ -x + 2y - z &= 6 \\ 2x - 3y - 2z &= 9 \end{aligned}$$

[Example 2]

3. **Solve linear systems in three variables that are inconsistent or dependent?**
Solve the system.

$$\begin{aligned} x - y + 2z &= 0 \\ -x + 4y + z &= 0 \\ -x + 2y - z &= 0 \end{aligned}$$

[Example 5]

6.3 Can you:

1. **Evaluate a second-order determinant?**

Find the value of $\begin{vmatrix} -9 & -2 \\ 11 & 7 \end{vmatrix}$.

[Example 1b]

2. **Solve a linear system in two variables using Cramer's rule?**
Use Cramer's rule to solve the system of equations.

$$\begin{aligned} 2x + 5y &= 3 \\ 6x - 10y &= -9 \end{aligned}$$

[Example 2]

3. **Evaluate a third-order determinant?**
Evaluate the determinant by expansion by minors about the first column.

$$\begin{vmatrix} 2 & -2 & 3 \\ 4 & 0 & 7 \\ 3 & 1 & -1 \end{vmatrix}$$

[Example 5]

4. **Solve a linear system in three variables using Cramer's rule?**
Use Cramer's rule to solve the system of equations.

$$\begin{aligned} -2x + y - z &= 3 \\ 2x + 3y + 2z &= 5 \\ 3x + 2y + z &= -3 \end{aligned}$$

[Example 7]

6.4 Can you:

1. **Solve a linear system by transforming the system to triangular form?**
Solve by transforming to triangular form.

$$\begin{aligned} 2x - 3y - 2z &= 9 \\ x - y + z &= -1 \\ -x + 2y + z &= -2 \end{aligned}$$

[Example 1]

2. **Solve a linear system using matrices?**
Use matrix form to solve the system.

$$\begin{aligned} 2x + 5y - 1 &= 0 \\ 3x - 2y - 11 &= 0 \end{aligned}$$

[Example 3]

6.5 Can you:

1. **Solve systems of linear inequalities graphically?**
   Graph the solution set of the system.

$$\begin{aligned} 5x + y &\le 10 \\ y &> 2x + 4 \end{aligned}$$

[Example 1]

2. **Solve word problems that translate to systems of linear inequalities?**
   For redecorating a room, you will buy some solid-color material at $9 per yard and some with a print design at $12 per yard. You can spend *no more than* $108 but you need *at least* 3 yd of the solid-color material and at least 4 yd of the print.
   a. Translate these requirements into a system of linear inequalities.
   b. Solve the system graphically.
   c. Specify and interpret one solution that is in the solution set.

[Example 3]

KEY TERMS

Addition-elimination method (6.1)
Augmented matrix (6.4)
Consistent system (6.1)
Dependent system (6.1)
Elementary operations (6.4)
Elementary row operations (6.4)
Entry (or element) of a matrix (6.4)
Equivalent systems (6.4)
Expansion by minors (6.3)
Gaussian elimination (6.4)
Inconsistent system (6.1)
Intersection of solution sets (6.5)
Linear (first-degree) equations in three variables (6.2)
Linear system of equations (6.1)
Matrix (6.4)
Minor (6.3)
Ordered triplet (6.2)
Second-order determinant (6.3)
Solution of a system of inequalities (6.5)
Solution set of a system of linear equations (6.1)
Substitution method (6.1)
Third-order determinant (6.3)
Triangular form (6.4)

KEY CONCEPTS AND PROCEDURES

Section	Key Concepts or Procedures to Review
6.1	■ The solution set of a system of linear equations consists of all pairs that satisfy both equations at the same time. Graphically, this is equivalent to finding all points the two lines have in common. ■ Methods to solve a linear system by the addition-elimination method and by the substitution method
6.2	■ The solution set of a system of three linear equations in three variables consists of all ordered triples that satisfy all the equations at the same time. ■ Addition-elimination method of solving a linear system in three variables ■ Geometric interpretations of consistent, inconsistent, and dependent linear systems in three variables
6.3	■ The value of a second-order determinant is given by $\begin{vmatrix} a_1 & b_1 \\ a_2 & b_2 \end{vmatrix} = a_1b_2 - b_1a_2.$ ■ The value of a third-order determinant is given by $\begin{vmatrix} a_1 & b_1 & c_1 \\ a_2 & b_2 & c_2 \\ a_3 & b_3 & c_3 \end{vmatrix} = a_1\begin{vmatrix} b_2 & c_2 \\ b_3 & c_3 \end{vmatrix} - a_2\begin{vmatrix} b_1 & c_1 \\ b_3 & c_3 \end{vmatrix} + a_3\begin{vmatrix} b_1 & c_1 \\ b_2 & c_2 \end{vmatrix}.$ ■ Method to evaluate a third-order determinant by expansion by minors ■ Cramer's rule for 2 by 2 and 3 by 3 systems

Section	Key Concepts or Procedures to Review
6.4	■ The following operations are used in Gaussian elimination to solve a linear system by transforming it to triangular form. **Elementary Operations on Equations** 1. Interchange two equations. 2. Multiply both sides of an equation by a nonzero number. 3. Add a multiple of one equation to another. **Elementary Row Operations on Matrices** 1. Interchange two rows. 2. Multiply each entry in a row by a nonzero number. 3. Add a multiple of the entries in one row to another row.
6.5	■ Methods to graph inequalities in two variables ■ Methods to solve a system of inequalities by graphing

CHAPTER 6 REVIEW EXERCISES

6.1

1. Determine if $(5,-2)$ is a solution of the system. Yes
$$x - y = 7$$
$$2x + 3y = 4$$
2. Solve by graphing: $2x + y = 5$, $y = x + 2$.

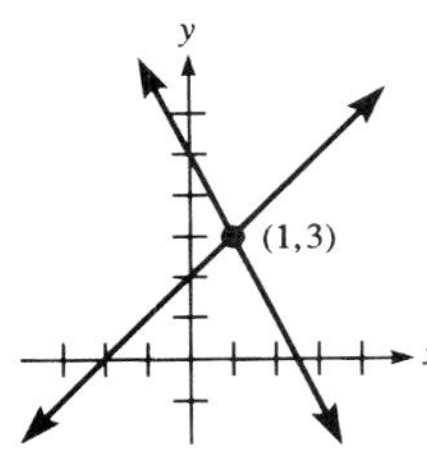

3. Solve by addition-elimination: $5x - 3y = 1$, $-2x + 2y = -2$. $(-1,-2)$
4. Solve by substitution: $3x + 4y = -1$, $-5x = y + 13$. $(-3,2)$
5. How should a $40,000 investment be split so that the total annual earnings are $2,700 if one portion is invested at 7 percent annual interest and the rest at 5 percent?
$35,000 at 7 percent; $5,000 at 5 percent

6.2

6. Determine whether $(3,-2,4)$ is a solution of the system.
$$\begin{aligned} 2x - y - z &= 4 \\ -3x + 2y + 3z &= 1 \\ 5x + 3y - 2z &= 1 \end{aligned}$$
No
7. Solve the system.
$$\begin{aligned} 2x + y + 3z &= 5 \\ -x - 2y + z &= -8 \\ -3x + 3y - z &= -2 \end{aligned}$$
$(3,2,-1)$
8. Solve the system.
$$\begin{aligned} 2B - 3C &= 21 \\ A + B + C &= 0 \\ 5A - 3C &= -6 \end{aligned}$$
$(-3,6,-3)$
9. Solve the system.
$$\begin{aligned} -x + 2y - z &= 5 \\ 3x - y + 2z &= -8 \\ -2x - y - z &= 3 \end{aligned}$$
Dependent system; infinitely many solutions
10. Solve the system.
$$\begin{aligned} 5x + 6y + 3z &= 6 \\ x + y + z &= 6 \\ x + 2y - z &= 0 \end{aligned}$$
Inconsistent system; ∅

6.3

11. Find the value of $\begin{vmatrix} -4 & 5 \\ -8 & 3 \end{vmatrix}$. 28
12. Use Cramer's rule to solve the system.
$$\begin{aligned} 3x + 7y &= -1 \\ -4x - 9y &= 2 \end{aligned}$$
$(-5,2)$
13. Solve by Cramer's rule.
$$\begin{aligned} -8y - 3 &= -5x \\ 10x - 1 &= -24y \end{aligned}$$
$(\frac{2}{5},-\frac{1}{8})$
14. Evaluate the determinant by expansion by minors about the first column.
$$\begin{vmatrix} 2 & -3 & 4 \\ 1 & 0 & -2 \\ 2 & 1 & -1 \end{vmatrix}$$
17
15. Solve by Cramer's rule.
$$\begin{aligned} 2x + 3y + 2z &= 7 \\ -2x + y - z &= -5 \\ 3x + 2y + z &= 12 \end{aligned}$$
$(4,1,-2)$

6.4

16. Solve by transforming to triangular form.
$$\begin{aligned} 3x + y + z &= 0 \\ x + 3y - z &= -4 \\ x - 4y + 4z &= 5 \end{aligned}$$
$(1,-2,-1)$

17. Use matrix form (Gaussian elimination) to solve the system.

$$3x + 4y - 4 = 0$$
$$-2x - 5y - 2 = 0 \quad (4,-2)$$

18. Use matrix form (Gaussian elimination) to solve the system.

$$x - y + z = 2$$
$$2x - 3y + 2z = 6$$
$$3x + y + z = 2 \quad (2,-2,-2)$$

19. A 3,000-seat theater was sold out for a weekend concert. Total receipts on Friday night were $45,300 when seats in the orchestra, mezzanine, and balcony sold for $18, $15, and $12, respectively. On Saturday, prices were raised to $25, $20, and $14, and total receipts increased to $59,800. How many seats are in the balcony? 700

6.5

20. Graph the solution set of the system.

$$3x + y \leq 6$$
$$y > 2x + 1$$

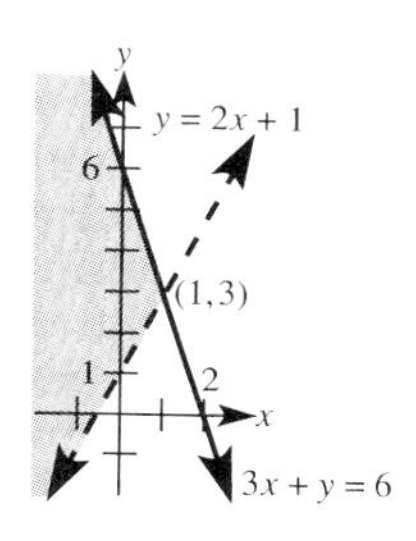

21. Graph the solution set of the system.

$$x - 3y > 6$$
$$x + y > 2$$

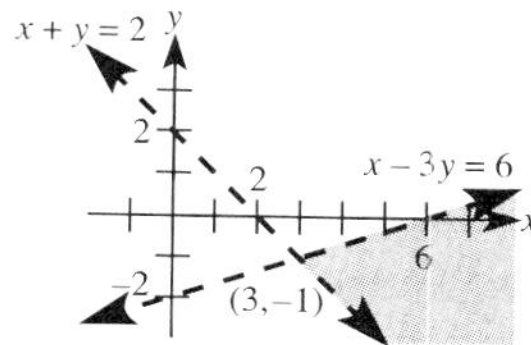

22. For a party you need to buy some roast beef at $6 per pound and cold cuts at $4 per pound. You will need at least 3 lb of roast beef and at least 4 lb of cold cuts, and you wish to spend no more than $48 on these items.

a. Translate these requirements into a system of linear inequalities.

b. Solve the system graphically.

c. Specify and interpret one solution that is in the solution set.

a. $6x + 4y \leq 48$
$x \geq 3$
$y \geq 4$

b.

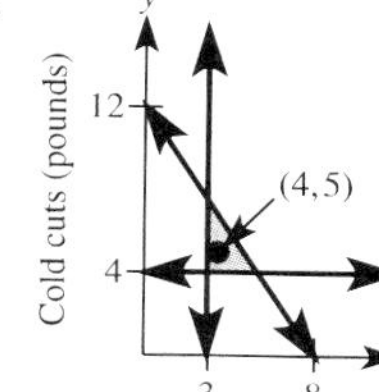

c. (4,5); 4 lb of roast beef and 5 lb of cold cuts may be purchased.

ADDITIONAL REVIEW EXERCISES

Evaluate.

23. $\begin{vmatrix} -6 & -5 \\ 1 & 0 \end{vmatrix}$ 5

24. $\begin{vmatrix} 2 & -1 & 4 \\ 1 & 0 & 2 \\ 0 & 3 & -1 \end{vmatrix}$ −1

Determine if the given ordered pair or triple is a solution of the system.

25. (4,−3); $2x - 5 = y$
$x + y = 1$ No

26. (8,0,−2); $x - 2y + z = 6$
$-x + y - 3z = -2$
$-2x - 3y + 4z = -24$ Yes

Solve. Use the indicated method.

27. $4x - y = -2$
$2x - y = -5$ (substitution) $(\frac{3}{2}, 8)$

28. $3x + 2y = 4$
$-4x - y = 3$ (addition-elimination) (−2,5)

29. $5x + 2y = 1$
$4x + 3y = 5$ (addition-elimination) (−1,3)

30. $y = x + 7$
$y = 2x + 6$ (graphing)

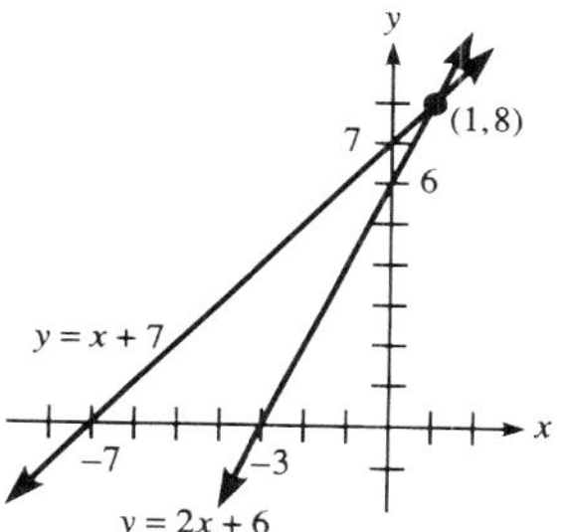

31. $4x - 3y = -26$
$-5x + 3y = 7$ (Cramer's rule) (19,34)

32. $x + y - 2z = 0$
$2x + 2y - z = 1$
$3x + 2y - 3z = 3$ (Cramer's rule) $(\frac{8}{3}, -2, \frac{1}{3})$

33. $3a - 7b = -2$
$5a + 2b = -17$ (matrix form) (−3,−1)

34. $-x - y + 2z = 7$
$x + 2y - 2z = -7$
$2x - y + z = -4$ (matrix form) (−3,0,2)

35. $-5x - 2y - 1 = 0$
$8x + 4y + 4 = 0$ (any method) (1,−3)

36. $x + y + 3z = 1$
$2x + 5y + 2z = 0$
$3x - 2y - z = 3$ (triangular form) $(\frac{4}{5}, -\frac{2}{5}, \frac{1}{5})$

37. $2x - y + 3z = 1$
$-x + y - z = -1$
$3x - 2y + 4z = 2$ (any method) Infinitely many solutions

38. $x + y \leq 7$
$x - 2y \leq -8$ (graphing)

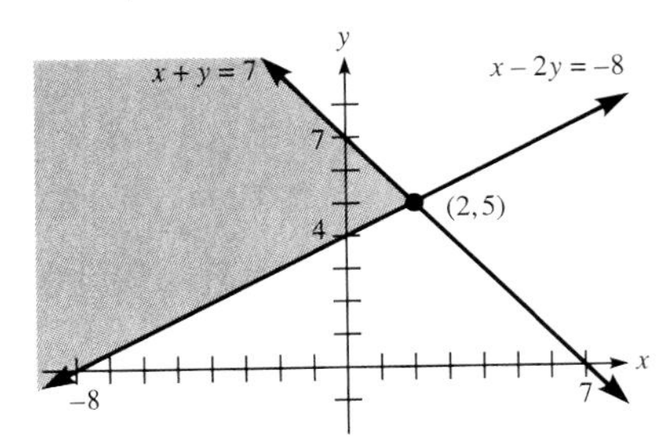

39. Admission to a concert was \$5 for students and \$7 for nonstudents. There were 525 paid admissions, and total receipts were \$3,053. How many students attended the concert? 311

40. For planting in your garden, you will buy some flowering plants at \$6 each and some small shrubs at \$12 each. You can spend no more than \$180, but you need at least 8 of the flowering plants and at least 9 of the shrubs.

a. Translate these requirements into a system of linear inequalities.

b. Solve the system graphically.

c. Specify and interpret one solution that is in the solution set.

a. $6x + 12y \leq 180$; $x \geq 8$; $y \geq 9$ b.

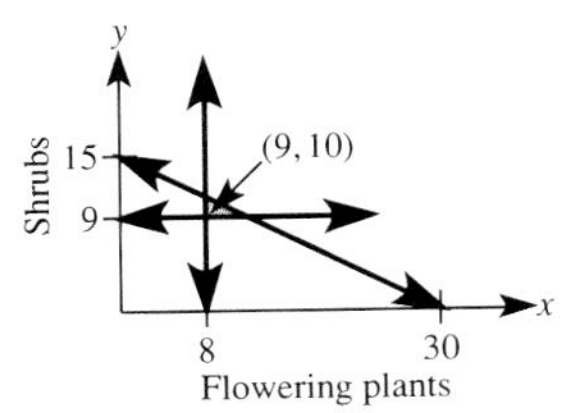

c. 10 flowering plants and 9 shrubs may be purchased within budget.

CHAPTER 6 TEST

1. Evaluate $\begin{vmatrix} -2 & -1 \\ 5 & 1 \end{vmatrix}$. 3

2. Evaluate $\begin{vmatrix} 1 & 2 & 0 \\ 1 & 0 & 3 \\ 0 & 1 & -1 \end{vmatrix}$. Expand by minors around the first column. −1

3. The ordered pair $(-5, -2)$ is a solution of which of the following systems?

a. $-x + 2y = -1$
$2x + y = 8$
b. $x + y + 7 = 0$
$-x + 2y - 1 = 0$
c. $x = y - 3$
$2x = y + 8$ b

4. Which of the following is a solution of the system?

$$-4x + y + 2z = 3$$
$$x + 2y - z = -7$$
$$2x - y + 3z = -4$$

a. $(-2, -3, -1)$ **b.** $(0, 1, 1)$ **c.** $(2, -1, -5)$ a

In Questions 5–15, solve the given system using the method indicated.

5. $2x + 3y = 0$
$4x + 9y = 1$ (addition-elimination) $(-\frac{1}{2}, \frac{1}{3})$

6. $y = -3x + 4$
$y = 2x - 6$ (substitution) $(2, -2)$

7. $x - y + z = 5$
$-x + 2y - 2z = -5$
$2x - 2y + 3z = 9$ (triangular form) $(5, -1, 1)$

8. $3x + y = 2$
$-3x + y = -2$ (Cramer's rule) $(\frac{2}{3}, 0)$

9. $x + z = 10$
$y + 2z = 3$
$-x + y = -2$ (Cramer's rule) $(15, 13, -5)$

10. $5x + 3y = 1$
$3x + 2y = 0$ (matrix form) $(2, -3)$

11. $-x + 3y - 5 = 0$
$4x - y - 2 = 0$ (matrix form) $(1, 2)$

12. $5x - 4y = -3$
$2x + 7y = -27$ (addition-elimination) $(-3, -3)$

13. $x + 3y + 3z = 5$
$x + 4y + 3z = 2$
$x + 3y + 4z = 8$ (matrix form) $(5, -3, 3)$

14. $x + y - z = 2$
$2x - y + 5z = 3$
$-x - y + z = -1$ (addition-elimination) Inconsistent system; $\emptyset$

15. $x + y = 6$
$y = 2x$ (graphing)

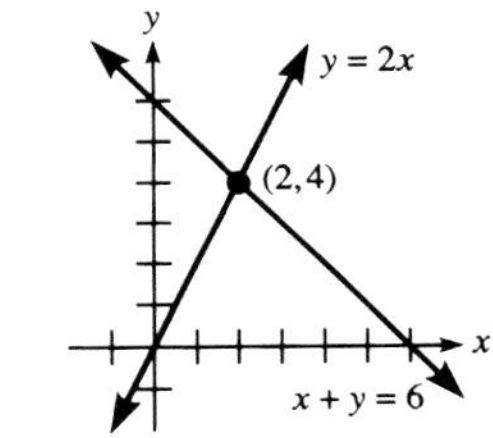

16. Graph $2x - y \leq 8$.

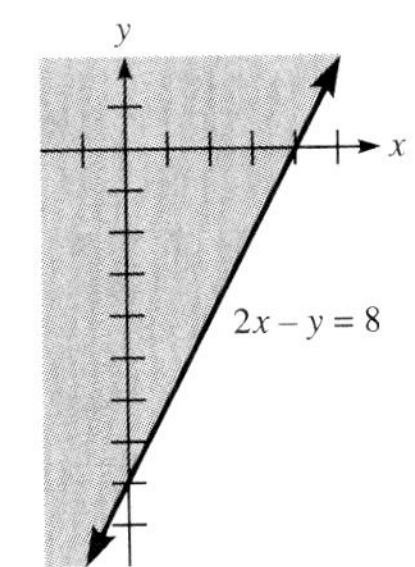

17. Graph the solution set of the system.

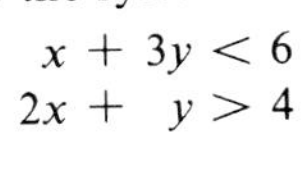

$x + 3y < 6$
$2x + y > 4$

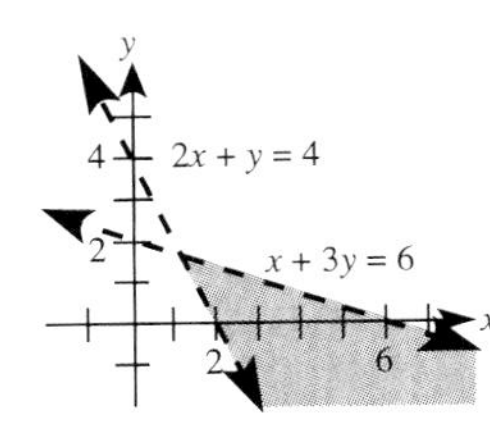

18. Graph the solution set of the system.

$$x + y \leq 5$$
$$x - y < -5$$

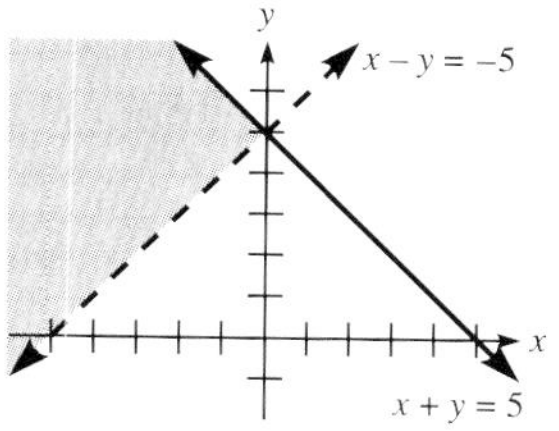

19. A manager invests a total of \$150,000. The investment is split between a bank CD yielding 6 percent interest and a stock that pays a 9 percent dividend. If the total annual income from the investment is \$12,300, how much is invested in the bank CD? \$40,000

20. Translate the following requirements into a system of inequalities: For a decorating project, you need to purchase some red ribbon at \$1.25 per yard and some white trim at \$1.50 per yard. You need at least 3 yd of the red ribbon and at least 5 yd of the white trim, but you can spend no more than \$14.

$$1.25r + 1.50w \leq 14.00$$
$$r \geq 3$$
$$w \geq 3$$

CUMULATIVE TEST 6

1. Which of the following illustrates the associative property?
a. $a(x - y) = ax - ay$ **b.** $(ax)y = y(ax)$
c. $x + y = y + x$ **d.** $(x + y) + z = x + (y + z)$ d

2. Solve $-5x - 2y = -10$ for y. $y = -\frac{5}{2}x + 5$

3. Express the solution set in interval notation: $|-x + 1| \leq 5$. $[-4,6]$

4. Graph the solution set: $-3(x - 1) > -6x$.

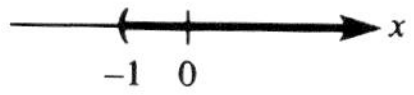

5. Solve: $\frac{2}{3}x - \frac{1}{5} = \frac{1}{3}$. $\{\frac{4}{5}\}$

6. Simplify $(2x^2y^3)^{-1}(3xy^4)$, and write the result using only positive exponents. $\frac{3y}{2x}$

7. Simplify: $(y^2 - 3x + 1) - (3y^2 + 4x + 1) + (y^2 + x + 4)$. $-y^2 - 6x + 4$

8. Factor completely: $2x^3 + 6x^2 - 20x$. $2x(x - 2)(x + 5)$

9. Solve: $x^2 + x - 22 = x + 3$. $\{-5,5\}$

10. Multiply and express the result in lowest terms:

$$\frac{-5x^2 - 10x}{x^2 + 3x} \cdot \frac{x^2 - x - 12}{x + 2}. \quad -5(x - 4)$$

11. Divide $x^2 - 5x + 7$ by $x - 4$. $x - 1 + \frac{3}{x^2 - 5x + 7}$

12. Simplify: $\dfrac{\frac{2}{x} - \frac{1}{y}}{\frac{1}{3xy}}$. $6y - 3x$

13. One person can paint a ceiling in 45 minutes. Another person can paint the same ceiling in 30 minutes. How long would it take them to paint the ceiling working together? 18 minutes

14. Find an equation for the line through the points $(-1,-1)$ and $(2,5)$. $y = 2x + 1$

15. Find the distance between the points $(-1,-1)$ and $(2,4)$. $\sqrt{34}$

16. If y varies directly as x, and $y = -4$ when $x = 10$, find y when $x = -15$. 6

17. Evaluate $\begin{vmatrix} 0 & -2 & 2 \\ 5 & 1 & -4 \\ 6 & 3 & 1 \end{vmatrix}$. 76

18. Solve the system.

$$10x - 3y + 7 = 0$$
$$-6x - 9y + 3 = 0 \quad (-\tfrac{1}{2}, \tfrac{2}{3})$$

19. Solve the system.

$$x + 3y - z = 0$$
$$-x + y + 2z = -5$$
$$x - 3y + 2z = 3 \quad (2,-1,-1)$$

20. Graph the solution set of the system.

$$y \leq x - 2$$
$$x + y < 3$$

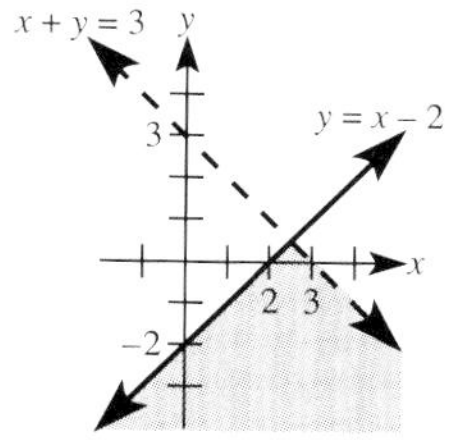

CHAPTER

7 Radicals and Complex Numbers

The **geometric mean** of n positive numbers is defined as the nth root of their product, so for positive numbers $a_1, a_2, \ldots, a_n$,

$$\text{geometric mean} = \sqrt[n]{(a_1)(a_2)\cdots(a_n)}.$$

a. Find the geometric mean of 2 and 18.

b. The geometric mean is used to average rates of change, ratios, or indexes. For instance, if sales at a biotechnology company over a three-year period had annual growth rates of 45 percent, 16 percent, and 73 percent, then annual sales increased by factors of 1.45, 1.16, and 1.73, respectively. Find the geometric mean of these three numbers, and interpret the result. (See Example 4 of Section 7.1.)

TO ANALYZE equations in one or two variables that extend beyond linear equations, we first need to develop some algebraic techniques for extracting roots. In this chapter we introduce radical notation and then establish some properties that enable us to simplify, add, subtract, multiply, and divide radicals. Methods for solving radical equations are then considered. The concluding section of this chapter shows how we may extend the number system beyond real numbers to the set of complex numbers.

7.1 Radicals and Rational Exponents

OBJECTIVES

1. **Find square roots and principal square roots.**
2. **Find principal *n*th roots.**
3. **Evaluate expressions containing rational exponents.**
4. **Use exponent properties to simplify expressions with rational exponents.**

1 In many applications we need to reverse the process of raising a number to a power, and we ask

$$\text{if } x^n = a, \text{ what is } x?$$

To answer this question when $n = 2$, it is useful to define a square root.

Definition of Square Root

The number b is a square root of a if $b^2 = a$.

When a is a positive number, then a always has two square roots, as illustrated next.

EXAMPLE 1 Find the square roots of 16.

Solution Because $4^2 = 16$ and $(-4)^2 = 16$, the square roots of 16 are 4 and -4.

PROGRESS CHECK 1 Find the square roots of 49.

To avoid ambiguity, we define the **principal square roots** of a positive number to be its *positive* square root, and we denote this root by using the symbol $\sqrt{}$. Thus,

$$\sqrt{16} = 4.$$

We write $-\sqrt{16}$ to symbolize the negative square root of 16, so

$$-\sqrt{16} = -4.$$

In general, note the following ideas concerning $\sqrt{a}$.

1. When $\sqrt{a}$ is a rational number, then a is called a **perfect square.**
2. When a is positive and not a perfect square, then $\sqrt{a}$ is an irrational number that may be approximated using the key $\boxed{\sqrt{}}$ on a calculator. Recall from Section 1.1 that irrational numbers are real numbers that are not rational.
3. When a is negative, then $\sqrt{a}$ is not a real number, because the product of two equal real numbers is never negative.
4. The number 0 has exactly one square root, and $\sqrt{0} = 0$.

The symbol $\sqrt{}$ is due to Christoff Rudolff (German, 1525). Other notations date back to Egyptian papyri. Rudolff's algebra text was called *The Art of the Coss* (from the Italian word for "the thing"), the name for the unknown.

Progress Check Answer

1. 7, -7

EXAMPLE 2 Find each square root that is a real number. Approximate irrational numbers to the nearest hundredth, and identify all numbers that are not real numbers.

a. $\sqrt{144}$ **b.** $-\sqrt{144}$ **c.** $\sqrt{-144}$ **d.** $\sqrt{14}$

Solution

a. $\sqrt{144}$ denotes the positive square root of 144. Because $12^2 = 144$, $\sqrt{144} = 12$.
b. $-\sqrt{144}$ denotes the negative square root of 144, so $-\sqrt{144} = -12$.
c. Square roots of negative numbers are never real numbers, so $\sqrt{-144}$ is not a real number.
d. 14 is not a perfect square, so $\sqrt{14}$ is irrational. By calculator,

$$14 \boxed{\sqrt{\ }} \boxed{3.7416574} .$$

Rounding off to the nearest hundredth, $\sqrt{14} \approx 3.74$ where $\approx$ is read, "is approximately equal to."

PROGRESS CHECK 2 Find each square root that is a real number. Approximate irrational numbers to the nearest hundredth, and identify all numbers that are not real numbers.

a. $\sqrt{200}$ **b.** $\sqrt{225}$ **c.** $\sqrt{-225}$ **d.** $-\sqrt{225}$

2 Our ideas to this point may be generalized with the following definition of an nth root. Note that square roots are specific cases of nth roots for the case when $n = 2$.

Definition of nth Root

For any positive integer n, the number b is an nth root of a if $b^n = a$.

The current way of writing the index in the hook of the radical sign was popularized in the early eighteenth century. Many other styles had been tried.

The **principal nth root** of a is denoted $\sqrt[n]{a}$ and is defined by

$$\sqrt[n]{a} = b \quad \text{if and only if} \quad b^n = a \quad \begin{cases} \text{for } a \geq 0, b \geq 0, \text{ if } n \text{ is even,} \\ \text{for any real number } a, \text{ if } n \text{ is odd.} \end{cases}$$

In the expression $\sqrt[n]{a}$, which is called a **radical,** we say $\sqrt{\ }$ is the **radical sign,** a is the **radicand,** and n is the **index.** The index is usually omitted for the square root radical, and $\sqrt[3]{a}$ is called the **cube root** of a.

EXAMPLE 3 Find each root that is a real number.

a. $\sqrt[3]{8}$ **b.** $\sqrt[3]{-8}$ **c.** $\sqrt[4]{81}$ **d.** $\sqrt[4]{-81}$

Solution

a. $\sqrt[3]{8} = 2$, because $2^3 = 8$. Read $\sqrt[3]{8} = 2$ as "the cube root of 8 is 2."
b. $\sqrt[3]{-8} = -2$, because $(-2)^3 = -8$. Note that cube roots of negative numbers are negative numbers; and in general, odd roots of negative numbers are negative.
c. $\sqrt[4]{81}$ denotes the positive fourth root of 81. Since $3^4 = 81$, $\sqrt[4]{81} = 3$.
d. $\sqrt[4]{-81}$ is not a real number. In general, when n is even and a is negative, then $\sqrt[n]{a}$ is not a real number.

PROGRESS CHECK 3 Find each root that is a real number.

a. $\sqrt[4]{16}$ **b.** $\sqrt[3]{-125}$ **c.** $\sqrt[4]{-16}$ **d.** $-\sqrt[3]{27}$

Use of the root key on a calculator is considered in the next example.

Progress Check Answers

2. (a) 14.14 (b) 15 (c) Not real (d) -15
3. (a) 2 (b) -5 (c) Not real (d) -3

EXAMPLE 4 Solve the problem in the section introduction on page 318.

Solution Use the definition of geometric mean given in the question.

a. The geometric mean of two numbers is the square root of their product. Therefore,

$$\text{geometric mean} = \sqrt{(2)(18)} = \sqrt{36} = 6.$$

The geometric mean of 2 and 18 is 6.

b. The geometric mean of *three* numbers is the *cube root* of their product. Thus,

$$\text{geometric mean} = \sqrt[3]{(1.45)(1.16)(1.73)}.$$

On a calculator, the root key $\boxed{\sqrt[x]{y}}$ is used to take the xth root of the number y, and a keystroke sequence to compute the requested geometric mean is

$$1.45 \boxed{\times} 1.16 \boxed{\times} 1.73 \boxed{=} \boxed{\sqrt[x]{y}}\ 3 \boxed{=} \boxed{1.4276575}.$$

Over the three years, annual sales increased by an average factor of about 1.4276575, which translates to an average annual sales increase of about 42.77 percent.

Note Some calculators have a cube root key $\boxed{\sqrt[3]{\ }}$ that is used just like the square root key $\boxed{\sqrt{\ }}$. Also, the root key sometimes looks like $\boxed{y^{1/x}}$, and the rationale for this labeling is discussed next.

On many graphing calculators the $\boxed{\wedge}$ key is used with fractional exponents.

PROGRESS CHECK 4 The geometric mean of n positive numbers is the nth root of their product.

a. Find the geometric mean of 3, 1, and 9.

b. If sales at a company over a two-year period had annual growth rates of 84 percent and 27 percent, then annual sales increased by factors of 1.84 and 1.27, respectively. Find the geometric mean of these two numbers, and interpret the result.

3 It is a goal of algebra to be able to use any real number as an exponent, and we can use our work with radicals to give meaning to expressions with rational number exponents such as

$$16^{1/2},\ 16^{-3/4},\ \text{and}\ 8^{2/3}.$$

Recall that the power-to-a-power property for integral exponents is

$$(a^m)^n = a^{mn}.$$

If this law is to hold for rational exponents, then consider

$$(16^{1/2})^2 = 16^{(1/2)2} = 16^1 = 16.$$

We see that squaring $16^{1/2}$ results in 16, so $16^{1/2}$ is a square root of 16. We choose to define $16^{1/2}$ as the positive square root of 16, so

$$16^{1/2} = \sqrt{16} = 4.$$

In general, our previous laws of exponents may be extended by defining $a^{1/n}$ as the principal nth root of a.

The first references to fractional powers are in the work of Nicholas Oresme (France, 1320–1382), who also indicated that irrational powers must exist, but he had no notation for them. Our current symbolism is due to Wallis (1655) and Newton (1669). Imaginary exponents are due to Euler (1740).

Definition of $a^{1/n}$

If n is a positive integer and $\sqrt[n]{a}$ is a real number, then

$$a^{1/n} = \sqrt[n]{a}.$$

Progress Check Answers

4. (a) 3 (b) 1.5287; average annual increase in sales is about 52.9 percent.

EXAMPLE 5 Evaluate each expression.

a. $8^{1/3}$ **b.** $81^{1/2}$ **c.** $81^{1/4}$ **d.** $(-8)^{1/3}$

Solution Convert to radical form and then simplify.

a. $8^{1/3} = \sqrt[3]{8} = 2$
b. $81^{1/2} = \sqrt{81} = 9$
c. $81^{1/4} = \sqrt[4]{81} = 3$
d. $(-8)^{1/3} = \sqrt[3]{-8} = -2$

PROGRESS CHECK 5 Evaluate each expression.

a. $100^{1/2}$ **b.** $(-27)^{1/3}$ **c.** $32^{1/5}$ **d.** $625^{1/4}$

The power-to-a-power property, $a^{mn} = (a^m)^n$, is the basis for extending the definition of rational exponent to the case when the numerator is not 1. For instance, to evaluate $8^{2/3}$, we reason that

$$8^{2/3} = (8^{1/3})^2 = (\sqrt[3]{8})^2 = 2^2 = 4,$$

or

$$8^{2/3} = (8^2)^{1/3} = 64^{1/3} = \sqrt[3]{64} = 4.$$

By both methods $8^{2/3} = 4$, and this example suggests how a rational number exponent should be defined.

Rational Exponent

If m and n are integers with $n > 0$, and if m/n represents a reduced fraction such that $a^{1/n}$ is a real number, then

$$a^{m/n} = (\sqrt[n]{a})^m = \sqrt[n]{a^m}.$$

When $a^{m/n}$ is a rational number, it is usually easier to find the root first by using $a^{m/n} = (\sqrt[n]{a})^m$.

EXAMPLE 6 Evaluate each expression.

a. $27^{4/3}$ **b.** $16^{5/4}$ **c.** $9^{3/2}$ **d.** $9^{-1/2}$

Solution Use $a^{m/n} = (\sqrt[n]{a})^m$, since each root is a rational number.

a. $27^{4/3} = (\sqrt[3]{27})^4 = 3^4 = 81$
b. $16^{5/4} = (\sqrt[4]{16})^5 = 2^5 = 32$
c. $9^{3/2} = (\sqrt{9})^3 = 3^3 = 27$
d. $9^{-1/2} = (\sqrt{9})^{-1} = 3^{-1} = \frac{1}{3}$

PROGRESS CHECK 6 Evaluate each expression.

a. $4^{5/2}$ **b.** $4^{-5/2}$ **c.** $(-27)^{2/3}$ **d.** $32^{4/5}$

In this section the determining principle in our work is that properties that apply for integer exponents should continue to apply for rational number exponents. Therefore, an alternative method in Example 6d for evaluating $9^{-1/2}$ is

$$9^{-1/2} = \frac{1}{9^{1/2}} = \frac{1}{\sqrt{9}} = \frac{1}{3}.$$

Progress Check Answers

5. (a) 10 (b) -3 (c) 2 (d) 5
6. (a) 32 (b) $\frac{1}{32}$ (c) 9 (d) 16

And in general, if all powers are real numbers,

$$a^{-r} = \frac{1}{a^r}, \quad \text{where } a \neq 0 \text{ and } r \text{ is a rational number.}$$

The simplifications in Example 7 also use this extended definition of negative exponent.

EXAMPLE 7 Evaluate each expression.

a. $8^{-2/3}$ b. $(-32)^{-3/5}$ c. $\left(\frac{49}{4}\right)^{-1/2}$

Solution

a. $8^{-2/3} = \frac{1}{8^{2/3}} = \frac{1}{(\sqrt[3]{8})^2} = \frac{1}{2^2} = \frac{1}{4}$

b. $(-32)^{-3/5} = \frac{1}{(-32)^{3/5}} = \frac{1}{(\sqrt[5]{-32})^3} = \frac{1}{(-2)^3} = -\frac{1}{8}$

c. $\left(\frac{49}{4}\right)^{-1/2} = \frac{1}{(\frac{49}{4})^{1/2}} = \frac{1}{\sqrt{\frac{49}{4}}} = \frac{1}{\frac{7}{2}} = \frac{2}{7}$

Note The most straightforward general way to find rational powers on a calculator is to use the power key $\boxed{y^x}$ and enclose the exponent within parentheses. By this method, $8^{-2/3}$ is computed as

8 $\boxed{y^x}$ $\boxed{(}$ 2 $\boxed{+/-}$ $\boxed{\div}$ 3 $\boxed{)}$ $\boxed{=}$ $\boxed{0.25}$.

PROGRESS CHECK 7 Evaluate each expression.

a. $25^{-3/2}$ b. $(-64)^{-2/3}$ c. $(\frac{27}{1,000})^{-1/3}$

4 Because all previous laws of exponents hold for rational number exponents, we may use these extended properties to simplify expressions with rational exponents.

EXAMPLE 8 Perform the indicated operations, and write the result with only positive exponents. Assume all variables represent positive real numbers.

a. $2^{1/2} \cdot 2^{3/2}$ b. $y^{2/3} \cdot y^{1/2}$ c. $(8x^9)^{1/3}$

d. $\frac{5}{5^{1/2}}$ e. $\frac{x^{4/5}y^{-3/2}}{x^{-1}y}$ f. $\left(\frac{4x^3}{100x^{-3}}\right)^{1/2}$

Solution

a. $2^{1/2} \cdot 2^{3/2} = 2^{1/2 + 3/2} = 2^2 = 4$

b. $y^{2/3} \cdot y^{1/2} = y^{2/3 + 1/2} = y^{4/6 + 3/6} = y^{7/6}$

c. $(8x^9)^{1/3} = 8^{(1)(1/3)}x^{(9)(1/3)} = 8^{1/3}x^3 = 2x^3$

d. $\frac{5}{5^{1/2}} = 5^{1 - 1/2} = 5^{1/2}$

e. $\frac{x^{4/5}y^{-3/2}}{x^{-1}y} = x^{4/5 - (-1)}y^{-3/2 - 1} = x^{9/5}y^{-5/2} = \frac{x^{9/5}}{y^{5/2}}$

f. $\left(\frac{4x^3}{100x^{-3}}\right)^{1/2} = \left(\frac{x^{3-(-3)}}{25}\right)^{1/2} = \left(\frac{x^6}{25}\right)^{1/2} = \frac{x^{(6)(1/2)}}{25^{(1)(1/2)}} = \frac{x^3}{5}$

PROGRESS CHECK 8 Perform the indicated operations, and write the result with only positive exponents. Assume all variables represent positive real numbers.

a. $5^{1/2} \cdot 5^{1/2}$ b. $x^{3/2} \cdot x^{1/3}$ c. $(9y^6)^{1/2}$

d. $\frac{2^{1/3}}{2}$ e. $\frac{a^{1/4}b^{1/2}}{a^{3/4}b^{-1}}$ f. $\left(\frac{3y^2}{24y^{-4}}\right)^{2/3}$

Progress Check Answers

7. (a) $\frac{1}{125}$ (b) $\frac{1}{16}$ (c) $\frac{10}{3}$

8. (a) 5 (b) $x^{11/6}$ (c) $3y^3$ (d) $\frac{1}{2^{2/3}}$ (e) $\frac{b^{3/2}}{a^{1/2}}$ (f) $\frac{y^4}{4}$

EXERCISES 7.1

In Exercises 1–6, find the square roots.

1. 121 11, −11
2. 196 14, −14
3. 361 19, −19
4. 441 21, −21
5. 576 24, −24
6. 729 27, −27

In Exercises 7–24, find each square root that is a real number. Approximate irrational numbers to the nearest hundredth, and identify all numbers that are *not* real numbers.

7. $\sqrt{289}$ 17
8. $\sqrt{625}$ 25
9. $\sqrt{12}$ 3.46
10. $\sqrt{28}$ 5.29
11. $-\sqrt{676}$ −26
12. $-\sqrt{324}$ −18
13. $\sqrt{-16}$ Not real
14. $\sqrt{-64}$ Not real
15. $\sqrt{34}$ 5.83
16. $\sqrt{90}$ 9.49
17. $-\sqrt{400}$ −20
18. $-\sqrt{529}$ −23
19. $\sqrt{-169}$ Not real
20. $\sqrt{-841}$ Not real
21. $-\sqrt{900}$ −30
22. $-\sqrt{1{,}024}$ −32
23. $\sqrt{-196}$ Not real
24. $\sqrt{-361}$ Not real

In Exercises 25–42, find each root that is a real number.

25. $\sqrt[3]{64}$ 4
26. $\sqrt[3]{125}$ 5
27. $-\sqrt[3]{216}$ −6
28. $-\sqrt[3]{729}$ −9
29. $\sqrt[3]{-27}$ −3
30. $\sqrt[3]{-64}$ −4
31. $\sqrt[4]{16}$ 2
32. $\sqrt[4]{256}$ 4
33. $-\sqrt[4]{16}$ −2
34. $-\sqrt[4]{625}$ −5
35. $\sqrt[4]{-64}$ Not real
36. $\sqrt[4]{-8}$ Not real
37. $\sqrt[3]{-512}$ −8
38. $\sqrt[3]{-1{,}331}$ −11
39. $\sqrt[4]{-225}$ Not real
40. $\sqrt[4]{-484}$ Not real
41. $\sqrt[5]{-32}$ −2
42. $\sqrt[6]{64}$ 2

In Exercises 43–66, evaluate each expression.

43. $49^{1/2}$ 7
44. $256^{1/4}$ 4
45. $(-216)^{1/3}$ −6
46. $(-1{,}000)^{1/3}$ −10
47. $27^{5/3}$ 243
48. $16^{3/2}$ 64
49. $16^{3/4}$ 8
50. $81^{5/4}$ 243
51. $81^{-1/4}$ $\frac{1}{3}$
52. $16^{-1/2}$ $\frac{1}{4}$
53. $4^{-3/2}$ $\frac{1}{8}$
54. $81^{-3/2}$ $\frac{1}{729}$
55. $-64^{4/3}$ −256
56. $-125^{2/3}$ −25
57. $(-216)^{-2/3}$ $\frac{1}{36}$
58. $(-343)^{-2/3}$ $\frac{1}{49}$
59. $(-32)^{-3/5}$ $-\frac{1}{8}$
60. $(-243)^{-4/5}$ $\frac{1}{81}$
61. $(\frac{25}{16})^{3/2}$ $\frac{125}{64}$
62. $(\frac{8}{27})^{2/3}$ $\frac{4}{9}$
63. $(\frac{25}{16})^{-1/2}$ $\frac{4}{5}$
64. $(\frac{8}{27})^{-1/3}$ $\frac{3}{2}$
65. $(\frac{16}{81})^{-1/4}$ $\frac{3}{2}$
66. $(\frac{32}{243})^{-1/5}$ $\frac{3}{2}$

In Exercises 67–96, perform the indicated operations, and write the result with only positive exponents. Assume all variables represent positive real numbers.

67. $3^{1/4} \cdot 3^{3/2}$ $3^{7/4}$
68. $4^{1/3} \cdot 4^{1/2}$ $4^{5/6}$
69. $(3^{1/2})^6$ 27
70. $(16^3)^{1/2}$ 64
71. $\frac{7}{7^{1/3}}$ $7^{2/3}$
72. $\frac{3^{1/4}}{3^{3/4}}$ $\frac{1}{3^{1/2}}$
73. $a^{1/4} \cdot a^{7/4}$ a^2
74. $x^{2/3} \cdot x^{4/3}$ x^2
75. $b^{-1/2} \cdot b^{3/4}$ $b^{1/4}$
76. $y^{4/3} \cdot y^{-1/2}$ $y^{5/6}$
77. $(4b^8)^{1/2}$ $2b^4$
78. $(27a^3)^{1/3}$ $3a$
79. $(8x^6)^{2/3}$ $4x^4$
80. $(27y^3)^{4/3}$ $81y^4$
81. $\frac{a^{1/2}b^{3/8}}{a^{1/4}b^{1/8}}$ $a^{1/4}b^{1/4}$
82. $\frac{x^{2/3}y^{5/6}}{x^{-1/3}y^{1/2}}$ $xy^{1/3}$
83. $\frac{a^{1/2}b}{a^{3/2}b^{1/2}}$ $\frac{b^{1/2}}{a}$
84. $\frac{x^{-5/2}y^{3/4}}{xy^{-1/4}}$ $\frac{y}{x^{7/2}}$
85. $\frac{a^{3/4}b^{-4/3}}{a^{1/2}b^{1/3}}$ $\frac{a^{1/4}}{b^{5/3}}$
86. $\frac{x^{2/3}y^{7/3}}{x^{4/3}y^{-1/3}}$ $\frac{y^{8/3}}{x^{2/3}}$
87. $\left(\frac{81x^2}{9x^{-6}}\right)^{1/2}$ $3x^4$
88. $\left(\frac{12y^5}{3y}\right)^{1/2}$ $2y^2$
89. $\left(\frac{3a^6}{81a^{-6}}\right)^{1/3}$ $\frac{a^4}{3}$
90. $\left(\frac{2b^7}{128b^{-2}}\right)^{1/3}$ $\frac{b^3}{4}$
91. $\left(\frac{4a}{a^{-5}}\right)^{3/2}$ $8a^9$
92. $\left(\frac{72b^6}{2b^{-2}}\right)^{3/2}$ $216b^{12}$
93. $\left(\frac{32a^{-4}}{a^{-9}}\right)^{4/5}$ $16a^4$
94. $\left(\frac{81b^{-1}}{b^7}\right)^{3/4}$ $\frac{27}{b^6}$
95. $\left(\frac{x^2}{64x^{-1/2}}\right)^{2/3}$ $\frac{x^{5/3}}{16}$
96. $\left(\frac{125y^{-3/4}}{y^{-3}}\right)^{2/3}$ $25y^{3/2}$

In Exercises 97–100, refer to Example 4 of this section for the definition of the geometric mean of a set of positive numbers.

97. a. Find the geometric mean of 3 and 27. 9
b. It can be shown that the geometric mean of two positive numbers can never be larger than their arithmetic mean (the ordinary average). Show that this is correct for 3 and 27. $9 < 15$

98. a. Find the geometric mean of 1, 2, 8, and 16. 4
b. Is the geometric mean smaller than the arithmetic mean? Yes; $4 < 6.75$

99. If sales at a company over a two-year period had annual growth rates of 40 percent and 10 percent, then annual growth increased by factors of 1.40 and 1.10, respectively. Find and interpret the geometric mean of these two numbers. Give the answer to the nearest tenth of a percent. Geometric mean = 1.24097; average annual increase in sales is about 24.1 percent.

100. The value of a house increased by factors of 1.10, 1.06, and 1.05 each year over a three-year period. Find and interpret the geometric mean of these three numbers. Give your answer to the nearest tenth of a percent. Geometric mean = 1.0698; average annual increase in value is about 7.0 percent.

101. a. A rectangle has length 16 in. and width 4 in. What size square has the same area? Side = 8 in.
b. A rectangle has length a and width b. What size square has the same area? Side = $\sqrt{ab}$

102. a. A rectangular solid has length = 12, width = 9, and height = 2 cm. What size cube has the same volume? Side = 6 cm
b. A rectangular solid has length = a, width = b, and height = c. What size cube has the same volume? Side = $\sqrt[3]{abc}$

THINK ABOUT IT

1. In this section we defined rational exponents but not irrational exponents. For instance 2^{π} and $5^{\sqrt{2}}$ were not defined. We shall assume that it is possible to write irrational exponents and that the regular rules apply to them.
   a. Explain why $(5^{\sqrt{2}})^{\sqrt{2}}$ must be equal to 25.
   b. Explain why 2^{π} should be between $2^{3.13}$ and $2^{3.15}$. Approximate π by 3.14159 and compute 2^{π} to three decimal places.
2. The Pythagoreans of ancient Greece studied many types of means. In general, a mean of a and b is some number between them. The three earliest recorded types are the arithmetic mean, $A = (a + b)/2$; the geometric mean, $G = \sqrt{ab}$; and the harmonic mean, $H = 2ab/(a + b)$. Compute each of these means when $a = 4$ and $b = 9$. Numerically, it is always true that A is largest and H is the smallest. Verify that this is true in this particular example.
3. Simplify. **a.** $3^{1/2}(3^{1/2} + 3^{1/2})$
   b. $a^{1/2}(a^{1/2} + a^{1/2})$
4. Show that $a^{-1/n}$ and $a^{1/n}$ are reciprocals by showing that their product is 1. Use a calculator to find both $2^{-1/3}$ and its reciprocal to the nearest thousandth.
5. Use FOIL to multiply $(x^{1/2} + y^{1/2})(x^{1/2} - y^{1/2})$.

REMEMBER THIS

1. True or false? $\sqrt[3]{8 \cdot 27} = (\sqrt[3]{8})(\sqrt[3]{27})$. True
2. True or false? $\sqrt[3]{8} + \sqrt[3]{27} = \sqrt[3]{8 + 27}$. False
3. Rewrite 45 as the product of a perfect square (other than 1) and another number. $9 \cdot 5$
4. Write the next two numbers in this sequence of perfect cubes: 1, 8, 27, 64, ___, ___. 125, 216
5. In the expression $\sqrt[3]{5^6}$, which number is called the index of the radical? 3
6. Solve the system. $$\begin{aligned} x + y - z &= 4 \\ x - 2y + z &= 0 \\ 2x - y + 3z &= 7 \end{aligned}$$ (3,2,1)
7. If y varies inversely as the square of x and $y = 1$ when $x = \frac{1}{2}$, find y when $x = 1$. $\frac{1}{4}$
8. Simplify $\dfrac{-30x^4y^2z}{18xy^2z^2}$. $\dfrac{-5x^3}{3z}$
9. Find the product of $3x^{-4}$ and $\frac{1}{3}x^4$. 1
10. Factor $2x^2 + 3x - 2$. $(2x - 1)(x + 2)$

7.2 Product and Quotient Properties of Radicals

The period T of a pendulum is the time required for the pendulum to complete one round-trip of motion, that is, one complete cycle. When the period is measured in seconds, then a formula for the period is

$$T = 2\pi\sqrt{\frac{\ell}{32}},$$

where ℓ is the length in feet of the pendulum. In simplified radical form, what is the period for a pendulum that is 1 ft long? Approximate this number to the nearest hundredth of a second. (See Example 5.)

OBJECTIVES

1 **Simplify radicals.**

2 **Multiply and divide radicals, and simplify where possible.**

Properties of radicals are used often to simplify radicals and to operate on radicals. To illustrate two such properties, consider

$$\sqrt[3]{8 \cdot 27} = \sqrt[3]{216} = 6 \quad \text{and} \quad \sqrt[3]{8} \cdot \sqrt[3]{27} = 2 \cdot 3 = 6,$$

and

$$\sqrt{\frac{36}{4}} = \sqrt{9} = 3 \quad \text{and} \quad \frac{\sqrt{36}}{\sqrt{4}} = \frac{6}{2} = 3.$$

We see that $\sqrt[3]{8 \cdot 27} = \sqrt[3]{8} \cdot \sqrt[3]{27}$ and $\sqrt{36/4} = \sqrt{36}/\sqrt{4}$, which illustrates product and quotient properties of radicals, respectively.

Note that these properties do not hold if both a and b are negative and n is even—thus the insistence that $\sqrt[n]{a}$ and $\sqrt[n]{b}$ be real.

Product and Quotient Properties of Radicals

For real numbers a, b, $\sqrt[n]{a}$, and $\sqrt[n]{b}$:

1. $\sqrt[n]{a \cdot b} = \sqrt[n]{a} \cdot \sqrt[n]{b}$

2. $\sqrt[n]{\frac{a}{b}} = \frac{\sqrt[n]{a}}{\sqrt[n]{b}} \; (b \neq 0)$

The product property may be proved by converting between radical form and exponential form and using the product-to-a-power property of exponents.

$$\sqrt[n]{ab} = (ab)^{1/n} = a^{1/n} \cdot b^{1/n} = \sqrt[n]{a} \cdot \sqrt[n]{b}$$

The quotient property is proved in a similar way, and this proof is requested in the exercises. The first application of these properties that we consider is their role in simplifying radicals.

1 Simplifying radicals

One condition that must be met for a radical to be expressed in simplified form is that we remove all factors of the radicand whose indicated root can be taken exactly.

EXAMPLE 1 Simplify each radical.

a. $\sqrt{45}$ **b.** $\sqrt{32}$ **c.** $\sqrt[3]{54}$ **d.** $\sqrt[4]{80}$

Solution

a. Rewrite 45 as the product of a perfect square and another factor and simplify.

$$\begin{aligned} \sqrt{45} &= \sqrt{9 \cdot 5} && \text{9 is a perfect square and } 45 = 9 \cdot 5. \\ &= \sqrt{9} \cdot \sqrt{5} && \text{Product property of radicals.} \\ &= 3\sqrt{5} && \text{Simplify.} \end{aligned}$$

Simplification of radicals is valuable for algebraic manipulation but not necessary for calculator approximation. It is quicker to approximate $\sqrt{45}$ on a calculator than to first convert it to $3\sqrt{5}$.

b. Both 4 and 16 are perfect square factors of 32. Choosing the *larger* perfect square factor is more efficient, so

$$\sqrt{32} = \sqrt{16 \cdot 2} = \sqrt{16} \cdot \sqrt{2} = 4\sqrt{2}.$$

Check that using $\sqrt{32} = \sqrt{4 \cdot 8}$ leads to the same result but with more work.

c. To simplify cube roots, look for factors of the radicand from the perfect cubes 8, 27, 64, 125, and so on. Seeing $54 = 27 \cdot 2$ yields

$$\sqrt[3]{54} = \sqrt[3]{27 \cdot 2} = \sqrt[3]{27} \cdot \sqrt[3]{2} = 3\sqrt[3]{2}.$$

d. Consider fourth powers of 2, 3, 4, and so on, to simplify fourth roots. Since $2^4 = 16$, and 16 is a factor of 80, we have

$$\sqrt[4]{80} = \sqrt[4]{16 \cdot 5} = \sqrt[4]{16} \cdot \sqrt[4]{5} = 2\sqrt[4]{5}.$$

PROGRESS CHECK 1 Simplify each radical.

a. $\sqrt{75}$ **b.** $\sqrt{80}$ **c.** $\sqrt[3]{32}$ **d.** $\sqrt[4]{162}$

Example 2 discusses how to simplify $\sqrt[n]{a^m}$ when m and n have common factors (other than 1) and a is nonnegative. In parts **c** and **d**, note that the index for the radical has been reduced. A second consideration for expressing a radical in simplified form is that the index of the radical be as small as possible.

EXAMPLE 2 Simplify each radical. Assume $x \geq 0$, $y \geq 0$.

a. $\sqrt[3]{x^6}$ **b.** $\sqrt{7^4}$ **c.** $\sqrt[6]{5^3}$ **d.** $\sqrt[8]{y^6}$

Solution Use $\sqrt[n]{a^m} = a^{m/n}$ and reduce the rational exponent. Then convert back to radical form where necessary.

a. $\sqrt[3]{x^6} = x^{6/3} = x^2$
b. $\sqrt{7^4} = 7^{4/2} = 7^2 = 49$
c. $\sqrt[6]{5^3} = 5^{3/6} = 5^{1/2} = \sqrt{5}$
d. $\sqrt[8]{y^6} = y^{6/8} = y^{3/4} = \sqrt[4]{y^3}$

Note In this example we assumed $x \geq 0$ and $y \geq 0$ so that expressions like $\sqrt{x^2}$ simplify to x. Without this assumption it is necessary to use absolute value and write

$$\sqrt{x^2} = |x|.$$

For instance, $\sqrt{(-7)^2} \neq -7$. Instead, $\sqrt{(-7)^2} = \sqrt{49} = 7$, so $\sqrt{(-7)^2} = |-7| = 7$. The general rule is

$$\sqrt[n]{a^n} = \begin{cases} a, \text{ if } n \text{ is odd,} \\ |a|, \text{ if } n \text{ is even.} \end{cases}$$

Throughout this chapter we will restrict radicands involving variables to nonnegative real numbers so that simplifications involving absolute value will not be necessary.

PROGRESS CHECK 2 Simplify each radical. Assume $x \geq 0$, $y \geq 0$.

a. $\sqrt{11^2}$ **b.** $\sqrt[4]{y^{12}}$ **c.** $\sqrt[12]{5^4}$ **d.** $\sqrt[9]{x^6}$

Parts **a** and **b** in Example 2 showed how to simplify a radical when the radicand contains a power of a variable that is a multiple of the index. This method is also used in the simplifications in the next example.

EXAMPLE 3 Simplify each radical. Assume $x \geq 0$, $y \geq 0$.

a. $\sqrt{y^9}$ **b.** $\sqrt{24x^9y^6}$ **c.** $\sqrt[3]{x^5y^8}$

Solution

a. The largest power in which the exponent is a multiple of the index of 2 is y^8, so

$$\sqrt{y^9} = \sqrt{y^8 \cdot y} = \sqrt{y^8} \cdot \sqrt{y} = y^4\sqrt{y}.$$

b. Rewrite $24x^9y^6$ as a product of its largest perfect square factor and another factor, and simplify.

$$\sqrt{24x^9y^6} = \sqrt{(4x^8y^6)(6x)} = \sqrt{4x^8y^6} \cdot \sqrt{6x} = 2x^4y^3\sqrt{6x}$$

c. The largest powers in which exponents are multiples of the index of 3 are x^3 and y^6, so

$$\sqrt[3]{x^5y^8} = \sqrt[3]{(x^3y^6)(x^2y^2)} = \sqrt[3]{x^3y^6} \cdot \sqrt[3]{x^2y^2} = xy^2\sqrt[3]{x^2y^2}.$$

Progress Check Answers

1. (a) $5\sqrt{3}$ (b) $4\sqrt{5}$ (c) $2\sqrt[3]{4}$ (d) $3\sqrt[4]{2}$
2. (a) 11 (b) y^3 (c) $\sqrt[3]{5}$ (d) $\sqrt[3]{x^2}$

PROGRESS CHECK 3 Simplify each radical. Assume $x \geq 0$, $y \geq 0$.

a. $\sqrt{x^{11}}$ **b.** $\sqrt{125x^5y^8}$ **c.** $\sqrt[4]{x^7y^9}$

A third consideration for expressing a radical in simplified form is to eliminate any fractions in the radicand. The quotient property of radicals is used in such simplifications, as shown in Example 4.

EXAMPLE 4 Simplify each radical. Assume $x > 0$.

a. $\sqrt{\frac{9}{64}}$ **b.** $\sqrt{\frac{2}{5}}$ **c.** $\sqrt{\frac{25}{x}}$ **d.** $\sqrt[3]{\frac{2}{9x}}$

Solution

a. Both 9 and 64 are perfect squares, so

$$\sqrt{\frac{9}{64}} = \frac{\sqrt{9}}{\sqrt{64}} = \frac{3}{8}.$$

b. Rewrite $\frac{2}{5}$ as an equivalent fraction whose denominator is a perfect square, and simplify.

$$\sqrt{\frac{2}{5}} = \sqrt{\frac{2}{5} \cdot \frac{5}{5}} = \sqrt{\frac{10}{25}} = \frac{\sqrt{10}}{\sqrt{25}} = \frac{\sqrt{10}}{5}$$

c. Since $\sqrt{x^2} = x$ for $x > 0$,

$$\sqrt{\frac{25}{x}} = \sqrt{\frac{25}{x} \cdot \frac{x}{x}} = \sqrt{\frac{25x}{x^2}} = \frac{\sqrt{25}\sqrt{x}}{\sqrt{x^2}} = \frac{5\sqrt{x}}{x}.$$

d. Rewrite $\frac{2}{9x}$ as an equivalent fraction whose denominator is a perfect cube, and simplify.

$$\sqrt[3]{\frac{2}{9x}} = \sqrt[3]{\frac{2}{9x} \cdot \frac{3x^2}{3x^2}} = \sqrt[3]{\frac{6x^2}{27x^3}} = \frac{\sqrt[3]{6x^2}}{\sqrt[3]{27x^3}} = \frac{\sqrt[3]{6x^2}}{3x}$$

PROGRESS CHECK 4 Simplify each radical. Assume $y > 0$.

a. $\sqrt{\frac{4}{81}}$ **b.** $\sqrt{\frac{7}{11}}$ **c.** $\sqrt{\frac{8}{y}}$ **d.** $\sqrt[3]{\frac{5}{4y}}$

The discovery that the period of a pendulum depends only on its length is due to Galileo (1581). The actual formula came later, after Newton's work on gravity.

EXAMPLE 5 Solve the problem in the section introduction on page 325.

Solution Replacing ℓ by 1 in

$$T = 2\pi\sqrt{\frac{\ell}{32}} \quad \text{gives} \quad T = 2\pi\sqrt{\frac{1}{32}}.$$

To express the period T in simplified radical form, rewrite $\frac{1}{32}$ as an equivalent fraction whose denominator is a perfect square, and then simplify.

$$T = 2\pi\sqrt{\frac{1}{32}} = 2\pi\sqrt{\frac{1}{32} \cdot \frac{2}{2}} = 2\pi\frac{\sqrt{2}}{\sqrt{64}} = 2\pi\frac{\sqrt{2}}{8} = \frac{\pi\sqrt{2}}{4}$$

The period is $\pi\sqrt{2}/4$ seconds. By calculator,

$$\pi \boxed{\times} 2 \boxed{\sqrt{\ }} \boxed{\div} 4 \boxed{=} \boxed{1.1107207}.$$

Thus, to the nearest hundredth of a second, the period is 1.11 seconds.

PROGRESS CHECK 5 In simplified radical form, what is the period of a pendulum that is 4 ft long? Approximate this answer to the nearest hundredth of a second.

Progress Check Answers

3. (a) $x^5\sqrt{x}$ (b) $5x^2y^4\sqrt{5x}$ (c) $xy^2\sqrt[4]{x^3y}$
4. (a) $\frac{2}{9}$ (b) $\frac{\sqrt{77}}{11}$ (c) $\frac{2\sqrt{2y}}{y}$ (d) $\frac{\sqrt[3]{10y^2}}{2y}$
5. $\frac{\pi\sqrt{2}}{2}$ seconds; 2.22 seconds

2 To multiply and divide expressions containing radicals, we use the product and quotient properties of radicals in the forms

$$\sqrt[n]{a} \cdot \sqrt[n]{b} = \sqrt[n]{ab},$$
$$\frac{\sqrt[n]{a}}{\sqrt[n]{b}} = \sqrt[n]{\frac{a}{b}}, \qquad b \neq 0.$$

Applying these properties sometimes results in expressions that can be simplified, and final answers must be stated in simplified radical form.

EXAMPLE 6 Multiply and simplify where possible. Assume $x > 0$, $y > 0$.

a. $\sqrt{5} \cdot \sqrt{10}$ **b.** $5\sqrt[3]{7} \cdot 4\sqrt[3]{2}$ **c.** $\sqrt[4]{8x^3y^2} \cdot \sqrt[4]{4x^5y^5}$

Solution

a. First, multiply to get

$$\sqrt{5} \cdot \sqrt{10} = \sqrt{5 \cdot 10} = \sqrt{50}.$$

Now simplify.

$$\sqrt{50} = \sqrt{25 \cdot 2} = \sqrt{25} \cdot \sqrt{2} = 5\sqrt{2}$$

Thus, $\sqrt{5}\sqrt{10} = 5\sqrt{2}$.

b. Reorder and regroup as shown. Then multiply.

$$5\sqrt[3]{7} \cdot 4\sqrt[3]{2} = (5 \cdot 4)(\sqrt[3]{7} \cdot \sqrt[3]{2}) = 20\sqrt[3]{14}$$

No factor of 14 is a perfect cube, so the result does not simplify.

c. Multiply first.

$$\sqrt[4]{8x^3y^2} \cdot \sqrt[4]{4x^5y^5} = \sqrt[4]{8x^3y^2 \cdot 4x^5y^5} = \sqrt[4]{32x^8y^7}$$

Then simplify.

$$\sqrt[4]{32x^8y^7} = \sqrt[4]{16x^8y^4 \cdot 2y^3} = \sqrt[4]{16x^8y^4} \cdot \sqrt[4]{2y^3} = 2x^2y\sqrt[4]{2y^3}$$

Thus, $\sqrt[4]{8x^3y^2} \cdot \sqrt[4]{4x^5y^5} = 2x^2y\sqrt[4]{2y^3}$.

PROGRESS CHECK 6 Multiply and simplify where possible. Assume $x > 0$, $y > 0$.

a. $\sqrt{15} \cdot \sqrt{5}$ **b.** $8\sqrt[4]{9} \cdot 5\sqrt[4]{9}$ **c.** $\sqrt[3]{4x^2y} \cdot \sqrt[3]{6xy^7}$

EXAMPLE 7 Divide and simplify where possible. Assume $x > 0$, $y > 0$.

a. $\dfrac{\sqrt[3]{54}}{\sqrt[3]{2}}$ **b.** $\dfrac{\sqrt{3}}{\sqrt{10}}$ **c.** $\dfrac{\sqrt[3]{21x^2}}{\sqrt[3]{12x}}$

Solution

a.
$$\frac{\sqrt[3]{54}}{\sqrt[3]{2}} = \sqrt[3]{\frac{54}{2}} \quad \text{Quotient property of radicals.}$$
$$= \sqrt[3]{27} \quad \text{Divide.}$$
$$= 3 \quad \text{Simplify.}$$

b. Apply the quotient property and then simplify.

$$\frac{\sqrt{3}}{\sqrt{10}} = \sqrt{\frac{3}{10}} = \sqrt{\frac{3}{10} \cdot \frac{10}{10}} = \sqrt{\frac{30}{100}} = \frac{\sqrt{30}}{\sqrt{100}} = \frac{\sqrt{30}}{10}$$

c. Divide first.

$$\frac{\sqrt[3]{21x^2}}{\sqrt[3]{12x}} = \sqrt[3]{\frac{21x^2}{12x}} = \sqrt[3]{\frac{7x}{4}}$$

Progress Check Answers

6. (a) $5\sqrt{3}$ (b) 120 (c) $2xy^2\sqrt[3]{3y^2}$

To simplify, we see that $4 \cdot 2$ gives the perfect cube 8, so

$$\sqrt[3]{\frac{7x}{4}} = \sqrt[3]{\frac{7x}{4} \cdot \frac{2}{2}} = \sqrt[3]{\frac{14x}{8}} = \frac{\sqrt[3]{14x}}{\sqrt[3]{8}} = \frac{\sqrt[3]{14x}}{2}.$$

Thus, $\sqrt[3]{21x^2}/\sqrt[3]{12x} = \sqrt[3]{14x}/2$.

PROGRESS CHECK 7 Divide and simplify where possible. Assume $x > 0$.

a. $\dfrac{\sqrt[4]{405}}{\sqrt[4]{5}}$ **b.** $\dfrac{\sqrt{2}}{\sqrt{7}}$ **c.** $\dfrac{\sqrt[3]{4y^7}}{\sqrt[3]{18y^5}}$

In Example 7b the quotient $\sqrt{3}/\sqrt{10}$ led to a division that did not reduce. In this case it is easier to obtain an equivalent fraction that eliminates the radical in the denominator by multiplying by 1 in the form $\sqrt{10}/\sqrt{10}$, as shown next.

$$\frac{\sqrt{3}}{\sqrt{10}} = \frac{\sqrt{3}}{\sqrt{10}} \cdot \frac{\sqrt{10}}{\sqrt{10}} = \frac{\sqrt{30}}{\sqrt{100}} = \frac{\sqrt{30}}{10}$$

The process of obtaining a radical-free denominator is called **rationalizing the denominator.** The method above is also used to rationalize the denominator when the numerator does not contain a radical. And as a final condition we adopt the requirement that a simplified radical expression cannot have radicals in the denominator.

EXAMPLE 8 Rationalize each denominator. Assume $x > 0$.

a. $\dfrac{5}{\sqrt{3}}$ **b.** $\dfrac{6}{\sqrt{20}}$ **c.** $\dfrac{\sqrt{6}}{\sqrt{7x}}$ **d.** $\dfrac{2}{\sqrt[3]{3}}$

Solution

a. $\dfrac{5}{\sqrt{3}} = \dfrac{5}{\sqrt{3}} \cdot \dfrac{\sqrt{3}}{\sqrt{3}} = \dfrac{5\sqrt{3}}{\sqrt{9}} = \dfrac{5\sqrt{3}}{3}$

b. $\dfrac{6}{\sqrt{20}} = \dfrac{6}{\sqrt{20}} \cdot \dfrac{\sqrt{5}}{\sqrt{5}} = \dfrac{6\sqrt{5}}{\sqrt{100}} = \dfrac{6\sqrt{5}}{10} = \dfrac{3\sqrt{5}}{5}$

c. $\dfrac{\sqrt{6}}{\sqrt{7x}} = \dfrac{\sqrt{6}}{\sqrt{7x}} \cdot \dfrac{\sqrt{7x}}{\sqrt{7x}} = \dfrac{\sqrt{42x}}{\sqrt{49x^2}} = \dfrac{\sqrt{42x}}{7x}$

d. $\dfrac{2}{\sqrt[3]{3}} = \dfrac{2}{\sqrt[3]{3}} \cdot \dfrac{\sqrt[3]{9}}{\sqrt[3]{9}} = \dfrac{2\sqrt[3]{9}}{\sqrt[3]{27}} = \dfrac{2\sqrt[3]{9}}{3}$

PROGRESS CHECK 8 Rationalize each denominator. Assume $x > 0$, $y > 0$.

a. $\dfrac{1}{\sqrt{2}}$ **b.** $\dfrac{\sqrt{x}}{\sqrt{y}}$ **c.** $\dfrac{8}{\sqrt{12}}$ **d.** $\dfrac{7}{\sqrt[3]{25}}$

In summary, note that the following conditions have been given in this section for writing a simplified radical.

> **Simplified Radical**
>
> To write a radical in simplified form:
>
> 1. Remove all factors of the radicand whose indicated root can be taken exactly.
> 2. Write the radical so that the index is as small as possible.
> 3. Eliminate all fractions in the radicand and all radicals in the denominator (which is called rationalizing the denominator).

Progress Check Answers

7. (a) 3 (b) $\dfrac{\sqrt{14}}{7}$ (c) $\dfrac{\sqrt[3]{6y^2}}{3}$

8. (a) $\dfrac{\sqrt{2}}{2}$ (b) $\dfrac{\sqrt{xy}}{y}$ (c) $\dfrac{4\sqrt{3}}{3}$ (d) $\dfrac{7\sqrt[3]{5}}{5}$

EXERCISES 7.2

In Exercises 1–62, simplify each radical. Assume $x > 0$, $y > 0$.

1. $\sqrt{28}$ $2\sqrt{7}$
2. $\sqrt{54}$ $3\sqrt{6}$
3. $\sqrt{147}$ $7\sqrt{3}$
4. $\sqrt{99}$ $3\sqrt{11}$
5. $\sqrt{72}$ $6\sqrt{2}$
6. $\sqrt{405}$ $9\sqrt{5}$
7. $\sqrt[3]{40}$ $2\sqrt[3]{5}$
8. $\sqrt[3]{108}$ $3\sqrt[3]{4}$
9. $\sqrt[3]{375}$ $5\sqrt[3]{3}$
10. $\sqrt[3]{128}$ $4\sqrt[3]{2}$
11. $\sqrt[4]{243}$ $3\sqrt[4]{3}$
12. $\sqrt[4]{96}$ $2\sqrt[4]{6}$
13. $\sqrt{4^6}$ 64
14. $\sqrt{5^4}$ 25
15. $\sqrt{x^4}$ x^2
16. $\sqrt{y^8}$ y^4
17. $\sqrt[3]{9^6}$ 81
18. $\sqrt[3]{2^9}$ 8
19. $\sqrt[3]{14^3}$ 14
20. $\sqrt[4]{13^4}$ 13
21. $\sqrt[3]{y^9}$ y^3
22. $\sqrt[3]{x^{12}}$ x^4
23. $\sqrt[8]{x^2}$ $\sqrt[4]{x}$
24. $\sqrt[12]{y^2}$ $\sqrt[6]{y}$
25. $\sqrt[12]{y^{10}}$ $\sqrt[6]{y^5}$
26. $\sqrt[10]{x^4}$ $\sqrt[5]{x^2}$
27. $\sqrt{x^7}$ $x^3\sqrt{x}$
28. $\sqrt{y^5}$ $y^2\sqrt{y}$
29. $\sqrt{72y^3}$ $6y\sqrt{2y}$
30. $\sqrt{243x^{11}}$ $9x^5\sqrt{3x}$
31. $\sqrt[3]{y^{11}}$ $y^3\sqrt[3]{y^2}$
32. $\sqrt[3]{x^{16}}$ $x^5\sqrt[3]{x}$
33. $\sqrt{x^4y^9}$ $x^2y^4\sqrt{y}$
34. $\sqrt{x^3y^6}$ $xy^3\sqrt{x}$
35. $\sqrt{32x^2y^2}$ $4xy\sqrt{2}$
36. $\sqrt{147x^4y^6}$ $7x^2y^3\sqrt{3}$
37. $\sqrt{40x^5y^6}$ $2x^2y^3\sqrt{10x}$
38. $\sqrt{48x^8y^9}$ $4x^4y^4\sqrt{3y}$
39. $\sqrt{98x^3y^5}$ $7xy^2\sqrt{2xy}$
40. $\sqrt{162x^{11}y^9}$ $9x^5y^4\sqrt{2xy}$
41. $\sqrt[3]{24x^3y}$ $2x\sqrt[3]{3y}$
42. $\sqrt[3]{108x^9y^7}$ $3x^3y^2\sqrt[3]{4y}$
43. $\sqrt[3]{81x^5y^4}$ $3xy\sqrt[3]{3x^2y}$
44. $\sqrt[3]{125x^{11}y^7}$ $5x^3y^2\sqrt[3]{x^2y}$
45. $\sqrt[4]{x^{11}y^5}$ $x^2y\sqrt[4]{x^3y}$
46. $\sqrt[5]{x^7y^{12}}$ $xy^2\sqrt[5]{x^2y^2}$
47. $\sqrt{\frac{16}{121}}$ $\frac{4}{11}$
48. $\sqrt{\frac{36}{81}}$ $\frac{6}{9}$
49. $\sqrt{\frac{32}{49}}$ $\frac{4\sqrt{2}}{7}$
50. $\sqrt{\frac{147}{25}}$ $\frac{7\sqrt{3}}{5}$
51. $\sqrt{\frac{36}{7}}$ $\frac{6\sqrt{7}}{7}$
52. $\sqrt{\frac{9}{13}}$ $\frac{3\sqrt{13}}{13}$
53. $\sqrt[3]{\frac{2}{3}}$ $\frac{\sqrt[3]{18}}{3}$
54. $\sqrt[3]{\frac{3}{4}}$ $\frac{\sqrt[3]{6}}{2}$
55. $\sqrt{\frac{49}{y}}$ $\frac{7\sqrt{y}}{y}$
56. $\sqrt{\frac{121}{x}}$ $\frac{11\sqrt{x}}{x}$
57. $\sqrt{\frac{24}{x}}$ $\frac{2\sqrt{6x}}{x}$
58. $\sqrt{\frac{50}{y}}$ $\frac{5\sqrt{2y}}{y}$
59. $\sqrt[3]{\frac{3}{4y}}$ $\frac{\sqrt[3]{6y^2}}{2y}$
60. $\sqrt[3]{\frac{2}{25x^2}}$ $\frac{\sqrt[3]{10x}}{5x}$
61. $\sqrt[3]{\frac{7}{108y^2}}$ $\frac{\sqrt[3]{14y}}{6y}$
62. $\sqrt[3]{\frac{15}{32x}}$ $\frac{\sqrt[3]{30x^2}}{4x}$

63. In simplified radical form, what is the period of a pendulum that is 16 ft long? Approximate the answer to the nearest hundredth of a second. Use the formula from Example 5.
$\pi\sqrt{2} \approx 4.44$ seconds

64. In simplified radical form, what is the period of a pendulum that is 9 ft long? Approximate the answer to the nearest hundredth of a second. Use the formula from Example 5.
$\frac{3}{4}\pi\sqrt{2} \approx 3.33$ seconds

65. When a piano string is struck, it vibrates at a certain frequency, which gives it its particular sound. Higher frequencies correspond to higher-pitched tones. The frequency f depends on the length of the string L, the tension in the string T, and the mass of the string m, as given by

$$f = \frac{1}{2L}\sqrt{\frac{T}{m}}.$$

Find the frequency when $L = 0.5$ m, $T = 400$ newtons, and $m = 10^{-6}$ kg. With these units the frequency is expressed in cycles per second. 20,000 cycles/second

66. Repeat Exercise 65, but change the length of the string to 2 m.
5,000 cycles/second

In Exercises 67–82, multiply and simplify where possible. Assume $x \geq 0$, $y \geq 0$.

67. $\sqrt{3} \cdot \sqrt{27}$ 9
68. $\sqrt{20} \cdot \sqrt{5}$ 10
69. $2\sqrt{3} \cdot 5\sqrt{2}$ $10\sqrt{6}$
70. $7\sqrt{2} \cdot 2\sqrt{5}$ $14\sqrt{10}$
71. $\sqrt{14} \cdot \sqrt{2}$ $2\sqrt{7}$
72. $\sqrt{6} \cdot \sqrt{30}$ $6\sqrt{5}$
73. $\sqrt{63} \cdot \sqrt{9}$ $9\sqrt{7}$
74. $\sqrt{8} \cdot \sqrt{24}$ $8\sqrt{3}$
75. $3\sqrt[3]{4} \cdot 2\sqrt[3]{6}$ $12\sqrt[3]{3}$
76. $5\sqrt[3]{3} \cdot 6\sqrt[3]{18}$ $90\sqrt[3]{2}$
77. $6\sqrt[4]{9} \cdot 4\sqrt[4]{27}$ $72\sqrt[4]{3}$
78. $7\sqrt[4]{8} \cdot 3\sqrt[4]{4}$ $42\sqrt[4]{2}$
79. $3\sqrt{8x^3y} \cdot 2\sqrt{5x^2y}$ $12x^2y\sqrt{10x}$
80. $4\sqrt{2xy^3} \cdot \sqrt{8x^3y^4}$ $16x^2y^3\sqrt{y}$
81. $\sqrt[4]{6x^3y^2} \cdot \sqrt[4]{8x^2y^5}$ $2xy\sqrt[4]{3xy^3}$
82. $\sqrt[4]{27x^3y} \cdot \sqrt[4]{9x^3y^4}$ $3xy\sqrt[4]{3x^2y}$

In Exercises 83–102, divide and simplify where possible. Assume $x > 0$, $y > 0$.

83. $\frac{\sqrt{56}}{\sqrt{14}}$ 2
84. $\frac{\sqrt{250}}{\sqrt{10}}$ 5
85. $\frac{\sqrt[3]{88}}{\sqrt[3]{11}}$ 2
86. $\frac{\sqrt[3]{135}}{\sqrt[3]{5}}$ 3
87. $\frac{\sqrt[4]{192}}{\sqrt[4]{6}}$ $2\sqrt[4]{2}$
88. $\frac{\sqrt[4]{256}}{\sqrt[4]{2}}$ $2\sqrt[4]{8}$
89. $\frac{\sqrt{2}}{\sqrt{11}}$ $\frac{\sqrt{22}}{11}$
90. $\frac{\sqrt{5}}{\sqrt{26}}$ $\frac{\sqrt{130}}{26}$
91. $\frac{\sqrt[3]{32}}{\sqrt[3]{6}}$ $\frac{2\sqrt[3]{18}}{3}$
92. $\frac{\sqrt[3]{108}}{\sqrt[3]{8}}$ $\frac{3\sqrt[3]{4}}{2}$
93. $\frac{\sqrt{2x}}{\sqrt{5x}}$ $\frac{\sqrt{10}}{5}$
94. $\frac{\sqrt{5x^3}}{\sqrt{2x}}$ $\frac{x\sqrt{10}}{2}$
95. $\frac{\sqrt{27y^2}}{\sqrt{6y}}$ $\frac{3\sqrt{2y}}{2}$
96. $\frac{\sqrt{25y^5}}{\sqrt{15y^4}}$ $\frac{\sqrt{15y}}{3}$
97. $\frac{\sqrt[3]{3x^4}}{\sqrt[3]{4x^2}}$ $\frac{\sqrt[3]{6x^2}}{2}$
98. $\frac{\sqrt[3]{6y^6}}{\sqrt[3]{49y^4}}$ $\frac{\sqrt[3]{42y^2}}{7}$
99. $\frac{\sqrt{64x^2y^3}}{\sqrt{10x^2y^2}}$ $\frac{4\sqrt{10y}}{5}$
100. $\frac{\sqrt{56x^8y^4}}{\sqrt{6x^7y^3}}$ $\frac{2\sqrt{21xy}}{3}$
101. $\frac{\sqrt[4]{45x^5}}{\sqrt[4]{8x^2}}$ $\frac{\sqrt[4]{90x^3}}{2}$
102. $\frac{\sqrt[4]{7y^9}}{\sqrt[4]{125y^6}}$ $\frac{\sqrt[4]{35y^3}}{5}$

In Exercises 103–122, rationalize each denominator. Assume $x > 0$, $y > 0$.

103. $\frac{7}{\sqrt{7}}$ $\sqrt{7}$
104. $\frac{3}{\sqrt{3}}$ $\sqrt{3}$
105. $\frac{12}{\sqrt{y}}$ $\frac{12\sqrt{y}}{y}$
106. $\frac{14}{\sqrt{x}}$ $\frac{14\sqrt{x}}{x}$
107. $\frac{\sqrt{3}}{\sqrt{2}}$ $\frac{\sqrt{6}}{2}$
108. $\frac{\sqrt{5}}{\sqrt{3}}$ $\frac{\sqrt{15}}{3}$

109. $\dfrac{4}{\sqrt{32}}$ $\frac{\sqrt{2}}{2}$

110. $\dfrac{6}{\sqrt{27}}$ $\frac{2\sqrt{3}}{3}$

111. $\dfrac{\sqrt{3}}{\sqrt{2y}}$ $\frac{\sqrt{6y}}{2y}$

112. $\dfrac{\sqrt{5}}{\sqrt{3x}}$ $\frac{\sqrt{15x}}{3x}$

113. $\dfrac{\sqrt{10}}{\sqrt{5y}}$ $\frac{\sqrt{2y}}{y}$

114. $\dfrac{\sqrt{18}}{\sqrt{6x}}$ $\frac{\sqrt{3x}}{x}$

115. $\dfrac{\sqrt{2}}{\sqrt{98x}}$ $\frac{\sqrt{x}}{7x}$

116. $\dfrac{\sqrt{6}}{\sqrt{24y}}$ $\frac{\sqrt{y}}{2y}$

117. $\dfrac{3}{\sqrt[3]{49}}$ $\frac{3\sqrt[3]{7}}{7}$

118. $\dfrac{4}{\sqrt[3]{9}}$ $\frac{4\sqrt[3]{3}}{3}$

119. $\dfrac{5}{\sqrt[4]{4}}$ $\frac{5\sqrt[4]{4}}{2}$

120. $\dfrac{6}{\sqrt[4]{9}}$ $2\sqrt[4]{9}$

121. $\dfrac{6}{\sqrt{45y}}$ $\frac{2\sqrt{5y}}{5y}$

122. $\dfrac{4}{\sqrt{28x}}$ $\frac{2\sqrt{7x}}{7x}$

123. **a.** Find the solution set for $x = \sqrt{16}$. {4}; principal root only
b. Find the solution set for $x^2 = 16$. {4, −4}; both square roots
c. Are your answers to parts **a** and **b** the same? No

124. **a.** Find the solution set for $x = \sqrt[3]{8}$. {2}
b. Find the solution set for $x^3 = 8$. {2}
c. Are your answers to parts **a** and **b** the same? Yes

125. Consider (i) $\sqrt{7^2}$ and (ii) $(\sqrt{7})^2$.
a. How do these two expressions compare in order of operations? (i) Square first; then extract square root. (ii) Get square root; then square.
b. How do these two expressions compare in value? Same value

126. Compare (i) $\sqrt[3]{(-5)^3}$ and (ii) $(\sqrt[3]{-5})^3$ in terms of value.
Both equal −5

127. Prove $\sqrt[n]{\dfrac{a}{b}} = \dfrac{\sqrt[n]{a}}{\sqrt[n]{b}}$ $(b \neq 0)$.
(*Hint:* The quotient property may be proved by converting between radical form and exponential form and using the quotient-to-a-power property of exponents.)
$\sqrt[n]{\frac{a}{b}} = \left(\frac{a}{b}\right)^{1/n} = \frac{a^{1/n}}{b^{1/n}} = \frac{\sqrt[n]{a}}{\sqrt[n]{b}}$

THINK ABOUT IT

1. The formula for the period of a pendulum as illustrated in Example 5 is $T = 2\pi\sqrt{\frac{\ell}{32}}$. The formula shows how the period depends on the length.
a. Does the period increase or decrease if you make the pendulum longer?
b. Explain why the period does not double when you double the length.
c. The 32 in the formula represents the force of gravity near the surface of the earth. On the moon the force of gravity would be less than 32. If you moved a pendulum from the earth to the moon, would its period be longer or shorter there?

2. Simplifying radicals by rationalizing the denominator is an important algebraic technique because it puts results in standard form, but for calculator evaluation of radicals it may not be crucial. Evaluate the following expression in two ways: (i) First rationalize the denominator; then use the calculator. (ii) Use the calculator immediately. The answer should be the same both ways, but one way may seem more convenient.
a. $\dfrac{1}{\sqrt{2}}$ **b.** $\dfrac{3 + \sqrt{2}}{\sqrt{2}}$

3. It is easy to show that $\sqrt[3]{-8} \times \sqrt[3]{-8}$ equals $\sqrt[3]{(-8)(-8)}$. Do this. In contrast, it is not possible to show (unless you know about imaginary numbers) whether $\sqrt{-8} \times \sqrt{-8}$ is equal to $\sqrt{(-8)(-8)}$. What is the difficulty with the second problem?

4. Rewrite the following expression by rationalizing the *numerator:* $\dfrac{\sqrt{3}}{\sqrt{5}}$.

5. **a.** Show by FOIL that $(\sqrt{3} + \sqrt{2})(\sqrt{3} - \sqrt{2})$ equals an integer.
b. Use the result from part **a** to rationalize the denominator in $\dfrac{4}{\sqrt{3} + \sqrt{2}}$.
c. If a and b are integers, explain why $(\sqrt{a} + \sqrt{b})(\sqrt{a} - \sqrt{b})$ must be an integer.

REMEMBER THIS

1. True or false? $5\sqrt{9} + 6\sqrt{9} = 11\sqrt{9}$. True
2. Write a true statement by selecting the correct relationship symbol (<, >, =): $\sqrt{3} + \sqrt{2}$ ____ $\sqrt{3 + 2}$. >
3. Simplify $\sqrt{40}$. $2\sqrt{10}$
4. Find the length of the diagonal of a square of side 5 in. $5\sqrt{2}$ in.
5. Rationalize the denominator for $\dfrac{\sqrt{2}}{\sqrt{7}}$. $\frac{\sqrt{14}}{7}$
6. Evaluate $\begin{vmatrix} 2 & \frac{1}{3} \\ 3 & \frac{1}{2} \end{vmatrix}$. 0
7. Graph $y = \frac{1}{2}x + 2$, and mark the intercepts.

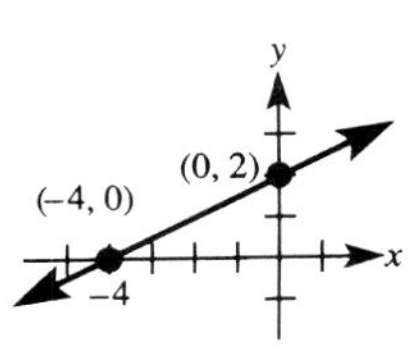

8. Subtract: $\dfrac{3x + 1}{2x - 5} - \dfrac{x}{5 - 2x}$. $\frac{4x + 1}{2x - 5}$
9. Expand $(x + 1)^3$. $x^3 + 3x^2 + 3x + 1$
10. Solve $m = \dfrac{a + b}{2}$ for a. $a = 2m - b$

7.3 Addition and Subtraction of Radicals

If a beam emerging from a laser travels as shown in the diagram, find the distance traveled by this beam in simplest radical form. (See Example 3.)

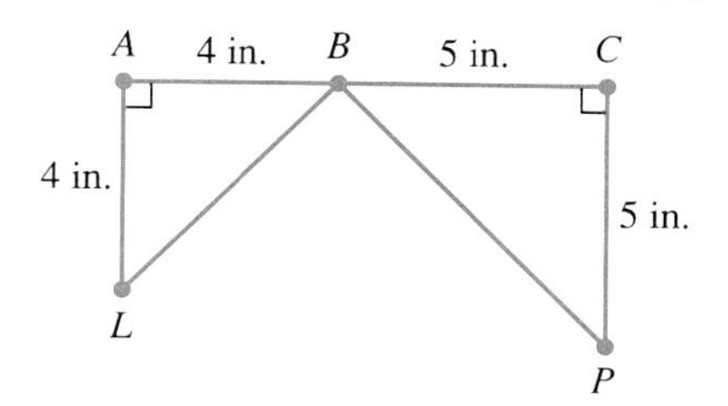

OBJECTIVES

1. **Add and subtract radicals by using the distributive property.**
2. **Add and subtract radicals that involve rationalizing denominators.**

1 We have seen that to add or subtract algebraic expressions, we combine like terms using the distributive property. For instance,

$$5x + 3x - x = (5 + 3 - 1)x = 7x.$$

Similarly, if x is replaced by $\sqrt{3}$,

$$5\sqrt{3} + 3\sqrt{3} - \sqrt{3} = (5 + 3 - 1)\sqrt{3} = 7\sqrt{3},$$

or if x is replaced by $\sqrt[3]{2x}$,

$$5\sqrt[3]{2x} + 3\sqrt[3]{2x} - \sqrt[3]{2x} = (5 + 3 - 1)\sqrt[3]{2x} = 7\sqrt[3]{2x}.$$

Thus, to add or subtract radicals, we combine like radicals using the distributive property. By definition, **like radicals** are radicals that have the same radicand and the same index. Note that only like radicals may be combined.

EXAMPLE 1 Simplify where possible. Assume $x > 0$, $y > 0$.

a. $\sqrt{11} + 14\sqrt{11}$ **b.** $\sqrt{7} + \sqrt{3}$ **c.** $4\sqrt[3]{5} - 6\sqrt[3]{5}$
d. $x\sqrt{xy} + 9\sqrt{xy}$ **e.** $\sqrt{3} + \sqrt[3]{3}$

Solution Use the distribution property (where possible) and simplify.

a. $\sqrt{11} + 14\sqrt{11} = (1 + 14)\sqrt{11} = 15\sqrt{11}$
b. $\sqrt{7}$ and $\sqrt{3}$ are not like radicals, because the radicands are different, and $\sqrt{7} + \sqrt{3}$ does not simplify.
c. $4\sqrt[3]{5} - 6\sqrt[3]{5} = (4 - 6)\sqrt[3]{5} = -2\sqrt[3]{5}$
d. $x\sqrt{xy} + 9\sqrt{xy} = (x + 9)\sqrt{xy}$
e. $\sqrt{3}$ and $\sqrt[3]{3}$ are not like radicals, because the indexes are different, and $\sqrt{3} + \sqrt[3]{3}$ does not simplify.

Caution Although $\sqrt[n]{a} \cdot \sqrt[n]{b} = \sqrt[n]{ab}$ and $\sqrt[n]{a}/\sqrt[n]{b} = \sqrt[n]{a/b}$ are properties of radicals, note that

$$\sqrt[n]{a} + \sqrt[n]{b} \quad \text{does not equal} \quad \sqrt[n]{a + b},$$

except for certain instances. For example,

$$\sqrt{16} + \sqrt{9} = 4 + 3 = 7, \quad \text{while} \quad \sqrt{16 + 9} = \sqrt{25} = 5,$$

so $\sqrt{16} + \sqrt{9} \neq \sqrt{16 + 9}$.

Since this is a common mistake, it pays to give several illustrations.

PROGRESS CHECK 1 Simplify where possible. Assume $x > 0$, $y > 0$.

a. $7\sqrt{5} - 19\sqrt{5}$ **b.** $8\sqrt[4]{2} + \sqrt[4]{2}$ **c.** $\sqrt[3]{x} + \sqrt[4]{x}$
d. $x\sqrt{y} + x\sqrt{y}$ **e.** $y\sqrt{x} + x\sqrt{y}$

Sometimes, simplifying radicals in a sum or difference results in like radicals, which can then be combined.

EXAMPLE 2 Simplify where possible. Assume $x > 0$, $y > 0$.

a. $\sqrt{40} + \sqrt{90}$ **b.** $\sqrt[3]{-16} + \sqrt[3]{250}$ **c.** $2\sqrt{27x^2y} - 5\sqrt{12x^2y}$

Solution

a. First, simplify each square root.

$$\sqrt{40} = \sqrt{4 \cdot 10} = \sqrt{4} \cdot \sqrt{10} = 2\sqrt{10}$$
$$\sqrt{90} = \sqrt{9 \cdot 10} = \sqrt{9} \cdot \sqrt{10} = 3\sqrt{10}$$

Then, $\sqrt{40} + \sqrt{90} = 2\sqrt{10} + 3\sqrt{10} = (2 + 3)\sqrt{10} = 5\sqrt{10}$.

b. First, simplify each cube root.

$$\sqrt[3]{-16} = \sqrt[3]{-8 \cdot 2} = \sqrt[3]{-8} \cdot \sqrt[3]{2} = -2\sqrt[3]{2}$$
$$\sqrt[3]{250} = \sqrt[3]{125 \cdot 2} = \sqrt[3]{125} \cdot \sqrt[3]{2} = 5\sqrt[3]{2}$$

Then, $\sqrt[3]{-16} + \sqrt[3]{250} = -2\sqrt[3]{2} + 5\sqrt[3]{2} = (-2 + 5)\sqrt[3]{2} = 3\sqrt[3]{2}$.

c. Simplify each radical, and then combine like radicals.

$$\begin{aligned} 2\sqrt{27x^2y} - 5\sqrt{12x^2y} &= 2\sqrt{9x^2} \cdot \sqrt{3y} - 5\sqrt{4x^2} \cdot \sqrt{3y} \\ &= 2 \cdot 3x\sqrt{3y} - 5 \cdot 2x\sqrt{3y} \\ &= 6x\sqrt{3y} - 10x\sqrt{3y} \\ &= (6x - 10x)\sqrt{3y} \\ &= -4x\sqrt{3y} \end{aligned}$$

PROGRESS CHECK 2 Simplify where possible. Assume $x > 0$, $y > 0$.

a. $\sqrt{80} + \sqrt{45}$ **b.** $\sqrt[3]{-54} - \sqrt[3]{128}$ **c.** $y\sqrt{63x^2y} + x\sqrt{28y^3}$

EXAMPLE 3 Solve the problem in the section introduction on page 333.

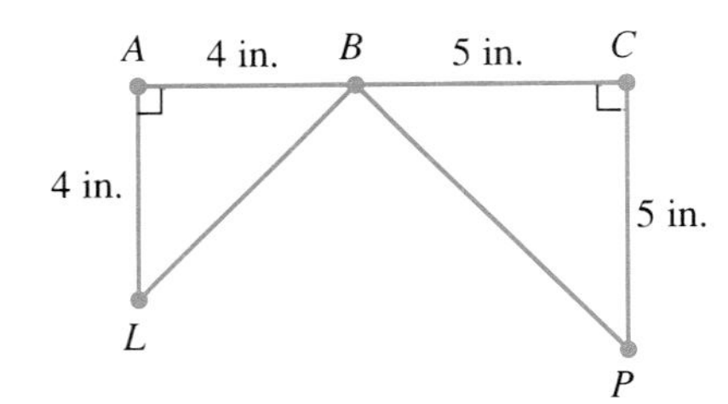

Figure 7.1

Solution Consider the sketch of the problem in Figure 7.1. Using the Pythagorean relation in right triangles LAB and PCB gives

$$\begin{aligned} (LB)^2 &= 4^2 + 4^2 \\ (LB)^2 &= 32 \\ LB &= \sqrt{32} \end{aligned} \quad \text{and} \quad \begin{aligned} (PB)^2 &= 5^2 + 5^2. \\ (PB)^2 &= 50 \\ PB &= \sqrt{50} \end{aligned}$$

The total distance $LB + PB$ is then

$$\begin{aligned} \sqrt{32} + \sqrt{50} &= \sqrt{16}\sqrt{2} + \sqrt{25}\sqrt{2} \\ &= 4\sqrt{2} + 5\sqrt{2} \\ &= 9\sqrt{2}. \end{aligned}$$

Thus, the beam from the laser travels $9\sqrt{2}$ in. Check by calculator that this distance is about 12.7 in.

PROGRESS CHECK 3 Redo Example 3 but assume that $LA = 10$, $AB = 5$, $CB = 4$, and $PC = 8$ cm.

2 Rationalizing denominators may also simplify radical expressions to a form that is more useful in addition and subtraction problems, as shown next.

Progress Check Answers

1. (a) $-12\sqrt{5}$ (b) $9\sqrt[4]{2}$ (c) Does not simplify (d) $2x\sqrt{y}$ (e) Does not simplify
2. (a) $7\sqrt{5}$ (b) $-7\sqrt[3]{2}$ (c) $5xy\sqrt{7y}$
3. $9\sqrt{5}$ cm $\approx$ 20.12 cm

EXAMPLE 4 Simplify where possible. Assume $x > 0$, $y > 0$.

a. $\sqrt{56} + \frac{1}{2}\sqrt{\frac{2}{7}}$ **b.** $\sqrt{3} + \frac{2}{\sqrt{3}}$ **c.** $2\sqrt{\frac{y}{x}} - 3\sqrt{\frac{x}{y}}$

Solution

a. First, simplify.

$$\sqrt{56} = \sqrt{4 \cdot 14} = \sqrt{4}\sqrt{14} = 2\sqrt{14}$$

$$\frac{1}{2}\sqrt{\frac{2}{7}} = \frac{1}{2}\sqrt{\frac{2 \cdot 7}{7 \cdot 7}} = \frac{1}{2}\,\frac{\sqrt{14}}{\sqrt{49}} = \frac{1}{2}\,\frac{\sqrt{14}}{7} = \frac{1}{14}\sqrt{14}$$

Then,

$$\sqrt{56} + \frac{1}{2}\sqrt{\frac{2}{7}} = 2\sqrt{14} + \frac{1}{14}\sqrt{14} = \left(2 + \frac{1}{14}\right)\sqrt{14} = \frac{29}{14}\sqrt{14}.$$

b. Rationalizing the denominator in $2/\sqrt{3}$ gives

$$\frac{2}{\sqrt{3}} = \frac{2}{\sqrt{3}} \cdot \frac{\sqrt{3}}{\sqrt{3}} = \frac{2\sqrt{3}}{3}.$$

Then,

$$\sqrt{3} + \frac{2}{\sqrt{3}} = \sqrt{3} + \frac{2\sqrt{3}}{3} = \left(1 + \frac{2}{3}\right)\sqrt{3} = \frac{5}{3}\sqrt{3}.$$

c. Simplify each radical, and then combine like radicals.

$$\begin{aligned} 2\sqrt{\frac{y}{x}} - 3\sqrt{\frac{x}{y}} &= 2\sqrt{\frac{y}{x} \cdot \frac{x}{x}} - 3\sqrt{\frac{x}{y} \cdot \frac{y}{y}} \\ &= 2\sqrt{\frac{xy}{x^2}} - 3\sqrt{\frac{xy}{y^2}} \\ &= 2\frac{\sqrt{xy}}{x} - 3\frac{\sqrt{xy}}{y} \\ &= \left(\frac{2}{x} - \frac{3}{y}\right)\sqrt{xy} \\ &= \frac{2y - 3x}{xy}\sqrt{xy} \end{aligned}$$

PROGRESS CHECK 4 Simplify where possible. Assume $x > 0$, $y > 0$.

a. $\sqrt{75} - \frac{1}{5}\sqrt{\frac{1}{3}}$ **b.** $\sqrt{2} + \frac{1}{\sqrt{2}}$ **c.** $\sqrt{\frac{2x}{y}} - \sqrt{\frac{2y}{x}}$

Progress Check Answers

4. (a) $\frac{74\sqrt{3}}{15}$ (b) $\frac{3\sqrt{2}}{2}$ (c) $\frac{x - y}{xy}\sqrt{2xy}$

EXERCISES 7.3

In Exercises 1–32, simplify where possible. Assume $x > 0$, $y > 0$.

1. $2\sqrt{5} + 4\sqrt{5}$ $6\sqrt{5}$
2. $3\sqrt{7} + 8\sqrt{7}$ $11\sqrt{7}$
3. $6\sqrt[3]{17} + 3\sqrt[3]{17}$ $9\sqrt[3]{17}$
4. $7\sqrt[3]{21} + 3\sqrt[3]{21}$ $10\sqrt[3]{21}$
5. $\sqrt{6} + 8\sqrt{6}$ $9\sqrt{6}$
6. $\sqrt{15} + 7\sqrt{15}$ $8\sqrt{15}$
7. $9\sqrt{13} + \sqrt{13}$ $10\sqrt{13}$
8. $12\sqrt{2} + \sqrt{2}$ $13\sqrt{2}$
9. $6\sqrt{6} - 3\sqrt{6}$ $3\sqrt{6}$
10. $10\sqrt{19} - 4\sqrt{19}$ $6\sqrt{19}$
11. $12\sqrt{10} - 20\sqrt{10}$ $-8\sqrt{10}$
12. $5\sqrt{23} - 72\sqrt{23}$ $-67\sqrt{23}$
13. $\sqrt[4]{2} - 9\sqrt[4]{2}$ $-8\sqrt[4]{2}$
14. $12\sqrt[4]{6} - 18\sqrt[4]{6}$ $-6\sqrt[4]{6}$
15. $3\sqrt{2} + 2\sqrt{3}$ Does not simplify.
16. $5\sqrt{8} + 8\sqrt{5}$ Does not simplify.
17. $\sqrt{14} + \sqrt{15}$ Does not simplify.
18. $\sqrt{11} + \sqrt{17}$ Does not simplify.
19. $\sqrt{14} + \sqrt[3]{14}$ Does not simplify.
20. $\sqrt[4]{125} + \sqrt[3]{125}$ $\sqrt[4]{125} + 5$
21. $4\sqrt{x} + y\sqrt{x}$ $(4 + y)\sqrt{x}$
22. $x\sqrt{3} + 7\sqrt{3}$ $(x + 7)\sqrt{3}$
23. $15\sqrt{13} - x\sqrt{13}$ $(15 - x)\sqrt{13}$
24. $20\sqrt{10} - y\sqrt{10}$ $(20 - y)\sqrt{10}$
25. $\sqrt[4]{7} + 5\sqrt[4]{7}$ $6\sqrt[4]{7}$
26. $\sqrt[3]{12} + 7\sqrt[3]{12}$ $8\sqrt[3]{12}$
27. $5\sqrt{x} + 5\sqrt{x}$ $10\sqrt{x}$
28. $6\sqrt{y} + 8\sqrt{y}$ $14\sqrt{y}$
29. $2y\sqrt{x} + y\sqrt{x}$ $3y\sqrt{x}$
30. $23x\sqrt{y} + x\sqrt{y}$ $24x\sqrt{y}$
31. $\sqrt[4]{3xy} + \sqrt[4]{3xy}$ $2\sqrt[4]{3xy}$
32. $\sqrt{2y} - \sqrt{2y}$ 0

In Exercises 33–54, simplify where possible. Assume $x > 0$, $y > 0$.

33. $3\sqrt{2} + \sqrt{8}$ $5\sqrt{2}$
34. $\sqrt{128} + 4\sqrt{2}$ $12\sqrt{2}$
35. $\sqrt{75} + \sqrt{12}$ $7\sqrt{3}$
36. $\sqrt{18} + \sqrt{50}$ $8\sqrt{2}$
37. $\sqrt{32} + \sqrt{98}$ $11\sqrt{2}$
38. $\sqrt{500} + \sqrt{180}$ $16\sqrt{5}$
39. $\sqrt{48} - \sqrt{27}$ $\sqrt{3}$
40. $-\sqrt{108} + \sqrt{192}$ $2\sqrt{3}$
41. $-3\sqrt[3]{3} + \sqrt[3]{81}$ 0
42. $-\sqrt[3]{3} + \sqrt[3]{24}$ $\sqrt[3]{3}$
43. $\sqrt[3]{-192} + 3\sqrt[3]{81}$ $5\sqrt[3]{3}$
44. $\sqrt[3]{-375} + \sqrt[3]{81}$ $-2\sqrt[3]{3}$
45. $-\sqrt[3]{54} - 2\sqrt[3]{16}$ $-7\sqrt[3]{2}$
46. $-\sqrt[3]{128} - \sqrt[3]{432}$ $-10\sqrt[3]{2}$
47. $\sqrt{24xy^2} + y\sqrt{54x}$ $5y\sqrt{6x}$
48. $\sqrt{98xy^2} + \sqrt{50xy^2}$ $12y\sqrt{2x}$
49. $-\sqrt{49x^3y} + \sqrt{16x^3y}$ $-3x\sqrt{xy}$
50. $-y\sqrt{75x} + \sqrt{12y^2x}$ $-3y\sqrt{3x}$
51. $y\sqrt{27x^3} + x\sqrt{48y^2x}$ $7xy\sqrt{3x}$
52. $\sqrt{40x^2y^3} + xy\sqrt{250y}$ $7xy\sqrt{10y}$
53. $\sqrt{18x} - \sqrt{72xy^2}$ $(3 - 6y)\sqrt{2x}$
54. $\sqrt{180y} - \sqrt{80x^2y}$ $(6 - 4x)\sqrt{5y}$

In Exercises 55–60, give exact answers using radicals in simplest form.

55. If a beam emerging from a laser travels as shown in the diagram, find the distance traveled by the beam in simplified radical form. $3\sqrt{17}$

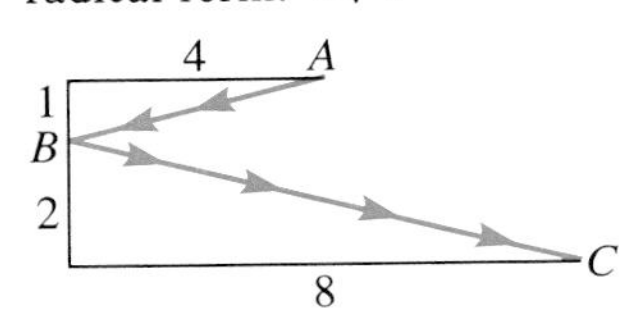

56. If a beam emerging from a laser travels as shown in the diagram, find the distance traveled by the beam in simplified radical form. $5\sqrt{10}$

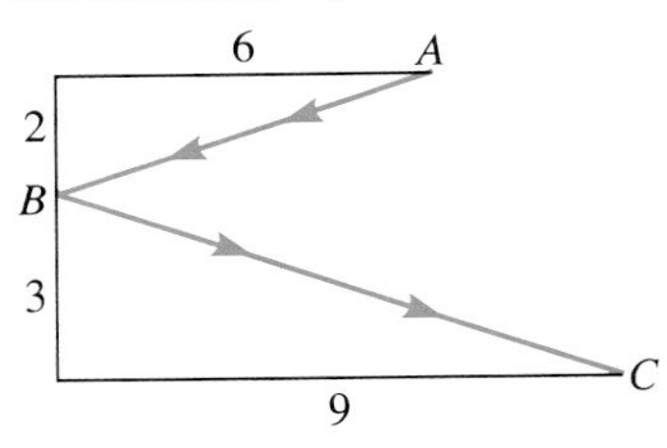

57. Square $ABCD$ is inscribed in a larger square as shown.
a. Find the perimeter of $ABCD$. $4\sqrt{10}$
b. Find the area of $ABCD$. 10

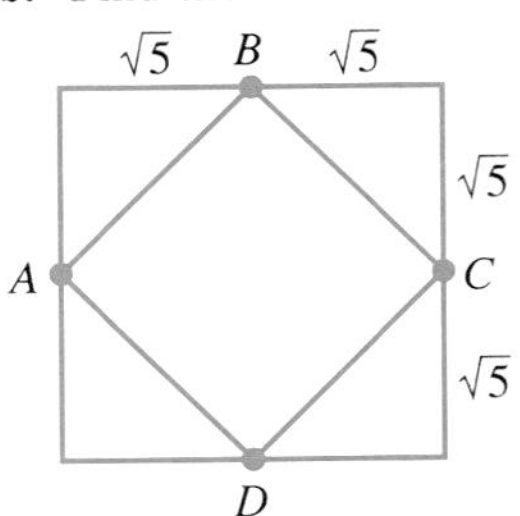

58. Square $ABCD$ is inscribed in a larger square as shown. The side of the large square is 4.
a. Find the perimeter of $ABCD$. $8\sqrt{2}$
b. Find the area of $ABCD$. 8

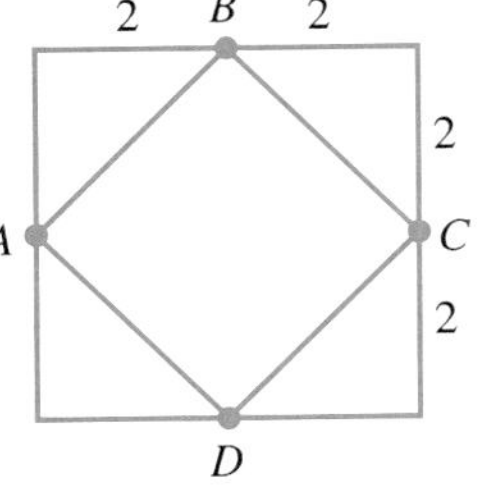

59. Use the triangle pictured.
a. Find the hypotenuse of the small triangle. $\sqrt{5}$
b. Find the hypotenuse of the large triangle. $2\sqrt{5}$
c. Find the length of the piece marked x. $\sqrt{5}$

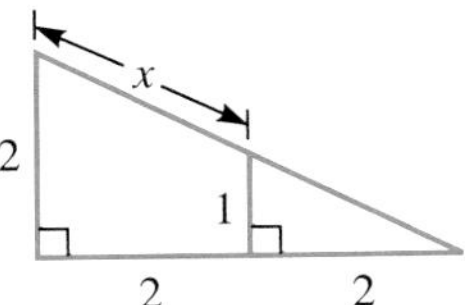

60. Use the triangle pictured.
a. Find the hypotenuse of the small triangle. $\sqrt{5}$
b. Find the hypotenuse of the large triangle. $\frac{3\sqrt{5}}{2}$
c. Find the length of the piece marked x. $\frac{\sqrt{5}}{2}$

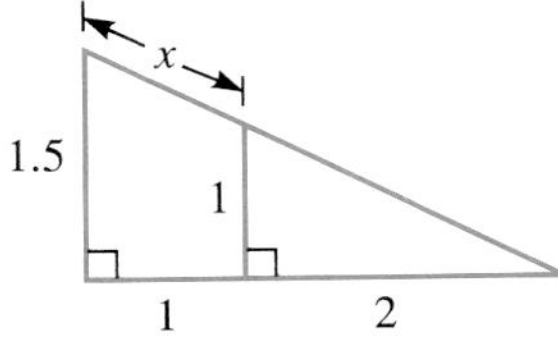

In Exercises 61–80, simplify when possible. Assume $x > 0$, $y > 0$.

61. $4\sqrt{50} + 6\sqrt{\frac{1}{2}}$ $23\sqrt{2}$
62. $\frac{1}{2}\sqrt{24} - \frac{12}{\sqrt{6}}$ $-\sqrt{6}$
63. $\frac{2}{3}\sqrt{6} - \frac{8}{\sqrt{6}}$ $-\frac{2}{3}\sqrt{6}$
64. $5\sqrt{54} - \frac{6}{\sqrt{6}}$ $14\sqrt{6}$
65. $\frac{\sqrt{32}}{12} + \frac{2}{\sqrt{8}}$ $\frac{5}{6}\sqrt{2}$
66. $\frac{3}{\sqrt{5}} + 2\sqrt{45}$ $\frac{33}{5}\sqrt{5}$
67. $\sqrt{12} + \frac{1}{\sqrt{3}}$ $\frac{7}{3}\sqrt{3}$
68. $\frac{3}{2}\sqrt{80} + \frac{1}{\sqrt{5}}$ $\frac{31}{5}\sqrt{5}$
69. $\sqrt{\frac{5}{8}} + \sqrt{\frac{5}{2}}$ $\frac{3}{4}\sqrt{10}$
70. $\sqrt{\frac{4}{3}} + \sqrt{\frac{50}{6}}$ $\frac{7}{3}\sqrt{3}$
71. $\sqrt{\frac{2}{3}} - 3\sqrt{\frac{3}{2}}$ $-\frac{7}{6}\sqrt{6}$
72. $\sqrt{\frac{3}{4}} - 3\sqrt{\frac{1}{3}}$ $-\frac{1}{2}\sqrt{3}$
73. $-\frac{4}{\sqrt{3}} + \frac{2\sqrt{3}}{5}$ $-\frac{14}{15}\sqrt{3}$
74. $4\sqrt{\frac{2}{3}} - \frac{3\sqrt{3}}{2\sqrt{2}}$ $\frac{7}{12}\sqrt{6}$
75. $3\sqrt{\frac{1}{6}} - \frac{\sqrt{150}}{2}$ $-2\sqrt{6}$
76. $\frac{2}{3}\sqrt{108} - 4\sqrt{\frac{3}{4}}$ $2\sqrt{3}$
77. $x\sqrt{48x} - 12x\sqrt{\frac{x}{3}}$ 0
78. $\sqrt{50y} - \sqrt{\frac{y}{2}}$ $\frac{9}{2}\sqrt{2y}$
79. $3\sqrt{\frac{x}{y}} + 4\sqrt{\frac{y}{x}}$ $\left(\frac{3x + 4y}{xy}\right)\sqrt{xy}$
80. $\sqrt{\frac{3x}{y}} - \sqrt{\frac{3y}{x}}$ $\left(\frac{x - y}{xy}\right)\sqrt{3xy}$

THINK ABOUT IT

1. This problem demonstrates that visual patterns are not always helpful when adding and subtracting radicals.
   a. Is it true that $1 + 49 = 25 + 25$? Will the equality still be correct if you just write a square root symbol in front of each term? In other words, is it true that $\sqrt{1} + \sqrt{49} = \sqrt{25} + \sqrt{25}$?
   b. Is it true that $512 + 216 + 1 = 729$? Will the equality still be correct if you just write a cube root symbol in front of each term? In other words, is it true that $\sqrt[3]{512} + \sqrt[3]{216} + \sqrt[3]{1} = \sqrt[3]{729}$?
   c. Is it true that $75 + 12 = 48 + 27$? Is it true that $\sqrt{75} + \sqrt{12} = \sqrt{48} + \sqrt{27}$?
   d. Given that $a + b = c + d$, which of these statements must also be correct?
   i. $\sqrt[n]{a + b} = \sqrt[n]{c + d}$ **ii.** $\sqrt[n]{a} + \sqrt[n]{b} = \sqrt[n]{c} + \sqrt[n]{d}$
2. **a.** Using the figure, find the perimeter and area of the inscribed square.
   b. Find the area of the larger square, and then explain (using the result from part **a**) why $(a + b)^2$ does not equal $a^2 + b^2$.

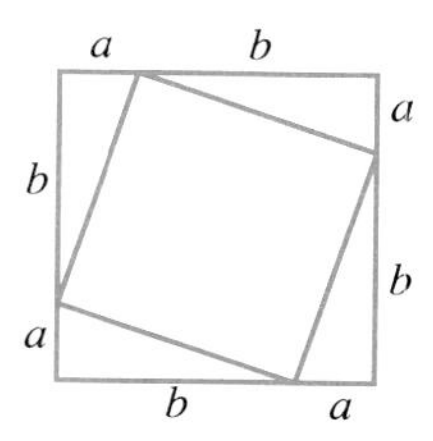

3. Write as a single fraction, and rationalize the denominator.
   a. $1 + \dfrac{1}{\sqrt{x}}$ **b.** $1 + \dfrac{1}{\sqrt[3]{x}}$ **c.** $1 + \dfrac{1}{\sqrt[n]{x}}$
4. Assume that n represents a positive integer. Explain why this statement is true for some values of n but not for others: $\sqrt[n]{-5} = -\sqrt[n]{5}$.
5. If you allow factoring "over the real numbers" instead of just "over the integers," then you can factor $\sqrt{3}$ from both terms of $\sqrt{3} + \sqrt{6}$.
   a. What will the factored expression look like?
   b. Factor $\sqrt[3]{5}$ from both terms of $2\sqrt[3]{5} + \sqrt[3]{15}$.

REMEMBER THIS

1. Simplify $(\sqrt[3]{7})(\sqrt[3]{7})(\sqrt[3]{7})$. 7
2. Which property of real numbers is illustrated by $a(x + 1) = ax + a$? Distributive
3. Factor $a^2 - b^2$. $(a + b)(a - b)$
4. Subtract $\sqrt{7} - 1$ from $\sqrt{7} + 1$. 2
5. Use FOIL to multiply $(\sqrt{3} + x)(\sqrt{3} - x)$. $3 - x^2$
6. Find the point of intersection of the lines determined by $y = 3x + 2$ and $y = 2x + 3$. (1,5)
7. Which is steeper, a line with slope -5 or one with slope $\frac{1}{5}$? Slope -5
8. Find the distance between $(1,-1)$ and $(-3,3)$. $4\sqrt{2}$
9. Divide: $\dfrac{2 - y}{x^2 - x} \div \dfrac{y^2 - 4}{x}$. $\dfrac{-1}{(x - 1)(y + 2)}$
10. Evaluate $E = \dfrac{Pr}{1 - (1 + r)^{-n}}$ if $r = 0.01$, $P = 40{,}000$, and $n = 180$. This gives the monthly payment for a \$40,000 loan at 12 percent annual interest over 15 years. \$480.07

7.4 Further Radical Simplification

The design for a rectangular solar panel specifies that the area is to be 6 m² and the length is to be 2 m greater than the width. Verify that a rectangular panel that measures $\sqrt{7} + 1$ m by $\sqrt{7} - 1$ m satisfies these specifications. (See Example 4.)

OBJECTIVES

1. **Use $(\sqrt[n]{a})^n = a$.**
2. **Multiply radical expressions involving more than one term.**
3. **Use a special product formula to multiply certain radical expressions.**
4. **Divide radical expressions involving more than one term in the numerator and a single term in the denominator.**
5. **Use a conjugate to rationalize a denominator.**

1 A property that is used often when simplifying radicals stems from the definition of $\sqrt[n]{a}$ and states

$$(\sqrt[n]{a})^n = a$$

if $\sqrt[n]{a}$ is a real number. For instance, $(\sqrt[3]{8})^3 = 8$, and this result checks, since $\sqrt[3]{8} = 2$ and $2^3 = 8$. Example 1 shows some simplifications that use this property.

EXAMPLE 1 Simplify each radical.

a. $(7\sqrt{5})^2$ **b.** $\sqrt{41} \cdot \sqrt{41}$ **c.** $(\sqrt[3]{x^2 - 5})^3$ **d.** $(4\sqrt{3x})^2, x \geq 0$

Solution

a. $(7\sqrt{5})^2 = 7^2(\sqrt{5})^2 = 49 \cdot 5 = 245$
b. $\sqrt{41} \cdot \sqrt{41} = (\sqrt{41})^2 = 41$
c. $(\sqrt[3]{x^2 - 5})^3 = x^2 - 5$
d. $(4\sqrt{3x})^2 = 4^2(\sqrt{3x})^2 = 16 \cdot 3x = 48x$

PROGRESS CHECK 1 Simplify each radical.

a. $(-3\sqrt{2})^2$ **b.** $(\sqrt[3]{4x - 1})^3$ **c.** $2\sqrt{19} \cdot \sqrt{19}$ **d.** $(6\sqrt{10y})^2, y \geq 0$

2 When a product involves a radical expression with more than one term, then we apply the distributive property, as shown in Example 2.

EXAMPLE 2 Simplify each expression.

a. $\sqrt{3}(\sqrt{6} - \sqrt{3})$ **b.** $(\sqrt{7} + \sqrt{2})(\sqrt{7} - \sqrt{2})$ **c.** $(\sqrt[3]{x} + 2)(\sqrt[3]{x} + 5)$

Solution

a. First, multiply.

$$\begin{aligned}\sqrt{3}(\sqrt{6} - \sqrt{3}) &= \sqrt{3} \cdot \sqrt{6} - \sqrt{3} \cdot \sqrt{3} \quad \text{Distributive property.}\\ &= \sqrt{18} - 3\end{aligned}$$

Now simplify. Since $\sqrt{18} = \sqrt{9} \cdot \sqrt{2} = 3\sqrt{2}$,

$$\sqrt{3}(\sqrt{6} - \sqrt{3}) = 3\sqrt{2} - 3.$$

Progress Check Answers

1. (a) 18 (b) $4x - 1$ (c) 38 (d) $360y$

b. Based on the distributive property, multiply each term of $\sqrt{7} + \sqrt{2}$ by each term of $\sqrt{7} - \sqrt{2}$, and then simplify.

$$\begin{aligned}(\sqrt{7} + \sqrt{2})(\sqrt{7} - \sqrt{2}) &= \sqrt{7}\cdot\sqrt{7} + \sqrt{7}(-\sqrt{2}) + \sqrt{2}\cdot\sqrt{7} + \sqrt{2}(-\sqrt{2})\\ &= 7 - \sqrt{14} + \sqrt{14} - 2\\ &= 5\end{aligned}$$

c. In this multiplication, note that $5\sqrt[3]{x}$ and $2\sqrt[3]{x}$ are like radicals that should be combined.

$$\begin{aligned}(\sqrt[3]{x} + 2)(\sqrt[3]{x} + 5) &= \sqrt[3]{x}\sqrt[3]{x} + \sqrt[3]{x}\cdot 5 + 2\cdot\sqrt[3]{x} + 2\cdot 5\\ &= \sqrt[3]{x^2} + 7\sqrt[3]{x} + 10\end{aligned}$$

PROGRESS CHECK 2 Simplify each expression.

a. $\sqrt{5}(\sqrt{10} - \sqrt{5})$ **b.** $(\sqrt{10} - \sqrt{3})(\sqrt{10} + \sqrt{3})$ **c.** $(\sqrt[5]{y} + 3)(\sqrt[5]{y} + 1)$

3 The special product formulas from Section 3.4 are often used to simplify certain products involving square roots. For instance, the product in Example 2b may be found more easily by using

$$(a + b)(a - b) = a^2 - b^2,$$

with a replaced by $\sqrt{7}$ and b replaced by $\sqrt{2}$, to obtain

$$\begin{aligned}(\sqrt{7} + \sqrt{2})(\sqrt{7} - \sqrt{2}) &= (\sqrt{7})^2 - (\sqrt{2})^2\\ &= 7 - 2 = 5.\end{aligned}$$

Other useful special product formulas from Section 3.4 are

$$\begin{aligned}(a + b)^2 &= a^2 + 2ab + b^2\\ (a - b)^2 &= a^2 - 2ab + b^2.\end{aligned}$$

EXAMPLE 3 Simplify each expression. Assume $x \geq 0$.

a. $(3\sqrt{x} - 4)^2$ **b.** $(5 + 2\sqrt{3})(5 - 2\sqrt{3})$

Solution

a. Replace a with $3\sqrt{x}$ and b with 4 in the formula for $(a - b)^2$.

$$\begin{aligned}(a - b)^2 &= a^2 - 2ab + b^2\\ (3\sqrt{x} - 4)^2 &= (3\sqrt{x})^2 - 2(3\sqrt{x})(4) + (4)^2\\ &= 9x - 24\sqrt{x} + 16\end{aligned}$$

b. Use $(a + b)(a - b) = a^2 - b^2$ with $a = 5$ and $b = 2\sqrt{3}$.

$$\begin{aligned}(5 + 2\sqrt{3})(5 - 2\sqrt{3}) &= (5)^2 - (2\sqrt{3})^2\\ &= 25 - 12\\ &= 13\end{aligned}$$

PROGRESS CHECK 3 Simplify each expression. Assume $x \geq 0$.

a. $(2\sqrt{3} + 5)^2$ **b.** $(4 + \sqrt{x})(4 - \sqrt{x})$

EXAMPLE 4 Solve the problem in the section introduction on page 338.

Solution If the length is $\sqrt{7} + 1$ m and the width is $\sqrt{7} - 1$ m, then the length is 2 m greater than the width, since

$$(\sqrt{7} + 1) - (\sqrt{7} - 1) = 2.$$

We verify the area is 6 m² for the rectangular solar panel next.

$$A = \ell w = (\sqrt{7} + 1)(\sqrt{7} - 1) = (\sqrt{7})^2 - (1)^2 = 7 - 1 = 6$$

Thus, a panel with the given measurement satisfies both specifications.

Progress Check Answers

2. (a) $5\sqrt{2} - 5$ (b) 7 (c) $\sqrt[5]{y^2} + 4\sqrt[5]{y} + 3$
3. (a) $37 + 20\sqrt{3}$ (b) $16 - x$

PROGRESS CHECK 4 The design for a rectangular computer chip specifies that the area is to be 18 mm^2 and the length is to be 4 mm greater than the width. Verify that a rectangular computer chip that measures $\sqrt{22} + 2$ mm by $\sqrt{22} - 2$ mm satisfies these specifications.

4 Methods for simplifying quotients involving radical expressions that have more than one term in the numerator and a single term in the denominator are shown in the next example.

EXAMPLE 5 Simplify each expression. Assume $r > 0$.

a. $\dfrac{3 - 9\sqrt{2}}{6}$ **b.** $\dfrac{2 + \sqrt{r}}{\sqrt{r}}$

Solution

a. Factor out the common factor 3 in the numerator, and then simplify.

$$\frac{3 - 9\sqrt{2}}{6} = \frac{3(1 - 3\sqrt{2})}{6} = \frac{1 - 3\sqrt{2}}{2}$$

b. Since $\sqrt{r} \cdot \sqrt{r} = r$ if $r > 0$, multiply by $\sqrt{r}/\sqrt{r}$ to rationalize the denominator.

$$\frac{2 + \sqrt{r}}{\sqrt{r}} = \frac{2 + \sqrt{r}}{\sqrt{r}} \cdot \frac{\sqrt{r}}{\sqrt{r}} = \frac{2\sqrt{r} + r}{r}$$

PROGRESS CHECK 5 Simplify each expression. Assume $x > 0$.

a. $\dfrac{8 + 4\sqrt{3}}{12}$ **b.** $\dfrac{\sqrt{x} - 3}{\sqrt{x}}$

5 To eliminate radicals in denominators that contain square roots and two terms, consider that for nonnegative a and b

$$(\sqrt{a} + \sqrt{b})(\sqrt{a} - \sqrt{b}) = (\sqrt{a})^2 - (\sqrt{b})^2 = a - b.$$

In general, the sum and the difference of the same two terms are called **conjugates** of each other. Binomial denominators involving square roots are rationalized by multiplying the numerator and the denominator by the conjugate of the denominator.

EXAMPLE 6 Rationalize the denominator of $\dfrac{8}{3 - \sqrt{5}}$.

Solution The conjugate of $3 - \sqrt{5}$ is $3 + \sqrt{5}$. Then,

$$\frac{8}{3 - \sqrt{5}} = \frac{8}{3 - \sqrt{5}} \cdot \frac{3 + \sqrt{5}}{3 + \sqrt{5}}$$ Multiply using the conjugate of the denominator.

$$= \frac{8(3 + \sqrt{5})}{(3)^2 - (\sqrt{5})^2}$$ Multiply fractions, and use $(a + b)(a - b) = a^2 - b^2$.

$$= \frac{8(3 + \sqrt{5})}{4}$$ Simplify the denominator.

$$= 2(3 + \sqrt{5})$$ Express in lowest terms.

$$= 6 + 2\sqrt{5}.$$ Remove parentheses.

Progress Check Answers

4. $(\sqrt{22} + 2)(\sqrt{22} - 2) = 18$; $(\sqrt{22} + 2) - (\sqrt{22} - 2) = 4$

5. (a) $\dfrac{2 + \sqrt{3}}{3}$ (b) $\dfrac{x - 3\sqrt{x}}{x}$

Note To express the result in lowest terms, it is recommended that fractions be simplified where possible before removing parentheses in the numerator, as was done in this example.

PROGRESS CHECK 6 Rationalize the denominator of $\frac{10}{3+\sqrt{7}}$.

EXAMPLE 7 Rationalize the denominator of $\frac{x\sqrt{2x}}{x+\sqrt{2x}}$. Assume $x > 0$.

Solution Multiply the numerator and the denominator by $x - \sqrt{2x}$, which is the conjugate of the denominator.

$$\frac{x\sqrt{2x}}{x+\sqrt{2x}} = \frac{x\sqrt{2x}}{x+\sqrt{2x}} \cdot \frac{x-\sqrt{2x}}{x-\sqrt{2x}}$$ Multiply using the conjugate of the denominator.

$$= \frac{x\sqrt{2x}(x-\sqrt{2x})}{(x)^2-(\sqrt{2x})^2}$$ Multiply fractions, and use $(a+b)(a-b)=a^2-b^2$.

$$= \frac{x\sqrt{2x}(x-\sqrt{2x})}{x^2-2x}$$ Simplify the denominator.

$$= \frac{x\sqrt{2x}(x-\sqrt{2x})}{x(x-2)}$$ Factor the denominator.

$$= \frac{\sqrt{2x}(x-\sqrt{2x})}{x-2}$$ Divide out the common factor x.

$$= \frac{x\sqrt{2x}-2x}{x-2}$$ Remove parentheses.

PROGRESS CHECK 7 Rationalize the denominator of $\frac{y\sqrt{5y}}{y-\sqrt{5y}}$. Assume $y > 0$.

Progress Check Answers

6. $15 - 5\sqrt{7}$
7. $\frac{y\sqrt{5y}+5y}{y-5}$

EXERCISES 7.4

In Exercises 1–22, simplify each radical. Assume $x \geq 0$, $y \geq 0$.

1. $(3\sqrt{7})^2$ 63
2. $(5\sqrt{3})^2$ 75
3. $(8\sqrt{10})^2$ 640
4. $(2\sqrt{15})^2$ 60
5. $\sqrt{57}\cdot\sqrt{57}$ 57
6. $\sqrt{29}\cdot\sqrt{29}$ 29
7. $3\sqrt{21}\cdot\sqrt{21}$ 63
8. $8\sqrt{2}\cdot\sqrt{2}$ 16
9. $(-7\sqrt{6})^2$ 294
10. $(-10\sqrt{5})^2$ 500
11. $(-3\sqrt{17})^2$ 153
12. $(-12\sqrt{2})^2$ 288
13. $(\sqrt[3]{4+2x})^3$ $4+2x$
14. $(\sqrt[3]{x^2+8})^3$ x^2+8
15. $(\sqrt[3]{3x^2-1})^3$ $3x^2-1$
16. $(\sqrt[3]{5x-3})^3$ $5x-3$
17. $(3\sqrt{5x})^2$ $45x$
18. $(6\sqrt{7y})^2$ $252y$
19. $(7\sqrt{2y})^2$ $98y$
20. $(10\sqrt{12x})^2$ $1{,}200x$
21. $(-9\sqrt{11x})^2$ $891x$
22. $(-3\sqrt{15y})^2$ $135y$

In Exercises 23–44, simplify each expression.

23. $\sqrt{3}(\sqrt{6}+\sqrt{3})$ $3\sqrt{2}+3$
24. $\sqrt{7}(\sqrt{7}+\sqrt{14})$ $7+7\sqrt{2}$
25. $\sqrt{10}(\sqrt{20}-\sqrt{10})$ $10\sqrt{2}-10$
26. $\sqrt{11}(\sqrt{22}-\sqrt{11})$ $11\sqrt{2}-11$
27. $\sqrt{15}(\sqrt{5}-\sqrt{3})$ $5\sqrt{3}-3\sqrt{5}$
28. $\sqrt{21}(\sqrt{3}-\sqrt{7})$ $3\sqrt{7}-7\sqrt{3}$
29. $\sqrt{2}(\sqrt{3}+\sqrt{5})$ $\sqrt{6}+\sqrt{10}$
30. $\sqrt{7}(\sqrt{5}-\sqrt{11})$ $\sqrt{35}-\sqrt{77}$
31. $(\sqrt{12}+\sqrt{5})(\sqrt{12}-\sqrt{5})$ 7
32. $(\sqrt{8}+\sqrt{6})(\sqrt{8}-\sqrt{6})$ 2
33. $(\sqrt{40}-\sqrt{30})(\sqrt{40}+\sqrt{30})$ 10
34. $(\sqrt{15}-\sqrt{11})(\sqrt{15}+\sqrt{11})$ 4
35. $(\sqrt{5}+\sqrt{7})(\sqrt{5}-\sqrt{7})$ -2
36. $(\sqrt{7}+\sqrt{8})(\sqrt{7}-\sqrt{8})$ -1
37. $(\sqrt{15}-\sqrt{16})(\sqrt{15}+\sqrt{16})$ -1
38. $(\sqrt{6}-\sqrt{12})(\sqrt{6}+\sqrt{12})$ -6
39. $(\sqrt[3]{x}+4)(\sqrt[3]{x}+1)$ $\sqrt[3]{x^2}+5\sqrt[3]{x}+4$
40. $(\sqrt[3]{y}+1)(\sqrt[3]{y}+3)$ $\sqrt[3]{y^2}+4\sqrt[3]{y}+3$
41. $(\sqrt[3]{x}-1)(\sqrt[3]{x}+4)$ $\sqrt[3]{x^2}+3\sqrt[3]{x}-4$
42. $(\sqrt[5]{y}+7)(\sqrt[5]{y}-3)$ $\sqrt[5]{y^2}+4\sqrt[5]{y}-21$
43. $(\sqrt[5]{y}-6)(\sqrt[5]{y}-4)$ $\sqrt[5]{y^2}-10\sqrt[5]{y}+24$
44. $(\sqrt[5]{y}-4)(\sqrt[5]{y}-5)$ $\sqrt[5]{y^2}-9\sqrt[5]{y}+20$

In Exercises 45–62, simplify each expression. Assume $x \geq 0$.

45. $(2\sqrt{5}+1)^2$ $21+4\sqrt{5}$
46. $(7\sqrt{6}+5)^2$ $319+70\sqrt{6}$
47. $(3\sqrt{7}-2)^2$ $67-12\sqrt{7}$
48. $(10\sqrt{3}-4)^2$ $316-80\sqrt{3}$
49. $(3+4\sqrt{x})^2$ $9+24\sqrt{x}+16x$

50. $(2 + 6\sqrt{4})^2$ $4 + 24\sqrt{x} + 36x$
51. $(5\sqrt{x} - 5)^2$ $25x - 50\sqrt{x} + 25$
52. $(9\sqrt{x} - 7)^2$ $81x - 126\sqrt{x} + 49$
53. $(3 + 2\sqrt{5})(3 - 2\sqrt{5})$ -11
54. $(4 + 3\sqrt{2})(4 - 3\sqrt{2})$ -2
55. $(7\sqrt{6} - 4)(7\sqrt{6} + 4)$ 278
56. $(11\sqrt{7} - 3)(11\sqrt{7} + 3)$ 838
57. $(5 + \sqrt{x})(5 - \sqrt{x})$ $25 - x$
58. $(12 - \sqrt{x})(12 + \sqrt{x})$ $144 - x$
59. $(\sqrt{x} + 3\sqrt{5})(\sqrt{x} - 3\sqrt{5})$ $x - 45$
60. $(\sqrt{x} + 2\sqrt{7})(\sqrt{x} - 2\sqrt{7})$ $x - 28$
61. $(2\sqrt{x} - 4\sqrt{3})(2\sqrt{x} + 4\sqrt{3})$ $4x - 48$
62. $(3\sqrt{x} - 5\sqrt{2})(3\sqrt{x} + 5\sqrt{2})$ $9x - 50$

63. Competitors in a contest were asked to find the dimensions of a rectangular plate where the length is 4 in. greater than the width and the area of the plate is 4 in.^{2}. Verify that $(2\sqrt{2} + 2)$ and $(2\sqrt{2} - 2)$ are the correct dimensions.
$(2\sqrt{2} + 2) - (2\sqrt{2} - 2) = 4$; $(2\sqrt{2} + 2)(2\sqrt{2} - 2) = 4$

64. An architect requires the length and width of a rectangular shed to be such that the area of the shed floor is 12 ft^2 and the length of the shed is 6 ft longer than the width.
a. Verify that the exact values of the length and width are $(\sqrt{21} + 3)$ and $(\sqrt{21} - 3)$, respectively.
$(\sqrt{21} + 3) - (\sqrt{21} - 3) = 6$; $(\sqrt{21} + 3)(\sqrt{21} - 3) = 12$
b. Give the dimensions to the nearest hundredth of a foot.
Length = 7.58 ft, width = 1.58 ft

65. An important expression in statistics is $\sqrt{\dfrac{p(1 - p)}{n}}$, which is used in computing the margin of error for a survey of n people, where p is the proportion who agree on some opinion.
a. Assume $p = 0.5$ and simplify the formula. $\dfrac{0.5\sqrt{n}}{n}$
b. Evaluate the formula to the nearest thousandth if $p = 0.5$ and $n = 1{,}000$. 0.016

66. Repeat Exercise 65, but take $p = 0.1$. a. $\dfrac{0.3\sqrt{n}}{n}$ b. 0.009

67. The harmonic mean of two numbers a and b is given by $\dfrac{2ab}{a + b}$.
a. What is the harmonic mean of n and $\sqrt{n}$? $\dfrac{2n(\sqrt{n} - 1)}{n - 1}$
b. Compute the harmonic mean of 3 and 9 by both formulas. 4.5

68. The contraharmonic mean of a and b is given by $\dfrac{a^2 + b^2}{a + b}$.
a. What is the contraharmonic mean of n and $\sqrt{n}$?
$(n + 1)(n - \sqrt{n})/(n - 1)$
b. Compute the contraharmonic mean of 3 and 9 by both formulas. 7.5
c. Show that the sum of the harmonic and contraharmonic means of a and b is $a + b$. See Exercise 67 for harmonic mean.

69. According to Ohm's law, if three resistors of R_1, R_2, and R_3 ohms are connected in parallel, then the total resistance of the circuit is given by

$$R = \frac{R_1R_2R_3}{R_1R_2 + R_1R_3 + R_2R_3}.$$

Suppose $R_1 = c$ ohms while R_2 and R_3 equal $\sqrt{c}$ ohms. Find an expression for R in simplest radical form. $R = \dfrac{c(2\sqrt{c} - 1)}{4c - 1}$

70. Repeat Exercise 69, but assume that R_1 and R_2 equal c ohms while $R_3 = \sqrt{c}$ ohms. $R = \dfrac{c(\sqrt{c} - 2)}{c - 4}$

71. What is the volume of a cube with side length $\dfrac{1}{\sqrt{a}}$ cm?
$\sqrt{a}/a^2$ cm^3

72. What is the volume of a sphere with radius $\sqrt{3}$ ft? Give the answer in simplified radical form and also to the nearest hundredth. $4\sqrt{3}\pi$ ft^3 ≈ 21.77 ft^3

In Exercises 73–84, simplify each expression. Assume $x > 0$.

73. $\dfrac{2 + 2\sqrt{3}}{12}$ $\dfrac{1 + \sqrt{3}}{6}$
74. $\dfrac{6 + 6\sqrt{6}}{36}$ $\dfrac{1 + \sqrt{6}}{6}$
75. $\dfrac{5 - 15\sqrt{15}}{10}$ $\dfrac{1 - 3\sqrt{15}}{2}$
76. $\dfrac{14 - 7\sqrt{5}}{21}$ $\dfrac{2 - \sqrt{5}}{3}$
77. $\dfrac{6 + 12\sqrt{7}}{18}$ $\dfrac{1 + 2\sqrt{7}}{3}$
78. $\dfrac{4 + 2\sqrt{3}}{8}$ $\dfrac{2 + \sqrt{3}}{4}$
79. $\dfrac{3 + \sqrt{5}}{\sqrt{5}}$ $\dfrac{3\sqrt{5} + 5}{5}$
80. $\dfrac{7 - \sqrt{3}}{\sqrt{3}}$ $\dfrac{7\sqrt{3} - 3}{3}$
81. $\dfrac{5 + \sqrt{x}}{\sqrt{x}}$ $\dfrac{5\sqrt{x} + x}{x}$
82. $\dfrac{\sqrt{x} + 4}{\sqrt{x}}$ $\dfrac{x + 4\sqrt{x}}{x}$
83. $\dfrac{\sqrt{x} - 6}{\sqrt{x}}$ $\dfrac{x - 6\sqrt{x}}{x}$
84. $\dfrac{7 - \sqrt{x}}{\sqrt{x}}$ $\dfrac{7\sqrt{x} - x}{x}$

In Exercises 85–94, rationalize the denominator. Assume $x > 0$, $y > 0$.

85. $\dfrac{4}{1 + \sqrt{3}}$ $2\sqrt{3} - 2$
86. $\dfrac{20}{4 + \sqrt{6}}$ $8 - 2\sqrt{6}$
87. $\dfrac{2}{4 - \sqrt{2}}$ $\dfrac{4 + \sqrt{2}}{7}$
88. $\dfrac{5}{6 - \sqrt{6}}$ $\dfrac{6 + \sqrt{6}}{6}$
89. $\dfrac{9}{8 - \sqrt{10}}$ $\dfrac{8 + \sqrt{10}}{6}$
90. $\dfrac{10}{6 - \sqrt{11}}$ $\dfrac{12 + 2\sqrt{11}}{5}$
91. $\dfrac{x\sqrt{x}}{x + \sqrt{x}}$ $\dfrac{x\sqrt{x} - x}{x - 1}$
92. $\dfrac{y\sqrt{3y}}{y + \sqrt{3y}}$ $\dfrac{y\sqrt{3y} - 3y}{y - 3}$
93. $\dfrac{y\sqrt{6y}}{y - \sqrt{6y}}$ $\dfrac{y\sqrt{6y} + 6y}{y - 6}$
94. $\dfrac{x\sqrt{7x}}{x + \sqrt{7x}}$ $\dfrac{x\sqrt{7x} - 7x}{x - 7}$

THINK ABOUT IT

1. For the rectangle shown, find the following.

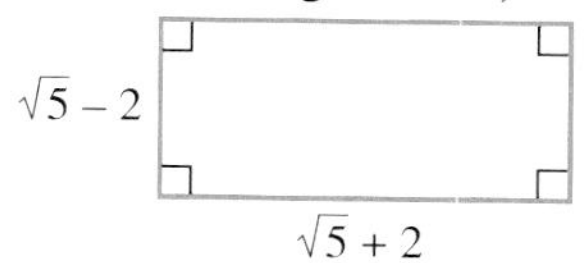

   a. The area
   b. The length of the diagonal

2. a. What is the volume of a cube with side length $\sqrt[3]{5}$ units?
   b. What is the volume of a cube with side length $\sqrt[3]{5} + 1$ units?

3. a. Show that $(\sqrt{6} - \sqrt{5})(\sqrt{6} + \sqrt{5})$ equals 1.
   b. Show that $(\sqrt[3]{6} - \sqrt[3]{5})(\sqrt[3]{36} + \sqrt[3]{30} + \sqrt[3]{25})$ equals 1.
   c. What factor times $\sqrt[4]{6} - \sqrt[4]{5}$ will give a product of 1? [*Hint:* $a^4 - b^4 = (a^2 - b^2)(a^2 + b^2)$ $= (a - b)(a^3 + a^2b + ab^2 + b^3)$.]

4. a. Show that $(1 + \sqrt[3]{2})(1 - \sqrt[3]{2} + \sqrt[3]{4})$ yields an integer product.
   b. Use the result of part **a** to rationalize the denominator: $\dfrac{1}{1 + \sqrt[3]{2}}$.

5. Rationalize the denominator: $\dfrac{1}{1 + \sqrt[3]{a}}$.

REMEMBER THIS

1. What is the result of squaring $\sqrt{2x + 3}$? $2x + 3$
2. True or false? 11 is a solution of $\sqrt{2x + 3} = 5$. True
3. If $n = \sqrt[3]{-5}$, what is the value of n^3? -5
4. True or false? -3 is a solution of both $3x + 18 = x^2$ and $\sqrt{3x + 18} = x$. False
5. Use FOIL to multiply $(1 + \sqrt{x})(1 + \sqrt{x})$. $1 + 2\sqrt{x} + x$
6. Solve. $6x + 10y = 4$
   $3x - 5y = 0$ $\left(\frac{1}{3}, \frac{1}{5}\right)$
7. Graph this line: $x = -2$.

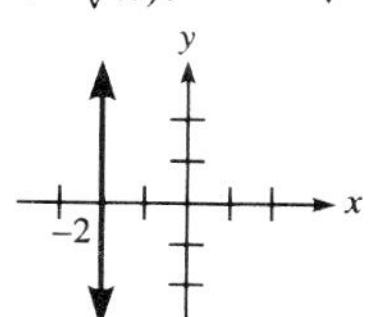

8. Express in lowest terms: $\dfrac{x^2 + cx}{c^2 + cx}$. $\dfrac{x}{c}$
9. Evaluate $-2x^{-1}$ if $x = \frac{1}{3}$. -6
10. Which of these are rational numbers? $\{2.\overline{3}, 2.3, 23, \frac{2}{3}\}$. All

7.5 Radical Equations

The formula $t = \sqrt{d}/4$ relates the distance d in feet traveled by a free-falling object to the time t of the fall in seconds, neglecting air resistance. If a stone is dropped from a bridge that spans a river and hits the water in 3.4 seconds, then to the nearest foot how far above the water is that particular point on the bridge? (See Example 3.)

OBJECTIVES

1 **Solve radical equations using the principle of powers once.**

2 **Solve radical equations using the principle of powers twice.**

1 To solve the problem that opens this section, we must solve an equation in which the unknown appears in a radicand. This type of equation is called a radical equation, and such equations are solved as follows.

To Solve Radical Equations

1. If necessary, isolate a radical term on one side of the equation.
2. Raise both sides of the equation to a power that matches the index of the isolated radical.
3. Solve the resulting equation, and check all solutions in the *original* equation.

This procedure is illustrated in Examples 1–3.

EXAMPLE 1 Solve $\sqrt{2x + 3} - 5 = 0$.

Solution First, isolate the radical on one side of the equation.

$$\begin{aligned} \sqrt{2x+3} - 5 &= 0 \\ \sqrt{2x+3} &= 5 \end{aligned}$$

Now, square both sides of the equation and solve for x.

$$\begin{aligned} (\sqrt{2x+3})^2 &= 5^2 && \text{Square both sides.} \\ 2x + 3 &= 25 \\ 2x &= 22 \\ x &= 11 \end{aligned}$$

To check, replace x by 11 in the original equation.

$$\begin{aligned} \sqrt{2x+3} - 5 &= 0 \\ \sqrt{2(11)+3} - 5 &\stackrel{?}{=} 0 && \text{Replace } x \text{ by 11.} \\ 0 &\stackrel{\checkmark}{=} 0 \end{aligned}$$

The proposed solution checks, and the solution set is $\{11\}$.

PROGRESS CHECK 1 Solve $\sqrt{3x - 2} - 4 = 3$.

EXAMPLE 2 Solve $\sqrt[3]{y - 5} = \sqrt[3]{2y - 3}$.

Solution A cube root term is isolated on each side of the equation, so begin by raising both sides to the third power, since the index of the radical is 3.

$$\begin{aligned} \sqrt[3]{y-5} &= \sqrt[3]{2y-3} \\ (\sqrt[3]{y-5})^3 &= (\sqrt[3]{2y-3})^3 && \text{Cube both sides.} \\ y - 5 &= 2y - 3 \\ -2 &= y \end{aligned}$$

Check

$$\begin{aligned} \sqrt[3]{y-5} &= \sqrt[3]{2y-3} \\ \sqrt[3]{(-2)-5} &\stackrel{?}{=} \sqrt[3]{2(-2)-3} && \text{Replace } x \text{ by } -2. \\ \sqrt[3]{-7} &\stackrel{\checkmark}{=} \sqrt[3]{-7} \end{aligned}$$

Thus, the solution set is $\{-2\}$.

PROGRESS CHECK 2 Solve $\sqrt[4]{3x + 3} = \sqrt[4]{5x - 35}$.

EXAMPLE 3 Solve the problem in the section introduction on page 343.

Solution Replacing t by 3.4 in

$$t = \frac{\sqrt{d}}{4} \quad \text{gives} \quad 3.4 = \frac{\sqrt{d}}{4}.$$

Progress Check Answers

1. $\{17\}$
2. $\{19\}$

The square root term is already isolated, so square both sides of the equation and solve for d.

$$(3.4)^2 = \left(\frac{\sqrt{d}}{4}\right)^2 \quad \text{Square both sides.}$$

$$(3.4)^2 = \frac{d}{16} \quad \text{Quotient-to-a-power property.}$$

$$16(3.4)^2 = d \quad \text{Multiply both sides by 16.}$$

Now compute $16(3.4)^2$ using a calculator.

16 [×] 3.4 [x^2] [=] [184.96]

To the nearest foot, the stone was dropped from a point on the bridge that is 185 ft above the water.

To check this answer, replace d by 185 in the given formula and compute $\sqrt{185}/4$.

185 [$\sqrt{\ }$] [÷] 4 [=] [3.4003676]

When $d \approx 185$ ft, $t \approx 3.4$ seconds, as requested.

PROGRESS CHECK 3 The bridge over the Royal Gorge of the Arkansas River in Colorado is the highest bridge in the world. A stone dropped from the recorded height above water level of this bridge takes about 8.112 seconds to hit the water. To the nearest foot, what is this recorded height?

The check in step 3 of the procedure to solve radical equations is not optional, because this procedure is based on the following principle.

Principle of Powers

If P and Q are algebraic expressions, then the solution set of the equation $P = Q$ is a subset of the solution set of $P^n = Q^n$ for any positive integer n.

Thus, every solution of $P = Q$ is a solution of $P^n = Q^n$; but solutions of $P^n = Q^n$ may or may not be solutions of $P = Q$, so checking is necessary. Solutions of $P^n = Q^n$ that do not satisfy the original equation are called **extraneous solutions,** and Examples 4 and 5 illustrate this possibility.

EXAMPLE 4 Solve $\sqrt{3x + 18} = x$.

Solution A radical is already isolated, so square both equation members and solve the resulting quadratic equation by the methods of Section 3.9.

$$\sqrt{3x + 18} = x$$

$$(\sqrt{3x + 18})^2 = x^2 \quad \text{Square both sides.}$$

$$3x + 18 = x^2 \quad \text{Simplify.}$$

$$0 = x^2 - 3x - 18 \quad \text{Rewrite so one side is 0.}$$

$$0 = (x + 3)(x - 6) \quad \text{Factor the nonzero side.}$$

$$x + 3 = 0 \quad \text{or} \quad x - 6 = 0 \quad \text{Set each factor equal to 0.}$$

$$x = -3 \qquad x = 6 \quad \text{Solve each linear equation.}$$

Now check.

$$\sqrt{3x + 18} = x$$
$$\sqrt{3(-3) + 18} \stackrel{?}{=} -3$$
$$\sqrt{9} \stackrel{?}{=} -3$$
$$3 \neq -3 \quad \text{Extraneous solution}$$

$$\sqrt{3x + 18} = x$$
$$\sqrt{3(6) + 18} \stackrel{?}{=} 6$$
$$\sqrt{36} \stackrel{?}{=} 6$$
$$6 \stackrel{\checkmark}{=} 6$$

Only 6 is a solution of the original equation, so the solution set is $\{6\}$.

Progress Check Answer

3. 1,053 ft

PROGRESS CHECK 4 Solve $\sqrt{20 - 8x} = x$.

In the remaining examples it is necessary to square both sides of an equation where one side of the equation contains two terms. To square a binomial, use

$$(a + b)^2 = a^2 + 2ab + b^2 \quad \text{or} \quad (a - b)^2 = a^2 - 2ab + b^2.$$

EXAMPLE 5 Solve $5 + \sqrt{2x + 5} = x$.

Solution First, isolate the radical. Then square both sides of the equation and solve for x.

$$\begin{aligned} 5 + \sqrt{2x + 5} &= x \\ \sqrt{2x + 5} &= x - 5 \\ (\sqrt{2x + 5})^2 &= (x - 5)^2 \qquad \text{Square both sides.} \\ 2x + 5 &= x^2 - 10x + 25 \\ 0 &= x^2 - 12x + 20 \\ 0 &= (x - 10)(x - 2) \end{aligned}$$

$$x - 10 = 0 \quad \text{or} \quad x - 2 = 0$$
$$x = 10 \qquad\qquad x = 2$$

Check

$$5 + \sqrt{2(10) + 5} = 10 \qquad 5 + \sqrt{2(2) + 5} = 2$$
$$5 + \sqrt{25} \stackrel{?}{=} 10 \qquad 5 + \sqrt{9} \stackrel{?}{=} 2$$
$$10 \stackrel{\checkmark}{=} 10 \qquad 8 \neq 2 \quad \text{Extraneous solution}$$

The check shows that 10 is a solution, while 2 is extraneous. Thus, the solution set is $\{10\}$.

PROGRESS CHECK 5 Solve $1 + \sqrt{x + 11} = x$.

2 In the next example we must square both sides of an equation twice in order to solve the equation.

EXAMPLE 6 Solve $\sqrt{2x - 3} - \sqrt{x - 2} = 1$.

Solution First, isolate a radical, and square both sides of the resulting equation.

$$\begin{aligned} \sqrt{2x - 3} - \sqrt{x - 2} &= 1 \\ \sqrt{2x - 3} &= 1 + \sqrt{x - 2} \\ (\sqrt{2x - 3})^2 &= (1 + \sqrt{x - 2})^2 \\ 2x - 3 &= 1^2 + 2(1)\sqrt{x - 2} + (\sqrt{x - 2})^2 \\ 2x - 3 &= 1 + 2\sqrt{x - 2} + x - 2 \end{aligned}$$

A radical remains, so isolate this radical term and square both sides again.

$$\begin{aligned} x - 2 &= 2\sqrt{x - 2} \\ (x - 2)^2 &= (2\sqrt{x - 2})^2 \\ x^2 - 4x + 4 &= 4(x - 2) \\ x^2 - 8x + 12 &= 0 \\ (x - 6)(x - 2) &= 0 \end{aligned}$$

$$x - 6 = 0 \quad \text{or} \quad x - 2 = 0$$
$$x = 6 \qquad\qquad x = 2$$

Progress Check Answers

4. $\{2\}$
5. $\{5\}$

Checking these proposed solutions in the original equation, we find

$$\sqrt{2(6)-3}-\sqrt{6-2}=\sqrt{9}-\sqrt{4}=3-2=1$$

and

$$\sqrt{2(2)-3}-\sqrt{2-2}=\sqrt{1}-\sqrt{0}=1-0=1.$$

Both solutions check, and the solution set is $\{6,2\}$.

PROGRESS CHECK 6 Solve $\sqrt{4x+1}-\sqrt{2x}=1$.

Progress Check Answer

6. $\{0,2\}$

EXERCISES 7.5

In Exercises 1–16, solve the given equation.

1. $\sqrt{3x-9}-6=0$ $\{15\}$
2. $\sqrt{4x+9}-5=0$ $\{4\}$
3. $\sqrt{2x-4}-2=4$ $\{20\}$
4. $\sqrt{5x+14}+6=13$ $\{7\}$
5. $\sqrt{6x-2}+6=10$ $\{3\}$
6. $\sqrt{3x-5}+15=22$ $\{18\}$
7. $10-\sqrt{7x+11}=5$ $\{2\}$
8. $8-\sqrt{2x+18}=2$ $\{9\}$
9. $\sqrt{x+1}+5=0$ $\emptyset$
10. $\sqrt{x-3}+4=0$ $\emptyset$
11. $\sqrt[3]{y+10}=\sqrt[3]{6y-20}$ $\{6\}$
12. $\sqrt[3]{2x+3}=\sqrt[3]{5x-18}$ $\{7\}$
13. $\sqrt[3]{6x+10}=\sqrt[3]{3x+1}$ $\{-3\}$
14. $\sqrt[3]{9x-3}=\sqrt[3]{5x-27}$ $\{-6\}$
15. $\sqrt[4]{12x+3}=\sqrt[4]{9x+9}$ $\{2\}$
16. $\sqrt[4]{9x+13}=\sqrt[4]{5x+9}$ $\{-1\}$

The formula $t=\sqrt{d}/4$ relates the distance d in feet traveled by a free-falling object to the time t of the fall in seconds, neglecting air resistance. Use this relationship to solve Exercises 17 and 18.

17. As you travel along the rim of the Grand Canyon, you stop at a place called Mojave Cliffs. Using a stopwatch, you calculate the length of time it takes for a rock dropped off the edge to hit the bottom of the canyon as 13.7 seconds. Approximate, to the nearest foot, the depth of the Grand Canyon at this point. 3,003 ft
18. On a farm there's an old well you'd like to use. You have a bucket, and you're going to buy some rope, but you don't know how much rope to get. You decide to measure the depth of the well by dropping a small stone. If it takes about 1.3 seconds for the stone to hit the water, how deep is the well (to the nearest foot)? 27 ft
19. The period of a pendulum is the time required for the pendulum to complete one round-trip of motion, that is, one complete cycle. When the period is measured in seconds and the pendulum length ℓ is measured in feet, then a formula for the period is
$$T=2\pi\sqrt{\frac{\ell}{32}}.$$
To the nearest hundredth of a foot, what is the length of a pendulum whose period is 1 second? 0.81 ft
20. Using the formula from Exercise 19, find the length of a pendulum whose period is 2 seconds. 3.24 ft
21. A kind of average called the **root-mean-square** is defined as the square root of the mean of a collection of squares. The formula $A=\sqrt{\frac{m^2+n^2}{2}}$ gives the root-mean-square of two positive numbers, m and n. Find n if $m=1$ and $A=5$. $n=7$
22. Use the formula from Exercise 21 and find n if $m=2$ and $A=2\sqrt{5}$. $n=6$

In Exercises 23–50, solve the given equation.

23. $x=\sqrt{5x-6}$ $\{3,2\}$
24. $x=\sqrt{6x-5}$ $\{1,5\}$
25. $\sqrt{4x+5}=x$ $\{5\}$
26. $\sqrt{3x+10}=x$ $\{5\}$
27. $\sqrt{x+6}=x$ $\{3\}$
28. $\sqrt{x+42}=x$ $\{7\}$
29. $x=\sqrt{27-6x}$ $\{3\}$
30. $\sqrt{2x+8}=x$ $\{4\}$
31. $\sqrt{x+3}-3=x$ $\{-3,-2\}$
32. $\sqrt{x-1}+1=x$ $\{1,2\}$
33. $3+\sqrt{21-2x}=x$ $\{6\}$
34. $\sqrt{13-4x}-2=x$ $\{1\}$
35. $2\sqrt{12-x}-3=x$ $\{3\}$
36. $5+\sqrt{2x-2}=x$ $\{9\}$
37. $\sqrt{2x+3}-\sqrt{x-2}=2$ $\{3,11\}$
38. $\sqrt{2x-1}-\sqrt{x-4}=2$ $\{5,13\}$
39. $1=\sqrt{3x-5}-\sqrt{2x-5}$ $\{3,7\}$
40. $\sqrt{3x+10}-\sqrt{2x+6}=1$ $\{5,-3\}$
41. $\sqrt{3x-11}-\sqrt{2x-9}=1$ $\{5,9\}$
42. $\sqrt{3x+4}-\sqrt{2x+1}=1$ $\{0,4\}$
43. $\sqrt{2x+3}+\sqrt{x-2}=2$ $\emptyset$
44. $\sqrt{3x-6}+\sqrt{2x-6}=1$ $\emptyset$
45. $\sqrt{x+4}=2-\sqrt{x+16}$ $\emptyset$
46. $\sqrt{x-5}+\sqrt{x+10}=-3$ $\emptyset$
47. $3=\sqrt{x+11}-\sqrt{x-4}$ $\{5\}$
48. $\sqrt{3x-2}+\sqrt{2x-2}=1$ $\{1\}$
49. $\sqrt{x}=3-\sqrt{x-5}$ $\{\frac{49}{9}\}$
50. $\sqrt{x}-\sqrt{x-5}=2$ $\{\frac{81}{16}\}$

THINK ABOUT IT

1. a. We have seen that $\sqrt{x^2 + 9}$ is not identical to $x + 3$. For almost all values of x, these two expressions are unequal. Find any values of x for which they are equal.
   b. We have seen that $\sqrt{x} + \sqrt{x}$ is not identical to $\sqrt{2x}$. Find any values of x for which they are equal.
   c. We have seen that $(x + 1)^2$ is not identical to $x^2 + 1$. Find any values of x for which they are equal.
2. a. A graphical approach to solving the equation $\sqrt{x} = 2$ uses the two graphs which result from setting each side of the equation equal to y. Graph $y = \sqrt{x}$ and $y = 2$ on the same set of axes. The solution to $\sqrt{x} = 2$ will be the x-coordinate of the point of intersection of the two graphs.
   b. Explain how the graphs of $y = \sqrt{x}$ and $y = -2$ show that $\sqrt{x} = -2$ has no solution.
   c. Graph $y = \sqrt{x} + 3$ and $y = \sqrt{x}$ on the same set of axes. Discuss why $\sqrt{x} + 3 = \sqrt{x}$ has no solution.
3. Solve these equations.
   a. $\sqrt{(5x - 2)^2} = 5x - 2$
   (*Hint:* The solutions must satisfy the original equation.)
   b. $\dfrac{10}{\sqrt{x - 5}} = \sqrt{x - 5} - 3$
   c. $\dfrac{x}{\sqrt{x + 16}} - \dfrac{2}{\sqrt{3}} = 0$
   d. $\sqrt{1 + 4\sqrt{y}} = 1 + \sqrt{y}$
4. The following geometry problems lead to radical equations.
   a. The perimeter of a right triangle is 3 in., and one of the legs is 1 in. Find the other leg and the hypotenuse.
   b. An isosceles triangle has perimeter 8 cm and height 2 cm. Find its base and its area.
5. Solve these equations by using the principle of powers.
   a. $\sqrt[3]{x} = x$ b. $\sqrt[3]{x^2} = x$
   c. $\sqrt[3]{x^3} = x$ d. $\sqrt[3]{x^4} = x$

REMEMBER THIS

1. True or false? Every number on the real number line must be either positive or negative or zero. True
2. True or false? There is no real number x for which $x^2 = -4$. True
3. Multiply $(3 + \sqrt{2})(3 - \sqrt{2})$. 7
4. Choose the relationship symbol ($<$, $>$, $=$) which makes $(-1)^{13}$ ____ $(-1)^{12}$ a true statement. $<$
5. If $n = -1$, what is the value of $n(-n)$? -1
6. Use Cramer's rule to find the value of x in the solution to
$$3x + 2y + 4z = 6$$
$$3x + 4y - 3z = 0$$
$$-2y + 3z = 2.$$
$\frac{1}{3}$
7. True or false? The graphs of $y = 2x + 1$ and $y = -2x + 1$ are perpendicular lines. False
8. Simplify $\dfrac{\frac{1}{a} + \frac{1}{b}}{\frac{a + b}{5}}$. $\frac{5}{ab}$
9. Multiply and express in scientific notation: $(4.5 \times 10^{22})(2.0 \times 10^{-9})$. 9×10^{13}
10. Find the volume of the circular cylinder shown. Use 3.14 for π. $V = 282.6 \text{ cm}^3$

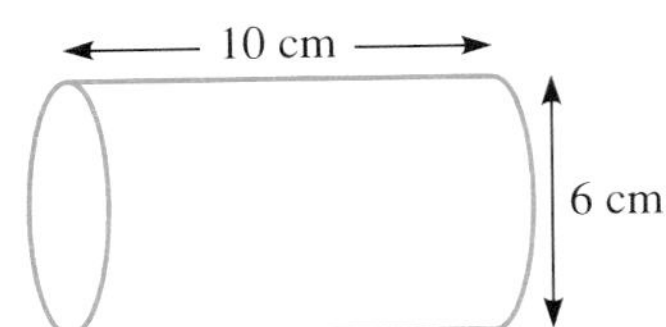

7.6 Complex Numbers

Ohm's law for alternating current circuits states

$$V = IZ,$$

where V is the voltage in volts, I is the current in amperes, and Z is the impedance in ohms. If the voltage in a particular circuit is $18 - 21i$ volts, and the current is $3 - 6i$ amperes, find the complex number that measures the impedance. (See Example 6.)

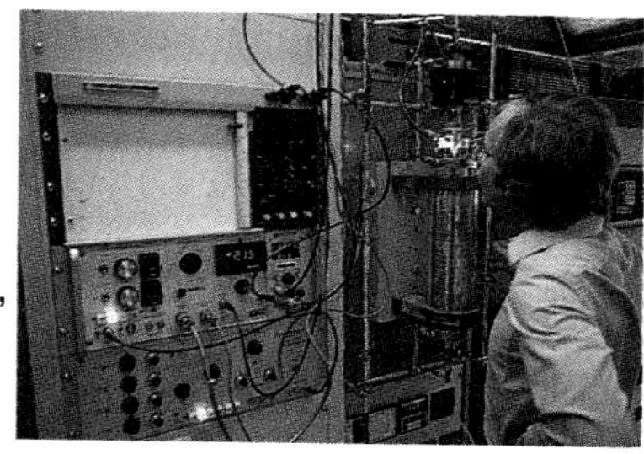

OBJECTIVES

1. **Express square roots of negative numbers in terms of *i*.**
2. **Add, subtract, and multiply complex numbers.**
3. **Divide complex numbers.**
4. **Find powers of *i*.**

In this section we extend the number system beyond real numbers to complex numbers which have significant applications in mathematics, physics, and engineering. For instance, in the section-opening problem complex numbers are used to describe voltage and other electrical quantities, because such a numerical representation indicates both the strength and time (or phase) relationships of the quantities.

1 To define a complex number, we first introduce a new set of numbers called **imaginary numbers** in which square roots of negative numbers are defined. The basic unit in imaginary numbers is $\sqrt{-1}$, and it is designated by i. Thus, by definition,

Descartes (1637) coined the names *real* and *imaginary*. Euler (1748) was the first to use i for $\sqrt{-1}$.

$$i = \sqrt{-1} \quad \text{and} \quad i^2 = -1.$$

Square roots of negative numbers may now be written in terms of i by defining the principal square root of a negative number as follows.

Principal Square Root of a Negative Number

If a is a positive number, then

$$\sqrt{-a} = i\sqrt{a}.$$

In this definition we choose to write i in front of any radicals so that expressions like $\sqrt{a}\,i$ are not confused with $\sqrt{ai}$.

EXAMPLE 1 Express each number in terms of i.

a. $\sqrt{-3}$ **b.** $-\sqrt{-3}$ **c.** $\sqrt{-4}$ **d.** $\sqrt{-50}$

Solution

a. $\sqrt{-3} = i\sqrt{3}$

b. $-\sqrt{-3} = -i\sqrt{3}$

c. $\sqrt{-4} = i\sqrt{4} = 2i$

d. $\sqrt{-50} = i\sqrt{50} = i\sqrt{25}\sqrt{2} = 5i\sqrt{2}$

PROGRESS CHECK 1 Express each number in terms of i.

a. $\sqrt{-9}$ **b.** $-\sqrt{-9}$ **c.** $\sqrt{-7}$ **d.** $\sqrt{-8}$

2 Using real numbers and imaginary numbers, we may extend the number system to include a number like $3 - 6i$ that was used to describe the current in the section-opening problem. This type of number is called a complex number.

Gauss (1832) suggested the word *complex* to distinguish $a + bi$ from ai.

Definition of Complex Number

A number of the form $a + bi$, where a and b are real numbers and $i = \sqrt{-1}$, is called a complex number.

The number a is called the **real part** of $a + bi$, and b is called the **imaginary part** of $a + bi$. Note that both the real part and the imaginary part of a complex number are real numbers. Two complex numbers are **equal** if and only if their real parts are equal and their imaginary parts are equal. That is,

$$a + bi = c + di \text{ if and only if } a = c \text{ and } b = d.$$

If we let a and/or b equal zero, both real numbers and imaginary numbers may be expressed in $a + bi$ form. For instance:

Real number → $2 = 2 + 0i$

Imaginary number → $5i = 0 + 5i$

Complex number in $a + bi$ form

Thus, the complex numbers include the real numbers and the imaginary numbers. Figure 7.2 illustrates the relationship among the various sets of numbers.

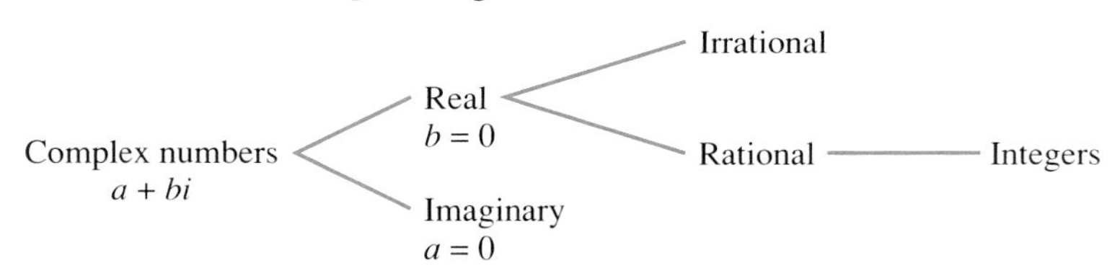

Figure 7.2

The rules for operating with complex numbers were first laid out clearly by Raphael Bombelli (1526–1572).

Operations with complex numbers are defined so that the properties of real numbers (like the commutative properties) continue to apply, and in many ways computations with complex numbers are similar to computations with polynomials. Consider carefully Examples 2–4, which illustrate how to add, subtract, and multiply with complex numbers.

EXAMPLE 2 Combine the complex numbers.

a. $\sqrt{-36} + \sqrt{-100}$ **b.** $(7 - 3i) + (1 + 6i)$ **c.** $(5 + 4i) - (9 - i)$

Solution

a.
$$\begin{aligned}\sqrt{-36} + \sqrt{-100} &= i\sqrt{36} + i\sqrt{100}\\ &= 6i + 10i\\ &= 16i\end{aligned}$$

b.
$$\begin{aligned}(7 - 3i) + (1 + 6i) &= (7 + 1) + (-3 + 6)i\\ &= 8 + 3i\end{aligned}$$

Progress Check Answers

1. (a) $3i$ (b) $-3i$ (c) $i\sqrt{7}$ (d) $2i\sqrt{2}$

c. $(5 + 4i) - (9 - i) = 5 + 4i - 9 + i$
$= (5 - 9) + (4 + 1)i$
$= -4 + 5i$

PROGRESS CHECK 2 Combine the complex numbers.

a. $\sqrt{-9} - \sqrt{-25}$ **b.** $(-1 - 2i) + (8 + 5i)$ **c.** $(9 + 3i) - (16 - i)$

EXAMPLE 3 Multiply the complex numbers.

a. $\sqrt{-25}\sqrt{-4}$ **b.** $(3 + 4i)(2 - 6i)$ **c.** $(5 + 2i)^2$

Solution

a. $\sqrt{-25}\sqrt{-4} = i\sqrt{25}\, i\sqrt{4}$
$= 5i \cdot 2i$
$= 10i^2$
$= 10(-1)$ Replace i^2 by -1.
$= -10$

b. $(3 + 4i)(2 - 6i)$
$= 3(2) + 3(-6i) + 4i(2) + 4i(-6i)$ Use FOIL.
$= 6 - 18i + 8i - 24i^2$
$= 6 - 18i + 8i - 24(-1)$ Replace i^2 by -1.
$= 30 - 10i$

c. $(5 + 2i)^2 = 5^2 + 2(5)(2i) + (2i)^2$ Use $(a + b)^2 = a^2 + 2ab + b^2$
$= 25 + 20i + 4i^2$
$= 25 + 20i + 4(-1)$ Replace i^2 by -1.
$= 21 + 20i$

Caution The property of radicals $\sqrt{a}\sqrt{b} = \sqrt{ab}$ does not hold when a and b are both negative. For instance, the solution in part **a** shows

$$\sqrt{-25} \cdot \sqrt{-4} = -10.$$

Therefore,

$$\sqrt{-25} \cdot \sqrt{-4} = \sqrt{(-25)(-4)} = \sqrt{100} = 10 \text{ is } \textbf{wrong.}$$

Always remember to express complex numbers in terms of i before performing computations.

You might point out as an aid in checking calculations that the product of the square roots of two negative numbers is itself a negative number which lies between them. For example, $-25 < (\sqrt{-25})(\sqrt{-4}) = -10 < -4$. For positive values the middle number is called the geometric mean.

PROGRESS CHECK 3 Multiply the complex numbers.

a. $\sqrt{-16}\sqrt{-9}$ **b.** $(2 + 3i)(10 - i)$ **c.** $(4 + 7i)^2$

EXAMPLE 4 Evaluate $x^2 + 2x + 3$ if $x = -1 - i\sqrt{2}$.

Solution Substitute $-1 - i\sqrt{2}$ for x and simplify.

$$\begin{aligned} x^2 + 2x + 3 &= (-1 - i\sqrt{2})^2 + 2(-1 - i\sqrt{2}) + 3 \\ &= (-1)^2 - 2(-1)(i\sqrt{2}) + (i\sqrt{2})^2 - 2 - 2i\sqrt{2} + 3 \\ &= 1 + 2i\sqrt{2} + 2i^2 - 2 - 2i\sqrt{2} + 3 \\ &= 2i^2 + 2 \\ &= 0 \qquad (\text{since } i^2 = -1) \end{aligned}$$

PROGRESS CHECK 4 Evaluate $x^2 + 2x + 3$ if $x = -1 + i\sqrt{2}$.

Use Examples 2–4 as a basis for understanding the definitions of the following operations on complex numbers.

Progress Check Answers

2. (a) $-2i$ (b) $7 + 3i$ (c) $-7 + 4i$
3. (a) -12 (b) $23 + 28i$ (c) $-33 + 56i$
4. 0

Addition, Subtraction, and Multiplication of Complex Numbers

Addition Two complex numbers are added by adding separately their real parts and their imaginary parts.

$$(a + bi) + (c + di) = (a + c) + (b + d)i$$

Subtraction Two complex numbers are subtracted by subtracting separately their real parts and their imaginary parts.

$$(a + bi) - (c + di) = (a - c) + (b - d)i$$

Multiplication Two complex numbers are multiplied as two binomials are multiplied, with i^2 being replaced by -1.

$$\begin{aligned}(a + bi)(c + di) &= ac + adi + bci + bdi^2 \\ &= (ac - bd) + (ad + bc)i\end{aligned}$$

3 Conjugates and division

To understand the procedure for dividing two complex numbers, consider

$$\begin{aligned}(a + bi)(a - bi) &= a^2 - abi + abi - b^2i^2 \\ &= a^2 - b^2(-1) \\ &= a^2 + b^2.\end{aligned}$$

The name *conjugate* is due to Cauchy (1821).

The complex numbers $a + bi$ and $a - bi$ are called **conjugates** of each other, and since a and b are real numbers, the above product shows that the product of a complex number and its conjugate is always a real number. Based on this property of conjugates, the quotient of two complex numbers can be written in the form $a + bi$ by multiplying both the numerator and the denominator by the conjugate of the denominator. Examples 5 and 6 illustrate this procedure.

EXAMPLE 5 Write the quotient $\dfrac{2 + i}{3 + i}$ in the form $a + bi$.

Solution As discussed, we divide two complex numbers by multiplying the numerator and the denominator by the conjugate of the denominator. The conjugate of $3 + i$ is $3 - i$, so

$$\begin{aligned}\frac{2 + i}{3 + i} &= \frac{2 + i}{3 + i} \cdot \frac{3 - i}{3 - i} \\ &= \frac{6 - 2i + 3i - i^2}{9 - 3i + 3i - i^2} \\ &= \frac{6 + i - (-1)}{9 - (-1)} \\ &= \frac{7 + i}{10} = \frac{7}{10} + \frac{1}{10}i.\end{aligned}$$

Progress Check Answer

5. $\frac{9}{10} + \frac{4}{5}i$

PROGRESS CHECK 5 Write the quotient $\dfrac{2 + 5i}{4 + 2i}$ in the form $a + bi$.

Ohm's law for alternating current has impedance, and in direct current has resistance.

EXAMPLE 6 Solve the problem in the section introduction on page 349.

Solution To find the impedance, which is symbolized by Z, first solve $V = IZ$ for Z, to obtain

$$Z = \frac{V}{I}.$$

Now replace V by $18 - 21i$ and I by $3 - 6i$ and divide, using the conjugate of the denominator.

$$Z = \frac{18 - 21i}{3 - 6i} = \frac{18 - 21i}{3 - 6i} \cdot \frac{3 + 6i}{3 + 6i} = \frac{54 + 108i - 63i - 126i^2}{9 + 18i - 18i - 36i^2}$$
$$= \frac{54 + 45i - 126(-1)}{9 - 36(-1)} = \frac{180 + 45i}{45} = 4 + i$$

The impedance is therefore $4 + i$ ohms.

PROGRESS CHECK 6 Find the impedance in an alternating current circuit in which the voltage is $23 - 14i$ volts and the current is $3 - 4i$ amperes.

4 Powers of i

To simplify i^n, where n is a positive integer, consider the cyclic pattern contained in the following simplifications.

$$\begin{array}{ll} i^1 = i & i^5 = i^4 i = 1i = i \\ i^2 = -1 & i^6 = i^4 i^2 = 1(-1) = -1 \\ i^3 = i^2 i = (-1)i = -i & i^7 = i^4 i^3 = 1(-i) = -i \\ i^4 = i^2 i^2 = (-1)(-1) = 1 & i^8 = (i^4)^2 = 1^2 = 1 \end{array}$$

Continuing this pattern to higher powers of i gives the cyclic property shown in Figure 7.3. Note in particular that $i^n = 1$ if n is a multiple of 4, because we may use this property to simplify large powers of i, as shown in Example 7.

Point out that the equivalent power in the $n = 0$ to 3 range is the remainder after the original power is divided by 4.

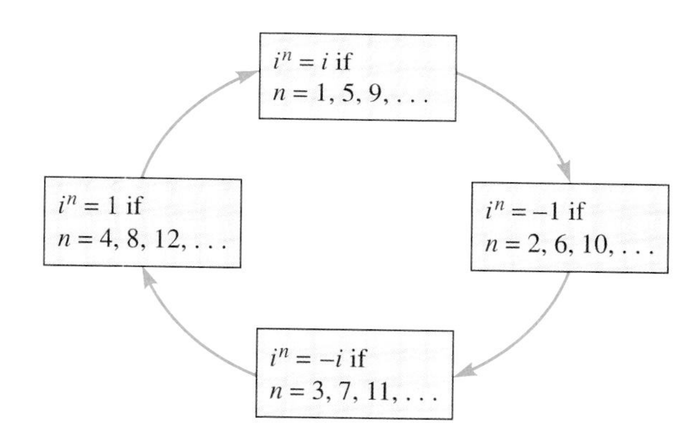

Figure 7.3

EXAMPLE 7 Simplify each power of i.

a. i^{15} **b.** i^{50}

Solution

a. The largest multiple of 4 that is less than or equal to the exponent is 12. Therefore, rewrite i^{15} with i^{12} as one factor, and then simplify.

$$i^{15} = i^{12}i^3 = (i^4)^3 i^3 = (1)^3(-i) = -i$$

b. Because $i^{48} = (i^4)^{12} = 1$, we have

$$i^{50} = i^{48}i^2 = 1(-1) = -1.$$

PROGRESS CHECK 7 Simplify each power of i.

a. i^{18} **b.** i^{81}

Progress Check Answers

6. $5 + 2i$ ohms
7. (a) -1 (b) i

EXERCISES 7.6

In Exercises 1–18, express each number in terms of i.

1. $\sqrt{-1}$ i
2. $\sqrt{-25}$ $5i$
3. $\sqrt{-16}$ $4i$
4. $\sqrt{-100}$ $10i$
5. $-\sqrt{-4}$ $-2i$
6. $-\sqrt{-1}$ $-i$
7. $-\sqrt{-49}$ $-7i$
8. $-\sqrt{-36}$ $-6i$
9. $\sqrt{-12}$ $2i\sqrt{3}$
10. $\sqrt{-18}$ $3i\sqrt{2}$
11. $\sqrt{-24}$ $2i\sqrt{6}$
12. $\sqrt{-72}$ $6i\sqrt{2}$
13. $\sqrt{-125}$ $5i\sqrt{5}$
14. $\sqrt{-48}$ $4i\sqrt{3}$
15. $-\sqrt{-27}$ $-3i\sqrt{3}$
16. $-\sqrt{-98}$ $-7i\sqrt{2}$
17. $-\sqrt{-75}$ $-5i\sqrt{3}$
18. $-\sqrt{-242}$ $-11i\sqrt{2}$

In Exercises 19–44, combine the complex numbers.

19. $\sqrt{-49} + \sqrt{-121}$ $18i$
20. $\sqrt{-256} + \sqrt{-1}$ $17i$
21. $\sqrt{-64} + \sqrt{-81}$ $17i$
22. $\sqrt{-400} + \sqrt{-144}$ $32i$
23. $\sqrt{-25} - \sqrt{-16}$ i
24. $\sqrt{-196} - \sqrt{-169}$ i
25. $\sqrt{-49} - \sqrt{-9}$ $4i$
26. $\sqrt{-400} - \sqrt{-100}$ $10i$
27. $(2 + i) + (3 + 2i)$ $5 + 3i$
28. $(5 + 3i) + (7 + 4i)$ $12 + 7i$
29. $(2 - 4i) + (5 + i)$ $7 - 3i$
30. $(8 - 6i) + (4 + 5i)$ $12 - i$
31. $(8 + 7i) + (8 - 3i)$ $16 + 4i$
32. $(4 + 3i) + (9 - 7i)$ $13 - 4i$
33. $(4 - 4i) + (6 + 4i)$ 10
34. $(3 - 7i) + (7 + 7i)$ 10
35. $(8 - i) + (7 - 2i)$ $15 - 3i$
36. $(8 - 4i) + (3 - 8i)$ $11 - 12i$
37. $(4 - 4i) - (6 + 4i)$ $-2 - 8i$
38. $(8 - 6i) - (4 + 5i)$ $4 - 11i$
39. $(2 - 4i) - (5 - i)$ $-3 - 3i$
40. $(5 + 4i) - (7 + 3i)$ $-2 + i$
41. $(8 - 3i) - (8 + 7i)$ $-10i$
42. $(4 - 6i) - (4 + 4i)$ $-10i$
43. $(8 - i) - (-7 - i)$ 15
44. $(-6 - 2i) - (4 - 2i)$ -10

In Exercises 45–66, multiply the complex numbers.

45. $\sqrt{-64} \cdot \sqrt{-81}$ -72
46. $\sqrt{-49} \cdot \sqrt{-121}$ -77
47. $\sqrt{-36} \cdot \sqrt{-100}$ -60
48. $\sqrt{-400} \cdot \sqrt{-144}$ -240
49. $\sqrt{-25} \cdot \sqrt{-16}$ -20
50. $\sqrt{-4} \cdot \sqrt{-196}$ -28
51. $(2 + i)(3 + 2i)$ $4 + 7i$
52. $(5 + 3i)(7 + 4i)$ $23 + 41i$
53. $(2 - 4i)(5 + i)$ $14 - 18i$
54. $(8 - 6i)(4 + 5i)$ $62 + 16i$
55. $(8 + 7i)(8 - 3i)$ $85 - 32i$
56. $(4 + 3i)(9 - 7i)$ $57 - i$
57. $(2 - 4i)(5 - i)$ $6 - 22i$
58. $(5 + 4i)(7 - 3i)$ $47 + 13i$
59. $(3 + 4i)^2$ $-7 + 24i$
60. $(3 + 8i)^2$ $-55 + 48i$
61. $(2 + 6i)^2$ $-32 + 24i$
62. $(6 + 5i)^2$ $11 + 60i$
63. $(5 - 2i)^2$ $21 - 20i$
64. $(2 - 9i)^2$ $-77 - 36i$
65. $(7 - i)^2$ $48 - 14i$
66. $(10 - 3i)^2$ $91 - 60i$

In Exercises 67–72, evaluate the given polynomial.

67. Evaluate $x^2 + 4x + 5$ if $x = -2 + i$. 0
68. Evaluate $x^2 + 4x + 5$ if $x = -2 - i$. 0
69. Evaluate $x^2 + 6x + 12$ if $x = -3 - i\sqrt{3}$. 0
70. Evaluate $x^2 + 6x + 12$ if $x = -3 + i\sqrt{3}$. 0
71. Evaluate $x^2 + 2x + 9$ if $x = -1 + i\sqrt{5}$. 3
72. Evaluate $x^2 + 2x + 9$ if $x = -1 - i\sqrt{5}$. 3

In Exercises 73–78, write the indicated quotient in the form $a + bi$.

73. $\dfrac{1 + 2i}{1 + i}$ $\frac{3}{2} + \frac{1}{2}i$
74. $\dfrac{1 + 3i}{2 + i}$ $1 + i$
75. $\dfrac{6 + 2i}{7 + i}$ $\frac{22}{25} + \frac{4}{25}i$
76. $\dfrac{1 + 5i}{4 + 3i}$ $\frac{19}{25} + \frac{17}{25}i$
77. $\dfrac{7 + 2i}{6 - 2i}$ $\frac{19}{20} + \frac{13}{20}i$
78. $\dfrac{5 - 4i}{3 - 5i}$ $\frac{35}{34} + \frac{13}{34}i$

Ohm's law for alternating current circuits states that $V = IZ$, where V is the voltage in volts, I is the current in amperes, and Z is the impedance in ohms. Use this law to solve Exercises 79–84.

79. Find the complex number that represents the impedance in a particular circuit if the voltage is $14 + 8i$ volts and the current is $1 + 2i$ amperes. $6 - 4i$ ohms
80. Find the complex number that represents the impedance in a given circuit if the voltage is $12 - 5i$ volts and the current is $2 - 3i$ amperes. $3 + 2i$ ohms
81. Find the complex number that represents the voltage in a given circuit if the impedance is $240 - 50i$ ohms and the current is $0.4 + 0.3i$ amperes. $111 + 52i$ volts
82. Find the complex number that represents the voltage in a given circuit if the impedance is $10 + i$ ohms and the current is $5 - 0.5i$ amperes. 50.5 volts
83. Find the complex number that represents the current in a given circuit if the voltage is 12 volts and the impedance is $1 - i$ ohms. $6 + 6i$ amperes
84. Find the complex number that represents the current in a given circuit if the voltage is 120 volts and the impedance is $-3i$ amperes. $40i$ amperes

In Exercises 85–100, simplify each power of i.

85. i^5 i
86. i^7 $-i$
87. i^8 1
88. i^{10} -1
89. i^{57} i
90. i^{101} i
91. i^{82} -1
92. $i^{1,002}$ -1

THINK ABOUT IT

1. Not only real numbers but *all* complex numbers (except zero) have two square roots. Show that $\frac{\sqrt{2}}{2} + \frac{\sqrt{2}}{2}i$ and $-\frac{\sqrt{2}}{2} - \frac{\sqrt{2}}{2}i$ are both square roots of i. To do this, show that the square of each number is i.
2. The complex numbers are said to be "closed" under the four basic arithmetic operations. That is, combining $a + bi$ and $c + di$ by any of the operations $+, -, \times, \div$ will always yield a number of that same form.
   a. Illustrate this by finding the sum, difference, product, and quotient of $1 + i$ and $2 - i$.
   b. Repeat part **a** in the general case for $a + bi$ and $c + di$.
3. Complex numbers are not ordered like real numbers. The relations "less than" and "greater than" do not apply to complex numbers. But they can be assigned a magnitude, similar to the concept of the absolute value of a real number. The **absolute value** of the complex number $a + bi$ is defined as $\sqrt{a^2 + b^2}$. Find the absolute value of these complex numbers.
   a. $1 + i$ **b.** $1 - i$ **c.** i **d.** $\frac{\sqrt{2}}{2} + \frac{\sqrt{2}}{2}i$
4. Just as a real number can be represented by a point on a line, a complex number can be represented by a point in a plane. The number $a + bi$ is represented by the point (a,b). Find the point represented by each of these. Verify that the absolute value of each number (see Exercise 3) equals its distance from the origin.
   a. $4 + i$ **b.** $3 - 2i$ **c.** $(4 + i) + (3 - 2i)$
   d. $(4 + i)(3 - 2i)$
5. Because complex numbers can be represented by ordered pairs, it is possible to define the operations for them entirely in terms of ordered pairs without any reference to i. For instance, the sum of (a,b) and (c,d) is defined as $(a + c, b + d)$. This takes the place of saying that the sum of $a + bi$ and $c + di$ equals $(a + c) + (b + d)i$.
   a. Show that the product of (a,b) and (c,d) should be defined as $(ac - bd, ad + bc)$.
   b. What should the definition be of $(a,b) \div (c,d)$?

REMEMBER THIS

1. Which of these equations is a quadratic equation?
   a. $2x - 4 = 0$ **b.** $x^2 - 4 = 0$ b
2. Solve $(x - 6)(x + 6) = 0$. $\{6, -6\}$
3. Simplify $[(-4 + \sqrt{7}) + 4]^2$. 7
4. Which of these is a perfect square trinomial?
   a. $x^2 + 6x + 6$ **b.** $x^2 + 6x + 9$ **c.** $x^2 + 6x + 12$ b
5. Evaluate $\frac{1 + \sqrt{10}}{3}$ on a calculator. Round to the nearest thousandth. 1.387
6. Solve graphically. $y > x - 2$
   $x + y < 1$

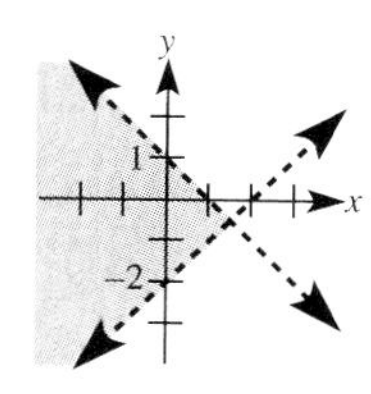

7. Find the slope and y-intercept for $3x + y - 5 = 0$.
   $m = -3$, $(0,5)$
8. Simplify $\frac{x^{-2} - 1}{1 + x^{-1}}$. $\frac{1 - x}{x}$
9. Simplify $\frac{2^n}{2^{n-2}}$. 4
10. Rewrite $a - (3 - x)$ without parentheses. $a - 3 + x$

Estimating Solutions of Equations and Testing the Equivalence of Algebraic Expressions

An advantage of using a graphing calculator is the ability to transform questions about equations into questions about graphs, which can then be answered (if only approximately) by looking at the screen.

EXAMPLE 1 Use the graphing calculator to solve the equation $7(x-2)=2x-1$. Determine the solution to the nearest tenth.

Solution Solving this equation means finding all values of x that make both sides of the equation have the same value. But this is equivalent to finding the x-coordinate of the point where the graphs of $y=7(x-2)$ and $y=2x-1$ intersect. We therefore draw both graphs on the same set of axes and estimate the point of intersection. See Figure 7.4.

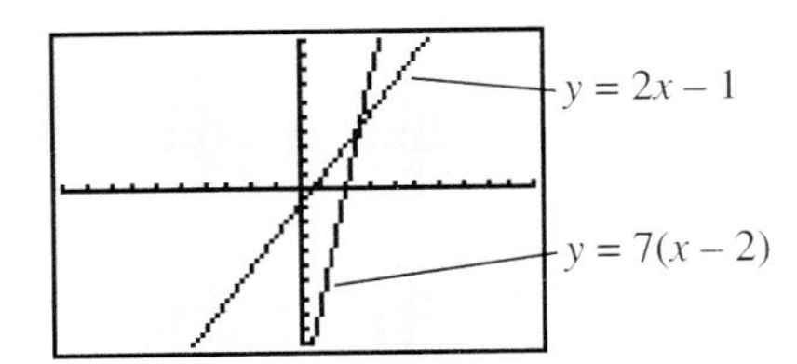

Figure 7.4

Zooming in on the point of intersection shows that the solution is about $x=2.6$. An algebraic check confirms that this is the exact solution.

Example 2 illustrates another way to find values of x for which two expressions are equal. We depend on the fact that if two expressions have the same value, then their *difference* must equal zero. Thus, we plot the graph of $Y_1 - Y_2$ and look to see where it crosses the x-axis.

EXAMPLE 2 Use the difference method to solve the equation $7(x-2)=2x-1$. Determine the solution to the nearest tenth.

Solution

Step 1 As in Example 1, use [Y=] to key in $7(X-2)$ for Y_1 and $2X-1$ for Y_2.

Step 2 For Y_3 key in $Y_1 - Y_2$. Recall that you must use the [2nd] [Y-VARS] key to produce Y_1 and Y_2 on the screen. The screen should look like Figure 7.5.

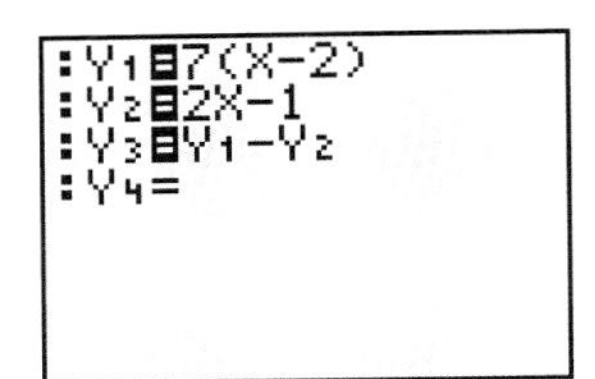

Figure 7.5

Step 3 Turn off the highlighting on the equals sign for Y_1 and Y_2 by putting the cursor on the equals sign and pressing [ENTER]. At this point, only the equals sign for Y_3 is highlighted. (Recall that the calculator draws graphs only for the highlighted expressions.)
Press [ZOOM] Standard to get a screen like Figure 7.6.

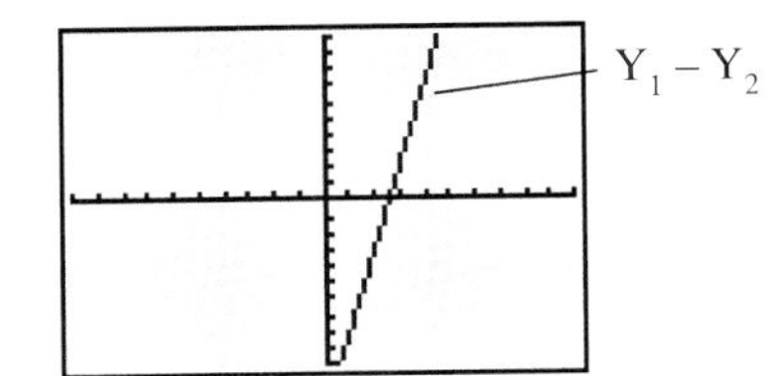

Figure 7.6

This is the graph of $Y_1 - Y_2$.
Use [TRACE] and [ZOOM] to estimate the value of X where it crosses the x-axis. As in Example 1 the solution is $X = 2.6$.

Testing the Equivalence of Two Algebraic Expressions

If two algebraic expressions are equivalent, they will produce the same graph. This property can be used to check algebraic simplifications.

EXAMPLE 3 Use the graphing calculator.

a. Show that $3x - (2x - 3)$ simplifies to $x + 3$.

b. Show that $\dfrac{\sqrt[3]{6x^2}}{\sqrt[3]{2x}}$ simplifies to $\sqrt[3]{3x}$.

Solution Using [Y=], key in $Y_1 = 3X - (2X - 3)$, $Y_2 = X + 3$, and $Y_3 = Y_1 - Y_2 + 4$. (This expression for Y_3 is explained below.)

a. Step 1 (Plot both graphs.) Plot the graphs of Y_1 and Y_2 separately, and observe that the two graphs do appear to be the same. Step 2 will confirm that the graphs are identical.

Step 2 (Plot $Y_1 - Y_2$ + constant.) The graph of the difference of two equivalent expressions is identically zero and therefore graphs as the x-axis (and so cannot be seen). A more visible display is made by adding a constant to the difference; that is why we added 4 in defining Y_3. Thus if Y_1 and Y_2 are equivalent, then the graph of Y_3 will be a horizontal line 4 units above the x-axis and will be very easy to see. In this problem the graph of Y_3 is such a horizontal line, which confirms that the expressions $3x - (2x - 3)$ and $x + 3$ are equivalent.

b. This exercise involves cube roots, which can be entered two different ways on the TI-81. They can be entered as an expression to the 1/3 power, such as [X|T] [^] [(] 1 [÷] 3 [)], or by putting the cube root symbol $\sqrt[3]{\ }$ on the screen. To display the $\sqrt[3]{\ }$ symbol, press the [MATH] key followed by 4, which we symbolize as [MATH] $<\sqrt[3]{\ }>$.

Step 1 Press [Y=].
For Y_1 key in [MATH] $<\sqrt[3]{\ }>$ 6 [X|T] [x^2] [÷] [MATH] $<\sqrt[3]{\ }>$ 2 [X|T].
For Y_2 key in [MATH] $<\sqrt[3]{\ }>$ 3 [X|T].
For Y_3 key in $Y_1 - Y_2 + 4$.
The screen appears as shown in Figure 7.7.

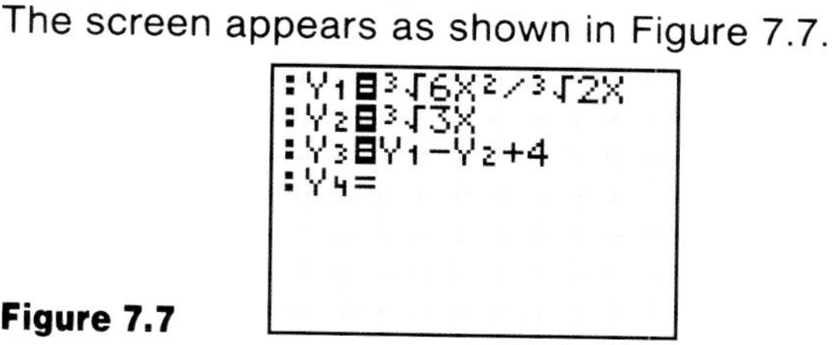

Figure 7.7

Step 2 Graph Y_1 and Y_2 separately to observe that they do look alike. To confirm the equality of Y_1 and Y_2, graph Y_3 by itself. The display of Y_3 will show a horizontal line 4 units above the x-axis.

EXERCISES

1. Use the graphing calculator to solve $6(x + 1) = 4x - 4$. Determine the solution to the nearest tenth. Ans. {−5.0}
2. Show that $\dfrac{3x + 1}{3}$ simplifies to $x + \frac{1}{3}$.
   Ans. The graphs are identical.
3. Show that $2x^{3/4} \cdot x^{1/4}$ simplifies to $2x$ (assume $x \geq 0$). Note that the TI-81 will not graph x raised to certain rational powers when x is negative. Ans. The graphs are identical.

Use the graphing calculator to solve or check these exercises from Chapter 7.

Section	*Exercises*
7.1	9, 27, 47, 91
7.2	27, 43, 85, 103, 119
7.3	37
7.4	11, 31, 57
7.5	13

Chapter 7 SUMMARY

OBJECTIVES CHECKLIST Specific chapter objectives are summarized below along with numbered example problems from the text that should clarify the objectives. If you do not understand any objectives or do not know how to do the selected problems, then restudy the material.

7.1 Can you:

1. **Find square roots and principal square roots?**
   Find $\sqrt{144}$. [Example 2a]
2. **Find principal *n*th roots?**
   Find $\sqrt[3]{-8}$. [Example 3b]
3. **Evaluate expressions containing rational exponents?**
   Evaluate $(-32)^{-3/5}$. [Example 7b]
4. **Use exponent properties to simplify expressions with rational exponents?**
   Perform the indicated operation(s), and write the result with only positive exponents:
   $\dfrac{x^{4/5}y^{-3/2}}{x^{-1}y}$. [Example 8e]

7.2 Can you:

1. **Simplify radicals?**
   Simplify $\sqrt{24x^9y^6}$. [Example 3b]
2. **Multiply and divide radicals, and simplify where possible?**
   Divide and simplify: $\dfrac{\sqrt[3]{21x^2}}{\sqrt[3]{12x}}$. [Example 7c]

7.3 Can you:

1. **Add and subtract radicals by using the distributive property?**
   Simplify $\sqrt{40} + \sqrt{90}$. [Example 2a]
2. **Add and subtract radicals that involve rationalizing denominators?**
   Simplify $\sqrt{3} + \dfrac{2}{\sqrt{3}}$. [Example 4b]

7.4 Can you:

1. **Use $(\sqrt[n]{a})^n = a$?**
   Simplify $(7\sqrt{5})^2$. [Example 1a]
2. **Multiply radical expressions involving more than one term?**
   Simplify $(\sqrt[3]{x} + 2)(\sqrt[3]{x} + 5)$. [Example 2c]
3. **Use a special product formula to multiply certain radical expressions?**
   Simplify $(5 + 2\sqrt{3})(5 - 2\sqrt{3})$. [Example 3b]
4. **Divide radical expressions involving more than one term in the numerator and a single term in the denominator?**
   Simplify $\dfrac{3 - 9\sqrt{2}}{6}$. [Example 5a]
5. **Use a conjugate to rationalize a denominator?**
   Rationalize the denominator of $\dfrac{8}{3 - \sqrt{5}}$. [Example 6]

7.5 Can you:

1. **Solve radical equations using the principle of powers once?**
   Solve $\sqrt{3x + 18} = x$. [Example 4]
2. **Solve radical equations using the principle of powers twice?**
   Solve $\sqrt{2x - 3} - \sqrt{x - 2} = 1$. [Example 6]

7.6 Can you:

1. **Express square roots of negative numbers in terms of *i*?**
   Express $-\sqrt{-3}$ in terms of i. [Example 1b]
2. **Add, subtract, and multiply complex numbers?**
   Multiply $\sqrt{-25}\sqrt{-4}$. [Example 3a]
3. **Divide complex numbers?**
   Write the quotient $\frac{2 + i}{3 + i}$ in the form $a + bi$. [Example 5]
4. **Find powers of *i*?**
   Simplify i^{15}. [Example 7a]

KEY TERMS

Complex number (7.6)
Conjugate of a complex number (7.6)
Conjugates (7.4)
Cube root (7.1)
Extraneous solution (7.5)
Geometric mean (7.1)
Imaginary number (7.6)
Imaginary part of a complex number (7.6)
Index of radical (7.1)
Like radicals (7.3)
Perfect square (7.1)
Principal *n*th root (7.1)
Principal square root (7.1)
Radical (7.1)
Radical equation (7.5)
Radical sign (7.1)
Radicand (7.1)
Rationalizing the denominator (7.2)
Real part of a complex number (7.6)
Square root (7.1)

KEY CONCEPTS AND PROCEDURES

Section	Key Concepts or Procedures to Review
7.1	■ Definitions of $a^{1/n}$, square root and nth root of a number ■ If m/n represents a reduced fraction such that $a^{1/n}$ is a real number, then $a^{m/n} = (\sqrt[n]{a})^m = \sqrt[n]{a^m}$.
7.2	■ Product and quotient properties of radicals (a, b, $\sqrt[n]{a}$, $\sqrt[n]{b}$ denote real numbers) 1. $\sqrt[n]{a \cdot b} = \sqrt[n]{a} \cdot \sqrt[n]{b}$ 2. $\sqrt[n]{\frac{a}{b}} = \frac{\sqrt[n]{a}}{\sqrt[n]{b}}$ $(b \neq 0)$ ■ Methods to simplify a radical by removing any factor of the radicand whose indicated root can be taken exactly, by eliminating any fractions in the radicand, or by writing the radical so that the index is as small as possible ■ Methods to multiply and divide radicals ■ Methods to rationalize the denominator
7.3	■ Methods to add and subtract radicals
7.4	■ $(\sqrt[n]{a})^n = a$ if $\sqrt[n]{a}$ is a real number. ■ Methods to multiply and divide radical expressions involving more than one term ■ Methods to rationalize a binomial denominator using conjugates

Section	Key Concepts or Procedures to Review
7.5	■ Methods to solve radical equations ■ Principle of powers
7.6	■ Definitions of principal square root of a negative number, imaginary number, complex number, and equality of complex numbers ■ $i = \sqrt{-1}$ and $i^2 = -1$ ■ Relationships among the various sets of numbers ■ Methods to add, subtract, multiply, and divide complex numbers ■ Property of conjugates: $(a + bi)(a - bi) = a^2 + b^2$ ■ Powers of i

CHAPTER 7 REVIEW EXERCISES

7.1

1. Find the square roots of 100. $-10, 10$
2. Find $\sqrt[3]{-64}$. -4
3. Evaluate $16^{1/4}$. 2
4. Evaluate $(-27)^{-2/3}$. $\frac{1}{9}$
5. Perform the indicated operation, and write the result using only positive exponents: $\left(\dfrac{3x^5}{12x^{-3}}\right)^{3/2}$. $\dfrac{x^{12}}{8}$

7.2

6. Simplify $\sqrt{125x^7y^4}$. $5x^3y^2\sqrt{5x}$
7. Simplify $\sqrt[8]{x^4}$. Assume $x \geq 0$. $\sqrt{x}$
8. Simplify $\sqrt{\dfrac{12}{x}}$. Assume $x > 0$. $\dfrac{2\sqrt{3x}}{x}$
9. Multiply and simplify $6\sqrt[3]{16} \cdot 3\sqrt[3]{2}$. $36\sqrt[3]{4}$
10. Rationalize the denominator: $\dfrac{8}{\sqrt{12}}$. $\dfrac{4\sqrt{3}}{3}$

7.3

Simplify where possible. Assume $x > 0$, $y > 0$.

11. $8\sqrt{7} - 12\sqrt{7}$. $-4\sqrt{7}$
12. $\sqrt[3]{-24} + 2\sqrt[3]{375}$. $8\sqrt[3]{3}$
13. $\sqrt{\dfrac{x}{y}} - \sqrt{\dfrac{y}{x}}$ $\dfrac{x-y}{xy}\sqrt{xy}$
14. $\sqrt{60} + \frac{1}{2}\sqrt{\frac{3}{5}}$. $\frac{21}{10}\sqrt{15}$

7.4

Simplify. Assume $x > 0$, $y > 0$.

15. $(\sqrt[4]{x+7})^4$ $x + 7$
16. $\sqrt{10}(\sqrt{5} - \sqrt{10})$ $5\sqrt{2} - 10$
17. $(2\sqrt{y} - 3)^2$ $4y - 12\sqrt{y} + 9$
18. $\dfrac{4 - \sqrt{x}}{\sqrt{x}}$ $\dfrac{4\sqrt{x} - x}{x}$
19. Rationalize the denominator: $\dfrac{6}{1 + \sqrt{11}}$. $\dfrac{-3 + 3\sqrt{11}}{5}$

7.5

Solve each equation.

20. $\sqrt{3x+1} - 4 = 0$ $\{5\}$
21. $\sqrt[3]{2x+1} = \sqrt[3]{x-2}$ -3
22. $\sqrt{3x+4} = x$ $\{4\}$
23. $\sqrt{3x+16} - 6 = x$ $\{-5, -4\}$
24. $\sqrt{x+1} - \sqrt{x-4} = 1$ $\{8\}$

7.6

25. Combine $\sqrt{-16} + \sqrt{-100}$. $14i$
26. Multiply $(2 + 4i)(5 - 2i)$. $18 + 16i$
27. Evaluate $x^2 + 2x - 1$ if $x = 1 + i\sqrt{2}$. $4i\sqrt{2}$
28. Write the quotient $\dfrac{1 - 3i}{2 + 3i}$ in the form $a + bi$. $\frac{-7}{13} - \frac{9}{13}i$
29. Simplify i^{23}. $-i$

ADDITIONAL REVIEW EXERCISES

30. Express $-\sqrt{-25}$ in terms of i. $-5i$
31. Evaluate $x^2 + 3x + 1$ if $x = 1 + i\sqrt{3}$. $2 + 5i\sqrt{3}$
32. Express $\dfrac{4 + 5i}{3 + 2i}$ in the form $a + bi$. $\frac{22}{13} + \frac{7}{13}i$
33. Find the geometric mean of 60 and 15. 30
34. What is the period in simplified radical form of a pendulum that is 9 ft long? Use $T = 2\pi\sqrt{\dfrac{\ell}{32}}$. $\dfrac{3\pi\sqrt{2}}{4}$ seconds

Rationalize the denominator.

35. $\dfrac{8}{3 + \sqrt{5}}$ $6 - 2\sqrt{5}$
36. $\dfrac{5}{\sqrt{12x}}$ $\dfrac{5\sqrt{3x}}{6x}$
37. $\dfrac{x\sqrt{3x}}{x - \sqrt{3x}}$ $\dfrac{x\sqrt{3x} + 3x}{x - 3}$

Simplify where possible. Assume $x > 0$, $y > 0$.

38. $\sqrt[4]{48}$ $2\sqrt[4]{3}$
39. $7\sqrt{20xy^2} - \sqrt{45xy^2}$ $11y\sqrt{5x}$
40. $(4\sqrt{5x})^2$ $80x$
41. $(8 + 2\sqrt{5})(8 - 2\sqrt{5})$ 44
42. $\sqrt[4]{x^{12}}$ x^3
43. $5\sqrt{2xy} + y\sqrt{2xy}$ $(5 + y)\sqrt{2xy}$
44. $\sqrt{-100}$ $10i$
45. $(-32)^{-2/5}$ $\frac{1}{4}$

46. $\sqrt[3]{16x^7y^5}$ $2x^2y\sqrt[3]{2xy^2}$

47. $3\sqrt{6} + \dfrac{9}{\sqrt{6}}$ $\dfrac{9\sqrt{6}}{2}$

48. $(\sqrt{3} + \sqrt{2})(\sqrt{3} - \sqrt{2})$ 1

49. $\sqrt{\frac{5}{7}}$ $\dfrac{\sqrt{35}}{7}$

50. $\dfrac{5 + 10\sqrt{3}}{10}$ $\dfrac{1 + 2\sqrt{3}}{2}$

51. i^{19} $-i$

52. $(-125)^{1/3}$ -5

53. $\dfrac{5^{1/4}}{5}$ $\dfrac{1}{5^{3/4}}$

54. $\dfrac{x^{-3} \cdot x \cdot y^{3/4}}{x^{-2}y^{1/2}}$ $y^{1/4}$

55. $6\sqrt[3]{18} \cdot 3\sqrt[3]{4}$ $36\sqrt[3]{9}$

56. $\dfrac{\sqrt[3]{15x^2}}{\sqrt[3]{10x}}$ $\dfrac{\sqrt[3]{12x}}{2}$

57. $\sqrt{-16}\sqrt{-4}$ -8

58. $(4 - 3i) - (2 + 6i)$ $2 - 9i$

Solve.

59. $\sqrt{4x + 5} - 3 = 2$ $\{5\}$

60. $\sqrt[4]{2x - 6} = \sqrt[4]{3x - 24}$ $\{18\}$

61. $\sqrt{-4x - 3} = x$ $\emptyset$

62. $3 + \sqrt{-2x + 6} = x$ $\{3\}$

63. $\sqrt{8x + 1} - \sqrt{4x} = 1$ $\{0,1\}$

CHAPTER 7 TEST

1. Simplify $\dfrac{x^4 \cdot y^2 \cdot y^{-3}}{xy}$, and write the results using only positive exponents. $\dfrac{x^3}{y^2}$
2. Rationalize the denominator: $\dfrac{6}{2 - \sqrt{7}}$. $-4 - 2\sqrt{7}$
3. Evaluate $x^2 - 2x - 2$ for $x = 1 - i\sqrt{5}$. -8
4. Evaluate $(32)^{-1/5}$. $\frac{1}{2}$
5. Simplify i^{51}. $-i$

In Questions 6–8, solve each equation.

6. $\sqrt[3]{4x + 1} = \sqrt[3]{5x - 12}$ $\{13\}$
7. $\sqrt{2x + 3} = x$ $\{3\}$
8. $\sqrt{5x + 6} + 6 = 0$ $\emptyset$

In Questions 9–12, perform the indicated operation, and write the results in the form $a + bi$.

9. $\sqrt{-15}\sqrt{-5}$ $-5\sqrt{3} + 0i$
10. $(8 + 6i) - (-3 + 2i)$ $11 + 4i$
11. $(7 + 2i)(8 - 3i)$ $62 - 5i$
12. $\dfrac{2 - 3i}{1 + 2i}$ $-\frac{4}{5} - \frac{7}{5}i$

In Questions 13–20, simplify where possible. Assume $x > 0$, $y > 0$.

13. $(8\sqrt{6x})^2$ $384x$
14. $\dfrac{\sqrt[3]{6y^2}}{\sqrt[3]{2y}}$ $\sqrt[3]{3y}$
15. $(\sqrt{5} + \sqrt{7})(\sqrt{5} - \sqrt{7})$ -2
16. $8\sqrt{3xy} - 7\sqrt{27x^3y^3}$ $(8 - 21xy)\sqrt{3xy}$
17. $\sqrt{\dfrac{5}{3x}}$ $\dfrac{\sqrt{15x}}{3x}$
18. $\sqrt[6]{x^8}$ $x\sqrt[3]{x}$
19. $(6\sqrt{2x})(3\sqrt{10x})$ $36x\sqrt{5}$
20. $(\sqrt[3]{y + 2})^3$ $y + 2$

CUMULATIVE TEST 7

1. List the numbers in the set of negative integers. $\{-1, -2, -3, -4, \ldots\}$
2. Simplify $1 - 2[3 - 4(5 - 6)]$. -13
3. Solve $|-2x + 1| = 5$. $\{-2, 3\}$
4. Solve $-4x + 1 \leq -x - 11$. $[4, \infty)$
5. A book and a toy together cost \$45. The toy costs \$3 more than half the price of the book. How much does each item cost? Toy, \$17; book, \$28
6. Express the number 405,000,000 in scientific notation. 4.05×10^8
7. Multiply and simplify: $(4x - 2)(5x - 3)$. $20x^2 - 22x + 6$
8. Solve $y^2 + 13y - 30 = 0$. $\{-15, 2\}$
9. Simplify $\dfrac{x^{-1} + 1}{x^{-2} - 1}$. $\dfrac{x}{1 - x}$
10. Divide and express the result in lowest terms: $\dfrac{6a}{(a + 6)^4} \div \dfrac{a^3}{(a + 6)^6}$. $\dfrac{6(a + 6)^2}{a^2}$
11. Combine and express the results in lowest terms: $1 + \dfrac{2}{n + 1} - \dfrac{8}{(n + 1)^2}$. $\dfrac{n^2 + 4n - 5}{(n + 1)^2}$
12. Graph the equation $y = -x + 2$.

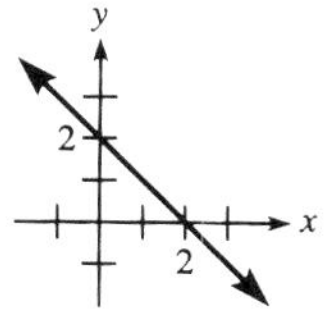

13. Find the slope of the line through the points $(1, -2)$ and $(5, 2)$. 1
14. Write an equation of the line with y-intercept $(0, 2)$ that is perpendicular to the line whose equation is $y = -3x + 1$. $y = \frac{1}{3}x + 2$
15. Solve the system. $-x + 2y = 8$, $3x - y = -9$ $(-2, 3)$
16. Evaluate $\begin{vmatrix} -2 & 3 & 2 \\ 0 & 7 & 4 \\ 1 & -1 & 3 \end{vmatrix}$. -52
17. Solve the system. $x + y + z = 1$, $x + y - z = 3$, $x - y - z = 5$ $(3, -1, -1)$
18. Simplify $3\sqrt{50x^4y^3} - 2\sqrt{128x^4y^3}$. Assume $x > 0$, $y > 0$. $-x^2y\sqrt{2y}$
19. Solve: $11 - \sqrt{-2x + 5} = 2$. $\{-38\}$
20. Multiply the complex numbers $(2 - 3i)(3 + 4i)$. Express the product in the form $a + bi$. $18 - i$

CHAPTER 8

Second-Degree Equations in One and Two Variables

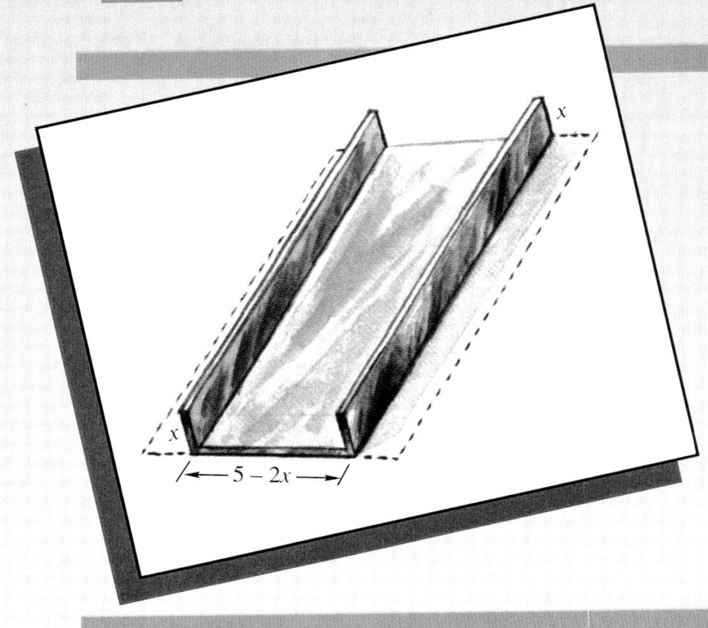

For use in the construction of a computer case, a strip of metal 5 in. wide is to be shaped into an open channel with cross-sectional area of 1 in.². (See sketch.) For the channel to be useful, the height and the width must both be at least $\frac{1}{4}$ in. Find the height and width of the channel to the nearest hundredth of an inch. (See Example 6 of Section 8.1.)

FIRST-DEGREE (or linear) equations in one variable and in two variables were considered in Chapters 2 and 5, respectively. Moving up one degree, this chapter first shows how to solve any second-degree (or quadratic) equation and then considers second-degree equations in two variables. In this progression, the coverage of inequalities and systems of equations will be extended.

8.1 Solving Quadratic Equations by Completing the Square

Chapter Outline

OBJECTIVES

1. **Solve quadratic equations using the square root property.**
2. **Complete the square for $x^2 + bx$, and express the resulting trinomial in factored form.**
3. **Solve $ax^2 + bx + c = 0$ by completing the square when $a = 1$.**
4. **Solve $ax^2 + bx + c = 0$ by completing the square when $a \neq 1$.**

1 As stated in Chapter 3, an equation that can be written in the general form $ax^2 + bx + c = 0$, where a, b, and c are real numbers with $a \neq 0$, is called a **second-degree** or **quadratic equation.** The simplest equation of this type has the form $x^2 + c = 0$, and it can be solved directly by using the following **square root property.**

See "Think About It" Exercises 1 and 2 for extensions to cube root and fourth root properties.

Square Root Property

If a is any real number, then

$$x^2 = a \quad \text{implies} \quad x = \sqrt{a} \quad \text{or} \quad x = -\sqrt{a}.$$

To see the origin of this property in terms of the factoring method of Section 3.9, consider the following steps.

$$x^2 = a$$
$$x^2 - a = 0$$
$$(x - \sqrt{a})(x + \sqrt{a}) = 0 \qquad \text{Factoring over the complex numbers}$$
$$x - \sqrt{a} = 0 \quad \text{or} \quad x + \sqrt{a} = 0 \qquad \text{Zero product principle}$$
$$x = \sqrt{a} \qquad\qquad x = -\sqrt{a}$$

Often, for convenience, the phrase "$x = \sqrt{a}$ or $x = -\sqrt{a}$" is abbreviated by writing $x = \pm\sqrt{a}$, which may be informally read as "plus or minus the square root of a."

EXAMPLE 1 Solve using the square root property.

a. $x^2 - 18 = 0$ **b.** $x^2 = -1$

Solution

a. $x^2 - 18 = 0$ is equivalent to $x^2 = 18$, which implies

$$x = \sqrt{18} \quad \text{or} \quad x = -\sqrt{18}.$$

Because $\sqrt{18} = \sqrt{9}\sqrt{2} = 3\sqrt{2}$, the solution set is $\{3\sqrt{2}, -3\sqrt{2}\}$, or $\{\pm 3\sqrt{2}\}$. Check this answer by substituting both solutions in the original equation.

b. $x^2 = -1$ implies $x = \sqrt{-1}$ or $x = -\sqrt{-1}$. Since $\sqrt{-1} = i$, the solution set is $\{i, -i\}$, or $\{\pm i\}$. Check that both solutions are correct.

Remind students that $i^2 = -1$. A common confusion is to write $i = -1$.

PROGRESS CHECK 1 Solve using the square root property.

a. $y^2 = 32$ **b.** $x^2 + 16 = 0$

The square root property is also efficient for solving quadratic equations of the form $(px + q)^2 = a$, as illustrated next in Example 2.

Progress Check Answers

1. (a) $\{4\sqrt{2}, -4\sqrt{2}\}$ (b) $\{4i, -4i\}$

EXAMPLE 2 Solve $(x + 4)^2 - 7 = 0$.

Solution The first step is to add 7 to both sides so that the equation has the form $X^2 = a$, where X stands for $x + 4$.

$$(x + 4)^2 - 7 = 0$$
$$(x + 4)^2 = 7 \quad \text{Add 7 to both sides.}$$

By the square root property, $(x + 4)^2 = 7$ implies that

$$x + 4 = \sqrt{7} \quad \text{or} \quad x + 4 = -\sqrt{7}.$$

This is a good place to point out that solutions of this form come in conjugate pairs.

Subtracting 4 from both sides will isolate the variable, giving

$$x = -4 + \sqrt{7} \quad \text{or} \quad x = -4 - \sqrt{7}.$$

Check Replacing x by $-4 + \sqrt{7}$ on the left in the original equation gives

$$[(-4 + \sqrt{7}) + 4]^2 - 7 = (\sqrt{7})^2 - 7 = 0,$$

which matches the member on the right. Similarly, $-4 - \sqrt{7}$ checks, so the solution set is $\{-4 + \sqrt{7}, -4 - \sqrt{7}\}$, or $\{-4 \pm \sqrt{7}\}$.

PROGRESS CHECK 2 Solve $(x - 3)^2 = 2$.

2 Example 2 showed that the square root property can be used to solve quadratic equations of a particular form. It is remarkable to note that by creative algebra it is possible to rewrite *any* quadratic equation in this desirable form. This ability means it is possible to derive a general method for solving all quadratic equations. As a first step, we develop a technique called **completing the square,** whose name dates back to the earliest Greek methods for solving quadratic equations geometrically. Consider an expression like

$$x^2 + 6x.$$

What constant needs to be added to make this expression a perfect square trinomial? Since the factoring model for a perfect square trinomial may be written as

$$x^2 + 2kx + k^2 = (x + k)^2,$$

we set $2k = 6$, so $k = 3$ and $k^2 = 9$. Thus, adding 9 gives

$$x^2 + 6x + 9 = (x + 3)^2.$$

More generally, to complete the square for

$$x^2 + bx,$$

we set $2k = b$, so $k = b/2$ and $k^2 = (b/2)^2$. Therefore, adding $(b/2)^2$ completes the square.

Completing the Square

To complete the square for $x^2 + bx$ with $b \neq 0$, add $(b/2)^2$, which is the square of one-half the coefficient of x.

In the diagram in Figure 8.1, note that the area of the figure is $x^2 + bx$ and that the area of the missing corner is $(b/2)^2$. This shows geometrically why adding $(b/2)^2$ turns the expression $x^2 + bx$ into a "complete" square.

Progress Check Answer

2. $\{3 + \sqrt{2}, 3 - \sqrt{2}\}$

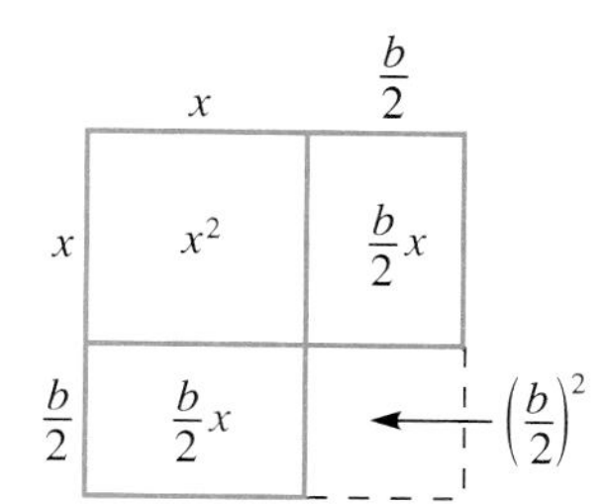

Figure 8.1

It is instructive to have students confirm that the area of the original figure is $x^2 + bx$.

EXAMPLE 3 Determine the number that should be added to make the given expression a perfect square. Then add this number, and factor the resulting trinomial.

a. $x^2 + 12x$ **b.** $y^2 - y$

Solution

a. The coefficient of x is 12, so $b = 12$. To complete the square, we add $(b/2)^2$, which is $(12/2)^2 = 6^2 = 36$. Adding 36 to the expression and factoring gives

$$x^2 + 12x + 36 = (x + 6)^2.$$

b. The coefficient of y is -1. Thus $b = -1$, and $(b/2)^2 = (-1/2)^2 = 1/4$. To complete the square, add $1/4$. Doing this and factoring yields

$$y^2 - y + \tfrac{1}{4} = (y - \tfrac{1}{2})^2.$$

PROGRESS CHECK 3 Determine the number that should be added to make the expression a perfect square. Then add this number, and factor the resulting trinomial.

a. $x^2 + x$ **b.** $y^2 - 10y$

3 The next example shows how completing the square can be used to solve a quadratic equation.

EXAMPLE 4 Solve $x^2 - 8x + 3 = 0$.

Solution First, note that the factoring method does not work for this equation. To solve by completing the square, begin by rearranging the equation so that the x terms are on one side of the equation and the constant term is on the other.

$$x^2 - 8x + 3 = 0$$
$$x^2 - 8x = -3 \quad \text{Subtract 3 from both sides.}$$

Next, complete the square on the left. Half of -8 is -4, and $(-4)^2 = 16$. By adding 16 to both sides of the equation, we then have

$$x^2 - 8x + 16 = -3 + 16 \quad \text{Add 16 to both sides.}$$
$$(x - 4)^2 = 13. \quad \text{Factor and simplify.}$$

Now apply the square root property and solve each equation.

$$x - 4 = \sqrt{13} \quad \text{or} \quad x - 4 = -\sqrt{13}$$
$$x = 4 + \sqrt{13} \qquad\qquad x = 4 - \sqrt{13}$$

Thus, the solution set is $\{4 + \sqrt{13}, 4 - \sqrt{13}\}$. Check this answer.

PROGRESS CHECK 4 Solve $x^2 + 6x - 1 = 0$.

4 The procedure for completing the square requires that the coefficient of x^2 be one. When the coefficient is not one, then as a first step, divide both sides of the equation by the coefficient of x^2, as shown next.

Progress Check Answers

3. (a) $x^2 + x + \frac{1}{4} = (x + \frac{1}{2})^2$
(b) $y^2 - 10y + 25 = (y - 5)^2$
4. $\{-3 + \sqrt{10}, -3 - \sqrt{10}\}$

"Think About It" Exercise 3 gives another approach for handling $a \neq 1$.

EXAMPLE 5 Solve $3x^2 - 2x - 3 = 0$.

Solution First, divide both sides of the equation by 3. Then write the x terms to the left of the equal sign and the constant term to the right.

$$3x^2 - 2x - 3 = 0$$
$$x^2 - \tfrac{2}{3}x - 1 = 0$$
$$x^2 - \tfrac{2}{3}x = 1$$

Now complete the square. Half of $-\frac{2}{3}$ is $-\frac{1}{3}$, and $(-\frac{1}{3})^2 = \frac{1}{9}$. Add $\frac{1}{9}$ to both sides of the equation and simplify.

$$x^2 - \tfrac{2}{3}x + \tfrac{1}{9} = 1 + \tfrac{1}{9}$$
$$(x - \tfrac{1}{3})^2 = \tfrac{10}{9}$$

By the square root property,

$$x - \tfrac{1}{3} = \sqrt{\tfrac{10}{9}} \quad \text{or} \quad x - \tfrac{1}{3} = -\sqrt{\tfrac{10}{9}}.$$

Finally, simplify $\sqrt{10/9}$ to $\sqrt{10}/3$, and solve both equations.

$$x - \frac{1}{3} = \frac{\sqrt{10}}{3} \quad \text{or} \quad x - \frac{1}{3} = -\frac{\sqrt{10}}{3}$$
$$x = \frac{1}{3} + \frac{\sqrt{10}}{3} \qquad x = \frac{1}{3} - \frac{\sqrt{10}}{3}$$
$$x = \frac{1 + \sqrt{10}}{3} \qquad x = \frac{1 - \sqrt{10}}{3}$$

Check to confirm that the solution set is $\left\{\frac{1 + \sqrt{10}}{3}, \frac{1 - \sqrt{10}}{3}\right\}$.

"Think About It" Exercise 5 asks for an exact algebraic check.

Note The solution can be checked by substituting these exact radical expressions in the original equation or by using the calculator, where $\sqrt{10}$ will be given a rational approximation. For instance, to check if $(1 + \sqrt{10})/3$ is a solution, first store the computed value of this number.

1 [+] 10 [√] [=] [÷] 3 [=] [1.3874259] [STO]

Now evaluate $3x^2 - 2x - 3$, where the stored value is x.

3 [×] [RCL] [x^2] [−] 2 [×] [RCL] [−] 3 [=] [0]

For this sequence, the end result in the display for some calculators may look like

[−1 −10]

which is scientific notation for -0.0000000001. This discrepancy from 0 is due to round-off errors and should be ignored.

PROGRESS CHECK 5 Solve $3x^2 - x - 5 = 0$.

Progress Check Answer

5. $\left\{\frac{1 + \sqrt{61}}{6}, \frac{1 - \sqrt{61}}{6}\right\}$

Recall that $x = \pm\sqrt{a}$ is the abbreviation for "$x = \sqrt{a}$ or $x = -\sqrt{a}$." In the next example we use such an abbreviation because it allows for a more compact method of solution.

EXAMPLE 6 Solve the problem in the chapter introduction on page 362.

Solution Referring to Figure 8.2, if we label the height of the channel as x, then the width can be represented by $5 - 2x$. The cross-sectional area is rectangular, and using $A = \ell w$ leads to the equation

$$\text{area} = 1$$
$$x(5 - 2x) = 1.$$

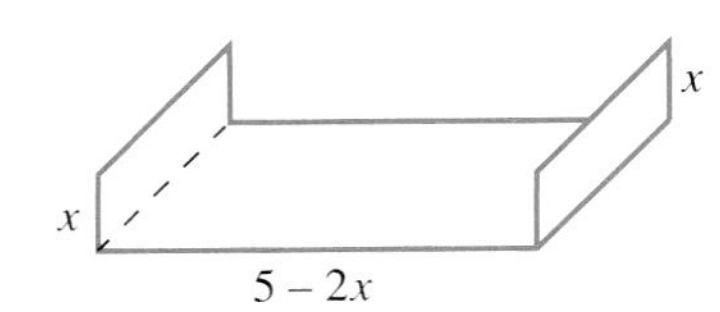

Figure 8.2

Now apply the method of completing the square.

$$5x - 2x^2 = 1 \quad \text{Remove parentheses.}$$
$$-2x^2 + 5x = 1 \quad \text{Reorder terms.}$$
$$x^2 - \frac{5}{2}x = -\frac{1}{2} \quad \text{Divide both sides by } -2.$$
$$x^2 - \frac{5}{2}x + \frac{25}{16} = -\frac{1}{2} + \frac{25}{16} \quad \text{Add } \left[\frac{1}{2}\left(-\frac{5}{2}\right)\right]^2 \text{ to both sides.}$$
$$\left(x - \frac{5}{4}\right)^2 = \frac{17}{16} \quad \text{Factor and simplify.}$$
$$x - \frac{5}{4} = \pm\sqrt{\frac{17}{16}} \quad \text{Square root property.}$$
$$x - \frac{5}{4} = \pm\frac{\sqrt{17}}{4} \quad \text{Simplify the radical.}$$
$$x = \frac{5}{4} \pm \frac{\sqrt{17}}{4} \quad \text{Add } \frac{5}{4} \text{ to both sides.}$$
$$x = \frac{5 \pm \sqrt{17}}{4} \quad \text{Combine fractions.}$$

The solution set for the equation is $\left\{\frac{5 + \sqrt{17}}{4}, \frac{5 - \sqrt{17}}{4}\right\}$.

Finally, we check to see if both solutions meet the requirements of the problem that the height and width of the channel measure at least $\frac{1}{4}$ in. By calculator (to the nearest hundredth),

$$\frac{5 + \sqrt{17}}{4} = 2.28 \quad \text{and} \quad \frac{5 - \sqrt{17}}{4} = 0.22.$$

Therefore, only the first answer is useful, and the approximate dimensions of the channel are a height equal to 2.28 in. and a width equal to $5 - 2(2.28) = 0.44$ in.

PROGRESS CHECK 6 The design for a rectangular solar panel specifies that the area is to be 18 m^2 and the length is to be 4 m greater than the width. To the nearest hundredth of a meter, find the length and the width of this panel.

Progress Check Answer

6. Width = 2.69 m, length = 6.69 m

EXERCISES 8.1

In Exercises 1–34, solve using the square root property.

1. $x^2 = 4$ $\{\pm 2\}$
2. $x^2 = 9$ $\{\pm 3\}$
3. $3x^2 = 48$ $\{\pm 4\}$
4. $2x^2 = 72$ $\{\pm 6\}$
5. $y^2 = 12$ $\{\pm 2\sqrt{3}\}$
6. $y^2 = 20$ $\{\pm 2\sqrt{5}\}$
7. $x^2 + 4 = 0$ $\{\pm 2i\}$
8. $x^2 + 9 = 0$ $\{\pm 3i\}$
9. $y^2 + 28 = 0$ $\{\pm 2i\sqrt{7}\}$
10. $y^2 + 50 = 0$ $\{\pm 5i\sqrt{2}\}$
11. $(x + 2)^2 = 16$ $\{-6, 2\}$
12. $(x + 1)^2 = 25$ $\{-6, 4\}$
13. $(x + 3)^2 = -16$ $\{-3 \pm 4i\}$
14. $(y + 4)^2 = -27$ $\{-4 \pm 3i\sqrt{3}\}$
15. $(x + 2)^2 - 6 = 0$ $\{-2 \pm \sqrt{6}\}$
16. $(x + 4)^2 - 12 = 0$ $\{-4 \pm 2\sqrt{3}\}$
17. $(y + 5)^2 + 5 = 0$ $\{-5 \pm i\sqrt{5}\}$
18. $(y + 2)^2 + 20 = 0$ $\{-2 \pm 2i\sqrt{5}\}$
19. $(x - 3)^2 = -25$ $\{3 \pm 5i\}$
20. $(x - 5)^2 = -36$ $\{5 \pm 6i\}$
21. $(y - 4)^2 = -45$ $\{4 \pm 3i\sqrt{5}\}$
22. $(y - 7)^2 = -50$ $\{7 \pm 5i\sqrt{2}\}$
23. $(x - 5)^2 - 36 = 0$ $\{-1, 11\}$
24. $(x - 2)^2 - 60 = 0$ $\{2 \pm 2\sqrt{15}\}$
25. $(y - 3)^2 + 7 = 0$ $\{3 \pm i\sqrt{7}\}$
26. $(y - 4)^2 + 6 = 0$ $\{4 \pm i\sqrt{6}\}$
27. $(3x + 2)^2 = 9$ $\{\frac{1}{3}, -\frac{5}{3}\}$
28. $(2x - 3)^2 = 4$ $\{\frac{1}{2}, \frac{5}{2}\}$
29. $(5x - 2)^2 = 10$ $\left\{\frac{2 \pm \sqrt{10}}{5}\right\}$
30. $(4x + 1)^2 = 7$ $\left\{\frac{-1 \pm \sqrt{7}}{4}\right\}$
31. $(3x - 1)^2 + 4 = 0$ $\left\{\frac{1 \pm 2i}{3}\right\}$
32. $(2x + 3)^2 + 1 = 0$ $\left\{\frac{-3 \pm i}{2}\right\}$
33. $(2x + 3)^2 = -12$ $\{-\frac{3}{2} \pm i\sqrt{3}\}$
34. $(3x - 4)^2 = -18$ $\{\frac{4}{3} \pm i\sqrt{2}\}$

In Exercises 35–46, determine the number that should be added to make the given expression a perfect square. Then add this number, and factor the resulting trinomial.

35. $x^2 + 10x$ 25; $(x + 5)^2$
36. $x^2 - 20x$ 100; $(x - 10)^2$
37. $y^2 - 8y$ 16; $(y - 4)^2$
38. $y^2 + 14y$ 49; $(y + 7)^2$
39. $x^2 - 5x$ $\frac{25}{4}$; $(x - \frac{5}{2})^2$
40. $x^2 + 7x$ $\frac{49}{4}$; $(x + \frac{7}{2})^2$
41. $y^2 - \frac{1}{4}y$ $\frac{1}{64}$; $(y - \frac{1}{8})^2$
42. $y^2 + \frac{1}{3}y$ $\frac{1}{36}$; $(y + \frac{1}{6})^2$
43. $x^2 - \frac{2}{3}x$ $\frac{1}{9}$; $(x - \frac{1}{3})^2$
44. $x^2 + \frac{2}{5}x$ $\frac{1}{25}$; $(x + \frac{1}{5})^2$
45. $y^2 + \frac{3}{5}y$ $\frac{9}{100}$; $(y + \frac{3}{10})^2$
46. $y^2 - \frac{5}{6}y$ $\frac{25}{144}$; $(y - \frac{5}{12})^2$

In Exercises 47–76, solve the given quadratic equations by completing the square.

47. $x^2 + 6x = 0$ $\{0, -6\}$
48. $x^2 - 10x = 0$ $\{0, 10\}$
49. $x^2 - 2x - 2 = 0$ $\{-1 \pm \sqrt{3}\}$
50. $x^2 + 2x - 4 = 0$ $\{-1 \pm \sqrt{5}\}$
51. $y^2 - 12y + 6 = 0$ $\{6 \pm \sqrt{30}\}$
52. $y^2 + 14y - 6 = 0$ $\{-7 \pm \sqrt{55}\}$
53. $x^2 + 10x - 4 = 0$ $\{-5 \pm \sqrt{29}\}$
54. $x^2 - 16x + 8 = 0$ $\{8 \pm 2\sqrt{14}\}$
55. $x^2 + 2x + 2 = 0$ $\{-1 \pm i\}$
56. $x^2 + 2x + 4 = 0$ $\{-1 \pm i\sqrt{3}\}$
57. $x^2 + 4x = -2$ $\{-2 \pm \sqrt{2}\}$
58. $x^2 - 10x = 4$ $\{5 \pm \sqrt{29}\}$
59. $y^2 - 8y = 2$ $\{4 \pm 3\sqrt{2}\}$
60. $y^2 + 12y = -6$ $\{-6 \pm \sqrt{30}\}$
61. $y^2 - y = -4$ $\left\{\frac{1 \pm i\sqrt{15}}{2}\right\}$
62. $y^2 + y = -6$ $\left\{\frac{-1 \pm i\sqrt{23}}{2}\right\}$
63. $x^2 + 6 = -2x$ $\{-1 \pm i\sqrt{5}\}$
64. $x^2 - 8 = 8x$ $\{4 \pm 2\sqrt{6}\}$
65. $y^2 - 12 = -6y$ $\{-3 \pm \sqrt{21}\}$
66. $y^2 + 4 = -6y$ $\{-3 \pm \sqrt{5}\}$
67. $2y^2 - y - 4 = 0$ $\left\{\frac{1 \pm \sqrt{33}}{4}\right\}$
68. $3x^2 + 4x - 6 = 0$ $\left\{\frac{-2 \pm \sqrt{22}}{3}\right\}$
69. $2y^2 - 4y - 5 = 0$ $\left\{\frac{2 \pm \sqrt{14}}{2}\right\}$
70. $4y^2 - 2y - 5 = 0$ $\left\{\frac{1 \pm \sqrt{21}}{4}\right\}$
71. $3x^2 + 9x - 1 = 0$ $\left\{\frac{-9 \pm \sqrt{93}}{6}\right\}$
72. $5x^2 - x - 1 = 0$ $\left\{\frac{1 \pm \sqrt{21}}{10}\right\}$
73. $2x^2 + 2x + 1 = 0$ $\left\{\frac{-1 \pm i}{2}\right\}$
74. $3x^2 + 3x + 2 = 0$ $\left\{\frac{-3 \pm i\sqrt{15}}{6}\right\}$
75. $3y^2 - 5y - 2 = 0$ $\{2, -\frac{1}{3}\}$
76. $-2y^2 + 5y + 3 = 0$ $\{-\frac{1}{2}, 3\}$

In Exercises 77–84, solve the equation by completing the square or by using the square root property. Give solutions to the nearest hundredth.

77. A 14-in. piece of wire must be bent as shown in the figure to make a three-sided frame. The area is to be 10 in.². Show that there are two different solutions to the problem.

x x
$14 - 2x$

Height = 6.19 in., base = 1.62 in.; height = 0.81 in., base = 12.38 in.

78. Refer to Exercise 77. A student claims that she can achieve the required area of 10 in.² using a shorter piece of wire (just 10 in. long). What are her two solutions? Height = 3.62 in., base = 2.76 in.; height = 1.38 in., base = 7.24 in.

79. Find the length of the side of the regular octagon that will fit between two studs set 30 in. apart. (See sketch.) Note that the corner right triangles have sides x, $\frac{30 - x}{2}$, and $\frac{30 - x}{2}$; and use the Pythagorean theorem to get a quadratic equation that you can solve for x. 12.43 in.

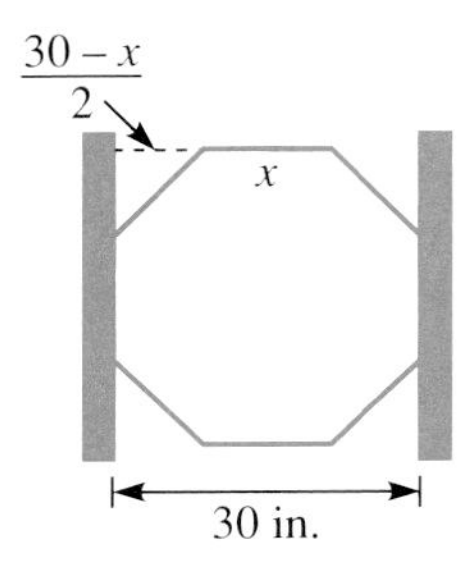

80. Refer to Exercise 79. Find the length of the sides of the octagon if the studs are 18 in. apart. 7.46 in.

81. On a sailboat the wind causes a wind pressure gauge to register 4.2 lb/ft². If the pressure p in pounds per square foot of a wind blowing at v mi/hour is given by

$$p = 0.003v^2,$$

find the wind speed at that moment. (Round to the nearest tenth.) 37.42 mi/hour

82. The scientist Galileo (1564–1642) discovered, when he rolled balls down an inclined plane (see figure), that the equation $v^2 = 64h$ relates the velocity of the ball and the *vertical* distance it has covered. Starting from rest, a ball which has dropped h ft has a velocity of v ft/second. It is remarkable that this velocity has nothing to do with the steepness of the plane. Use the given equation to find the velocity of a rolling ball when its vertical height is 2 ft below its starting height. 11.31 ft/second

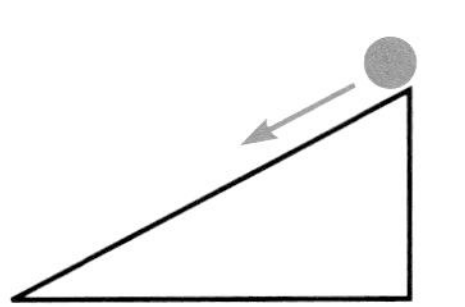

83. When P dollars is invested at an annual compound interest rate r for 2 years, then the compounded amount A is given by

$$A = P(1 + r)^2.$$

To the nearest hundredth of a percent, what compounded annual interest rate is needed for an investment to grow from \$1,000 to \$1,200 in 2 years? 9.54 percent

84. Refer to Exercise 83. To the nearest hundredth of a percent, what compounded interest rate is needed for an investment to grow from \$5,000 to \$5,200 in 2 years? 1.98 percent

THINK ABOUT IT

1. By analogy to the square root property, you can construct a "cube root property," useful in solving equations with the form $x^3 = a$. It would look like this: If a is any real number, then $x^3 = a$ implies $x = \sqrt[3]{a}$.

a. Use this property to solve $x^3 = -27$.

b. In more advanced work it is shown that every nonzero real number has three distinct cube roots, one real and two which have an imaginary component. Thus the cube root property in part **a** does not find *all* the solutions of $x^3 = -27$. Confirm that the other two solutions are $\dfrac{3 \pm 3i\sqrt{3}}{2}$ by showing that cubing each yields -27.

2. a. As in Exercise 1, write and use a "fourth root property" to solve $x^4 = 81$.

b. Confirm that there are four distinct fourth roots of 81 by showing that 3, -3, $3i$, and $-3i$ are all solutions to $x^4 = 81$.

3. To complete the square in $ax^2 + bx$, we divide by a to make the coefficient of x^2 equal to one. Another approach multiplies by some value to make the coefficient of x^2 a perfect square (other than one). For instance, to complete the square for $3x^2 + 2x$, we first multiply by 3, to get $9x^2 + 6x$. Then to get the desired constant to make $9x^2 + 6x$ a perfect square, we divide the coefficient of x by twice 3, which gives 1 in this problem. In general, divide by twice the square root of the new coefficient of x^2.

$$9x^2 + 6x + \frac{6}{2(3)} = 9x^2 + 6x + 1 = (3x + 1)^2$$

Use this method to solve $3x^2 + 2x + 1 = 0$.

4. Some quadratic equations of the form $P(x) = 0$ can be solved by factoring $P(x)$ into a product of the form $(x - a)(x - b)$, thus showing that a and b are the solutions. You can reverse this process to find a quadratic equation when you are given its solutions: Just set $(x - \text{solution}_1)(x - \text{solution}_2)$ equal to zero. Use this approach to find equations with these given solution sets.

a. $\{\pm\sqrt{7}\}$ **b.** $\{\pm 2i\}$ **c.** $\{5 \pm \sqrt{7}\}$ **d.** $\{-3 \pm i\}$

5. In Example 5 of this section completing the square gave $\dfrac{1 + \sqrt{10}}{3}$ and $\dfrac{1 - \sqrt{10}}{3}$ as the exact solutions of the equation $3x^2 - 2x - 3 = 0$.

a. By substitution, show that both solutions check in the original equation.

b. By substitution, show that $\dfrac{1 \pm \sqrt{61}}{6}$ both check as solutions to "Progress Check" Exercise 5: $3x^2 - x - 5 = 0$.

REMEMBER THIS

1. What are the two values of $\dfrac{-3 \pm \sqrt{9}}{2}$? $-3, 0$

2. Simplify $\sqrt{8}$. $2\sqrt{2}$

3. Simplify $\dfrac{c + ac}{c}$. $1 + a$

4. Simplify $\dfrac{2 + 2\sqrt{2}}{2}$. $1 + \sqrt{2}$

5. Multiply $(3 + i\sqrt{7})^2$. $2 + 6i\sqrt{7}$

6. Solve $\sqrt{2x} - 8 = 0$. $\{32\}$

7. Solve this system. $2x + 3y = -13$
$3x + 2y = -12$ $(-2, -3)$

8. If x varies directly as y and $x = 2$ when $y = 7$, find x when $y = 10$. $\frac{20}{7}$

9. Find the slope and y-intercept of the graph of $2x + 3y = 4$. $m = -\frac{2}{3}$, $(0, \frac{4}{3})$

10. Divide $3x^3 + x - 5$ by $x + 1$. $3x^2 - 3x + 4 - \dfrac{9}{x + 1}$

8.2 The Quadratic Formula

A motorist sees a dangerous situation straight ahead and immediately applies the brakes. Under certain conditions, a formula that approximates the braking distance of a car is

$$d = 0.045s^2 + 1.1s,$$

where d is the distance in feet and s is the speed in miles per hour. If these conditions apply and the braking distance cannot exceed 200 ft, what is the highest (integer) speed at which this car can be traveling to stop in time? (See Example 6.)

OBJECTIVES

1. **Solve quadratic equations using the quadratic formula.**
2. **Use the discriminant to determine the nature of the solutions of a quadratic equation.**
3. **Solve quadratic equations after choosing an efficient method.**

It wasn't until the seventeenth century that it was clear that all quadratic equations could be solved by one algebraic method. The earliest solutions were based on geometric constructions or arithmetical algorithms, which varied according to the form of the equation and avoided imaginary numbers.

1 The key feature of solving quadratic equations by completing the square is that this method applies to *all* quadratic equations. Therefore, if we apply this method to the general quadratic equation $ax^2 + bx + c = 0$ with $a \neq 0$, then we obtain a formula for solving quadratic equations that always works. In the following derivation of this formula, we also display the corresponding steps to solve the particular equation $2x^2 + 5x + 1 = 0$ to illustrate in specific terms what is happening.

General Equation	Particular Equation	Comment
$ax^2 + bx + c = 0, a \neq 0$	$2x^2 + 5x + 1 = 0$	Given equation.
$x^2 + \frac{b}{a}x + \frac{c}{a} = 0$	$x^2 + \frac{5}{2}x + \frac{1}{2} = 0$	Divide on both sides by the coefficient of x^2.
$x^2 + \frac{b}{a}x = -\frac{c}{a}$	$x^2 + \frac{5}{2}x = -\frac{1}{2}$	Subtract the constant term from both sides.
$x^2 + \frac{b}{a}x + \frac{b^2}{4a^2} = -\frac{c}{a} + \frac{b^2}{4a^2}$	$x^2 + \frac{5}{2}x + \frac{25}{16} = -\frac{1}{2} + \frac{25}{16}$	Add the square of one-half of the coefficient of x to both sides.
$\left(x + \frac{b}{2a}\right)^2 = \frac{b^2 - 4ac}{4a^2}$	$\left(x + \frac{5}{4}\right)^2 = \frac{17}{16}$	Factor on the left and add fractions on the right.
$x + \frac{b}{2a} = \pm\sqrt{\frac{b^2 - 4ac}{4a^2}}$	$x + \frac{5}{4} = \pm\sqrt{\frac{17}{16}}$	Apply the square root property.
$x + \frac{b}{2a} = \pm\frac{\sqrt{b^2 - 4ac}}{2a}$	$x + \frac{5}{4} = \frac{\pm\sqrt{17}}{4}$	Simplify the radical.
$x = \frac{-b}{2a} + \frac{\pm\sqrt{b^2 - 4ac}}{2a}$	$x = \frac{-5}{4} + \frac{\pm\sqrt{17}}{4}$	Isolate x on the left.
$x = \frac{-b \pm \sqrt{b^2 - 4ac}}{2a}$	$x = \frac{-5 \pm \sqrt{17}}{4}$	Combine fractions on the right.

In the particular equation, the two solutions are

$$x = \frac{-5 + \sqrt{17}}{4} \quad \text{and} \quad x = \frac{-5 - \sqrt{17}}{4},$$

and in the general equation, the two solutions are

$$x = \frac{-b + \sqrt{b^2 - 4ac}}{2a} \quad \text{and} \quad x = \frac{-b - \sqrt{b^2 - 4ac}}{2a}.$$

The work with the general equation results in the quadratic formula.

Quadratic Formula

If $ax^2 + bx + c = 0$, and $a \neq 0$, then

$$x = \frac{-b \pm \sqrt{b^2 - 4ac}}{2a}.$$

Any quadratic equation may be solved with this formula. The idea is to substitute appropriate values for a, b, and c in the formula and then simplify.

EXAMPLE 1 Solve $x^2 - 2x - 1 = 0$ using the quadratic formula.

Solution By comparing the given equation with the general equation,

$$\begin{array}{ccccccc} 1x^2 & - & 2x & - & 1 & = & 0 \\ \updownarrow & & \updownarrow & & \updownarrow & & \\ ax^2 & + & bx & + & c & = & 0 \end{array}$$

we see that $a = 1$, $b = -2$, and $c = -1$. Substituting these values into the quadratic formula and simplifying gives the solution.

$$\begin{aligned} x = \frac{-b \pm \sqrt{b^2 - 4ac}}{2a} &= \frac{-(-2) \pm \sqrt{(-2)^2 - 4(1)(-1)}}{2(1)} \\ &= \frac{2 \pm \sqrt{4 + 4}}{2} \\ &= \frac{2 \pm \sqrt{8}}{2} \\ &= \frac{2 \pm 2\sqrt{2}}{2} && \text{Simplify the radical.} \\ &= \frac{2(1 \pm \sqrt{2})}{2} && \text{Factor in the numerator.} \\ &= 1 \pm \sqrt{2} && \text{Divide out 2.} \end{aligned}$$

Now convert from abbreviated form to show both solutions.

$$x = 1 + \sqrt{2} \quad \text{or} \quad x = 1 - \sqrt{2}$$

Thus, the solution set is $\{1 + \sqrt{2}, 1 - \sqrt{2}\}$. Check this answer.

Caution A common student error is to interpret the quadratic formula as

$$\textbf{Wrong} \qquad\qquad\qquad\qquad \textbf{Wrong}$$

$$x = -b \pm \frac{\sqrt{b^2 - 4ac}}{2a} \quad \text{or as} \quad x = \frac{-b}{2a} \pm \sqrt{b^2 - 4ac}.$$

Remember to divide the *entire* expression $-b \pm \sqrt{b^2 - 4ac}$ by $2a$.

Other common errors are represented by these incorrect "simplifications."

$$\frac{1 + \sqrt{6}}{2} \neq 1 + \sqrt{3}; \quad \frac{1 + 2\sqrt{3}}{2} \neq 1 + \sqrt{3};$$

$$\frac{2 + \sqrt{3}}{2} \neq 1 + \sqrt{3}$$

PROGRESS CHECK 1 Solve $x^2 + 8x + 3 = 0$ using the quadratic formula.

For the next example, note that the quadratic formula is easily adjusted when the variable in the quadratic equation is not symbolized by x.

EXAMPLE 2 Solve $3y^2 - 2y = 5$ using the quadratic formula.

Solution First, put the equation in standard form to identify a, b, and c.

$$3y^2 - 2y = 5$$
$$3y^2 - 2y - 5 = 0 \quad \text{Subtract 5 from both sides.}$$

In this equation the variable is y, with $a = 3$, $b = -2$, and $c = -5$. Therefore,

$$y = \frac{-b \pm \sqrt{b^2 - 4ac}}{2a} = \frac{-(-2) \pm \sqrt{(-2)^2 - 4(3)(-5)}}{2(3)}$$
$$= \frac{2 \pm \sqrt{4 + 60}}{6}$$
$$= \frac{2 \pm \sqrt{64}}{6}$$
$$= \frac{2 \pm 8}{6}. \quad \text{Simplify the radical.}$$

Then simplify each solution separately.

$$y = \frac{2 + 8}{6} = \frac{10}{6} = \frac{5}{3} \quad \text{or} \quad y = \frac{2 - 8}{6} = \frac{-6}{6} = -1$$

Thus, the solution set is $\{\frac{5}{3}, -1\}$. Check this answer.

Note To solve quadratic equations using a calculator and the quadratic formula, it is usually best to start with the radical and store its value by substituting for a, b, and c in the following sequence.

b [x^2] [−] 4 [×] a [×] c [=] [√] [STO]

Then one solution adds the stored number.

b [+/−] [+] [RCL] [=] [÷] 2 [÷] a [=]

The other solution subtracts the stored number.

b [+/−] [−] [RCL] [=] [÷] 2 [÷] a [=]

Check this method using the equation in this example.

Note that division by $2a$ is done by consecutive divisions rather than by using parentheses.
A quadratic formula program for the TI-81 calculator is given in the graphing calculator section at the end of this chapter.

PROGRESS CHECK 2 Solve $5t^2 + 3t - 2 = 0$ using the quadratic formula.

In the next example the solution has an imaginary component.

EXAMPLE 3 Solve $x^2 + 4 = 3x$.

Solution To obtain the form $ax^2 + bx + c = 0$, first subtract $3x$ from both sides of the equation.

$$x^2 + 4 = 3x$$
$$x^2 - 3x + 4 = 0$$

Progress Check Answers

1. $\{-4 \pm \sqrt{13}\}$
2. $\{-1, \frac{2}{5}\}$

From this equation, we see that $a = 1$, $b = -3$, and $c = 4$, so

$$x = \frac{-b \pm \sqrt{b^2 - 4ac}}{2a} = \frac{-(-3) \pm \sqrt{(-3)^2 - 4(1)(4)}}{2(1)}$$

$$= \frac{3 \pm \sqrt{9 - 16}}{2}$$

$$= \frac{3 \pm \sqrt{-7}}{2}$$

$$= \frac{3 \pm i\sqrt{7}}{2}. \quad \text{Simplify the radical.}$$

The solution set is $\left\{\frac{3 + i\sqrt{7}}{2}, \frac{3 - i\sqrt{7}}{2}\right\}$.

Check We show the check for $\frac{3 + i\sqrt{7}}{2}$. Because the solutions must come in conjugate pairs, it is sufficient to check just one of them.

$$x^2 + 4 = 3x \quad \text{Original equation.}$$

$$\left(\frac{3 + i\sqrt{7}}{2}\right)^2 + 4 \stackrel{?}{=} 3\left(\frac{3 + i\sqrt{7}}{2}\right) \quad \text{Replace } x \text{ by } \frac{3 + i\sqrt{7}}{2}.$$

$$\frac{9 + 6i\sqrt{7} - 7}{4} + 4 \stackrel{?}{=} \frac{9 + 3i\sqrt{7}}{2}$$

$$\frac{2 + 6i\sqrt{7}}{4} + 4 \stackrel{?}{=} \frac{9 + 3i\sqrt{7}}{2}$$

$$\frac{1 + 3i\sqrt{7}}{2} + \frac{8}{2} \stackrel{?}{=} \frac{9 + 3i\sqrt{7}}{2}$$

$$\frac{9 + 3i\sqrt{7}}{2} \stackrel{\checkmark}{=} \frac{9 + 3i\sqrt{7}}{2}$$

The solution checks and the solution set is correct.

PROGRESS CHECK 3 Solve $2x^2 + 3 = x$.

2 Examples 1–3 have illustrated quadratic equations with solutions that are irrational numbers, rational numbers, and complex conjugate numbers, respectively. Based on the quadratic formula, note that the nature of the solutions depends on the value of $b^2 - 4ac$, which is called the **discriminant.** This expression appears under the radical in the quadratic formula and reveals the following about the solutions of $ax^2 + bx + c = 0$.

When a, b, c Are Rational and	The Solutions Are
$b^2 - 4ac < 0$	Conjugate complex numbers
$b^2 - 4ac = 0$	Real, rational, equal (1 solution)
$b^2 - 4ac > 0$ and a perfect square	Real, rational, unequal (2 solutions)
$b^2 - 4ac > 0$ but not a perfect square	Real, irrational, unequal (2 solutions)

Note the importance of saying a, b, c are rational. For instance, this analysis fails for $x^2 + \sqrt{8}x + 1 = 0$.

To illustrate, we use the equation from Example 1: $x^2 - 2x - 1 = 0$. Replacing a by 1, b by -2, and c by -1 in the expression for the discriminant gives

$$b^2 - 4ac = (-2)^2 - 4(1)(-1) = 8.$$

Because 8 is greater than 0 and is not a perfect square, we know (without solving the equation) that there are two different real number solutions that are irrational.

Progress Check Answer

3. $\left\{\frac{1 + i\sqrt{23}}{4}, \frac{1 - i\sqrt{23}}{4}\right\}$

EXAMPLE 4 Use the discriminant to determine the nature of the solutions of the equation $2x^2 + 5 = 4x$.

Solution This equation is equivalent to $2x^2 - 4x + 5 = 0$, in which $a = 2$, $b = -4$, and $c = 5$. The discriminant is

$$b^2 - 4ac = (-4)^2 - 4(2)(5) = -24.$$

Since -24 is less than 0, the solutions are a pair of complex conjugate numbers.

PROGRESS CHECK 4 Use the discriminant to determine the nature of the solutions of the equation $25y^2 + 16 = 40y$.

3 In this chapter we have shown several additional methods besides the factoring method of Section 3.9 for solving quadratic equations. Selecting an efficient method depends on the particular equation to be solved, and the following guidelines will help in your choice of methods.

Guidelines to Solve a Quadratic Equation

Equation Type	Recommended Method
$ax^2 + c = 0$	Use the square root property. If $ax^2 + c$ is a difference of squares, consider the factoring method.
$ax^2 + bx = 0$	Use the factoring method.
$(px + q)^2 = k$	Use the square root property.
$ax^2 + bx + c = 0$	First, try the factoring method, which applies when the discriminant reveals rational number solutions. If $ax^2 + bx + c$ does not factor or is hard to factor, use the quadratic formula.

For instance, completing the square is useful in determining the center of a circle. See Section 8.6.

Note that solving by completing the square is not recommended as an efficient method. Instead, this method is important because it is the basis for the quadratic formula and because in other problem-solving situations, completing the square is used to convert an expression to a standard form that is easier to analyze.

EXAMPLE 5 Solve the equation $25s^2 - 40s + 16 = 0$. Choose an efficient method.

Solution The equation is of the form $ax^2 + bx + c = 0$, where $a = 25$, $b = -40$, and $c = 16$. First, we try the factoring method, and we find that it can be used.

$$25s^2 - 40s + 16 = 0$$
$$(5s - 4)(5s - 4) = 0$$
$$5s - 4 = 0 \quad \text{or} \quad 5s - 4 = 0$$
$$s = \tfrac{4}{5} \qquad\qquad s = \tfrac{4}{5}$$

If $(x - a)^n$ is a factor of $f(x) = 0$, then a is called a root (or solution) with *multiplicity* n. For $n = 1$, a is called a *simple* root. In this example $\frac{4}{5}$ is a root of multiplicity 2.

Thus, there is exactly one solution, and the solution set is $\{\frac{4}{5}\}$.

If the factoring step in this solution seemed hard, then an alternative solution method is to use the quadratic formula.

$$x = \frac{-b \pm \sqrt{b^2 - 4ac}}{2a} = \frac{-(-40) \pm \sqrt{(-40)^2 - 4(25)(16)}}{2(25)}$$
$$= \frac{40 \pm \sqrt{1{,}600 - 1{,}600}}{50} = \frac{40 \pm \sqrt{0}}{50} = \frac{40 \pm 0}{50} = \frac{40}{50} = \frac{4}{5}$$

By either method, the solution set is $\{\frac{4}{5}\}$.

Progress Check Answer

4. One real number that is rational

PROGRESS CHECK 5 Solve $9x^2 = 100$. Choose an efficient method.

EXAMPLE 6 Solve the problem in the section introduction on page 370.

Solution Replacing d by 200 in the given formula leads to

$$200 = 0.045s^2 + 1.1s$$
$$0 = 0.045s^2 + 1.1s - 200.$$

Then applying the quadratic formula to the resulting equation in which $a = 0.045$, $b = 1.1$, and $c = -200$ gives

$$s = \frac{-1.1 \pm \sqrt{1.1^2 - 4(0.045)(-200)}}{2(0.045)} = \frac{-1.1 \pm \sqrt{1.21 + 36}}{0.09}$$
$$= \frac{-1.1 \pm \sqrt{37.21}}{0.09} = \frac{-1.1 \pm 6.1}{0.09}.$$

Because the speed cannot be negative, reject $s = (-1.1 - 6.1)/0.09$, which leaves

$$s = \frac{-1.1 + 6.1}{0.09} = \frac{5}{0.09} \approx 55.555556 \quad \text{(by calculator)}.$$

To the nearest integer, $s = 56$. But the braking distance is over 200 ft in this case, and the required value for s is 55, as confirmed in the following check.

$$s = 56: \quad d = 0.045(56)^2 + 1.1(56) = 202.72$$
$$s = 55: \quad d = 0.045(55)^2 + 1.1(55) = 196.625$$

Thus, if the car is to stop in time, the highest (integer) speed at which it can be traveling is 55 mi/hour.

PROGRESS CHECK 6 Redo the problem in Example 6, but assume that the braking distance cannot exceed 100 ft.

Progress Check Answers

5. $\{\frac{10}{3}, -\frac{10}{3}\}$
6. 36 mi/hour

EXERCISES 8.2

In Exercises 1–16, solve the given equations using the quadratic formula.

1. $x^2 - 2x - 3 = 0$ $\{-1, 3\}$
2. $y^2 + 3y - 4 = 0$ $\{1, -4\}$
3. $y^2 - 4y + 4 = 0$ $\{2\}$
4. $x^2 - 4x + 3 = 0$ $\{1, 3\}$
5. $t^2 + 5t - 4 = 0$ $\left\{\frac{-5 \pm \sqrt{41}}{2}\right\}$
6. $x^2 + 3x - 2 = 0$ $\left\{\frac{-3 \pm \sqrt{17}}{2}\right\}$
7. $y^2 - 4y - 3 = 0$ $\{2 \pm \sqrt{7}\}$
8. $t^2 + 4t = -3$ $\{-3, -1\}$
9. $x^2 + 5x = -4$ $\{-4, -1\}$
10. $y^2 - 2y = -5$ $\{1 \pm 2i\}$
11. $x^2 + 2x + 3 = 0$ $\{-1 \pm i\sqrt{2}\}$
12. $4x^2 - 4x - 15 = 0$ $\{-\frac{3}{2}, \frac{5}{2}\}$
13. $2x^2 - 5x - 3 = 0$ $\{-\frac{1}{2}, 3\}$
14. $5t^2 + 33t = 14$ $\{-7, \frac{2}{5}\}$
15. $6x^2 - 13x = -6$ $\{\frac{2}{3}, \frac{3}{2}\}$
16. $4x^2 - 25 = 0$ $\{-\frac{5}{2}, \frac{5}{2}\}$

In Exercises 17–36, use the discriminant to determine the nature of the solutions.

17. $x^2 - 3x + 2 = 0$ Real, rational, unequal (2 solutions)
18. $x^2 + x - 12 = 0$ Real, rational, unequal (2 solutions)
19. $x^2 + 8x + 16 = 0$ Real, rational, equal (1 solution)
20. $y^2 + 12y + 36 = 0$ Real, rational, equal (1 solution)
21. $t^2 - 3t + 7 = 0$ Conjugate complex numbers
22. $t^2 - 7t + 14 = 0$ Conjugate complex numbers
23. $y^2 - 3y = 0$ Real, rational, unequal (2 solutions)
24. $x^2 + 4x = 0$ Real, rational, unequal (2 solutions)
25. $x^2 - 2 = -3x$ Real, irrational, unequal (2 solutions)
26. $x^2 - 3 = 5x$ Real, irrational, unequal (2 solutions)
27. $3t^2 - 7t + 4 = 0$ Real, rational, unequal (2 solutions)
28. $2t^2 - 5t - 3 = 0$ Real, rational, unequal (2 solutions)
29. $y^2 - 9 = 0$ Real, rational, unequal (2 solutions)
30. $y^2 - 16 = 0$ Real, rational, unequal (2 solutions)
31. $x^2 + 49 = 0$ Conjugate complex numbers
32. $x^2 + 64 = 0$ Conjugate complex numbers
33. $5y^2 - 3y + 4 = 0$ Conjugate complex numbers
34. $2y^2 + 5y - 3 = 0$ Real, rational, unequal (2 solutions)
35. $\frac{1}{2}t^2 - \frac{3}{8}t + \frac{1}{4} = 0$ Conjugate complex numbers
36. $\frac{1}{8}t^2 + \frac{3}{2}t - \frac{1}{4} = 0$ Real, irrational, unequal (2 solutions)

In Exercises 37–60, solve the quadratic equations after choosing an efficient method.

37. $(x - 2)^2 = 16$ $\{-2, 6\}$
38. $(x + 3)^2 = 9$ $\{-6, 0\}$
39. $x^2 = 9$ $\{\pm 3\}$
40. $x^2 = 36$ $\{\pm 6\}$
41. $x^2 + 5x = 0$ $\{-5, 0\}$
42. $x^2 - 6x = 0$ $\{0, 6\}$

43. $2t^2 - 5t - 3 = 0$ $\{-\frac{1}{2}, 3\}$
44. $5t^2 - 14 = -33t$ $\{-7, \frac{2}{5}\}$
45. $y^2 - 2y - 15 = 0$ $\{-3, 5\}$
46. $y^2 + 7y + 12 = 0$ $\{-4, -3\}$
47. $t^2 - 16 = 0$ $\{\pm 4\}$
48. $t^2 - 121 = 0$ $\{\pm 11\}$
49. $x^2 + 5x - 4 = 0$ $\left\{\frac{-5 \pm \sqrt{41}}{2}\right\}$
50. $x^2 + 3x - 1 = 0$ $\left\{\frac{-3 \pm \sqrt{13}}{2}\right\}$
51. $9t^2 - 25 = 0$ $\{-\frac{5}{3}, \frac{5}{3}\}$
52. $4t^2 - 1 = 0$ $\{-\frac{1}{2}, \frac{1}{2}\}$
53. $2y^2 - 9y = 0$ $\{0, \frac{9}{2}\}$
54. $3y^2 + 7y = 0$ $\{0, -\frac{7}{3}\}$
55. $3x^2 - 7x + 4 = 0$ $\{\frac{4}{3}, 1\}$
56. $2t^2 - 5t - 3 = 0$ $\{-\frac{1}{2}, 3\}$
57. $x^2 + 64 = 0$ $\{\pm 8i\}$
58. $x^2 = -25$ $\{\pm 5i\}$
59. $5y^2 - 3y + 4 = 0$ $\left\{\frac{3 \pm i\sqrt{71}}{10}\right\}$
60. $7y^2 - 3y + 5 = 0$ $\left\{\frac{3 \pm i\sqrt{131}}{14}\right\}$

61. Two adjacent building lots are each square with a total area of 14,600 ft². The total street frontage for the two lots is 160 ft. (See the figure.) Find the dimensions of each square lot.

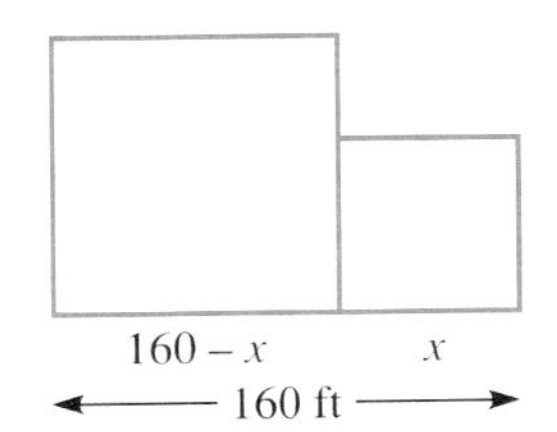

Side of smaller lot = 50 ft; side of larger lot = 110 ft

62. Refer to the figure for Exercise 61. Two adjacent building lots have a total area of 10,952 ft². The total street frontage is 148 ft. Find the dimensions of each square lot. Each has side 74 ft.

63. An ancient problem leading to a quadratic equation is given on an Egyptian papyrus from about 2000 B.C. The problem is as follows: Divide 100 square units into two squares such that the side of one of the squares is three-fourths the side of the other. Small square has side 6; large square has side 8.

64. Divide 10 square units into two squares such that the side of one of the squares is half the side of the other. Small square has side $\sqrt{2}$; large square has side $2\sqrt{2}$.

65. An object is thrown upward from a platform 20 ft above the ground with an initial velocity of 40 ft/second. The formula $y = 20 + 40t - 16t^2$ gives its height above the ground (y) t seconds after it is released. After how many seconds will the object be 40 ft above the ground? Why are there two answers? (Round to the nearest hundredth of a second.) 0.69 and 1.81 seconds; one is for the way up and one is for the way down.

66. Refer to Exercise 65. After how many seconds will the object be 10 ft above the ground? Why is there only one answer? (Round to the nearest hundredth of a second.) 2.73 seconds; because the object is more than 10 ft above the ground to begin with

67. Babylonian mathematics texts from about 3000 B.C. contain problems asking for a number which added to its reciprocal gives a specified sum. They worked out a general procedure for such problems. Try this one. What number when added to its reciprocal equals 5? Give the answers in radical form and in decimal approximation. $\frac{5 \pm \sqrt{21}}{2}$; 4.791 and 0.209

68. What number when added to its reciprocal equals 6? Give the answer in radical form and in decimal approximation. $3 \pm 2\sqrt{2}$; 5.828 and 0.172

THINK ABOUT IT

1. The great equation solver of the sixteenth century, Gerolamo Cardano, perhaps the most bizarre character in the whole history of mathematics, used a technique called "depressing" to put equations in simpler form. When you "depress" an equation, you get rid of the term which has the next to the highest power of the variable. Through this technique Cardano was able to solve cubic equations. You can use it to solve quadratic equations.

a. Given the equation $ax^2 + bx + c = 0$, rewrite it by replacing each occurrence of x by $y - b/2a$. This will give you a simpler (depressed) equation in y which you can solve by the square root property. You will see that this is another way to arrive at the quadratic formula. [For more on Cardano see *Journey Through Genius* by William Dunham (Wiley, 1990) or any history of mathematics text.]

b. Solve the equation $x^2 + 4x + 1 = 0$ by the method of depression.

2. a. Show that the sum of the solutions of $ax^2 + bx + c = 0$ with $a \neq 0$ is $-b/a$.

b. Find a simple expression for the product of the solutions.

c. For the equation $3x^2 + 2x - 4 = 0$, find the sum and product of the solutions.

3. In general, there are n solutions (including all complex solutions) to $x^n = 1$; they are called the nth roots of 1. Find the six solutions of $x^6 = 1$. Rewrite as $x^6 - 1 = 0$; then use the zero product principle and the quadratic formula.

4. A fascinating number which appears in many areas of mathematics is called the golden ratio. For example, in an isosceles triangle where the base angles are double the vertex angle, it is the ratio of the side to the base. An early appearance is in Euclid's *Elements*, where an exercise says: "Divide a line segment such that the ratio of the large part to the whole is equal to the ratio of the small part to the large." Refer to the sketch, from which we derive golden ratio $= \frac{\ell}{\ell + s} = \frac{s}{\ell}$. If you set $s = 1$ and solve the resulting quadratic equation, you will determine the exact value of the golden ratio.

5. In Example 4 of Section 8.2 the discriminant was used to determine the nature of the solutions of the equation $2x^2 - 4x + 5 = 0$.
   a. Show that the discriminant is not changed if the equation is rewritten as $-2x^2 + 4x - 5 = 0$, which results when both sides are multiplied by -1.
   b. Show that, in general, the discriminant $b^2 - 4ac$ is not changed if a, b, and c are replaced by their opposites.
   c. Show that in the quadratic formula if a, b, and c are replaced by their opposites, then the solutions do not change.

REMEMBER THIS

1. What is the LCD for $\frac{10}{x}$ and $\frac{18}{x^2}$? x^2
2. Multiply $x^2\left(\frac{10}{x} - \frac{18}{x^2}\right)$. $10x - 18$
3. Simplify $\sqrt{(-10)^2 - 4(1)(18)}$. $2\sqrt{7}$
4. Factor $t^2 - 5t + 4$. $(t - 4)(t - 1)$
5. Solve $\sqrt{2x - 4} = \sqrt{x + 1}$. $\{5\}$
6. Determine the number that should be added to make this expression a perfect square: $y^2 - y$. Then add this number, and factor the resulting trinomial. $\frac{1}{4}$; $(y - \frac{1}{2})^2$
7. Simplify $(1 + \sqrt{3}i)(1 - \sqrt{3}i)$. 4
8. Evaluate $(\frac{1}{4})^{1/2} + (\frac{1}{4})^{-1/2}$. $\frac{5}{2}$
9. Evaluate the determinant. $\begin{vmatrix} 3 & 4 \\ 5 & 6 \end{vmatrix}$ -2
10. Graph the inequality $y \leq x + 1$.

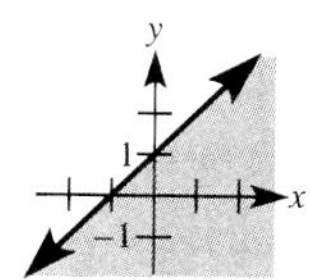

8.3 Other Equations Which Lead to Quadratic Equations

The cost C of producing x units of a product is often approximated by formulas of the form

$$C = px^2 + qx + r.$$

Solve this formula for x in terms of p, q, r, and C. (See Example 6.)

OBJECTIVES

1 **Solve fractional equations and radical equations that lead to quadratic equations.**

2 **Solve equations with quadratic form.**

3 **Solve literal equations for a variable with highest power 2.**

1 Equations containing rational expressions and equations containing radicals may lead to quadratic equations, as was shown in Sections 4.7 and 7.5, respectively. With the aid of the additional methods of this chapter, we can now solve a wider variety of such equations.

EXAMPLE 1 Solve $\frac{10}{x} - \frac{18}{x^2} = 1$.

Solution First, we remove fractions by multiplying both sides of the equation by the LCD, which is x^2. Thus, if $x \neq 0$,

$$x^2\left(\frac{10}{x} - \frac{18}{x^2}\right) = x^2 \cdot 1$$
$$10x - 18 = x^2$$
$$0 = x^2 - 10x + 18.$$

Then by the quadratic formula,

$$x = \frac{-(-10) \pm \sqrt{(-10)^2 - 4(1)(18)}}{2(1)} = \frac{10 \pm \sqrt{28}}{2}$$
$$= \frac{10 \pm 2\sqrt{7}}{2} = 5 \pm \sqrt{7}.$$

The restriction $x \neq 0$ does not affect the proposed solution, and both solutions check in the original equation. Thus, the solution set is $\{5 + \sqrt{7}, 5 - \sqrt{7}\}$.

PROGRESS CHECK 1 Solve $1 - \frac{2}{x} = \frac{17}{x^2}$.

Remind students that fractional equations and radical equations may have extraneous roots.

EXAMPLE 2 Solve $x + \sqrt{14 - x^2} = 0$.

Solution First, isolate the square root term.

$$x + \sqrt{14 - x^2} = 0$$
$$\sqrt{14 - x^2} = -x$$

Then square both sides of the equation and solve for x.

$$(\sqrt{14 - x^2})^2 = (-x)^2 \qquad \text{Square both sides.}$$
$$14 - x^2 = x^2$$
$$14 = 2x^2$$
$$7 = x^2$$
$$\pm\sqrt{7} = x \qquad \text{Square root property.}$$

Now check both solutions in the original equation.

$$x + \sqrt{14 - x^2} = 0$$
$$\sqrt{7} + \sqrt{14 - (\sqrt{7})^2} \stackrel{?}{=} 0$$
$$\sqrt{7} + \sqrt{7} \stackrel{?}{=} 0$$
$$2\sqrt{7} \neq 0 \quad \text{Extraneous}$$

$$x + \sqrt{14 - x^2} = 0$$
$$-\sqrt{7} + \sqrt{14 - (-\sqrt{7})^2} \stackrel{?}{=} 0$$
$$-\sqrt{7} + \sqrt{7} \stackrel{?}{=} 0$$
$$0 \stackrel{\checkmark}{=} 0$$

The check shows that $-\sqrt{7}$ is a solution, while $\sqrt{7}$ is extraneous. Thus, the solution set is $\{-\sqrt{7}\}$.

PROGRESS CHECK 2 Solve $\sqrt{16 - x^2} - x = 0$.

2 By a well-chosen substitution, it is sometimes possible to convert equations that are not originally quadratic into the form

$$at^2 + bt + c = 0 \qquad (a \neq 0),$$

where t is some variable expression. Such equations are called **equations with quadratic form.** For example, the fourth-degree equation

$$x^4 - 5x^2 + 4 = 0$$

Progress Check Answers

1. $\{1 \pm 3\sqrt{2}\}$
2. $\{2\sqrt{2}\}$

is not a quadratic (or second-degree) equation. But if we let $t = x^2$, then $t^2 = x^4$; and this equation becomes

$$t^2 - 5t + 4 = 0.$$

By the methods for solving a quadratic equation, t may be found; and back substitution using $t = x^2$ then leads to the solution of the original equation, as shown in Example 3.

EXAMPLE 3 Solve $x^4 - 5x^2 + 4 = 0$.

Solution As discussed, first let $t = x^2$ and solve for t.

$$\begin{aligned} x^4 - 5x^2 + 4 &= 0 \\ t^2 - 5t + 4 &= 0 && \text{Let } t = x^2\text{; then } t^2 = x^4. \\ (t-4)(t-1) &= 0 \\ t - 4 = 0 \quad \text{or} \quad t - 1 &= 0 && \text{Use the factoring method.} \\ t = 4 \qquad\quad t &= 1 \end{aligned}$$

Now resubstitute x^2 for t and solve for x.

$$\begin{aligned} x^2 = 4 \quad &\text{or} \quad x^2 = 1 \\ x = \pm 2 \quad &\qquad x = \pm 1 && \text{Square root property.} \end{aligned}$$

Check all four solutions in the original equation to confirm that the solution set is $\{2,-2,1,-1\}$.

Stress that the goal is to find the solutions of the *original* equation, not of the equation for the substitute variable t.

Note The equation in this example is intended as a simple illustration of the substitution technique associated with equations that are quadratic in form. This equation is also easily solved without substitution, as follows.

$$\begin{aligned} x^4 - 5x^2 + 4 &= 0 \\ (x^2-4)(x^2-1) &= 0 && \text{Factor.} \\ x^2 - 4 = 0 \quad \text{or} \quad x^2 - 1 &= 0 && \text{Zero product principle.} \\ x^2 = 4 \qquad\quad x^2 &= 1 && \text{Isolate } x^2. \\ x = \pm 2 \qquad\quad x &= \pm 1 && \text{Square root property.} \end{aligned}$$

However, the substitution method will be easier in some cases and required in others. For instance, try solving $x^4 - 4x^2 + 1 = 0$ (as requested in "Think About It" Exercise 2).

PROGRESS CHECK 3 Solve $x^4 - 11x^2 + 18 = 0$.

To spot equations with quadratic form, look for the exponent in one term to be double the exponent in another term.

EXAMPLE 4 Solve $5x^{2/3} - 8x^{1/3} - 4 = 0$.

Solution In the exponents in the variable terms, note that 2/3 is double 1/3. If we let $t = x^{1/3}$, then $t^2 = x^{2/3}$; and this substitution reveals an equation with quadratic form that we can solve for t.

$$\begin{aligned} 5x^{2/3} - 8x^{1/3} - 4 &= 0 \\ 5t^2 - 8t - 4 &= 0 && \text{Let } t = x^{1/3}\text{; then } t^2 = x^{2/3}. \\ (5t+2)(t-2) &= 0 \\ 5t + 2 = 0 \quad \text{or} \quad t - 2 &= 0 && \text{Use the factoring method.} \\ t = -\tfrac{2}{5} \qquad\quad t &= 2 \end{aligned}$$

Progress Check Answer

3. $\{\pm 3, \pm\sqrt{2}\}$

Remind students of the principle of powers stated in Section 7.5, and remind them that $x^{1/3} = \sqrt[3]{x}$.

Now replace t by $x^{1/3}$ and solve for x.

$$x^{1/3} = -\tfrac{2}{5} \quad \text{or} \quad x^{1/3} = 2$$
$$x = -\tfrac{8}{125} \qquad\qquad x = 8 \qquad \text{Cube both sides.}$$

Both solutions check, and the solution set is $\{-\frac{8}{125}, 8\}$.

PROGRESS CHECK 4 Solve $x + 2x^{1/2} - 3 = 0$.

3 The methods for solving quadratic equations may be applied to solve a literal equation or a formula for a variable with highest power 2.

EXAMPLE 5 Solve $x = y^2 - 4$ for y.

Solution Isolate y^2 and then solve for y using the square root property.

$$x = y^2 - 4$$
$$x + 4 = y^2 \qquad \text{Add 4 to both sides.}$$
$$\pm\sqrt{x + 4} = y \qquad \text{Square root property.}$$

The graphing calculator section at the end of Chapter 6 discusses drawing several graphs on one screen.

Note To graph an equation on a graphing calculator, you must enter an expression that is solved for y. Using the result in this example, $x = y^2 - 4$ may be graphed by displaying the graphs of $y = \sqrt{x + 4}$ and $y = -\sqrt{x + 4}$ in the same calculator screen.

PROGRESS CHECK 5 Solve $x^2 + y^2 = 1$ for y.

EXAMPLE 6 Solve the problem in the section introduction on page 377.

Solution We use the quadratic formula to solve the equation

$$C = px^2 + qx + r$$

for x. To match the form $0 = ax^2 + bx + c$, subtract C on both sides to obtain

$$0 = px^2 + qx + (r - C).$$

In this form the coefficient of x^2 represents a, the coefficient of x represents b, and the remaining expression represents c. Replacing a by p, b by q, and c by $r - C$ in the quadratic formula gives

$$x = \frac{-q \pm \sqrt{q^2 - 4p(r - C)}}{2p} \quad \text{or} \quad x = \frac{-q \pm \sqrt{q^2 - 4pr + 4pC}}{2p}.$$

PROGRESS CHECK 6 Solve $y = -16t^2 + v_0t + y_0$ for t.

Progress Check Answers

4. $\{1\}$; 9 is extraneous.
5. $y = \pm\sqrt{1 - x^2}$
6. $t = \dfrac{-v_0 \pm \sqrt{v_0^2 + 64y_0 - 64y}}{-32}$

EXERCISES 8.3

In Exercises 1–30, solve the given equation.

1. $\frac{4}{x} + \frac{12}{x^2} = 1$ $\{-2, 6\}$
2. $\frac{8}{x} - \frac{15}{x^2} = 1$ $\{3, 5\}$
3. $-\frac{13}{x} - \frac{7}{x^2} = -2$ $\{-\frac{1}{2}, 7\}$
4. $3 = -\frac{4}{x} + \frac{15}{x^2}$ $\{-3, \frac{5}{3}\}$
5. $\frac{1}{x} - \frac{12}{x^2} = -1$ $\{-4, 3\}$
6. $1 = \frac{5}{x} + \frac{6}{x^2}$ $\{-1, 6\}$
7. $3 = \frac{-2}{x} + \frac{2}{x^2}$ $\left\{\frac{-1 \pm \sqrt{7}}{3}\right\}$
8. $2 = \frac{3}{x} + \frac{1}{x^2}$ $\left\{\frac{3 \pm \sqrt{17}}{4}\right\}$
9. $1 = \frac{2}{x} - \frac{2}{x^2}$ $\{1 \pm i\}$
10. $1 = -\frac{4}{x} - \frac{5}{x^2}$ $\{2 \pm i\}$
11. $1 = \frac{-9}{x^2}$ $\emptyset$
12. $1 = \frac{-16}{x^2}$ $\emptyset$
13. $2 + \frac{5}{x^2} = \frac{6}{x}$ $\left\{\frac{3 \pm i}{2}\right\}$
14. $9 + \frac{5}{x^2} = \frac{12}{x}$ $\left\{\frac{2 \pm i}{3}\right\}$
15. $2x - \frac{20}{x} = 0$ $\{\pm\sqrt{10}\}$
16. $2 = \frac{60}{x^2}$ $\{\pm\sqrt{30}\}$
17. $x + \sqrt{12 - x^2} = 0$ $\{-\sqrt{6}\}$
18. $x - \sqrt{12 - x^2} = 0$ $\{\sqrt{6}\}$

19. $\sqrt{24 - x^2} = x$ $\{2\sqrt{3}\}$
20. $\sqrt{10 - x^2} = x$ $\{\sqrt{5}\}$
21. $x + \sqrt{4 - x^2} = 0$ $\{-\sqrt{2}\}$
22. $\sqrt{6 - x^2} = -x$ $\{-\sqrt{3}\}$
23. $x - \sqrt{x^2 - 8} = 0$ $\emptyset$
24. $x - \sqrt{x^2 - 9} = 0$ $\emptyset$
25. $x = \sqrt{x^2}$ $\{x: x \geq 0\}$
26. $x + \sqrt{x^2} = 0$ $\{x: x \leq 0\}$
27. $\sqrt{14 - 3x^2} = 2x$ $\{\sqrt{2}\}$
28. $\sqrt{45 - 6x^2} = 3x$ $\{\sqrt{3}\}$
29. $\sqrt{1 + x + x^2} = 3x + 1$ $\{0\}$
30. $\sqrt{9 - x - x^2} = 2x + 3$ $\{0\}$

In Exercises 31–48, solve the given equations, which all have quadratic form.

31. $y^4 + 3y^2 - 4 = 0$ $\{\pm 1, \pm 2i\}$
32. $y^4 - 4y^2 + 3 = 0$ $\{\pm 1, \pm \sqrt{3}\}$
33. $x^4 - 11x^2 + 30 = 0$ $\{\pm \sqrt{5}, \pm \sqrt{6}\}$
34. $6x^4 - 13x^2 + 6 = 0$ $\{\pm \sqrt{\frac{2}{3}}, \pm \sqrt{\frac{3}{2}}\}$
35. $y^4 - 2y^2 - 3 = 0$ $\{\pm \sqrt{3}\}$
36. $y^4 - 2y^2 - 8 = 0$ $\{\pm 2\}$
37. $x^4 - 25x^2 + 144 = 0$ $\{\pm 3, \pm 4\}$
38. $x^4 - 26x^2 + 25 = 0$ $\{\pm 1, \pm 5\}$
39. $x^4 - 3x^2 - 4 = 0$ $\{\pm 2, \pm i\}$
40. $x^4 + 8x^2 - 9 = 0$ $\{\pm 1, \pm 3i\}$
41. $y^4 - y^2 - 6 = 0$ $\{\pm \sqrt{3}, \pm i\sqrt{2}\}$
42. $y^4 + y^2 - 6 = 0$ $\{\pm \sqrt{2}, \pm i\sqrt{3}\}$
43. $x^{2/3} - 2x^{1/3} - 3 = 0$ $\{-1, 27\}$
44. $x^{2/3} + 3x^{1/3} - 4 = 0$ $\{-64, 1\}$
45. $x + 4x^{1/2} + 3 = 0$ $\emptyset$
46. $x + 5x^{1/2} + 4 = 0$ $\emptyset$
47. $x^{1/2} - 2x^{1/4} - 8 = 0$ $\{256\}$
48. $x^{1/2} - 4x^{1/4} + 3 = 0$ $\{1, 81\}$

In Exercises 49–58, solve for x.

49. $x^2 = y + 3$ $x = \pm\sqrt{y + 3}$
50. $x^2 = y - 7$ $x = \pm\sqrt{y - 7}$
51. $y = x^2 - 8$ $x = \pm\sqrt{y + 8}$
52. $y = x^2 + 7$ $x = \pm\sqrt{y - 7}$
53. $y^2 = x^2 + 6$ $x = \pm\sqrt{y^2 - 6}$
54. $y^2 = x^2 - 8$ $x = \pm\sqrt{y^2 + 8}$
55. $t = a_1x^2 + a_2x + a_3$ $x = \dfrac{-a_2 \pm \sqrt{a_2^2 - 4a_1(a_3 - t)}}{2a_1}$
56. $y = -16x^2 - 5x + y_0$ $x = \dfrac{5 \pm \sqrt{25 + 64(y_0 - y)}}{-32}$
57. $t = dx^2 + rx$ $x = \dfrac{-r \pm \sqrt{r^2 + 4dt}}{2d}$
58. $y = -16x^2 + 16x$ $x = \dfrac{2 \pm \sqrt{4 - y}}{4}$

Exercises 59–62 are based on those found in American algebra texts of 100 years ago. The wording is quite dated, but you can still solve them by using quadratic equations, as algebra students did then.

59. A gentleman has two square rooms whose sides are to each other as 2 to 3. He finds that it will require 20 yd² more of carpeting to cover the floor of the larger than of the smaller room. What is the length of one side of each room? (*Hint:* The side of the smaller square is $\frac{2}{3}$ the size of the larger one.) 6 yd, 4 yd
60. A man purchased a rectangular field whose length was $\frac{10}{9}$ times its breadth. It contained 9 acres. What was the length in rods of each side? (*Note:* 1 acre is 160 square rods, so 9 acres is 1,440 square rods. You don't need it for this exercise, but a rod is 6.5 yd, and an acre originally was supposed to indicate the size of a field that a team of oxen could plow in one day.) 40 rods, 36 rods
61. An orchard containing 2,000 trees had 10 rows more than it had trees in a row. How many rows were there? How many trees were there in each row? 50 rows of 40 trees each
62. A person purchased a flock of sheep for $100. If he had purchased 5 more for the same sum, they would have cost $1 less per head. How many did he buy? [*Hint:* If he bought x sheep, then each one cost $100/x$ dollars. If he bought $x + 5$ sheep, then each cost $100/(x + 5)$ dollars.] 20 sheep

THINK ABOUT IT

1. Give three examples of equations with quadratic form.
2. Solve $x^4 - 4x^2 + 1 = 0$.
3. If the perimeter and the area of a rectangle are given by p and a, respectively, then show that the length ℓ is given by
$$\ell = \frac{p \pm \sqrt{p^2 - 16a}}{4}.$$
Note that since $a = \ell w$, you can use a/ℓ in place of w.
4. The curved surface area S for a right circular cone (see figure) is given by $S = \pi r\sqrt{r^2 + h^2}$, where r is the base radius and h is the altitude.

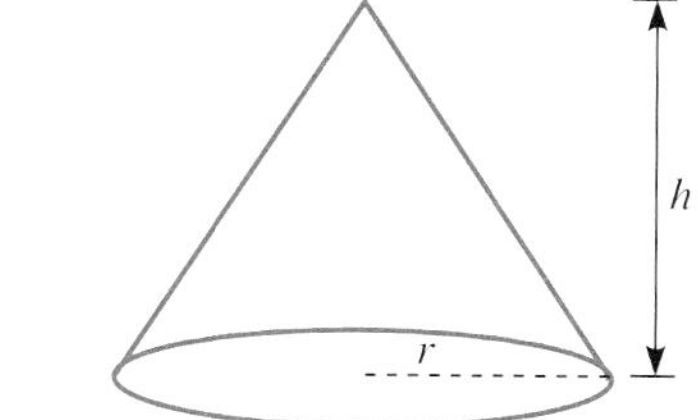

   a. Solve this formula for r.
   b. If $h = 1$, what radius (to the nearest hundredth) is necessary so that the surface area is 10 in.²?
5. a. Solve $x^{-2} - 4x^{-1} - 3 = 0$ by letting $t = x^{-1}$.
   b. Solve the same equation by multiplying both sides by x^2.

REMEMBER THIS

1. For what values of x is the expression $x + 1$ positive? $\{x{:}x > -1\}$
2. When $x = 0$, is $(x + 1)(x - 2)(x - 4)$ positive, negative, or zero? Positive
3. Are both solutions of $t^2 - 6t + 4$ positive? Yes
4. Solve $x(x - 2)(3x + 5) = 0$. $\{0, 2, -\frac{5}{3}\}$
5. Solve $\dfrac{x + 2}{x - 3} = 0$. $\{-2\}$
6. Simplify $\sqrt{63} - \sqrt{28}$. $\sqrt{7}$
7. Find the distance between the points (1,1) and (5,0). $\sqrt{17}$
8. Simplify $\dfrac{3x^{1/2}y^{-3}}{6x^{-1/2}y^3}$. $\dfrac{x}{2y^6}$
9. Solve the system. $\begin{aligned} x + y + z &= 0 \\ y + 2z &= -4 \\ 3z &= -3 \end{aligned}$ $(3, -2, -1)$
10. If a radio sells for x dollars and there is a 6 percent sales tax, give an expression for the total cost (radio plus tax). $1.06x$

8.4 Solving Inequalities Involving Polynomials

The height (y) of a projectile that is shot directly up from the ground with an initial velocity of 96 ft/second is given by

$$y = -16t^2 + 96t,$$

where y is measured in feet and t is the elapsed time in seconds. To the nearest hundredth of a second, for what values of t is the projectile more than 64 ft off the ground? (See Example 3.)

OBJECTIVES

1. **Solve quadratic inequalities.**
2. **Solve polynomial inequalities of degree higher than 2.**
3. **Solve inequalities involving quotients of polynomials.**

1 **Quadratic inequalities** are inequalities that may be expressed in the standard forms

$$ax^2 + bx + c > 0, \qquad ax^2 + bx + c \geq 0,$$
$$\text{or}$$
$$ax^2 + bx + c < 0, \qquad ax^2 + bx + c \leq 0,$$

where a, b, and c are real numbers with $a \neq 0$. There are several methods for solving such inequalities, and Example 1 shows a method that is based on creating a **sign graph.**

EXAMPLE 1 Solve $x^2 - 2 > x$ using the sign graph method.

Solution First, rewrite the inequality so that one side is zero. Then factor.

$$\begin{aligned} x^2 - 2 &> x \\ x^2 - x - 2 &> 0 \\ (x + 1)(x - 2) &> 0 \end{aligned}$$

Now display graphically where each of these factors is zero, positive, or negative, as shown in the top two number lines in Figure 8.3. To determine when $(x + 1)(x - 2)$

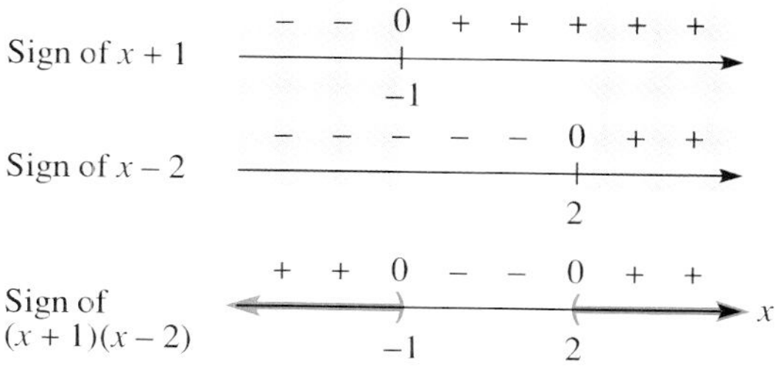

Figure 8.3

is greater than zero, look for intervals where both factors are positive or both factors are negative. We see that the product is positive if x is to the left of -1 or to the right of 2, as shown in the bottom row of the sign graph in Figure 8.3. Thus, the solution set is $\{x: x < -1 \text{ or } x > 2\}$, in set-builder notation, or $(-\infty,-1) \cup (2,\infty)$, in interval notation; and it is specified graphically as in Figure 8.4.

Figure 8.4

PROGRESS CHECK 1 Solve $x^2 + 2x > 3$ using the sign graph method.

By observing the sign graph in Figure 8.3, we can formulate an alternative method for solving quadratic inequalities that is often easier to use in practice. First, real numbers that make $ax^2 + bx + c$ equal to zero are called **critical numbers.** Note in Example 1 that the critical numbers are -1 and 2 and that these two critical numbers separate the number line into three intervals, namely, $(-\infty,-1)$, $(-1,2)$, and $(2,\infty)$. Also, observe that throughout each interval the sign of $x^2 - x - 2$ remains the same. Thus an efficient method, called the **test point method,** for solving quadratic inequalities is to find the sign of a convenient number in each of the intervals determined by the critical numbers. By comparing the resulting sign with the inequality in question, we may determine the solution set, as shown in Example 2.

EXAMPLE 2 Solve $2x^2 - 7x - 4 \leq 0$ using the test point method.

Solution Find the critical numbers by using the factoring method to determine when $2x^2 - 7x - 4$ equals zero.

$$2x^2 - 7x - 4 = 0$$
$$(2x + 1)(x - 4) = 0$$
$$2x + 1 = 0 \quad \text{or} \quad x - 4 = 0$$
$$2x = -1 \qquad\qquad x = 4$$
$$x = -\tfrac{1}{2}$$

The critical numbers $-\frac{1}{2}$ and 4 separate the number line into the intervals $(-\infty,-\frac{1}{2})$, $(-\frac{1}{2},4)$ and $(4,\infty)$, and Figure 8.5 shows whether a true statement results when a specific number in each of these intervals is tested.

Point out that specific numbers may be tested in $2x^2 - 7x - 4 \leq 0$ or in $(2x + 1)(x - 4) \leq 0$. Although it is necessary to determine only the sign that results from the substitution, some students prefer to find the actual number.

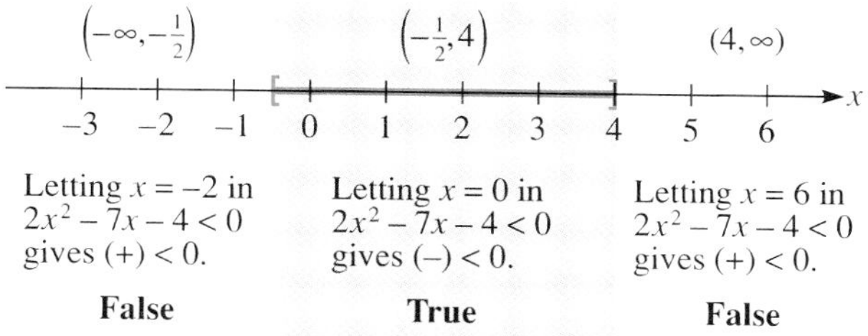

Figure 8.5

Progress Check Answer

1. $(-\infty,-3) \cup (1,\infty)$

A common student error is to write $\{-\frac{1}{2},4\}$ for the solution set. This error results when students forget that the original problem is an inequality (not an equation) or when they do not distinguish the different meanings of $\{-\frac{1}{2},4\}$, $(-\frac{1}{2},4)$, and $[-\frac{1}{2},4]$.

All numbers in each interval lead to the same result as the specific number tested, so $2x^2 - 7x - 4 < 0$ is true on the interval $(-\frac{1}{2},4)$. Because the inequality in question is less than or equal to zero, $-\frac{1}{2}$ and 4 are also solutions; so the solution set is the closed interval $[-\frac{1}{2},4]$.

PROGRESS CHECK 2 Solve $3x^2 + 2 \leq 5x$ using the test point method.

When critical numbers are irrational, these numbers may be found by using the quadratic formula, as considered in Example 3.

EXAMPLE 3 Solve the problem in the section introduction on page 382.

Solution The height (y) of the projectile is given by $-16t^2 + 96t$, so the projectile is more than 64 ft off the ground for all values of t satisfying

$$-16t^2 + 96t > 64.$$

We solve the inequality by first making the right side equal to zero and then simplifying the inequality.

$$-16t^2 + 96t - 64 > 0 \quad \text{Subtract 64 from both sides.}$$
$$t^2 - 6t + 4 < 0 \quad \text{Divide both sides by } -16 \text{ and reverse the inequality sign.}$$

Now determine the critical numbers by using the quadratic formula to find when $t^2 - 6t + 4$ equals zero.

$$t = \frac{-(-6) \pm \sqrt{(-6)^2 - 4(1)(4)}}{2(1)}$$
$$= \frac{6 \pm \sqrt{20}}{2} = \frac{6 \pm 2\sqrt{5}}{2} = 3 \pm \sqrt{5}$$

The critical numbers are $3 + \sqrt{5}$ and $3 - \sqrt{5}$. The question requests answers to the nearest hundredth, so we approximate these numbers by 5.24 and 0.76, respectively, and then test in the intervals determined by these critical numbers, as shown in Figure 8.6.

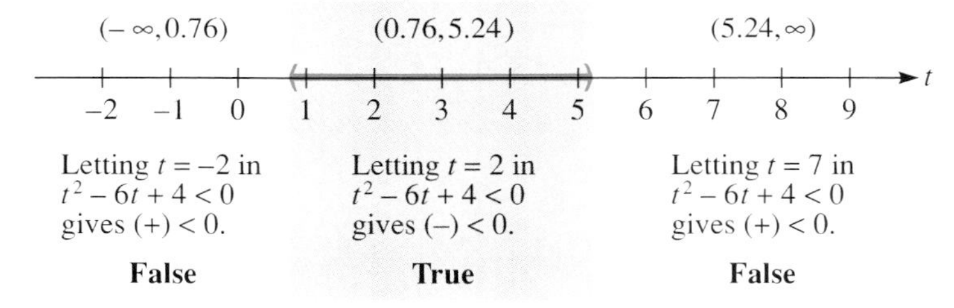

Figure 8.6

The tests show that an approximate solution set is the interval (0.76 second,5.24 seconds).

Note If an exact solution had been requested in this problem, then we would leave the answers in radical form, and the solution set would be written as the interval $(3 - \sqrt{5}$ second, $3 + \sqrt{5}$ seconds).

PROGRESS CHECK 3 To the nearest hundredth, for what values of t is the projectile considered in Example 3 more than 32 ft off the ground?

Special cases of quadratic inequalities may result in which the inequality is always true, never true, or true only for one number. The next example considers one of these cases.

Progress Check Answers

2. $[\frac{2}{3},1]$
3. (0.35 second, 5.65 seconds)

EXAMPLE 4 Solve $x^2 + 4 < 0$.

Solution We first attempt to find critical numbers by finding when $x^2 + 4$ equals zero.

$$\begin{aligned} x^2 + 4 &= 0 \\ x^2 &= -4 \\ x &= \pm\sqrt{-4} \\ x &= \pm 2i \end{aligned}$$

The solutions are not real numbers, so no critical numbers exist. This result means that $x^2 + 4$ never changes sign. Replacing x by a convenient number, say 0, in the original inequality gives $4 < 0$, which is false. Thus, no number satisfies the inequality, and the solution set is $\emptyset$.

Remind students that the number line means the *real* number line, so imaginary numbers are not associated with points on the line.

Note Because $x^2 + 4$ is always positive, an inequality like $x^2 + 4 > 0$ illustrates a quadratic inequality that is always true. In this case the solution is the set of all real numbers, or $(-\infty,\infty)$. An inequality like $(x - 2)^2 \leq 0$ illustrates a quadratic inequality that is true for only one number, namely, 2.

Ask students to create examples of quadratic inequalities that illustrate special cases. See "Think About It" Exercise 1.

PROGRESS CHECK 4 Solve $(x - 4)^2 > -2$.

2 Our current methods may be applied to solve polynomial inequalities of degree higher than 2. For such problems, critical numbers are defined as real numbers that make the *polynomial* in question zero. The key principle is that throughout each of the intervals determined by the critical numbers, the sign of the *polynomial* remains the same.

EXAMPLE 5 Solve $x(x - 2)(3x + 5) > 0$.

Solution The inequality is given with zero on one side and the polynomial in factored form on the other side, so the critical numbers are easily found using the zero product principle.

$$x(x - 2)(3x + 5) = 0$$

$$x = 0 \quad \text{or} \quad x - 2 = 0 \quad \text{or} \quad 3x + 5 = 0$$
$$x = 2 \qquad\qquad x = -\tfrac{5}{3}$$

The critical numbers are 0, 2, and $-\frac{5}{3}$, and we test in the intervals determined by these critical numbers, as shown in Figure 8.7.

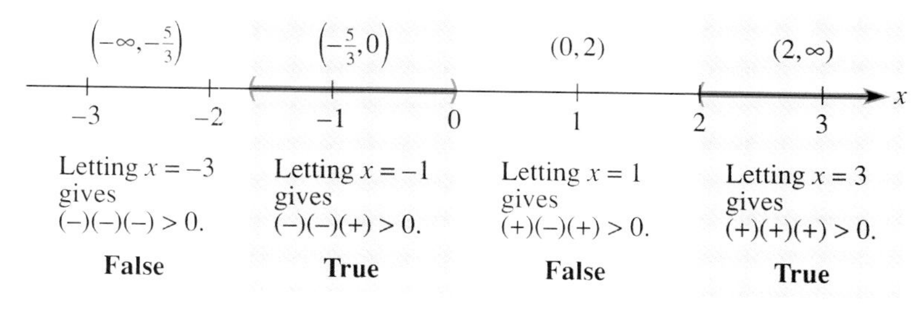

Figure 8.7

Thus, $x(x - 2)(3x + 5) > 0$ is true if $-\frac{5}{3} < x < 0$ or $x > 2$, and the solution set is $(-\frac{5}{3},0) \cup (2,\infty)$.

PROGRESS CHECK 5 Solve $x(2x - 5)(x + 1) < 0$.

Progress Check Answers

4. $(-\infty,\infty)$
5. $(-\infty,-1) \cup (0,\frac{5}{2})$

3 To solve inequalities involving quotients of polynomials, first rewrite (if necessary) the inequality so that one side is zero and the other side is a single rational expression. Then apply our current methods, where critical numbers are defined as real numbers that either make the numerator zero or make the denominator zero. When analyzing quotients, remember that division by zero is undefined, so critical numbers that make the denominator zero are never included in the solution set.

EXAMPLE 6 Solve $\frac{x + 2}{x - 3} \geq 0$.

Mention that multiplying both sides of an inequality by a variable LCD is not recommended because two cases result. In one case the LCD is positive and the direction of the inequality is preserved. In the other the LCD is negative and the direction of the inequality is reversed. See "Think About It" Exercise 4.

Solution The numerator $x + 2$ is zero when $x = -2$, and the denominator $x - 3$ is zero when $x = 3$. So the critical numbers are -2 and 3. Figure 8.8 shows whether a true statement results when a specific number is tested in each of the intervals $(-\infty,-2)$, $(-2,3)$, and $(3,\infty)$.

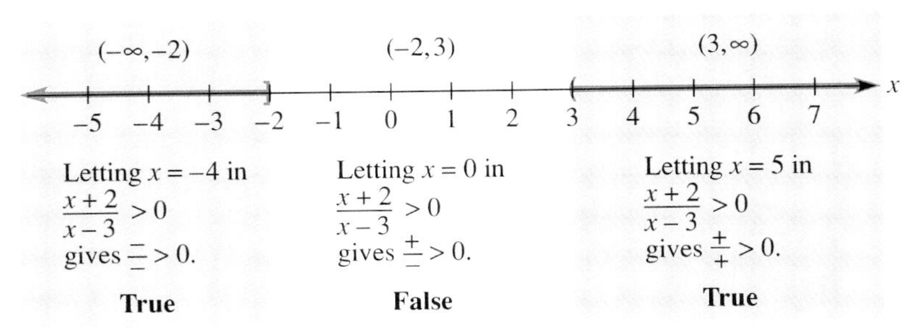

Figure 8.8

We see that $(x + 2)/(x - 3) > 0$ is true if $x < -2$ or $x > 3$. Because the inequality in question is $\geq$, the solution set is $(-\infty,-2] \cup (3,\infty)$. Note that -2 is included and 3 is excluded from the solution set, since the quotient is 0 when $x = -2$ and is undefined when $x = 3$.

PROGRESS CHECK 6 Solve $\frac{x - 5}{x} \leq 0$.

EXAMPLE 7 Solve $\frac{3}{x} > 2$.

Solution First, rewrite the inequality so that one side is zero and the other side is a single rational expression.

$$\frac{3}{x} > 2$$

$$\frac{3}{x} - 2 > 0$$

$$\frac{3 - 2x}{x} > 0$$

The critical numbers are $\frac{3}{2}$ and 0, because the numerator is zero when $x = \frac{3}{2}$ and the denominator is zero when $x = 0$. In Figure 8.9 we test in the intervals determined by these two numbers. Thus, the given inequality is true if $0 < x < \frac{3}{2}$, and the solution set is $(0,\frac{3}{2})$.

Progress Check Answer

6. $(0,5]$

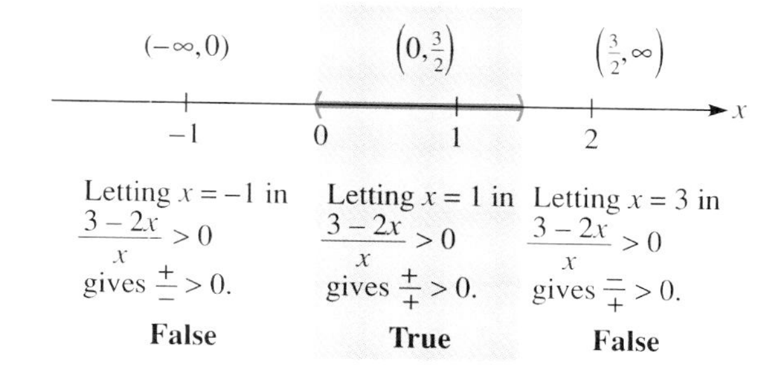

Figure 8.9

PROGRESS CHECK 7 Solve $\frac{2}{x}<5$.

Progress Check Answer

7. $(-\infty,0)\cup(\frac{2}{5},\infty)$

EXERCISES 8.4

In Exercises 1–20, solve the quadratic inequality using the sign graph method or the test point method.

1. $x^2-2x-8\le 0$
2. $x^2+x-12>0$
3. $y^2-5y-6\le 0$
4. $x^2-x-20<0$
5. $2x^2-13x-7>0$
6. $3x^2+4x-15\ge 0$
7. $9x^2+15x-14<0$
8. $5x^2-9x-2\ge 0$
9. $3x^2+2x-2\le 0$ (Give the endpoints to the nearest hundredth.)

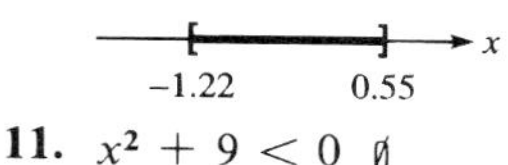

10. $2x^2-3x-1\ge 0$ (Give the endpoints to the nearest hundredth.)
11. $x^2+9<0$ $\emptyset$
12. $y^2+16\le 0$ $\emptyset$
13. $(x-5)^2\le 0$ $\{5\}$
14. $(y+8)^2\le 0$ $\{-8\}$
15. $x^2\ge -1$ $(-\infty,\infty)$
16. $y^2+3\ge 0$ $(-\infty,\infty)$
17. $y^2-25<0$
18. $t^2-36\ge 0$
19. $2x^2+3x\le 7$
20. $5x^2+4>3x$ $\emptyset$

In Exercises 21–44, solve the given inequality.

21. $x(x+2)(x-3)<0$
22. $x(x-4)(x-6)\le 0$
23. $x(x+3)(x-5)\ge 0$
24. $x(x+7)(x-4)>0$
25. $x(x+3)(x-1)\le 0$
26. $(x+3)(x-2)(x-5)\le 0$
27. $(2x+1)(x-1)(x+4)<0$
28. $(3x-1)(x+4)(2x+3)\ge 0$
29. $(x+1)(x+2)(x-3)(x-5)>0$
30. $(x+3)(x+5)(x-4)(x-7)<0$
31. $x(x+2)(x+5)(x+7)\le 0$
32. $x(x-3)(x-5)(x-9)\ge 0$
33. $\frac{x+2}{x-1}<0$
34. $\frac{x+3}{x-2}\ge 0$
35. $\frac{x-3}{x+4}\le 0$
36. $\frac{x-5}{x-2}>0$
37. $\frac{x-6}{x}\le 0$
38. $\frac{x+7}{x}<0$

39. $\frac{2x - 1}{x + 4} > 0$

-4 $\frac{1}{2}$ x

40. $\frac{3x - 2}{x - 3} < 0$

$\frac{2}{3}$ 3 x

41. $\frac{5}{x} < 2$

0 $\frac{5}{2}$ x

42. $\frac{4}{x} > 6$

0 $\frac{2}{3}$ x

43. $\frac{x}{5} < 2$

10 x

44. $\frac{x}{7} > -5$

-35 x

45. The height of a projectile that is shot directly up from the ground with an initial velocity of 144 ft/second is given by

$$y = -16t^2 + 144t,$$

where y is measured in feet and t is the elapsed time in seconds. To the nearest hundredth of a second, for what values of t is the projectile more than 250 ft off the ground? (2.35,6.65)

46. Refer to Exercise 45. To the nearest hundredth of a second, for what values of t is the projectile more than 304 ft off the ground? (3.38,5.62)

47. A rectangle has perimeter 16 in. If the length is denoted by x, then the width is denoted by $8 - x$. (See the figure.) Assume that the length is not shorter than the width.

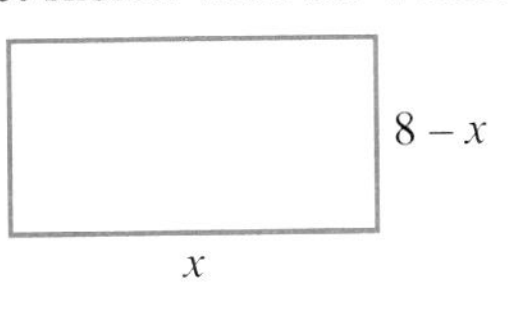

a. For what lengths is the area of the rectangle greater than 10 in.²? (Give the answer exactly using radical notation and also as a decimal to the nearest hundredth.) Lengths from 4 to $4 + \sqrt{6}$ in., or 4 to 6.45 in.

b. What length yields the maximum area? 4 in.

48. Repeat Exercise 47, but use a perimeter equal to 20 in.
a. Lengths from 5 to $5 + \sqrt{15}$ in., or 5 to 8.87 in. b. 5 in.

THINK ABOUT IT

1. Create a quadratic inequality with the given solution set.
   a. $[1,3]$ b. $(-\infty,-1) \cup (2,\infty)$ c. $(-\sqrt{2},\sqrt{2})$
   d. $(-\infty,\infty)$ e. $\{0\}$ f. $\emptyset$
2. Create an inequality involving a quotient of two polynomials that is greater than or equal to zero with the given solution set.
   a. $(-\infty,0] \cup (3,\infty)$ b. $(0,3]$
3. For what values of x will $\sqrt{x^2 - 3x - 10}$ be a real number?
4. Examine the following line of reasoning.

$$\frac{3}{x} > 1$$

$3 > x$ Multiply both sides by x.

$x < 3$ $a > b$ is equivalent to $b < a$.

The end result suggests that the solution set is $(-\infty,3)$, but this answer is incorrect. What is wrong?

5. a. Graph $y = x^2 - 4$.
   b. Solve $x^2 - 4 < 0$ using the graph in part **a.** Explain your method.
   c. Solve $x^2 - 4 \geq 0$ using the graph in part **a.** Explain your method.

REMEMBER THIS

1. Complete these ordered pairs so that they are solutions to $y = x^2$: (2,), (−1,), (0,). (2,4), (−1,1), (0,0)
2. What is the y-intercept of this graph? (0,1)

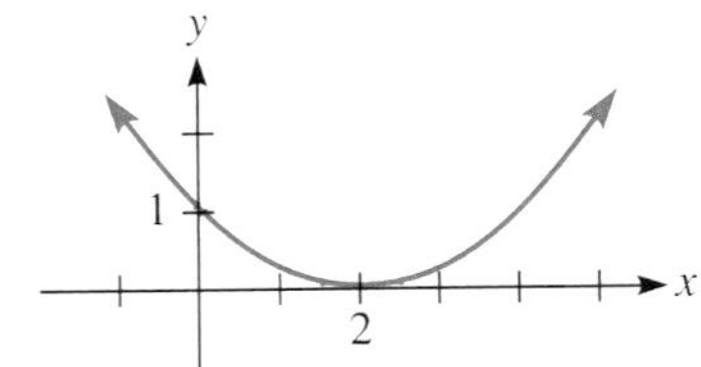

3. Find the average of 0 and x. $x/2$
4. If $y = ax^2 + bx + c$, find y when $x = 0$. c
5. Evaluate $-x^2 + 3x - 4$ when $x = \frac{3}{2}$. $-\frac{7}{4}$
6. The height above the ground (y) of a projectile is given by $y = -16t^2 + 96t$, where y is measured in feet and t is time elapsed in seconds. For what values of t is the projectile at ground level? 0, 6
7. Solve $x^4 - 16 = 0$. $\{2,-2,2i,-2i\}$
8. Simplify $(3^{1/2})^{-2} + (4^{1/3})^{-3}$. $\frac{7}{12}$
9. Solve the system. $x + 3y = 5$
   $2x + 6y = 9$ $\emptyset$
10. Simplify $\frac{a + 1/b}{b - 1/a}$. $\frac{a^2b + a}{ab^2 - b}$

8.5 The Parabola

As stated in the opening to Section 8.4, the height (y) of a projectile that is shot directly up from the ground with an initial velocity of 96 ft/second is given by

$$y = -16t^2 + 96t,$$

where y is measured in feet and t is the elapsed time in seconds.

a. When does the projectile attain its maximum height?
b. What is the maximum height?
c. When does the projectile hit the ground?
d. Graph the equation. (See Example 7.)

OBJECTIVES

1. **Graph $y = ax^2 + bx + c$ by finding ordered-pair solutions.**
2. **Graph $y = ax^2 + bx + c$ by using the vertex and intercepts of the graph.**
3. **Find the number of x-intercepts in the graph of $y = ax^2 + bx + c$ by using the discriminant.**
4. **Graph $x = ay^2 + by + c$ by using the vertex and intercepts of the graph.**
5. **Solve applied problems involving parabolas.**

Graphs of first-degree (or linear) equations in two variables were considered in Chapter 5. Recall that the lines shown in such graphs provide a useful picture of all ordered-pair solutions of the equations. The next three sections extend our coverage of graphing to obtain pictures of the solutions of second-degree equations in two variables with general form

$$Ax^2 + Bxy + Cy^2 + Dx + Ey + F = 0,$$

where A, B, and C are not all zero. The graphs we consider are called parabolas, circles, ellipses, or hyperbolas, and such curves are called **conic sections.** This name was first used by Greek mathematicians who discovered that these curves result from the intersection of a cone with an appropriate plane, as shown in Figure 8.10.

It is not clear when the ancient Greeks first analyzed the conic sections. They are mentioned by Euclid and others in the fourth century B.C. But Apollonius of Perga wrote a definitive text (*Conics*) on them about 220 B.C., which defined the sections basically as we do now.

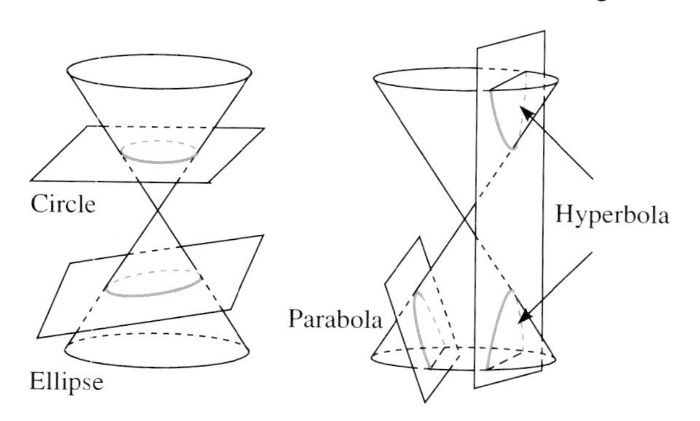

Figure 8.10

1 The conic sections may be analyzed from different viewpoints, and in this text we primarily start with an equation for a conic section and then draw its graph. We begin by graphing second-degree equations of the form

$$y = ax^2 + bx + c,$$

where a, b, and c are real numbers with $a \neq 0$. One method for graphing such equations is to generate a list of ordered-pair solutions, as shown in Examples 1 and 2.

EXAMPLE 1 Graph $y = x^2$.

Solution Make a list of ordered-pair solutions by replacing x with integer values from, say, 2 to -2.

If $x =$	Then $y = x^2$	Thus, the Ordered-Pair Solutions Are
2	$y = 2^2 = 4$	(2,4)
1	$y = 1^2 = 1$	(1,1)
0	$y = 0^2 = 0$	(0,0)
-1	$y = (-1)^2 = 1$	$(-1,1)$
-2	$y = (-2)^2 = 4$	$(-2,4)$

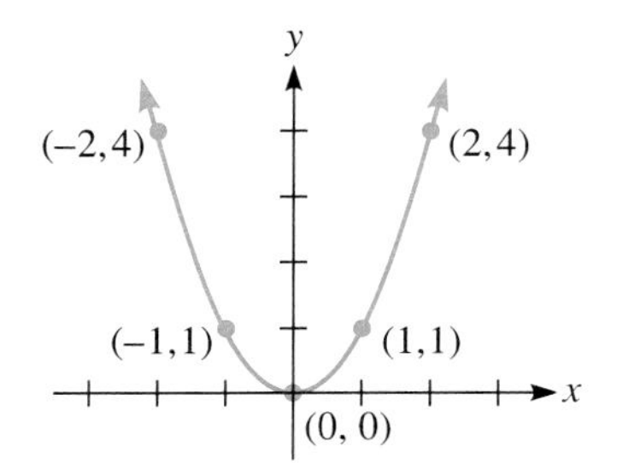

Figure 8.11

By graphing these ordered pairs and drawing a smooth curve through them, we obtain a graph of $y = x^2$, as shown in Figure 8.11.

PROGRESS CHECK 1 Graph $y = x^2 - 4$.

EXAMPLE 2 Graph $y = -x^2 + 4x$.

Solution Generate a list of ordered-pair solutions, as in the chart below, and then graph the equation, as shown in Figure 8.12. (Although you may have difficulty deciding on substitutions for x that are efficient for drawing the graph, we show how to overcome this difficulty following this example.)

If $x =$	Then $y = -x^2 + 4x$	Thus, the Ordered-Pair Solutions
-1	$y = -(-1)^2 + 4(-1) = -5$	$(-1,-5)$
0	$y = -(0)^2 + 4(0) = 0$	(0,0)
1	$y = -(1)^2 + 4(1) = 3$	(1,3)
2	$y = -(2)^2 + 4(2) = 4$	(2,4)
3	$y = -(3)^2 + 4(3) = 3$	(3,3)
4	$y = -(4)^2 + 4(4) = 0$	(4,0)

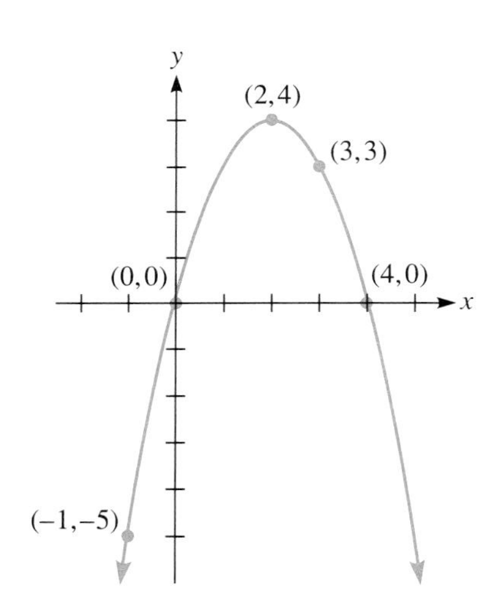

Figure 8.12

PROGRESS CHECK 2 Graph $y = 4x - 2x^2$.

Progress Check Answers

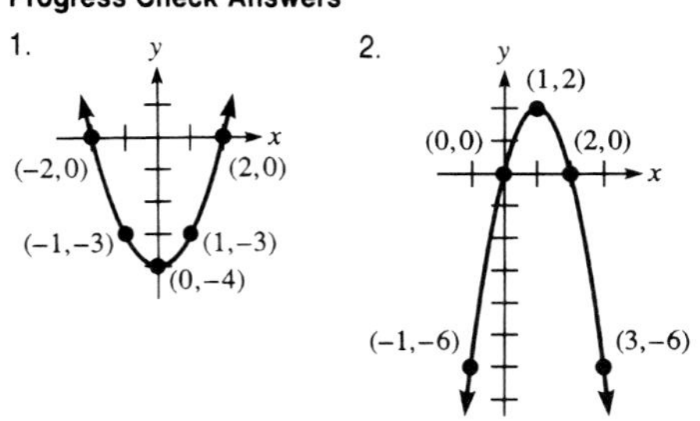

2 The graphs in Examples 1 and 2 are curves called **parabolas.** Every equation of the form $y = ax^2 + bx + c$ with $a \neq 0$ graphs as a parabola, and we may use some features of parabolas to graph such equations efficiently. First, Figure 8.11 shows that

the graph of $y = x^2$ has a minimum turning point and opens up like a cup, while Figure 8.12 shows that the graph of $y = -x^2 + 4x$ turns at its highest point and opens down. These examples illustrate that the sign of the coefficient of x^2 determines whether the parabola opens up or down, as follows:

1. If $a > 0$, the parabola opens upward and turns at the lowest point on the graph.
2. If $a < 0$, the parabola opens downward and turns at the highest point on the graph.

The names *parabola, ellipse,* and *hyperbola* came from Apollonius. In his construction of the sections the names respectively meant that a certain line came out even with, was short of, or was longer than a fixed reference length. See "Think About It" Exercise 4 in Section 8.7.

A second feature of a parabola is that a vertical line that passes through the turning point, as shown in Figure 8.13, divides the parabola into two segments such that if we fold on this line, the two halves will coincide. This line is called the **axis of symmetry** of the parabola, and the turning point is called the **vertex** of the parabola. To find the equation of the axis of symmetry, we note that the line is halfway between any pair of points on the parabola that have the same y-coordinate. The easiest pair of points to analyze is the pair of points on the graph of $y = ax^2 + bx + c$ whose y-coordinate is c. Therefore, we replace y by c in this equation and solve for x.

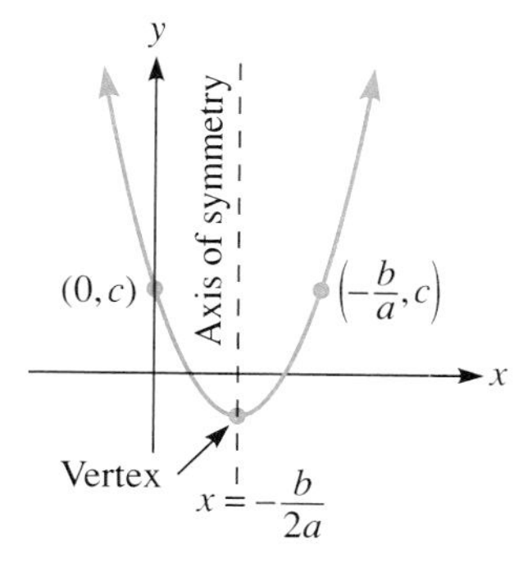

Figure 8.13

$$
\begin{aligned}
y &= ax^2 + bx + c \\
c &= ax^2 + bx + c \\
0 &= ax^2 + bx \\
0 &= x(ax + b) \\
x &= 0 \quad \text{or} \quad ax + b = 0 \\
& \qquad\qquad\quad ax = -b \\
& \qquad\qquad\quad x = -\frac{b}{a}
\end{aligned}
$$

The x-coordinate halfway between 0 and $-b/a$ is given by

$$x = \frac{0 + (-b/a)}{2},$$

so the equation of the axis of symmetry is

$$x = \frac{-b}{2a}$$

The method of finding the vertex by completing the square will be considered in Section 9.4 on quadratic functions, following a general discussion of graphing functions using translation and reflection.

Because the vertex lies on the axis of symmetry, the x-coordinate of the vertex is $-b/(2a)$. Replacing x by the value of this coordinate in the equation to be graphed produces the y-coordinate of the vertex.

A third consideration for graphing and analyzing parabolas is to determine the location of all intercepts. In general, to find y-intercepts, we let $x = 0$ and solve for y. Replacing x by 0 in $y = ax^2 + bx + c$ gives

$$
\begin{aligned}
y &= a(0)^2 + b(0) + c \\
&= c.
\end{aligned}
$$

Thus, the y-intercept for the graph of this equation is always $(0,c)$.

In general, to find x-intercepts, let $y = 0$ and solve for x. Replacing y by 0 in the equation $y = ax^2 + bx + c$ shows that the x-coordinate of any x-intercepts are found by solving the quadratic equation $ax^2 + bx + c = 0$.

EXAMPLE 3 Consider the equation $y = x^2 - 2x - 3$.

a. Find the equation of the axis of symmetry.
b. Find the coordinates of the vertex.
c. Find the coordinates of any x- and y-intercepts.
d. Graph the equation.

Solution

a. y has the form $ax^2 + bx + c$, with $a = 1$, $b = -2$, and $c = -3$. Replacing a by 1 and b by -2 in the axis of symmetry formula gives

$$x = \frac{-b}{2a} = \frac{-(-2)}{2(1)} = 1.$$

Thus, $x = 1$ is the equation of the axis of symmetry.

b. The x-coordinate of the vertex is 1, since the vertex lies on the axis of symmetry. To find the y-coordinate, replace x by 1 in the given equation.

$$\begin{aligned} y &= x^2 - 2x - 3 \\ y &= (1)^2 - 2(1) - 3 \quad \text{Replace } x \text{ by } 1. \\ &= -4 \end{aligned}$$

The vertex is located at $(1,-4)$. Since $a > 0$, this vertex is a minimum point.

c. The y-intercept is at $(0,-3)$, since $c = -3$ in the given equation. To find any x-intercepts, set $y = 0$ and solve the resulting quadratic equation.

$$\begin{aligned} x^2 - 2x - 3 &= 0 \\ (x - 3)(x + 1) &= 0 \end{aligned}$$

$$x - 3 = 0 \quad \text{or} \quad x + 1 = 0$$
$$x = 3 \qquad\qquad x = -1$$

The x-intercepts are at $(3,0)$ and $(-1,0)$.

d. By drawing a parabola through the vertex and the intercepts, we obtain the graph of the equation, which is shown in Figure 8.14.

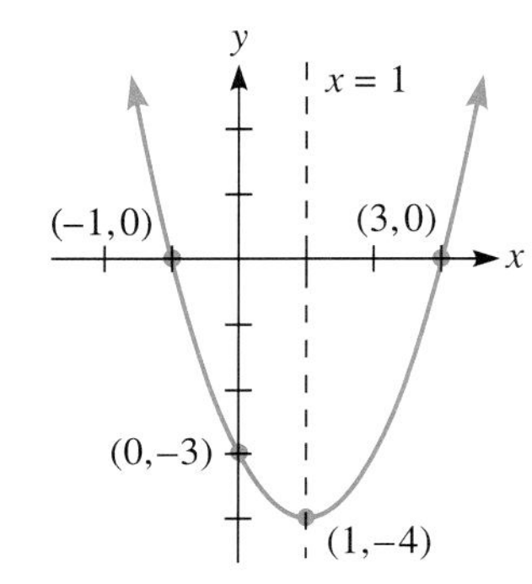

Figure 8.14

Note If the objective in this example had been only to graph the equation, then finding the vertex and plotting additional points on the parabola on both sides of $x = 1$ (which is the axis of symmetry) would be sufficient.

PROGRESS CHECK 3 Answer the questions in Example 3 for the equation $y = x^2 + 2x - 8$.

Progress Check Answer

3.

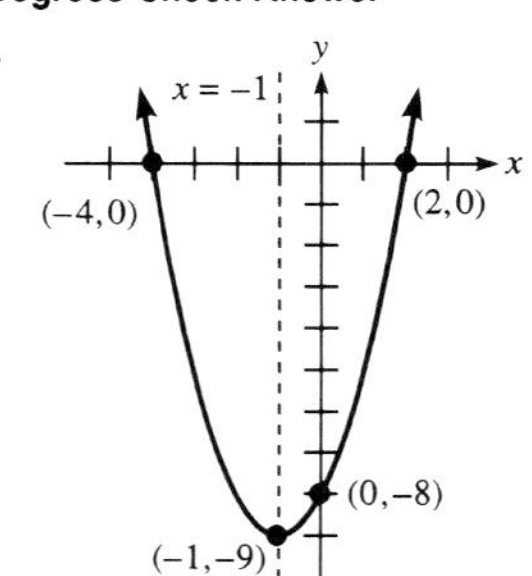

EXAMPLE 4 Consider the equation $y = -x^2 + 3x - 4$.

a. Find the equation of the axis of symmetry.
b. Find the coordinates of the vertex.
c. Find the coordinates of any x- and y-intercepts.
d. Graph the equation.

Solution

a. In the given equation $a = -1$ and $b = 3$, so

$$x = \frac{-b}{2a} = \frac{-(3)}{2(-1)} = \frac{3}{2}.$$

The equation of the axis of symmetry is $x = \frac{3}{2}$.

b. The x-coordinate of the vertex is $\frac{3}{2}$, and we use this result to find the y-coordinate as follows.

$$\begin{aligned} y &= -x^2 + 3x - 4 \\ y &= -(\tfrac{3}{2})^2 + 3(\tfrac{3}{2}) - 4 \quad \text{Replace } x \text{ by } \tfrac{3}{2}. \\ &= -\tfrac{9}{4} + \tfrac{9}{2} - 4 \\ &= -\tfrac{7}{4} \end{aligned}$$

The vertex is located at $(\frac{3}{2}, -\frac{7}{4})$. Since $a < 0$, this vertex is a maximum point.

c. Since $c = -4$, the y-intercept is $(0,-4)$. We let $y = 0$ and solve

$$0 = -x^2 + 3x - 4,$$

using the quadratic formula to find any x-intercepts.

$$x = \frac{-3 \pm \sqrt{3^2 - 4(-1)(-4)}}{2(-1)} = \frac{-3 \pm \sqrt{-7}}{-2}$$

There are no x-intercepts because the solutions are not real numbers.

d. Using the results of parts **a–c** and plotting two additional points at $(1,-2)$ and $(3,-4)$ gives the graph of the equation, which is shown in Figure 8.15.

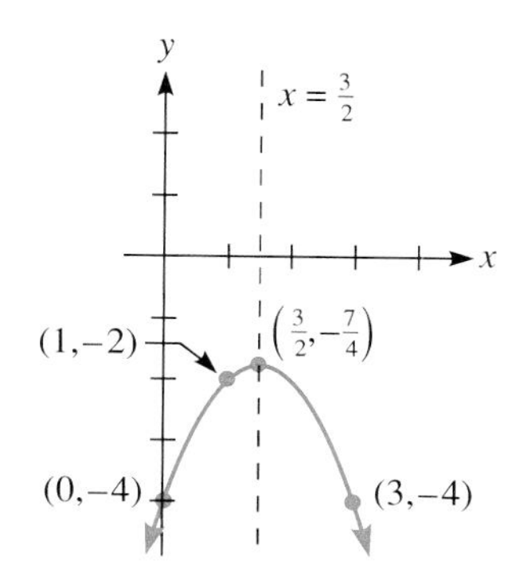

Figure 8.15

PROGRESS CHECK 4 Answer the questions in Example 4 for the equation $y = -x^2 + 5x - 7$.

3 Equations of the form $y = ax^2 + bx + c$ have exactly one y-intercept at $(0,c)$, but Examples 3 and 4 illustrate that the number of x-intercepts may vary. We can predict the number of x-intercepts by recalling from Section 8.2 that the discriminant $b^2 - 4ac$ tells us the nature of the solutions of the equation $ax^2 + bx + c = 0$ with $a \neq 0$. For instance, when the discriminant indicates real and unequal solutions, then there are two x-intercepts. One x-intercept results when the solutions are real and equal, and there are no x-intercepts when the solutions are not real numbers. Figure 8.16 summarizes these three cases.

Remind students that the discriminant is the expression under the radical in the quadratic formula.

Discriminant	Number of x-intercepts	Possible Graph
$b^2 - 4ac > 0$	Two	
$b^2 - 4ac = 0$	One	
$b^2 - 4ac < 0$	None	

Figure 8.16

Progress Check Answer

4.

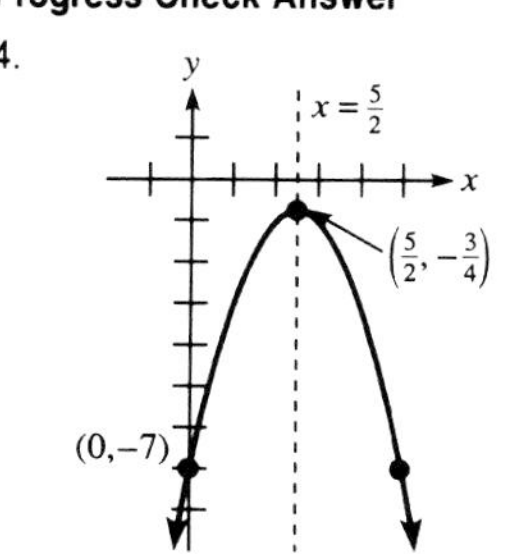

EXAMPLE 5 Use the discriminant to determine the number of x-intercepts of the graph of $y = 4x^2 - 4x + 1$.

Solution In the given equation $a = 4$, $b = -4$, and $c = 1$, so

$$b^2 - 4ac = (-4)^2 - 4(4)(1) = 0.$$

Since the discriminant is 0, the graph has one x-intercept.

PROGRESS CHECK 5 Use the discriminant to determine the number of x-intercepts of the graph of $y = x^2 + 2x + 3$.

4 Reversing the roles of x and y in the equation $y = ax^2 + bx + c$ produces a second-degree equation of the form

$$x = ay^2 + by + c.$$

Every equation of this form also graphs as a parabola. However, the parabola in such graphs will open to the right (if $a > 0$) or the left (if $a < 0$), instead of opening up or down. Because the equation of the axis of symmetry in this case is the horizontal line $y = -b/(2a)$, we refer to such parabolas as **horizontal parabolas.**

EXAMPLE 6 Consider the equation $x = y^2 + 4y$.

a. Find the equation of the axis of symmetry.
b. Find the coordinates of the vertex.
c. Find the coordinates of any x- and y-intercepts.
d. Graph the equation.

Solution

a. In the given equation $a = 1$ and $b = 4$, so

$$y = \frac{-b}{2a} = \frac{-4}{2(1)} = -2.$$

The equation of the axis of symmetry is $y = -2$.

b. The y-coordinate of the vertex is -2 since the vertex lies on the axis of symmetry. To find the x-coordinate, replace y by -2 in the given equation.

$$x = y^2 + 4y$$
$$x = (-2)^2 + 4(-2) \quad \text{Replace } y \text{ by } -2.$$
$$= -4$$

The vertex is located at $(-4, -2)$.

c. To find y-intercepts, let $x = 0$ and solve the resulting equation for y.

$$y^2 + 4y = 0$$
$$y(y + 4) = 0$$
$$y = 0 \quad \text{or} \quad y + 4 = 0$$
$$y = 0 \qquad\qquad y = -4$$

The y-intercepts are (0,0) and $(0,-4)$. Letting $y = 0$ in the equation to be graphed shows that (0,0) is also an x-intercept.

d. By drawing a parabola through the vertex and the intercepts, we obtain the graph of $x = y^2 + 4y$, which is shown in Figure 8.17. Notice that this graph is in agreement with the general principle that a horizontal parabola opens to the right when $a > 0$.

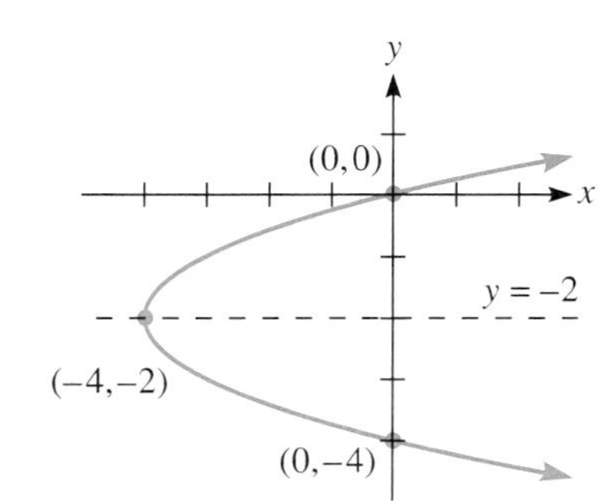

Figure 8.17

PROGRESS CHECK 6 Answer the questions in Example 6 for the equation $x = 3y^2 - 6y$.

Progress Check Answers

5. No x-intercept 6.

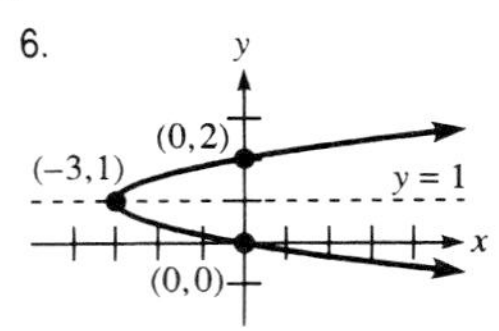

5 The section-opening problem considers an applied problem that may be solved by the methods we have developed for analyzing parabolas.

EXAMPLE 7 Solve the problem in the section introduction on page 389.

Solution

a. In the formula $y = -16t^2 + 96t$, $a = -16$ and $b = 96$. Since $a < 0$, the vertex is the highest point in the graph. Use the axis of symmetry formula to find when this highest point is reached.

$$t = \frac{-b}{2a} = \frac{-96}{2(-16)} = 3$$

The projectile attains its maximum height when $t = 3$ seconds.

b. To find the maximum height, replace t by 3 in the given formula.

$$\begin{aligned} y &= -16t^2 + 96t \\ y &= -16(3)^2 + 96(3) \quad \text{Replace } t \text{ by 3.} \\ &= -144 + 288 \\ &= 144 \end{aligned}$$

The maximum height is 144 ft.

c. The height (y) of the projectile is zero when the projectile hits the ground. Replacing y by 0 in the formula and solving for t gives

$$-16t^2 + 96t = 0$$
$$-16t(t - 6) = 0$$
$$-16t = 0 \quad \text{or} \quad t - 6 = 0$$
$$t = 0 \qquad\qquad t = 6.$$

The projectile is on the ground initially, and it hits the ground 6 seconds after launching.

d. Parts **a** and **b** show that the vertex is at (3,144), while part **c** indicates the intercepts are (0,0) and (6,0). Drawing a parabola using these points gives the graph in Figure 8.18. Note that the graph is restricted to values for t between 0 and 6, inclusive, because the formula is not meaningful outside this interval.

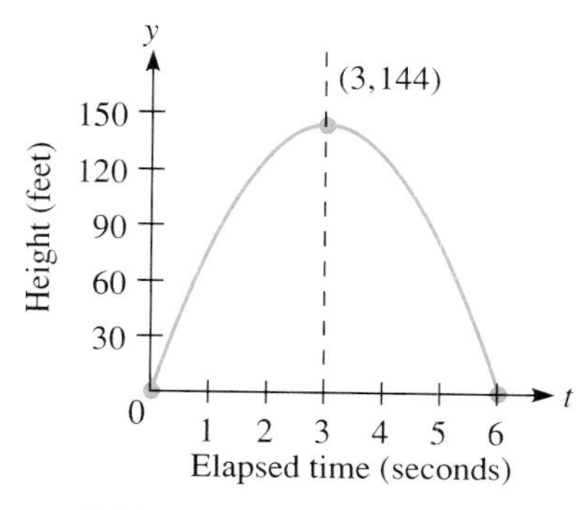

Figure 8.18

PROGRESS CHECK 7 The height (y) in feet of a projectile shot vertically up from the ground with an initial velocity of 128 ft/second is given by $y = -16t^2 + 128t$, where t is the elapsed time in seconds. Answer the questions in Example 7 with respect to this formula.

In Example 7 a parabola helps to visualize the relation between two variables, which is a main use of parabolas. Parabolas also have many applications that are based on geometric properties of parabolas. For instance, searchlights and radar dishes are shaped like a parabola that is rotated about its axis of symmetry because of reflection properties of parabolas. This type of application is often considered in a course in college algebra.

Progress Check Answers

7. (a) 4 seconds (b) 256 ft (c) 8 seconds
(d)

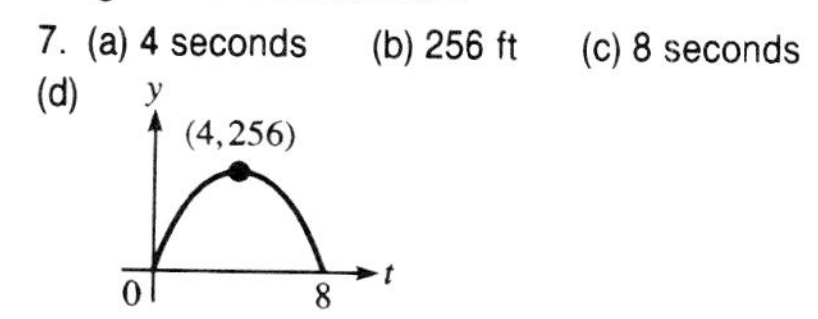

EXERCISES 8.5

In Exercises 1–44 for each equation, find the following:

a. The equation of the axis of symmetry
b. The coordinates of the vertex
c. The coordinates of any x- and y-intercepts
d. The graph

1. $y = x^2 + 2x$

a. $x = -1$
b. $(-1, -1)$
c. $(-2, 0)$, $(0, 0)$
d.

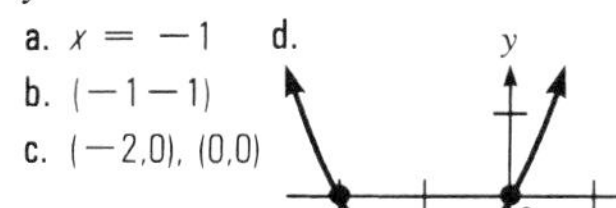

2. $y = x^2 + 9x$

a. $x = -\frac{9}{2}$
b. $(-\frac{9}{2}, -\frac{81}{4})$
c. $(-9, 0)$, $(0, 0)$
d.

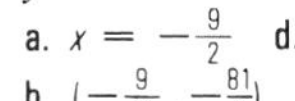

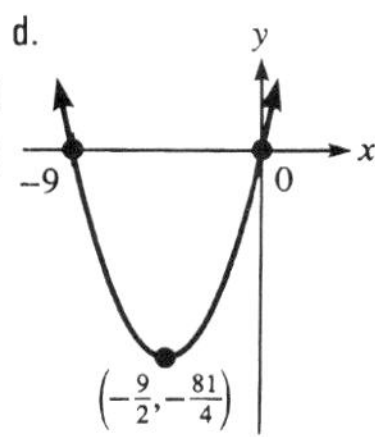

3. $y = -x^2 + 3x$

a. $x = \frac{3}{2}$
b. $(\frac{3}{2}, \frac{9}{4})$
c. $(3, 0)$, $(0, 0)$
d.

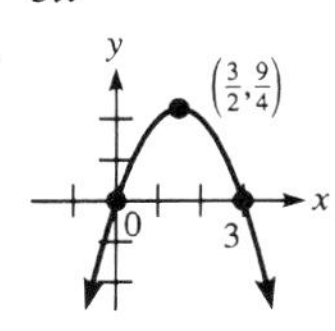

4. $y = -x^2 + 5x$

a. $x = \frac{5}{2}$
b. $(\frac{5}{2}, \frac{25}{4})$
c. $(5, 0)$, $(0, 0)$
d.

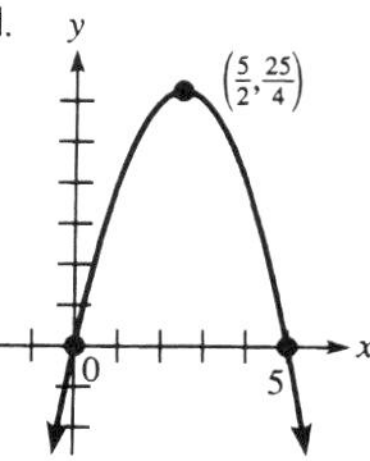

5. $y = x^2 - 5x$

a. $x = \frac{5}{2}$
b. $(\frac{5}{2}, -\frac{25}{4})$
c. $(0, 0)$, $(5, 0)$
d.

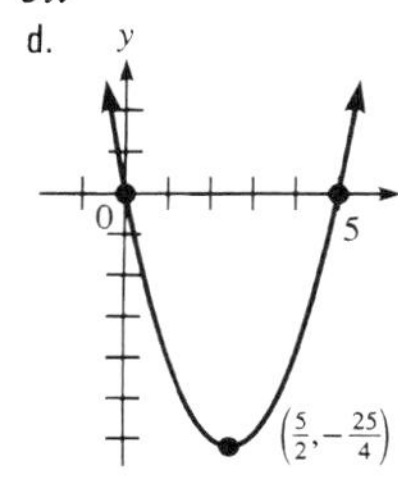

6. $y = x^2 - 7x$

a. $x = \frac{7}{2}$
b. $(\frac{7}{2}, -\frac{49}{4})$
c. $(0, 0)$, $(7, 0)$
d.

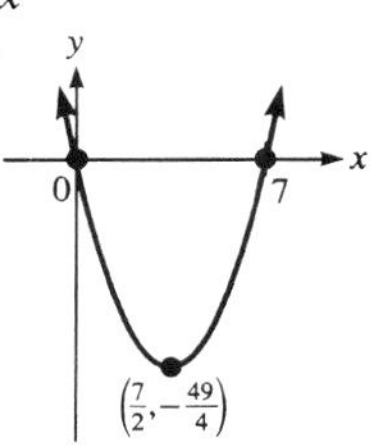

7. $y = x^2 - 4x + 4$

a. $x = 2$
b. $(2, 0)$
c. $(2, 0)$, $(0, 4)$
d.

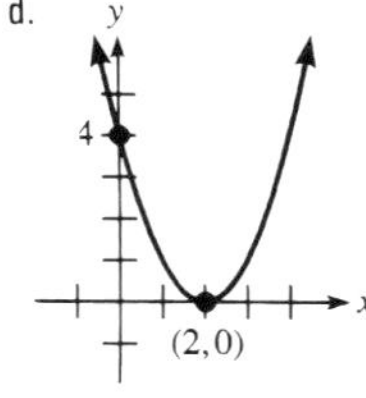

8. $y = x^2 + 6x + 9$

a. $x = -3$
b. $(-3, 0)$
c. $(-3, 0)$, $(0, 9)$
d.

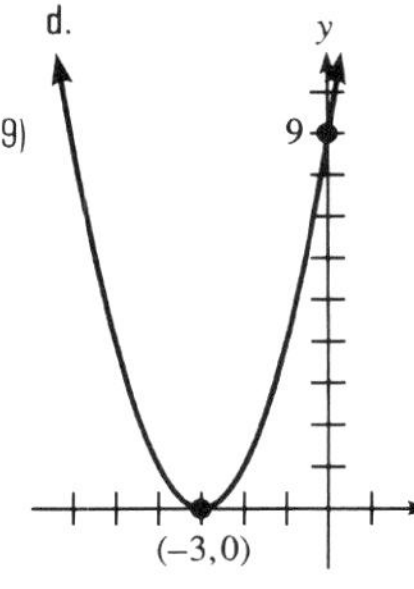

9. $y = -x^2 - 6x - 9$

a. $x = -3$
b. $(-3, 0)$
c. $(-3, 0)$, $(0, -9)$
d.

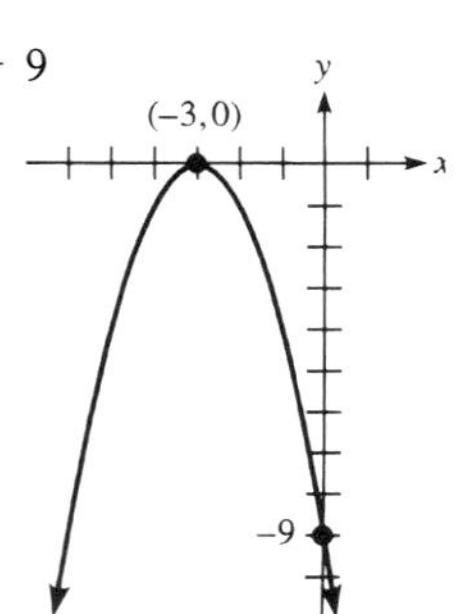

10. $y = -x^2 + 4x - 4$

a. $x = 2$
b. $(2, 0)$
c. $(2, 0)$, $(0, -4)$
d.

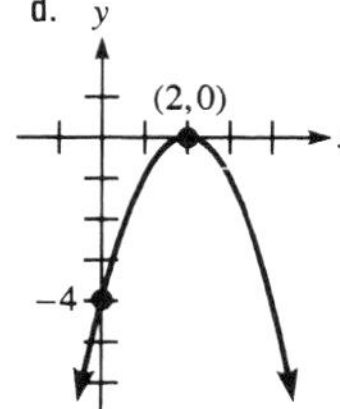

11. $y = x^2 + 9$

a. $x = 0$
b. $(0, 9)$
c. $(0, 9)$
d.

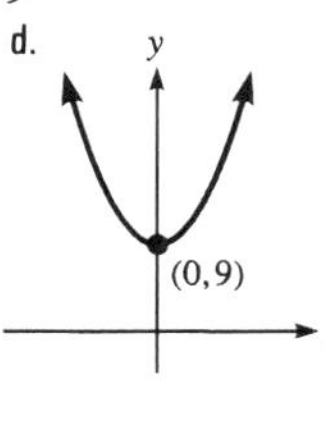

12. $y = -x^2 - 5$

a. $x = 0$
b. $(0, -5)$
c. $(0, -5)$
d.

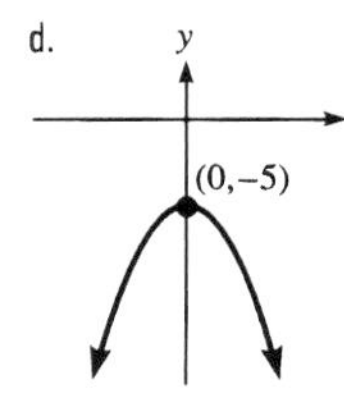

13. $y = x^2 - 2x - 8$

a. $x = 1$
b. $(1, -9)$
c. $(-2, 0)$, $(4, 0)$, $(0, -8)$
d.

y
x
−2
4
−8
(1,−9)

14. $y = x^2 + x - 6$

a. $x = -\frac{1}{2}$
b. $(-\frac{1}{2}, -\frac{25}{4})$
c. $(-3, 0)$, $(2, 0)$, $(0, -6)$
d.

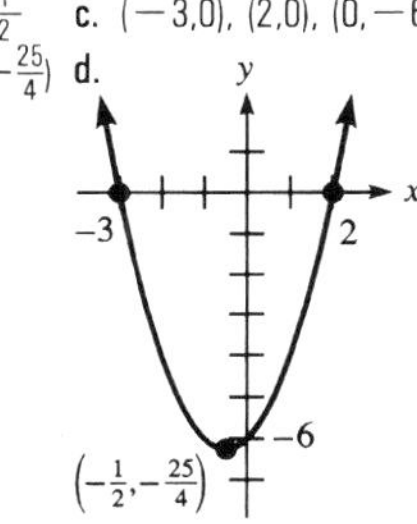

15. $y = -x^2 + 2x + 8$

a. $x = 1$
b. $(1, 9)$
c. $(-2, 0)$, $(4, 0)$, $(0, 8)$
d.

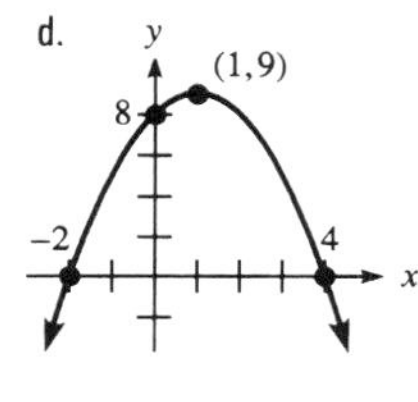

16. $y = -x^2 - x + 6$

a. $x = -\frac{1}{2}$
b. $(-\frac{1}{2}, \frac{25}{4})$
c. $(-3, 0)$, $(2, 0)$, $(0, 6)$
d.

y
x
$(-\frac{1}{2}, \frac{25}{4})$
6
−3
2

17. $y = x^2 + x - 20$

a. $x = -\frac{1}{2}$
b. $(-\frac{1}{2}, -\frac{81}{4})$
c. $(4, 0)$, $(-5, 0)$, $(0, -20)$
d.

y
x
−5
4
−20
$(-\frac{1}{2}, -\frac{81}{4})$

18. $y = x^2 - 3x - 28$

a. $x = \frac{3}{2}$
b. $(\frac{3}{2}, -\frac{121}{4})$
c. $(-4, 0)$, $(7, 0)$, $(0, -28)$
d.

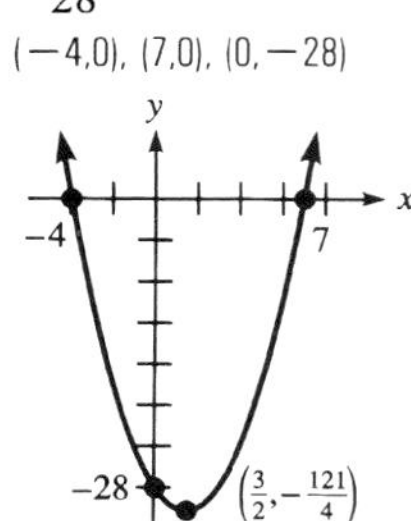

19. $y = -x^2 + x + 20$

a. $x = \frac{1}{2}$ c. $(-4,0)$, $(5,0)$, $(0,20)$

b. $(\frac{1}{2}, \frac{81}{4})$ d.

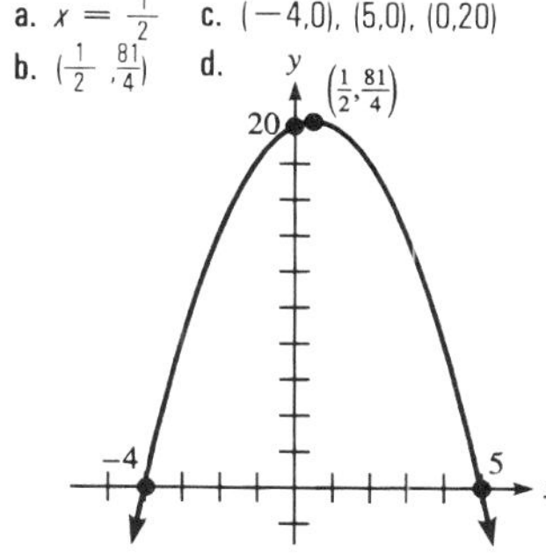

20. $y = -x^2 + 3x + 28$

a. $x = \frac{3}{2}$ c. $(-4,0)$, $(7,0)$, $(0,28)$

b. $(\frac{3}{2}, \frac{121}{4})$ d.

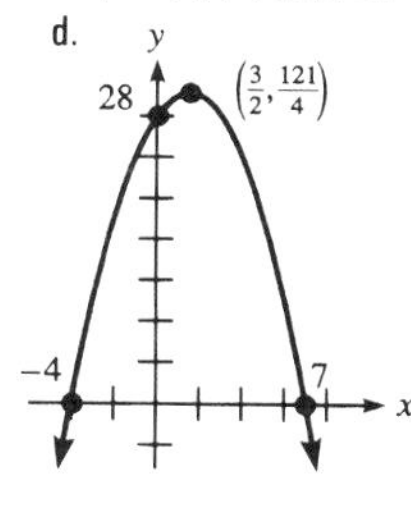

21. $y = -x^2 - 9x - 18$

a. $x = -\frac{9}{2}$ c. $(-6,0)$, $(-3,0)$, $(0,-18)$

b. $(-\frac{9}{2}, \frac{9}{4})$ d.

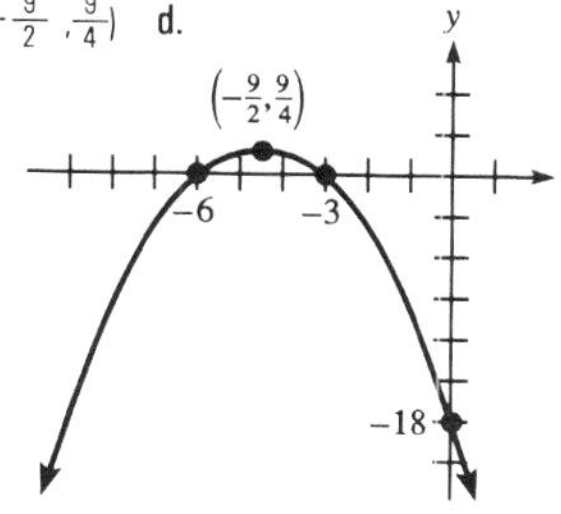

22. $y = -x^2 - 12x - 35$

a. $x = -6$ b. $(-6,1)$

c. $(-7,0)$, $(-5,0)$, $(0,-35)$

d.

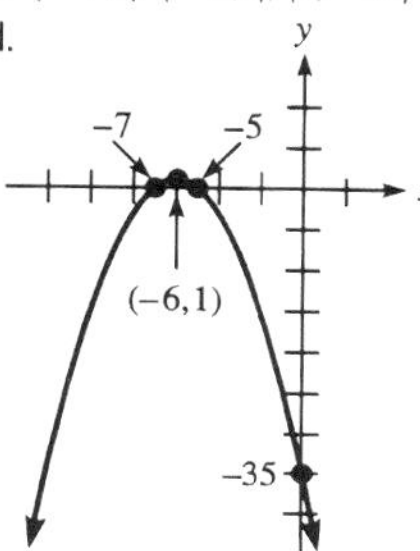

23. $y = x^2 + 9x + 18$

a. $x = -\frac{9}{2}$ b. $(-\frac{9}{2}, -\frac{9}{4})$

c. $(-6,0)$, $(-3,0)$, $(0,18)$

d.

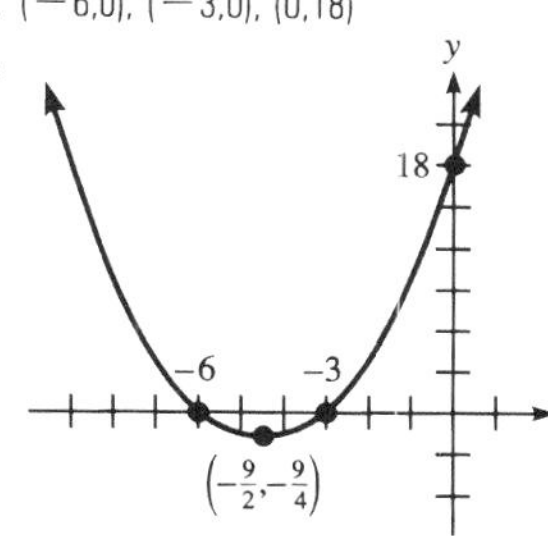

24. $y = x^2 + 12x + 35$

a. $x = -6$ b. $(-6,-1)$

c. $(-7,0)$, $(-5,0)$, $(0,35)$

d.

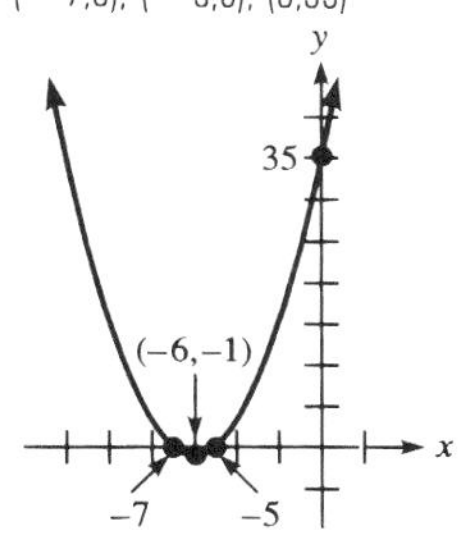

25. $y = -6x^2 + x + 2$

a. $x = \frac{1}{12}$

b. $(\frac{1}{12}, \frac{49}{24})$

c. $(-\frac{1}{2}, 0)$, $(\frac{2}{3}, 0)$, $(0,2)$

d.

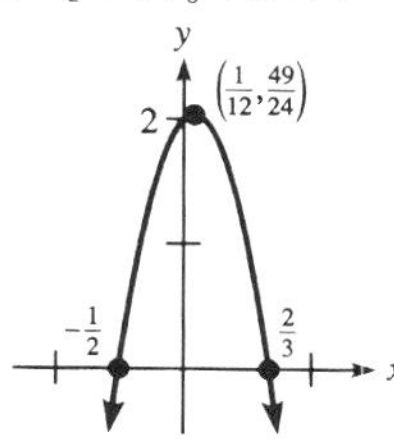

26. $y = 12x^2 + 5x - 3$

a. $x = -\frac{5}{24}$

b. $(-\frac{5}{24}, -\frac{169}{48})$

c. $(-\frac{3}{4}, 0)$, $(\frac{1}{3}, 0)$, $(0,-3)$

d.

y x $-\frac{3}{4}$ $\frac{1}{3}$ −3 $(-\frac{5}{24}, -\frac{169}{48})$

27. $y = 6x^2 - x - 2$

a. $x = \frac{1}{12}$

b. $(\frac{1}{12}, -\frac{49}{24})$

c. $(-\frac{1}{2}, 0)$, $(\frac{2}{3}, 0)$, $(0,-2)$

d.

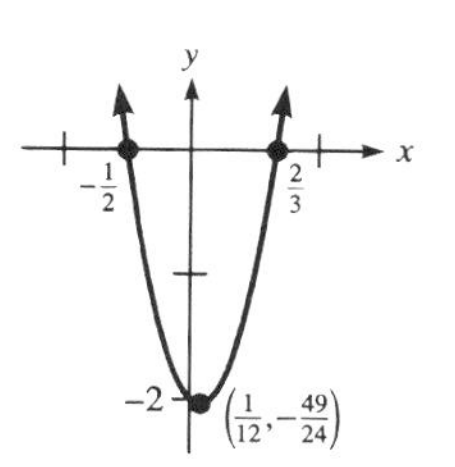

28. $y = -12x^2 - 5x + 3$

a. $x = -\frac{5}{24}$ b. $(-\frac{5}{24}, \frac{169}{48})$

c. $(-\frac{3}{4}, 0)$, $(\frac{1}{3}, 0)$, $(0,3)$

d.

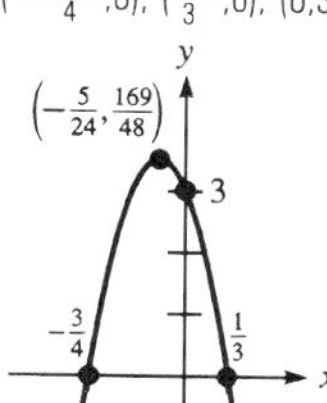

29. $y = x^2 - 20$

a. $x = 0$ b. $(0,-20)$

c. $(\pm 2\sqrt{5}, 0)$, $(0,-20)$

d.

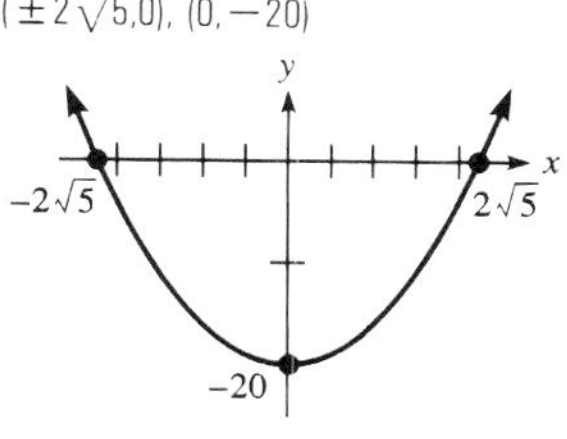

30. $y = x^2 - 27$

a. $x = 0$ b. $(0,-27)$

c. $(\pm 3\sqrt{3}, 0)$, $(0,-27)$

d.

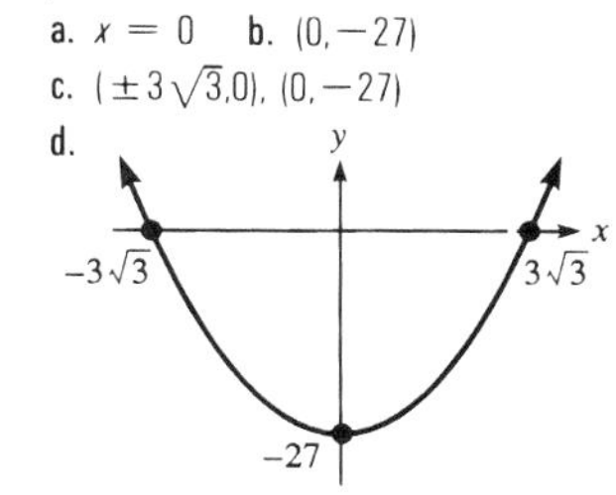

31. $y = x^2 - 4x - 3$

a. $x = 2$ b. $(2,-7)$

c. $(2 \pm \sqrt{7}, 0)$, $(0,-3)$

d.

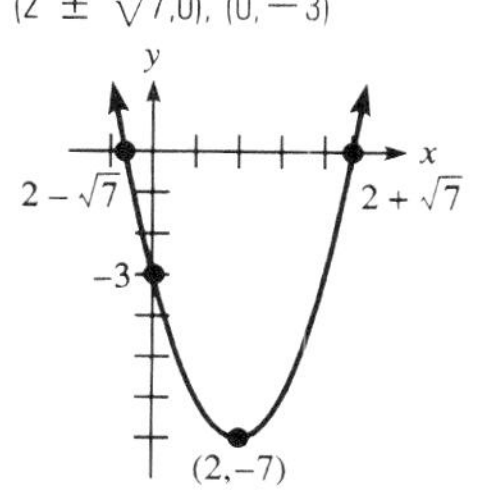

32. $y = x^2 - 8x + 12$

a. $x = 4$ b. $(4,-4)$

c. $(2,0)$, $(6,0)$, $(0,12)$

d.

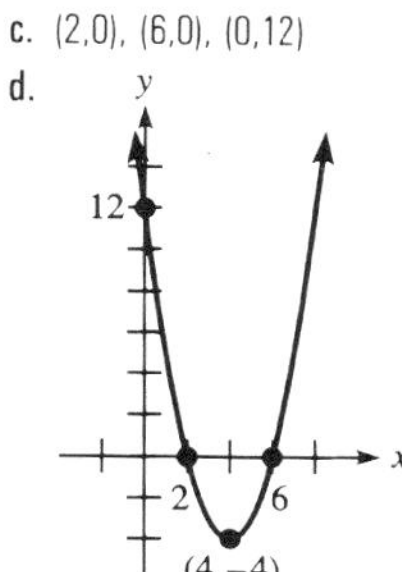

33. $y = x^2 - 2x + 5$

a. $x = 1$ d.

b. $(1,4)$

c. $(0,5)$

y 5 (1,4) x

34. $y = x^2 + 2x + 3$

a. $x = -1$ d.

b. $(-1,2)$

c. $(0,3)$

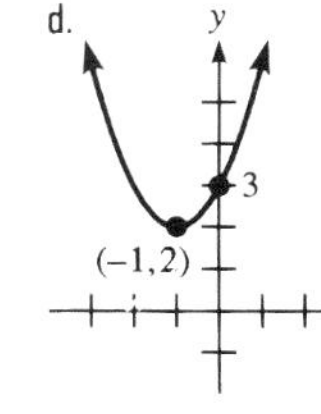

35. $y = -x^2 + 2x - 5$

a. $x = 1$ d.

b. $(1,-4)$

c. $(0,-5)$

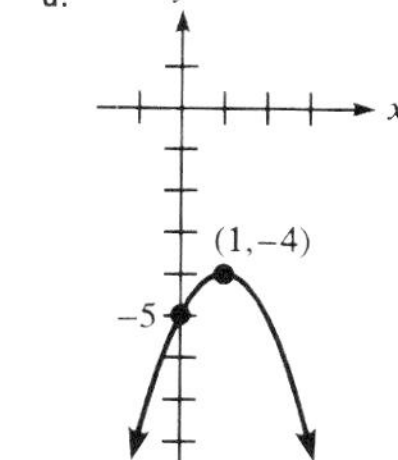

36. $y = -x^2 - 2x - 3$

a. $x = -1$ d.

b. $(-1,-2)$

c. $(0,-3)$

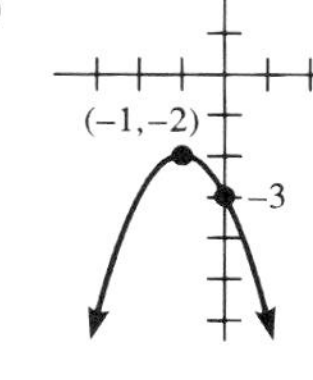

37. $x = y^2 + 5y$

a. $y = -\frac{5}{2}$ b. $(-\frac{25}{4}, -\frac{5}{2})$

c. $(0,0)$, $(0,-5)$

d.

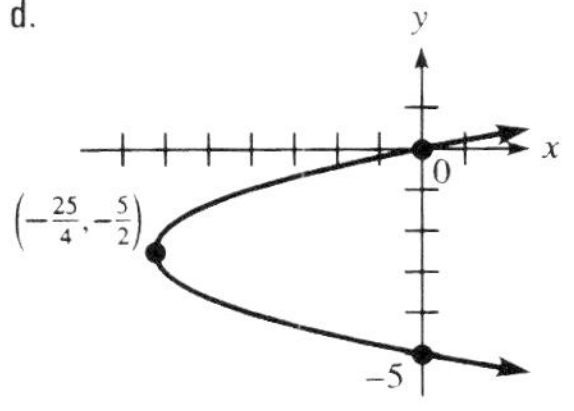

38. $x = y^2 + 9y$
a. $y = -\frac{9}{2}$ b. $(-\frac{81}{4}, -\frac{9}{2})$
c. $(0,0)$, $(0,-9)$
d.

39. $x = -y^2 + 3y$
a. $y = \frac{3}{2}$ b. $(\frac{9}{4}, \frac{3}{2})$
c. $(0,0)$, $(0,3)$
d.

40. $x = y^2 - 5y$
a. $y = \frac{5}{2}$ b. $(-\frac{25}{4}, \frac{5}{2})$
c. $(0,0)$, $(0,5)$
d.

41. $x = -y^2 - 4y$
a. $y = -2$ b. $(4,-2)$
c. $(0,-4)$, $(0,0)$
d.

42. $x = y^2 - 2y - 8$
a. $y = 1$ b. $(-9,1)$
c. $(-8,0)$, $(0,-2)$, $(0,4)$
d.

43. $x = -y^2 + 2y + 8$
a. $y = 1$ b. $(9,1)$
c. $(0,-2)$, $(0,4)$, $(8,0)$
d.

44. $x = y^2 - 20$
a. $y = 0$
b. $(-20,0)$
c. $(-20,0)$, $(0, \pm 2\sqrt{5})$
d.

In Exercises 45–54, use the discriminant to determine the number of x-intercepts.

45. $y = x^2 - 7x$ 2
46. $y = -x^2 - 4x$ 2
47. $y = -x^2 + 25$ 2
48. $y = x^2 - 2x - 8$ 2
49. $y = x^2$ 1
50. $y = -x^2$ 1
51. $y = x^2 - 2x + 5$ 0
52. $y = x^2 + 2x + 3$ 0
53. $y = x^2 - 2x + 1$ 1
54. $y = 9x^2 - 12x + 4$ 1

55. The height of a projectile shot directly from the ground with an initial velocity of 32 ft/second is given by $y = -16t^2 + 32t$.
a. Graph the equation.
b. When does the projectile attain its maximum height?
c. What is the maximum height?
d. When does it hit the ground?

a. b. 1 second
c. 16 ft
d. 2 seconds

56. The projectile of Exercise 55 is now launched with twice the old initial velocity, so that the equation becomes $y = -16t^2 + 64t$. Does it now go twice as high? Is it now in the air twice as long before hitting the ground? Answer all the questions from Exercise 55 for this new equation. No (64 ft); yes (2 seconds)

a.

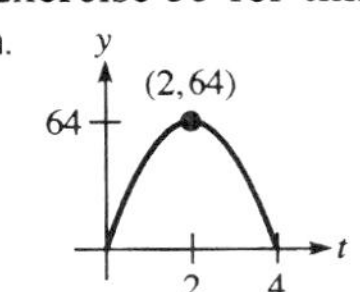

b. 2 seconds
c. 64 ft
d. 4 seconds

57. An object (like a ball or an apple) thrown from a height of 10 ft straight up into the air with an initial velocity of 78 ft/second has a distance (in feet) above the ground after t seconds given by the formula $y = 10 + 78t - 16t^2$.
a. Graph the equation.
b. When does the object attain its maximum height?
c. What is the maximum height?
d. When does it hit the ground?

a.

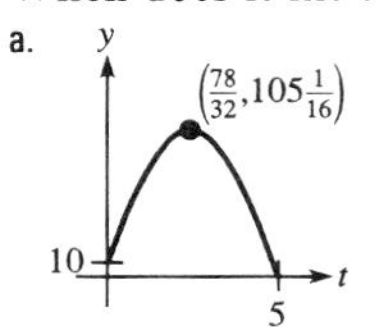

b. $\frac{78}{32} = 2.4375$ seconds
c. $105\frac{1}{16} = 105.0625$ ft
d. 5 seconds

58. If the object of Exercise 57 is thrown straight up from a height of 108 ft, then the equation becomes $y = 108 + 78t - 16t^2$. Answer the questions given in Exercise 57.

a.

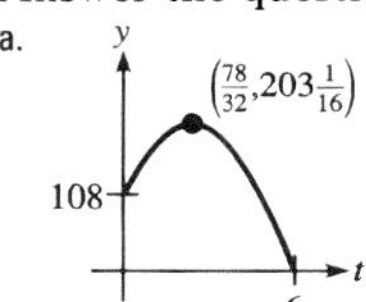

b. $\frac{78}{32} = 2.4375$ seconds
c. $203\frac{1}{16} = 203.0625$ ft
d. 6 seconds

59. On one episode of the TV show "Northern Exposure," a piano was catapulted through the air. Such an object would follow a parabolic arc through the air. If it is launched at a 45° angle, the distance it travels is given approximately by $d = V^2/32$, where V is the initial velocity. Graph the equation. Then find out approximately what initial velocity is needed for it to travel 100 ft. 56.6 ft/second

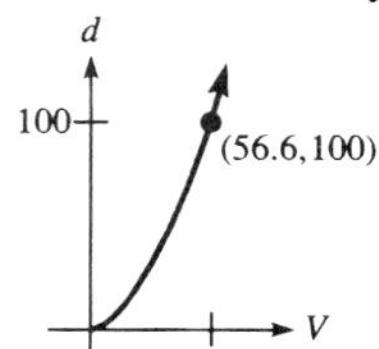

60. If the piano of Exercise 59 is launched at a 30° angle, the equation becomes $d = \sqrt{3}V^2/64$. Answer the questions of Exercise 59 for this equation. 60.8 ft/second

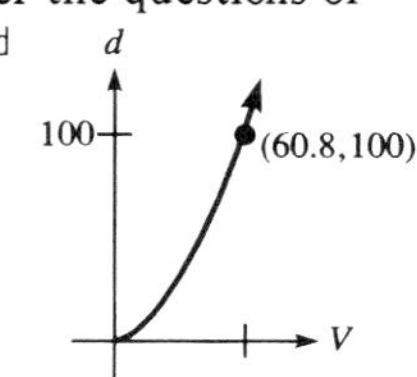

61. In estimating the margin of error for a statistical survey, one must find the value of x which maximizes the value of the expression $x(1 - x)$. Solve this problem by analyzing the graph of $y = x(1 - x)$. What is the maximum value?
The maximum value is $\frac{1}{4}$, when $x = \frac{1}{2}$.

62. The table below consists of readings from a parabola. Use it to find the equation of the axis of symmetry, the equation of the parabola, and the coordinates of the vertex. (*Hint:* Assume that the equation has the form $y = ax^2 + bx + c$; then use the formula for the axis of symmetry to determine a relationship between a and b.)

x	-3	-2	-1	0	1	2	3
y	15	7	3	3	7	15	27

$x = -\frac{1}{2}$; $y = 2x^2 + 2x + 3$; $\left(-\frac{1}{2}, \frac{5}{2}\right)$

THINK ABOUT IT

1. This section of the text showed that the x-coordinate of the vertex of the graph of $y = ax^2 + bx + c$ is given by $-b/(2a)$. Find a general expression for the y-coordinate of the vertex.
2. Find an equation for the parabola whose intercepts are $(4,0)$, $(-6,0)$, and $(0,-12)$. Assume that the form of the equation is $y = ax^2 + bx + c$, and use the given points to arrive at a system of three equations with variables a, b, and c. (*Note:* It is also possible to find an equation for a circle through these points.)
3. First, graph $y = x^2$, and then graph each of the following equations. Explain what effect the constant has in moving the graph of $y = x^2$.
   a. $y = x^2 + 1$ **b.** $y = x^2 - 1$
   c. $y = (x - 1)^2$ **d.** $y = (x + 1)^2$
4. The great Greek mathematician Archimedes (225 B.C.) showed that the area inside a parabola (see the figure) equals $\frac{4}{3}$ the area of the triangle. Use his discovery to find the area enclosed by the parabola $y = x^2$ and the line $y = 4$.

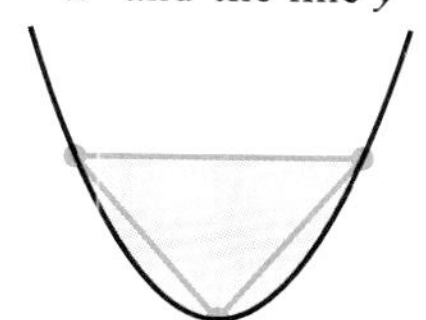

5. If you draw a line and a point not on the line, and then you mark all other points that are equidistant from the original line and point, these new points will form a parabola. For example, if you draw the point $(0,1)$ and the line $y = -1$, then the collection of points equidistant from them forms the parabola whose equation is $y = \frac{1}{4}x^2$.
   a. Draw the point $(0,1)$ and the line $y = -1$ on the same set of axes.
   b. Now add the graph of $y = \frac{1}{4}x^2$ to the picture.
   c. The point $(2,1)$ is on the parabola. Show that its distance from $(0,1)$ is the same as its distance from the line $y = -1$.
   d. The point $(4,4)$ is on the parabola. Show that its distance from $(0,1)$ is the same as its distance from the line $y = -1$.

REMEMBER THIS

1. True or false? All points on a circle are the same distance from the center. True
2. True or false? The center of a circle is not a point on the circle. True
3. If the center of a circle is $(1,1)$ and one point of the circle is $(4,5)$, what is the radius of the circle? 5
4. Is the ordered pair $(-5,2)$ a solution to $(x + 3)^2 + y^2 = 8$? Yes
5. What is the formula for the area of a circle of radius r? $A = \pi r^2$
6. Solve $x^2 - 4 \leq 0$. Give the solution graphically and in interval notation. $[-2,2]$ (number line shaded from -2 to 2)
7. Solve $x^2 + x - 1 = 0$. $\left\{\frac{-1 \pm \sqrt{5}}{2}\right\}$
8. Find an equation for the line through $(0,0)$ and $(-2,5)$. $y = -\frac{5}{2}x$
9. Factor $6x^3 + 5x^2 - 6x$. $x(2x + 3)(3x - 2)$
10. What property is used to say that $3(x - 4)$ and $3x - 12$ are equivalent? Distributive

8.6 The Circle

In addition to its great practical value, the circle is also an aesthetically pleasing figure. For instance, in architecture circles are basic to the construction of many beautiful designs, as shown in the bridge in the photo. If the diameter in such a bridge is 100 m, how high above the water is the bridge 30 m on each side of the center? (See Example 6.)

OBJECTIVES

1. **Find the equation in standard form of a circle given its center and either the radius of the circle or a point on the circle.**
2. **Find the center and radius of a circle given its equation, and then graph it.**
3. **Graph and write equations for circles and semicircles with centers at the origin.**

The word *circle* comes from the Latin word *circus,* which means "ring"; *circle* means "a little ring."

1 A **circle** is the set of all points in a plane at a given distance from a fixed point. To derive an equation for a circle, we recall from Section 5.2 that the distance between any two points (x_1,y_1) and (x_2,y_2) is given by the formula

$$d = \sqrt{(x_2 - x_1)^2 + (y_2 - y_1)^2}.$$

If we let (h,k) be the fixed point (the center of the circle), r be the given distance (the radius of the circle), and (x,y) be any point on the circle, as shown in Figure 8.19, then by the distance formula,

$$\sqrt{(x - h)^2 + (y - k)^2} = r.$$

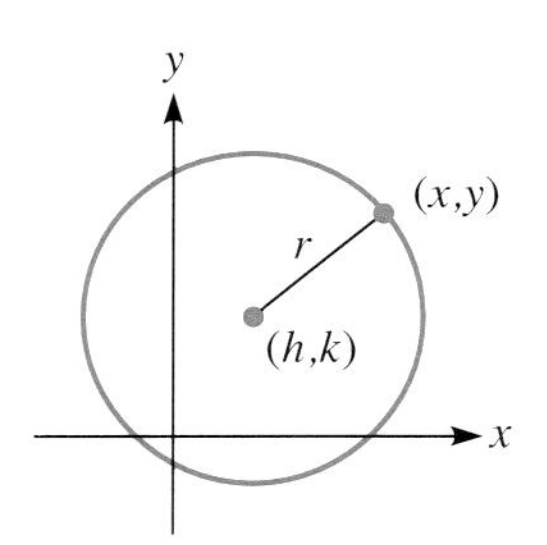

Figure 8.19

Squaring both sides produces an equation for a circle that is called the standard form.

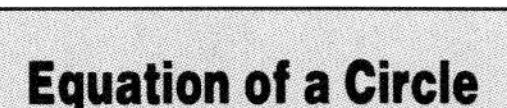

The standard form of the equation of a circle of radius r with center (h,k) is

$$(x - h)^2 + (y - k)^2 = r^2.$$

EXAMPLE 1 Find the equation in standard form of the circle with center at $(5,-2)$ and radius 3.

Solution Replacing h by 5, k by -2, and r by 3 in the equation

$$(x - h)^2 + (y - k)^2 = r^2$$

yields

$$(x - 5)^2 + [y - (-2)]^2 = (3)^2$$
$$(x - 5)^2 + (y + 2)^2 = 9.$$

PROGRESS CHECK 1 Find the equation in standard form of the circle with center at $(-1,6)$ and radius 4.

Progress Check Answer

1. $(x + 1)^2 + (y - 6)^2 = 16$

EXAMPLE 2 Find the equation in standard form of the circle with center at $(-3,0)$ that passes through $(-5,2)$.

Solution Substituting $h = -3$ and $k = 0$ into $(x - h)^2 + (y - k)^2 = r^2$ gives

$$[x - (-3)]^2 + (y - 0)^2 = r^2$$
$$(x + 3)^2 + y^2 = r^2.$$

Since the circle passes through $(-5,2)$, we find r^2 by substituting the coordinates of this point in the equation.

$$(-5 + 3)^2 + 2^2 = r^2$$
$$8 = r^2$$

The standard form of the equation of this circle is then

$$(x + 3)^2 + y^2 = 8.$$

PROGRESS CHECK 2 Find the equation in standard form of the circle with center at $(2,-5)$ that passes through $(1,0)$.

2 Expressing an equation for a circle in standard form is beneficial because the center and the radius of the circle can be identified by inspection of the equation. It is then easy to graph the circle once these features are known.

EXAMPLE 3 Find the center and the radius of the circle given by

$$x^2 + (y + 2)^2 = 9.$$

Also, sketch the circle.

Solution To match standard form, view the given equation as

$$(x - 0)^2 + [y - (-2)]^2 = (3)^2.$$

Then $h = 0$, $k = -2$, and $r = 3$, so the center is $(0,-2)$ and the radius is 3. Figure 8.20 shows the graph of this circle.

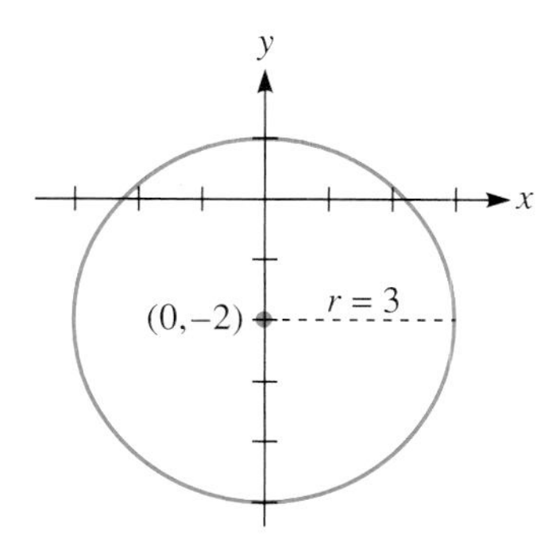

Figure 8.20

PROGRESS CHECK 3 Find the center and radius of the circle given by $(x + 3)^2 + y^2 = 4$. Also, sketch the circle.

EXAMPLE 4 Find the center and the radius of the circle given by

$$x^2 + y^2 - 10x + 6y - 15 = 0.$$

Solution The initial objective is to transform the given equation to the standard form

$$(x - h)^2 + (y - k)^2 = r^2.$$

To obtain this form, we complete the square in the x terms and the y terms by first rewriting the given equation as

$$(x^2 - 10x) + (y^2 + 6y) = 15.$$

Recall from Section 8.1 that we complete the square for $x^2 + bx$ by adding $(b/2)^2$. So we complete the square for $x^2 - 10x$ by adding $(-10/2)^2$, which is 25, and this number must be added on both sides of the equation to preserve equality. Similarly, one-half of the coefficient of y is 3 and $3^2 = 9$, so we complete the square on the y terms by adding 9, and this number is added on both sides of the equation. Thus, we have

$$(x^2 - 10x + 25) + (y^2 + 6y + 9) = 15 + 25 + 9$$
$$(x - 5)^2 + (y + 3)^2 = 49.$$

By comparing this equation with the standard form, we determine that the center of the circle is at $(5,-3)$, and the radius is $\sqrt{49}$, or 7.

Progress Check Answers

2. $(x - 2)^2 + (y + 5)^2 = 26$

3.

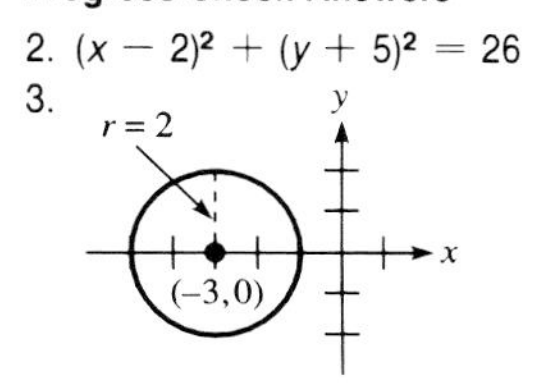

PROGRESS CHECK 4 Find the center and radius of the circle given by $x^2 + y^2 + 4x - 6y - 12 = 0$.

3 For circles centered at (0,0) the standard equation becomes

$$(x - 0)^2 + (y - 0)^2 = r^2,$$

so the standard equation of a circle with center at the origin and radius r is

$$x^2 + y^2 = r^2.$$

This result is used to graph the type of equations given in the next example.

EXAMPLE 5 Graph each equation.

a. $x^2 + y^2 = 4$ **b.** $y = \sqrt{4 - x^2}$ **c.** $y = -\sqrt{4 - x^2}$

Solution

a. The equation $x^2 + y^2 = 4$ fits the form $x^2 + y^2 = r^2$, so the graph is a circle centered at the origin. Because $r^2 = 4$, the radius is $\sqrt{4}$, or 2. This circle is graphed in Figure 8.21(a).

b. Squaring both sides of $y = \sqrt{4 - x^2}$ gives

$$y^2 = 4 - x^2$$
$$x^2 + y^2 = 4,$$

which is the equation from part **a.** However, because y represents a principal square root, y is restricted to positive numbers or zero in the equation $y = \sqrt{4 - x^2}$, so the graph is the semicircle shown in Figure 8.21(b).

c. As in part **b,** the graph of $y = -\sqrt{4 - x^2}$ is a semicircle. However, y is restricted to negative numbers or zero in this equation, so the graph is the semicircle in Figure 8.21(c).

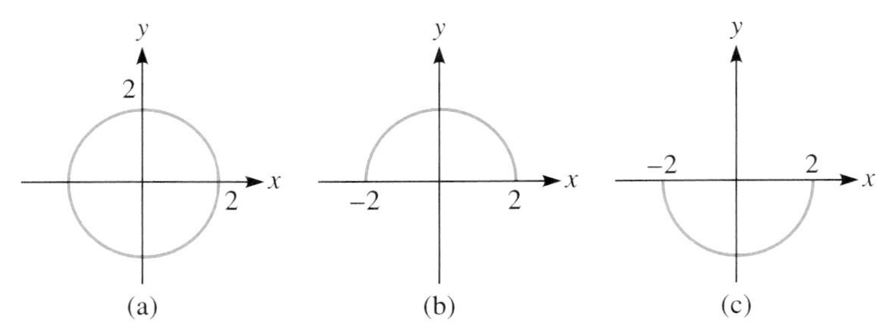

Figure 8.21

On the TI-81 the Zoom Square feature must be invoked or the circle will not appear round.

Note To graph $x^2 + y^2 = 4$ on a graphing calculator, we graph $y = \sqrt{4 - x^2}$ and $y = -\sqrt{4 - x^2}$ in the same calculator screen. This method is required because it is necessary to enter an expression that is solved for y on graphing calculators.

PROGRESS CHECK 5 Graph each equation.

a. $x^2 + y^2 = 1$ **b.** $y = -\sqrt{1 - x^2}$ **c.** $y = \sqrt{1 - x^2}$

EXAMPLE 6 Solve the problem in the section introduction on page 400.

Solution We are free to place the specific figure in any convenient position in a coordinate system, so we start by diagraming the problem as shown in Figure 8.22.

Progress Check Answers

4. (−2,3); 5 5. (a)

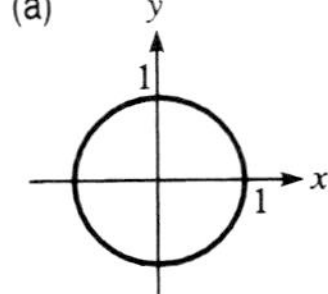

(b)

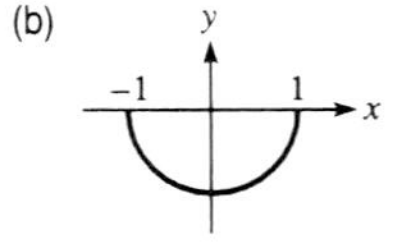

(c)

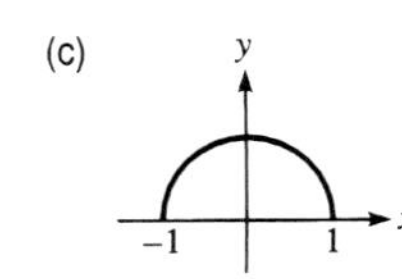

The equation for the entire circle is

$$x^2 + y^2 = (50)^2,$$

so

$$y^2 = 2{,}500 - x^2$$
$$y = \pm\sqrt{2{,}500 - x^2},$$

and the equation for the semicircle in the diagram is

$$y = \sqrt{2{,}500 - x^2}.$$

Now we find y when $x = 30$ to determine the requested height.

$$y = \sqrt{2{,}500 - (30)^2} = \sqrt{1{,}600} = 40$$

The bridge is 40 m above the water at the specified point.

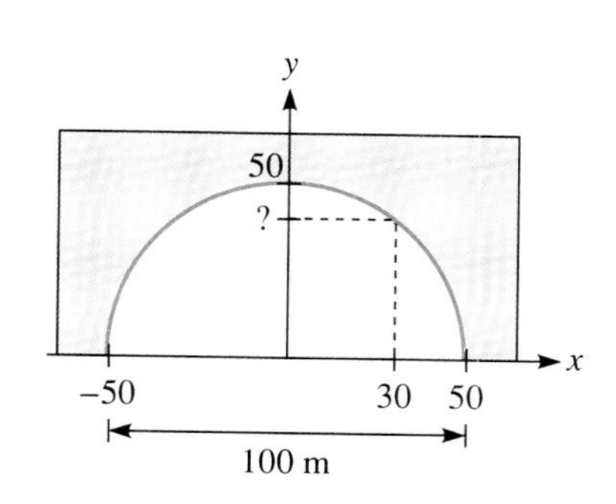

Figure 8.22

PROGRESS CHECK 6 If the diameter of a bridge as described in Example 6 is 200 ft, how high above the water is the bridge 80 ft on each side of the center?

Progress Check Answer
6. 60 ft

EXERCISES 8.6

In Exercises 1–24, find the equation in standard form of a circle given its center and either its radius or a point on the circle.

1. Center (0,0), radius 5 $x^2 + y^2 = 25$
2. Center (0,0), radius 4 $x^2 + y^2 = 16$
3. Center (0,0), point $(-3,0)$ $x^2 + y^2 = 9$
4. Center (0,0), point $(0,-2)$ $x^2 + y^2 = 4$
5. Center $(-2,3)$, radius 7 $(x + 2)^2 + (y - 3)^2 = 49$
6. Center $(4,-7)$, radius 9 $(x - 4)^2 + (y + 7)^2 = 81$
7. Center $(-5,-8)$, radius 3 $(x + 5)^2 + (y + 8)^2 = 9$
8. Center $(-3,-2)$, radius 2 $(x + 3)^2 + (y + 2)^2 = 4$
9. Center (4,4), radius 5 $(x - 4)^2 + (y - 4)^2 = 25$
10. Center (7,7), radius 12 $(x - 7)^2 + (y - 7)^2 = 144$
11. Center $(0,-5)$, radius 6 $x^2 + (y + 5)^2 = 36$
12. Center $(-4,0)$, radius 9 $(x + 4)^2 + y^2 = 81$
13. Center $(-5,0)$, point (4,0) $(x + 5)^2 + y^2 = 81$
14. Center $(0,-2)$, point (0,6) $x^2 + (y + 2)^2 = 64$
15. Center (2,3), point (5,3) $(x - 2)^2 + (y - 3)^2 = 9$
16. Center $(-5,2)$, point $(-5,7)$ $(x + 5)^2 + (y - 2)^2 = 25$
17. Center (6,3), point (10,6) $(x - 6)^2 + (y - 3)^2 = 25$
18. Center $(-5,12)$, point $(-8,16)$ $(x + 5)^2 + (y - 12)^2 = 25$
19. Center $(-3,-5)$, point $(-7,-1)$ $(x + 3)^2 + (y + 5)^2 = 32$
20. Center $(-2,-1)$, point (3,11) $(x + 2)^2 + (y + 1)^2 = 169$
21. Center (3,2), point $(3\frac{5}{8},2)$ $(x - 3)^2 + (y - 2)^2 = \frac{25}{64}$
22. Center $(-3,7)$, point $(-3,8\frac{1}{6})$ $(x + 3)^2 + (y - 7)^2 = \frac{49}{36}$
23. Center $(2,-3)$, radius $\frac{2}{3}$ $(x - 2)^2 + (y + 3)^2 = \frac{4}{9}$
24. Center $(-5,-4)$, radius $\frac{3}{4}$ $(x + 5)^2 + (y + 4)^2 = \frac{9}{16}$

In Exercises 25–40, find the center and radius of a circle given its equation. Then graph it.

25. $(x - 5)^2 + (y - 3)^2 = 16$

Center (5,3); radius 4

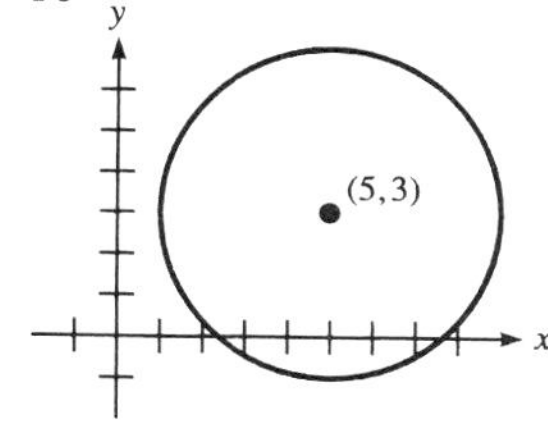

26. $(x + 3)^2 + (y - 6)^2 = 25$

Center $(-3,6)$; radius 5

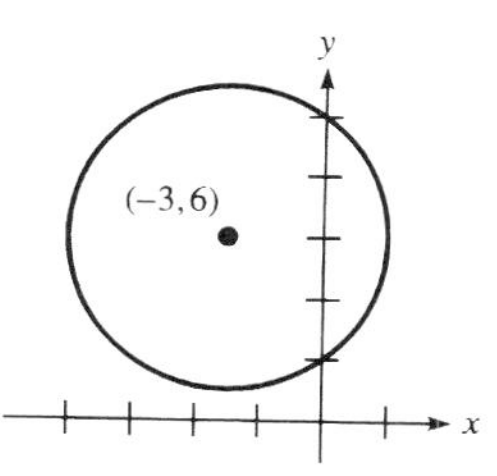

27. $(x + 4)^2 + (y + 3)^2 = 36$

Center $(-4,-3)$; radius 6

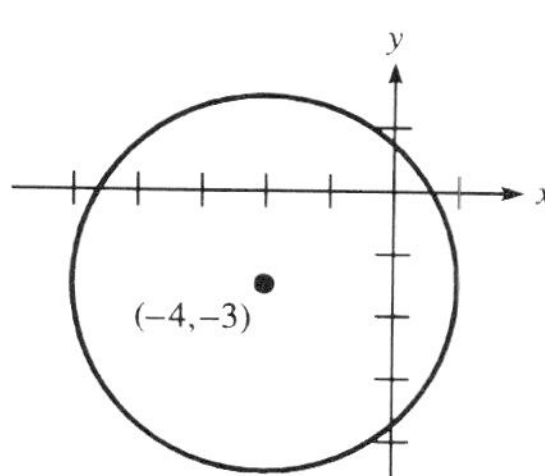

28. $(x - 6)^2 + (y - 3)^2 = 9$

Center (6,3); radius 3

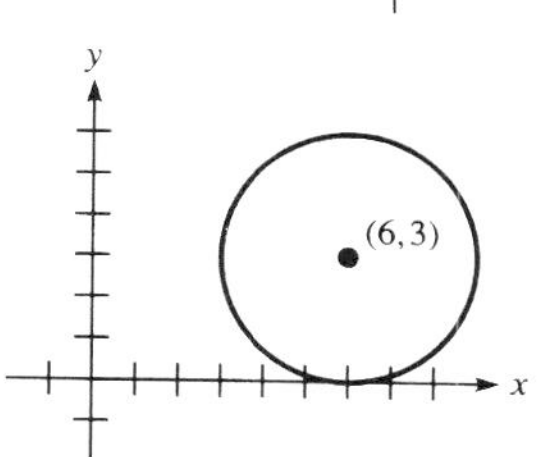

29. $x^2 + y^2 = 81$

Center (0,0); radius 9

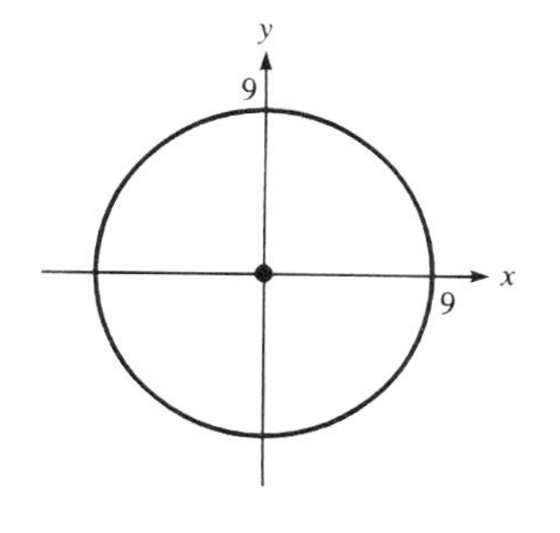

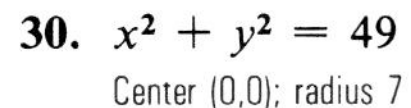

30. $x^2 + y^2 = 49$
Center (0,0); radius 7

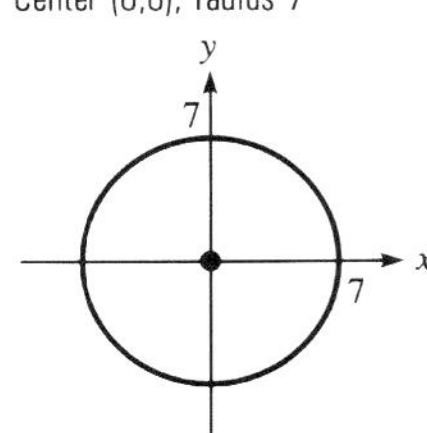

31. $x^2 + (y - 5)^2 = 4$
Center (0,5); radius 2

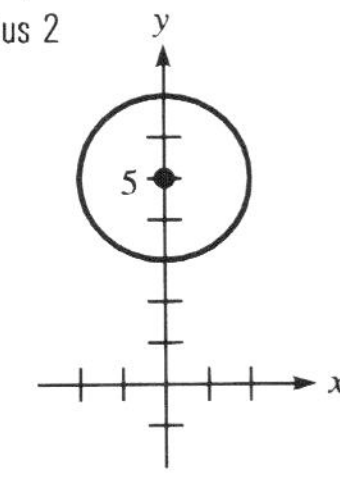

32. $(x + 4)^2 + y^2 = 12$
Center (−4,0); radius $2\sqrt{3}$

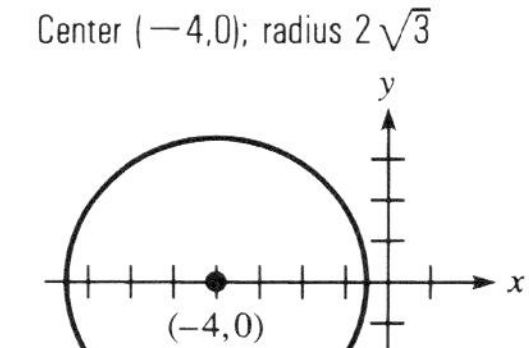

33. $(x - 2)^2 + (y + 1)^2 = 20$
Center (2, −1); radius $2\sqrt{5}$

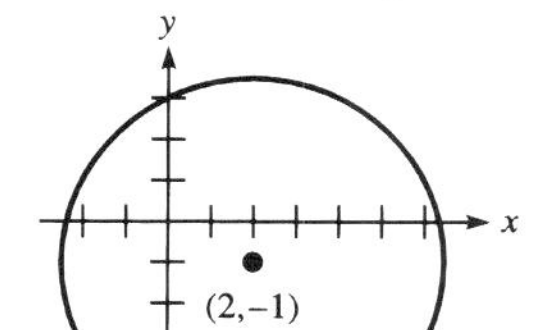

34. $(x + 3)^2 + (y - 1)^2 = 32$
Center (−3,1); radius $4\sqrt{2}$

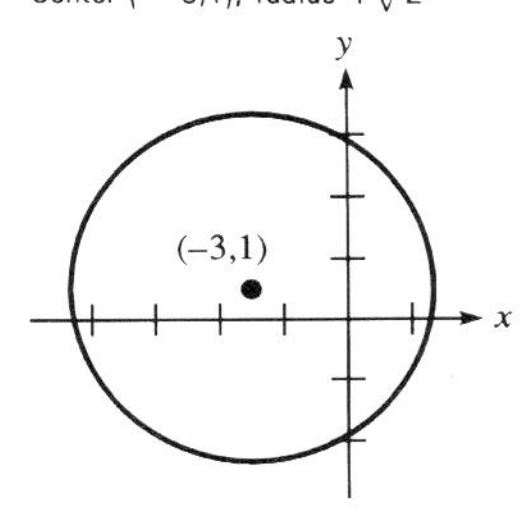

35. $x^2 + 4x + y^2 = 5$
Center (−2,0); radius 3

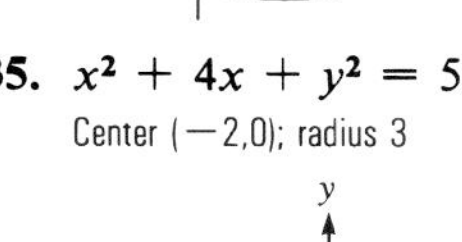

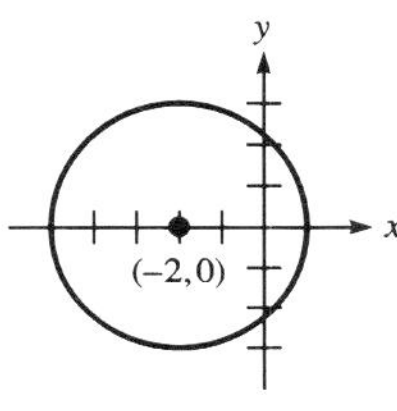

36. $x^2 + y^2 + 10y = 11$
Center (0, −5); radius 6

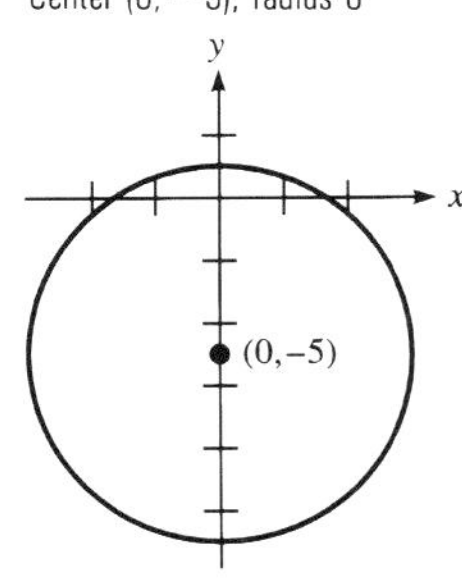

37. $x^2 + 12x + y^2 - 6y = 4$
Center (−6,3); radius 7

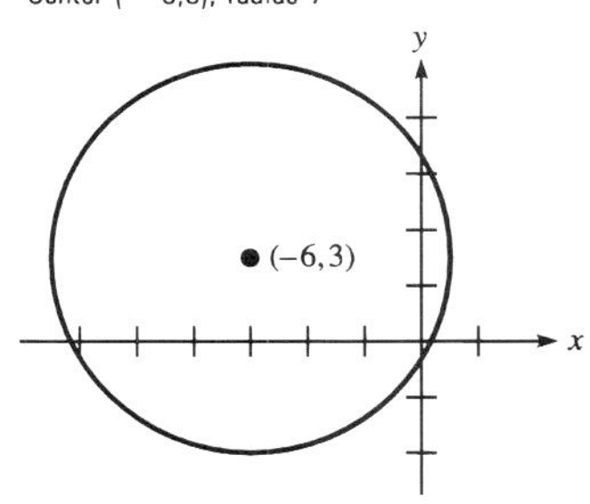

38. $x^2 - 8x + y^2 - 4y = 5$
Center (4,2); radius 5

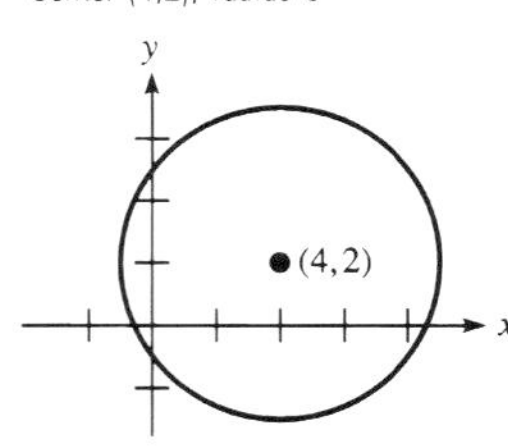

39. $x^2 - 14x + y^2 - 10y = 6$
Center (7,5); radius $4\sqrt{5}$

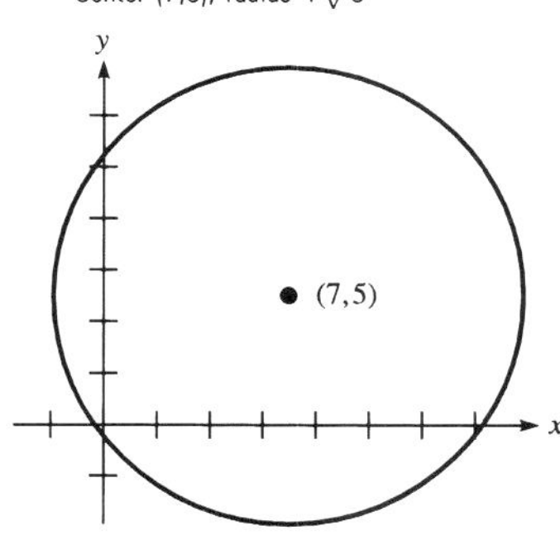

40. $x^2 - x + y^2 + 2y = \frac{3}{4}$
Center $(\frac{1}{2}, -1)$; radius $\sqrt{2}$

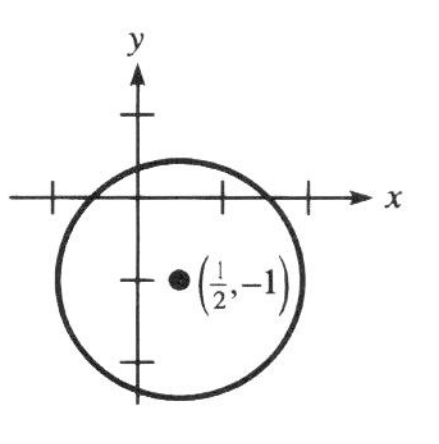

In Exercises 41–48, graph the given equations.

41. **a.** $x^2 + y^2 = 9$ **b.** $y = \sqrt{9 - x^2}$ **c.** $y = -\sqrt{9 - x^2}$

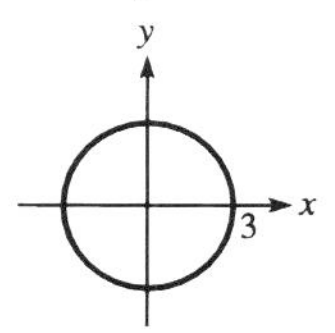

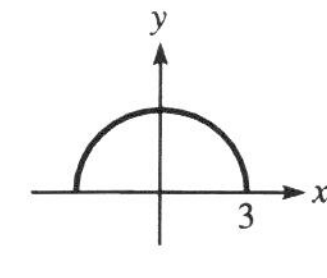

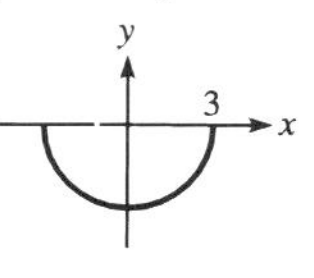

42. **a.** $x^2 + y^2 = 16$ **b.** $y = \sqrt{16 - x^2}$ **c.** $y = -\sqrt{16 - x^2}$

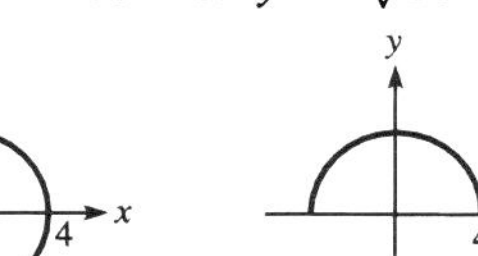

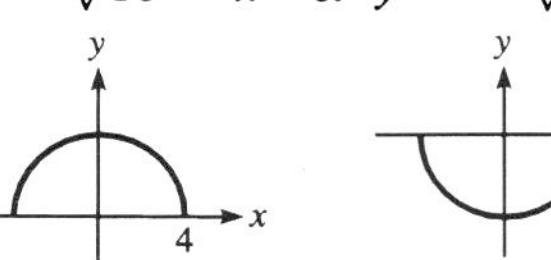

43. **a.** $x^2 + y^2 = 25$ **b.** $y = \sqrt{25 - x^2}$ **c.** $y = -\sqrt{25 - x^2}$

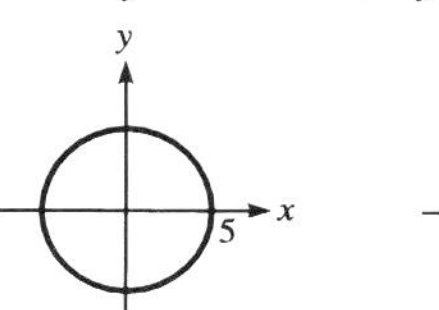

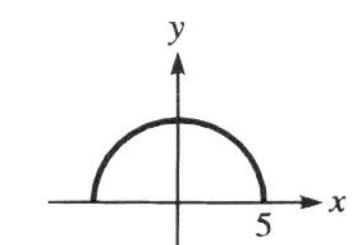

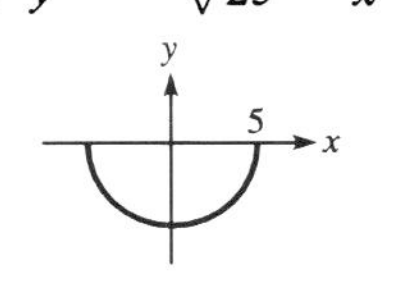

44. **a.** $x^2 + y^2 = 49$ **b.** $y = \sqrt{49 - x^2}$ **c.** $y = -\sqrt{49 - x^2}$

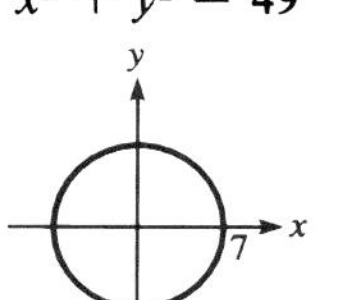

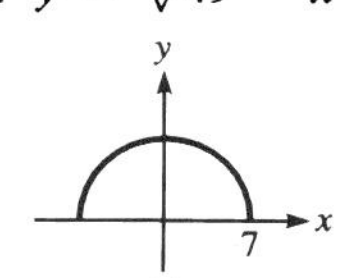

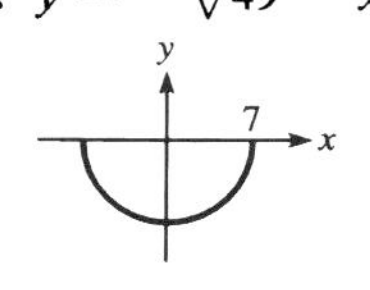

45. **a.** $x^2 + y^2 = 20$ **b.** $y = \sqrt{20 - x^2}$ **c.** $y = -\sqrt{20 - x^2}$

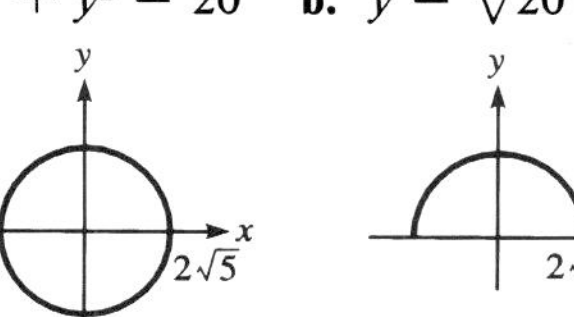

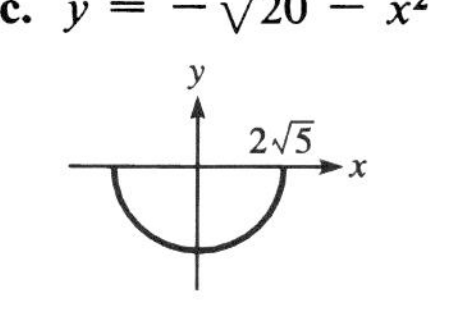

46. **a.** $x^2 + y^2 = 50$ **b.** $y = \sqrt{50 - x^2}$ **c.** $y = -\sqrt{50 - x^2}$

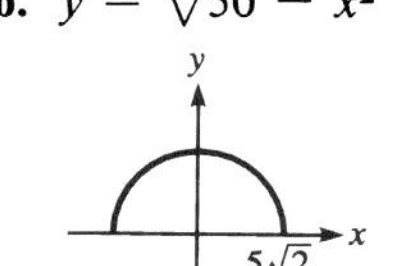

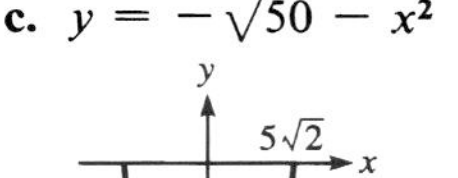

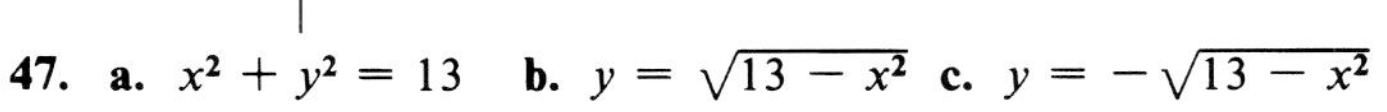

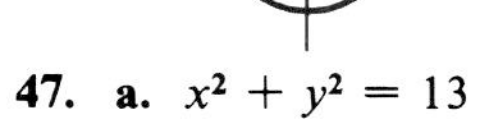

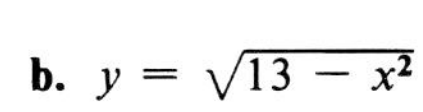

47. **a.** $x^2 + y^2 = 13$ **b.** $y = \sqrt{13 - x^2}$ **c.** $y = -\sqrt{13 - x^2}$

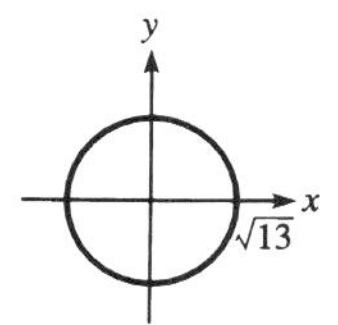

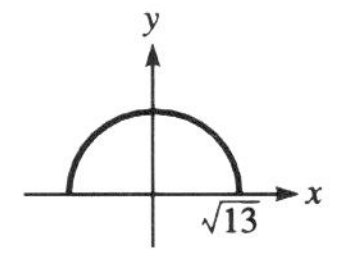

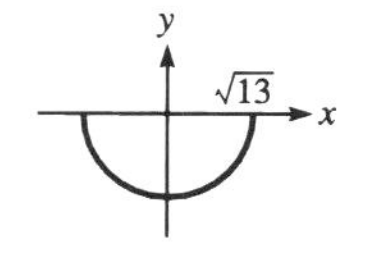

48. **a.** $x^2 + y^2 = 17$ **b.** $y = \sqrt{17 - x^2}$ **c.** $y = -\sqrt{17 - x^2}$

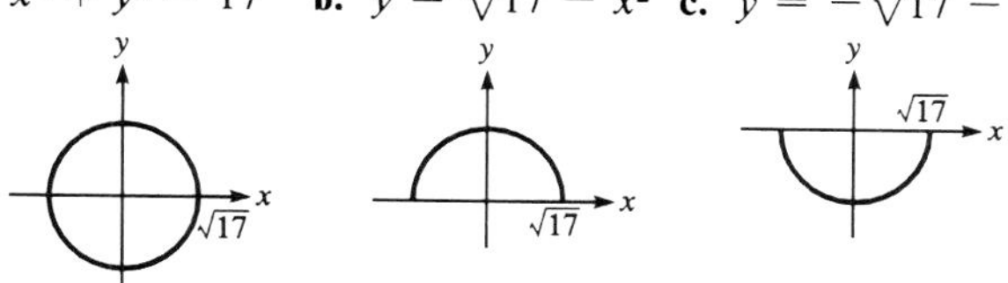

49. The cross section of a tunnel is the semicircle shown in the figure. How high above the road is the tunnel ceiling 30 ft on each side of the center? 40 ft

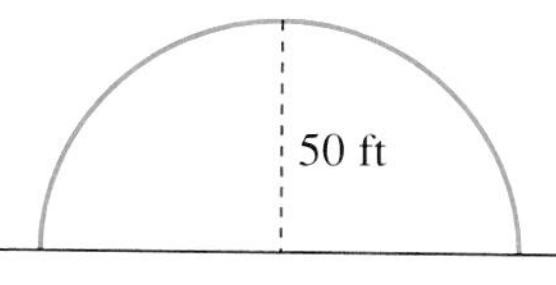

50. Refer to the figure in Exercise 49. How high above the road is the tunnel 25 ft on each side of the center? Give the answer in exact radical form and also approximated to the nearest tenth of a foot. $25\sqrt{3} \approx 43.3$ ft

51. The figure for this exercise shows two circles. Point A is on the larger circle and it is directly above C, the center of the smaller circle.

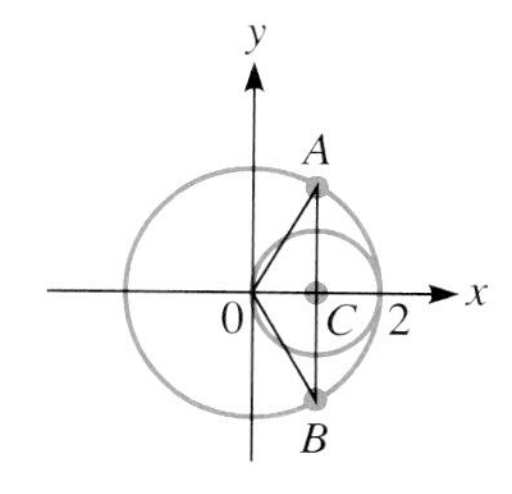

a. Find the coordinates of point C. (1,0)

b. Find the equation of the larger circle. $x^2 + y^2 = 4$

c. Find the coordinates of point A. $(1, \sqrt{3})$

d. Find the distance between A and B. $2\sqrt{3}$

e. Find the area of triangle ABO. $\sqrt{3}$

52. For the given figure, answer the questions in Exercise 51.

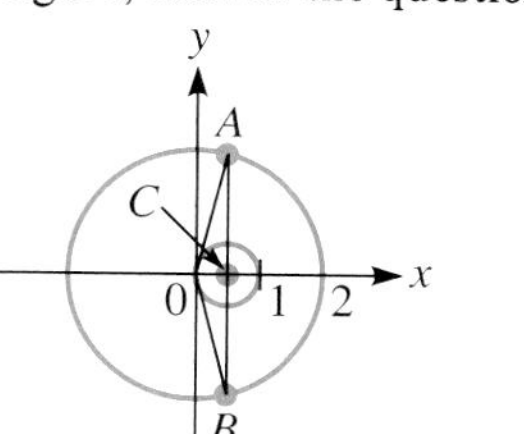

a. $(1/2, 0)$
b. $x^2 + y^2 = 4$
c. $(1/2, \sqrt{15}/2)$
d. $\sqrt{15}$
e. $\sqrt{15}/4$

53. The circle shown has equation $x^2 + y^2 = 2$. Find the area of the shaded square. 1

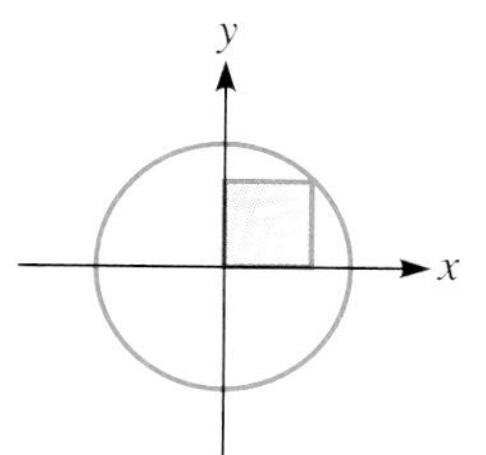

54. The circle shown has equation $x^2 + y^2 = 1$. Find the area of the shaded square. $\frac{1}{2}$

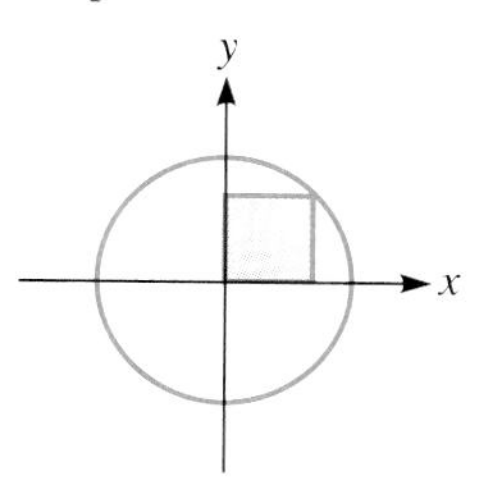

THINK ABOUT IT

1. The semicircular cross section of a one-lane tunnel through a mountain is shown in the figure.

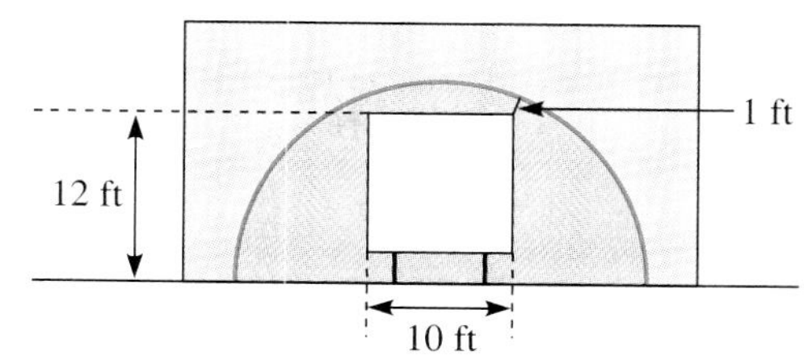

a. What is the minimum radius for the tunnel that will allow a truck that is 10 ft wide and 12 ft high to pass through with 1 ft to spare (as shown) by driving exactly down the center?

b. Will a truck that is 10 ft wide and 13 ft tall be able to make it through the tunnel?

2. A line is tangent to a circle if the two intersect at exactly one point. Such tangent lines are perpendicular to the radius at the point of intersection.

a. Find the equation in standard form for the circle with center (1,4) that is tangent to the line $y = -3$. Find the area and the circumference of the circle.

b. Find the slope of the line which is tangent to the circle at the point in quadrant 1 where $x = 4$.

3. Use the given figure and the concept of slope to show that if a triangle is inscribed in a circle with the diameter as one of its sides, then the triangle is a right triangle.

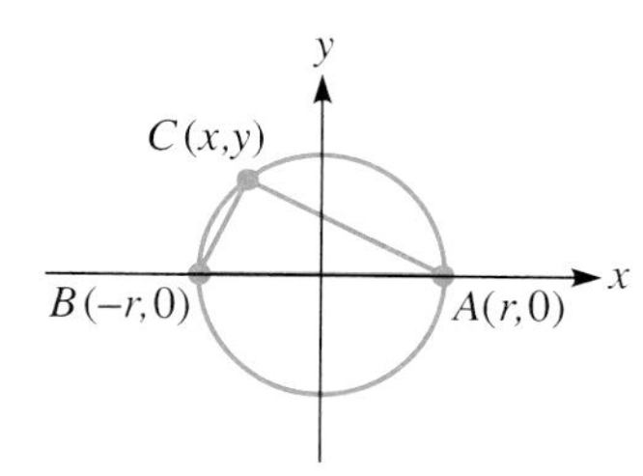

4. Do the circles given by $(x - 5)^2 + (y - 12)^2 = 121$ and $x^2 + y^2 = 4$ intersect? Explain in terms of the centers and radii of the circles.

5. Write each equation in standard form for the equation of a circle. Then describe its graph.

a. $x^2 + y^2 + 4x - 6y + 13 = 0$

b. $x^2 + y^2 + 4x - 6y + 14 = 0$

REMEMBER THIS

1. By letting $y = 0$, find the x-intercepts for the graph of the equation $\frac{x^2}{16} + \frac{y^2}{9} = 1$. $(-4,0), (4,0)$
2. Solve $\frac{x^2}{16} - \frac{y^2}{9} = 1$ for y when $x = 0$. $\{\pm 3i\}$
3. A rectangle has vertices at $(-3,2)$, $(-3,-2)$, $(3,2)$, and $(3,-2)$. Find the slope of both of its diagonals. Are they perpendicular? $m_1 = \frac{2}{3}$, $m_2 = -\frac{2}{3}$; no
4. Does the graph of $y = \frac{2}{x}$ have any x- or y-intercepts? No
5. If A varies inversely with p, represent this relationship in an equation. $A = k/p$
6. Find the vertex of the parabola whose equation is $y = (x - 3)^2 + 2$. $(3,2)$
7. Solve $x^2 + 1 = 0$. $\{\pm i\}$
8. Find any points of intersection for the lines $y = 3x - 2$ and $y = 2x - 3$. $(-1,-5)$
9. Solve $\frac{x-2}{5} = \frac{x+2}{10}$. $\{6\}$
10. Solve $y = \frac{ax - b}{b}$ for b. $b = \frac{ax}{y+1}$

8.7 The Ellipse and the Hyperbola

Because of uncertainties in world events, an investor decides to take a position in precious metals and buy \$10,000 worth of gold. The amount (A) of gold in troy ounces that can be bought with \$10,000 varies inversely as the price (P) per troy ounce of gold. Write a variation equation that expresses the relationship between A and P, and then graph the equation. (See Example 6.)

OBJECTIVES

1. Graph an ellipse.
2. Graph a hyperbola.
3. Graph $xy = c$.
4. Classify certain equations as defining either a circle, an ellipse, a hyperbola, or a parabola.

1 The ellipse, which is an oval-shaped curve, is defined geometrically as follows.

Students should see how an ellipse may be drawn by using a pencil, string, and tacks. Either show them or assign "Think About It" Exercise 1.

> **Definition of an Ellipse**
>
> An **ellipse** is the set of all points in a plane the sum of whose distances from two fixed points is a constant.

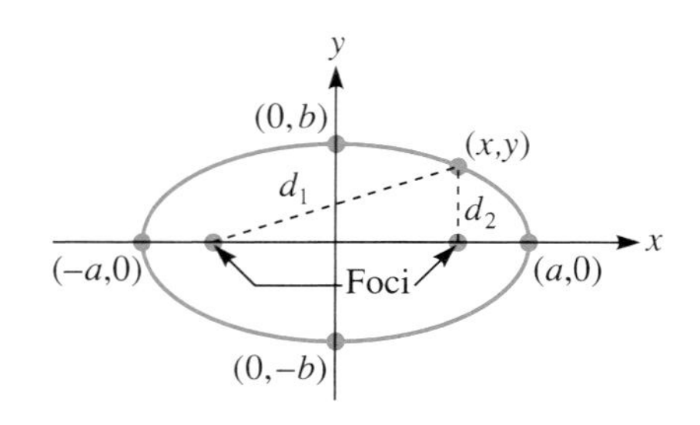

Figure 8.23

Each fixed point is called a **focus** of the ellipse, and the two fixed points are called the **foci.**

To understand this definition, consider Figure 8.23, which shows an ellipse centered at the origin with foci on the x-axis. If we let (x,y) be any point on the ellipse as shown, then by definition,

$$d_1 + d_2 = \text{positive constant.}$$

Although we do not determine the coordinates of the foci in this text, their location is important in many applications of the ellipse. For instance, one of the well-known applications of the ellipse is that the earth moves in an elliptical orbit with the sun at one focus, as shown in Figure 8.24.

From the given definition it can be shown that an ellipse with intercepts as shown in Figure 8.23 has the following equation.

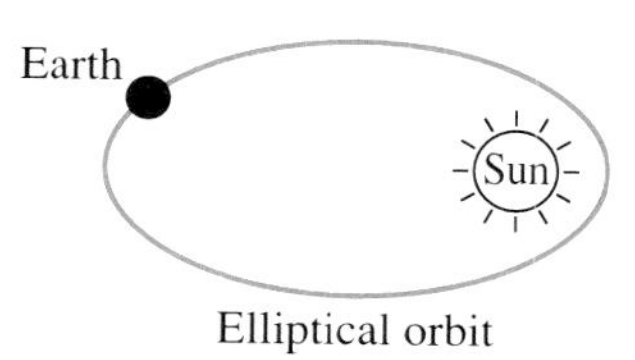

Figure 8.24

Consider mentioning that if $a = b$, then the graph is a circle, which is a special case of an ellipse. Also, note that the area of this ellipse is $A = \pi ab$, which becomes $A = \pi r^2$ if $a = b = r$.

Equation of an Ellipse

The **standard form** of the equation of an ellipse centered at the origin with x-intercepts $(a,0)$ and $(-a,0)$ and y-intercepts $(0,b)$ and $(0,-b)$ is

$$\frac{x^2}{a^2} + \frac{y^2}{b^2} = 1,$$

where a and b are positive numbers.

When an equation for an ellipse is given in this standard form, then the equation may be graphed by drawing an ellipse through the four intercepts after determining the values of a and b.

EXAMPLE 1 Graph $\frac{x^2}{16} + \frac{y^2}{9} = 1$.

Solution The equation fits the form for an ellipse, where

$$a^2 = 16, \quad \text{so} \quad a = 4; \qquad \text{and} \qquad b^2 = 9, \quad \text{so} \quad b = 3.$$

Thus the x-intercepts are $(4,0)$ and $(-4,0)$, while the y-intercepts are $(0,3)$ and $(0,-3)$. Drawing an ellipse through these four intercepts gives the graph in Figure 8.25.

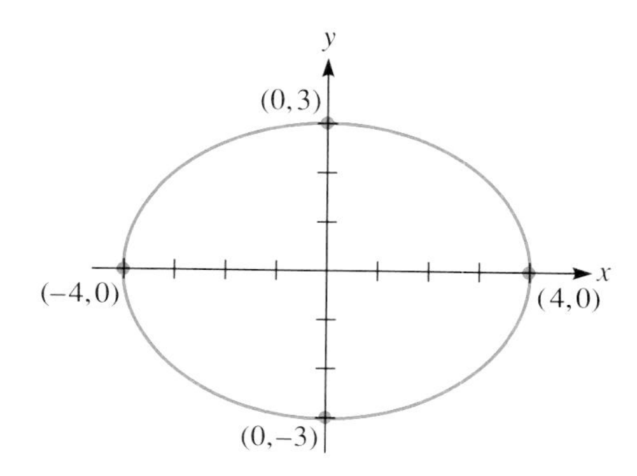

Figure 8.25

PROGRESS CHECK 1 Graph $\frac{x^2}{4} + \frac{y^2}{25} = 1$.

EXAMPLE 2 Graph $4x^2 + y^2 = 4$.

Solution First, put the equation in standard form by dividing both sides of the equation by 4.

$$4x^2 + y^2 = 4$$
$$\frac{4x^2}{4} + \frac{y^2}{4} = \frac{4}{4}$$
$$\frac{x^2}{1} + \frac{y^2}{4} = 1$$

The rewritten equation fits the form for an ellipse, where

$$a^2 = 1, \quad \text{so} \quad a = 1; \qquad \text{and} \qquad b^2 = 4, \quad \text{so} \quad b = 2.$$

Drawing an ellipse through the intercepts, which are $(1,0)$, $(-1,0)$, $(0,2)$, and $(0,-2)$, gives the graph of the equation in Figure 8.26.

Students may want to find x- and y-intercepts by letting $y = 0$ and $x = 0$, respectively. This method works well here, but conversion to standard form should also be emphasized, because other features of the ellipse (like the location of the foci) are determined from the standard forms.

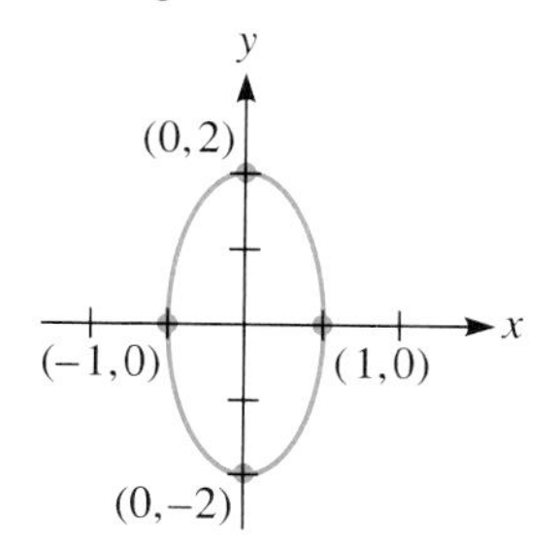

Figure 8.26

Progress Check Answer

1.

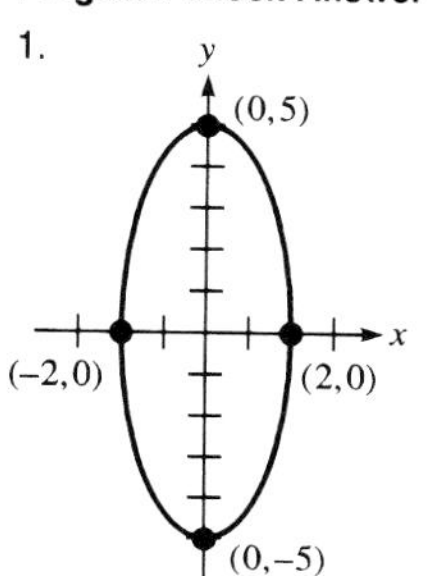

PROGRESS CHECK 2 Graph $x^2 + 9y^2 = 36$.

2 The next conic section we consider is the hyperbola, and its geometric definition resembles the definition of the ellipse.

> **Definition of a Hyperbola**
>
> A **hyperbola** is the set of all points in a plane the difference of whose distances from two fixed points (foci) is a constant.

Note that the distances between the foci and a point in the figure maintain a *constant difference* for a hyperbola and a *constant sum* for an ellipse. Equations in standard forms for hyperbolas may be derived from the geometric definition, and the standard forms for the hyperbolas shown in Figure 8.27 are stated next.

Standard form: $\frac{x^2}{a^2} - \frac{y^2}{b^2} = 1$

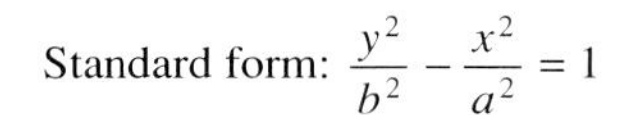

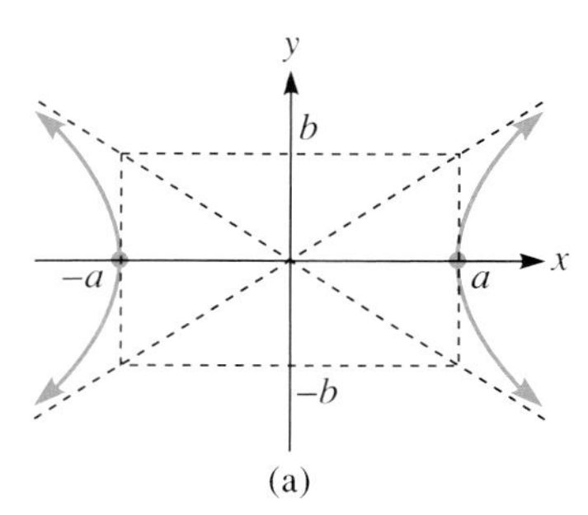

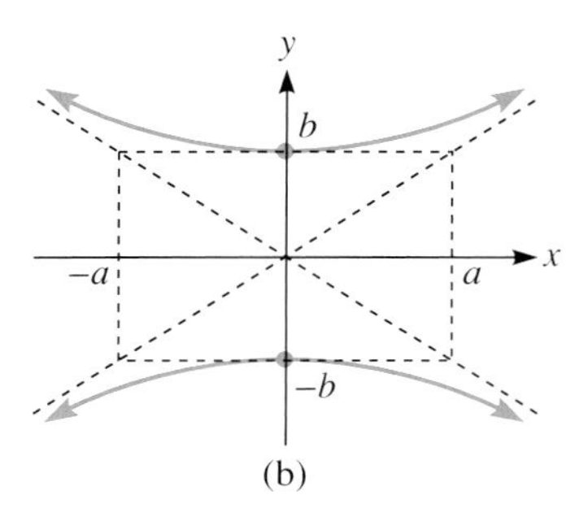

Figure 8.27

> **Equations of a Hyperbola**
>
> The **standard form** of the hyperbola centered at the origin with x-intercepts at $(a,0)$ and $(-a,0)$ and no y-intercepts is
>
> $$\frac{x^2}{a^2} - \frac{y^2}{b^2} = 1.$$
>
> The **standard form** of the hyperbola centered at the origin with y-intercepts at $(0,b)$ and $(0,-b)$ and no x-intercepts is
>
> $$\frac{y^2}{b^2} - \frac{x^2}{a^2} = 1.$$

Progress Check Answer

2.

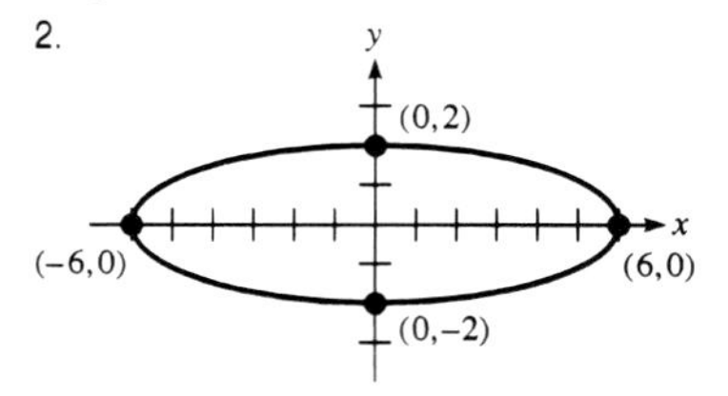

In Figure 8.27, observe that a hyperbola consists of two disconnected curves, known as **branches.** In this figure note also the dashed lines that are the extended diagonals of a rectangle, called the **fundamental rectangle,** with corners at (a,b), $(a,-b)$, $(-a,b)$, and $(-a,-b)$. These diagonal lines are called the **asymptotes of the hyperbola.** They are a great aid in sketching the curve because the hyperbola approaches these lines as x increases in absolute value. In the next two examples we graph hyperbolas by drawing curves that pass through the intercepts on one of the axes and then approach asymptotes as the graph moves further from the origin in both the positive and the negative direction.

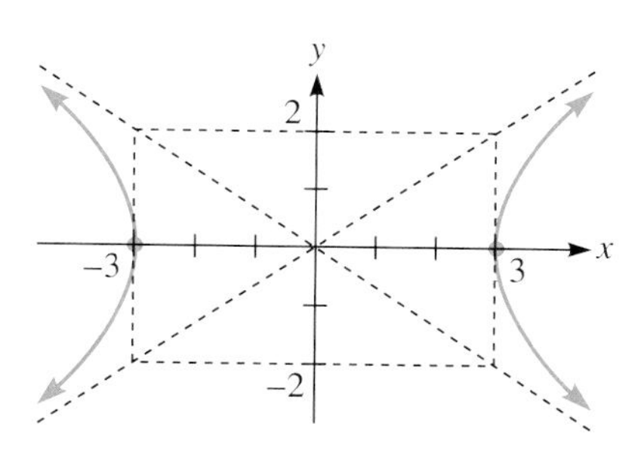

Figure 8.28

EXAMPLE 3 Graph $\frac{x^2}{9} - \frac{y^2}{4} = 1$.

Solution The equation fits the form for a hyperbola, where

$$a^2 = 9, \quad \text{so} \quad a = 3; \qquad \text{and} \qquad b^2 = 4, \quad \text{so} \quad b = 2.$$

The graph has x-intercepts because the x^2 term is positive, and these intercepts are (3,0) and (−3,0), since $a = 3$. Now we graph the hyperbola, as shown in Figure 8.28, by drawing branches that pass through the intercepts and approach the asymptotes that are the extended diagonals of the rectangle with corners (3,2), (3,−2), (−3,2), and (−3,−2).

It is instructive to ask students for the equations of the asymptotes (even though they are not needed to draw the hyperbola). See "Think About It" Exercise 2.

Note The fundamental rectangle and the dashed lines associated with it are not part of the hyperbola but, rather, are graphing aids for sketching the hyperbola.

PROGRESS CHECK 3 Graph $\frac{x^2}{4} - \frac{y^2}{25} = 1$.

EXAMPLE 4 Graph $4y^2 - 9x^2 = 144$.

Solution Divide both sides of the given equation by 144 to obtain the standard form

$$\frac{y^2}{36} - \frac{x^2}{16} = 1.$$

This equation fits the form $(y^2/b^2) - (x^2/a^2) = 1$, with

$$b^2 = 36, \quad \text{so} \quad b = 6; \qquad \text{and} \qquad a^2 = 16, \quad \text{so} \quad a = 4.$$

Thus, the graph is a hyperbola with y-intercepts at (0,6) and (0,−6) and no x-intercepts. We graph the hyperbola as shown in Figure 8.29, using the fundamental rectangle with corners at (4,6), (4,−6), (−4,6), and (−4,−6).

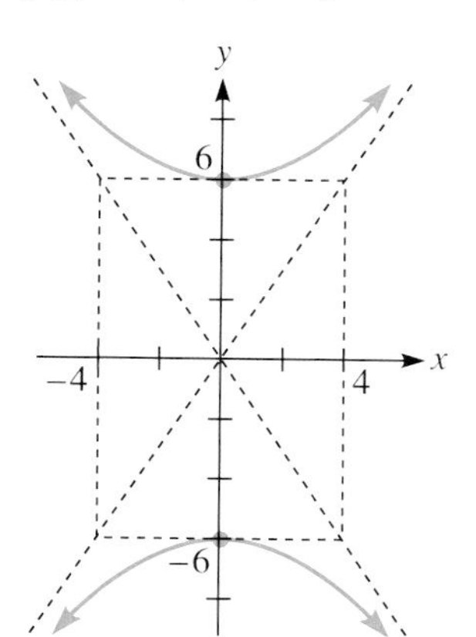

Figure 8.29

PROGRESS CHECK 4 Graph $y^2 - x^2 = 9$.

3 An equation of the form

$$xy = c,$$

where c is a nonzero constant, is an important case of another type of second-degree equation that graphs as a hyperbola. The asymptotes in these graphs are the x-axis and the y-axis, and knowing these asymptotes along with some points on each branch enables us to draw these hyperbolas.

Progress Check Answers

3.

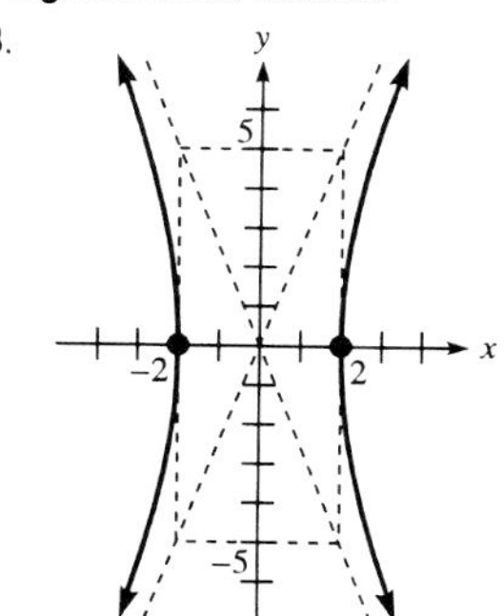

4.

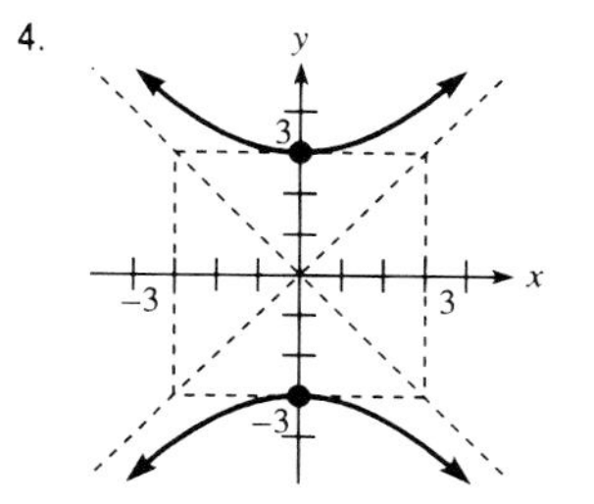

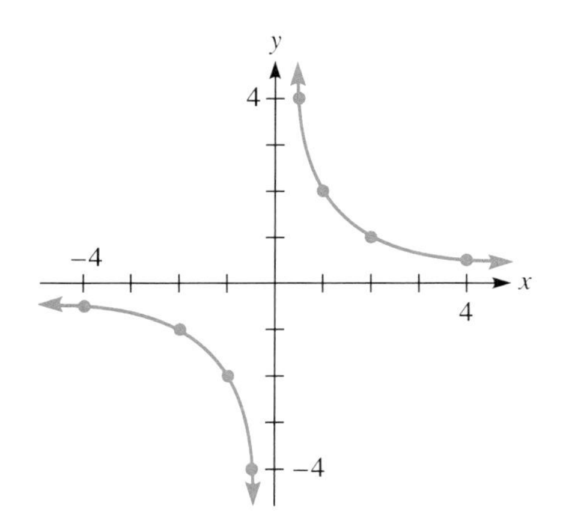

Figure 8.30

EXAMPLE 5 Graph $xy = 2$.

Solution The equation $xy = 2$, or $y = 2/x$, graphs as a hyperbola with both axes as asymptotes. Using these facts and plotting a few points, say (1,2), (−1,−2), $(4,\frac{1}{2})$, and $(-4,-\frac{1}{2})$, gives the graph in Figure 8.30.

PROGRESS CHECK 5 Graph $xy = -2$.

To solve the section-opening problem, you may need to review the concept of inverse variation considered in Section 5.5.

EXAMPLE 6 Solve the problem in the section introduction on page 406.

Solution Since the amount (A) of gold in troy ounces that may be bought varies inversely with the price (P) per troy ounce of gold,

$$AP = k, \qquad \text{or} \qquad A = \frac{k}{P}.$$

Because \$10,000 is to be invested, the variation constant k is 10,000, and A and P are related by

$$A = \frac{10{,}000}{P}.$$

This equation graphs as a hyperbola with both axes as asymptotes. However, the graph is limited to the branch in quadrant 1, since both A and P must be positive. With the aid of the points specified by the following table, we can draw the graph as shown in Figure 8.31.

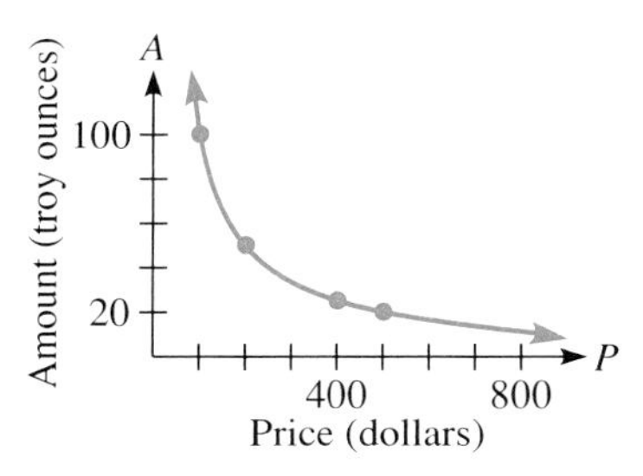

Figure 8.31

Price (in dollars), P	100	200	400	500
Amount (in troy ounces), A	100	50	25	20

PROGRESS CHECK 6 If a car travels at a constant rate r, then the time t required to drive 60 mi varies inversely with r. Write an equation that expresses the relationship between r and t, and then graph this equation.

4 With the exception of the equation $xy = c$, the equations considered in Sections 8.5–8.7 have all been of the general form

$$Ax^2 + Cy^2 + Dx + Ey + F = 0,$$

where A and C are not both zero. If a graph is a circle, an ellipse, a hyperbola, or a parabola, then A and C must satisfy certain conditions that follow from the forms given for each conic section.

The general form shown here omits the Bxy term. When Bxy is present the graph (for the nondegenerate case) represents a conic section which has been rotated. The graph is a parabola, ellipse, or hyperbola according as $B^2 - 4AC$ is zero, negative, or positive.

Conditions on A and C	Conic Section	Degenerate Possibilities
$A = C \neq 0$	Circle	A point or no graph at all
$A \neq C$; A and C have the same sign	Ellipse	A point or no graph at all
A and C have opposite signs	Hyperbola	Two intersecting straight lines
$A = 0$ or $C = 0$ (but not both)	Parabola	A line, a pair of parallel lines, or no graph at all

In this text we do not consider equations that lead to the degenerate possibilities (but see "Think About It" Exercise 3). However, it is interesting to note that many of

Progress Check Answers

5.

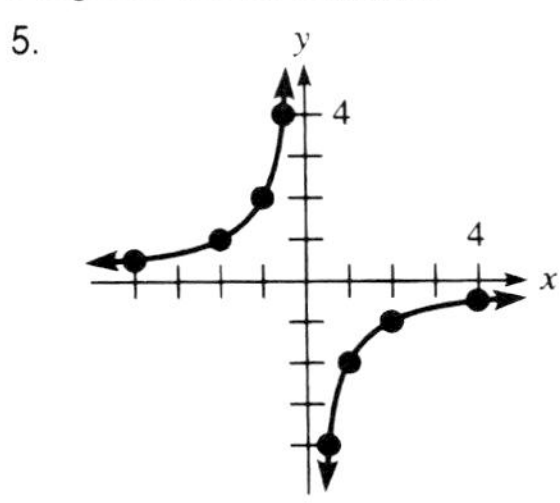

6. $rt = 60$, or $t = \dfrac{60}{r}$

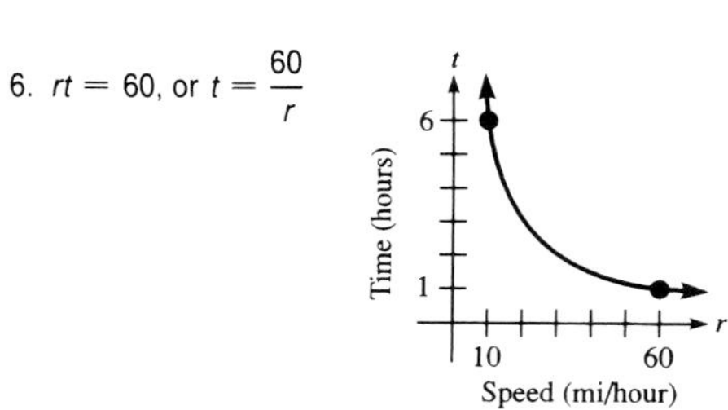

these cases also result from the intersection of a cone with an appropriate plane, and the degenerate possibilities are called **degenerate conic sections.** It can be shown that the graph of every second-degree equation in two variables is a conic section or a degenerate conic section.

See "Think About It" Exercise 3 for illustrations of some degenerate cases.

EXAMPLE 7 Identify the graph of each equation as a circle, an ellipse, a hyperbola, or a parabola.

a. $2x^2 + 2y^2 - 10x - 2y - 31 = 0$
b. $2x^2 - y + 5 = 0$
c. $4x^2 = 9y^2 + 36$

Solution Since the degenerate possibilities have been eliminated, we proceed as follows.

a. The given equation is in general form with $A = 2$ and $C = 2$. Since A and C are nonzero and equal, the graph is a circle.
b. In this equation $A = 2$ and there is no y^2 term, so $C = 0$. Because one (but not both) of these values is zero, the graph is a parabola.
c. Rewrite $4x^2 = 9y^2 + 36$ in general form as

$$4x^2 - 9y^2 - 36 = 0.$$

Then $A = 4$ and $C = -9$. Since A and C have opposite signs, the graph is a hyperbola.

PROGRESS CHECK 7 Identify the graph of each equation as a circle, an ellipse, a hyperbola, or a parabola.

a. $y^2 + 4y + 3x - 8 = 0$
b. $4y^2 - x^2 = 36$
c. $9y^2 = 36 - 4x^2$

Progress Check Answers
7. (a) Parabola (b) Hyperbola (c) Ellipse

EXERCISES 8.7

In Exercises 1–10, graph the given ellipse.

1. $\frac{x^2}{9} + \frac{y^2}{16} = 1$

2. $\frac{x^2}{4} + \frac{y^2}{1} = 1$

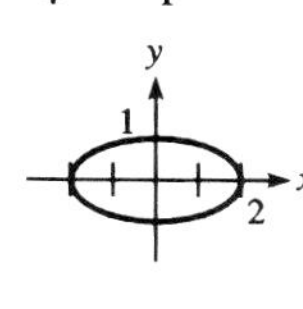

5. $4x^2 + y^2 = 4$

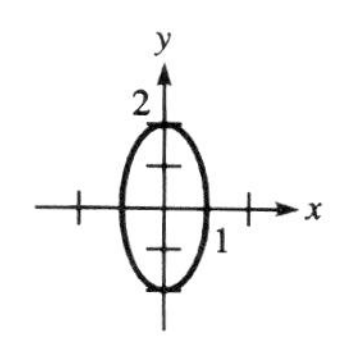

6. $x^2 + 5y^2 = 20$

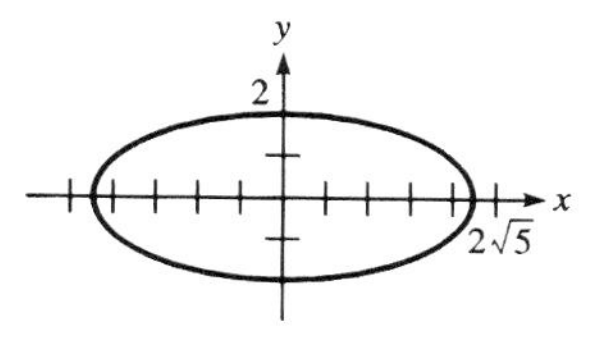

3. $\frac{x^2}{4} + \frac{y^2}{25} = 1$

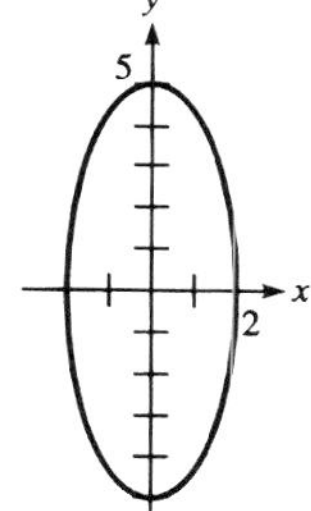

4. $\frac{x^2}{36} + \frac{y^2}{9} = 1$

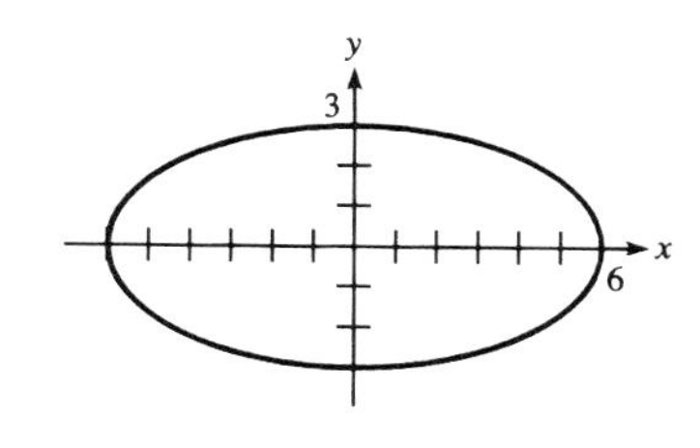

7. $9x^2 + 16y^2 = 144$

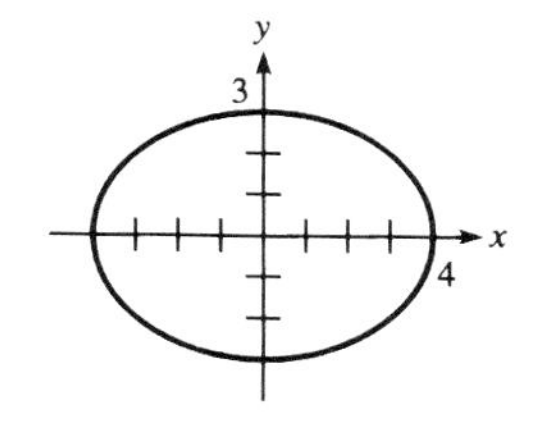

8. $4x^2 + 9y^2 = 36$

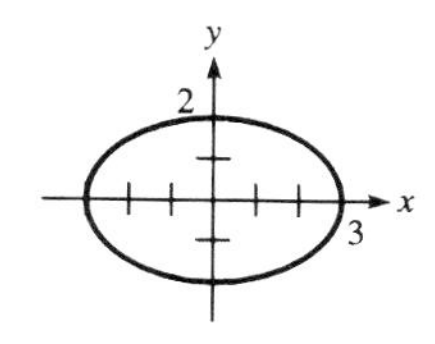

9. $100x^2 + 9y^2 = 900$

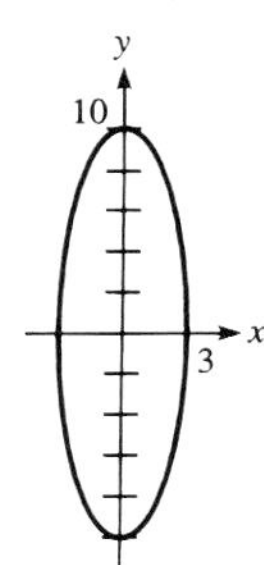

10. $64x^2 + 8y^2 = 16$

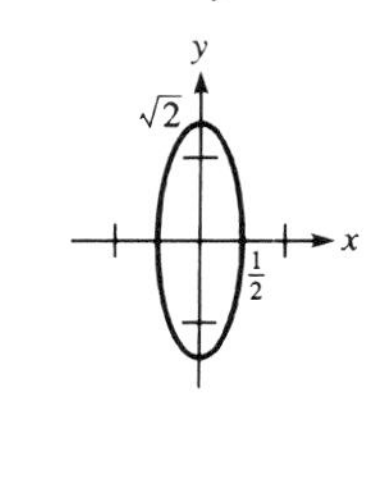

In Exercises 11–30, graph the given hyperbolas.

11. $\dfrac{x^2}{36} - \dfrac{y^2}{9} = 1$

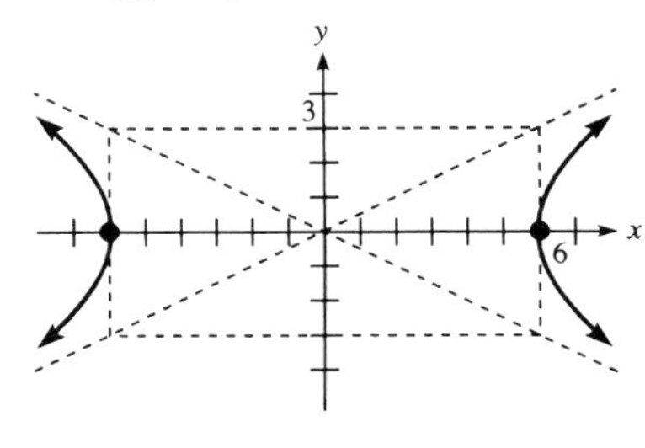

12. $\dfrac{x^2}{4} - \dfrac{y^2}{25} = 1$

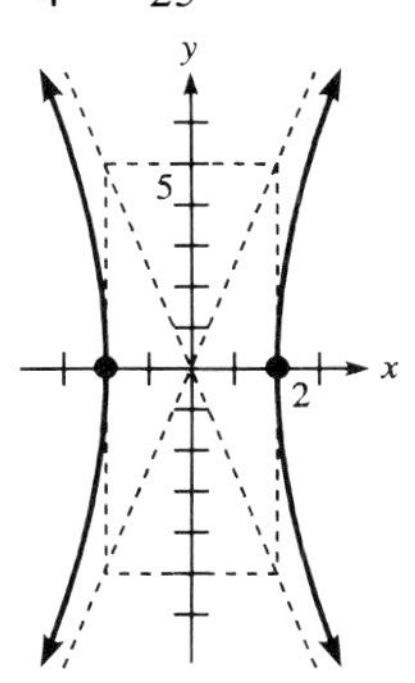

13. $\dfrac{x^2}{4} - y^2 = 1$

14. $\dfrac{x^2}{9} - \dfrac{y^2}{16} = 1$

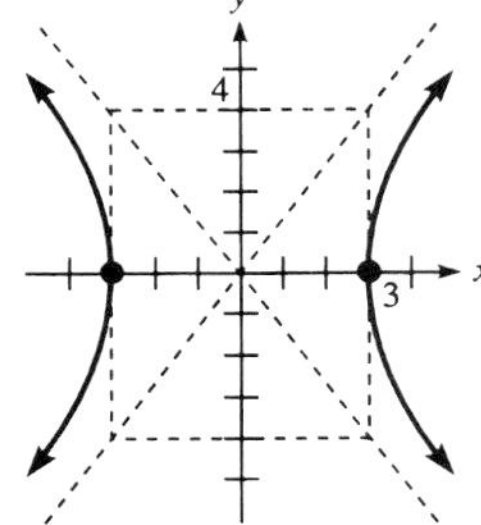

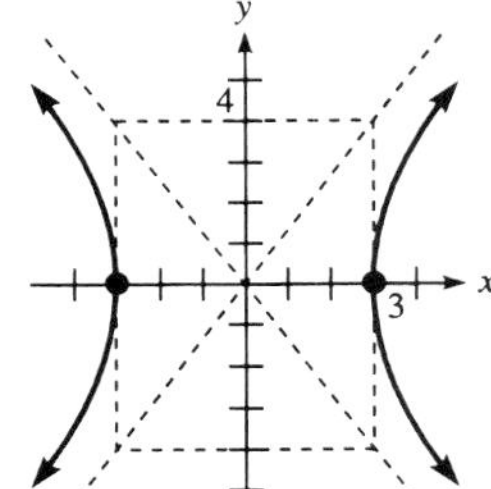

15. $\dfrac{y^2}{9} - \dfrac{x^2}{25} = 1$

16. $\dfrac{y^2}{36} - \dfrac{x^2}{4} = 1$

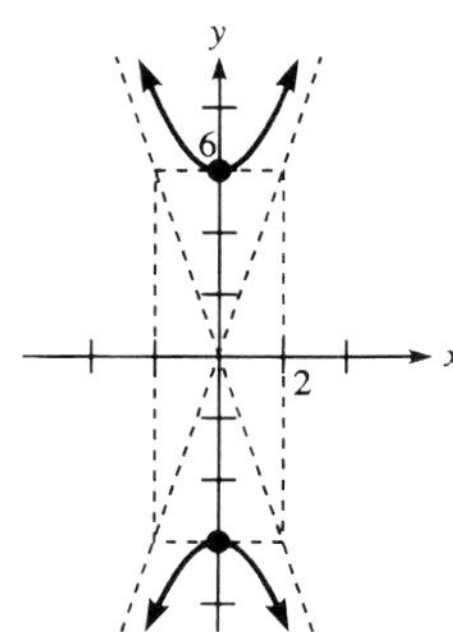

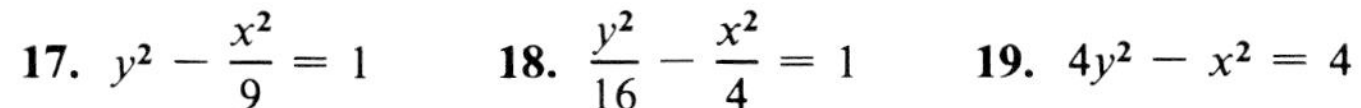

17. $y^2 - \dfrac{x^2}{9} = 1$

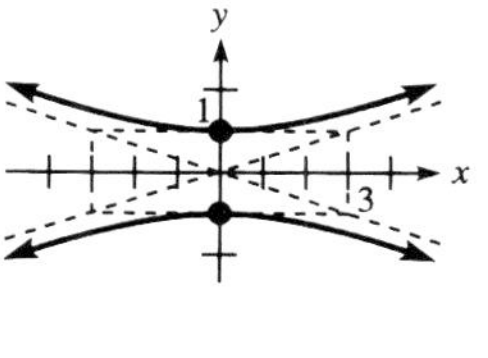

18. $\dfrac{y^2}{16} - \dfrac{x^2}{4} = 1$

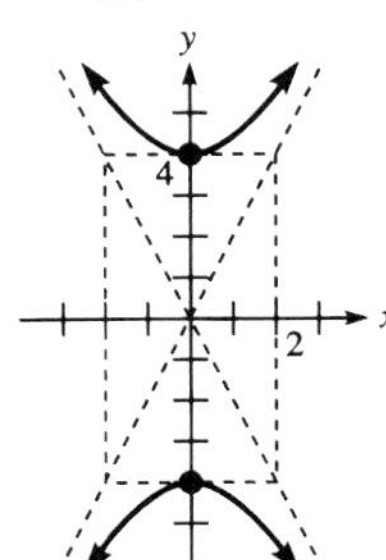

19. $4y^2 - x^2 = 4$

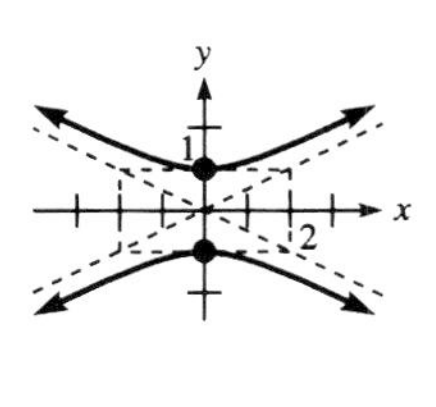

20. $y^2 - 5x^2 = 20$

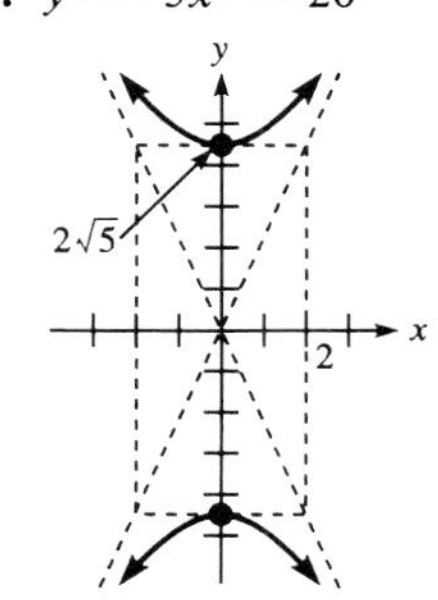

21. $9y^2 - 16x^2 = 144$

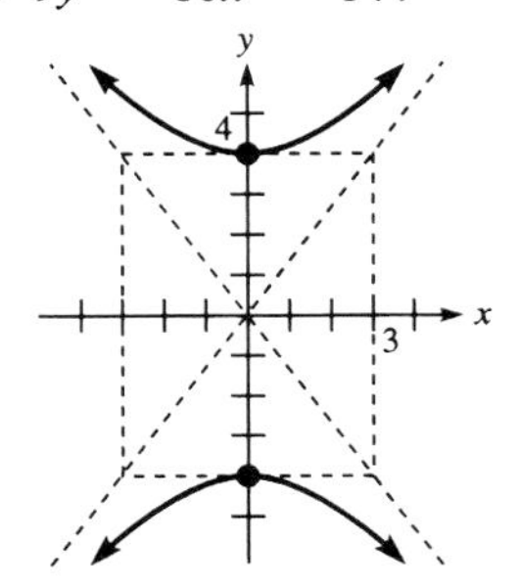

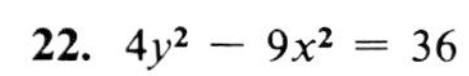

22. $4y^2 - 9x^2 = 36$

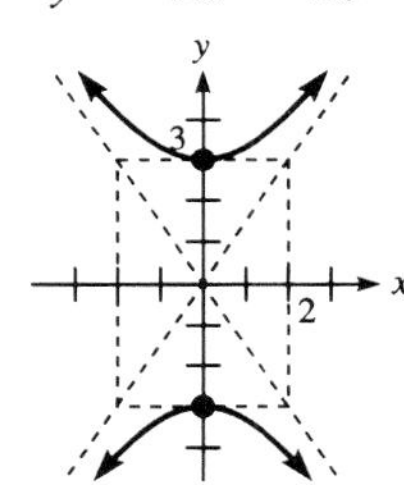

23. $64y^2 - 8x^2 = 16$

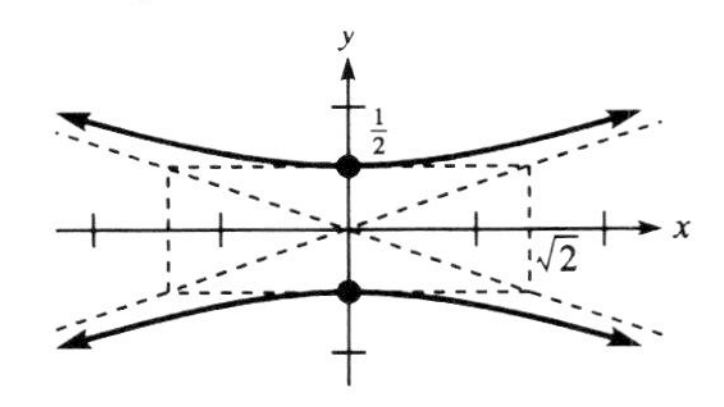

24. $4x^2 - y^2 = 4$

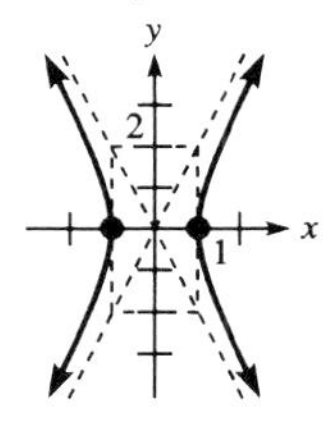

25. $100x^2 - 9y^2 = 900$

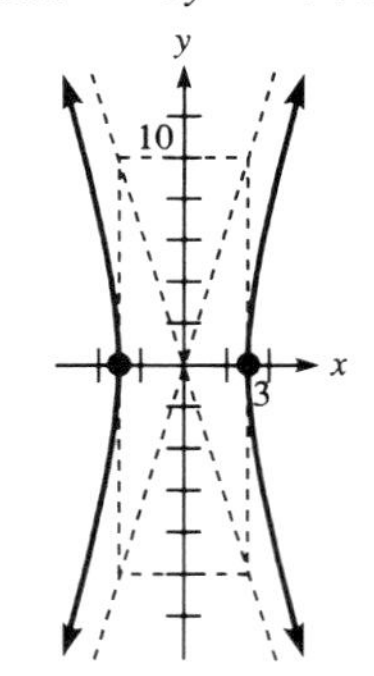

26. $x^2 - 5y^2 = 20$

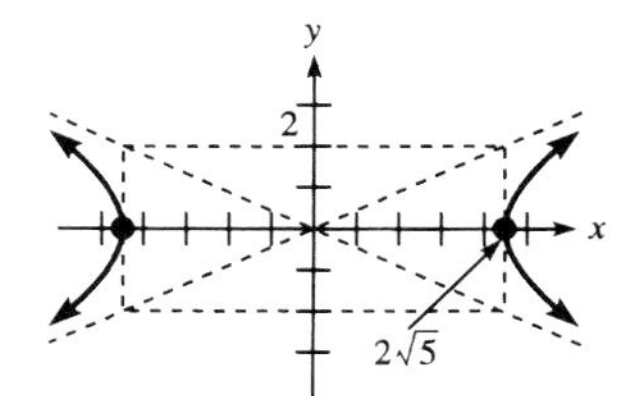

27. $xy = 3$

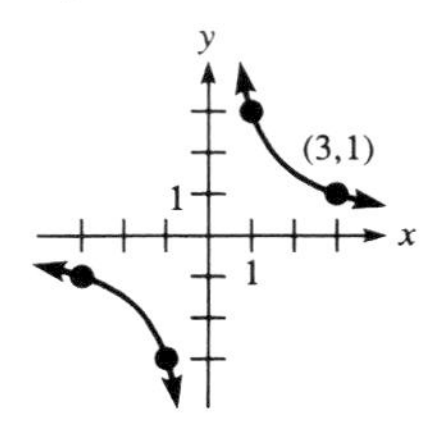

28. $xy = 9$

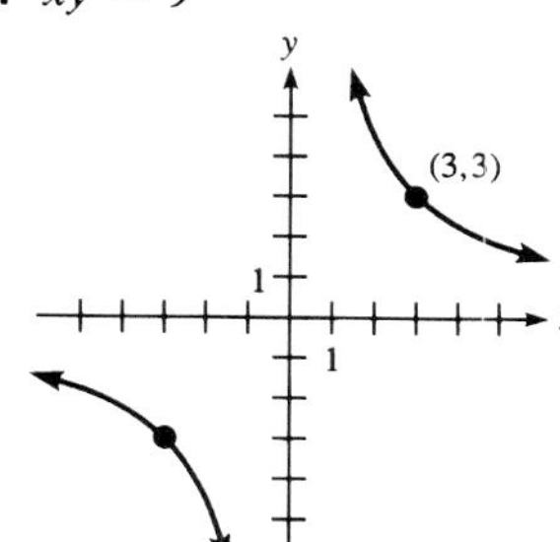

29. $xy = -5$

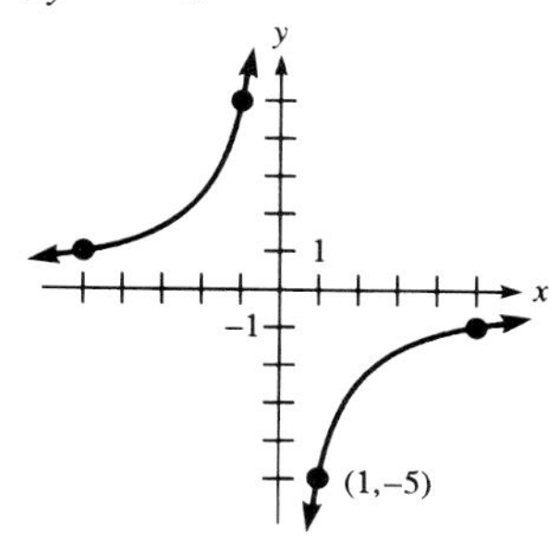

30. $xy = -7$

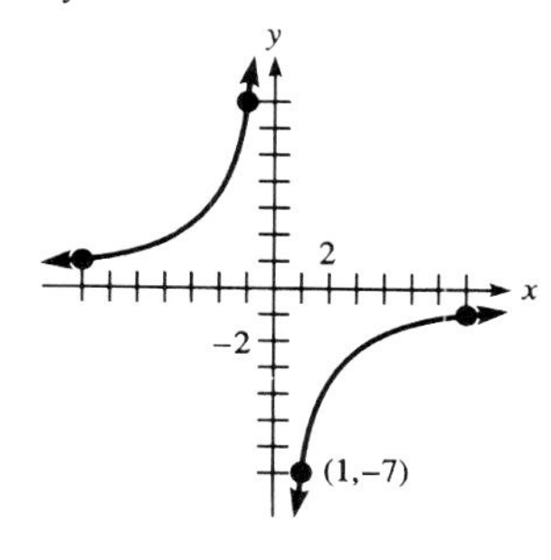

In Exercises 31–46, identify the graph of each equation as a circle, an ellipse, a hyperbola, or a parabola.

31. $-4x^2 - y + 3 = 0$ Parabola
32. $9x^2 + 9y^2 - 81 = 0$ Circle
33. $9x^2 = 9y^2 + 81$ Hyperbola
34. $2y^2 - x + 6 = 0$ Parabola
35. $3x^2 + 3y^2 - 6x + 30y + 30 = 0$ Circle
36. $5x^2 - 4y^2 + 6x - 3y + 10 = 0$ Hyperbola
37. $5x^2 + 9y^2 - 45 = 0$ Ellipse
38. $y^2 + 4 = x$ Parabola
39. $6x^2 - 7y^2 - 42 = 0$ Hyperbola
40. $4x^2 + 4y^2 + x - 4y + 1 = 0$ Circle
41. $5x^2 + 3y^2 + 10x - 9y - 3 = 0$ Ellipse
42. $9x^2 - 3y - 3 = 0$ Parabola
43. $3x^2 + 7y^2 - 2x + y + 5 = 0$ Ellipse
44. $9x^2 - 4y^2 - 6x + 5y + 2 = 0$ Hyperbola
45. $3x^2 + 3y^2 + 12x - 18y + 4 = 0$ Circle
46. $10x^2 + 7y^2 - 100x + 28y + 8 = 0$ Ellipse
47. **a.** Given that y varies inversely as x, and $y = 5$ when $x = 3$, find the variation constant, and write an equation which shows y in terms of x. $k = 15$; $y = 15/x$
    b. Graph the equation from part **a.** $xy = 15$

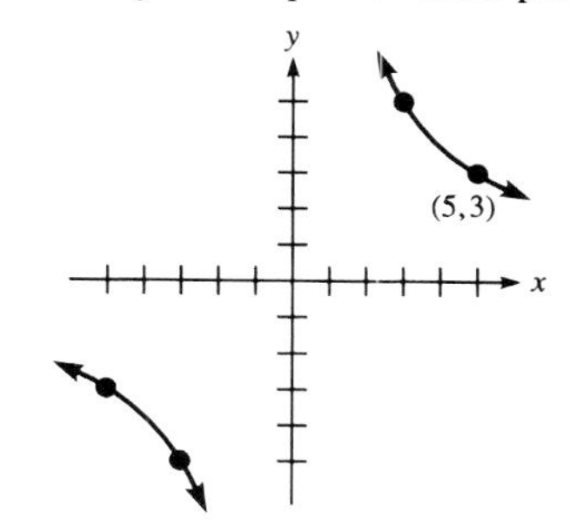

48. **a.** Given that y varies inversely as x, and $y = 1$ when $x = -1$, find the variation constant, and write an equation which shows y in terms of x. $k = -1$; $y = -1/x$
    b. Graph the equation from part **a.** $xy = -1$

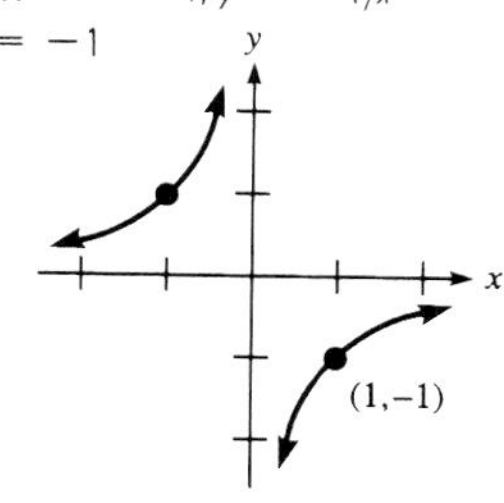

In Exercises 49–50, use the relationship that for something traveling a fixed distance, the time elapsed varies inversely with the average speed.

49. **a.** Write an equation showing that the time t to get from A to B varies inversely with speed s. $t = k/s$
    b. Use the fact that a car averaging 55 mi/hour along an interstate highway takes 2 hours to get from point A to point B to find the variation constant for the equation in part **a.** $k = 110$
    c. Graph this relation, with time on the vertical axis and speed on the horizontal axis. Indicate the appropriate units.

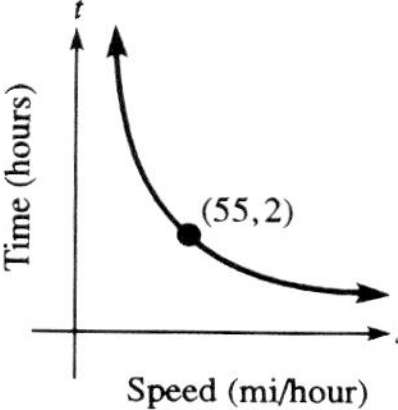

    d. About how long will it take to get from A to B if the speed is reduced to 45 mi/hour? $t = 2.444$ hours $= 2$ hours 27 minutes
50. **a.** Use the fact that a car averaging 55 mi/hour along an interstate highway takes 3 hours to get from point A to point C to find an equation expressing the time needed to get from A to C in terms of the average speed of the car. $t = 165/s$
    b. Graph this relation, with time on the vertical axis and speed on the horizontal axis. Indicate the appropriate units.

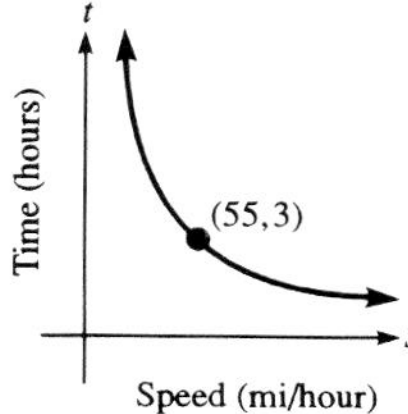

    c. About how long will it take to get from A to C if the speed is reduced to 45 mi/hour? $t = 3.667$ hours $= 3$ hours 40 minutes

THINK ABOUT IT

1. An ellipse may be drawn using pencil, string, and tacks, as shown here. Explain the mathematical basis for this method, referring to the definition of the ellipse.

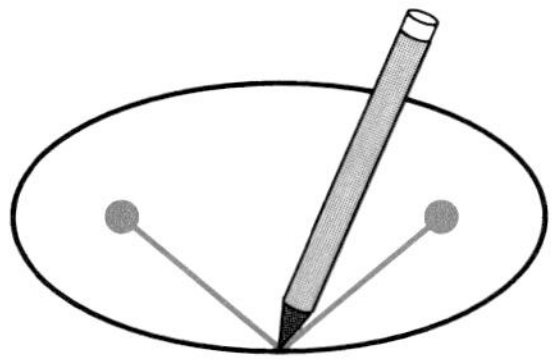

2. **a.** Find the equations of the asymptotes of the hyperbola in Example 3.
   b. In general form, what are the equations of the asymptotes of $(x^2/a^2) - (y^2/b^2) = 1$ or $(y^2/b^2) - (x^2/a^2) = 1$?
3. Match each graph with the equation that illustrates that case.

Graph	Equation
1. Two distinct parallel lines	a. $y^2 = 0$
2. One line through the origin	b. $y^2 = 1$
3. Two distinct lines through the origin	c. $x^2 + y^2 = -1$
4. No graph	d. $x^2 + y^2 = 0$
5. A point	e. $x^2 - y^2 = 0$

4. The names *ellipse, parabola,* and *hyperbola* were given to the conic sections by the Greek mathematician Apollonius (c. 262–190 B.C.). His work was entirely geometric but is equivalent to algebraically noting three types of equations for conics.

$y^2 = px - \dfrac{px^2}{d}$	The case of "ellipsis," which means to fall short
$y^2 = px$	The case of "parabole," which means to coincide
$y^2 = px + \dfrac{px^2}{d}$	The case of "hyperbole," which means to exceed

The letters p and d refer to lengths of line segments used in the geometric construction of these shapes.
   a. Let $p = 4$ and $d = 1$; then by plotting points, draw each of the resulting graphs and confirm that they do give the "correct" shape.
   b. Let $p = 1$ and $d = 4$, and repeat part **a.**
   c. What values for p and d will make the "ellipse" into a circle?
5. Refer to the given figure of an ellipse. It can be shown that $a^2 = b^2 + c^2$ (the Pythagorean relationship). The ratio c/a is called the *eccentricity* of the ellipse. Eccentricity means "out of roundness." If $c = 0$, then the focus is at the center, a equals b, the eccentricity is 0, and the figure is a circle. Recall that the orbits of the planets around the sun are ellipses with the sun at one focus. For instance, the orbit of Venus has eccentricity 0.0068, while earth's eccentricity is 0.0168 and Mercury's is 0.206.

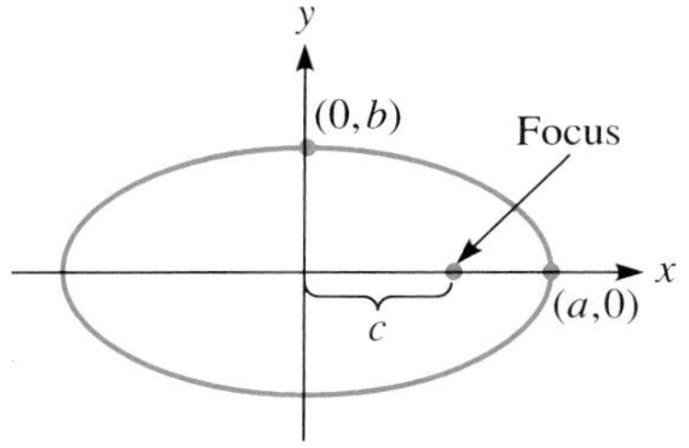

   a. Use the Pythagorean relationship above to find the eccentricity of the ellipse whose x-intercepts are (2,0) and (−2,0) and whose y-intercepts are (0,1) and (0,−1).
   b. If the x-intercepts are twice as far from the origin as the y-intercepts, will the eccentricity be the same as that in part **a?**

REMEMBER THIS

1. By plotting both graphs, determine the number of points of intersection of $y = 2x + 2$ and $y = (x - 1)^2 - 4$. 2
2. What is the maximum number of points of intersection between a circle and a hyperbola? The minimum number? 4; 0
3. Solve for x: $x^2 + (x + 1)^2 = 1$. $\{0, -1\}$
4. Solve for x: $x^2 + \left(\dfrac{4}{x}\right)^2 = 8$. $\{2, -2\}$
5. Is (2,−2) a solution of the system $y = 2 - x^2$, $y = -x$? Yes
6. Graph $(x - 1)^2 + (y - 2)^2 = 4$.

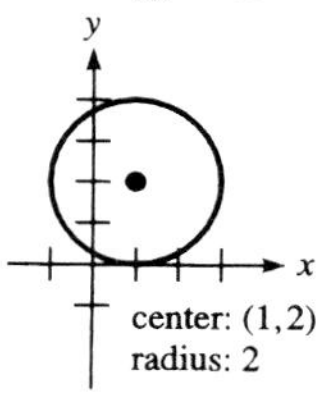

7. Find the x-intercepts of the parabola $y = (x - 2)^2 - 3$. $2 \pm \sqrt{3}$
8. Evaluate $27^{2/3}$. 9
9. Evaluate the determinant. $\begin{vmatrix} 1 & 1 & 0 \\ 1 & 0 & 1 \\ 0 & 1 & 1 \end{vmatrix}$ −2
10. Solve $-5 \le 2x + 1 \le 5$. $[-3, 2]$

8.8 Nonlinear Systems of Equations

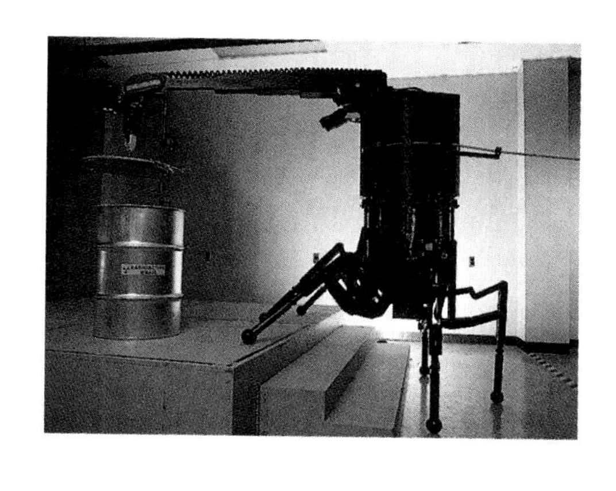

The design for a rectangular component in an industrial robot specifies that the area must measure 120 cm² and the diagonal must measure 17 cm. Find the dimensions of the component. (See Example 3.)

OBJECTIVES

1. **Solve a nonlinear system by the substitution method.**
2. **Solve a nonlinear system by the addition-elimination method.**

1 Systems of linear equations in two variables were solved algebraically and graphically in Section 6.1. On the basis of the work with second-degree equations in this chapter, we can now solve certain **nonlinear systems of equations,** which are systems with at least one equation that is not linear. Graphically, solving the systems considered in this section will require finding all intersection points of one of the conic sections (parabola, circle, ellipse, hyperbola) with a line or another conic section. The algebraic approach for finding such solutions will rely once again on either the substitution method or the addition-elimination method. As with linear systems when at least one equation in the system is solved for one of the variables, then the substitution method is easy to apply, as illustrated in Example 1.

EXAMPLE 1 Solve the system:

$$y = x^2 - 2x - 3 \quad (1)$$
$$y = 2x + 2. \quad (2)$$

Solution Both equations are solved for y. We choose to substitute $x^2 - 2x - 3$ for y in equation (2) and solve for x.

$$x^2 - 2x - 3 = 2x + 2$$
$$x^2 - 4x - 5 = 0$$
$$(x - 5)(x + 1) = 0$$
$$x - 5 = 0 \quad \text{or} \quad x + 1 = 0$$
$$x = 5 \qquad\qquad x = -1$$

Next, find y by substituting these numbers for x in equation (2), which is the simpler equation.

$$\begin{aligned} y &= 2x + 2 \\ &= 2(5) + 2 \\ &= 12 \end{aligned} \quad \text{or} \quad \begin{aligned} y &= 2x + 2 \\ &= 2(-1) + 2 \\ &= 0 \end{aligned}$$

Thus, the solution set is $\{(5,12),(-1,0)\}$. Figure 8.32 confirms that the parabola and line of the system intersect at these points. Check these solutions algebraically in the usual way.

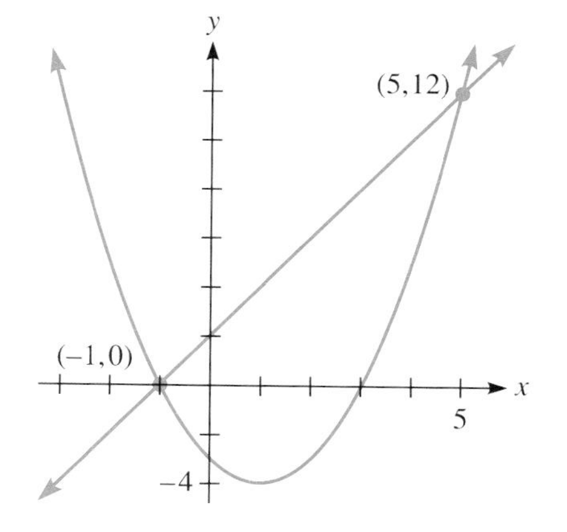

Figure 8.32

Caution In the solution of systems of equations, a common student error is to solve for only one variable and write answers like $x = 5$ or $x = -1$. Remember that it is important to find *both* coordinates of a solution. Then use these two coordinates to check solutions in the original equations of the system.

PROGRESS CHECK 1 Solve the system:

$$y = 2 - x^2$$
$$y = -x.$$

Progress Check Answer

1. $\{(2,-2),(-1,1)\}$

The substitution method is also a useful method when one equation in a system is linear or one equation in a system is of the form $xy = c$, as shown in Examples 2 and 3.

EXAMPLE 2 Solve the system: $x^2 + y^2 = 1$ (1)
$-x + y = 1.$ (2)

Solution Start with the linear equation and solve for one of the variables. Solving for y in equation (2) gives

$$-x + y = 1$$
$$y = x + 1. \quad (3)$$

Now substitute $x + 1$ for y in equation (1) and solve for x.

$$x^2 + (x + 1)^2 = 1$$
$$x^2 + x^2 + 2x + 1 = 1$$
$$2x^2 + 2x = 0$$
$$2x(x + 1) = 0$$
$$2x = 0 \quad \text{or} \quad x + 1 = 0$$
$$x = 0 \qquad\qquad x = -1$$

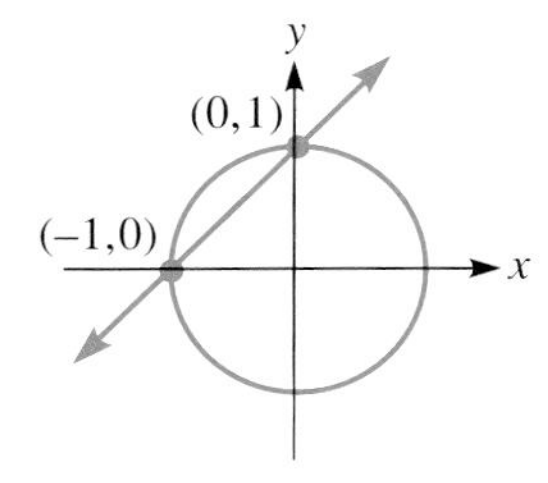

Figure 8.33

Then replacing x by 0 and -1 in equation (3) gives

$$y = x + 1 \quad \text{or} \quad y = x + 1$$
$$= 0 + 1 \qquad\qquad = -1 + 1$$
$$= 1 \qquad\qquad = 0.$$

The solution set is $\{(0,1),(-1,0)\}$. Figure 8.33 shows the circle and line of this system intersecting at these points.

Note A line and a conic section may intersect in two, one, or no points, as illustrated in Figure 8.34. Therefore, we can anticipate that a system with one first-degree equation and one second-degree equation will have two, one, or no real solutions.

Ask students to create specific systems in equation form that illustrate the case in Figure 8.34(c). See "Think About It" Exercise 2.

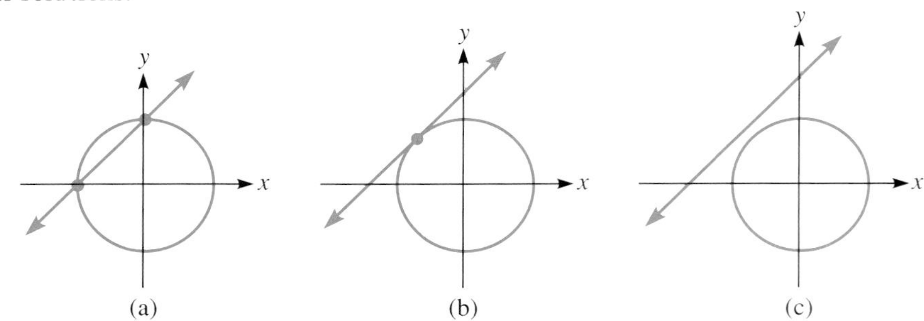

Figure 8.34

PROGRESS CHECK 2 Solve the system: $x^2 + y^2 = 25$
$x - y = 1.$

EXAMPLE 3 Solve the problem in the section introduction on page 415.

Solution Let x represent the width and y represent the length, and sketch the problem situation as in Figure 8.35. Then set up a system of equations.

By the area formula: $xy = 120$ (1)
By the Pythagorean theorem: $x^2 + y^2 = 289$ (2)

Progress Check Answer
2. $\{(4,3),(-3,-4)\}$

Start with the equation of the form $xy = c$ and solve for one of the variables. Solving for y in equation (1) gives

$$xy = 120$$
$$y = \frac{120}{x}. \qquad (3)$$

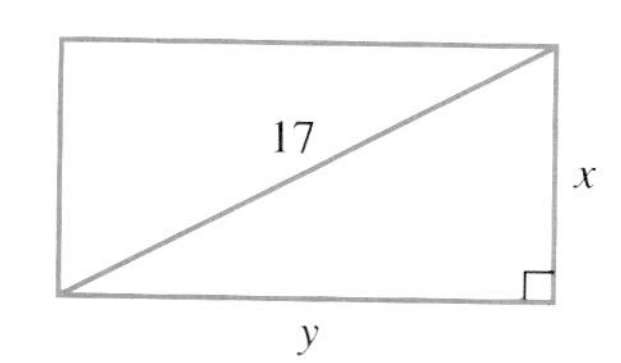

Figure 8.35

Now substitute $120/x$ for y in equation (2) and solve. (The solution will involve the methods of Section 8.3 for solving an equation in quadratic form.)

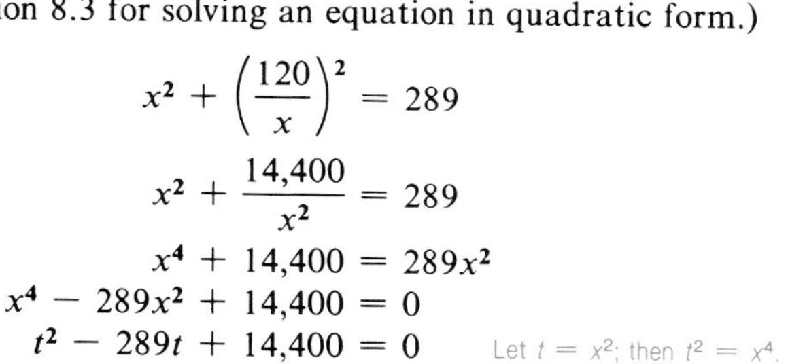

$$x^2 + \left(\frac{120}{x}\right)^2 = 289$$
$$x^2 + \frac{14{,}400}{x^2} = 289$$
$$x^4 + 14{,}400 = 289x^2$$
$$x^4 - 289x^2 + 14{,}400 = 0$$
$$t^2 - 289t + 14{,}400 = 0 \qquad \text{Let } t = x^2\text{; then } t^2 = x^4.$$
$$t = 225 \quad \text{or} \quad t = 64 \qquad \text{By the quadratic formula.}$$

Then, since $t = x^2$,

$$x^2 = 225 \quad \text{or} \quad x^2 = 64$$
$$x = \pm 15 \qquad\quad x = \pm 8.$$

In the context of the problem, negative solutions are rejected. Substituting the two positive solutions into equation (3) yields

$$y = 8 \quad \text{when} \quad x = 15$$
$$\text{and} \quad y = 15 \quad \text{when} \quad x = 8.$$

Because x represents the width (which must be smaller than the length), we conclude that the rectangular component must be 8 cm wide and 15 cm long.

PROGRESS CHECK 3 If the area of a rectangular component must measure 60 cm² and the diagonal must measure 13 cm, find the dimensions of the component.

2 When the two equations in a system both contain an x^2 term and a y^2 term, then the addition-elimination method is often useful, as shown next.

EXAMPLE 4 Find all intersection points of the ellipse $3x^2 + y^2 = 35$ and the hyperbola $5y^2 - 2x^2 = 2$.

Solution The system we need to solve is

$$3x^2 + 2y^2 = 35 \qquad (1)$$
$$-2x^2 + 5y^2 = 2. \qquad (2)$$

If equivalent equations are formed by multiplying both sides of equation (1) by 2, and both sides of equation (2) by 3, then the x variable can be eliminated.

$$6x^2 + 4y^2 = 70 \qquad (3)$$
$$-6x^2 + 15y^2 = 6 \qquad (4)$$
$$19y^2 = 76 \qquad \text{Add equations (3) and (4).}$$
$$y^2 = 4$$
$$y = \pm 2$$

Progress Check Answer

3. 5 cm wide; 12 cm long

Two conic sections may intersect in four or fewer points, so a system with two second-degree equations may have four, three, two, one, or no real solutions. Ask students to illustrate this concept using an ellipse and a hyperbola, as requested in "Think About It" Exercise 1.

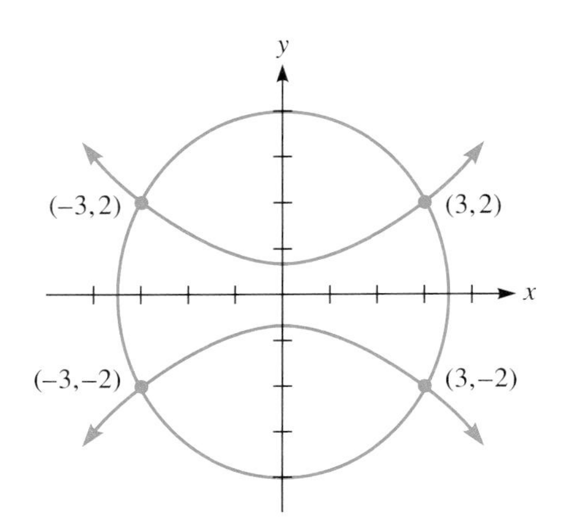

Figure 8.36

Then substitute $y = \pm 2$ into equation (1) to find x. Because 2^2 and $(-2)^2$ are both 4, the solution may be continued in consolidated form.

$$\begin{aligned} 3x^2 + 2(\pm 2)^2 &= 35 \\ 3x^2 + 8 &= 35 \\ 3x^2 &= 27 \\ x^2 &= 9 \\ x &= \pm 3 \end{aligned}$$

Associating each solution for y with each solution for x gives the solution set $\{(3,2),(3,-2),(-3,2),(-3,-2)\}$. Figure 8.36 shows the ellipse and the hyperbola of this system intersecting at these points.

Progress Check Answer

4. $(1,1),(1,-1),(-1,1),(-1,-1)$

PROGRESS CHECK 4 Find all intersection points of the ellipse $2x^2 + 3y^2 = 5$ and the hyperbola $4y^2 - 3x^2 = 1$.

EXERCISES 8.8

In Exercises 1–14, solve the given system by the substitution method.

1. $y = x^2 - x + 2$
   $y = 2x + 2$ $\{(0,2),(3,8)\}$
2. $y = x^2 - 4x + 3$
   $y = -5x + 9$ $\{(-3,24),(2,-1)\}$
3. $y = -x^2 - x + 13$
   $y = -4x + 3$
   $\{(-2,11),(5,-17)\}$
4. $y = -x^2 - 2x + 20$
   $y = 2x - 1$ $\{(-7,-15),(3,5)\}$
5. $y = x^2 - 4x + 12$
   $y = 2x + 3$ $\{(3,9)\}$
6. $y = -x^2 + 3x - 1$
   $y = -x + 3$ $\{(2,1)\}$
7. $y = x^2 + 6$
   $y = x - 3$ $\emptyset$
8. $y = -3x^2 - 5$
   $y = x + 3$ $\emptyset$
9. $x^2 + y^2 = 36$
   $y = x - 6$ $\{(0,-6),(6,0)\}$
10. $x^2 + y^2 = 9$
    $y = x + 3$ $\{(0,3),(-3,0)\}$
11. $x^2 + y^2 = 9$
    $y = 2x - 3$ $\{(0,-3),(\frac{12}{5},\frac{9}{5})\}$
12. $x^2 + y^2 = 16$
    $y = -3x + 4$ $\{(0,4),(\frac{12}{5},-\frac{16}{5})\}$
13. $x^2 + y^2 = 25$
    $y = -x + 8$ $\emptyset$
14. $x^2 + y^2 = 1$
    $y = x + 2$ $\emptyset$

In Exercises 15–20, find all the intersection points of the given ellipses and hyperbolas. Use the elimination method.

15. $2x^2 + 3y^2 = 93$
    $2y^2 - x^2 = 41$ $(3,5), (3,-5), (-3,5), (-3,-5)$
16. $3x^2 + 4y^2 = 19$
    $-2x^2 + 5y^2 = 18$ $(1,2), (1,-2), (-1,2), (-1,-2)$
17. $6x^2 + 7y^2 = 159$
    $-3x^2 + y^2 = -39$ $(4,3), (4,-3), (-4,3), (-4,-3)$
18. $x^2 + 2y^2 = 57$
    $-x^2 + 3y^2 = 23$ $(5,4), (5,-4), (-5,4), (-5,-4)$
19. $x^2 + 2y^2 = 33$
    $2y^2 - x^2 = 3$ $(\sqrt{15},-3), (\sqrt{15},3), (-\sqrt{15},3), (-\sqrt{15},-3)$
20. $3x^2 + 4y^2 = 34$
    $2y^2 - x^2 = 2$ $(\sqrt{6},2), (\sqrt{6},-2), (-\sqrt{6},2), (-\sqrt{6},-2)$
21. The area of a rectangle is 48 cm², and the diagonal is 10 cm. Find the dimensions of the rectangle. 6 by 8 cm
22. The area of a right triangle is 120 in.² and the hypotenuse is 26 in. Find the lengths of both legs. 10 and 24 in.
23. To the nearest thousandth, find the dimensions of a rectangle whose diagonal is 10 in. and whose area is 10 in.². 1.005 by 9.949 in.
24. To the nearest thousandth, find the dimensions of a rectangle whose diagonal is 10 cm and whose area is 1 cm³. 0.100 by 9.999 cm

25. **a.** Find the equations of the line and the circle in the figure. $y = \sqrt{3}x$; $x^2 + y^2 = 4$

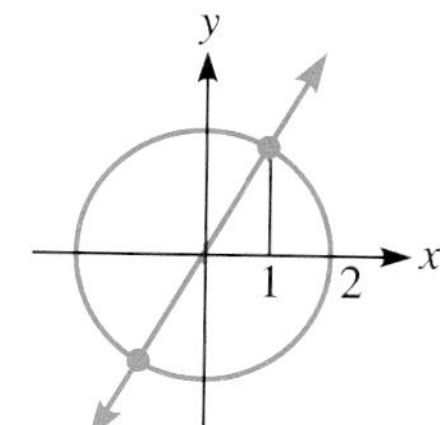

b. Find both points of intersection. $(1, \sqrt{3}), (-1, -\sqrt{3})$

26. Find all points on the ellipse $\frac{x^2}{9} + \frac{y^2}{25} = 1$ that are at a distance of 4 units from the origin. $\left(\frac{9}{4}, \frac{5\sqrt{7}}{4}\right), \left(-\frac{9}{4}, \frac{5\sqrt{7}}{4}\right), \left(\frac{9}{4}, -\frac{5\sqrt{7}}{4}\right), \left(-\frac{9}{4}, -\frac{5\sqrt{7}}{4}\right)$

THINK ABOUT IT

1. Show in a series of sketches that an ellipse and a hyperbola may intersect in four or less points. What does this say about the solution of a system that contains an equation for an ellipse and an equation for a hyperbola?
2. Create specific systems by writing equations that illustrate the given case.
   a. A circle and a line that do not intersect
   b. An ellipse and a hyperbola that intersect in two points
   c. Any two conic sections that intersect in one point
3. Greek legend has it that Apollo was responsible for a plague in Delos. The oracle there said that Apollo would be appeased if the people would double the size of the altar, keeping its shape the same. The shape was cubical. The mathematical question then is, "How much longer should each side of the new cube be so that its volume is double the original volume?" This was a difficult problem for Greek geometry. One solution, often attributed to Menaechmus (c. 350 B.C.), depended on finding the intersection of a parabola and a hyperbola. The side of the new altar is given by the x-coordinate of the intersection of the graphs of $y = x^2$ and $xy = 2$. Solve this system and show that multiplying the side (s) of a cube by this solution gives a new cube with double the volume of the old one.
4. When two resistors are connected in series, their combined resistance is given by $R = R_1 + R_2$. If the resistors are connected in parallel, their combined resistance is given by $R = R_1R_2/(R_1 + R_2)$. Find the value of resistors R_1 and R_2 if their combined resistance is 32 ohms when connected in series and 6 ohms when connected in parallel.
5. Find the radius of each circle in the illustration if their combined area is 68π cm².

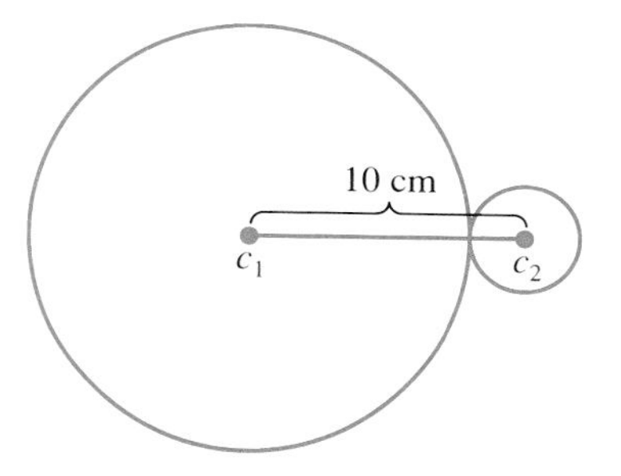

REMEMBER THIS

1. Solve for y: $x^2 + y^2 = 25$. $y = \pm\sqrt{25 - x^2}$
2. By referring to a graph, determine the maximum real value of x that satisfies $x^2 + y^2 = 25$. 5
3. For what values of x is $\frac{1}{x - 2}$ not defined? 2
4. For what values of x is $\sqrt{x + 4}$ a real number? $\{x: x \geq -4\}$
5. How many ordered pairs are there in the solution set of $y = x^2 + 1$? Infinitely many
6. Express this relationship by an equation: One number (y) equals 3 more than twice another (x). $y = 2x + 3$
7. Graph $x^2 + \frac{y^2}{4} = 1$.

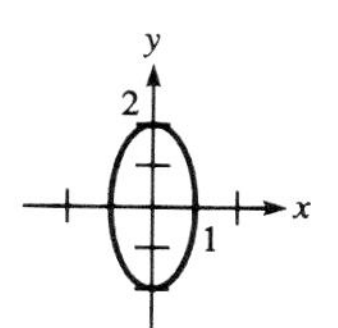

8. Solve $3x^2 = 4x$. $\{0, \frac{4}{3}\}$
9. Find an equation for the line through the origin which is perpendicular to $y = 3x - 4$. $y = -\frac{1}{3}x$
10. The total price of an item including the 8 percent sales tax is \$42.66. Find the before-tax price. \$39.50

Graphing Conic Sections and Solving Quadratic Equations

As shown in Example 1, graphing a parabola of the form $y = ax^2 + bx + c$ is straightforward. Just enter the formula using the [Y=] key.

EXAMPLE 1 Graph the parabola $y = 3x^2 + x - 1$, and estimate the coordinates of the vertex to the nearest hundredth.

Solution

Step 1 Press [Y=] and enter $3X^2 + X - 1$ for Y_1.

Step 2 Press [ZOOM] Standard.
The graph appears as in Figure 8.37.

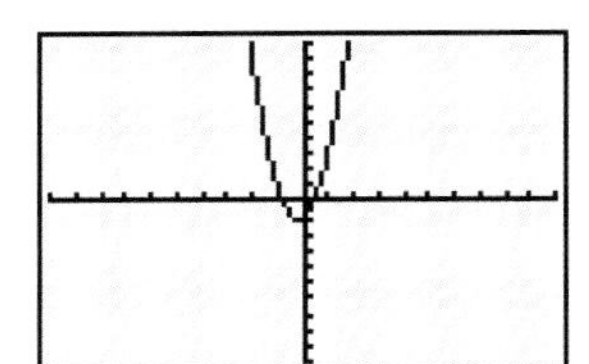

Figure 8.37

Step 3 By repeated use of [TRACE] and [ZOOM], the vertex to the nearest hundredth is $(-0.17, -1.08)$.

Example 2 deals with curves whose equations involve y^2 and shows that such curves are graphed in two pieces.

EXAMPLE 2 Graph the circle $x^2 + y^2 = 25$.

Solution

Step 1 Solve for y to get $y = \pm\sqrt{25 - x^2}$.

Step 2 Press [Y=].
For Y_1, key in $\sqrt{(25 - X^2)}$.
For Y_2, key in $-Y_1$.
The screen should look like Figure 8.38(a).

Step 3 Press [ZOOM] Square. This will put equal spacing on both axes so that the graph will appear round. See Figure 8.38(b).

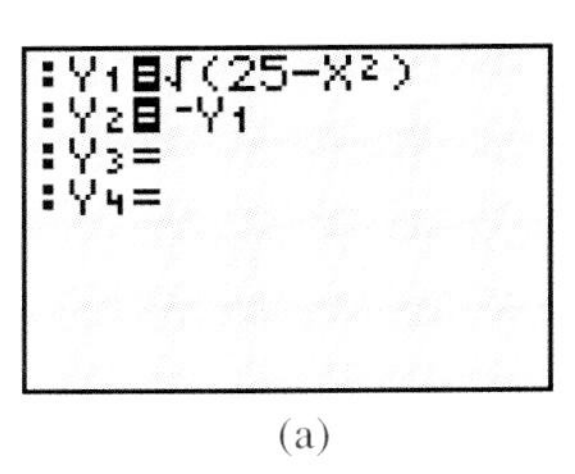

(a)

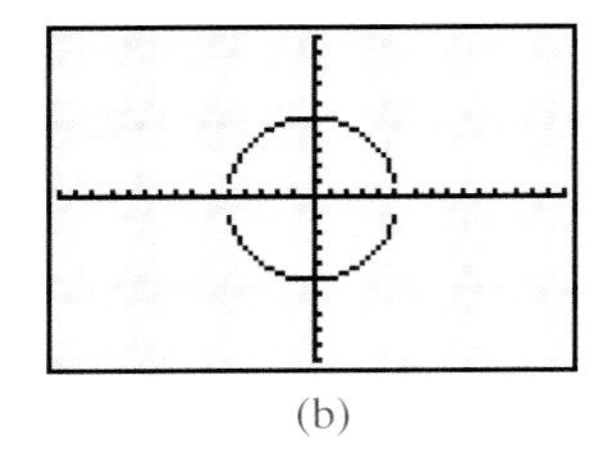

(b)

Figure 8.38

The technique used to graph the circle can also be used to graph ellipses and hyperbolas after their equations are solved for y.

In Example 3 the calculator is used to solve an equation involving x^2. The method of solution illustrated is to plot the two sides of the equation and look for the point of intersection. You could also use the method of plotting their difference.

EXAMPLE 3 Use the graphing calculator to solve $\sqrt{6 - x^2} = -x$. Estimate any solutions to the nearest hundredth.

Solution

Step 1 Press [Y=].
For Y_1, key in $\sqrt{(6 - X^2)}$.
For Y_2, key in $-X$.

Step 2 Plot both graphs using [ZOOM] Square, as shown in Figure 8.39.
Use [TRACE] and [ZOOM] to find the point of intersection. To the nearest hundredth the only solution is -1.73.

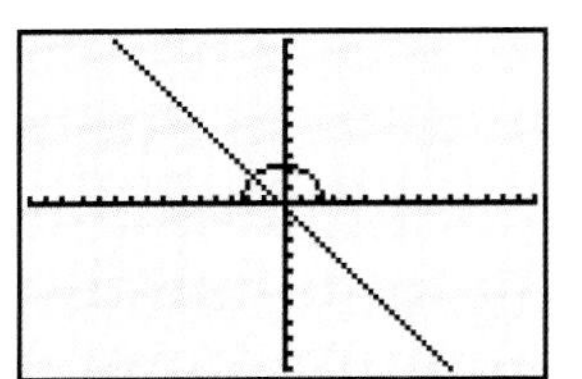

Figure 8.39

A Program for the Quadratic Formula

The following program can be copied into your calculator for automatic evaluation of the quadratic formula.

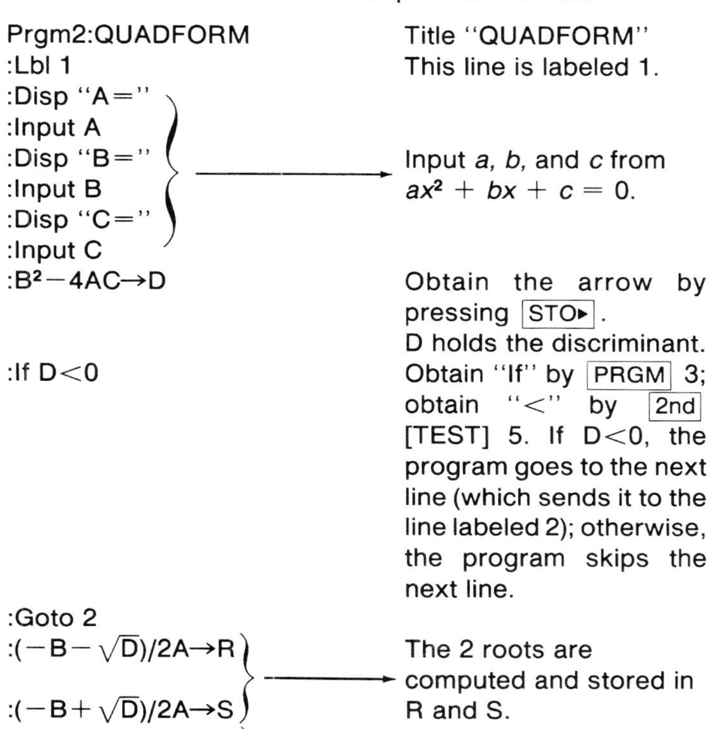

Program	Comment
Prgm2:QUADFORM	Title "QUADFORM"
:Lbl 1	This line is labeled 1.
:Disp "A=" :Input A :Disp "B=" :Input B :Disp "C=" :Input C	Input *a, b,* and *c* from $ax^2 + bx + c = 0$.
:B²−4AC→D	Obtain the arrow by pressing [STO▸]. D holds the discriminant.
:If D<0	Obtain "If" by [PRGM] 3; obtain "<" by [2nd] [TEST] 5. If D<0, the program goes to the next line (which sends it to the line labeled 2); otherwise, the program skips the next line.
:Goto 2	
:(−B−√D)/2A→R :(−B+√D)/2A→S	The 2 roots are computed and stored in R and S.
:Disp "ROOTS =" :Disp R :Disp S	The 2 roots are displayed.
:Goto 1	The program starts again at label 1.
:Lbl 2	This line is labeled 2.
:Disp "NO REAL SOLUTIONS"	
:Goto 1	The program starts again at label 1.

After entering the program, press [2nd] [QUIT] to return to the home screen. To use the program, press [PRGM] and the number for QUADFORM.

EXERCISES

1. Graph this parabola, and approximate the coordinates of the vertex to the nearest hundredth:
   $y = \frac{1}{2}x^2 - \frac{2}{5}x + 3$. Ans. (0.40,2.92)
2. Graph this circle, and approximate the x-intercepts to the nearest hundredth: $x^2 + 4x + y^2 - 30 = 0$.
   Ans. (3.83,0), (−7.83,0)

Use the graphing calculator to solve and check these Exercises from Chapter 8.

Section	*Exercises*
8.1	11
8.2	7, 49
8.3	5, 21
8.5	1, 15, 37
8.6	25, 35
8.7	1, 11
8.8	1, 15

Chapter 8 SUMMARY

OBJECTIVES CHECKLIST Specific chapter objectives are summarized below along with numbered example problems from the text that should clarify the objectives. If you do not understand any objectives or do not know how to do the selected problems, then restudy the material.

8.1 Can you:

1. **Solve quadratic equations using the square root property?**
   Solve $x^2 - 18 = 0$ using the square root property. [Example 1a]

2. **Complete the square for $x^2 + bx$, and express the resulting trinomial in factored form?**
   Determine the number that should be added to make $y^2 - y$ a perfect square. Then add this number, and factor the resulting trinomial. [Example 3b]

3. **Solve $ax^2 + bx + c = 0$ by completing the square when $a = 1$?**
   Solve $x^2 - 8x + 3 = 0$. [Example 4]

4. **Solve $ax^2 + bx + c = 0$ by completing the square when $a \neq 1$?**
   Solve $3x^2 - 2x - 3 = 0$. [Example 5]

8.2 Can you:

1. **Solve quadratic equations using the quadratic formula?**
   Solve $x^2 - 2x - 1 = 0$ using the quadratic formula. [Example 1]

2. **Use the discriminant to determine the nature of the solutions of a quadratic equation?**
   Use the discriminant to determine the nature of the solutions of the equation $2x^2 + 5 = 4x$. [Example 4]

3. **Solve quadratic equations after choosing an efficient method?**
   Solve the equation $25s^2 - 40s + 16 = 0$. Choose an efficient method. [Example 5]

8.3 Can you:

1. **Solve fractional equations and radical equations that lead to quadratic equations?**
   Solve $x + \sqrt{14 - x^2} = 0$. [Example 2]

2. **Solve equations with quadratic form?**
   Solve $x^4 - 5x^2 + 4 = 0$. [Example 3]

3. **Solve literal equations for a variable with highest power 2?**
   Solve $x = y^2 - 4$ for y. [Example 5]

8.4 Can you:

1. **Solve quadratic inequalities?**
   Solve $2x^2 - 7x - 4 \leq 0$ using the test point method. [Example 2]

2. **Solve polynomial inequalities of degree higher than 2?**
   Solve $x(x - 2)(3x + 5) > 0$. [Example 5]

3. **Solve inequalities involving quotients of polynomials?**
   Solve $\dfrac{x + 2}{x - 3} \geq 0$. [Example 6]

8.5 Can you:

1. **Graph $y = ax^2 + bx + c$ by finding ordered-pair solutions?**
   Graph $y = -x^2 + 4x$. [Example 2]

2. **Graph $y = ax^2 + bx + c$ by using the vertex and intercepts of the graph?**
   For the equation $y = x^2 - 2x - 3$, find the coordinates of the vertex and of the x- and y-intercepts, and graph the equation. [Example 3b–d]

3. **Find the number of x-intercepts in the graph of $y = ax^2 + bx + c$ by using the discriminant?**
Use the discriminant to determine the number of x-intercepts of the graph of $y = 4x^2 - 4x + 1$. [Example 5]

4. **Graph $x = ay^2 + by + c$ by using the vertex and intercepts of the graph?**
For the equation $x = y^2 + 4y$, find the coordinates of the vertex and of the x- and y-intercepts, and graph the equation. [Example 6b–d]

5. **Solve applied problems involving parabolas?**
Solve the problem in the section introduction on page 389. [Example 7]

8.6 Can you:

1. **Find the equation in standard form of a circle given its center and either the radius of the circle or a point on the circle?**
Find the equation in standard form of the circle with center at $(-3,0)$ that passes through $(-5,2)$. [Example 2]

2. **Find the center and radius of a circle given its equation, and then graph it?**
Find the center and the radius of the circle given by $x^2 + (y + 2)^2 = 9$. Also, sketch the circle. [Example 3]

3. **Graph and write equations for circles and semicircles with centers at the origin?**
Graph $y = -\sqrt{4 - x^2}$. [Example 5c]

8.7 Can you:

1. **Graph an ellipse?**
Graph $4x^2 + y^2 = 4$. [Example 2]

2. **Graph a hyperbola?**
Graph $4y^2 - 9x^2 = 144$. [Example 4]

3. **Graph $xy = c$?**
Graph $xy = 2$. [Example 5]

4. **Classify certain equations as defining either a circle, an ellipse, a hyperbola, or a parabola?**
Identify the graph of $2x^2 + 2y^2 - 10x - 2y - 31 = 0$ as a circle, an ellipse, a hyperbola or a parabola. [Example 7a]

8.8 Can you:

1. **Solve a nonlinear system by the substitution method?**
Solve the system: $y = x^2 - 2x - 3$
$y = 2x + 2$. [Example 1]

2. **Solve a nonlinear system by the addition-elimination method?**
Find all intersection points of the ellipse $3x^2 + y^2 = 35$ and the hyperbola $5y^2 - 2x^2 = 2$. [Example 4]

KEY TERMS

Asymptotes of a hyperbola (8.7)
Axis of symmetry (8.5)
Branches (of a hyperbola) (8.7)
Circle (8.6)
Completing the square (8.1)
Conic sections (8.5)
Critical numbers (8.4)
Degenerate conic sections (8.7)
Discriminant (8.2)
Ellipse (8.7)
Equation with quadratic form (8.3)
Fundamental rectangle (8.7)
Horizontal parabola (8.5)
Hyperbola (8.7)
Nonlinear system of equations (8.8)
Parabola (8.5)
Quadratic inequality (8.4)
Second-degree (or quadratic) equation (8.1)
Sign graph (8.4)
Square root property (8.1)
Test point method (8.4)
Vertex (of a parabola) (8.5)

KEY CONCEPTS AND PROCEDURES

Section	Key Concepts or Procedures to Review
8.1	■ Methods to solve a quadratic equation by using the square root property and by completing the square
8.2	■ Quadratic formula: If $ax^2 + bx + c = 0$ and $a \neq 0$, then $x = \dfrac{-b \pm \sqrt{b^2 - 4ac}}{2a}$. ■ Methods to solve quadratic equations by using the quadratic formula ■ Methods to determine the nature of the solutions to a quadratic equation from the discriminant $(b^2 - 4ac)$ ■ Guidelines for selecting an efficient method for solving a quadratic equation
8.3	■ Methods to solve fractional equations and radical equations that lead to quadratic equations and to solve equations with quadratic form ■ Methods to solve literal equations for a variable with highest power 2
8.4	■ Methods to solve quadratic inequalities, polynomial inequalities of degree higher than 2, and inequalities involving quotients of polynomials by using the test point method
8.5	■ The graph of an equation of the form $y = ax^2 + bx + c$ with $a \neq 0$ is a parabola. If $a > 0$, the parabola opens upward. If $a < 0$, the parabola opens downward. ■ The graph of an equation of the form $x = ay^2 + by + c$ with $a \neq 0$ is a parabola. If $a > 0$, the parabola opens to the right. If $a < 0$, the parabola opens to the left. ■ Procedure for graphing a parabola by determining the axis of symmetry, the vertex, and the coordinates of any x- and y-intercepts ■ Methods to determine the number of x-intercepts of the graph of an equation of the form $y = ax^2 + bx + c$ from the discriminant $(b^2 - 4ac)$
8.6	■ Standard form of the equation of a circle of radius r with center (h,k): $(x - h)^2 + (y - k)^2 = r^2$ ■ Standard equation of a circle with center at the origin and radius r: $x^2 + y^2 = r^2$ ■ The graph of an equation of the form $y = \sqrt{r^2 - x^2}$ or $y = -\sqrt{r^2 - x^2}$ is a semicircle.
8.7	■ Definitions of an ellipse and a hyperbola ■ Standard form of the equation of an ellipse centered at the origin with x-intercepts $(a,0)$ and $(-a,0)$ and y-intercepts $(0,b)$ and $(0,-b)$: $\dfrac{x^2}{a^2} + \dfrac{y^2}{b^2} = 1$, where $a > 0, b > 0$. ■ Standard form of the hyperbola centered at the origin, with (**1**) x-intercepts at $(a,0)$ and $(-a,0)$ and no y-intercepts: $\dfrac{x^2}{a^2} - \dfrac{y^2}{b^2} = 1$ (**2**) y-intercepts at $(0,b)$ and $(0,-b)$ and no x-intercepts: $\dfrac{y^2}{b^2} - \dfrac{x^2}{a^2} = 1$ ■ An equation of the form $xy = c$, with $c \neq 0$, graphs as a hyperbola with both axes as asymptotes. ■ Chart summarizing the graphing possibilities for $Ax^2 + Cy^2 + Dx + Ey + F = 0$, where A and C are not both zero
8.8	■ Methods to solve a nonlinear system of equations by the substitution method and by the addition-elimination method

CHAPTER 8 REVIEW EXERCISES

8.1

1. Solve $x^2 = -3$ using the square root property. $\{i\sqrt{3}, -i\sqrt{3}\}$
2. Solve $(x - 2)^2 - 9 = 0$. $\{-1,5\}$
3. Determine the number that should be added to make the expression $x^2 + 8x$ a perfect square. Then add this number and factor the resulting trinomial. $x^2 + 8x + 16 = (x + 4)^2$
4. Solve $x^2 + 4x - 15 = 0$ by completing the square. $\{-2 - \sqrt{19}, -2 + \sqrt{19}\}$
5. Solve $3x^2 - x - 3 = 0$ by completing the square. $\left\{\frac{1 + \sqrt{37}}{6}, \frac{1 - \sqrt{37}}{6}\right\}$

8.2

6. Solve $x^2 + 5x + 3 = 0$ using the quadratic formula. $\left\{\frac{-5 \pm \sqrt{13}}{2}\right\}$
7. Solve $4x^2 + 5x - 6 = 0$ using the quadratic formula. $\{-2, \frac{3}{4}\}$
8. Solve $x^2 + 5 = 2x$ using the quadratic formula. $\{1 \pm 2i\}$
9. Use the discriminant to determine the nature of the solutions of the equation $3x^2 - 4 = 5x$. 2 solutions; real, irrational, unequal
10. Solve the equation $4x^2 - 16 = -7$. Choose an efficient method. $\{-\frac{3}{2}, \frac{3}{2}\}$

8.3

11. Solve $\frac{5}{x} - \frac{6}{x^2} = 1$. $\{2,3\}$
12. Solve $x + \sqrt{6 - x^2} = 0$. $\{-\sqrt{3}\}$
13. Solve $x^4 - 10x^2 + 9 = 0$. $\{\pm 1, \pm 3\}$
14. Solve $3x^{1/2} - 7x^{1/4} + 2 = 0$. $\{\frac{1}{81}, 16\}$
15. Solve $x^2 = 9 - y^2$ for y. $y = \pm\sqrt{9 - x^2}$

8.4

16. Solve $2x^2 + x - 3 \le 0$ using the test point method. $[-\frac{3}{2}, 1]$
17. Solve $(x - 1)^2 > -3$. $(-\infty, \infty)$
18. Solve $x(2x + 7)(x - 2) < 0$. $(-\infty, -\frac{7}{2}) \cup (0,2)$
19. Solve $\frac{5}{x} > 3$. $(0, \frac{5}{3})$
20. The height of a projectile that is shot directly up from the ground with an initial velocity of 128 ft/second is given by $y = -16t^2 + 128t$, where y is measured in feet and t is the elapsed time in seconds. To the nearest hundredth of a second, for what values of t is the projectile more than 96 ft off the ground? (0.84 second, 7.16 seconds)

8.5

21. Graph $y = x^2 + 4$ by finding ordered-pair solutions.

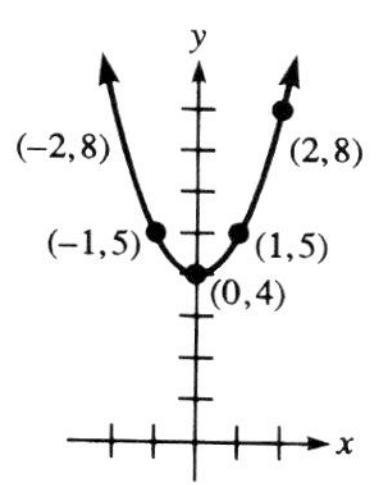

22. Use the discriminant to determine the number of x-intercepts of the graph of $y = 2x^2 - 3x + 2$. No x-intercept

In Exercises 23 and 24 for the given equation, do the following:

a. Find the equation of the axis of symmetry.
b. Find the coordinates of the vertex.
c. Find the coordinates of any x- and y-intercepts.
d. Graph the equation.

23. $y = x^2 - x - 6$
a. $x = \frac{1}{2}$ b. $(\frac{1}{2}, -\frac{25}{4})$ c. $(-2,0)$, $(3,0)$, $(0,-6)$ d.

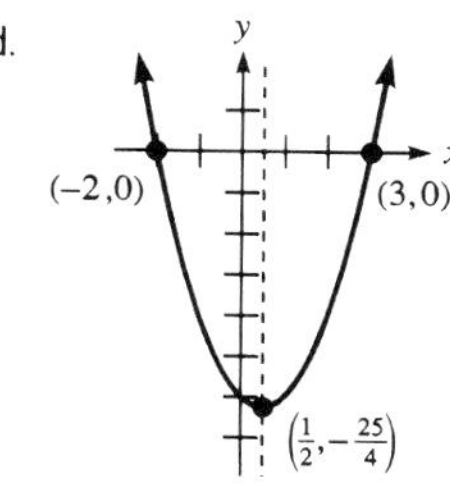

24. $x = 2y^2 - 4y$ a. $y = 1$ b. $(-2,1)$ c. $(0,0)$, $(0,2)$ d.

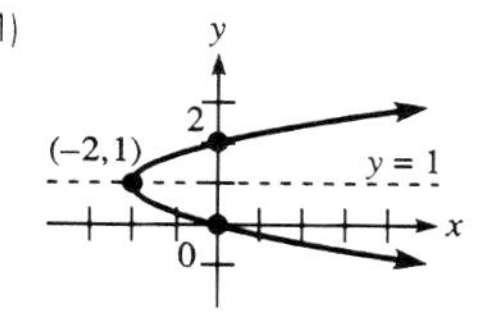

25. The height (y) of a projectile that is shot directly up from the ground with an initial velocity of 144 ft/second is given by $y = -16t^2 + 144t$, where y is measured in feet and t is the elapsed time in seconds.
a. When does the projectile attain its maximum height? 4.5 seconds
b. What is the maximum height? 324 ft
c. When does the projectile hit the ground? 9 seconds

8.6

26. Find the equation in standard form of the circle with center at $(2,-5)$ and radius 6. $(x - 2)^2 + (y + 5)^2 = 36$
27. Find the equation in standard form of the circle with center at $(-2,6)$ that passes through $(1,3)$. $(x + 2)^2 + (y - 6)^2 = 18$
28. Find the center and radius of the circle given by $(x - 1)^2 + y^2 = 16$. Center $(1,0)$; radius 4
29. Find the center and radius of the circle given by $x^2 - 6x + y^2 - 4y + 2 = 0$. Center $(3,2)$; radius $\sqrt{11}$
30. Graph the equation $y = \sqrt{9 - x^2}$.

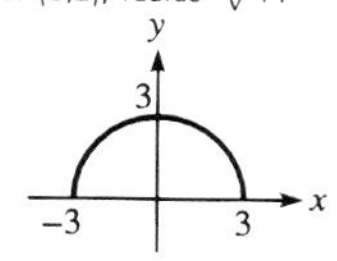

8.7

31. Graph $\frac{x^2}{25} + \frac{y^2}{9} = 1$.

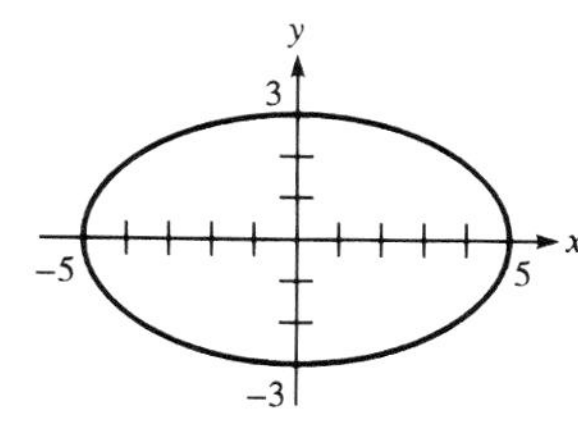

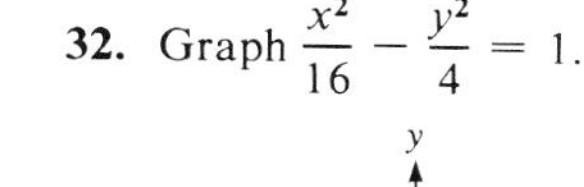

32. Graph $\frac{x^2}{16} - \frac{y^2}{4} = 1$.

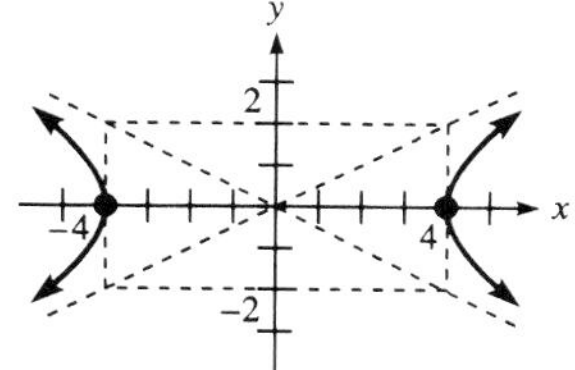

33. Graph $9y^2 - 16x^2 = 144$.

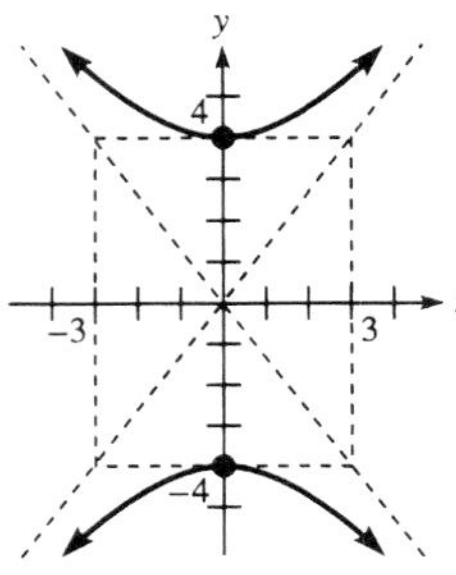

34. If a car travels at a constant rate r, then the time t required to travel 40 mi varies inversely with r. Write an equation that expresses the relationship between r and t, and graph this equation.

$rt = 40$, or $t = \frac{40}{r}$

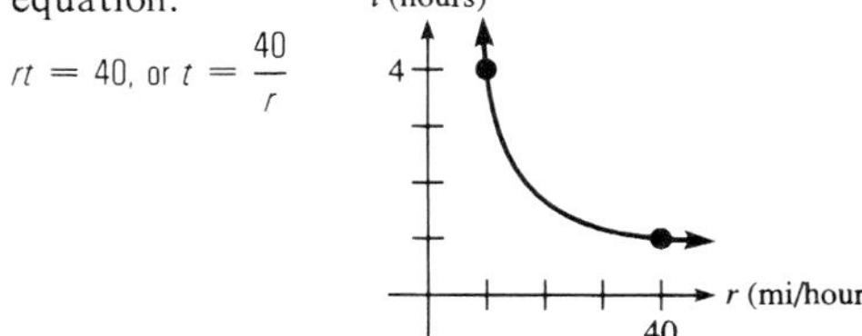

35. Identify the graph of the equation $4y^2 = -25x^2 + 16$ as a circle, an ellipse, a hyperbola or a parabola. Ellipse

8.8

36. Solve the system. $y = x^2 + x - 4$
$y = 3x - 1$ $\{(3,8),(-1,-4)\}$

37. Solve the system. $x^2 + y^2 = 4$
$-3x + y = -6$ $\{(2,0),(\frac{8}{5},-\frac{6}{5})\}$

38. Find all intersection points of the ellipse $2x^2 + y^2 = 31$ and the hyperbola $3x^2 - 2y^2 = 1$. $(3,\sqrt{13}),(3,-\sqrt{13}),(-3,\sqrt{13}),(-3,-\sqrt{13})$

39. Solve the system. $xy = 80$
$x^2 + y^2 = 281$ $\{(16,5),(-16,-5),(5,16),(-5,-16)\}$

40. If the area of a rectangle measures 108 cm² and the diagonal measures 15 cm, find the dimensions of the rectangle.
9 cm wide; 12 cm long

Additional Review Exercises

41. Find all intersection points of the ellipse $3x^2 + 2y^2 = 5$ and the hyperbola $4x^2 - 3y^2 = 1$. $(1,1),(1,-1),(-1,1),(-1,-1)$

42. Find the center and radius of the circle given by $x^2 + y^2 - 4x - 2y + 1 = 0$. $(2,1)$; $r = 2$

43. Find the center and radius of the circle given by $(x + 1)^2 + y^2 = 6$. $(-1,0)$; $r = \sqrt{6}$

44. Use the discriminant to determine the number of x-intercepts of the graph of $y = x^2 - 8x + 10$. 2

45. Find the equation in standard form of the circle with center at $(-1,4)$ and the radius 7. $(x + 1)^2 + (y - 4)^2 = 49$

46. Find the equation in standard form of the circle with center at $(-2,8)$ that passes through $(-1,-2)$. $(x + 2)^2 + (y - 8)^2 = 101$

47. Complete the square for $y^2 + 10y$, and express the resulting trinomial in factored form. $y^2 + 10y + 25 = (y + 5)^2$

48. Use the discriminant to determine the nature of the solutions of the equation $3x^2 + 6 = 2x$. Conjugate complex numbers

49. The amount (A) of silver in troy ounces that can be bought with \$500 varies inversely as the price (p) per troy ounce of silver. Write a variation equation that expresses the relationship between A and p, and then graph the equation.

$Ap = 500$, or $A = \frac{500}{p}$

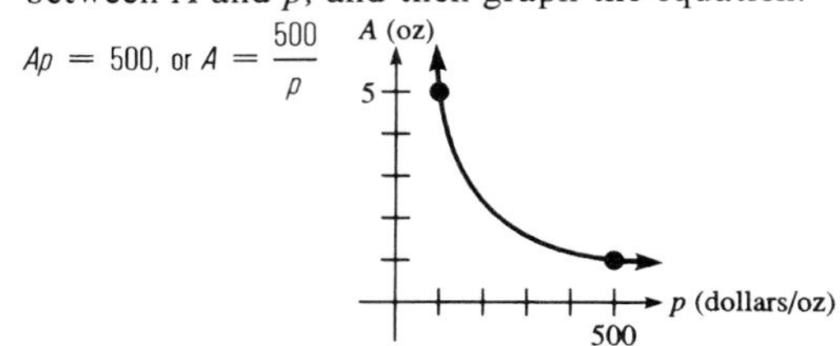

50. The area of a rectangle measures 240 cm² and the diagonal measures 26 cm. Find the dimensions of the rectangle.
Width 10 cm; length 24 cm

51. The length of a rectangle is 5 in. greater than its width, and its area is 22 in.². To the nearest thousandth of an inch, find the length and width of the rectangle. Width 2.815 in.; length 7.815 in.

52. The height (y) in feet of a projectile shot vertically up from the ground with an initial velocity of 32 ft/second is given by $y = -16t^2 + 32t$, where t is the elapsed time in seconds.

a. When does the projectile attain its maximum height?
1 second

b. What is the maximum height? 16 ft

c. When does the projectile hit the ground? 2 seconds

d. Graph the equation.

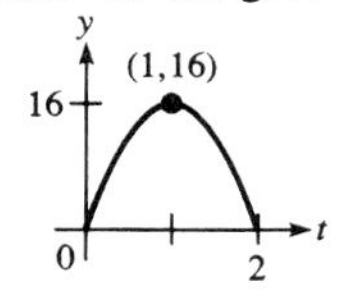

In Exercises 53–55 for the given equation, do the following.

a. Find the equation of the axis of symmetry.
b. Find the coordinates of the vertex.
c. Find the coordinates of the x- and y-intercepts.
d. Graph the equation.

53. $y = x^2 - 4x + 4$
a. $x = 2$ b. $(2,0)$ c. $(2,0)$, $(0,4)$
d.

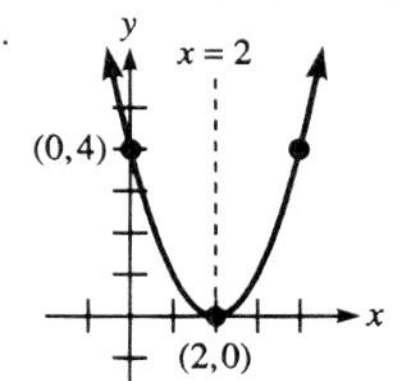

54. $y = -x^2 + 2x - 4$
a. $x = 1$ b. $(1,-3)$ c. $(0,-4)$
d.

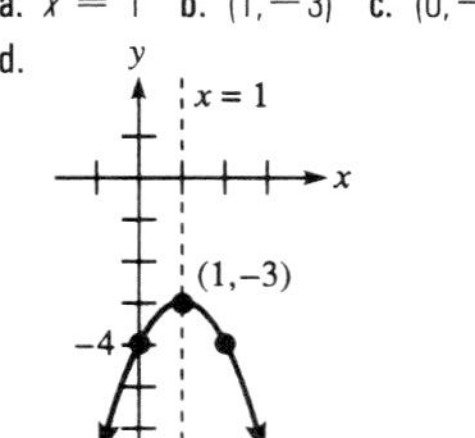

55. $x = y^2 + 8y$ a. $y = -4$ b. $(-16,-4)$
c. $(0,0)$, $(0,-8)$ d.

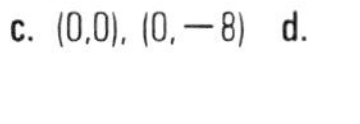
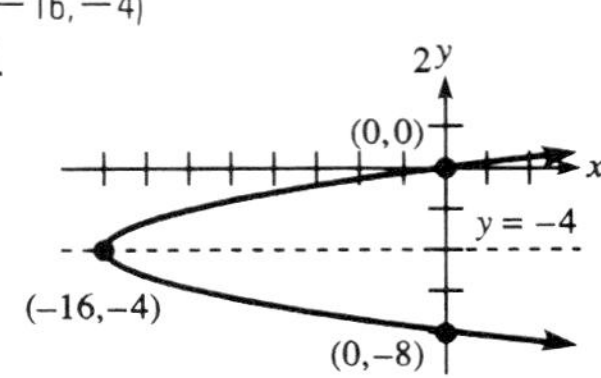

56. Solve for y: $x = y^2 + 6$. $y = \pm\sqrt{x - 6}$

57. Solve for t: $y = -32t^2 + pt + q$. $t = \frac{-p \pm \sqrt{p^2 + 128q - 128y}}{-64}$

Solve each system.

58. $y = x^2 + 2x - 4$
$y = x - 4$ $\{(-1,-5),(0,-4)\}$

59. $x^2 + y^2 = 34$
$y - x = 2$ $\{(3,5),(-5,-3)\}$

Solve using the indicated method.

60. $x^2 - 20 = 0$ (square root property) $\{\pm 2\sqrt{5}\}$

61. $x^2 - 3x - 5 = 0$ (quadratic formula) $\left\{\frac{3-\sqrt{29}}{2}, \frac{3+\sqrt{29}}{2}\right\}$

62. $x^2 - 10x + 16 = 0$ (completing the square) $\{2,8\}$

63. $3x^2 + 7x = 6$ (quadratic formula) $\{-3, \frac{2}{3}\}$

64. $2x^2 + x - 2 = 0$ (completing the square)
$\left\{\frac{-1+\sqrt{17}}{4}, \frac{-1-\sqrt{17}}{4}\right\}$

Solve using any efficient method.

65. $x^2 + 5 = 4x$ $\{2+i, 2-i\}$

66. $x - \sqrt{8 - x^2} = 0$ $\{2\}$

67. $9y^2 - 6y + 1 = 0$ $\{\frac{1}{3}\}$

68. $y^4 - 6y^2 = -5$ $\{-1, 1, -\sqrt{5}, \sqrt{5}\}$

69. $y^2 + 25 = 0$ $\{5i, -5i\}$

70. $3x^{2/3} - 2x^{1/3} - 1 = 0$ $\{-\frac{1}{27}, 1\}$

71. $(y - 4)^2 = 5$ $\{4-\sqrt{5}, 4+\sqrt{5}\}$

72. $1 - \frac{3}{x} = \frac{10}{x^2}$ $\{-2, 5\}$

73. $\frac{x - 3}{x} \leq 0$ $(0,3]$

74. $2x^2 - x - 10 < 0$ $(-2, \frac{5}{2})$

75. $x(x + 1)(2x - 1) > 0$ $(-1,0) \cup (\frac{1}{2}, \infty)$

Classify each equation as determining either a circle, a semicircle, an ellipse, a hyperbola, or a parabola. Graph the equation.

76. $y = 6x - 3x^2$ Parabola

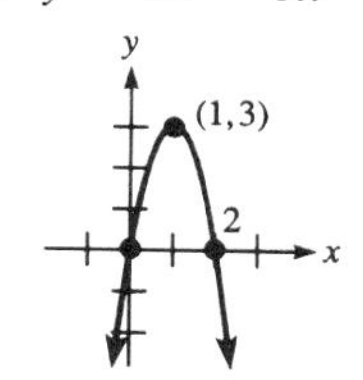

77. $x^2 + y^2 = 9$ Circle

78. $\frac{x^2}{4} - \frac{y^2}{9} = 1$ Hyperbola

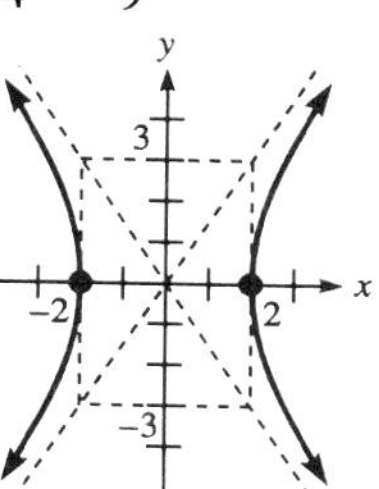

79. $y = x^2 - 25$ Parabola

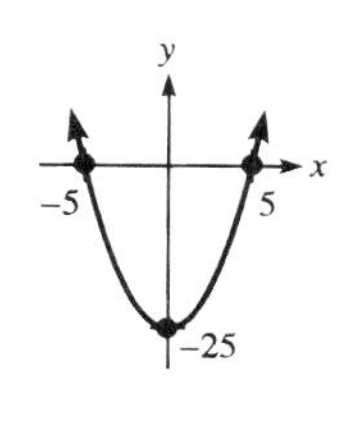

80. $xy = 10$ Hyperbola

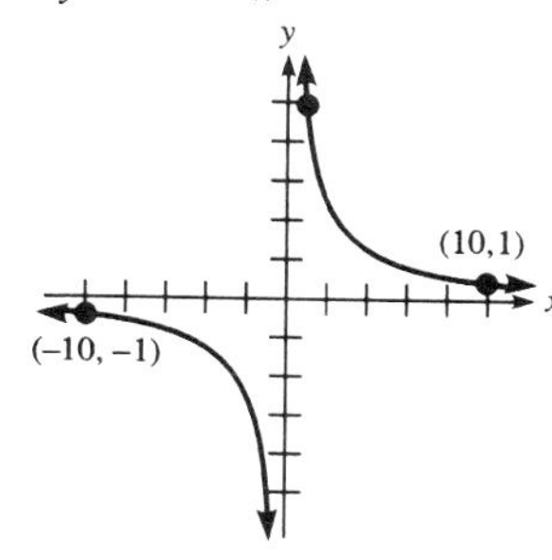

81. $\frac{x^2}{100} + \frac{y^2}{25} = 1$ Ellipse

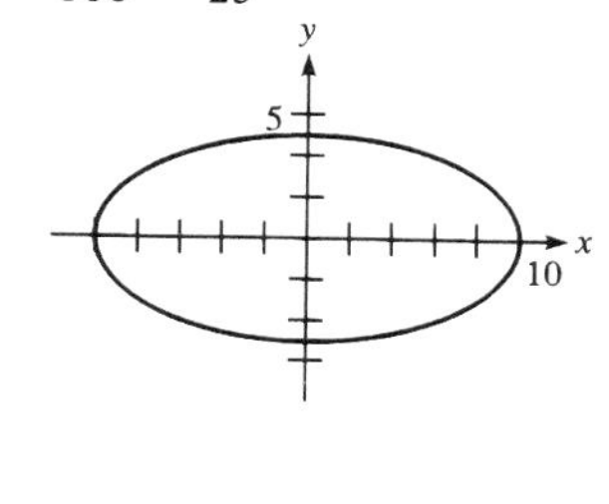

82. $x^2 + 4y^2 = 36$ Ellipse

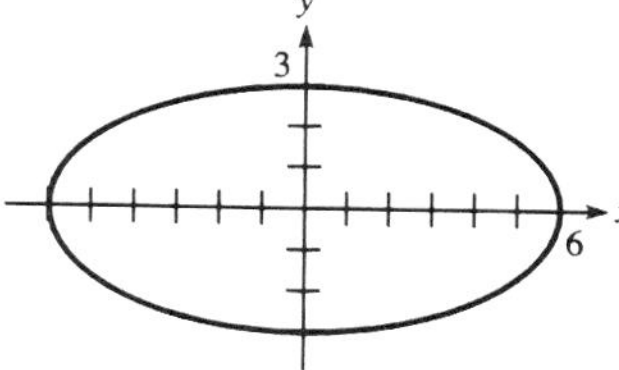

83. $x + 2y^2 + 4y = 0$ Parabola

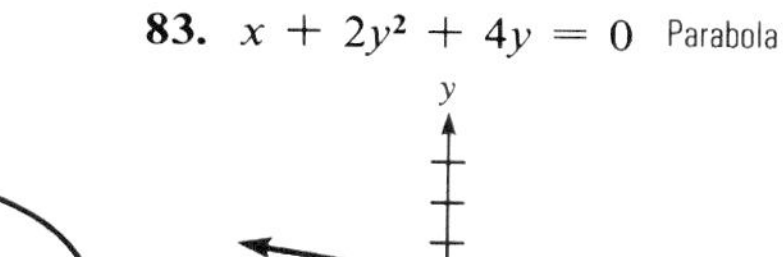

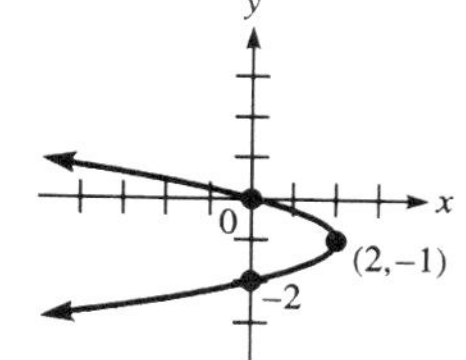

84. $y = \sqrt{16 - x^2}$ Semicircle

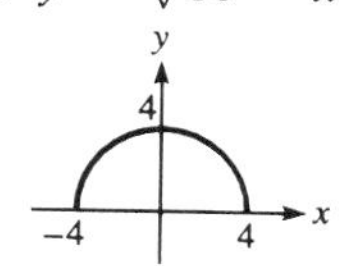

85. $y^2 - x^2 = 16$ Hyperbola

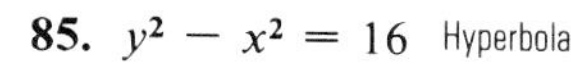

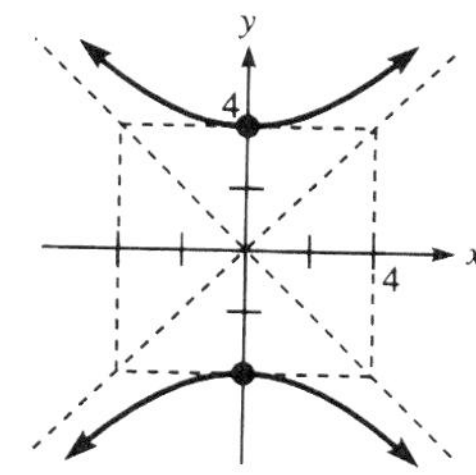

CHAPTER 8 TEST

1. Complete the square for $x^2 - 8x$, and express the resulting trinomial in factored form. $x^2 - 8x + 16 = (x - 4)^2$

2. Use the discriminant to determine the nature of the solutions of the equation $16x^2 + 16 = 32x$. One rational solution

3. Find the center and radius of the circle given by $x^2 + y^2 + 2x - 2y - 3 = 0$. $(-1,1); r = \sqrt{5}$

4. Find the equation in standard form of the circle with center at $(5,1)$ that passes through $(2,-3)$. $(x - 5)^2 + (y - 1)^2 = 25$

5. Consider the equation $y = x^2 - 4x + 3$.

a. Find the equation of the axis of symmetry. $x = 2$

b. Find the coordinates of the vertex. $(2,-1)$

c. Find the coordinates of any x- and y-intercepts. $(1,0), (3,0), (0,3)$

d. Graph the equation.

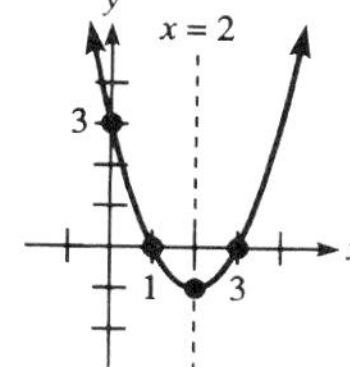

6. Solve $x^2 = y^2 + 7$ for y. $y = \pm\sqrt{x^2 - 7}$
7. Solve the system. $y = 2x + 1$
   $y = x^2 + 3x - 1$ $\{(-2,-3),(1,3)\}$
8. Graph $\frac{x^2}{4} - \frac{y^2}{16} = 1$.

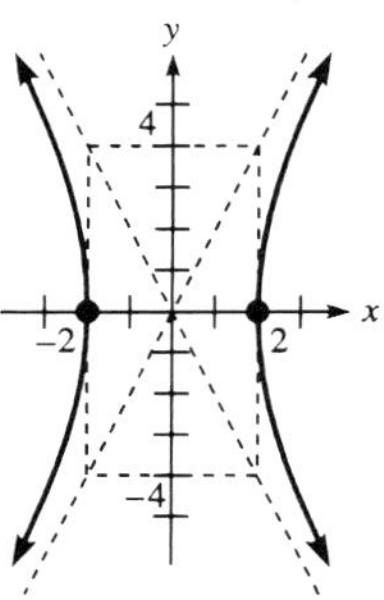

9. Graph $9x^2 + 4y^2 = 36$.

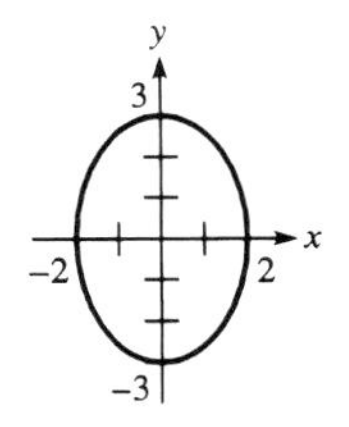

In Questions 10–19, solve the given equation or inequality. Use the method indicated, if any.

10. $x^2 - 4x - 1 = 0$ (quadratic formula) $\{2 + \sqrt{5}, 2 - \sqrt{5}\}$
11. $2y^2 + y - 4 = 0$ (completing the square) $\left\{\frac{-1+\sqrt{33}}{4}, \frac{-1-\sqrt{33}}{4}\right\}$
12. $(x - 6)^2 + 4 = 0$ (square root property) $\{6 + 2i, 6 - 2i\}$
13. $1 - \frac{3}{x} = \frac{4}{x^2}$ $\{-1, 4\}$
14. $3y^2 + 4y - 4 = 0$ $\{\frac{2}{3}, -2\}$
15. $\sqrt{4 - x^2} - x = 0$ $\{\sqrt{2}\}$
16. $y^4 - 8y^2 + 12 = 0$ $\{\sqrt{6}, -\sqrt{6}, \sqrt{2}, -\sqrt{2}\}$
17. $x^2 + 6x > 7$ $(-\infty, -7) \cup (1, \infty)$
18. $x(x + 4)(3x - 1) < 0$ $(-\infty, -4) \cup (0, \frac{1}{3})$
19. $\frac{x + 5}{x - 6} \leq 0$ $[-5, 6)$
20. The height (y) of a projectile that is shot directly up from the ground with an initial velocity of 112 ft/second is given by $y = -16t^2 + 112t$, where y is measured in feet and t is the elapsed time in seconds.
    a. When does the projectile attain its maximum height? 3.5 seconds
    b. What is the maximum height? 196 ft
    c. When does the projectile hit the ground? 7 seconds

CUMULATIVE TEST 8

1. Evaluate $-3^2 + 4|-1 - 2|^3$. 99
2. Solve for T: $R = \frac{1}{S + T}$. $T = \frac{1}{R} - S$, or $T = \frac{1 - RS}{R}$
3. Solve and write the results using interval notation: $-10 < -4t + 2 < 10$. $(-2, 3)$
4. Multiply and simplify: $(2x^2 - 2x + 3)(3x - 4)$. $6x^3 - 14x^2 + 17x - 12$
5. Factor out the greatest common factor: $24x^4y^2 + 18x^3y^3 + 6x^2y^2$. $6x^2y^2(4x^2 + 3xy + 1)$
6. A man weighs 40 lb less than twice the weight of his son. How much do they each weigh if their combined weight is 335 lb? Man 210 lb; son 125 lb
7. Subtract and express in lowest terms: $\frac{4x + 1}{3x - 1} - \frac{2x - 3}{1 - 3x}$. 2
8. Divide and express in lowest terms: $\frac{3 - 3x}{x^2 + x - 2} \div \frac{4}{x^2 + 4x + 4}$. $\frac{-3(x + 2)}{4}$
9. Divide $2x^2 - 7x + 3$ by $x + 4$. Use synthetic division. $2x - 15 + \frac{63}{x + 4}$
10. Write an equation of the line through the origin which is perpendicular to $2x - y = 5$. $y = -\frac{1}{2}x$
11. Suppose that y varies directly as x^2 and inversely as z and that $y = 8$ when $x = 4$ and $z = \frac{1}{2}$. Find y when $x = 2$ and $z = 2$. $y = \frac{1}{2}$
12. The gardening club sold flowering plants for \$5 each and cacti for \$3 each. A total of 80 plants were sold, and total receipts were \$350. How many of each type of plant were sold? 55 flowering plants; 25 cacti
13. Solve by graphing. $y = x - 1$
    $2x - y = 4$

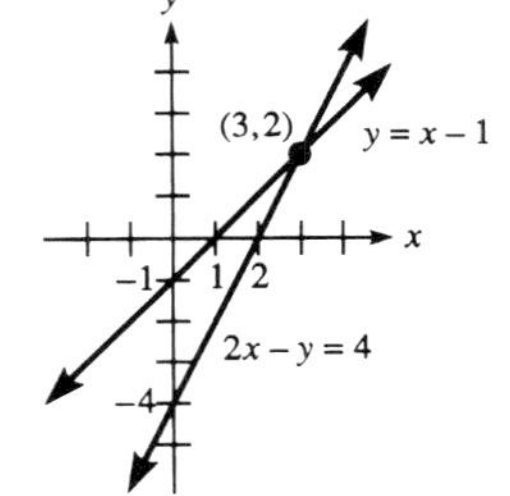

14. Solve the system. $x + 2y + z = 5$
    $-2x + 6y - z = -4$
    $3x - 2y + 2z = 10$ $(3, \frac{1}{2}, 1)$
15. Rationalize the denominator: $\frac{6}{2 - \sqrt{3}}$. $12 + 6\sqrt{3}$
16. Divide and simplify: $\frac{\sqrt[3]{15x^2}}{\sqrt[3]{9x}}$. $\frac{\sqrt[3]{45x}}{3}$
17. Write the quotient $\frac{1 + i}{4 + i}$ in the form $a + bi$. $\frac{5}{17} + \frac{3}{17}i$
18. Solve $x^2 + x - 12 > 0$. $(-\infty, -4) \cup (3, \infty)$
19. Write an equation in standard form of the circle with center at $(1, -5)$ that passes through the point $(-2, -3)$. $(x - 1)^2 + (y + 5)^2 = 13$
20. Identify the graph of the equation $y^2 = -9x^2 + 25$ as a circle, an ellipse, a hyperbola, or a parabola. Ellipse

CHAPTER

9 Functions

A designer makes and sells round tablecloths for \$3 per square foot.

a. Express the cost C of such a tablecloth as a function of its diameter d.

b. What is the cost of a tablecloth that is 6 ft in diameter?

c. Specify the domain of the cost function.

(See Example 10 of Section 9.1.)

IN EARLIER chapters equations, inequalities, and graphs have been used to describe the relationship between two variables. For instance, Chapter 5 opened with a problem where the equation $C = 3n + 4$ revealed how the cost of a certain mail order was related to the number n of items ordered. Analysis of this equation then showed that the graph of this relationship is a straight line. In this chapter we will discuss in more generality the relationship between two variables while concentrating on the language and notation of functions, which are special kinds of mathematical relationships. The function concept will then become a central theme of the remaining chapters.

9.1 Functions

OBJECTIVES

1. **Find the domain and range given a set of ordered pairs.**
2. **Determine if a set of ordered pairs is a function.**
3. **Determine if an equation or graph defines *y* as a function of *x*.**
4. **Find the domain and range given an equation or a graph.**
5. **Find a formula that defines the functional relationship between two variables.**

1 The equation $C = 3n + 4$ provides a rule that reveals the correspondence between values of n and values of C. By using the equation, we can determine a set of ordered pairs that behaves according to the given rule. But it is not necessary for the relationship between two variables to be given by an equation. We will now take a more general approach which begins with the definition of a mathematical relation.

Relation Definition

A **relation** is a set of ordered pairs. The set of all first components of the ordered pairs is called the **domain** of the relation. The set of all second components is called the **range** of the relation.

EXAMPLE 1 In a certain class the correspondence between the final averages and final grades for four students is given by $\{(88,B),(92,A),(71,C),(87,B)\}$. Find the domain and range of this relation.

Solution The domain, which is the set of all first components, is $\{88,92,71,87\}$. The range, which is the set of all second components, is $\{A,B,C\}$.

PROGRESS CHECK 1 Find the domain and the range of the relation $\{(65,D),(100,A),(43,F),(94,A)\}$.

EXAMPLE 2 The relation "less than" in the set $\{2,3,5\}$ is defined by $\{(2,3),(2,5),(3,5)\}$. Find the domain and range of this relation.

Solution The domain of this relation is $\{2,3\}$, and the range is $\{3,5\}$. Note that an ordered pair like (2,5) belongs to this relation because 2 is less than 5.

PROGRESS CHECK 2 The relation "greater than" in the set $\{1,4,9\}$ is defined by $\{(4,1),(9,1),(9,4)\}$. Find the domain and range of this relation.

2 Compare the relations in the first two examples and observe that in Example 1 none of the pairs have the same first component (which customarily represents x). But in Example 2 the pairs (2,3) and (2,5) do have the same first component. Relations in which there is no duplication in the x values are particularly convenient to work with, because each x value leads to only one y value. In other words, the rule for determining y for a specific x always produces exactly one answer. This convenient type of relation is called a function.

The concept of function originated in studies of motion; in words (without algebraic symbolism) the ideas are found in the work of Nicholas Oresmé (1350) and Galileo (1638). Isaac Newton (1670) called relations "fluents." The name *function* was used by Gottfried Leibniz and Johann Bernoulli in the 1690s and was picked up by others.

Progress Check Answers

1. Domain = $\{65,100,43,94\}$; range = $\{A,D,F\}$
2. Domain = $\{4,9\}$; range = $\{1,4\}$

Function

A **function** is a relation in which no two different ordered pairs have the same first component.

Defining functions formally in terms of independent and dependent variables is due to Dirichlet (c. 1850). The definition in terms of ordered pairs (not necessarily numbers) and the distinction between relations and functions are modern developments of abstract algebra.

EXAMPLE 3 Determine if the given relation is a function.

a. $\{(5,3),(6,3),(7,3)\}$ **b.** $\{(4,2),(0,0),(4,-2)\}$

Solution

a. This relation is a function because the first component in the ordered pairs is always different. Note that the definition of a function does not require the second components to be different.

b. This relation is not a function because the number 4 is the first component in more than one ordered pair.

PROGRESS CHECK 3 Determine if the given relation is a function.

a. $\{(-1,1),(0,0),(1,1)\}$ **b.** $\{(-1,5),(-1,6),(-1,7)\}$

3 The examples to this point have only considered relations with a few ordered pairs. However, relations and functions that consist of an infinite set of ordered pairs are common. For example, the solution set of the equation $y = x^2$ is an infinite set of ordered pairs that define the relation

$$\{(x,y): y = x^2\}.$$

Furthermore, this relation is a function, since to each real number input for x there corresponds exactly one y value (which is the square of x). When y is a function of x as just described, the value of y depends on the choice of x, so x is called the **independent variable,** and y is called the **dependent variable.** To work with a relation or function that is an infinite set of ordered pairs, we usually focus on the rule (like $y = x^2$ or its graph) that is used to define the relation, as shown in the remaining examples.

EXAMPLE 4 Does the given rule determine y as a function of x?

a. $y - x = 1$ **b.** $x = y^2$

Solution

a. $y - x = 1$ implies $y = x + 1$. The assignment of a real number to x results in exactly one output for y, so the equation determines y as a function of x. To describe this function verbally, say "for each real number, add 1 to it."

b. $x = y^2$ implies $y = \pm\sqrt{x}$. The equation does not determine y as a function of x because it is possible for an x value to correspond to two different y values. For instance, if $x = 4$, then $y = 2$ or $y = -2$. Since square roots of negative numbers are not real numbers, a verbal description for this rule is "for each nonnegative number, take its positive and its negative square root."

PROGRESS CHECK 4 Does the given rule determine y as a function of x?

a. $x^2 + y^2 = 1$ **b.** $y = x^2$

Because relations are sets of ordered pairs, it is natural to study them by looking at their graphs. The **graph** of a relation is the set of points that correspond to ordered pairs in the relation. Furthermore, it is easy to recognize the graph of a function because none of its points can have the same x-coordinate. Thus, no point in the graph of a function can be directly above any other point. This feature is often summarized in the vertical line test.

Progress Check Answers

3. (a) Yes (b) No
4. (a) No (b) Yes

Remind students that the vertical line test decides whether y is a function of x, not whether x is a function of y.

Vertical Line Test

Imagine a vertical line sweeping across a graph. If the vertical line at any position intersects the graph in more than one point, then the graph is not the graph of a function.

EXAMPLE 5 Use the vertical line test to determine which graphs in Figure 9.1 represent the graph of a function.

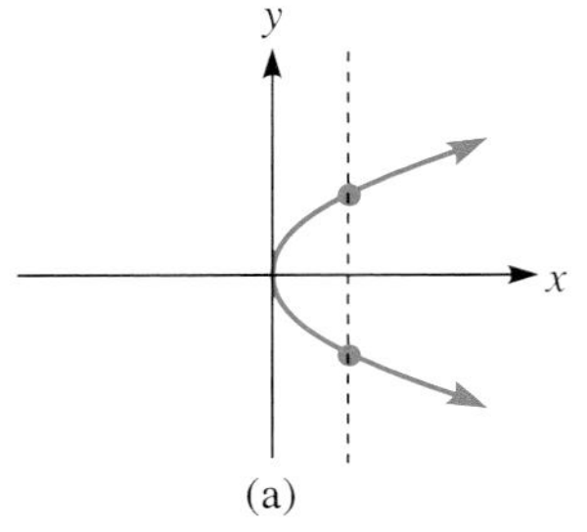

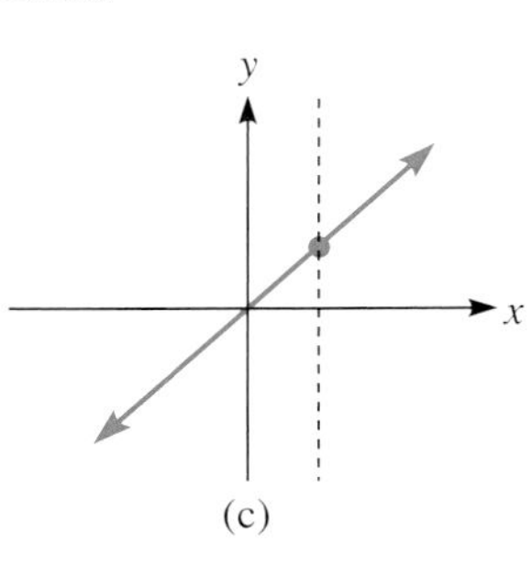

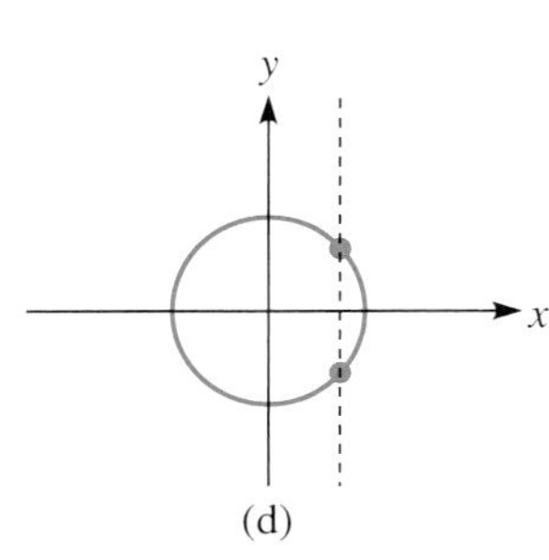

Figure 9.1

Solution By the vertical line test graphs (b) and (c) represent functions, whereas (a) and (d) do not.

PROGRESS CHECK 5 Use the vertical line test to determine which graphs in Figure 9.2 represent the graph of a function.

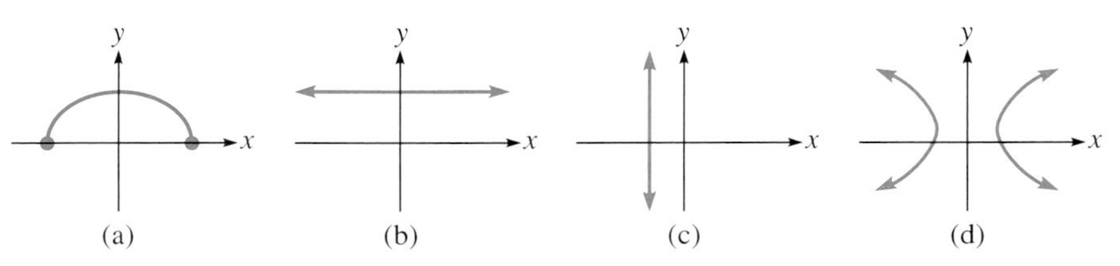

Figure 9.2

The next example combines the methods of Examples 4 and 5 and considers whether certain relations are functions from both the algebraic and the geometric viewpoint.

EXAMPLE 6 Does the given rule determine y as a function of x? Use both algebraic and geometric methods.

a. $y < x$ **b.** $x^2 + y^2 = 25$ **c.** $y = 2$

Solution

a. The inequality $y < x$ does not determine y as a function of x, because to each real number x there corresponds an infinite number of y values that are less than x. In Figure 9.3(a) this result is confirmed graphically by the vertical line test. For example, (3,2), (3,1), and (3,0) all specify points in the graph.

b. The equation $x^2 + y^2 = 25$ does not determine y as a function of x. Its graph is the circle in Figure 9.3(b), and the vertical line test indicates that a circle is never the graph of a function. From an algebraic viewpoint, $x^2 + y^2 = 25$ implies $y = \pm\sqrt{25 - x^2}$, so it is possible for an x value to correspond to two different y values.

Progress Check Answers

5. Function: (a), (b); not a function: (c), (d)

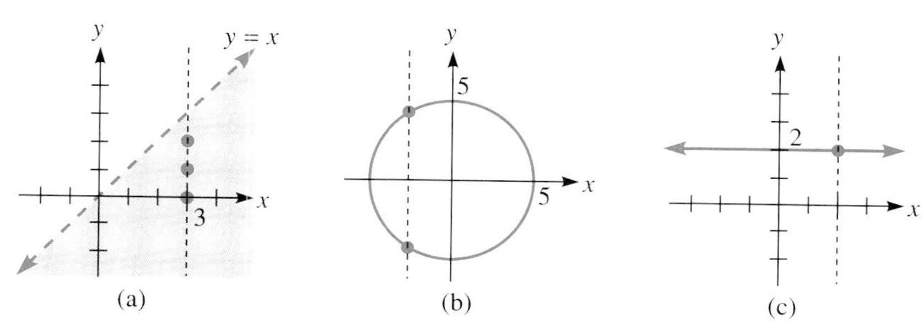

Figure 9.3

c. The equation $y = 2$ graphs as a horizontal line, as shown in Figure 9.3(c). Applying the vertical line test to this graph shows that the equation does determine y as a function of x. This equation is a rule that assigns to each real number x exactly one y value, namely, 2.

PROGRESS CHECK 6 Does the given equation determine y as a function of x? Use both algebraic and geometric methods.

a. $xy = 1$ b. $x = -y^2$ c. $x = 2$

4 Combining algebraic and geometric methods is also useful for finding the domain and range of a relation. Beginning from the geometric viewpoint, the domain is given by the variation of the graph in a horizontal direction, because the domain is the set of all first components in the ordered pairs. The range (which is the set of all second components) is given by the variation of the graph in the vertical direction.

Determining the domain and range from a graph may be new for some students. This skill will be useful throughout Chapters 9 and 10 and is worth stressing at this point.

EXAMPLE 7 Find the domain and range of the relation in Figure 9.4. Is this relation a function?

Solution The x values in the semicircle start at -5, and the graph extends to the right and ends when x equals 5. Thus, the domain is

$$\{x: -5 \le x \le 5\} \quad \text{or} \quad [-5,5] \text{ in interval notation.}$$

The minimum y value is 0, and the graph extends up to a maximum y value of 5. Thus, the range is

$$\{y: 0 \le y \le 5\} \quad \text{or} \quad [0,5] \text{ in interval notation.}$$

Figure 9.5 specifies these results and also indicates that this relation is a function according to the vertical line test.

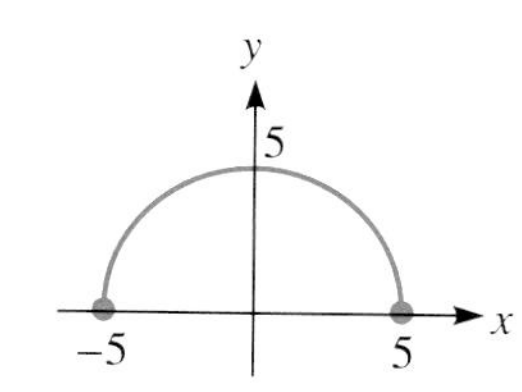

Figure 9.4

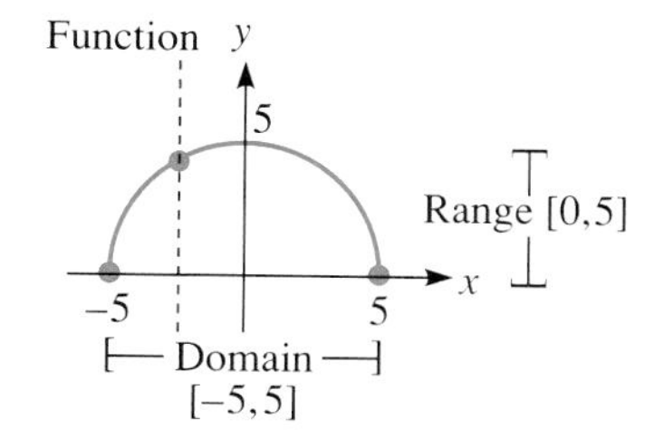

Figure 9.5

PROGRESS CHECK 7 Find the domain and range of the relation in Figure 9.6. Is this relation a function?

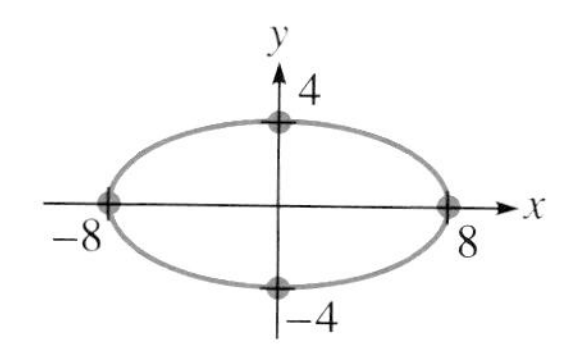

Figure 9.6

Progress Check Answers

6. (a) Yes (b) No (c) No

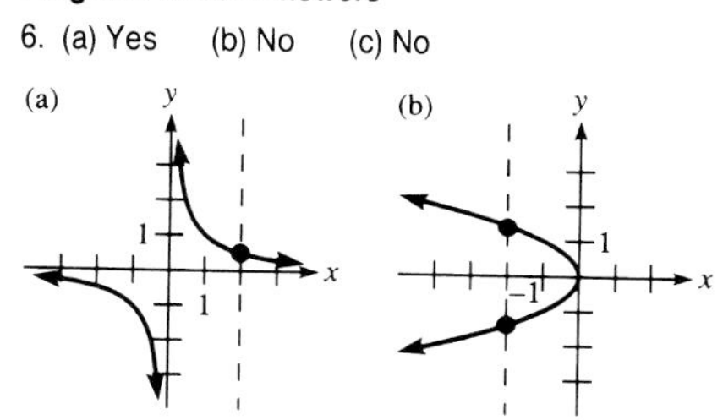

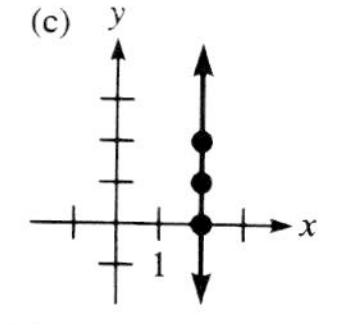

7. Domain: $[-8,8]$; range: $[-4,4]$; no

When a relation is defined by an algebraic equation like $y = 1/x$, then the domain is the set of all real numbers for which a real number exists in the range. Thus, we exclude from the domain values for the independent variable (x) that result in division by zero or in an even root of a negative number, as considered in Example 8.

EXAMPLE 8 Determine the domain and range of each relation.

a. $y = \dfrac{1}{x - 2}$ **b.** $y = \sqrt{x + 4}$

Solution

a. Algebraic Analysis When $x = 2$, $y = 1/0$, which is undefined. Otherwise, the assignment of any real number for x results in a real number output for y. Thus, the domain is the set of all real numbers except 2, which may be written $\{x: x \neq 2\}$. The corresponding y values are all numbers except zero, so the range is $\{y: y \neq 0\}$. Note that y can never be zero because the numerator of the fraction can never be zero.

Geometric Analysis The domain and range may be read from the graph of $y = 1/(x - 2)$ shown in Figure 9.7. This graph is a hyperbola and is a translation of the graph of $y = 1/x$ as considered in Section 8.7. Compare the graphs of $y = 1/x$ and $y = 1/(x - 2)$, and observe that the line $y = 0$ is a horizontal asymptote in both cases, while the vertical asymptotes are the lines $x = 0$ and $x = 2$, respectively.

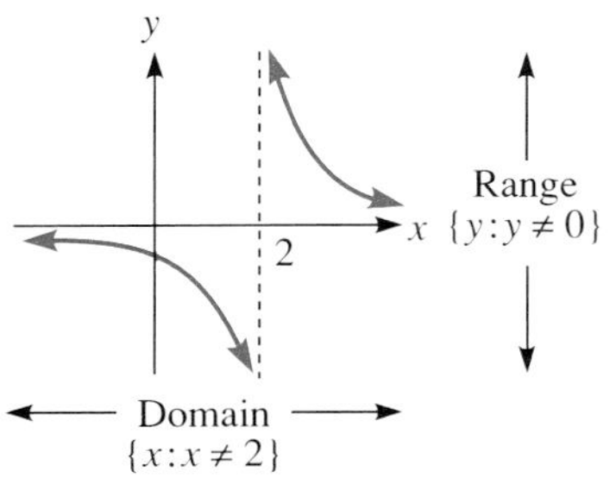

y
2
x
Range $\{y: y \neq 0\}$
Domain $\{x: x \neq 2\}$

Figure 9.7

b. Algebraic Analysis For the output of $y = \sqrt{x + 4}$ to be a real number, x must satisfy

$$\begin{aligned} x + 4 &\geq 0 \\ x &\geq -4. \end{aligned}$$

Thus, the domain is $\{x: x \geq -4\}$, or $[-4,\infty)$ in interval notation. The radical sign $\sqrt{\ }$ denotes the principal square root, so if $x \geq -4$, then $\sqrt{x + 4}$ is greater than or equal to zero. Therefore, the range is $\{y: y \geq 0\}$, or alternatively, $[0,\infty)$.

Geometric Analysis We graph $y = \sqrt{x + 4}$ as shown in Figure 9.8, and we can see from this graph that the domain is $[-4,\infty)$ and the range is $[0,\infty)$. A simple method for obtaining this graph will be shown in Section 9.3. At this time, constructing a table of ordered-pair solutions will produce this graph.

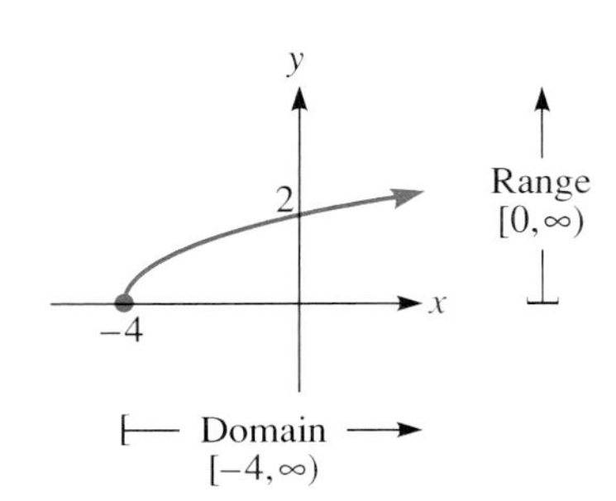

Figure 9.8

PROGRESS CHECK 8 Determine the domain and range of each relation.

a. $y = \dfrac{1}{x + 3}$ **b.** $y = \sqrt{x - 2}$

5 In the solution of a problem it is often necessary to write a formula that shows the relationship between two variables. From the context of the problem the domain is then assumed to be the set of numbers that are meaningful replacements for the independent variable. Example 9 uses the typical wording of such problems.

EXAMPLE 9 Write a rule in equation form that describes the given function. In each case, specify the domain of the function.

a. Express the distance y that a car going 55 mi/hour will travel as a function of t, the hours spent traveling.

b. Express the width w of a rectangle with length 8 units as a function of its area A.

Solution

a. Using $d = rt$, the equation $y = 55t$ expresses the distance y as a function of the time t when $r = 55$. Because t cannot be negative, the domain is the interval $[0,\infty)$.

Progress Check Answers

8. (a) $D: \{x: x \neq -3\}$; $R: \{y: y \neq 0\}$ (b) $D: [2,\infty)$; $R: [0,\infty)$

b. $A = \ell w$ implies $w = A/\ell$. When ℓ equals 8, the formula for w in terms of A is $w = A/8$. Since the area of a rectangle must be positive, the domain is the interval $(0,\infty)$.

PROGRESS CHECK 9 Use the directions given in Example 9.

a. Express the cost C of x gal of gasoline if the gas costs \$1.35/gal.
b. Express the side length s of a square as a function of the area A of the square.

EXAMPLE 10 Solve the problem in the chapter introduction on page 429.

Solution

a. The tablecloth is round, so its area is given by $A = \pi r^2$. To express A as a function of the diameter d, replace r by $d/2$, to obtain

$$A = \pi\left(\frac{d}{2}\right)^2 = \frac{\pi d^2}{4}.$$

Since the cost C is the product of the area and the price per square foot (which is \$3), the cost equation is

$$C = \frac{3\pi d^2}{4}.$$

b. When $d = 6$,

$$C = \frac{3\pi(6)^2}{4} = \$84.82 \text{ (to the nearest cent).}$$

c. The domain of the function is the set of numbers that are sensible replacements for the diameter of the tablecloth. From the mathematical viewpoint, the diameter of a circle must be positive, so the domain is the interval $(0,\infty)$. Practically speaking, other limits may need to be imposed on how large or how small the tablecloth can measure.

PROGRESS CHECK 10 A designer makes and sells square tablecloths for \$4 per square foot.

a. Express the cost C of such a tablecloth as a function of its perimeter P.
b. What is the cost of a tablecloth whose perimeter is 12 ft?
c. Specify the domain of the cost function.

Progress Check Answers

9. (a) $C = 1.35x$; $[0,\infty)$ (b) $s = \sqrt{A}$; $(0,\infty)$

10. (a) $C = \frac{P^2}{4}$ (b) \$36 (c) $(0,\infty)$

EXERCISES 9.1

In Exercises 1–6, find the domain and range for the given set of ordered pairs.

1. Final course averages and course grades for five students: $\{(80,B),(89,B),(85,B),(79,C),(78,C)\}$
Domain = $\{80,89,85,79,78\}$; range = $\{B,C\}$

2. Scores on a quality control checklist and final rating for four appliances: $\{(9.2,E),(8.7,G),(8.2,G),(9.4,E)\}$
Domain = $\{9.2,8.7,8.2,9.4\}$; range = $\{E,G\}$

3. Four points which make a graph: $\{(1,2),(1,3),(1,4),(2,2)\}$
Domain = $\{1,2\}$; range = $\{2,3,4\}$

4. Heights (inches) and weights (pounds) for four children: $\{(36,50),(48,75),(48,80),(35,50)\}$
Domain = $\{36,48,35\}$; range = $\{50,75,80\}$

5. The relation "is less than" in the set $\{-1,0,1\}$ is defined by $\{(-1,0),(-1,1),(0,1)\}$. Domain = $\{-1,0\}$; range = $\{0,1\}$

6. The relation "is less than or equal to" in the set $\{-1,0,1\}$ is defined by $\{(-1,-1),(-1,0),(-1,1),(0,0),(0,1),(1,1)\}$.
Domain = $\{-1,0,1\}$; range = $\{-1,0,1\}$

In Exercises 7–16, determine whether the given relation is a function.

7. $\{(1,1),(2,2),(3,3),(4,4),(5,5)\}$ Yes
8. $\{(-1,-1),(-2,-2),(-3,-3)\}$ Yes
9. $\{(2,0),(2,2),(2,4),(2,5)\}$ No
10. $\{(1,3),(1,5),(1,7),(1,9)\}$ No
11. $\{(-1,-1),(0,-1),(1,-1)\}$ Yes
12. $\{(-2,-\frac{1}{2}),(-\frac{1}{2},2),(-3,-\frac{1}{3})\}$ Yes
13. The relation "is less than" on the set $\{2,3,4\}$ No
14. The relation "is greater than" on the set $\{5,10,15\}$ No
15. The relation "is equal to" on the set $\{1,2\}$ Yes
16. The relation "is not equal to" on the set $\{1,2\}$ Yes

In Exercises 17–46, determine if the given graph or equation determines y as a function of x.

17.
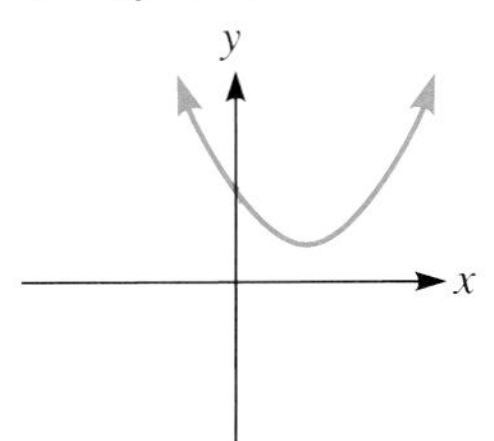
Yes

18.
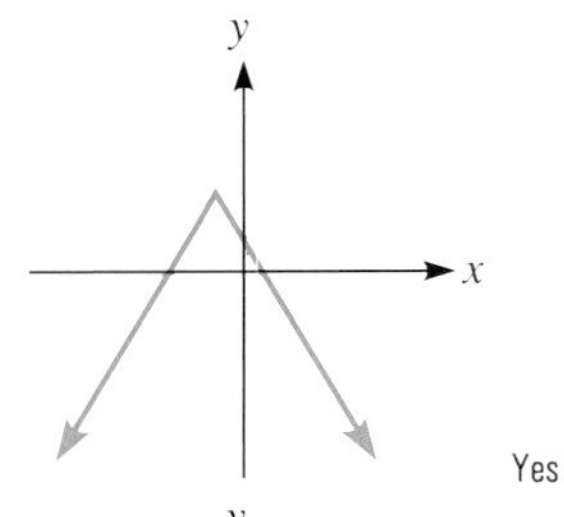
Yes

19.
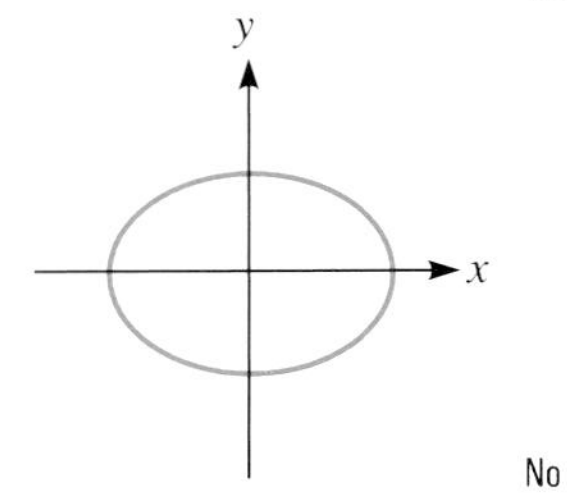
No

20.
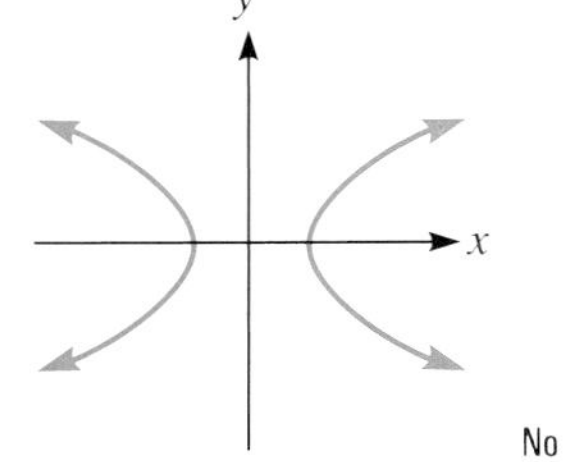
No

21.
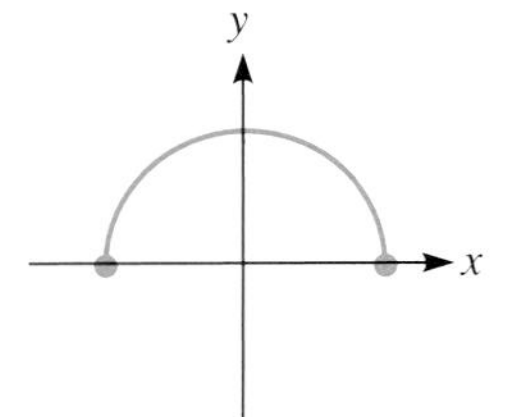
Yes

22.
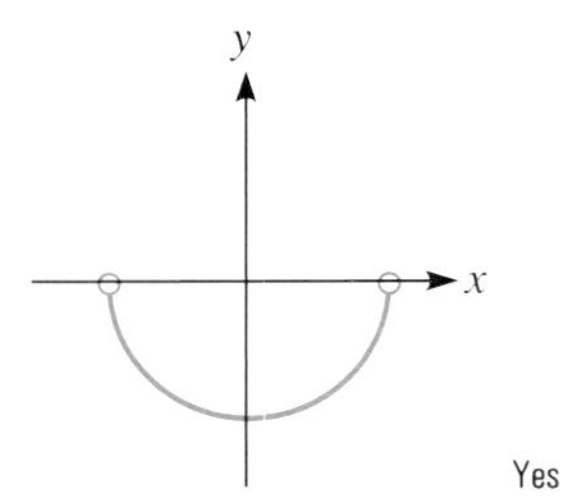
Yes

23.
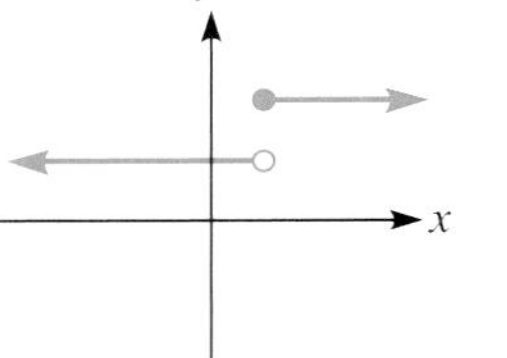
Yes

24.
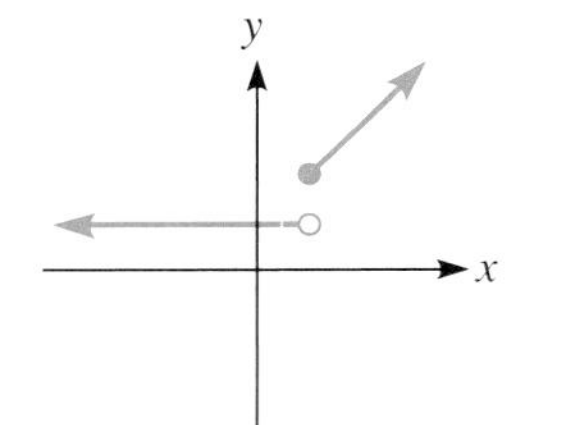
Yes

25.
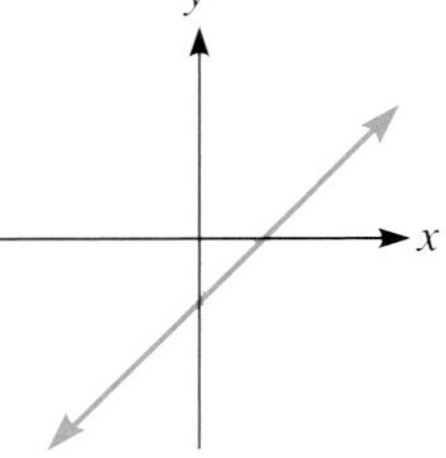
Yes

26.
Yes

27.
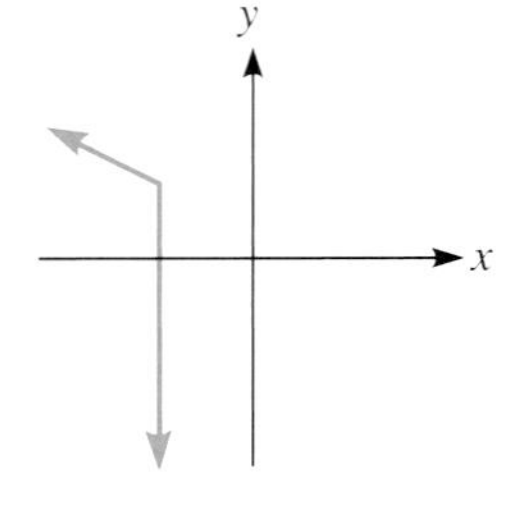
No

28.
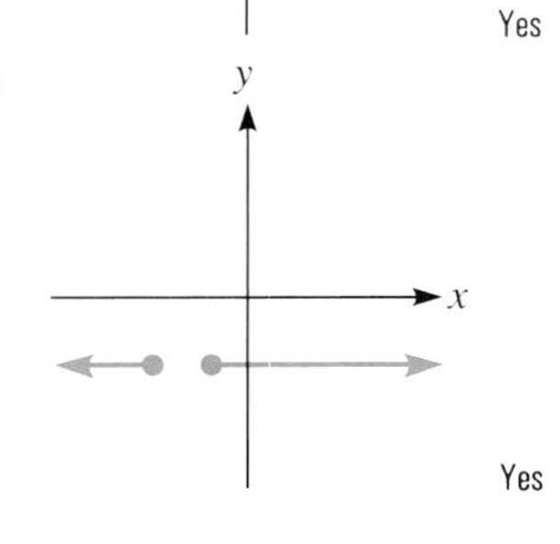
Yes

29. $y = x^2 + 2$ Yes

30. $y = x^2 - 3$ Yes

31. $x = y^2 - 1$ No

32. $x = y^2 + 4$ No

33. $y < x + 1$ No

34. $y > x + 1$ No

35. $y = x - 1$ Yes

36. $x = y - 1$ Yes

37. $x^2 + y^2 = 25$ No

38. $x^2 - y^2 = 5$ No

39. $\dfrac{x^2}{9} + \dfrac{y^2}{16} = 1$ No

40. $\dfrac{x^2}{9} - \dfrac{y^2}{16} = 1$ No

41. $xy = 12$ Yes

42. $xy = -12$ Yes

43. $y = \sqrt{1 - x^2}$ Yes

44. $y = -\sqrt{9 - x^2}$ Yes

45. $x = -\sqrt{4 - y^2}$ No

46. $x = \sqrt{1 - y^2}$ No

In Exercises 47–76, determine the domain and the range of the given relation.

47.
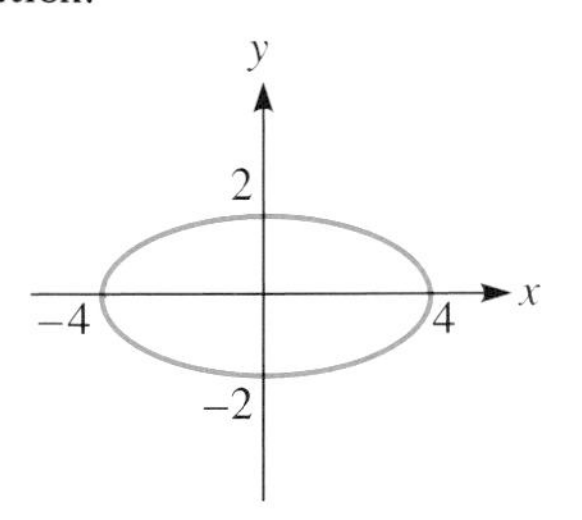
Domain = $[-4,4]$; range = $[-2,2]$

48.
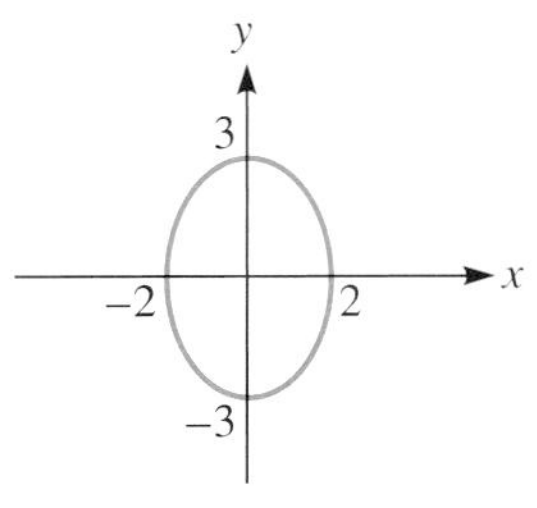
Domain = $[-2,2]$; range = $[-3,3]$

49.
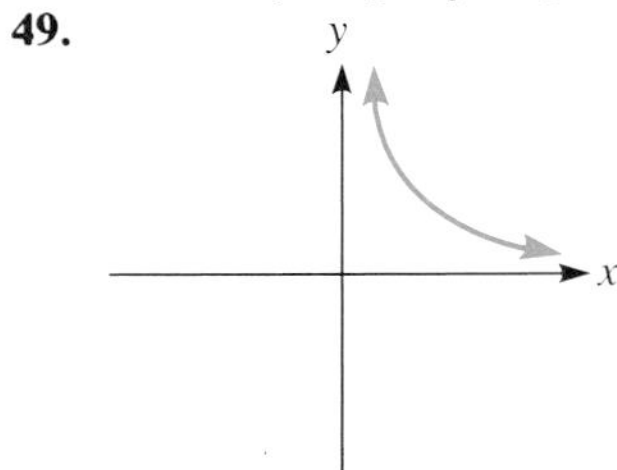
Domain = $(0,\infty)$; range = $(0,\infty)$

50.
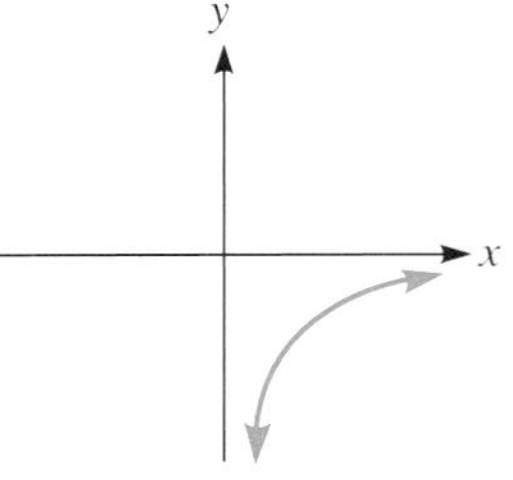
Domain = $(0,\infty)$; range = $(-\infty,0)$

51.
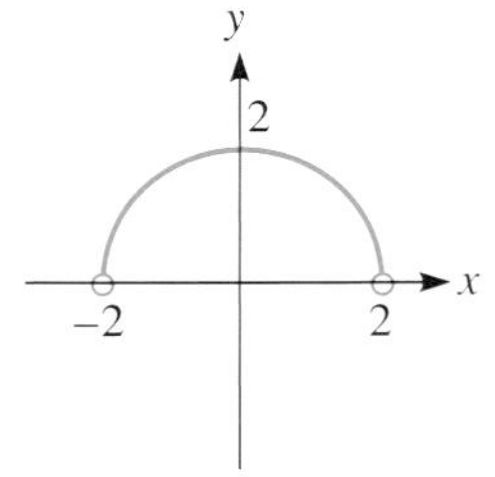
Domain = $(-2,2)$; range = $(0,2]$

52.
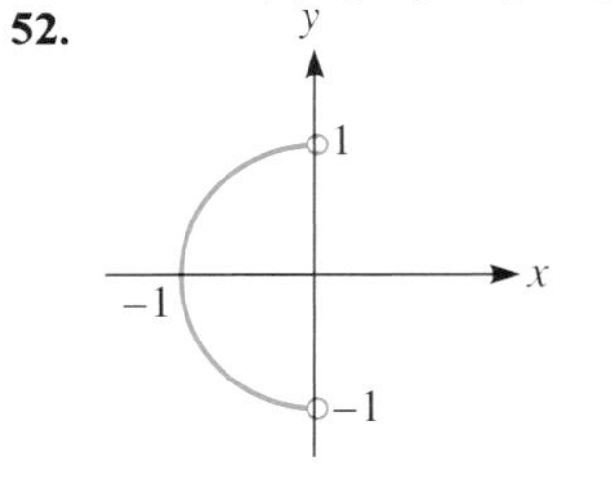
Domain = $[-1,0)$; range = $(-1,1)$

53.
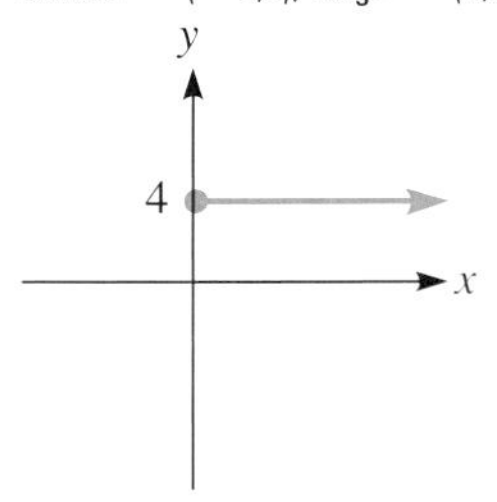
Domain = $[0,\infty)$; range = $\{4\}$

54.
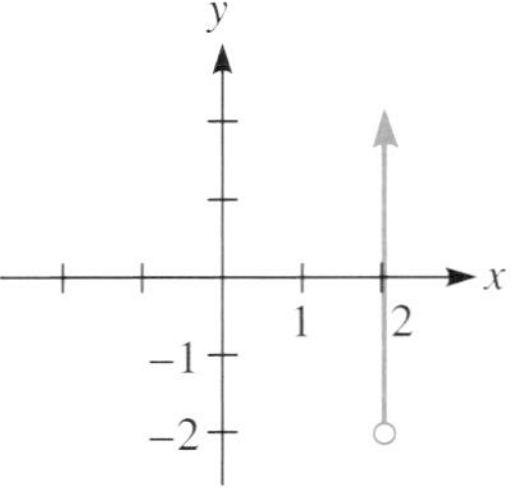
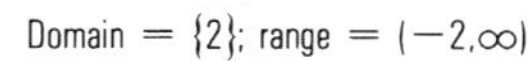
Domain = $\{2\}$; range = $(-2,\infty)$

55.

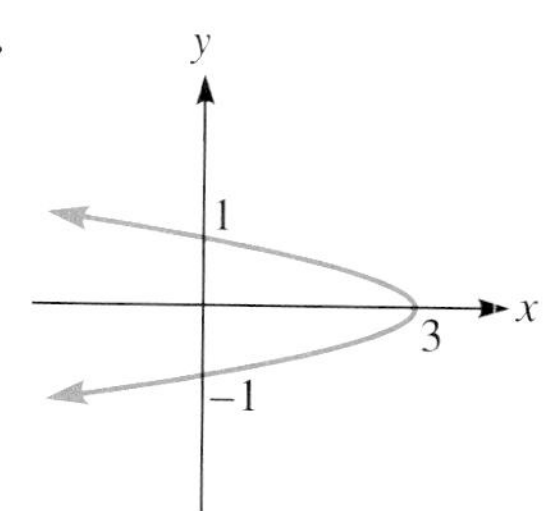

Domain = $(-\infty,3]$; range = $(-\infty,\infty)$

56.

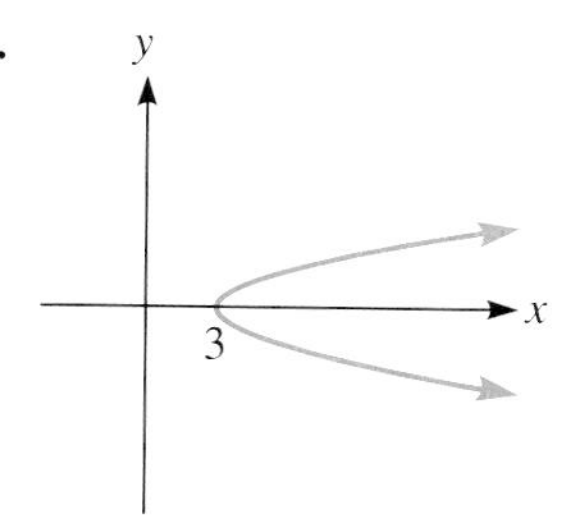

Domain = $[3,\infty)$; range = $(-\infty,\infty)$

57. $y = 2$ Domain = $(-\infty,\infty)$; range = $\{2\}$

58. $y = -3$ Domain = $(-\infty,\infty)$; range = $\{-3\}$

59. $y = 3x + 1$ Domain = $(-\infty,\infty)$; range = $(-\infty,\infty)$

60. $y = -2x + 3$ Domain = $(-\infty,\infty)$; range = $(-\infty,\infty)$

61. $y = \dfrac{x^2}{5}$ Domain = $(-\infty,\infty)$; range = $[0,\infty)$

62. $y = \dfrac{x^2}{3}$ Domain = $(-\infty,\infty)$; range = $[0,\infty)$

63. $y = \dfrac{3}{x^2}$ Domain = $\{x: x \neq 0\}$; range = $(0,\infty)$

64. $y = \dfrac{-4}{x^2}$ Domain = $\{x: x \neq 0\}$; range = $(-\infty,0)$

65. $y = \sqrt{x-2}$ Domain = $[2,\infty)$; range = $[0,\infty)$

66. $y = \sqrt{x+2}$ Domain = $[-2,\infty)$; range = $[0,\infty)$

67. $y = \sqrt{x^2}$ Domain = $(-\infty,\infty)$; range = $[0,\infty)$

68. $y = -\sqrt{x^2}$ Domain = $(-\infty,\infty)$; range = $(-\infty,0]$

69. $y = \dfrac{1}{x+4}$ Domain = $\{x: x \neq -4\}$; range = $\{y: y \neq 0\}$

70. $y = \dfrac{-2}{x-1}$ Domain = $\{x: x \neq 1\}$; range = $\{y: y \neq 0\}$

71. $\dfrac{x^2}{9} - \dfrac{y^2}{4} = 1$ Domain = $(-\infty,-3] \cup [3,\infty)$; range = $(-\infty,\infty)$

72. $\dfrac{y^2}{16} - \dfrac{x^2}{25} = 1$ Domain = $(-\infty,\infty)$; range = $(-\infty,-4] \cup [4,\infty)$

73. $x = \dfrac{1}{y} + 1$ Domain = $\{x: x \neq 1\}$; range = $\{y: y \neq 0\}$

74. $x = \dfrac{1}{y} - 3$ Domain = $\{x: x \neq -3\}$; range = $\{y: y \neq 0\}$

75. $x^2 + y^2 < 1$ Domain = $(-1,1)$; range = $(-1,1)$

76. $x^2 + y^2 \leq 9$ Domain = $[-3,3]$; range = $[-3,3]$

In Exercises 77–84, find a rule in equation form that defines the functional relationship described. For each function, specify the domain.

77. Light travels at about 186,000 mi/second through space.

a. Express the distance in miles (d) traveled by light in t seconds as a function of time (t). $d = 186{,}000t$; domain: $[0,\infty)$

b. It takes about 1.3 seconds for light to reach the moon from the earth. Find the approximate distance from the earth to the moon. 242,000 mi

c. Express the distance in miles (d) traveled by light in t *minutes* as a function of time (t).
$d = 11{,}160{,}000t$; domain: $[0,\infty)$

d. It takes about 8.3 minutes for light to travel from the sun to the earth. Find the approximate distance from the sun to the earth. 93 million mi

78. Sound travels at about 1,100 ft/second in air. There are 5,280 ft per mile.

a. Express the distance in feet (d) traveled by sound in t seconds as a function of time (t). $d = 1{,}100t$; domain: $[0,\infty)$

b. About how far (in feet) are you from the source of thunder if it takes 5 seconds for the sound to reach you?
5,500 ft

c. Express the distance in *miles* (d) traveled by sound in t seconds as a function of time (t). Round to two decimal places. $d = 0.21t$; domain: $[0,\infty)$

d. About how far (in miles) are you from the source of thunder if it takes 6 seconds for the sound to reach you? Round to one decimal place. 1.3 mi

79. A length of 1 m is about equal to 39 in.

a. Express the length of an object in inches (i) as a function of its length in meters (m). $i = 39m$; domain: $[0,\infty)$

b. Find the length in inches of a line which is 0.6 m long.
23.4 in.

c. Express the length of an object in meters (m) as a function of its length in inches (i). Round to three decimal places. $m = \dfrac{i}{39} = 0.026i$; domain: $[0,\infty)$

d. Find the length in meters of a line which is 50 in. long. Round to one decimal place. 1.3 m

80. A weight of 1 lb is about equal to 0.454 kg.

a. Express the weight of an object in kilograms (k) as a function of its weight in pounds (p). $k = 0.454p$; domain: $[0,\infty)$

b. Find the weight in kilograms of a 5-lb box of sugar.
2.270 kg

c. Express the weight of an object in pounds (p) as a function of its weight in kilograms (k). Round to three decimal places. $p = \dfrac{k}{0.454} = 2.203k$; domain: $[0,\infty)$

d. Find the weight in pounds of 5 kg of tobacco. Round to three decimal places. 11.015 lb

81. Ribbon sells for \$3.69 per yard.

a. Express the cost (C) of this ribbon as a function of its length in *feet* (f). $C = 1.23f$; domain: $[0,\infty)$

b. Find the cost of 20 ft of ribbon. \$24.60

82. According to the 1989 *Guinness Book of World Records*, in 1977 wild ginseng from the Chan Pak Mountain area of China sold for \$24,000 per ounce in Hong Kong. (It was thought to have aphrodisiacal properties.)

a. Express the cost (C) of this plant as a function of its weight in *pounds* (p). $C = 384{,}000p$; domain: $[0,\infty)$

b. Find the cost of 5 lb of wild ginseng. \$1,920,000

c. How much ginseng could you buy for \$100?
0.00026 lb = 0.0042 oz

83. **a.** Express the length of the diagonal (d) of a square as a function of the length of the side (s). $d = \sqrt{2}s$; domain: $(0,\infty)$

b. To the nearest tenth, find the length of the diagonal of a square with side 524 mm. 741.0 mm

84. For a certain triangle the altitude is double the base.

a. Express the area (A) of this triangle as a function of its base (b). $A = b^2$

b. Express the area (A) of this triangle as a function of its altitude (a). $A = \dfrac{a^2}{4}$

THINK ABOUT IT

1. Explain the difference between a relation and a function.
2. The relation "is west of" on the set {Denver, San Diego, Boston} is given by what set of ordered pairs? Is this set a function?
3. Write a different function which has the same domain and the same range as the function {(1,5),(2,6)}.
4. For homework a student is assigned *every other odd-numbered* section exercise (starting with 1) in a set of 100 problems.
   a. Is Exercise 67 assigned? What about Exercise 93? Write a formula that shows the student the numbers of the exercises that are assigned. State the domain of the function. Give a verbal description of the rule.
   b. What formula will assign every other *even* exercise starting with 2?
5. A function can be described numerically by a table of ordered pairs. There need not be a formula to show how one variable depends on another, but often a table is derived from some formula. Try to find a formula that expresses y as a function of x that would give the ordered pairs in these tables. There may be more than one correct formula.

a.

x	-2	-1	0	1	2
y	4	1	0	1	4

b.

x	-2	-1	0	1	2
y	-1	1	3	5	7

c.

x	-2	-1	0	1	2
y	3	2	1	2	3

REMEMBER THIS

1. Find the value of y when $x = 3$ in $y = 4x^2 + 1$. 37
2. Evaluate $\dfrac{x^2 - 2x + 10}{x^2 + 1}$ when $x = -4$. 2
3. Approximate $95(1.05)^{10}$ to the nearest integer. 155
4. True or false? $(9 + 16)^{1/2} = 9^{1/2} + 16^{1/2}$. False
5. For what values of x is the y-coordinate in the graph of $y = (x - 3)^2 - 4$ negative? (1,5)

   [number line: open interval from 1 to 5]
6. Solve for a: $b = \dfrac{1}{c + a}$. $a = \dfrac{1}{b} - c$
7. The total mailing weight for two packages is 12.8 lb. One package is 3 times as heavy as the other. What is the difference in their weights? 6.4 lb
8. Divide: $\dfrac{x - 3}{x + 3} \div \dfrac{x^2 - 9}{2x + 6}$. $\dfrac{2}{x + 3}$
9. Rationalize the denominator: $\dfrac{1}{1 - \sqrt{2}}$. $-1 - \sqrt{2}$
10. Is $3x^2 + 4y^2 = 25$ the equation of an ellipse, a hyperbola, a parabola, or none of these? Ellipse

9.2 Functional Notation

A function which approximates the value V of a home purchased for \$95,000 and which increases in value 5 percent per year is

$$V = f(x) = 95{,}000(1.05)^x,$$

where x is the number of years since the house was purchased. Find $f(10)$ and interpret its meaning. (See Example 2.)

OBJECTIVES

1. **Evaluate expressions using functional notation.**
2. **Find function values and graphs of functions defined by more than one equation.**
3. **Read from a graph of function f the domain, range, function values, and values of x for which $f(x) = 0$, $f(x) < 0$, and $f(x) > 0$.**

1 Functional notation is used extensively in mathematics because it provides a very convenient way to refer to the value of the dependent variable that corresponds to a particular value of the independent variable. In this notation a letter like f is used to name a function, and then an equation like

$$y = 4x^2 + 1 \quad \text{is written as} \quad f(x) = 4x^2 + 1.$$

The notation $f(x)$ originated with Euler (1734).

The dependent variable y is replaced by $f(x)$, with the independent variable x appearing in parentheses. The notation $f(x)$ is read "f of x" or "f at x" and means the value of the function (the y value) corresponding to the value of x. Similarly, $f(2)$ means the value of the function when $x = 2$. If $f(x) = 4x^2 + 1$, $f(3)$ is found by substituting 3 for x in the equation.

$$\begin{aligned} f(x) &= 4x^2 + 1 && \text{Given equation.} \\ f(3) &= 4(3)^2 + 1 && \text{Replace } x \text{ by 3.} \\ &= 37 && \text{Simplify.} \end{aligned}$$

The result $f(3) = 37$ says that $y = 37$ when $x = 3$. Note that the notation $f(x)$ does *not* mean f times x.

Reassure students that they will be able to tell from the context if an expression like $g(x)$ refers to function notation or multiplication.

EXAMPLE 1 If $y = f(x) = 5x^2 - x - 1$, find $f(0)$ and $f(-1)$.

Solution Replace x by the number inside the parentheses and then simplify.

$$y_{\text{when } x = 0} = f(0) = 5(0)^2 - 0 - 1 = -1$$
$$y_{\text{when } x = -1} = f(-1) = 5(-1)^2 - (-1) - 1 = 5$$

Thus, $f(0) = -1$, and $f(-1) = 5$.

PROGRESS CHECK 1 If $y = f(x) = 3x^2 - 2$, find $f(4)$ and $f(-1)$.

EXAMPLE 2 Solve the problem in the section introduction on page 438.

Solution $f(10)$ gives V when $x = 10$ years. Using $V = f(x) = 95{,}000(1.05)^x$ gives

$$V_{\text{when } x = 10} = f(10) = 95{,}000(1.05)^{10} = 154{,}744.99.$$

Thus, the value of the house in 10 years is about $155,000 (to the nearest thousand dollars).

PROGRESS CHECK 2 Use $V = f(x) = 95{,}000(1.05)^x$ and find $f(5)$. Interpret the meaning of $f(5)$ in the context of Example 2.

In functional notation it is customary to use the symbols f and x, but other symbols work just as well and are often useful. The notations $f(x) = 3x$, $g(t) = 3t$, and $h(r) = 3r$ all define exactly the same functions if x, t, and r are replaced by the same numbers.

EXAMPLE 3 If $g(x) = \dfrac{x^2 - 2x + 10}{x^2 + 1}$, find $g(-4)$.

Solution To find $g(-4)$, replace all occurrences of x by -4 and simplify.

$$\begin{aligned} g(x) &= \frac{x^2 - 2x + 10}{x^2 + 1} \\ g(-4) &= \frac{(-4)^2 - 2(-4) + 10}{(-4)^2 + 1} && \text{Replace } x \text{ by } -4. \\ &= \frac{16 + 8 + 10}{17} \\ &= 2 \end{aligned}$$

Thus, $g(-4) = 2$.

Progress Check Answers

1. $f(4) = 46$; $f(-1) = 1$
2. $f(5) = 121{,}246.75$; in five years the value of the house is about $121,000.

PROGRESS CHECK 3 If $h(x) = \dfrac{4x^2 + 9x - 5}{x^2 - 5}$, find $h(-2)$.

Students often have trouble interpreting expressions like $3f(-5) + 2g(3)$. Verbal descriptions (as in this example) are helpful.

EXAMPLE 4 If $f(x) = x - 1$ and $g(x) = x^2 + 1$, find $3f(-5) + 2g(3)$.

Solution The given expression means to find the sum of 3 times "f of -5" and 2 times "g of 3." First, we find $f(-5)$ and $g(3)$.

$$\begin{aligned} f(x) &= x - 1 \\ f(-5) &= (-5) - 1 \\ &= -6 \end{aligned} \qquad \begin{aligned} g(x) &= x^2 + 1 \\ g(3) &= (3)^2 + 1 \\ &= 10 \end{aligned}$$

Then,

$$\begin{aligned} 3f(-5) + 2g(3) &= 3(-6) + 2(10) \\ &= 2. \end{aligned}$$

PROGRESS CHECK 4 If $f(x) = x - 1$ and $g(x) = x^2 + 1$, find $4g(2) - 3f(-1)$.

Functional notation is useful for analyzing whether functions have certain properties, as illustrated next.

Stress that $f(a)$ is a y value and a is an x value. Ask students to create examples of functions with certain properties, as in "Think About It" Exercises 1 and 2.

EXAMPLE 5 If $f(x) = x + 2$, show that $f(a + b)$ does not equal $f(a) + f(b)$ for all a and b.

Solution To determine $f(a + b)$, $f(a)$, and $f(b)$, replace all occurrences of x in the function $f(x) = x + 2$ by $a + b$, a, and b, respectively.

$$\begin{aligned} f(a + b) &= a + b + 2 \\ f(a) &= a + 2 \\ f(b) &= b + 2 \end{aligned}$$

Since $a + b + 2 \neq (a + 2) + (b + 2)$, $f(a + b) \neq f(a) + f(b)$.

PROGRESS CHECK 5 If $f(x) = 2x$, show that $f(a + b)$ *does* equal $f(a) + f(b)$ for all a and b.

2 In some problems it is necessary to give the rule for the function by using more than one equation. Different equations are used for different values of x. An everyday example of this type of function is used by the telephone company when determining the charge for certain telephone calls, since different formulas are used depending on the time of day the call is made.

EXAMPLE 6 If $y = f(x) = \begin{cases} 3 \text{ if } x < 1 \\ 2x \text{ if } x \geq 1, \end{cases}$ find the following.

a. $f(-2)$ **b.** $f(1)$ **c.** $f(3)$ **d.** Graph the function.

Solution The top equation is used when x is less than 1, while the bottom equation is used when x is greater than or equal to 1.

a. Since -2 is less than 1, use $f(x) = 3$, so $f(-2) = 3$.
b. Since 1 is greater than or equal to 1, use $f(x) = 2x$, so $f(1) = 2$.
c. Since 3 is greater than or equal to 1, use $f(x) = 2x$, so $f(3) = 6$.
d. If $x < 1$, $y = f(x) = 3$, which graphs as a horizontal line with a constant y value of 3. If $x \geq 1$, $y = f(x) = 2x$, which graphs as a line with ordered pairs (1,2), (2,4), (3,6), and so on. Combining these results gives the graph of the function in Figure 9.9. Note that we draw a solid circle at (1,2) and an open circle at (1,3) to show that (1,2) is part of the graph while (1,3) is not.

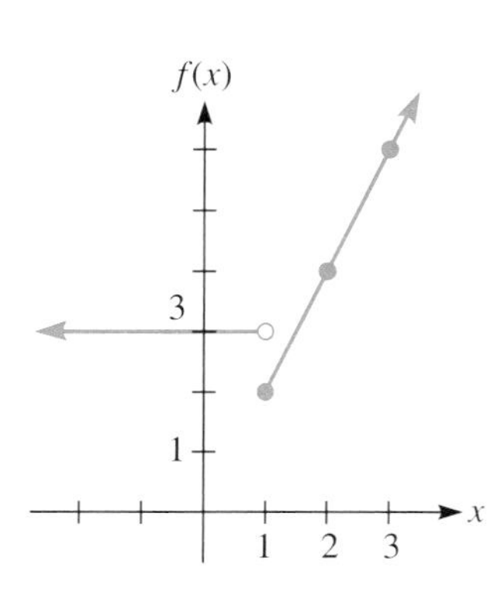

Figure 9.9

Progress Check Answers

3. $h(-2) = 7$
4. 26
5. $2(a + b) = 2a + 2b$

PROGRESS CHECK 6 If $y = f(x) = \begin{cases} -2x \text{ if } x \leq 0 \\ 1 \quad \text{if } x > 0, \end{cases}$ find the following.

a. $f(-3)$ **b.** $f(0)$ **c.** $f(3)$ **d.** Graph the function.

3 From a graph of a function f it is important to be able to read not only the domain and range (as considered in Section 9.1) but also function values and values of x for which $f(x) = 0$, $f(x) < 0$, and $f(x) > 0$.

EXAMPLE 7 Consider the graph of $y = f(x)$ in Figure 9.10.

a. What is the domain of f?
b. What is the range of f?
c. Determine $f(0)$.
d. For what values of x does $f(x) = 0$?
e. For what values of x is $f(x) < 0$?
f. Solve $f(x) > 0$.

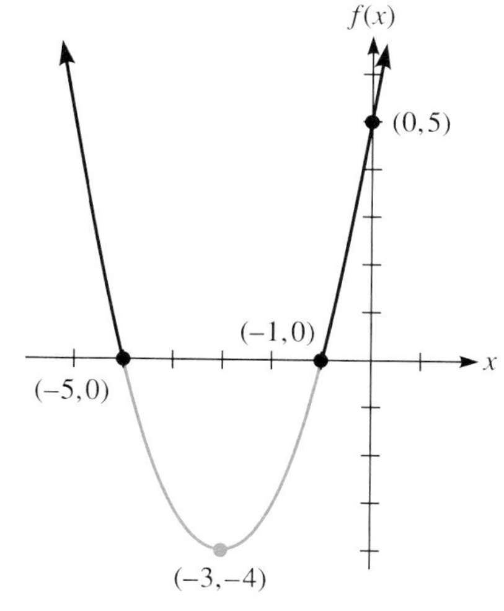

Figure 9.10

Solution

a. The graph is unbounded to the left and to the right, so the domain is the set of all real numbers, or $(-\infty,\infty)$ in interval notation.
b. The minimum y value is -4, and the graph extends indefinitely in the positive y direction, so there is no maximum value. Thus, the range is $\{y: y \geq -4\}$, or $[-4,\infty)$ in interval notation.
c. To determine $f(0)$ requires finding the y value where $x = 0$. Using the ordered pair (0,5) gives $f(0) = 5$.
d. From the ordered pairs $(-5,0)$ and $(-1,0)$, we know $f(x)$ or y equals 0 when $x = -5$ or $x = -1$. The solution set is therefore $\{-5,-1\}$.
e. The y values are less than zero when the graph is below the x-axis. As indicated in color, $f(x) < 0$ for $-5 < x < -1$, so the solution set is the interval $(-5,-1)$.
f. The y values are greater than zero when the graph is above the x-axis. Thus, $f(x) > 0$ when $x < -5$ or $x > -1$, so the solution set in interval notation is $(-\infty,-5) \cup (-1,\infty)$.

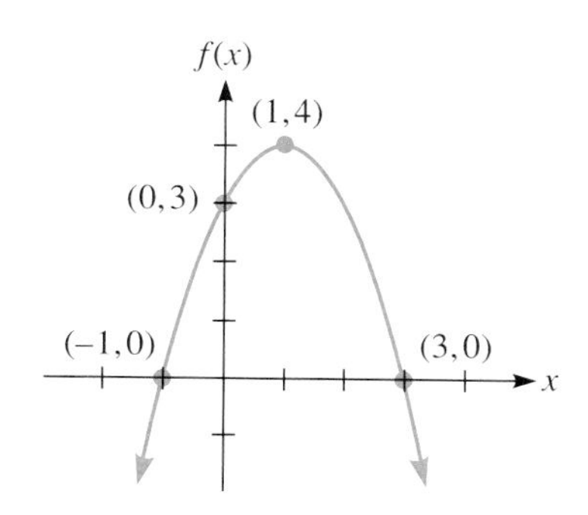

Figure 9.11

PROGRESS CHECK 7 Answer the questions in Example 7 for the graph of $y = f(x)$ in Figure 9.11.

Progress Check Answers

6. (a) 6 (b) 0 (c) 1 (d)

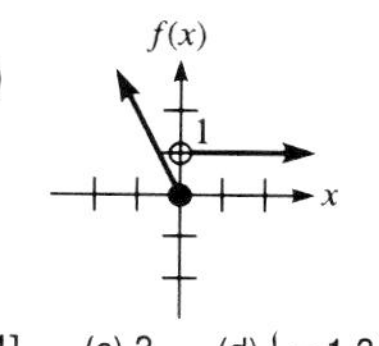

7. (a) $(-\infty,\infty)$ (b) $(-\infty,4]$ (c) 3 (d) $\{-1,3\}$ (e) $(-\infty,-1) \cup (3,\infty)$ (f) $(-1,3)$

EXERCISES 9.2

1. If $y = f(x) = 3x + 2$, find $f(0)$ and $f(-2)$. 2, −4
2. If $y = f(x) = -2x + 5$, find $f(-1)$ and $f(3)$. 7, −1
3. If $y = f(x) = 3x^2 - 5x + 1$, find $f(1)$ and $f(-1)$. −1, 9
4. If $y = f(x) = 7x^2 - 4$, find $f(4)$ and $f(0)$. 108, −4
5. If $y = f(x) = x^2 + 11x + 12$, find $f(0)$ and $f(-1)$. 12, 2
6. If $y = f(x) = -2x^2 - 2x - 2$, find $f(2)$ and $f(-2)$. −14, −6
7. If $y = f(x) = -3x^2 + 4x - 5$, find $f(-1)$ and $f(3)$. −12, −20
8. If $y = f(x) = -x^2 - 10x + 24$, find $f(2)$ and $f(3)$. 0 −15
9. If $y = f(x) = 8x^2 - 7x - 5$, find $f(5)$ and $f(-6)$. 160, 325
10. If $y = f(x) = 56x^2 - 5x - 6$, find $f(1)$ and $f(-2)$. 45, 228
11. If $y = f(x) = 10x^2 - 27x + 18$, find $f(1.5)$ and $f(1.2)$. 0, 0
12. If $y = f(x) = 15x^2 - 29x + 12$, find $f(\frac{4}{3})$ and $f(\frac{3}{5})$. 0, 0
13. A function which approximates the value V of a machine purchased for \$20,000 and which decreases in value 10 percent per year is $V = f(x) = 20{,}000(0.90)^x$, where x is the number of years since the machine was purchased. Find $f(4)$ and $f(7)$ and interpret their meaning. $f(4) = 13{,}122$; in 4 years the value will be about \$13,100; $f(7) = 9{,}565.94$; in 7 years the value will be about \$9,600.
14. If a machine that costs \$17,000 initially depreciates 6 percent per year thereafter, then its value V after x years is approximated by the function $V = f(x) = 17{,}000(0.94)^x$, where x is the number of years since the original purchase. Find $f(5)$ and $f(10)$ and interpret their meaning. $f(5) = 12{,}476.37$; in 5 years the value will be about \$12,500; $f(10) = 9{,}156.46$; in 5 years the value will be about \$9,200.

15. You are about to throw out an old lamp when a friend tells you it is now "highly collectible." An appraiser says the value of the lamp has increased about 15 percent per year and gives you the function $V = f(x) = (\text{original cost})(1.15)^x$ for its value, where x is the number of years since it was purchased. If your grandmother originally paid \$45 for it, what is it worth today (50 years later)? That is, find and interpret $f(50)$.
$f(50) = 48{,}764.58$; the current value is about \$48,800.

16. If the value of a particular work of art is given by the function $V = f(x) = 40{,}000(1.08)^x$, where x is the number of years since its purchase at \$40,000, find and interpret $f(12)$.
$f(12) = 100{,}726.80$; 12 years after purchase the value is about \$100,700.

17. The height y above the water in meters of a diver t seconds after stepping off a diving tower 10 m high is given by the function $y = f(t) = -4.9t^2 + 10$. Find and interpret $f(1)$ and $f(1.42)$. $f(1) = 5.1$; 1 second after stepping off, the diver is 5.1 m above the water; $f(1.42) = 0.12$; 1.42 seconds after stepping off, the diver is 0.12 m above the water (about to hit the water).

18. The height y above the water in meters of a diver t seconds after stepping off a diving tower 20 m high is given by the function $y = f(t) = -4.9t^2 + 20$. Find and interpret $f(1)$ and $f(2.02)$. $f(1) = 15.1$; 1 second after stepping off, the diver is 15.1 m above the water; $f(2.02) = 0.006$; 2.02 seconds after stepping off, the diver is 0.006 m above the water (about to hit the water).

19. During exercise a person's maximum target heart rate is a function of age. The recommended number of beats per minute is given by $y = f(x) = -0.85x + 187$, where x represents age in years and y represents the recommended number of beats per minute. Find and interpret $f(20)$ and $f(40)$. $f(20) = 170$; the recommended maximum heart rate for a 20-year-old is 170 beats per minute; $f(40) = 153$; the recommended maximum heart rate for a 40-year-old is 153 beats per minute.

20. During exercise a person's minimum target heart rate to have a training effect is a function of age. The recommended number of beats per minute is given by $y = f(x) = -0.7x + 154$, where x represents age in years and y represents the recommended number of beats per minute. Find and interpret $f(20)$ and $f(40)$. $f(20) = 140$; the recommended minimum heart rate for a 20-year-old is 140 beats per minute; $f(40) = 126$; the recommended minimum heart rate for a 40-year-old is 126 beats per minute.

21. If $g(x) = \dfrac{x^2 + 4x + 13}{x^2 + 1}$, find $g(-3)$. 1

22. If $h(x) = \dfrac{2x^2 - x - 6}{x^2 - 6}$, find $h(-2)$. -2

23. If $h(x) = \dfrac{x^2 - 8x - 26}{x^2 + 5}$, find $h(4)$. -2

24. If $g(x) = \dfrac{x^2 + 2x - 17}{x^2 - 3}$, find $g(7)$. 1

25. If $g(x) = \dfrac{x^2 + 2x + 3}{x^2 + x + 1}$, find $g(1)$. 2

26. If $g(x) = \dfrac{x^2 + 2x + 3}{x^2 + x + 1}$, find $g(-1)$. 2

27. If $f(x) = \dfrac{3x^2 - 7x + 2}{2x^2 - 2x + 1}$, find $f(-3)$. 2

28. If $f(x) = \dfrac{7x^2 - 10x - 8}{3x^2 - 12}$, find $f(5)$. $\frac{13}{7}$

29. If $h(x) = \dfrac{x^2 + 3x - 2}{x^2 - 1}$, find $h\left(\dfrac{1}{3}\right)$. 1

30. If $g(x) = \dfrac{x^2 + x + 1}{x^2 - 1}$, find $g\left(-\dfrac{1}{2}\right)$. -1

31. If $f(x) = x + 3$ and $g(x) = x^2$, find $2f(-4) + 3g(2)$. 10

32. If $f(x) = x + 3$ and $g(x) = x^2$, find $3f(5) + 4g(-3)$. 60

33. If $f(x) = x - 6$ and $g(x) = x^2 - 3$, find $4f(7) - 3g(-5)$. -62

34. If $f(x) = x - 6$ and $g(x) = x^2 - 3$, find $8f(-8) + 5g(-6)$. 53

35. If $f(x) = 2x + 3$ and $g(x) = 3x^2 - 2$, find $5f(-4) + 2g(3)$. 25

36. If $f(x) = 3x + 2$ and $g(x) = 2x^2 - 3$, find $6f(-2) + 7g(1)$. -31

37. If $f(x) = 3x^2 - 7$ and $g(x) = 5x + 9$, find $5f(4) - 8g(3)$. 13

38. If $f(x) = 3x^2 - 7$ and $g(x) = 5x + 9$, find $-2f(-6) - 9g(-6)$. -13

39. If $f(x) = x^2 - 2x + 10$ and $g(x) = x^2 + 1$, find $3f(-4) - 2g(3)$. 82

40. If $f(x) = x^2 - 2x + 10$ and $g(x) = x^2 + 1$, find $7f(5) - 5g(7)$. -75

41. If $f(x) = x - 3$, show that $f(a - b) \neq f(a) - f(b)$ for all values of a and b. $a - b - 3 \neq (a - 3) - (b - 3)$

42. If $f(x) = x + 4$, show that $f(a + b) \neq f(a) + f(b)$ for all values of a and b. $a + b + 4 \neq (a + 4) + (b + 4)$

43. If $f(x) = 3x$, show that $f(a - b) = f(a) - f(b)$ for all values of a and b. $3(a - b) = 3a - 3b$

44. If $f(x) = -12x$, show that $f(a + b) = f(a) + f(b)$ for all values of a and b. $-12(a + b) = -12a + (-12b)$

45. If $f(x) = x^2$, show that $f(a + b) \neq f(a) + f(b)$ for all values of a and b. $(a + b)^2 \neq a^2 + b^2$

46. If $f(x) = x^2$, show that $f\left(\dfrac{a}{b}\right) = \dfrac{f(a)}{f(b)}$ for all values of a and b except $b = 0$. $\left(\dfrac{a}{b}\right)^2 = \dfrac{a^2}{b^2}$, $b \neq 0$

47. If $y = f(x) = \begin{cases} -1 \text{ if } x < 2 \\ 3x \text{ if } x \geq 2, \end{cases}$ find the following.

a. $f(-2)$ **b.** $f(2)$ **c.** $f(3)$ **d.** Graph the function.

a. -1 b. 6 c. 9 d.

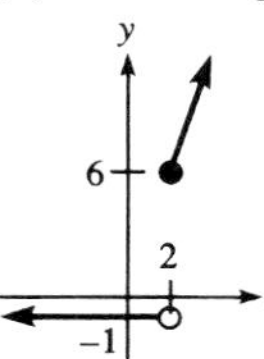

48. If $y = f(x) = \begin{cases} 2 \text{ if } x \geq 3 \\ -x \text{ if } x < 3, \end{cases}$ find the following.

a. $f(1)$ **b.** $f(3)$ **c.** $f(4)$ **d.** Graph the function.

a. -1 b. 2 c. 2 d.

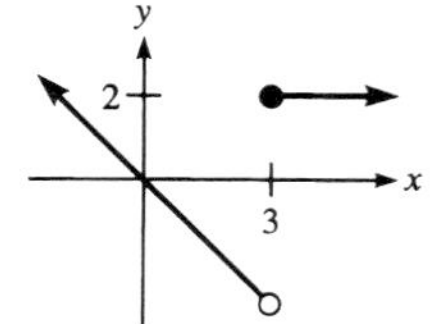

49. If $y = f(x) = \begin{cases} 3 \text{ if } x \geq 0 \\ 4x \text{ if } x < 0, \end{cases}$ find the following.

a. $f(2)$ **b.** $f(0)$ **c.** $f(-2)$ **d.** Graph the function.

a. 3 b. 3 c. -8 d.

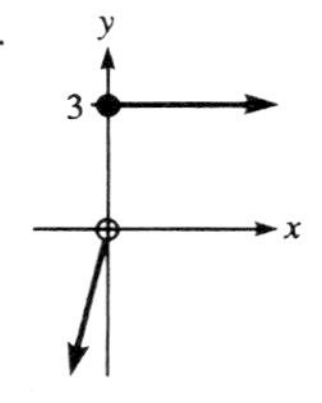

50. If $y = f(x) = \begin{cases} \frac{1}{2}x \text{ if } x \geq 0 \\ 0 \text{ if } x < 0, \end{cases}$ find the following.

a. $f(\frac{1}{2})$ **b.** $f(0)$ **c.** $f(-\frac{1}{2})$ **d.** Graph the function.

a. $\frac{1}{4}$ b. 0 c. 0 d.

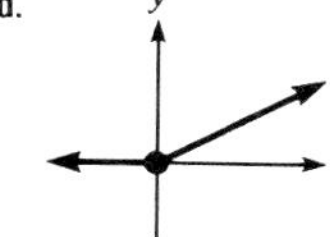

51. If $y = f(x) = \begin{cases} x \text{ if } x \geq 2 \\ 1 \text{ if } x < 2, \end{cases}$ find the following.

a. $f(1)$ **b.** $f(2)$ **c.** $f(3)$ **d.** Graph the function.

a. 1 b. 2 c. 3 d.

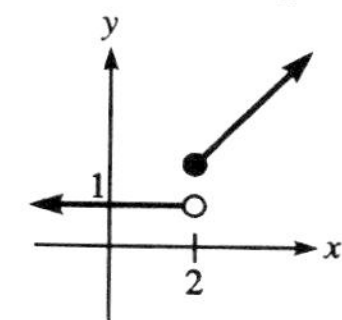

52. If $y = f(x) = \begin{cases} -\frac{1}{2}x \text{ if } x \leq -2 \\ 4 \text{ if } x > -2, \end{cases}$ find the following.

a. $f(-3)$ **b.** $f(-2)$ **c.** $f(-1)$ **d.** Graph the function.

a. 1.5 b. 1 c. 4 d.

53. If $y = f(x) = \begin{cases} -\frac{1}{2}x \text{ if } x \leq -2 \\ 1 \text{ if } x > -2, \end{cases}$ find the following.

a. $f(-4)$ **b.** $f(-2)$ **c.** $f(1)$ **d.** Graph the function.

a. 2 b. 1 c. 1 d.

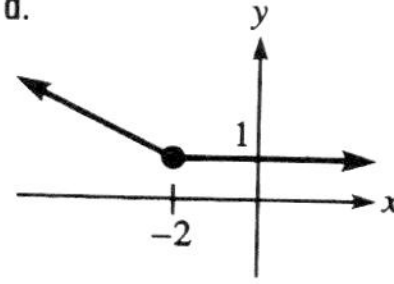

54. If $y = f(x) = \begin{cases} -x + 3 \text{ if } x \geq 3 \\ -x \text{ if } x < 3, \end{cases}$ find the following.

a. $f(2)$ **b.** $f(3)$ **c.** $f(4)$ **d.** Graph the function.

a. −2 b. 0 c. −1 d.

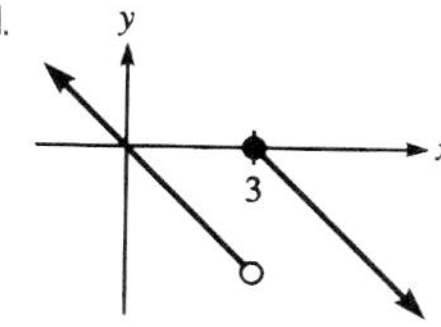

55. If $y = f(x) = \begin{cases} -x \text{ if } x < -3 \\ x \text{ if } x \geq 3, \end{cases}$ find the following.

a. $f(-4)$ **b.** $f(0)$ **c.** $f(3)$ **d.** $f(4)$

e. Graph the function.

a. 4 b. Undefined c. 3 d. 4 e.

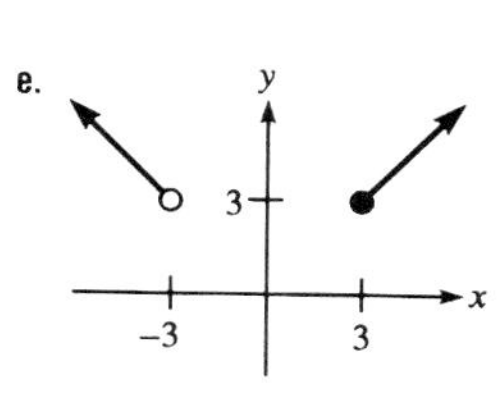

56. If $y = f(x) = \begin{cases} 2x \text{ if } x > 3 \\ 2 \text{ if } x \leq 2, \end{cases}$ find the following.

a. $f(1)$ **b.** $f(2.5)$ **c.** $f(3)$ **d.** $f(4)$

e. Graph the function.

a. 2 b. Undefined c. Undefined d. 8 e.

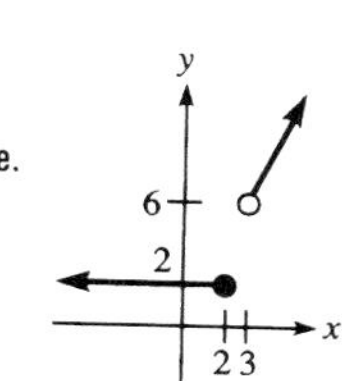

For each graph given in Exercises 57–64, answer questions **a** through **f.**

a. What is the domain of f?
b. What is the range of f?
c. Determine $f(0)$.
d. For what values of x does $f(x) = 0$?
e. For what values of x is $f(x) < 0$?
f. Solve $f(x) > 0$.

57.

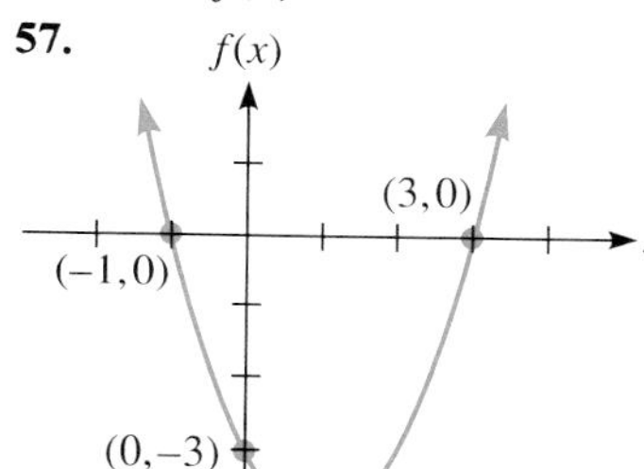

a. $(-\infty,\infty)$
b. $[-4,\infty)$
c. -3
d. $\{-1,3\}$
e. $(-1,3)$
f. $(-\infty,-1) \cup (3,\infty)$

58.

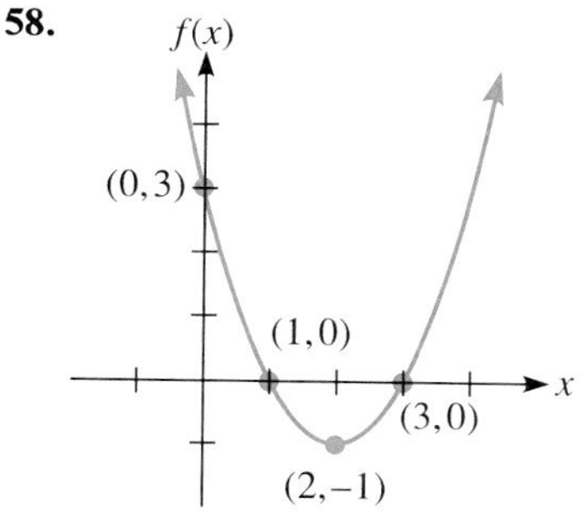

a. $(-\infty,\infty)$
b. $[-1,\infty)$
c. 3
d. $\{1,3\}$
e. $(1,3)$
f. $(-\infty,1) \cup (3,\infty)$

59.

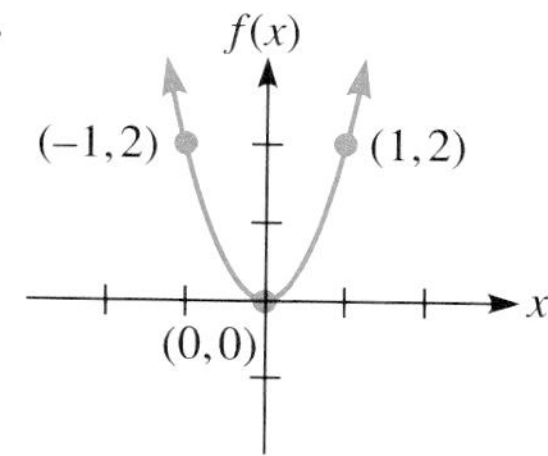

a. $(-\infty,\infty)$
b. $[0,\infty)$
c. 0
d. $\{0\}$
e. $\emptyset$
f. $(-\infty,0) \cup (0,\infty)$

60.

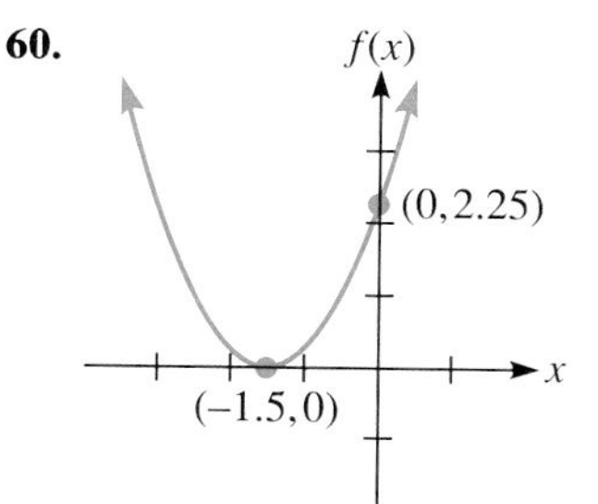

a. $(-\infty,\infty)$
b. $[0,\infty)$
c. 2.25
d. $\{-1.5\}$
e. $\emptyset$
f. $(-\infty,-1.5) \cup (-1.5,\infty)$

61.

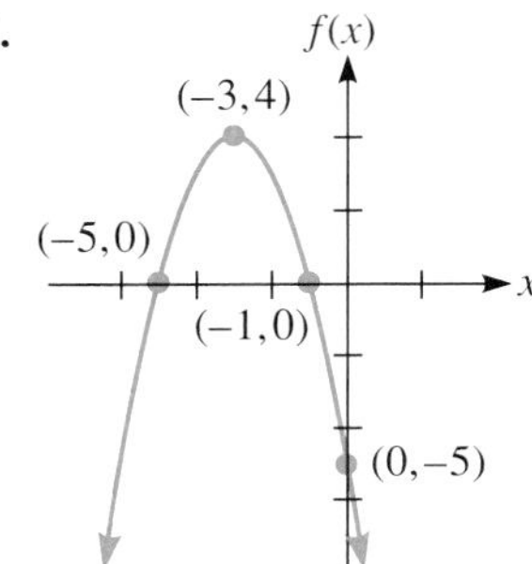

a. $(-\infty,\infty)$
b. $(-\infty,4]$
c. -5
d. $\{-5,-1\}$
e. $(-\infty,-5) \cup (-1,\infty)$
f. $(-5,-1)$

62.

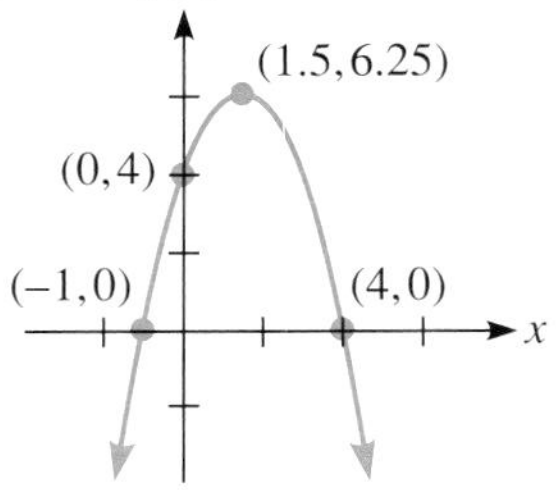

f(x)
(1.5,6.25)
(0,4)
(−1,0)
(4,0)
x

a. $(-\infty,\infty)$
b. $(-\infty,6.25]$
c. 4
d. $\{-1,4\}$
e. $(-\infty,-1) \cup (4,\infty)$
f. $(-1,4)$

63.

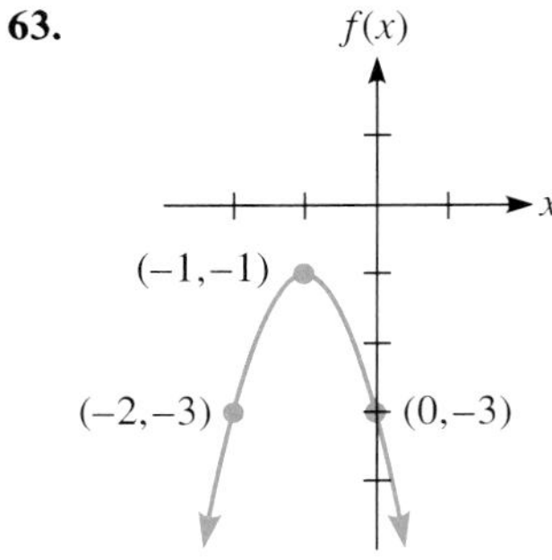

a. $(-\infty,\infty)$
b. $(-\infty,-1]$
c. -3
d. $\emptyset$
e. $(-\infty,\infty)$
f. $\emptyset$

64.

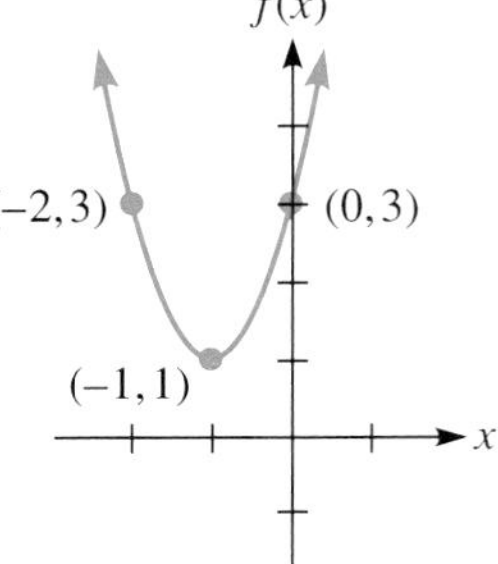

a. $(-\infty,\infty)$
b. $[1,\infty)$
c. 3
d. $\emptyset$
e. $\emptyset$
f. $(-\infty,\infty)$

THINK ABOUT IT

1. **a.** Give two examples of functions where $f(a + b)$ equals $f(a) + f(b)$ for all a and b.
   b. Give two examples of functions where $f(a + b)$ does not equal $f(a) + f(b)$ for all a and b.
2. **a.** Give two examples of functions where $f(-a) = f(a)$ for all a.
   b. Give two examples of functions where $f(-a) = -f(a)$ for all a.
3. Denote $f[f(a)]$ by $f^{(2)}(a)$, and $f(f[f(a)])$ by $f^{(3)}(a)$.
   a. If $f(x) = x^2$, what is $f^{(3)}(a)$?
   b. If $f(x) = -x$, what is $f^{(3)}(a)$?
   c. If $f(n) = (-1)^n$, what is $f^{(2)}(100)$?
4. **a.** Let $f(x) = 3x + 2$ and let $g(x) = (x - 2)/3$. Find $g[f(5)]$; $g[f(-8)]$; $g[f(a)]$; $f[g(5)]$; $f[g(-8)]$; $f[g(a)]$.
   b. If $f(x) = 4x + 7$, find $g(x)$ so that $f[g(a)] = a$ and $g[f(a)] = a$ for all a.
5. Consider the given graph of $y = f(x)$.
   a. What is the domain of f?
   b. What is the range of f?
   c. Determine $f(c)$.
   d. Solve $f(x) = 0$.
   e. Solve $f(x) < 0$.
   f. Solve $f(x) \geq 0$.
   g. Solve $f(x) = b$.

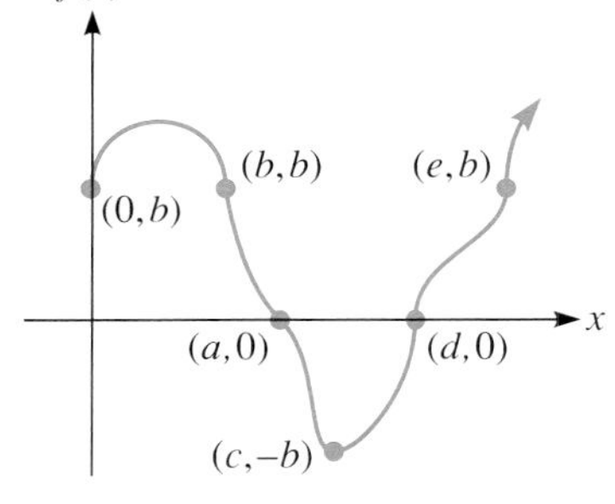

REMEMBER THIS

1. Graph $y = x^2$.

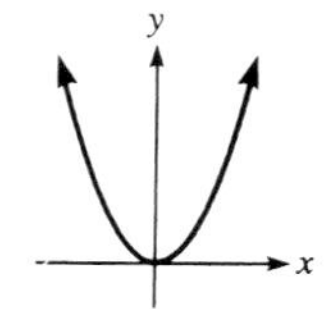

2. Graph $y = |x|$.

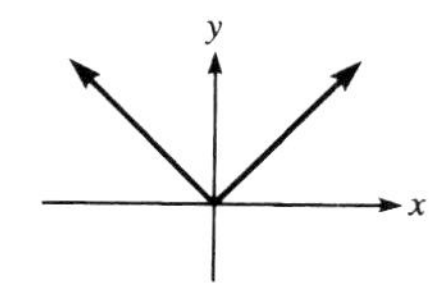

3. Graph $y = \sqrt{x}$.

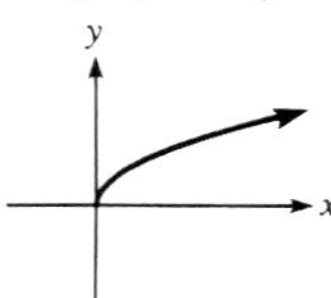

4. Graph $y = x$.

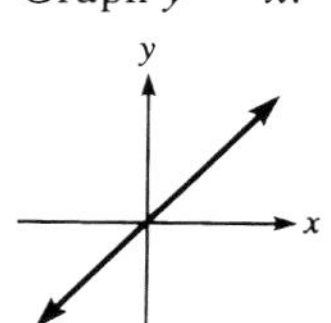

5. Find all points of intersection of the parabola $y = x^2$ and the line $y = 2x$. $(0,0)$, $(2,4)$
6. Find all points of intersection of the ellipse $\frac{x^2}{9} + \frac{y^2}{4} = 1$ and the hyperbola $\frac{x^2}{9} - \frac{y^2}{4} = 1$. $(3,0)$, $(-3,0)$
7. Simplify $\frac{3i}{1 + i}$. Write the answer in the form $a + bi$. $\frac{3}{2} + \frac{3}{2}i$
8. In solving the system $\begin{array}{r} 3x - 4y = 1 \\ -2x + 5y = 6 \end{array}$ by Cramer's rule, what is the determinant for the denominator of x and y? Evaluate it. $\begin{vmatrix} 3 & -4 \\ -2 & 5 \end{vmatrix}$; 7
9. Add $\frac{1}{x^2} + \frac{2}{x}$ and express as a single fraction. $\frac{1 + 2x}{x^2}$
10. Solve $|2x + 1| < 4$. $\left(-\frac{5}{2}, \frac{3}{2}\right)$ [number line from $-\frac{5}{2}$ to $\frac{3}{2}$, x]

9.3 Graphing Techniques

The height y above water in meters of a diver t seconds after stepping off a diving tower 10 m high is given by

$$y = -4.9t^2 + 10.$$

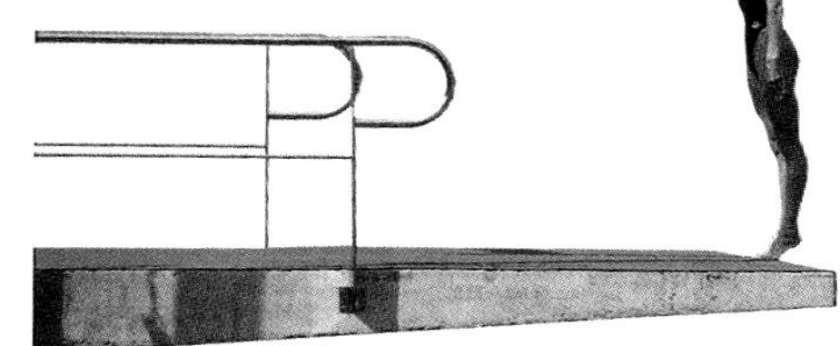

a. To the nearest tenth of a second, when does the diver hit the water?
b. What is the domain of the function?
c. Graph this function. (See Example 7.)

OBJECTIVES

1. **Graph a function using translations.**
2. **Graph a function using reflecting, stretching, or flattening.**
3. **Graph a function using combinations of the above methods.**

Many graphs may be sketched quickly if you learn to graph variations of familiar functions by properly adjusting a known curve. To draw the graphs in this section, you must memorize the graphs of the special functions in Figure 9.12. In each case some ordered pairs that were plotted as aids for drawing the graph are shown. Note that the graph of the squaring function is a curve that is called a parabola (as considered in Section 8.5).

Stress that the methods of this section apply to any function (not just the ones in Figure 9.12). For instance, these methods are useful in the next chapter for graphing exponential and logarithmic functions.

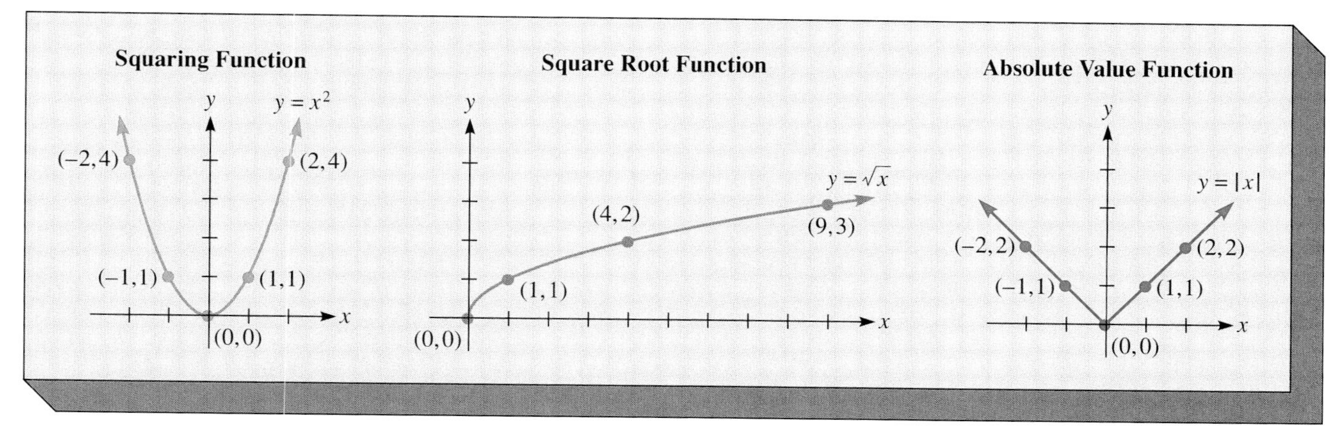

Figure 9.12

1 Consider Figure 9.13, which shows the graphs of $y = x^2$, $y = x^2 + 2$, and $y = x^2 - 2$ on the same coordinate system. Observe that the graph of $y = x^2 + 2$ is the graph of $y = x^2$ translated 2 units up, while the graph of $y = x^2 - 2$ is the graph of $y = x^2$ translated 2 units down. In other words, the curve that characterizes the squaring functions is used in all three graphs, and graphing each equation amounts to just positioning this parabola correctly. In Figure 9.13 we see that when 2 is added or subtracted *after* applying the squaring rule, the effect is to shift the parabola up (if adding) or down (if subtracting) a distance of 2 units. These results extend to give a general procedure for graphing $y = f(x) + c$ and $y = f(x) - c$ using the graph of $y = f(x)$.

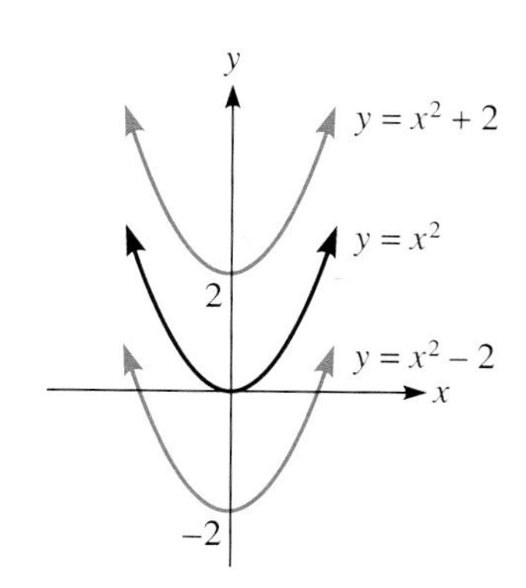

Figure 9.13

The graphing calculator section at the end of this chapter has illustrative exercises on translations and reflection.

Vertical Translations

Let c be a positive constant.

1. The graph of $y = f(x) + c$ is the graph of f shifted c units up.
2. The graph of $y = f(x) - c$ is the graph of f shifted c units down.

EXAMPLE 1 Use the graph of $y = |x|$ to graph each function.

a. $y = |x| - 3$ b. $y = |x| + 1$

Solution The graph of the absolute value function has a V shape, as shown in Figure 9.12. Translate this basic shape as follows.

a. The constant 3 is subtracted *after* the absolute value rule, so the graph of $y = |x| - 3$ is the graph of $y = |x|$ shifted 3 units down, as shown in Figure 9.14(a).
b. Shift the graph of $y = |x|$ up 1 unit to graph $y = |x| + 1$. See Figure 9.14(b).

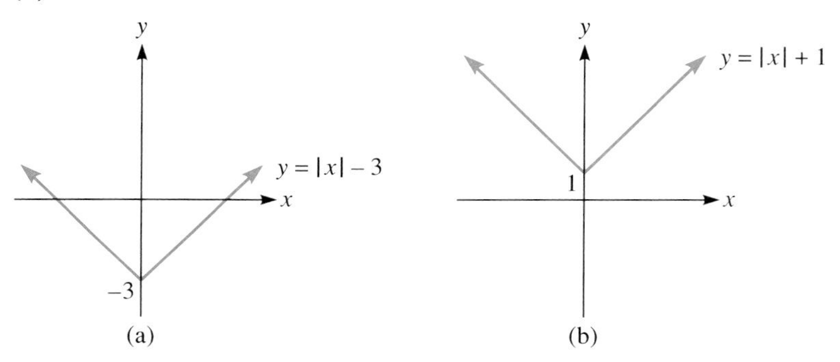

Figure 9.14

PROGRESS CHECK 1 Use the graph of $y = |x|$ to graph each function.

a. $y = |x| + 2$ b. $y = |x| - 4$

To determine what happens graphically when a constant is added or subtracted *before* applying a function rule, consider the graphs of $y = x^2$, $y = (x + 2)^2$, and $y = (x - 2)^2$ in Figure 9.15. Once again, all three graphs have the same shape. However, the parabola shifts horizontally in this case. Observe that the graph of $y = (x + 2)^2$ is the graph of $y = x^2$ shifted 2 units to the left, while the graph of $y = (x - 2)^2$ is the graph of $y = x^2$ shifted 2 units to the right.

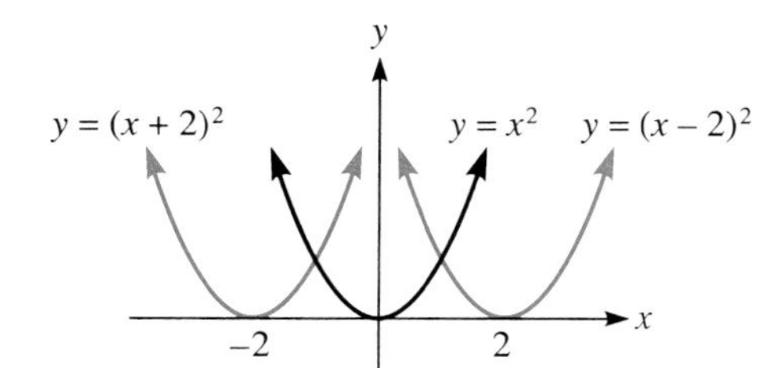

Figure 9.15

For horizontal shifts the direction of the translation is not supported by intuition, so be careful here. The graph of $y = (x - 2)^2$ is 2 units to the *right* (not the left) of $y = x^2$, because x must be 2 units more in $y = (x - 2)^2$ than in $y = x^2$ to produce the same y value. For instance, $y = 0$ when $x = 0$ in $y = x^2$, while $y = 0$ when $x = 2$ in $y = (x - 2)^2$. Similar reasoning applies for any function f and any constant, and the following statements summarize the rules for graphing using horizontal shifts.

Many students will find the direction of the shift puzzling; it is worth taking time for several illustrations.

Progress Check Answers

1. (a) (b)

y
2
x
y
x
−4

Horizontal Translations

Let c be a positive constant.

1. The graph of $y = f(x + c)$ is the graph of f shifted c units to the left.
2. The graph of $y = f(x - c)$ is the graph of f shifted c units to the righ1

EXAMPLE 2 Use the graph of $y = \sqrt{x}$ to graph each function.

a. $y = \sqrt{x + 4}$ **b.** $y = \sqrt{x - 1}$

Solution

a. The constant 4 is added *before* the square root rule. Therefore, the graph of $y = \sqrt{x + 4}$ is the graph of $y = \sqrt{x}$ (see Figure 9.12) shifted 4 units horizontally. Because a positive constant is added, the graph of $y = \sqrt{x}$ is shifted to the left to obtain the graph in Figure 9.16(a).

b. To graph $y = \sqrt{x - 1}$, start with the graph of $y = \sqrt{x}$ and shift this graph 1 unit to the right, as shown in Figure 9.16(b).

Constructing a small table of ordered-pair solutions can help students check that the graphing methods are being applied correctly.

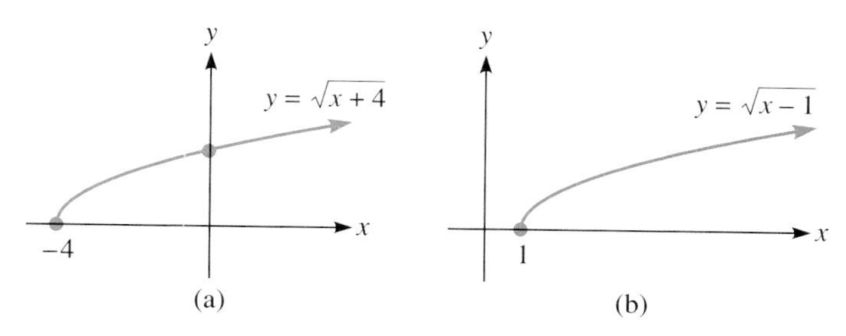

Figure 9.16

PROGRESS CHECK 2 Use the graph of $y = |x|$ to graph each function.

a. $y = |x - 3|$ **b.** $y = |x + 1|$

The graph of the function in the next example uses both a horizontal translation and a vertical translation.

EXAMPLE 3 Graph $y = (x + 2)^2 - 3$. Identify the vertex and the y-intercept of the graph.

Solution To graph this function, first shift the graph of $y = x^2$ in Figure 9.17(a) to the left 2 units, to obtain the graph of $y = (x + 2)^2$ in Figure 9.17(b). Then shift this graph down 3 units, since the constant 3 is subtracted after the squared term. The completed graph is shown in Figure 9.17(c). The vertex in this graph is at $(-2,-3)$. When $x = 0$,

$$y = (0 + 2)^2 - 3 = 1,$$

so the y-intercept is $(0,1)$.

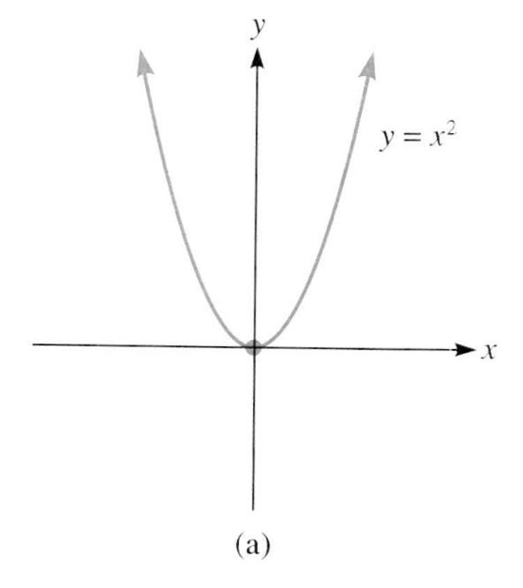

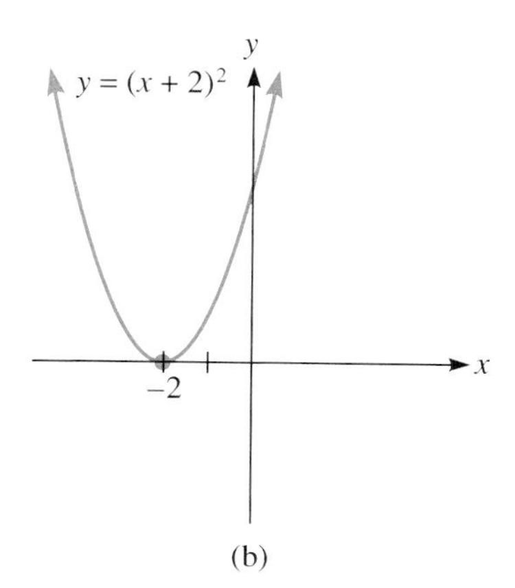

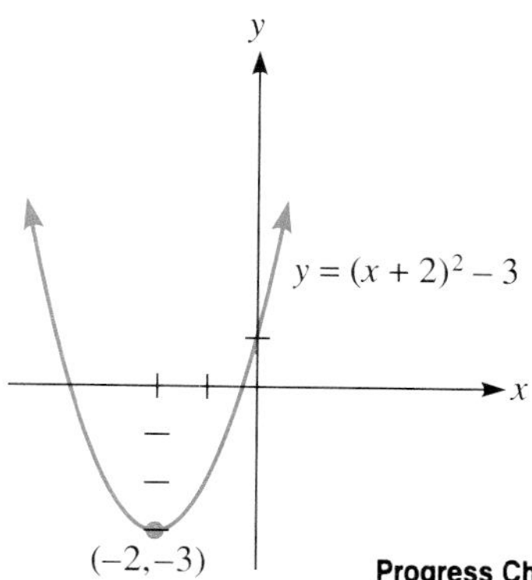

Figure 9.17

Progress Check Answers

2. (a)

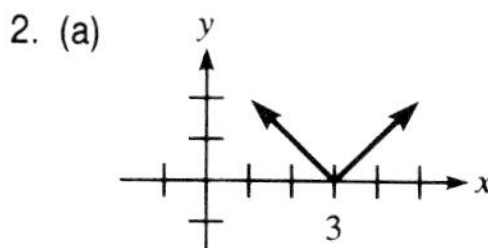

(b)

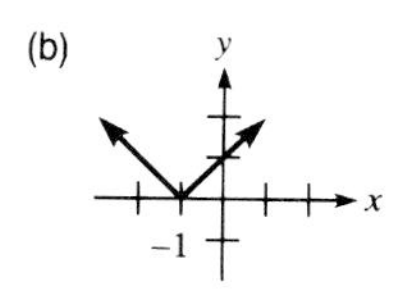

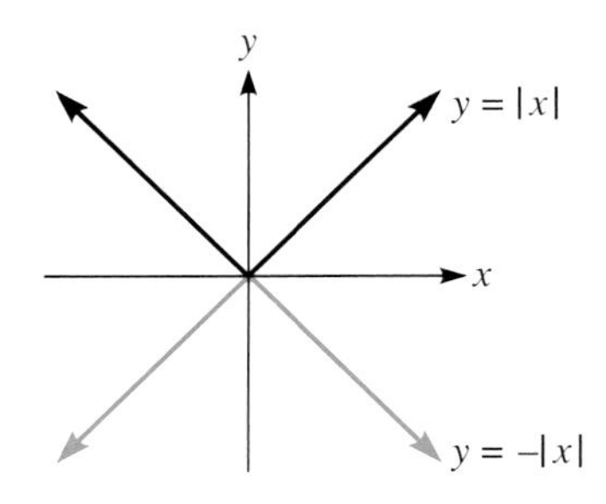

Figure 9.18

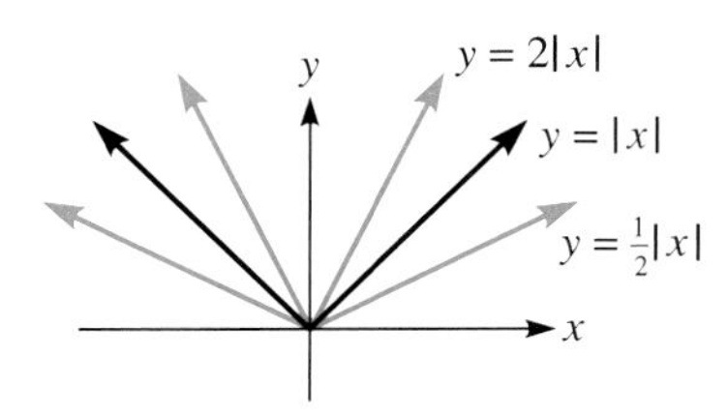

Figure 9.19

It may help to point out that "large" values of c make the graph grow faster, while "small" values of c make the graph grow slower.

PROGRESS CHECK 3 Graph $y = (x - 2)^2 + 1$. Identify the vertex and the y-intercept of the graph.

2 To see how to graph $y = cf(x)$ using the graph of $y = f(x)$, first consider the graphs of $y = |x|$ and $y = -|x|$ shown in Figure 9.18. Each graph is the reflection of the other about the x-axis, and the graphs of $y = f(x)$ and $y = -f(x)$ are always related in this way.

Next consider Figure 9.19, which shows the graphs of $y = |x|$, $y = 2|x|$, and $y = \frac{1}{2}|x|$. To graph $y = 2|x|$, we double each y value in $y = |x|$, which causes the graph of $y = |x|$ to stretch by a factor of 2. Likewise, we graph $y = \frac{1}{2}|x|$ by halving the y values in $y = |x|$, and we say the graph of $y = \frac{1}{2}|x|$ is the graph of $y = |x|$ flattened out by a factor of $\frac{1}{2}$. These observations generalize to the following rules for graphing $y = cf(x)$.

> **To Graph $y = cf(x)$**
>
> **Reflecting** The graph of $y = -f(x)$ is the graph of f reflected about the x-axis.
> **Stretching** If $c > 1$, the graph of $y = cf(x)$ is the graph of f stretched by a factor of c.
> **Flattening** If $0 < c < 1$, the graph of $y = cf(x)$ is the graph of f flattened out by a factor of c.

EXAMPLE 4 Graph each function.

a. $y = -\sqrt{x}$ **b.** $y = \frac{1}{4}x^2$

Solution

a. Start with the graph of $y = \sqrt{x}$. Then reflect this graph about the x-axis to obtain the graph of $y = -\sqrt{x}$, as shown in Figure 9.20(a).

b. The y values in $y = \frac{1}{4}x^2$ are one-fourth the y values in $y = x^2$. Thus, the graph of $y = \frac{1}{4}x^2$ is the graph of $y = x^2$ flattened out by a factor of $\frac{1}{4}$, as shown in Figure 9.20(b).

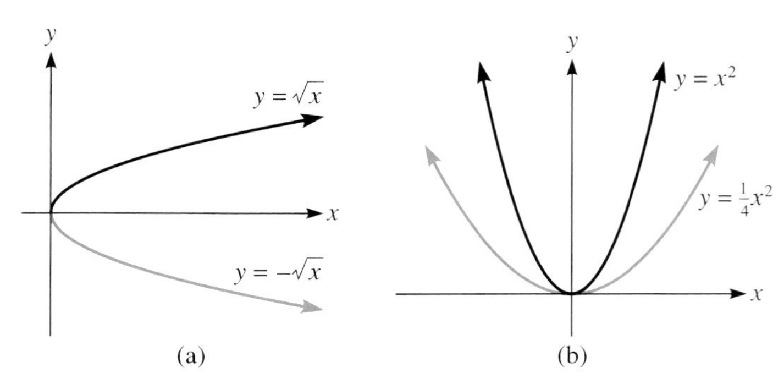

Figure 9.20

PROGRESS CHECK 4 Graph each function.

a. $y = 3\sqrt{x}$ **b.** $y = -x^2$

3 In the remaining examples graphing the function will require combinations of the graphing methods developed in this section.

Progress Check Answers

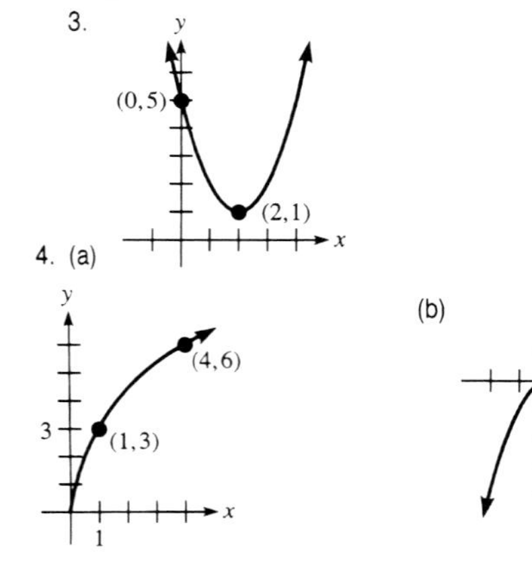

EXAMPLE 5 Graph $y = 4 - x^2$ using the graph of $y = x^2$.

Solution The equation of $y = 4 - x^2$ is equivalent to $y = -x^2 + 4$. To graph this function, first reflect the graph of $y = x^2$ about the x-axis to obtain the graph of $y = -x^2$ in Figure 9.21(a). Raising the graph of $y = -x^2$ up 4 units gives the graph of $y = -x^2 + 4$, shown in Figure 9.21(b).

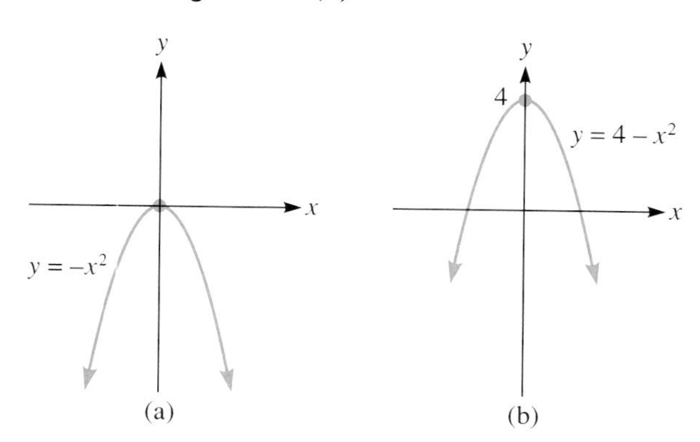

Figure 9.21

PROGRESS CHECK 5 Graph $y = 2 - |x|$ using the graph of $y = |x|$.

EXAMPLE 6 Graph $y = -|x - 4| + 5$ using the graph of $y = |x|$.

Solution Follow this sequence of graphing steps.

Step 1 Shift the graph of $y = |x|$ to the right 4 units to graph $y = |x - 4|$, as shown in Figure 9.22(a).

Step 2 Reflect the graph from the previous step about the x-axis to graph $y = -|x - 4|$, as shown in Figure 9.22(b).

Step 3 Shift the graph from the previous step up 5 units to graph $y = -|x - 4| + 5$, as shown in Figure 9.22(c).

Note In practice, it is usually possible to draw the completed graph in Figure 9.22(c) without actually sketching the other preliminary graphs. The graphs in Figures 9.22(a) and 9.22(b) are just part of the thought process that leads to the answer.

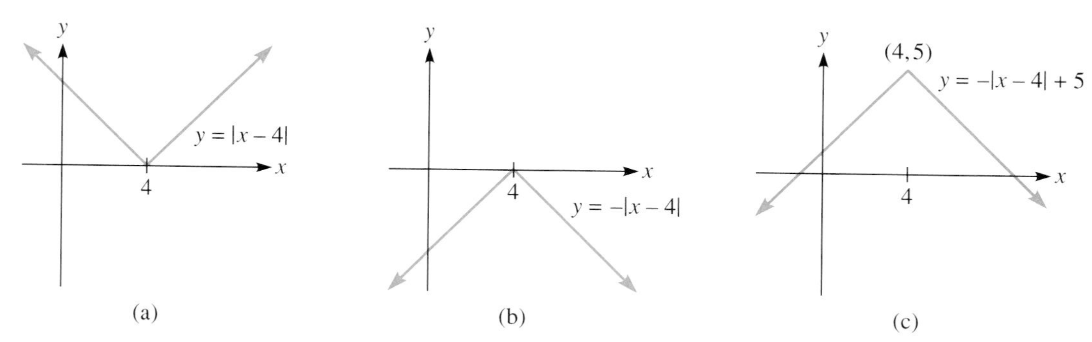

Figure 9.22

PROGRESS CHECK 6 Graph $y = 2 - |x - 1|$ using the graph of $y = |x|$.

Progress Check Answers

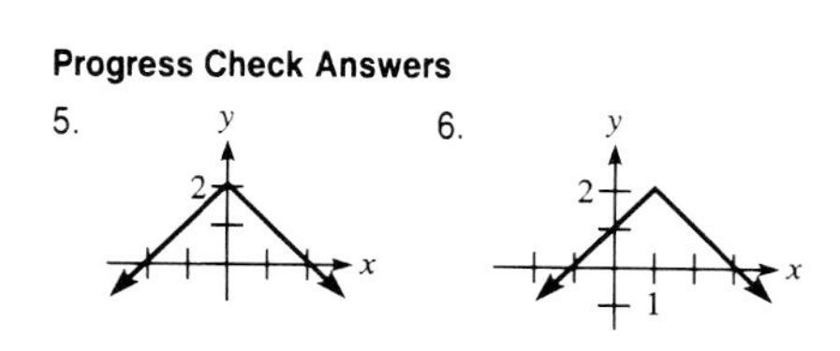

Consider pointing out that the value 4.9 represents one-half of the acceleration due to the force of gravity near the earth's surface when distance and time are given in meters and seconds, and that the corresponding value is 16 when the units are feet and seconds. Other constants are appropriate at different altitudes or on different bodies (e.g., 137 instead of 4.9 at the surface of the sun).

EXAMPLE 7 Solve the problem in the section introduction on page 445.

Solution

a. When the diver hits the water, the height y above water is 0. So replace y by 0 in the given formula and solve for t.

$$\begin{aligned} y &= -4.9t^2 + 10 \\ 0 &= -4.9t^2 + 10 \qquad \text{Replace } y \text{ by } 0. \\ 4.9t^2 &= 10 \\ t^2 &= \frac{10}{4.9} \\ t &= \pm\sqrt{\frac{10}{4.9}} \\ &= \pm 1.4 \quad \text{(to the nearest tenth)} \end{aligned}$$

Since t must be positive, the diver hits the water about 1.4 seconds after stepping off the diving tower.

b. The formula is meaningful starting at $t = 0$ seconds and ending at $t = 1.4$ seconds (when the diver hits the water), so the domain is the interval [0 seconds, 1.4 seconds].

c. To graph this function, stretch the graph of $y = t^2$ by a factor of 4.9, reflect this parabola about the x-axis, and then shift the resulting graph 10 units up. This graph is shown in Figure 9.23(a). Finally, use the results from parts **a** and **b** for the time at which the diver hits the water and the domain of the function to draw the graph of the formula, which is shown in Figure 9.23(b).

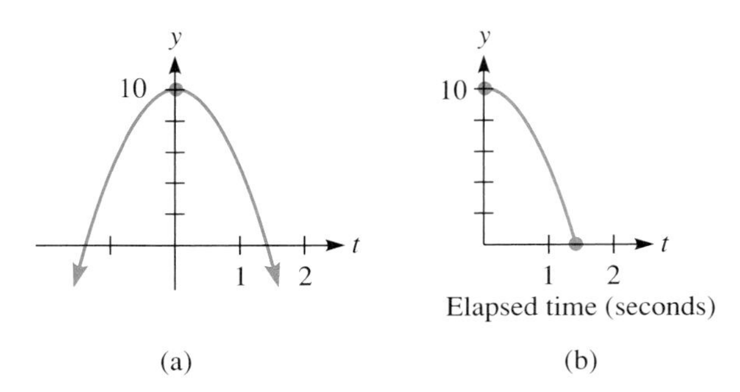

Figure 9.23

PROGRESS CHECK 7 If a stone is dropped from a bridge at a point that is 200 ft above water, then $y = -16t^2 + 200$ gives the height y above water in feet after t seconds have elapsed.

a. To the nearest tenth of a second, when does the stone hit the water?
b. What is the domain of the function?
c. Graph this function.

Progress Check Answers

7. (a) 3.5 seconds (b) [0 seconds, 3.5 seconds]
(c)

y
200
Height (feet)
1 2 3 4
t
Elapsed time (seconds)

EXERCISES 9.3

In all exercises, use the graphs of $y = |x|$, $y = x^2$, or $y = \sqrt{x}$ as appropriate to graph each of the given functions.

1. $y = |x| - 1$

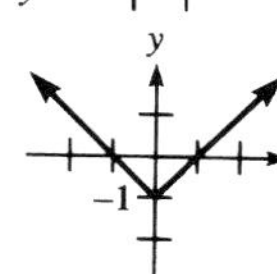

2. $y = |x| + 3$

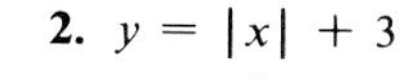

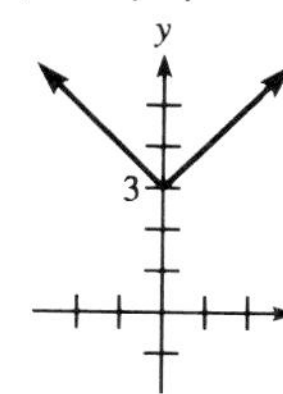

3. $y = x^2 + 3$

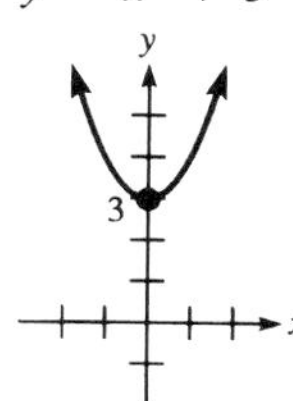

4. $y = x^2 - 5$

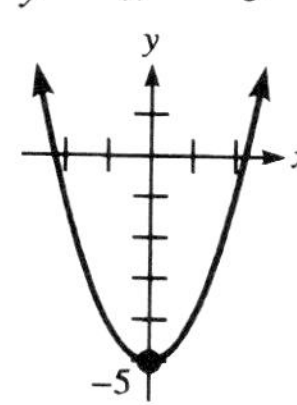

5. $y = \sqrt{x} + 2$

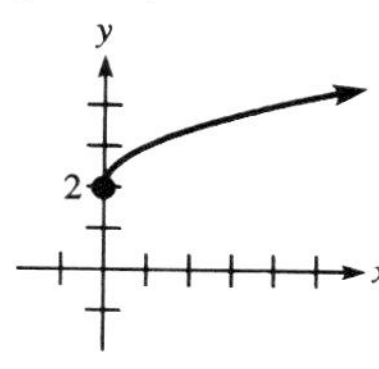

6. $y = \sqrt{x} - 2$

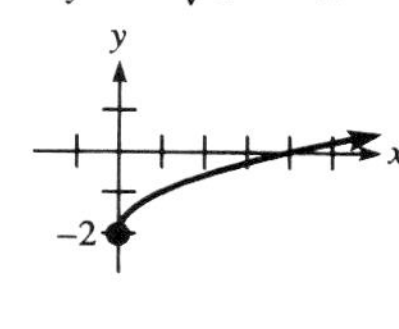

7. $y = |x| - 5$

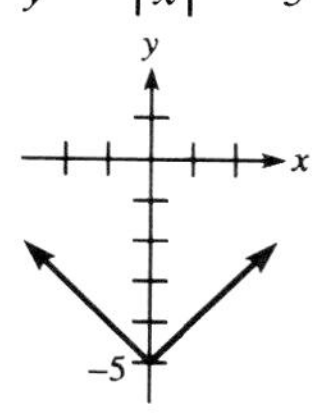

8. $y = x^2 + 5$

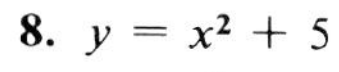

9. $y = \sqrt{x} - 1$

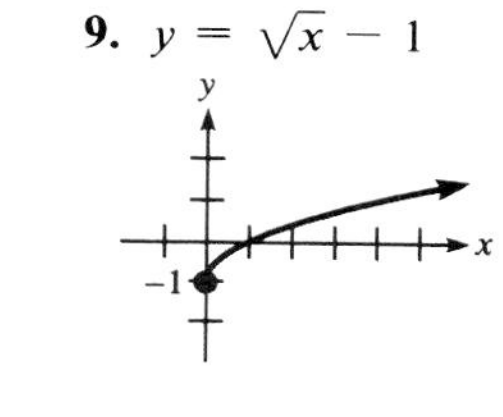

10. $y = \sqrt{x} + 3$

11. $y = |x - 1|$

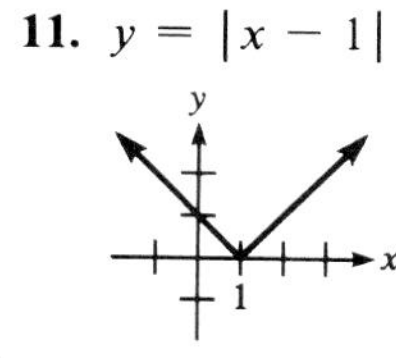

12. $y = |x + 3|$

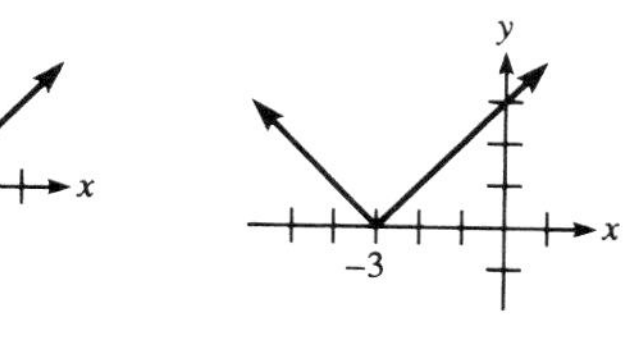

13. $y = (x + 3)^2$

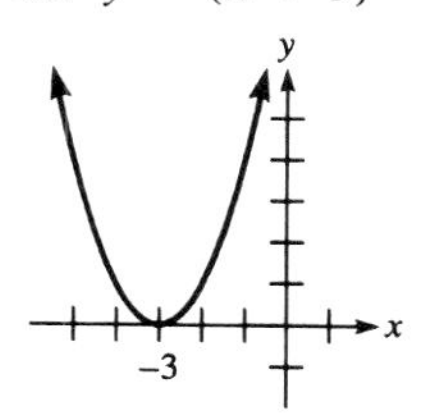

14. $y = (x - 5)^2$

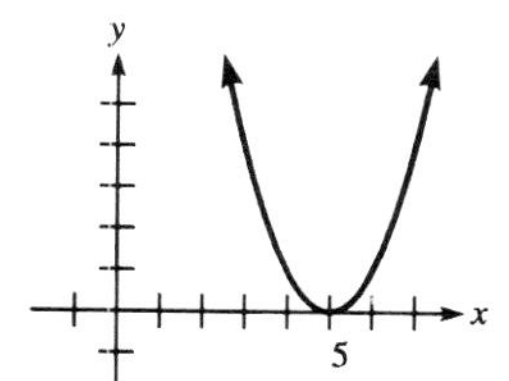

15. $y = \sqrt{x + 2}$

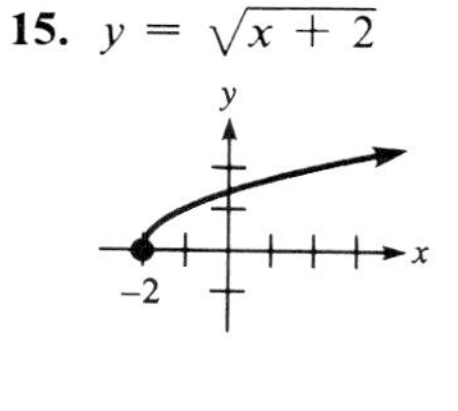

16. $y = \sqrt{x - 2}$

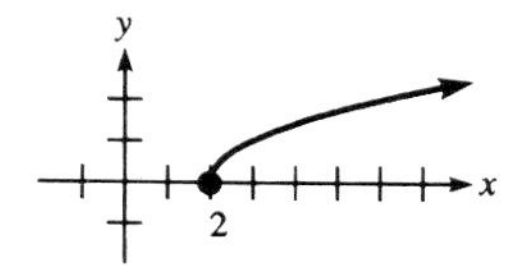

17. $y = |x + 5|$

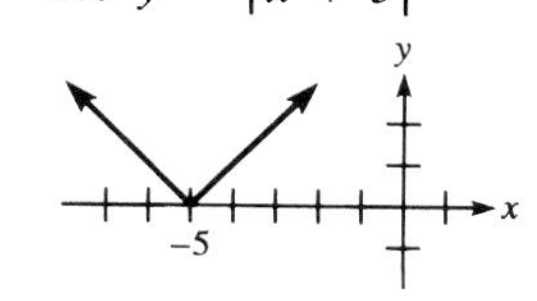

18. $y = (x + 5)^2$

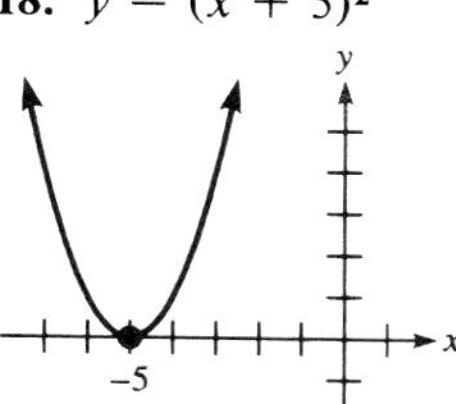

19. $y = \sqrt{x + 1}$

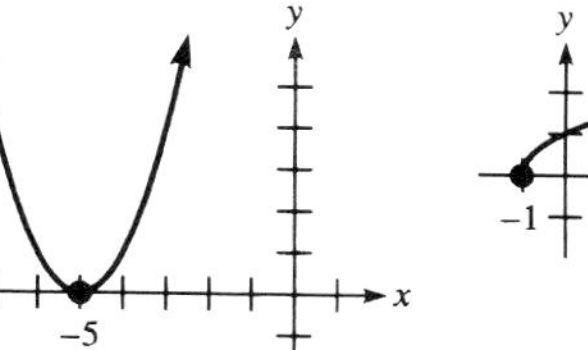

20. $y = \sqrt{x - 3}$

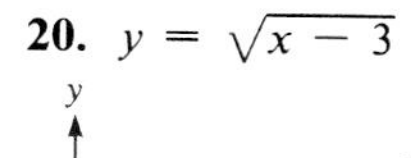

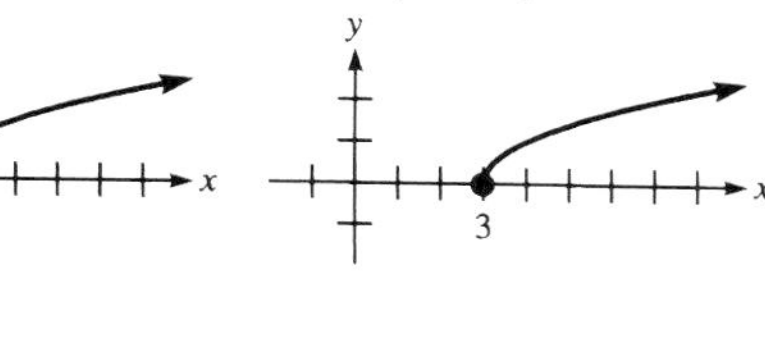

In Exercises 21–30, graph each function. Identify the minimum point and the y-intercept of the graph.

21. $y = (x - 1)^2 + 2$

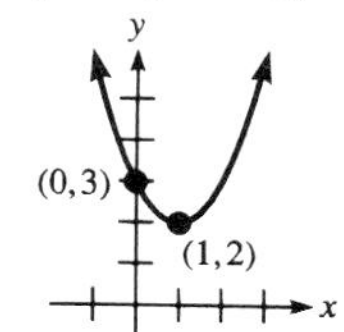

22. $y = (x + 1)^2 - 2$

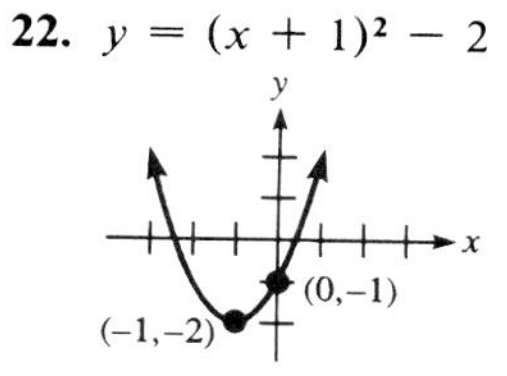

23. $y = |x + 1| - 1$

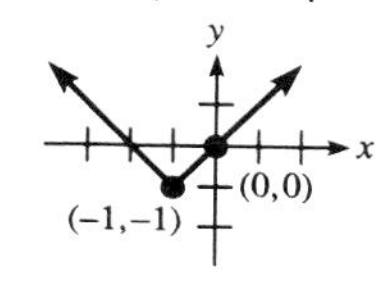

24. $y = |x - 1| + 2$

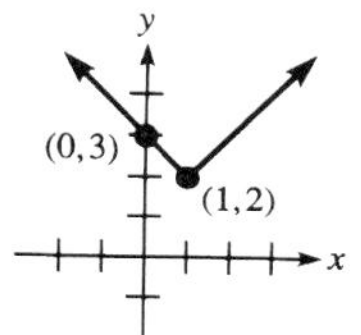

25. $y = \sqrt{x - 3} + 2$

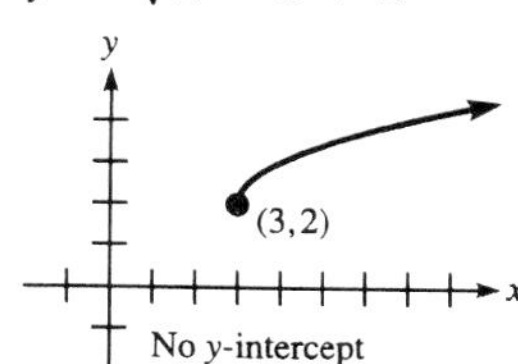

26. $y = \sqrt{x + 1} - 3$

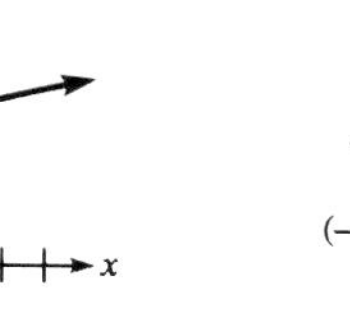

27. $y = (x - 3)^2 - 2$

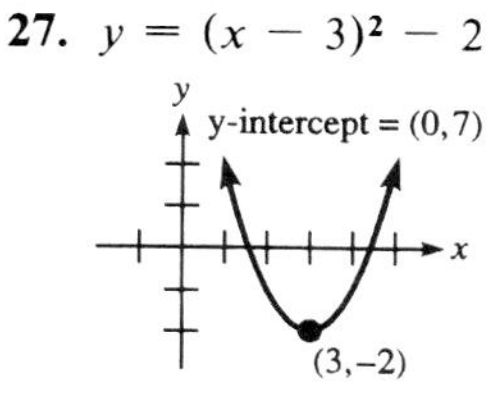

28. $y = (x - 1)^2 - 1$

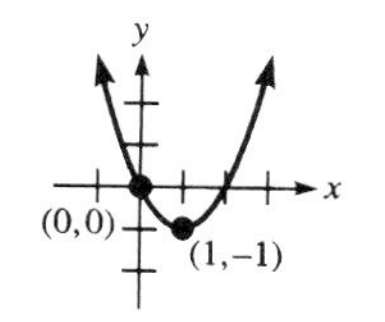

29. $y = |x + 4| + 1$

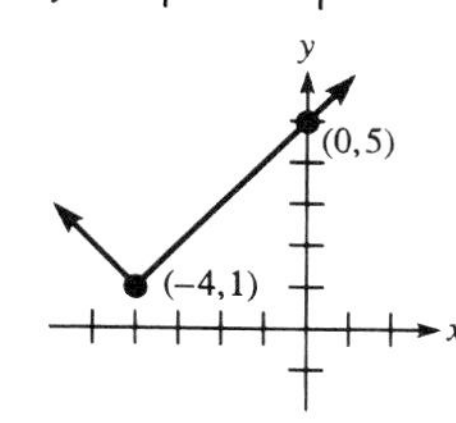

30. $y = |x - 4| - 1$

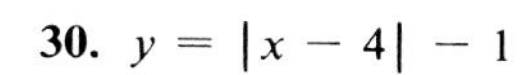

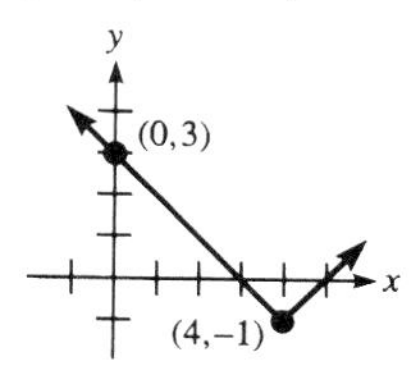

In Exercises 31–60, graph the given function.

31. $y = -|x|$

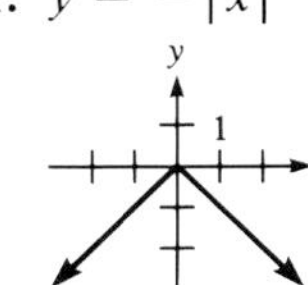

32. $y = \frac{1}{2}|x|$

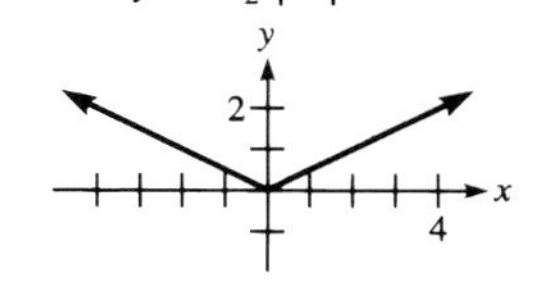

33. $y = 2x^2$

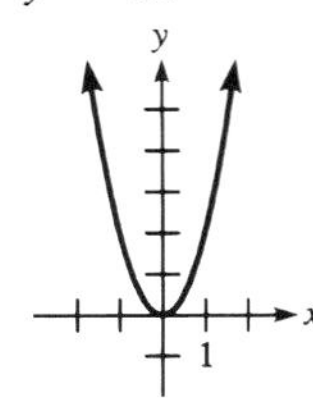

34. $y = \frac{1}{3}x^2$

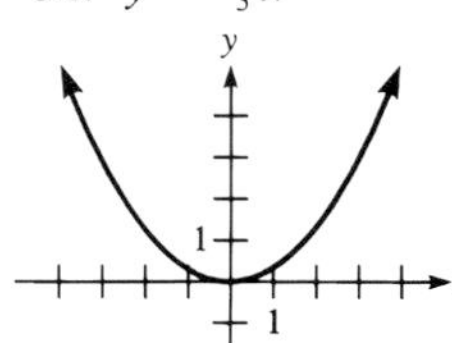

35. $y = \frac{1}{2}\sqrt{x}$

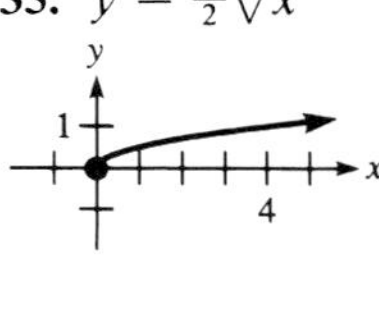

36. $y = 2\sqrt{x}$

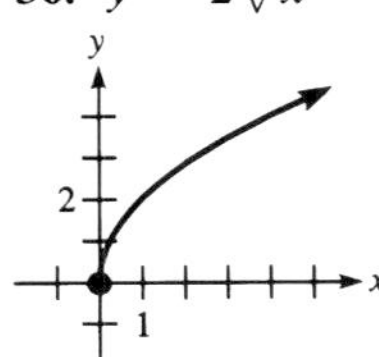

37. $y = 5|x|$

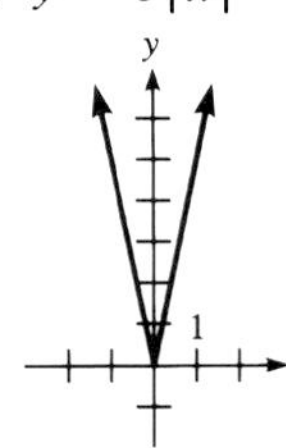

38. $y = -\frac{1}{4}|x|$

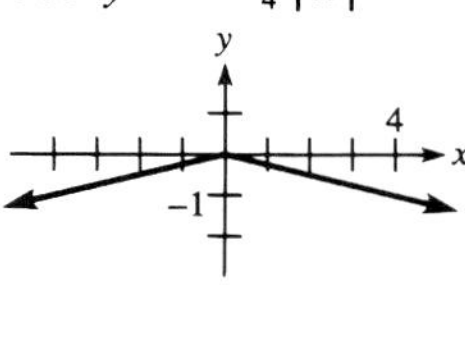

39. $y = -3x^2$

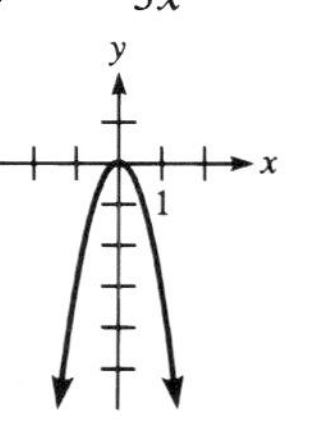

40. $y = -\frac{1}{2}x^2$

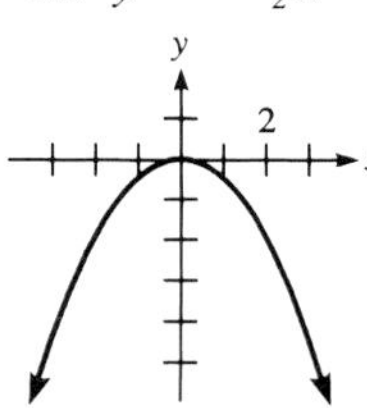

41. $y = 2 + |x|$

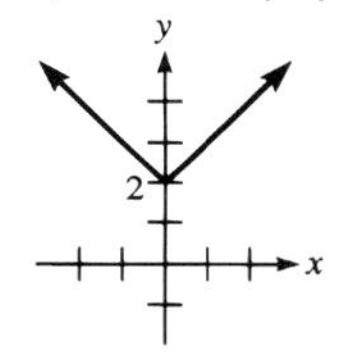

42. $y = 3 - |x|$

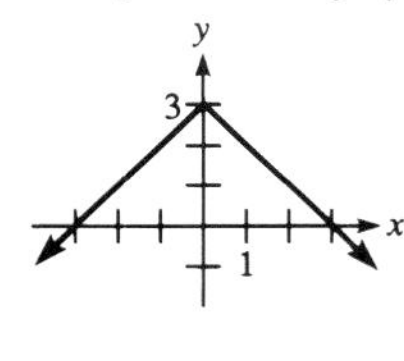

43. $y = 4 + \sqrt{x}$

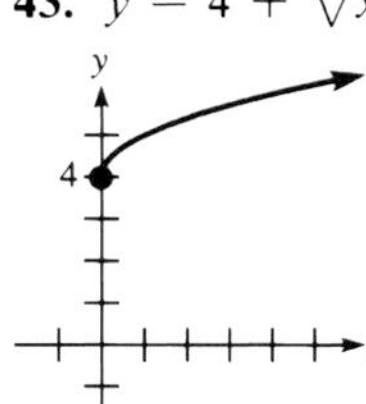

44. $y = 3 - \sqrt{x}$

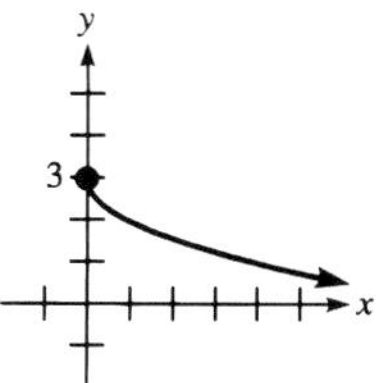

45. $y = -x^2 + 1$

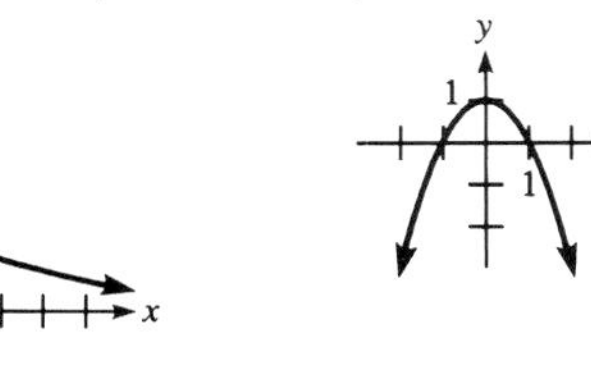

46. $y = 4 - x^2$

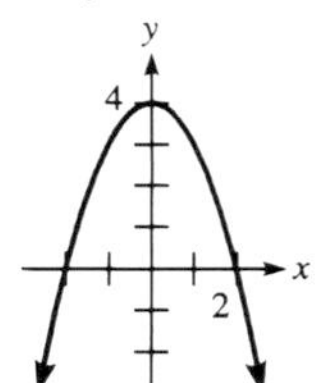

47. $y = -\sqrt{x} - 2$

48. $y = -|x| - 1$

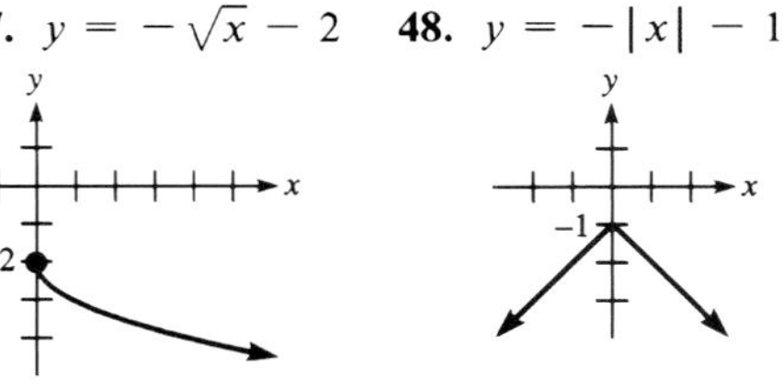

49. $y = -x^2 - 1$

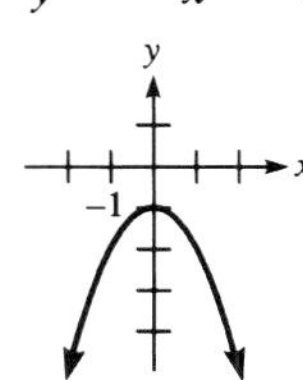

50. $y = -\sqrt{x} + 4$

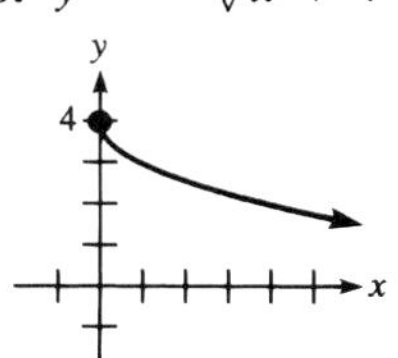

51. $y = -|x + 1|$

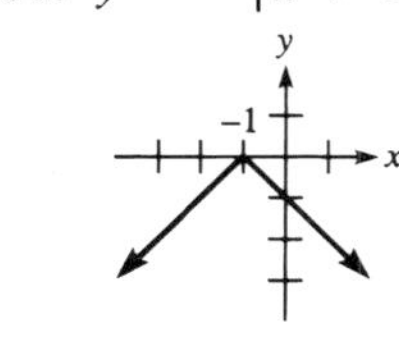

52. $y = -(x - 2)^2$

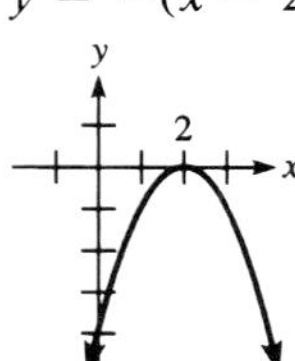

53. $y = -\sqrt{x} - 1$

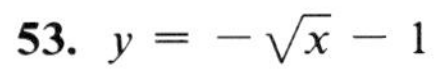

54. $y = -|x - 3|$

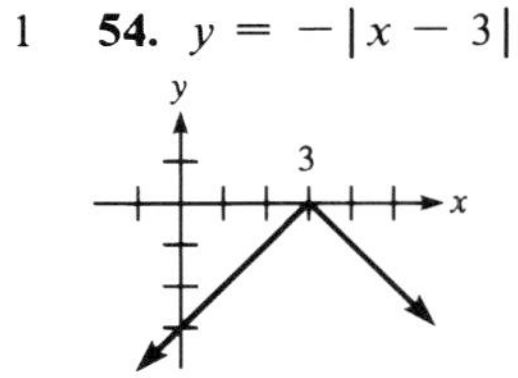

55. $y = -2 - (x + 3)^2$

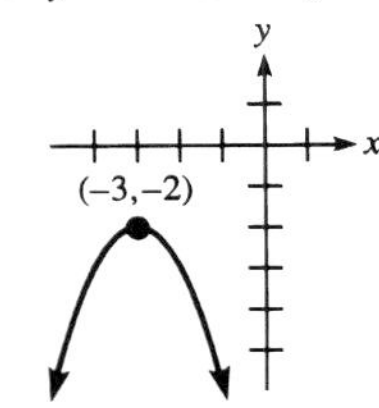

56. $y = -1 + (x - 3)^2$

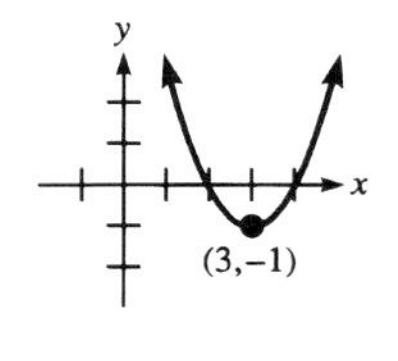

57. $y = 3 - |x - 1|$

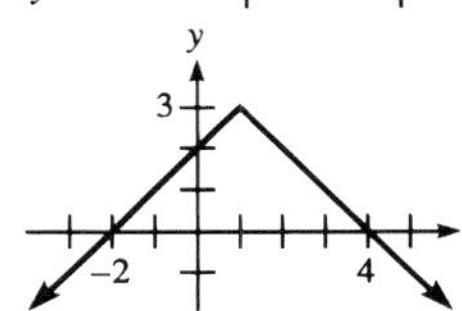

58. $y = 2 - |x + 1|$

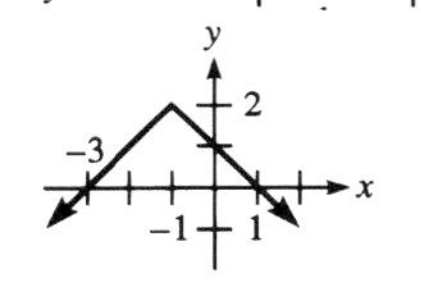

59. $y = -\sqrt{x + 1} - 5$

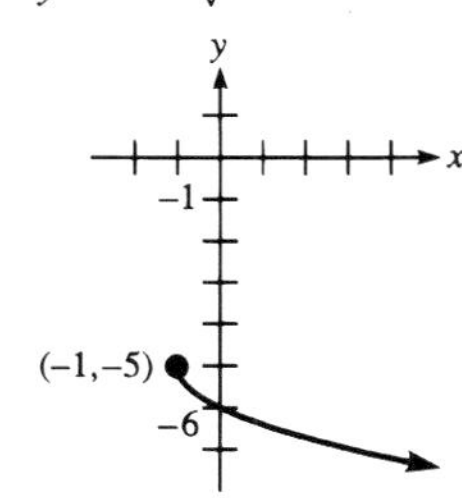

60. $y = -\sqrt{x - 1} + 2$

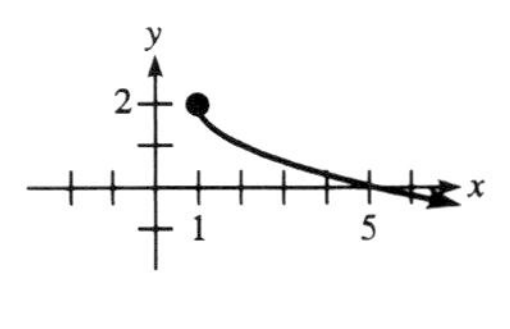

61. The height y in meters of a tightrope walker above a safety net t seconds after falling from the tightrope 40 m above the net is given by $y = -4.9t^2 + 40$.

a. To the nearest tenth of a second, when does the performer hit the safety net?

b. What is the domain of this function?

c. Graph the function.

a. 2.9 seconds

b. [0 seconds, 2.9 seconds]

c.

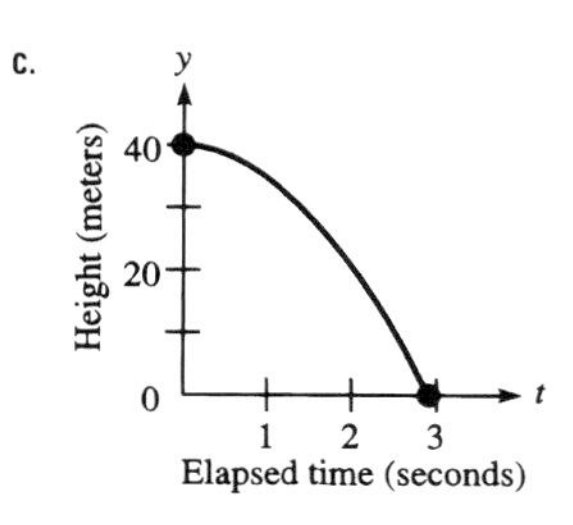

62. The conditions of Exercise 61 remain essentially the same, but the tightrope is lowered to 30 m above the safety net. Answer the same questions. Note that in this case, the function becomes $y = -4.9t^2 + 30$.

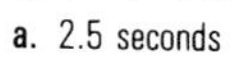

a. 2.5 seconds
b. [0 seconds, 2.5 seconds]
c.

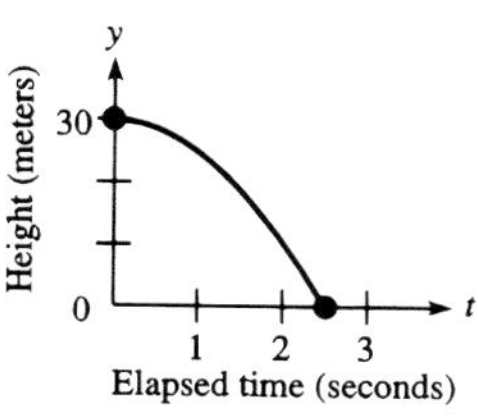

63. A lifeboat on a certain freighter can be lowered to the surface of the water by a cable system extending from the top deck. The height of the top deck above the waterline is 75 ft, and the height of the lifeboat t seconds after the lowering process begins is given by $h = -0.005t^2 + 75$.

a. How many seconds does it take (to the nearest second) to lower the lifeboat to the water?

b. Approximately how many minutes is this?

c. What is the domain of this function?

d. Graph the function.

a. 122 seconds
b. 2.0 minutes
c. [0 seconds, 122 seconds]
d.

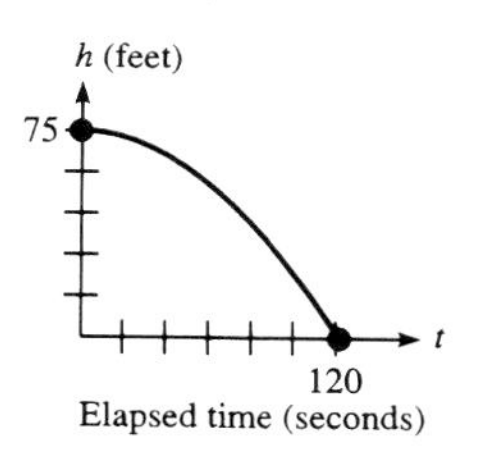

64. The conditions in Exercise 63 remain essentially the same, but the top deck is 100 ft above the waterline and the height function is accordingly $h = -0.005t^2 + 100$. Answer the same questions.

a. 141 seconds
b. 2.4 minutes
c. [0 seconds, 141 seconds]
d.

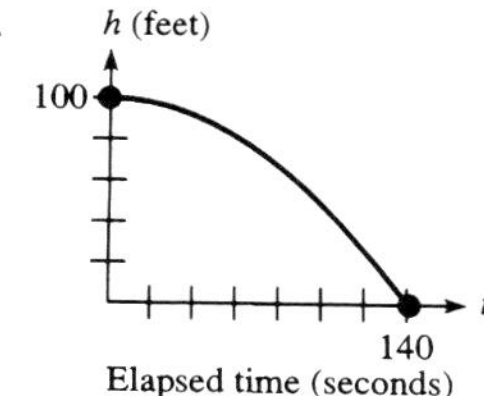

THINK ABOUT IT

1. Draw the graph of $y = x^3$ by constructing a table of ordered-pair solutions. Then use this graph and the methods of this section to graph the following functions.
   a. $y = -x^3$ **b.** $y = (x - 1)^3$
   c. $y = x^3 + 1$ **d.** $y = (x + 1)^3 - 2$
2. Use the given graph of $y = f(x)$ to graph each of these functions.
   a. $y = 3f(x)$ **b.** $y = -f(x)$
   c. $y = f(x) + 1$ **d.** $y = f(x + 1)$

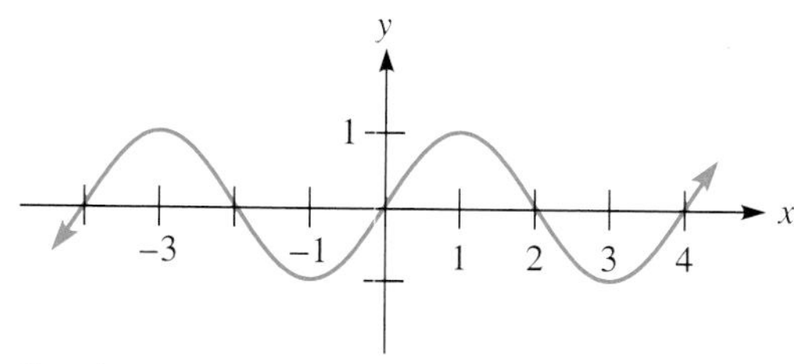

3. Graph $y = x^2 + 2x$ by completing the square on the right-hand side of the equation and then using the methods of this section.
4. Graph $y = \sqrt{x}$ and $y = \sqrt{-x}$ on the same coordinate system. (*Note:* They have different domains.) How do the graphs relate to each other?
5. In translating a graph, the shape does not change, so it is reasonable to talk about points in the translated graph corresponding to points in the original graph.
   a. What point on the graph of $y = x^2 + 3$ corresponds to the point (a,b) on the graph of $y = x^2$?
   b. What point on the graph of $y = x^2 - 5$ corresponds to the point (a,b) on the graph of $y = x^2$?
   c. What point on the graph of $y = (x - 2)^2$ corresponds to the point (a,b) on the graph of $y = x^2$?
   d. What point on the graph of $y = (x + 7)^2$ corresponds to the point (a,b) on the graph of $y = x^2$?
   e. What point on the graph of $y = -(x - 1)^2$ corresponds to the point (a,b) on the graph of $y = x^2$?

REMEMBER THIS

1. Express these in the form x^a, where a is a real number: $\sqrt{x}, \frac{1}{x}, \frac{1}{x^2}$. $x^{1/2}, x^{-1}, x^{-2}$
2. Graph $f(x) = 2$.

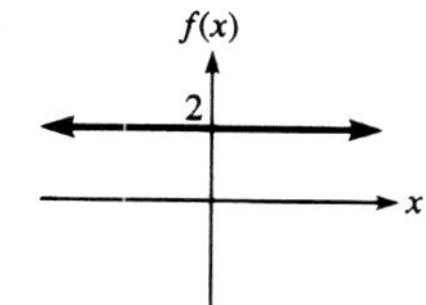

3. Identify those equations which graph as a straight line.
   a. $y = 3x - 2$ **b.** $y = \frac{1}{3}$ **c.** $y = x^2 + 3$ a, b
4. What is the slope of the line which connects (3,1) and (5,−3)? −2
5. Simplify $ax^2 + bx + c$ if $x = -b/a$. c
6. If $f(x) = 2$ and $g(x) = 5$, find $3f(0) - 5g(0)$. −19
7. Does the vertex of the parabola $y = 5 - (x + 3)^2$ mark a maximum point on the graph or a minimum point? Maximum
8. Multiply and then simplify the result: $\sqrt{xy} \cdot \sqrt{9xy}$. Assume that x and y represent positive real numbers. $3xy$

9. One job pays \$200 per week plus 15 percent commission on each sale. Another pays \$180 per week plus 20 percent commission on each sale. Represent each pay scheme by a linear function, and show the graphs on the same set of axes. At what point do the jobs provide the same pay?

$y = 200 + 0.15x$; $y = 180 + 0.20x$. They pay the same when the amount of sales is \$400.

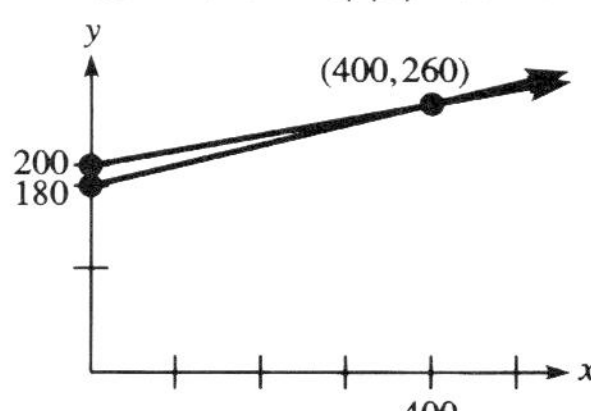

10. Solve $2x(2x - 1)(2x + 2) = 0$. $\{0, \frac{1}{2}, -1\}$

9.4 Polynomial Functions

During exercise a person's maximum target heart rate is a function of age. This relation is specified by a linear function, and the recommended maximum number of beats per minute is 153 at age 40 and 136 at age 60.

a. Find the equation that defines this linear function.

b. What is the maximum target heart rate for an 18-year-old? (See Example 4.)

OBJECTIVES

1. **Graph a constant function.**
2. **Graph a linear function.**
3. **Find the equation of a linear function f given two ordered pairs in the function.**
4. **Solve applied problems involving linear functions.**
5. **Graph a quadratic function and specify the range of the function.**
6. **Determine the vertex of the graph of a quadratic function by matching the function to the form $f(x) = a(x - h)^2 + k$.**
7. **Solve applied problems involving quadratic functions.**

One of the important types of elementary functions is a polynomial function. Informally, an equation for a polynomial function is characterized by terms of the form

$$(\text{real number})x^{\text{nonnegative integer}},$$

so typical terms look like 2, $-x$, $5x^2$, $\frac{1}{2}x^3$, and so on. More formally, a function of the form

$$y = f(x) = a_n x^n + a_{n-1}x^{n-1} + \cdots + a_1 x + a_0 \qquad (a_n \neq 0),$$

where n is a nonnegative integer and the a's are real numbers, is called a **polynomial function of degree n.** For instance, $f(x) = 3x^2 - 5x + 1$ defines a polynomial function of degree 2 in which $a_2 = 3$, $a_1 = -5$, and $a_0 = 1$. This section considers polynomial functions of degree 0, 1, and 2. Analyzing these functions requires that previous work with lines and parabolas be reconsidered in terms of the function concept.

The functions of this section are analyzed primarily by the methods of Sections 5.1–5.3 and 8.5. However, adjustments in terminology and notation are required.

1 An equation like $f(x) = 2$ may be written as $f(x) = 2x^0$, so this equation defines a zero-degree polynomial function. Such functions are called constant functions because the function value remains fixed for all values of x.

Constant Function

A function of the form

$$f(x) = c,$$

where c is a real number, is called a **constant function.**

Since a constant function f is defined by an equation of the form $y = f(x) = c$, it follows from Section 5.1 that the graph of f is a horizontal line that contains the point $(0,c)$. Note that for technical reasons, the constant function $f(x) = 0$ is called the zero polynomial function and is not assigned a degree.

Informally, $f(x) = 0$ is not assigned a degree because we cannot distinguish $0x^n$ from $0x^m$ for $m \neq n$.

EXAMPLE 1 Graph $f(x) = 2$.

Solution $f(x) = 2$ is a constant function that graphs as a horizontal line through $(0,2)$, as shown in Figure 9.24.

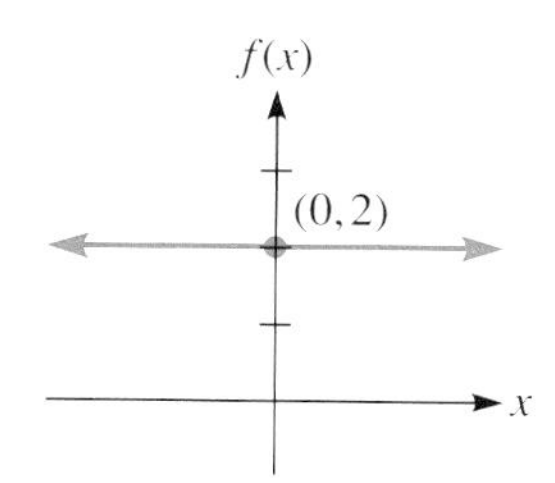

Figure 9.24

PROGRESS CHECK 1 Graph $f(x) = -3$.

2 First-degree polynomial functions are considered next.

Linear Function

A function of the form

$$f(x) = mx + b,$$

where m and b are real numbers with $m \neq 0$, is called a **linear function.**

Recall from Section 5.3 that the graph of the equation $y = mx + b$ is a line with slope m and y-intercept $(0,b)$. Since a linear function f may be defined by an equation of the form $y = f(x) = mx + b$, with $m \neq 0$, it follows that the graph of a linear function is a line (as its name suggests).

EXAMPLE 2 Graph $f(x) = -\frac{3}{2}x + 5$ using the slope and y-intercept of the graph.

Solution The equation fits the form $f(x) = mx + b$, where

$$m = -\tfrac{3}{2} \qquad \text{and} \qquad b = 5,$$

so the graph is a line with slope $-\frac{3}{2}$ and y-intercept $(0,5)$. To use the slope to find a second point, interpret $m = -\frac{3}{2}$ to mean y decreases 3 units as x increases 2 units. Starting at $(0,5)$ and going 2 units to the right and 3 units down gives another point at $(2,2)$. Drawing a line through these two points gives the graph of f in Figure 9.25.

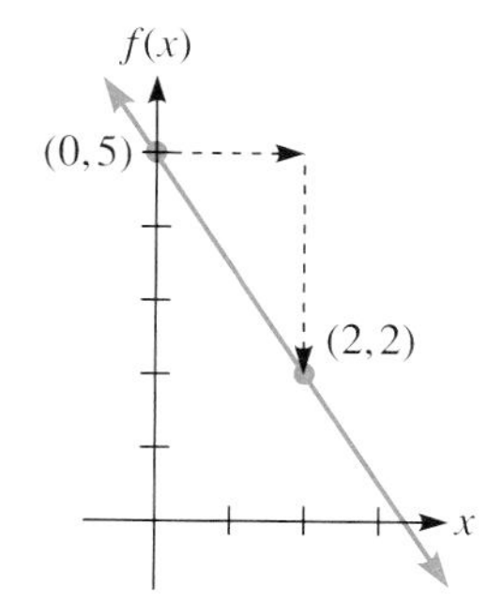

Figure 9.25

You could mention that it would also work to let y increase 3 units as x decreases 2 units.

PROGRESS CHECK 2 Graph $f(x) = \frac{2}{3}x - 4$.

Progress Check Answers

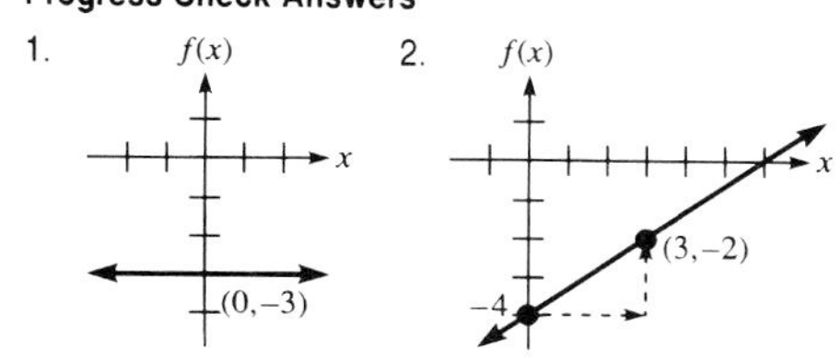

3 When two ordered pairs in a linear function are known, then the equation for the linear function may be found as shown in Examples 3 and 4.

EXAMPLE 3 Find the equation that defines the linear function f if $f(3) = 1$ and $f(5) = -3$.

Solution If $f(3) = 1$, then when $x = 3$, $y = 1$. Similarly, if $f(5) = -3$, then when $x = 5$, $y = -3$. Thus, (3,1) and (5,−3) are points on the graph. First, calculate the slope.

$$m = \frac{y_2 - y_1}{x_2 - x_1} = \frac{-3 - 1}{5 - 3} = \frac{-4}{2} = -2$$

It is instructive to have students show that the other point would work just as well.

Now use the point-slope equation with one of the points, say (3,1), as follows.

$$\begin{aligned} y - y_1 &= m(x - x_1) && \text{Point-slope equation} \\ y - 1 &= -2(x - 3) \\ y - 1 &= -2x + 6 \\ y &= -2x + 7 \end{aligned}$$

The equation that defines the function f is

$$f(x) = -2x + 7.$$

PROGRESS CHECK 3 Find the equation that defines the linear function f if $f(2) = 7$ and $f(4) = -1$.

4 An application of linear functions is illustrated by the section-opening problem.

EXAMPLE 4 Solve the problem in the section introduction on page 454.

Solution

a. Heart rate is a function of age, so let y represent the maximum target heart rate and x represent the person's age. Then we are asked to find the equation for the linear function f such that (40,153) and (60,136) belong to f. As in Example 3, first find the slope.

$$m = \frac{y_2 - y_1}{x_2 - x_1} = \frac{136 - 153}{60 - 40} = \frac{-17}{20} = -0.85$$

Now using one of the points, say (40,153), and the point-slope equation gives

$$\begin{aligned} y - y_1 &= m(x - x_1) && \text{Point-slope equation} \\ y - 153 &= -0.85(x - 40) \\ y - 153 &= -0.85x + 34 \\ y &= -0.85x + 187. \end{aligned}$$

The equation that defines the linear function f is

$$f(x) = -0.85x + 187.$$

b. When $x = 18$,

$$\begin{aligned} f(18) &= -0.85(18) + 187 \\ &= 171.7. \end{aligned}$$

Thus, the maximum target heart rate for an 18-year-old is about 172 beats per minute.

PROGRESS CHECK 4 During exercise a person's minimum target heart rate to have a training effect is a function of age. This relation is specified by a linear function, and the recommended minimum number of beats per minute is 126 at age 40 and 112 at age 60. (a) Find the equation that defines this linear function. (b) What is the minimum target heart rate for a 19-year-old?

Progress Check Answers

3. $f(x) = -4x + 15$
4. (a) $f(x) = -0.7x + 154$ (b) 141

5 Second-degree polynomial functions are called quadratic functions, so the following definition applies.

> **Quadratic Functions**
>
> A function of the form
>
> $$f(x) = ax^2 + bx + c,$$
>
> where a, b, and c are real numbers with $a \neq 0$, is called a **quadratic function.**

It follows from Section 8.5 that the graph of $y = f(x) = ax^2 + bx + c$ is a parabola with axis of symmetry $x = -b/(2a)$. The parabola opens upward when $a > 0$ and downward when $a < 0$.

Consider mentioning that equations of the form $x = ay^2 + by + c$, which graph as horizontal parabolas, do not define y as a function of x.

EXAMPLE 5 Consider the function defined by $f(x) = -x^2 + x + 2$.

a. Find the equation of the axis of symmetry.
b. Find the coordinates of the vertex.
c. Find the coordinates of the y-intercept.
d. Graph the function and specify the range using this graph.

Solution

a. In the given equation $a = -1$ and $b = 1$, so

$$x = \frac{-b}{2a} = \frac{-(1)}{2(-1)} = \frac{1}{2}.$$

The equation of the axis of symmetry is $x = \frac{1}{2}$.

b. The x-coordinate of the vertex is $\frac{1}{2}$, since the vertex lies on the axis of symmetry. We find the y-coordinate of the vertex by finding $f(\frac{1}{2})$, the value of the function when $x = \frac{1}{2}$.

$$\begin{aligned} f(x) &= -x^2 + x + 2 \\ f(\tfrac{1}{2}) &= -(\tfrac{1}{2})^2 + \tfrac{1}{2} + 2 \\ &= -\tfrac{1}{4} + \tfrac{1}{2} + 2 \\ &= \tfrac{9}{4} \end{aligned}$$

The vertex is at $(\frac{1}{2}, \frac{9}{4})$.

c. When $x = 0$, $y = 2$, so the y-intercept is $(0,2)$.

d. From the form of the equation, the graph is a parabola that opens downward (since $a < 0$). Combining this observation with the results from parts **a–c** gives the graph of f in Figure 9.26. From the graph we read that the range is the interval $(-\infty, \frac{9}{4}]$.

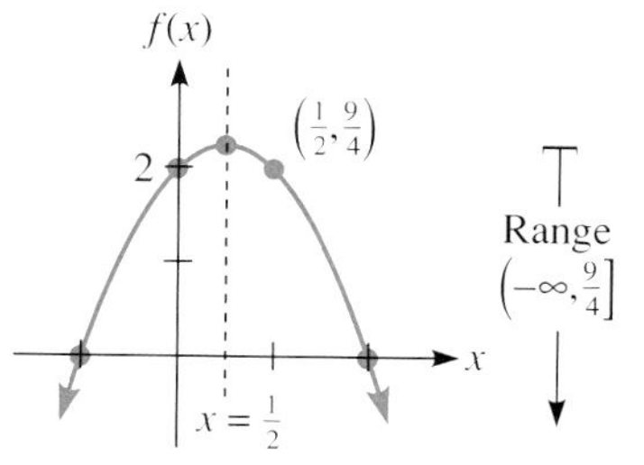

Figure 9.26

PROGRESS CHECK 5 Answer the questions in Example 5 for the function defined by $f(x) = -x^2 + 3x - 4$.

Progress Check Answers

5. (a) $x = \frac{3}{2}$ (b) $(\frac{3}{2}, -\frac{7}{4})$ (c) $(0, -4)$

(d)

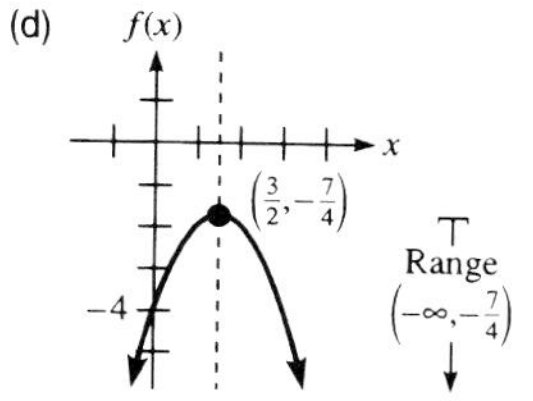

6 The vertex of a parabola can be found more efficiently in some cases by using the graphing techniques of the previous section. For instance, in Example 3 of Section 9.3 the graph of $y = (x + 2)^2 - 3$ was drawn by translating the graph of $y = x^2$ to the left 2 units and down 3 units, which positioned the vertex of the parabola at $(-2, -3)$. More generally, any quadratic function may be expressed in the form

$$f(x) = a(x - h)^2 + k \qquad (\text{with } a \neq 0)$$

by completing the square (as considered in Section 8.1). Graphing techniques applied

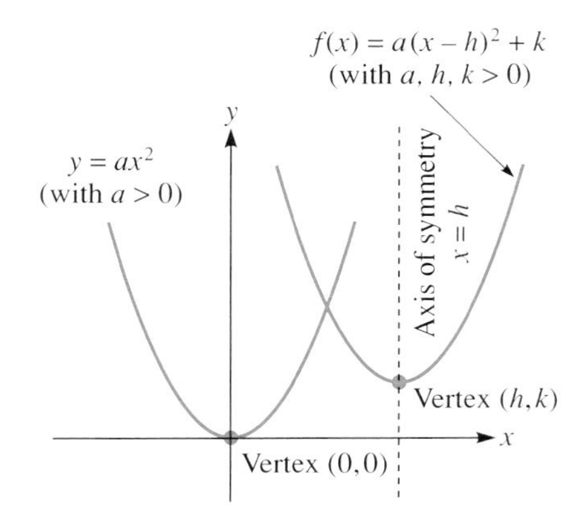

Figure 9.27

to this form indicate that the graph of f is the graph of $y = ax^2$ translated so that the vertex is the point (h,k) and the axis of symmetry is the line $x = h$, as diagramed in Figure 9.27. Example 6 illustrates this approach using the function analyzed in the previous example.

EXAMPLE 6 If $f(x) = -x^2 + x + 2$, find the vertex and axis of symmetry by matching the equation to the form $f(x) = a(x - h)^2 + k$.

Solution To complete the square, the coefficient of x^2 must be 1, so first factor out -1 from the x terms.

$$\begin{aligned} f(x) &= -x^2 + x + 2 \\ &= -1(x^2 - x) + 2 \end{aligned}$$

The square of one-half of the coefficient of x is $(-\frac{1}{2})^2$, or $1/4$. So inside the parentheses both add $\frac{1}{4}$ (to complete the square) and subtract $\frac{1}{4}$ [to keep $f(x)$ unchanged], and proceed as follows.

$$\begin{aligned} f(x) &= -1(x^2 - x + \tfrac{1}{4} - \tfrac{1}{4}) + 2 \\ &= -1(x^2 - x + \tfrac{1}{4}) + (-1)(-\tfrac{1}{4}) + 2 \\ &= -1(x - \tfrac{1}{2})^2 + \tfrac{9}{4} \end{aligned}$$

Matching this equation to the form $f(x) = a(x - h)^2 + k$ gives $h = \frac{1}{2}$ and $k = \frac{9}{4}$. Thus, the vertex is $(\frac{1}{2}, \frac{9}{4})$, and the axis of symmetry is $x = \frac{1}{2}$.

PROGRESS CHECK 6 If $f(x) = 2x^2 + 4x - 5$, find the vertex and axis of symmetry by matching the equation to the form $f(x) = a(x - h)^2 + k$.

7 Applications of quadratic functions often involve finding the maximum or minimum function value associated with the vertex, as illustrated next.

EXAMPLE 7 Find the maximum possible area that can be enclosed with 100 m of fencing shaped in the form of a rectangle.

Solution The area and perimeter formulas for a rectangle are

$$A = \ell w \qquad \text{and} \qquad P = 2\ell + 2w.$$

Since $P = 100$ m, we know that $100 = 2\ell + 2w$, so $\ell = 50 - w$. Substituting $50 - w$ for ℓ in the area formula gives

$$A = (50 - w)w = 50w - w^2.$$

The area is a maximum when

$$w = \frac{-b}{2a} = \frac{-50}{2(-1)} = 25.$$

And the corresponding value for A is

$$A = (50 - 25)25 = 625.$$

Thus, the maximum value for the area is 625 m^2. Note that the maximum area results from a rectangle that is a square.

PROGRESS CHECK 7 If 200 ft of fencing are available, find the area of the largest rectangular region that can be enclosed with the available fencing.

Progress Check Answers

6. $(-1, -7)$; $x = -1$
7. 2,500 ft^2

EXERCISES 9.4

In Exercises 1–4, graph the given constant function.

1. $f(x) = 4$

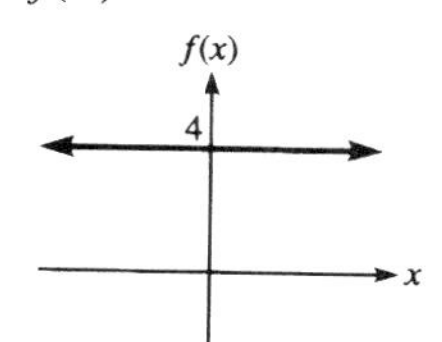

2. $f(x) = -2$

f(x)
−2
x

3. $f(x) = -5$

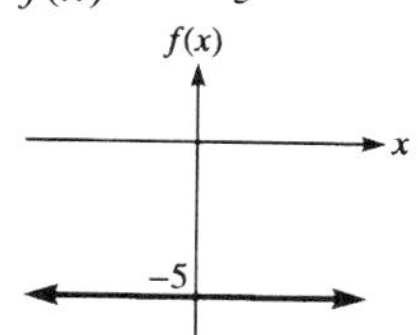

4. $f(x) = 7$

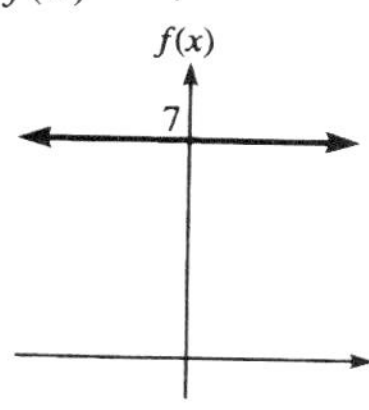

In Exercises 5–14, graph using the slope and y-intercept of the line.

5. $y = \frac{2}{3}x + 4$

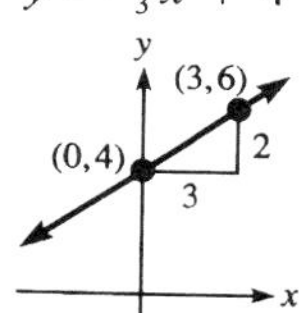

6. $y = \frac{3}{2}x + 5$

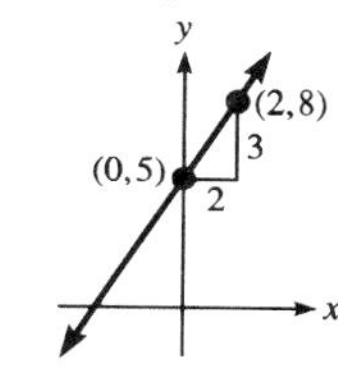

7. $y = -\frac{4}{3}x + 7$

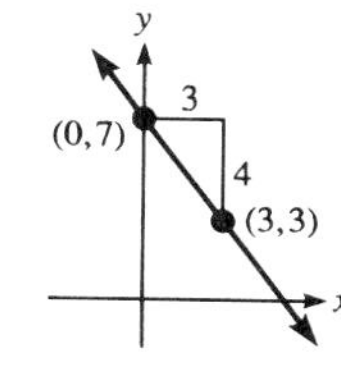

8. $y = -\frac{3}{4}x + 2$

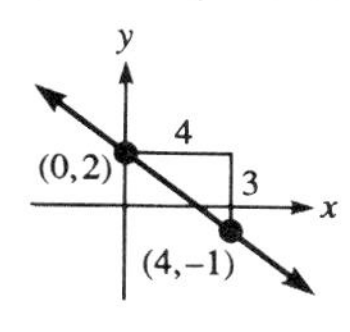

9. $y = -\frac{1}{3}x - 1$

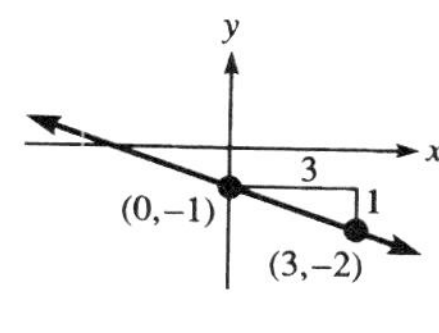

10. $y = -\frac{1}{4}x - 3$

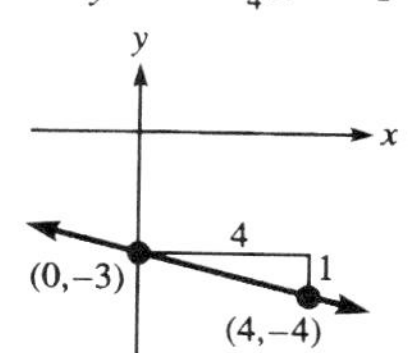

11. $y = \frac{4}{5}x - 3$

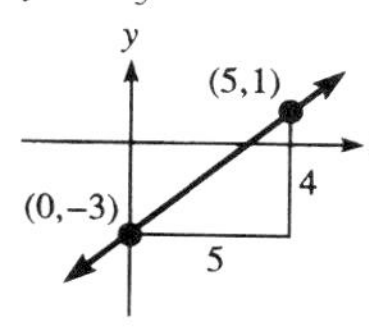

12. $y = \frac{7}{5}x - 2$

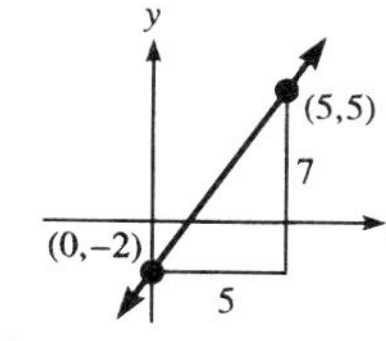

13. $y = -\frac{2}{3}x + 3$

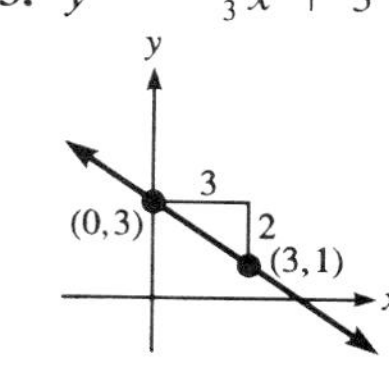

14. $y = -\frac{2}{3}x - 3$

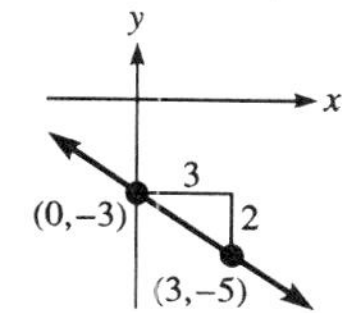

In Exercises 15–30, find an equation that defines the linear function f for each set of ordered pairs.

15. $f(2) = 0, f(3) = -2$ $f(x) = -2x + 4$

16. $f(-3) = 4, f(0) = -2$ $f(x) = -2x - 2$

17. $f(2) = 1, f(4) = 4$ $f(x) = \frac{3}{2}x - 2$

18. $f(3) = 2, f(6) = 1$ $f(x) = -\frac{1}{3}x + 3$

19. $f(5) = 2, f(-5) = 0$ $f(x) = \frac{1}{5}x + 1$

20. $f(5) = 2, f(-10) = -1$ $f(x) = \frac{1}{5}x + 1$

21. $f(3) = 5, f(6) = 3$ $f(x) = -\frac{2}{3}x + 7$

22. $f(3) = 1, f(6) = 3$ $f(x) = \frac{2}{3}x - 1$

23. $f(3) = -4, f(6) = -2$ $f(x) = \frac{2}{3}x - 6$

24. $f(-1) = 7, f(1) = -5$ $f(x) = -6x + 1$

25. $f(2) = -2, f(4) = -7$ $f(x) = -\frac{5}{2}x + 3$

26. $f(3) = -2, f(6) = -4$ $f(x) = -\frac{2}{3}x$

27. $f(-3) = 3, f(-6) = 4$ $f(x) = -\frac{1}{3}x + 2$

28. $f(4) = 1, f(2) = -2$ $f(x) = \frac{3}{2}x - 5$

29. $f(-5) = -4, f(-10) = -7$ $f(x) = \frac{3}{5}x - 1$

30. $f(-1) = -5, f(-2) = -3$ $f(x) = -2x - 7$

31. During exercise a person's maximum target heart rate is a function of age. This relation is specified by a linear function. The recommended maximum number of beats per minute is 170 at age 20 and 136 at age 60.

a. Find the equation that defines this linear function.
$y = -0.85x + 187$, where x = age and y = heart rate

b. What is the maximum target heart rate for a 30-year-old? 161.5, or about 162 beats per minute

32. During exercise a person's minimum target heart rate to have a training effect is a function of age. This relation is specified by a linear function, and the recommended number of beats per minute is 140 at age 20 and 112 at age 60.

a. Find the equation that defines this linear function.
$y = -0.7x + 154$

b. What is the minimum target heart rate for a 30-year-old? 133 beats per minute

In Exercises 33–44 for the given function, find the following.

a. The equation of the axis of symmetry
b. The coordinates of the vertex
c. The coordinates of the y-intercept
d. Then graph the function and specify the range using the graph.

33. $f(x) = 2x^2 - 4x + 5$ a. $x = 1$

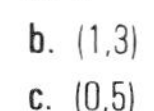

d.

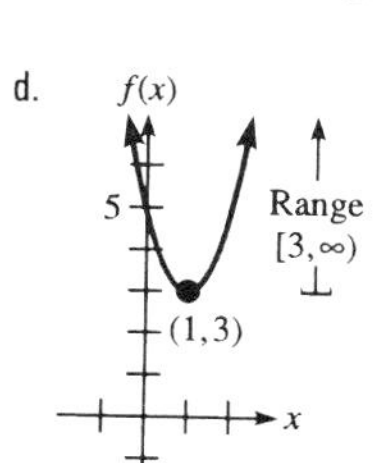

34. $f(x) = x^2 - 4x + 8$ a. $x = 2$
b. (2,4)
c. (0,8)

d.

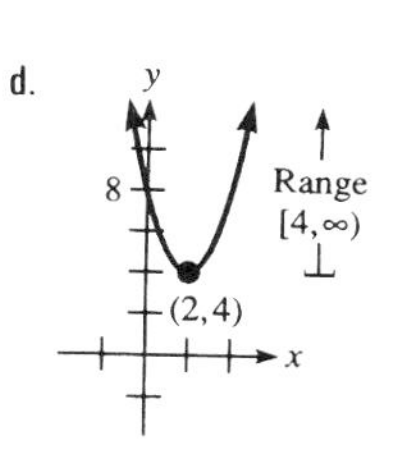

35. $f(x) = x^2 + 2x + 3$
a. $x = -1$
b. $(-1,2)$
c. $(0,3)$
d. Range $[2,\infty)$

36. $f(x) = x^2 + 6x + 10$
a. $x = -3$
b. $(-3,1)$
c. $(0,10)$
d. Range $[1,\infty)$

37. $f(x) = -x^2 + 6x - 14$
a. $x = 3$
b. $(3,-5)$
c. $(0,-14)$
d. Range $(-\infty,-5]$

38. $f(x) = -x^2 - 4x - 9$
a. $x = -2$
b. $(-2,-5)$
c. $(0,-9)$
d. Range $(-\infty,-5]$

39. $f(x) = -x^2 - 8x - 23$
a. $x = -4$
b. $(-4,-7)$
c. $(0,-23)$
d. Range $(-\infty,-7]$

40. $f(x) = -x^2 + 4x - 7$
a. $x = 2$
b. $(2,-3)$
c. $(0,-7)$
d. Range $(-\infty,-3]$

41. $f(x) = -x^2 + 4x - 1$
a. $x = 2$
b. $(2,3)$
c. $(0,-1)$
d. Range $(-\infty,3]$

42. $f(x) = -x^2 + 6x - 14$
a. $x = 3$
b. $(3,-5)$
c. $(0,-14)$
d. Range $(-\infty,-5]$

43. $f(x) = x^2 - x - 1$
a. $x = \frac{1}{2}$
b. $(\frac{1}{2},-\frac{5}{4})$
c. $(0,-1)$
d. Range $[-\frac{5}{4},\infty)$

44. $f(x) = x^2 + x - 2$
a. $x = -\frac{1}{2}$
b. $(-\frac{1}{2},-\frac{9}{4})$
c. $(0,-2)$
d. Range $[-\frac{9}{4},\infty)$

In Exercises 45–54 for the given quadratic function, find the equation of the axis of symmetry and the vertex by matching the equation to the form $f(x) = a(x - h)^2 + k$.

45. $y = x^2 - 2x - 3$ $x = 1$; $(1,-4)$
46. $y = x^2 - 4x + 4$ $x = 2$; $(2,0)$
47. $y = -x^2 - 6x - 14$ $x = -3$; $(-3,-5)$
48. $y = -x^2 + 4x - 1$ $x = 2$; $(2,3)$
49. $y = -x^2 + 3x + 4$ $x = \frac{3}{2}$; $(\frac{3}{2},\frac{25}{4})$
50. $y = -x^2 - 3x - 4$ $x = -\frac{3}{2}$; $(-\frac{3}{2},-\frac{7}{4})$
51. $y = 2x^2 - 4x + 5$ $x = 1$; $(1,3)$
52. $y = 2x^2 - 4x + 6$ $x = 1$; $(1,4)$
53. $y = 2x^2 - 6x - 3$ $x = \frac{3}{2}$; $(\frac{3}{2},-\frac{15}{2})$
54. $y = 2x^2 - 6x + 8$ $x = \frac{3}{2}$; $(\frac{3}{2},\frac{7}{2})$

In Exercises 55–66 the graph of an appropriate function will help you see why the solutions are reasonable.

55. Find the maximum possible area of a rectangular patio that can be bordered with 300 ft of decorative edging. 5,625 ft²

56. If you have 80 ft of decorative edging, what's the largest rectangular concrete area it can border? 400 ft²

57. To keep the rabbits out of your two-section garden (see the figure), you purchase 60 yd of chicken-wire fencing to enclose it. You want the largest possible growing area for your garden. What is the maximum number of square yards your 60 yd of fencing can enclose? (Confirm that the total fencing used is 60 yd.) 150 yd²

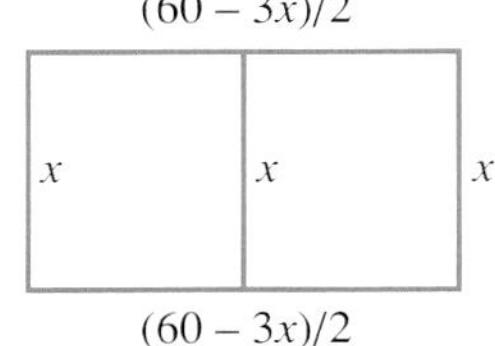

58. A farmer raises sheep and goats. If 1,200 m of fencing are available to surround the grazing areas shown in the figure, what's the largest area that can be enclosed? (Confirm that the total fencing used is 1,200 m.) 60,000 m²

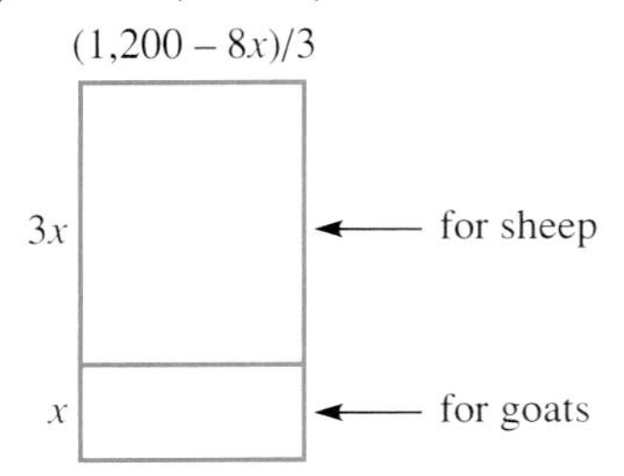

59. A bus tour charges a fare of \$10 per person and carries 200 people per day. The manager estimates that they will lose 10 passengers for each \$1 increase in fare. Find the most profitable fare to charge, and give the expected income. [*Hint:* Let x = the number of \$1 increases made in the fare. Then the number of passengers will be given by $200 - 10x$. So you need to find the maximum value of $(10 + x)(200 - 10x)$.]
Charge \$15 per person; income = \$2,250.

60. Redo Exercise 59, but assume that in the beginning the fare for 200 people is \$20 per person. Charge \$20 per person; income = \$4,000.

61. What positive number exceeds its square by the largest amount? By how much does the number exceed its square? (*Hint:* Maximize $x - x^2$.) $\frac{1}{2}$ exceeds $(\frac{1}{2})^2$ by $\frac{1}{4}$.

62. Find two positive numbers whose sum is 20 such that the sum of their squares is a minimum. What is that minimum sum? Both numbers are 10; minimum sum = 200.

63. You have 100 ft of fencing to use for three sides of a rectangular pen. The fourth side will be an existing stone wall. (See the figure.) What dimensions will surround the maximum area? What is the maximum area? The pen should be 25 by 50 ft; maximum area = 1,250 ft².

x x
$100 - 2x$

64. Redo Exercise 63, but assume that you have 120 ft of fence. 30 by 60 ft; maximum area = 1,800 ft²

65. Scientists often have to find equations for functions which "fit" their observations fairly well. In analyzing some lab data, a statistician uses a variant of a procedure called the "method of least squares" to find the slope of the straight line through the origin that comes closest to both points (2,1) and (3,4). (See the figure.) The equation of a line through the origin is $y = mx$, which leads to the problem of minimizing the expression $(2m - 1)^2 + (3m - 4)^2$. Find the value of m which minimizes this polynomial, and write the equation of the desired line through the origin. $m = \frac{14}{13}$; $y = \frac{14}{13}x$

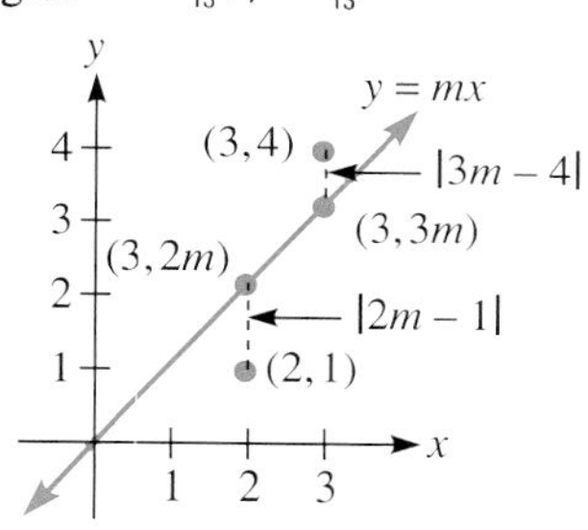

66. In the given figure the side of the large square is 10 units. (Check this.) The figure shows that there are many different-size smaller squares that can be inscribed in the larger one. Use the given labels to answer these questions.

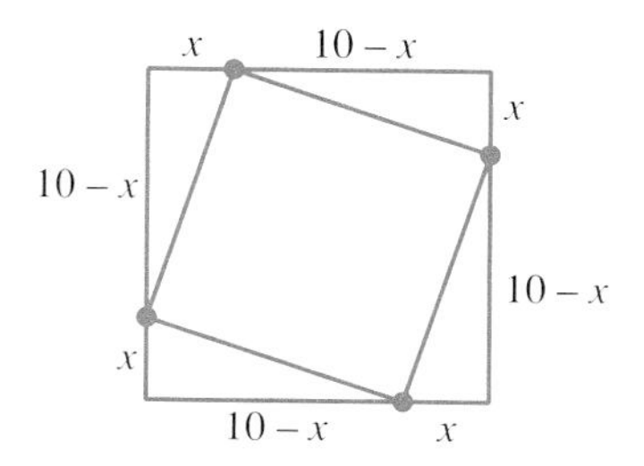

a. What expression represents the length of the side of the smaller square? $\sqrt{x^2 + (10 - x)^2} = \sqrt{2x^2 - 20x + 100}$

b. What expression represents the area of the smaller square? $2x^2 - 20x + 100$

c. What value of x gives the inscribed square with the smallest possible area? What is that minimum area? 5; 50 square units

THINK ABOUT IT

1. Is the given function a polynomial function? If yes, state the degree. If no, explain why.

a. $f(x) = \sqrt{2}\,x$ **b.** $f(x) = 2\sqrt{x}$

c. $f(x) = 2/x$ **d.** $f(x) = 2^{-1}$

2. Construct a table of values and graph the third-degree polynomial function defined by $f(x) = x^3 - x^2 - 10x - 8$.

3. The following table shows corresponding values on two temperature scales.

Degrees Celsius, x	0	5	10	15	20
Degrees Fahrenheit, y	32	41	50	59	68

This table was generated using a linear function $y = f(x)$. How can we recognize that the relation is linear from inspection of the table?

4. Given any three points that are not in a straight line, it is possible to find a parabola whose graph includes all three points.

a. Find an equation for a parabola which goes through (0,1), (1,0), and (2,1).

b. Find a polynomial function with degree 2 for which $f(0) = 1$, $f(1) = 0$, and $f(2) = 1$.

5. A line goes through the vertex (2,2) of a parabola and intersects it again at (3,4). Find the equations of the line and the parabola.

REMEMBER THIS

1. If $f(x) = 10x$ and $g(x) = 0.01x^2 + x + 200$, find $f(10) - g(10)$. -111
2. If $f(x) = x + 2$ and $g(x) = x^2$, find $g(x) - f(x)$. $x^2 - x - 2$
3. Express the circumference (C) of a circle as a function of its diameter (d). $C = \pi d$
4. Multiply $(4x - 3)(x^2 - x - 5)$. $4x^3 - 7x^2 - 17x + 15$
5. Use this graph to determine $f(-2)$. $f(-2) = 0$

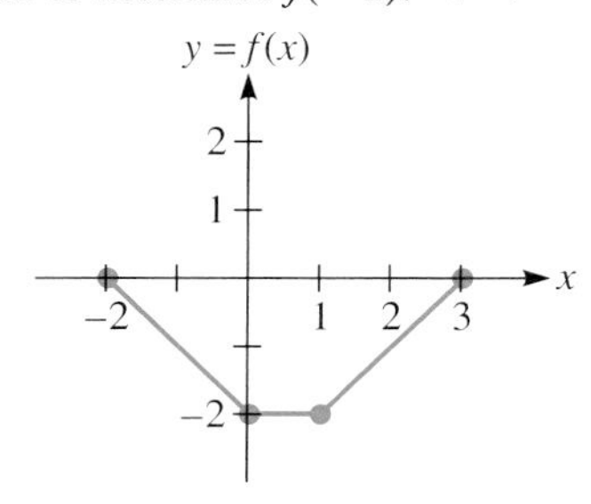

6. Use the given graph of $y = f(x)$ to draw the graph of $y = f(x) - 2$.

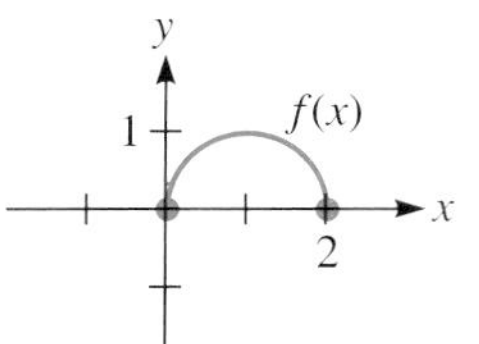

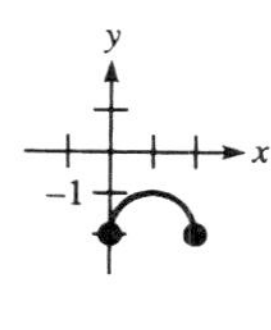

7. Write the equation of a circle with radius 3 and center at $(-2,2)$. $(x + 2)^2 + (y - 2)^2 = 9$
8. If the center of a circle is at the origin and one point of the circle is $(-12,5)$, what is the diameter of the circle? 26
9. Find the equation of the line through $(1,1)$ with slope -1. $y = -x + 2$
10. Factor $x^2 - 4x + 4$. $(x - 2)^2$

9.5 Operations with Functions

An oil leak is spreading over a plane surface in the shape of a circle. The radius of the circle is increasing at a rate of 7 cm/second, so the radius of this spill t seconds after the start of the leak may be expressed by $r = g(t) = 7t$. If function f expresses the area of this circular spill as a function of r so that $A = f(r) = \pi r^2$, find and interpret $(f \circ g)(t)$. (See Example 6.)

OBJECTIVES

1. **Add, subtract, multiply, and divide two functions.**
2. **Find the composite function of two functions.**

1 New functions are often formed by adding, subtracting, multiplying, or dividing two functions. For example,

$$\text{profit} = \text{revenue} - \text{cost}.$$

So if the revenue function for manufacturing and selling x units of a certain product is

$$R = f(x) = 10x$$

and the cost function is

$$C = g(x) = 0.01x^2 + x + 200,$$

then the profit function is

$$\begin{aligned} P &= f(x) - g(x) \\ &= 10x - (0.01x^2 + x + 200) \\ &= -0.01x^2 + 9x - 200. \end{aligned}$$

This example illustrates how to find a difference of two functions, and in general, $f + g$, $f - g$, $f \cdot g$, and f/g are defined as follows.

Sum, Difference, Product, and Quotient of Two Functions

If f and g are functions, then:

Sum of f and g: $(f + g)(x) = f(x) + g(x)$

Difference of f and g: $(f - g)(x) = f(x) - g(x)$

Product of f and g: $(f \cdot g)(x) = f(x) \cdot g(x)$

Quotient of f and g: $\left(\frac{f}{g}\right)(x) = \frac{f(x)}{g(x)}$, $g(x) \neq 0$

EXAMPLE 1 If $f(x) = 4x - 3$ and $g(x) = x^2 - x - 5$, find the following.

a. $(f + g)(x)$ **b.** $(f - g)(x)$ **c.** $(f \cdot g)(x)$ **d.** $\left(\frac{f}{g}\right)(x)$

Solution

a.
$$\begin{aligned}(f + g)(x) &= f(x) + g(x)\\ &= (4x - 3) + (x^2 - x - 5)\\ &= x^2 + 3x - 8\end{aligned}$$

b.
$$\begin{aligned}(f - g)(x) &= f(x) - g(x)\\ &= (4x - 3) - (x^2 - x - 5)\\ &= 4x - 3 - x^2 + x + 5\\ &= -x^2 + 5x + 2\end{aligned}$$

c.
$$\begin{aligned}(f \cdot g)(x) &= f(x) \cdot g(x)\\ &= (4x - 3)(x^2 - x - 5)\\ &= 4x^3 - 4x^2 - 20x - 3x^2 + 3x + 15\\ &= 4x^3 - 7x^2 - 17x + 15\end{aligned}$$

d.
$$\begin{aligned}\left(\frac{f}{g}\right)(x) &= \frac{f(x)}{g(x)}\\ &= \frac{4x - 3}{x^2 - x - 5}\end{aligned}$$

PROGRESS CHECK 1 If $f(x) = 3x^2 - 2x - 6$ and $g(x) = 2x - 1$, find the following.

a. $(f + g)(x)$ **b.** $(f - g)(x)$ **c.** $(f \cdot g)(x)$ **d.** $\left(\frac{f}{g}\right)(x)$

EXAMPLE 2 If $f(x) = x^2 - 5x + 3$ and $g(x) = 3x - 1$, find $(f \cdot g)(2)$.

Solution $(f \cdot g)(2) = f(2) \cdot g(2)$, so find $f(2)$ and $g(2)$.

$$\begin{aligned}f(x) &= x^2 - 5x + 3\\ f(2) &= (2)^2 - 5(2) + 3\\ &= -3\end{aligned} \qquad \begin{aligned}g(x) &= 3x - 1\\ g(2) &= 3(2) - 1\\ &= 5\end{aligned}$$

Then,

$$(f \cdot g)(2) = f(2) \cdot g(2) = (-3)(5) = -15.$$

An alternative solution method is to find $(f \cdot g)(2)$ by first finding $(f \cdot g)(x)$.

PROGRESS CHECK 2 If $f(x) = x^2 - 5x + 3$ and $g(x) = 3x - 1$, find $(f - g)(3)$.

Progress Check Answers

1. (a) $(f + g)(x) = 3x^2 - 7$
(b) $(f - g)(x) = 3x^2 - 4x - 5$
(c) $(f \cdot g)(x) = 6x^3 - 7x^2 - 10x + 6$
(d) $\left(\frac{f}{g}\right)(x) = \frac{3x^2 - 2x - 6}{2x - 1}$
2. -11

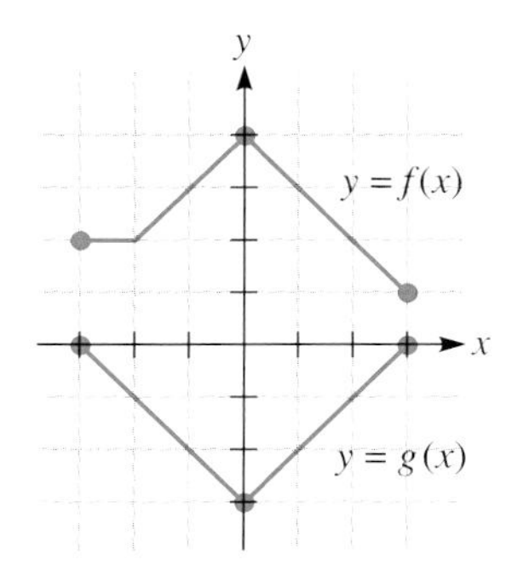

Figure 9.28

EXAMPLE 3 Use the graphs of f and g in Figure 9.28 to evaluate each expression.

a. $(f + g)(0)$ **b.** $\left(\frac{f}{g}\right)(-2)$

Solution

a. To find $f(0)$ and $g(0)$ requires finding the y value in each function when $x = 0$. Since $(0,4)$ is in f and $(0,-3)$ is in g, $f(0) = 4$ and $g(0) = -3$. Then,

$$(f + g)(0) = f(0) + g(0) = 4 + (-3) = 1.$$

b. From the graph we read that $f(-2) = 2$ and $g(-2) = -1$. Then,

$$\left(\frac{f}{g}\right)(-2) = \frac{f(-2)}{g(-2)} = \frac{2}{-1} = -2.$$

PROGRESS CHECK 3 Evaluate each expression using Figure 9.28.

a. $(f - g)(1)$ **b.** $(f \cdot g)(3)$

2 Functions may also be combined by an operation called composition. To illustrate, consider the problem of determining the cost of a square tablecloth that sells for $5 per square foot in terms of the side length s. The cost is a function of the area, as expressed by

$$C = f(A) = 5A.$$

Meanwhile, the area is a function of the side length, as expressed by

$$A = g(s) = s^2.$$

To find C as a function of s, we replace A by s^2 in function f to obtain

$$C = 5s^2,$$

and this substitution may be viewed in functional notation by writing

$$C = f[g(s)] = f(s^2) = 5s^2.$$

A function like $f[g(s)]$ is said to be a composition of f and g. In general, the symbol $\circ$ is used to denote the operation of composition, and the **composite functions** of f and g are defined as follows:

$$(f \circ g)(x) = f[g(x)]$$
$$(g \circ f)(x) = g[f(x)].$$

Basically, $f[g(x)]$ can be viewed as a "chain reaction" in which the output of the g rule becomes the input for the f rule, so that two functions are applied in succession.

EXAMPLE 4 If $f(x) = x^2$ and $g(x) = 2x - 5$, find the following.

a. $(f \circ g)(x)$ **b.** $(g \circ f)(x)$

Solution

a.
$$\begin{aligned} (f \circ g)(x) &= f[g(x)] && \text{By definition.} \\ &= f(2x - 5) && \text{Replace } g(x) \text{ by } 2x - 5. \\ &= (2x - 5)^2 && \text{Apply the } f \text{ rule.} \\ &= 4x^2 - 20x + 25 && \text{Expand.} \end{aligned}$$

b.
$$\begin{aligned} (g \circ f)(x) &= g[f(x)] && \text{By definition.} \\ &= g(x^2) && \text{Replace } f(x) \text{ by } x^2. \\ &= 2x^2 - 5 && \text{Apply the } g \text{ rule.} \end{aligned}$$

Note In this example $(f \circ g)(x) \neq (g \circ f)(x)$, and this result holds except for special classes of functions. Thus, it is *not* true that $(f \circ g)(x)$ equals $(g \circ f)(x)$ for all functions f and g, so composition of functions is not a commutative operation.

Progress Check Answers

3. (a) 5 (b) 0

Consider asking students which operations on functions are commutative and which are not. See "Think About It" Exercises 4 and 5 for some cases where $(f \circ g)(x) = (g \circ f)(x)$. Further consideration of inverse functions is given in Section 10.2.

PROGRESS CHECK 4 If $f(x) = x^2 - 1$ and $g(x) = 3x + 2$, find the following.

a. $(f \circ g)(x)$ **b.** $(g \circ f)(x)$

EXAMPLE 5 Use the graph of f and g in Figure 9.28 to evaluate $(f \circ g)(1)$.

Solution $(f \circ g)(1) = f[g(1)] = f(-2) = 2$, where $g(1) = -2$ and $f(-2) = 2$ are read from the graphs of g and f, respectively.

PROGRESS CHECK 5 Use the graphs of f and g in Figure 9.28 to evaluate $(g \circ f)(1)$.

EXAMPLE 6 Solve the problem in the section introduction on page 462.

Solution By definition, $(f \circ g)(t) = f[g(t)]$, and in this problem $g(t) = 7t$. Thus,

$$(f \circ g)(t) = f[g(t)] = f(7t) = \pi(7t)^2 = 49\pi t^2.$$

An interpretation of this result is as follows: If function f expresses A in terms of r, and function g expresses r in terms of t, then the function $f \circ g$ expresses A in terms of t. Thus, the area of this circular spill t seconds after the start of the leak is given by

$$A = (f \circ g)(t) = 49\pi t^2.$$

PROGRESS CHECK 6 An oil leak is spreading over a plane surface in the shape of a circle. The radius of this circle is increasing at a rate of 5 cm/second, so the radius of this spill t seconds after the start of the leak may be expressed by $r = g(t) = 5t$. If function f expresses the circumference of this circular spill as a function of r so that $C = f(r) = 2\pi r$, find and interpret $(f \circ g)(t)$.

Progress Check Answers

4. (a) $9x^2 + 12x + 3$ (b) $3x^2 - 1$
5. 0
6. $C = (f \circ g)(t) = 10\pi t$ is the formula for the circumference of the spill t seconds after the start of the leak.

EXERCISES 9.5

For Exercises 1–6, use the two functions $f(x)$ and $g(x)$ to find $(f + g)(x)$, $(f - g)(x)$, $(f \cdot g)(x)$, and $(f/g)(x)$.

1. $f(x) = 3x$, $g(x) = x^2$ $(f+g)(x) = 3x + x^2$; $(f-g)(x) = 3x - x^2$; $(f \cdot g)(x) = 3x^3$; $\left(\frac{f}{g}\right)(x) = \frac{3}{x}$

2. $f(x) = 2x$, $g(x) = 1 - x$ $(f+g)(x) = x + 1$; $(f-g)(x) = 3x - 1$; $(f \cdot g)(x) = 2x - 2x^2$; $\left(\frac{f}{g}\right)(x) = \frac{2x}{1-x}$

3. $f(x) = 5x - 2$, $g(x) = x^2 - 3x + 1$ $(f+g)(x) = x^2 + 2x - 1$; $(f-g)(x) = -x^2 + 8x - 3$; $(f \cdot g)(x) = 5x^3 - 17x^2 + 11x - 2$; $\left(\frac{f}{g}\right)(x) = \frac{5x - 2}{x^2 - 3x + 1}$

4. $f(x) = x^2 + 3x + 2$, $g(x) = 7 - 2x$ $(f+g)(x) = x^2 + x + 9$; $(f-g)(x) = x^2 + 5x - 5$; $(f \cdot g)(x) = -2x^3 + x^2 + 17x + 14$; $\left(\frac{f}{g}\right)(x) = \frac{x^2 + 3x + 2}{7 - 2x}$

5. $f(x) = 2x^2 - x + 5$, $g(x) = 3x + 7$ $(f+g)(x) = 2x^2 + 2x + 12$; $(f-g)(x) = 2x^2 - 4x - 2$; $(f \cdot g)(x) = 6x^3 + 11x^2 + 8x + 35$; $\left(\frac{f}{g}\right)(x) = \frac{2x^2 - x + 5}{3x + 7}$

6. $f(x) = 3x^2 - 4x + 2$, $g(x) = 4x - 3$ $(f+g)(x) = 3x^2 - 1$; $(f-g)(x) = 3x^2 - 8x + 5$; $(f \cdot g)(x) = 12x^3 - 25x^2 + 20x - 6$; $\left(\frac{f}{g}\right)(x) = \frac{3x^2 - 4x + 2}{4x - 3}$

Given $f(x) = x^2 - 4x + 5$ and $g(x) = 2x - 3$.

7. Find $(f \cdot g)(3)$. 6

8. Find $(f - g)(2)$. 0

Given $f(x) = 2x^2 - 3x + 2$ and $g(x) = 5 - 2x$.

9. Find $(f + g)(-1)$. 14

10. Find $(f/g)(1)$. $\frac{1}{3}$

Given $f(x) = 3x^2 - x - 1$ and $g(x) = 5x - 14$.

11. Find $(f/g)(2)$. $-\frac{9}{4}$

12. Find $(f + g)(4)$. 49

Given $f(x) = x^2 + 2$ and $g(x) = x^2 - 4x + 5$.

13. Find $(f + g)(3)$. 13

14. Find $(f - g)(-3)$. -15

15. Find $(f \cdot g)(3)$. 22

16. Find $(f/g)(-3)$. $\frac{11}{26}$

Use this graph for Exercises 17 and 18.

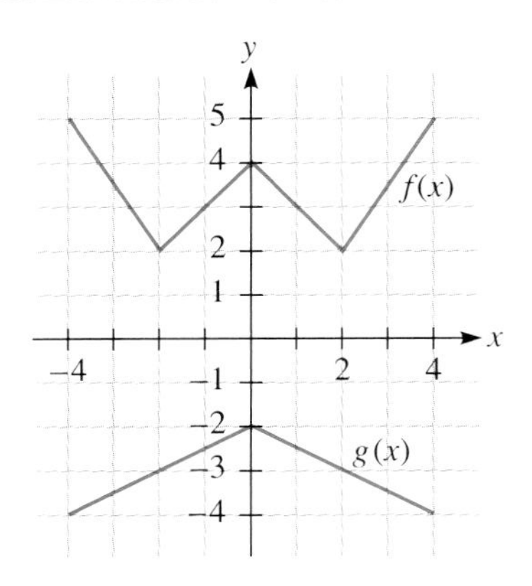

17. **a.** Find $(f + g)(0)$. 2
b. Find $(f - g)(0)$. 6
c. Find $(f \cdot g)(0)$. -8
d. Find $(f/g)(0)$. -2

18. **a.** Find $(f + g)(-4)$. 1
b. Find $(f - g)(-4)$. 9
c. Find $(f \cdot g)(-4)$. -20
d. Find $(f/g)(-4)$. $-\frac{5}{4}$

Use the given graph for Exercises 19 and 20.

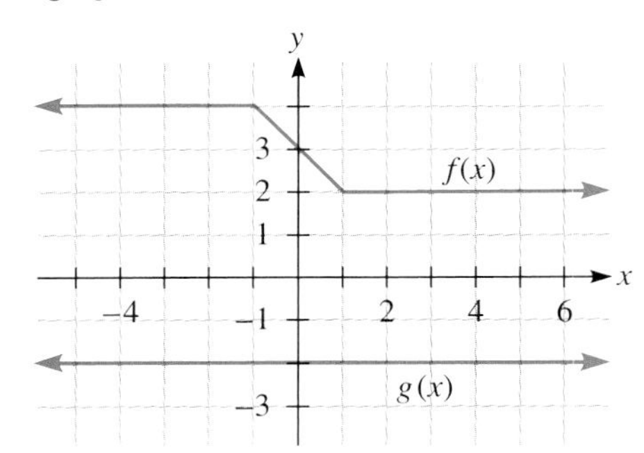

19. **a.** Find $(f + g)(1)$. 0
b. Find $(f - g)(1)$. 4
c. Find $(f \cdot g)(0)$. -6
d. Find $(f/g)(-1)$. -2

20. **a.** Find $(f + g)(-1)$. 2
b. Find $(f - g)(-1)$. 6
c. Find $(f \cdot g)(-2)$. -8
d. Find $(f/g)(2)$. -1

In Exercises 21–26, find (a) $(f \circ g)(x)$ and (b) $(g \circ f)(x)$.

21. $f(x) = 2x$, $g(x) = 3x - 1$ a. $6x - 2$ b. $6x - 1$

22. $f(x) = 5x - 2$, $g(x) = 3x - 3$ a. $15x - 17$ b. $15x - 9$

23. $f(x) = x^2 + 1$, $g(x) = 2 - 3x$ a. $5 - 12x + 9x^2$ b. $-3x^2 - 1$

24. $f(x) = x^2 - 2$, $g(x) = 7 - 2x$ a. $47 - 28x + 4x^2$ b. $11 - 2x^2$

25. $f(x) = 4x + 3$, $g(x) = x^2 - 4$ a. $4x^2 - 13$ b. $16x^2 + 24x + 5$

26. $f(x) = x^2 + 7$, $g(x) = 4 - 5x$ a. $23 - 40x + 25x^2$ b. $-31 - 5x^2$

For Exercises 27–32, use the graph to evaluate the given expression.

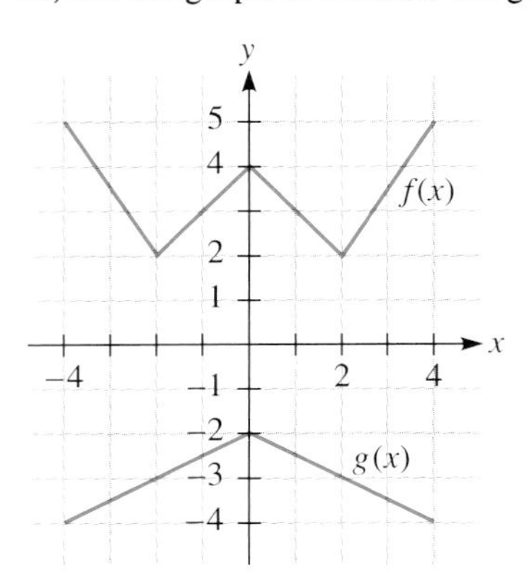

27. Find $(f \circ g)(0)$. 2

28. Find $(g \circ f)(0)$. -4

29. Find $(f \circ g)(4)$. 5

30. Find $(g \circ f)(4)$. Undefined

31. Find $(g \circ f)(2)$. -3

32. Find $(g \circ f)(-4)$. Undefined

For Exercises 33–40, use the graph to evaluate the given expression.

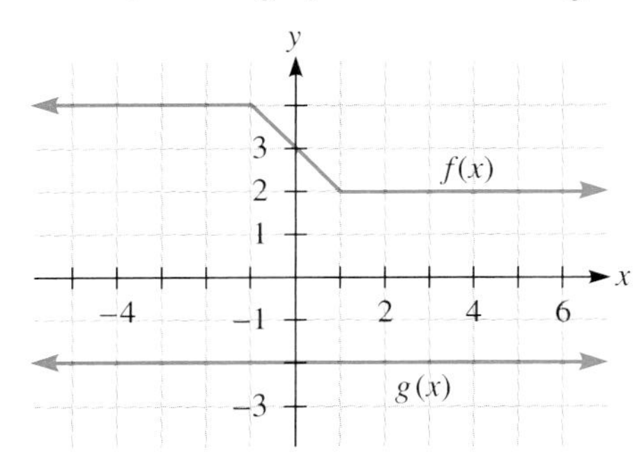

33. Find $(f \circ g)(0)$. 4

34. Find $(g \circ f)(0)$. -2

35. Find $(f \circ g)(1)$. 4

36. Find $(g \circ f)(1)$. -2

37. Find $(f \circ g)(-1)$. 4

38. Find $(g \circ f)(-1)$. -2

39. Find $(g \circ f)(4)$. -2

40. Find $(f \circ g)(4)$. 4

41. As soon as a stand of extremely fast-growing bamboo is planted, it begins growing outward in the shape of a circle. The radius of this circle is increasing at a rate of 6 in. per month, so the radius of the stand m months after planting may be expressed by $r = g(m) = 6m$. If the function f expresses the area of this circular stand of bamboo as a function of the radius r, so that $A = f(r) = \pi r^2$, find and interpret $(f \circ g)(m)$. $(f \circ g)(m) = 36\pi m^2$. If f expresses area A in terms of radius r, and g expresses r in terms of months m, then the composition function $f \circ g$ expresses area A in terms of months m.

42. Continuation of Exercise 41: Function r is the same, but now function f expresses the circumference of this circular stand of bamboo as a function of the radius, so that $C = f(r) = 2\pi r$. Find and interpret $(f \circ g)(m)$. $(f \circ g)(m) = 12\pi m$. If f expresses circumference C in terms of radius r, and g expresses r in terms of months m, then the composition function $f \circ g$ expresses circumference in terms of months m.

43. The sides (s) of an inflatable cube are 1 in. but then expand at the rate of 0.2 in./second, which gives $s = f(t) = 1 + 0.2t$, where t is time in seconds. The function $V = g(s) = s^3$ gives the volume of a cube as a function of its side.
a. Find and interpret $(g \circ f)(t)$.
b. Find and interpret $(g \circ f)(10)$.

a. $(g \circ f)(t) = (1 + 0.2t)^3$; this gives the volume after t seconds.
b. $(g \circ f)(10) = 27$; after 10 seconds the volume is 27 in.^{3}.

44. Refer to Exercise 43, and answer parts **a** and **b**, but assume that now the sides expand at the rate of 0.4 in./second.

a. $(g \circ f)(t) = (1 + 0.4t)^3$

b. $(g \circ f)(10) = 125$; after 10 seconds the volume is 125 in.³.

45. The cost of producing a full shipment of widgits is given by $C = f(h) = 340h + 200$, where h is the current hourly pay rate at the factory. For the next five years the hourly pay rate will go up 5 percent per year. Thus, $g(t) = h(1.05)^t$ gives the hourly pay rate after t years.

a. Find and interpret $(f \circ g)(t)$.
b. Find and interpret $(f \circ g)(4)$.
c. Find and interpret $(f \circ g)(4)$ if $h = 10$.

a. $(f \circ g)(t) = 340h(1.05)^t + 200$; this gives the cost of a full shipment of widgits t years from now.

b. $(f \circ g)(4) = 413.27h + 200$; this gives the cost of a full shipment of widgits 4 years from now.

c. $(f \circ g)(4) = 4{,}332.7$; 4 years from now it will cost $4,332.70 to produce a full shipment of widgets.

46. The cost of producing a full shipment of gimgiks is given by $C = f(h) = 500h + 300$, where h is the current hourly pay rate at the factory. For the next five years the hourly pay rate will go up 3 percent per year. Thus, $g(t) = h(1.03)^t$ gives the hourly pay rate after t years.

a. Find and interpret $(f \circ g)(t)$.
b. Find and interpret $(f \circ g)(5)$.
c. Find and interpret $(f \circ g)(5)$ if $h = 12$.

a. $(f \circ g)(t) = 500(1.03)^t h + 300$; this gives the cost of a full shipment of gimgiks t years from now.

b. $(f \circ g)(5) = 579.64h + 300$; this gives the cost of a full shipment of gimgiks 5 years from now.

c. $(f \circ g)(5) = 7{,}255.68$; 5 years from now it will cost $7,255.68 to produce a full shipment of gimgiks.

THINK ABOUT IT

1. The domain of the function $f + g$ is the intersection of the domains of f and g. If $f = \{(0,1),(1,2),(2,3)\}$ and $g = \{(1,-1),(2,-2),(3,-3)\}$, find the domain of $f + g$.

2. Find $f + g$ using the function in Exercise 1.

3. Use the graphs below to graph $y = (f + g)(x)$.

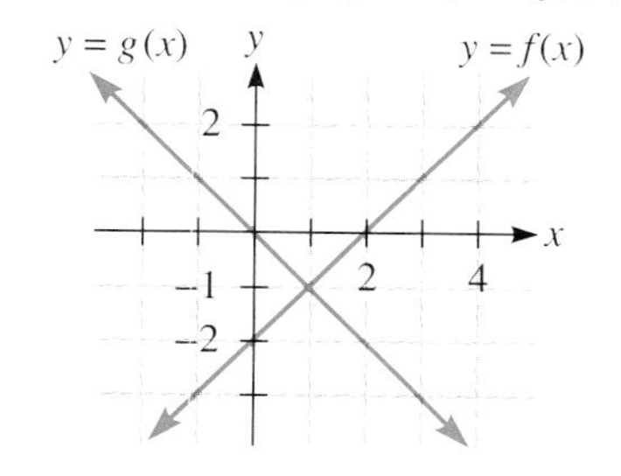

4. Let $f(x) = 2x - 5$ and $g(x) = \frac{x + 5}{2}$.

a. Show that $(f \circ g)(1) = 1$.
b. Show that $(f \circ g)(2) = 2$.
c. Show that $(f \circ g)(x) = x$ for all values of x.
d. Show that $(g \circ f)(x) = x$ for all values of x.

5. If $f(x) = 3x - 4$ and $(f \circ g)(x) = x$ for all values of x, find $g(x)$.

REMEMBER THIS

1. Use a calculator to evaluate $y = 2^x$ when $x = 15$. 32,768

2. Express this number in scientific notation: 1,073,700,000. 1.0737×10^9

3. Evaluate $y = 9(14.92)^x$ when $x = 0$. 9

4. Which is greater, $(\frac{1}{2})^5$ or $(\frac{1}{2})^3$? $(\frac{1}{2})^3$

5. Is $\frac{2}{3}$ a solution of the equation $8^x = 4$? Yes

6. If $f(x) = x^3$ and $g(x) = x^{1/3}$, find $(f \circ g)(11)$. 11

7. Graph $xy = 24$. Is this a hyperbola, a parabola, or neither? Hyperbola

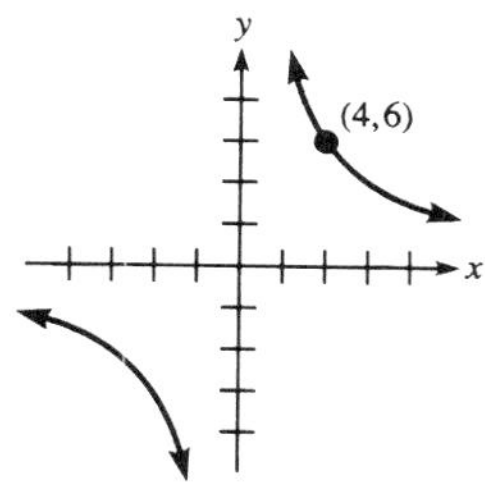

8. Solve $3x^2 - 2x = 3$. $\left\{\frac{1 \pm \sqrt{10}}{3}\right\}$

9. Solve $\sqrt{x} + 1 = \sqrt{x + 5}$. $\{4\}$

10. Solve $3(x - 5) + 10 = 3x - 5$. Identity; all real numbers are solutions.

Identifying Vertical Asymptotes and Investigating Graphing Techniques

In connection with the material of this chapter, it is worth pointing out that the [Y=] key records equations of functions. Thus, for instance, the graph for Y_1 must be the graph of a function. Example 1 illustrates what the TI-81 does when a graph has a vertical asymptote. You will see that the graph is an aid to but not a substitute for careful algebraic analysis.

EXAMPLE 1 Graph $y = \dfrac{1}{x-2}$.

Solution

Step 1 Use the [Y=] key and set $Y_1 = 1/(X - 2)$.

Step 2 Press [ZOOM] Standard to get the graph shown in Figure 9.29. The appearance of a break in the graph (or a sudden spike) may indicate the presence of a vertical asymptote, as it does in this case.

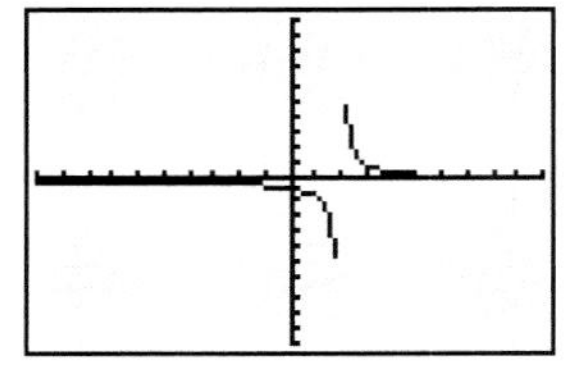

Figure 9.29

Step 3 Use [TRACE] to move the cursor to $X = 2$, and note that the y value is blank. This is further evidence that $x = 2$ is not in the domain.

Step 4 Press [ZOOM] In to get the graph shown in Figure 9.30. The (almost) vertical line that now appears is another clue that there is an asymptote at $x = 2$. This line is not actually the asymptote but results when the calculator connects a point just to the left of 2 with a point just to the right of 2. (Using [TRACE], you can see that when $x \approx 1.97$, for example, then $y \approx -38$, and when $x \approx 2.03$, then $y \approx 38$.)

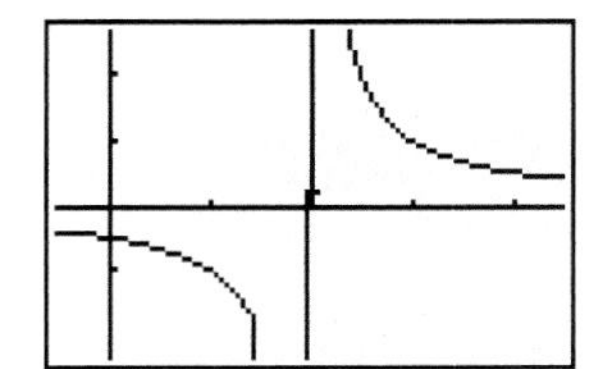

Figure 9.30

Note You can use the [MODE] key to select "Dot" instead of the default "Connected," and then the graphs will just be individual points, not connected. If the "Dot" mode is used for the graph of $y = 1/(x - 2)$ then the "asymptote" will not appear.

In the next two examples we demonstrate the use of the graphing calculator to investigate the graphing techniques discussed in Chapter 9. Example 2 demonstrates vertical and horizontal translations, while Example 3 exhibits reflection, stretching, and flattening.

EXAMPLE 2 Draw these graphs on the same screen to confirm the effect of changing $y = f(x)$ to $y = f(x) + c$ or to $y = f(x + c)$.

a. $y = \sqrt{x}$ **b.** $y = \sqrt{x} + 3$ **c.** $y = \sqrt{x + 5}$

Solution Use the [Y=] key and enter these expressions for Y_1, Y_2, and Y_3, respectively. With [ZOOM] Standard the screen should appear as shown in Figure 9.31.

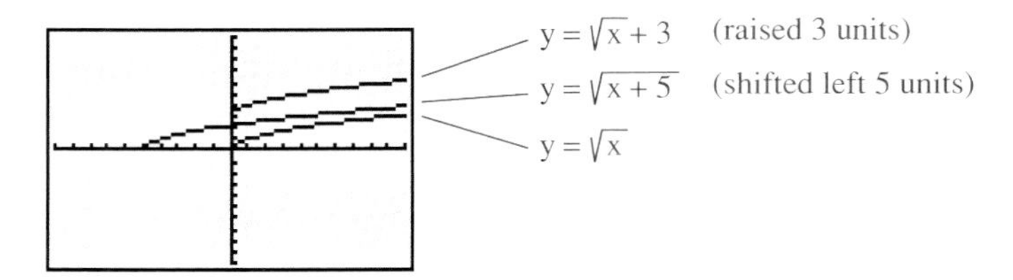

Figure 9.31

EXAMPLE 3 Draw these graphs on the same screen to confirm the effect of changing $y = f(x)$ to $y = cf(x)$.

a. $y = x^2$ **b.** $y = -x^2$
c. $y = 10x^2$ **d.** $y = \frac{1}{5}x^2$

Solution After entering the expressions using the Y= key and ZOOM Standard, we see the screen shown in Figure 9.32.

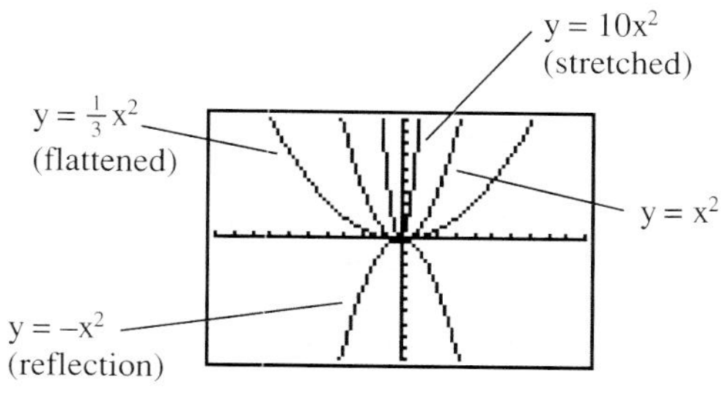

Figure 9.32

Example 4 shows how the graphing calculator can be used to visualize a composite function.

EXAMPLE 4 Consider the functions $f(x) = x^2 - 3$ and $g(x) = 2x - 5$.

a. Use the calculator to graph $y = f(x)$, $y = g(x)$, and $y = (f \circ g)(x)$.
b. Find the value of $(f \circ g)(9)$.

Solution

a. Use the Y= key and let $Y_1 = F(X) = X^2 - 3$, $Y_2 = G(X) = 2X - 5$, and $Y_3 = (F \circ G)(X) = Y_2{}^2 - 3$, as shown in Figure 9.33(a). Press ZOOM Standard to see all three graphs, as shown in Figure 9.33(b).

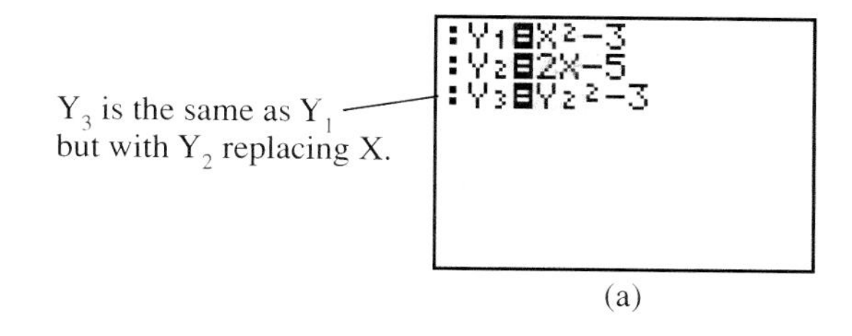

(a)

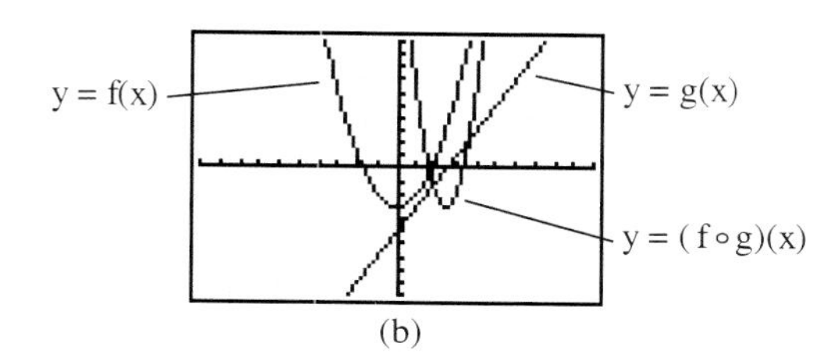

(b)

Figure 9.33

b. Assign the value 9 to X and then evaluate Y_3. The keystroke sequence follows.
9 STO▸ X ENTER 2nd [Y-VARS] 3 ENTER
The screen appears as shown in Figure 9.34, and the answer is $(f \circ g)(9) = 166$.

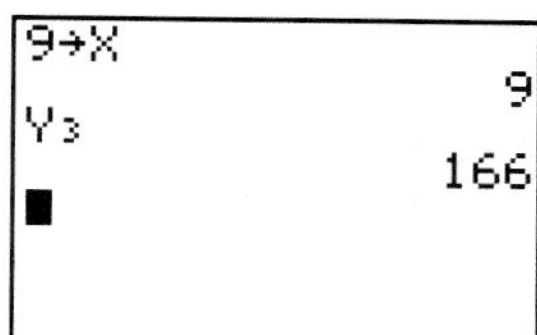

Figure 9.34

EXERCISES

1. Use the graphing calculator to graph $y = 2/(x + 5)$, and demonstrate that $x = -5$ is a vertical asymptote. Ans.

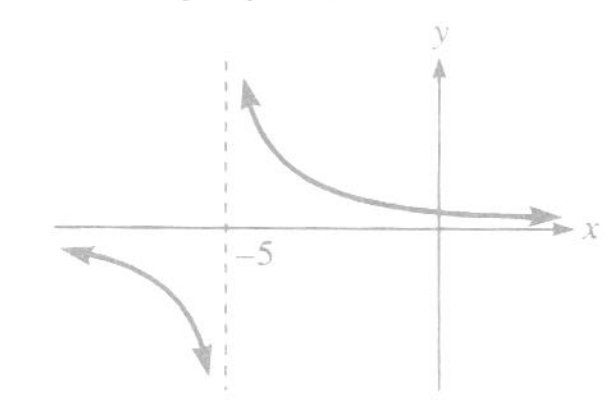

2. Predict how these graphs will look; then check your prediction by letting the calculator draw the graphs.
a. $y = \sqrt{x}$ **b.** $y = -3 + \sqrt{x}$ **c.** $y = \sqrt{x - 5}$
d. $y = -\sqrt{x}$ Ans.

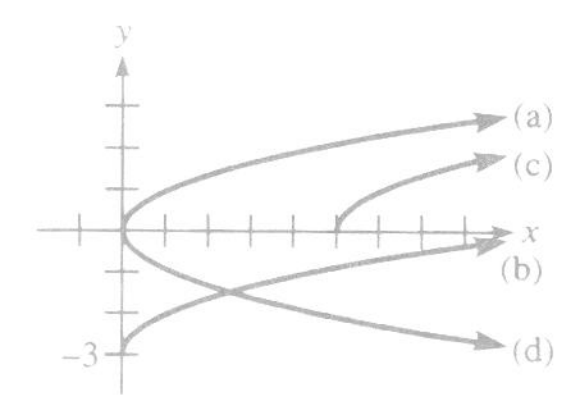

3. For $f(x) = 2x - 1$ and $g(x) = (x + 1)/2$, graph $y = (f \circ g)(x)$ and evaluate $(f \circ g)(7)$. Ans. $(f \circ g)(7) = 7$

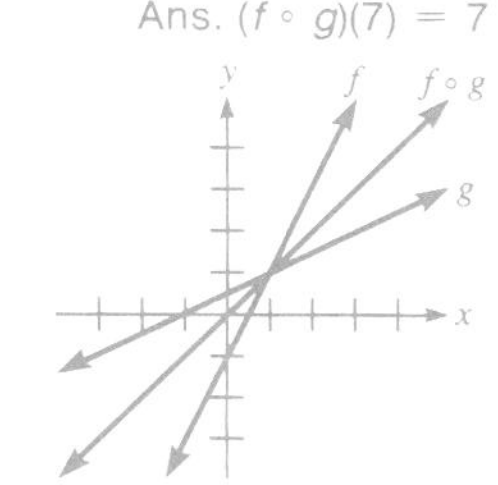

Use the graphing calculator to help solve the following exercises from Chapter 9. You can use the program EVAL (given in the graphing calculator section at the end of Chapter 4) for some of the exercises.

Section	*Exercises*
9.2	7, 11, 21, 39
9.3	1, 5, 13, 27, 47, 57
9.4	5, 9, 21, 33, 39, 47

Chapter 9 SUMMARY

OBJECTIVES CHECKLIST Specific chapter objectives are summarized below along with numbered example problems from the text that should clarify the objectives. If you do not understand any objectives or do not know how to do the selected problems, then restudy the material.

9.1 Can you:

1. **Find the domain and range given a set of ordered pairs?**
   The relation "less than" in the set $\{2,3,5\}$ is defined by $\{(2,3),(2,5),(3,5)\}$. Find the domain and range of this relation. [Example 2]

2. **Determine if a set of ordered pairs is a function?**
   Determine if the relation $\{(5,3),(6,3),(7,3)\}$ is a function. [Example 3a]

3. **Determine if an equation or graph defines y as a function of x?**
   Does the rule $x = y^2$ define y as a function of x? [Example 4b]

4. **Find the domain and range given an equation or a graph?**
   Determine the domain and range of the relation $y = \sqrt{x + 4}$. [Example 8b]

5. **Find a formula that defines the functional relationship between two variables?**
   Express the distance y that a car going 55 mi/hour will travel as a function of t, the hours spent traveling. [Example 9a]

9.2 Can you:

1. **Evaluate expressions using functional notation?**
   If $y = f(x) = 5x^2 - x - 1$, find $f(0)$ and $f(-1)$. [Example 1]

2. **Find function values and graphs of functions defined by more than one equation?**
   If $y = f(x) = \begin{cases} 3 & \text{if } x < 1 \\ 2x & \text{if } x \geq 1, \end{cases}$ find $f(-2)$. [Example 6a]

3. **Read from a graph of function f the domain, range, function values, and values of x for which $f(x) = 0$, $f(x) < 0$, and $f(x) > 0$?**
   Consider the graph of $y = f(x)$ in Figure 9.35. For what values of x is $f(x) < 0$? [Example 7e]

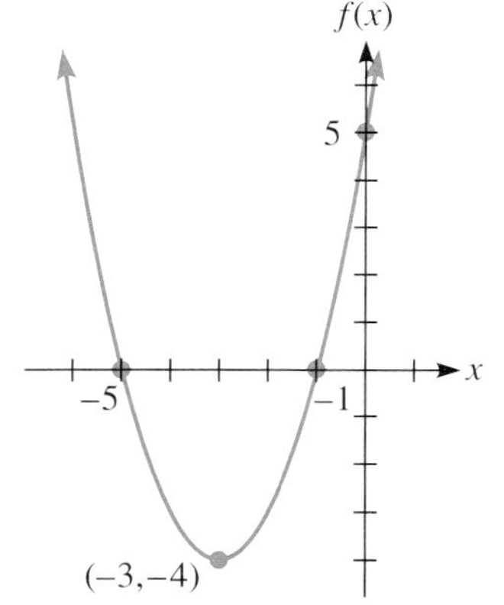

Figure 9.35

9.3 Can you:

1. **Graph a function using translations?**
   Use the graph of $y = |x|$ to graph $y = |x| - 3$. [Example 1a]

2. **Graph a function using reflecting, stretching, or flattening?**
   Graph the function $y = \frac{1}{4}x^2$. [Example 4b]

3. **Graph a function using combinations of the above methods?**
   Graph $y = -|x - 4| + 5$ using the graph of $y = |x|$. [Example 6]

9.4 Can you:

1. **Graph a constant function?**
   Graph $f(x) = 2$. [Example 1]
2. **Graph a linear function?**
   Graph $f(x) = -\frac{3}{2}x + 5$ using the slope and y-intercept of the graph. [Example 2]
3. **Find the equation of a linear function f given two ordered pairs in the function?**
   Find the equation that defines the linear function f if $f(3) = 1$ and $f(5) = -3$. [Example 3]
4. **Solve applied problems involving linear functions?**
   During exercise a person's maximum target heart rate is a function of age. This relation is specified by a linear function, and the recommended maximum number of beats per minute is 153 at age 40 and 136 at age 60. Find the equation that defines this linear function. [Example 4a]
5. **Graph a quadratic function and specify the range of the function?**
   For the function defined by $f(x) = -x^2 + x + 2$, graph the function and specify the range using this graph. [Example 5d]
6. **Determine the vertex of the graph of a quadratic function by matching the function to the form $f(x) = a(x - h)^2 + k$?**
   If $f(x) = -x^2 + x + 2$, find the vertex and axis of symmetry by matching the equation to the form $f(x) = a(x - h)^2 + k$. [Example 6]
7. **Solve applied problems involving quadratic functions?**
   Find the maximum possible area that can be enclosed with 100 m of fencing shaped in the form of a rectangle. [Example 7]

9.5 Can you:

1. **Add, subtract, multiply, and divide two functions?**
   If $f(x) = 4x - 3$ and $g(x) = x^2 - x - 5$, find $(f - g)(x)$. [Example 1b]
2. **Find the composite function of two functions?**
   If $f(x) = x^2$ and $g(x) = 2x - 5$, find $(f \circ g)(x)$. [Example 4a]

KEY TERMS

Composite functions (9.5)
Constant function (9.4)
Dependent variable (9.1)
Domain (9.1)
Function (9.1)
Graph (of a relation) (9.1)
Independent variable (9.1)
Linear function (9.4)
Polynomial function of degree n (9.4)
Quadratic function (9.4)
Range (9.1)
Relation (9.1)

KEY CONCEPTS AND PROCEDURES

Section	Key Concepts or Procedures to Review
9.1	■ Definitions of relation, domain of a relation, range of a relation, and function ■ Vertical line test ■ Methods for finding the domain and range of a relation
9.2	■ The notation $f(x)$ is read "f of x" or "f at x" and means the value of the function (the y value) corresponding to the value of x.
9.3	■ Graphs of the squaring function ($y = x^2$), square root function ($y = \sqrt{x}$), and absolute value function ($y = \lvert x \rvert$) ■ Methods to graph variations of a familiar function by using vertical and horizontal translations, reflecting, stretching, and flattening

Section	Key Concepts or Procedures to Review
9.4	▪ Definitions of constant function, linear function, and quadratic function ▪ The graph of a constant function [$f(x) = c$] is a horizontal line containing the point $(0,c)$. ▪ The graph of a linear function [$f(x) = mx + b$] is a straight line with slope m and y-intercept $(0,b)$. ▪ The graph of $f(x) = ax^2 + bx + c$ is a parabola with axis of symmetry $x = -b/(2a)$. The parabola opens upward when $a > 0$ and downward when $a < 0$. ▪ The graph of $f(x) = a(x - h)^2 + k$ (with $a \neq 0$) is a parabola with vertex (h,k) and axis of symmetry $x = h$. ▪ Methods to graph linear and quadratic functions
9.5	▪ Definitions of $f + g$, $f - g$, $f \cdot g$, and f/g for two functions f and g ▪ The symbol $\circ$ denotes the operation of composition. The composite functions are $(f \circ g)(x) = f[g(x)]$ and $(g \circ f)(x) = g[f(x)]$.

CHAPTER 9 REVIEW EXERCISES

9.1

1. Find the domain and range of the relation $\{(2,3),(10,3),(21,7),(70,4)\}$. Domain $= \{2,10,21,70\}$; range $= \{3,4,7\}$
2. Determine if the given relation is a function: $\{(2,3),(10,3),(21,7),(70,4)\}$. Yes
3. Does the graph below represent a function? Yes

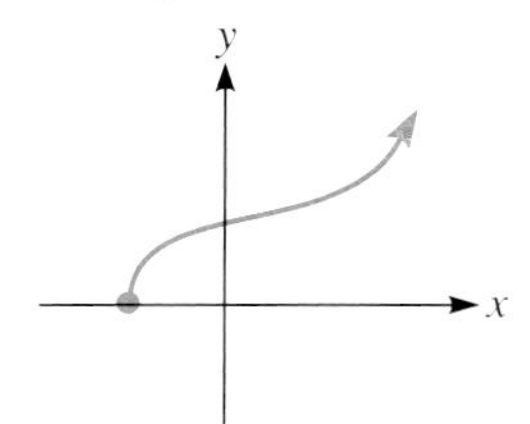

4. Determine the domain and range of the following function. Domain $= [-3,3]$; range $= [-2,0]$

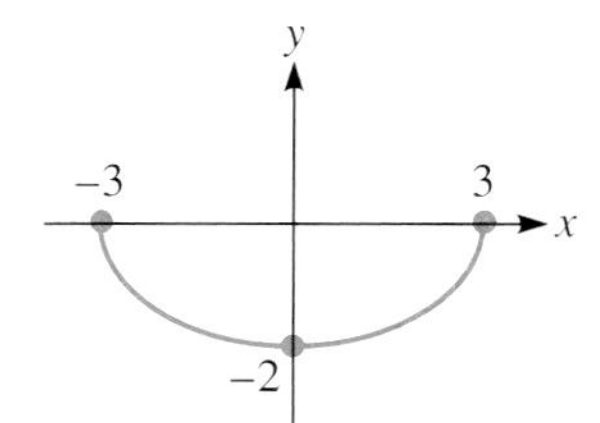

5. a. Write an equation to express the cost C before tax of n books if the books cost \$35 each.
   b. What is the domain of the function? a. $C = 35n$
   b. Domain $= \{n: n$ is a nonnegative integer$\}$

9.2

6. If $f(x) = 2x^2 + 3x - 4$, find $f(0)$ and $f(-2)$. $f(0) = -4$, $f(-2) = -2$
7. If $g(x) = x^2 - 1$ and $h(x) = 2x$, find $2g(-1) - 3h(2)$. -12
8. If $f(x) = \begin{cases} -1 \text{ if } x < 1 \\ x \text{ if } x \geq 1, \end{cases}$ find the following.
   a. $f(3)$ b. $f(1)$ c. $f(-1)$ d. Graph the function.
   a. 3 b. 1 c. -1 d.

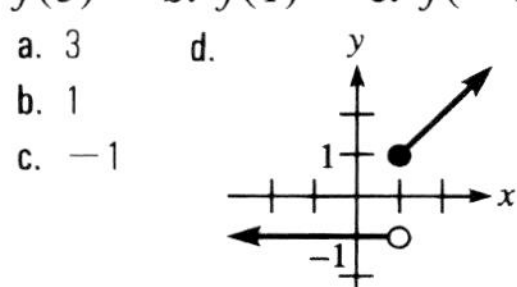

9. Consider the graph of $y = f(x)$ that is shown.

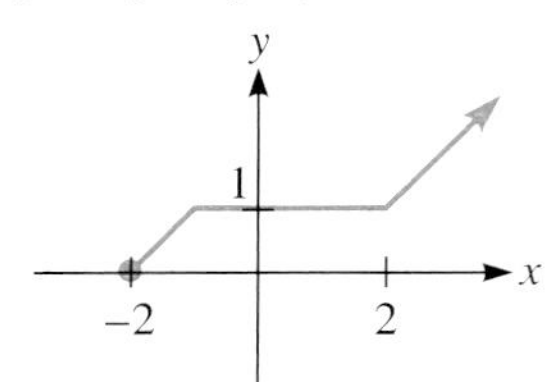

   a. What is the domain of f? $[-2,\infty)$
   b. What is the range of f? $[0,\infty)$
   c. Determine $f(0)$. 1
   d. For what values of x does $f(x) = 0$? -2
   e. For what values of x is $f(x) < 0$? None
   f. Solve $f(x) > 0$. $(-2,\infty)$
10. If $f(x) = 3x$, show that $f(a + b)$ does equal $f(a) + f(b)$ for all a and b. $3(a + b) = 3a + 3b$

9.3

11. Use the graph of $y = |x|$ to graph $y = |x| - 2$.

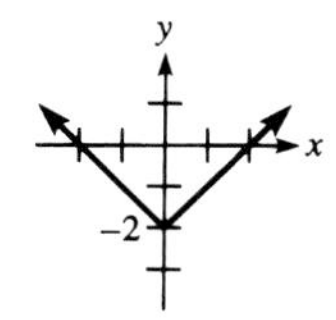

12. Use the graph of $y = x^2$ to graph $y = (x + 1)^2$.

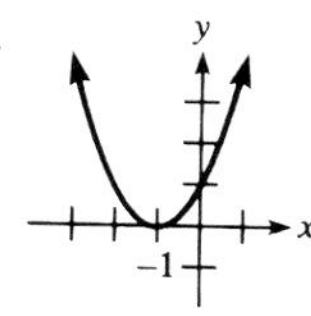

13. Use the graph of $y = |x|$ to graph $y = -2|x|$.

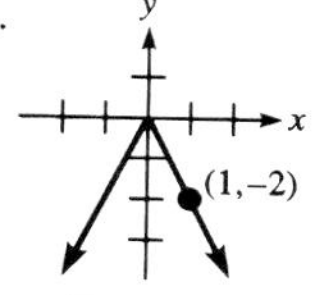

14. Graph $y = 2 - \sqrt{x - 1}$ using the graph of $y = \sqrt{x}$.

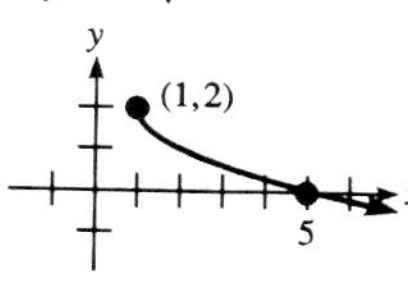

15. If a stone is dropped from a bridge at a point that is 384 ft above water, then $y = -16t^2 + 384$ gives the height y above water in feet after t seconds have elapsed.
a. To the nearest tenth of a second, when does the stone hit the water?
b. What is the domain of the function?
c. Graph this function. a. 4.9 seconds
b. [0 seconds, 4.9 seconds]
c.

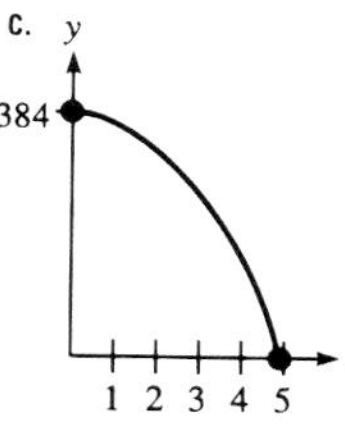

9.4

16. Graph $f(x) = -4$.

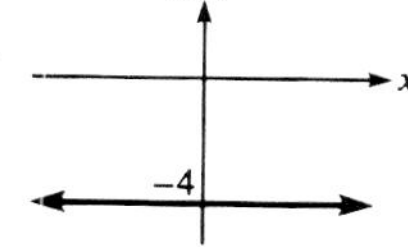

17. Graph $f(x) = \frac{3}{2}x - 1$ using the slope and y-intercept of the graph.

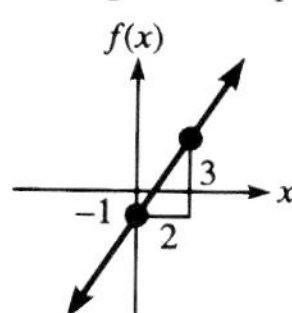

18. Find the equation that defines the linear function f if $f(2) = -1$ and $f(4) = -7$. $f(x) = -3x + 5$

19. Consider the function defined by $f(x) = x^2 - 4x + 3$.
a. Find the equation of the axis of symmetry.
b. Find the coordinates of the vertex.
c. Find the coordinates of the y-intercept.
d. Graph the function and specify the range using this graph. a. $x = 2$
b. $(2, -1)$
c. $(0,3)$
d.

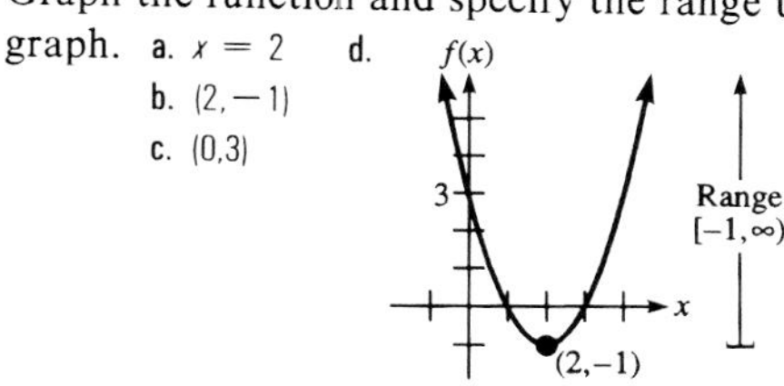

20. If $f(x) = -x^2 + 6x + 1$, find the vertex and axis of symmetry by matching the equation to the form $f(x) = a(x - h)^2 + k$.
$(3,10)$; $x = 3$

9.5

21. If $f(x) = 3x + 2$ and $g(x) = x^2 + x + 3$, find the following.
a. $(f + g)(x)$ **b.** $(f - g)(x)$ **c.** $(f \cdot g)(x)$ **d.** $(f/g)(x)$
a. $(f + g)(x) = x^2 + 4x + 5$
b. $(f - g)(x) = -x^2 + 2x - 1$
c. $(f \cdot g)(x) = 3x^3 + 5x^2 + 11x + 6$
d. $\left(\frac{f}{g}\right)(x) = \frac{3x + 2}{x^2 + x + 3}$

22. If $f(x) = x^2 - 3x + 2$ and $g(x) = 4x - 3$, find $(f/g)(2)$. 0

23. If $f(x) = -x^2$ and $g(x) = 2x - 4$, find the following.
a. $(f \circ g)(x)$ a. $-4x^2 + 16x - 16$
b. $(g \circ f)(x)$ b. $-2x^2 - 4$

24. Use the graphs of f and g to evaluate each expression.
a. $(f - g)(3)$
b. $(f \cdot g)(2)$
c. $(f \circ g)(2)$
a. 1 b. 0 c. 2

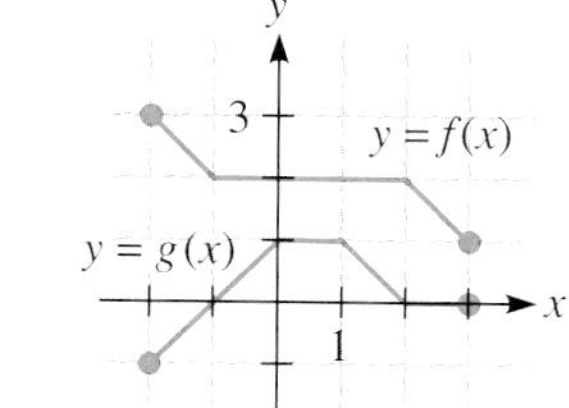

25. An oil leak is spreading over a plane surface in the shape of a circle. The diameter of this circle is increasing at a rate of 8 cm/second, so the diameter of this spill t seconds after the start of this leak may be expressed by $d = g(t) = 8t$. If function f expresses the circumference of this circular spill as a function of d, so that $C = f(d) = \pi d$, find and interpret $(f \circ g)(t)$. $C = (f \circ g)(t) = 8\pi t$ is the formula for the circumference of the spill t seconds after the start of the leak.

ADDITIONAL REVIEW EXERCISES

Determine if the following represent functions.

26. $\{(0,8),(1,8),(2,8),(3,8)\}$ Yes

27. $\{(-6,-6),(-7,-7),(-8,-8),(-9,-9)\}$ Yes

28. $\{(4,1),(4,2),(4,3),(4,4)\}$ No

29. Yes

30. No

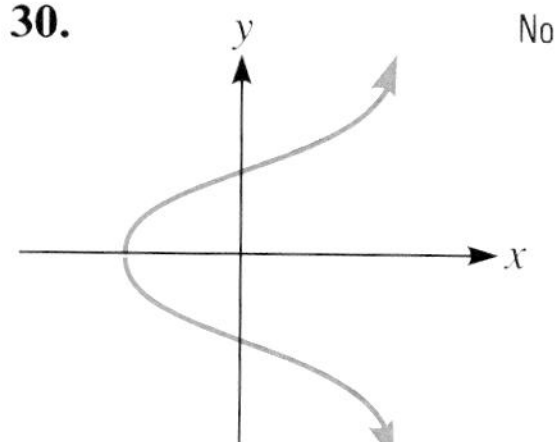

31. Yes

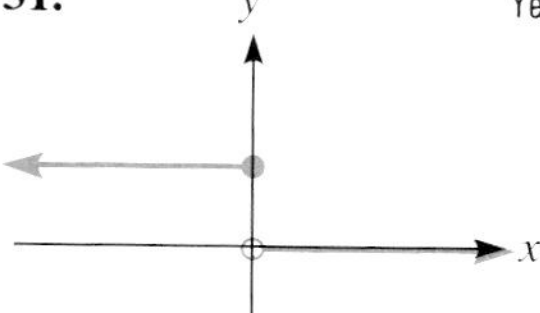

32. $2x + 3y = 6$ Yes

33. $y + x^2 = 9$ Yes

34. $x + y^2 = 9$ No

Determine the domain and range of each function.

35. $\{(9,95),(8,85),(6,65),(7,75)\}$ Domain $= \{6,7,8,9\}$; range $= \{65,75,85,95\}$

36. $y = \sqrt{x + 6}$ Domain $= [-6,\infty)$; range $= [0,\infty)$

37. $y = \frac{1}{x - 1}$ Domain $= \{x: x \neq 1\}$; range $= \{y: y \neq 0\}$

38. $y = \begin{cases} 1 \text{ if } x > 0 \\ -1 \text{ if } x \leq 0 \end{cases}$ Domain $= (-\infty, \infty)$; range $= \{1, -1\}$

Let $f(x) = x^2 + 3x - 5$ and $g(x) = 3x - 3$. Find each of the following.

39. $(f + g)(x)$ $x^2 + 6x - 8$

40. $(f - g)(x)$ $x^2 - 2$

41. $(f/g)(x)$ $\dfrac{x^2 + 3x - 5}{3x - 3}$

42. $(f \cdot g)(x)$ $3x^3 + 6x^2 - 24x + 15$

43. $(f \circ g)(x)$ $9x^2 - 9x - 5$

44. $(g \circ f)(x)$ $3x^2 + 9x - 18$

45. $f(0)$ -5

46. $g(0)$ -3

47. $3f(1) + g(1)$ -3

48. $(f + g)(2)$ 8

49. $(f/g)(2)$ $\frac{5}{3}$

50. $(f \circ g)(-1)$ 13

51. $(g \circ f)(-1)$ -24

52. Use the graphs of f and g to evaluate each expression.

a. $(f + g)(1)$
b. $(f/g)(-2)$
c. $(f \circ g)(3)$

a. 4 b. 2 c. 2

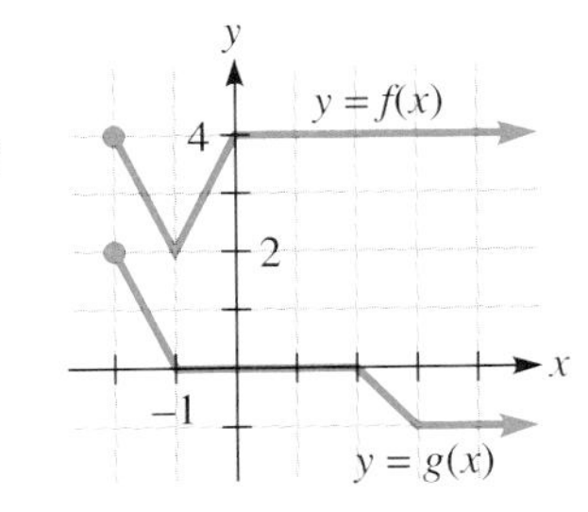

53. Consider the graph of $y = f(x)$ shown below.

a. What is the domain of f?
b. What is the range of f?
c. Determine the value of $f(0)$.
d. For what value(s) of x does $f(x) = 0$?
e. For what value(s) is $f(x) > 0$?
f. Solve $f(x) < 0$.

a. $[-3, \infty)$ b. $(-\infty, 5]$ c. 5
d. 3 e. $[-3, 3)$ f. $(3, \infty)$

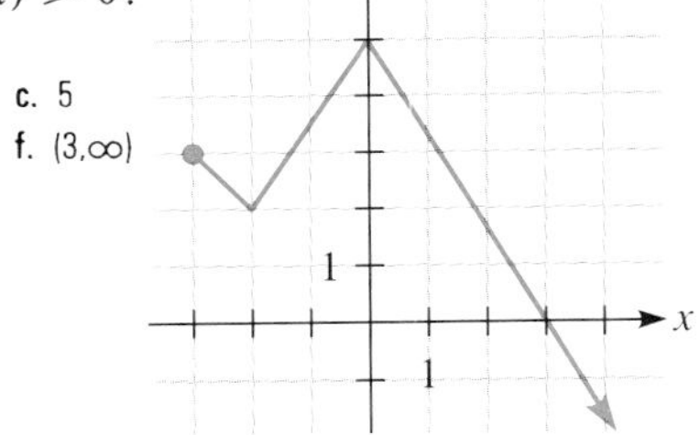

54. Write an equation that defines the linear function f if $f(-2) = -3$ and $f(2) = -1$. $f(x) = \frac{1}{2}x - 2$

55. Consider the function defined by $f(x) = -x^2 + 2x - 2$.

a. Find the axis of symmetry.
b. Find the coordinates of the vertex.
c. Find the coordinates of the y-intercept.
d. Graph the function.
e. Determine the domain of $f(x)$.
f. Determine the range of $f(x)$.

a. $x = 1$ b. $(1, -1)$
c. $(0, -2)$
d.

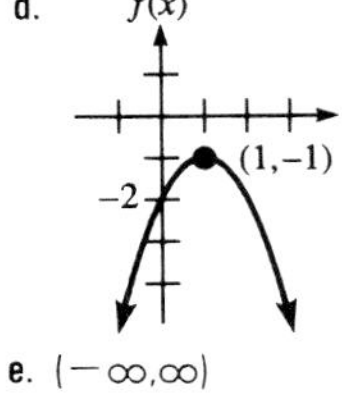

e. $(-\infty, \infty)$
f. $(-\infty, -1]$

56. If $f(x) = 2x^2 + 12x + 11$, find the vertex and axis of symmetry by matching the equation to the form $f(x) = a(x - h)^2 + k$. $(-3, -7)$; $x = -3$

Graph each of the following.

57. $f(x) = 5$

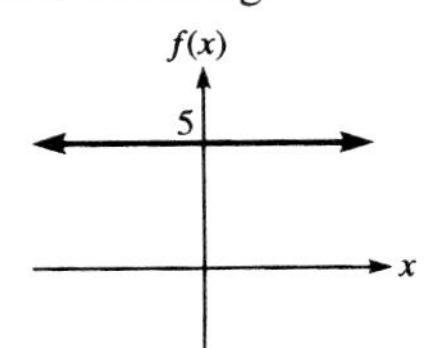

58. $f(x) = \frac{3}{4}x - 2$ (use slope and y-intercept)

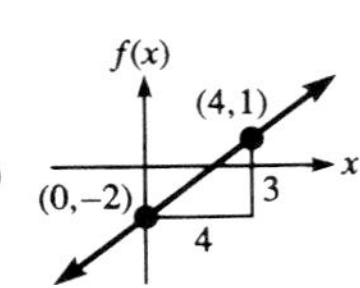

59. $f(x) = \begin{cases} 2x \text{ if } x \leq 1 \\ 2 \text{ if } x > 1 \end{cases}$

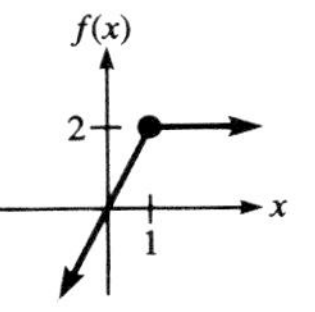

Use the graph of $y = |x|$, $y = x^2$, or $y = \sqrt{x}$ to graph each equation.

60. $y = x^2 - 3$

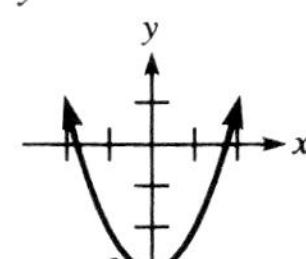

61. $y = |x + 2|$

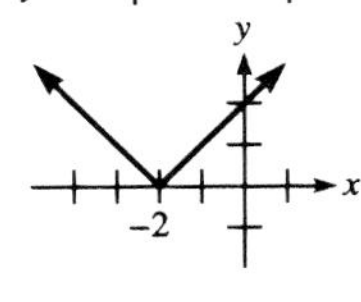

62. $y = 2\sqrt{x}$

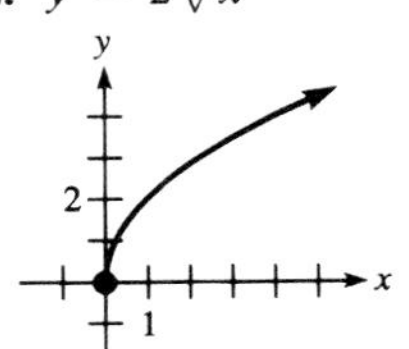

63. $y = (x - 1)^2 + 2$

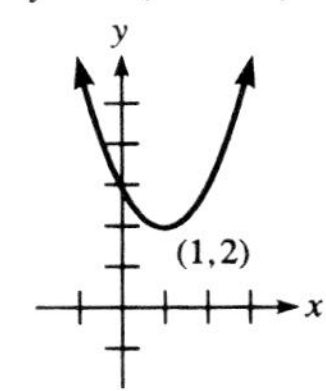

64. $y = 3 - |x|$

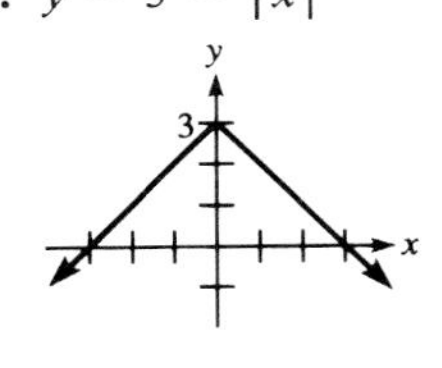

65. Write an equation to express the distance d traveled by a car in t hours at 45 mi/hour. Determine the domain of this function. $d = 45t$; $[0, \infty)$

66. The height y above water in meters of a diver t seconds after stepping off a diving tower 20 m high is given by the formula $y = -4.9t^2 + 20$.

a. To the nearest tenth of a second, when does the diver hit the water?
b. What is the domain of this function?
c. Graph this function.

a. 2.0 seconds b. [0 seconds, 2.0 seconds] c.

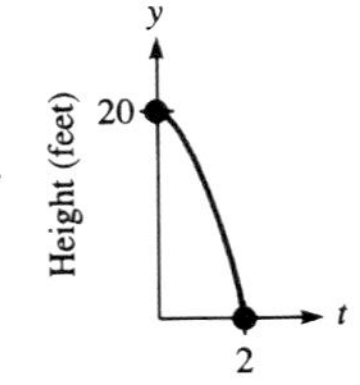

67. An oil leak is spreading over a plane surface in the shape of a circle. The radius of this circle is increasing at a rate of 12 cm/second, so the radius of this spill t seconds after the start of the leak may be expressed by $r = g(t) = 12t$. If function f expresses the area of this circular spill as a function of r, so that $A = f(r) = \pi r^2$, find and interpret $(f \circ g)(t)$. $A = 144\pi t^2$ is the formula for the area of the circular spill t seconds after the start of the leak.

CHAPTER 9 TEST

1. Find the domain and range of the relation $\{(5,3),(3,2),(5,2)\}$.
Domain = $\{3,5\}$; range = $\{3,2\}$
2. Determine if the given relation is a function: $\{(-3,4),(-3,5),(-2,2)\}$. No
3. Does the given rule determine y as a function of x?
   a. $x^2 + y = 4$ Yes
   b. $y + x = 3$ Yes
   c. $y^2 + x^2 = 5$ No
4. Determine the domain and range of the function $y = \dfrac{2}{x+4}$
Domain = $\{x: x \neq -4\}$; range = $\{y: y \neq 0\}$
5. Express the length ℓ of a rectangle with width 6 cm as a function of its perimeter P. $\ell = \dfrac{P}{2} - 6$
6. If $f(x) = -x^2 + x - 1$ and $g(x) = x + 1$, find $3g(2) - f(-2)$. 16
7. If $y = f(x) = \begin{cases} -x \text{ if } x \leq 0 \\ 2 \text{ if } x > 0, \end{cases}$ find the following.
   a. $f(-3)$ 3
   b. $f(0)$ 0
   c. $f(3)$ 2
   d. Graph the function.

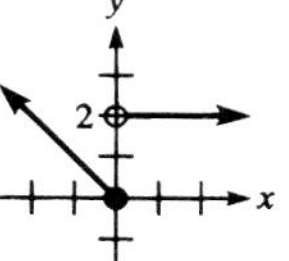

8. Consider the graph of $y = f(x)$ shown below.
   a. What is the domain of f?
   b. What is the range of f?
   c. For what values of x does $f(x) = 0$?
   d. For what values of x is $f(x) < 0$?
   e. Solve $f(x) > 0$.
   a. $(-\infty,\infty)$ b. $(-\infty,1]$ c. $\{-3,-1\}$
   d. $(-\infty,-3) \cup (-1,\infty)$ e. $(-3,-1)$

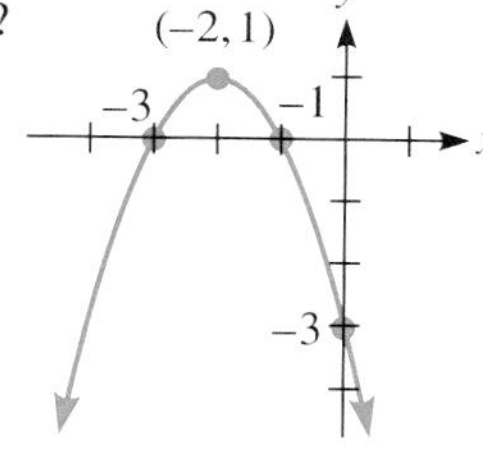

9. Use the graph of $y = x^2$ to graph $y = (x - 1)^2 + 2$. Identify the vertex and y-intercept of the graph.

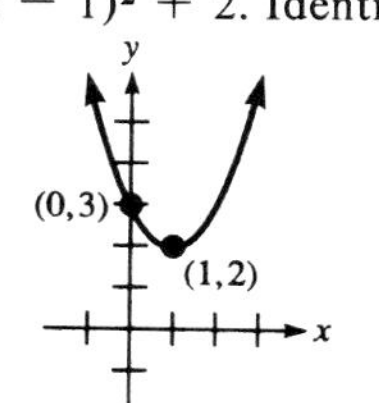

10. Graph $y = 2 - \sqrt{x}$ using the graph of $y = \sqrt{x}$.

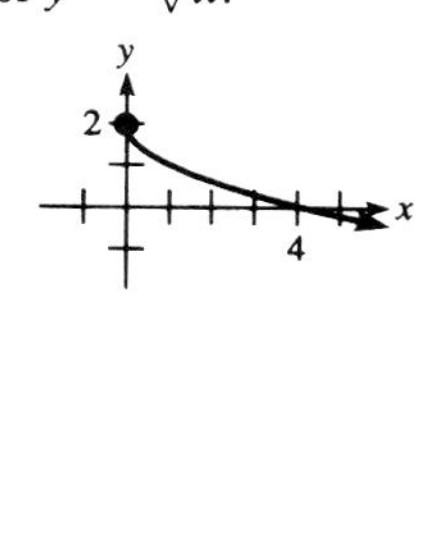

11. Graph $y = 3|x|$.

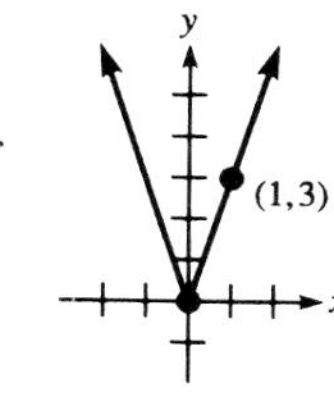

12. Graph $f(x) = -2$.

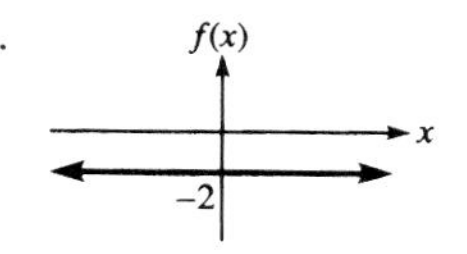

13. Graph $f(x) = \frac{1}{2}x - 3$ using the slope and y-intercept.

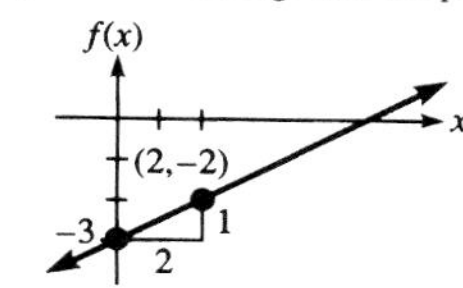

14. Find the equation that defines the linear function f if $f(2) = -1$ and $f(-1) = 8$. $f(x) = -3x + 5$
15. Consider the function defined by $f(x) = x^2 - 2x + 2$.
   a. Find the axis of symmetry.
   b. Find the coordinates of the vertex.
   c. Find the coordinates of the y-intercept.
   d. Graph the function and specify the range using this graph.
   a. $x = 1$ b. $(1,1)$ c. $(0,2)$ d.

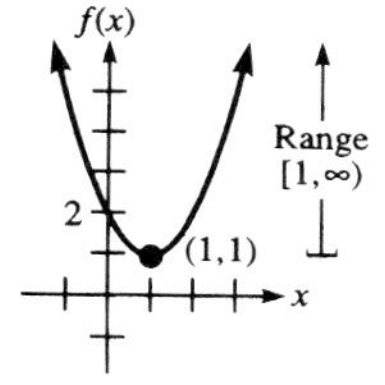

16. If $f(x) = 3x^2 - 12x + 8$, find the vertex and the axis of symmetry by matching the equation to the form $f(x) = a(x - h)^2 + k$. $(2,-4)$; $x = 2$
17. If 60 ft of fencing is available, find the area of the largest rectangular region that can be enclosed with the available fencing. 225 ft²
18. If $f(x) = 2x + 1$ and $g(x) = -x^2 + x - 3$, find the following.
   a. $(f + g)(x)$ $-x^2 + 3x - 2$
   b. $(f - g)(x)$ $x^2 + x + 4$
   c. $(f \cdot g)(x)$ $-2x^3 + x^2 - 5x - 3$
   d. $\left(\dfrac{f}{g}\right)(x)$ $\dfrac{2x+1}{-x^2+x-3}$
19. Use the graph of f and g to evaluate each expression.
   a. $(f + g)(0)$ 0
   b. $(f/g)(-2)$ $-\frac{2}{3}$

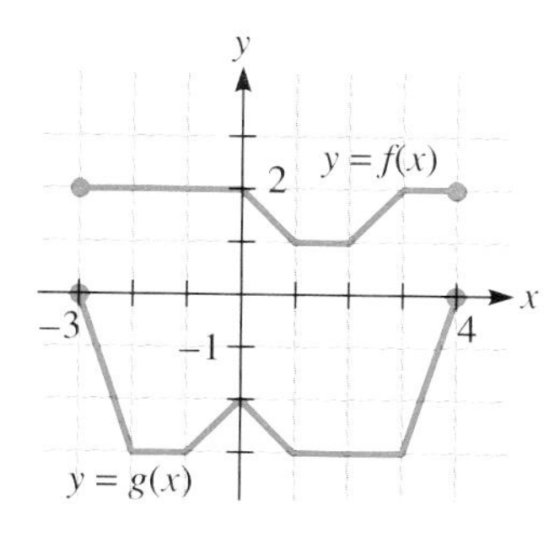

20. If $f(x) = x^2 + 2$ and $g(x) = -3x + 1$, find the following.
   a. $(f \circ g)(x)$ $9x^2 - 6x + 3$
   b. $(g \circ f)(x)$ $-3x^2 - 5$

CUMULATIVE TEST 9

1. From the set $\{1.\overline{3}, -\frac{5}{6}, 0, \pi, \sqrt{3}, 2 + 3i\}$, list all numbers that are real.
   All are real except $2 + 3i$.
2. Solve $x - 7 = \dfrac{3x - 33}{9}$. $\{5\}$
3. Solve $\left|\frac{1}{2}x + 4\right| = 9$. $\{-26, 10\}$
4. Simplify $5^{n+1} \cdot 5^n$, where n is an integer. 5^{2n+1}
5. Simplify $\dfrac{x^3 + 8}{x^2 - 4} \cdot \dfrac{x^2 - 2x + 4}{x - 2}$
6. Solve $(x - 1)(x + 1) + 8 = 0$. $\{\pm i\sqrt{7}\}$
7. Solve $a = \dfrac{b + b_0}{b}$ for b. $b = \dfrac{b_0}{a - 1}$
8. If $P(x) = 3x^4 - 2x^3 + x^2 - 5x - 8$, find $P(3)$ by using the remainder theorem. 175
9. Find the slope and y-intercept of the line $6x - 2y = 8$.
   Slope 3; y-intercept $(0, -4)$
10. Show that the points $A(0,0)$, $B(2,5)$, and $C(10,-4)$ are vertices of a right triangle.
    $m_{AB} = \frac{5}{2}$ and $m_{AC} = -\frac{2}{5}$; $m_{AB} \cdot m_{AC} = -1$
11. Solve by graphing: $x - 2y = -4$
    $y = x + 2$.

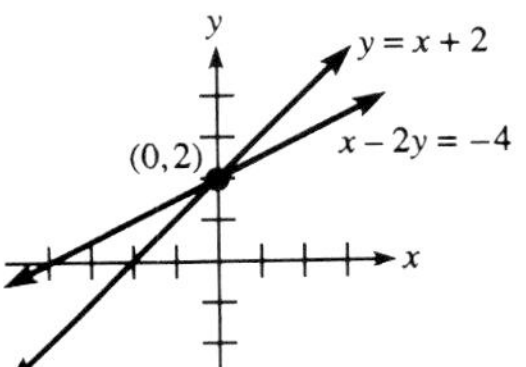

12. Solve by addition-elimination: $5x + 8y = 2$
    $3x + 5y = 1$. $(2, -1)$
13. Multiply and write the results in the form $a + bi$:
    $\sqrt{-8} \cdot \sqrt{-6}$. $-4\sqrt{3} + 0i$
14. Simplify i^{27}. $-i$
15. Simplify; assume $x > 0$, $y > 0$: $\sqrt{125x^5y^5} + \sqrt{20xy}$.
    $(5x^2y^2 + 2)\sqrt{5xy}$
16. Solve $6x^2 - 7x + 2 = 0$. $\{\frac{1}{2}, \frac{2}{3}\}$
17. Solve $(x - 1)(2x + 3) \leq 0$. $[-\frac{3}{2}, 1]$
18. Graph $4x^2 + y^2 = 16$.

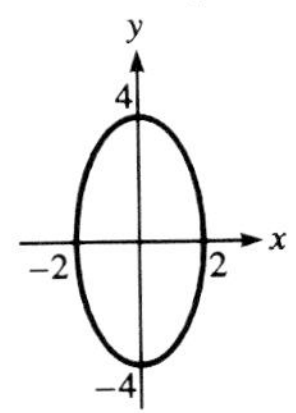

19. If $f(x) = 2x - 3$ and $g(x) = x^2 + 4$, find the following.
    a. $f(4)$ 5
    b. $g(-1)$ 5
    c. $(f \cdot g)(0)$ -12
    d. $(f \circ g)(x)$ $2x^2 + 5$
20. a. Use the graph of $y = |x|$ to graph $y = |x + 1| - 3$.
    b. What is the domain of this function?
    c. What is the range of this function?

a.

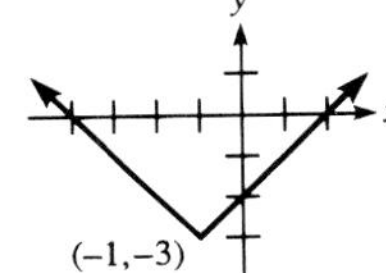

b. $(-\infty, \infty)$
c. $[-3, \infty)$

CHAPTER

10 Exponential and Logarithmic Functions

Financial planners advise that an individual retirement account (IRA) be started as early as possible, because the power of compounding an investment is realized over long periods of time. If just one \$2,000 investment is made in an IRA by a college student on her 17th birthday and the account grows at 8.5 percent compounded annually, what will be the value of this account on her 65th birthday? (See Example 9 of Section 10.1.)

JOINING THE polynomial functions as important types of elementary functions are the exponential functions and their inverses, the logarithmic functions. In contrast to the polynomial function, where the exponent is fixed and the base varies, for exponential functions the exponent varies. For instance, $y = x^2$ is a polynomial function, while $y = 2^x$ is an exponential function. A wide variety of relationships are analyzed using exponential and logarithmic functions; in this chapter population growth, radioactive decay, and compound interest are some of the applications that are considered.

Chapter Outline

Consider introducing the topic of exponential functions using the problem in "Think About It" Exercise 1.

10.1 Exponential Functions

OBJECTIVES

1. **Determine function values for an exponential function.**
2. **Graph an exponential function.**
3. **Solve exponential equations using $b^x = b^y$ implies $x = y$.**
4. **Find the base in the exponential function $y = b^x$ given an ordered pair in the function.**
5. **Solve applied problems involving exponential functions.**

1 To introduce the concept of an exponential function, Example 1 considers the growth of a population that starts from a single type of cell that continually reproduces itself by division into two cells of the same type after a certain period of time.

EXAMPLE 1 A biologist has one cell in a culture at the start of an experiment. Careful observations reveal that the number of cells is doubling every day.

a. Find a formula showing the number of cells present after x days.
b. Approximately how many cells are present at the end of 30 days?

Solution

a. Let y represent the number of cells in the culture. Then,

$$\begin{aligned} y &= 1 && \text{when } x = 0 \text{ days} \\ y &= 1 \cdot 2 = 2 && \text{when } x = 1 \text{ day} \\ y &= 2 \cdot 2 = 2^2 = 4 && \text{when } x = 2 \text{ days} \\ y &= 2^2 \cdot 2 = 2^3 = 8 && \text{when } x = 3 \text{ days} \\ y &= 2^3 \cdot 2 = 2^4 = 16 && \text{when } x = 4 \text{ days} \end{aligned}$$

and in general,

$$y = 2^x,$$

where x represents the number of complete days which have elapsed from the start of the experiment.

b. When $x = 30$,

$$y = 2^{30} \approx 1.0737 \times 10^9, \text{ or } 1{,}073{,}700{,}000.$$

Thus, there are about 1.0737 billion cells present after 30 days. Observe that the relation defined by $y = 2^x$ grows quickly to produce very large values of y for relatively small values of x.

Note Throughout this chapter a scientific calculator is an essential aid. In this example 2^{30} was calculated using the power key $\boxed{y^x}$ as follows.

$$2 \; \boxed{y^x} \; 30 \; \boxed{=} \; \boxed{1.0737 \;\; 09}$$

Refer students to Section 3.2 for a reminder about scientific notation.

Evaluating exponential expressions often leads to large numbers (or small numbers) that are displayed in a scientific notation format, as in this example.

Ernest Rutherford, a great physicist at the turn of the century, wrote the first major text about radioactivity (1904). He coined the term *half-life.* He was born in New Zealand but worked in Canada and England.

PROGRESS CHECK 1 A scientist has 1 g of a radioactive element which is disappearing through radioactive decay. Careful observations reveal that for every hour that passes, the quantity of this element which still remains is one-half the amount that was present at the beginning of that hour. (Scientists say the "half-life" of the element is 1 hour.)

a. Find a formula showing the amount of the element present after x hours.
b. Approximately how much of this element is left at the end of 1 day?

Progress Check Answers

1. (a) $y = (\frac{1}{2})^x$, x in hours (b) 5.96×10^{-8} g

Example 1 requires that 2^x have meaning only for nonnegative integer values of x because no change is considered to have occurred until an entire time period has elapsed. However, for the exponential function defined by $y = 2^x$, we wish 2^x to be meaningful for *all* real values of x. Up to this point in the text, there is no problem interpreting 2^x where x is a rational number. For instance,

$$\begin{aligned} 2^4 &= 16, \\ 2^0 &= 1, \\ 2^{-3} &= \frac{1}{2^3} = \frac{1}{8}, \\ 2^{2/3} &= \sqrt[3]{2^2} = \sqrt[3]{4} \approx 1.59. \end{aligned}$$

A precise definition for 2^x, where x is irrational, is given in higher mathematics, and the power key on a calculator may be used to approximate such expressions. For example, the following keystrokes show $2^\pi \approx 8.82$.

$$2 \; \boxed{y^x} \; \pi \; \boxed{=} \; \boxed{8.8249778}$$

Thus, an expression like 2^x is meaningful for all real numbers x, and an exponential function with a positive base other than 1 may now be defined.

Exponential Function

The function f defined by

$$f(x) = b^x,$$

with $b > 0$ and $b \neq 1$, is called the **exponential function with base *b*.**

The restriction $b \neq 1$ is made because $1^x = 1$ for all values of x, so f is a constant function in this case. A nonpositive base is not used in the defining equation because expressions like $(-4)^{1/2}$ and 0^{-2} are not real numbers.

EXAMPLE 2 If $f(x) = 4^x$, find $f(3)$, $f(-2)$, and $f(\frac{5}{2})$. Also, approximate $f(\sqrt{2})$ to the nearest hundredth.

Solution Using $f(x) = 4^x$ gives

$$\begin{aligned} f(3) &= 4^3 = 64, \\ f(-2) &= 4^{-2} = \frac{1}{4^2} = \frac{1}{16}, \\ f(\tfrac{5}{2}) &= 4^{5/2} = (\sqrt{4})^5 = 2^5 = 32. \end{aligned}$$

By calculator,

$$f(\sqrt{2}) = 4^{\sqrt{2}} = 7.10 \text{ (to the nearest hundredth).}$$

PROGRESS CHECK 2 If $f(x) = 8^x$, find $f(2)$, $f(-1)$, and $f(\frac{2}{3})$. Also, approximate $f(\sqrt{3})$ to the nearest hundredth.

2 One method for graphing an exponential function is to generate a list of ordered-pair solutions and then draw a smooth curve through the points given by these solutions. The next example graphs two exponential functions where $b > 1$ to illustrate the case of exponential growth.

Progress Check Answer

2. 64, $\frac{1}{8}$, 4, 36.66

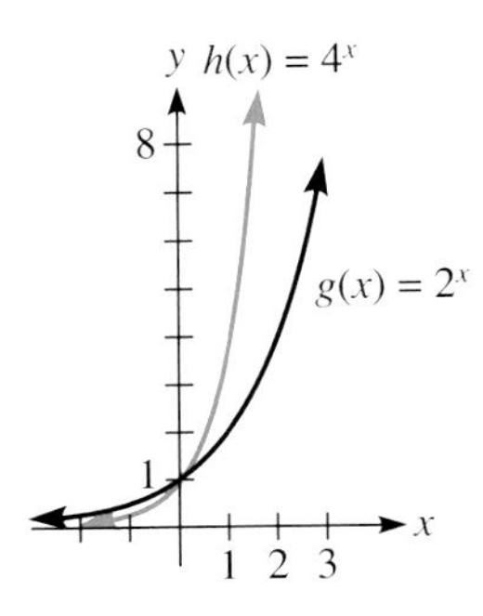

Figure 10.1

You might mention the use of special graph paper (semilogarithmic) to overcome the difficulty that y increases rapidly in equations like $y = 2^x$.

EXAMPLE 3 Sketch the graphs of $g(x) = 2^x$ and $h(x) = 4^x$ on the same coordinate system.

Solution Generate a table of values by replacing x with integer values from, say, -3 to 3.

x	-3	-2	-1	0	1	2	3
2^x	$\frac{1}{8}$	$\frac{1}{4}$	$\frac{1}{2}$	1	2	4	8
4^x	$\frac{1}{64}$	$\frac{1}{16}$	$\frac{1}{4}$	1	4	16	64

By graphing these solutions and drawing a smooth curve through them, we sketch the graphs of g and h, which are shown in Figure 10.1. Note that the graph of $y = b^x$ with $b > 1$ displays faster growth as b gets larger.

PROGRESS CHECK 3 Sketch the graphs of $g(x) = 3^x$ and $h(x) = 8^x$ on the same coordinate system.

When $0 < b < 1$, then the graph illustrates exponential decay. Comparison of the two graphs in the next example shows that the decay is faster when b is closer to zero.

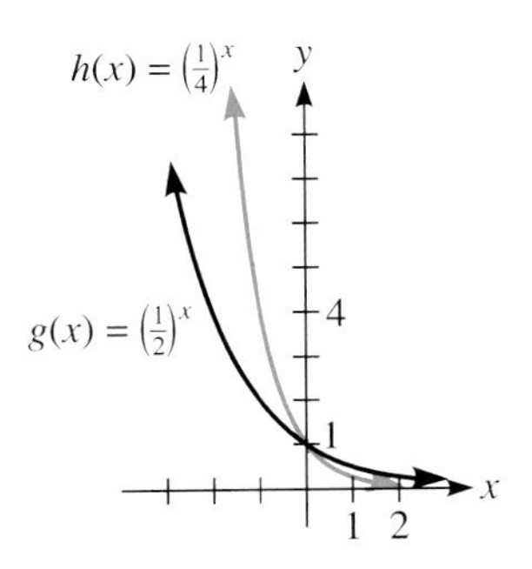

Figure 10.2

EXAMPLE 4 Sketch the graphs of $g(x) = (\frac{1}{2})^x$ and $h(x) = (\frac{1}{4})^x$ on the same coordinate system.

Solution Construct a table of values, as follows, and then graph the equations, as shown in Figure 10.2.

x	-3	-2	-1	0	1	2	3
$(\frac{1}{2})^x$	8	4	2	1	$\frac{1}{2}$	$\frac{1}{4}$	$\frac{1}{8}$
$(\frac{1}{4})^x$	64	16	4	1	$\frac{1}{4}$	$\frac{1}{16}$	$\frac{1}{64}$

Note The equation $y = (\frac{1}{2})^x$ is equivalent to $y = 2^{-x}$, since

$$\left(\frac{1}{2}\right)^x = \frac{1}{2^x} = 2^{-x}.$$

Therefore, exponential decay may also be described by equations of the form $y = b^{-x}$ with $b > 1$.

PROGRESS CHECK 4 Sketch the graphs of $g(x) = (\frac{1}{3})^x$ and $h(x) = (0.8)^x$ on the same coordinate system.

Important features of the exponential function with base b are illustrated in the graphs in Examples 3 and 4, and some of these properties are summarized next.

Properties of $f(x) = b^x$ (with $b > 0$, $b \neq 1$) and Its Graph

1. If $b > 1$, then as x increases, y increases; and a quantity is growing exponentially.
2. If $0 < b < 1$, then as x increases, y decreases; and a quantity is decaying exponentially.
3. The y-intercept is always $(0,1)$, and there are no x-intercepts.
4. The x-axis is a horizontal asymptote for the graph of f.
5. The domain of f is $(-\infty,\infty)$, and the range of f is $(0,\infty)$.

Progress Check Answers

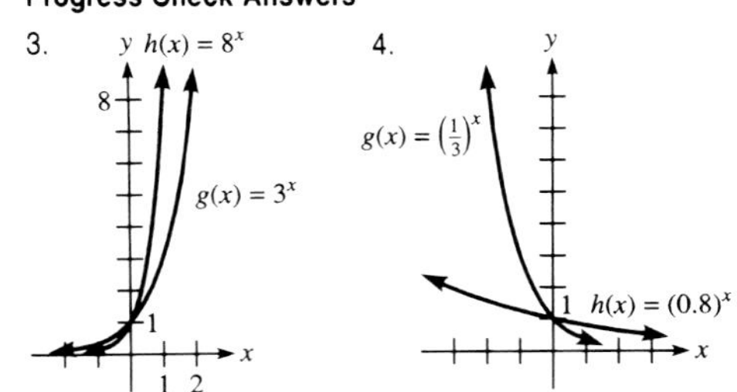

The graphing techniques of Section 9.3 may be used to graph certain variations of the function $y = b^x$, as discussed in Example 5.

EXAMPLE 5 Sketch the graph of each function.

a. $y = 4^x - 3$

b. $y = 1 - (\frac{1}{2})^x$

Solution

a. The graph of $y = 4^x - 3$ is identical in shape to the graph of $y = 4^x$, shown in Figure 10.1. To position this curve correctly, observe that the constant 3 is subtracted after the exponential term, so the graph of $y = 4^x - 3$ is the graph of $y = 4^x$ shifted 3 units down, as shown in Figure 10.3.

b. The equation $y = 1 - (\frac{1}{2})^x$ is equivalent to $y = -(\frac{1}{2})^x + 1$. To graph this function, first reflect the graph of $y = (\frac{1}{2})^x$ about the x-axis to obtain the graph of $y = -(\frac{1}{2})^x$, shown in Figure 10.4(a). Then raising the graph of $y = -(\frac{1}{2})^x$ up 1 unit gives the graph of $y = -(\frac{1}{2})^x + 1$, shown in Figure 10.4(b).

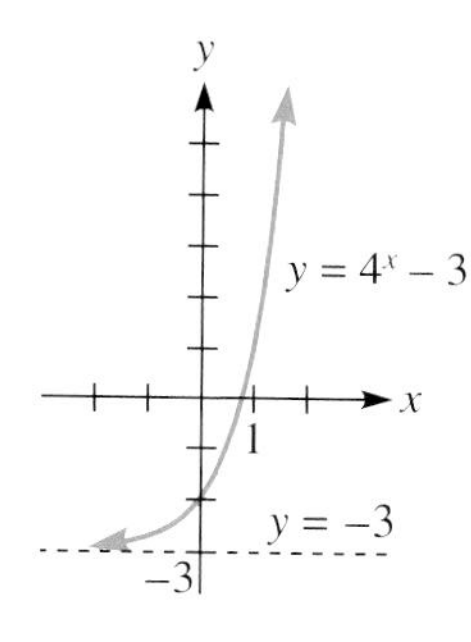

Figure 10.3

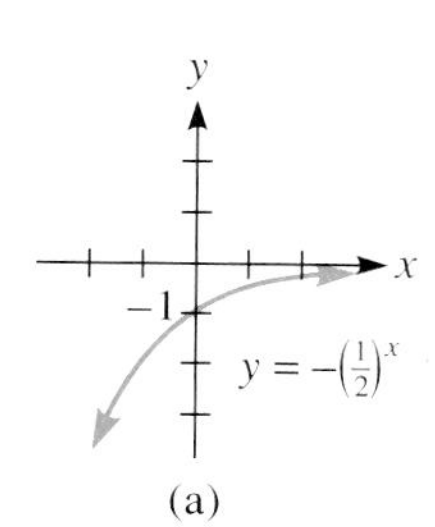

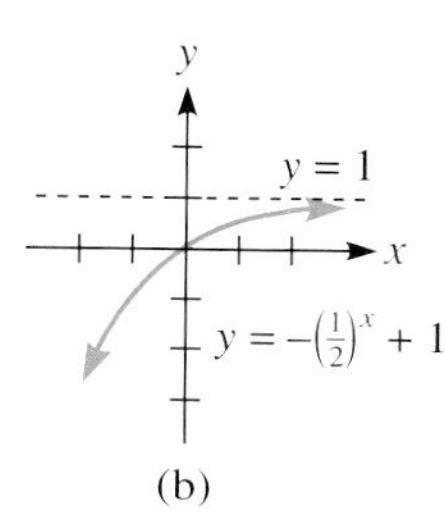

Figure 10.4

Note This example focused on sketching graphs using translations or reflection. Plotting convenient points on the graph as needed is also recommended in order to draw the graph accurately and to confirm that the graphing techniques are being applied correctly.

PROGRESS CHECK 5 Sketch the graph of each function.

a. $y = -4^x$

b. $y = (\frac{1}{2})^x + 1$

3 To analyze an exponential function, we often need to solve an **exponential equation,** which is an equation that has a variable in an exponent. In certain cases it is not difficult to write both sides of the equation in terms of the same base. Then because an exponential function takes on each value in its range exactly once, the following principle applies.

Equation-Solving Principle

If b is a positive number other than 1, then

$$b^x = b^y \quad \text{implies} \quad x = y.$$

This principle indicates that when certain exponential expressions with the same base are equal, then the equation can be solved by equating exponents, as illustrated in Example 6.

Progress Check Answers

5. (a)

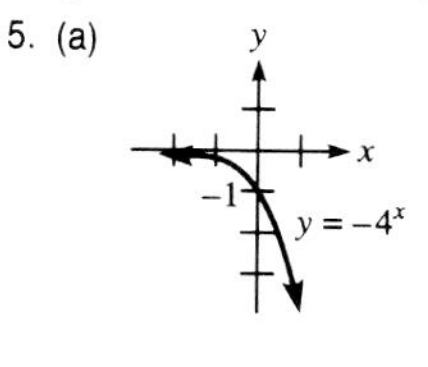

(b)

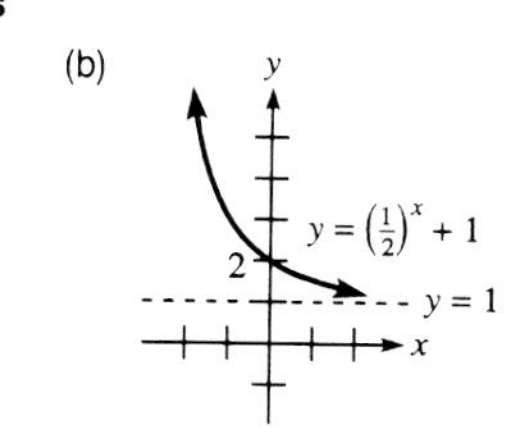

EXAMPLE 6 Solve $8^x = 4$.

Solution Both sides of the equation can be written as powers of 2 by recognizing that $8 = 2^3$ and $4 = 2^2$.

$$\begin{aligned} 8^x &= 4 \\ (2^3)^x &= 2^2 \\ 2^{3x} &= 2^2 \end{aligned}$$

Then $b^x = b^y$ implies $x = y$, so we equate exponents and solve.

$$\begin{aligned} 3x &= 2 \\ x &= \tfrac{2}{3} \end{aligned}$$

Thus, the solution set is $\{\tfrac{2}{3}\}$.

Note Section 10.5 discusses methods for solving equations like $3^x = 4$ in which it is difficult to rewrite the expressions in terms of a common base.

PROGRESS CHECK 6 Solve $9^x = \frac{1}{27}$.

4 Another type of problem associated with analyzing exponential functions involves finding the base of an exponential function when given one ordered pair in the function.

EXAMPLE 7 Find the base in the exponential function $y = b^x$ that contains the point (4,2).

Solution Replacing x by 4 and y by 2 in the equation $y = b^x$ gives

$$2 = b^4.$$

The principle of powers was first discussed in Section 7.5 with respect to radical equations.

To find b, use an extension of the principle of powers and raise both sides of the equation to the reciprocal power of 4, which is $\frac{1}{4}$, and then simplify.

$$\begin{aligned} 2^{1/4} &= (b^4)^{1/4} \quad \text{Raise both sides to the reciprocal power.} \\ 2^{1/4} &= b \end{aligned}$$

Thus, the base of the exponential function containing the given point is $2^{1/4}$, or $\sqrt[4]{2}$.

PROGRESS CHECK 7 Find the base of the exponential function $y = b^x$ that contains the point $(\frac{3}{2}, 27)$.

5 The next three examples illustrate applications of exponential functions. In particular, the examples focus on the topic of compound interest, where the growth of an investment can be seen as exponential growth.

EXAMPLE 8 \$5,000 is invested at 7 percent compounded annually.

a. Find a formula showing the value of the investment at the end of t years.
b. How much is the investment worth after 8 years?

Solution

a. Let A represent the compounded amount; then at the end of 1 year the \$5,000 amounts to

$$A = \$5{,}000(1 + 0.07) \text{ or } \$5{,}350.$$

The principal during the second year is \$5,000(1.07) and thus principal grows by a factor of 1.07 by the end of the year. Thus, the amount at the end of 2 years is

$$A = \$5{,}000(1.07)(1.07) = \$5{,}000(1.07)^2 = \$5{,}724.50.$$

Progress Check Answers

6. $\{-\frac{3}{2}\}$

7. 9

Continuing in this manner indicates that the value of the investment after t years is

$$A = \$5,000(1.07)^t.$$

b. When $t = 8$,

$$A = \$5,000(1.07)^8 = \$8,590.93.$$

PROGRESS CHECK 8 \$9,000 is invested at 5 percent compounded annually.

a. Find a formula showing the value of this investment at the end of t years.
b. How much is the investment worth after 12 years?

The procedure from Example 8 can be generalized to obtain the following formula for the compounded amount A when an original principal P is compounded annually for t years at annual interest rate r:

$$A = P(1 + r)^t.$$

A more general discussion of compound interest is given in Section 10.6.

The next two examples make use of this formula.

EXAMPLE 9 Solve the problem in the chapter introduction on page 477.

Solution Observe that 48 years will elapse from the student's 17th birthday to her 65th birthday. Then substituting $P = 2,000$, $r = 0.085$, and $t = 48$ in the formula above gives

$$\begin{aligned} A &= 2,000(1 + 0.085)^{48} \\ &= \$100,382.37. \end{aligned}$$

PROGRESS CHECK 9 If the student makes just one \$2,000 deposit in an IRA and delays this deposit until her 27th birthday, then what will be the value of her account on her 65th birthday, assuming 8.5 percent compounded annually?

EXAMPLE 10 At what interest rate compounded annually must a sum of money be invested if it is to double in 5 years?

Solution When an original principal P has doubled, then $A = 2P$. Replace A by $2P$ and t by 5 in the formula for interest compounded annually and solve for r.

$$\begin{aligned} 2P &= P(1 + r)^5 \\ 2 &= (1 + r)^5 \\ 2^{1/5} &= 1 + r \quad \text{Raise both sides to the reciprocal power.} \\ 2^{1/5} - 1 &= r \end{aligned}$$

By calculator, $2^{1/5} - 1 \approx 0.1486984$, so the required interest rate is about 14.87 percent.

PROGRESS CHECK 10 At what interest rate compounded annually must a sum of money be invested if it is to triple in 12 years?

Progress Check Answers

8. (a) $A = 9,000(1.05)^t$ (b) \$16,162.71
9. \$44,397.66
10. 9.59 percent

EXERCISES 10.1

In Exercises 1–20 for each of the given functions, find $f(0)$, $f(3)$, $f(-2)$, $f(\frac{1}{2})$, and $f(\sqrt{2})$. Approximate decimals to the nearest hundredth.

1. $f(x) = 5^x$ 1, 125, $\frac{1}{25}$, 2.24, 9.74
2. $f(x) = 7^x$ 1, 343, $\frac{1}{49}$, 2.65, 15.67
3. $f(x) = 3^x$ 1, 27, $\frac{1}{9}$, 1.73, 4.73
4. $f(x) = 9^x$ 1, 729, $\frac{1}{81}$, 3, 22.36
5. $f(x) = 4^{-x}$ 1, $\frac{1}{64}$, 16, $\frac{1}{2}$, 0.14
6. $f(x) = 9^{-x}$ 1, $\frac{1}{729}$, 81, $\frac{1}{3}$, 0.04
7. $f(x) = 6^{-x}$ 1, $\frac{1}{216}$, 36, 0.41, 0.08
8. $f(x) = 5^{-x}$ 1, $\frac{1}{125}$, 25, 0.45, 0.10
9. $f(x) = -9^x$ −1, −729, $-\frac{1}{81}$, −3, −22.36
10. $f(x) = -9^{-x}$ −1, $-\frac{1}{729}$, −81, $-\frac{1}{3}$, −0.04
11. $f(x) = -5^x$ −1, −125, $-\frac{1}{25}$, −2.24, −9.74
12. $f(x) = -5^{-x}$ −1, $-\frac{1}{125}$, −25, −0.45, −0.10
13. $f(x) = (\frac{1}{2})^x$ 1, $\frac{1}{8}$, 4, 0.71, 0.38
14. $f(x) = (\frac{1}{2})^{-x}$ 1, 8, $\frac{1}{4}$, 1.41, 2.67
15. $f(x) = (\frac{1}{4})^x$ 1, $\frac{1}{64}$, 16, $\frac{1}{2}$, 0.14

16. $f(x) = (\frac{1}{4})^{-x}$ 1, 64, $\frac{1}{16}$, 2, 7.10

17. $f(x) = -(\frac{1}{2})^{x}$ −1, $-\frac{1}{8}$, −4, −0.71, −0.38

18. $f(x) = -(\frac{1}{2})^{-x}$ −1, −8, $-\frac{1}{4}$, −1.41, −2.67

19. $f(x) = -(\frac{1}{4})^{x}$ −1, $-\frac{1}{64}$, −16, $-\frac{1}{2}$, −0.14

20. $f(x) = -(\frac{1}{4})^{-x}$ −1, −64, $-\frac{1}{16}$, −2, −7.10

In Exercises 21–34, sketch the graphs of the given function.

21. $y = 5^x$

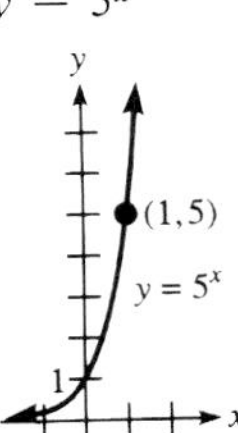

22. $y = 3^x$

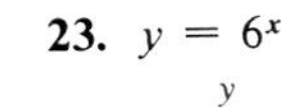

23. $y = 6^x$

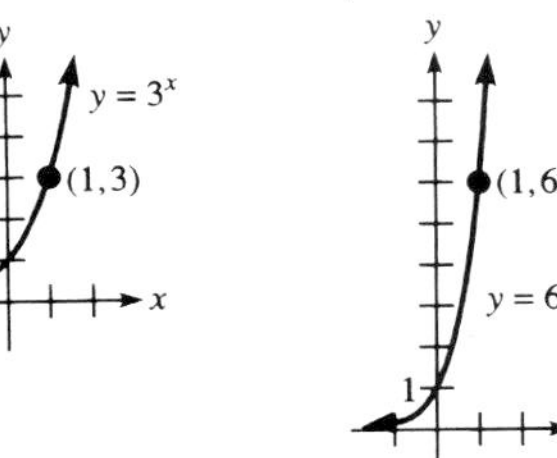

24. $y = 9^x$

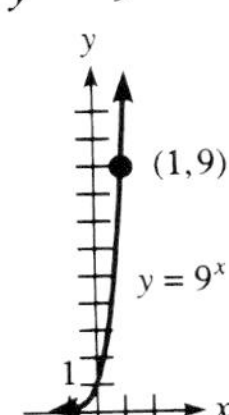

25. $y = 5^{-x}$

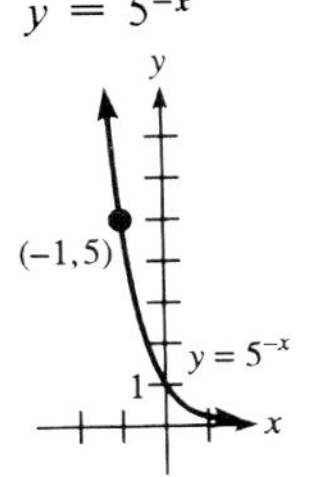

26. $y = 3^{-x}$

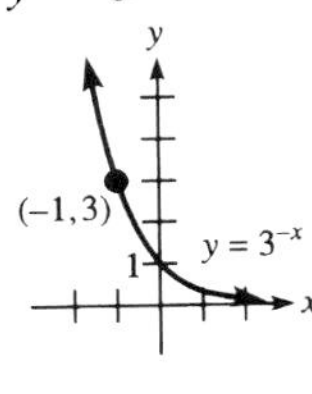

27. $y = 6^{-x}$

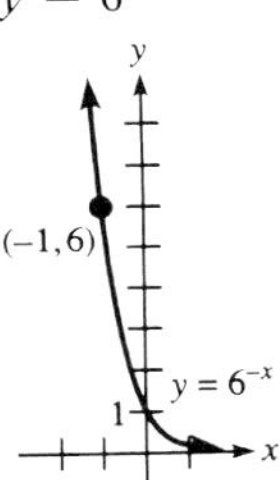

28. $y = 9^{-x}$

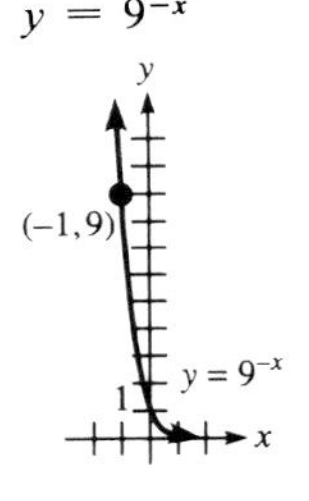

29. $y = -5^x$

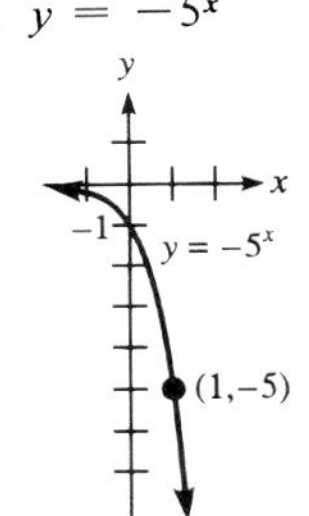

30. $y = -3^{-x}$

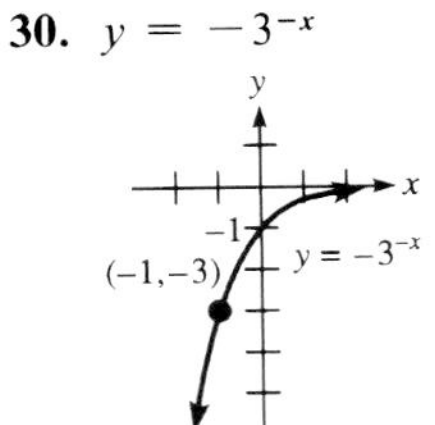

31. $y = (\frac{1}{2})^x$

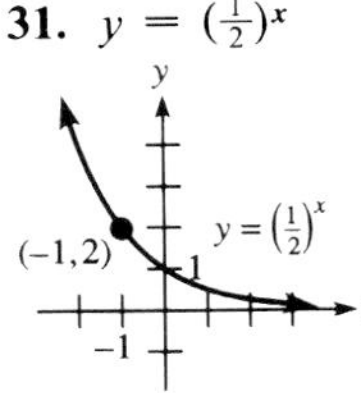

32. $y = (\frac{1}{2})^{-x}$

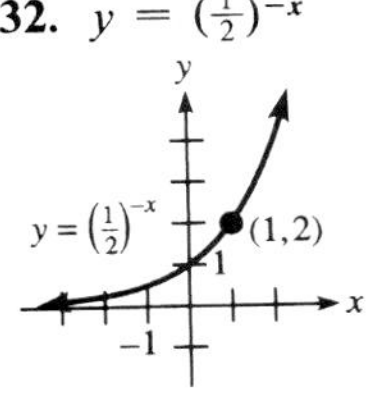

33. $y = -(\frac{1}{2})^x$

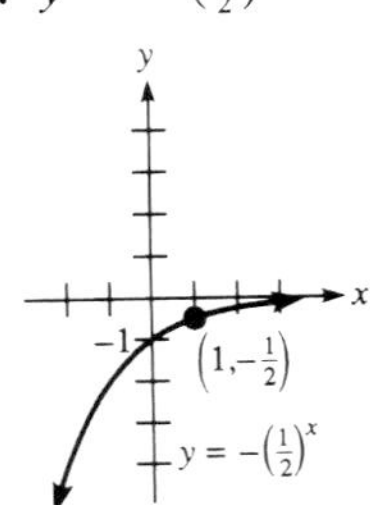

34. $y = -(\frac{1}{2})^{-x}$

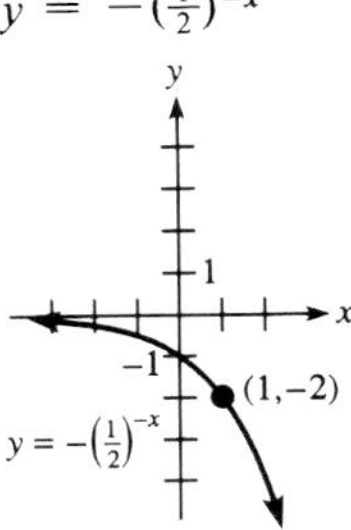

In Exercises 35–42, sketch the graph of f and g on the same coordinate system.

35. $f(x) = 3^x$, $g(x) = 5^x$

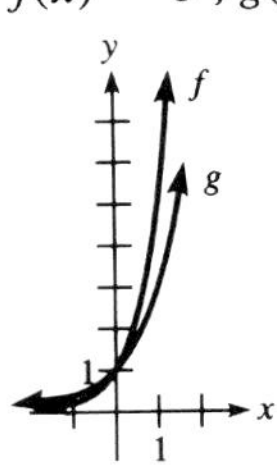

36. $f(x) = 3^{-x}$, $g(x) = 5^{-x}$

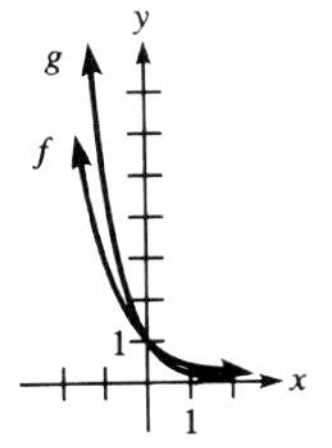

37. $f(x) = (\frac{1}{3})^x$, $g(x) = (\frac{1}{5})^x$

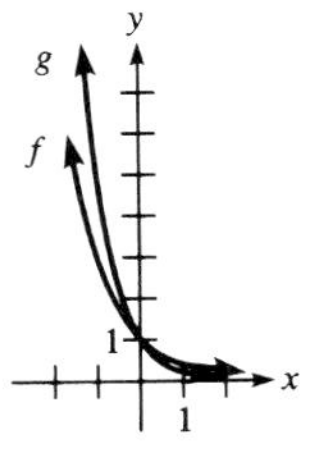

38. $f(x) = (\frac{1}{5})^x$, $g(x) = (\frac{1}{5})^{-x}$

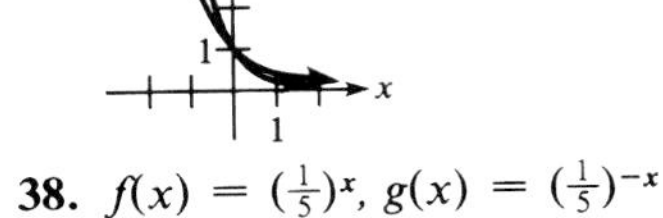

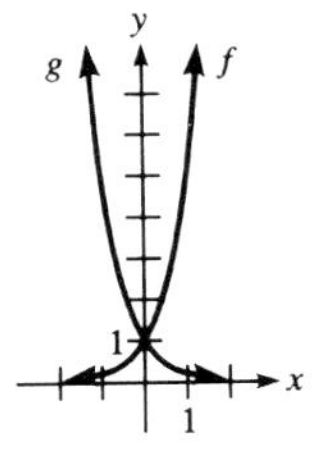

39. $f(x) = -3^x$, $g(x) = -5^x$

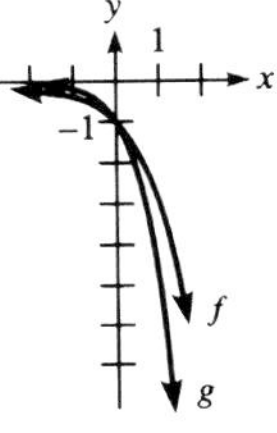

40. $f(x) = -3^{-x}$, $g(x) = -5^{-x}$

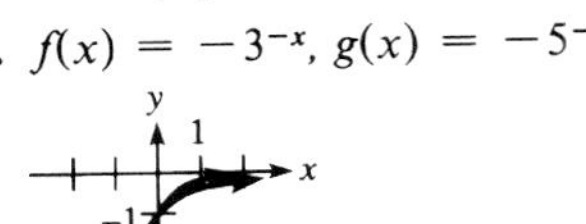

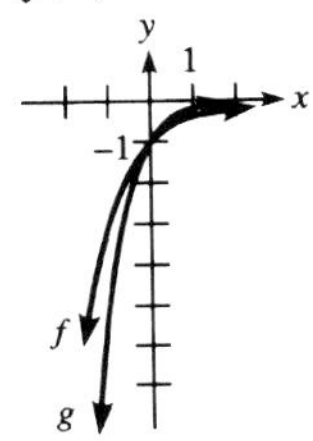

41. $f(x) = -(\frac{1}{3})^x$, $g(x) = -(\frac{1}{5})^x$

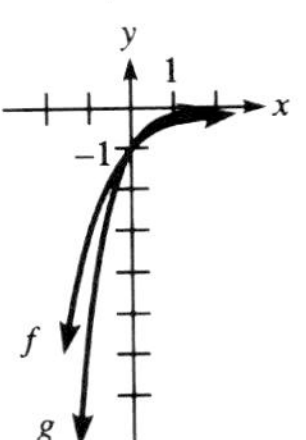

42. $f(x) = -(\frac{1}{3})^{-x}$, $g(x) = -3^x$

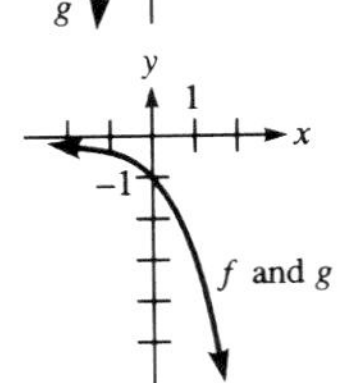

In Exercises 43–48, solve for x.

43. $9^x = \frac{1}{27}$ $-\frac{3}{2}$

44. $9^x = 27$ $\frac{3}{2}$

45. $16^x = 8$ $\frac{3}{4}$

46. $25^x = 125$ $\frac{3}{2}$

47. $36^x = \frac{1}{6}$ $-\frac{1}{2}$

48. $8^x = \frac{1}{64}$ −2

In Exercises 49–58, find the base b of the exponential function $y = b^x$ which contains the given point (x,y).

49. $(2,9)$ 3

50. $(3,64)$ 4

51. $(-2,\frac{1}{25})$ 5

52. $(-1,\frac{1}{7})$ 7

53. $(-2,9)$ $\frac{1}{3}$

54. $(-5,32)$ $\frac{1}{2}$

55. $(3,2)$ $\sqrt[3]{2}$

56. $(2,3)$ $\sqrt{3}$

57. $(2,\sqrt[3]{5})$ $\sqrt[6]{5}$

58. $(5,\sqrt[3]{2})$ $\sqrt[15]{2}$

In Exercises 59–62, sketch the graph of f and g on the same coordinate system.

59. $f(x) = 5^x + 3,\ g(x) = 5^x - 2$

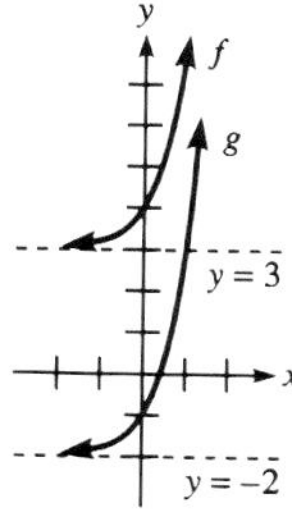

60. $f(x) = 3^x + 2,\ g(x) = 3^x - 1$

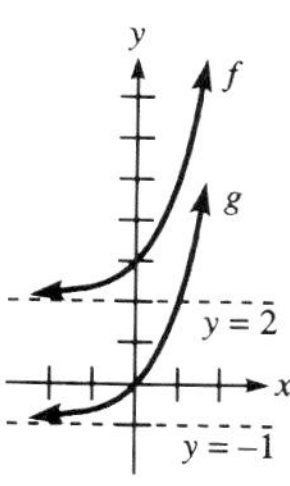

61. $f(x) = 5^{-x} + 3,\ g(x) = 5^{-x} - 2$

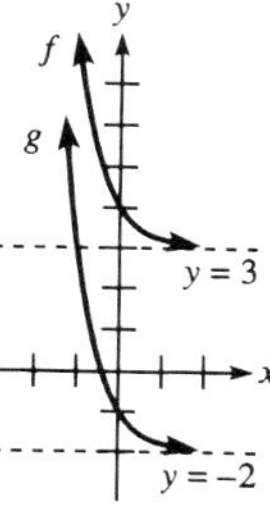

62. $f(x) = (\frac{1}{3})^x - 2,\ g(x) = (\frac{1}{3})^x + 2$

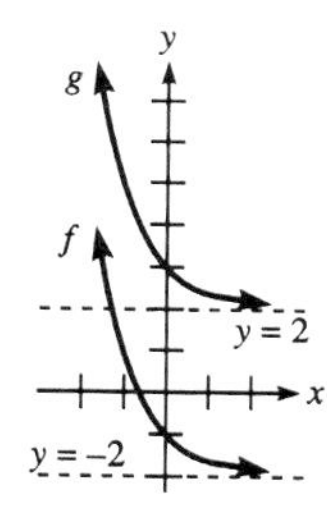

63. In 1990 the descendants of a man who lent the American government \$450,000 at the time of the revolution asked to have the loan repaid. Suppose the loan is compounded annually at 6 percent. To the nearest thousand dollars, how much was the family owed if the loan was made in 1776 and repaid in 1990? \$117,129,875,000

64. Which is worth more after compounding for 20 years: \$10 at 5 percent, or \$5 at 10 percent? \$5 at 10 percent

65. What is the value of a \$10 investment after 10 years if it is compounded at an annual interest of 10 percent? \$25.94

66. What is the value of an investment of \$300 after 30 years if it is compounded at an annual interest rate of 6 percent? \$1,723.05

67. The value of an investment which depreciates at a fixed rate every year can be evaluated by using negative values for r in the formula for compound interest. The value of a \$10,000 investment which depreciates by 8 percent a year is given by $A = 10{,}000(1 - 0.08)^t$. What is its value after 4 years? \$7,163.93

68. Use the definitions in Exercise 67 to find the value of a \$1 million piece of machinery after 14 years of depreciation at the rate of 5 percent annually. Give the answer to the nearest dollar. \$487,675

69. At what interest rate compounded annually must a sum of money be invested if it is to double in 10 years? 7.18 percent

70. At what interest rate compounded annually must a sum of money be invested if it is to triple in 10 years? 11.61 percent

71. What annual growth rate will cause a population to double in 50 years? 1.40 percent

72. What annual growth rate will cause a population to double in 25 years? 2.81 percent

THINK ABOUT IT

1. If you win the state lottery and are offered \$1 million or a penny that doubles in value on each day of the month of May, which option would you accept? What is the value of the penny by the end of the month?
2. The following table was generated using an exponential function.

x	0	1	2	3	4
y	10,000	8,000	6,400	5,120	4,096

   a. For such a table a fixed change in x results in a constant ratio between the corresponding y values. What is that ratio when the change in x is 1?

   b. Find an equation that will generate this table. (*Hint:* Think of it as the depreciating value of a \$10,000 investment.)

3. There is often more than one equation whose graph will contain certain given points.

   a. Given the points (0,1) and (1,4), find a linear function, a quadratic function, and an exponential function whose graphs contain both points. Sketch each graph.

   b. For each function in part **a,** find y when $x = 2$.

4. Based on the graph of $y = 2^x$, place the following numbers in their correct numerical order:

$$2^{\sqrt{3}},\ 2^{1/3},\ 2^{\pi},\ 2^{-\sqrt{2}},\ 2^{0}.$$

   Explain your method.

5. If $f(x) = b^x$ is the exponential function with base b, then show that $f(u + v) = f(u) \cdot f(v)$.

REMEMBER THIS

1. In the ordered pair (3,5), which value is called the x-coordinate? 3
2. Is this set of ordered pairs a function: $\{(1,2),(2,3),(3,4)\}$? Yes
3. Does the equation $y = 2$ determine a horizontal or a vertical line? Horizontal
4. Graph $y = x^3$.

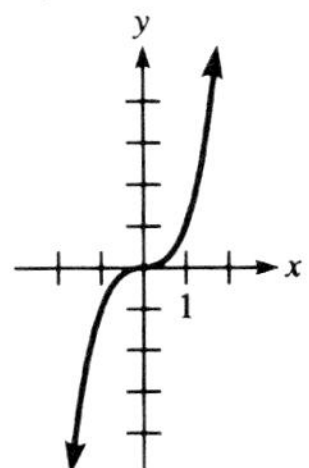

5. If $f(x) = x^{1/3}$ and $g(x) = x^3$, find $f[g(5)]$. 5
6. What algebraic expression represents 80 percent of a number x? $0.80x$
7. Which of these expressions determines a linear function?
   a. $y = 3x - 4$ **b.** $y = x^2 + 3x - 4$ a
8. What is the equation of a circle with radius 3 and center at the origin? $x^2 + y^2 = 9$
9. True or false? $\dfrac{x + 1/y}{y + 1/x} = \dfrac{x}{y}$. True

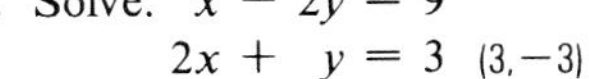

10. Solve. $x - 2y = 9$
    $2x + y = 3$ (3, −3)

10.2 Inverse Functions

The function $y = f(x) = \frac{5}{9}(x - 32)$ converts degrees Fahrenheit (x) to degrees Celsius (y). Find an equation for the inverse function. What formula does the inverse function represent? (See Example 7.)

OBJECTIVES

1. **Find the inverse of a function and its domain and range.**
2. **Determine if a function has an inverse function.**
3. **Graph $y = f^{-1}(x)$ from the graph of $y = f(x)$.**
4. **Find an equation that defines f^{-1}.**
5. **Determine whether two functions are inverses of each other.**
6. **Use inverse function concepts in applications.**

1 Consider the following tables, which illustrate ordered pairs that belong to the function $f(x) = x^3$ and ordered pairs that belong to the function $g(x) = \sqrt[3]{x}$.

x	$f(x) = x^3$	Ordered Pairs in f
1	$1^3 = 1$	(1,1)
2	$2^3 = 8$	(2,8)
−2	$(-2)^3 = -8$	(−2,−8)
3	$3^3 = 27$	(3,27)
−3	$(-3)^3 = -27$	(−3,−27)

x	$g(x) = \sqrt[3]{x}$	Ordered Pairs in g
1	$\sqrt[3]{1} = 1$	(1,1)
8	$\sqrt[3]{8} = 2$	(8,2)
−8	$\sqrt[3]{-8} = -2$	(−8,−2)
27	$\sqrt[3]{27} = 3$	(27,3)
−27	$\sqrt[3]{-27} = 3$	(−27,−3)

Observe that functions f and g are related and have reverse assignments. For example, $f(3) = 27$ and $g(27) = 3$. Two functions with exactly reverse assignments are called **inverse functions** of each other, and $f(x) = x^3$ and $g(x) = \sqrt[3]{x}$ are examples of inverse functions. The special symbol f^{-1} is used to denote the inverse of function f, so

$$\begin{array}{llll} \text{if} & f(x) = x^3, & \text{then} & f^{-1}(x) = \sqrt[3]{x}, \\ \text{and if} & f(x) = \sqrt[3]{x}, & \text{then} & f^{-1}(x) = x^3. \end{array}$$

The reverse assignments of f and f^{-1} mean that when (a,b) belongs to a function, then (b,a) belongs to the inverse function. Because the components in the ordered pairs are reversed, the domain of f equals the range of f^{-1} and the range of f equals the domain of f^{-1}.

At some point, caution students about the special use of the -1 superscript for denoting an inverse function. This warning is especially needed for problems like Example 5 (where such a caution is printed). The notation was developed for mappings in abstract algebra, where expressions like $(a\alpha)\alpha^{-1} = a$ are convenient. In this notation α represents a mapping, α^{-1} represents the inverse mapping, and a is the element being mapped.

EXAMPLE 1 If $f = \{(1,5),(2,6),(3,7)\}$, find f^{-1}. Find and compare the domain and the range of the two functions.

Solution Reversing assignments gives

$$f^{-1} = \{(5,1),(6,2),(7,3)\}.$$

The set of all first components gives the domain of a function, and the set of all second components gives the range. Therefore,

$$\begin{aligned} \text{domain of } f = \text{range of } f^{-1} &= \{1,2,3\}, \\ \text{range of } f = \text{domain of } f^{-1} &= \{5,6,7\}. \end{aligned}$$

As expected, f and f^{-1} interchange their domain and range.

PROGRESS CHECK 1 If $f = \{(-1,1),(-2,2),(-3,3)\}$, find f^{-1}. Find and compare the domain and the range of the two functions.

2 Not all functions have an inverse function. For instance, if a function f is given by

$$\{(1,5),(2,5),(3,5)\}$$

then reversing assignments produces

$$\{(5,1),(5,2),(5,3)\}.$$

The resulting relation is not a function because the number 5 is the first component in more than one ordered pair. Because the second component in f becomes the first component in the inverse relation, the following method may be used in determining whether f has an inverse function.

One-to-One Function

A function is **one-to-one** when each x value in the domain is assigned a different y value so that no two ordered pairs have the same second component. If f is one-to-one, then f has an inverse function; and if f is not one-to-one, then f does not have an inverse function.

EXAMPLE 2 If $f = \{(-5,5),(0,0),(5,5)\}$, does f have an inverse function?

Solution Because 5 appears as the second component in two ordered pairs, f is not a one-to-one function and f does not have an inverse function.

PROGRESS CHECK 2 If $f = \{(-3,3),(0,0),(3,-3)\}$, does f have an inverse function?

Progress Check Answers

1. $f^{-1} = \{(1,-1), (2,-2), (3,-3)\}$; $D_f = R_{f^{-1}} = \{-1,-2,-3\}$; $R_f = D_{f^{-1}} = \{1,2,3\}$
2. Yes

It is easy to recognize the graph of a one-to-one function because none of its points can have the same y-coordinate. Therefore, the graph of a one-to-one function cannot contain two or more points that lie on the same horizontal line. This feature is often summarized in the horizontal line test.

Ask students to articulate the different purposes of the horizontal line test and the vertical line test of Section 9.1. See "Think About It" Exercise 4.

> **Horizontal Line Test**
>
> Imagine a horizontal line sweeping down the graph of a function f. If the horizontal line at any position intersects the graph in more than one point, then f is not a one-to-one function and f does not have an inverse function.

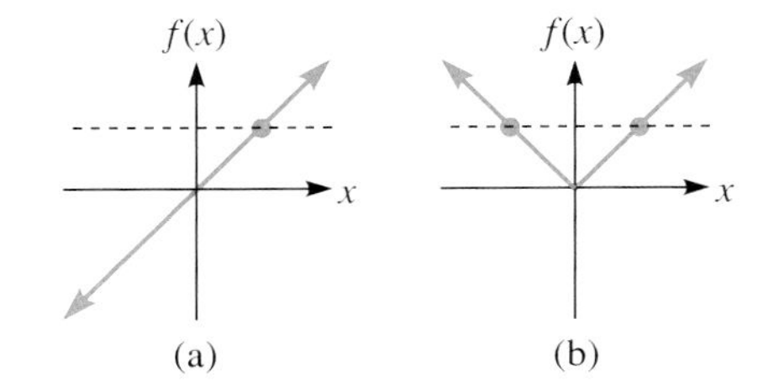

Figure 10.5

EXAMPLE 3 Which functions graphed in Figure 10.5 have an inverse function?

Solution

a. This function has an inverse function since no horizontal line intersects the graph at more than one point.
b. This function does not have an inverse function because a horizontal line may intersect the graph at two points.

PROGRESS CHECK 3 Which functions graphed in Figure 10.6 have an inverse function?

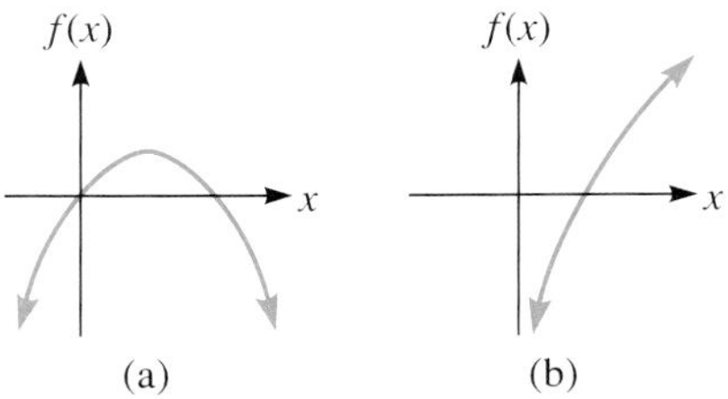

Figure 10.6

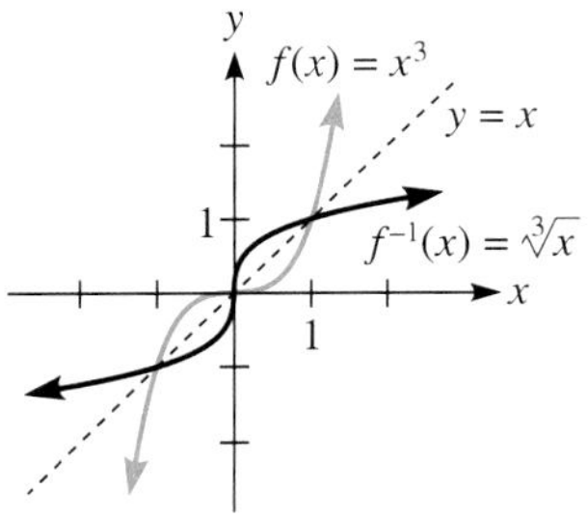

Figure 10.7

[3] When the horizontal line test indicates that a function f has an inverse function, then there is a simple geometric method for drawing f^{-1}. Because the x- and y-coordinates change places in inverse functions, the graphs of f and f^{-1} are related in that each one is the reflection of the other about the line $y = x$. For instance, this relationship between the graphs of $f(x) = x^3$ and $f^{-1}(x) = \sqrt[3]{x}$ is shown in Figure 10.7. In the next example the graph of f^{-1} is obtained from the graph of f by using this reflection method.

Consider mentioning that f in Example 4 is an exponential function and its inverse is a type of function (logarithmic) that will be studied in the next section.

EXAMPLE 4 Use the graph of $y = f(x)$ in Figure 10.8 to graph $y = f^{-1}(x)$.

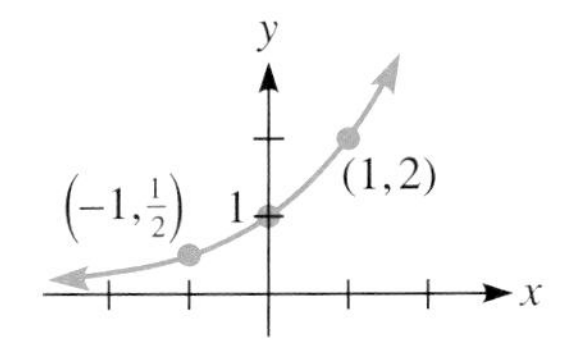

Figure 10.8

Solution First, observe that f is a one-to-one function (by the horizontal line test), so f has an inverse function. To graph $y = f^{-1}(x)$, we reflect the graph of $y = f(x)$ about

Progress Check Answers

3. (a) No (b) Yes

the line $y = x$. Note the ordered pairs $(-1,\frac{1}{2})$, $(0,1)$, and $(1,2)$ from f become $(\frac{1}{2},-1)$, $(1,0)$, and $(2,1)$ in f^{-1}. Both $y = f(x)$ and $y = f^{-1}(x)$ are graphed in Figure 10.9.

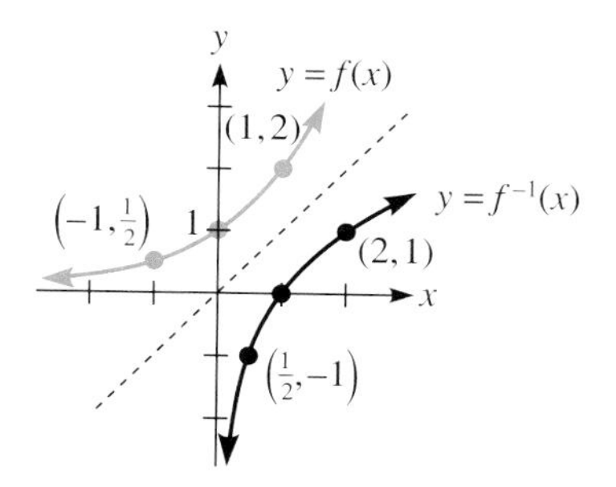

Figure 10.9

PROGRESS CHECK 4 Use the graph of $y = f(x)$ in Figure 10.10 to graph $y = f^{-1}(x)$.

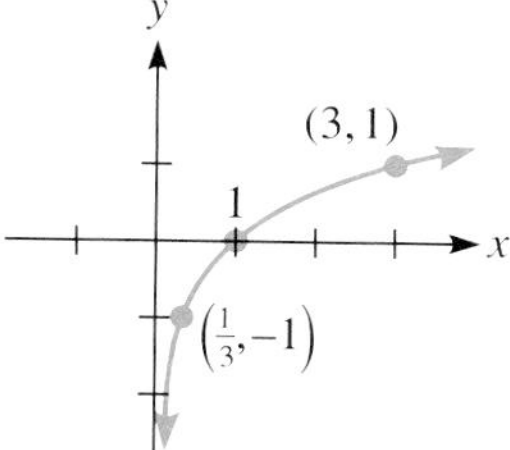

Figure 10.10

4 When f is a one-to-one function that is defined by an equation, then it is sometimes possible to find an equation that defines f^{-1}. The method is based on the fact that f and f^{-1} make reverse assignments, so we will reverse the roles of x and y in the equation that defines f and try to solve for y to obtain an equation that defines f^{-1}. For example, the inverse of the function defined by

$$y = x^3$$

is defined by

$$x = y^3.$$

Then taking the cube root of both sides of the resulting equation gives

$$y = \sqrt[3]{x}.$$

Thus, if $f(x) = x^3$, then $f^{-1}(x) = \sqrt[3]{x}$. The method just discussed is incorporated in the detailed procedure that follows for finding an equation that defines an inverse function.

It may be easy to show that f is one-to-one so that f^{-1} exists, but finding an equation that defines f^{-1} may be difficult or impossible because of the algebra involved. To illustrate, consider finding an equation defining f^{-1} if $f(x) = x^3 + x$.

To Find an Equation Defining f^{-1}

1. Start with a one-to-one function $y = f(x)$ and interchange x and y in this equation.
2. Solve the resulting equation for y, and then replace y by $f^{-1}(x)$.
3. Define the domain of f^{-1} to be equal to the range of f.

Progress Check Answer

4.

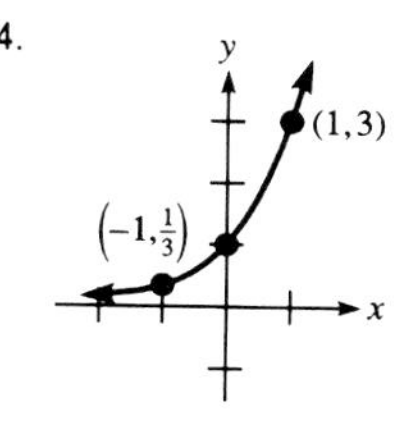

EXAMPLE 5 Find $f^{-1}(x)$ for each function. If the given function is not one-to-one, so that no inverse function exists, state this.

a. $f(x) = 4x - 1$ **b.** $f(x) = \sqrt{x}$ **c.** $f(x) = x^2$

Solution As a visual aid, the graphs of all functions considered in this example are shown in Figures 10.11–10.13.

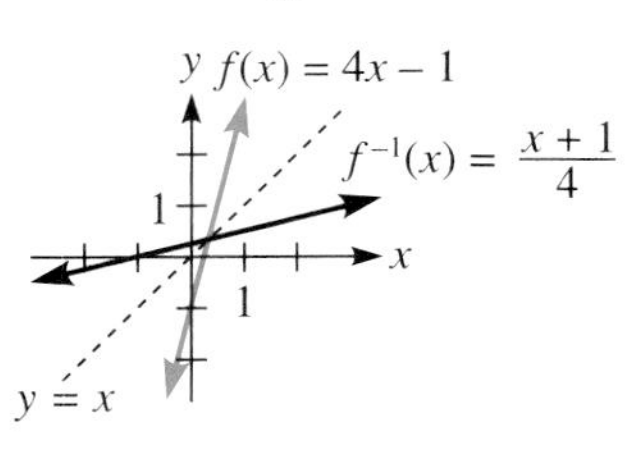

Figure 10.11

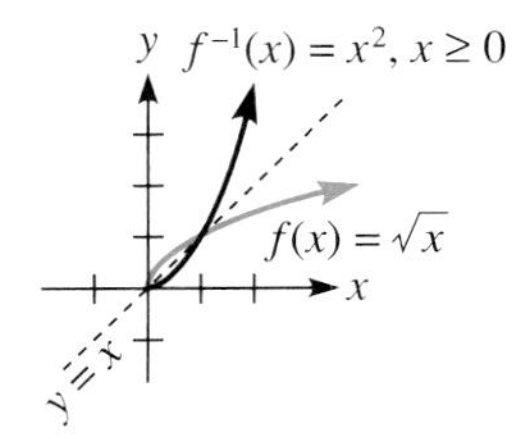

Figure 10.12

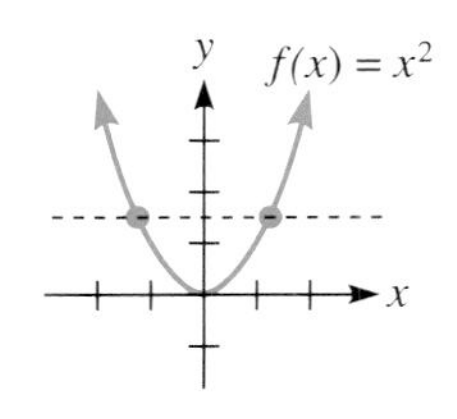

Figure 10.13

a. The function $f(x) = 4x - 1$ is a linear function. All such functions are one-to-one, so an inverse function exists. To find an equation defining f^{-1}, write the equation defining f as

$$y = 4x - 1$$

and then proceed as follows.

$$\begin{aligned} x &= 4y - 1 && \text{Interchange } x \text{ and } y. \\ x + 1 &= 4y && \text{Solve for } y. \\ \frac{x+1}{4} &= y && \\ \frac{x+1}{4} &= f^{-1}(x) && \text{Replace } y \text{ by } f^{-1}(x). \end{aligned}$$

Tell students to note in Example 5a that $f(0) = -1$ and $f^{-1}(-1) = 0$. Checking that f and f^{-1} make reverse assignments for a convenient specific input is a quick way to test that the answer for $f^{-1}(x)$ is reasonable.

Thus, if $f(x) = 4x - 1$, then $f^{-1}(x) = (x + 1)/4$. The domain of f^{-1} is equal to the range of f, which is the set of all real numbers. (See Figure 10.11.)

b. The square root function is a one-to-one function with range $[0,\infty)$, as can be seen in Figure 10.12. Thus, f has an inverse function whose equation is found as follows.

$$\begin{aligned} y &= \sqrt{x} && \text{Start with } y = f(x). \\ x &= \sqrt{y} && \text{Interchange } x \text{ and } y. \\ x^2 &= y && \text{Solve for } y. \\ x^2 &= f^{-1}(x) && \text{Replace } y \text{ by } f^{-1}(x). \end{aligned}$$

Although $y = x^2$ is defined for all real values of x, the domain of f^{-1} must match the range of f to specify inverse functions. Therefore, x is limited to the nonnegative real numbers, so f^{-1} is defined by

$$f^{-1}(x) = x^2, \quad \text{where} \quad x \geq 0.$$

c. The squaring function $f(x) = x^2$ is not a one-to-one function (as can be seen by applying the horizontal line test to its graph in Figure 10.13). Thus, f does not have an inverse function.

Caution Remember that f^{-1} is the special symbol that denotes the inverse function of function f. Do *not* interpret the -1 in this symbol as an exponent.

PROGRESS CHECK 5 Find $f^{-1}(x)$ for each function. If no inverse function exists, state this.

a. $f(x) = 3x + 2$ **b.** $f(x) = |x|$ **c.** $f(x) = \sqrt{x - 4}$

Progress Check Answers

5. (a) $f^{-1}(x) = \dfrac{x-2}{3}$ (b) No inverse function exists. (c) $f^{-1}(x) = x^2 + 4, x \geq 0$

5 An important concept associated with inverse functions is that each function "undoes" the other. Therefore, applying f and f^{-1}, one after the other, to a meaningful input x produces x as the output. For instance, if $f(x) = x^3$, then 2 is a meaningful input and $f^{-1}(x) = \sqrt[3]{x}$. Observe that

$$f(2) = 8 \quad \text{and} \quad f^{-1}(8) = 2, \quad \text{so} \quad f^{-1}[f(2)] = 2.$$

Recall from Section 9.5 that applying two functions in succession is called a composition of the functions, and inverse functions may be defined in terms of this operation as follows.

Definition of Inverse Functions

Two functions f and g are said to be inverses of each other provided that

$$(f \circ g)(x) = f[g(x)] = x \quad \text{for all } x \text{ in the domain of } g$$

and

$$(g \circ f)(x) = g[f(x)] = x \quad \text{for all } x \text{ in the domain of } f.$$

This definition may be used to prove that two functions are inverse functions, as shown next.

EXAMPLE 6 Verify that $f(x) = \sqrt{x}$ and $g(x) = x^2$, $x \geq 0$, are inverses of each other.

Solution First, show that $f[g(x)] = x$ for all nonnegative real numbers, which is the domain of g.

$$\begin{aligned} f[g(x)] &= f(x^2) \\ &= \sqrt{x^2} \\ &= x \quad (\text{since } x \geq 0) \end{aligned}$$

Next, show that $g[f(x)] = x$ for all nonnegative real numbers, which is the domain of f.

$$\begin{aligned} g[f(x)] &= g(\sqrt{x}) \\ &= (\sqrt{x})^2 \\ &= x \quad (\text{since } x \geq 0) \end{aligned}$$

Thus, f and g are inverses of each other.

PROGRESS CHECK 6 Verify that $f(x) = 4x - 1$ and $g(x) = \dfrac{x+1}{4}$ are inverse functions.

6 One type of application involving inverse functions is based on the fact that inverse functions contain the same information in different forms. Which function is more useful depends on what is given and what is to be found. In this context certain problems are viewed as inverse problems, as illustrated below.

Inverse problems	*Problem:*	Find the Celsius temperature for a given Fahrenheit temperature.
	Problem:	Find the Fahrenheit temperature for a given Celsius temperature.

The inverse problems of converting between two temperature scales are analyzed further in the section-opening problem.

Progress Check Answers

6. $f[g(x)] = 4\left(\dfrac{x+1}{4}\right) - 1 = x + 1 - 1 = x$; $g[f(x)] = \dfrac{(4x-1)+1}{4} = \dfrac{4x}{4} = x$

EXAMPLE 7 Solve the problem in the section introduction on page 486.

Solution The equation $y = \frac{5}{9}(x - 32)$ defines a linear function, so an inverse function exists and an equation defining f^{-1} may be found as follows.

$$\begin{aligned} y &= \tfrac{5}{9}(x - 32) && \text{Start with } y = f(x). \\ x &= \tfrac{5}{9}(y - 32) && \text{Interchange } x \text{ and } y. \\ \tfrac{9}{5}x &= y - 32 && \text{Solve for } y. \\ \tfrac{9}{5}x + 32 &= y && \\ \tfrac{9}{5}x + 32 &= f^{-1}(x) && \text{Replace } y \text{ by } f^{-1}(x). \end{aligned}$$

Thus, if $y = f(x) = \frac{5}{9}(x - 32)$, then $y = f^{-1}(x) = \frac{9}{5}x + 32$. In the context of the problem, it is stated that the formula given in function f converts degrees Fahrenheit (x) to degrees Celsius (y). Therefore, the formula in the inverse function converts degrees Celsius (x) to degrees Fahrenheit (y).

PROGRESS CHECK 7 The function $y = f(x) = x^2$, $x > 0$, gives the formula for the area (y) of a square in terms of the side length (x). Find an equation for the inverse function. What formula does the inverse function represent?

The fact that inverse functions "undo" each other is the basis for another type of application considered in Example 8.

EXAMPLE 8 Certain airline employees accept a 20 percent cut in hourly wage to avoid layoffs. What percent raise will then be needed to return these employees to their original hourly wage?

Solution A 20 percent cut in hourly wage means the new wage is 80 percent of the original wage. Therefore, the reduced wage is given by

$$y = f(x) = 0.8x.$$

The rule for f^{-1} gives the formula to "undo f," and the inverse of the function defined by $y = 0.8x$ is defined by

$$x = 0.8y.$$

Solving this equation for y then yields

$$y = \frac{x}{0.8} \quad \text{or} \quad y = 1.25x.$$

The offsetting formula $y = f^{-1}(x) = 1.25x$ indicates that a 25 percent increase is needed to return the workers to their original hourly wage.

Note To check this result, observe that

$$f^{-1}[f(x)] = f^{-1}(0.8x) = x.$$

Thus, f and f^{-1} undo each other to produce the original hourly wage.

PROGRESS CHECK 8 An investor's portfolio loses one-third of its value during the first year. For the following year, what percent increase is needed for the portfolio to return to its original value?

Progress Check Answers

7. $y = f^{-1}(x) = \sqrt{x}$, $x > 0$; formula for the side length (y) of a square in terms of the area (x).
8. 50 percent

EXERCISES 10.2

In Exercises 1–8, find the inverse for each of the following functions, if it exists. Compare the domain and range of the two functions.

1. $\{(1,2),(3,4),(5,6)\}$ $f^{-1} = \{(2,1),(4,3),(6,5)\}$; $D_f = R_{f^{-1}} = \{1,3,5\}$; $R_f = D_{f^{-1}} = \{2,4,6\}$
2. $\{(-2,1),(-3,0),(5,4)\}$ $f^{-1} = \{(1,-2),(0,-3),(4,5)\}$; $D_f = R_{f^{-1}} = \{-3,-2,5\}$; $R_f = D_{f^{-1}} = \{0,1,4\}$
3. $\{(2,1),(-3,4),(5,8)\}$ $f^{-1} = \{(1,2),(4,-3),(8,5)\}$; $D_f = R_{f^{-1}} = \{-3,2,5\}$; $R_f = D_{f^{-1}} = \{1,4,8\}$
4. $\{(-3,0),(0,-3),(7,12)\}$ $f^{-1} = \{(0,-3),(-3,0),(12,7)\}$; $D_f = R_{f^{-1}} = \{-3,0,7\}$; $R_f = D_{f^{-1}} = \{-3,0,12\}$
5. $\{(-2,-3),(-4,2),(6,-3)\}$ No inverse function
6. $\{(5,3),(2,2),(-5,3)\}$ No inverse function
7. $\{(3,2),(4,2),(5,2)\}$ No inverse function
8. $\{(1,-4),(2,-4),(3,-4)\}$ No inverse function

In Exercises 9–20, determine if the function whose graph is shown has an inverse function.

9.

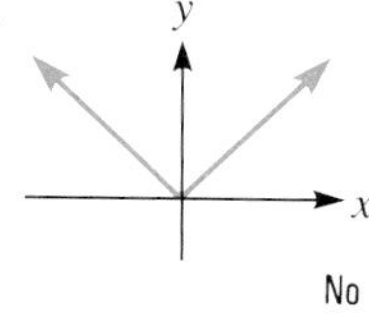

No

10.

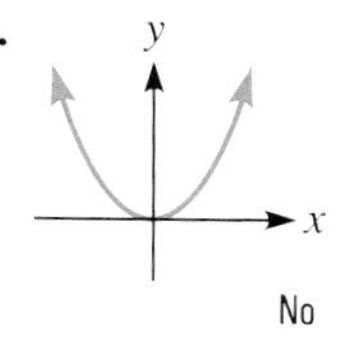

No

11.

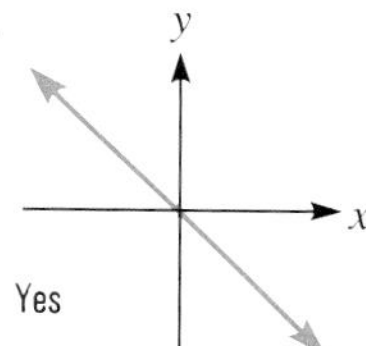

Yes

12.

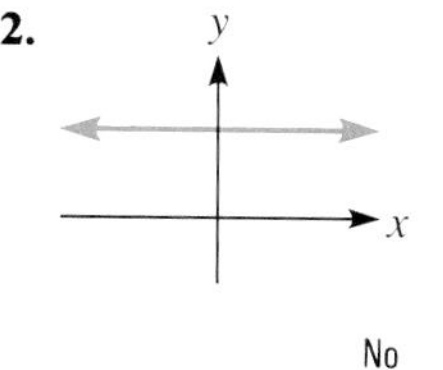

No

13.

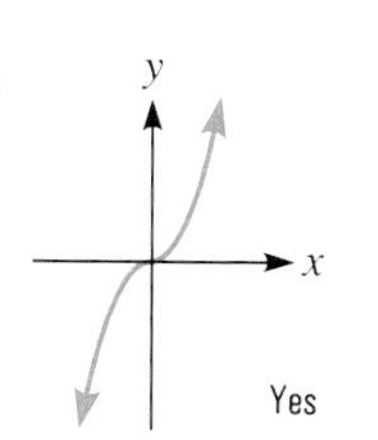

Yes

14.

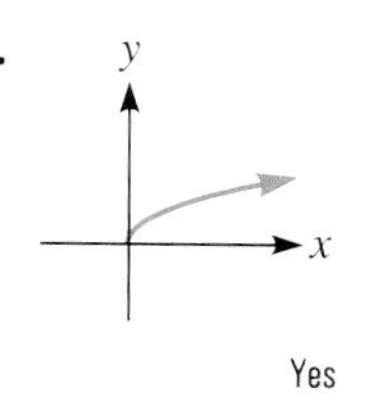

Yes

15.

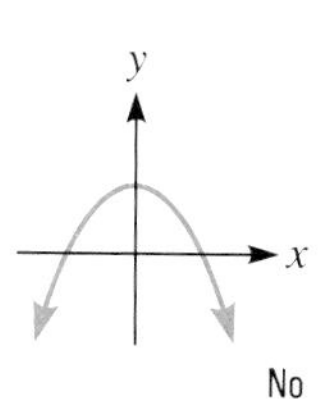

No

16.

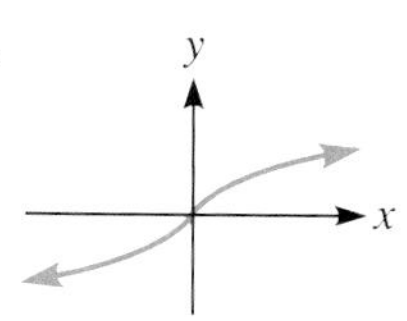

Yes

17.

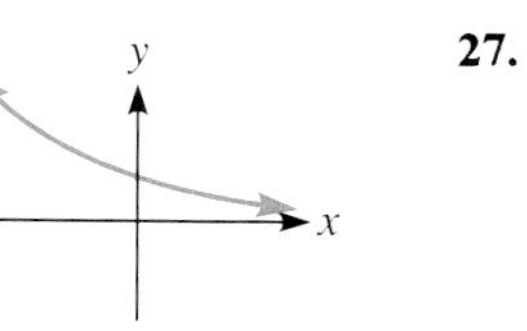

Yes

18.

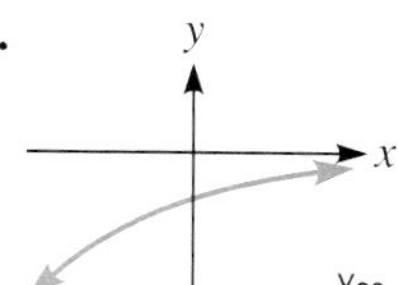

Yes

19.

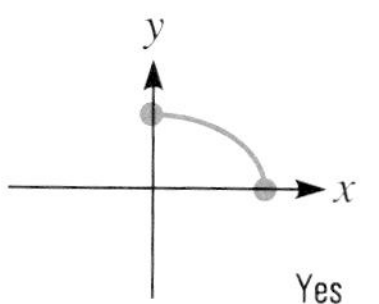

Yes

20.

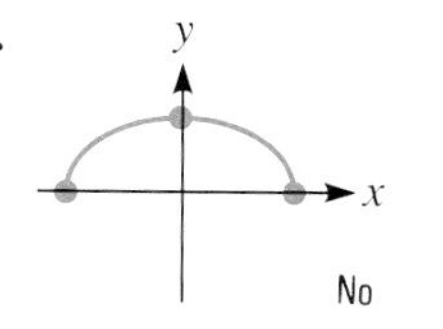

No

In Exercises 21–28, use the graph of f to graph f^{-1}.

21.

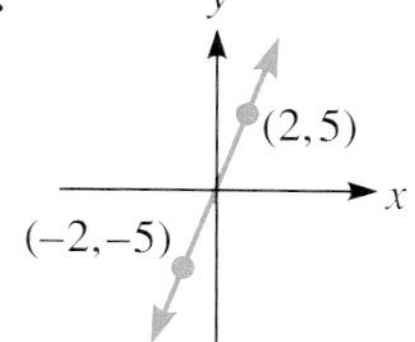

(5,2)
(−5,−2)

22.

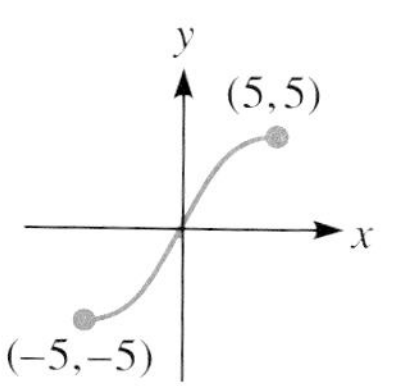

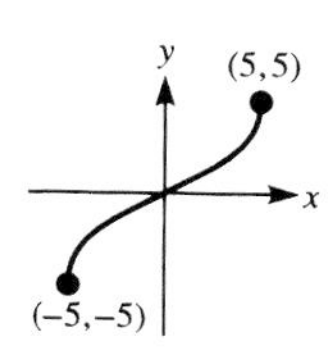

23.

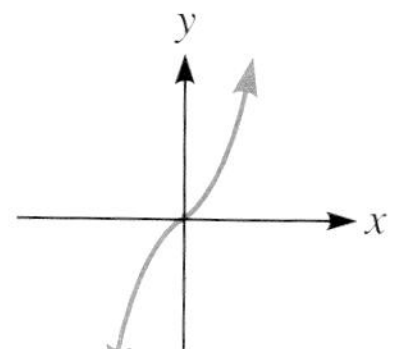

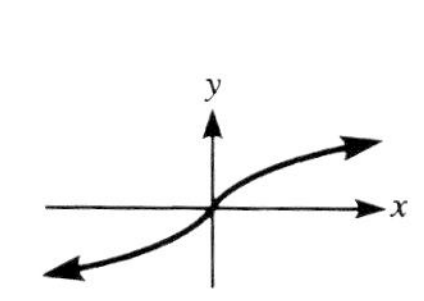

24.

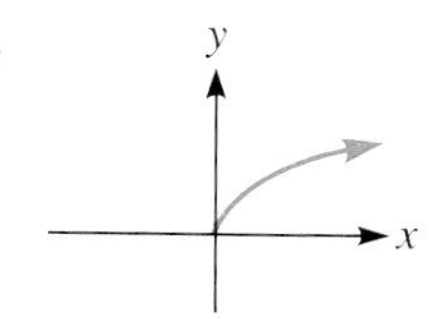

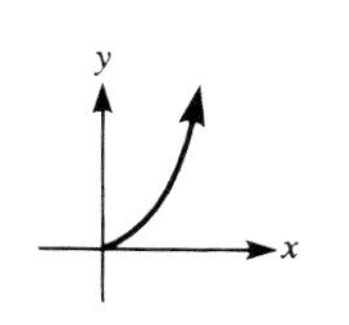

25.

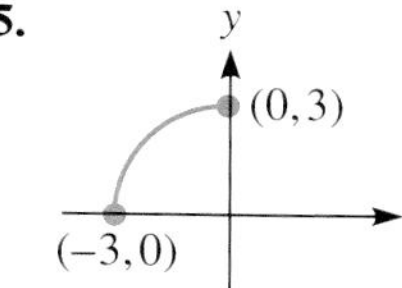

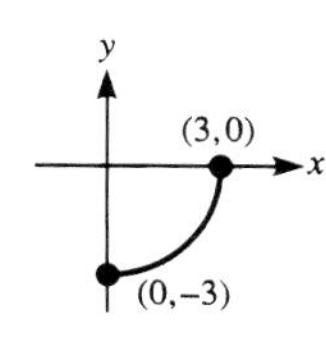

26.

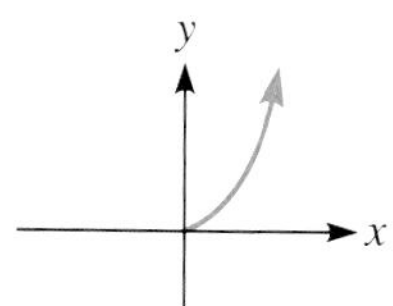

27.

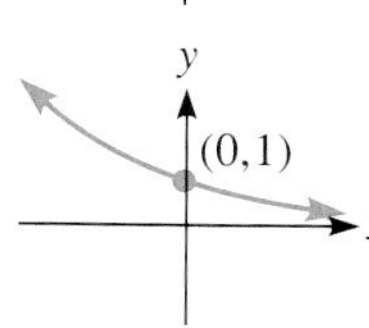

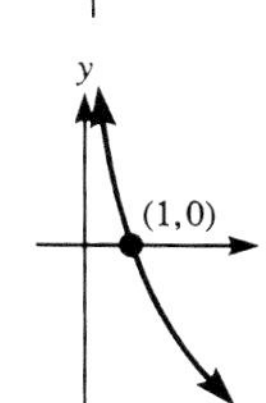

28.

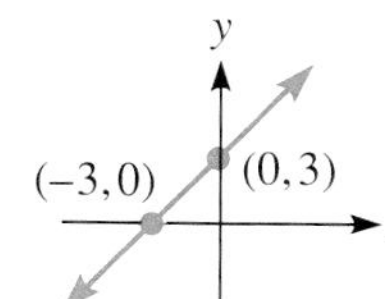

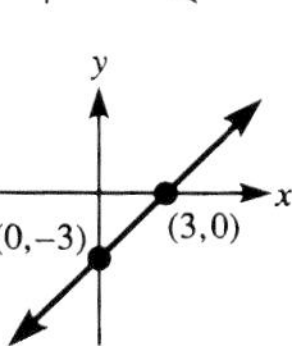

In Exercises 29–39, find $f^{-1}(x)$ for the given function. If the given function is not one-to-one, so that no inverse exists, state this.

29. $f(x) = x - 3$ $f^{-1}(x) = x + 3$
30. $f(x) = x + 2$ $f^{-1}(x) = x - 2$
31. $f(x) = 3x + 5$ $f^{-1}(x) = \dfrac{x-5}{3}$
32. $f(x) = -3x + 7$ $f^{-1}(x) = \dfrac{x-7}{-3}$
33. $f(x) = -\sqrt{x},\ x \geq 0$ $f^{-1}(x) = x^2,\ x \leq 0$
34. $f(x) = -\sqrt{x+2},\ x \geq -2$ $f^{-1}(x) = x^2 - 2,\ x \leq 0$

35. $f(x) = x^2 + 3$ No inverse function exists.
36. $f(x) = -x^2 - 2$ No inverse function exists.
37. $f(x) = |x + 3|$ No inverse function exists.
38. $f(x) = x^3$ $f^{-1}(x) = \sqrt[3]{x}$, or $x^{1/3}$
39. $f(x) = x^4, x \geq 0$ $f^{-1}(x) = \sqrt[4]{x}$, or $x^{1/4}$, $x \geq 0$

In Exercises 40–50, determine if f and g are inverses of each other.

40. $f(x) = x + 2, g(x) = x - 2$ Yes
41. $f(x) = \sqrt{x + 1}, g(x) = x^2 - 1, x \geq 0$ Yes
42. $f(x) = \sqrt[3]{x}, g(x) = x^3$ Yes
43. $f(x) = \dfrac{x - 5}{3}, g(x) = 3x + 5$ Yes
44. $f(x) = x^2 - 2, g(x) = x^2 + 2$ No
45. $f(x) = |x|, g(x) = |x|$ No
46. $f(x) = \dfrac{x}{3}, g(x) = 3x$ Yes
47. $f(x) = x^{1/4}, x \geq 0, g(x) = x^4, x \geq 0$ Yes
48. $f(x) = x, g(x) = x$ Yes
49. $f(x) = \dfrac{1}{x}, g(x) = \dfrac{1}{x}$ Yes
50. $f(x) = -\dfrac{1}{x}, g(x) = -\dfrac{1}{x}$ Yes
51. The function $y = f(x) = \frac{22}{15}x$ converts miles per hour (x) to feet per second (y). Find an equation for the inverse function. What formula does the inverse function represent? $f^{-1}(x) = \frac{15}{22}x$; this converts feet/second to miles/hour.
52. The function $y = f(x) = 1.609x$ converts miles per hour (x) to kilometers per hour (y). Find an equation for the inverse function. What formula does the inverse function represent? $f^{-1}(x) = 0.6215x$; this converts kilometers/hour to miles/hour.
53. A music store manager marks up his purchase price on guitars by 100 percent to determine his usual selling price. Thus $y = f(x) = 2x$ gives the usual selling price (y) as a function of the purchase price (x). How much of a discount could he offer before he loses money on the sale? $y = f^{-1}(x) = 0.50x$; 50 percent
54. If an investment portfolio loses 25 percent of its value during its first year, what percent increase is needed the next year to return it to its original value? $33\frac{1}{3}$ percent
55. Mountain campers can use the formula $y = f(a)$ which gives the temperature in degrees Fahrenheit (y) at which water boils as a function of altitude in feet (a) above sea level.
    a. What is the meaning of $f(1{,}000) = 200$? At 1,000 ft above sea level, water boils at 200° F.
    b. What is the meaning of $f^{-1}(190) = 2{,}000$? At 2,000 ft above sea level, water boils at 190° F.
    c. What is the meaning of $f^{-1}(j) = k$? At k ft above sea level, water boils at j° F.
56. The cost in dollars for a printing job is given by the formula $C = f(x) = 20 + 0.04x$, where x is the number of copies printed.
    a. What is the meaning of $f(1{,}000) = 60$? The cost of 1,000 copies is \$60.
    b. What is the meaning of $f^{-1}(100) = 2{,}000$? The cost of 2,000 copies is \$100.
    c. What is the meaning of $f^{-1}(m) = n$? The cost of n copies is m dollars.
    d. Find an equation for $f^{-1}(x)$. $f^{-1}(x) = \dfrac{x - 20}{0.04}$

THINK ABOUT IT

1. It is possible for the inverse of a function to be the same as the function. One such example is $f(x) = x$. Find two other such functions, and draw their graphs.
2. a. Graph the function $f(x) = 2^x$ and use the graph to explain why the inverse of f is a function.
   b. Use the graph in part **a** to graph $y = f^{-1}(x)$.
   c. Find the domain and range of f.
   d. Find the domain and range of f^{-1}.
3. a. If $f(x) = 10^x$, find $f^{-1}(100)$.
   b. If $f(x) = 8^x$, find $f^{-1}(4)$.
4. Explain the different purposes of the horizontal and vertical line tests.
5. Verify that $f(x) = \dfrac{1}{x + 2}$ and $g(x) = \dfrac{1 - 2x}{x}$ are inverses of each other using the composition definition of inverse functions. Give the domain and range of each, and sketch their graphs.

REMEMBER THIS

1. 10 must be raised to what power to equal 100? 2
2. Solve $7^{2x} = 7^3$. $\{\frac{3}{2}\}$
3. Simplify $4^{-1/2}$. $\frac{1}{2}$
4. Graph $y = 2^x$.

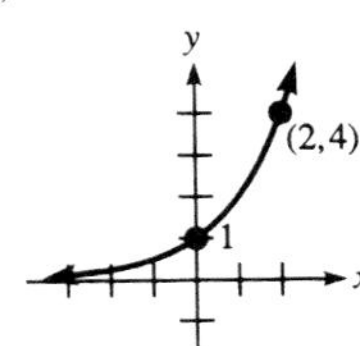

5. Simplify $2^m \cdot 2^n$. 2^{m+n}
6. Given $f(x) = x^{-1} + x^{-2}$, find $f(10)$. $\frac{11}{100}$
7. Graph $\dfrac{x^2}{4} + \dfrac{y^2}{9} = 1$.

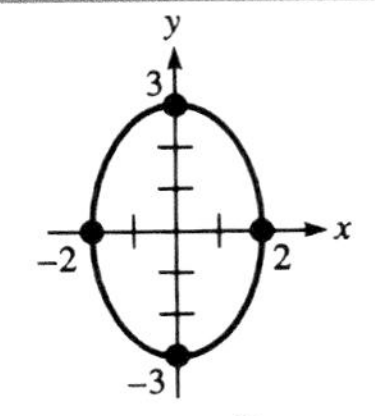

8. Solve $10\sqrt{x - 1} = 3x + 5$. $\{5, \frac{25}{9}\}$
9. Find the distance between $(-1,-1)$ and $(1,4)$. $\sqrt{29}$
10. Solve $\dfrac{x}{5} - \dfrac{1}{3} = x$. $\{-\frac{5}{12}\}$

10.3 Logarithmic Functions

To analyze chemical reactions involving liquids, one often must know the pH of the solution, which measures its hydrogen ion concentration (symbolized $[H^+]$). The formula for pH involves a logarithmic function, as discussed in the following problem.

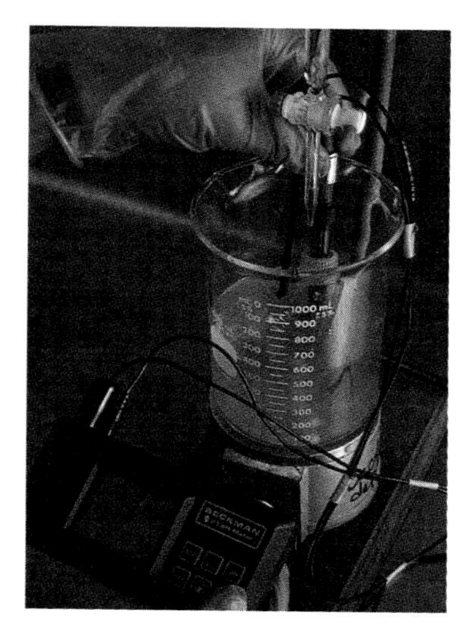

a. Because hydrogen ion concentrations are small numbers, it is useful to express such concentrations in exponential form. For instance, the concentration (measured in moles per liter) of hydrogen ions in distilled water is 1 part H^+ in 10,000,000, or 1/10,000,000, or 10^{-7}. To simplify further, pH is defined in exponential form by

$$[H^+] = 10^{-pH}.$$

Thus, the pH of distilled water is 7. Since pH is defined in terms of an exponent, it is more convenient to write a formula for pH in terms of a logarithm. What is this formula?

b. Use the answer in part **a** to find the pH (to the nearest tenth) of a sample of household ammonia whose hydrogen ion concentration is 5.2×10^{-12}. (See Example 8.)

OBJECTIVES

1. **Convert from the exponential form $b^L = N$ to the logarithmic form $\log_b N = L$, and vice versa.**
2. **Determine the value of the unknown in expressions of the form $\log_b N = L$.**
3. **Graph logarithmic functions.**
4. **Determine the common logarithm and antilogarithm of a number by using a calculator.**
5. **Solve applied problems involving logarithmic functions.**

1 To analyze inverse problems that involve an exponential function f, it is useful to create a rule that undoes f by defining the logarithm. The **logarithm** (abbreviated **log**) of a number is the *exponent* to which a fixed base is raised to obtain the number. The exponential statement $2^3 = 8$ is written in logarithmic form as $\log_2 8 = 3$, and we say that 3 is the logarithm to the base 2 of 8. In general, the key relation between exponential form and logarithmic form is expressed in the following definition.

John Napier (Scotland) published the first work on logarithms (1614); he invented the word *logarithm.* In his original system the log of 10 million was 0; and numbers greater than 10 million had negative logs. His goal was to make a table of the logs of the sines of angles which would be useful to astronomers. In the course of this work he invented our modern notation for decimal fractions (the use of the decimal point). Henry Briggs, after meeting with Napier, reworked Napier's ideas so that the log of 1 is 0, and he published the first extensive table of common logs (1624).

> **Definition of Logarithm**
>
> If b and N are positive numbers with $b \neq 1$, then
>
> $$\log_b N = L \quad \text{is equivalent to} \quad b^L = N.$$

In this definition it is important to observe that a logarithm is an exponent, as the following diagram emphasizes.

$$\log_b N = L \quad \text{is equivalent to} \quad b^L = N$$

(logarithm or exponent: L)

Examples 1 and 2 show how this definition is used to convert between exponential form and logarithmic form.

EXAMPLE 1 Write $10^2 = 100$ and $a^x = 4$ in logarithmic form.

Solution Since $b^L = N$ implies $\log_b N = L$,

$$10^2 = 100 \text{ may be written as } \log_{10} 100 = 2,$$
$$a^x = 4 \text{ may be written as } \log_a 4 = x.$$

PROGRESS CHECK 1 Write $3^4 = 81$ and $b^0 = 1$ in logarithmic form.

EXAMPLE 2 Write $\log_9 3 = \frac{1}{2}$ and $\log_b x = y$ in exponential form.

Solution Since $\log_b N = L$ implies $b^L = N$,

$$\log_9 3 = \tfrac{1}{2} \text{ may be written as } 9^{1/2} = 3,$$
$$\log_b x = y \text{ may be written as } b^y = x.$$

PROGRESS CHECK 2 Write $\log_8 2 = \frac{1}{3}$ and $\log_a y = x$ in exponential form.

2 The next example discusses how to find the value of an unknown in expressions of the form $\log_b N = L$.

EXAMPLE 3 Determine the value of the unknown in each expression.

a. $\log_5 5 = y$
b. $\log_4 8 = x$
c. $\log_{10} x = -1$
d. $\log_b 4 = -2$

Solution In each case it is helpful to first convert the expression from logarithmic form to exponential form.

a. $\log_5 5 = y$ is equivalent to $5^y = 5$. From the exponential form it is apparent that

$$y = \log_5 5 = 1.$$

Section 10.5 shows how to evaluate expressions like $\log_4 8$ using the change-of-base formula.

b. $\log_4 8 = x$ is equivalent to $4^x = 8$. Recall from Section 10.1 that this type of exponential equation may be solved by writing both sides of the equation in terms of the same base, as shown below.

$$\begin{aligned} 4^x &= 8 \\ (2^2)^x &= 2^3 \\ 2^{2x} &= 2^3 \\ 2x &= 3 \quad \text{Since } b^x = b^y \text{ implies } x = y \\ x &= \tfrac{3}{2} \end{aligned}$$

Thus, $x = \log_4 8 = \frac{3}{2}$.

c. $\log_{10} x = -1$ is equivalent to $10^{-1} = x$. Thus,

$$x = 10^{-1} = \tfrac{1}{10}.$$

d. $\log_b 4 = -2$ is equivalent to $b^{-2} = 4$. To find b, raise both sides of the equation to the reciprocal power of -2 (which is $-1/2$) to get

$$(b^{-2})^{-1/2} = 4^{-1/2},$$

and then simplify to obtain

$$b = 4^{-1/2} = \frac{1}{4^{1/2}} = \frac{1}{\sqrt{4}} = \frac{1}{2}.$$

Thus, $b = \frac{1}{2}$.

Progress Check Answers

1. $\log_3 81 = 4$, $\log_b 1 = 0$
2. $8^{1/3} = 2$, $a^x = y$

PROGRESS CHECK 3 Determine the value of the unknown in each expression.

a. $\log_{10} 1 = y$ **b.** $\log_8 32 = x$

c. $\log_2 x = -4$ **d.** $\log_b 64 = -3$

3 Inverse functions for exponential functions may now be defined using logarithms. The exponential function defined by $f(x) = b^x$ (with $b > 0$, $b \neq 1$) is one-to-one, so f has an inverse, and an equation defining f^{-1} may be found as follows.

$$y = b^x \quad \text{Start with } y = f(x).$$
$$x = b^y \quad \text{Interchange } x \text{ and } y.$$
$$y = \log_b x \quad \text{Solve for } y \text{ using the definition of logarithm.}$$
$$f^{-1}(x) = \log_b x \quad \text{Replace } y \text{ by } f^{-1}(x).$$

Thus, $y = b^x$ and $y = \log_b x$ are inverse functions. Because the base in an exponential function must be a positive number other than 1, this same restriction applies to a logarithmic function.

To see another important restriction, consider that

$$y = \log_b x \quad \text{is equivalent to} \quad x = b^y.$$

Since a positive base raised to any power is positive, it is also necessary to require that x be positive. In other words, we must incorporate in the definition of a logarithmic function that we may only take the logarithms of positive numbers.

Logarithmic Function

If b and x are positive numbers with $b \neq 1$, then the function f defined by

$$f(x) = \log_b x$$

is called the **logarithmic function with base *b*.**

EXAMPLE 4 Graph $y = \log_2 x$. Then sketch the graphs of $y = \log_2 x$ and $y = 2^x$ on the same coordinate system, and describe how the graphs are related.

See the graphing calculator section at the end of this chapter for one way to graph $y = \log_b x$.

Solution To graph $y = \log_2 x$, first construct a table of values, as shown below. To generate this table, rewrite

$$y = \log_2 x \quad \text{as} \quad x = 2^y,$$

and then replace y with integer values from -3 to 3.

x	$\frac{1}{8}$	$\frac{1}{4}$	$\frac{1}{2}$	1	2	4	8
y	-3	-2	-1	0	1	2	3

By graphing these solutions and drawing a smooth curve through them, we obtain the graph of $y = \log_2 x$ shown in Figure 10.14.

Consider mentioning that if $b > 1$, then both $y = b^x$ and $y = \log_b x$ are increasing functions. However, y increases very rapidly in $y = b^x$ and very slowly in $y = \log_b x$.

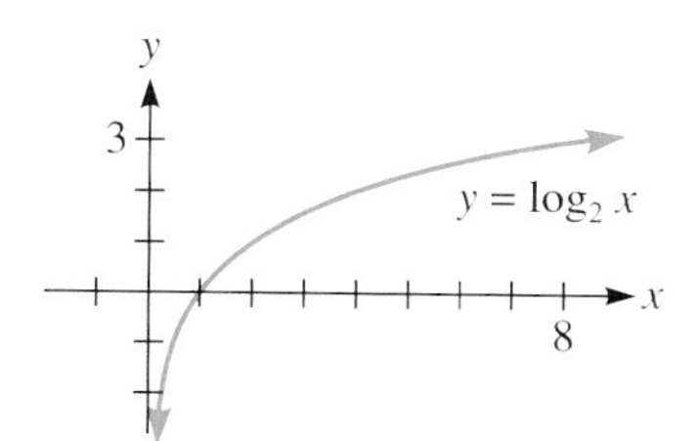

Figure 10.14

Progress Check Answers

3. (a) 0 (b) $\frac{5}{3}$ (c) $\frac{1}{16}$ (d) $\frac{1}{4}$

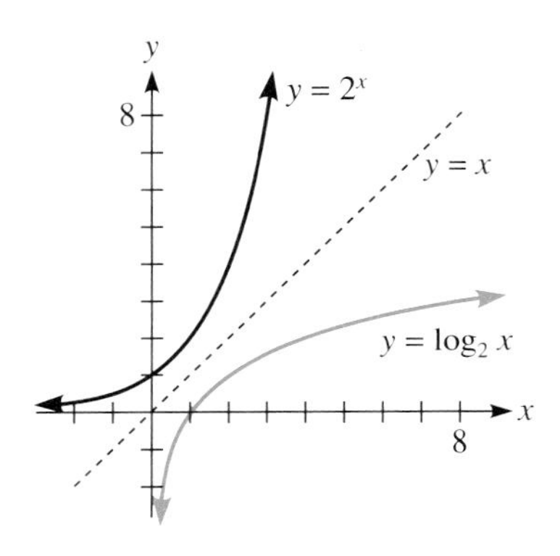

Figure 10.15

To relate this graph to the graph of its inverse $y = 2^x$, sketch both graphs on the same coordinate system, as in Figure 10.15. As expected, the graphs are related in that each is the reflection of the other about the line $y = x$.

PROGRESS CHECK 4 Graph $y = \log_{1/2} x$. Then sketch the graphs of the functions $y = \log_{1/2} x$ and $y = (\frac{1}{2})^x$ on the same coordinate system, and describe how the graphs are related.

Two typical logarithmic functions were graphed in Example 4 and "Progress Check" Exercise 4:

$$y = \log_2 x \text{ in which } b > 1,$$
$$y = \log_{1/2} x \text{ in which } 0 < b < 1.$$

In general, the graphs of $y = \log_b x$ and $y = b^x$ for these two cases are as shown in Figure 10.16. From these graphs some properties of the logarithmic function with base b are apparent.

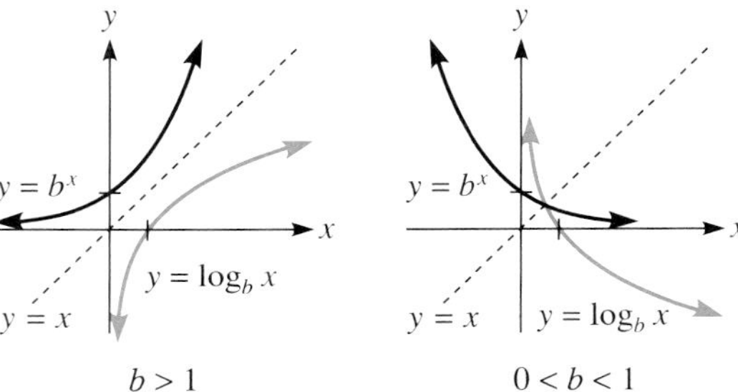

Figure 10.16

Properties of $f(x) = \log_b x$ (with $b > 0$, $b \neq 1$) and Its Graph

1. The domain of $y = \log_b x$ (which is the range of $y = b^x$) is $(0,\infty)$.
2. The range of $y = \log_b x$ (which is the domain of $y = b^x$) is $(-\infty,\infty)$.
3. The graph of $y = \log_b x$ has the x-intercept $(1,0)$, and the y-axis is a vertical asymptote.
4. The graph of $y = \log_b x$ is the reflection of the graph of $y = b^x$ about the line $y = x$.
5. If $b > 1$, then as x increases, y increases. If $0 < b < 1$, then as x increases, y decreases.

Progress Check Answer

4. Each graph is the reflection of the other about the line $y = x$.

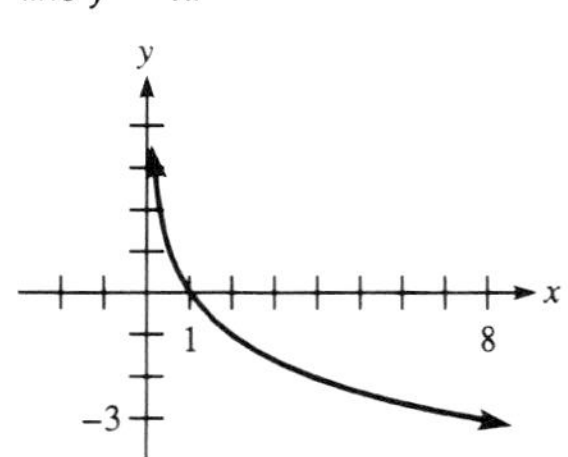

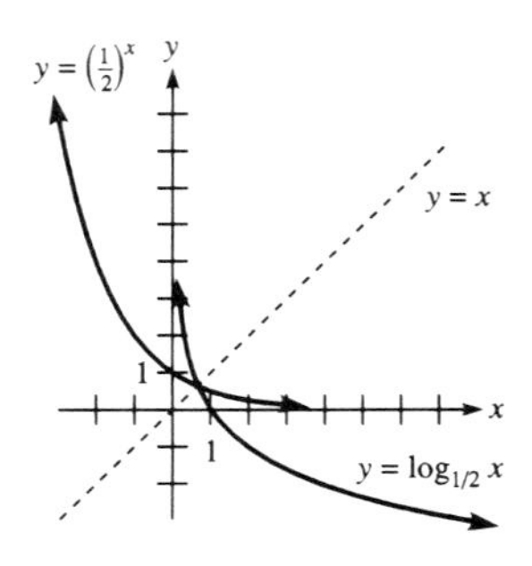

The domain of a log function is easily read from its graph. Ask students to do this here, and follow up by assigning "Think About It" Exercise 4.

The graphing techniques of Section 9.3 may be used to graph certain variations of $y = \log_b x$; Example 5 illustrates the case of a horizontal shift.

EXAMPLE 5 Graph $y = \log_2(x + 4)$.

Solution The graph of the function $y = \log_2(x + 4)$ is identical in shape to the graph of $y = \log_2 x$. Since the constant 4 is added to x before the log rule is applied, shift the graph of $y = \log_2 x$ to the left 4 units to position this curve correctly. Convenient points to plot for the graph are $(-3,0)$, $(0,2)$, and $(4,3)$. The completed graph is shown in Figure 10.17.

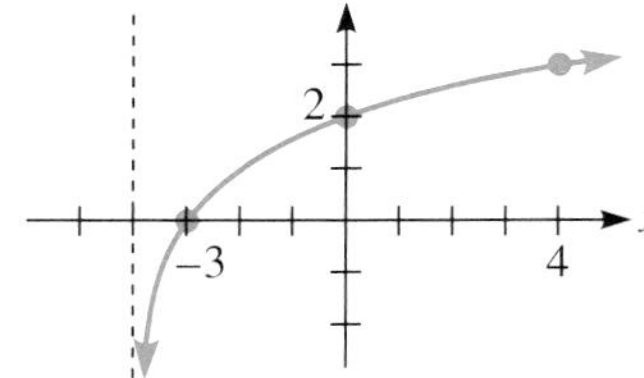

Figure 10.17

PROGRESS CHECK 5 Graph $y = -\log_2 x$.

4 Many formulas in science that involve logarithms are written in terms of base 10 logarithms, which are called **common logarithms.** It is standard notation to abbreviate $\log_{10} N$ as $\log N$, with base 10 being understood. Scientific calculators have a [log] key. To find $\log N$, simply enter N and press [log], as illustrated in Example 6.

EXAMPLE 6 Evaluate log 750 to four decimal places.

Solution From the calculator sequence below, log 750 = 2.8751, to four decimal places.

750 [log] [2.8750613]

PROGRESS CHECK 6 Evaluate log 325 to four decimal places.

To undo the common logarithm rule and find N when $\log N$ is known, keep in mind that $y = \log_{10} x$ and $y = 10^x$ define inverse functions. Therefore, calculators are usually programmed so that

[INV] [log] is equivalent to [10^x].

Many calculators write 10^x above the [log] key to specify that it is the inverse or second function associated with this key. In the following example we find N by calculator and round off the answer to the requested number of significant digits. Note that all digits, except the zeros that are written to indicate the position of the decimal point, are **significant digits.**

To suggest that logs are defined only for positive numbers, ask students to evaluate log 0 and log(−1) on their calculator so that an error message results. Keystroke sequences for a graphing calculator are given in the graphing calculator section at the end of this chapter. Some advanced calculators, like the TI-85, return complex number logarithms of negative numbers.

EXAMPLE 7 If $\log N = -1.4$, find N to three significant digits.

Solution If $\log N = -1.4$, then $N = 10^{-1.4}$. This expression is evaluated by the following calculator sequence:

1.4 [+/−] [INV] [log] [0.0398107].

Thus, $N = 0.0398$ to three significant digits. Observe that the zeros in this answer are not significant digits, because the number is 398 ten-thousandths and the zeros are written to indicate the correct position of the decimal point.

Students may wish to confirm that calculator evaluation of $10^{-1.4}$ using the [y^x] key gives the same result.

Note In science it is common to refer to an inverse logarithm as an antilogarithm. In this convention,

$$\text{antilog } x = 10^x.$$

PROGRESS CHECK 7 If $\log N = -2.3$, find N to three significant digits.

5 Calculators replace log *evaluations,* not log *functions*. To illustrate, we now solve the section-opening problem in which a relationship is analyzed by using a logarithmic function.

EXAMPLE 8 Solve the problem in the section introduction on page 495.

Solution

a. The problem states that pH is defined in exponential form by

$$[H^+] = 10^{-pH}.$$

Converting this expression to logarithmic form yields

$$\begin{aligned} -pH &= \log_{10}[H^+], \\ pH &= -\log[H^+]. \end{aligned}$$

Thus, pH is defined as the negative of the common logarithm of the hydrogen ion concentration.

Progress Check Answers

5.

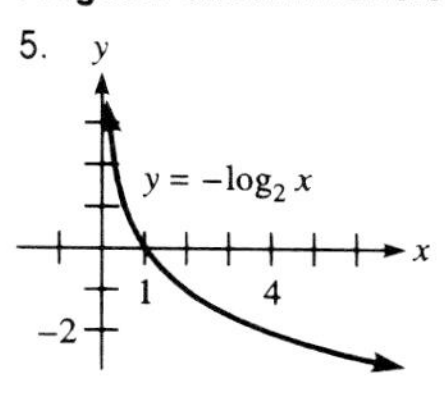

6. 2.5119
7. 0.00501

Remind students of the scientific notation capabilities of their calculators.

b. For household ammonia, $[H^+] = 5.2 \times 10^{-12}$, so

$$\begin{aligned} pH &= -\log[H^+] \\ &= -\log(5.2 \times 10^{-12}) \\ &= 11.3 \quad \text{(to the nearest tenth).} \end{aligned}$$

By calculator, a keystroke sequence is 5.2 [EE] 12 [+/−] [log] [+/−].

Note For reference, a solution with pH = 7 is called neutral. In acids, pH < 7; bases or alkalies have pH > 7.

PROGRESS CHECK 8 Find the pH (to the nearest tenth) of a sample of nitric acid whose hydrogen ion concentration is 5.1×10^{-4}.

EXAMPLE 9 Determine $[H^+]$ for a sample of blood whose pH is 7.4.

Solution From Example 8, the formula for pH is

$$pH = -\log[H^+].$$

To find $[H^+]$, first replace pH by 7.4 and solve for $\log[H^+]$.

$$\begin{aligned} 7.4 &= -\log[H^+] \\ -7.4 &= \log[H^+] \end{aligned}$$

To emphasize the principle of $f^{-1}[f(x)] = x$ you might point out that in the calculator sequence we find $[H^+]$ by taking the inverse log of both sides of the equation.

$$\begin{aligned} \text{INV}\log(-7.4) &= \text{INV}\log(\log[H^+]) \\ &= [H^+] \end{aligned}$$

Then compute an inverse logarithm by the following calculator sequence:

7.4 [+/−] [INV] [log] [3.9811 − 08].

Thus, $[H^+] \approx 4.0 \times 10^{-8}$.

PROGRESS CHECK 9 Determine $[H^+]$ for a sample of seawater whose pH is 8.9.

Progress Check Answers

8. 3.3
9. 1.3×10^{-9}

EXERCISES 10.3

In Exercises 1–16, write the given relation in logarithmic form.

1. $2^3 = 8$ $\log_2 8 = 3$
2. $3^4 = 81$ $\log_3 81 = 4$
3. $4^2 = 16$ $\log_4 16 = 2$
4. $5^4 = 625$ $\log_5 625 = 4$
5. $(\frac{1}{2})^3 = \frac{1}{8}$ $\log_{1/2} \frac{1}{8} = 3$
6. $(\frac{1}{3})^4 = \frac{1}{81}$ $\log_{1/3} \frac{1}{81} = 4$
7. $b^y = x$ $\log_b x = y$
8. $a^r = m$ $\log_a m = r$
9. $10^3 = 1{,}000$ $\log_{10} 1{,}000 = 3$
10. $10^4 = 10{,}000$ $\log_{10} 10{,}000 = 4$
11. $10^{-2} = \dfrac{1}{100}$ $\log_{10} \dfrac{1}{100} = -2$
12. $10^{-5} = \dfrac{1}{100{,}000}$ $\log_{10} \dfrac{1}{100{,}000} = -5$
13. $(0.2)^3 = 0.008$ $\log_{0.2} 0.008 = 3$
14. $(0.1)^3 = 0.001$ $\log_{0.1} 0.001 = 3$
15. $(0.1)^{-2} = 100$ $\log_{0.1} 100 = -2$
16. $(0.5)^{-3} = 8$ $\log_{0.5} 8 = -3$

In Exercises 17–30, write the given relation in exponential form.

17. $\log_4 64 = 3$ $4^3 = 64$
18. $\log_5 25 = 2$ $5^2 = 25$
19. $\log_4 2 = \frac{1}{2}$ $4^{1/2} = 2$
20. $\log_{16} 2 = \frac{1}{4}$ $16^{1/4} = 2$
21. $\log_{10} 1{,}000 = 3$ $10^3 = 1{,}000$
22. $\log_{10} 10{,}000 = 4$ $10^4 = 10{,}000$
23. $\log_8 \frac{1}{8} = -1$ $8^{-1} = \frac{1}{8}$
24. $\log_6 \frac{1}{36} = -2$ $6^{-2} = \frac{1}{36}$
25. $\log_2 \frac{1}{32} = -5$ $2^{-5} = \frac{1}{32}$
26. $\log_3 \frac{1}{27} = -3$ $3^{-3} = \frac{1}{27}$
27. $\log_b y = x$ $b^x = y$
28. $\log_{1/2} y = x$ $(\frac{1}{2})^x = y$
29. $\log_{1/2} 8 = -3$ $(\frac{1}{2})^{-3} = 8$
30. $\log_{1/3} 9 = -2$ $(\frac{1}{3})^{-2} = 9$

In Exercises 31–50, determine the value of the unknown in each equation.

31. $\log_3 3 = y$ 1
32. $\log_7 1 = y$ 0
33. $\log_{1/2} 1 = y$ 0
34. $\log_{1/2} 4 = y$ −2
35. $\log_{1/2} \frac{1}{16} = y$ 4
36. $\log_{1/3} 9 = y$ −2
37. $\log_b 27 = 3$ 3
38. $\log_b \frac{1}{4} = 2$ $\frac{1}{2}$
39. $\log_b \frac{1}{25} = -2$ 5
40. $\log_b \frac{1}{729} = -3$ 9
41. $\log_b 8 = -3$ $\frac{1}{2}$
42. $\log_b 81 = -4$ $\frac{1}{3}$
43. $\log_6 x = 2$ 36
44. $\log_9 x = 3$ 729
45. $\log_4 x = -2$ $\frac{1}{16}$
46. $\log_5 x = -3$ $\frac{1}{125}$
47. $\log_{1/5} x = -2$ 25
48. $\log_{1/3} x = -3$ 27
49. $\log_2 x = -3$ $\frac{1}{8}$
50. $\log_{1/2} x = 5$ $\frac{1}{32}$

In Exercises 51–58, graph the given logarithmic functions.

51. $y = \log_{10} x$

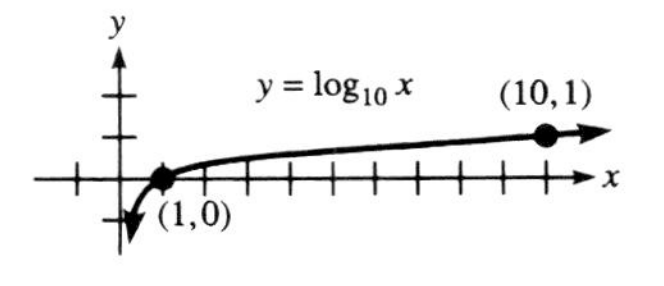

52. $y = \log_5 x$

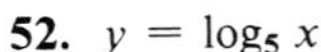
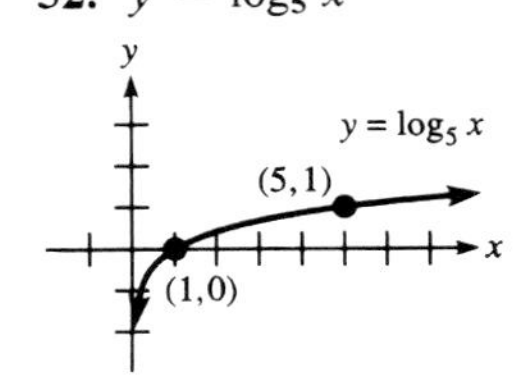

53. $y = \log_3 x$

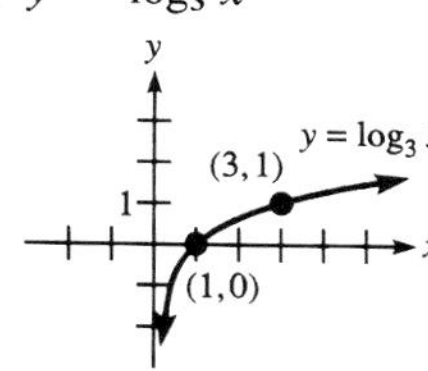

54. $y = \log_2 x$

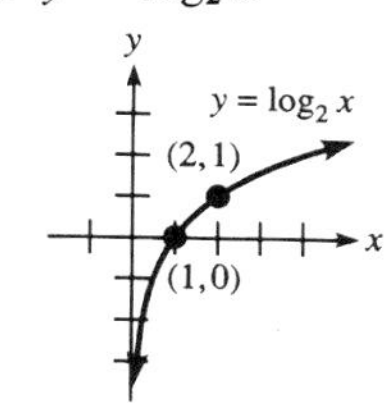

55. $y = \log_{1/3} x$

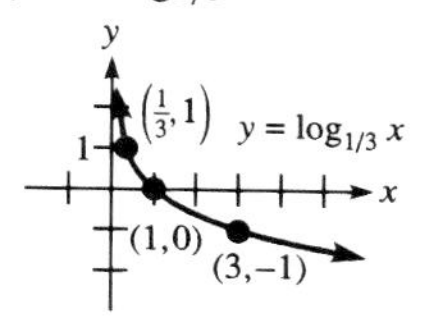

56. $y = \log_{1/10} x$

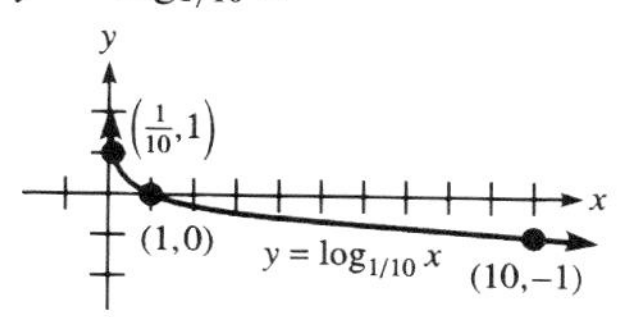

57. $y = \log_b x,\ b > 1$

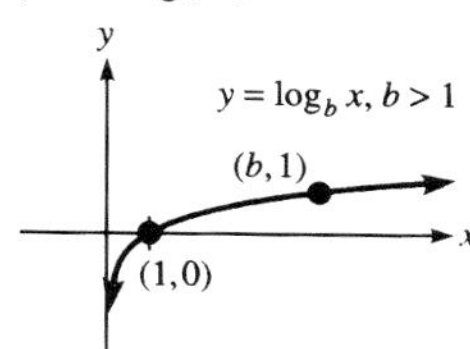

58. $y = \log_b x,\ 0 < b < 1$

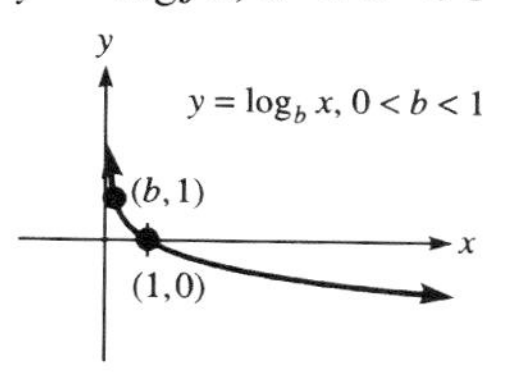

In Exercises 59–64, graph the given pair of functions on the same coordinate system, and describe how the graphs are related.

59–64. In each case one graph is the reflection of the other about the line $y = x$.

59. $y = \log_{10} x,\ y = 10^x$ **60.** $y = \log_{1/3} x,\ y = (\frac{1}{3})^x$

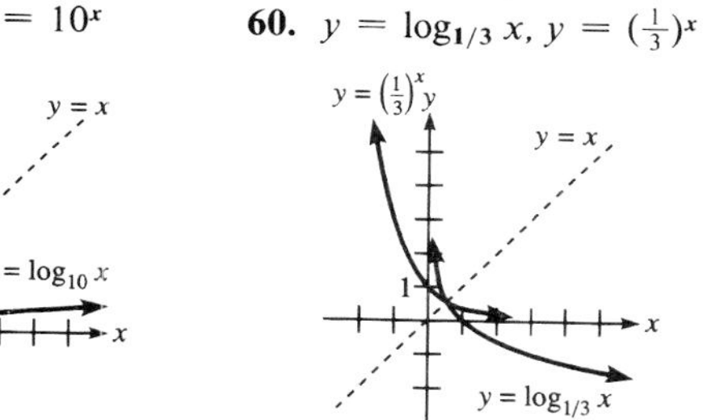

61. $y = \log_{1/4} x,\ y = (\frac{1}{4})^x$

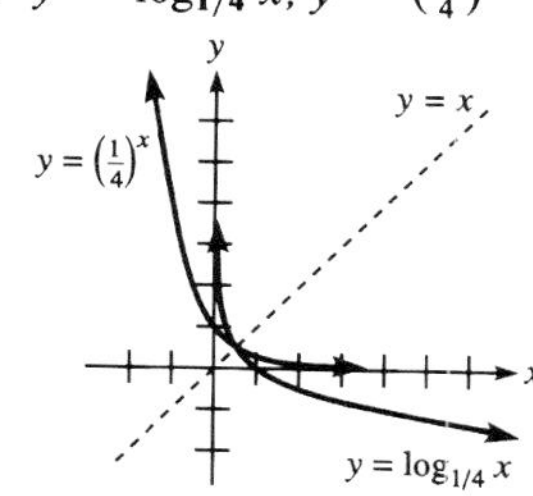

62. $y = \log_3 x,\ y = 3^x$

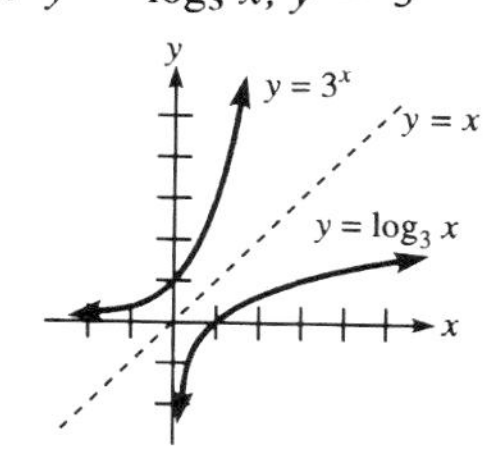

63. $y = \log_b x,\ b > 1;\ y = b^x,\ b > 1$

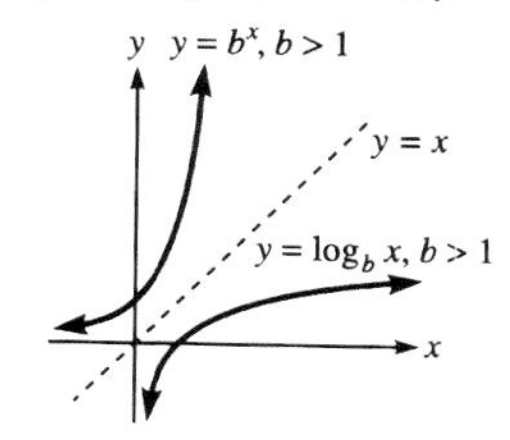

64. $y = \log_b x,\ 0 < b < 1;\ y = b^x,\ 0 < b < 1$

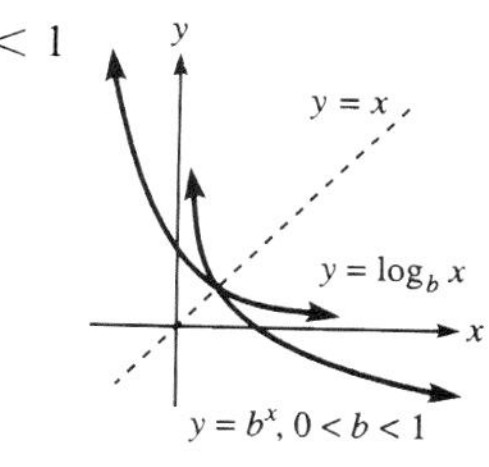

In Exercises 65–74, use graphing techniques for translation and reflection to graph the given function.

65. $y = \log_5(x + 4)$

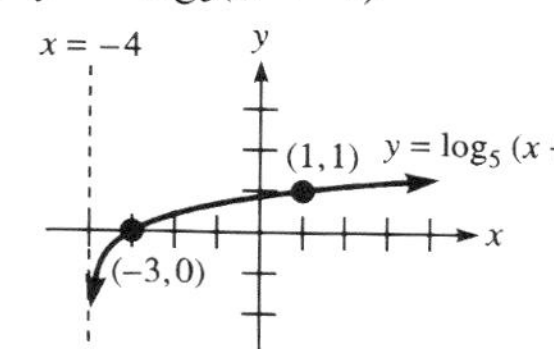

66. $y = \log_3(x - 2)$

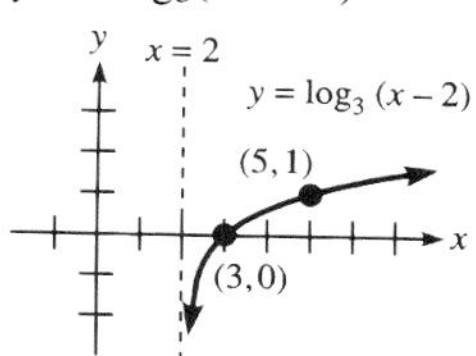

67. $y = \log_{1/2}(x + 4)$

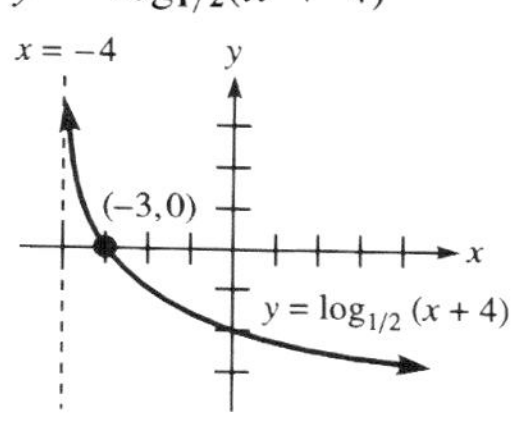

68. $y = \log_{1/2}(x - 3)$

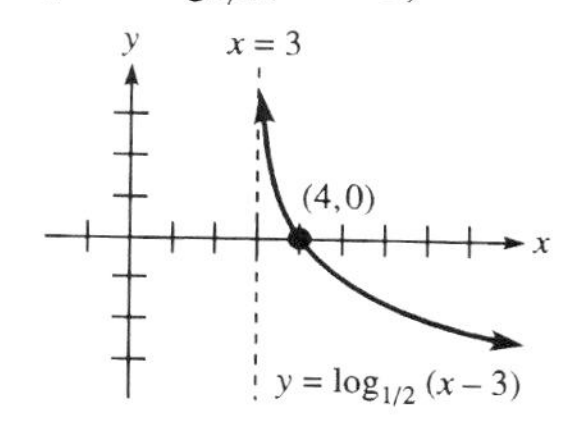

69. $y = -\log_3 x$

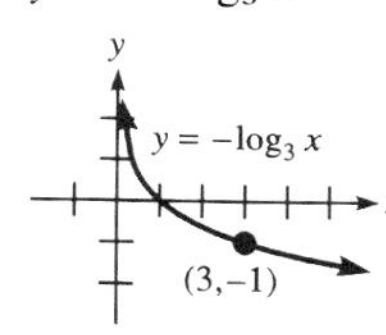

70. $y = -\log_5 x$

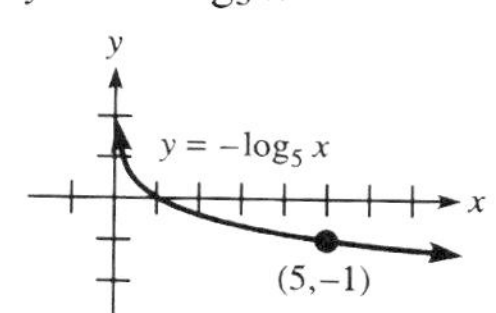

71. $y = -\log_{1/2} x$

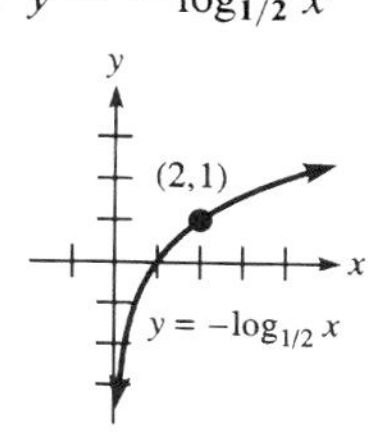

72. $y = -\log_{1/3} x$

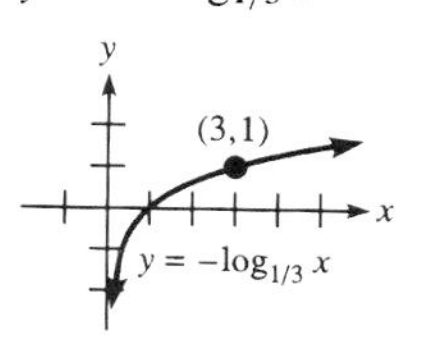

73. $y = -\log_5(x + 2)$

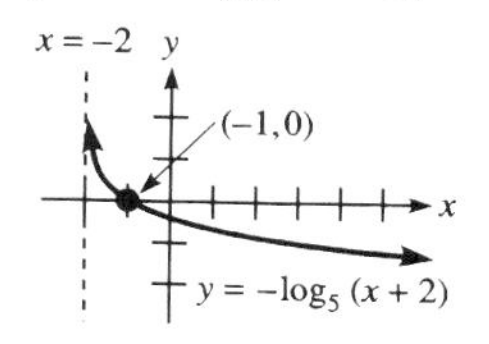

74. $y = -\log_5(x - 6)$

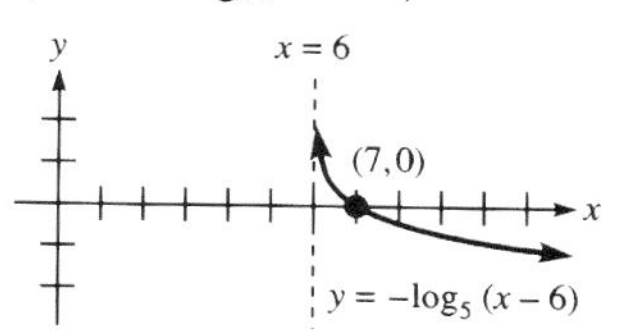

In Exercises 75–80, evaluate the given logarithms to four decimal places.

75. log 25 1.3979

76. log 75 1.8751

77. log 0.37 −0.4318

78. log 0.54 −0.2676

79. log 2.36 0.3729

80. log 0.004 −2.3979

In Exercises 81–88, find N to three significant digits.

81. $\log N = 2.2$ 158
82. $\log N = 1.7$ 50.1
83. $\log N = 0.042$ 1.10
84. $\log N = 0.057$ 1.14
85. $\log N = -2.6$ 0.00251
86. $\log N = -1.41$ 0.0389
87. $\log N = -0.31$ 0.490
88. $\log N = -0.46$ 0.347

In Exercises 89–94, use the relationship between pH and $[H^+]$ given in Example 8.

89. The concentration of hydrogen ions in a solution is 4×10^{-5}. Find the pH to the nearest tenth. Is this an acid or a base? 4.4; acid

90. The concentration of hydrogen ions in a solution is 1.8×10^{-9}. Find the pH to the nearest tenth. Is this an acid or a base? 8.7; base

91. Acid rain decreases the pH in rivers, which can endanger fish. For instance, salmon die when the pH falls to about 5.5. Determine the concentration of hydrogen ions $[H^+]$ for such river water. 3.16×10^{-6}

92. Acid-base indicators are highly colored dyes which change color when the pH of a solution changes. One such indicator is red when the pH is above 7.5, yellow when the pH is below 6.5, and orange otherwise. Find the values of $[H^+]$ which correspond to these pH values. pH = 7.5: $[H^+] = 3.16 \times 10^{-8}$; pH = 6.5: $[H^+] = 3.16 \times 10^{-7}$

93. A chemist must solve the equation $\frac{x^2}{0.05} = \frac{1 \times 10^{-14}}{1.8 \times 10^{-5}}$, where x gives the hydrogen ion concentration of a certain solution.
a. Find $[H^+]$. 5.27×10^{-6}
b. Find the corresponding value for the pH. 5.28

94. A chemist must solve the equation $\frac{x^2}{0.15} = 2 \times 10^{-6}$, where x gives the hydrogen ion concentration of a certain solution.
a. Find $[H^+]$. 5.5×10^{-4}
b. Find the corresponding value for the pH. 3.26

95. One version of the Richter scale for measuring the magnitude of earthquakes is given by

$$R = \tfrac{2}{3}(\log E - 4.4),$$

where E is the energy of the quake in joules and R is the Richter rating.
a. Find the energy (E) released by an earthquake which measures 6 on the Richter scale. 2.51×10^{13} joules
b. What is the energy released when R is 5? 7.94×10^{11} joules
c. How many times more energy is released by a quake which measures 6 on the Richter scale than by one which measures 5? 31.6

96. Refer to Exercise 95. Compare the energy released by earthquakes which register 7.5 and 7 on the Richter scale. (Answer parts **a**, **b**, and **c** as in Exercise 95.) a. 4.47×10^{15} joules b. 7.94×10^{14} joules c. 5.63

THINK ABOUT IT

1. Recall that the equation $[H^+] = 10^{-pH}$ defines the pH of a solution.
   a. Show that when the pH is doubled, the hydrogen ion concentration is squared.
   b. In general, what happens to the hydrogen ion concentration when the pH is multiplied by n?
2. The logarithm function with $b > 1$ is one which increases very slowly. This is in marked contrast to the exponential function, which grows quickly. By making tables of ordered pairs for $x = 0, 1, 2, 4, 8, 16$, and 32 and then drawing accurate graphs, compare the graphs of $y = \log_2 x$ and $y = x^{1/2}$, both of which grow slowly. Give function values to the nearest tenth. Where do the graphs intersect?
3. For the logarithmic function defined by $y = \log_b x$, explain why the restriction $b \neq 1$ is necessary for the equation to define a function.
4. Find the domain of $y = \log_2(x + 2)$. Use both a geometric and an algebraic approach.
5. **a.** If $\log_b 3 = m$ and $\log_b 4 = n$, find b^{m+n}.
   b. By calculator, find $10^{\log_{10} 3}$ and $10^{\log_{10} 5}$. Generalize to $b^{\log_b x}$, and explain why these answers make sense.

REMEMBER THIS

1. Simplify $\frac{b^m}{b^n}$. b^{m-n}
2. True or false? $(\sqrt{x} + \sqrt{y})^2 = x + y$. False
3. Express $\sqrt[3]{5^2}$ using fractional exponents. $5^{2/3}$
4. Simplify $2^x \cdot 2^7$. 2^{x+7}
5. Simplify $(8x^6y^9)^{1/3}$. $2x^2y^3$
6. Graph the solution set of $x + y \le 5$.

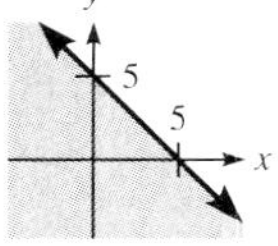

7. Solve $4^x = 2^{x+3}$. {3}
8. Multiply $(x^2 + x + 1)(x - 1)$. $x^3 - 1$
9. Solve $n = \frac{A + B}{A}$ for A. $A = \frac{B}{n - 1}$
10. Solve $|x + 2| \le 8$. $[-10,6]$

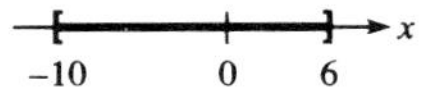

10.4 Properties of Logarithms

An alternative formula for pH is

$$\text{pH} = \log \frac{1}{[\text{H}^+]}.$$

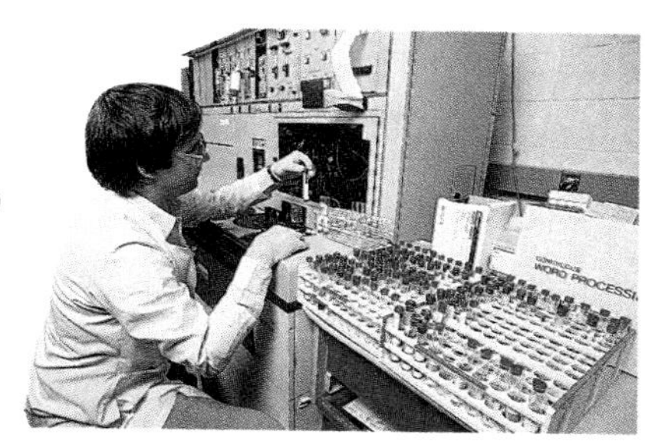

a. Show that the above formula is equivalent to the formula stated in the previous section. That is, show that

$$\log \frac{1}{[\text{H}^+]} = -\log[\text{H}^+].$$

b. Use the alternative formula above and find the pH (to the nearest tenth) of a sample of acid rain whose hydrogen ion concentration is 3.8×10^{-4}. (See Example 3.)

OBJECTIVES

1 **Use properties of logarithms to express certain log statements in terms of simpler logarithms or expressions.**

2 **Use properties of logarithms to convert certain statements involving logarithms to a single logarithm with coefficient 1.**

1 Because a logarithm is an exponent, properties of logarithms follow from exponent properties. Three key exponent properties from Section 3.1 that are the basis for the product, quotient, and power rules for logarithms may be stated as follows.

1. $b^m \cdot b^n = b^{m+n}$ **Product property**
2. $\dfrac{b^m}{b^n} = b^{m-n}$ **Quotient property**
3. $(b^m)^n = b^{mn}$ **Power-to-a-power property**

To each of these exponent properties, there corresponds a logarithm property, as stated next.

Product, Quotient, and Power Properties of Logarithms

If b, x, and y are positive numbers with $b \neq 1$, and k is any real number, then

1. $\log_b xy = \log_b x + \log_b y$ **Product property**
2. $\log_b \dfrac{x}{y} = \log_b x - \log_b y$ **Quotient property**
3. $\log_b x^k = k \log x.$ **Power property**

The historical use of logarithms to perform calculations is based on these properties. Point out that these properties allow problems involving products, quotients, and powers to be converted to problems involving sums, differences, and products, respectively.

To use the product property of exponents to prove the product property of logarithms, let

$$x = b^m \quad \text{and} \quad y = b^n,$$

and observe that the respective logarithmic forms of these statements are

$$\log_b x = m \quad \text{and} \quad \log_b y = n.$$

By the product rule of exponents,

$$x \cdot y = b^m \cdot b^n = b^{m+n},$$

and converting $xy = b^{m+n}$ to logarithmic form gives

$$\log_b xy = m + n.$$

Finally, substituting $\log_b x$ for m and $\log_b y$ for n yields the property

$$\log_b xy = \log_b x + \log_b y.$$

The quotient and power properties of logarithms may be established in similar ways (and these proofs are requested in Exercises 75 and 76).

One use of these properties is to convert certain logarithms to a sum, difference, or product involving simpler logarithms, as illustrated in Example 1.

EXAMPLE 1 Express each logarithm as a sum, difference, or product involving simpler logarithms.

a. $\log_b 5x$
b. $\log_4 \frac{3}{5}$
c. $\log_{10} x^4$
d. $\log_b \sqrt[3]{5}$
e. $\log_2 7x^2$
f. $\log_b \sqrt{\frac{x}{3y}}$

Solution In parts **d** and **f** note that radical expressions are converted to rational exponent form so that the power property can be applied.

a. $\log_b 5x = \log_b 5 + \log_b x$ Product property

b. $\log_4 \frac{3}{5} = \log_4 3 - \log_4 5$ Quotient property

c. $\log_{10} x^4 = 4 \log_{10} x$ Power property

d. $\log_b \sqrt[3]{5} = \log_b 5^{1/3}$ $\sqrt[3]{5} = 5^{1/3}$

$= \frac{1}{3} \log_b 5$ Power property

e. $\log_2 7x^2 = \log_2 7 + \log_2 x^2$ Product property

$= \log_2 7 + 2 \log_2 x$ Power property

f. $\log_b \sqrt{\frac{x}{3y}} = \log_b\left(\frac{x}{3y}\right)^{1/2}$ $\sqrt{\frac{x}{3y}} = \left(\frac{x}{3y}\right)^{1/2}$

$= \frac{1}{2} \log_b \frac{x}{3y}$ Power property

$= \frac{1}{2}(\log_b x - \log_b 3y)$ Quotient property

$= \frac{1}{2}[\log_b x - (\log_b 3 + \log_b y)]$ Product property

Letting $x = b^m$ and $y = b^n$ reveals there is no general property for $\log_b(x + y)$, because there is no general property for $b^m + b^n$.

Caution Properties have been stated for the logarithm of a product, a quotient, and a power. Logarithms of a *sum* or *difference* may not be converted using logarithm properties. In particular,

$\log_b(x + y)$ may *not* be replaced by $\log_b x + \log_b y$,
and $\log_b(x - y)$ may *not* be replaced by $\log_b x - \log_b y$.

PROGRESS CHECK 1 Express each logarithm as a sum, difference, or product involving simpler logarithms.

a. $\log_b\left(\frac{x}{4}\right)$
b. $\log_2 10x$
c. $\log_{10} 4^x$
d. $\log_b \sqrt{7}$
e. $\log_b 5(x + 2)^3$
f. $\log_5 \sqrt[3]{\frac{2}{9N}}$

Progress Check Answers

1. (a) $\log_b x - \log_b 4$ (b) $\log_2 10 + \log_2 x$ (c) $x \log_{10} 4$ (d) $\frac{1}{2} \log_b 7$ (e) $\log_b 5 + 3 \log_b(x + 2)$ (f) $\frac{1}{3}[\log_5 2 - (\log_5 9 + \log_5 N)]$

Two additional properties that are often used to simplify logarithmic expressions are the direct result of the exponent laws $b^1 = b$ and $b^0 = 1$. Converting these expressions to logarithmic form gives the result that if b is a positive number other than 1, then

$$\log_b b = 1 \quad \text{and} \quad \log_b 1 = 0.$$

For instance, $\log_{10} 10 = 1$ and $\log_2 1 = 0$. In the remaining examples these two properties will be used to further simplify logarithmic expressions whenever applicable.

EXAMPLE 2 Express each logarithm as a sum, difference, or product involving simpler logarithms.

a. $\log_{10} 10x$ **b.** $\log_4 4^5$ **c.** $\log_b \frac{1}{3}$

Solution

a. $\log_{10} 10x = \log_{10} 10 + \log_{10} x$ Product property
$= 1 + \log_{10} x$ $\log_b b = 1$

b. $\log_4 4^5 = 5 \log_4 4$ Power property
$= 5(1)$ $\log_b b = 1$
$= 5$

c. $\log_b \frac{1}{3} = \log_b 1 - \log_b 3$ Quotient property
$= 0 - \log_b 3$ $\log_b 1 = 0$
$= -\log_b 3$

The problem in Example 2b may be used to introduce the inverse property $\log_b b^x = x$ discussed later in this section.

PROGRESS CHECK 2 Express each logarithm as a sum, difference, or product involving simpler logarithms.

a. $\log_{10} 10^{-1}$ **b.** $\log_6 \frac{6}{5}$ **c.** $\log_6 \frac{1}{5}$

EXAMPLE 3 Solve the problem in the section introduction on page 503.

Solution

a. If the expression in the alternative formula for pH is viewed as the logarithm of a quotient, then we may show that the expressions in the formulas are equivalent, as follows.

$$\log \frac{1}{[H^+]} = \log 1 - \log[H^+] \quad \text{Quotient property}$$
$$= 0 - \log[H^+] \quad \log_b 1 = 0$$
$$= -\log[H^+]$$

b. For the sample of acid rain, $[H^+] = 3.8 \times 10^{-4}$, so

$$pH = \log \frac{1}{3.8 \times 10^{-4}}$$
$$= 3.4 \text{ (to the nearest tenth).}$$

By calculator, a keystroke sequence is 3.8 [EE] 4 [+/−] [1/x] [log].

Have students find the pH of the sample of acid rain using both pH formulas to show that the formulas are equivalent.

PROGRESS CHECK 3 The expression in the alternative formula for pH may also be viewed as the logarithm of the reciprocal of $[H^+]$.

a. Show that $\log \frac{1}{[H^+]} = -\log[H^+]$ by first rewriting $\frac{1}{[H^+]}$ as $[H^+]^{-1}$.

b. Use the alternative formula to find the pH (to the nearest tenth) of a sample of swimming pool water whose hydrogen ion concentration is 2.9×10^{-8}.

Progress Check Answers

2. (a) -1 (b) $1 - \log_6 5$ (c) $-\log_6 5$

3. (a) $\log \frac{1}{[H^+]} = \log [H^+]^{-1} = -1 \log [H^+] = -\log[H^+]$ (b) 7.5

EXAMPLE 4 If $\log_b 2 = m$ and $\log_b 3 = n$, express each of the following in terms of m and/or n.

a. $\log_b \frac{1}{2}$ **b.** $\log_b 72$

Solution

a. By the quotient property,

$$\log_b \tfrac{1}{2} = \log_b 1 - \log_b 2.$$

Then $\log_b 1 = 0$ and $\log_b 2 = m$, so

$$\log_b \tfrac{1}{2} = 0 - m = -m.$$

In Example 4a students sometimes replace $\log_b 2$ with $\log_b m$ and answer $-\log_b m$. Consider warning them about this type of error.

b. Using factors of 2 and 3, we write 72 as $2^3 \cdot 3^2$. Then,

$$\log_b 72 = \log_b(2^3 \cdot 3^2) = \log_b 2^3 + \log_b 3^2 = 3 \log_b 2 + 2 \log_b 3.$$

Replacing $\log_b 2$ by m and $\log_b 3$ by n gives

$$\log_b 72 = 3m + 2n.$$

PROGRESS CHECK 4 If $\log_b 2 = m$ and $\log_b 3 = n$, express each of the following in terms of m and/or n.

a. $\log_b 24$ **b.** $\log_b \frac{1}{9}$

Remind students that for inverse functions $f^{-1}[f(x)] = x$ and $f[f^{-1}(x)] = x$.

Two other useful properties of logarithms are referred to as inverse properties because they are a direct consequence of the inverse relation between exponential and logarithmic functions. That is, since $y = b^x$ and $y = \log_b x$ "undo" each other, applying the two rules one after the other to a meaningful input x produces x as the output.

Inverse Properties

If b is a positive number with $b \neq 1$, then

1. $\log_b b^x = x$,
2. $b^{\log_b x} = x$, for $x > 0$.

From a different viewpoint, $\log_b b^x = x$ because $b^x = b^x$, and $b^{\log_b x} = x$ (for $x > 0$) because $\log_b x$ is the power of b that results in x.

EXAMPLE 5 Simplify each expression.

a. $\log_{10}(m \times 10^k)$ **b.** $10^{\log_{10} 100}$

Solution

The simplification in Example 5a is the basis for finding common logarithms with the aid of a table of values for $\log m$, where $1.00 \leq m \leq 9.99$. See Exercises 67 and 68.

a. By the product property,

$$\log_{10}(m \times 10^k) = \log_{10} m + \log_{10} 10^k.$$

Then the inverse property $\log_b b^x = x$ indicates that $\log_{10} 10^k = k$, so

$$\log_{10}(m \times 10^k) = \log_{10} m + k.$$

b. Using $b^{\log_b x} = x$ gives

$$10^{\log_{10} 100} = 100.$$

To confirm this answer, observe that $\log_{10} 100 = 2$, so

$$10^{\log_{10} 100} = 10^2 = 100.$$

Progress Check Answers

4. (a) $3m + n$ (b) $-2n$

PROGRESS CHECK 5 Simplify each expression.

a. $\log_{10}(a \times 10^3)$ b. $4^{\log_4 64}$

2 To this point, logarithm properties have been used mainly to convert a simple log statement to a sum, difference, or product that involved simpler logarithms. Depending on the application, it may be more useful to convert sums and differences of logarithms to a single logarithm with coefficient 1. This type of conversion is considered in the next example.

Converting to single log with coefficient 1 is particularly useful in the next section for solving logarithmic equations.

EXAMPLE 6 Express as a single logarithm with coefficient 1.

a. $\log_3 5 + \log_3 y$ b. $\log_{10} 2 - \log_{10} x$

c. $2 \log_b x + 3 \log_b y$ d. $\frac{1}{2}[\log_b(x + 1) - \log_b 3]$

Solution

a. $\log_3 5 + \log_3 y = \log_3 5y$ Product property

b. $\log_{10} 2 - \log_{10} x = \log_{10} \dfrac{2}{x}$ Quotient property

c. $2 \log_b x + 3 \log_b y = \log_b x^2 + \log_b y^3$ Power property

$= \log_b x^2y^3$ Product property

d. $\dfrac{1}{2}[\log_b(x + 1) - \log_b 3] = \dfrac{1}{2} \log_b \dfrac{x + 1}{3}$ Quotient property

$= \log_b\left(\dfrac{x + 1}{3}\right)^{1/2}$ or $\log_b \sqrt{\dfrac{x + 1}{3}}$ Power property

PROGRESS CHECK 6 Express as a single logarithm with coefficient 1.

a. $\log_b 12 - \log_b 4$ b. $\log_2 x + \log_2(x - 1)$

c. $2 \log_{10} x + \frac{1}{2} \log_{10} 9$ d. $\frac{1}{2}[\log_{10} L - \log_{10} g]$

Progress Check Answers

5. (a) $\log_{10} a + 3$ (b) 64
6. (a) $\log_b 3$ (b) $\log_2(x^2 - x)$ (c) $\log_{10} 3x^2$ (d) $\log_{10} \sqrt{\dfrac{L}{g}}$

EXERCISES 10.4

In Exercises 1–36, express the given logarithm as a sum, difference, or product of simpler logarithms.

1. $\log_5 \frac{7}{9}$ $\log_5 7 - \log_5 9$
2. $\log_8 \frac{2}{3}$ $\log_8 2 - \log_8 3$
3. $\log_8 6x$ $\log_8 6 + \log_8 x$
4. $\log_5 2x$ $\log_5 2 + \log_5 x$
5. $\log_3 x^5$ $5 \log_3 x$
6. $\log_5 x^4$ $4 \log_5 x$
7. $\log_b \sqrt[3]{5}$ $\frac{1}{3} \log_b 5$
8. $\log_b \sqrt[7]{3}$ $\frac{1}{7} \log_b 3$
9. $\log_3 5x^3$ $\log_3 5 + 3 \log_3 x$
10. $\log_7 6x^4$ $\log_7 6 + 4 \log_7 x$
11. $\log_b \sqrt{\dfrac{x}{2y}}$ $\frac{1}{2}[\log_b x - (\log_b 2 + \log_b y)]$
12. $\log_b \sqrt{\dfrac{3}{xy}}$ $\frac{1}{2}[\log_b 3 - (\log_b x + \log_b y)]$
13. $\log_b \sqrt{\dfrac{xy}{z}}$ $\frac{1}{2}[\log_b x + \log_b y - \log_b z]$
14. $\log_b \sqrt[3]{\dfrac{2x}{yz^2}}$ $\frac{1}{3}[\log_b 2 + \log_b x - (\log_b y + 2 \log_b z)]$
15. $\log_b \sqrt{\dfrac{x + 5}{y}}$ $\frac{1}{2}[\log_b(x + 5) - \log_b y]$
16. $\log_b \sqrt{\dfrac{x}{y - 3}}$ $\frac{1}{2}[\log_b x - \log_b(y - 3)]$
17. $\log_4 4$ 1
18. $\log_9 9$ 1
19. $\log_3 3^5$ 5
20. $\log_5 5^9$ 9
21. $\log_5 1$ 0
22. $\log_6 1$ 0
23. $\log_{10} \frac{1}{10}$ -1
24. $\log_{10} \frac{1}{100}$ -2
25. $\log_{10} 10^{-3}$ -3
26. $\log_{10} 10^{-4}$ -4
27. $\log_5 5^{-3}$ -3
28. $\log_7 7^{-2}$ -2
29. $\log_6 6^3$ 3
30. $\log_8 8^5$ 5
31. $\log_b \frac{1}{3}$ $-\log_b 3$
32. $\log_b \frac{1}{7}$ $-\log_b 7$
33. $\log_b \dfrac{b}{7}$ $1 - \log_b 7$
34. $\log_b \dfrac{b}{5}$ $1 - \log_b 5$
35. $\log_b \dfrac{3}{b^2}$ $\log_b 3 - 2$
36. $\log_b \dfrac{5}{b^3}$ $\log_b 5 - 3$

In Exercises 37–46, let $\log_b 2 = m$ and $\log_b 3 = n$. Express each of the following in terms of m and/or n.

37. $\log_b 108$ $2m + 3n$
38. $\log_b 144$ $4m + 2n$
39. $\log_b 48$ $4m + n$
40. $\log_b 162$ $m + 4n$
41. $\log_b \frac{1}{27}$ $-3n$
42. $\log_b \frac{1}{64}$ $-6m$
43. $\log_b \frac{1}{32}$ $-5m$
44. $\log_b \frac{1}{81}$ $-4n$
45. $\log_b \frac{27}{64}$ $3n - 6m$
46. $\log_b \frac{32}{81}$ $5m - 4n$

In Exercises 47–54, simplify the given expression.

47. $\log_{10}(t \times 10^k)$ $\log_{10} t + k$
48. $\log_{10}(2y \times 10^k)$ $\log_{10} 2 + \log_{10} y + k$
49. $\log_{10}(y \times 100^k)$ $\log_{10} y + 2k$
50. $\log_{10}(y \times 1{,}000^k)$ $\log_{10} y + 3k$
51. $5^{\log_5 32}$ 32
52. $5^{\log_5 64}$ 64
53. $7^{\log_7 49}$ 49
54. $8^{\log_8 23}$ 23

In Exercises 55–66, express the given expression as a single log with coefficient 1.

55. $\log_2 8 + \log_2 y$ $\log_2 8y$

56. $\log_5 x + \log_5 y$ $\log_5 xy$

57. $\log_b 3 - \log_b y$ $\log_b \frac{3}{y}$

58. $\log_b 7 - \log_b y$ $\log_b \frac{7}{y}$

59. $3 \log_{10} y - 4 \log_{10} x$ $\log_{10} \frac{y^3}{x^4}$

60. $2 \log_{10} x - 9 \log_{10} y$ $\log_{10} \frac{x^2}{y^9}$

61. $\frac{1}{3}[\log_{10}(x + 3) - \log_{10}(y - 1)]$ $\log_{10} \sqrt[3]{\frac{x+3}{y-1}}$

62. $\frac{1}{2}[\log_{10}(x - 2) - \log_{10}(y + 5)]$ $\log_{10} \sqrt{\frac{x-2}{y+5}}$

63. $\frac{1}{2}[\log_{10} x - (\log_{10} y + \log_{10} z)]$ $\log_{10} \sqrt{\frac{x}{yz}}$

64. $\frac{1}{3}\{\log_{10} y - [\log_{10}(x + 2) + \log_{10} z]\}$ $\log_{10} \sqrt[3]{\frac{y}{(x+2)z}}$

65. $\frac{1}{2}(\log_{10} y - \log_{10} z - \log_{10} x)$ $\log_{10} \sqrt{\frac{y}{xz}}$

66. $\frac{1}{3}(\log_{10} x - \log_{10} y - \log_{10} z)$ $\log_{10} \sqrt[3]{\frac{x}{yz}}$

67. Without using a calculator, fill in the missing entries. Make use of the properties of logarithms and the given entries. (*Hint:* $\log 20 = \log(2 \times 10) = \log 2 + \log 10$.)

x	1	2	5	10	20	50	100
$y = \log x$	0	0.3010	0.6990	1	1.3010	1.6990	2
$y = \log x^2$	0	0.6020	1.3980	2	2.6020	3.3980	4
$y = \log \sqrt{x}$	0	0.1505	0.3495	0.5	0.6505	0.8495	1

68. Without using a calculator, fill in the missing entries. Make use of the properties of logarithms and the given entries.

x	1	2	5	10	20	50	100
$y = \log_2 x$	0	1	2.3219	3.3219	4.3219	5.6438	6.6438
$y = \log_2 x^2$	0	2	4.6438	6.6438	8.6438	11.2876	13.2876
$y = \log_2 \sqrt{x}$	0	0.5	1.16095	1.66095	2.16095	2.8219	3.3219

69. **a.** Assume that x is any positive real number. Use the graph of $y = \log x$ to draw the graph of $y = \log x^2$. Explain why each y value on the graph of $y = \log x^2$ is twice the corresponding value of y on the graph of $y = \log x$.

b. Assume that x is any positive real number. Use the graph of $y = \log x$ to draw the graph of $y = \log \sqrt{x}$. Explain why each y value on the graph of $y = \log \sqrt{x}$ is half the corresponding value of y on the graph of $y = \log x$.

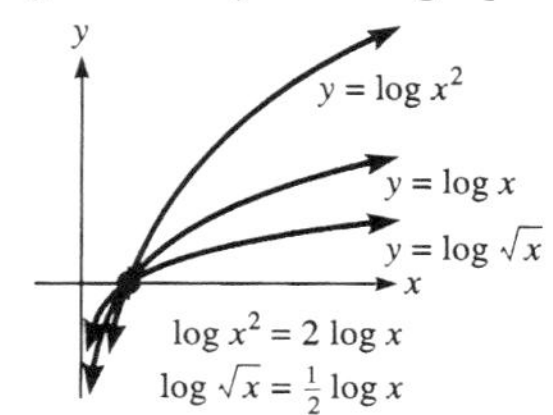

70. The graph of $y = \log x$ is shown. Use this graph and the properties of logarithms to sketch the graph of $y = \log(1/x)$.

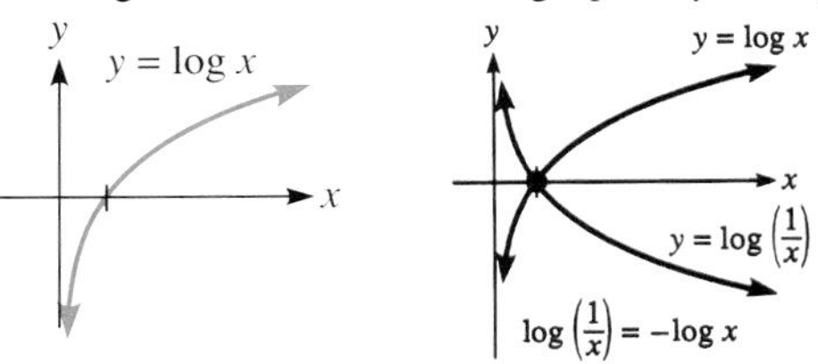

71. A useful equation in statistics is called the *log-linear model* because the logarithm of one variable is expressed as a linear function of the other variable. Such models are often used to investigate the effects of various health hazards on the survival time of patients. An example of a log-linear model is $\log y = ax + b$.

a. Solve this equation for x. $x = (\log y - b)/a$

b. Find x when $a = 0.4$, $b = 0$, and $y = 16$. $x = 3.0103$

c. Solve the given equation for y. $y = 10^{ax+b}$

d. Find y when $a = 0.4$, $b = 0$, and $x = 3$. $y = 15.85$

72. In the statistical study of life expectancy, a simple but useful mathematical model is $\log p = -mt$, where p represents the fraction of some original population that is still alive after time t. The constant m is called the "hazard rate" or the "force of mortality."

a. Solve this equation for p. $p = 10^{-mt}$

b. Solve this equation for t. $t = (\log p)/(-m)$

c. Assume that $m = 0.1$ and $t = 4$. Find p to the nearest hundredth. 0.40

d. Assume that $m = 0.1$ and $p = 0.5$. Find t to the nearest hundredth. 3.01

73. In the scientific study of sound, several properties of sound are measured on the decibel scale. (Decibel $= \frac{1}{10}$ bel, a unit named after Alexander Graham Bell.) One such property is called sound intensity, and a formula for calculating the intensity level (L) in decibels (dB) is

$$L = 10 \log(I) + 90,$$

where I is the intensity measured in ergs per square centimeter per second.

a. Show that the given formula is equivalent to

$$L = 10 \log\left(\frac{I}{10^{-9}}\right),$$

which is another common version of this formula.

$L = 10 \log(I) + 90 = 10[\log(I) + \log(10^9)] = 10 \log(I \cdot 10^9) = 10 \log \frac{I}{10^{-9}}$

b. Find L when $I = 100$. 110 dB

c. Find I when $L = 120$ decibels (this is the threshold of pain). 1,000 ergs/cm²/second

74. Another characteristic of sound is called the sound pressure. One formula for the pressure level is

$$L = 20 \log(p) + 74,$$

where p is measured in dynes per square centimeter.

a. Show that the formula is equivalent to

$$L = 20 \log\left(\frac{p}{10^{-3.7}}\right).$$

$20 \log(p) + 74 = 20 \log(p) + 20(3.7) = 20[\log(p) + \log(10^{3.7})] = 20 \log\left(\frac{p}{10^{-3.7}}\right)$

b. Find L when $p = 0.2$ dynes/cm². 60.02 dB

c. Find p when $L = 0$ decibels. 0.0002 dynes/cm²

75. Prove the quotient property of logarithms in a way similar to that used in Section 10.4 to prove the product property.
Let $x = b^m$ and $y = b^n$. $\frac{x}{y} = \frac{b^m}{b^n} = b^{m-n}$. $\log_b \frac{x}{y} = m - n = \log_b x - \log_b y$

76. Prove the power property of logarithms in a way similar to that used in Section 10.4 to prove the product property.
Let $x = b^m$. $x^k = b^{km}$. $\log_b x^k = km = k \log_b x$

THINK ABOUT IT

1. In words, the product property of logarithms states that the logarithm of a product is equal to the sum of the logarithms of the factors. Give a verbal description for the quotient and power properties of logarithms.

2. Give some specific examples to *disprove* both of these statements.

a. $\log_b(x + y) = \log_b x + \log_b y$

b. $\frac{\log_b x}{\log_b y} = \log_b x - \log_b y$

3. If $\log_{10} 2 = m$ and $\log_{10} 3 = n$, express these logarithms in terms of m and/or n.

a. $\log_{10} 5$ **b.** $\log_{10} 15$ **c.** $\log_{10} 6\frac{2}{3}$

4. If $\log_{12} 2 = m$ and $\log_{12} 3 = n$, show that $n = 1 - 2m$.

5. Use the fact that $x = b^{\log_b x}$ and $y = b^{\log_b y}$ to prove the product property of logarithms.

REMEMBER THIS

1. Solve $xm - xn = k$ for x. $x = \frac{k}{m - n}$

2. Solve $x(x - 2) = 2^3$. $\{-2, 4\}$

3. Which one of these is called an exponential equatio

a. $y = x^2$ **b.** $y = 2^x$ b

4. Graph $y = (\frac{1}{3})^x$.

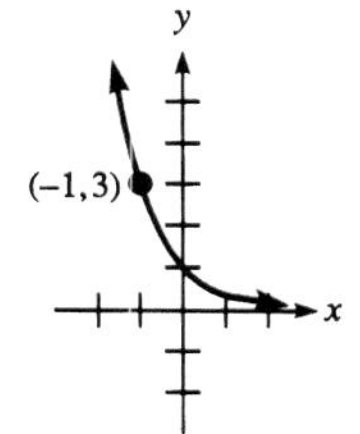

5. Which is larger, $(-\frac{1}{2})^4$ or $-\frac{1}{2}^4$? $(-\frac{1}{2})^4$

6. If $f(x) = 3x - 2$, find the inverse $f^{-1}(x)$. $f^{-1}(x) = \frac{x + 2}{3}$

7. What property is illustrated by $3(x - a) = 3x - 3a$? Distributive

8. Evaluate $1 + 3(4 - 5^2)$. -62

9. If the price of an item including the 8 percent sales tax comes to \$21.06, what was the price before the tax was added? \$19.50

10. Rationalize the denominator: $\frac{1}{\sqrt{x} + \sqrt{y}}$. Assume $x,y > 0$.
$\frac{\sqrt{x} - \sqrt{y}}{x - y}$

10.5 Exponential and Logarithmic Equations

The **doubling time** of an exponentially increasing quantity is the time required for the quantity to double its size or value. Use the fact that the world's population is currently growing at a rate of about 1.85 percent per year to determine the current doubling time (to the nearest year) for the world's population. The required formula is

$$P = P_0(1 + r)^t,$$

where P is the population t years from now, P_0 is the current world population, and r is the annual growth rate. (See Example 3.)

OBJECTIVES

1 **Solve exponential equations by using logarithms.**

2 **Apply the change-of-base formula.**

3 **Solve logarithmic equations.**

1 An **exponential equation** is an equation that has a variable in an exponent. Section 10.1 considered how to solve such equations when it is not difficult to rewrite both sides of the equation in terms of the same base. For instance, $8^x = 4$ was solved in Example 6 of that section as shown next.

$$\begin{aligned} 8^x &= 4 \\ 2^{3x} &= 2^2 \\ 3x &= 2 \quad && b^x = b^y \text{ implies } x = y. \\ x &= \tfrac{2}{3} \end{aligned}$$

With the aid of logarithms a more general method for solving exponential equations may be stated. When logarithms are involved in equation solving, then the following principle is important.

Equation-Solving Principle

If x, y, and b are positive real numbers with $b \neq 1$, then
1. $x = y$ implies $\log_b x = \log_b y$, and conversely,
2. $\log_b x = \log_b y$ implies $x = y$.

Consider mentioning that another equation-solving principle based on the fact that an exponential correspondence is a one-to-one function is that
1. $x = y$ implies $b^x = b^y$, and conversely,
2. $b^x = b^y$ implies $x = y$.

This principle is based on the fact that a logarithmic correspondence is a one-to-one function, and we may use it to solve exponential equations by taking the common logarithm of both sides of the equation, as shown in Example 1.

EXAMPLE 1 Solve $3^x = 6$. Give the solution to four significant digits.

Solution

$$\begin{aligned} 3^x &= 6 \\ \log 3^x &= \log 6 \quad && \text{Apply common logarithms to each side.} \\ x \log 3 &= \log 6 \quad && \text{Power property.} \\ x &= \frac{\log 6}{\log 3} \quad && \text{Divide both sides by log 3.} \end{aligned}$$

A warning to students that $\dfrac{\log 6}{\log 3} \neq \log \dfrac{6}{3}$ and $\dfrac{\log 6}{\log 3} \neq \log 6 - \log 3$ may be necessary.

To approximate this solution, compute

6 [log] [÷] 3 [log] [=] [1.6309298].

Thus, to four significant digits the solution set is $\{1.631\}$.

Note In the solution above, introducing logarithms is helpful because the power property may then be used to convert from a problem involving a power to a problem involving a product.

PROGRESS CHECK 1 Solve $4^x = 7$. Give the solution to four significant digits.

Example 1 illustrates a general approach that may be used to solve exponential equations.

To Solve Exponential Equations Using Logarithms

1. Take the logarithm to the same base of both sides of the equation.
2. Simplify by applying the property $\log_b x^k = k \log_b x$.
3. Solve the resulting equation using previous equation-solving methods.

Progress Check Answer

1. $\{1.404\}$

EXAMPLE 2 Solve $5^x = 3^{x+1}$. Give the solution to four significant digits.

Solution

$$\begin{aligned}
5^x &= 3^{x+1} \\
\log 5^x &= \log 3^{x+1} && \text{Apply common logarithms to each side.} \\
x \log 5 &= (x+1)\log 3 && \text{Power property.} \\
x \log 5 &= x \log 3 + \log 3 && \text{Distributive property.} \\
x \log 5 - x \log 3 &= \log 3 && \text{Subtract } x \log 3 \text{ from both sides.} \\
x(\log 5 - \log 3) &= \log 3 && \text{Factor.} \\
x &= \frac{\log 3}{\log 5 - \log 3} && \text{Divide both sides by } \log 5 - \log 3.
\end{aligned}$$

By calculator,

3 [log] [÷] [(] 5 [log] [−] 3 [log] [)] [=] [2.1506601].

To four significant digits, the solution set is {2.151}.

PROGRESS CHECK 2 Solve $4^x = 6^{x-1}$. Give the solution to four significant digits.

EXAMPLE 3 Solve the problem in the section introduction on page 509.

Solution When the current world population has doubled, then $P = 2P_0$. Replace P by $2P_0$ and r by 1.85 percent (or 0.0185) in the given formula and solve for t.

$$\begin{aligned}
P &= P_0(1+r)^t \\
2P_0 &= P_0(1+0.0185)^t && \text{Replace } P \text{ by } 2P_0 \text{ and } r \text{ by } 0.0185. \\
2 &= (1.0185)^t \\
\log 2 &= \log(1.0185)^t && \text{Apply common logarithms to each side.} \\
\log 2 &= t \log 1.0185 && \text{Power property.} \\
\frac{\log 2}{\log 1.0185} &= t
\end{aligned}$$

By calculator, $t \approx 37.81293$, so the current doubling time for the world's population is about 38 years.

On the basis of a doubling time of 38 years and current mortality tables, the world's population will roughly quadruple during an average American's lifetime.

PROGRESS CHECK 3 What would be the doubling time (to the nearest year) for the world's population if the annual growth rate were to increase to 2.35 percent?

2 Depending on the application, it may be more convenient to write a logarithmic statement in a certain base. A change-of-base formula may be derived using our current methods for solving exponential equations. To express $\log_b x$ in terms of a different base, say a, first recall that

$$y = \log_b x \quad \text{is equivalent to} \quad b^y = x.$$

By taking the logarithm to the base a of both sides of $b^y = x$, we have

$$\begin{aligned}
\log_a b^y &= \log_a x \\
y \log_a b &= \log_a x \\
y &= \frac{\log_a x}{\log_a b}.
\end{aligned}$$

Then replacing y by $\log_b x$ yields the formula

$$\log_b x = \frac{\log_a x}{\log_a b}.$$

Because calculators have a log key, converting to base 10 logarithms is often useful, as illustrated next.

Progress Check Answers

2. {4.419}
3. 30 years

EXAMPLE 4 Evaluate each logarithm to four significant digits.

a. $\log_3 6$ **b.** $\log_{1/2} 0.8$

Solution Convert to common logarithms using the change-of-base formula.

a. $\log_3 6 = \dfrac{\log 6}{\log 3} \approx 1.631$ **b.** $\log_{1/2} 0.8 = \dfrac{\log 0.8}{\log \frac{1}{2}} \approx 0.3219$

Note When solving equations like $3^x = 6$, students sometimes begin by writing

$$x = \log_3 6$$

but then are stumped. The continuation of this line of reasoning uses the change-of-base formula, so

$$x = \log_3 6 = \frac{\log 6}{\log 3} \approx 1.631.$$

Compare this alternative method with the solution shown in Example 1.

PROGRESS CHECK 4 Evaluate each logarithm to four significant digits.

a. $\log_4 7$ **b.** $\log_{1/3} 0.4$

3 A **logarithmic equation** is an equation that involves a logarithm of a variable expression. In some cases such equations may be solved by applying the principle that $\log_b x = \log_b y$ implies $x = y$. It is important to remember that logarithms are not defined for negative numbers or zero. Therefore, in the solution of logarithmic equations, it is necessary to check answers in the original equation and accept only solutions that result in the logarithms of positive numbers.

EXAMPLE 5 Solve $\log x = \log (1 - x)$.

Solution

$$\begin{aligned} \log x &= \log(1 - x) \\ x &= 1 - x \qquad \log_b x = \log_b y \text{ implies } x = y. \\ 2x &= 1 \\ x &= \tfrac{1}{2} \end{aligned}$$

Replacing x by $\frac{1}{2}$ in the original equation leads to $\log \frac{1}{2} = \log \frac{1}{2}$, so the proposed solution checks, and the solution set is $\{\frac{1}{2}\}$.

PROGRESS CHECK 5 Solve $\log (2x - 7) = \log x$.

When an equation contains only one log statement, then we may solve by converting the equation from logarithmic form to exponential form, as shown next.

The special case of solving log equations of the form $\log_b N = k$ was covered in Section 10.3. Consider reviewing this special case before discussing Example 6.

EXAMPLE 6 Solve $\log_5(3x - 8) = 2$.

Solution We convert to exponential form and then solve for x.

$$\begin{aligned} \log_5(3x - 8) &= 2 \\ 3x - 8 &= 5^2 \qquad \log_b N = L \text{ implies } N = b^L. \\ 3x &= 33 \\ x &= 11 \end{aligned}$$

To check, replace x by 11 in the original equation.

$$\begin{aligned} \log_5[3(11) - 8] &\stackrel{?}{=} 2 \\ \log_5 25 &\stackrel{?}{=} 2 \\ 2 &\stackrel{\checkmark}{=} 2 \end{aligned}$$

The proposed solution checks, and the solution set is $\{11\}$.

PROGRESS CHECK 6 Solve $\log_2(5x + 1) = 4$.

Progress Check Answers

4. (a) 1.404 (b) 0.8340
5. {7}
6. {3}

Before you apply the methods of the last two examples, it may be necessary to use logarithm properties on one or both sides of the equation. Such a case is illustrated in Example 7.

EXAMPLE 7 Solve $\log_2(x - 2) = 3 - \log_2 x$.

Solution First, rewrite the given equation as

$$\log_2 x + \log_2(x - 2) = 3$$

so that the product property of logarithms may be applied to obtain

$$\log_2[x(x - 2)] = 3.$$

Then convert from logarithmic form to exponential form and solve.

$$\begin{aligned} x(x - 2) &= 2^3 \\ x^2 - 2x &= 8 \\ x^2 - 2x - 8 &= 0 \\ (x - 4)(x + 2) &= 0 \\ x - 4 = 0 \quad &\text{or} \quad x + 2 = 0 \\ x = 4 \quad & \qquad\quad x = -2 \end{aligned}$$

Now check.

$$\begin{aligned} \log_2(x - 2) &= 3 - \log_2 x \\ \log_2(4 - 2) &\overset{?}{=} 3 - \log_2 4 \\ 1 &\overset{?}{=} 3 - 2 \\ 1 &\overset{\checkmark}{=} 1 \end{aligned}$$

$$\begin{aligned} \log_2(x - 2) &= 3 - \log_2 x \\ \log_2(-2 - 2) &\overset{?}{=} 3 - \log_2(-2) \end{aligned}$$

No solution because $\log_2(-4)$ and $\log_2(-2)$ are not defined.

Only 4 is a solution of the original equation, so the solution set is $\{4\}$.

PROGRESS CHECK 7 Solve $\log_2(x + 4) = 2 - \log_2(x + 1)$.

A solution to a log equation may be a negative number or zero, as illustrated in "Progress Check" Exercise 7. Students sometimes misinterpret the phrase "accept only solutions that *result* in the *logarithm* of positive numbers" and incorrectly reject all nonpositive solutions. Assign some exercises that emphasize this point.

Progress Check Answer

7. $\{0\}$

EXERCISES 10.5

In Exercises 1–16, solve the given equation. Give solutions to four significant digits.

1. $2^x = 12$ $\{3.585\}$
2. $5^x = 75$ $\{2.683\}$
3. $7^x = 22$ $\{1.588\}$
4. $3^x = 19$ $\{2.680\}$
5. $4^x = 32$ $\{2.500\}$
6. $6^x = 23$ $\{1.750\}$
7. $(\frac{1}{2})^x = 24$ $\{-4.585\}$
8. $(\frac{1}{4})^x = 9$ $\{-1.585\}$
9. $6^x = 2^{x-2}$ $\{-1.262\}$
10. $5^x = 3^{x+1}$ $\{2.151\}$
11. $7^x = 5^{x-1}$ $\{-4.783\}$
12. $8^x = 2^{x+3}$ $\{1.500\}$
13. $3^x = 7^{x-2}$ $\{4.593\}$
14. $5^x = 3^{x+2}$ $\{4.301\}$
15. $4^x = 9^{x-4}$ $\{10.84\}$
16. $2^x = 7^{x+3}$ $\{-4.660\}$

In Exercises 17–28, evaluate each logarithm to four significant digits.

17. $\log_2 7$ 2.807
18. $\log_5 3$ 0.6826
19. $\log_4 12$ 1.792
20. $\log_6 8$ 1.161
21. $\log_2 7$ 2.807
22. $\log_3 10$ 2.096
23. $\log_{1/8} 12$ −1.195
24. $\log_{1/5} 14$ −1.640
25. $\log_{1/2} 21$ −4.392
26. $\log_{3/5} 12$ −4.864
27. $\log_{1/4} 14$ −1.904
28. $\log_{3/8} 24$ −3.240

In Exercises 29–50, solve the given equation for x.

29. $\log(3x - 2) = \log x$ $\{1\}$
30. $\log(5x - 8) = \log x$ $\{2\}$
31. $\log_3(x + 6) = \log_3(3x - 2)$ $\{4\}$
32. $\log_7(3x - 7) = \log_7(2x - 3)$ $\{4\}$
33. $\log_2(x - 5) = \log_2(7x + 7)$ $\emptyset$
34. $\log_5(3x + 4) = \log_5(6x + 5)$ $\{-\frac{1}{3}\}$
35. $\log_3(1 - 2x) = 4$ $\{-40\}$
36. $\log_5(3x + 2) = 3$ $\{41\}$
37. $\log_9(2x - 3) = 2$ $\{42\}$
38. $\log_2(11x - 1) = 5$ $\{3\}$
39. $\log_5(-2x + 5) = 3$ $\{-60\}$
40. $\log_7(5x + 4) = 2$ $\{9\}$
41. $\log(3x + 100) = 5$ $\{33{,}300\}$
42. $\log(4x + 8) = 3$ $\{248\}$
43. $\log_2(x + 3) = 2 - \log_2 x$ $\{1\}$
44. $\log_2(x - 2) = 3 - \log_2 x$ $\{4\}$
45. $\log_3(x - 2) = 1 - \log_3 x$ $\{3\}$
46. $\log_3(x + 6) = 3 - \log_3 x$ $\{3\}$
47. $\log_2(x + 2) = 2 - \log_2(x - 1)$ $\{2\}$
48. $\log_6(x + 3) = 1 - \log_6(x - 2)$ $\{3\}$
49. $\log_2(x + 3) = 1 + \log_2(x + 2)$ $\{-1\}$
50. $\log_3(x - 1) = 1 + \log_3(x - 3)$ $\{4\}$

In Exercises 51–56, use the formula for population growth given in Example 3.

51. Find the doubling time for a deer population with an annual growth rate of 10 percent. Round to the nearest year. 7 years

52. Find the doubling time for a mouse population with an annual growth rate of 20 percent. Round to the nearest year. 4 years

53. The U.S. population has been growing recently at about 1 percent annually. In 1990 the population was about 248.7 million.
   a. To the nearest year, when will the population reach 300 million? 2009
   b. In about what year will the U.S. population be double the 1990 number? 2060
54. The population of Canada has been growing recently at about 0.8 percent annually. In 1990 the population was about 26.6 million.
   a. To the nearest year, when will the population reach 30 million? 2005
   b. In about what year will the Canadian population be double the 1990 number? 2077
55. What annual growth rate (to the nearest tenth) will cause a population to double in 5 years? 14.9 percent
56. What annual growth rate (to the nearest tenth) will cause a population to double in 25 years? 2.8 percent
57. Some filters remove a fixed percentage of a substance each time a liquid is passed through the filter. For instance, if a filter removes 90 percent of some pollutant from a quantity of water each time the water goes through the filter, then the amount of pollutant remaining after n passes is given by $A = A_0(1 - 0.90)^n$, where A_0 is the original amount of pollutant.
   a. How many passes are needed to reduce the pollutant to 1 percent of its original value? (That is, A must equal $0.01A_0$.) 2
   b. How many passes are needed to reduce the pollutant to one-thousandth of its original value? 3
58. Suppose the filter described in Exercise 57 is only 60 percent effective.
   a. How many passes are needed to reduce the pollutant to below 1 percent of its original value? 6
   b. How many passes are needed to reduce the pollutant to below one-thousandth of its original value? 8

THINK ABOUT IT

1. Just as squaring both sides of an equation can introduce extraneous roots, certain operations with logarithms can alter the solution set of an equation. Solve $\log_3(x - 5)^2 = 2$ by two methods and compare solution sets.
   Method 1: Use the definition of logarithm to get $3^2 = (x - 5)^2$. Then solve this quadratic equation.
   Method 2: Use the power property of logarithms to get $2 \log_3(x - 5) = 2$. Divide both sides by 2, and solve the resulting logarithmic equation.
   Which method gives the correct solution set to the original equation?
2. In the equation for population growth, $P = P_0(1 + r)^t$, does doubling the annual growth rate (r) cut the doubling time in half?
   a. Compare the doubling time for $r = 5$ and $r = 10$ percent.
   b. Compare the doubling time for $r = 10$ and $r = 20$ percent.
   c. What value of r has exactly half the doubling time of $r = 5$ percent?
3. a. Graph $y = \log_{1/2} x$.
   b. Graph $y = -\log_2 x$.
   c. How are the graphs in parts **a** and **b** related? Explain this relationship in terms of the change-of-base formulas.
4. a. Graph $y = \log_2 x^2$. What is the domain?
   b. Graph $y = 2 \log_2 x$. What is the domain?
   c. What restrictions are necessary to say that $\log_2 x^2$ is equal to $2 \log_2 x$?
   d. What restrictions are necessary to say that $\log_b x^2$ is equal to $2 \log_b x$?
5. a. Solve $4^x - 2^x = 30$, which is an equation with quadratic form. (*Hint:* $4^x = 2^{2x}$.)
   b. Solve $9^x - 5 \cdot 3^x + 6 = 0$.

REMEMBER THIS

1. Evaluate $A = P(1 + r)^t$ for $t = 8$, $r = 0.07$, and $P = 5{,}000$ to find the value of a \$5,000 investment compounded once a year for 7 years at an 8 percent annual interest rate. Round to the nearest dollar. \$8,591
2. To four significant digits evaluate $\left(1 + \frac{1}{10{,}000}\right)^{10{,}000}$. 2.718
3. Evaluate $5 \cdot 2^{-4} - (5 \cdot 2)^{-4}$. 0.3124
4. Evaluate $10^{\log 41}$. 41
5. *Semiannual* refers to how many times per year? 2
6. Graph $y = \log x$.

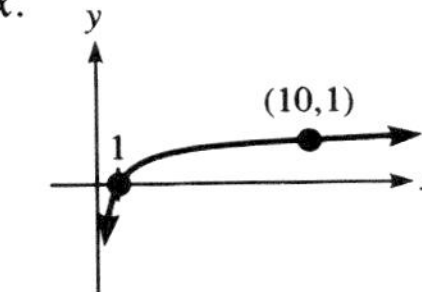

7. What function is the inverse of $y = \frac{3x - 4}{9}$? $y = \frac{9x + 4}{3}$

8. Graph $\frac{x^2}{4} - \frac{y^2}{9} = 1$.

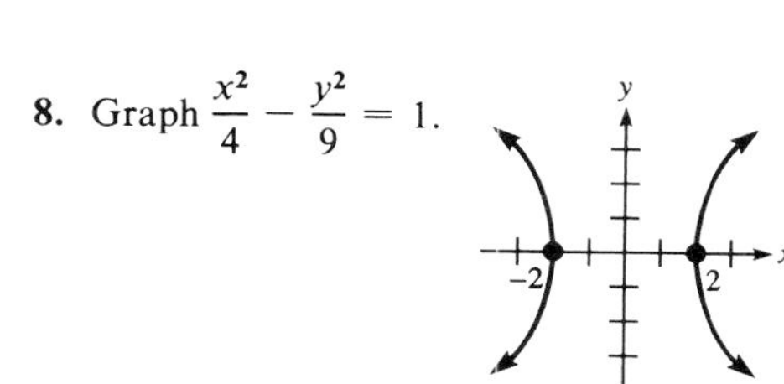

9. Find the point of intersection of the lines $3x + y = 4$ and $6x - y = 4$. $\left(\frac{8}{9}, \frac{4}{3}\right)$

10. Express the sum of 5×10^{-3} and 3×10^{-5} in scientific notation. 5.03×10^{-3}

10.6 More Applications and the Number *e*

In 1991 the oldest intact human body was discovered in an Alpine glacier. He became known as the Iceman, and radiocarbon-dating methods established his age at about 5,300 years. This dating method is based on the fact that there is a fixed ratio of radioactive carbon 14 to ordinary stable carbon in the cells of all living plants and animals. However, when the plant or animal dies, the carbon 14 decays exponentially according to the formula

$$A = A_0 e^{-0.000121t},$$

where A_0 is the initial amount present and A is the amount present after t years have elapsed. Use 5,300 years for the Iceman's age and find the percent of the initial carbon 14 that remains in a sample of bone from Iceman. Express the answer to the nearest percent. (See Example 4.)

OBJECTIVES

1. **Use the compound-interest formula for an investment compounded *n* times per year.**
2. **Solve applied problems that require computing a power of *e*.**
3. **Solve exponential and logarithmic equations involving the number *e*.**

1 The formula $A = P(1 + r)^t$ was used in Section 10.1 to analyze investments that are compounded once a year. However, if the interest is computed more frequently and added to the principal, then the amount grows at a faster rate as the additional interest earns interest. For instance, Example 8 of Section 10.1 showed that \$5,000 amounts to \$8,590.93 when compounded annually at 7 percent for 8 years. The effect of changing this investment to compounding twice per year is to increase the compounded amount by \$79, as shown in Example 1.

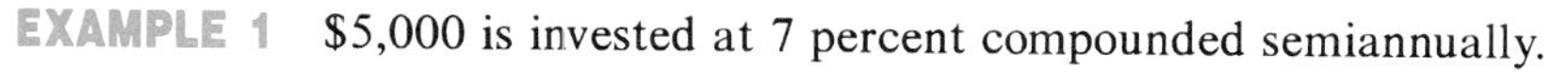

EXAMPLE 1 \$5,000 is invested at 7 percent compounded semiannually.

a. Find a formula showing the value of the investment at the end of t years.
b. How much is the investment worth after 8 years?

Solution

a. Since the investment is compounded semiannually, the interest is determined two times per year, and the interest rate for each of these periods is 7 percent/2, or 0.035. Then,

$$\begin{aligned} A &= 5{,}000(1.035) && \text{when} \quad t = \tfrac{1}{2} \text{ year}, \\ A &= 5{,}000(1.035)^2 && \text{when} \quad t = 1 \text{ year}, \\ A &= 5{,}000(1.035)^4 && \text{when} \quad t = 2 \text{ years}, \end{aligned}$$

and in general, the compounded amount after t years is given by

$$A = 5{,}000(1.035)^{2t}.$$

b. When $t = 8$.

$$\begin{aligned} A &= 5{,}000(1.035)^{2(8)} \\ &= 5{,}000(1.035)^{16} \\ &= \$8{,}669.93. \end{aligned}$$

PROGRESS CHECK 1 $9,000 is invested at 5 percent compounded semiannually.

a. Find a formula showing the value of the investment at the end of t years.
b. How much is the investment worth after 12 years?

The procedure for Example 1 can be generalized to obtain the following compound-interest formula.

Compound-Interest Formula

The compounded amount A when an original principal P is compounded n times per year for t years at annual interest rate r is given by

$$A = P\left(1 + \frac{r}{n}\right)^{nt}.$$

Observe that the above formula simplifies to $A = P(1 + r)^t$ for interest compounded annually, since $n = 1$ in this case.

Have students compare the result in this example with the result of Example 9 of Section 10.1. They will see that the effect of changing the investment from compounding *annually* to compounding *daily* is to increase the compounded amount by $17,852.39!

EXAMPLE 2 $2,000 is invested in an IRA account by a college student on her 17th birthday. If the account grows at 8.5 percent compounded daily, what will be the value of this account on her 65th birthday? Use $n = 365$, and round to the nearest dollar.

Solution 48 years will elapse from the student's 17th birthday to her 65th birthday. Then substituting $P = 2{,}000$, $r = 0.085$, $n = 365$, and $t = 48$ in the compound-interest formula gives

$$\begin{aligned} A &= 2{,}000\left(1 + \frac{0.085}{365}\right)^{365(48)} \\ &= 2{,}000\left(1 + \frac{0.085}{365}\right)^{17{,}520} \end{aligned}$$

To evaluate the resulting expression, compute

2,000 [×] [(] 1 [+] 0.085 [÷] 365 [)] [y^x] 17,520 [=] [118,234.76].

Thus, to the nearest dollar the compounded amount on her 65th birthday is $118,235.

Caution In this example students sometimes introduce a large round-off error by approximating $1 + 0.085/365$ with 1.0002 and computing

$$A = 2{,}000(1.0002)^{17{,}520} = \$66{,}473.07.$$

To avoid such errors, use a keystroke sequence (like the one displayed in Example 2) that allows intermediate computations to be carried forward with as much accuracy as possible.

PROGRESS CHECK 2 $1,500 is invested in an IRA account by a college student on her 27th birthday. If the account grows at 7.8 percent compounded daily, to the nearest dollar what will be the value of this account on her 65th birthday? Use $n = 365$.

Progress Check Answers

1. (a) $A = 9{,}000(1.025)^{2t}$ (b) $16,278.53
2. $29,054

2 When an investment is compounded n times per year, then we have observed that the compounded amount increases as n increases. However, there is a limit to this growth. To illustrate, fix P, r, and t at 1 in the compound-interest formula to obtain

$$A = 1\left(1 + \frac{1}{n}\right)^{n(1)} = \left(1 + \frac{1}{n}\right)^n.$$

The result gives the compounded amount when \$1 is invested at 100 percent interest for 1 year and is compounded n times. Now consider how A changes as the frequency of compounding is increased, as shown in the following table.

Type of Compounding	Number of Conversions (n)	Compounded Amount
Annually	1	$A = \left(1 + \frac{1}{1}\right)^1 = 2$
Semiannually	2	$A = \left(1 + \frac{1}{2}\right)^2 = 2.25$
Quarterly	4	$A = \left(1 + \frac{1}{4}\right)^4 \approx 2.441\ldots$
Monthly	12	$A = \left(1 + \frac{1}{12}\right)^{12} \approx 2.613\ldots$
Daily	365	$A = \left(1 + \frac{1}{365}\right)^{365} \approx 2.714\ldots$
Hourly	8,760	$A = \left(1 + \frac{1}{8{,}760}\right)^{8{,}760} \approx 2.718\ldots$

Students should confirm the compounded amounts displayed in this table using their calculators.

Notice that A increases by a small amount as the conversion period changes from daily to hourly, and more frequent conversions lead to even smaller changes in A. In higher mathematics it is shown that as n gets larger, $(1 + 1/n)^n$ gets closer to an irrational number that is denoted by the letter e. To six significant digits,

$$e \approx 2.71828\ldots$$

The symbol e for the base of natural logarithms was introduced by Leonhard Euler (Swiss) in 1742. Other symbols were suggested from time to time, but e is now universally used.

When n increases without bound, we say that the investment is **compounded continuously.** If the \$1 investment analyzed above is compounded in this manner, then it grows to \$$e$ by the end of the year. A base e exponential function therefore describes investments that are compounded continuously, and a general formula that allows for interest rates and principals that are more practical than 100 percent and \$1 is stated next.

Continuous-Compounding Formula

The compounded amount A when an original principal P is compounded continuously for t years at annual interest rate r is given by

$$A = Pe^{rt}.$$

EXAMPLE 3 \$6,000 is invested at 5 percent compounded continuously. How much is the investment worth in 7 years?

Solution Substituting $P = 6{,}000$, $r = 0.05$, and $t = 7$ in the formula $A = Pe^{rt}$ gives

$$A = 6{,}000e^{(0.05)(7)}$$
$$= 6{,}000e^{0.35}$$

For the use of e^x and $\ln x$ on the graphing calculator, see the graphing calculator section at the end of the chapter.

To evaluate powers of e, most calculators write e^x above the [ln] key, so

[INV] [ln] is equivalent to [e^x].

In this case we compute $6{,}000e^{0.35}$ by pressing

6000 [×] 0.35 [INV] [ln] [=] [8514.4053].

Thus, the compounded amount is \$8,514.41.

PROGRESS CHECK 3 \$2,000 is invested at 11 percent compounded continuously. How much is the investment worth in 10 years?

For analysis of exponential growth or decay at a continuous rate for physical quantities, the formula $A = Pe^{rt}$ is expressed more generally as

$$A = A_0e^{kt},$$

where A is the amount at time t, A_0 is the initial amount, and k is the growth or decay constant. The constant k is positive when describing growth and negative when describing decay. A specific formula that fits this form was given in the section-opening problem, which may now be solved.

The radiocarbon-dating method was developed in 1947 by Dr. Willard Libby, who received the Nobel Prize in chemistry for his discovery.

EXAMPLE 4 Solve the problem in the section introduction on page 515.

Solution The percent of the initial amount of carbon 14 remaining today is given by the current amount divided by the initial amount. Symbolically, this ratio is given by A/A_0, so begin by dividing both sides of the given formula by A_0.

$$A = A_0e^{-0.000121t}$$
$$\frac{A}{A_0} = e^{-0.000121t}$$

Then replacing t by 5,300 yields

$$\frac{A}{A_0} = e^{-0.000121(5{,}300)}.$$

To evaluate the resulting expression, one possible calculation sequence is

0.000121 [+/−] [×] 5,300 [=] [INV] [ln] [0.5266074].

To the nearest percent, about 53 percent of the initial carbon 14 remains in the sample from Iceman.

PROGRESS CHECK 4 Egyptian mummies have been found that are older than Iceman, but their bodies were not fully intact because of embalming. What percent (to the nearest percent) of the initial carbon 14 would be present in a sample from a mummy that is 5,900 years old?

Progress Check Answer

3. \$6,008.33
4. 49 percent

3 The calculator evaluation of powers of e in Examples 3 and 4 once again suggests the inverse relation between exponential and logarithmic functions. The inverse of the function defined by $y = e^x$ is defined by $x = e^y$, which is equivalent to $y = \log_e x$. Logarithms to the base e are called **natural logarithms,** and $\log_e x$ is usually abbreviated as $\ln x$. Thus,

$$y = e^x \text{ and } y = \ln x \text{ are inverse functions.}$$

The graphs of $y = e^x$ and $y = \ln x$ are shown in Figure 10.18. Observe that the graphs are symmetric about the line $y = x$ (as are all pairs of inverse functions), and they behave as exponential and logarithmic functions with base b that satisfy $b > 1$ (since $e > 1$).

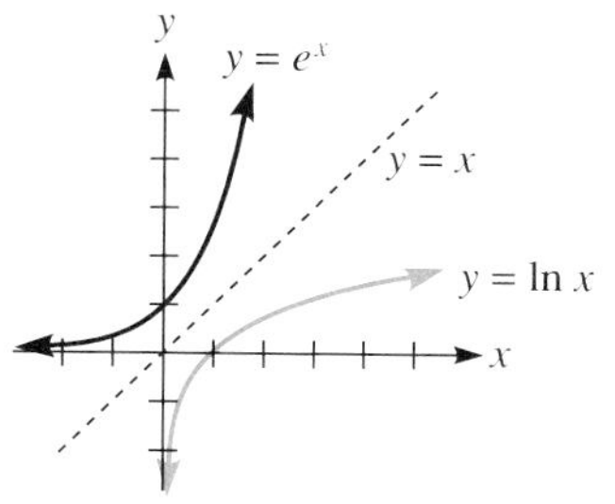

Properties of natural logarithms follow from the logarithm properties given in Section 10.4. For instance, stating the product, quotient, and power properties in terms of base e logarithms results in the following properties.

Figure 10.18

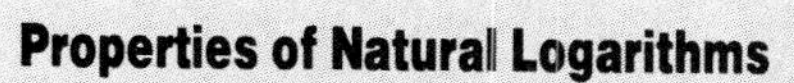

Properties of Natural Logarithms

If x and y are positive numbers and k is any real number, then

1. $\ln xy = \ln x + \ln y$ **Product property**
2. $\ln \frac{x}{y} = \ln x - \ln y$ **Quotient property**
3. $\ln x^k = k \ln x$. **Power property**

Furthermore, the inverse properties $\log_b b^x = x$ and $b^{\log_b x} = x$ (if $x > 0$) become

$$\ln e^x = x \quad \text{and} \quad e^{\ln x} = x \quad (\text{if } x > 0)$$

when stated in terms of base e, while

$$\ln e = 1 \quad \text{and} \quad \ln 1 = 0.$$

In the remaining examples, properties of natural logarithms are used together with the equation-solving methods of Section 10.5 to solve exponential and logarithmic equations that involve the number e.

EXAMPLE 5 Solve $e^{-0.12t} = 0.8$. Give the solution to four significant digits.

Solution When the unknown is in an exponent above base e, solve by taking the natural logarithm of both sides of the equation and then using the inverse property $\ln e^x = x$, as illustrated below.

$$\begin{aligned} e^{-0.12t} &= 0.8 \\ \ln e^{-0.12t} &= \ln 0.8 && \text{Apply natural logarithms to each side.} \\ -0.12t &= \ln 0.8 && \text{Inverse property } \ln e^x = x. \\ t &= \frac{\ln 0.8}{-0.12} \end{aligned}$$

By calculator,

0.8 [ln] [÷] 0.12 [+/−] [=] [1.8595296].

Thus, to four significant digits, the solution set is $\{1.860\}$.

PROGRESS CHECK 5 Solve $e^{0.15t} = 2$. Give the solution to four significant digits.

Progress Check Answer

5. $\{4.621\}$

EXAMPLE 6 Solve each equation. Give exact answers and also approximate solutions to four significant digits.

a. $\ln x = 2.3$ **b.** $\ln x = 1 - \ln 2$

Solution

a. If $\ln x = 2.3$, then $x = e^{2.3}$, and the solution set is $\{e^{2.3}\}$. A calculator sequence to approximate this power of e is

2.3 [INV] [ln] [9.9741825].

The solution set to four significant digits is therefore $\{9.974\}$.

b. First, rewrite the given equation as

$$\ln x + \ln 2 = 1$$

so that the product property of natural logarithms yields

$$\ln 2x = 1.$$

Then convert to exponential form and solve for x.

$$2x = e^1$$
$$x = \frac{e}{2}$$

Confirm that the solution set is $\{e/2\}$ by substituting in the original equation. To approximate this solution, compute

1 [INV] [ln] [÷] 2 [=] [1.3591409].

Thus, to four significant digits, the solution set is $\{1.359\}$.

PROGRESS CHECK 6 Solve each equation. Give exact answers and also approximate solutions to four significant digits.

a. $\ln x = -1.8$ **b.** $\ln x = 2 + \ln 3$

EXAMPLE 7 At what rate of interest compounded continuously must money be deposited, if the amount is to double in 9 years? Answer to the nearest hundredth of a percent.

Solution When the original principal P has doubled, then $A = 2P$. Substitute $2P$ for A and 9 for t in the formula $A = Pe^{rt}$ and solve for r.

$$A = Pe^{rt}$$
$$2P = Pe^{r(9)} \quad \text{Replace } A \text{ by } 2P \text{ and then } t \text{ by } 9.$$
$$2 = e^{9r}$$
$$\ln 2 = \ln e^{9r} \quad \text{Apply natural logarithms to each side.}$$
$$\ln 2 = 9r \quad \text{Inverse property } \ln e^x = x.$$
$$\frac{\ln 2}{9} = r$$

By calculator, $r \approx 0.0770164$, so the required interest rate is 7.70 percent, to the nearest hundredth of a percent.

PROGRESS CHECK 7 At what rate of interest compounded continuously must money be deposited if the amount is to double in 6 years? Answer to the nearest hundredth of a percent.

Progress Check Answers

6. (a) $\{e^{-1.8}\}$; $\{0.1653\}$ (b) $\{3e^2\}$; $\{22.17\}$

7. 11.55 percent

EXERCISES 10.6

In Exercises 1–14, use the given amounts, rates, compounding periods, and years to find (a) a formula showing the value of the investment at the end of t years, and (b) how much the investment is worth at the end of the number of years given.

1. \$10,000, 8 percent, quarterly, 5 years
a. $A = 10{,}000(1.02)^{4t}$ b. \$14,859.47

2. \$5,000, 12 percent, quarterly, 6 years
a. $A = 5{,}000(1.03)^{4t}$ b. \$10,163.97

3. \$20,000, 10 percent, semiannually, 8 years
a. $A = 20{,}000(1.05)^{2t}$ b. \$43,657.49

4. \$15,000, 6 percent, semiannually, 7 years
a. $A = 15{,}000(1.03)^{2t}$ b. \$22,688.85

5. \$8,000, 13 percent, annually, 10 years
a. $A = 8{,}000(1.13)^{t}$ b. \$27,156.54

6. \$12,000, 16 percent, annually, 9 years
a. $A = 12{,}000(1.16)^{t}$ b. \$45,635.54

7. \$30,000, 12 percent, monthly, 10 years
a. $A = 30{,}000(1.01)^{12t}$ b. \$99,011.61

8. \$6,000, 6 percent, monthly, 8 years
a. $A = 6{,}000(1.005)^{12t}$ b. \$9,684.86

9. \$10,000, 10 percent, daily, 7 years
a. $A = 10{,}000\left(1 + \dfrac{0.1}{365}\right)^{365t}$ b. \$20,135.60

10. \$25,000, 5 percent, daily, 10 years
a. $A = 25{,}000\left(1 + \dfrac{0.05}{365}\right)^{365t}$ b. \$41,216.62

11. \$15,000, 7 percent, continuously, 12 years
a. $A = 15{,}000e^{0.07t}$ b. \$37,745.50

12. \$8,000, 9 percent, continuously, 15 years
a. $A = 9{,}000e^{0.09t}$ b. \$30,859.40

13. \$10,000, 10 percent, continuously, 7 years
a. $A = 10{,}000e^{0.10t}$ b. \$20,137.53

14. \$25,000, 5 percent, continuously, 10 years
a. $A = 25{,}000e^{0.05t}$ b. \$41,218.03

In Exercises 15–30, solve the given equation for t. Give the solution to four significant digits.

15. $e^{-0.21t} = 0.2231$ {7.143}
16. $e^{-0.36t} = 0.4628$ {2.140}
17. $e^{-0.14t} = 0.2134$ {11.03}
18. $e^{-0.43t} = 0.7341$ {0.7189}
19. $e^{-3.52t} = 0.3753$ {0.2784}
20. $e^{-4.71t} = 0.1246$ {0.4422}
21. $e^{-1.26t} = 0.0315$ {2.744}
22. $e^{-2.34t} = 0.0684$ {1.146}
23. $e^{0.13t} = 1.364$ {2.388}
24. $e^{0.47t} = 3.413$ {2.612}
25. $e^{0.06t} = 2.175$ {12.95}
26. $e^{0.05t} = 1.831$ {12.10}
27. $e^{1.01t} = 0.821$ {−0.1953}
28. $e^{2.21t} = 0.532$ {−0.2856}
29. $e^{3.04t} = 0.041$ {−1.051}
30. $e^{4.78t} = 0.035$ {−0.7013}

In Exercises 31–56, solve the given equation for x, and (a) give the exact value and (b) give the approximate value to four significant digits.

31. $\ln x = 7.4$ a. $\{e^{7.4}\}$ b. {1,636}
32. $\ln x = 2.8$ a. $\{e^{2.8}\}$ b. {16.44}
33. $\ln x = 0.21$ a. $\{e^{0.21}\}$ b. {1.234}
34. $\ln x = 0.56$ a. $\{e^{0.56}\}$ b. {1.751}
35. $\ln x = 5.2$ a. $\{e^{5.2}\}$ b. {181.3}
36. $\ln x = 6.4$ a. $\{e^{6.4}\}$ b. {601.8}
37. $\ln x = 0.02$ a. $\{e^{0.02}\}$ b. {1.020}
38. $\ln x = 0.05$ a. $\{e^{0.05}\}$ b. {1.051}
39. $\ln x = -3.8$ a. $\{e^{-3.8}\}$ b. {0.0224}
40. $\ln x = -6.2$ a. $\{e^{-6.2}\}$ b. {0.002029}
41. $\ln x = -0.2$ a. $\{e^{-0.2}\}$ b. {0.8187}
42. $\ln x = -0.4$ a. $\{e^{-0.4}\}$ b. {0.6703}
43. $\ln x + \ln 3 = 6$ a. $\{e^{6}/3\}$ b. {134.5}
44. $\ln x + \ln 5 = 8$ a. $\{e^{8}/5\}$ b. {596.2}
45. $\ln x + \ln 2 = 5$ a. $\{e^{5}/2\}$ b. {74.21}
46. $\ln x + \ln 8 = 3$ a. $\{e^{3}/8\}$ b. {2.511}
47. $\ln x = 1 - \ln 3$ a. $\{e/3\}$ b. {0.9061}
48. $\ln x = 2 - \ln 5$ a. $\{e^{2}/5\}$ b. {1.478}
49. $\ln x = 4 - \ln 7$ a. $\{e^{4}/7\}$ b. {7.800}
50. $\ln x = 3 - \ln 4$ a. $\{e^{3}/4\}$ b. {5.021}
51. $\ln x = 2 + \ln 5$ a. $\{5e^{2}\}$ b. {36.95}
52. $\ln x = 3 + \ln 2$ a. $\{2e^{3}\}$ b. {40.17}
53. $\ln x = 5 + \ln 3$ a. $\{3e^{5}\}$ b. {445.2}
54. $\ln x = 1 + \ln 7$ a. $\{7e\}$ b. {19.03}
55. $2 \ln x = 3 + 2 \ln 5$ a. $\{5e^{3/2}\}$ b. {22.41}
56. $3 \ln x = 2 + 3 \ln 2$ a. $\{2e^{2/3}\}$ b. {3.895}

In Exercises 57–64, at what rate of interest compounded continuously must money be invested to grow as indicated? (Round to the nearest hundredth of a percent.)

57. Double in 15 years 4.62 percent
58. Double in 5 years 13.86 percent
59. Double in 7 years 9.90 percent
60. Double in 10 years 6.93 percent
61. Triple in 10 years 10.99 percent
62. Triple in 7 years 15.69 percent
63. Quadruple in 10 years 13.86 percent
64. Quadruple in 7 years 19.80 percent

65. Which becomes more valuable after 10 years, \$1 invested continuously at 10 percent or \$2 invested continuously at 5 percent? \$2 at 5 percent

66. Which becomes more valuable after 20 years, \$1 invested continuously at 10 percent or \$2 invested continuously at 5 percent? \$1 at 10 percent

67. About how long does it take an investment to double in value if it is invested at 5 percent interest compounded continuously? 13.9 years

68. About how long does it take an investment to double in value if it is invested at 4 percent interest compounded continuously? 17.3 years

69. In 1920 two scientists, Pearl and Reed, published a formula which described the growth of the U.S. population over the years 1790 to 1910. Their formula is $N = \dfrac{197{,}273{,}000}{1 + e^{-0.0314t}}$, where N is the population and t is the number of years elapsed since 1914 (t is negative for years before 1914). Their formula was almost exactly on target for the years 1790, 1850, and 1910. What population does their formula yield for each of these years? Round to the nearest thousand. 3,939,000; 23,317,000; 92,450,000

70. A formula published in 1922 by H. G. Thornton gives the area (A) in square centimeters of a growing colony of bacteria as a function of the number of days (x) elapsed. His formula was

$$A = \frac{0.2524}{0.005125 + e^{-2.13x}} \qquad (0 \leq x \leq 5).$$

What are the areas predicted by his formula after 1, 3, and 5 days? 2.04 cm²; 37.10 cm²; 49.02 cm²

71. The decay constant for carbon 14 is -0.000121. Approximately what percent of the initial carbon 14 remains in a fossil that is 10,000 years old? 29.8 percent

72. The decay constant for carbon 14 is -0.000121. Approximately what percent of the initial carbon 14 remains in a fossil that is 100 years old? 98.8 percent

73. The great physicist Ernest Rutherford was among the first to use radioactive decay to date the age of the earth. After analyzing the decay of radium and uranium in a pitchblende rock, he dramatically declared (about 100 years ago) to a geology professor at Cambridge University he was sure the rock in his hand was 700 million years old. The decay constant for uranium 238 is -1.55×10^{-10}. What percentage of an initial amount of uranium 238 remains after 700 million years? 89.7 percent

74. The most widely used method for radiometric dating employed by geologists today involves the radioactive decay of potassium 40, which has a decay constant of -5.55×10^{-10}. What percentage of an initial amount of potassium 40 remains after 700 million years? 67.8 percent

75. A nuclear accident at Chernobyl in Ukraine in 1986 released dangerous amounts of the radioactive isotope cesium 137, which has a decay constant -0.023. Approximately what percent of the cesium released in 1986 remains in the year 2000? 72.5 percent

76. The Chernobyl accident in 1986 also released dangerous amounts of strontium 90, which has decay constant -0.024. Approximately what percent of the strontium released in 1986 remains in the year 2000? 71.5 percent

THINK ABOUT IT

1. a. Graph $y = e^{-x}$ and $y = -\ln x$ on the same coordinate system. How are the graphs related?

b. Show algebraically that if $f(x) = e^{-x}$, then $f^{-1}(x) = -\ln x$. Specify the domain and range of f^{-1}.

2. a. If the half-life of radium is 1,620 years, find the decay constant.

b. Find a formula that gives the decay constant C as a function of the half-life T, given that $A = A_0e^{Ct}$.

3. Solve $\ln x = 1 - \ln(x + 2)$. Give the exact solution set and also approximate solutions to four significant digits.

4. After what amount of time are these two investments equally valuable, a \$1 investment at 10 percent, compounded continuously, and a \$2 investment at 5 percent, compounded continuously?

5. The analysis of growth given limited resources often involves equations containing e. A common form is called the **logistic growth function.** The equations given in Exercises 69 and 70 both can be written in one of the characteristic forms of a logistic growth function:

$$y = \frac{L}{1 + be^{-kt}}.$$

To see what a typical logistic growth curve looks like, assume that $L = 100$, $b = 99$, and $k = 2$, and plot the graph of y. Construct a table for $t = 0, 1, 2, 3, 4, 5, 6$. In the given equation y represents the size of the growing entity and t represents elapsed time.

REMEMBER THIS

1. What number is missing in this sequence: $1, \frac{1}{2}, \frac{1}{4},$ ______ $, \frac{1}{16}$? $\frac{1}{8}$

2. Evaluate $f(n) = n^2 - 1$ for $n = 1, 2,$ and 3. 0, 3, 8

3. Evaluate $f(n) = 3n^2$ for $n = 1, 2,$ and 3. 3, 12, 27

4. What is the sixth number in the sequence that begins $\frac{1}{1}, \frac{3}{2}, \frac{7}{4}, \frac{15}{8}, \ldots$? $\frac{63}{32}$

5. Solve $3^x = 5$ to the nearest hundredth. $\{1.46\}$

6. Find the coordinate of the vertex of the parabola $y = (x - 1)^2 + 5$. (1,5)

7. Multiply $(\sqrt{3x} - \sqrt{x})(\sqrt{3x} + \sqrt{x})$. Assume $x > 0$. $2x$

8. Evaluate the determinant $\begin{vmatrix} 5 & 5 \\ 5 & 5 \end{vmatrix}$. 0

9. Find the slope and the y-intercept of the line given by $2x - 3y = 6$. $m = \frac{2}{3}$; $(0, -2)$

10. Solve $x^3 - x = 0$. $\{0, 1, -1\}$

Inverse Functions; Exponential and Logarithmic Functions

Inverse Functions

If two functions f and g are inverses, then both compositions $f \circ g$ and $g \circ f$ must be identically equal to x; and so the graph of $f \circ g$ and the graph of $g \circ f$ will each be the straight line $y = x$. The composite functions are found as described in the graphing calculator section of Chapter 9.

EXAMPLE 1 Determine graphically that $f(x) = x^3 + 4$ and $g(x) = \sqrt[3]{x - 4}$ are inverse functions.

Solution The entries for the [Y=] key are shown in Figure 10.19 (note that the cube and the cube root are obtained from the [MATH] key). The resulting graphs, plotted with [ZOOM] Square, are shown in Figure 10.20. The graphs of Y_3 and Y_4 do appear to be the line $y = x$. To check this equation, trace along the line and note that the x- and y-coordinates are always equal. Thus, f and g are inverses. Further evidence that they are inverses may be seen in the symmetry of the graphs of f and g about the line $y = x$.

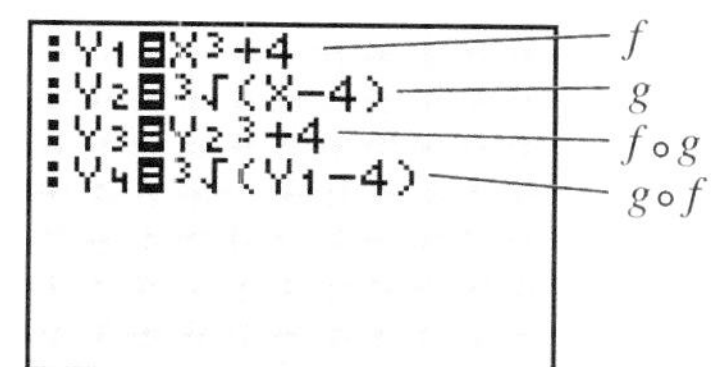

Figure 10.19

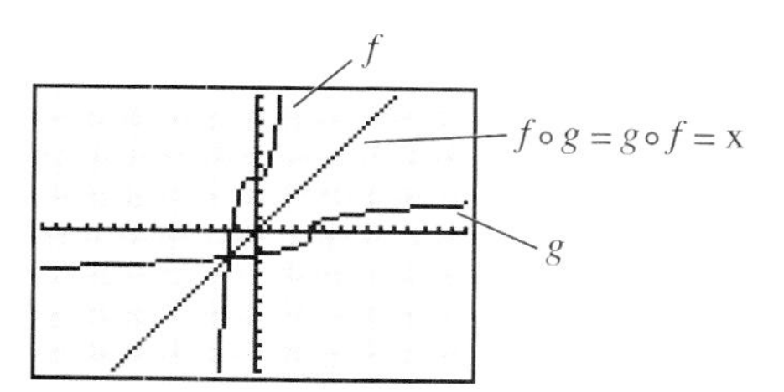

Figure 10.20

Exponential Functions

The power key [∧] permits easy evaluation and graphing of exponential functions, as demonstrated in Examples 2 and 3. For expressions involving e^x, use [2nd] [e^x], which appears above the black [LN] key.

EXAMPLE 2 Evaluate to three decimal places.

a. e^3 **b.** e

Solution The keystroke sequences are shown.

a. [2nd] [e^x] 3 [ENTER] The answer is 20.086.

b. It is necessary to use the fact that e is the same as e^1.

[2nd] [e^x] 1 [ENTER] The answer is 2.718.

EXAMPLE 3 Graph.

a. $y = 2^x + 1$ **b.** $y = (\frac{1}{2})^x$ **c.** $y = -e^x - 2$

Solution These exponential graphs can be entered as shown in Figure 10.21, and the resulting screen is shown in Figure 10.22.

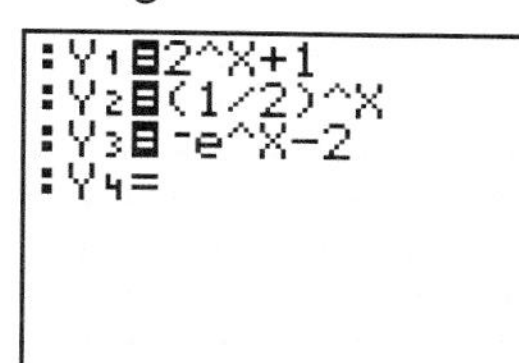

Figure 10.21

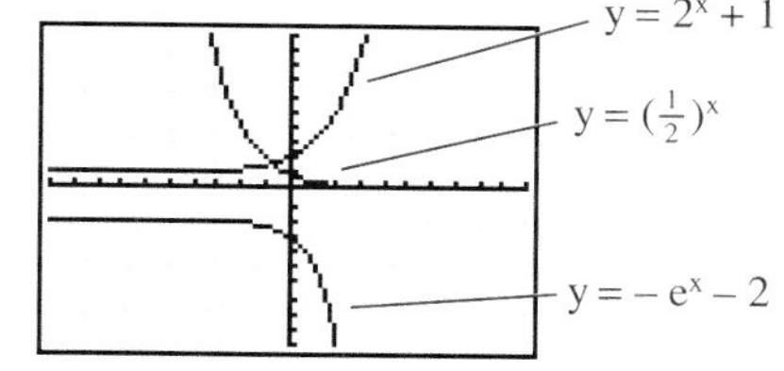

Figure 10.22

Logarithmic Functions

Base 10 and base e logarithmic functions are built into the calculator. For other bases, apply the change-of-base formula given in Section 10.5.

EXAMPLE 4 Evaluate each of these terms to four decimal places.

a. log 35 **b.** ln 20 **c.** $\log_2 14$

Solution

a. [LOG] 35 [ENTER] The answer is 1.5441.

b. [LN] 20 [ENTER] The answer is 2.9957.

c. The change-of-base formula gives

$$\log_2 14 = \log 14/\log 2.$$

[LOG] 14 [÷] [LOG] 2 [ENTER] The answer is 3.8074.

EXAMPLE 5 Graph each of these functions.

a. $y = \log x$ **b.** $y = \ln x$ **c.** $y = \log_2 x$

Solution Since X cannot be negative, we set the RANGE so that $X_{min} = 0$ and $X_{max} = 10$. For clarity, we set $Y_{min} = -5$ and $Y_{max} = 5$. The Y= entries and the resulting graphs are shown in Figures 10.23 and 10.24.

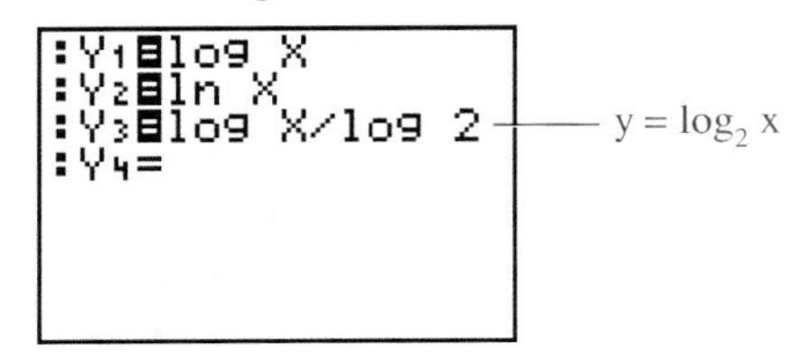

Figure 10.23

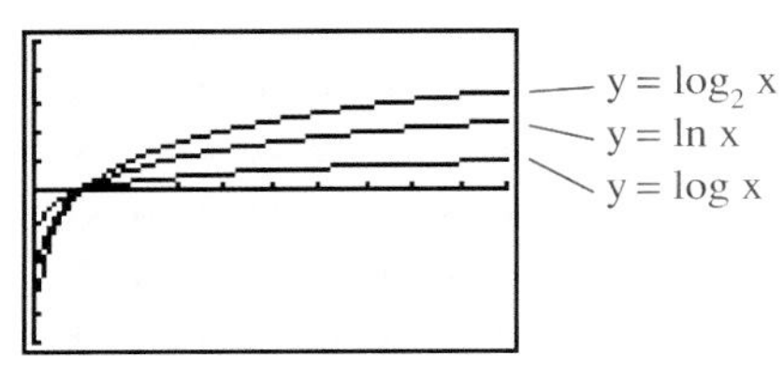

Figure 10.24

Solving Logarithmic and Exponential Equations

Graphical methods for estimating solutions of equations were described in the graphing calculator section of Chapter 7. These methods are applicable for solving logarithmic and exponential equations.

EXAMPLE 6 Solve to two decimal places.

a. $5^x = 3^{x+2}$ **b.** $\log_3(x - 2) = 1 - \log_3 x$

Solution

a. Let $Y_1 = 5^X$, $Y_2 = 3^{X+2}$, and $Y_3 = Y_1 - Y_2$. Then plot Y_3 and find any x-intercepts. Observe that the graph of Y_3 crosses the x-axis very steeply at about $x = 4$, so several rounds of TRACE and ZOOM will be necessary to get the desired accuracy. The solution set is $\{4.30\}$.

b. Let $Y_1 = \log(X - 2)/\log 3$, $Y_2 = 1 - \log X/\log 3$, and $Y_3 = Y_1 - Y_2$. With use of TRACE and ZOOM the x-intercept of Y_3 can be estimated as 3.00. (Check by algebra that the exact solution is 3.)

EXERCISES

1. Determine graphically that $f(x) = \log x$ and $g(x) = 10^x$ are inverse functions. Ans. The composition is $y = x$.
2. Evaluate each of the given expressions to three decimal places.
   a. $\sqrt{e}$ **b.** $\log 121$ **c.** $\ln 121$ **d.** $\log_5 121$
   Ans. **a.** 1.649 **b.** 2.083 **c.** 4.796 **d.** 2.980
3. Graph each given expression.
   a. $y = (\frac{1}{4})^x$ Ans.

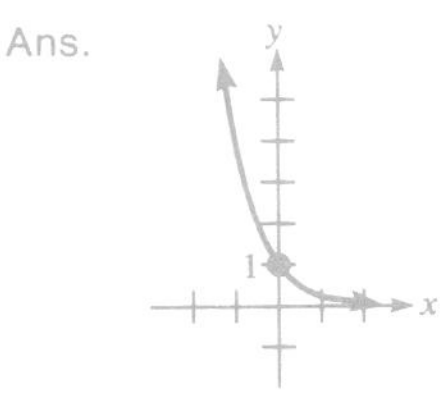

   b. $y = e^{-x}$ Ans.

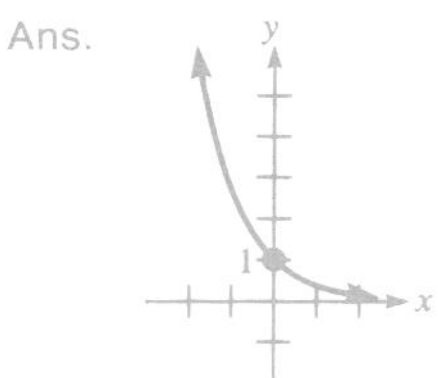

   c. $y = -\log x$ Ans.

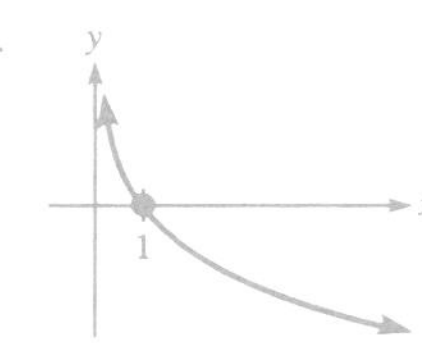

   d. $y = 5 \ln(2x)$ Ans.

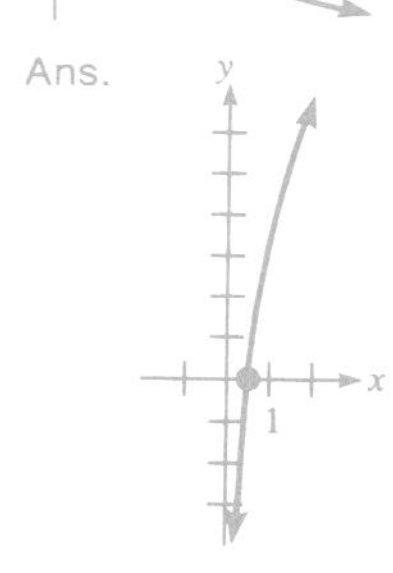

4. Solve to two decimal places.
   a. $e^{-0.21x} = 0.2231$ **b.** $\log_5(2x + 5) = 3.12$
   Ans. **a.** $\{7.14\}$ **b.** $\{73.32\}$
5. This exercise explores the formula given in Section 10.6 in the discussion of continuous growth. Graph $y = (1 + 1/x)^x$ and verify that as x increases, the value of y approaches e. What is the smallest integer value of x for which y exceeds 2.71? Ans. 164

Use the graphing calculator to help solve these problems from Chapter 10. Use the program EVAL when helpful.

Section	*Exercises*
10.1	1, 11, 21, 27, 29, 43, 59
10.2	41
10.3	51, 55, 59, 65, 75, 81
10.5	1, 11, 17, 29, 43
10.6	1, 15, 31

Chapter 10 SUMMARY

OBJECTIVES CHECKLIST Specific chapter objectives are summarized below along with numbered example problems from the text that should clarify the objectives. If you do not understand any objectives or do not know how to do the selected problems, then restudy the materials.

10.1 Can you:

1. **Determine function values for an exponential function?**
   If $f(x) = 4^x$, find $f(3)$, $f(-2)$, and $f(\frac{5}{2})$. Also, approximate $f(\sqrt{2})$ to the nearest hundredth. [Example 2]

2. **Graph an exponential function?**
   Sketch the graphs of $g(x) = 2^x$ and $h(x) = 4^x$ on the same coordinate system. [Example 3]

3. **Solve exponential equations using $b^x = b^y$ implies $x = y$?**
   Solve $8^x = 4$. [Example 6]

4. **Find the base in the exponential function $y = b^x$ given an ordered pair in the function?**
   Find the base in the exponential function $y = b^x$ that contains the point (4,2). [Example 7]

5. **Solve applied problems involving exponential functions?**
   $5,000 is invested at 7 percent compounded annually.
   a. Find a formula showing the value of this investment at the end of t years.
   b. How much is the investment worth after 8 years? [Example 8]

10.2 Can you:

1. **Find the inverse of a function and its domain and range?**
   If $f = \{(1,5),(2,5),(3,7)\}$, find f^{-1}. Find and compare the domain and the range of the two functions. [Example 1]

2. **Determine if a function has an inverse function?**
   Which functions graphed in Figure 10.25 have an inverse function? [Example 3]

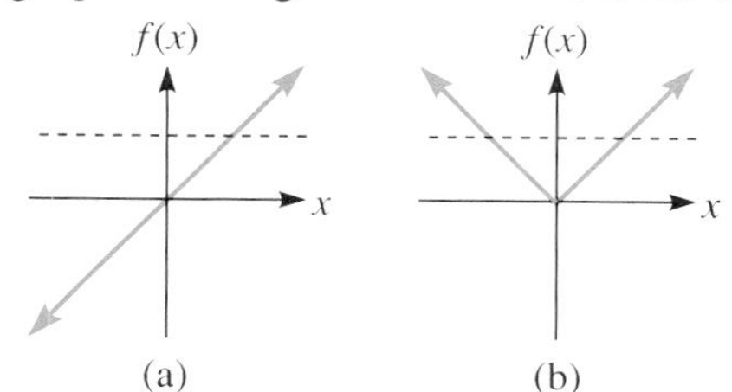

Figure 10.25

3. **Graph $y = f^{-1}(x)$ from the graph of $y = f(x)$?**
   Use the graph of $y = f(x)$ in Figure 10.26 to graph $y = f^{-1}(x)$. [Example 4]

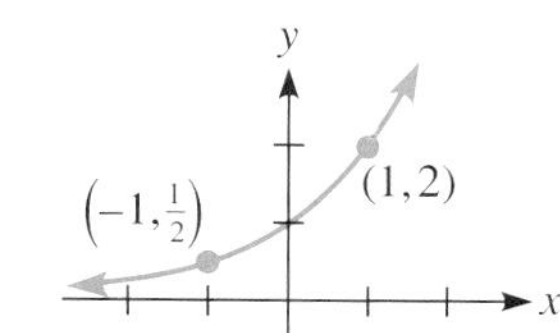

Figure 10.26

4. **Find an equation that defines f^{-1}?**
   Find $f^{-1}(x)$ for the function $f(x) = \sqrt{x}$. If f is not one-to-one, so that no inverse function exists, state this. [Example 5b]

5. **Determine whether two functions are inverses of each other?**
   Verify that $f(x) = \sqrt{x}$ and $g(x) = x^2$, $x \geq 0$, are inverses of each other. [Example 6]

6. **Use inverse function concepts in applications?**
   The function $y = f(x) = \frac{5}{9}(x - 32)$ converts degrees Fahrenheit (x) to degrees Celsius (y). Find an equation for the inverse function. What formula does the inverse function represent? [Example 7]

10.3 **Can you:**

1. **Convert from the exponential form $b^L = N$ to the logarithmic form $\log_b N = L$, and vice versa?**
   Write $10^2 = 100$ and $a^x = 4$ in logarithmic form. [Example 1]

2. **Determine the value of an unknown in expressions of the form $\log_b N = L$?**
   Determine the value of x: $\log_4 8 = x$. [Example 3b]

3. **Graph logarithmic functions?**
   Graph $y = \log_2(x + 4)$. [Example 5]

4. **Determine the common logarithm and antilogarithm of a number by using a calculator?**
   If $\log N = -1.4$, find N to three significant digits. [Example 7]

5. **Solve applied problems involving logarithmic functions?**
   The pH of a solution measures its hydrogen ion concentration (symbolized by $[H^+]$). In exponential form, pH is defined by $[H^+] = 10^{-pH}$. Determine $[H^+]$ for a sample of blood whose pH is 7.4. [Example 9]

10.4 **Can you:**

1. **Use properties of logarithms to express certain log statements in terms of simpler logarithms or expressions?**
   Express $\log_2 7x^2$ as a sum, difference, or product involving simpler logarithms. [Example 1e]

2. **Use properties of logarithms to convert certain statements involving logarithms to a single logarithm with coefficient 1?**
   Express $2 \log_b x + 3 \log_b y$ as a single logarithm with coefficient 1. [Example 6c]

10.5 **Can you:**

1. **Solve exponential equations by using logarithms?**
   Solve $5^x = 3^{x+1}$. Give the solution to four significant digits. [Example 2]

2. **Apply the change-of-base formula?**
   Evaluate $\log_3 6$ to four significant digits. [Example 4a]

3. **Solve logarithmic equations?**
   Solve $\log_5(3x - 8) = 2$. [Example 6]

10.6 **Can you:**

1. **Use the compound-interest formula for an investment compounded *n* times per year?**
   $2,000 is invested in an IRA account by a college student on her 17th birthday. If the account grows at 8.5 percent compounded daily, what will be the value of this account on her 65th birthday? Use $n = 365$, and round to the nearest dollar. [Example 2]

2. **Solve applied problems that require computing a power of *e*?**
   $6,000 is invested at 5 percent compounded continuously. How much is the investment worth in 7 years? [Example 3]

3. **Solve exponential and logarithmic equations involving the number *e*?**
   Solve $e^{-0.12t} = 0.8$. Give the solution to four significant digits. [Example 5]

KEY TERMS

Common logarithm (10.3)
Compounded continuously (10.6)
Exponential equation (10.1)
Exponential function (10.1)
Inverse functions (10.2)
Logarithm (10.3)
Logarithmic equation (10.5)
Logarithmic function (10.3)
Natural logarithm (10.6)
One-to-one function (10.2)
Significant digits (10.3)

KEY CONCEPTS AND PROCEDURES

Section	Key Concepts or Procedures to Review
10.1	■ Definition of the exponential function with base b ■ For $f(x) = b^x$ with $b > 0$, $b \neq 1$: Domain: $(-\infty,\infty)$ Horizontal asymptote: x-axis Range: $(0,\infty)$ y-intercept: $(0,1)$ If $b > 1$, f represents exponential growth. If $0 < b < 1$, f represents exponential decay. ■ Methods from Section 9.3 to graph variations of $f(x) = b^x$ ■ If $b > 0$, $b \neq 1$, then $b^x = b^y$ implies $x = y$. ■ Formula for the compounded amount A when a principal P is compounded annually for t years at annual interest rate r: $$A = P(1 + r)^t$$
10.2	■ Definitions of one-to-one functions and inverse functions ■ The special symbol f^{-1} is used to denote the inverse of function f. ■ Methods to determine if the inverse of a function exists and to find f^{-1}, if it exists ■ f and f^{-1} interchange their domain and range. ■ f and f^{-1} are reflections of each other about the line $y = x$.
10.3	■ Definitions of logarithm and logarithmic function with base b ■ $\log_b N = L$ is equivalent to $b^L = N$. ■ The logarithmic function $y = \log_b x$ and the exponential function $y = b^x$ are inverse functions. Therefore: **(a)** their domain and range are interchanged; **(b)** their graphs are reflections of each other about the line $y = x$. ■ For $f(x) = \log_b x$ (with $b > 0$, $b \neq 1$): Domain: $(0,\infty)$ Range: $(-\infty,\infty)$ x-intercept: $(1,0)$ Vertical asymptote: y-axis If $b > 1$, then as x increases, y increases. If $0 < b < 1$, then as x increases, y decreases. ■ Methods from Section 9.3 to graph variations of $y = \log_b x$. ■ $\log_{10} N$ is usually abbreviated as $\log N$.
10.4	■ Properties of logarithms (for b, x, $y > 0$, $b \neq 1$, and k any real number): **1.** $\log_b xy = \log_b x + \log_b y$ **2.** $\log_b \frac{x}{y} = \log_b x - \log_b y$ **3.** $\log_b x^k = k \log_b x$ **4.** $\log_b b = 1$ **5.** $\log_b 1 = 0$ **6.** $\log_b b^x = x$ **7.** $b^{\log_b x} = x$

Section	Key Concepts or Procedures to Review
10.5	■ If $x, y, b > 0$ with $b \neq 1$, then $x = y$ implies $\log_b x = \log_b y$; and $\log_b x = \log_b y$ implies $x = y$. ■ Change-of-base formula: $$\log_b x = \frac{\log_a x}{\log_a b}$$ ■ Methods to solve exponential and logarithmic equations
10.6	■ Compound-interest formula: $$A = P\left(1 + \frac{r}{n}\right)^{nt}$$ ■ As n gets larger, $\left(1 + \frac{1}{n}\right)^n$ gets closer to an irrational number that is denoted by the letter e. To six significant digits, $e \approx 2.71828$. ■ Continuous-compounding formula: $A = Pe^{rt}$. ■ The general formula for continuous growth or decay is $A = A_0e^{kt}$. ■ $\log_e x$ is usually abbreviated as $\ln x$. ■ $y = \ln x$ and $y = e^x$ are inverse functions. ■ Graphs of $y = \ln x$ and $y = e^x$ ■ Properties of natural logarithms: **1.** $\ln xy = \ln x + \ln y$ **2.** $\ln \frac{x}{y} = \ln x - \ln y$ **3.** $\ln x^k = k \ln x$ **4.** $\ln e^x = x$ **5.** $e^{\ln x} = x$, if $x > 0$

CHAPTER 10 REVIEW EXERCISES

10.1

1. If $f(x) = 9^x$, find $f(2)$, $f(-3)$, and $f(\frac{3}{2})$. Also, approximate $f(\sqrt{2})$ to the nearest hundredth. $81, \frac{1}{729}, 27, 22.36$

2. Sketch the graph of $f(x) = (\frac{1}{3})^x$.

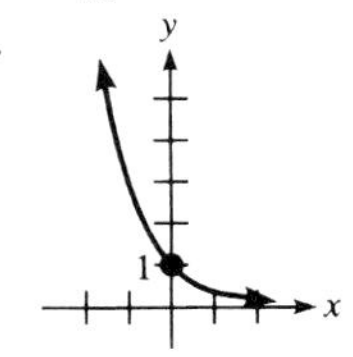

3. Solve $16^x = 64$. $\{\frac{3}{2}\}$

4. Find the base of the exponential function $y = b^x$ that contains the point $(-\frac{1}{2}, \frac{1}{2})$. 4

5. At what interest rate compounded annually must a sum of money be invested if it is to double in 6 years? 12.25 percent

10.2

6. If $f = \{(-2,3),(-1,5),(0,7)\}$, find f^{-1} and determine its domain and range. $f^{-1} = \{(3,-2),(5,-1),(7,0)\}$; $D = \{3,5,7\}$; $R = \{-2,-1,0\}$

7. Which functions graphed in this figure have inverse functions?

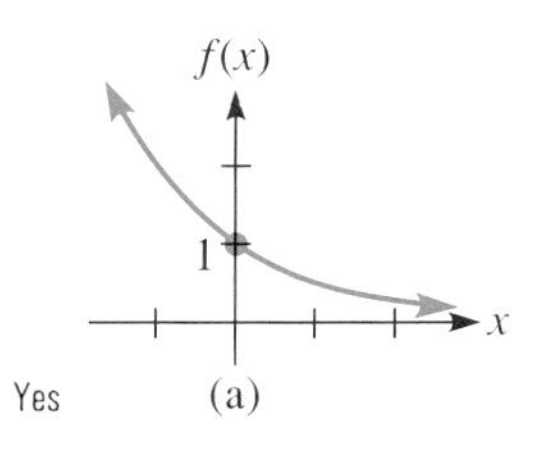

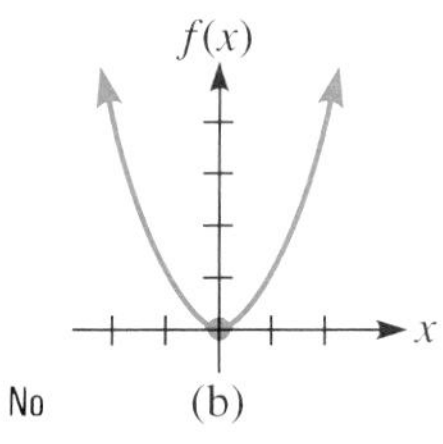

8. Use the graph of $y = f(x)$ in the figure to graph $y = f^{-1}(x)$.

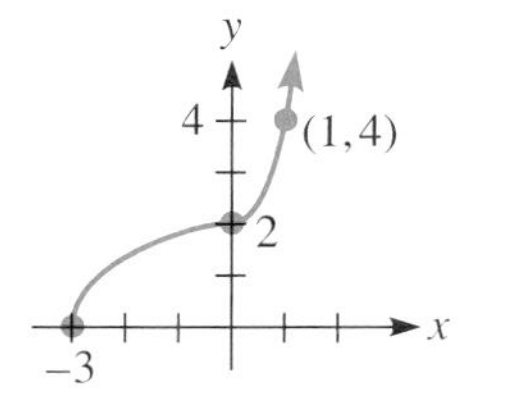

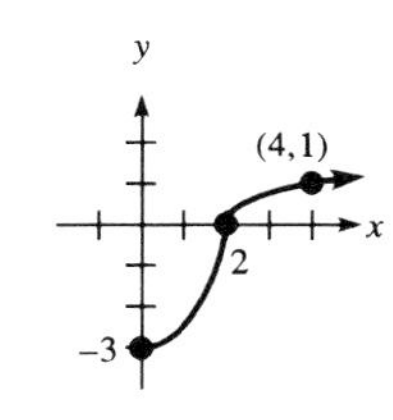

9. Find $f^{-1}(x)$ for the function $f(x) = -3x + 2$. $f^{-1}(x) = \frac{-x+2}{3}$

10. Verify that $f(x) = \sqrt{x-1}$ and $g(x) = x^2 + 1$, $x \geq 1$, are inverses of each other. $f[g(x)] = \sqrt{(x^2+1) - 1} = \sqrt{x^2} = x$; $g[f(x)] = (\sqrt{x-1})^2 + 1 = (x-1) + 1 = x$

11. An inventor's portfolio loses one-sixth of its value during its first year. For the following year, what percent increase is needed for the portfolio to return to its original value? 20 percent

10.3

12. Write $\log_b y = x$ in exponential form. $b^x = y$

13. Determine the value of a: $\log_a \frac{1}{16} = -2$. 4

14. Graph $y = -\log_{1/2} x$.

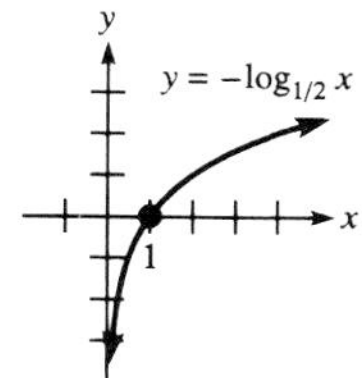

15. Evaluate log 650 to four decimal places. 2.8129

16. Determine $[H^+]$ for a sample of unknown solution whose pH is 8.2. 6.3×10^{-9}

10.4

17. Express as a sum, difference, or product involving simpler logarithms: $\log_b \sqrt{\frac{2x}{y}}$. $\frac{1}{2}(\log_b 2 + \log_b x - \log_b y)$

18. Simplify the expression $3^{\log_3 81}$. 81

19. Express as a single logarithm with coefficient 1: $5 \log_{10} x + \frac{1}{3} \log_{10} 8$. $\log_{10} 2x^5$

20. If $\log_b 2 = m$ and $\log_b 3 = n$, then express $\log_b \frac{1}{27}$ in terms of m and/or n. $-3n$

10.5

21. Solve $5^x = 18$. Give the solution to five significant digits. $\{1.7959\}$

22. Evaluate $\log_5 8$ to four significant digits. 1.292

23. Solve $\log(3x + 1) = \log x$. $\emptyset$

24. Solve $\log_2(5x + 3) = 3$. $\{1\}$

10.6

25. \$8,000 is invested at 6 percent compounded monthly. How much is the investment worth in 10 years? \$14,555.17

26. \$8,000 is invested at 7 percent compounded continuously. How much is the investment worth in 15 years? \$22,861.21

27. Solve $e^{0.12t} = 0.6$. Give the solution to four significant digits. $\{-4.257\}$

28. Solve $\ln 5 = 3 - \ln x$. Give the exact answer and also an approximate solution to four significant digits. $\left\{\frac{e^3}{5}\right\}$; $\{4.017\}$

29. At what rate of interest compounded continuously must money be deposited if the amount is to double in 12 years? Answer to the nearest hundredth of a percent. 5.78 percent

ADDITIONAL REVIEW EXERCISES

30. Find the base of the exponential function $y = b^x$ that contains the point (4,3). $\sqrt[4]{3}$

31. If $f(x) = (\frac{1}{2})^x$, find $f(3)$ and $f(-2)$. Also, approximate $f\sqrt{2}$ to the nearest hundredth. $\frac{1}{8}$, 4, 0.38

32. At what interest rate (to the nearest tenth of a percent) compounded annually must a sum of money be invested to triple in 18 years? 6.3 percent

33. \$9,000 is invested at 5.5 percent compounded continuously. How much is the money worth in 8 years? \$13,974.36

34. How long (to the nearest year) would it take the world's population to double if the annual growth rate was 2.65 percent? 27 years

35. Certain municipal workers accept a 15 percent cut in annual salary to avoid layoffs. What percent raise (to the nearest hundredth of a percent) will then be needed to return these employees to their original annual salary? 17.65 percent

36. At what interest rate compounded continuously (to the nearest hundredth of a percent) must money be deposited if the amount is to double in 8 years? 8.66 percent

37. \$10,000 is invested at 6 percent compounded annually. How much is the investment worth after 7 years? \$15,036.30

38. \$1,000 is invested in an IRA account by a graduate student on her 25th birthday. If the account grows at 6.8 percent compounded daily, what will the value of the account be on her 65th birthday? \$15,176.48

39. Find the pH (to the nearest tenth) of a solution whose hydrogen ion concentration $[H^+]$ is 4.5×10^{-8}. 7.3

40. A scientist has 1 g of a radioactive element. This element is decaying to one-half its amount at a given time after 1 hour has elapsed. Approximately how much of this element is left at the end of 1 day? 5.96×10^{-8} g

Determine the value of the unknown in each expression.

41. $\log_4 32 = x$ $\frac{5}{2}$

42. $\log_6 1 = y$ 0

43. $\log_{10} x = -2$ $\frac{1}{100}$

44. $\log_b 3 = \frac{1}{2}$ 9

Express in logarithmic form.

45. $3^0 = 1$ $\log_3 1 = 0$

46. $10^1 = 10$ $\log_{10} 10 = 1$

47. $5^x = 125$ $\log_5 125 = x$

Express in exponential form.

48. $\log_{10} 1 = 0$ $10^0 = 1$

49. $\log_{10} 3 = x$ $10^x = 3$

50. $\log_b 2 = -3$ $b^{-3} = 2$

Express as a sum, difference, or product involving simpler logarithms.

51. $\log_{10} x^6$ $6 \log_{10} x$

52. $\log_b \sqrt{7}$ $\frac{1}{2} \log_b 7$

53. $\log_b 8(x + 1)^3$ $3[\log_b 8 + \log_b(x + 1)]$

54. $\log_{10} \frac{1}{5}$ $-\log_{10} 5$

Express as a single logarithm with coefficient 1.

55. $\log_{10} x + \log_{10}(x + 2)$ $\log_{10}(x^2 + 2x)$

56. $\frac{1}{3}[\log_2 3 - \log_2 x]$ $\log_2 \sqrt[3]{\frac{3}{x}}$

57. $3 \log_b x + \frac{1}{2} \log_b 4$ $\log_b 2x^3$

Simplify.

58. $\log_6 6^{-2}$ -2

59. $2^{\log_2 32}$ 32

Solve. Give exact answers unless otherwise indicated.

60. $25^x = 125$ $\{\frac{3}{2}\}$

61. $\log N = -3.2$ (three significant digits) $\{0.000631\}$

62. $\log_4(2x - 10) = 3$ $\{37\}$

63. $\ln x = 2.5$ (four significant digits) $\{12.18\}$

64. $e^{-0.05t} = 6$ (four significant digits) $\{-35.84\}$

65. $4^x = 3^{x+1}$ (four significant digits) $\{3.819\}$
66. $\log x = \log(2 - x)$ $\{1\}$
67. $-\ln 2 + 4 = \ln x$ $\left\{\frac{e^4}{2}\right\}$
68. $\log 500 = x$ (four significant digits) $\{2.699\}$
69. $6^x = 9$ (four significant digits) $\{1.226\}$
70. $\log_2(x - 6) = 4 - \log_2 x$ $\{8\}$
71. $\log_{1/2} 0.6 = x$ (four significant digits) $\{0.7370\}$

Find the inverse of each function. If no inverse function exists, state this.

72. $f = \{(-2,2),(-1,3),(0,4)\}$ $\{(2,-2),(3,-1),(4,0)\}$
73. $f = \{(3,1),(0,2),(-3,1)\}$ f^{-1} does not exist.
74. $f = \{(-4,4),(0,0),(4,-4)\}$ $\{(4,-4),(0,0),(-4,4)\}$
75. $f(x) = 4x - 3$ $f^{-1}(x) = \frac{x+3}{4}$
76. $f(x) = 2x^2$ f^{-1} does not exist.
77. $f(x) = \frac{x-1}{2}$ $f^{-1}(x) = 2x + 1$
78.

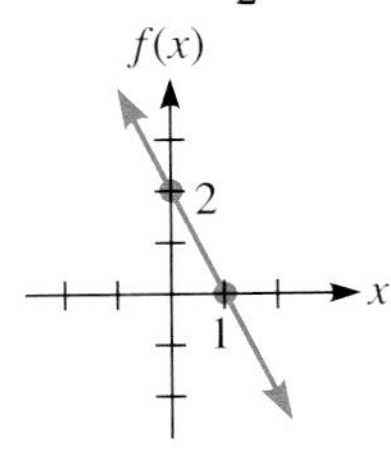

79.

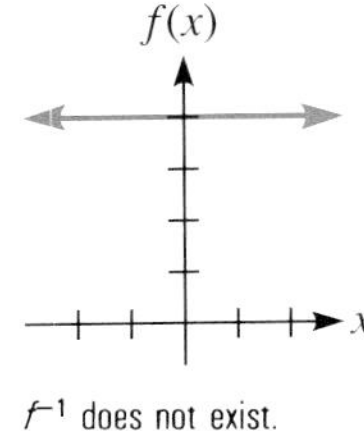

f^{-1} does not exist.

Sketch the graph.

80. $y = -(\frac{1}{3})^x$

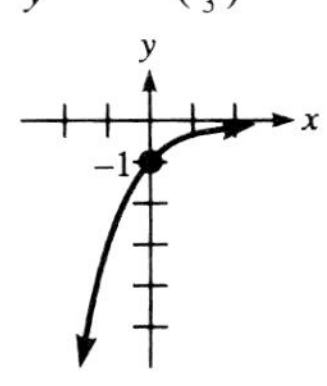

81. $y = 2^x - 3$

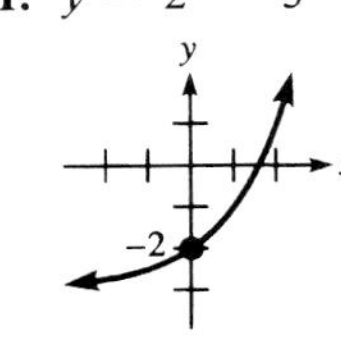

82. $y = \log_{1/3} x$

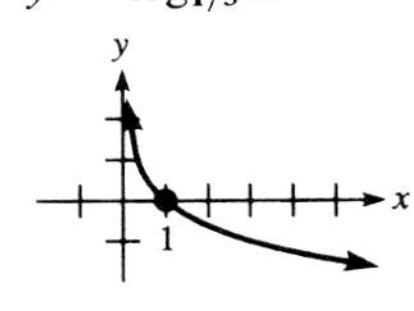

83. $y = \log_3(x + 5)$

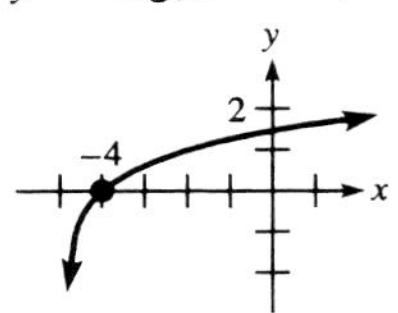

CHAPTER 10 TEST

1. Express in logarithmic form: $b^3 = 8$. $\log_b 8 = 3$
2. Express in exponential form: $\ln x = \frac{1}{2}$. $e^{1/2} = x$
3. Evaluate $\log_5 \frac{1}{125}$. -3
4. Simplify $7^{2 \log_7 6}$. 36
5. Find the base of the exponential function $y = b^x$ that contains the point $(-4,16)$. $\frac{1}{2}$
6. Express as a single logarithm with coefficient 1: $4 \log_2 x - \log_2 3$. $\log_2 \frac{x^4}{3}$
7. Express as a sum, difference, or product involving simpler logarithms: $\log_b 8\sqrt{x}$. $\log_b 8 + \frac{1}{2} \log_b x$
8. Evaluate $\log_6 11$ to four significant digits. 1.338
9. If $f(x) = \frac{5x - 6}{4}$, determine $f^{-1}(x)$. $f^{-1}(x) = \frac{4x + 6}{5}$
10. Verify that $f(x) = x^2 - 2$ and $g(x) = \sqrt{x + 2}$, $x \geq 0$, are inverses of each other. $f[g(x)] = (\sqrt{x+2})^2 - 2 = x$; $g[f(x)] = \sqrt{(x^2 - 2) + 2} = x$
11. \$5,000 is invested at 7 percent compounded monthly. How much is the investment worth in 30 years? \$40,582.49
12. At what rate of interest compounded continuously must money be deposited if the amount is to triple in 20 years? Answer to the nearest hundredth of a percent. 5.49 percent
13. Solve $7^x = 0.9$. Give the solution to four significant digits. $\{-0.05414\}$
14. If $\log N = -0.6$, find N to three significant digits. 0.251
15. Solve $2^{x-2} = 6^x$. Give the solution to four significant digits. $\{-1.262\}$
16. Solve $\log_4(10x + 4) = 3$. $\{6\}$
17. Solve $e^{-0.8t} = 0.5$. Give the solution to four significant digits. $\{0.8664\}$
18. Graph $f(x) = -2^x$.

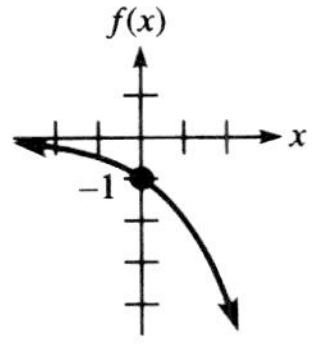

19. Graph $f(x) = \log_2 x + 1$.

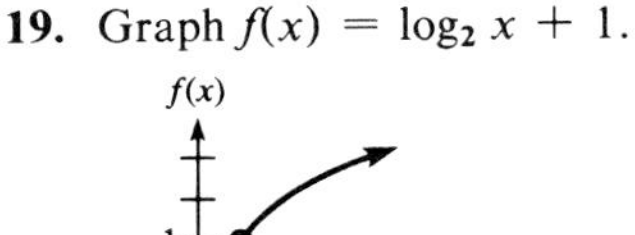

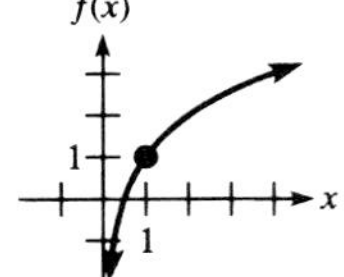

20. Use the graph of $y = f(x)$ in the figure to graph $y = f^{-1}(x)$.

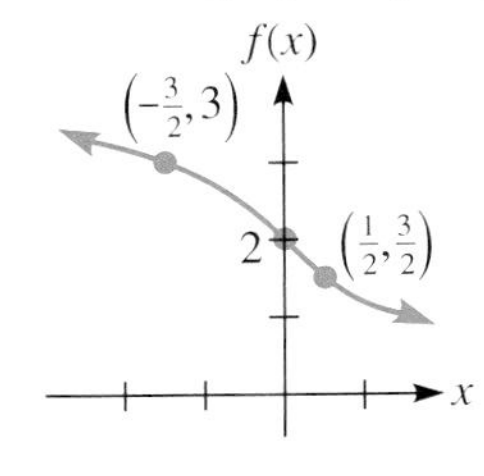

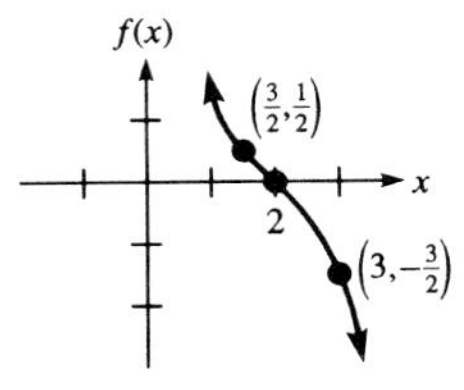

CUMULATIVE TEST 10

1. Solve $A = \frac{1}{2}bh$ for b. $b = \frac{2A}{h}$
2. Solve $-2(x - 4) \leq 3(x + 1)$, and graph the solution set. $[1,\infty)$

3. Simplify $3(4x^2 - 2) - (x^2 - 3)$. $11x^2 - 3$
4. Simplify $\left(\frac{8a^2b^5}{-4a^4b}\right)^{-2}$, and write the result using only positive exponents. $\frac{a^4}{4b^8}$
5. The height (y) of a projectile that is shot directly up from the ground with an initial velocity of 40 ft/second is given by the formula $y = -16t^2 + 40t$, where t is the elapsed time in seconds. When does the projectile hit the ground? In 2.5 seconds
6. Express as a single fraction in lowest terms.
$$\frac{x}{x^2 - 9} - \frac{x + 1}{2x^2 - 5x - 3} \qquad \frac{x^2 - 3x - 3}{(x - 3)(x + 3)(2x + 1)}$$
7. Solve $\frac{4}{x - 1} + 3 = -\frac{6}{x}$. $\{-3, \frac{2}{3}\}$
8. Find an equation in general form of the line through the points $(-8,5)$ and $(-2,-3)$. $4x + 3y = -17$
9. Graph $x + 2y = -8$.

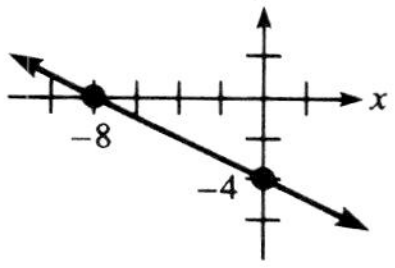

10. Solve the system.
$$-8x + 6y = 7$$
$$10x - 4y = -7 \quad \left(-\frac{1}{2}, \frac{1}{2}\right)$$
11. Solve the system.
$$2x - y + z = -5$$
$$x + 3y - z = 6$$
$$x - 2y - 2z = -3 \quad (-1,2,-1)$$
12. Simplify $\sqrt{48x^5y^3}$. Assume $x,y \geq 0$. $4x^2y\sqrt{3xy}$
13. Combine: $\sqrt{-36} + \sqrt{-25}$. $11i$
14. Rationalize the denominator: $\frac{1}{\sqrt{3}}$. $\frac{\sqrt{3}}{3}$
15. Solve $x^2 - 5x + 4 < 0$, and graph the solution set. $(1,4)$

16. Identify the graph of the equation $4x^2 - 9y^2 = 144$ as a circle, a parabola, an ellipse, or a hyperbola. Hyperbola
17. Determine the domain and range of the function whose graph is shown in the figure. $D: [-1,\infty)$; $R: [0,\infty)$

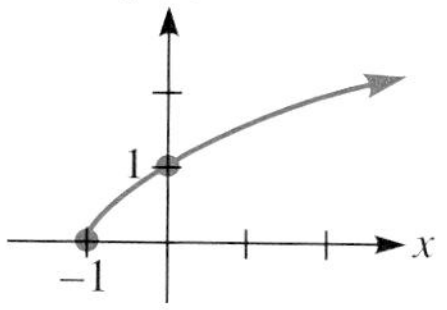

18. If $f(x) = x^2 - 1$ and $g(x) = -x$, find $2f(3) - 3g(-2)$. 10
19. Evaluate $\log_2 10$ to four significant digits. 3.322
20. Solve $9^x = \frac{1}{27}$. $\{-\frac{3}{2}\}$

CHAPTER 11

Sequences, Series, and the Binomial Theorem

Business equipment is often depreciated by using the **double-declining balance method.** Under this method, the value of the equipment at the beginning of each year (called the **book value**) is multiplied by a fixed percent to determine its value at the end of that year. When the estimated useful life for equipment is five years, then the equipment may be depreciated by this method at a rate of 40 percent each year. Use the double-declining balance method to find the depreciated book value at the end of five years for office equipment with a useful life of five years that is purchased for $20,000. (See Example 5 of Section 11.1.)

WHEN linear functions and exponential functions are restricted so that the independent variable is limited to positive integers, then the resulting functions are called arithmetic sequences and geometric sequences, respectively. This chapter focuses primarily on such sequences and the series (or sums of sequences) that are associated with them. Arithmetic and geometric series are just two examples of the many important types of series, and this chapter concludes by discussing the series contained in the binomial theorem.

11.1 Sequences

OBJECTIVES

1. **Find any term in a sequence when given a formula for the *n*th term of the sequence.**
2. **Find a formula for the general term a_n in a given arithmetic sequence.**
3. **Find a formula for the general term a_n in a given geometric sequence.**
4. **Determine if a sequence is an arithmetic sequence, a geometric sequence, or neither.**
5. **Solve applied problems involving sequences.**

1 A sequence of numbers is generated by listing numbers in a definite order. For example, the sequence of positive odd numbers is given by

$$1, 3, 5, 7, 9, \ldots .$$

Additionally, the sequence of year-end book values for the office equipment that is depreciating as described in the chapter-opening problem is given by

$$\$12{,}000, \$7{,}200, \$4{,}320, \$2{,}592, \$1{,}555.20$$

as shown in Example 5. Each number in a sequence is called a **term** of the sequence. A sequence with an infinite number of terms is called an **infinite sequence,** as illustrated by the sequence of positive odd numbers. When a sequence has a first and last term, like the sequence of book values above, then it is called a **finite sequence.**

In general form, we usually write a sequence as

$$a_1, a_2, a_3, \ldots, a_n, \ldots ,$$

where the subscript gives the term number and a_n represents the general or nth term. The concept of a function applies to a sequence of numbers because a correspondence exists that assigns to each term number exactly one term, as illustrated in Figure 11.1.

For students to interpret a sequence as a function, it is helpful to show a correspondence explicitly as in Figure 11.1.

Term Number	Term	Ordered Pair
1	a_1	$(1,a_1)$
2	a_2	$(2,a_2)$
.	.	.
.	.	.
.	.	.
n	a_n	(n,a_n)
.	.	.
.	.	.
.	.	.

Figure 11.1

To analyze a sequence, we usually work from a formula that defines the function. For example,

$$a(n) = 2n, \qquad n = 1, 2, 3, \ldots,$$

specifies the sequence of even positive integers. Note that a is a function name (just like f), and the domain of a is the set of positive integers. By substituting the positive integers for n in the given formula, we can generate the terms of the sequence, as shown in the following chart.

		Subscript Notation		Functional Notation		Terms of Sequence
1st term	=	a_1	=	$a(1)$	$= 2(1) =$	2
2nd term	=	a_2	=	$a(2)$	$= 2(2) =$	4
3rd term	=	a_3	=	$a(3)$	$= 2(3) =$	6
.		.		.	.	.
.		.		.	.	.
.		.		.	.	.
nth term	=	a_n	=	$a(n)$	$= 2(n) =$	$2n$
.		.		.	.	.
.		.		.	.	.
.		.		.	.	.

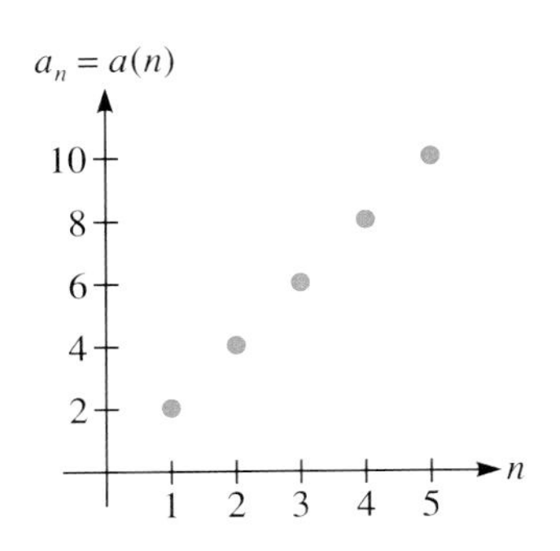

Figure 11.2

The discrete or unattached nature of a sequence is emphasized by the graph of function a shown in Figure 11.2. Observe that the graph consists of the isolated points

$$(1,2), (2,4), (3,6), (4,8), (5,10), \ldots$$

and that only a portion of the graph may be drawn since only a finite number of points may be plotted. With the aid of the function concept we can now state more formally the definition of a sequence.

> **Definition of a Sequence**
>
> A **sequence** is a function whose domain is the set of positive integers 1, 2, 3, ... The functional values or range elements are called the **terms** of the sequence.

Note that this definition follows the standard practice of interpreting *sequence* to mean *infinite sequence*. It is also conventional to write a formula that defines a sequence in terms of subscript notation, and this practice will be used throughout this chapter.

EXAMPLE 1 Write the first four terms of the sequence given by $a_n = 5n - 2$; also, find a_{50}.

Solution Substituting $n = 1, 2, 3, 4$ in the formula for a_n gives

$$a_1 = 5(1) - 2 = 3, \qquad a_2 = 5(2) - 2 = 8,$$
$$a_3 = 5(3) - 2 = 13, \qquad a_4 = 5(4) - 2 = 18.$$

To find the 50th term a_{50}, replace n by 50 to obtain

$$a_{50} = 5(50) - 2 = 248.$$

PROGRESS CHECK 1 Write the first four terms of the sequence given by $a_n = 3n + 5$; also, find a_{75}.

2 Two special types of sequences that have many applications are called arithmetic sequences and geometric sequences. To illustrate an arithmetic sequence, consider the sequence

$$4, 7, 10, 13, \ldots .$$

Observe that each term after the first may be found by adding 3 to the previous term, as shown below.

$$4, \underbrace{\ \ }_{3} 7, \underbrace{\ \ }_{3} 10, \underbrace{\ \ }_{3} 13, \ldots .$$

Therefore, the sequence is called an arithmetic sequence, and the number 3 is called the common difference in this sequence, as specified in the following definitions.

Arithmetic Sequence

An **arithmetic sequence** is a sequence of numbers in which each number after the first is found by adding a constant to the preceding term. This constant is called the **common difference** and is symbolized by d.

A formula for the general term in an arithmetic sequence with first term a_1 and common difference d may be found by observing the following pattern in such a sequence.

1st term,	2nd term,	3rd term,	. . . ,	nth term,	. . .
a_1,	$a_1 + 1d$,	$a_1 + 2d$,	. . . ,	$a_1 + (n - 1)d$,	. . .

Thus, the nth term is given by $a_1 + (n - 1)d$, and we have the following formula for a_n.

Formula for nth Term of an Arithmetic Sequence

The nth term of an arithmetic sequence is given by

$$a_n = a_1 + (n - 1)d,$$

where a_1 is the first term and d is the common difference.

EXAMPLE 2 Find a formula for the general term a_n in the arithmetic sequence

$$4, 7, 10, 13, \ldots .$$

What is the 28th term in the sequence?

Solution The first term is 4, so $a_1 = 4$. By subtracting any term from the next term, say a_1 from a_2, we obtain the common difference.

$$d = 7 - 4 = 3$$

Then a formula for a_n may be found as follows.

$$\begin{aligned} a_n &= a_1 + (n - 1)d \\ &= 4 + (n - 1)3 && \text{Replace } a_1 \text{ by 4 and } d \text{ by 3.} \\ &= 4 + 3n - 3 \\ &= 3n + 1 \end{aligned}$$

Progress Check Answer

1. 8, 11, 14, 17; 230

To find the 28th term, substitute 28 for n in this formula.

$$\begin{aligned} a_n &= 3n + 1 \\ a_{28} &= 3(28) + 1 \quad \text{Replace } n \text{ by 28.} \\ &= 85 \end{aligned}$$

PROGRESS CHECK 2 Find a formula for the general term a_n in the arithmetic sequence 6, 13, 20, 27, ... What is the 45th term in this sequence?

3 In an arithmetic sequence, the terms in the sequence maintain a *common difference.* When the terms in a sequence maintain a *common ratio,* then the sequence is called a geometric sequence, as given in the following definitions.

Geometric Sequence

A **geometric sequence** is a sequence of numbers in which each number after the first number is found by multiplying the preceding term by a constant. This constant is called the **common ratio** and is symbolized by r.

To illustrate, the following sequences are geometric sequences with common ratios r as shown.

$3, 9, 27, 81, \ldots \qquad r = \frac{9}{3} = \frac{27}{9} = \frac{81}{27} = 3$

$1, 1.05, (1.05)^2, (1.05)^3, \ldots \qquad r = \frac{1.05}{1} = \frac{(1.05)^2}{1.05} = \frac{(1.05)^3}{(1.05)^2} = 1.05$

A formula for the general term in a geometric sequence with first term a_1 and common ratio r may be found by observing the following pattern in such a sequence.

1st term,	2nd term,	3rd term,	. . . ,	nth term,	. . .
a_1,	a_1r^1,	a_1r^2,	. . . ,	a_1r^{n-1}	. . .

Thus, a_1r^{n-1} specifies the general term, and we have the following formula for a_n in a geometric sequence.

Formula for *n*th Term in a Geometric Sequence

The nth term in a geometric sequence is given by

$$a_n = a_1r^{n-1},$$

where a_1 is the first term and r is the common ratio.

EXAMPLE 3 Find a formula for the general term a_n in the geometric sequence

$$2, -1, \tfrac{1}{2}, -\tfrac{1}{4}, \ldots .$$

What is the 10th term in the sequence?

Solution The first term is 2, so $a_1 = 2$. By dividing any term into the next term, say a_2 into a_3, the common ratio is

$$r = \frac{\frac{1}{2}}{-1} = -\frac{1}{2}.$$

Progress Check Answer

2. $a_n = 7n - 1$; 314

It is instructive to have students confirm that the common ratio is $-\frac{1}{2}$ between all pairs of consecutive terms in this example.

Then replacing a_1 by 2 and r by $-\frac{1}{2}$ in $a_n = a_1 r^{n-1}$ yields

$$a_n = 2(-\tfrac{1}{2})^{n-1}.$$

To find the 10th term, substitute 10 for n in this formula to obtain

$$a_{10} = 2(-\tfrac{1}{2})^{10-1} = 2(-\tfrac{1}{2})^9 = 2(-\tfrac{1}{512}) = -\tfrac{1}{256}.$$

Note In this example, to check that the 10th term is $-\frac{1}{256}$, just generate successive terms by multiplying by the common ratio of $-\frac{1}{2}$, to obtain

$$2, -1, \tfrac{1}{2}, -\tfrac{1}{4}, \tfrac{1}{8}, -\tfrac{1}{16}, \tfrac{1}{32}, -\tfrac{1}{64}, \tfrac{1}{128}, -\tfrac{1}{256}, \ldots.$$

Writing the sequence in this form confirms that the 10th term is $-\frac{1}{256}$.

PROGRESS CHECK 3 Find a formula for the general term a_n in the geometric sequence $-9, 3, -1, \frac{1}{3}, \ldots$. What is the 8th term in the sequence?

4 When given a sequence of numbers, we must be able to classify the sequence as arithmetic, geometric, or neither, as considered in the next example.

The English economist Thomas Malthus in 1798 wrote a famous essay on population in which he claimed that population increases geometrically while food increases arithmetically. Ask students to explain why this prediction is alarming.

EXAMPLE 4 State whether the sequence is an arithmetic sequence, a geometric sequence, or neither. For any arithmetic sequence, state the common difference and write the next two terms. For any geometric sequence, state the common ratio and write the next two terms.

a. 2, 6, 18, 54, . . .
b. 1, 8, 15, 22, . . .
c. 1, 4, 9, 16, . . .

Solution

a. Each term after the first is found by multiplying the preceding term by 3, as illustrated below.

$$2, \quad 6, \quad 18, \quad 54, \ldots$$

$$2 \cdot 3 = 6 \quad 6 \cdot 3 = 18 \quad 18 \cdot 3 = 54$$

Therefore, the sequence is geometric with common ratio 3. The next two terms in the sequence are 162 and 486.

b. Each term after the first is found by adding 7 to the preceding term, as shown next.

$$1, \quad 8, \quad 15, \quad 22, \ldots$$

$$1 + 7 = 8 \quad 8 + 7 = 15 \quad 15 + 7 = 22$$

Therefore, the sequence is arithmetic with common difference 7. The next two terms in the sequence are 29 and 36.

c. The terms in this sequence do not have a common difference and they do not have a common ratio. Therefore, the sequence is neither arithmetic nor geometric.

Consider asking students to write a formula for the general term in each sequence. In this example, $a_n = 2(3)^{n-1}$, $7n - 6$, and n^2, respectively.

PROGRESS CHECK 4 Answer the questions in Example 4 for each sequence.

a. 1.125, 1.25, 1.375, 1.5, . . .
b. $1, \frac{1}{2}, \frac{1}{3}, \frac{1}{4}, \ldots$
c. 27, 9, 3, 1, . . .

5 The mathematics of finance contains many applications of sequences. One such application is considered in the chapter-opening problem.

Progress Check Answers

3. $a_n = -9(-\frac{1}{3})^{n-1}$; $\frac{1}{243}$

4. (a) Arithmetic; 0.125; 1.625, 1.75 (b) Neither (c) Geometric; $\frac{1}{3}$; $\frac{1}{3}, \frac{1}{9}$

The allowable depreciation rate is equal to double the straight-line rate. For this example, (100 percent ÷ 5) · 2 = 40 percent. In actual practice, the depreciation schedule would be modified based on the Tax Reform Act of 1986 to incorporate the *half-year convention*, which assumes that the property was purchased in midyear. For more details, see a business math text.

EXAMPLE 5 Solve the problem in the chapter introduction on page 532.

Solution If the office equipment depreciates 40 percent each year, at the end of the year this equipment is worth 60 percent of the book value at which it began the year. Thus, the sequence of year-end book values forms a geometric sequence with common ratio 0.6, and the terms in this sequence are computed as shown in the following table.

Year	Book Value at Beginning of Year	Book Value at End of Year
1	20,000	20,000(0.6) = 12,000
2	12,000	12,000(0.6) = 7,200
3	7,200	7,200(0.6) = 4,320
4	4,320	4,320(0.6) = 2,592
5	2,592	2,592(0.6) = 1,555.20

This depreciation schedule shows that the book value of the equipment at the end of five years is $1,555.20.

Note The book value at the end of five years may also be found using $a_n = a_1r^{n-1}$. Substituting $a_1 = 12{,}000$, $r = 0.6$, and $n = 5$ in this formula gives

$$a_5 = 12{,}000(0.6)^{5-1} = 1{,}555.20.$$

PROGRESS CHECK 5 Use the double-declining balance method to find the depreciated book value at the end of 10 years for property with a useful life of 10 years that is purchased for $100,000. The allowable depreciation rate in this case is 20 percent each year.

Progress Check Answer

5. $10,737.42

EXERCISES 11.1

In Exercises 1–8, use the formula for the nth term to write the first four terms of the given arithmetic sequence; also, find the term indicated.

1. $a_n = 2n - 5$; a_{10} $-3, -1, 1, 3; 15$
2. $a_n = 3n + 2$; a_{20} $5, 8, 11, 14; 62$
3. $a_n = 6n + 8$; a_{15} $14, 20, 26, 32; 98$
4. $a_n = 7n - 3$; a_{25} $4, 11, 18, 25; 172$
5. $a_n = \frac{1}{2}n + 2$; a_7 $\frac{5}{2}, 3, \frac{7}{2}, 4; \frac{11}{2}$
6. $a_n = \frac{2}{3}n - 5$; a_{12} $-\frac{13}{3}, -\frac{11}{3}, -3, -\frac{7}{3}; 3$
7. $a_n = \frac{3}{5}n - \frac{2}{5}$; a_{25} $\frac{1}{5}, \frac{4}{5}, \frac{7}{5}, 2; \frac{73}{5}$
8. $a_n = \frac{2}{3}n - \frac{1}{6}$; a_{18} $\frac{1}{2}, \frac{7}{6}, \frac{11}{6}, \frac{5}{2}; \frac{71}{6}$

In Exercises 9–18, find a formula for the general term a_n in each of the following arithmetic sequences. Also, find the indicated term for each sequence.

9. 1, 5, 9, 13, . . . ; a_{12} $4n - 3; 45$
10. 1, 6, 11, 16, . . . ; a_{15} $5n - 4; 71$
11. 2, 8, 14 ,20, . . . ; a_{20} $6n - 4; 116$
12. 3, 7, 11, 15, . . . ; a_{17} $4n - 1; 67$
13. 5, 12, 19, 26, . . . ; a_{27} $7n - 2; 187$
14. 3, 13, 23, 33, . . . ; a_{16} $10n - 7; 153$
15. 4, 13, 22, 31, . . . ; a_{40} $9n - 5; 355$
16. 8, $9\frac{1}{2}$, 11, $12\frac{1}{2}$, . . . ; a_{11} $\frac{3}{2}n + \frac{13}{2}; 23$
17. 12, $13\frac{1}{3}$, $14\frac{2}{3}$, 16, . . . ; a_{16} $\frac{4}{3}n + \frac{32}{3}; 32$
18. 6, 17, 28, 39, . . . ; a_{20} $11n - 5; 215$

In Exercises 19–28, find a formula for the general term a_n in the given geometric sequence. Also, find the indicated term in each sequence.

19. 2, 4, 8, 16, . . . ; a_{10} $a_n = 2(2)^{n-1}; 1{,}024$
20. 5, 10, 20, 40, . . . ; a_{10} $a_n = 5(2)^{n-1}; 2{,}560$
21. 3, 15, 75, 375, . . . ; a_7 $a_n = 3(5)^{n-1}; 46{,}875$
22. 2, 8, 32, 128, . . . ; a_8 $a_n = 2(4)^{n-1}; 32{,}768$
23. 7, 21, 63, 189, . . . ; a_9 $a_n = 7(3)^{n-1}; 45{,}927$
24. 3, −1, $\frac{1}{3}$, $-\frac{1}{9}$, . . . ; a_{10} $a_n = 3(-\frac{1}{3})^{n-1}; -\frac{1}{6{,}561}$
25. 5, −1, $\frac{1}{5}$, $-\frac{1}{25}$, . . . ; a_7 $a_n = 5(-\frac{1}{5})^{n-1}; \frac{1}{3{,}125}$
26. $\frac{1}{3}$, $\frac{1}{9}$, $\frac{1}{27}$, $\frac{1}{81}$, . . . ; a_8 $a_n = \frac{1}{3}(\frac{1}{3})^{n-1}; \frac{1}{6{,}561}$
27. $\frac{1}{2}$, $-\frac{1}{4}$, $\frac{1}{8}$, $-\frac{1}{16}$, . . . ; a_{10} $a_n = \frac{1}{2}(-\frac{1}{2})^{n-1}; -\frac{1}{1{,}024}$
28. $\frac{1}{5}$, $-\frac{2}{15}$, $\frac{4}{45}$, $-\frac{8}{135}$, . . . ; a_6 $a_n = \frac{1}{5}(-\frac{2}{3})^{n-1}; -\frac{32}{1{,}215}$

In Exercises 29–38, state whether the sequence is an arithmetic sequence, a geometric sequence, or neither. For any arithmetic sequence, state the common difference and write the next two terms. For any geometric sequence, state the common ratio and write the next two terms.

29. 1, 4, 16, 64, . . . Geometric, $r = 4$; 256, 1,024
30. 2, 7, 12, 17, . . . Arithmetic, $d = 5$; 22, 27
31. 3, 5, 8, 11, . . . Neither
32. 3, 12, 48, 192, . . . Geometric, $r = 4$; 768, 3,072
33. 5, 15, 45, 135, . . . Geometric, $r = 3$; 405, 1,215

34. 7, 21, 28, 35, . . . Neither

35. 7, 13, 19, 25, . . . Arithmetic, $d = 6$; 31, 37

36. $2, \frac{2}{3}, \frac{2}{9}, \frac{2}{27}, \ldots$ Geometric, $r = \frac{1}{3}$; $\frac{2}{81}, \frac{2}{243}$

37. $\frac{1}{5}, \frac{1}{10}, \frac{1}{15}, \frac{1}{20}, \ldots$ Neither

38. $\frac{2}{3}, \frac{5}{3}, \frac{8}{3}, \frac{11}{3}, \ldots$ Arithmetic, $d = 1$; $\frac{14}{3}, \frac{17}{3}$

In Exercises 39–40, use the double-declining balance method from the problem which opens the chapter.

39. Find the depreciated book value at the end of five years for office equipment with a useful life of five years that is purchased originally for $8,000. It depreciates 40 percent per year. What kind of sequence is the yearly book value? $622.08; geometric

40. Find the depreciated book value at the end of 10 years for office equipment with a useful life of 10 years that is purchased originally for $60,000. It depreciates 20 percent per year. What kind of sequence is the yearly book value? $6,442.45; geometric

41. An amount of $1,000 is invested at 8 percent annual interest, payable on the anniversary of the deposit. The formula $a_n = 1{,}000(1.08)^n$ gives the value of the deposit after n complete years. Find the value of a_n for each of years 1, 2, 3, 4, 5. What kind of sequence is this? 1,080, 1,166.40, 1,259.71, 1,360.49, 1,469.33; geometric

42. An amount of $2,000 is invested at 6 percent annual interest, payable on the anniversary of the deposit. The formula $a_n = 2{,}000(1.06)^n$ gives the value of the deposit after n complete years. Find the value of a_n for each of years 1, 2, 3, 4, 5. What kind of sequence is this? 2,120, 2,247.2, 2,382.03, 2,524.95, 2,676.45; geometric

43. Sequences are important in the theory of limits in calculus, where the object is to see if the terms of a sequence are approaching some fixed value. When this happens, that value is called the *limit* of the sequence. In these examples, write the first five terms of the given sequence, and guess the limit.

a. $a_n = \dfrac{n}{n+1}$ **b.** $a_n = \dfrac{1}{n}$

c. $a_n = \dfrac{1}{n^3 - 5n^2 + 11n - 6}$ **d.** $a_n = \dfrac{1}{n^2}$

e. $a_n = 2 - \dfrac{1}{3^n}$ **f.** $a_n = 1 - 0.6^n$

a. $\frac{1}{2}, \frac{2}{3}, \frac{3}{4}, \frac{4}{5}, \frac{5}{6}$; 1 b. $1, \frac{1}{2}, \frac{1}{3}, \frac{1}{4}, \frac{1}{5}$; 0 c. $1, \frac{1}{4}, \frac{1}{9}, \frac{1}{22}, \frac{1}{49}$; 0

d. $1, \frac{1}{4}, \frac{1}{9}, \frac{1}{16}, \frac{1}{25}$; 0 e. $\frac{5}{3}, \frac{17}{9}, \frac{53}{27}, \frac{161}{81}, \frac{485}{243}$; 2 f. 0.4, 0.64, 0.784, 0.8704, 0.92224; 1

44. Here is a sequence, called the Galileo sequence, with a clear pattern.

$$\frac{1}{3}, \frac{1+3}{5+7}, \frac{1+3+5}{7+9+11}, \ldots$$

a. Follow the given pattern and write the next term.

b. Simplify each term as much as possible to notice a remarkable result.

a. $\dfrac{1+3+5+7}{9+11+13+15}$

b. All terms equal $\frac{1}{3}$.

45. Some sequences are clearly defined even though there is no obvious formula. For instance, what are the first five terms of the sequence of prime numbers? That is, a_n = nth prime number. Is this sequence arithmetic, geometric, or neither? 2, 3, 5, 7, 11; neither

46. Prime numbers are positive integers that have exactly two factors. Thus, 3 is prime because it has exactly two factors, namely, 3 and 1. In a similar way, we can construct a sequence of positive integers that have exactly three factors.

	Factors
$a_1 = 4$	1, 2, 4
$a_2 = 9$	1, 3, 9
$a_3 = 25$	1, 5, 25

What is the next number in this sequence? Is this sequence arithmetic, geometric, or neither? 49; neither

47. A sequence whose next term is found by adding the two preceding terms is called a Fibonacci sequence. For instance, if $a_1 = 1$ and $a_2 = 1$, then $a_3 = 2$ and $a_4 = 3$. Find a_5 and a_6. Is this sequence arithmetic, geometric, or neither? 5, 8; neither

48. Find the next two terms of this sequence of logarithms. Is the sequence arithmetic, geometric, or neither? $a_1 = \log_2 2$, $a_2 = \log_2 4$, $a_3 = \log_2 8$, $a_4 = \log_2 16$ $\log_2 32$, $\log_2 64$; arithmetic ($d = 1$)

THINK ABOUT IT

1. Many sequences have been discovered whose terms approach irrational numbers. Here are two which you can investigate. For each sequence, (a) write out the fourth term, and (b) find the decimal value of the fourth term to the nearest thousandth, and compare it with the value of the irrational number as approximated by your calculator.

π: $\quad 4 \cdot 1,\ 4 \cdot (1 - \frac{1}{3}),\ 4 \cdot (1 - \frac{1}{3} + \frac{1}{5}), \ldots$

$\sqrt{2}$: $\quad 1 + \frac{1}{2},\ 1 + \dfrac{1}{2 + \frac{1}{2}},\ 1 + \dfrac{1}{2 + \dfrac{1}{2 + \frac{1}{2}}}$

2. If the first three terms in a geometric sequence are b^2, b^x, and b^8, find x.

3. Is 4,000 a term in the arithmetic sequence 2, 5, 8, . . . ? Explain your answer.

4. a. If a, b, and c are the first three terms in an arithmetic sequence, express b in terms of a and c.

b. If a, b, and c are the first three terms in a geometric sequence, express c in terms of a and b.

5. In an arithmetic sequence $a_6 = 15$ and $a_{12} = 24$. Find a_1 and d.

REMEMBER THIS

1. Find $p + q$ in simplest form, where $p = a_1 + (a_1 + d)$ and $q = a_1 + (a_1 - d)$. $4a_1$
2. Simplify $a_1 + [a_1 + (n - 1)d]$. $2a_1 + (n - 1)d$
3. Solve $S_n(1 - r) = a_1(1 - r^n)$ for S_n. $S_n = \frac{a_1(1 - r^n)}{1 - r}$
4. For what value of r is $\frac{1 - r^n}{1 - r}$ undefined? 1
5. Evaluate $\frac{4{,}320(1 - 1.08^4)}{1 - 1.08}$ to two decimal places. 19,466.40
6. Express in logarithmic form: $b^4 = 10$. $\log_b 10 = 4$
7. Express as a single logarithm with coefficient 1: $3 \log_4 x + \log_4 5$. $\log_4(5x^3)$
8. Solve $e^{0.3t} = 0.75$. Give the solution to four significant digits. $\{-0.9589\}$
9. Solve $b = a + at$ for a. $a = \frac{b}{1 + t}$
10. Graph $2x - 3y = 6$.

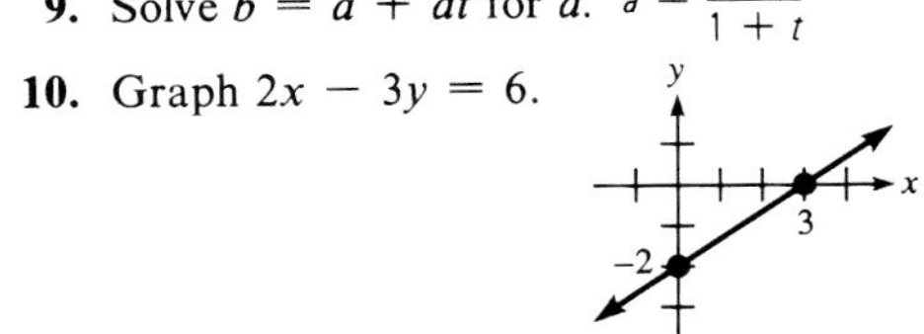

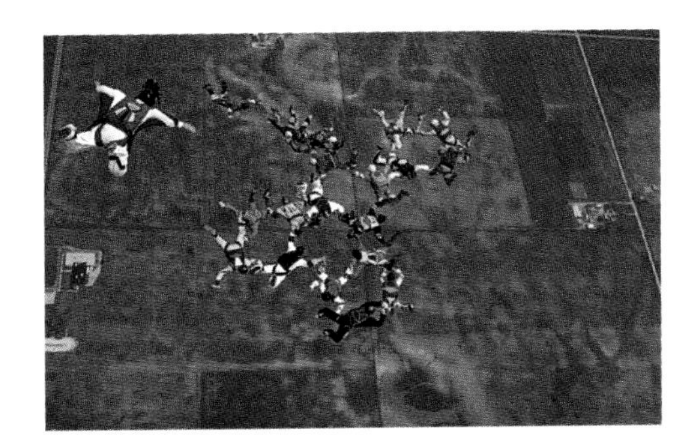

11.2 Series

A free-falling body that starts from rest drops about 16 ft the first second, 48 ft the second second, 80 ft the third second, and so on. About how many feet does a parachutist drop during the first 10 seconds of free-fall? (See Example 7.)

OBJECTIVES

1. **Find the sum of an indicated number of terms in an arithmetic sequence.**
2. **Find the sum of an indicated number of terms in a geometric sequence.**
3. **Write a series given in sigma notation in its expanded form, and determine the sum.**
4. **Write a series given in expanded form using sigma notation.**
5. **Solve applied problems involving series.**

Students may enjoy the following anecdote about the mathematician Carl Friedrich Gauss (1777–1855). When Gauss was nine years old, his teacher assigned this problem as "busy work" and Gauss wrote the answer (5,050) on his slate practically as soon as the teacher finished explaining the problem. His method was to mentally arrange the 100 values in 50 pairs that all add up to 101 from which the sum is easily seen to be 50(101) = 5,050. The teacher was so impressed that he arranged for Gauss to have special textbooks and tutoring and to then be admitted to a secondary school.

1 Associated with any sequence

$$a_1, a_2, \ldots, a_n, \ldots$$

is a series

$$a_1 + a_2 + \cdots + a_n + \cdots,$$

which is the sum of all the terms in the sequence. In this section we consider only series that are associated with finite sequences. In Section 11.3 a special type of infinite series will be discussed.

The series associated with an arithmetic sequence is called an **arithmetic series.** To illustrate a method for finding such a sum, consider the problem of finding the sum of the first 100 positive integers. This series can be written as

$$S = 1 + 2 + 3 + \cdots + 98 + 99 + 100.$$

By reversing the order of the terms, we can also write this series as

$$S = 100 + 99 + 98 + \cdots + 3 + 2 + 1.$$

If we now add, term by term, the two equivalent expressions for S, we have 100 pairs, which all add up to 101. So

$$2S = 100(101) \quad \text{and} \quad S = \frac{100(101)}{2} = 5{,}050.$$

Thus, the sum of the first 100 positive integers is 5,050.

By applying the above method to an arithmetic series in general form, we can derive a formula for the sum of an arithmetic series. The sum of the first n terms in an arithmetic sequence can be written as

$$S_n = a_1 + (a_1 + d) + (a_1 + 2d) + \cdots + (a_n - 2d) + (a_n - d) + a_n.$$

By reversing the order of the terms, we can also write this series as

$$S_n = a_n + (a_n - d) + (a_n - 2d) + \cdots + (a_1 + 2d) + (a_1 + d) + a_1.$$

Adding term by term the two equivalent expressions for S_n gives n pairs, which all add up to $a_1 + a_n$. So

$$2S_n = n(a_1 + a_n) \quad \text{and} \quad S_n = \frac{n}{2}(a_1 + a_n).$$

Another version is $S_n = \left(\frac{a_1 + a_n}{2}\right)n$, which points out that the series is the sum of n terms all equal to the average of the first and last terms.

This formula is used when n, a_1, and a_n are known. However, d is often known in place of a_n. To derive an alternative formula, replace a_n in the above formula by $a_1 + (n - 1)d$ to get

$$S_n = \frac{n}{2}\{a_1 + [a_1 + (n - 1)d]\}$$
$$= \frac{n}{2}[2a_1 + (n - 1)d].$$

In summary, the following formulas have been derived.

Arithmetic Series Formulas

The sum of the first n terms of an arithmetic series is given by

$$S_n = \frac{n}{2}(a_1 + a_n)$$

or

$$S_n = \frac{n}{2}[2a_1 + (n - 1)d].$$

EXAMPLE 1 Find the sum of the first 250 positive integers.

Solution The problem is to evaluate

$$S_{250} = 1 + 2 + 3 + \cdots + 250.$$

Because a_1, a_n, and n are known, use the top formula in the box above.

$$S_n = \frac{n}{2}(a_1 + a_n)$$
$$S_{250} = \frac{250}{2}(1 + 250) \quad \text{Let } n = 250,\ a_1 = 1, \text{ and } a_n = 250.$$
$$= 31{,}375$$

The sum of the first 250 positive integers is 31,375.

PROGRESS CHECK 1 Find the sum of the first 180 positive integers.

EXAMPLE 2 Find the sum of the first 25 terms of the arithmetic sequence 6, 13, 20, ...

Solution For the given arithmetic sequence, $a_1 = 6$ and $d = 7$. To find the sum of the first 25 terms, let $n = 25$ and use the second formula in the box above.

$$S_n = \frac{n}{2}[2a_1 + (n - 1)d]$$

$$S_{25} = \frac{25}{2}[2(6) + (25 - 1)7] \quad \text{Let } n = 25, a_1 = 6, \text{ and } d = 7.$$

$$= \frac{25}{2}(180) = 2{,}250$$

PROGRESS CHECK 2 Find the sum of the first 35 terms of the arithmetic sequence 4, 7, 10, ...

2 The series associated with a geometric sequence is called a **geometric series,** and in general form, a geometric series with n terms can be written as

$$S_n = a_1 + a_1r + a_1r^2 + \cdots + a_1r^{n-1}. \tag{1}$$

To derive a formula for the sum in a geometric series, first multiply both sides of equation (1) by r to obtain

$$rS_n = a_1r + a_1r^2 + \cdots + a_1r^{n-1} + a_1r^n. \tag{2}$$

Then subtracting equation (2) from equation (1) yields

$$S_n - rS_n = a_1 - a_1r^n$$

$$S_n(1 - r) = a_1(1 - r^n)$$

$$S_n = \frac{a_1(1 - r^n)}{1 - r}, \quad \text{for} \quad r \neq 1.$$

The result is a formula for S_n when a_1, r, and n are known.

It is also common to write this formula as $S_n = \dfrac{a_1(r^n - 1)}{r - 1}$. Consider asking students to show that these formulas are equivalent. Also consider assigning "Think About It" Exercise 4, which asks students to develop a formula for S_n in terms of a_1, a_n, and r.

Geometric Series Formula

The sum of the first n terms of a geometric series with $r \neq 1$ is given by

$$S_n = \frac{a_1(1 - r^n)}{1 - r}.$$

EXAMPLE 3 Find the sum of the first eight terms of the geometric sequence 2, 6, 18, ...

Solution Here $a_1 = 2$, $r = 3$, and $n = 8$. Then,

$$S_n = \frac{a_1(1 - r^n)}{1 - r}$$

$$S_8 = \frac{2(1 - 3^8)}{1 - 3} \quad \text{Let } a_1 = 2, r = 3, \text{ and } n = 8.$$

$$= \frac{2(1 - 6{,}561)}{-2}$$

$$= 6{,}560.$$

Progress Check Answers

1. 16,290
2. 1,925

PROGRESS CHECK 3 Find the sum of the first seven terms of the geometric sequence 3, 6, 12, ...

3 The Greek letter Σ, read "sigma," is commonly used to write a series in compact form. By this convention, called **sigma notation,**

$$S_n = a_1 + a_2 + \cdots + a_n$$

is abbreviated as

$$S_n = \sum_{i=1}^{n} a_i,$$

so that $\sum_{i=1}^{n} a_i$ means to add the terms that result from replacing i by 1, then 2, . . . , then n. The letter i in this notation is called the **index of summation,** and the choice of this letter is arbitrary. For instance,

$$\sum_{i=1}^{n} a_i, \qquad \sum_{j=1}^{n} a_j, \qquad \sum_{k=1}^{n} a_k$$

all represent the same series.

Advise students that a series can be written in sigma notation in more than one way. See also the note accompanying Example 6. Some students may need to be cautioned that i is an arbitrary index symbol here, not a complex number.

EXAMPLE 4 Write the series $\sum_{i=1}^{4}(i^2 - 1)$ in expanded form, and determine the sum.

Solution

$$\begin{aligned}\sum_{i=1}^{4}(i^2 - 1) &= (1^2 - 1) + (2^2 - 1) + (3^2 - 1) + (4^2 - 1)\\ &= 0 + 3 + 8 + 15\\ &= 26\end{aligned}$$

PROGRESS CHECK 4 Write the series $\sum_{i=1}^{5}(3i + 2)$ in expanded form, and determine the sum.

EXAMPLE 5 Write the series $\sum_{i=1}^{5} 4(\frac{1}{2})^i$ in expanded form, and determine the sum.

Solution In expanded form the series is

$$\sum_{i=1}^{5} 4(\tfrac{1}{2})^i = 4(\tfrac{1}{2})^1 + 4(\tfrac{1}{2})^2 + 4(\tfrac{1}{2})^3 + 4(\tfrac{1}{2})^4 + 4(\tfrac{1}{2})^5.$$

This series is a geometric series with $a_1 = 4(\frac{1}{2}) = 2$, $r = \frac{1}{2}$, and $n = 5$. So

$$S_n = \frac{a_1(1 - r^n)}{1 - r}$$

$$S_5 = \frac{2[1 - (\frac{1}{2})^5]}{1 - \frac{1}{2}} \qquad \text{Let } a_1 = 2,\ r = \tfrac{1}{2},\ \text{and } n = 5.$$

$$= \frac{2[1 - \frac{1}{32}]}{\frac{1}{2}} = \frac{\frac{31}{16}}{\frac{1}{2}} = \frac{31}{8}.$$

Therefore, $\sum_{i=1}^{5} 4(\frac{1}{2})^i = \frac{31}{8}$.

PROGRESS CHECK 5 Write the series $\sum_{i=1}^{4} 6(\frac{1}{2})^i$ in expanded form, and determine the sum.

Progress Check Answers

3. 381
4. $5 + 8 + 11 + 14 + 17 = 55$
5. $6(\frac{1}{2}) + 6(\frac{1}{2})^2 + 6(\frac{1}{2})^3 + 6(\frac{1}{2})^4 = \frac{45}{8}$

4 When we need to convert a series from expanded form to compact form using sigma notation, the main obstacle is finding a formula for the general term, as illustrated in the next example.

EXAMPLE 6 Find an expression for the general term and write the series $6 + 11 + 16 + 21 + 26 + 31 + 36$ in sigma notation.

Solution In many cases we hope to be able to predict the general term by inspection. For this series a_i is given by

$$5i + 1$$

and since there are seven terms, we have

$$\sum_{i=1}^{7}(5i + 1) = 6 + 11 + 16 + 21 + 26 + 31 + 36.$$

If you had difficulty predicting the general term, you could note in this case that the series is arithmetic with $a_1 = 6$ and $d = 5$ so that

$$\begin{aligned} a_n &= a_1 + (n - 1)d \\ &= 6 + (n - 1)5 \\ &= 5n + 1. \end{aligned}$$

Note In sigma notation it is sometimes convenient to begin the index of summation at a number other than 1. For instance, the series in this example can also be represented by

$$\sum_{i=0}^{6}(5i + 6).$$

Keep in mind that a series can be represented in sigma notation in more than one way.

PROGRESS CHECK 6 Find an expression for the general term and write the series $7 + 13 + 19 + \cdots + 49$ in sigma notation.

5 The problem that opens this section illustrates an application of an arithmetic series.

EXAMPLE 7 Solve the problem in the section introduction on page 540.

Solution The drop in feet for the parachutist during the first 10 seconds of the free-fall is given by the sum of the first 10 terms of the sequence

$$16, 48, 80, \ldots .$$

This sequence is arithmetic with $a_1 = 16$ and $d = 32$. Let $n = 10$ and use the formula that gives S_n in terms of a_1, d, and n.

$$\begin{aligned} S_n &= \frac{n}{2}[2a_1 + (n - 1)d] \\ &= \frac{10}{2}[2(16) + (10 - 1)32] \quad \text{Let } n = 10,\ a_1 = 16, \text{ and } d = 32. \\ &= \frac{10}{2}(320) = 1{,}600 \end{aligned}$$

Thus, a parachutist drops about 1,600 ft during the first 10 seconds of free-fall.

Note also that this answer is $f(10)$, where $f(t) = 16t^2$, which is the formula for the distance covered by a freely falling body with an initial velocity of zero.

PROGRESS CHECK 7 From the sequence given in Example 7, about how many feet does a parachutist drop during the first 15 seconds of free-fall?

Progress Check Answers

6. $a_i = 6i + 1$; $\sum_{i=1}^{8}(6i + 1)$ 7. 3,600 ft

The next example shows an important type of application associated with geometric series.

EXAMPLE 8 To help finance their daughter's college education, a couple invests \$4,000 on her birthday each year, starting with her 13th birthday. If the money is placed in an account that pays 8 percent interest compounded annually, how much is in this account on her 17th birthday? Assume that the last deposit is made on her 16th birthday.

Solution Observe that four deposits will be made corresponding to the daughter's 13th, 14th, 15th, and 16th birthdays. Determine separately the compounded amount of each \$4,000 deposit using the formula $A = P(1 + r)^t$ from Section 10.1.

Birthday	Deposit	Interest Period	Compounded Amount
13th	1st	4 years	$4{,}000(1 + 0.08)^4$
14th	2nd	3 years	$4{,}000(1 + 0.08)^3$
15th	3rd	2 years	$4{,}000(1 + 0.08)^2$
16th	4th	1 year	$4{,}000(1 + 0.08)^1$

The amount S in the account at the end of four years is the sum of the four compounded amounts, so

$$S = 4{,}000(1.08) + 4{,}000(1.08)^2 + 4{,}000(1.08)^3 + 4{,}000(1.08)^4.$$

This series is a geometric series with $a_1 = 4{,}000(1.08) = 4{,}320$, $r = 1.08$, and $n = 4$. Thus,

$$S_n = \frac{a_1(1 - r^n)}{1 - r}$$

$$S_4 = \frac{4{,}320(1 - 1.08^4)}{1 - 1.08} \approx 19{,}466.40 \text{ (by calculator)}.$$

The daughter's college fund is worth \$19,466.40 on her 17th birthday.

An **annuity** is any series of equal payments made at equal time intervals. Example 8 illustrates an **annuity due,** in which payments are made at the *beginning* of each time period that coincides with a point at which interest is converted. The sum S is called the **future value** of the annuity or the **amount of the annuity.** The future value of an **ordinary annuity** (payments are made at the *end* of each interval) is considered in "Think About It" Exercise 5.

PROGRESS CHECK 8 Redo the problem in Example 8, but assume that \$3,000 is deposited each year in an account that pays 7 percent interest compounded annually, and that deposits are started with the daughter's 7th birthday.

Progress Check Answer

8. \$44,350.80

EXERCISES 11.2

In Exercises 1–10, find the sum of the first n positive integers.

1. $n = 6$ 21
2. $n = 8$ 36
3. $n = 12$ 78
4. $n = 15$ 120
5. $n = 18$ 171
6. $n = 20$ 210
7. $n = 125$ 7,875
8. $n = 250$ 31,375
9. $n = 300$ 45,150
10. $n = 500$ 125,250

In Exercises 11–20, find the sum of the first n terms of the given arithmetic sequence.

11. $n = 8$; 1, 4, 7, . . . 92
12. $n = 12$; 2, 6, 10, . . . 288
13. $n = 15$; 5, 10, 15, . . . 600
14. $n = 20$; 3, 9, 15, . . . 1,200
15. $n = 22$; 1, 8, 15, . . . 1,639
16. $n = 37$; 6, 9, 12, . . . 2,220
17. $n = 48$; 14, 17, 20, . . . 4,056
18. $n = 78$; 1, 5, 9, . . . 12,090
19. $n = 138$; 2, 7, 12, . . . 47,541
20. $n = 537$; 1, 3, 5, . . . 288,369

In Exercises 21–30, find the sum of the first n terms of the given geometric sequence.

21. $n = 5$; 3, 9, 27 363
22. $n = 6$; 3, 15, 75 11,718
23. $n = 10$; 2, 8, 32 699,050
24. $n = 9$; 2, 10, 50 976,562
25. $n = 7$; 3, 6, 12 381
26. $n = 8$; 3, 12, 48 65,535
27. $n = 11$; 1, 3, 9 88,573
28. $n = 12$; 1, 5, 25 61,035,156
29. $n = 15$; 1, 2, 4 32,767
30. $n = 17$; 1, 4, 16 5,726,623,061

In Exercises 31–42, write the given series in expanded form and determine the sum.

31. $\sum_{i=1}^{5} 3i$ $3(1) + 3(2) + 3(3) + 3(4) + 3(5) = 45$

32. $\sum_{i=1}^{9} 2i$ $2(1) + 2(2) + 2(3) + 2(4) + 2(5) + 2(6) + 2(7) + 2(8) + 2(9) = 90$

33. $\sum_{i=0}^{4}(2i + 1)$

$[2(0) + 1] + [2(1) + 1] + [2(2) + 1] + [2(3) + 1] + [2(4) + 1] = 25$

34. $\sum_{i=0}^{3}(3i - 1)$ $[3(0) - 1] + [3(1) - 1] + [3(2) - 1] + [3(3) - 1] = 14$

35. $\sum_{i=1}^{6}(i^2 + 2)$

$(1^2 + 2) + (2^2 + 2) + (3^2 + 2) + (4^2 + 2) + (5^2 + 2) + (6^2 + 2) = 103$

36. $\sum_{i=1}^{5}(i^2 - 3)$ $(1^2 - 3) + (2^2 - 3) + (3^2 - 3) + (4^2 - 3) + (5^2 - 3) = 40$

37. $\sum_{i=2}^{4} 3^i$ $3^2 + 3^3 + 3^4 = 117$

38. $\sum_{i=3}^{8} 2^i$ $2^3 + 2^4 + 2^5 + 2^6 + 2^7 + 2^8 = 504$

39. $\sum_{i=1}^{4} 2^{i-2}$ $2^{1-2} + 2^{2-2} + 2^{3-2} + 2^{4-2} = 7\frac{1}{2}$

40. $\sum_{i=1}^{5} 3^{i+1}$ $3^{1+1} + 3^{2+1} + 3^{3+1} + 3^{4+1} + 3^{5+1} = 1{,}089$

41. $\sum_{i=1}^{4}(\frac{1}{3})^i$ $(\frac{1}{3})^1 + (\frac{1}{3})^2 + (\frac{1}{3})^3 + (\frac{1}{3})^4 = \frac{40}{81}$

42. $\sum_{i=1}^{5} 3(\frac{1}{2})^i$ $3(\frac{1}{2})^1 + 3(\frac{1}{2})^2 + 3(\frac{1}{2})^3 + 3(\frac{1}{2})^4 + 3(\frac{1}{2})^5 = \frac{93}{32}$

In Exercises 43–52, find an expression for the general term, and write the given series in sigma notation. (More than one form is possible.)

43. $6 + 12 + 18 + 24 + 30 + 36$ $a_i = 6i$; $\sum_{i=1}^{6} 6i$

44. $4 + 8 + 12 + 16 + 20 + 24$ $a_i = 4i$; $\sum_{i=1}^{6} 4i$

45. $1 + 4 + 7 + 10 + 13 + 16$ $a_i = 3i - 2$; $\sum_{i=1}^{6}(3i - 2)$

46. $-1 + 1 + 3 + 5 + 7$ $a_i = 2i - 3$; $\sum_{i=1}^{5}(2i - 3)$

47. $6 + 11 + 18 + 27 + 38$

(*Hint:* i^2 is needed.) $a_i = (i + 1)^2 + 2$; $\sum_{i=1}^{5}[(i + 1)^2 + 2]$

48. $6 + 13 + 22 + 33 + 46$ (*Hint:* i^2 is needed.)

$a_i = (i + 2)^2 - 3$; $\sum_{i=1}^{5}(i + 2)^2 - 3$

49. $1 + 3 + 9 + 27 + 81$ $a_i = 3^{i-1}$; $\sum_{i=1}^{5} 3^{i-1}$

50. $1 + 2 + 4 + 8$ $a_i = 2^{i-1}$; $\sum_{i=1}^{4} 2^{i-1}$

51. $3 + \frac{3}{4} + \frac{3}{16} + \frac{3}{64} + \frac{3}{128}$ $a_i = 3(\frac{1}{4})^{i-1}$; $\sum_{i=1}^{5} 3(\frac{1}{4})^{i-1}$

52. $\frac{2}{3} + \frac{2}{9} + \frac{2}{27} + \frac{2}{81} + \frac{2}{243}$ $a_i = 2(\frac{1}{3})^i$; $\sum_{i=1}^{5} 2(\frac{1}{3})^i$

53. In many winter celebrations such as Kwanza or Hanukkah an increasing number of candles are burned each day for a period of time. For instance, over the 8 days of Hanukkah 2, 3, 4, 5, 6, 7, 8, and then 9 candles are used. Express the total number of candles as a series and find its sum. Is the series arithmetic, geometric, or neither? $2 + 3 + 4 + \cdots + 9 = 44$; arithmetic, $d = 1$

54. In this famous riddle the correct answer is 1 (why?), but how many were coming *from* St. Ives? Express the answer as a series. Is it arithmetic, geometric, or neither?

> As I was going to St. Ives,
> I met a man with seven wives.
> Each wife had seven sacks,
> Each sack had seven cats,
> Each cat had seven kits:
> Kits, cats, sacks and wives,
> How many were going to St. Ives?

Only "I" was going *to* St. Ives; $1 + 7 + 7^2 + 7^3 + 7^4 = 2{,}801$ were coming *from* St. Ives; geometric, $r = 7$. But should you count the sacks? If not, subtract 49.

55. A person makes a purchase for \$6,000, paying \$1,500 down and agreeing to pay at the end of each year 8 percent interest on the unpaid balance plus an additional \$900 to reduce the principal.

a. Show that the loan will be paid off after five years.

b. Write out the five-term series which gives the total of the interest payments. What kind of series is this? What is the total amount of interest paid?

a. $(6{,}000 - 1{,}500)/900 = 5$ years

b. Arithmetic series with $d = -72$; $360 + 288 + 216 + 144 + 72 = \$1{,}080$

56. A person makes a purchase for \$6,000, paying \$1,200 down and agreeing to pay at the end of each year 8 percent interest on the unpaid balance plus an additional \$800 to reduce the principal.

a. Show that the loan will be paid off after six years.

b. Write out the six-term series which gives the total of the interest payments. What kind of series is this? What is the total amount of interest paid?

a. $(6{,}000 - 1{,}200)/800 = 6$ years

b. Arithmetic series with $d = -64$; $384 + 320 + 256 + 192 + 128 + 64 = \$1{,}344$

57. Refer to the figure and express the total area as the sum of individual rectangular areas. Is this series arithmetic, geometric, or neither? $5 + 4.5 + 4 + 3.5 + 3 + 2.5 = 22.5$; arithmetic with $d = -0.5$

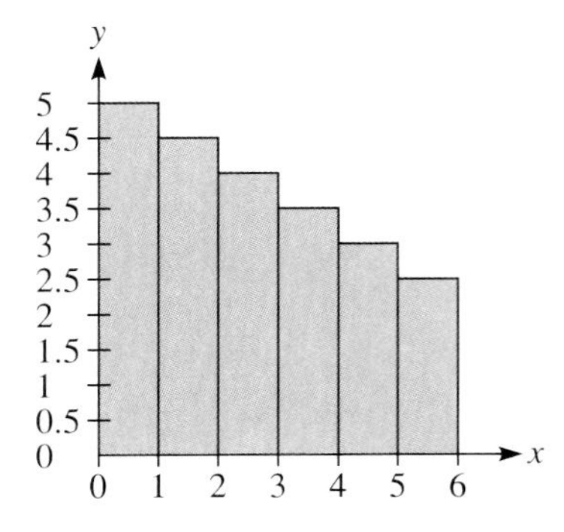

58. The area under the parabola $y = x^2$ may be estimated by the sum of the areas of the 10 rectangles shown in the figure. Write out the 10-term series which represents the sum of the areas of the rectangles and find the sum. (*Hint:* The base of each rectangle is 0.1; the height is given by x^2.) Is this series arithmetic, geometric, or neither?

$0.1(0.1^2) + 0.1(0.2^2) + \cdots + 0.1(1^2) = 0.385$; neither

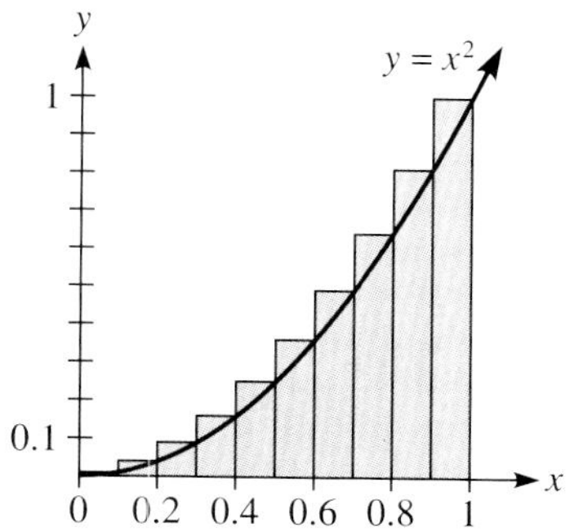

59. A young student is asked to memorize the "times" table for the digits from 0 to 9. How many products must be memorized? (Assume, for example, that 2×3 and 3×2 are just one item to be memorized.) Here is one way of listing the products. Write a series that gives the sum of the number of products in each row of the table.

$0 \times 0, 0 \times 1, 0 \times 2, 0 \times 3, 0 \times 4, \ldots, 0 \times 9$
$1 \times 1, 1 \times 2, 1 \times 3, 1 \times 4, \ldots, 1 \times 9$
$2 \times 2, 2 \times 3, 2 \times 4, \ldots, 2 \times 9$
$3 \times 3, 3 \times 4, \ldots, 3 \times 9$
.
.
.
9×9

$10 + 9 + 8 + 7 + \cdots + 1 = 55$

60. In the familiar base 10 number system we use 10 symbols for digits from 0 to 9. In the base 12 number system 12 symbols are used for digits: 0, 1, 2, 3, 4, 5, 6, 7, 8, 9, *, #. How many products must be memorized for the base 12 times table? Model the solution after the one in Exercise 59.

$12 + 11 + 10 + \cdots + 1 = 78$

61. A free-falling body that starts from rest drops about 16 ft the first second, 48 ft the second second, 80 ft the third second, and so on. About how many feet does an apple released from an airplane fall during the first 12 seconds of free-fall? 2,304 ft

62. Redo Exercise 61, but assume that the apple falls for 16 seconds. 4,096 ft

63. To help finance their son's college education, a couple invests \$1,000 on his birthday each year starting with his 5th birthday. If the money is placed in an account that pays 6 percent interest compounded annually, how much is in the account on his 18th birthday? Assume that the last payment is made on his 17th birthday. How much of the total is interest? \$20,015.07; \$7,015.07

64. Redo Exercise 63, but assume that they start investing on his first birthday. \$29,905.65; \$12,905.65

THINK ABOUT IT

1. **a.** Show that the sum of the first n positive odd integers is n^2.
   b. What is the sum of the first n positive even integers?
2. **a.** Find the sum of the first 10 terms of the sequence $3, 3\sqrt{3}, 9, \ldots$.
   b. Find the sum of the first 10 terms of the sequence $3^{-1}, 3^{-2}, 3^{-3}, \ldots$.
3. Show that $\sum_{i=1}^{n} ca_i = c \sum_{i=1}^{n} a_i$.
4. Start with the formula for the sum S_n of the first n terms of a geometric series given in this section and derive a formula for S_n in terms of a_1, a_n, and r.
5. The future value of an *ordinary annuity* is given by $S = p + p(1 + i) + p(1 + i)^2 + \cdots + p(1 + i)^{n-1}$. Find the sum of this series.

REMEMBER THIS

1. Calculate these sums: $\frac{1}{2} + \frac{1}{4}$; $\frac{1}{2} + \frac{1}{4} + \frac{1}{8}$; $\frac{1}{2} + \frac{1}{4} + \frac{1}{8} + \frac{1}{16}$. $\frac{3}{4}$; $\frac{7}{8}$; $\frac{15}{16}$
2. Graph the solution set of $|r| < 1$.

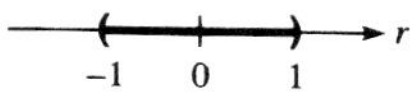

3. Simplify the complex fraction $\dfrac{5}{1 - \frac{1}{3}}$. $\frac{15}{2}$
4. Which is largest, $\dfrac{-3^0}{4}$, $\left(-\dfrac{3}{4}\right)^0$, or $-\left(\dfrac{3}{4}\right)^0$? $\left(-\dfrac{3}{4}\right)^0$
5. Put these in order of increasing magnitude: 0.46, 0.45, $0.\overline{45}$. 0.45, $0.\overline{45}$, 0.46
6. Find a formula for the general term of the sequence 5, 8, 11, 14, ... $a_n = 3n + 2$

7. Graph $y = \log_2 x + 2$.

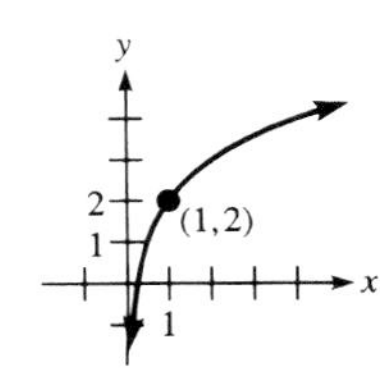

8. Solve $x^2 + x + 1 = 0$. $\frac{-1 \pm i\sqrt{3}}{2}$

9. Solve the system. $x + y = 10$
$x - y = 10$ (10,0)

10. Simplify $\left(\frac{6a^3b^4}{3a^5b}\right)^{-1}$. $\frac{a^2}{2b^3}$

11.3 Infinite Geometric Series

In a certain filtration system wastewater is collected in a holding tank so that each time a liquid is passed through this system only 75 percent as much wastewater is collected as in the previous pass. If 12 gal of wastewater are collected the first time a liquid is passed through this system, how many gallons of wastewater must the holding tank be able to accommodate? (See Example 4.)

OBJECTIVES

1. **Find the sum of certain infinite geometric series.**
2. **Use an infinite geometric series to express a repeating decimal as the ratio of two integers.**
3. **Solve applied problems involving infinite geometric series.**

1 A question that intrigued mathematicians for some time was whether it was possible to assign a number as the sum of certain infinite series in some meaningful way. To explore this problem, consider the series

$$\tfrac{1}{2} + \tfrac{1}{4} + \tfrac{1}{8} + \tfrac{1}{16} + \cdots + (\tfrac{1}{2})^n + \cdots$$

and examine the behavior of S_n (the sum of the first n terms in the series) as n increases from 1 to 6.

$$\begin{aligned}
S_1 &= \tfrac{1}{2}\\
S_2 &= \tfrac{1}{2} + \tfrac{1}{4} = \tfrac{3}{4}\\
S_3 &= \tfrac{1}{2} + \tfrac{1}{4} + \tfrac{1}{8} = \tfrac{7}{8}\\
S_4 &= \tfrac{1}{2} + \tfrac{1}{4} + \tfrac{1}{8} + \tfrac{1}{16} = \tfrac{15}{16}\\
S_5 &= \tfrac{1}{2} + \tfrac{1}{4} + \tfrac{1}{8} + \tfrac{1}{16} + \tfrac{1}{32} = \tfrac{31}{32}\\
S_6 &= \tfrac{1}{2} + \tfrac{1}{4} + \tfrac{1}{8} + \tfrac{1}{16} + \tfrac{1}{32} + \tfrac{1}{64} = \tfrac{63}{64}
\end{aligned}$$

It appears that as n gets larger, S_n approaches, but never equals, 1. In fact, we can get S_n as close to 1 as we wish merely by taking a sufficiently large value for n. It is in this sense that we are going to assign the number 1 as the "sum" of the series. That is, to say 1 is the sum of the above infinite series is to say that as n gets larger, S_n converges to or closes in on 1.

On an intuitive level, observe that if a series is to converge to a sum, then the terms in the series must approach 0 as n get larger. This condition occurs in infinite geometric series with $|r| < 1$ (or equivalently, with $-1 < r < 1$), so we will analyze this type of series. A formula for the sum of the first n terms of a geometric series is

$$S_n = \frac{a_1(1 - r^n)}{1 - r}.$$

Then if $|r| < 1$, r^n approaches 0 as n gets larger, so that

$$S_n \text{ converges to } \frac{a_1(1 - 0)}{1 - r} = \frac{a_1}{1 - r}.$$

This result establishes a useful formula.

Consider introducing limit notation and writing the condition as $\lim_{n \to \infty} a_n = 0$. This condition is a necessary, but not sufficient, condition for the convergence of an infinite series. Note that the harmonic series $1 + 1/2 + 1/3 + \cdots + 1/n + \cdots$ does not converge.

Sum of an Infinite Geometric Series

An infinite geometric series with $|r| < 1$ converges to the value or sum

$$S = \frac{a_1}{1 - r}.$$

EXAMPLE 1 Find the sum of the infinite geometric series

$$5 + \tfrac{5}{3} + \tfrac{5}{9} + \tfrac{5}{27} + \cdots.$$

Solution Here $a_1 = 5$ and $r = \frac{5}{3} \div 5 = \frac{1}{3}$. Since r is between -1 and 1, a sum can be assigned to this series. Substituting in the formula

$$S = \frac{a_1}{1 - r}$$

gives

$$S = \frac{5}{1 - \frac{1}{3}} = \frac{5}{\frac{2}{3}} = \frac{15}{2}.$$

You may wish to mention that geometric series are not the only infinite series which converge. For example, Leibniz (1672) proved that the sum of the reciprocals of the triangular numbers, $1 + 1/3 + 1/6 + 1/10 + 1/15 + \cdots$, converges to 2; Euler (1734) proved that sum of the reciprocals of the squares, $1 + 1/4 + 1/9 + 1/16 + \cdots$, converges to $\pi^2/6$.

PROGRESS CHECK 1 Find the sum of the infinite geometric series $3 + \frac{3}{2} + \frac{3}{4} + \frac{3}{8} + \cdots$.

EXAMPLE 2 Evaluate $\sum_{i=1}^{\infty} (-\frac{3}{4})^{i-1}$.

Solution In expanded form the series is

$$\sum_{i=1}^{\infty} (-\tfrac{3}{4})^{i-1} = (-\tfrac{3}{4})^0 + (-\tfrac{3}{4})^1 + (-\tfrac{3}{4})^2 + (-\tfrac{3}{4})^3 + \cdots$$
$$= 1 - \tfrac{3}{4} + \tfrac{9}{16} - \tfrac{27}{64} + \cdots.$$

This series is an infinite geometric series with $a_1 = 1$ and $r = -\frac{3}{4}$. A sum S can be assigned to this series since $|r| < 1$, and

$$S = \frac{a_1}{1 - r} = \frac{1}{1 - (-\frac{3}{4})} = \frac{1}{\frac{7}{4}} = \frac{4}{7}.$$

PROGRESS CHECK 2 Evaluate $\sum_{i=1}^{\infty} (-\frac{1}{2})^{i-1}$.

Progress Check Answers

1. 6
2. $\frac{2}{3}$

2 Recall from Section 1.1 that every repeating decimal is a rational number and can be expressed as the quotient of two integers. One method for finding such a fraction is to use infinite geometric series, as illustrated next.

Students are sometimes familiar with the following method for writing $0.\overline{45}$ as the ratio of two integers.

$$\begin{aligned} 100x &= 45.\overline{45} \\ x &= 0.\overline{45} \\ \hline 99x &= 45 \\ x &= \tfrac{45}{99} \text{ or } \tfrac{5}{11} \end{aligned}$$

The point here is the concept of an infinite geometric series, so this method should not be used now.

EXAMPLE 3 Express the repeating decimal $0.\overline{45}$ as the ratio of two integers.

Solution When we write repeating decimals, a bar is placed above the portion of the decimal that repeats. Therefore $0.\overline{45}$ is equivalent to 0.454545 . . . and may be written as

$$0.45 + 0.0045 + 0.000045 + \cdots.$$

This series is an infinite geometric series with $a_1 = 0.45$ and $r = 0.01$. Since $|r| < 1$, a sum S can be determined for this series as follows.

$$S = \frac{a_1}{1 - r} = \frac{0.45}{1 - 0.01} = \frac{0.45}{0.99} = \frac{45}{99} = \frac{5}{11}$$

Thus, $0.\overline{45} = \frac{5}{11}$. Check this result using your calculator.

PROGRESS CHECK 3 Express the repeating decimal $0.\overline{4}$ as the ratio of two integers.

3 Using the concepts developed in this section enables us to solve the problem that opens the section.

EXAMPLE 4 Solve the problem in the section introduction on page 548.

Solution The amount S of wastewater being collected is given by

$$S = 12 + 12(0.75) + 12(0.75)^2 + \cdots.$$

This series is an infinite geometric series with $r = 0.75$, so S can be determined since $|r| < 1$. Replacing a_1 by 12 and r by 0.75 in the formula for S gives

$$S = \frac{a_1}{1 - r} = \frac{12}{1 - 0.75} = \frac{12}{0.25} = 48.$$

Thus, the holding tank must be able to accommodate 48 gal of wastewater.

PROGRESS CHECK 4 Redo the problem in Example 4, but assume that 18 gal of wastewater are collected on the first pass and that on each subsequent pass only 60 percent as much wastewater is collected as on the pass before.

Progress Check Answers

3. $\frac{4}{9}$
4. 45 gal

EXERCISES 11.3

In Exercises 1–14, find the sum of the given infinite geometric series.

1. $1 + \frac{1}{10} + \frac{1}{100} + \cdots$ $\frac{10}{9}$
2. $1 + \frac{1}{5} + \frac{1}{25} + \cdots$ $\frac{5}{4}$
3. $3 + 1 + \frac{1}{3} + \cdots$ $\frac{9}{2}$
4. $3 + \frac{3}{7} + \frac{3}{49} + \cdots$ $\frac{7}{2}$
5. $6 + 3 + \frac{3}{2} + \cdots$ 12
6. $6 + \frac{3}{2} + \frac{3}{8} + \cdots$ 8
7. $\frac{1}{3} - \frac{1}{6} + \frac{1}{12} - \cdots$ $\frac{2}{9}$
8. $\frac{1}{3} - \frac{1}{9} + \frac{1}{27} - \cdots$ $\frac{1}{4}$
9. $\frac{1}{5} + \frac{2}{15} + \frac{4}{45} + \cdots$ $\frac{3}{5}$
10. $\frac{1}{5} + \frac{2}{25} + \frac{4}{125} + \cdots$ $\frac{1}{3}$
11. $3 + 0.3 + 0.03 + \cdots$ $\frac{10}{3}$
12. $2 + 0.4 + 0.08 + \cdots$ $\frac{5}{2}$
13. $2 + 0.6 + 0.18 + \cdots$ $\frac{20}{7}$
14. $3 + 1.5 + 0.75 + \cdots$ 6

In Exercises 15–30, evaluate the given sum.

15. $\sum_{i=1}^{\infty}(\frac{2}{3})^i$ 2
16. $\sum_{i=1}^{\infty}(\frac{1}{5})^i$ $\frac{1}{4}$
17. $\sum_{i=1}^{\infty}(-\frac{2}{5})^i$ $-\frac{2}{7}$
18. $\sum_{i=1}^{\infty}(-\frac{1}{3})^i$ $-\frac{1}{4}$
19. $\sum_{i=1}^{\infty}(\frac{1}{4})^{i-1}$ $\frac{4}{3}$
20. $\sum_{i=1}^{\infty}(\frac{2}{5})^{i-1}$ $\frac{5}{3}$
21. $\sum_{i=1}^{\infty}(\frac{1}{8})^{i+1}$ $\frac{1}{56}$
22. $\sum_{i=1}^{\infty}(\frac{1}{7})^{i+1}$ $\frac{1}{42}$
23. $\sum_{i=1}^{\infty}(-0.1)^{i-1}$ $\frac{10}{11}$
24. $\sum_{i=1}^{\infty}(-0.3)^{i-1}$ $\frac{10}{13}$
25. $\sum_{i=1}^{\infty}(0.5)^{i+1}$ 0.5
26. $\sum_{i=1}^{\infty}(0.2)^{i+1}$ 0.05
27. $\sum_{i=1}^{\infty}(-\frac{1}{5})^{i+2}$ $-\frac{1}{150}$
28. $\sum_{i=1}^{\infty}(-\frac{1}{9})^{i-2}$ $-\frac{81}{10}$
29. $\sum_{i=1}^{\infty}(-0.6)^{i-2}$ $-\frac{25}{24}$
30. $\sum_{i=1}^{\infty}(-0.7)^{i+3}$ 2,401/17,000

In Exercises 31–40, write the given repeating decimal as an infinite series (first three terms) and then as the quotient of two integers.

31. $0.\overline{2}$ $0.2 + 0.02 + 0.002 + \cdots; \frac{2}{9}$

32. $0.\overline{3}$ $0.3 + 0.03 + 0.003 + \cdots; \frac{1}{3}$

33. $0.\overline{7}$ $0.7 + 0.07 + 0.007 + \cdots; \frac{7}{9}$

34. $0.\overline{6}$ $0.6 + 0.06 + 0.006 + \cdots; \frac{2}{3}$

35. $0.\overline{13}$ $0.13 + 0.0013 + 0.000013 + \cdots; \frac{13}{99}$

36. $0.\overline{56}$ $0.56 + 0.0056 + 0.000056 + \cdots; \frac{56}{99}$

37. $0.\overline{375}$ $0.375 + 0.000375 + 0.000000375 + \cdots; \frac{375}{999}$

38. $0.\overline{241}$ $0.241 + 0.000241 + 0.000000241 + \cdots; \frac{241}{999}$

39. $0.0\overline{7}$ $0.07 + 0.007 + 0.0007 + \cdots; \frac{7}{90}$

40. $0.00\overline{5}$ $0.005 + 0.0005 + 0.00005 + \cdots; \frac{1}{180}$

41. A system for removing pollutants from kerosene involves passing the kerosene repeatedly through a filtering system. Suppose that each time the kerosene passes through the filters some waste is removed and stored in a special tank. In a certain application each pass through the filters removes 10 percent as much waste as the previous pass. Suppose the first pass produces 18 gal of waste. What size holding tank is large enough to hold all the waste this application will produce? 20 gal

42. Redo Exercise 41, but assume that each pass through the filters removes 20 percent as much waste as the previous pass. 22.5 gal

43. A game consists of tossing a coin until you get heads. The appearance of a head (H) is called a success (S). The appearance of a tail (T) is called a failure (F). Here are all the possible outcomes and their probabilities.

Outcome	*Probability*
Heads first appears on 1st toss: H	$\frac{1}{2}$
Heads first appears on 2nd toss: TH	$\frac{1}{2} \cdot \frac{1}{2} = \frac{1}{2^2}$
Heads first appears on 3rd toss: TTH	$\frac{1}{2} \cdot \frac{1}{2} \cdot \frac{1}{2} = \frac{1}{2^3}$
⋮	⋮
Heads first appears on nth toss: $\underbrace{\text{TTT . . . T}}_{n-1 \text{ T's}}\text{H}$	$\frac{1}{2^n}$
⋮	⋮

Show that the sum of the probabilities of all the possible outcomes is a geometric series whose sum is 1. (*Note:* In probability theory the sum of the probabilities of all possible outcomes of an experiment is always 1.) $S = \frac{\frac{1}{2}}{1 - \frac{1}{2}} = 1$

44. A game consists of rolling a die until you get a 6. The appearance of a 6 is called a success (S). The appearance of any other number is called a failure (F). Here are all the possible outcomes for such a game.

Outcome	*Probability*
6 first appears on 1st roll: S	$\frac{1}{6}$
6 first appears on 2nd roll: FS	$\frac{5}{6} \cdot \frac{1}{6} = \frac{5}{6^2}$
6 first appears on 3rd roll: FFS	$\frac{5}{6} \cdot \frac{5}{6} \cdot \frac{1}{6} = \frac{5^2}{6^3}$
⋮	⋮
6 first appears on nth roll: $\underbrace{FFF\ldots F}_{n-1\ F\text{'s}}S$	$\frac{5^{n-1}}{6^n}$
⋮	⋮

Show that the sum of the probabilities of all the possible outcomes is a geometric series whose sum is 1. $S = \frac{\frac{1}{6}}{1 - \frac{5}{6}} = 1$

45. The curve drawn in the figure is a segment of a parabola. The largest triangle has the same base and vertex as the curve. Each smaller triangle is constructed in the same way, resulting in a sequence of polygons with more and more sides, whose perimeter approaches the shape of the parabola. Archimedes showed that the areas of the successive polygons are given by this sequence, where A represents the area of the largest triangle.

Polygon	*Area*
1st polygon	A
2nd polygon	$A + \frac{1}{4}A$
3rd polygon	$A + \frac{1}{4}A + \frac{1}{4^2}A$
4th polygon	$A + \frac{1}{4}A + \frac{1}{4^2}A + \frac{1}{4^3}A$
⋮	⋮
nth polygon	$A + \frac{1}{4}A + \frac{1}{4^2}A + \cdots + \frac{1}{4^{n-1}}A$

Archimedes argued that the sum of the infinite series gives the area of the parabola in terms of the area (A) of the largest triangle. Discover Archimedes' formula by finding the sum of this geometric series. This is the earliest recorded example of the summation of an infinite series. (*Hint:* Factor out A from the series before finding the sum.) $S = \frac{4}{3}A$

46. Here is a classic problem as discussed in W. W. Sawyer's *Mathematician's Delight* (Penguin Books, 1943): "If a ton of seed potatoes will produce a crop of 3 tons, which can either be consumed or used again as seed, how much must a gardener buy, if his family want to consume a ton of potatoes every year forever?" [*Hint:* For the first year's harvest $\frac{1}{3}$ ton of seed planted now will be enough. For the second year's harvest $\frac{1}{9}$ ton planted now will be enough. (Why?)]
   a. Write a geometric series for the total amount of seed potatoes needed.
   b. Find the sum of the series to answer the question.
   a. $\frac{1}{3} + \frac{1}{9} + \frac{1}{27} + \cdots$ b. 0.5 ton of seed potatoes

47. A pendulum is swinging back and forth, but each diminishing swing takes 0.999 times as long as the previous one. If the first swing takes 2 seconds, what is the total time elapsed before the pendulum stops? Approximate this time as the sum of an infinite geometric series. 2,000 seconds = $33\frac{1}{3}$ minutes

48. A certain ball always rebounds $\frac{2}{3}$ as far as it falls. If the ball is dropped from a height of 9 ft, how far up and down has it traveled before it comes to rest? Approximate this distance as the sum of an infinite geometric series. 45 ft

THINK ABOUT IT

1. In the formula for the sum of an infinite geometric series $|r|$ is required to be less than 1. Try using the formula for the series $2 + 2^2 + 2^3 \cdots$ to see that nonsense results.
2. What is the common ratio in an infinite series whose first term is 4 and whose sum is $4\frac{1}{4}$?
3. Explain in terms of infinite geometric series why $\sum_{i=1}^{\infty} x^{i-1}$ converges to a sum if $|x| < 1$. What is this sum?
4. For what values of x does the series
$$(x + 1) + 2(x + 1)^2 + 4(x + 1)^3 + \cdots$$
converge to a sum? What is the sum?
5. Express the repeating decimal $2.1\overline{43}$ as the ratio of two integers.

REMEMBER THIS

1. Expand $(a + b)^2$. $a^2 + 2ab + b^2$
2. Expand $(a - b)^3$. $a^3 - 3a^2b + 3ab^2 - b^3$
3. Simplify: $\dfrac{1 \cdot 2 \cdot 3 \cdot 4 \cdot 5}{3 \cdot 2 \cdot 1}$. 20
4. Evaluate $n(n - 1)(n - 2)$ when $n = 3$. 6
5. What is the degree of the polynomial $3xy$? 2
6. Write this series in expanded form and find its sum: $\sum_{i=1}^{4} i^2$.
   $1 + 4 + 9 + 16 = 30$
7. Evaluate the determinant. $\begin{vmatrix} 3 & 5 \\ 2 & 4 \end{vmatrix}$ 2
8. Factor $x^2 - x - 12$. $(x - 4)(x + 3)$
9. Divide $\dfrac{22 \times 10^{500}}{2 \times 10^{499}}$. 110
10. Subtract $\dfrac{x}{x + 4} - \dfrac{x + 3}{x + 4}$. $\dfrac{-3}{x + 4}$

11.4 Binomial Theorem

A student guesses randomly at six multiple-choice questions. If each question has four possible choices, then the probability that the student gets exactly three correct is given by the fourth term in the expansion of $(\frac{1}{4} + \frac{3}{4})^6$. Find the probability. (See Example 8.)

OBJECTIVES

1. **Expand $(a + b)^n$ using Pascal's triangle.**
2. **Evaluate binomial coefficients in the form $\binom{n}{r}$ for given values of n and r.**
3. **Expand $(a + b)^n$ using the binomial theorem.**
4. **Find the rth term in the expansion of $(a + b)^n$.**

1 In Section 3.4 a special product formula was used to square a binomial and expand $(a + b)^2$. Other expressions of the form $(a + b)^n$, where n is a positive integer, may also be expanded in a systematic way, and we now discuss a method for finding such expansions.

Consider carefully the following expansions of the powers of $a + b$ and try to find some patterns. Note that direct multiplication can be used to verify these results.

$$(a + b)^1 = a + b \qquad \text{(2 terms)}$$
$$(a + b)^2 = a^2 + 2ab + b^2 \qquad \text{(3 terms)}$$
$$(a + b)^3 = a^3 + 3a^2b + 3ab^2 + b^3 \qquad \text{(4 terms)}$$
$$(a + b)^4 = a^4 + 4a^3b + 6a^2b^2 + 4ab^3 + b^4 \qquad \text{(5 terms)}$$
$$(a + b)^5 = a^5 + 5a^4b + 10a^3b^2 + 10a^2b^3 + 5ab^4 + b^5 \qquad \text{(6 terms)}$$

In each case, observe that the expansion of $(a + b)^n$ behaved as follows:

1. The number of terms in the expansion is $n + 1$.
2. The first term is a^n and the last term is b^n.
3. The second term is $na^{n-1}b$ and the nth term is nab^{n-1}.
4. The exponent of a decreases by 1 in each successive term, while the exponent of b increases by 1.
5. The sum of the exponents of a and b in any term is n.

It can be proved that the above patterns continue to hold if n is a positive integer greater than 5, so we now need only determine a method for finding the constant coefficients when we expand such expressions. For this purpose, we have arranged the constant coefficients in the above expansions of $(a + b)^n$ for $n \leq 5$ in the triangular array shown in the following chart.

Powers of $a + b$		**Constant Coefficients (Pascal's Triangle)**
$(a + b)^0$	Row 0	1
$(a + b)^1$	Row 1	1 1
$(a + b)^2$	Row 2	1 2 1
$(a + b)^3$	Row 3	1 3 3 1
$(a + b)^4$	Row 4	1 4 6 4 1
$(a + b)^5$	Row 5	1 5 10 10 5 1
⋮	⋮	⋮

The triangular array of numbers that specifies the constant coefficients in the expansions of $(a + b)^n$ for $n = 0, 1, 2, \ldots$ is called **Pascal's triangle.** Observe that except for the 1's, each entry in Pascal's triangle is the sum of the two numbers on either side of it in the preceding row, as diagramed in the chart. Based on our observations to this point, we can now expand positive integral powers of $a + b$ using

$$(a + b)^n = a^n + na^{n-1}b + (\text{constant})a^{n-2}b^2 + (\text{constant})a^{n-3}b^3 + \cdots + nab^{n-1} + b^n,$$

where Pascal's triangle may be used to find the constant coefficients.

In the Arabic world the array we call Pascal's triangle was used by Al-Karaji, who died in 1019. A little later, in the mid-eleventh century, the Chinese mathematician Jia Xian made use of the array for solving polynomial equations. The first appearances in print in Western Europe are about 500 years later in works by Michael Stifel (1544) and Petrus Apianus (1527). Blaise Pascal (1623–1662) developed the triangle (he called it the arithmetical triangle) for use in analyzing the probabilities connected with gambling problems.

EXAMPLE 1 Expand $(2x + y)^4$ using Pascal's triangle.

Solution Use the expansion formula above with $a = 2x$, $b = y$, and $n = 4$. From Pascal's triangle we find the coefficients of the five terms in the expansion are 1, 4, 6, 4, and 1, respectively. Therefore,

$$\begin{aligned}(2x + y)^4 &= (2x)^4 + 4(2x)^3y + 6(2x)^2y^2 + 4(2x)y^3 + y^4 \\ &= 16x^4 + 32x^3y + 24x^2y^2 + 8xy^3 + y^4.\end{aligned}$$

PROGRESS CHECK 1 Expand $(x + 4y)^3$ using Pascal's triangle.

EXAMPLE 2 Expand $(y - 5)^3$ using Pascal's triangle.

Solution First, rewrite $(y - 5)^3$ as $[y + (-5)]^3$. We now substitute y for a, -5 for b, and 3 for n in our expansion formula. The coefficients of the four terms in the expansion are determined from Pascal's triangle to be 1, 3, 3, and 1, respectively. Therefore,

$$\begin{aligned}(y - 5)^3 &= y^3 + 3y^2(-5) + 3y(-5)^2 + (-5)^3 \\ &= y^3 - 15y^2 + 75y - 125.\end{aligned}$$

Note The terms in the expansion alternate in sign when the binomial is the difference of two terms, as illustrated by this example.

PROGRESS CHECK 2 Expand $(x - 3)^5$ using Pascal's triangle.

EXAMPLE 3 Expand $(x + h)^6$ using Pascal's triangle.

Solution We substitute x for a, h for b, and 6 for n in our expansion formula. To determine the constant coefficients using Pascal's triangle, we must continue the pattern in the previous chart to specify the entries in row 6. Starting from row 5, the next row is obtained as follows:

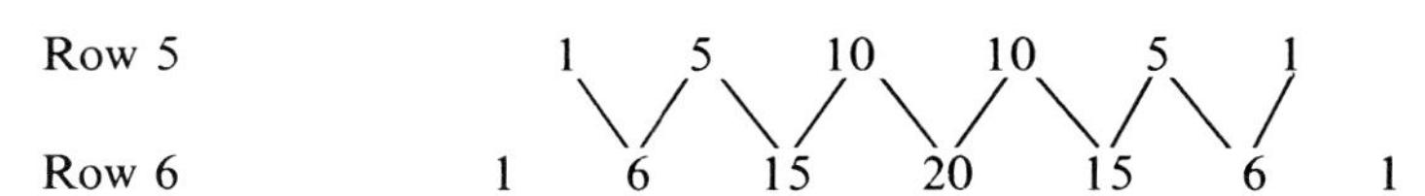

Thus, the expansion of $(x + h)^6$ is

$$(x + h)^6 = x^6 + 6x^5h + 15x^4h^2 + 20x^3h^3 + 15x^2h^4 + 6xh^5 + h^6.$$

PROGRESS CHECK 3 Expand $(y - m)^7$ using Pascal's triangle.

2 Using Pascal's triangle to determine the constant coefficients in a binomial expansion is sometimes impractical, since constructing the table can be time-consuming, particularly for large values of n. Therefore, it is useful to have a formula for these constant coefficients. Before such a formula can be stated, it is helpful to first introduce **factorial notation.**

n Factorial

For any positive integer n, the symbol $n!$ (read "n factorial") is defined by

$$n! = n(n - 1)(n - 2) \cdots (2)(1).$$

For example,

$$3! = 3 \cdot 2 \cdot 1 = 6 \qquad \text{and} \qquad 7! = 7 \cdot 6 \cdot 5 \cdot 4 \cdot 3 \cdot 2 \cdot 1 = 5{,}040.$$

Progress Check Answers

1. $x^3 + 12x^2y + 48xy^2 + 64y^3$
2. $x^5 - 15x^4 + 90x^3 - 270x^2 + 405x - 243$
3. $y^7 - 7y^6m + 21y^5m^2 - 35y^4m^3 + 35y^3m^4 - 21y^2m^5 + 7ym^6 - m^7$

If you use a scientific calculator with a factorial key $\boxed{x!}$, it is easy to evaluate factorials. For instance, the following keystrokes show $6! = 720$.

$$6 \; \boxed{x!} \; \boxed{\quad 720}$$

As a special case, it is also useful to define 0! so that

$$0! = 1.$$

For example, with $0! = 1$, it is easy to describe a row of Pascal's triangle with consistent notation, as in Example 4.

Using factorials, we can now write a formula that determines a constant coefficient in a binomial expansion. Such members are called binomial coefficients.

Binomial Coefficient

Let r and n be nonnegative integers with $r \leq n$. Then the symbol $\binom{n}{r}$ is defined by

$$\binom{n}{r} = \frac{n!}{r!(n-r)!}.$$

Each of the numbers $\binom{n}{r}$ is called a **binomial coefficient.**

As you might expect, binomial coefficients are entries in Pascal's triangle, and the next example illustrates this point.

EXAMPLE 4 Evaluate the following binomial coefficients and compare the results with the entries in Pascal's triangle.

$$\binom{4}{0}, \binom{4}{1}, \binom{4}{2}, \binom{4}{3}, \binom{4}{4}$$

Solution By applying the formula in the definition, we have

$$\binom{4}{0} = \frac{4!}{0!(4-0)!} = \frac{4!}{0!4!} = \frac{4 \cdot 3 \cdot 2 \cdot 1}{1 \cdot (4 \cdot 3 \cdot 2 \cdot 1)} = 1$$

$$\binom{4}{1} = \frac{4!}{1!(4-1)!} = \frac{4!}{1!3!} = \frac{4 \cdot 3 \cdot 2 \cdot 1}{1 \cdot (3 \cdot 2 \cdot 1)} = 4$$

$$\binom{4}{2} = \frac{4!}{2!(4-2)!} = \frac{4!}{2!2!} = \frac{4 \cdot 3 \cdot 2 \cdot 1}{(2 \cdot 1)(2 \cdot 1)} = 6$$

$$\binom{4}{3} = \frac{4!}{3!(4-3)!} = \frac{4!}{3!1!} = \frac{4 \cdot 3 \cdot 2 \cdot 1}{(3 \cdot 2 \cdot 1) \cdot 1} = 4$$

$$\binom{4}{4} = \frac{4!}{4!(4-4)!} = \frac{4!}{4!0!} = \frac{4 \cdot 3 \cdot 2 \cdot 1}{(4 \cdot 3 \cdot 2 \cdot 1) \cdot 1} = 1.$$

By direct comparison, observe that the binomial coefficients 1, 4, 6, 4, and 1 match the entries in row 4 of Pascal's triangle.

Note This example illustrates the general result that the binomial coefficients

$$\binom{n}{0}, \binom{n}{1}, \binom{n}{2}, \ldots, \binom{n}{n}$$

are precisely the numbers in the nth row of Pascal's triangle.

Point out the symmetry in the binomial coefficients here. Perhaps offer the proof

$$\binom{n}{n-r} = \frac{n!}{(n-r)![n-(n-r)]!} = \frac{n!}{(n-r)!r!} = \binom{n}{r}.$$

PROGRESS CHECK 4 Evaluate the following binomial coefficients and compare the results with the entries in Pascal's triangle.

$$\binom{5}{0}, \binom{5}{1}, \binom{5}{2}, \binom{5}{3}, \binom{5}{4}, \binom{5}{5}$$

EXAMPLE 5 Evaluate $\binom{10}{4}$.

Solution Replace n by 10 and r by 4 in the above formula to get

$$\binom{10}{4} = \frac{10!}{4!(10-4)!} = \frac{10!}{4!6!} = 210.$$

One possible keystroke sequence for evaluating $10!/(4!6!)$ is

10 [x!] [÷] 4 [x!] [÷] 6 [x!] [=] [210].

Consider showing students whose calculators have a [$_nC_r$] key how to use it to evaluate binomial coefficients.

PROGRESS CHECK 5 Evaluate $\binom{9}{5}$.

3 A formula for expanding $(a + b)^n$ can be stated very efficiently by using binomial coefficients and sigma notation. This expansion formula is called the **binomial theorem.**

For extensions of the binomial theorem to fractional and negative exponents, see "Think About It" Exercises 4 and 5.

Binomial Theorem

For any positive integer n,

$$(a + b)^n = \sum_{r=0}^{n} \binom{n}{r} a^{n-r} b^r$$
$$= \binom{n}{0}a^n + \binom{n}{1}a^{n-1}b + \binom{n}{2}a^{n-2}b^2 + \cdots + \binom{n}{n}b^n.$$

EXAMPLE 6 Expand $(y + d)^7$ using the binomial theorem.

Solution Applying the binomial theorem with $y = a$, $d = b$, and $n = 7$ yields

$$(y + d)^7 = \binom{7}{0}y^7 + \binom{7}{1}y^6d + \binom{7}{2}y^5d^2 + \binom{7}{3}y^4d^3$$
$$+ \binom{7}{4}y^3d^4 + \binom{7}{5}y^2d^5 + \binom{7}{6}yd^6 + \binom{7}{7}d^7.$$

Now evaluate the binomial coefficients. Use the symmetry in the coefficients and the binomial coefficient formula to obtain

$$\binom{7}{0} = \binom{7}{7} = \frac{7!}{0!7!} = 1, \qquad \binom{7}{1} = \binom{7}{6} = \frac{7!}{1!6!} = 7,$$
$$\binom{7}{2} = \binom{7}{5} = \frac{7!}{2!5!} = 21, \qquad \binom{7}{3} = \binom{7}{4} = \frac{7!}{3!4!} = 35.$$

Then, the desired expansion is

$$(y + d)^7 = y^7 + 7y^6d + 21y^5d^2 + 35y^4d^3 + 35y^3d^4 + 21y^2d^5 + 7yd^6 + d^7.$$

PROGRESS CHECK 6 Expand $(x + c)^8$ using the binomial theorem.

Progress Check Answers

4. $\binom{5}{0} = \binom{5}{5} = 1$, $\binom{5}{1} = \binom{5}{4} = 5$, $\binom{5}{2} = \binom{5}{3} = 10$; the binomial coefficients 1, 5, 10, 10, 5, 1 match the entries in row 5 of Pascal's triangle.
5. 126
6. $x^8 + 8x^7c + 28x^6c^2 + 56x^5c^3 + 70x^4c^4 + 56x^3c^5 + 28x^2c^6 + 8xc^7 + c^8$

In practice, the coefficients of y^7, y^6d, yd^6, and d^7 are found more easily on the basis of patterns already discussed. The associated binomial coefficients should be computed here to confirm that the binomial theorem works for these cases.

4 It is sometimes important to know only a single term in a binomial expansion, and this term may be found without producing the rest of the terms. Observe in the binomial theorem that the exponent of b is 1 in the second term, 2 in the third term, and in general, $r - 1$ in the rth term. Because the sum of the exponents of a and b is always n, the exponent above a in the rth term must be $n - (r - 1)$. Finally, the binomial coefficient of the term containing b^{r-1} is $\binom{n}{r-1}$. These observations lead to the following formula.

rth Term of the Binomial Expansion

The rth term of the binomial expansion of $(a + b)^n$ is

$$\binom{n}{r-1} a^{n-(r-1)} b^{r-1}.$$

EXAMPLE 7 Find the seventh term in the expansion of $(2x - y)^{10}$.

Solution Here $a = 2x$, $b = -y$, and $n = 10$. To find the seventh term, let $r = 7$, so $r - 1 = 6$. Then applying the above formula gives the seventh term as

$$\begin{aligned}\binom{10}{6}(2x)^4(-y)^6 &= 210(16x^4)(y^6)\\ &= 3{,}360x^4y^6.\end{aligned}$$

PROGRESS CHECK 7 Find the sixth term in the expansion of $(3x - y)^8$.

EXAMPLE 8 Solve the problem in the section introduction on page 552.

Solution The requested probability is the fourth term in the expansion of $(\frac{1}{4} + \frac{3}{4})^6$. In this case, $a = \frac{1}{4}$, $b = \frac{3}{4}$, $n = 6$, and $r = 4$ (so $r - 1 = 3$). By the above formula, the fourth term is

$$\begin{aligned}\binom{6}{3}\left(\frac{1}{4}\right)^3\left(\frac{3}{4}\right)^3 &= 20\left(\frac{1}{4}\right)^3\left(\frac{3}{4}\right)^3\\ &= \frac{540}{4{,}096} = \frac{135}{1{,}024}.\end{aligned}$$

The probability that this student gets exactly three correct is $135/1{,}024 \approx 13.18$ percent.

Note that the desired term contains p^m where p is the probability of an individual correct guess and m is the number of correct guesses. For further details, consult a statistics text.

PROGRESS CHECK 8 The probability that the student described in Example 8 gets exactly four correct is given by the third term in the expansion of $(\frac{1}{4} + \frac{3}{4})^6$. Find this probability.

Progress Check Answers

7. $-1{,}512x^3y^5$
8. $\frac{135}{4{,}096} \approx 3.30$ percent

EXERCISES 11.4

In Exercises 1–12, use Pascal's triangle to expand the given expression.

1. $(x + y)^3$ $x^3 + 3x^2y + 3xy^2 + y^3$
2. $(x + y)^4$ $x^4 + 4x^3y + 6x^2y^2 + 4xy^3 + y^4$
3. $(x - y)^3$ $x^3 - 3x^2y + 3xy^2 - y^3$
4. $(x - y)^4$ $x^4 - 4x^3y + 6x^2y^2 - 4xy^3 + y^4$
5. $(x - 3)^5$ $x^5 - 15x^4 + 90x^3 - 270x^2 + 405x - 243$
6. $(x + 2)^6$ $x^6 + 12x^5 + 60x^4 + 160x^3 + 240x^2 + 192x + 64$
7. $(x + 4)^3$ $x^3 + 12x^2 + 48x + 64$
8. $(x - 5)^4$ $x^4 - 20x^3 + 150x^2 - 500x + 625$
9. $(2x - n)^6$ $64x^6 - 192x^5n + 240x^4n^2 - 160x^3n^3 + 60x^2n^4 - 12xn^5 + n^6$
10. $(x + 3y)^5$ $x^5 + 15x^4y + 90x^3y^2 + 270x^2y^3 + 405xy^4 + 243y^5$
11. $(2x + 5y)^6$ $64x^6 + 960x^5y + 6{,}000x^4y^2 + 20{,}000x^3y^3 + 37{,}500x^2y^4 + 37{,}500xy^5 + 15{,}625y^6$
12. $(3x - 2y)^4$ $81x^4 - 216x^3y + 216x^2y^2 - 96xy^3 + 16y^4$

In Exercises 13 and 14, evaluate the given list of binomial coefficients, and match the list to a row in Pascal's triangle.

13. $\binom{7}{0}, \binom{7}{1}, \binom{7}{2}, \binom{7}{3}, \binom{7}{4}, \binom{7}{5}, \binom{7}{6}, \binom{7}{7}$
1, 7, 21, 35, 35, 21, 7, 1; matches seventh row of Pascal's triangle

14. $\binom{8}{0}, \binom{8}{1}, \binom{8}{2}, \binom{8}{3}, \binom{8}{4}, \binom{8}{5}, \binom{8}{6}, \binom{8}{7}, \binom{8}{8}$
1, 8, 28, 56, 70, 56, 28, 8, 1; matches eighth row of Pascal's triangle

In Exercises 15–22, evaluate the given binomial coefficients.

15. $\binom{5}{2}$ 10 **16.** $\binom{12}{4}$ 495 **17.** $\binom{10}{6}$ 210 **18.** $\binom{6}{1}$ 6

19. $\binom{88}{0}$ 1 **20.** $\binom{55}{0}$ 1 **21.** $\binom{66}{66}$ 1 **22.** $\binom{957}{957}$ 1

In Exercises 23–30, expand the given expression using the binomial theorem. Do not refer to Pascal's triangle.

23. $(x + h)^4$ $x^4 + 4x^3h + 6x^2h^2 + 4xh^3 + h^4$

24. $(x + h)^5$ $x^5 + 5x^4h + 10x^3h^2 + 10x^2h^3 + 5xh^4 + h^5$

25. $(x - y)^7$ $x^7 - 7x^6y + 21x^5y^2 - 35x^4y^3 + 35x^3y^4 - 21x^2y^5 + 7xy^6 - y^7$

26. $(x - y)^{10}$ $x^{10} - 10x^9y + 45x^8y^2 - 120x^7y^3 + 210x^6y^4 - 252x^5y^5 + 210x^4y^6 - 120x^3y^7 + 45x^2y^8 - 10xy^9 + y^{10}$

27. $(x + 3)^6$ $x^6 + 18x^5 + 135x^4 + 540x^3 + 1{,}215x^2 + 1{,}458x + 729$

28. $(x + 5)^8$ $x^8 + 40x^7 + 700x^6 + 7{,}000x^5 + 43{,}750x^4 + 175{,}000x^3 + 437{,}500x^2 + 625{,}000x + 390{,}625$

29. $(3x - 1)^3$ $27x^3 - 27x^2 + 9x - 1$

30. $(2x - 4)^5$ $32x^5 - 320x^4 + 1{,}280x^3 - 2{,}560x^2 + 2{,}560x - 1{,}024$

In Exercises 31–38, find the desired term in the given expansion.

31. $(x + y)^9$, sixth term $126x^4y^5$

32. $(x + y)^{12}$, seventh term $924x^6y^6$

33. $(x - y)^{10}$, eighth term $-120x^3y^7$

34. $(x - y)^7$, fourth term $-35x^4y^3$

35. $(x - 3)^8$, sixth term $-13{,}608x^3$

36. $(x - 4)^{10}$, seventh term $860{,}160x^4$

37. $(2x - 1)^7$, fifth term $280x^3$

38. $(3x + 2)^6$, fifth term $2{,}160x^2$

39. A student guesses randomly at 10 multiple-choice questions. If each question has 5 possible choices, then the probability that the student gets exactly three correct is given by the eighth term in the expansion of $(\frac{1}{5} + \frac{4}{5})^{10}$. Find this probability.
393,216/1,953,125 ≈ 20.13 percent

40. Refer to Exercise 39. The probability that the student gets no answers correct is given by the last term in the expansion of $(\frac{1}{5} + \frac{4}{5})^{10}$. Find this probability. 1,048,576/9,765,625 ≈ 10.7 percent

41. If you guess randomly at five questions on a multiple-choice test, where each question has three possible choices, then the probability that you get *at least* four answers correct is the sum of the first and second terms in the expansion of $(\frac{1}{3} + \frac{2}{3})^5$. Find this probability. $\frac{11}{243} = 4.53$ percent

42. If you guess randomly at six questions on a multiple-choice test, where each question has three possible choices, then the probability that you get *at least* four answers correct is the sum of the first, second, and third terms in the expansion of $(\frac{1}{3} + \frac{2}{3})^6$. Find this probability. $\frac{73}{729} = 10.01$ percent

43. A seminar has six students from which a committee of two will be chosen to report to the dean about the conditions in the classroom. How many different combinations of students are possible for the makeup of this committee? Answer this question two ways.

a. Call the students A, B, C, D, E, F; then write down all the distinct pairs you can find (AB, AC, and so on) and count them. 15

b. Evaluate $\binom{6}{2}$ to get the correct answer directly. 15

44. Redo Exercise 43, but now suppose that the committee consists of three members of the class. For part **b,** evaluate $\binom{6}{3}$.
a. 20 b. 20

THINK ABOUT IT

1. Expand $(\sqrt{x} + \sqrt{y})^4$. Assume $x, y \geq 0$.

2. Expand $(x + y + 1)^3$ by using the binomial theorem. [*Hint:* Treat $(x + y)$ as the first term.]

3. Simplify $\dfrac{(n + 1)!}{(n - 1)!}$.

4. A great early contribution by Isaac Newton was the extension of the binomial theorem to fractional and negative exponents.

a. Note that the binomial coefficient $\binom{5}{2} = \dfrac{5!}{2!3!}$ reduces to $\dfrac{5 \cdot 4}{1 \cdot 2}$ because 3! divides out in the numerator and denominator. In general, $\dfrac{n!}{r!(n - r)!}$ reduces to $\dfrac{n \cdot (n - 1) \cdot (n - 2) \cdots (n - r + 1)}{r!}$ because $(n - r)!$ divides out in the numerator and denominator. Using this approach, show that $\binom{7}{3} = \dfrac{7 \cdot 6 \cdot 5}{1 \cdot 2 \cdot 3}$.

b. Newton decided that he could follow the pattern in part **a** even with fractional and negative values in the binomial coefficients. For example,

$$\binom{\frac{1}{2}}{3} = \frac{\frac{1}{2}(\frac{1}{2} - 1)(\frac{1}{2} - 2)}{1 \cdot 2 \cdot 3} = \frac{(\frac{1}{2})(-\frac{1}{2})(-\frac{3}{2})}{6} = \frac{3}{48} = \frac{1}{16},$$

$$\binom{-1}{3} = \frac{-1(-1 - 1)(-1 - 2)}{1 \cdot 2 \cdot 3} = \frac{-1(-2)(-3)}{6} = \frac{-6}{6} = -1.$$

Use this technique to evaluate $\binom{\frac{1}{2}}{2}$ and $\binom{-2}{3}$.

5. a. Newton applied the method of Exercise 4 to the expansion of $(1 + x)^{1/2}$, producing a nonending expansion which he used to approximate irrational numbers. Write out the first four terms of this expansion; then let $x = 1$ to get an approximation for $\sqrt{2}$. Compare this with the calculator approximation for $\sqrt{2}$. The expansion is started below.

$$(1 + x)^{1/2} = \binom{\frac{1}{2}}{0}1^{1/2} + \binom{\frac{1}{2}}{1}1^{-1/2}x + \binom{\frac{1}{2}}{2}1^{-3/2}x^2 + \cdots$$

b. Using the expansion from part **a**, let $x = 0.1$ to get an approximation for $\sqrt{1.1}$. Compare the result with the calculator approximation for $\sqrt{1.1}$.

c. Expand $(1 + x)^{-1}$ for four terms; then let $x = 0.01$ to get an approximation for the reciprocal of 1.01.

REMEMBER THIS

1. Evaluate $\sum_{i=1}^{\infty}(\frac{1}{2})^{i-1}$. 2
2. Why will the formula for the sum of an infinite geometric series not work for this series? $1 + 2 + 4 + 8 + \cdots$.
   Because r is greater than 1
3. Express $0.\overline{345}$ as a ratio of two integers. $\frac{345}{999} = \frac{115}{333}$
4. Which label is correct for 2, 5, 8, 11, . . . ?
   a. Geometric sequence
   b. Geometric series
   c. Arithmetic sequence
   d. Arithmetic series
   c
5. Find the sum of the first 20 terms of the series whose general term is $a_n = 3n + 2$. 670
6. Solve $3(x - 1) = 3x - 1$. $\emptyset$
7. Solve $x^2 + 4 = 0$. $\pm 2i$
8. Solve $\log_3 x = 5$. $\{243\}$
9. Solve $e^{-0.1t} = 5$. Give the solution to four decimal places. $\{-16.0944\}$
10. Solve the system. $2x - y = 11$
    $2x + y = 11$ $(\frac{11}{2}, 0)$

GRAPHING CALCULATOR

Sequences, Series, and Binomial Coefficients

For solving problems of the type discussed in Chapter 11, we can make use of the programming capabilities of the TI-81 as well as a special feature for calculating binomial coefficients. Example 1 demonstrates a program which will evaluate individual terms in a sequence when a formula for the general term is known.

EXAMPLE 1 The nth term of a sequence is given by $a_n = \dfrac{2n - 3}{5}$. Evaluate a_n for $n = 1, 2,$ and 3.

Solution Enter the program SEQ shown in Figure 11.3 into the calculator. This program uses N as the variable for convenience. Each time this program is used for a new sequence, you must key in the formula for a_n at the starred line. Also, note that the calculator will produce answers in decimal form, not fractional form.

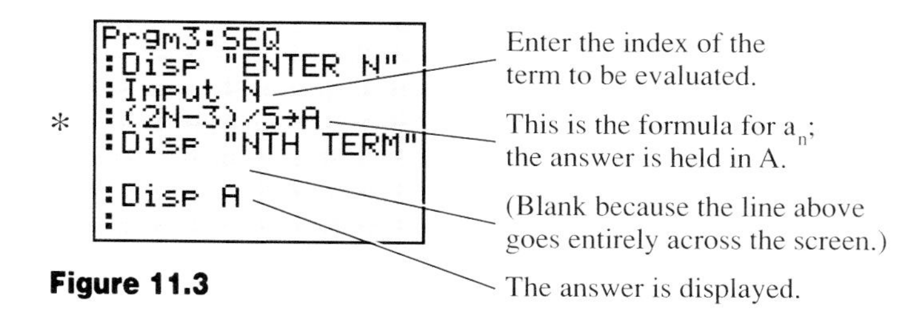

Prgm3:SEQ
:Disp "ENTER N"
:Input N
* :(2N-3)/5→A
:Disp "NTH TERM"
:Disp A
:
Enter the index of the term to be evaluated.
This is the formula for a_n; the answer is held in A.
(Blank because the line above goes entirely across the screen.)
The answer is displayed.

Figure 11.3

After the program is stored in the calculator, you can execute it to find the values required in this example. In response to "ENTER N," key in 1, then press ENTER. The answer, −.2, appears on the right side of the screen. To continue the program, just press ENTER, and the program will start again. (This is an alternative to putting a label on the first line and a Goto instruction on the last line.) The correct results are $a_1 = -0.2$, $a_2 = 0.2$, $a_3 = 0.6$. ◢

Example 2 illustrates a program for finding the sum of the first n terms of a series when a formula for the sum is known.

EXAMPLE 2 Find the sum of the first 25 terms of the arithmetic series $6 + 13 + 20 + \cdots$.

Solution The program SUMFORM shown in Figure 11.4 uses the formula $S_n = (n/2)[2a_1 + (n - 1)d]$ for the sum of the first n terms of an arithmetic series in terms of a_1, d, and n.

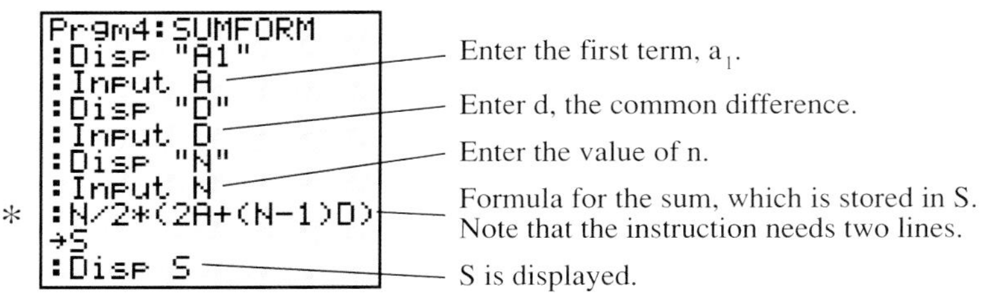

Prgm4:SUMFORM
:Disp "A1"
:Input A
:Disp "D"
:Input D
:Disp "N"
:Input N
* :N/2*(2A+(N-1)D)
→S
:Disp S
Enter the first term, a_1.
Enter d, the common difference.
Enter the value of n.
Formula for the sum, which is stored in S. Note that the instruction needs two lines.
S is displayed.

Figure 11.4

In this example $a_1 = 6$, $d = 7$, and $n = 25$; so to execute the program, enter these values as requested. The correct answer is 2,250. This program may be modified for the sum of a geometric series—or any other series where you have a formula for the sum. Just enter the appropriate formula at the starred line, and adjust the set of variables to display and input as needed.

In Example 3 we demonstrate a method for finding the sum of the first n terms of any series whose general term is known.

EXAMPLE 3 Find the sum of the squares of the integers from 1 to 25,

$$S = \sum_{i=1}^{25} i^2 = 1^2 + 2^2 + 3^2 + \cdots + 25^2,$$

and then find the sum of the squares of the integers from 1 to 50.

Solution This series is neither an arithmetic nor a geometric series, so the sum will be computed by actually adding all the individual terms in a recursive manner. Program SERSUM for finding this sum is given in Figure 11.5.

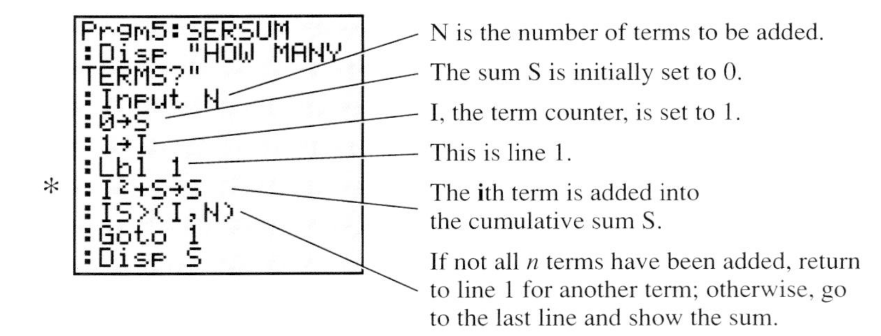

Prgm5:SERSUM
:Disp "HOW MANY
TERMS?"
:Input N
:0→S
:1→I
:Lbl 1
* :I²+S→S
:IS>(I,N)
:Goto 1
:Disp S
N is the number of terms to be added.
The sum S is initially set to 0.
I, the term counter, is set to 1.
This is line 1.
The ith term is added into the cumulative sum S.
If not all n terms have been added, return to line 1 for another term; otherwise, go to the last line and show the sum.

Figure 11.5

In executing the program, key in 25 for N when requested, and press ENTER. The answer 5,525 appears. By pressing ENTER, you start the program again, and you can key in 50 to answer the second question. The correct result is 42,925. To use this program for a different series, replace I^2 in the starred line by the general term of the new series.

Example 4 demonstrates the use of the nCr feature for evaluating binomial coefficients.

EXAMPLE 4 Use the calculator to evaluate the binomial coefficients $\binom{4}{2}$ and $\binom{13}{0}$.

Solution Built into many calculators is a special key or menu choice called nCr which evaluates binomial coefficients. In probability applications the binomial coefficient $\binom{n}{r}$ gives the number of **c**ombinations of $\boldsymbol{n}$ objects taken $\boldsymbol{r}$ at a time, and thus the key or menu choice is labeled with n, C, and r. To display nCr on the screen of the TI-81, you press the MATH key and then move the cursor to PRB and press 3. Here is the keystroke sequence for calculating $\binom{4}{2}$. (*Note:* ◀ means to press the left arrow key.)

4 MATH ◀ 3 2 ENTER

(MATH ◀ 3: These keys cause nCr to appear on the screen.)

The resulting screen is shown in Figure 11.6; the correct answer is 6. Also shown are the entry and the answer for $\binom{13}{0}$.

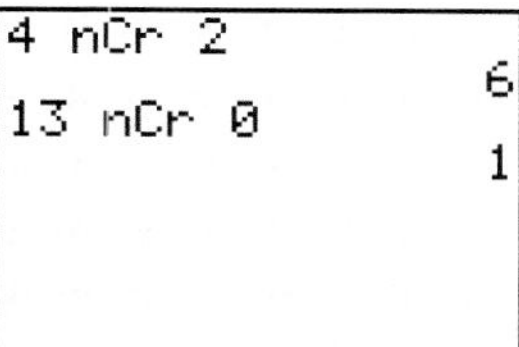

Figure 11.6

EXERCISES

1. Use the program SEQ to evaluate a_n (to three decimal places) for $n = 1, 2,$ and 3 in the sequence for which $a_n = e^n - \ln(10n)$. Ans. $a_1 = 0.416$, $a_2 = 4.393$, $a_3 = 16.684$
2. Use the program SUMFORM to find the sum of the first 50 terms of the series $-10 + (-5) + 0 + 5 + \cdots$. Ans. 5,625
3. Use the program SERSUM to find the sum of the first 15 terms of the series $1^3 + 2^3 + 3^3 + \cdots$. Ans. 14,400
4. Use the nCr feature to evaluate $\binom{10}{5}$ and $\binom{20}{10}$. Ans. 252; 184,756

Use the calculator and the programs of this section to help solve these exercises from Chapter 11.

Section	*Exercises*
11.1	1, 15
11.2	19, 21, 35
11.3	15
11.4	13, 25

Chapter 11 SUMMARY

OBJECTIVES CHECKLIST Specific chapter objectives are summarized below along with numbered example problems from the text that should clarify the objectives. If you do not understand any objectives or do not know how to do the selected problems, then restudy the material.

11.1 Can you:

1. **Find any term in a sequence when given a formula for the nth term of the sequence?**
   Write the first four terms of the sequence given by $a_n = 5n - 2$; also, find a_{50}. [Example 1]
2. **Find a formula for the general term a_n in a given arithmetic sequence?**
   Find a formula for the general term a_n in the arithmetic sequence
   $$4, 7, 10, 13, \ldots .$$
   [Example 2]
3. **Find a formula for the general term a_n in a given geometric sequence?**
   Find a formula for the general term a_n in the geometric sequence
   $$2, -1, \tfrac{1}{2}, -\tfrac{1}{4}, \ldots .$$
   [Example 3]
4. **Determine if a sequence is an arithmetic sequence, a geometric sequence, or neither?**
   State whether the sequence is an arithmetic sequence, a geometric sequence, or neither.
   $$1, 4, 9, 16, \ldots$$
   [Example 4c]
5. **Solve applied problems involving sequences?**
   Solve the problem in the chapter introduction on page 532. [Example 5]

11.2 Can you:

1. **Find the sum of an indicated number of terms in an arithmetic sequence?**
   Find the sum of the first 25 terms of the arithmetic sequence
   $$6, 13, 20, \ldots .$$
   [Example 2]
2. **Find the sum of an indicated number of terms in a geometric sequence?**
   Find the sum of the first eight terms of the geometric sequence
   $$2, 6, 18, \ldots .$$
   [Example 3]

3. **Write a series given in sigma notation in its expanded form, and determine the sum?**
   Write the series $\sum_{i=1}^{5} 4(\frac{1}{2})^i$ in expanded form, and determine the sum. [Example 5]

4. **Write a series given in expanded form using sigma notation?**
   Find an expression for the general term and write the series $6 + 11 + 16 + 21 + 26 + 31 + 36$ in sigma notation. [Example 6]

5. **Solve applied problems involving series?**
   To help finance their daughter's college education, a couple invests $4,000 on her birthday each year, starting with her 13th birthday. If the money is placed in an account that pays 8 percent interest compounded annually, how much is in the account on her 17th birthday? Assume that the last deposit is made on her 16th birthday. [Example 8]

11.3 Can you:

1. **Find the sum of certain infinite geometric series?**
   Find the sum of the infinite geometric series

$$5 + \frac{5}{3} + \frac{5}{9} + \frac{5}{27} + \ldots.$$

[Example 1]

2. **Use an infinite geometric series to express a repeating decimal as the ratio of two integers?**
   Express the repeating decimal $0.\overline{45}$ as the ratio of two integers. [Example 3]

3. **Solve applied problems involving geometric series?**
   In a certain filtration system wastewater is collected in a holding tank so that each time a liquid is passed through this system only 75 percent as much wastewater is collected as in the previous pass. If 12 gal of wastewater are collected the first time a liquid is passed through this system, how many gallons of wastewater must the holding tank be able to accommodate? [Example 4]

11.4 Can you:

1. **Expand $(a + b)^n$ using Pascal's triangle?**
   Expand $(2x + y)^4$ using Pascal's triangle. [Example 1]

2. **Evaluate binomial coefficients in the form $\binom{n}{r}$ for given values of n and r?**
   Evaluate $\binom{10}{4}$. [Example 5]

3. **Expand $(a + b)^n$ using the binomial theorem?**
   Expand $(y + d)^7$ using the binomial theorem. [Example 6]

4. **Find the rth term in the expansion of $(a + b)^n$?**
   Find the seventh term in the expansion of $(2x - y)^{10}$. [Example 7]

KEY TERMS

Arithmetic sequence (11.1)
Arithmetic series (11.2)
Binomial coefficient (11.4)
Binomial theorem (11.4)
Book value (11.1)
Common difference (11.1)
Common ratio (11.1)
Double-declining balance method (11.1)
Factorial notation (11.4)
Finite sequence (11.1)
Geometric sequence (11.1)
Geometric series (11.2)
Index of summation (11.2)
Infinite sequence (11.1)
Pascal's triangle (11.4)
Sequence (11.1)
Sigma notation (11.2)
Terms (of a sequence) (11.1)

KEY CONCEPTS AND PROCEDURES

Section	Key Concepts or Procedures to Review
11.1	■ Definition of sequence, arithmetic sequence, and geometric sequence ■ Formulas for the nth term of a sequence: Arithmetic sequence: $a_n = a_1 + (n - 1)d$ Geometric sequence: $a_n = a_1 r^{n-1}$
11.2	■ Formulas for the sum of the first n terms of a series: Arithmetic series: $S_n = \frac{n}{2}(a_1 + a_n)$ or $S_n = \frac{n}{2}[2a_1 + (n - 1)d]$ Geometric series: $S_n = \frac{a_1(1 - r^n)}{1 - r}$ ■ The Greek letter Σ, read "sigma," is used to write a series in compact form. In sigma notation, $\sum_{i=1}^{n} a_i = a_1 + a_2 + \cdots + a_n$.
11.3	■ An infinite geometric series with $\lvert r \rvert < 1$ converges to the value or sum $S = \frac{a_1}{1 - r}$.
11.4	■ Pascal's triangle ■ Method to expand $(a + b)^n$ using Pascal's triangle ■ The symbol $n!$ (read "n factorial") means the product $n(n - 1)(n - 2) \cdots (2)(1)$. ■ The binomial coefficient $\binom{n}{r}$ is defined by $\binom{n}{r} = \frac{n!}{r!(n - r)!}$. ■ The binomial coefficients $\binom{n}{0}, \binom{n}{1}, \binom{n}{2}, \ldots, \binom{n}{n}$ are the numbers in the nth row of Pascal's triangle. ■ Binomial theorem: For any positive integer n, $(a + b)^n = \sum_{r=0}^{n} \binom{n}{r} a^{n-r} b^r = \binom{n}{0}a^n + \binom{n}{1}a^{n-1}b + \binom{n}{2}a^{n-2}b^2 + \cdots + \binom{n}{n}b^n$. ■ The rth term of the binomial expansion of $(a + b)^n$ is $\binom{n}{r - 1}a^{n-(r-1)}b^{r-1}$.

CHAPTER 11 REVIEW EXERCISES

11.1

1. Write the first four terms of the sequence given by $a_n = 3n + 2$; also, find a_{60}. 5, 8, 11, 14; 182
2. Find a formula for the general term a_n in the arithmetic sequence 5, 11, 17, 23, ... What is the tenth term in the sequence? $a_n = 6n - 1$; 59
3. Find a formula for the general term in the geometric sequence 3, -6, 12, -24, ... What is the tenth term in the sequence? $a_n = 3(-2)^{n-1}$; $-1{,}536$
4. State whether the following sequence is an arithmetic sequence, a geometric sequence, or neither. If the sequence is arithmetic, find the common difference; if it is geometric, find the common ratio.
   2, -2, -6, -10, ... Arithmetic; -4

5. Use the double-declining balance method to find the depreciated book value at the end of 10 years for property with a useful life of 10 years that is purchased for $80,000. The allowable depreciation rate in this case is 20 percent each year. $8,589.93

11.2

6. Find the sum of the first 30 terms of the arithmetic sequence 5, 9, 13, 17, ... 1,890

7. Find the sum of the first 12 terms of the geometric sequence 2, 8, 32, 128, ... 11,184,810

8. Write the series $\sum_{i=1}^{5}(i^2 + 2)$ in expanded form, and determine the sum. $3 + 6 + 11 + 18 + 27 = 65$

9. Find an expression for the general term and write the series $7 + 11 + 15 + 19 + 23 + 27$ in sigma notation. $a_i = 4i + 3$; $\sum_{i=1}^{6}(4i + 3)$

10. A free-falling body that starts from rest drops about 16 ft the first second, 48 ft the second second, 80 ft the third second, and so on. About how many feet does a parachutist drop during the first 9 seconds of free-fall? 1,296 ft

11.3

11. Find the sum of the infinite geometric series $3 + \frac{3}{4} + \frac{3}{16} + \frac{3}{64} + \cdots$. 4

12. Evaluate $\sum_{i=1}^{\infty}(-\frac{2}{3})^{i-1}$. $\frac{3}{5}$

13. Express the repeating decimal $0.\overline{63}$ as the ratio of two integers. $\frac{7}{11}$

14. In a certain filtration system wastewater is collected in a holding tank so that each time a liquid is passed through the system only 80 percent as much wastewater is collected as on the pass before. If 20 gal of wastewater are collected on the first pass, how many gallons of wastewater must the holding tank be able to accommodate? 100 gal

11.4

15. Expand $(x + 2y)^3$ using Pascal's triangle. $x^3 + 6x^2y + 12xy^2 + 8y^3$

16. Expand $(y - a)^8$ using Pascal's triangle.
$y^8 - 8y^7a + 28y^6a^2 - 56y^5a^3 + 70y^4a^4 - 56y^3a^5 + 28y^2a^6 - 8ya^7 + a^8$

17. Evaluate $\binom{8}{6}$. 28

18. Expand $(x + m)^6$ using the binomial theorem.
$x^6 + 6x^5m + 15x^4m^2 + 20x^3m^3 + 15x^2m^4 + 6xm^5 + m^6$

19. Find the seventh term in the expansion of $(4x + y)^9$. $5{,}376x^3y^6$

ADDITIONAL REVIEW EXERCISES

State whether each of the following sequences is arithmetic, geometric, or neither. If it is an arithmetic sequence, find the common difference. If it is a geometric sequence, find the common ratio.

20. $6, 1, \frac{1}{6}, \frac{1}{36}, \ldots$ Geometric; $r = \frac{1}{6}$

21. 15, 18, 21, 24, . . . Arithmetic; $d = 3$

22. 1, 2, 4, 7, 11, . . . Neither

23. 1, −3, −7, −11, . . . Arithmetic; $d = -4$

Find an expression for the general term a_n in each of the following sequences. What is the tenth term in the sequence?

24. 8, 11, 14, 17, . . . $a_n = 3n + 5$; 35

25. −5, −11, −17, −23, . . . $a_n = -6n + 1$; −59

26. $-3, -\frac{3}{2}, -\frac{3}{4}, -\frac{3}{8}, \ldots$ $a_n = -3(\frac{1}{2})^{n-1}$; $-\frac{3}{512}$

Write the series in expanded form, and determine the sum.

27. $\sum_{i=1}^{6} 8(\frac{1}{2})^i$ $8(\frac{1}{2})^1 + 8(\frac{1}{2})^2 + 8(\frac{1}{2})^3 + 8(\frac{1}{2})^4 + 8(\frac{1}{2})^5 + 8(\frac{1}{2})^6 = \frac{63}{8}$

28. $\sum_{i=1}^{5}(3i - 1)$ $2 + 5 + 8 + 11 + 14 = 40$

Expand using Pascal's triangle.

29. $(4x + y)^3$ $64x^3 + 48x^2y + 12xy^2 + y^3$

30. $(y - 6)^4$ $y^4 - 24y^3 + 216y^2 - 864y + 1{,}296$

31. $(x + h)^5$ $x^5 + 5x^4h + 10x^3h^2 + 10x^2h^3 + 5xh^4 + h^5$

32. Evaluate $\binom{9}{3}$. 84

33. Find the fifth term in the expansion of $(2x + y)^8$. $1{,}120x^4y^4$

34. Find the sum of the first 20 terms of the arithmetic sequence 4, 10, 16, ... 1,220

35. Find the sum of the first nine terms of the geometric sequence 4, 12, 36, ... 39,364

36. Find the sum of the infinite geometric series $5 + \frac{5}{2} + \frac{5}{4} + \frac{5}{8} + \cdots$. 10

37. Find the sum of the first 200 positive integers. 20,100

38. Express $0.\overline{23}$ as the ratio of two integers. $\frac{23}{99}$

39. Find an expression for the general term and write the series $6 + 10 + 14 + \cdots + 42$ in sigma notation.
$a_i = 4i + 2$; $\sum_{i=1}^{10}(4i + 2)$

40. Evaluate $\sum_{i=1}^{\infty}(-\frac{1}{3})^{i-1}$. $\frac{3}{4}$

41. A student guesses randomly at four multiple-choice questions. If each question has five possible choices, then the probability that the student gets exactly three correct is given by the second term in the expansion of $(\frac{1}{5} + \frac{4}{5})^4$. Find this probability. $\frac{16}{625}$

42. A free-falling body that starts from rest drops about 16 ft the first second, 48 ft the second second, 80 ft the third second, and so on. If a ball is dropped off the roof of an office building, how far will it drop during the first 6 seconds of free-fall? 576 ft

43. Suppose that you place $200 in a savings account each year beginning with your 21st birthday. If the account pays 5 percent interest compounded annually, how much will the account be worth on your 26th birthday? (Assume that no deposit is made on your 26th birthday.) $1,160.38

CHAPTER 11 TEST

State whether the sequence is an arithmetic sequence, a geometric sequence, or neither. For any arithmetic or geometric sequence, write the next two terms.

1. 125, 25, 5, 1, . . . Geometric; $\frac{1}{5}$, $\frac{1}{25}$
2. 0.0001, 0.001, 0.01, 0.1, . . . Geometric; 1, 10
3. 1, 5, 9, 13, . . . Arithmetic; 17, 21
4. Find a formula for the general term a_n in the arithmetic sequence $-3, -2, -1, 0, \ldots$, and find a_{25}. $a_n = n - 4$; 21
5. Find a formula for the general term a_n in the geometric sequence $-2, 1, -\frac{1}{2}, \frac{1}{4}, \ldots$, and find a_8. $a_n = -2(-\frac{1}{2})^{n-1}$; $\frac{1}{64}$
6. Find the sum of the first 20 terms of the arithmetic sequence $-3, -1, 1, 3, \ldots$. 320
7. Find the sum of the first eight terms of the geometric sequence 2, 20, 200, ... 22,222,222
8. Find an expression for the general term and write the series $5 + 9 + 13 + \cdots + 41$ in sigma notation. $a_i = 4i + 1$; $\sum_{i=1}^{10}(4i + 1)$

Determine the sum of each series.

9. $\sum_{i=1}^{4}(1 - i^2)$ -26
10. $\sum_{i=1}^{5} 9(\frac{1}{3})^i$ $\frac{121}{27}$
11. $3 + \frac{3}{5} + \frac{3}{25} + \frac{3}{125} + \cdots$ $\frac{15}{4}$
12. $\sum_{i=1}^{\infty}(\frac{1}{2})^{i-1}$ 2
13. Express the repeating decimal $0.\overline{5}$ as the ratio of two integers. $\frac{5}{9}$
14. Determine the entries in the fifth row of Pascal's triangle. 1, 5, 10, 10, 5, 1
15. Expand $(x + 3y)^4$ using Pascal's triangle. $x^4 + 12x^3y + 54x^2y^2 + 108xy^3 + 81y^4$
16. Evaluate $\binom{9}{3}$. 84
17. Expand $(x + y)^7$ using the binomial theorem. $x^7 + 7x^6y + 21x^5y^2 + 35x^4y^3 + 35x^3y^4 + 21x^2y^5 + 7xy^6 + y^7$
18. Find the eighth term in the expansion of $(2x + y)^{10}$. $960x^3y^7$
19. A couple invests $2,000 each year on their son's birthday, starting with his 12th birthday and stopping with his 18th birthday (when no investment is made). If the money is placed in an account that pays 6 percent interest compounded annually, how much is in the account on the son's 18th birthday? $14,787.68
20. A free-falling body that starts from rest drops about 4.9 m the first second, 14.6 m the second second, 24.3 m the third second, and so on. About how many meters does a parachutist drop during the first 7 seconds of free-fall? 238 m

CUMULATIVE TEST 11 (FINAL EXAMINATION)

1. Evaluate $\frac{-x^2 + y^2}{-(x - y)}$ given that $x = -3$ and $y = -2$. -5
2. Solve for x: $|-3x + 1| = 25$. $\{-8, \frac{26}{3}\}$
3. Solve for y_2: $m = \frac{y_2 - y_1}{x_2 - x_1}$. $y_2 = m(x_2 - x_1) + y_1$
4. Perform the indicated operations and simplify. Write the result using only positive exponents: $\frac{(12a^3b^{-2})(3ab)^{-1}}{2a^4b^5}$. $\frac{2}{a^2b^8}$
5. Solve for x: $8x^2 - 2x - 3 = 0$. $\{\frac{3}{4}, -\frac{1}{2}\}$
6. Express $\frac{1}{x^2 - 4} + \frac{2}{x + 2}$ as a single fraction in lowest term. $\frac{2x - 3}{(x + 2)(x - 2)}$
7. Solve for x: $\frac{-2}{x^2} - \frac{3}{2x} = \frac{1}{4}$. $\{-4, -2\}$
8. Determine the x- and y-intercepts of the line whose equation is $-2x + 3y = 6$. x-intercept: $(-3,0)$; y-intercept: $(0,2)$
9. Graph $y \geq -2x - 4$.

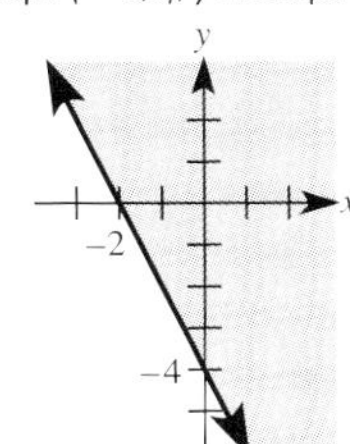

10. Solve the system. $y = 300 + 5x$
$y = 180 + 7x$ (60,600)
11. Solve the system.
$2A - B + 4C = -1$
$A + C = 2$
$B - C = 3$ (4,1,−2)
12. Simplify $\sqrt{45} + \sqrt{320} - \sqrt{180}$. $5\sqrt{5}$
13. Evaluate $x^2 - 2x + 1$ if $x = 3i$. $-8 - 6i$
14. Find the center and radius of the circle given by $(x - 6)^2 + (y + 3)^2 = 4$. Center: $(6, -3)$; radius: 2
15. Use the quadratic formula to solve $3x^2 - x - 1 = 0$. $\left\{\frac{1 \pm \sqrt{13}}{6}\right\}$
16. Determine the domain and range of $y = f(x)$ whose graph is shown in the figure. D: $(-\infty,\infty)$; R: $[-1,4]$

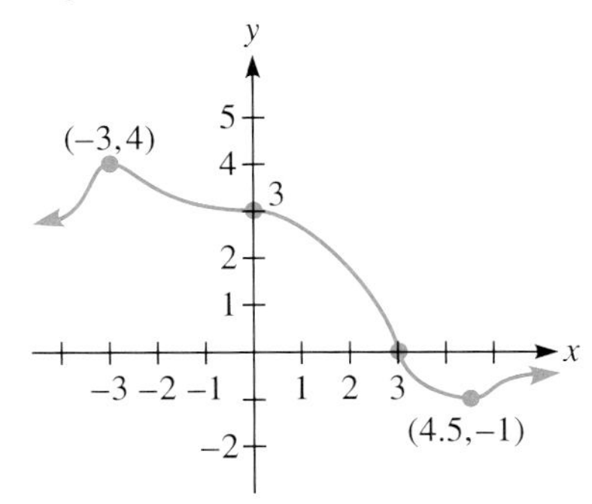

17. If $f(x) = x - 4$ and $g(x) = x^2 + 1$, find $f[g(2)]$. 1
18. Solve $2^{x+1} = 10^x$. Give the solution to four significant digits. $\{0.4307\}$
19. Find the fifth term in the expansion of $(x + 2)^6$. $240x^2$
20. Find the sum of the first 50 terms of the arithmetic sequence $1 + 7 + 13 + 19 + \cdots$. 7,400

Answers to Odd-Numbered Problems

Chapter 1

Exercises 1.1

1. **3.**

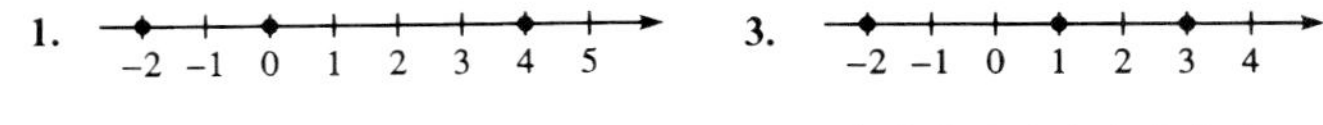

5. 0 $\frac{1}{4}$ $\frac{1}{2}$ 1 **7.** 0 $\frac{8}{9}$ 1 $\frac{9}{8}$

9. $\sqrt{2}$ $\sqrt{4}$ $\sqrt{8}$ 0 1 2 3 **11.** $\frac{\pi}{2}$ π 2π 0 1 2 3 4 5 6 7

13. True **15.** True **17.** True **19.** 5 **21.** −4

23. $-a$ a −1 $-\frac{2}{3}$ 0 $\frac{2}{3}$ 1 **25.** a $-a$ −1.2 −1 0 1 1.2

27. $-a$ a −2 $-1\frac{2}{3}$ −1 0 1 $1\frac{2}{3}$ 2

29. a. Left **b.** Right **31.** Yes; a and $-a$ are equidistant from 0. **33.** $-2, -1, |-1|, |-2|$ **35.** 1.7 **37.** $\frac{1}{5}$ **39.** −4 **41.** 6 **43. a** **45.** a can be any negative number. **47. a** **49.** −100 ft/second **51.** Drop from 3000 to 2910 **53.** True **55.** True **57.** False **59. a.** $\sqrt{9}$ **b.** All **c.** None **d.** All **61. a.** 1, 0 **b.** 3.4, 1, −5.3, 0 **c.** $\sqrt{5}$ **d.** All **63. a.** $\sqrt{1}$, $\sqrt{4}$ **b.** $\sqrt{1}$, $\sqrt{4}$ **c.** $\sqrt{2}$, $\sqrt{3}$, $\sqrt{5}$ **d.** All **65.** 1.01001000100001 . . . never repeats, so it is irrational; 1.010101 . . . does repeat and so is rational. **67.** True **69.** $\{-1,1,-2,2\}$ **71.** $\{-1,1\}$ **73.** True **75.** $\{0\}$ **77.** $\{1\}$ **79.** False **81.** $\{0,1,2,3,4,8,9\}$ **83.** $\{0,2,4,8\}$ **85.** True **87.** True **89.** $\emptyset$ **91.** $\{0\}$

Remember This (1.1)

1. a. iv **b.** ii **c.** i **d.** iii **2.** True **3.** True **4.** True **5.** $\frac{1}{2}, \frac{1}{3}, \frac{1}{4}, \frac{1}{5}, \frac{1}{6}$; other answers are possible **6.** 8 **7.** 5 **8.** True **9.** $\frac{7}{6}$ **10.** $\frac{1}{3}$

Exercises 1.2

1. Sum **3.** Difference **5. b** **7. a** **9.** Start −2 −1 0 −1

11. **13.**

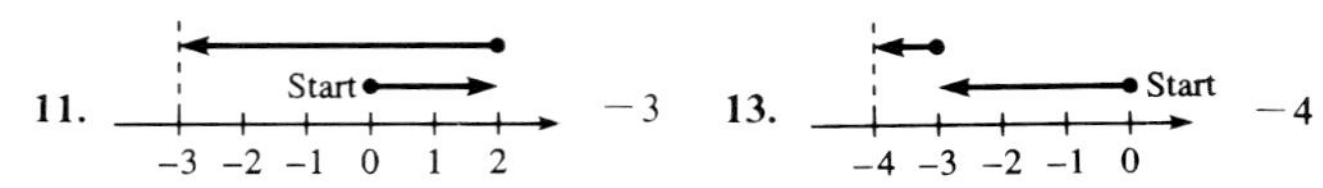

15. 1 minus 5 **17.** 3 minus negative 5 **19.** Negative a minus a **21.** $7 - (-3)$ **23.** $-x - (-4)$ **25.** 8 **27.** −46 **29.** −101 **31.** −52 **33.** −100 **35.** 56 **37.** −0.2 **39.** −1 **41.** −1 **43.** $-\frac{1}{6}$ **45.** $5 + 2$ **47.** $-1 + 2$ **49.** $x + (-y)$ **51.** −7 **53.** 2 **55.** 10 **57.** −3 **59.** −31 **61.** −34 **63.** 80 **65.** 3.2 **67.** $-\frac{7}{12}$ **69.** $-\frac{1}{12}$ **71.** $-\frac{1}{6}$ **73. d** **75. b** **77.** 6 **79.** −1 **81.** 6 **83.** −14 **85.** 0 **87.** Undefined **89.** $-\frac{1}{2}$ **91.** −1 **93.** 30(3) **95.** $\left(\frac{3}{5}\right)\left(\frac{1}{2}\right)$ **97.** −54 **99.** $\frac{1}{6}$ **101.** −14 **103.** 1 **105.** Negative **107.** Negative **109.** 9 **111.** −9 **113.** −1 **115.** −1 **117.** $\frac{1}{16}$ **119.** −1 **121.** 11 **123.** −131 **125.** $\frac{2}{3}$ **127.** −7.44 percent **129.** −6.90 percent

Remember This (1.2)

1. No **2.** 5 **3.** −3 **4.** $2 \cdot 3 - 4$ **5.** True **6.** $a(7 + y)$ **7.** −4 −3 −2 −1 0 1 2 **8.** $-1, 0, \frac{5}{3}, \sqrt{5}, |-3|$ **9.** True **10.** True

Exercises 1.3

1. a. $x + y = y + x = 4$ **b.** $xy = yx = -12$ **c.** $x - y = -8, y - x = 8$ **d.** $x \div y = -\frac{1}{3}, y \div x = -3$ **3. a.** $x + y = y + x = -\frac{5}{6}$ **b.** $xy = yx = \frac{1}{6}$ **c.** $x - y = -\frac{1}{6}, y - x = \frac{1}{6}$ **d.** $x \div y = \frac{3}{2}, y \div x = \frac{2}{3}$ **5. a.** $(a + b) + c = a + (b + c) = 11$ **b.** $(ab)c = a(bc) = -144$ **c.** $(a - b) - c = 13, a - (b - c) = 19$ **d.** $(a \div b) \div c = -1$, $a \div (b \div c) = -9$ **7. a.** $(a + b) + c = a + (b + c) = \frac{7}{12}$ **b.** $(ab)c = a(bc) = -\frac{1}{24}$ **c.** $(a - b) - c = \frac{5}{12}, a - (b - c) = -\frac{1}{12}$ **d.** $(a \div b) \div c = -6, a \div (b \div c) = -\frac{3}{8}$ **9.** Associative, multiplication **11.** Commutative, addition **13.** Commutative, addition **15.** Associative, addition **17.** 0 **19.** Negative or opposite of x **21.** n **23.** −7 **25.** −4; addition identity **27.** 1; multiplication inverse **29.** 0; addition inverse **31. a.** $1 = 1$ **b.** $-5 = -5$ **33. a.** $0 = 0$ **b.** $16 = 16$ **35.** $4x + 20$ **37.** $-6x - 2$ **39.** $-x + y - 2$ **41.** $x - \frac{1}{3}y$ **43.** $3(8 + 9)$ **45.** $(8 + 11)x$ **47.** $a(w + 1)$ **49.** $3(a + b + c)$ **51.** $(2 - 1)x$ **53.** Associative, multiplication **55.** Distributive **57.** Associative, multiplication; commutative, multiplication; associative, multiplication. **59.** $-a - b$ **61.** $-1 + x$ **63.** $7 - 3n$ **65.** $-2x + 3y - 4z$ **67.** $-x + 2y + 3z$

Remember This (1.3)

1. No **2.** Yes **3.** Yes; both are 70. **4.** Yes; both are −2. **5.** Area **6.** 86 **7.** −87 **8.** Reciprocals **9.** −121 **10.** −1

Exercises 1.4

1. $3n, 5w$ **3.** $2x, -4y$ **5.** $x^2, -2x, -4$ **7.** $\frac{1}{2}mn$ **9.** $1, -a$ **11.** $3abc$ **13.** $11x$ **15.** $6a$ **17.** $3w$ **19.** Cannot be simplified **21.** $4x - 6y$ **23.** $5x^2$ **25.** $3x - 3$ **27.** $-x + 5$ **29.** x^2 **31.** $x - 6$ **33.** $6x + 4$ **35.** $-5x + 5$ **37.** 8 **39.** −12 **41.** 22 **43.** 21 **45.** 0 **47.** −1 **49.** $-\frac{1}{4}$ **51.** 294 **53.** 196 **55.** $\frac{13}{14} = 0.93$ **57.** −13.29 **59.** −64,993 **61.** −399 **63.** 19,140 **65.** $20,400 **67.** $535 per month **69.** $9.14x$ **71.** $3.14 + 8x + 6.28x^2$ **73. a.** 707 ft³ **b.** 5,300 gal **75. a.** 197 in.³ **b.** 59 lb **77.** $0.05t$ **79.** $0.005v$ **81.** $v + 5$ **83.** $p + 0.06p$ **85.** $a + bc$ **87.** $\frac{x}{x + 2}$ **89.** The quotient of 2 more than x and a; the sum of x and the quotient of 2 and a **91.** The sum of 3 times x and y **93.** x divided by the product of y and z **95.** 5 percent of the sum of x and y

Remember This (1.4)

1. Yes **2.** Yes **3.** Variable **4.** A is not equal to negative 7. **5.** 5 **6.** $5x$ **7.** $(4 + 5)3$ **8.** $3x + 4$ **9.** True **10.** -5

Chapter 1 Review Exercises

1. (number line with point at $\frac{3}{2}$; marks -2, -1, 0, 1)

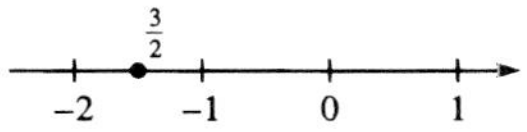

3. 6 **5.** $\sqrt{9}, -\sqrt{4}, 0, -7$ **7.** -1 **9.** -16 **11.** -10 percent **13.** 1; multiplication inverse property **15.** $7 - x$ **17.** $8y - 5$ **19.** 9,586 **21.** $\frac{x - 10}{5}$ **23.** $\emptyset$ **25.** Commutative property of multiplication **27.** -4; identity, addition **29.** -6 **31.** 0 **33.** $3s^2, -4s, 12$ **35.** -1 **37.** x^3 **39.** $\frac{5}{6}$ **41.** 64 **43.** $-\frac{16}{11}$ **45.** $15x - 10$ **47.** $-2y + 2$ **49.** The sum of a and 4 times b **51.** True **53.** $1,280

Chapter 1 Test

1. Commutative property of multiplication **2.** 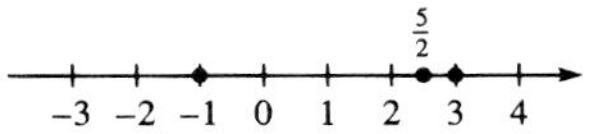 (number line with points marked, including $\frac{5}{2}$; marks -3, -2, -1, 0, 1, 2, 3, 4)

3. $\frac{1}{2}(x + y)$ **4.** **d** **5.** $2x - y + 1$ **6.** 0 **7.** $-8x + 9$ **8.** $2(x + y)$ **9.** The difference between a and two times b **10.** **b** **11.** $\{-3, -2, -1, 0\}$ **12.** 4,320 in.^{3} **13.** 5,695 **14.** $-a + 2b + c$ **15.** -125 **16.** 7 **17.** -12 **18.** -3 **19.** -6 **20.** 0

Chapter 2

Exercises 2.1

1. Identity **3.** Conditional equation **5.** Identity **7.** False equation **9.** False equation **11.** Conditional equation **13.** Identity **15.** No **17.** Yes **19.** No **21.** No **23.** Yes **25.** No **27.** No **29.** $\{-2\}$ **31.** $\{3\}$ **33.** $\{1\}$ **35.** $\{6\}$ **37.** $\{7\}$ **39.** $\{-1\}$ **41.** $\{9\}$ **43.** $\{1\}$ **45.** $\{-3\}$ **47.** $\{\frac{5}{2}\}$ **49.** False equation; $\emptyset$ **51.** Identity; all real numbers **53.** $\{-4\}$ **55.** $\{3\}$ **57.** $\{6\}$ **59.** $\{4.5\}$ **61.** $\{7\}$ **63.** False equation; $\emptyset$ **65.** Identity; all real numbers **67.** $\{14\}$ **69.** $\{5\}$ **71.** $\{24\}$ **73.** $\{72\}$ **75.** $\{2\}$ **77.** $\{3\}$ **79.** $\{\frac{1}{4}\}$ **81.** $\{24\}$ **83.** $159,574 **85.** $1,111,111 **91.** $6.50x - 5.00x - 50 = 100$; 100 items

Remember This (2.1)

1. $V = \pi r^2 h$ **2.** False **3.** 86 **4.** $-y$ **5.** $-b$ **6.** $-2x + 1$ **7.** 4.44 **8.** $x + 2$ **9.** $3x + 3y$ **10.** True

Exercises 2.2

1. 12 **3.** 0 **5.** 32.77 **7.** 14 **9.** -4 **11.** 100 **13.** 0.53 **15.** -2 **17.** 12 **19.** 10.19 **21.** 32 **23.** 20 **25.** 3 **27.** 89.57 **29.** 0.42 **31.** 7 **33.** $b = \frac{2A}{h}$ **35.** $r^3 = \frac{3V}{4\pi}$ **37.** $a = \ell - (n - 1)d$ **39.** $n = \frac{\ell + d - a}{d}$ **41.** $a = \frac{2S - n\ell}{n}$ **43.** $x = \frac{y - b}{m}$ **45.** $n = \frac{90A}{\pi r^2}$ **47.** $y = 3x$ **49.** $y = 3x - 7$ **51.** $y = x - 3$ **53.** $y = -\frac{6}{5}x - 4$ **55.** $y = x - 12$ **57.** $y = -\frac{8}{3}x + 16$ **59.** $y = x + 3$ **61.** $x = \frac{s}{r}$ **63.** $x = \frac{c + b}{a}$ **65.** **a.** $w = 5.5h - 220$ **b.** 143 **67.** **a.** $k = 1.609m$ **b.** 40,000 km **69.** **a.** $t = \frac{d}{1{,}100}$ **b.** 1.4 seconds

Remember This (2.2)

1. $\frac{1}{4}x$ **2.** $x + 1$ **3.** 180° **4.** 10,000 **5.** $0.30x$ **6.** $\{\frac{8}{3}\}$ **7.** $-\frac{3}{19}$ **8.** Distributive **9.** Negative **10.** $\sqrt{2}, \sqrt{200}$

Exercises 2.3

1. 240 for one family, 60 for the other **3.** $105 **5.** 15.5, 4.5 ft **7.** 42, 6 ft **9.** **a.** 16,000 ft^2, 4,000 ft^2 **b.** Yes, exactly **11.** $36,000, $24,000 **13.** Yes; 99 **15.** 1,500 **17.** 10 **19.** -16 **21.** 73, 810 **23.** $x + (x + 1) + (x + 2) = 3(x + 2) - 3$ **25.** $x + (x + 2) + (x + 4) + (x + 6) = 4(x + 2) + 4$ **27.** $x + (x + 1) + (x + 2) = 47$; no integer solution **29.** $3(x + 1) = 3(x + 2) - 3$ **31.** 67.5° **33.** 54°, 36°, 90° **35.** 56° **37.** 20°, 55°, 105° **39.** 58°, 60°, 62° **41.** 35°, 145° **43.** 41°, 49° **45.** 9, 18, 25 cm **47.** $1,250 at 8 percent, $13,750 at 6 percent **49.** $13,777.78 at 7 percent, $15,777.78 at 11 percent **51.** 7 percent **53.** $190,500 **55.** **a.** $271,000 at 5 percent, $104,000 at 12 percent **b.** $129,000 at 5 percent, $246,000 at 12 percent **57.** $24,706 **59.** In 3 hours **61.** 44 mi/hour and 48 mi/hour; 264 mi **63.** 660 mi **65.** 3, 6 gal **67.** 6 gal of 6 percent, 2 gal of 8 percent **69.** 0.125 oz of each **71.** 36.25 percent

Remember This (2.3)

1. Yes **2.** True **3.** Yes **4.** **a** **5.** $-5x + 2$ **6.** $x = \frac{y + 4}{3}$ **7.** False equation **8.** No, $0 \neq \emptyset$ **9.** $2x - 3$ **10.** $3ab + (2cd + 5ef)$

Exercises 2.4

1. $(-\infty, 4]$ (graph: 4) **3.** $[-1, \infty)$ (graph: -1) **5.** $(-\infty, 0)$ (graph: 0) **7.** $[\frac{1}{2}, \infty)$ (graph: $\frac{1}{2}$) **9.** $(-\infty, 3)$ (graph: 3) **11.** $[4, \infty)$ (graph: 4) **13.** $[\frac{1}{2}, \infty)$ (graph: $\frac{1}{2}$) **15.** $(-14, \infty)$ (graph: -14) **17.** $(-\infty, -9]$ (graph: -9) 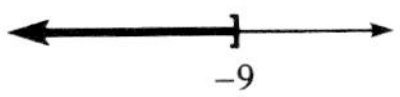 **19.** $(-\infty, -\frac{3}{4})$ (graph: $-\frac{3}{4}$) **21.** $(-3, \infty)$ (graph: -3) 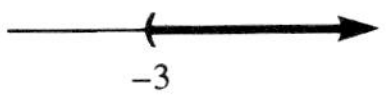 **23.** $(-\infty, 3)$ (graph: 3) **25.** $[4, \infty)$ (graph: 4) **27.** $(-\infty, -4]$ (graph: -4) **29.** $(4, \infty)$ (graph: 4) 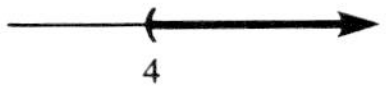 **31.** $(-5, \infty)$ (graph: -5) **33.** $(-\infty, 9)$ (graph: 9) 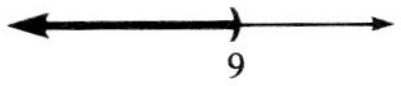 **35.** $(-\infty, 2]$ (graph: 2) **37.** $(-10, \infty)$ (graph: -10) 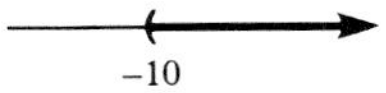 **39.** $[-2, \infty)$ (graph: -2) **41.** $(-\infty, 1]$ (graph: 1) 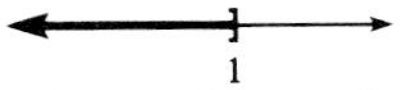 **43.** $[1, \infty)$ (graph: 1) **45.** $(-\infty, \infty)$ (graph: 0) **47.** $\emptyset$ **49.** $(-\infty, \infty)$ (graph: 0) **51.** More than 20 **53.** $k > 27.21$

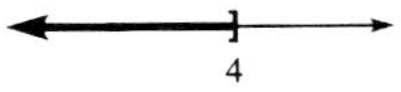

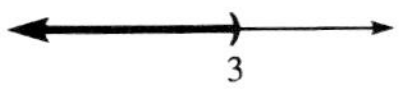

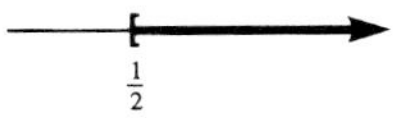

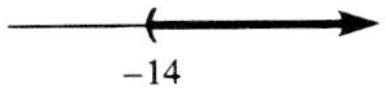

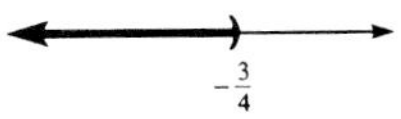

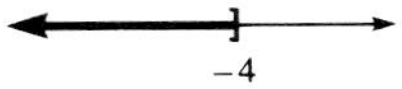

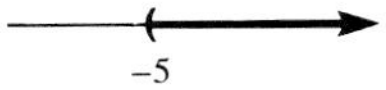

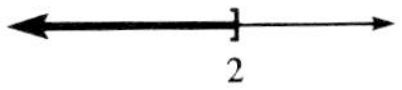

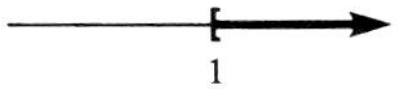

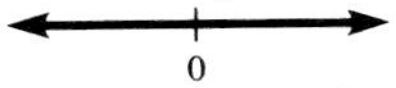

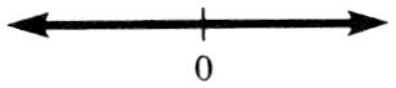

Remember This (2.4)

1. True **2.** $\{8, 9, 10\}$ **3.** $\{0, 1, 2, 3\}$ **4.** $x = -\frac{1}{2}$ is not an integer. **5.** $2x + 2(2x) = 9$; width $= \frac{3}{2}$ in., length $= 3$ in. **6.** $x = \frac{y - b}{m}$ **7.** $\{\frac{3}{4}\}$ **8.** $\frac{1}{n} + (-n)$ **9.** 0 **10.** $-\frac{1}{5}$

Exercises 2.5

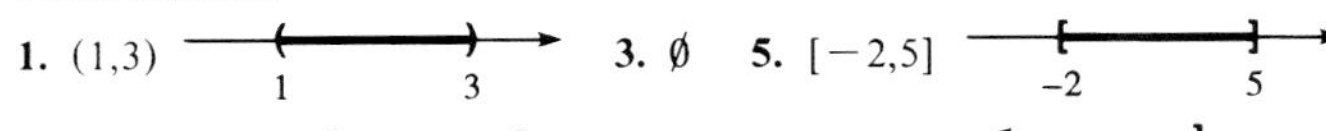

1. $(1,3)$ **3.** $\emptyset$ **5.** $[-2,5]$ **7.** $(-1,4]$ **9.** $(-\infty,-5]$ **11.** $[-5,3)$ **13.** $\emptyset$ **15.** $(-3,-\frac{3}{4}]$ **17.** $[-8,-2]$ **19.** $[-1,9)$ **21.** $(-\infty,-5)$ **23.** $(-1,2]$ **25.** $[-4,-1)$ **27.** $(2,3)$ **29.** $(5,31)$ **31.** $(2,14]$ **33.** $[-14,4]$ **35.** $[80,100]$ **37.** $2.78 < C < 3.89$ **39.** $(-\infty,5) \cup (7,\infty)$ **41.** $(-\infty,\infty)$ **43.** $(-\infty,7)$ **45.** $(-\infty,-5) \cup (1,\infty)$ **47.** $(-\infty,\infty)$ **49.** $(-\infty,8]$ **51.** $(-\infty,1)$ **53.** $(-\infty,-1) \cup [2,\infty)$ **55.** $(-\infty,\infty)$ **57.** $\emptyset$ **59.** $[-1,2]$ **61.** $(-\infty,7] \cup [12,\infty)$

Remember This (2.5)

1. No **2.** Yes **3.** Absolute value of x can never be negative. **4.** 384 ft **5.** (graph from $-\frac{4}{5}$ to 0) **6.** 4.1 qt of 2 percent oil, 3.9 qt of 100 percent oil **7.** 4.5 in. **8.** $\{\frac{24}{5}\}$ **9.** $x + 0.05x = 1.05x$ **10.** 1

Exercises 2.6

1. $\{1,-1\}$ **3.** $\{-7,7\}$ **5.** $\emptyset$ **7.** $\{0\}$ **9.** $\{-\frac{1}{3},3\}$ **11.** $\{9\}$ **13.** $\{7.5,-10\}$ **15.** $\{-6,12\}$ **17.** $\{4,-5\}$ **19.** $\emptyset$ **21.** 5, 10 seconds **23.** $\{-\frac{92}{3},36\}$ **25.** $\{0,2\}$ **27.** $\{0.75,-3.5\}$ **29.** $\{-10,-\frac{1}{4}\}$ **31.** $\{1,\frac{15}{16}\}$

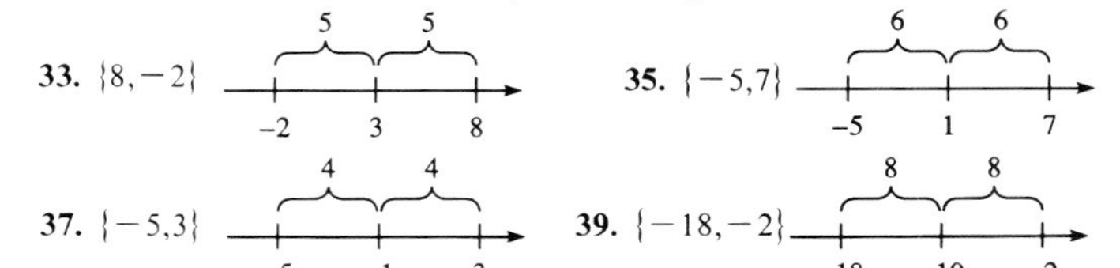

33. $\{8,-2\}$ **35.** $\{-5,7\}$ **37.** $\{-5,3\}$ **39.** $\{-18,-2\}$ **41.** $[0,12]$ **43.** $[-2,12]$ **45.** $(-18,-2)$ **47.** $(-6,8)$ **49.** $[-10,-6]$ **51.** $(-9,19)$ **53.** $(5,10)$ **55.** $(-\infty,-18.75) \cup (31.25,\infty)$ **57.** $(-\infty,-2) \cup (4,\infty)$ **59.** $(-\infty,5] \cup [19,\infty)$ **61.** $(-\infty,-5) \cup (1,\infty)$ **63.** $(-\infty,-\frac{4}{3}] \cup [2,\infty)$ **65.** $(-\infty,-2.5] \cup [4,\infty)$ **67.** $(-\infty,\frac{7}{3}] \cup [3,\infty)$ **69.** $(-\infty,-\frac{8}{7}) \cup (2,\infty)$ **71.** $[-4.5,5]$ **73.** $(-\infty,\frac{1}{2}) \cup (2,\infty)$

Remember This (2.6)

1. $(-5)(-5)(-5)(-5)$ **2.** Base **3.** a **4.** Both equal $-1{,}792$ **5.** True **6.** $(5,10]$ **7.** \$14.50 **8.** $a = 2A - b$ **9.** $\emptyset$ **10.** a

Chapter 2 Review Exercises

1. False equation **3.** $\{-\frac{1}{2}\}$ **5.** $\{3\}$ **7.** 82 **9.** $y = 3x - 8$ **11.** \$14,200, \$42,600 **13.** 25°, 65°, 90° **15.** 1 hour and 12 minutes **17.** $[2,\infty)$ **19.** $(-\infty,1)$ **21.** $(-\infty,\infty)$ **23.** $[1,3)$ **25.** $(-\infty,2) \cup (4,\infty)$ **27.** $\{-\frac{7}{3},5\}$ **29.** $\{-1,7\}$ **31.** $(-\infty,-1] \cup [-\frac{1}{3},\infty)$ **33.** b **35.** $\{-4\}$ **37.** $\{-18,21\}$ **39.** 15°, 75°, 90° **41.** 38 liters of 35 percent, 57 liters of 10 percent **43.** \$2,100 at 5 percent, \$2,500 at 6 percent **45.** $b = 3A - a - c$

47. 101 **49.** $(\infty,-2]$ −2

51. $(-2,\infty)$ −2

53. $(-\infty,-4) \cup (1,\infty)$ −4 1

55. $(-\infty,-2]$ −2

57. $(-\infty,-3] \cup [7,\infty)$ −3 7 **59.** $(-5,8)$ −5 8

Chapter 2 Test

1. a **2.** 32°, 58° **3.** 3:30 P.M. **4.** $\{-5\}$ **5.** $\{\frac{5}{6}\}$ **6.** $\{-4,3\}$
7. $\{-10,2\}$ **8.** $y = x - z - 10P$ **9.** $y = -\frac{5}{3}x + 2$ **10.** $x = \frac{9}{2}$
11. $(-\infty,-3]$ **12.** $(-\infty,\infty)$ **13.** $[1,7]$ **14.** $(-4,5)$ **15.** $\emptyset$
16. −3 15 **17.** $\frac{2}{3}$
18. 0 **19.** 2 3
20. −1 3

Cumulative Test 2

1. $w = \dfrac{P - 2\ell}{2}$ **2.** $\{1,2,3, \ldots\}$ **3.** $\frac{5}{6}$ **4.** \$440 **5.** −4
6. Associative property of multiplication **7.** $-2x + 3y - 5$
8. 25.12 in. **9.** $2(x + y)$ **10.** −4 **11.** $10a + 10$ **12.** 5,515.38
13. 47.5°, 42.5° **14.** $\{\frac{2}{3}\}$ **15.** $\{-\frac{13}{3},-3\}$ **16.** $\{\frac{5}{4}\}$ **17.** $\{-\frac{3}{5},-5\}$
18. $(-\infty,2]$ 0 2 **19.** $(1,5)$ 1 5
20. $(-15,3)$ −15 3

Chapter 3

Exercises 3.1

1. 6^3; 6, 3 **3.** $(-a)^4$; $-a$, 4 **5.** -3^4; 3, 4 **7.** m^1; m, 1 **9.** 4^8
11. x^5 **13.** y^4 **15.** m^6 **17.** $-6x^4$, x^2 **19.** x^{10}, $2x^5$ **21.** $-x^2$, 0
23. 3 **25.** 1 **27.** 3 **29.** 1 **31.** 1 **33.** $\frac{1}{2}$ **35.** $\frac{1}{16}$ **37.** $\frac{5}{3}$
39. $\frac{1}{72}$ **41.** $\dfrac{2}{x^3}$ **43.** $\dfrac{1}{8x^3}$ **45.** a^3 **47.** $\dfrac{1}{2^6}$ **49.** 4 **51.** $125x^3$
53. $\dfrac{27}{x^3}$ **55.** $\dfrac{c^5}{a}$ **57.** 2^{12} **59.** $\dfrac{1}{x^2}$ **61.** $\dfrac{1}{a^4}$ **63.** $\dfrac{1}{2^{12}}$ **65.** 1 **67.** $8x^6$
69. $-8x^9y^6$ **71.** $\dfrac{a^2}{25x^2}$ **73.** $\dfrac{x^6}{27}$ **75.** $\frac{9}{4}$ **77.** $\dfrac{27}{8x^6}$ **79.** 4.661
81. 474.310 **83.** \$848.16

Remember This (3.1)

1. 10^7 **2.** 1 **3.** True **4.** 5^{n+3} **5.** $27y^6$ **6.** $\{4,2\}$ **7.** Distributive
8. $\ell = \dfrac{P - 2w}{2}$ **9.** \$210 **10.** −1 1 x

Exercises 3.2

1. $9x^{12}$ **3.** $12x^4$ **5.** $\dfrac{9}{2x^4}$ **7.** y^6 **9.** $\dfrac{4}{x^4}$ **11.** $-\dfrac{3}{2y^3}$ **13.** $\dfrac{b}{a^4}$ **15.** $\frac{1}{18}$
17. $\dfrac{2}{3y^6}$ **19.** 2^{2n+1} **21.** 3^{n+2} **23.** 1 **25.** x^{2n} **27.** 2 **29.** x^{2n+2}
31. x^{-n+1} **33.** 2^{6n+3} **35.** $\dfrac{4}{n^2}$ **37.** 1.1×10^3 **39.** 9.3×10^7
41. 3.9×10^9 **43.** 5×10^{-11} **45.** -5.79×10^8 **47.** -3.2×10^{-4}
49. 3,400,000,000,000,000,000,000,000,000 **51.** 0.000 000 003
53. 0.000 000 000 000 000 000 16 **55.** 511,000 **57.** −0.01
59. −0.35 **61.** 7.2 **63.** $-7 \times 10^3 = -7{,}000$ **65.** 9×10^{16}
67. 1.671×10^{-27} kg **69.** 2.58×10^9
71. 3×10^9 seconds; about 95 years
73. $\dfrac{4.252 \times 10^{11}}{9.755 \times 10^{11}} \times 100 = 43.6$ percent
75. 1.8×10^{17} joules, 43.2 million tons of TNT

Remember This (3.2)

1. 3 **2.** True **3.** True **4.** $5x^3$ and $4x^3$ **5.** $9x^4$ **6.** 3,561.67
7. Base, exponent **8.** Negative integers **9.** $\frac{1}{5}$ **10.** $\{0\}$

Exercises 3.3

1. Y **3.** Y **5.** Y **7.** N **9.** N **11.** Y **13.** 5; trinomial
15. 2; monomial **17.** 2; trinomial **19.** 0; monomial **21.** 2; binomial
23. 3 **25.** 14 **27.** 1.7 **29.** $\frac{3}{4}$ **31.** −2 **33.** 9, −3 **35.** 6, 2
37. $3 + 4 = 7$ **39.** True **41.** 30 **43.** 41; 43; 47; $1{,}681 = 41^2$
45. a. $\frac{1}{2}(16 + 4) = 10$, $\frac{1}{2}(25 + 5) = 15$ **b.** 5,050 **47.** $4x^2 - 2x + 8$
49. $x + 2$ **51.** $x^3 + x^2 - 1$ **53.** $3x + 2$ **55.** $2x^2 + \dfrac{5x}{3} + \dfrac{1}{8}$
57. $x^4 + x^3 + x^2 - x - 1$ **59.** $x - 5$ **61.** $-2x + 2$
63. $x^2 + x - 1$ **65.** $x^3 - x^2 + 3x - 1$ **67.** 0 **69.** $\dfrac{x^2}{4} + \dfrac{x}{6} - \dfrac{3}{8}$
71. $3x^2 + x - 6$ **73.** $11x^2 - 13x + 10$ **75.** $-2x^3 + 4x + 2$
77. $2x$ **79.** $\dfrac{x^2}{4} + \dfrac{2x}{35} + \dfrac{1}{3}$ **81.** $-2x^2 + 5x - 5$ **83.** $4x^2 + 20x + 6$
85. $\{6\}$ **87.** $\{3\}$

Remember This (3.3)

1. $28x^6$ **2.** $xn + xm$ **3.** True **4.** $3x + (-4y)$ **5.** $-6x^6$
6. 5.86×10^{14} **7.** False **8.** $\{-7,13\}$ **9.** $\sqrt{2}$, $\sqrt{3}$ **10.** \$580

Exercises 3.4

1. $-14x^8$ **3.** x^5 **5.** $4x^3$ **7.** $20x^3y^4$ **9.** $12x^3 - 15x^2 + 18x$
11. $-2y^3 + 6y^2 - 10y$ **13.** $x^3y - 2x^2y^2 + xy^3$ **15.** $3x^6 + x^3 + 2$
17. $12x^2 + 17x + 6$ **19.** $4y^3 + 15y^2 + 7y - 6$ **21.** $x^3 - 1$
23. $x^4 + 4x^3 + 6x^2 + 4x + 1$ **25.** $x^3 + 3x^2y + 3xy^2 + y^3$
27. $x^2 + 8x + 15$ **29.** $6x^2 - 14x + 4$ **31.** $2x^2 - \frac{7}{2}x - 1$
33. $a^2 - ab - 6b^2$ **35.** $6a^3 + 4a^2 - 12a - 8$
37. $m^3 + m^2n + mn^2 + n^3$
39. Sum and difference of two expressions; $x^2 - 49$
41. Sum and difference of two expressions; $4y^2 - 25$
43. Sum and difference of two expressions; $25x^2 - 9y^2$
45. Square of binomial; $y^2 + 18y + 81$
47. Square of binomial; $4x^2 - 28x + 49$
49. Square of binomial; $x^2 + 6xy + 9y^2$
51. Square of binomial; $x^4 + 2x^3 + x^2$ **53.** $m^3 - 3m^2x + 3mx^2 - x^3$
55. $x^3 + 6x^2 + 12x + 8$ **57.** $8x^3 - 12x^2y + 6xy^2 - y^3$
59. Because $x^2 \geq 0$, $100 - x^2$ cannot be more than 100. **61. a.** $25 - x^2$
b. $x = 0$ **63.** $A = 63$ in.^{2}

Remember This (3.4)

1. Multiplication **2.** 1 **3.** $2^2 \cdot 3 \cdot 5$ **4.** $5x^2$ **5.** True **6.** 9 **7.** 2
8. $\dfrac{1}{y^8}$ **9.** Coefficient **10.** 5 mi

Exercises 3.5

1. 6 **3.** 6 **5.** $2x$ **7.** $7x$ **9.** $3(2x + 3)$ **11.** $4x(3x + 2)$
13. $xy(3x + 5y)$ **15.** $2x^2y(-2 + 3x)$ **17.** $m(m - 1)$
19. $3x^2(1 + 2x + 3x^2)$ **21.** $4x(3 - 4x - 2x^2)$ **23.** $n^2(-n^4 + n^2 - 1)$
25. $8xy(y^2 + 8xy + x^2)$ **27.** $9ab^2c^3(c^2 - 4abc - 9a^2b^2)$
29. $(x + 3)(x + 7)$ **31.** $(x - 1)(3y^2 + 2)$ **33.** $(x - 2)(2x + 4)$
35. $6(x + 4)$ **37.** $(x + 1)[(x + 1)^2 - 2(x + 1) + 3]$
39. $(x + 2)(2x + 3)$ **41.** $(3x + 1)(5x - 4)$ **43.** $(y - 5)^2$
45. $(3a + 4b)^2$ **47.** $(2m - n)(3m + n)$ **49.** $(1 + a)(1 + x)$
51. $0.91x$; less **53.** $1.32x$; a 32 percent increase of the original price

Remember This (3.5)

1. True **2.** $-7, -2$ **3.** $-2, 15$ **4.** True **5.** $\frac{3}{5}, \sqrt{5}$
6. $9x^2 - 24x + 16$ **7.** 15 **8.** $\frac{9}{4x^4y^6}$ **9.** \$444.89
10. 

Exercises 3.6

1. $(x + 3)(x + 5)$ **3.** $(y + 1)(y + 9)$ **5.** $(t + 6)(t + 2)$
7. $(x - 3)(x - 5)$ **9.** $(y - 24)(y - 1)$ **11.** $(c - 3)(c - 6)$
13. $(x - 3)(x + 8)$ **15.** $(y - 4)(y + 2)$ **17.** $(s - 3)(s + 2)$
19. $(x + a)(x + 2a)$ **21.** $(y - 4b)(y - 3b)$ **23.** $(d - b)(d + 2b)$
25. $2(x + 3)(x + 5)$ **27.** $4(y - 1)(y - 2)$ **29.** $6(z + 2)(z - 4)$
31. $x^2 + 2x + 2$ **33.** $y^2 - 3y + 3$ **35.** $(3x + 2)(5x + 7)$
37. $(2t - 1)(3t - 2)$ **39.** $(2y + 3)(2y - 5)$ **41.** $2x(2x + 1)(4x - 3)$
43. $3y^2(y - 1)(y - 2)$ **45.** $4ab(3a + 2b)(a + b)$ **47.** $(t + 5)(t + 7)$
49. $(2r - 1)(3r + 8)$ **51.** $(2x - 5)(2x - 7)$ **53.** $(y^2 + 3)(y^2 + 5)$
55. $(3t^2 - 1)(2t^2 - 9)$ **57.** $(xy + 1)(xy + 2)$
59. $(2x + 11)(x - 5) = 0$

Remember This (3.6)

1. $x^2 + 2ax + a^2$ **2.** c **3.** $\{-3,0\}$ **4.** $y^2 - c^2$ **5.** $a^3 - b^3$
6. $6x^2y(5x^2 + 4y^2)$ **7.** $(3x + 1)(x + 5)$
8. Associative property of multiplication **9.** -2 **10.** \$40,000

Exercises 3.7

1. $(x + 5)^2$ **3.** $(3x + 1)^2$ **5.** $(2x - 3)^2$ **7.** $(4a - 3b)^2$
9. $(8x + 3)(2x + 3)$ **11.** $(y - 9)(y - 4)$ **13.** $(m + 5)(m - 5)$
15. $(3 + x)(3 - x)$ **17.** $(2x + 3y)(2x - 3y)$
19. $(5ac + 6bd)(5ac - 6bd)$ **21.** $(4t^3 + 1)(4t^3 - 1)$
23. $(3x^2 + 5y^4)(3x^2 - 5y^4)$ **25.** $(x + y + 3)(x + y - 3)$
27. $(2x + 3y + 4z)(2x + 3y - 4z)$ **29.** $3x(x + 1)(x - 1)$
31. $2xy(x + 3y)(x - 3y)$ **33.** $(x^2 + 9)(x + 3)(x - 3)$
35. $(4x^2 + 9y^4)(2x + 3y^2)(2x - 3y^2)$ **37.** $(x + 1)(x^2 - x + 1)$
39. $(2m + 3n)(4m^2 - 6mn + 9n^2)$ **41.** $(xy + 2)(x^2y^2 - 2xy + 4)$
43. $(x^2 + 1)(x^4 - x^2 + 1)$ **45.** $2(m - 2)(m^2 + 2m + 4)$
47. $2x(y^3 + 2x^2)(y^6 - 2x^2y^3 + 4x^4)$ **49.** $4a(a + b)(a - b)$
51. a. $(c + d)^2$ **b.** 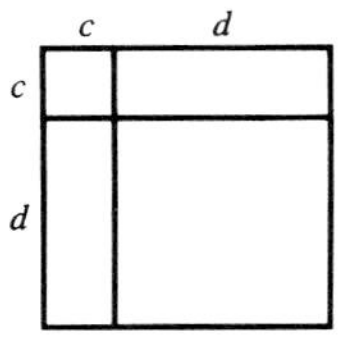

53. $\pi(R + r)(R - r)$

Remember This (3.7)

1. $a^2 - ab + b^2$ **2.** $(x + n)(x + m)$ **3.** $5xy(x^2 - 3x + 2)$
4. $(x + 2)(x + 8)$ **5.** $(x + 4)(x - 3)$ **6.** $(6x + 1)(x - 6)$ **7.** 0
8. 2^{m+1} **9.** 4.21×10^{-4} **10.** $h = \frac{S - 2\pi r^2}{2\pi r}$; 1

Exercises 3.8

1. $x(x + 1)$ **3.** $3(3x - 1)(2x + 1)$ **5.** $2y(y - 5)(y + 7)$
7. $(2a + 5)(5a + 2)$ **9.** Prime **11.** $(4d^2 + e^2)(2d + e)(2d - e)$
13. $(x + c)(x + 2)$ **15.** $(a + 1)(b + 2)c$
17. $(2x + 1)(4x^2 - 2x + 1)(2x - 1)(4x^2 + 2x + 1)$ **19.** $(a + b + 1)^2$
21. $s^2t^2(2st + 1)(2st - 1)$ **23.** $(x + 2)(x - 2)(x^2 + 1)$
25. $(x + y + 3)(x + y - 3)$ **27.** Prime **29.** $xy(y + 3x)^2$
31. $(6w + 1)(4w - 15)$ **33.** $(a - 3 + b)(a - 3 - b)$
35. $(n + 10)(n^2 - 10n + 100)$
37. $(a + b + c)[(a + b)^2 - (a + b)c + c^2]$ **39.** Prime
41. $(xy + 4)(xy - 3)$ **43.** $(x^3 - 2a)(x^3 + 2a)$ **45.** $(x + c)(x + 2)$
47. $(x - 6)(x - c)$ **49. a.** $y = 225 - 4t^2$
b. $y = (15 + 2t)(15 - 2t)$ **c.** When $t = 7.5$, $y = 0$
51. $P = 2(a + b)^2$ **53.** $A = (a + b)(a - b)$

Remember This (3.8)

1. 2 **2.** True **3.** False; $\frac{1}{2} \cdot 2 = 1$
4. $4 \cdot 0^2 \stackrel{?}{=} 20 \cdot 0, 0 \stackrel{\checkmark}{=} 0$; $4 \cdot 5^2 \stackrel{?}{=} 20 \cdot 5, 100 \stackrel{\checkmark}{=} 100$. **5.** d
6. $(x + 4y)(x - 4y)$ **7.** $2(2y + 1)(y - 3)$ **8.** 6×10^{14}
9. $\{x{:}x > -2\}$ 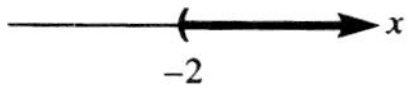

10. 2.5 gal

Exercises 3.9

1. $\{0,6\}$ **3.** $\{0,-3\}$ **5.** $\{0,\frac{18}{5}\}$ **7.** $\{7,-7\}$ **9.** $\{1,-1\}$ **11.** $\{2,5\}$
13. $\{-1,-6\}$ **15.** $\{1,-2\}$ **17.** $\{-\frac{1}{2},2\}$ **19.** $\{\frac{1}{3},\frac{1}{4}\}$ **21.** $\{2\}$
23. $\{0,\frac{1}{3}\}$ **25.** $\{1,-2\}$ **27.** $\{4,-6\}$ **29.** $\{0,\frac{13}{6}\}$ **31.** $\{7,-7\}$
33. $\{412,-6.02,-\pi\}$ **35.** $\{\frac{3}{2},\frac{4}{3},\frac{5}{4}\}$ **37.** $\{0,1,-1\}$ **39.** $\{0,1,2\}$
41. $\{-5,8\}$ **43.** $x^2 - 8x + 15 = 0$ **45.** $x^2 - 36 = 0$
47. $x^3 - 12x^2 + 44x - 48 = 0$ **49.** $t = \frac{3}{2}$ seconds **51.** 8 seconds
53. 7, 8 and $-7, -8$ **55.** 8 and -5
57. Side = 6; area = volume = 216 **59.** 49 ft²
61. Length = 12 in.; width = 2 in. **63.** 192 yd²
65. A to B takes 36 minutes; A to X to B takes 42 minutes.

Remember This (3.9)

1. 3/0 and 0/0 **2.** $\frac{2}{3}$ **3.** True **4.** $18 = 2 \cdot 3^2$ **5.** 0 **6.** True
7. $\{2,5\}$ **8.** $-x^2 + x$ **9.** 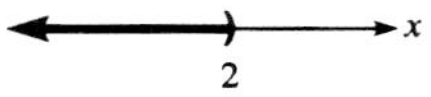

10. \$2.20

Chapter 3 Review Exercises

1. $(-b)^6$; base $= -b$; exponent $= 6$ **3.** 4 **5.** \$1,018.61 **7.** 3
9. -8×10^5; $-800,000$ **11.** No, because the exponents are negative
13. $P(-1) = 9$; $P(1) = 3$ **15.** $2x^2 - x + 3$ **17.** $x^3 - x^2 - 10x + 12$
19. $4x^2 + 4xy + y^2$ **21.** $8xy^2$ **23.** $(y - 3)(7 - y)$
25. $1.21x$; the result is a 21 percent increase of the original price.
27. $(y + 8)(y - 3)$ **29.** $(6x^2 - 1)(3x^2 + 1)$ **31.** $(3x - 4)^2$
33. $6y(y + 1)(y - 1)$ **35.** $(c + d)(c - 2d)$ **37.** $c(a + 2)(b + 1)$
39. $(x^2 + h^2)(x + h)(x - h)$ **41.** $\{\frac{1}{2},4\}$ **43.** $\{\frac{1}{2},-\frac{4}{3},-\frac{1}{4}\}$
45. 64 square units **47.** b **49.** $(x + y)^3$; base: $x + y$; exponent: 3
51. $x^2 - 4xy + 4y^2$ **53.** -15 and -16; 15 and 16
55. $4xy^2(2x^2 + 6xy + 3)$ **57.** $(x - 3)(2x + 5)$ **59.** 1 **61.** $\{-3,0,3\}$
63. $\{-\frac{5}{2},3\}$ **65.** $\frac{27x^3}{y^6}$ **67.** $(-x)^8$ or x^8 **69.** $\frac{49}{9a^4}$ **71.** $8y^4$
73. $3x^2 - 5x + 4$ **75.** $3y^3 - 11y^2 + 10y - 12$ **77.** $(x + 8)(x + 4)$
79. $(4x + 3)(3x - 10)$ **81.** $(s - 10)(s + 2)$
83. $2m(m + 10)(m + 3)$ **85.** $(2x + 3y)^2$

87. $(2y-1)(2y+1)(4y^2+1)$ **89.** $(5s^4-3t^2)(5s^4+3t^2)$
91. $s^2(t-3s^2)(t^2+3s^2t+9s^4)$

Chapter 3 Test

1. $15y^9$ **2.** 4 **3.** 3^{m+1} **4.** 1.4×10^3 **5.** All except **b** **6.** 5
7. $2x^3-7x^2+8x-3$ **8.** $5x(x+2)$ **9.** $(x+1)(x+9)$
10. $(2a-1)^2$ **11.** $(4y+5)(4y-5)$ **12.** $(n+1)(n^2-n+1)$
13. Prime polynomial **14.** $(2x-3)(3x+2)$ **15.** $3a(ab-1)(ab+2)$
16. $(3x+2)(x-2)$ **17.** $\{-\frac{3}{2}\}$; check: $\frac{81}{4}=\frac{81}{4}$
18. $\{-\frac{1}{2},\frac{5}{3},0\}$; all checks give $0=0$.
19. $\{-2\}$; check: $(-2)^2+4(-2)=-4$ **20.** 6 in.

Cumulative Test 3

1. The integers **2.** Commutative property of multiplication
3. a. 2 **b.** 2 **4.** 1 **5.** It is the ratio of two integers. **6.** $7x-8$
7. $\{2\}$ **8.** $r=A/(2\pi)$ **9.** 2.2, 3.4, and 4.4 ft
10. 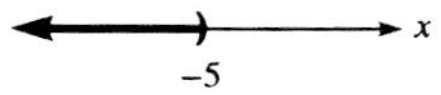

11. 

12. $\{x: -1<x<6\}$ or $(-1,6)$ **13.** $\frac{1}{2}+\frac{1}{4}=\frac{3}{4}$
14. 143^{-12}, 143×1.04^{-12}, $\dfrac{143}{1.04^{-12}}$ **15.** $\dfrac{-8}{x^6y^9}$ **16.** 30
17. Many correct answers; an example is x^5+2. **18.** 2
19. $(2x-1)(x+2)$ **20.** $\{\frac{1}{2},-4\}$

Chapter 4

Exercises 4.1

1. $\frac{4}{3}$ **3.** -2 **5.** No value **7.** 0 **9.** $3,-3$ **11.** 5 **13.** No value
15. Yes **17.** Yes **19.** No **21.** $\frac{1}{5}$ **23.** $\frac{5}{4}$ **25.** $\frac{3}{4}$
27. Already in lowest terms **29.** $x-1$ **31.** $\dfrac{w+1}{w+3}$ **33.** $\frac{3}{4}$
35. $\dfrac{2}{5(x+y)^3}$ **37.** $\dfrac{2y+1}{y-2}$ **39.** $\dfrac{1}{m+2}$ **41.** $\frac{1}{3}$ **43.** 1 **45.** -1
47. $\dfrac{-x}{7}$ **49.** $\dfrac{-1}{x+3}$ **51.** $\dfrac{x+1}{x^2+x+1}$ **53.** $\dfrac{a^2}{a^2-a+1}$ **55.** $\dfrac{6x}{10x^2y^2}$
57. $\dfrac{48(n+2)}{8m(n+2)}$ **59.** $\dfrac{4x(x-2)}{20x^2}$ **61.** $\dfrac{3(x+5)}{x^2-25}$ **63.** $\dfrac{y(y-5)}{y^2-4y-5}$
65. $\dfrac{(x-1)(x+2)}{x^2-4}$ **67. a.** About 2.5 seconds
b. $g=0$; it would not swing back and forth when released.
69. a. 25 percent **b.** 0 **71. a.** 100 percent **b.** 0
73. a. $\frac{3}{2}$; x must be greater than 3. **b.** 6

Remember This (4.1)

1. $-\frac{2}{3}$ **2.** $-\frac{2}{7}$ **3.** $3(x+2)$ **4.** $(x+4)^4$ **5.** False
6. $\{-1,2,-3\}$ **7.** $(3y+4x^2)(3y-4x^2)$ **8.** 15 **9.** $c=b/a$
10. $x=5$; length $=8$ cm, width $=3$ cm

Exercises 4.2

1. 12 **3.** $3y$ **5.** $\dfrac{10}{x}$ **7.** $\dfrac{5(2x+3)}{12(x+2)}$ **9.** $\dfrac{t-1}{t-2}$ **11.** $\dfrac{x+2}{x+3}$
13. $\dfrac{(y-4)(y+2)}{(y-5)(y-2)}$ **15.** $\dfrac{(t+1)^2}{(t+3)^2}$ **17.** 1 **19.** -1 **21.** $\dfrac{1}{a^2b^2(x+2)^2}$
23. $\dfrac{x-2}{2(x+1)}$ **25.** $\dfrac{x+2}{x+1}$ **27.** $\dfrac{2}{x^3}$ **29.** $\dfrac{1}{x+1}$ **31.** $\dfrac{x^3}{(y-3)^2}$
33. $\dfrac{w+3}{5w+2}$ **35.** $\dfrac{1-x}{x^2}$ **37.** $\dfrac{1}{(t+2)^2(t+4)}$ **39.** $\dfrac{x-y}{x+y}$ **41. a.** $\dfrac{y}{n}$
b. $\dfrac{n-x-a}{n}$ **43.** $\dfrac{x+w}{2}$ **45.** $\dfrac{m-y}{y}$

Remember This (4.2)

1. $\dfrac{2}{21}$ **2.** $\dfrac{6}{5}$ **3.** 300 **4.** $h(2x+h)$ **5.** $-3, 3$ **6.** Yes **7.** $\dfrac{4}{m-1}$
8. $\{-7,4\}$ **9.** $\{x: x>-1\}$; $(-1,\infty)$ 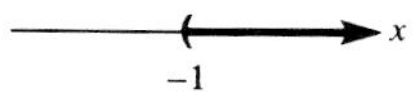

10. 15 and 16

Exercises 4.3

1. $2x$ **3.** $\dfrac{3-x}{x}$ **5.** 1 **7.** $x+2$ **9.** $\dfrac{1}{w-3}$ **11.** $\dfrac{x+2}{x+1}$ **13.** 0
15. $\dfrac{4}{x-4}$ **17.** $\dfrac{2y}{2y-3}$ **19.** $t+5$ **21.** $35x^2y^2$ **23.** $12x^3y^2z^3$
25. $x(x+2)$ **27.** $(y+1)^2(y-1)$ **29.** $6(n+1)(n-1)$
31. $6m(3m-1)$ **33.** $\dfrac{3+2x}{6x}$ **35.** $\dfrac{6y-3}{10y^2}$ **37.** $\dfrac{ax+by}{x^2y^2}$
39. $\dfrac{1-x^2}{x}$ **41.** $\dfrac{x^2+x-1}{x(x-1)}$ **43.** $\dfrac{-x^2+2x+6}{x(x+3)}$ **45.** $\dfrac{v-3}{v^2-4}$
47. $\dfrac{-9}{4x^2-9}$ **49.** $\dfrac{x+1}{x+2}$ **51.** $\dfrac{y^2+4y+2}{(y+1)(y+2)(y+3)}$
53. $\dfrac{x^2-2}{(x+1)^2(x+2)}$ **55.** $\dfrac{4x-4}{(x+3)^2}$ **57.** $\dfrac{x^2+4x-5}{(x+1)^2}$ **59.** $\dfrac{t+5}{t^2-1}$
61. $\dfrac{2}{a-b}$ **63.** $\dfrac{-3x^2}{(x+y)^2(2x+y)}$ **65.** 2
67. a. $\dfrac{n}{n(n+1)}+\dfrac{1}{n(n+1)}=\dfrac{n+1}{n(n+1)}=\dfrac{1}{n}$ **b.** $\frac{1}{4}+\frac{1}{12}=\frac{1}{3}$
69. $\dfrac{ax-bx+bn}{n}$

Remember This (4.3)

1. $\dfrac{t+1}{t}$ **2.** $\dfrac{20}{x}$ **3.** 21 **4.** $y-1$ **5.** $x+3$ **6.** No **7.** $\{\frac{2}{3},-5\}$
8. False **9.** No solution **10.** 12.5 percent

Exercises 4.4

1. $\dfrac{4}{29}$ **3.** $\dfrac{x}{2}$ **5.** $\dfrac{1+mn}{1-mn}$ **7.** $\dfrac{x+1}{2x-1}$ **9.** $\dfrac{b}{2b-3a}$ **11.** $\dfrac{xy}{y^2-x^2}$
13. $\dfrac{w}{3-w^2}$ **15.** $\dfrac{a^2bc+ac+ab}{abc^2+ac+bc}$ **17.** $\dfrac{x}{1+y}$ **19.** -2
21. $\dfrac{3hk+4k}{2hk-h-6}$ **23.** $-\dfrac{x^2-2xy-y^2}{x^2+y^2}$ **25.** $\dfrac{x^3-3x^2-2x-8}{3x}$
27. $\dfrac{(x-1)(4x^2+x-2)}{4x^2}$ **29.** $\dfrac{(x+y)xy}{x^2+xy+y^2}$ **31.** $\dfrac{1+x}{x}$ **33.** $\dfrac{1}{n^2}$
35. $\dfrac{y+2}{y-2}$ **37.** $f=\dfrac{ad}{a+d}$ **39. a.** $m=\dfrac{w_1a+w_2b}{w_1+w_2}$ **b.** $m=\dfrac{a+b}{2}$

Remember This (4.4)

1. an^2 **2.** $\dfrac{a}{c}+\dfrac{b}{c}$ **3.** **b** and **c** **4.** **c** **5. a.** 4 **b.** 2 **c.** 3 **d.** 14
6. $\dfrac{-x}{n(x+n)}$ **7.** $\dfrac{2}{n(n-1)}$ **8.** $\dfrac{3xy^3}{x^2y^2}$ **9.** $\{\frac{1}{3}\}$
10. $\{x: 95.5<x\le100\}$, or $(95.5,100]$

Exercises 4.5

1. $4x^6$ **3.** $\dfrac{x}{2y}$ **5.** $-\dfrac{3z^2}{y^2}$ **7.** $6n-2+\dfrac{1}{2n}$ **9.** $-\dfrac{4}{y}+3x$
11. $-\dfrac{3rs}{4t}+\dfrac{7r}{6t^2}$ **13.** $-\dfrac{2}{y}+\dfrac{3}{x}-\dfrac{4}{xy}$ **15.** $\dfrac{a}{b}+2+\dfrac{b}{a}$

17. $3x + 5 + \frac{6}{x - 1}$ **19.** $4y - 10 + \frac{19}{y + 2}$ **21.** $3y - 4$

23. $3x - 1 + \frac{1}{2x + 3}$ **25.** $x^2 - 2x + 4$ **27.** $n^2 - n + 1 - \frac{2}{n + 1}$

29. $2x - 3 + \frac{18}{2x + 3}$ **31.** $3x^3 + x^2 + 2x + 1 + \frac{4}{3x - 1}$

33. $x^2 + x + 1 + \frac{x - 5}{2x^2 + 1}$ **35.** $2c^2 + 3c - 4 - \frac{2c}{c^2 - c + 1}$

37. $x^3 - x + 3$ **39.** $m^2 + m + 1$

41. $\frac{2A - bh}{h} = \frac{2A}{h} - \frac{bh}{h} = \frac{2A}{h} - b$ **43.** $\frac{cy - ab}{a} = \frac{cy}{a} - \frac{ab}{a} = \frac{cy}{a} - b$

Remember This (4.5)

1. True **2.** Both answers are 123. **3.** Yes **4.** $x^3 - 5x^2 + 2x + 8$

5. $\frac{a}{b}$ **6.** $\frac{-2}{x - 4}$ **7.** $(y + 4)(y^2 - 4y + 16)$ **8.** $2 \times 10^1 = 20$

9. $\frac{43}{72}$ **10.** This is an identity; solution set is all real numbers.

Exercises 4.6

1. $4x^2 + 6x + 15 + \frac{29}{x - 2}$ **3.** $2y^2 - 6y + 12 - \frac{20}{x + 1}$

5. $-5t^3 - t^2 - 2t + 4$ **7.** $x^3 - x^2 + \frac{1}{x + 1}$

9. $n^2 - 2n + 4$ **11.** $8y^2 - 4y + 2 - \frac{1}{y + 1}$

13. $6x^4 - x + 8 = (x - 2)(6x^3 + 12x^2 + 24x + 47) + 102$

15. $2t^3 - 5t^2 + 7t - 30 = (t - 3)(2t^2 + t + 10) + 0$

17. $x^3 + 1 = (x - 1)(x^2 + x + 1) + 2$ **19.** 29 **21.** -20 **23.** 0

25. Yes **27.** Yes **29.** No **31.** $\{1,2,3\}$ **33.** $\{1,\frac{1}{2},\frac{1}{3}\}$ **35.** $\{-1\}$

37. There are 2; 2, 3, 4 and 4, 5, 6. **39. a.** $x^3 - 3x = 2$ **b.** $\{-1,2\}$

Remember This (4.6)

1. $5x + 2$ **2.** $5(y + 1)$ **3.** $\{5\}$ **4.** 3, -3 **5.** $\frac{1}{20}$ **6.** $P(1) = -2$

7. $(m + 7)(3m + 6)$ **8.** $5 + 3x^2$ **9.** \$14,000 **10.** $\{1,-1\}$

Exercises 4.7

1. $\{\frac{4}{5}\}$ **3.** $\{48\}$ **5.** $\{\frac{10}{3}\}$ **7.** No solution; false equation

9. Identity; all real numbers except 0 **11.** $\{\frac{1}{4}\}$ **13.** $\{-\frac{28}{9}\}$ **15.** $\{2\}$

17. $\{\frac{1}{8}\}$ **19.** $\{10\}$ **21.** $\{-\frac{22}{3}\}$ **23.** $\{\frac{21}{10}\}$ **25.** $\{0\}$ **27.** $\{1\}$ **29.** $\{\frac{1}{2},2\}$

31. $\{-\frac{1}{6},1\}$ **33.** $\{-\frac{1}{2},0\}$ **35.** $\{5\}$ **37.** $\{2\}$ **39.** $\{1,-1\}$

41. $\{-1,-3\}$ **43.** $b = 12$ **45.** $R_1 = 15$ ohms **47.** 100 units

49. $p = \frac{1}{1 + w}$ **51.** $d_1 = \frac{fd_2}{d_2 - f}$ **53.** $i = \frac{S - P}{Pn}$ **55.** 2.4 hours

57. 12 minutes **59.** $\frac{1}{100} + \frac{1}{100} = \frac{1}{x}$; $x = 50$ minutes

61. a. 6 mi/hour **b.** 4 hours **c.** $3\frac{1}{4}$ hours **63.** 2 hours **65.** 1 hour

67. a. 150 mi/hour **b.** 1.8 hours from A to B; 3.6 hours from B to A

c. 133.33 mi/hour

Remember This (4.7)

1. Yes **2.** No **3.** True **4.** $C = 13n + 3$ **5.** $x^2 + x + 4 + \frac{10}{x - 3}$

6. $\frac{ac - b}{c^2}$ **7.** $\frac{-2}{x^2 - 1}$ **8.** $2y(3x - 4)(4x + 5)$ **9.** $\{0,3\}$ **10.** \$2.50

Chapter 4 Review Exercises

1. -2, 2 **3.** $\frac{-3}{x^2}$ **5.** 2π; $g = 0$ **7.** $10y$ **9.** $2y^5$ **11.** $18x^3y^2z$

13. $\frac{a - b}{3a - b}$ **15.** $\frac{x^2 - x - 1}{(x - 3)^2(x + 2)}$ **17.** $\frac{5b + 2a}{ab}$ **19.** $\frac{1}{n^3}$

21. $\frac{x}{y^3} + \frac{2}{y} + \frac{2}{x} + \frac{y^2}{x^3}$ **23.** $n^2 + n + 1 + \frac{2}{n - 1}$

25. $4x^2 - 10x + 23 + \frac{-47}{x + 2}$ **27.** 43 **29.** $\{-1,1,2\}$

31. No solution; $x = 2$ is extraneous. **33.** $1\frac{7}{8}$ hours **35.** $\frac{5m}{m + 5}$

37. $\frac{20b - 3a^2}{24a^3b^3}$ **39.** $\frac{3}{4x^2}$ **41.** $\frac{2}{x - 3}$ **43.** $\frac{-4}{x - 1}$ **45.** $\frac{(y - 2)^2}{x^5}$

47. $\frac{23}{30}$ **49.** $\frac{3x^2 + x + 4}{(x - 3)(x - 4)^2}$ **51.** $\{-\frac{5}{2}\}$ **53.** $\{\frac{2}{5}\}$ **55.** $\{-1,-5\}$

57. $\{-3,1,5\}$ **59.** 6 mi/hour **61.** $40x^2y^2$ **63.** -2 **65.** $\frac{6}{7}$

67. $\frac{x - 1}{x - 2}$ **69.** $\frac{22}{21}$ **71.** $\frac{-1}{x + 4}$ **73.** $\frac{-(x + 5)}{5x}$

Chapter 4 Test

1. 2, -2 **2.** $\frac{2}{2a + 3}$ **3.** $a(a - 1)(a + 1)(a - 3)$ **4.** $\frac{2ab^4}{a^3b^3}$

5. $\frac{x - 1}{4}$ **6.** $\frac{y - 1}{2(y + 2)}$ **7.** $\frac{-x^3}{x + 2}$ **8.** $\frac{x + 1}{x - 4}$ **9.** $\frac{2t + 1}{2t - 1}$

10. $\frac{3a - 2b}{6a^2b^2}$ **11.** $\frac{xy + 1}{y + x}$ **12.** $\frac{1}{n^3}$ **13.** $\frac{-5x^3}{2z^4}$

14. $\frac{3x}{y} - \frac{4y}{x} + xy$ **15.** $2x - 3 + \frac{6}{2x + 1}$

16. $x^3 + 6x^2 + 5x - 10 = (x + 4)(x^2 + 2x - 3) + 2$ **17.** -21

18. $\{2,-1,\frac{1}{2}\}$ **19.** $\{28\}$ **20.** 36 minutes

Cumulative Test 4

1. Distributive property **2.** -2 **3.** b **4.** $-9x - 10$ **5.** $\{8\}$

6. $g = \frac{2v}{t^2}$ **7.** Width 15 in., length 43 in. **8.** $(-\infty,-5)$

9. 

10. $(-\infty,0) \cup (\frac{2}{3},\infty)$ **11.** $\frac{11}{8}$ **12.** $36x^5y^5$

13. 10 **14.** $(3x - 4)(x + 3)$ **15.** $\{\frac{3}{2},2\}$ **16.** $\frac{1}{3(x - 1)(x + 2)}$

17. $\frac{x^2 + 4x - 1}{(x + 3)(x + 1)^2}$ **18.** $\frac{x^3 + y^3}{x^3y^3}$ **19.** $2x^2 + 3x + 9 + \frac{31}{x - 3}$ **20.** $\{4\}$

Chapter 5

Exercises 5.1

1. Yes **3.** No **5.** Yes **7.**

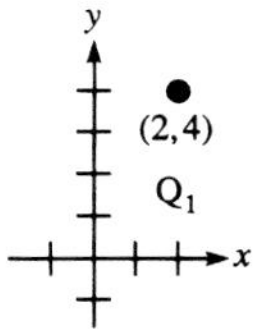

9.

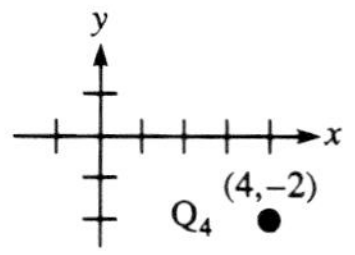

11.

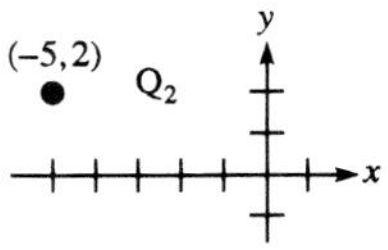

13.

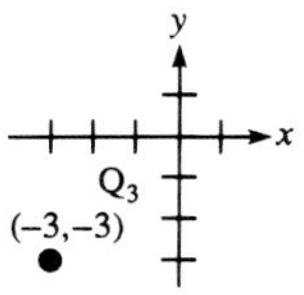

15. No quadrant

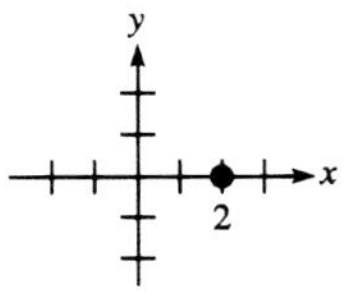

17. No quadrant

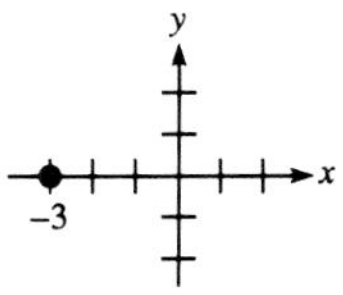

19. (0,1), (1,2), (2,3) **21.** (0,4), (1,4), (2,4)

23.

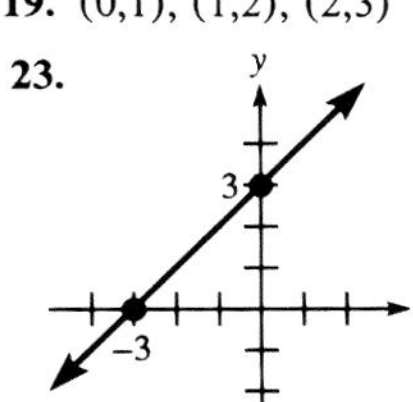

25.

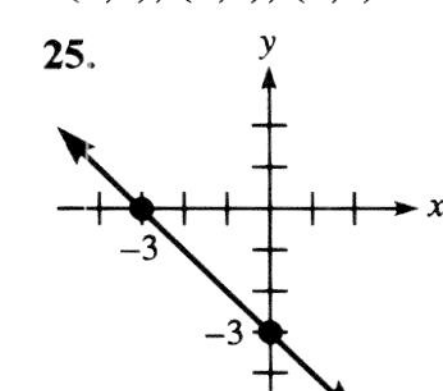

27.

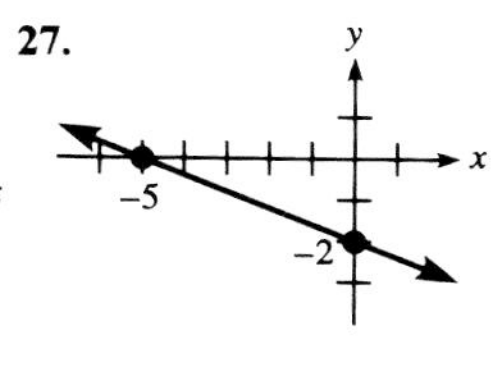

29.

31.

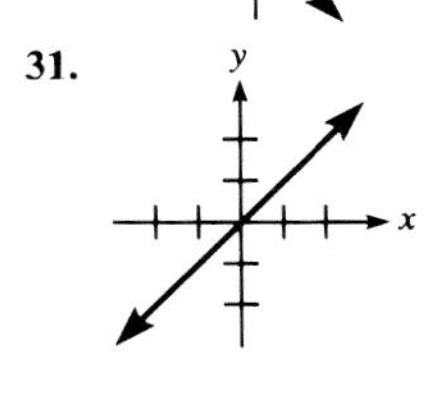

33.

35.

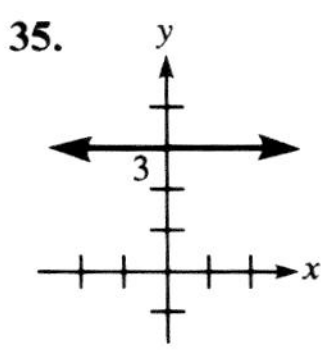

37. **39.**

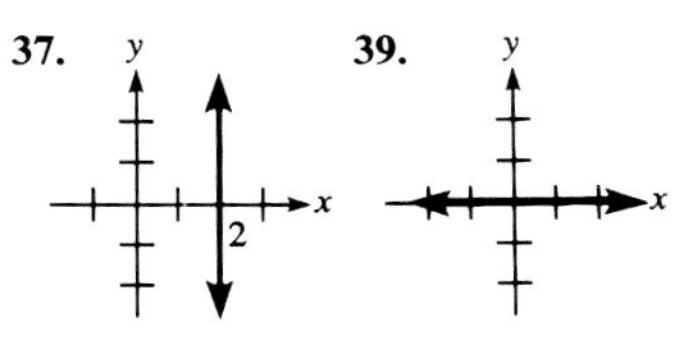

41.

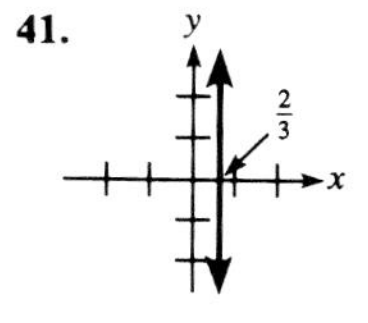

43.

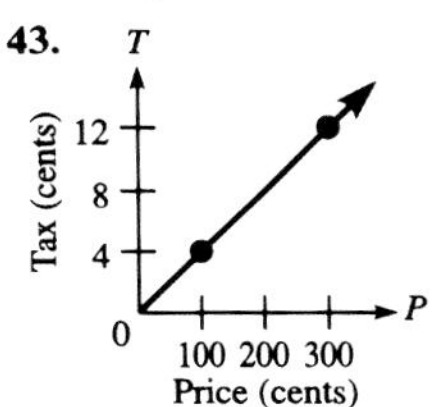

45. a. $C = 60 + 0.20m$ **b.**

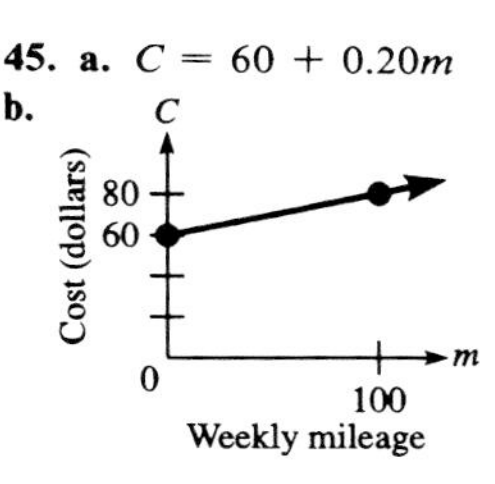

47. a. $C = 3 + 0.30n$ **b.**

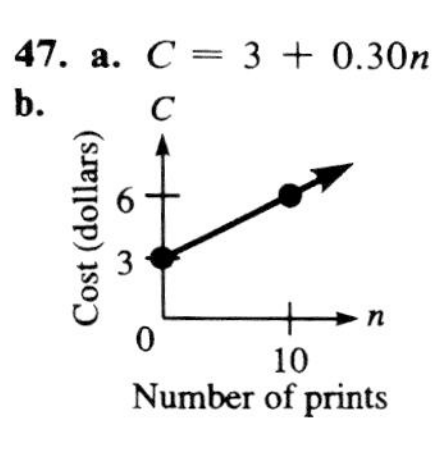

Remember This (5.1)

1. 20/500 **2.** Parallel **3.** Four **4.** $r = d/t$ **5.** $\{\frac{1}{2}, -2\}$

6. $\dfrac{10x + 1}{(x + 1)(2x - 1)}$ **7.** $(10a + 7b)(10a - 7b)$ **8.** $-x^2 - 8x$

9. $(-2,2)$ or $\{x: -2 < x < 2\}$ **10.** Width $= \frac{3}{2}$ in.; length $= \frac{7}{2}$ in.

Exercises 5.2

1. $\frac{4}{3}$; 5 **3.** -1; $\sqrt{18}$ **5.** $\frac{2}{5}$; $\sqrt{29}$ **7.** $-\frac{2}{5}$; $\sqrt{29}$ **9.** 0; 3
11. Undefined; 8 **13.** 0.8 **15.** -1.75
17. $m = 0.20$; mileage charge is 20¢ per mile.
19. $m = 0.06$; rate is 6¢ per ounce. **21.** $m = 6$; well provides 6 gal/minute.

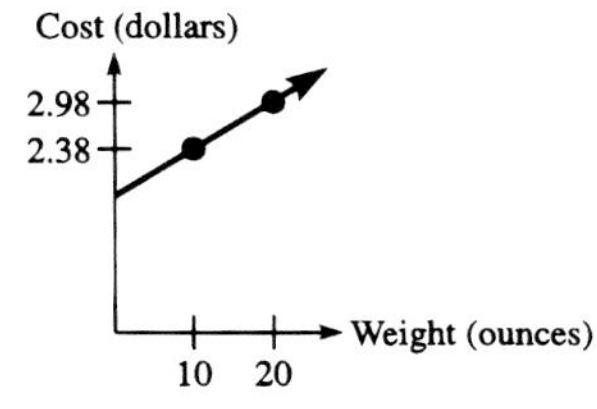

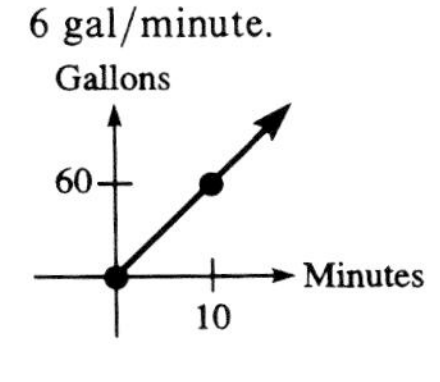

23. 4.8 percent per year **25. a.** $m_{AB} = m_{CD} = \frac{1}{4}$; $m_{AD} = m_{BC} = -4$
b. $\frac{1}{4}(-4) = -1$ **c.** $\sqrt{34} = \sqrt{34}$
27. $m_{AB} = m_{CD} = 3$; $m_{BC} = m_{AD} = 0$; $3 \cdot 0$ does not equal -1; $\sqrt{34} \neq \sqrt{18}$ **29. a.** $-\frac{1}{5}$ **b.** Undefined **c.** $-\frac{10}{7}$

31. Yes; $m_{AB} = m_{AC} = \frac{1}{6}$ **33.** No; $m_{AB} = \frac{1}{4}$, $m_{AC} = \frac{1}{6}$ **35. a.** 2962
b. 4 P.M. **c.** 2–3 P.M. **d.** 12 noon–1 P.M. **e.** 3–4 P.M.
37. a. 2; December **b.** 130; May
c. March and April; also April and May **d.** May and June

Remember This (5.2)

1. $y = \dfrac{c - ax}{b}$ **2.** 275 **3.** $\frac{3}{5}$ **4.** $m = \dfrac{y - b}{x}$

5.

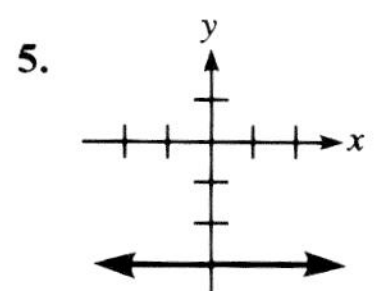

Horizontal

6. $\dfrac{9x^2}{y^6}$ **7.** $\{1,2\}$ **8.** $n^3 + 3n^2 + 3n + 1$

9. 6 **10.** 11 and 13

Exercises 5.3

1. $3x + y = 20$ **3.** $3x - 4y = 26$ **5.** $y = 1$ **7.** $x - 2y = 0$
9. $x - 3y = 5$ **11.** $4x + 5y = 2$ **13.** Slope = 2; y-intercept is (0,1).
15. Slope $= -\frac{2}{3}$; y-intercept is $(0,\frac{4}{3})$.
17. Slope = 0; y-intercept is (0,6). **19.** Slope undefined; no y-intercept
21. a. $y = -\frac{1}{2}x + 1$ **23. a.** $y = 6x + \frac{1}{2}$ **25. a.** $y = -2$

b.

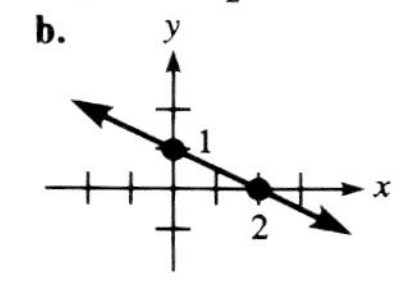

b.

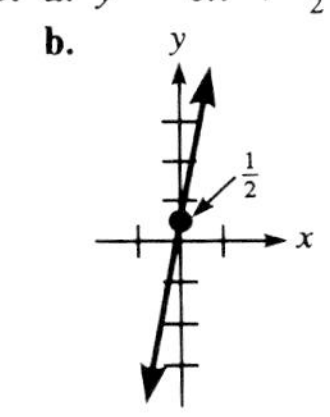

b.

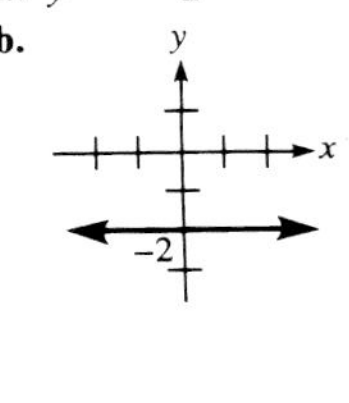

27. Yes **29.** Yes **31.** Same line **33.** Yes **35.** $x + 3y = 7$
37. $y = \frac{1}{3}x - \frac{2}{3}$ **39.** $x = 3$ **41.** $2x - 3y = -7$ **43.** $x - y = 0$
45. $x = 1$ **47. a.** $y = 0.11947x + 684.83$ **b.** 780 kilowatt-hours
49. a. $y = 0.7778x - 1{,}490.04$ **b.** 61.67 percent

Remember This (5.3)

1. d **2.** $0 \leq 2$ **3.** True **4.** True **5.**

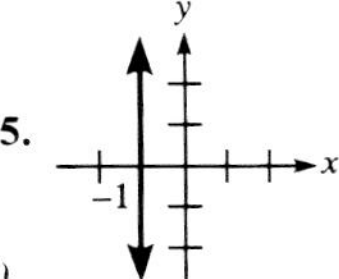

6. $\sqrt{2}$ **7.** $P(1) = 0$ **8.** $\dfrac{9}{xy}$ **9.** $\{0, \frac{7}{3}\}$

10. Two solutions: width $= \frac{3}{2}$, length $= 2$; width $= 2$, length $= \frac{8}{3}$

Exercises 5.4

1. No **3.** Yes **5.** Yes

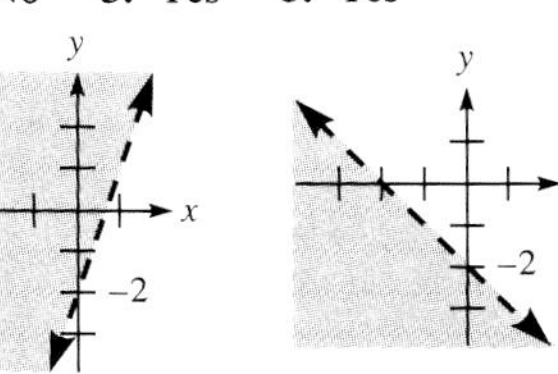

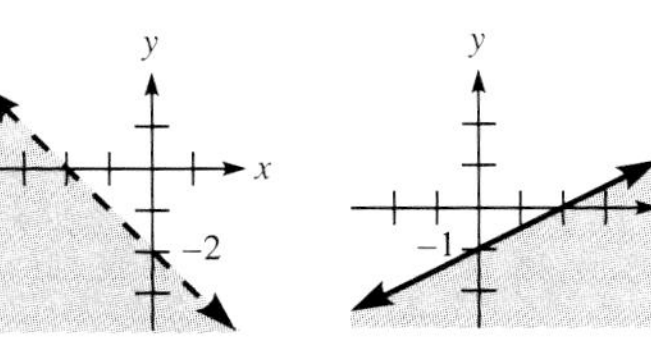

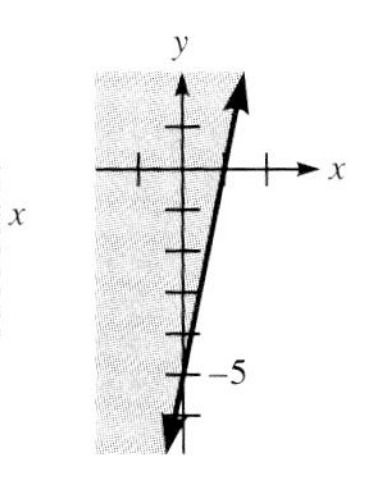

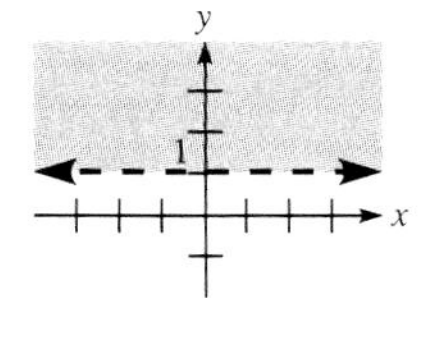

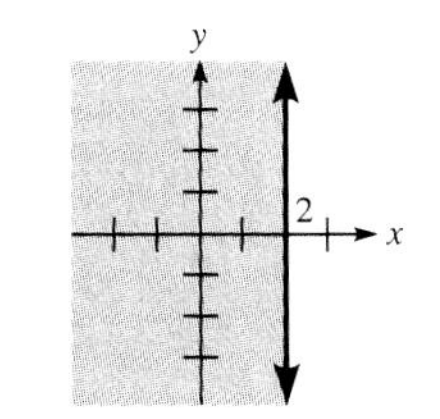

19.
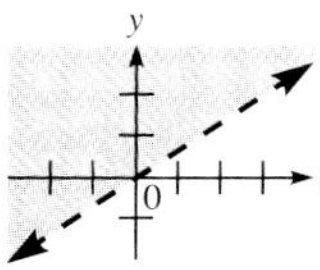

21.
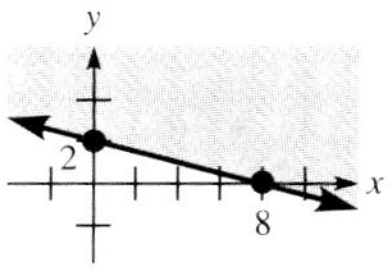

23.
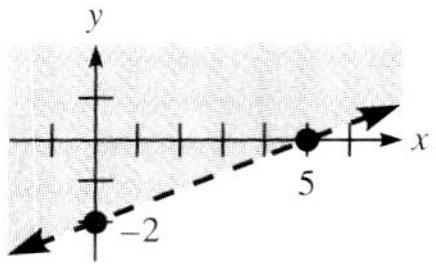

25.
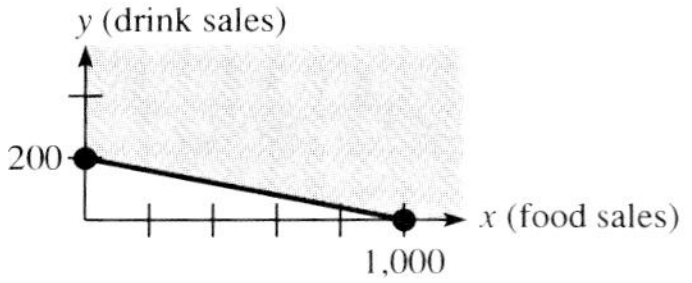

b. y-intercept, (0, 200): Selling \$200 worth of drinks and no food makes a profit of \$100. x-intercept, (1,000, 0): Selling \$1,000 worth of food and no drinks makes a profit of \$100. **c.** \$120 **27. a.** x is the number of acres of land; y is the number of square feet in the house.

b.
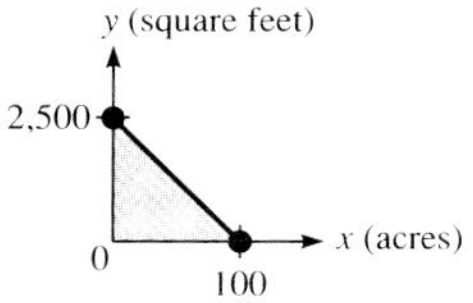

c. 28 acres at most **d.** 2,250 ft² at most

Remember This (5.4)

1. y increases also. **2.** $x = k/y$ **3.** $k = \dfrac{Fd^2}{m_1m_2}$ **4.** $y = 3x$

5. Slope $= -\frac{1}{2}$ **6.**
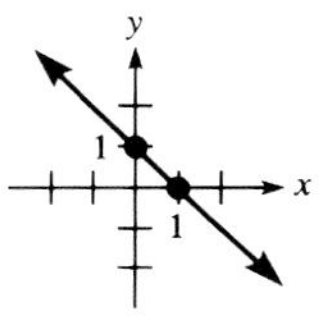

7. $\{\frac{3}{2}, 5\}$ **8.** 4.59×10^{14} **9.** $b = \dfrac{c}{a-1}$ **10.** \$469

Exercises 5.5

1. $P = ks;\ k = 4$ **3.** $d = kt;\ k = 60$ **5.** $p = kA$ **7.** $c = k/n$
9. $t = k/n$ **11.** $d = k/v$ **13.** $y = kx^2/z$ **15.** $f = kab/c^2$
17. $g = kxy/\sqrt{z}$ **19.** $\frac{20}{3}$ **21.** 1 **23.** $\frac{1}{4}$ **25.** $\frac{15}{4}$ **27.** 3 **29.** $-\frac{1}{64}$
31. $\frac{1}{2}$ **33.** 3 **35. a.** \$450 **b.** Side = 3 ft **37.** \$666.67 **39.** 64 in.
41. Kinetic energy becomes 36 times as great. **43.** 1.18×10^{18} kg/m³

Remember This (5.5)

1. (1,3) is a solution to both equations. **2.** $9x$ **3.** $15x - 28y = 70$
4. True; the point of intersection is on both lines.
5. True; different slopes mean they are not parallel. **6.**
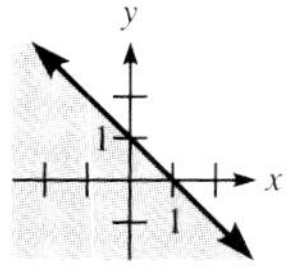

7. $3x - 7y = 12$ **8.** $\dfrac{x}{x+1}$; 0, 1, −1

9. $(x-3)(x^2+3x+9)$

10. Yes; $m_{AB} = 1$ and $m_{AC} = 1$; all three points lie on the line $y = x$.

Chapter 5 Review Exercises

1. No **3.**
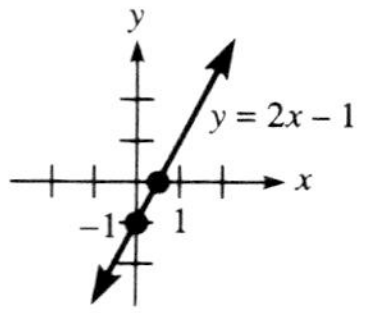

5.
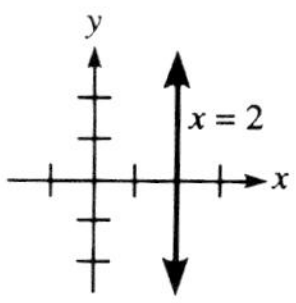

7. $m = 0.05$; rate is 5¢ per ounce. **9. a.** 1–2 P.M. **b.** 10–11 A.M.
11. $-2x + y = 5$ **13.** $m = -2,\ b = -1$ **15.** $y = \frac{2}{3}x + 5$

17.
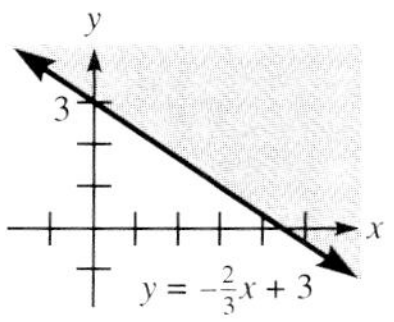

19.
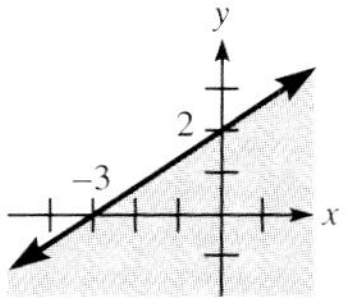

21. $C = kA$ **23.** 40 **25.** \$142.50

27.
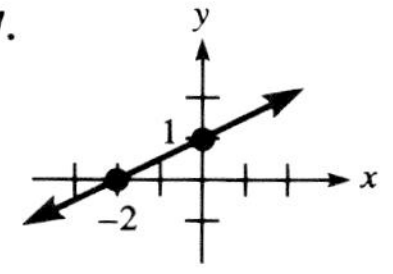

29.
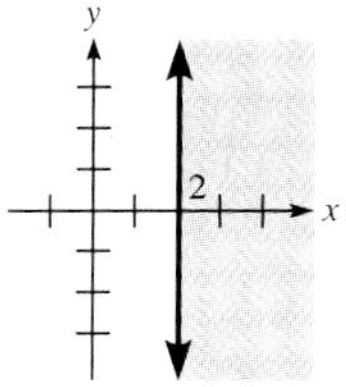

31.
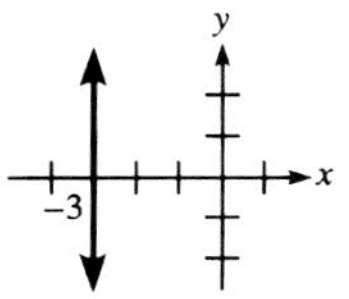

33. $x + 6y = 12$ **35.** $x + y = 0$ **37.** (0,1), (1,−4), (2,−9) **39.** 1
41. Same line **43.** 4 **45.** $-\frac{1}{4}$, (0,0)
47. $m_{AB} = m_{CD} = \frac{1}{3}$; $m_{BC} = m_{AD} = -3$; $\frac{1}{3}(-3) = -1$
49. $m = 5$; pump provides 5 gal/minute. **51. a.** 11 A.M.–12 noon
b. 1–2 P.M.

Chapter 5 Test

1. 10 **2.** $y = \frac{1}{2}x - 1$ **3.** $m = \frac{1}{2},\ b = \frac{3}{2}$ **4.** −4
5. $m = 0.10$; 10¢ per invitation **6.** Parallel
7. $w = \dfrac{kst}{r^2}$ **8.** (2,−6), (0,−4), (−2,−2)
9. a. $m_{AB} = m_{CD} = \frac{3}{2}$; $m_{AD} = m_{BC} = -\frac{2}{3}$ **b.** $\frac{3}{2}(-\frac{2}{3}) = -1$
10. $x + y = -4$ **11.** $y = 1$ **12.** $-2x + 3y = -3$
13. $-x + 3y = -6$

14.
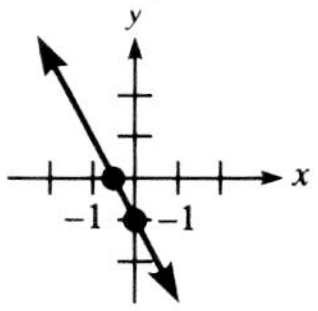

15.
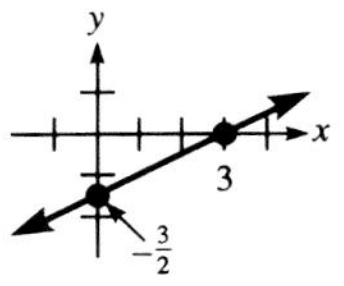

16.
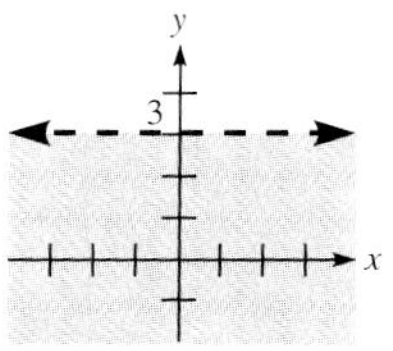

17.
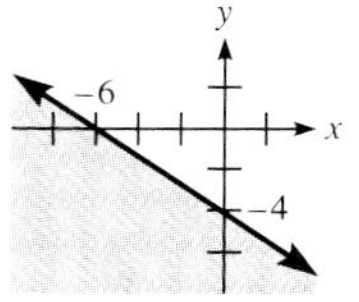

18.
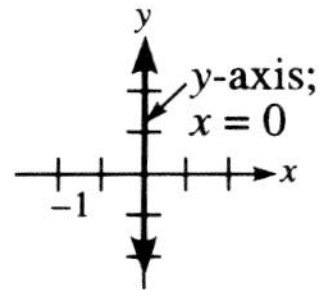

19. a. $C = 0.50n + 2$
b.
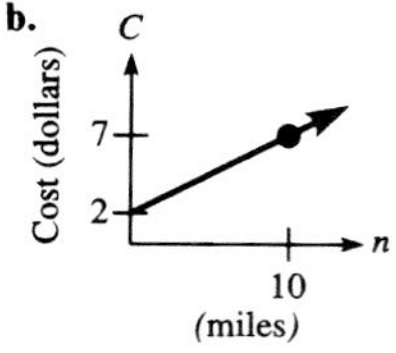

20. 400 ft

Cumulative Test 5

1. a **2.** b **3.** {2} **4.** 188 **5.** 1 **6.** $-2x^3 - 8x^2 + 12x + 4$
7. $-x + 2$ **8.** $x = 2y + 4$ **9.** 2 **10.** $[-\frac{3}{2}, 2]$ **11.** \$15,000
12. $(-\infty, -4)$
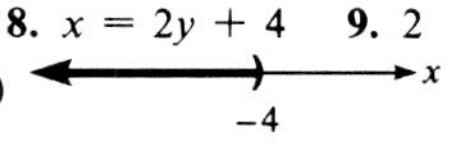

13. $(x-12)(x+1)$ **14.** $\dfrac{2x^2}{3}$ **15.** −2, 2 **16.** $\dfrac{1}{3x-1}$
17. {−2,1,3} **18.** 2 **19.** $y = -3x + 16$ **20.** 13 units

Chapter 6

Exercises 6.1

1. a. Yes **b.** No **3. a.** No **b.** Yes **5. a.** No **b.** Yes

7.
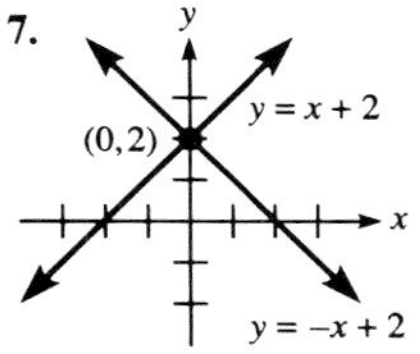

(0,2); consistent

9.
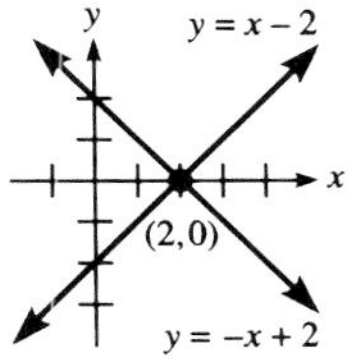

(2,0); consistent

11.
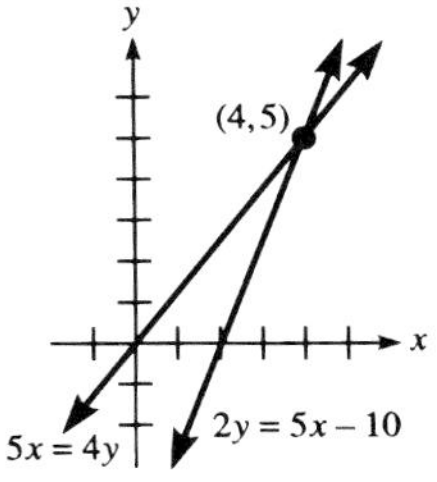

(4,5); consistent

13.
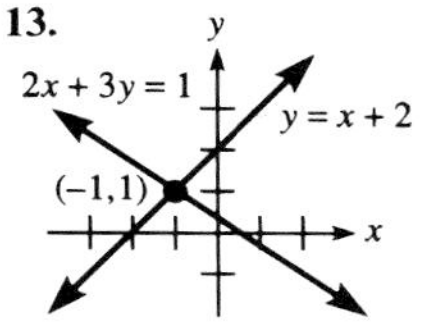

(−1,1); consistent

15.
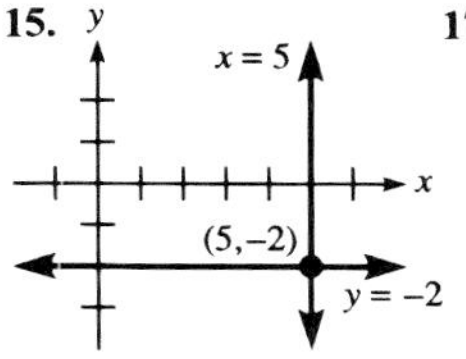

(5,−2); consistent

17.
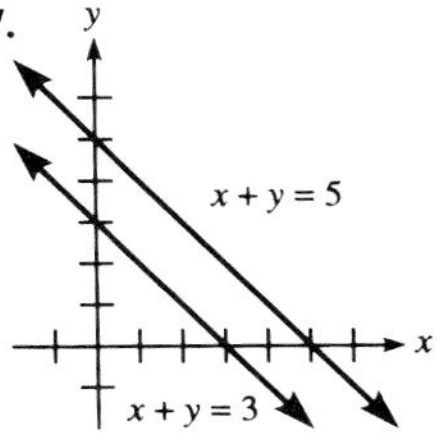

∅; inconsistent

19.
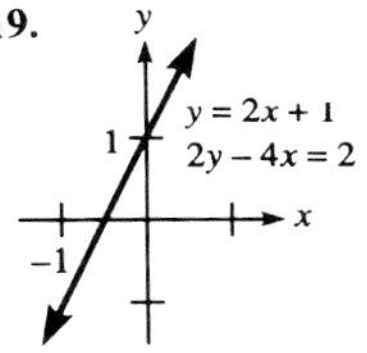

Dependent

21. $(2,-\frac{1}{3})$ **23.** $(-\frac{7}{3},-4)$ **25.** (0,5) **27.** $(\frac{7}{3},\frac{3}{2})$ **29.** $(-4,-\frac{13}{3})$
31. $(\frac{9}{8},\frac{1}{8})$ **33.** $(\frac{5}{11},\frac{2}{11})$ **35.** (1,1) **37.** (4,−1) **39.** $(\frac{34}{11},-\frac{7}{11})$
41. (7,−10) **43.** ∅ **45.** ∅
47. Dependent system; $\{(x,y)\colon 4x - 6y = 10\}$
49. Dependent system; $\{(x,y)\colon 10y = -6x + 14\}$ **51.** (100, 700)
53. $(-\frac{4}{5},34)$ **55.** (14,3) **57.** $(\frac{13}{7},\frac{31}{7})$ **59.** (5,5) **61.** ∅
63. Dependent system; $\{(x,y)\colon x = 3y - 1\}$
65. \$150,000 at 10 percent; \$50,000 at 6 percent
67. \$45,000 salary; \$15,000 investments **69.** \$45,094
71. State tax: \$28,340; federal tax: \$291,498
73. A's income: \$9,500; B's income: \$38,000 **75.** 6 minutes
77. 4 gal of 2 percent solution; 2 gal of 5 percent solution
79. 100,000 copies; 20 months **81.** 50 years **83.** 12.5 and −2.5
85. 57° and 33° **87.** 18° **89.** 6.75 cm² **91.** 8.5 in. by 2.5 in.
93. \$83.20 **95.** \$94.50 on paperbacks; \$119.76 on hardbacks
97. Plane speed = 425 mi/hour; wind speed = 25 mi/hour
99. $\frac{1}{30}$ mi/minute, or 2 mi/hour

Remember This (6.1)

1. b **2.** Yes **3.** $x + y = 13$ **4.** a **5.** $\frac{27}{2}$

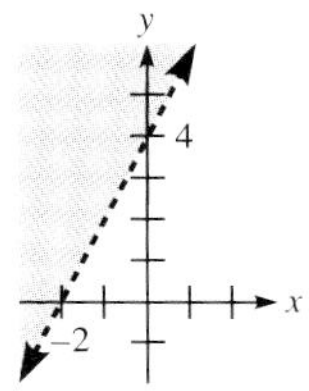

7. $x = 3$ **8.** $\sqrt{41}$ **9.** $\dfrac{1}{2t-3}$ **10.** 2 in.

Exercises 6.2

1. a. Yes **b.** No **3. a.** Yes **b.** Yes **5. a.** Yes **b.** No **7.** (1,2,3)
9. (−1,1,2) **11.** (−2,0,2) **13.** (2,2,2) **15.** (−1,−2,3) **17.** (5,3,1)
19. (1,−1,1) **21.** (3,−2,1) **23.** $(\frac{1}{2},\frac{1}{3},\frac{1}{4})$ **25.** (0,0,0)
27–33. Answers will vary. Solutions lead to $0 = a$, $a \neq 0$.
35–41. Answers will vary. Solutions lead to $0 = 0$.
43. $I_1 = \frac{2}{13}$ ampere, $I_2 = \frac{4}{65}$ ampere, $I_3 = \frac{14}{65}$ ampere
45. $4 = a + b + c$
$9 = 4a + 2b + c$
$6 = a - b + c$
Solution: (2,−1,3); polynomial: $P(x) = 2x^2 - x + 3$ **47.** 16, 24, 5, 80

Remember This (6.2)

1. $x = \dfrac{c}{a_1b_2 - b_1a_2}$ **2.** −41 **3.** −52
4. Yes **5.** $0.05x + 0.10y = 0.06n$, $x + y = n$ **6.** ∅ **7.** $y = -\frac{1}{3}x$
8. $F = \dfrac{km_1m_2}{d^2}$ **9.** $\dfrac{5a+10}{5ab+3a}$ **10.** $\frac{13}{3}$

Exercises 6.3

1. −2 **3.** −5 **5.** 0 **7.** $ad - bc$ **9.** 0 **11.** 1 **13.** a^2 **15.** $\frac{17}{6}$

17. $D = \begin{vmatrix} 1 & 2 \\ 2 & 3 \end{vmatrix} = -1,\ D_x = \begin{vmatrix} 3 & 2 \\ 4 & 3 \end{vmatrix} = 1,\ D_y = \begin{vmatrix} 1 & 3 \\ 2 & 4 \end{vmatrix} = -2$

19. $D = \begin{vmatrix} 3 & -2 \\ 1 & 3 \end{vmatrix} = 11,\ D_x = \begin{vmatrix} 5 & -2 \\ 1 & 3 \end{vmatrix} = 17,\ D_y = \begin{vmatrix} 3 & 5 \\ 1 & 1 \end{vmatrix} = -2$

21. $D = \begin{vmatrix} 1 & 2 \\ 2 & 4 \end{vmatrix} = 0,\ D_x = \begin{vmatrix} 5 & 2 \\ 10 & 4 \end{vmatrix} = 0,\ D_y = \begin{vmatrix} 1 & 5 \\ 2 & 10 \end{vmatrix} = 0$

23. $D = \begin{vmatrix} 1 & -3 \\ -7 & 1 \end{vmatrix} = -20,\ D_x = \begin{vmatrix} -5 & -3 \\ -1 & 1 \end{vmatrix} = -8,\ D_y = \begin{vmatrix} 1 & -5 \\ -7 & -1 \end{vmatrix} = -36$

25. (0,1) **27.** (0,0) **29.** (1,2) **31.** $(-\frac{1}{6},-\frac{5}{8})$
33. (−1,−1) **35.** $D = 0$; inconsistent system; ∅.
37. $D = 0$; inconsistent system; ∅
39. $D = 0$; dependent system; $\{(x,y)\colon x - y = 4\}$
41. $D = 0$; dependent system; $\{(x,y)\colon y = 3x + 4\}$ **43.** (−8,5) **45.** 0
47. 1 **49.** −49 **51.** 0 **53.** 6 **55.** −2 **57.** 0 **59.** 0

61. $D = \begin{vmatrix} -2 & 1 & -1 \\ 1 & 2 & -3 \\ 3 & -1 & 2 \end{vmatrix} = -6,\ D_x = \begin{vmatrix} 2 & 1 & -1 \\ 4 & 2 & -3 \\ -1 & -1 & 2 \end{vmatrix} = -1,$

$D_y = \begin{vmatrix} -2 & 2 & -1 \\ 1 & 4 & -3 \\ 3 & -1 & 2 \end{vmatrix} = -19,\ D_z = \begin{vmatrix} -2 & 1 & 2 \\ 1 & 2 & 4 \\ 3 & -1 & -1 \end{vmatrix} = -5$

63. $D = \begin{vmatrix} 3 & 1 & 4 \\ 1 & 5 & 9 \\ 2 & 6 & 5 \end{vmatrix} = -90 = D_x,\ D_y = \begin{vmatrix} 3 & 3 & 4 \\ 1 & 1 & 9 \\ 2 & 2 & 5 \end{vmatrix} = 0,\ D_z = \begin{vmatrix} 3 & 1 & 3 \\ 1 & 5 & 1 \\ 2 & 6 & 2 \end{vmatrix} = 0$

65. $(-\frac{1}{2},-\frac{20}{7},\frac{23}{14})$ **67.** $(-\frac{1}{2},\frac{1}{3},\frac{1}{4})$ **69.** $(0,5,\frac{1}{5})$ **71.** $(\frac{1}{4},\frac{1}{4},\frac{3}{4})$
73. ∅, inconsistent system **75.** Dependent system
77. 80 oz of 35 percent solution; 16 oz of 25 percent solution
79. \$120,000 at 12.5 percent; \$130,000 at 8.5 percent **81.** −4.5, 3.5
83. Width = 125 yd; length = 250 yd **85. a.** 100 days **b.** 200 liters
87. Yes; 30 oz alloy 1, 50 oz alloy 2, 20 oz alloy 3
89. $A = 80°$, $B = 70°$, $C = 30°$

Remember This (6.3)

1. (−1,4) **2.** $-y - 4z = 11$ **3. a.** g, h, i **b.** a, d, g
4. Dependent system **5.** True **6.** 3.85 **7.** $(1,\infty)$
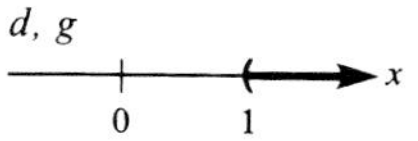

8. $(2x - 3)(6x + 1)$ **9.** $\{-1\}$ **10.** $y = x + 1$

Exercises 6.4

1. $(1,-2,1)$ **3.** $(2,0,1)$ **5.** $(1,-2,1)$ **7.** $(3,5,7)$ **9.** $(-3,1)$
11. $(-1,-2)$ **13.** $(0,\frac{1}{3},\frac{2}{3})$ **15.** $(\frac{1}{2},0,\frac{3}{4})$ **17.** $(\frac{1}{4},-\frac{1}{4})$
19. $(3,-1)$ **21.** $(2,2,3)$ **23.** $(1,-1,2)$
25. 300 orchestra and mezzanine; 400 balcony
27. 30 cassettes, 20 records, 100 CDs

Remember This (6.4)

1. $\{4,5,6\}$ **2.**

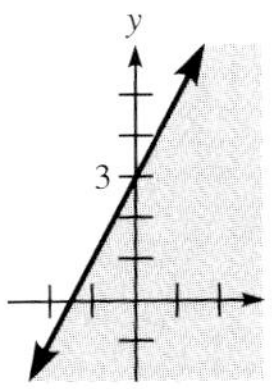

3. Yes **4.** $12x + 11y$ **5.** $(-\frac{1}{2},\frac{1}{2})$
6. 10 **7.** No; $P(-2) = 4 \neq 0$
8. $\{1,3,-\frac{4}{3}\}$ **9.** $b = \dfrac{c}{a-1}$
10. $5(b + a)$

Exercises 6.5

1.

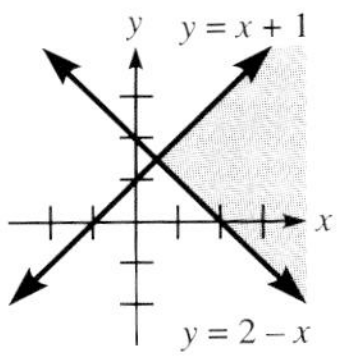

3.

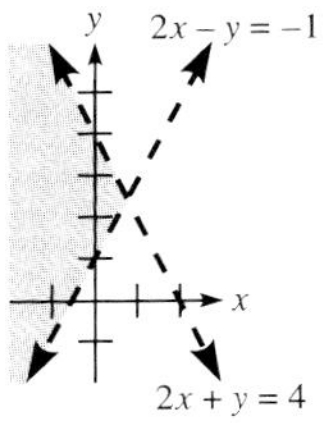

5.

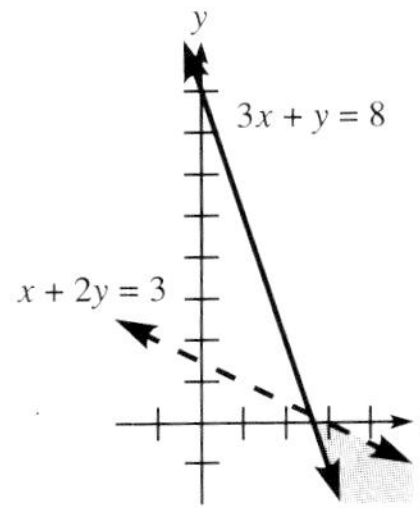

7.

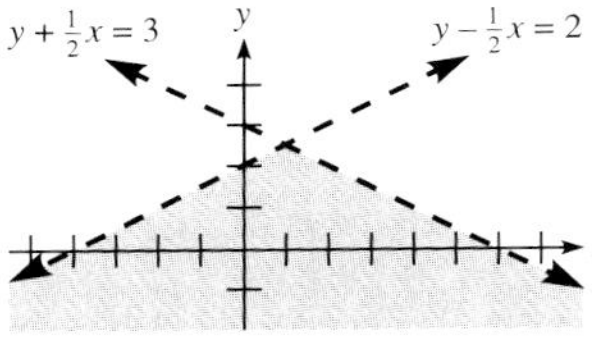

9.

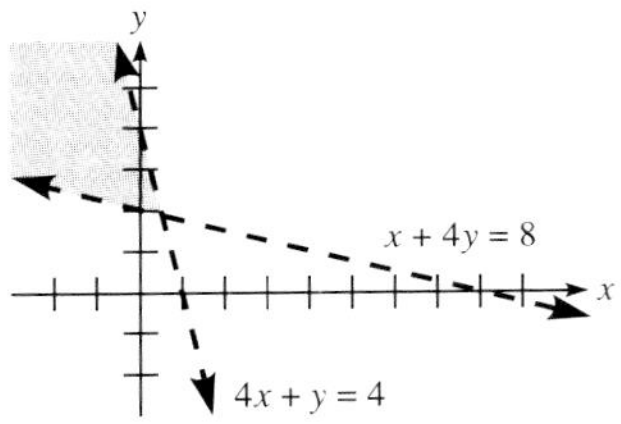

11.

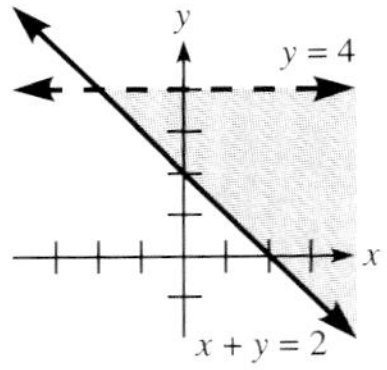

13.

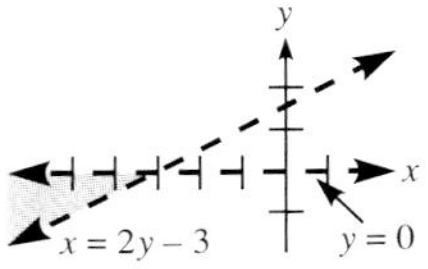

15. No solution; the solution sets do not overlap.

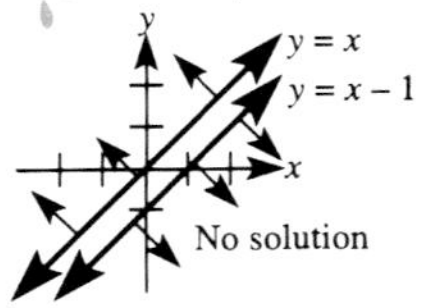

17.

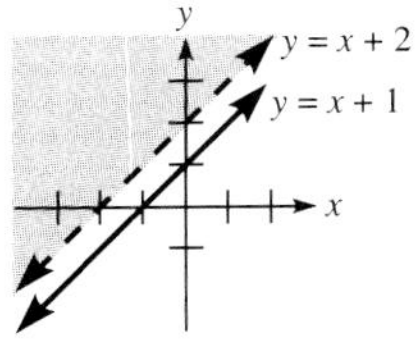

19.

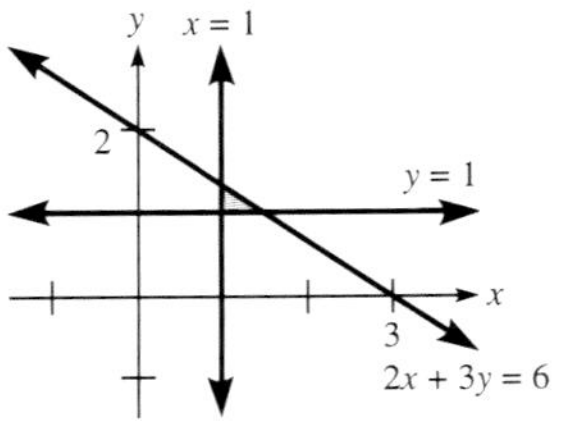

21.

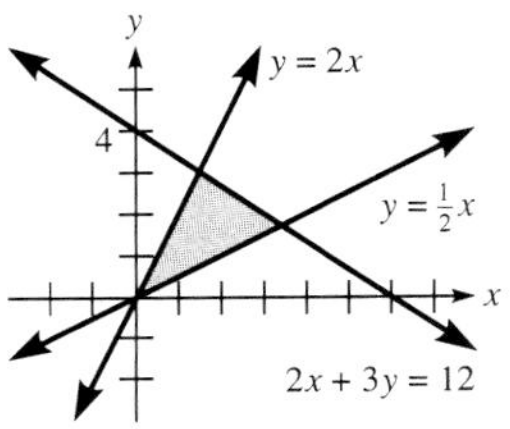

23. a. $3x + 4y \leq 180$
$x \geq 20$
$y \geq 20$

b.

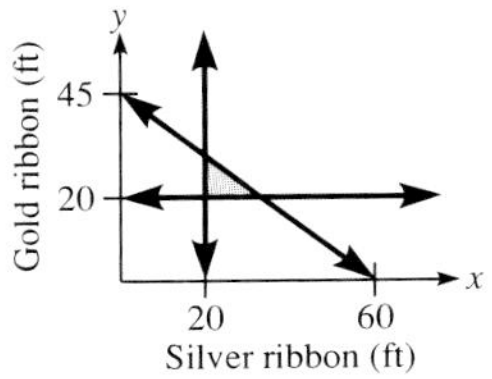

c. One solution is 25 ft of each; total cost: \$175.

25. a. $15x + 10y \leq 200$
$15x \geq 50$
$10y \geq 50$

b.

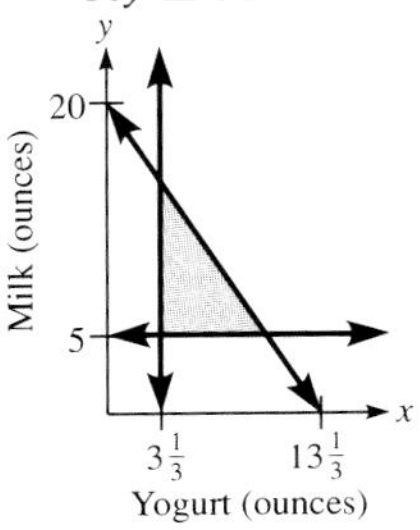

c. 6 oz of each provides about 150 calories.

27. a. $3x + 4y + 0.5(3x + 4y) \leq 500$
$3.15x + 4.20y \leq 500$
$x \geq 75$
$y \geq 50$

b.

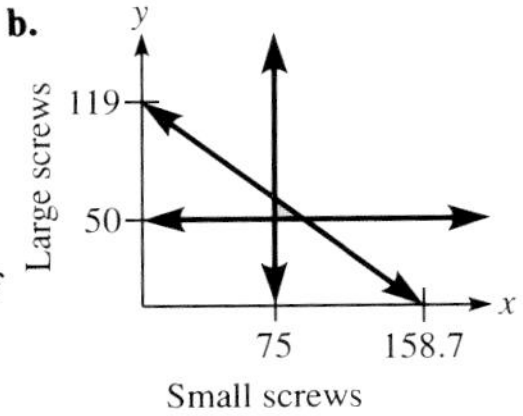

c. One solution is 80 small screws and 55 large screws, for a total cost of \$4.83.

Remember This (6.5)

1. True **2.** 2, 4, $\sqrt{4}$ **3.** True **4.** No **5.** $(-\frac{1}{3},\frac{1}{4})$ **6.** 28
7. $\dfrac{-1}{x-2}$ **8.** -5.5 and -2; 2 and 5.5 **9.** $\dfrac{1}{x^4}$
10.

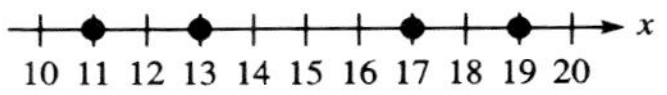

Chapter 6 Review Exercises

1. Yes **3.** $(-1,-2)$ **5.** \$35,000 at 7 percent; \$5,000 at 5 percent
7. $(3,2,-1)$ **9.** Dependent system; infinitely many solutions **11.** 28
13. $(\frac{2}{5},-\frac{1}{8})$ **15.** $(4,1,-2)$ **17.** $(4,-2)$ **19.** 700
21.

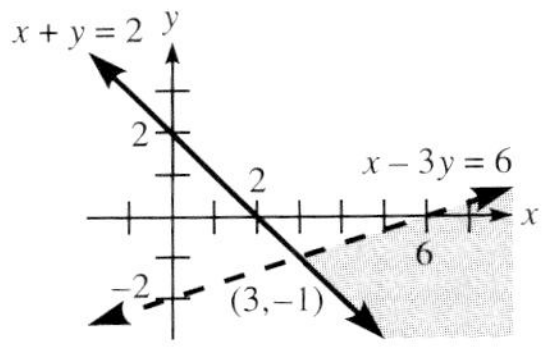

23. 5 **25.** No **27.** $(\frac{3}{2},8)$ **29.** $(-1,3)$ **31.** $(19,34)$
33. $(-3,-1)$ **35.** $(1,-3)$ **37.** Infinitely many solutions **39.** 311

Chapter 6 Test

1. 3 **2.** -1 **3.** b **4.** a **5.** $(-\frac{1}{2},\frac{1}{3})$ **6.** $(2,-2)$
7. $(5,-1,1)$ **8.** $(\frac{2}{3},0)$ **9.** $(15,13,-5)$ **10.** $(2,-3)$
11. $(1,2)$ **12.** $(-3,-3)$ **13.** $(5,-3,3)$ **14.** Inconsistent system; $\emptyset$
15.

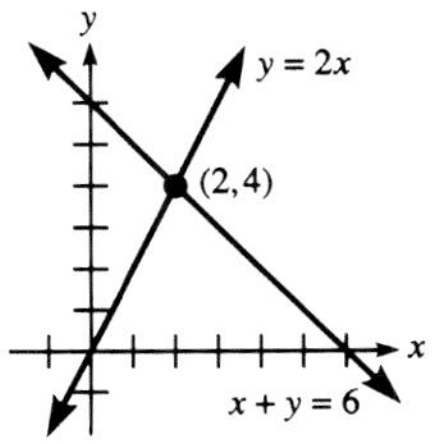

16.

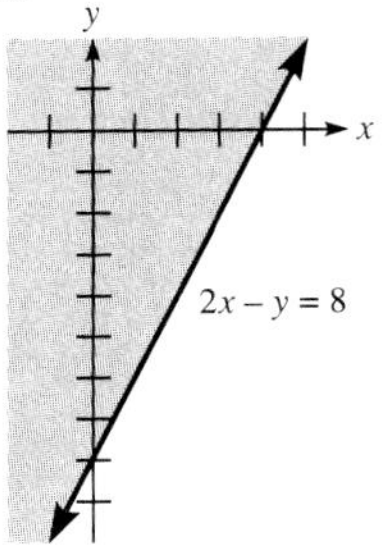

17.

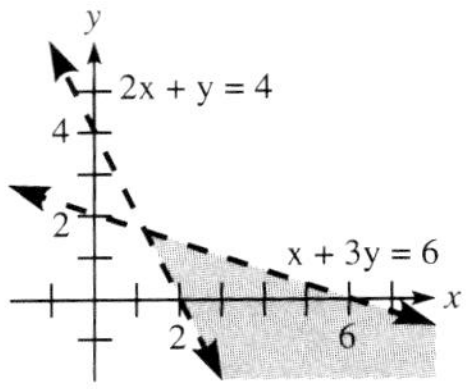

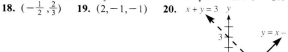

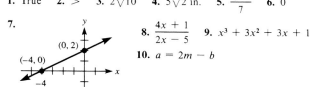

18.

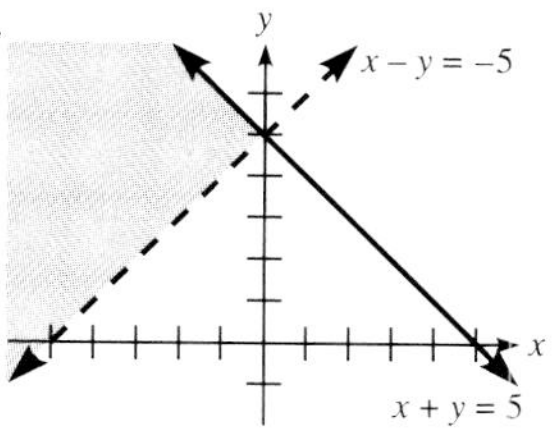

19. \$40,000

20. $1.25r + 1.50w \le 14.00$
$r \ge 3$
$w \ge 3$

Cumulative Test 6

1. d **2.** $y = -\frac{5}{2}x + 5$ **3.** $[-4,6]$ **4.** (number line: −1 0, x)

5. $\{\frac{4}{5}\}$ **6.** $\frac{3y}{2x}$ **7.** $-y^2 - 6x + 4$ **8.** $2x(x-2)(x+5)$ **9.** $\{-5,5\}$

10. $-5(x-4)$ **11.** $x - 1 + \dfrac{3}{x^2 - 5x + 7}$ **12.** $6y - 3x$

13. 18 minutes **14.** $y = 2x + 1$ **15.** $\sqrt{34}$ **16.** 6 **17.** 76

18. $(-\frac{1}{2}, \frac{2}{3})$ **19.** $(2,-1,-1)$ **20.**

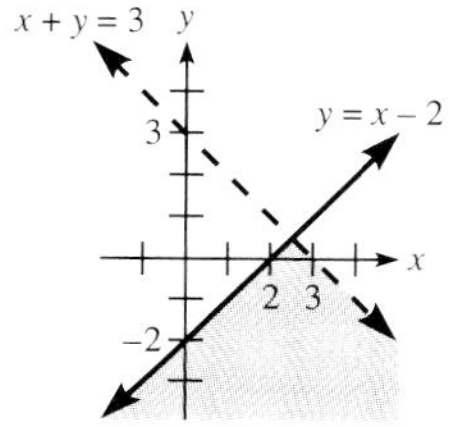

Chapter 7
Exercises 7.1

1. 11, −11 **3.** 19, −19 **5.** 24, −24 **7.** 17 **9.** 3.46 **11.** −26
13. Not real **15.** 5.83 **17.** −20 **19.** Not real **21.** −30
23. Not real **25.** 4 **27.** −6 **29.** −3 **31.** 2 **33.** −2
35. Not real **37.** −8 **39.** Not real **41.** −2 **43.** 7 **45.** −6
47. 243 **49.** 8 **51.** $\frac{1}{3}$ **53.** $\frac{1}{8}$ **55.** −256 **57.** $\frac{1}{36}$ **59.** $-\frac{1}{8}$
61. $\frac{125}{64}$ **63.** $\frac{4}{5}$ **65.** $\frac{3}{2}$ **67.** $3^{7/4}$ **69.** 27 **71.** $7^{2/3}$ **73.** a^2
75. $b^{1/4}$ **77.** $2b^4$ **79.** $4x^4$ **81.** $a^{1/4}b^{1/4}$ **83.** $\dfrac{b^{1/2}}{a}$ **85.** $\dfrac{a^{1/4}}{b^{5/3}}$
87. $3x^4$ **89.** $\dfrac{a^4}{3}$ **91.** $8a^9$ **93.** $16a^4$ **95.** $\dfrac{x^{5/3}}{16}$ **97. a.** 9 **b.** $9 < 15$
99. Geometric mean = 1.24097; average annual increase in sales is about 24.1 percent. **101. a.** Side = 8 in. **b.** Side = $\sqrt{ab}$

Remember This (7.1)

1. True **2.** False **3.** $9 \cdot 5$ **4.** 125, 216 **5.** 3 **6.** (3,2,1) **7.** $\frac{1}{4}$
8. $\dfrac{-5x^3}{3z}$ **9.** 1 **10.** $(2x-1)(x+2)$

Exercises 7.2

1. $2\sqrt{7}$ **3.** $7\sqrt{3}$ **5.** $6\sqrt{2}$ **7.** $2\sqrt[3]{5}$ **9.** $5\sqrt[3]{3}$ **11.** $3\sqrt[4]{3}$ **13.** 64
15. x^2 **17.** 81 **19.** 14 **21.** y^3 **23.** $\sqrt[4]{x}$ **25.** $\sqrt[6]{y^5}$ **27.** $x^3\sqrt{x}$
29. $6y\sqrt{2y}$ **31.** $y^3\sqrt[3]{y^2}$ **33.** $x^2y^4\sqrt{y}$ **35.** $4xy\sqrt{2}$ **37.** $2x^2y^3\sqrt{10x}$
39. $7xy^2\sqrt{2xy}$ **41.** $2x\sqrt[3]{3y}$ **43.** $3xy\sqrt[3]{3x^2y}$ **45.** $x^2y\sqrt[4]{x^3y}$ **47.** $\frac{4}{11}$
49. $\dfrac{4\sqrt{2}}{7}$ **51.** $\dfrac{6\sqrt{7}}{7}$ **53.** $\dfrac{\sqrt[3]{18}}{3}$ **55.** $\dfrac{7\sqrt{y}}{y}$ **57.** $\dfrac{2\sqrt{6x}}{x}$ **59.** $\dfrac{\sqrt[3]{6y^2}}{2y}$
61. $\dfrac{\sqrt[3]{14y}}{6y}$ **63.** $\pi\sqrt{2} \approx 4.44$ seconds **65.** 20,000 cycles/second **67.** 9
69. $10\sqrt{6}$ **71.** $2\sqrt{7}$ **73.** $9\sqrt{7}$ **75.** $12\sqrt[3]{3}$ **77.** $72\sqrt[4]{3}$
79. $12x^2y\sqrt{10x}$ **81.** $2xy\sqrt[4]{3xy^3}$ **83.** 2 **85.** 2 **87.** $2\sqrt[4]{2}$ **89.** $\dfrac{\sqrt{22}}{11}$
91. $\dfrac{2\sqrt[3]{18}}{3}$ **93.** $\dfrac{\sqrt{10}}{5}$ **95.** $\dfrac{3\sqrt{2y}}{2}$ **97.** $\dfrac{\sqrt[3]{6x^2}}{2}$ **99.** $\dfrac{4\sqrt{10y}}{5}$
101. $\dfrac{\sqrt[4]{90x^3}}{2}$ **103.** $\sqrt{7}$ **105.** $\dfrac{12\sqrt{y}}{y}$ **107.** $\dfrac{\sqrt{6}}{2}$ **109.** $\dfrac{\sqrt{2}}{2}$
111. $\dfrac{\sqrt{6y}}{2y}$ **113.** $\dfrac{\sqrt{2y}}{y}$ **115.** $\dfrac{\sqrt{x}}{7x}$ **117.** $\dfrac{3\sqrt[3]{7}}{7}$ **119.** $\dfrac{5\sqrt[4]{4}}{2}$
121. $\dfrac{2\sqrt{5y}}{5y}$ **123. a.** $\{4\}$; principal root only **b.** $\{4,-4\}$; both square roots **c.** No **125. a.** (i) Square first; then extract square root. (ii) Get square root; then square. **b.** Same value
127. $\sqrt[n]{\dfrac{a}{b}} = \left(\dfrac{a}{b}\right)^{1/n} = \dfrac{a^{1/n}}{b^{1/n}} = \dfrac{\sqrt[n]{a}}{\sqrt[n]{b}}$

Remember This (7.2)

1. True **2.** > **3.** $2\sqrt{10}$ **4.** $5\sqrt{2}$ in. **5.** $\dfrac{\sqrt{14}}{7}$ **6.** 0

7.

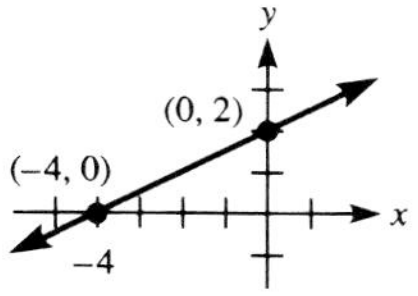

8. $\dfrac{4x+1}{2x-5}$ **9.** $x^3 + 3x^2 + 3x + 1$
10. $a = 2m - b$

Exercises 7.3

1. $6\sqrt{5}$ **3.** $9\sqrt[3]{17}$ **5.** $9\sqrt{6}$ **7.** $10\sqrt{13}$ **9.** $3\sqrt{6}$ **11.** $-8\sqrt{10}$
13. $-8\sqrt[4]{2}$ **15.** Does not simplify. **17.** Does not simplify.
19. Does not simplify. **21.** $(4+y)\sqrt{x}$ **23.** $(15-x)\sqrt{13}$ **25.** $6\sqrt[4]{7}$
27. $10\sqrt{x}$ **29.** $3y\sqrt{x}$ **31.** $2\sqrt[4]{3xy}$ **33.** $5\sqrt{2}$ **35.** $7\sqrt{3}$ **37.** $11\sqrt{2}$
39. $\sqrt{3}$ **41.** 0 **43.** $5\sqrt[3]{3}$ **45.** $-7\sqrt[3]{2}$ **47.** $5y\sqrt{6x}$ **49.** $-3x\sqrt{xy}$
51. $7xy\sqrt{3x}$ **53.** $(3-6y)\sqrt{2x}$ **55.** $3\sqrt{17}$ **57. a.** $4\sqrt{10}$ **b.** 10
59. a. $\sqrt{5}$ **b.** $2\sqrt{5}$ **c.** $\sqrt{5}$ **61.** $23\sqrt{2}$ **63.** $-\frac{2}{3}\sqrt{6}$ **65.** $\frac{5}{6}\sqrt{2}$
67. $\frac{7}{3}\sqrt{3}$ **69.** $\frac{3}{4}\sqrt{10}$ **71.** $-\frac{7}{6}\sqrt{6}$ **73.** $-\frac{14}{15}\sqrt{3}$ **75.** $-2\sqrt{6}$
77. 0
79. $\left(\dfrac{3x+4y}{xy}\right)\sqrt{xy}$

Remember This (7.3)

1. 7 **2.** Distributive **3.** $(a+b)(a-b)$ **4.** 2 **5.** $3 - x^2$ **6.** (1,5)
7. Slope −5 **8.** $4\sqrt{2}$ **9.** $\dfrac{-1}{(x-1)(y+2)}$ **10.** \$480.07

Exercises 7.4

1. 63 **3.** 640 **5.** 57 **7.** 63 **9.** 294 **11.** 153 **13.** $4 + 2x$
15. $3x^2 - 1$ **17.** $45x$ **19.** $98y$ **21.** $891x$ **23.** $3\sqrt{2} + 3$
25. $10\sqrt{2} - 10$ **27.** $5\sqrt{3} - 3\sqrt{5}$ **29.** $\sqrt{6} + \sqrt{10}$ **31.** 7 **33.** 10
35. −2 **37.** −1 **39.** $\sqrt[3]{x^2} + 5\sqrt[3]{x} + 4$ **41.** $\sqrt[3]{x^2} + 3\sqrt[3]{x} - 4$
43. $\sqrt[5]{y^2} - 10\sqrt[5]{y} + 24$ **45.** $21 + 4\sqrt{5}$ **47.** $67 - 12\sqrt{7}$
49. $9 + 24\sqrt{x} + 16x$ **51.** $25x - 50\sqrt{x} + 25$ **53.** −11 **55.** 278
57. $25 - x$ **59.** $x - 45$ **61.** $4x - 48$
63. $(2\sqrt{2}+2) - (2\sqrt{2}-2) = 4$; $(2\sqrt{2}+2)(2\sqrt{2}-2) = 4$
65. a. $\dfrac{0.5\sqrt{n}}{n}$ **b.** 0.016 **67. a.** $\dfrac{2n(\sqrt{n}-1)}{n-1}$ **b.** 4.5
69. $R = \dfrac{c(2\sqrt{c}-1)}{4c-1}$ **71.** $\sqrt{a}/a^2$ cm^3 **73.** $\dfrac{1+\sqrt{3}}{6}$ **75.** $\dfrac{1-3\sqrt{15}}{2}$
77. $\dfrac{1+2\sqrt{7}}{3}$ **79.** $\dfrac{3\sqrt{5}+5}{5}$ **81.** $\dfrac{5\sqrt{x}+x}{x}$ **83.** $\dfrac{x-6\sqrt{x}}{x}$
85. $2\sqrt{3} - 2$ **87.** $\dfrac{4+\sqrt{2}}{7}$ **89.** $\dfrac{8+\sqrt{10}}{6}$ **91.** $\dfrac{x\sqrt{x}-x}{x-1}$
93. $\dfrac{y\sqrt{6y}+6y}{y-6}$

Remember This (7.4)

1. $2x + 3$ **2.** True **3.** -5 **4.** False **5.** $1 + 2\sqrt{x} + x$ **6.** $(\frac{1}{3}, \frac{1}{5})$ **7.** **8.** $\frac{x}{c}$ **9.** -6 **10.** All

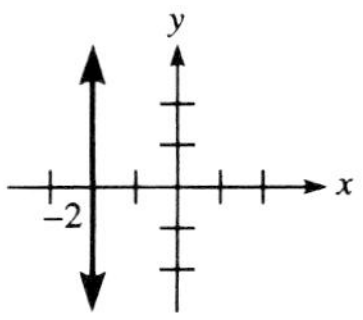

Exercises 7.5

1. {15} **3.** {20} **5.** {3} **7.** {2} **9.** $\emptyset$ **11.** {6} **13.** {−3} **15.** {2} **17.** 3,003 ft **19.** 0.81 ft **21.** $n = 7$ **23.** {3,2} **25.** {5} **27.** {3} **29.** {3} **31.** {−3,−2} **33.** {6} **35.** {3} **37.** {3,11} **39.** {3,7} **41.** {5,9} **43.** $\emptyset$ **45.** $\emptyset$ **47.** {5} **49.** $\{\frac{49}{9}\}$

Remember This (7.5)

1. True **2.** True **3.** 7 **4.** $<$ **5.** -1 **6.** $\frac{1}{3}$ **7.** False **8.** $\frac{5}{ab}$ **9.** 9×10^{13} **10.** $V = 282.6 \text{ cm}^3$

Exercises 7.6

1. i **3.** $4i$ **5.** $-2i$ **7.** $-7i$ **9.** $2i\sqrt{3}$ **11.** $2i\sqrt{6}$ **13.** $5i\sqrt{5}$ **15.** $-3i\sqrt{3}$ **17.** $-5i\sqrt{3}$ **19.** $18i$ **21.** $17i$ **23.** i **25.** $4i$ **27.** $5 + 3i$ **29.** $7 - 3i$ **31.** $16 + 4i$ **33.** 10 **35.** $15 - 3i$ **37.** $-2 - 8i$ **39.** $-3 - 3i$ **41.** $-10i$ **43.** 15 **45.** -72 **47.** -60 **49.** -20 **51.** $4 + 7i$ **53.** $14 - 18i$ **55.** $85 - 32i$ **57.** $6 - 22i$ **59.** $-7 + 24i$ **61.** $-32 + 24i$ **63.** $21 - 20i$ **65.** $48 - 14i$ **67.** 0 **69.** 0 **71.** 3 **73.** $\frac{3}{2} + \frac{1}{2}i$ **75.** $\frac{22}{25} + \frac{4}{25}i$ **77.** $\frac{19}{20} + \frac{13}{20}i$ **79.** $6 - 4i$ ohms **81.** $111 + 52i$ volts **83.** $6 + 6i$ amperes **85.** i **87.** 1 **89.** i **91.** -1

Remember This (7.6)

1. b **2.** {6,−6} **3.** 7 **4.** b **5.** 1.387 **6.** **7.** $m = -3$, (0,5) **8.** $\frac{1-x}{x}$ **9.** 4 **10.** $a - 3 + x$

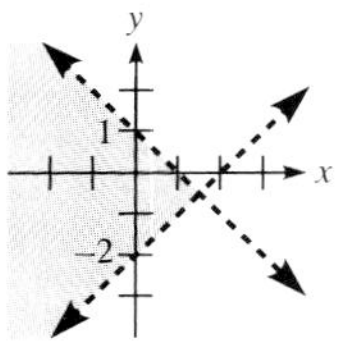

Chapter 7 Review Exercises

1. $-10, 10$ **3.** 2 **5.** $\frac{x^{12}}{8}$ **7.** $\sqrt{x}$ **9.** $36\sqrt[3]{4}$ **11.** $-4\sqrt{7}$ **13.** $\frac{x-y}{xy}\sqrt{xy}$ **15.** $x + 7$ **17.** $4y - 12\sqrt{y} + 9$ **19.** $\frac{-3 + 3\sqrt{11}}{5}$ **21.** -3 **23.** {−5,−4} **25.** $14i$ **27.** $4i\sqrt{2}$ **29.** $-i$ **31.** $2 + 5i\sqrt{3}$ **33.** 30 **35.** $6 - 2\sqrt{5}$ **37.** $\frac{x\sqrt{3x} + 3x}{x - 3}$ **39.** $11y\sqrt{5x}$ **41.** 44 **43.** $(5 + y)\sqrt{2xy}$ **45.** $\frac{1}{4}$ **47.** $\frac{9\sqrt{6}}{2}$ **49.** $\frac{\sqrt{35}}{7}$ **51.** $-i$ **53.** $\frac{1}{5^{3/4}}$ **55.** $36\sqrt[3]{9}$ **57.** -8 **59.** {5} **61.** $\emptyset$ **63.** {0,1}

Chapter 7 Test

1. $\frac{x^3}{y^2}$ **2.** $-4 - 2\sqrt{7}$ **3.** -8 **4.** $\frac{1}{2}$ **5.** $-i$ **6.** {13} **7.** {3} **8.** $\emptyset$ **9.** $-5\sqrt{3} + 0i$ **10.** $11 + 4i$ **11.** $62 - 5i$ **12.** $-\frac{4}{5} - \frac{7}{5}i$ **13.** $384x$ **14.** $\sqrt[3]{3y}$ **15.** -2 **16.** $(8 - 21xy)\sqrt{3xy}$ **17.** $\frac{\sqrt{15x}}{3x}$ **18.** $x\sqrt[3]{x}$ **19.** $36x\sqrt{5}$ **20.** $y + 2$

Cumulative Test 7

1. {−1,−2,−3,−4, . . .} **2.** -13 **3.** {−2,3} **4.** $[4,\infty)$ **5.** Toy, \$17; book, \$28 **6.** 4.05×10^8 **7.** $20x^2 - 22x + 6$ **8.** {−15,2} **9.** $\frac{x}{1-x}$ **10.** $\frac{6(a+6)^2}{a^2}$ **11.** $\frac{n^2 + 4n - 5}{(n+1)^2}$ **12.** **13.** 1 **14.** $y = -\frac{1}{3}x + 2$ **15.** $(-2,3)$ **16.** -52 **17.** $(3,-1,-1)$ **18.** $-x^2y\sqrt{2y}$ **19.** {−38} **20.** $18 - i$

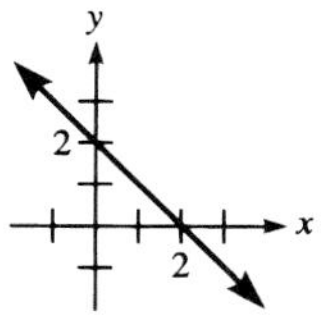

Chapter 8

Exercises 8.1

1. {±2} **3.** {±4} **5.** $\{\pm 2\sqrt{3}\}$ **7.** $\{\pm 2i\}$ **9.** $\{\pm 2i\sqrt{7}\}$ **11.** {−6,2} **13.** $\{-3 \pm 4i\}$ **15.** $\{-2 \pm \sqrt{6}\}$ **17.** $\{-5 \pm i\sqrt{5}\}$ **19.** $\{3 \pm 5i\}$ **21.** $\{4 \pm 3i\sqrt{5}\}$ **23.** {−1,11} **25.** $\{3 \pm i\sqrt{7}\}$ **27.** $\{\frac{1}{3}, -\frac{5}{3}\}$ **29.** $\left\{\frac{2 \pm \sqrt{10}}{5}\right\}$ **31.** $\left\{\frac{1 \pm 2i}{3}\right\}$ **33.** $\{-\frac{3}{2} \pm i\sqrt{3}\}$ **35.** 25; $(x + 5)^2$ **37.** 16; $(x - 4)^2$ **39.** $\frac{25}{4}$; $(x - \frac{5}{2})^2$ **41.** $\frac{1}{64}$; $(x - \frac{1}{8})^2$ **43.** $\frac{1}{9}$; $(x - \frac{1}{3})^2$ **45.** $\frac{9}{100}$; $(x + \frac{3}{10})^2$ **47.** {0,−6} **49.** $\{-1 \pm \sqrt{3}\}$ **51.** $\{6 \pm \sqrt{30}\}$ **53.** $\{-5 \pm \sqrt{29}\}$ **55.** $\{-1 \pm i\}$ **57.** $\{-2 \pm \sqrt{2}\}$ **59.** $\{4 \pm 3\sqrt{2}\}$ **61.** $\left\{\frac{1 \pm i\sqrt{15}}{2}\right\}$ **63.** $\{-1 \pm i\sqrt{5}\}$ **65.** $\{-3 \pm \sqrt{21}\}$ **67.** $\left\{\frac{1 \pm \sqrt{33}}{4}\right\}$ **69.** $\left\{\frac{2 \pm \sqrt{14}}{2}\right\}$ **71.** $\left\{\frac{-9 \pm \sqrt{93}}{6}\right\}$ **73.** $\left\{\frac{-1 \pm i}{2}\right\}$ **75.** $\left\{2, -\frac{1}{3}\right\}$ **77.** Height = 6.19 in., base = 1.62 in.; height = 0.81 in., base = 12.38 in. **79.** 12.43 in. **81.** 37.42 mi/hour **83.** 9.54 percent

Remember This (8.1)

1. $-3, 0$ **2.** $2\sqrt{2}$ **3.** $1 + a$ **4.** $1 + \sqrt{2}$ **5.** $2 + 6i\sqrt{7}$ **6.** {32} **7.** $(-2,-3)$ **8.** $\frac{20}{7}$ **9.** $m = -\frac{2}{3}$, $(0,\frac{4}{3})$ **10.** $3x^2 - 3x + 4 - \frac{9}{x+1}$

Exercises 8.2

1. {−1,3} **3.** {2} **5.** $\left\{\frac{-5 \pm \sqrt{41}}{2}\right\}$ **7.** $\{2 \pm \sqrt{7}\}$ **9.** {−4,−1} **11.** $\{-1 \pm i\sqrt{2}\}$ **13.** $\{-\frac{1}{2}, 3\}$ **15.** $\{\frac{2}{3}, \frac{3}{2}\}$ **17.** Real, rational, unequal (2 solutions) **19.** Real, rational, equal (1 solution) **21.** Conjugate complex numbers **23.** Real, rational, unequal (2 solutions) **25.** Real, irrational, unequal (2 solutions) **27.** Real, rational, unequal (2 solutions) **29.** Real, rational, unequal (2 solutions) **31.** Conjugate complex numbers **33.** Conjugate complex numbers **35.** Conjugate complex numbers **37.** {−2,6} **39.** {±3} **41.** {−5,0} **43.** $\{-\frac{1}{2}, 3\}$ **45.** {−3,5} **47.** {±4} **49.** $\left\{\frac{-5 \pm \sqrt{41}}{2}\right\}$ **51.** $\{-\frac{5}{3}, \frac{5}{3}\}$ **53.** $\{0, \frac{9}{2}\}$ **55.** $\{\frac{4}{3}, 1\}$ **57.** $\{\pm 8i\}$ **59.** $\left\{\frac{3 \pm i\sqrt{71}}{10}\right\}$ **61.** Side of smaller lot = 50 ft; side of larger lot = 110 feet **63.** Small square has side 6; large square has side 8. **65.** 0.69 and 1.81 seconds; one is for the way up and one is for the way down. **67.** $\frac{5 \pm \sqrt{21}}{2}$; 4.791 and 0.209

Remember This (8.2)

1. x^2 **2.** $10x - 18$ **3.** $2\sqrt{7}$ **4.** $(t-4)(t-1)$ **5.** $\{5\}$
6. $\frac{1}{4}$; $(y - \frac{1}{2})^2$ **7.** 4 **8.** $\frac{5}{2}$ **9.** -2 **10.**

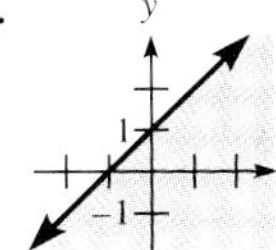

Exercises 8.3

1. $\{-2,6\}$ **3.** $\{-\frac{1}{2},7\}$ **5.** $\{-4,3\}$ **7.** $\left\{\frac{-1 \pm \sqrt{7}}{3}\right\}$ **9.** $\{1 \pm i\}$
11. $\emptyset$ **13.** $\left\{\frac{3 \pm i}{2}\right\}$ **15.** $\{\pm\sqrt{10}\}$ **17.** $\{-\sqrt{6}\}$ **19.** $\{2\sqrt{3}\}$
21. $\{-\sqrt{2}\}$ **23.** $\emptyset$ **25.** $\{x: x \geq 0\}$ **27.** $\{\sqrt{2}\}$ **29.** $\{0\}$
31. $\{\pm 1, \pm 2i\}$ **33.** $\{\pm\sqrt{5}, \pm\sqrt{6}\}$ **35.** $\{\pm\sqrt{3}\}$ **37.** $\{\pm 3, \pm 4\}$
39. $\{\pm 2, \pm i\}$ **41.** $\{\pm\sqrt{3}, \pm i\sqrt{2}\}$ **43.** $\{-1,27\}$ **45.** $\emptyset$ **47.** $\{256\}$
49. $x = \pm\sqrt{y+3}$ **51.** $x = \pm\sqrt{y+8}$ **53.** $x = \pm\sqrt{y^2-6}$
55. $x = \dfrac{-a_2 \pm \sqrt{a_2^2 - 4a_1(a_3 - t)}}{2a_1}$ **57.** $x = \dfrac{-r \pm \sqrt{r^2 + 4dt}}{2d}$
59. 6 yd, 4 yd **61.** 50 rows of 40 trees each

Remember This (8.3)

1. $\{x: x > -1\}$ **2.** Positive **3.** Yes **4.** $\{0, 2, -\frac{5}{3}\}$ **5.** $\{-2\}$
6. $\sqrt{7}$ **7.** $\sqrt{17}$ **8.** $\dfrac{x}{2y^6}$ **9.** $(3,-2,-1)$ **10.** $1.06x$

Exercises 8.4

1. −2 4 x **3.** −1 6 y **5.** $-\frac{1}{2}$ 7 x
7. $-\frac{7}{3}$ $\frac{2}{3}$ x **9.** −1.22 0.55 x
11. $\emptyset$ **13.** $\{5\}$ **15.** $(-\infty,\infty)$ **17.** −5 5 y
19. −2.77 1.27 x **21.** −2 0 3 x **23.** −3 0 5 x
25. −3 0 1 x **27.** −4 $-\frac{1}{2}$ 1 x **29.** −2 −1 3 5 x
31. −7 −5 −2 0 x **33.** −2 1 x **35.** −4 3 x
37. 0 6 x **39.** −4 $\frac{1}{2}$ x
41. 0 $\frac{5}{2}$ x **43.** 10 x

45. $(2.35, 6.65)$ **47. a.** Lengths from 4 to $4 + \sqrt{6}$ in., or 4 to 6.45 in.
b. 4 in.

Remember This (8.4)

1. $(2,4), (-1,1), (0,0)$ **2.** $(0,1)$ **3.** $x/2$ **4.** c **5.** $-\frac{7}{4}$ **6.** 0, 6
7. $\{2,-2,2i,-2i\}$ **8.** $\frac{7}{12}$ **9.** $\emptyset$ **10.** $\dfrac{a^2b + a}{ab^2 - b}$

Exercises 8.5

1. a. $x = -1$ **b.** $(-1,-1)$ **c.** $(-2,0), (0,0)$ **d.**

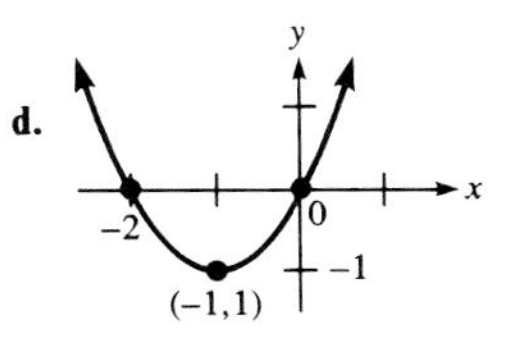

3. a. $x = \frac{3}{2}$ **b.** $(\frac{3}{2}, \frac{9}{4})$
c. $(3,0), (0,0)$
d.

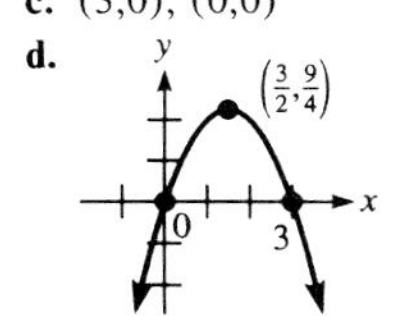

5. a. $x = \frac{5}{2}$ **b.** $(\frac{5}{2}, -\frac{25}{4})$
c. $(0,0), (5,0)$
d.

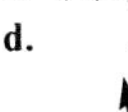
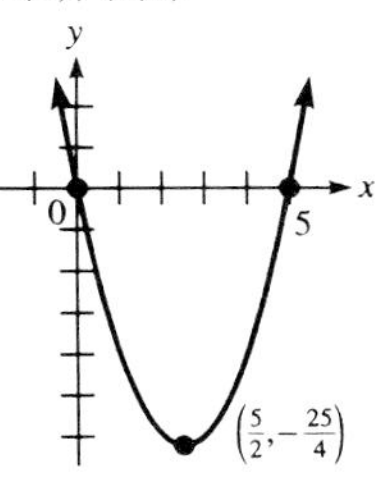

7. a. $x = 2$ **b.** $(2,0)$
c. $(2,0), (0,4)$
d.

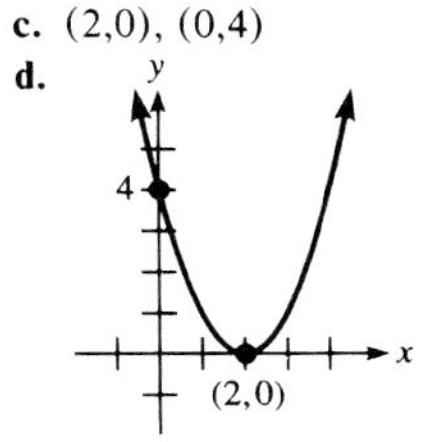

9. a. $x = -3$ **b.** $(-3,0)$
c. $(-3,0), (0,-9)$
d.

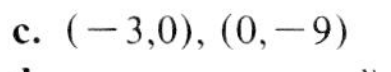
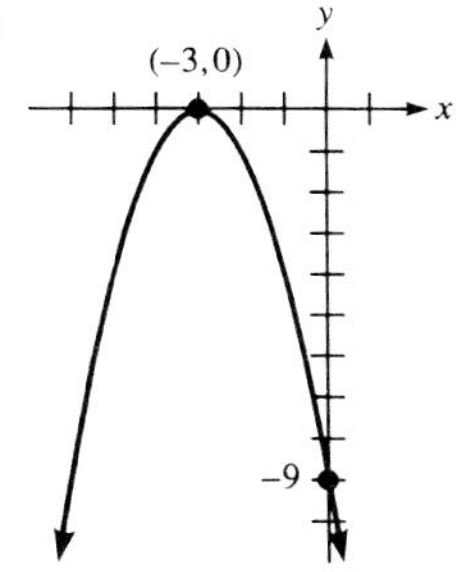

11. a. $x = 0$ **b.** $(0,9)$
c. $(0,9)$
d.

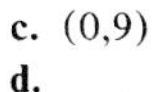
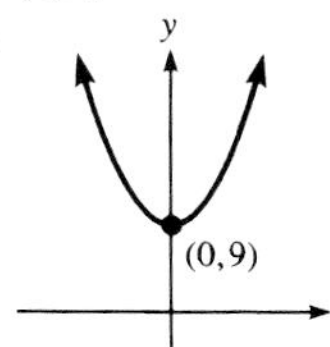

13. a. $x = 1$ **b.** $(1,-9)$
c. $(-2,0), (4,0), (0,-8)$
d.

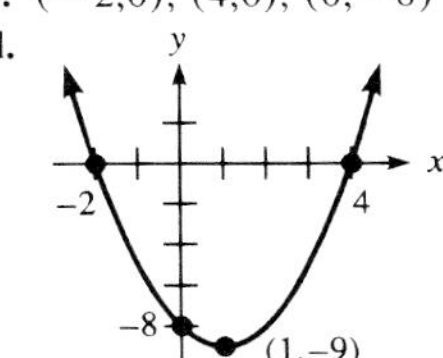

15. a. $x = 1$ **b.** $(1,9)$
c. $(-2,0), (4,0), (0,8)$
d.

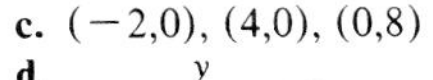
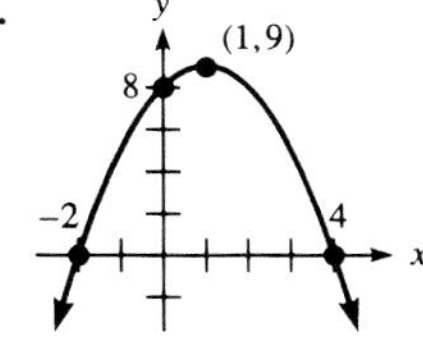

17. a. $x = \frac{1}{2}$ **b.** $(\frac{1}{2}, -\frac{81}{4})$
c. $(-4,0), (5,0), (0,-20)$
d.

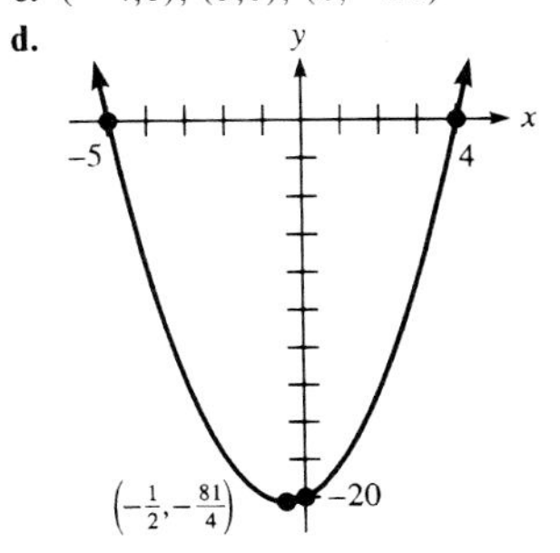

19. a. $x = \frac{1}{2}$ **b.** $(\frac{1}{2}, \frac{81}{4})$
c. $(-4,0), (5,0), (0,20)$
d.

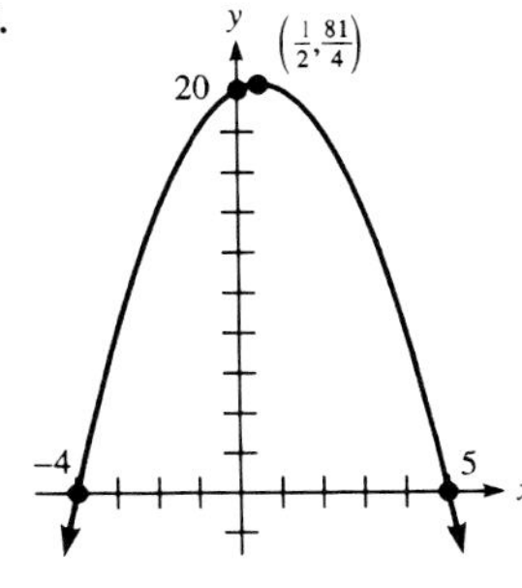

21. a. $x = -\frac{9}{2}$ **b.** $(-\frac{9}{2}, \frac{9}{4})$
c. $(-6,0), (-3,0), (0,-18)$
d.

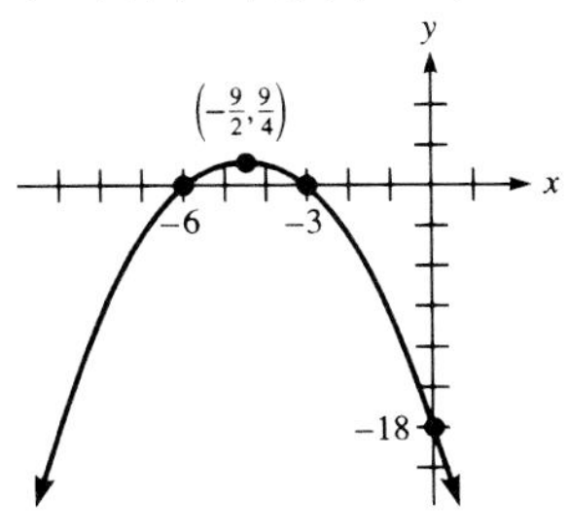

23. **a.** $x = -\frac{9}{2}$ **b.** $(-\frac{9}{2}, -\frac{9}{4})$
c. $(-6,0)$, $(-3,0)$, $(0,18)$
d.

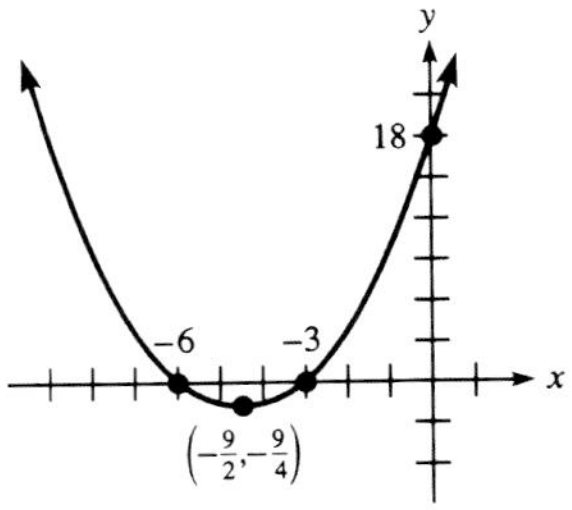

25. **a.** $x = \frac{1}{12}$ **b.** $(\frac{1}{12}, \frac{49}{24})$
c. $(-\frac{1}{2},0)$, $(\frac{2}{3},0)$, $(0,2)$
d.

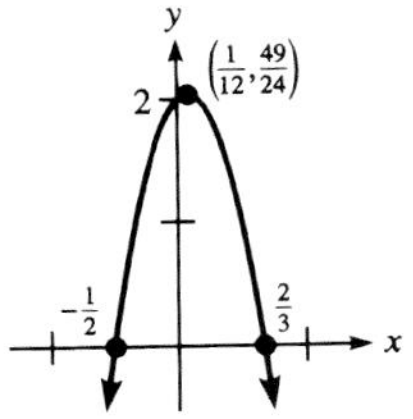

27. **a.** $x = \frac{1}{12}$ **b.** $(\frac{1}{12}, -\frac{49}{24})$
c. $(-\frac{1}{2},0)$, $(\frac{2}{3},0)$, $(0,-2)$
d.

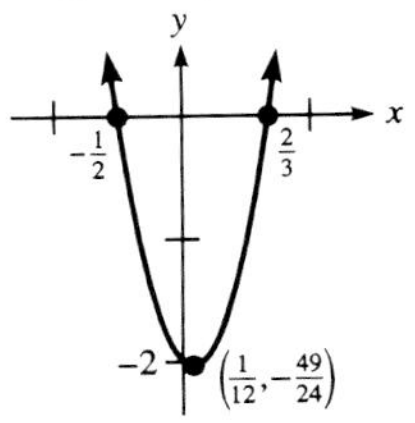

29. **a.** $x = 0$ **b.** $(0,-20)$
c. $(\pm 2\sqrt{5},0)$, $(0,-20)$
d.

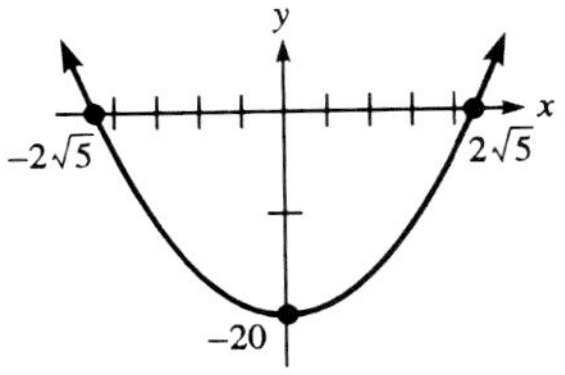

31. **a.** $x = 2$ **b.** $(2,-7)$
c. $(2 \pm \sqrt{7},0)$, $(0,-3)$
d.

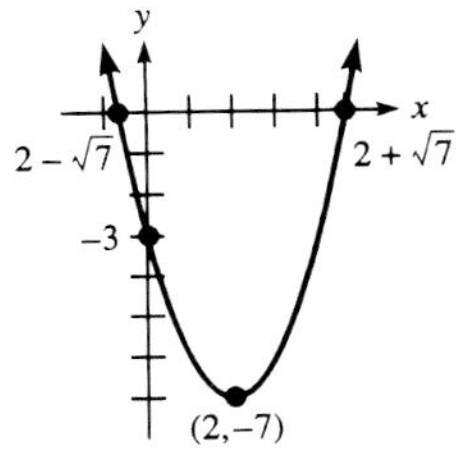

33. **a.** $x = 1$ **b.** $(1,4)$ **c.** $(0,5)$
d.

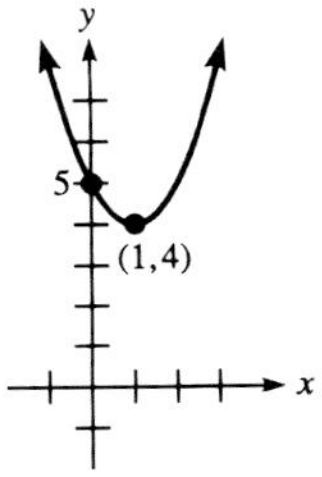

35. **a.** $x = 1$ **b.** $(1,-4)$
c. $(0,-5)$ **d.**

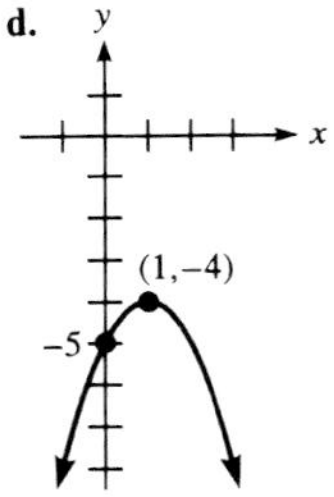

37. **a.** $y = -\frac{5}{2}$ **b.** $(-\frac{25}{4}, -\frac{5}{2})$
c. $(0,0)$, $(0,-5)$
d.

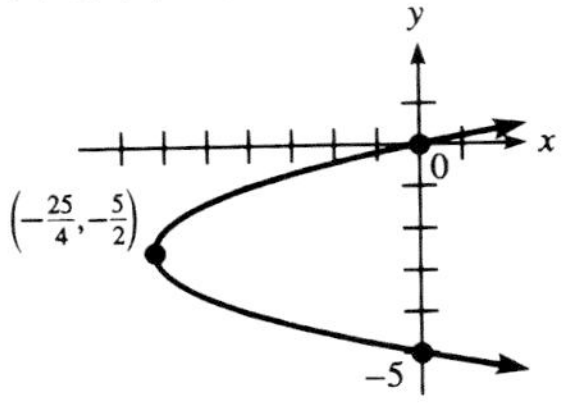

39. **a.** $y = \frac{3}{2}$ **b.** $(\frac{9}{4}, \frac{3}{2})$
c. $(0,0)$, $(0,3)$
d.

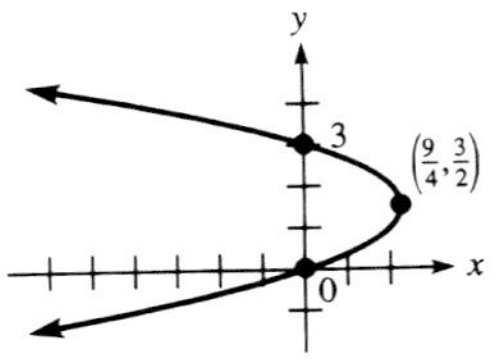

41. **a.** $y = -2$ **b.** $(4,-2)$
c. $(0,-4)$, $(0,0)$
d.

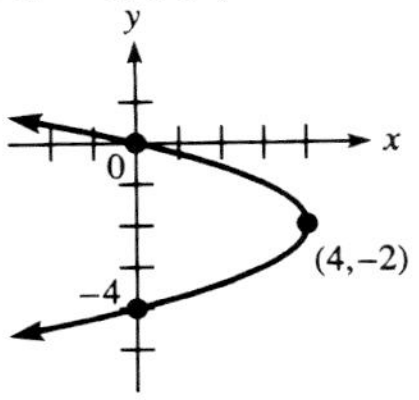

43. **a.** $y = 1$ **b.** $(9,1)$
c. $(0,-2)$, $(0,4)$, $(8,0)$ **d.**

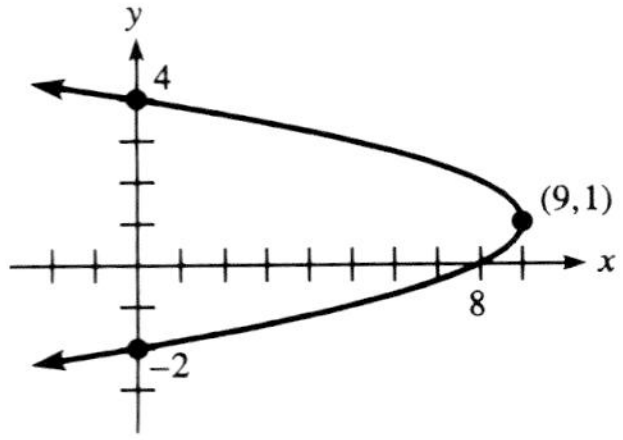

45. 2 **47.** 2 **49.** 1 **51.** 0 **53.** 1

55. **a.**

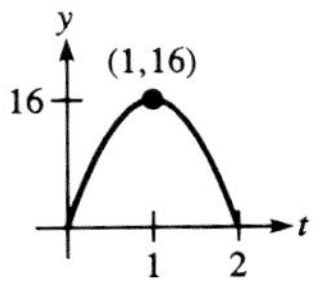

b. 1 second **c.** 16 ft
d. 2 seconds

57. **a.**

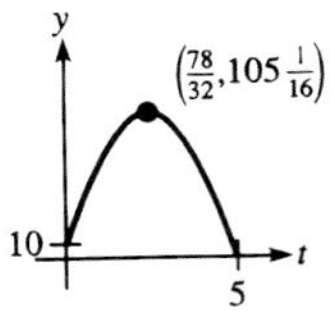

b. $\frac{78}{32} = 2.4375$ seconds
c. $105\frac{1}{16} = 105.0625$ ft
d. 5 seconds

59. 56.6 ft/second **61.** The maximum value is $\frac{1}{4}$, when $x = \frac{1}{2}$.

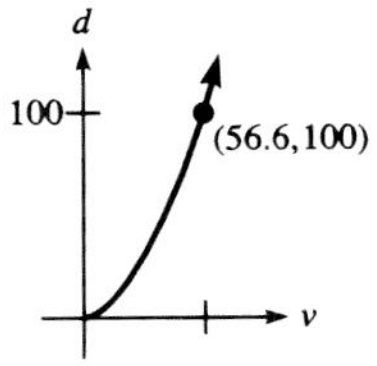

Remember This (8.5)

1. True **2.** True **3.** 5 **4.** Yes **5.** $A = \pi r^2$

6. $[-2,2]$ 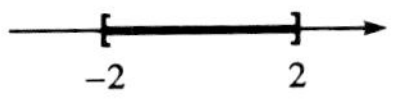

7. $\left\{\frac{-1 \pm \sqrt{5}}{2}\right\}$ **8.** $y = -\frac{5}{2}x$

9. $x(2x + 3)(3x - 2)$ **10.** Distributive

Exercises 8.6

1. $x^2 + y^2 = 25$ **3.** $x^2 + y^2 = 9$ **5.** $(x + 2)^2 + (y - 3)^2 = 49$
7. $(x + 5)^2 + (y + 8)^2 = 9$ **9.** $(x - 4)^2 + (y - 4)^2 = 25$
11. $x^2 + (y + 5)^2 = 36$ **13.** $(x + 5)^2 + y^2 = 81$
15. $(x - 2)^2 + (y - 3)^2 = 9$ **17.** $(x - 6)^2 + (y - 3)^2 = 25$
19. $(x + 3)^2 + (y + 5)^2 = 32$ **21.** $(x - 3)^2 + (y - 2)^2 = \frac{25}{64}$
23. $(x - 2)^2 + (y + 3)^2 = \frac{4}{9}$

25. Center (5,3); radius 4

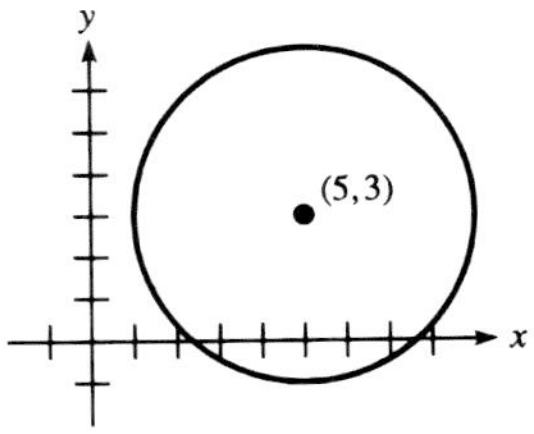

27. Center (−4,−3); radius 6

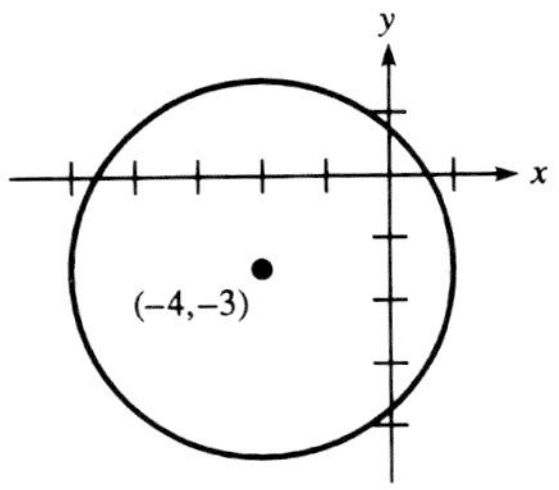

29. Center (0,0); radius 9

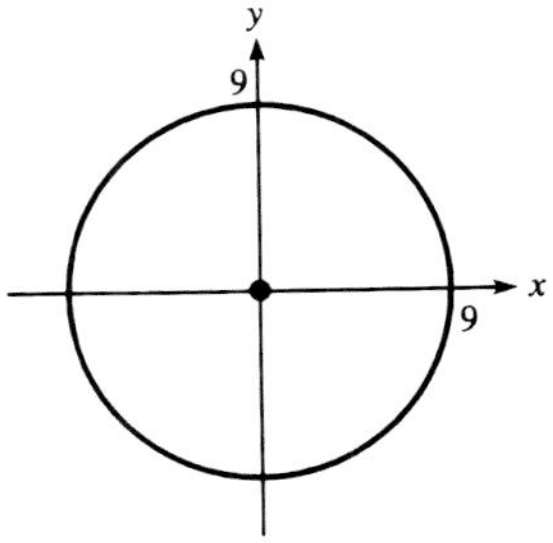

31. Center (0,5); radius 2

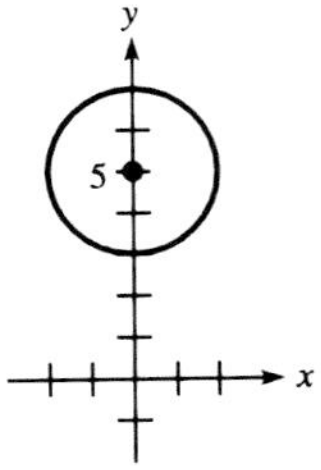

33. Center $(2,-1)$; radius $2\sqrt{5}$

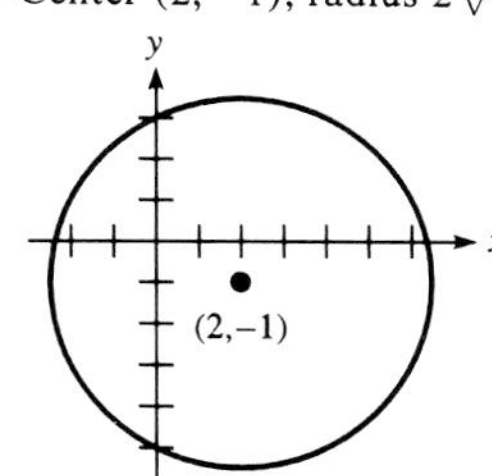

35. Center $(-2,0)$; radius 3

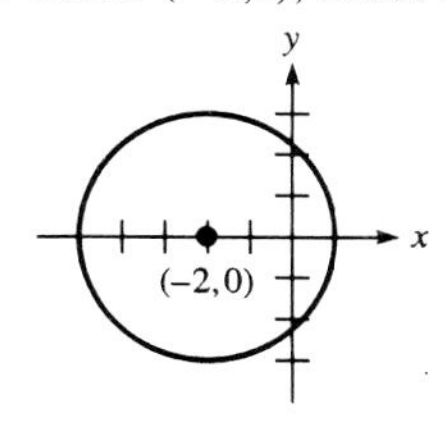

37. Center $(-6,3)$; radius 7

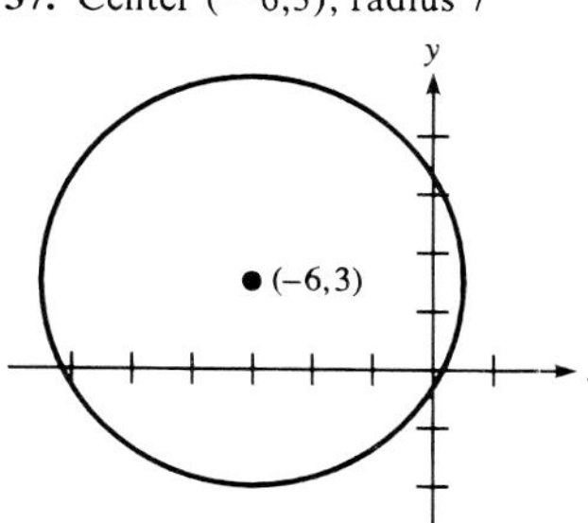

39. Center $(7,5)$; radius $4\sqrt{5}$

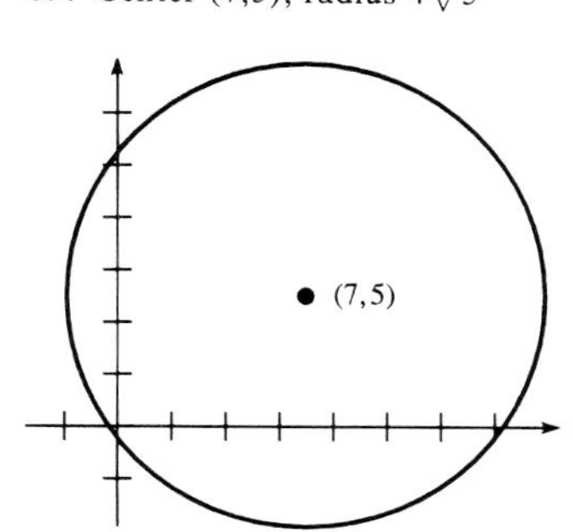

41. a.

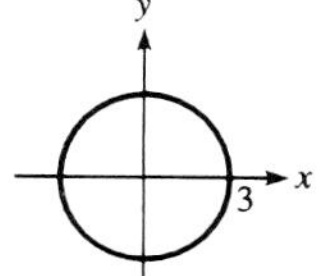

b. **c.**

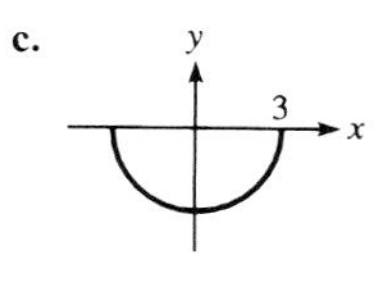

43. a.

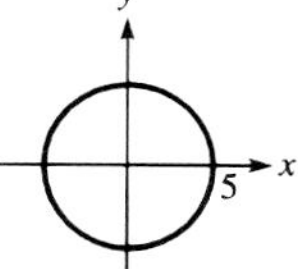

b. **c.**

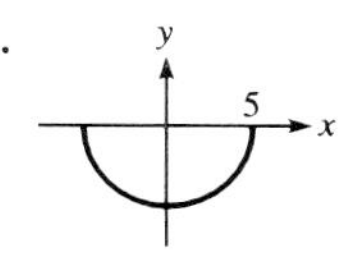

45. a.

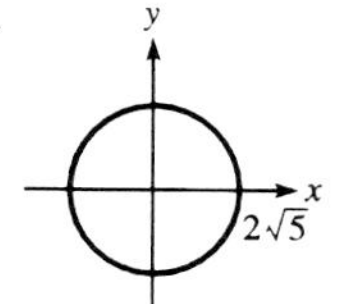

b. **c.**

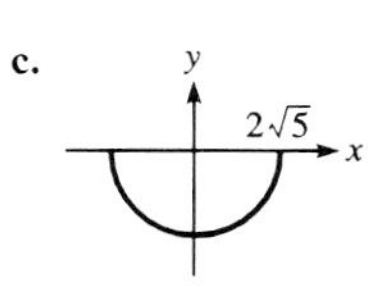

47. a.

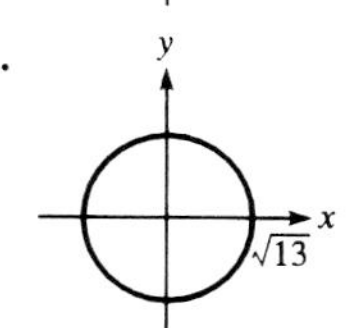

b. **c.**

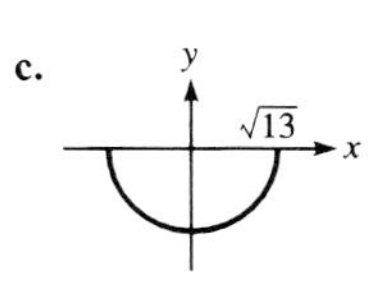

49. 40 ft **51. a.** $(1,0)$ **b.** $x^2 + y^2 = 4$ **c.** $(1,\sqrt{3})$ **d.** $2\sqrt{3}$ **e.** $\sqrt{3}$ **53.** 1

Remember This (8.6)

1. $(-4,0)$, $(4,0)$ **2.** $\{\pm 3i\}$ **3.** $m_1 = \frac{2}{3}$, $m_2 = -\frac{2}{3}$; no **4.** No **5.** $A = k/p$ **6.** $(3,2)$ **7.** $\{\pm i\}$ **8.** $(-1,-5)$ **9.** $\{6\}$ **10.** $b = \dfrac{ax}{y+1}$

Exercises 8.7

1.

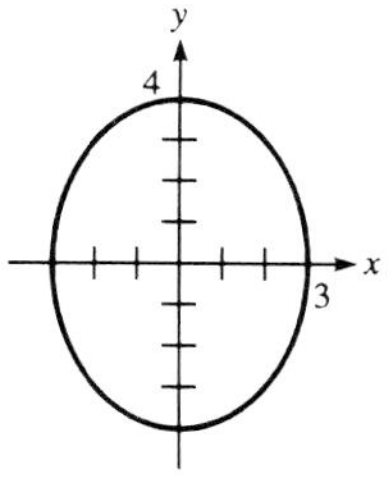

3.

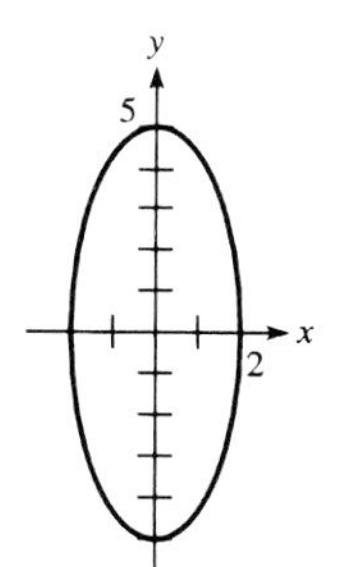

5.

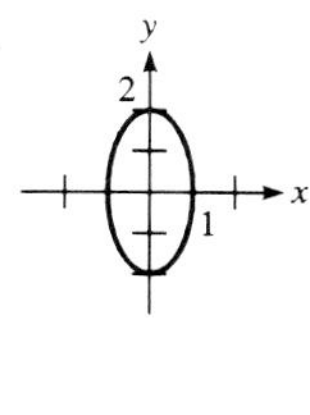

7. **9.**

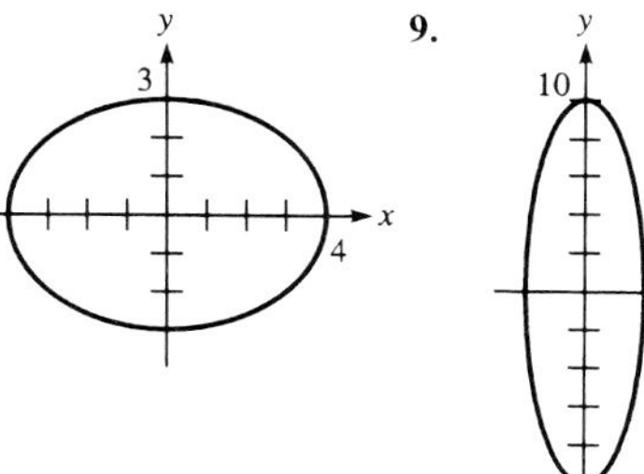

11.

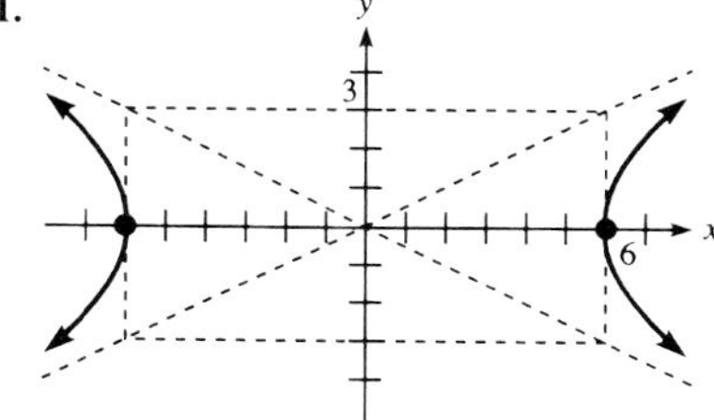

13.

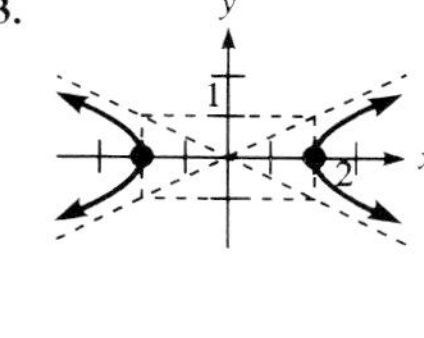

15.

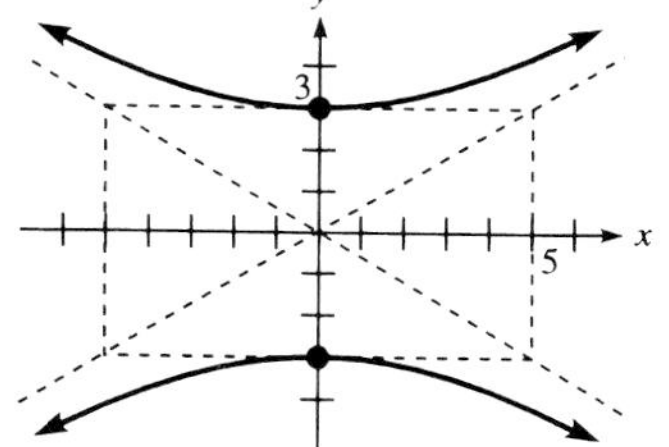

17.

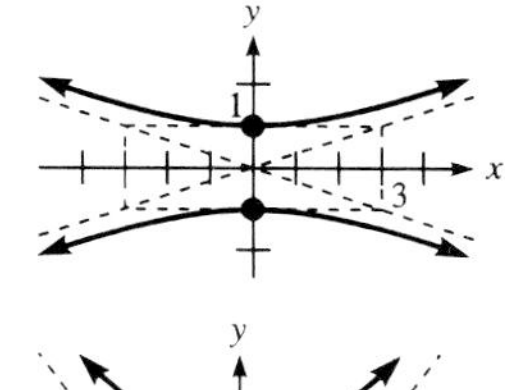

19.

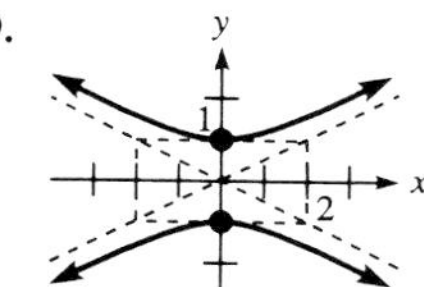

21.

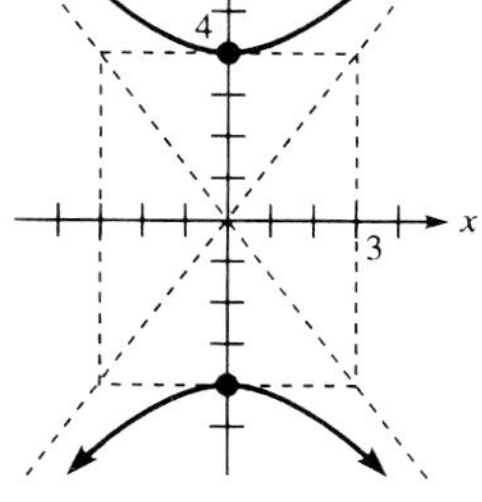

23.

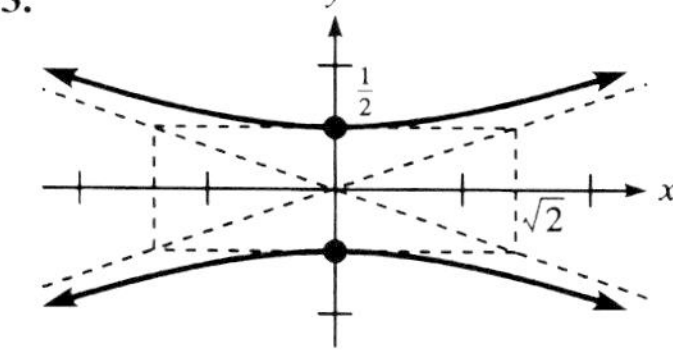

25.

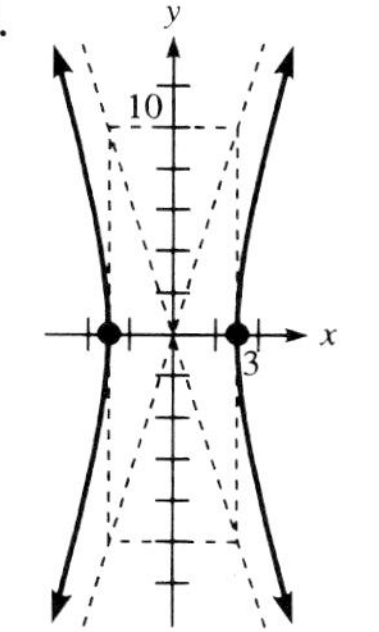

27.

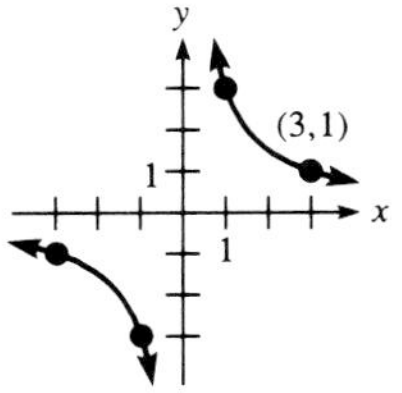

29.

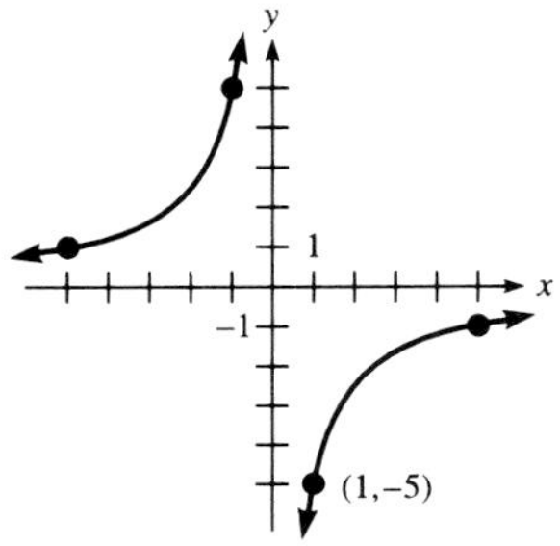

31. Parabola **33.** Hyperbola **35.** Circle **37.** Ellipse **39.** Hyperbola
41. Ellipse **43.** Ellipse **45.** Circle
47. a. $k = 15$; $y = 15/x$
b. $xy = 15$

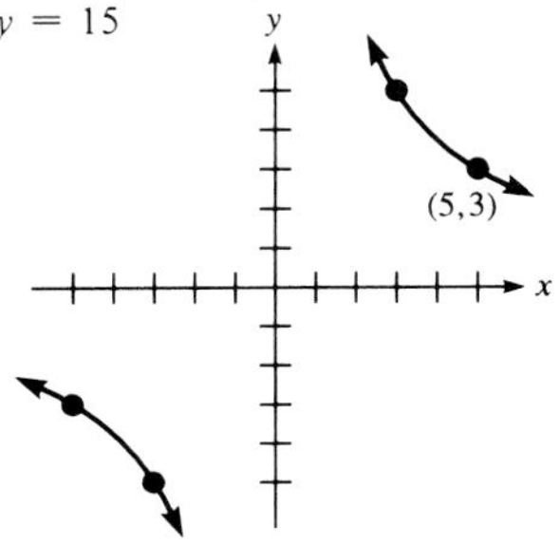

49. a. $t = k/s$ **b.** $k = 110$
c.

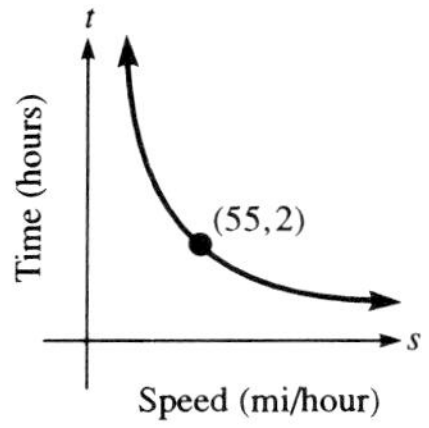

d. $t = 2.444$ hours
$= 2$ hours 27 minutes

Remember This (8.7)

1. 2 **2.** 4; 0 **3.** $\{0, -1\}$ **4.** $\{2, -2\}$ **5.** Yes **6.**

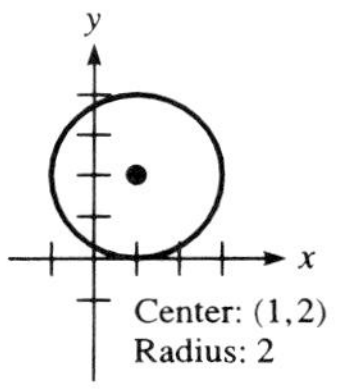

7. $2 \pm \sqrt{3}$ **8.** 9 **9.** -2
10. $[-3,2]$

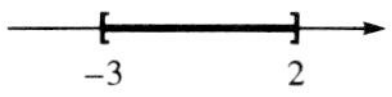

Exercises 8.8

1. $\{(0,2),(3,8)\}$ **3.** $\{(-2,11),(5,-17)\}$ **5.** $\{(3,9)\}$ **7.** $\emptyset$
9. $\{(0,-6),(6,0)\}$ **11.** $\{(0,-3),(\frac{12}{5},\frac{9}{5})\}$ **13.** $\emptyset$
15. $(3,5)$, $(3,-5)$, $(-3,5)$, $(-3,-5)$
17. $(4,3)$, $(4,-3)$, $(-4,3)$, $(-4,-3)$
19. $(\sqrt{15},-3)$, $(\sqrt{15},3)$, $(-\sqrt{15},3)$, $(-\sqrt{15},-3)$ **21.** 6 by 8 cm
23. 1.005 by 9.949 in. **25. a.** $y = \sqrt{3}x$; $x^2 + y^2 = 4$
b. $(1,\sqrt{3})$, $(-1,-\sqrt{3})$

Remember This (8.8)

1. $y = \pm\sqrt{25 - x^2}$ **2.** 5 **3.** 2 **4.** $\{x: x \geq -4\}$ **5.** Infinitely many
6. $y = 2x + 3$ **7.**

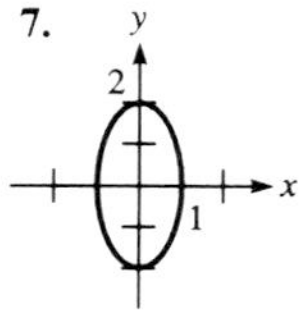

8. $\{0,\frac{4}{3}\}$ **9.** $y = -\frac{1}{3}x$ **10.** \$39.50

Chapter 8 Review Exercises

1. $\{i\sqrt{3}, -i\sqrt{3}\}$ **3.** $x^2 + 8x + 16 = (x + 4)^2$
5. $\left\{\frac{1 + \sqrt{37}}{6}, \frac{1 - \sqrt{37}}{6}\right\}$ **7.** $\{-2, \frac{3}{4}\}$
9. 2 solutions; real, irrational, unequal **11.** $\{2,3\}$ **13.** $\{\pm 1, \pm 3\}$
15. $y = \pm\sqrt{9 - x^2}$ **17.** $(-\infty,\infty)$ **19.** $(0,\frac{5}{3})$

21.

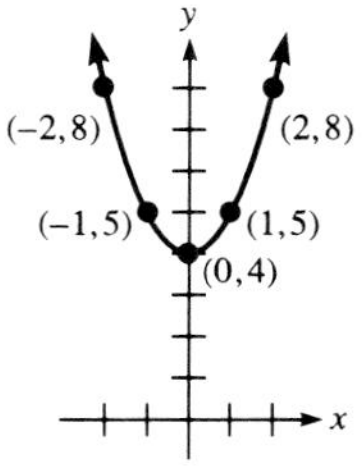

23. a. $x = \frac{1}{2}$ **b.** $(\frac{1}{2}, -\frac{25}{4})$
c. $(-2,0)$, $(3,0)$, $(0,-6)$
d.

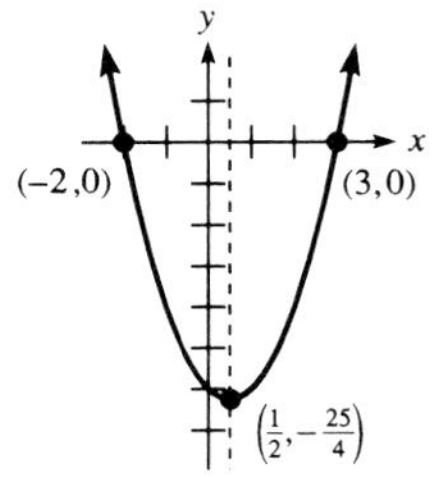

25. a. 4.5 seconds **b.** 324 ft **c.** 9 seconds
27. $(x + 2)^2 + (y - 6)^2 = 18$ **29.** Center (3,2); radius $\sqrt{11}$
31.

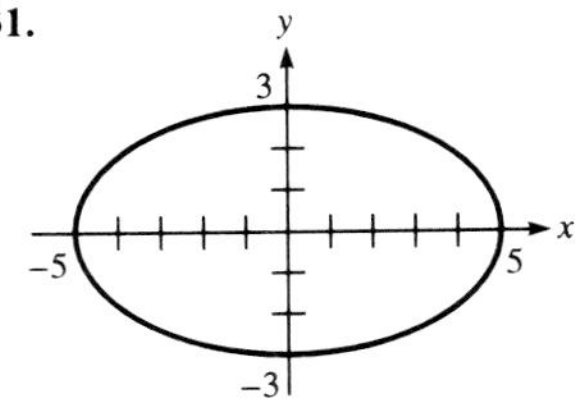

33.

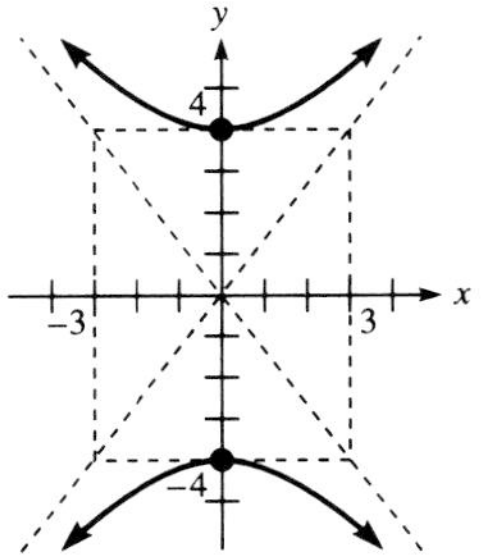

35. Ellipse **37.** $\{(2,0),(\frac{8}{5},-\frac{6}{5})\}$
39. $\{(16,5),(-16,-5),(5,16),(-5,-16)\}$
41. $(1,1),(1,-1),(-1,1),(-1,-1)$ **43.** $(-1,0)$; $r = \sqrt{6}$
45. $(x + 1)^2 + (y - 4)^2 = 49$ **47.** $y^2 + 10y + 25 = (y + 5)^2$
49. $Ap = 500$, or $A = \dfrac{500}{p}$ **51.** Width 2.815 in.; length 7.815 in.

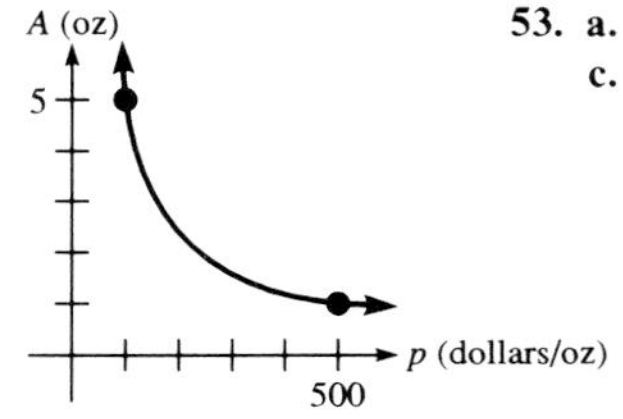

53. a. $x = 2$ **b.** $(2,0)$
c. $(2,0)$, $(0,4)$ **d.**

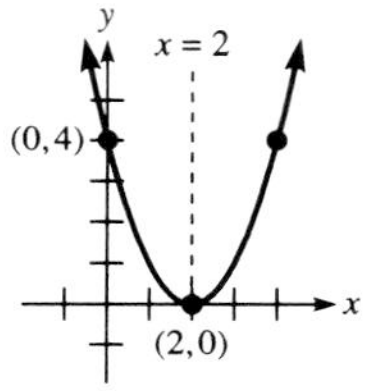

55. a. $y = -4$ **b.** $(-16,-4)$
c. $(0,0)$, $(0,-8)$
d.

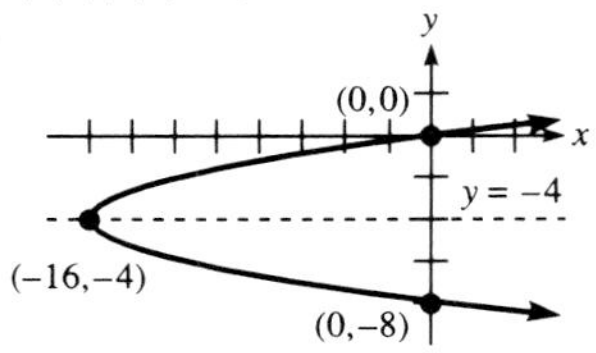

57. $t = \dfrac{-p \pm \sqrt{p^2 + 128q - 128y}}{-64}$ **59.** $\{(3,5),(-5,-3)\}$
61. $\left\{\frac{3 - \sqrt{29}}{2}, \frac{3 + \sqrt{29}}{2}\right\}$ **63.** $\{-3,\frac{2}{3}\}$ **65.** $\{2 + i, 2 - i\}$ **67.** $\{\frac{1}{3}\}$
69. $\{5i, -5i\}$ **71.** $\{4 - \sqrt{5}, 4 + \sqrt{5}\}$ **73.** $(0,3]$
75. $(-1,0) \cup (\frac{1}{2},\infty)$
77. Circle

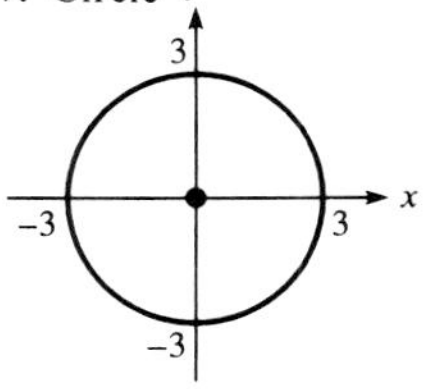

79. Parabola

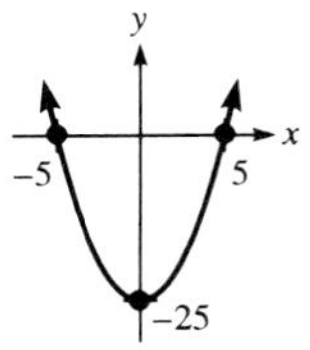

81. Ellipse

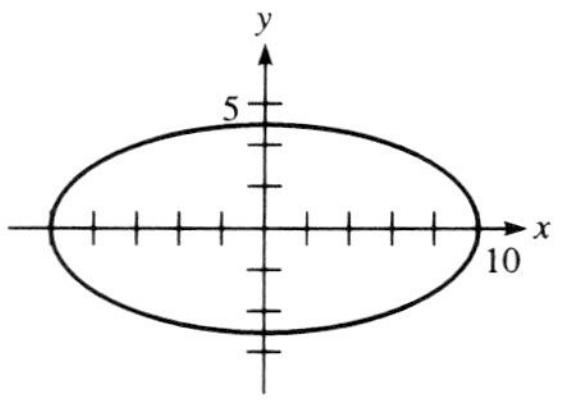

83. Parabola

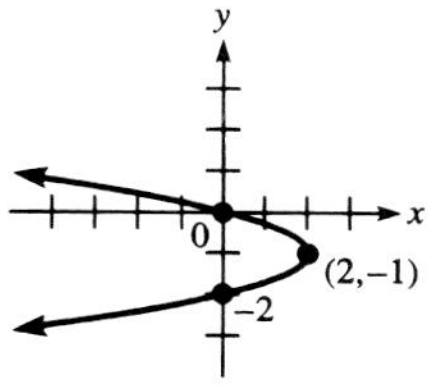

85. Hyperbola

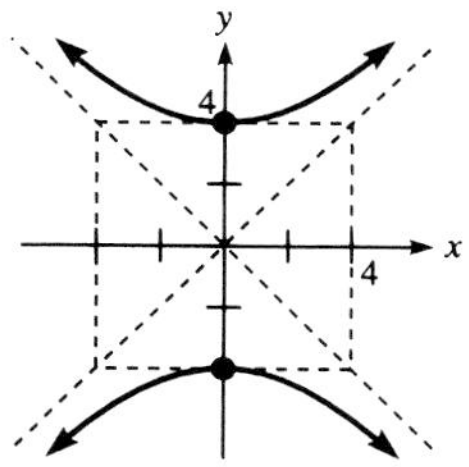

Chapter 8 Test

1. $x^2 - 8x + 16 = (x - 4)^2$ **2.** One rational solution
3. $(-1,1); r = \sqrt{5}$ **4.** $(x - 5)^2 + (y - 1)^2 = 25$
5. a. $x = 2$ **b.** $(2,-1)$ **6.** $y = \pm\sqrt{x^2 - 7}$ **7.** $\{(-2,-3),(1,3)\}$
c. (1,0), (3,0), (0,3)
d.

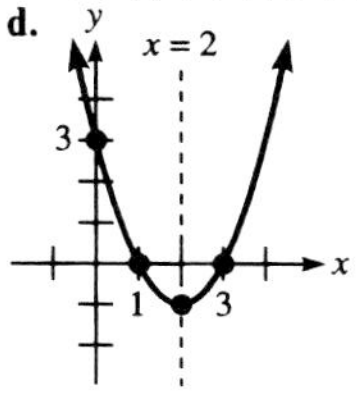

8.

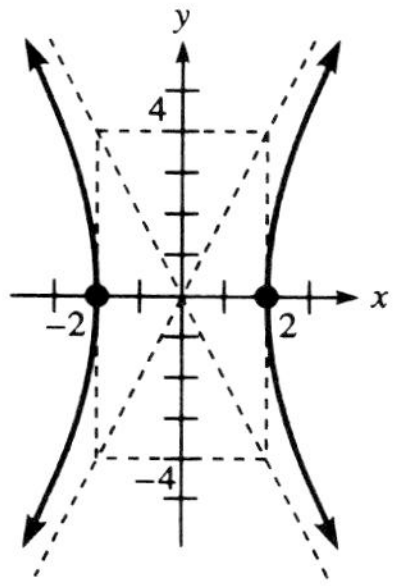

9.

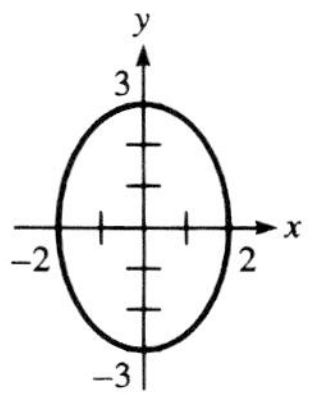

10. $\{2 + \sqrt{5}, 2 - \sqrt{5}\}$ **11.** $\left\{\frac{-1 + \sqrt{33}}{4}, \frac{-1 - \sqrt{33}}{4}\right\}$
12. $\{6 + 2i, 6 - 2i\}$ **13.** $\{-1,4\}$ **14.** $\{\frac{2}{3}, -2\}$ **15.** $\{\sqrt{2}\}$
16. $\{\sqrt{6}, -\sqrt{6}, \sqrt{2}, -\sqrt{2}\}$ **17.** $(-\infty,-7) \cup (1,\infty)$
18. $(-\infty,-4) \cup (0,\frac{1}{3})$ **19.** $[-5,6)$ **20. a.** 3.5 seconds
b. 196 ft **c.** 7 seconds

Cumulative Test 8

1. 99 **2.** $T = \frac{1}{R} - S$, or $T = \frac{1 - RS}{R}$ **3.** $(-2,3)$
4. $6x^3 - 14x^2 + 17x - 12$ **5.** $6x^2y^2(4x^2 + 3xy + 1)$
6. Man 210 lb; son 125 lb **7.** 2 **8.** $\frac{-3(x + 2)}{4}$ **9.** $2x - 15 + \frac{63}{x + 4}$
10. $y = -\frac{1}{2}x$ **11.** $y = \frac{1}{2}$ **12.** 55 flowering plants; 25 cacti
13.

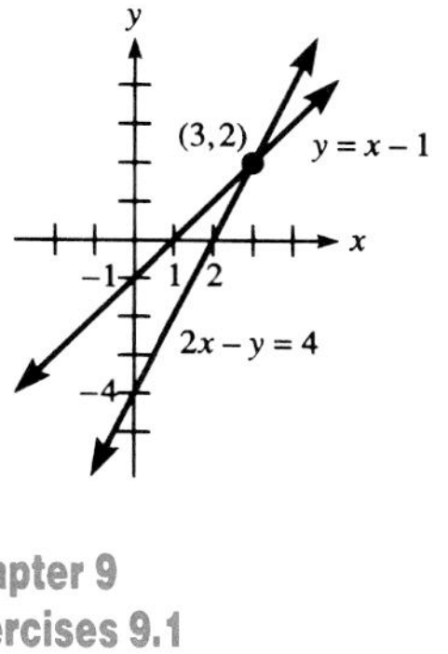

14. $(3,\frac{1}{2},1)$ **15.** $12 + 6\sqrt{3}$ **16.** $\frac{\sqrt[3]{45x}}{3}$
17. $\frac{5}{17} + \frac{3}{17}i$ **18.** $(-\infty,-4) \cup (3,\infty)$
19. $(x - 1)^2 + (y + 5)^2 = 13$ **20.** Ellipse

Chapter 9
Exercises 9.1

1. Domain = {80,89,85,79,78}; range = {B,C}
3. Domain = {1,2}; range = {2,3,4} **5.** Domain = {−1,0}; range = {0,1}
7. Yes **9.** No **11.** Yes **13.** No **15.** Yes **17.** Yes **19.** No
21. Yes **23.** Yes **25.** Yes **27.** No **29.** Yes **31.** No **33.** No
35. Yes **37.** No **39.** No **41.** Yes **43.** Yes **45.** No
47. Domain = $[-4,4]$; range = $[-2,2]$
49. Domain = $(0,\infty)$; range = $(0,\infty)$
51. Domain = $(-2,2)$; range = $(0,2]$ **53.** Domain = $[0,\infty)$; range = $\{4\}$
55. Domain = $(-\infty,3]$; range = $(-\infty,\infty)$
57. Domain = $(-\infty,\infty)$; range = $\{2\}$
59. Domain = $(-\infty,\infty)$; range = $(-\infty,\infty)$
61. Domain = $(-\infty,\infty)$; range = $[0,\infty)$
63. Domain = $\{x: x \neq 0\}$; range = $(0,\infty)$
65. Domain = $[2,\infty)$; range = $[0,\infty)$
67. Domain = $(-\infty,\infty)$; range = $[0,\infty)$
69. Domain = $\{x: x \neq -4\}$; range = $\{y: y \neq 0\}$
71. Domain = $(-\infty,-3] \cup [3,\infty)$; range = $(-\infty,\infty)$
73. Domain = $\{x: x \neq 1\}$; range = $\{y: y \neq 0\}$
75. Domain = $(-1,1)$; range = $(-1,1)$
77. a. $d = 186{,}000t$; domain: $[0,\infty)$ **b.** 242,000 mi
c. $d = 11{,}160{,}000t$; domain: $[0,\infty)$ **d.** 93 million mi
79. a. $i = 39m$; domain: $[0,\infty)$ **b.** 23.4 in.
c. $m = \frac{i}{39} = 0.026i$; domain $[0,\infty)$ **d.** 1.3 m
81. a. $C = 1.23f$; domain: $[0,\infty)$ **b.** \$24.60
83. a. $d = \sqrt{2s}$; domain: $(0,\infty)$ **b.** 741.0 mm

Remember This (9.1)

1. 37 **2.** 2 **3.** 155 **4.** False **5.** (1,5)

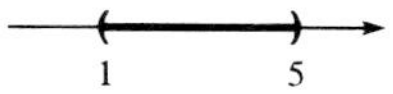

6. $a = \frac{1}{b} - c$ **7.** 6.4 lb **8.** $\frac{2}{x + 3}$ **9.** $-1 - \sqrt{2}$ **10.** Ellipse

Exercises 9.2

1. 2, −4 **3.** −1, 9 **5.** 12, 2 **7.** −12, −20 **9.** 160, 325 **11.** 0, 0
13. $f(4) = 13{,}122$; in 4 years the value will be about \$13,100; $f(7) = 9{,}565.94$; in 7 years the value will be about \$9,600.
15. $f(50) = 48{,}764.58$; the current value is about \$48,800.
17. $f(1) = 5.1$; 1 second after stepping off, the diver is 5.1 m above the water. $f(1.42) = 0.12$; 1.42 seconds after stepping off, the diver is 0.12 m above the water (about to hit the water).
19. $f(20) = 170$; the recommended maximum heart rate for a 20-year-old is 170 beats per minute; $f(40) = 153$; the recommended maximum heart rate for a 40-year-old is 153 beats per minute. **21.** 1 **23.** −2 **25.** 2
27. 2 **29.** 1 **31.** 10 **33.** −62 **35.** 25 **37.** 13 **39.** 82
41. $a - b - 3 \neq (a - 3) - (b - 3)$ **43.** $3(a - b) = 3a - 3b$
45. $(a + b)^2 \neq a^2 + b^2$
47. a. −1 **b.** 6 **c.** 9 **d.**

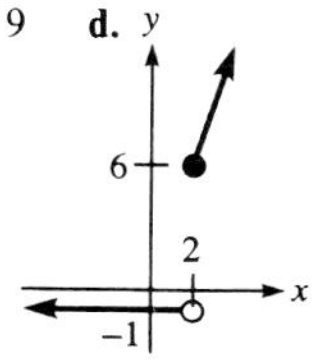

49. a. 3 **b.** 3 **c.** −8 **d.**

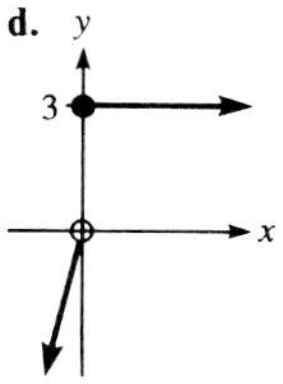

51. a. 1 **b.** 2 **c.** 3 **d.**

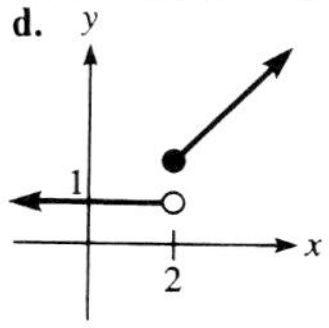

53. a. 2 **b.** 1 **c.** 1 **d.**

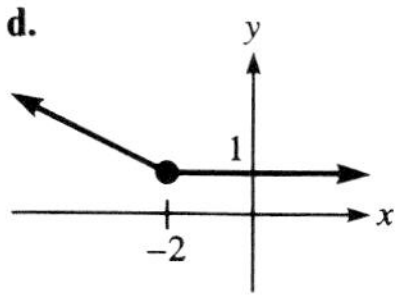

55. a. 4 **b.** Undefined **c.** 3 **d.** 4 **e.**

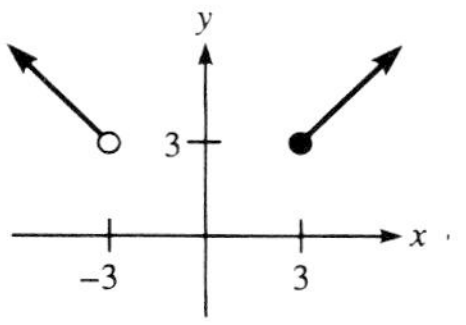

57. a. $(-\infty,\infty)$ **b.** $[-4,\infty)$ **c.** −3 **d.** $\{-1,3\}$ **e.** $(-1,3)$
f. $(-\infty,-1) \cup (3,\infty)$ **59. a.** $(-\infty,\infty)$ **b.** $[0,\infty)$ **c.** 0 **d.** $\{0\}$
e. $\emptyset$ **f.** $(-\infty,0) \cup (0,\infty)$ **61. a.** $(-\infty,\infty)$ **b.** $(-\infty,4]$ **c.** −5
d. $\{-5,-1\}$ **e.** $(-\infty,-5) \cup (-1,\infty)$ **f.** $(-5,-1)$ **63. a.** $(-\infty,\infty)$
b. $(-\infty,-1]$ **c.** −3 **d.** $\emptyset$ **e.** $(-\infty,\infty)$ **f.** $\emptyset$

Remember This (9.2)

1. 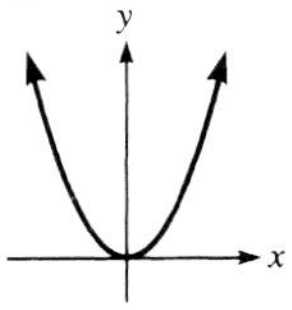

2. 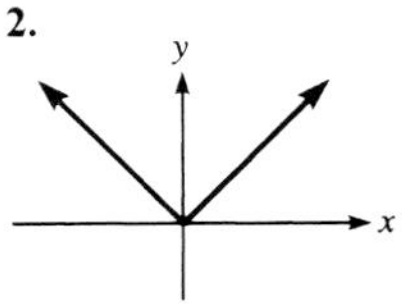

3. 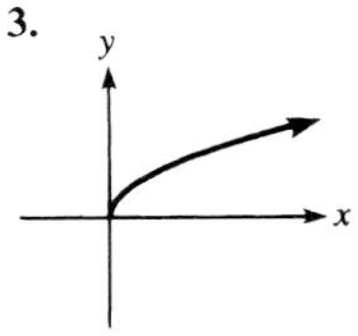

4. 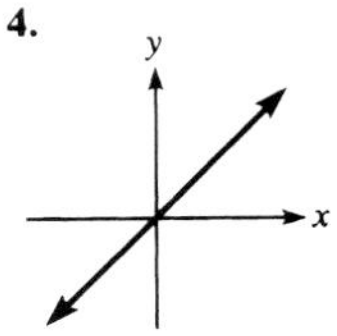

5. (0,0), (2,4) **6.** (3,0), (−3,0) **7.** $\frac{3}{2} + \frac{3}{2}i$ **8.** $\begin{vmatrix} 3 & -4 \\ -2 & 5 \end{vmatrix}$; 7

9. $\dfrac{1 + 2x}{x^2}$ **10.** $(-\frac{5}{2}, \frac{3}{2})$ 

Exercises 9.3

1. 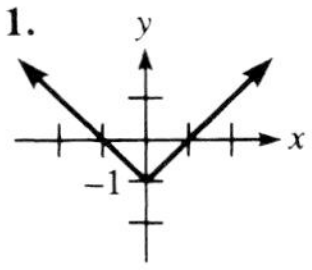

3. 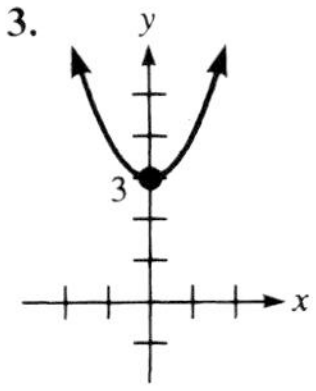

5. 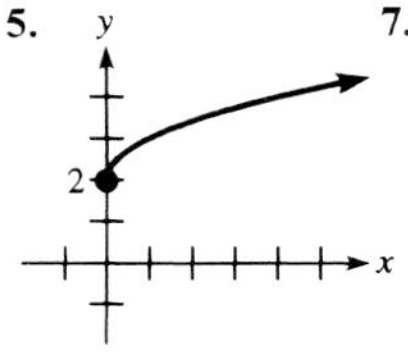

7. 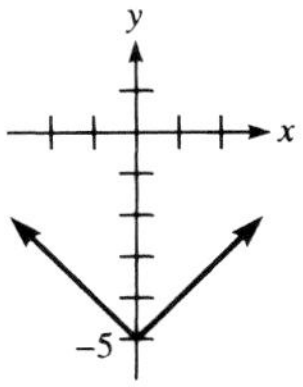

9. 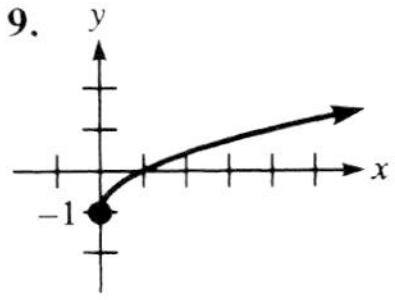

11. 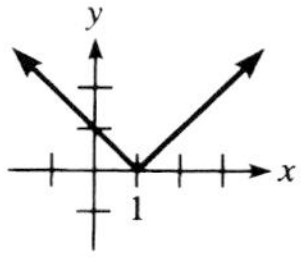

13. 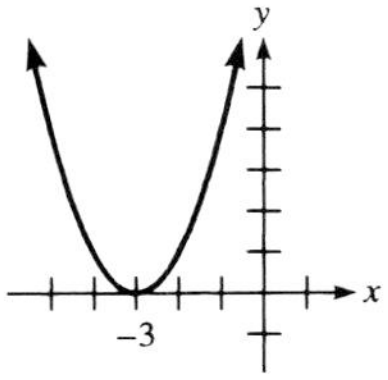

15. 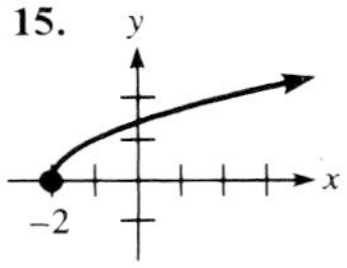

17. 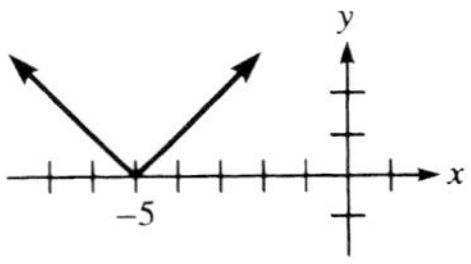

19. 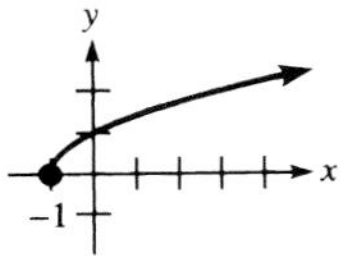

21. 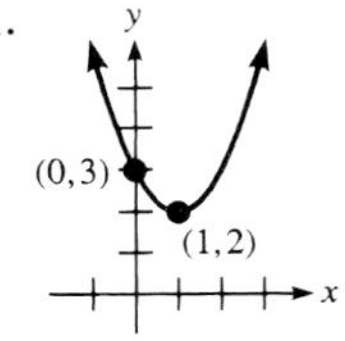

23. 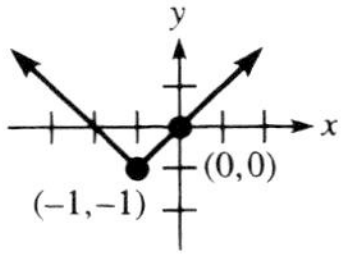

25. 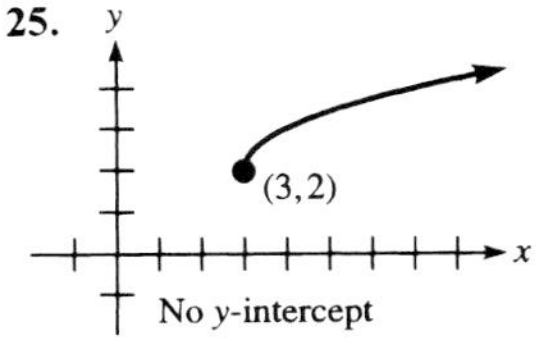

27. 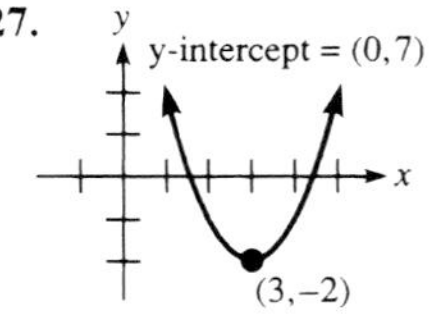

29. 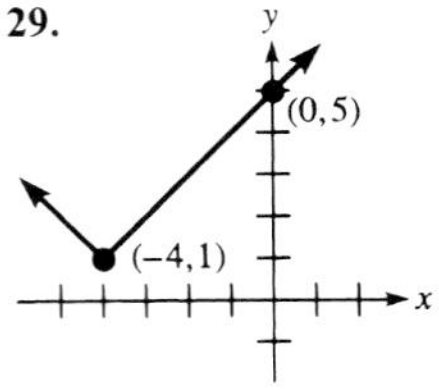

31. 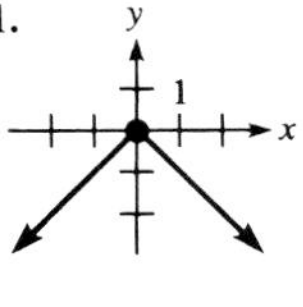

33. 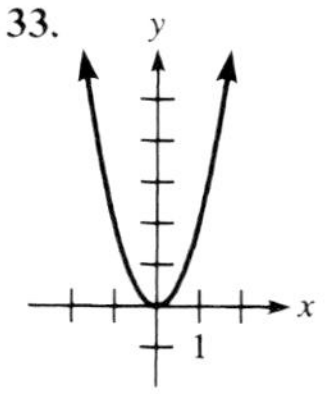

35. 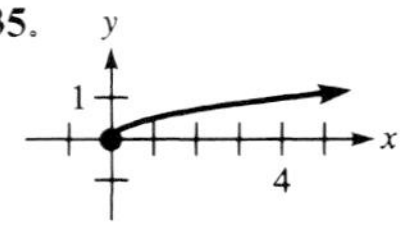

37. 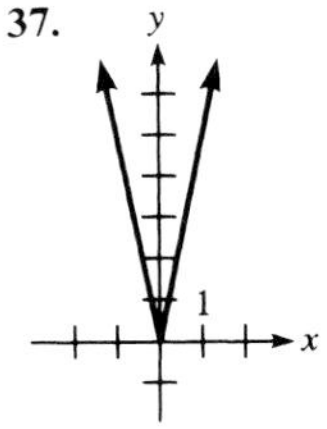

39. 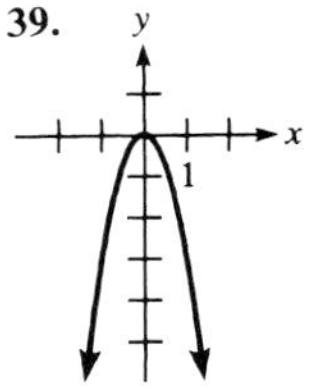

41. 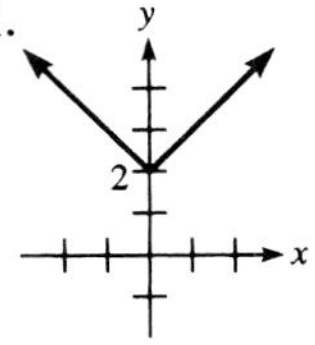

43. 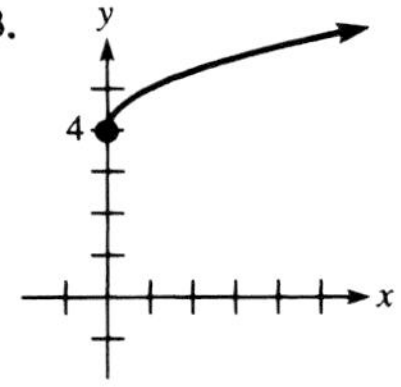

45. 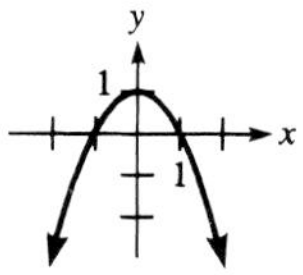

47. 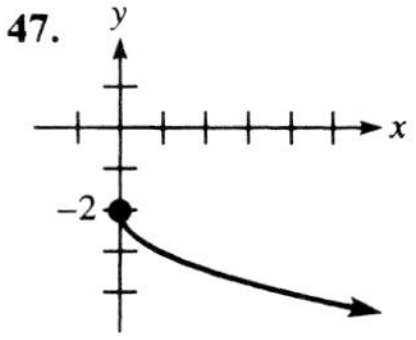

49. 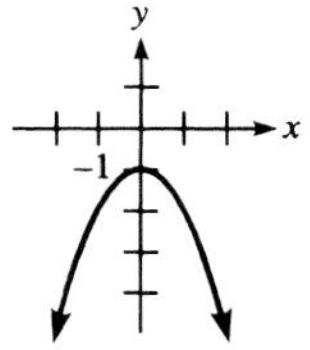

51. 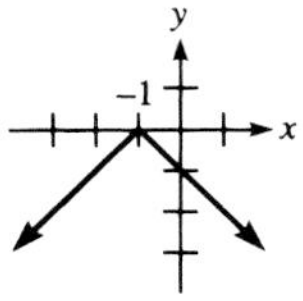

53. 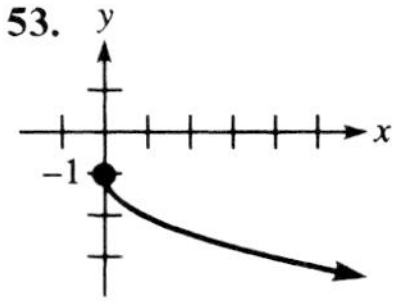

55. 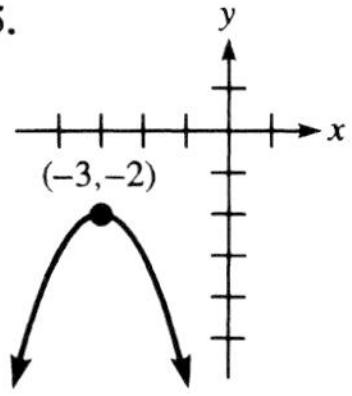

57. 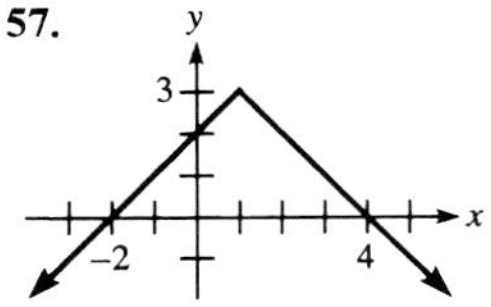

59. 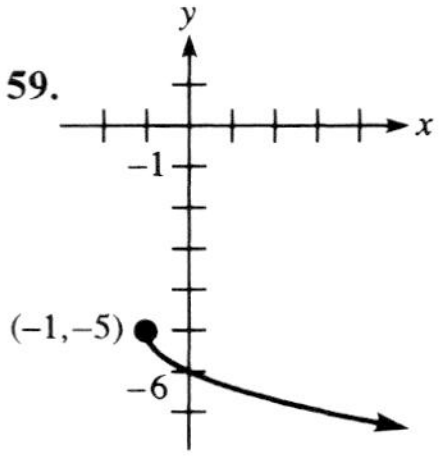

61. a. 2.9 seconds
b. [0 seconds, 2.9 seconds]
c. 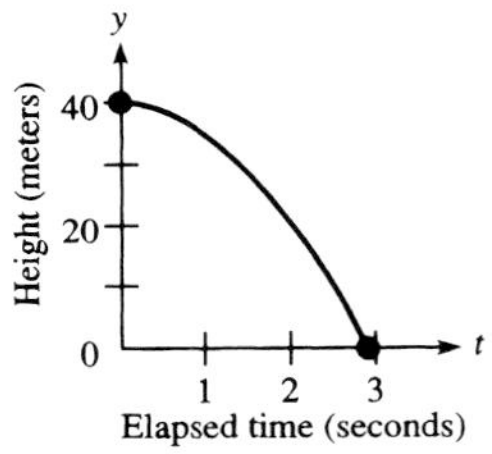

63. a. 122 seconds **b.** 2.0 minutes
c. [0 seconds, 122 seconds]
d. 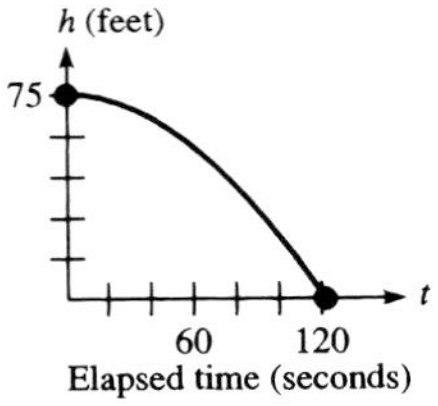

Remember This (9.3)

1. $x^{1/2}$, x^{-1}, x^{-2} **2.** 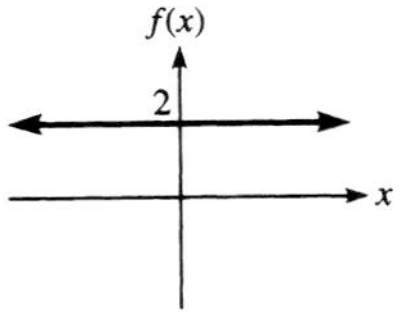

3. a, b **4.** −2 **5.** c **6.** −19 **7.** Maximum **8.** $3xy$
9. $y = 200 + 0.15x$; $y = 180 + 0.20x$. They pay the same when the amount of sales is $400.

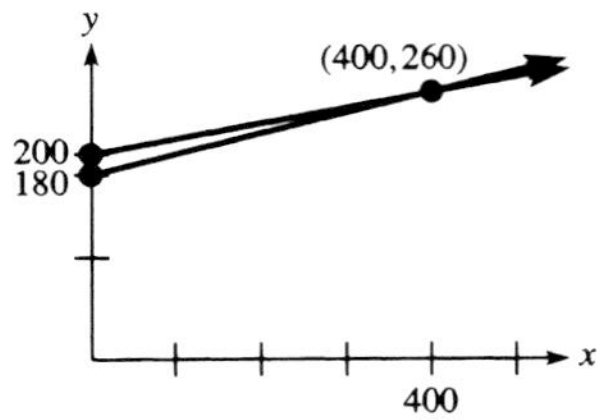

10. {0,1/2,−1}

Exercises 9.4

1.
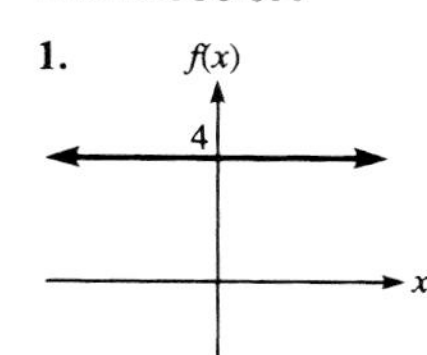

3.
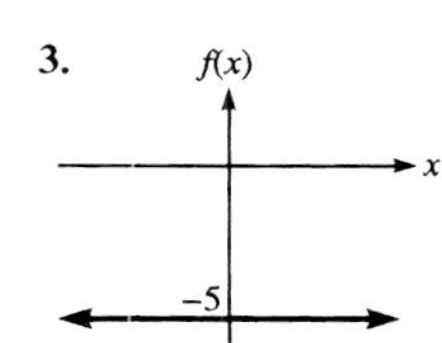

5.
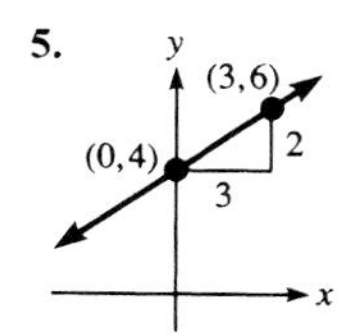

7.
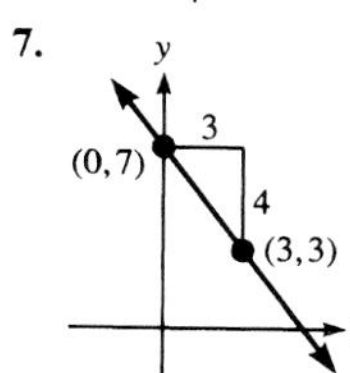

9.
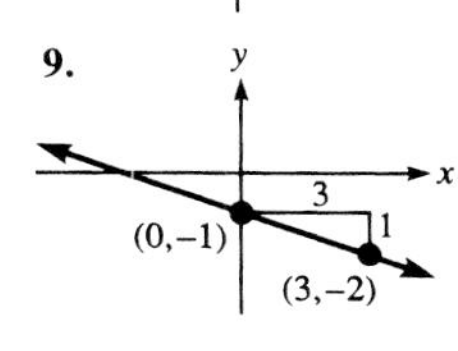

11.
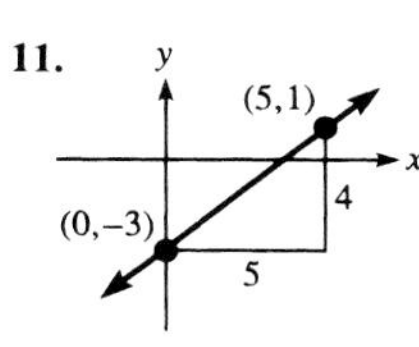

13.
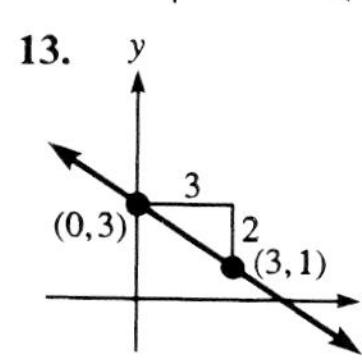

15. $f(x) = -2x + 4$ **17.** $f(x) = \frac{3}{2}x - 2$ **19.** $f(x) = \frac{1}{5}x + 1$ **21.** $f(x) = -\frac{2}{3}x + 7$ **23.** $f(x) = \frac{2}{3}x - 6$ **25.** $f(x) = -\frac{5}{2}x + 3$ **27.** $f(x) = -\frac{1}{3}x + 2$ **29.** $f(x) = \frac{3}{5}x - 1$ **31. a.** $y = -0.85x + 187$, where x = age and y = heart rate **b.** 161.5, or about 162 beats per minute

33. a. $x = 1$ **b.** (1,3) **c.** (0,5) **d.**
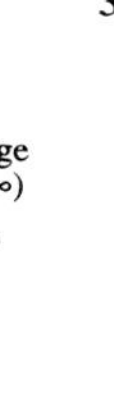

35. a. $x = -1$ **b.** $(-1,2)$ **c.** (0,3) **d.**
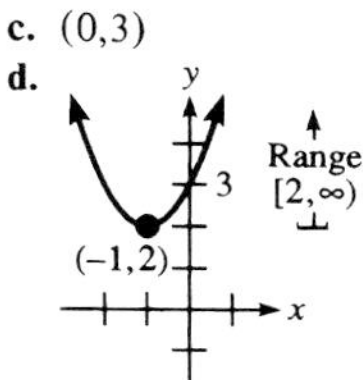

37. a. $x = 3$ **b.** $(3,-5)$ **c.** $(0,-14)$ **d.**
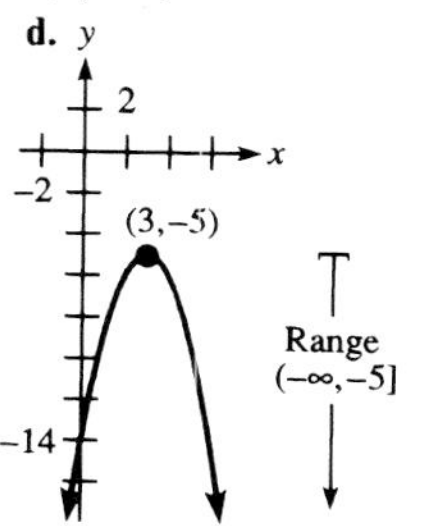

39. a. $x = -4$ **b.** $(-4,-7)$ **c.** $(0,-23)$ **d.**
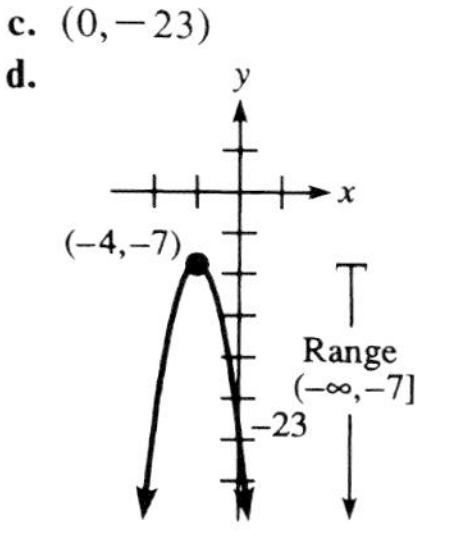

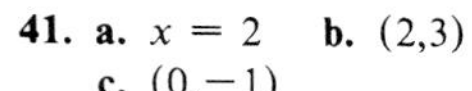
41. a. $x = 2$ **b.** (2,3) **c.** $(0,-1)$ **d.**
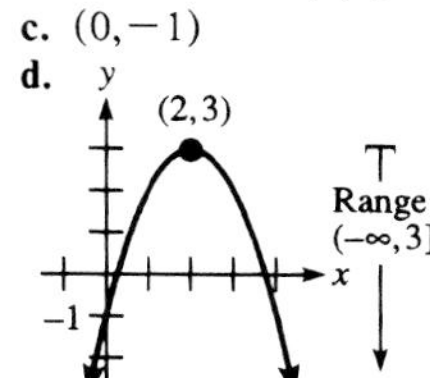

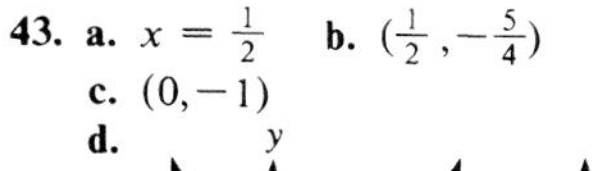
43. a. $x = \frac{1}{2}$ **b.** $(\frac{1}{2}, -\frac{5}{4})$ **c.** $(0,-1)$ **d.**
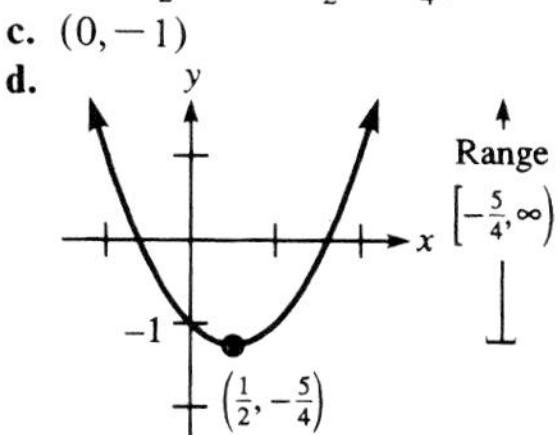

45. $x = 1$; $(1,-4)$ **47.** $x = -3$; $(-3,-5)$ **49.** $x = \frac{3}{2}$; $(\frac{3}{2}, \frac{25}{4})$ **51.** $x = 1$; (1,3) **53.** $x = \frac{3}{2}$; $(\frac{3}{2}, -\frac{15}{2})$ **55.** 5,625 ft² **57.** 150 yd² **59.** Charge $15 per person; income = $2,250. **61.** $\frac{1}{2}$ exceeds $(\frac{1}{2})^2$ by $\frac{1}{4}$. **63.** The pen should be 25 by 50 ft; maximum area = 1,250 ft². **65.** $m = \frac{14}{13}$; $y = \frac{14}{13}x$

Remember This (9.4)

1. -111 **2.** $x^2 - x - 2$ **3.** $C = \pi d$ **4.** $4x^3 - 7x^2 - 17x + 15$ **5.** $f(-2) = 0$ **6.** (graph) **7.** $(x + 2)^2 + (y - 2)^2 = 9$ **8.** 26 **9.** $y = -x + 2$ **10.** $(x - 2)^2$

Exercises 9.5

1. $(f + g)(x) = 3x + x^2$; $(f - g)(x) = 3x - x^2$; $(f \cdot g)(x) = 3x^3$; $\left(\frac{f}{g}\right)(x) = \frac{3}{x}$
3. $(f + g)(x) = x^2 + 2x - 1$; $(f - g)(x) = -x^2 + 8x - 3$; $(f \cdot g)(x) = 5x^3 - 17x^2 + 11x - 2$; $\left(\frac{f}{g}\right)(x) = \frac{5x - 2}{x^2 - 3x + 1}$
5. $(f + g)(x) = 2x^2 + 2x + 12$; $(f - g)(x) = 2x^2 - 4x - 2$; $(f \cdot g)(x) = 6x^3 + 11x^2 + 8x + 35$; $\left(\frac{f}{g}\right)(x) = \frac{2x^2 - x + 5}{3x + 7}$ **7.** 6
9. 14 **11.** $-\frac{9}{4}$ **13.** 13 **15.** 22 **17. a.** 2 **b.** 6 **c.** -8 **d.** -2
19. a. 0 **b.** 4 **c.** -6 **d.** -2 **21. a.** $6x - 2$ **b.** $6x - 1$
23. a. $5 - 12x + 9x^2$ **b.** $-3x^2 - 1$ **25. a.** $4x^2 - 13$ **b.** $16x^2 + 24x + 5$ **27.** 2 **29.** 5 **31.** -3 **33.** 4 **35.** 4 **37.** 4 **39.** -2
41. $(f \circ g)(m) = 36\pi m^2$. If f expresses area A in terms of radius r, and g expresses r in terms of months m, then the composition function $f \circ g$ expresses area A in terms of months m.
43. a. $(g \circ f)(t) = (1 + 0.2t)^3$; this gives the volume after t seconds. **b.** $(g \circ f)(10) = 27$; after 10 seconds the volume is 27 in.³.
45. a. $(f \circ g)(t) = 340h(1.05)^t + 200$; this gives the cost of a full shipment of widgits t years from now. **b.** $(f \circ g)(4) = 413.27h + 200$; this gives the cost of a full shipment of widgits 4 years from now. **c.** $(f \circ g)(4) = 4{,}332.7$; 4 years from now it will cost $4,322.70 to produce a full shipment of widgits.

Remember This (9.5)

1. 32,768 **2.** 1.0737×10^9 **3.** 9 **4.** $(\frac{1}{2})^3$ **5.** Yes **6.** 11
7. Hyperbola
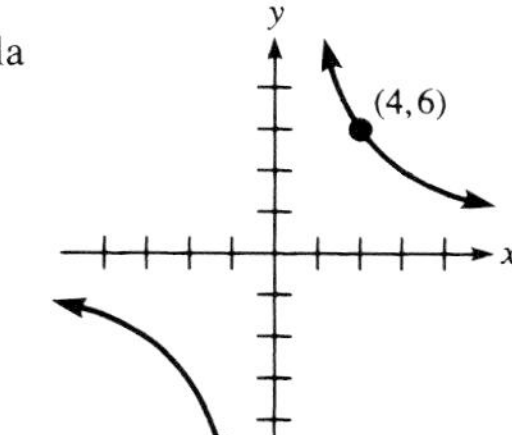

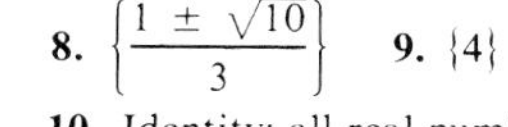
8. $\left\{\frac{1 \pm \sqrt{10}}{3}\right\}$ **9.** {4}
10. Identity; all real numbers are solutions.

Chapter 9 Review Exercises

1. Domain = {2,10,21,70}; range = {3,4,7} **b.** Domain = {n:n is a nonnegative integer} **3.** Yes **5. a.** $C = 35n$ **7.** -12 **9. a.** $[-2,\infty)$ **b.** $[0,\infty)$ **c.** 1 **d.** -2 **e.** None **f.** $(-2,\infty)$

11.
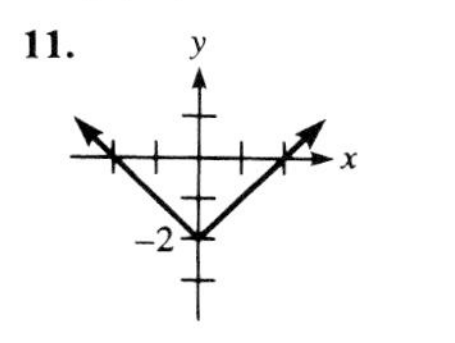

13.
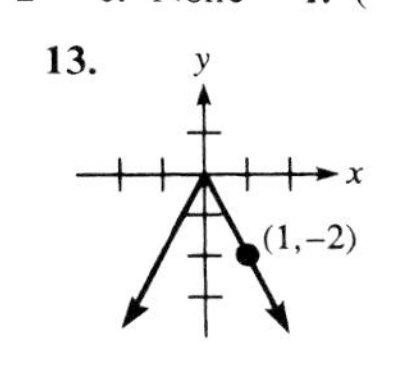

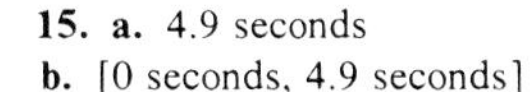
15. a. 4.9 seconds **b.** [0 seconds, 4.9 seconds] **c.**
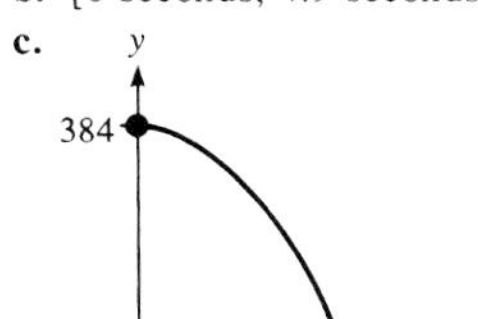

17.

19. a. $x = 2$ **b.** $(2,-1)$ **c.** $(0,3)$ **d.**

Range $[-1,\infty)$

21. a. $(f + g)(x) = x^2 + 4x + 5$ **b.** $(f - g)(x) = -x^2 + 2x - 1$
c. $(f \cdot g)(x) = 3x^3 + 5x^2 + 11x + 6$ **d.** $\left(\dfrac{f}{g}\right)(x) = \dfrac{3x + 2}{x^2 + x + 3}$
23. a. $-4x^2 + 16x - 16$ **b.** $-2x^2 - 4$
25. $C = (f \circ g)(t) = 8\pi t$ is the formula for the circumference of the spill t seconds after the start of the leak. **27.** Yes **29.** Yes **31.** Yes
33. Yes **35.** Domain $= \{6,7,8,9\}$; range $= \{65,75,85,95\}$
37. Domain $= \{x: x \neq 1\}$; range $= \{y: y \neq 0\}$ **39.** $x^2 + 6x - 8$
41. $\dfrac{x^2 + 3x - 5}{3x - 3}$ **43.** $9x^2 - 9x - 5$ **45.** -5 **47.** -3 **49.** $\frac{5}{3}$
51. -24 **53. a.** $[-3,\infty)$ **b.** $(-\infty,5]$ **c.** 5 **d.** 3 **e.** $[-3,3)$
f. $(3,\infty)$
55. a. $x = 1$ **b.** $(1,-1)$ **c.** $(0,-2)$ **d.**

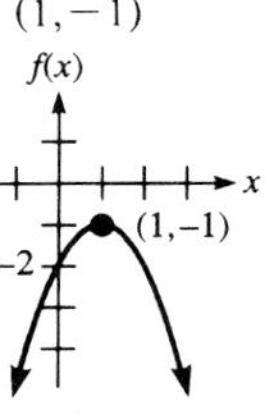

e. $(-\infty,\infty)$ **f.** $(-\infty,-1]$

57. **59.**

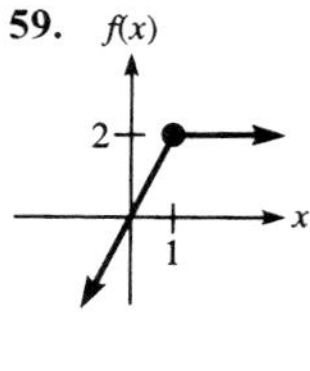

61.

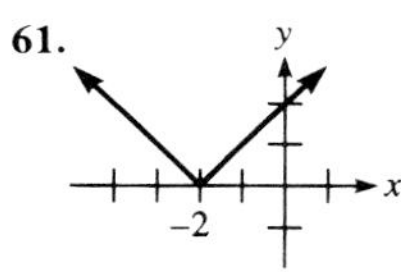

63.

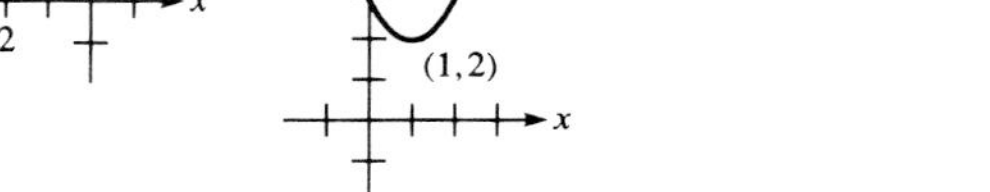

65. $d = 45t$; $[0,\infty)$

67. $A = 144\pi t^2$ is the formula for the area of the circular spill t seconds after the start of the leak.

Chapter 9 Test

1. Domain $= \{3,5\}$; range $= \{3,2\}$ **2.** No **3. a.** Yes **b.** Yes **c.** No
4. Domain $= \{x: x \neq -4\}$; range $= \{y: y \neq 0\}$ **5.** $\ell = \dfrac{P}{2} - 6$ **6.** 16
7. a. 3 **b.** 0 **c.** 2 **d.**

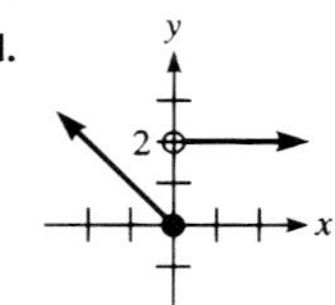

8. a. $(-\infty,\infty)$ **b.** $(-\infty,1]$ **c.** $-3,-1$ **d.** $(-\infty,-3) \cup (-1,\infty)$
e. $(-3,-1)$
9.

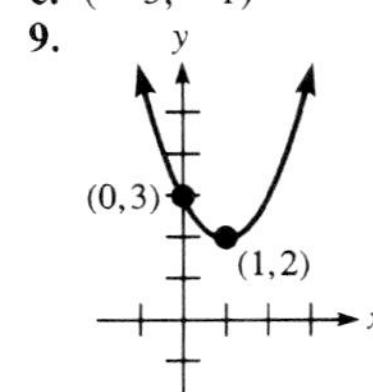

10.

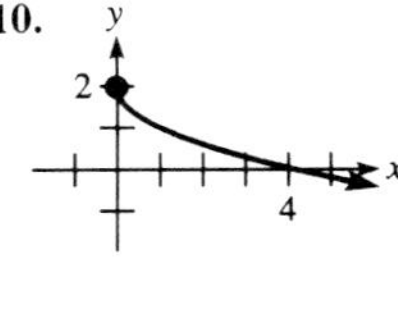

11.

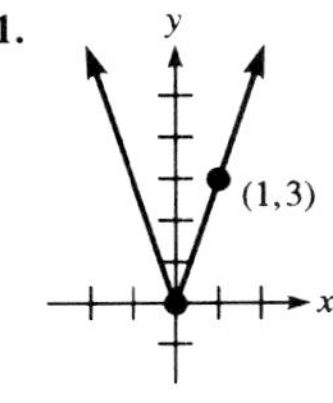

12. **13.**

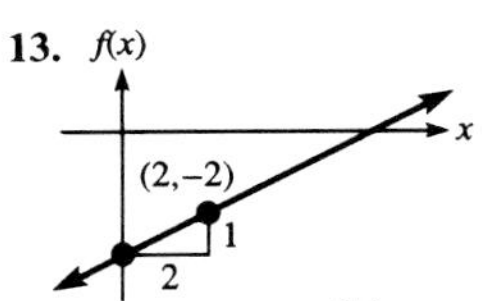

14. $f(x) = -3x + 5$

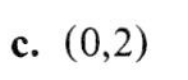

15. a. $x = 1$ **b.** $(1,1)$ **c.** $(0,2)$ **d.**

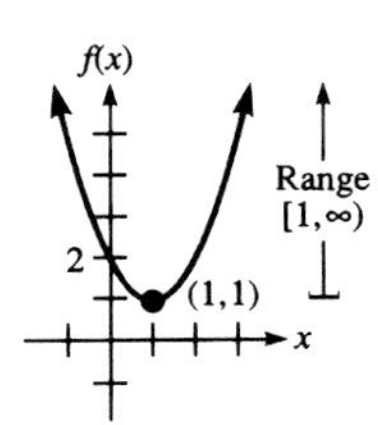

Range $[1,\infty)$

16. $(2,-4)$; $x = 2$ **17.** 225 ft^2 **18. a.** $-x^2 + 3x - 2$
b. $x^2 + x + 4$ **c.** $-2x^3 + x^2 - 5x - 3$ **d.** $\dfrac{2x + 1}{-x^2 + x - 3}$
19. a. 0 **b.** $-\frac{2}{3}$ **20. a.** $9x^2 - 6x + 3$ **b.** $-3x^2 - 5$

Cumulative Test 9

1. All are real except $2 + 3i$ **2.** $\{5\}$ **3.** $\{-26,10\}$ **4.** 5^{2n+1}
5. $\dfrac{x^2 - 2x + 4}{x - 2}$ **6.** $\{\pm i\sqrt{7}\}$ **7.** $b = \dfrac{b_0}{a - 1}$ **8.** 175
9. Slope 3; y-intercept $(0,-4)$
10. $m_{AB} = \frac{5}{2}$ and $m_{AC} = -\frac{2}{5}$; $m_{AB} \cdot m_{AC} = -1$
11.

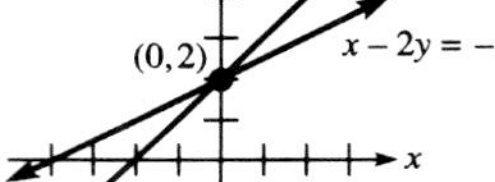

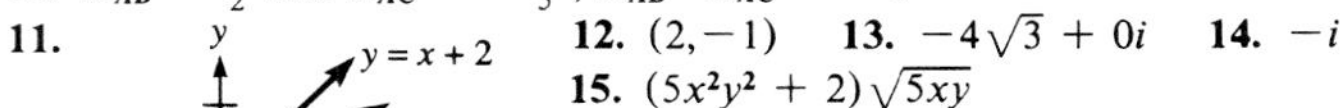

12. $(2,-1)$ **13.** $-4\sqrt{3} + 0i$ **14.** $-i$
15. $(5x^2y^2 + 2)\sqrt{5xy}$
16. $\{\frac{1}{2},\frac{2}{3}\}$ **17.** $[-\frac{3}{2},1]$
18.

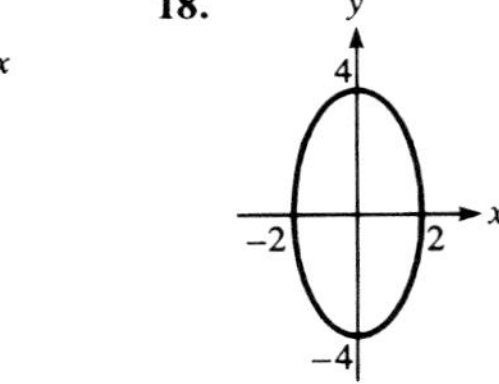

19. a. 5 **b.** 5 **c.** -12 **d.** $2x^2 + 5$
20. a.

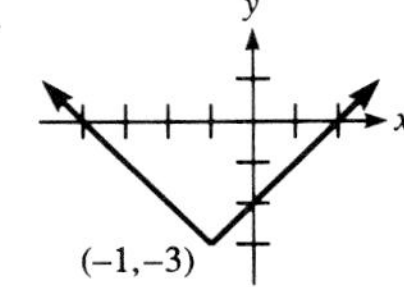

b. $(-\infty,\infty)$ **c.** $[-3,\infty)$

Chapter 10
Exercises 10.1

1. $1, 125, \frac{1}{25}, 2.24, 9.74$ **3.** $1, 27, \frac{1}{9}, 1.73, 4.73$ **5.** $1, \frac{1}{64}, 16, \frac{1}{2}, 0.14$
7. $1, \frac{1}{216}, 36, 0.41, 0.08$ **9.** $-1, -729, -\frac{1}{81}, -3, -22.36$
11. $-1, -125, -\frac{1}{25}, -2.24, -9.74$ **13.** $1, \frac{1}{8}, 4, 0.71, 0.38$
15. $1, \frac{1}{64}, 16, \frac{1}{2}, 0.14$ **17.** $-1, -\frac{1}{8}, -4, -0.71, -0.38$
19. $-1, -\frac{1}{64}, -16, -\frac{1}{2}, -0.14$
21. **23.**

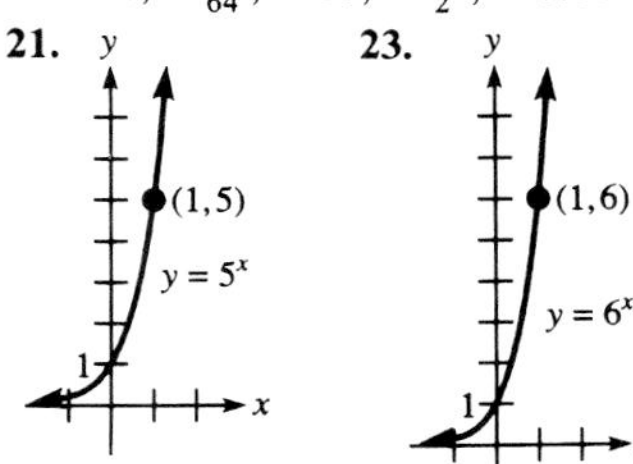

25.

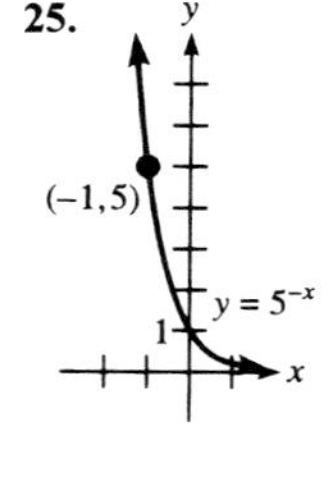

27.

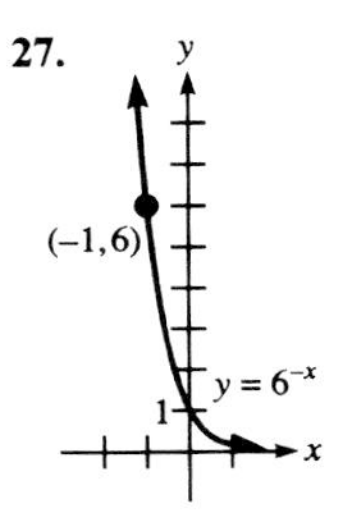

29. **31.** **33.** **35.**

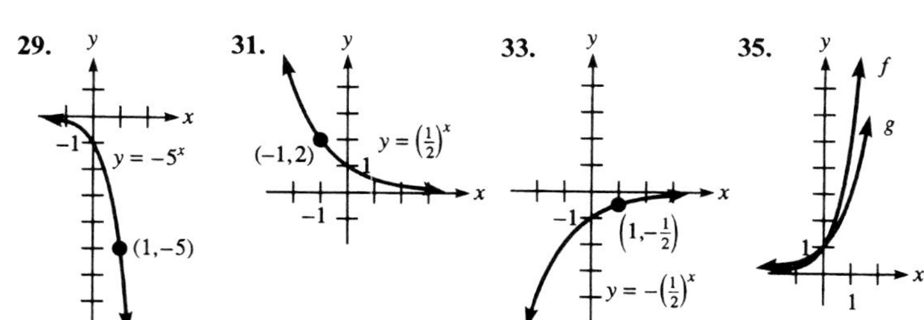

37.

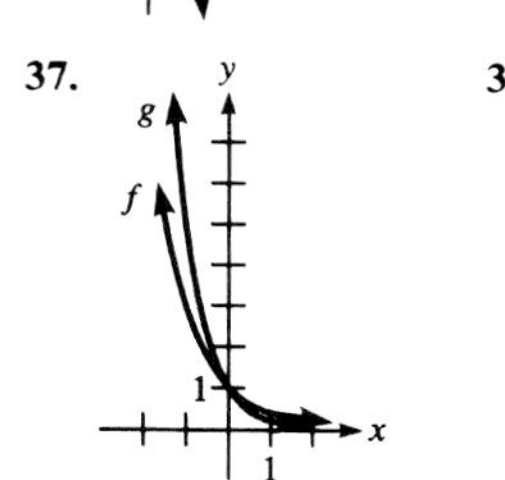

39.

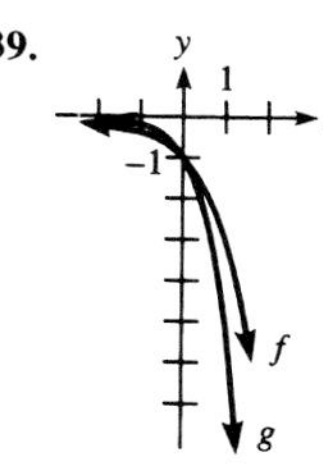

41.

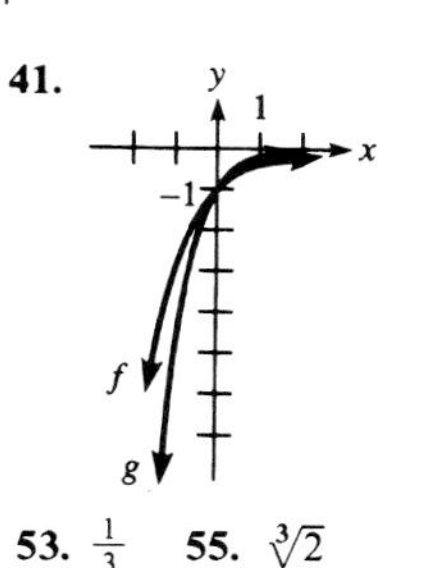

43. $-\frac{3}{2}$ **45.** $\frac{3}{4}$ **47.** $-\frac{1}{2}$ **49.** 3 **51.** 5 **53.** $\frac{1}{3}$ **55.** $\sqrt[3]{2}$
57. $\sqrt[6]{5}$

59. **61.**

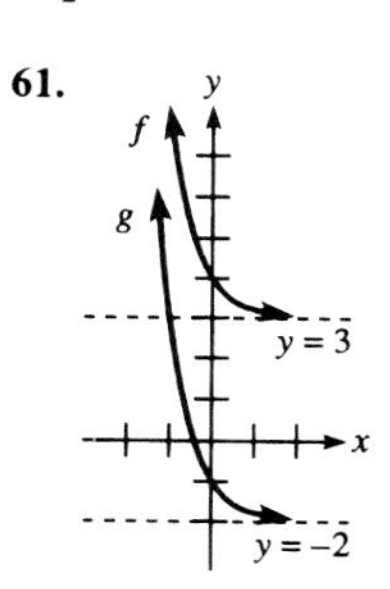

63. \$117,129,875,000 **65.** \$25.94 **67.** \$7,163.93 **69.** 7.18 percent
71. 1.40 percent

Remember This (10.1)

1. 3 **2.** Yes **3.** Horizontal **4.**

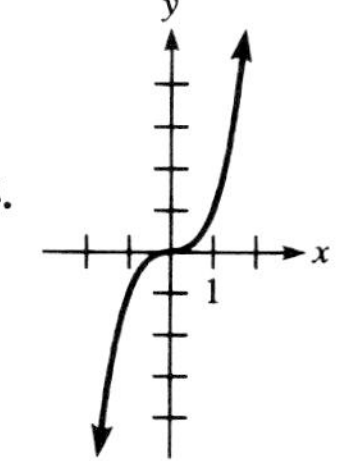

5. 5 **6.** $0.80x$ **7.** a **8.** $x^2 + y^2 = 9$ **9.** True **10.** $(3,-3)$

Exercises 10.2

1. $f^{-1} = \{(2,1),(4,3),(6,5)\}$; $D_f = R_{f^{-1}} = \{1,3,5\}$; $R_f = D_{f^{-1}} = \{2,4,6\}$
3. $f^{-1} = \{(1,2),(4,-3),(8,5)\}$; $D_f = R_{f^{-1}} = \{-3,2,5\}$; $R_f = D_{f^{-1}} = \{1,4,8\}$
5. No inverse function **7.** No inverse function **9.** No **11.** Yes
13. Yes **15.** No **17.** Yes **19.** Yes

21.

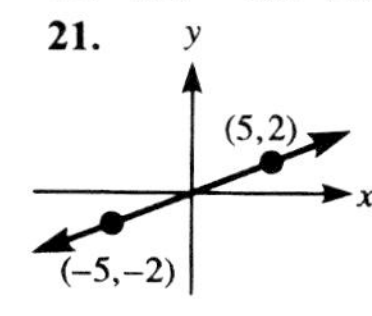

23.

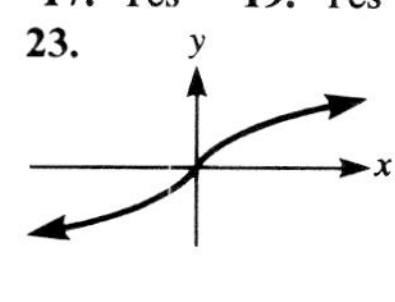

25.

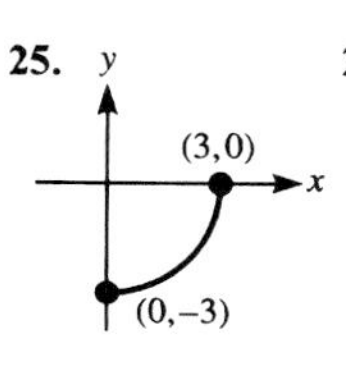

27.

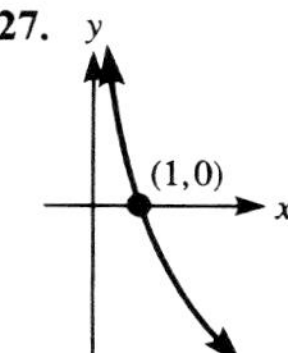

29. $f^{-1}(x) = x + 3$ **31.** $f^{-1}(x) = \dfrac{x-5}{3}$ **33.** $f^{-1}(x) = x^2, x \le 0$
35. No inverse function exists. **37.** No inverse function exists.
39. $f^{-1}(x) = \sqrt[4]{x}$, or $x^{1/4}, x \ge 0$ **41.** Yes **43.** Yes **45.** No **47.** Yes
49. Yes **51.** $f^{-1}(x) = \frac{15}{22}x$; this converts feet/second to miles/hour.

53. $y = f^{-1}(x) = 0.50x$; 50 percent
55. a. At 1,000 ft above sea level, water boils at 200° F.
b. At 2,000 ft above sea level, water boils at 190° F.
c. At k ft above sea level, water boils at j° F.

Remember This (10.2)

1. 2 **2.** $\{\frac{3}{2}\}$ **3.** $\frac{1}{2}$ **4.**

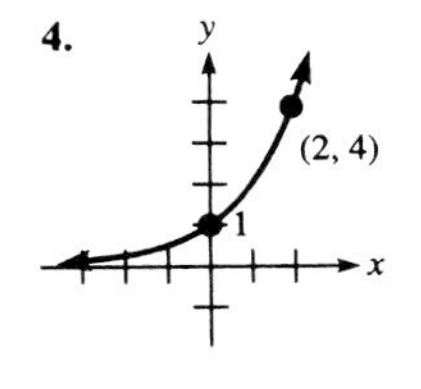

5. 2^{m+n} **6.** $\frac{11}{100}$
7.

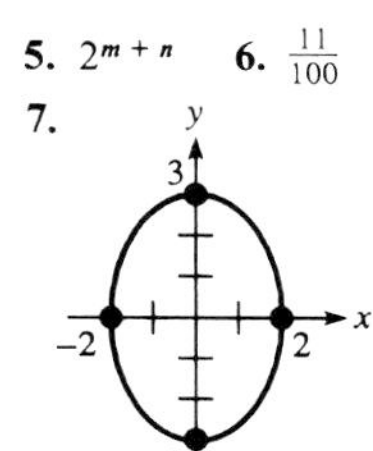

8. $\{5, \frac{25}{9}\}$ **9.** $\sqrt{29}$ **10.** $\{-\frac{5}{12}\}$

Exercises 10.3

1. $\log_2 8 = 3$ **3.** $\log_4 16 = 2$ **5.** $\log_{1/2} \frac{1}{8} = 3$ **7.** $\log_b x = y$
9. $\log_{10} 1{,}000 = 3$ **11.** $\log_{10} \dfrac{1}{100} = -2$ **13.** $\log_{0.2} 0.008 = 3$
15. $\log_{0.1} 100 = -2$ **17.** $4^3 = 64$ **19.** $4^{1/2} = 2$ **21.** $10^3 = 1{,}000$
23. $8^{-1} = \frac{1}{8}$ **25.** $2^{-5} = \frac{1}{32}$ **27.** $b^x = y$ **29.** $(\frac{1}{2})^{-3} = 8$ **31.** 1
33. 0 **35.** 4 **37.** 3 **39.** 5 **41.** $\frac{1}{2}$ **43.** 36 **45.** $\frac{1}{16}$ **47.** 25
49. $\frac{1}{8}$

51. **53.**

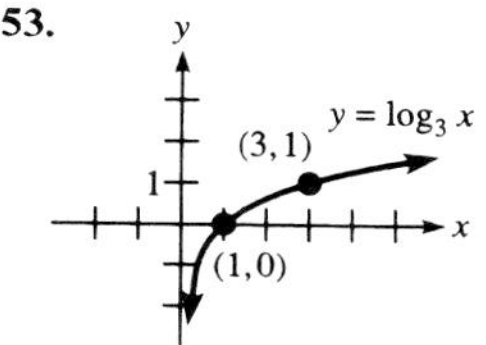

55.

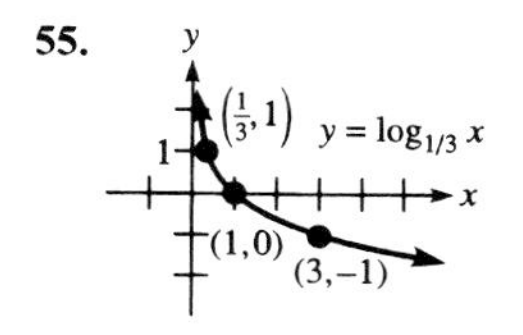

57.

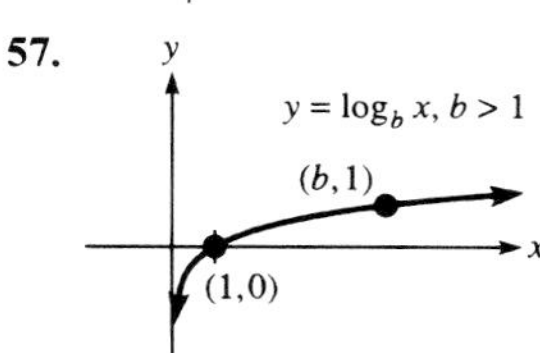

59–63. In each case one graph is the reflection of the other about the line $y = x$.

59.

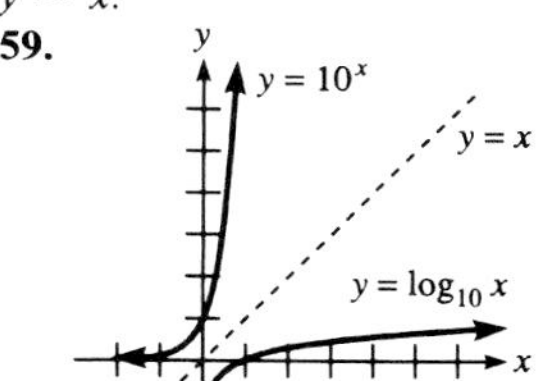

61.

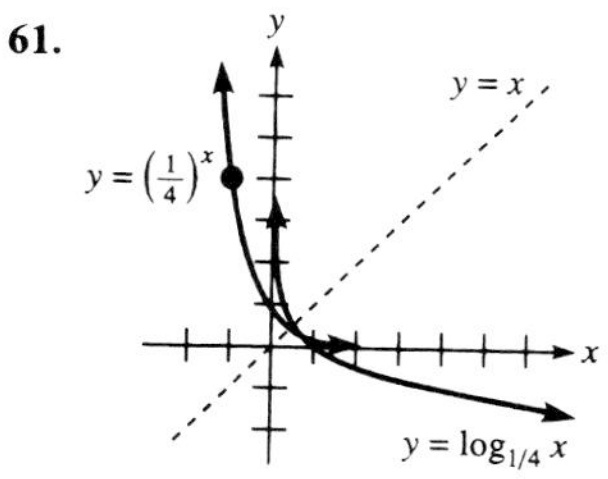

63.

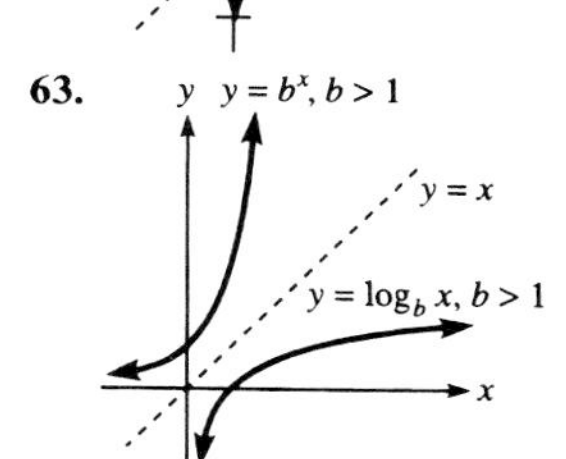

65.

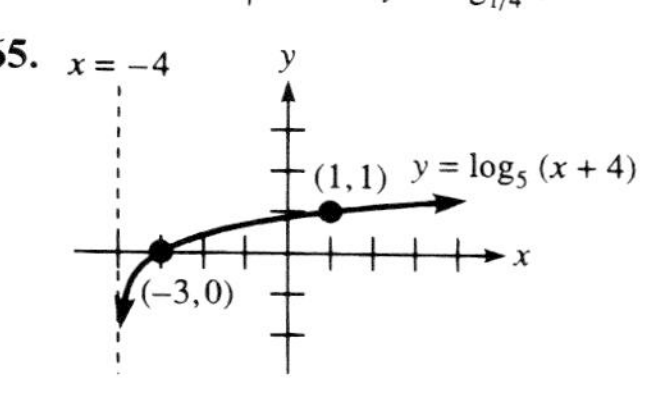

67.

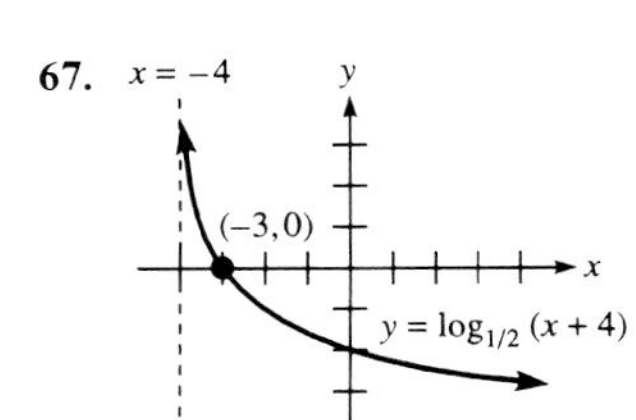

69.

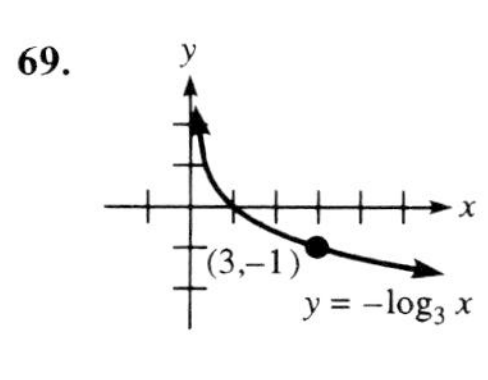

71.

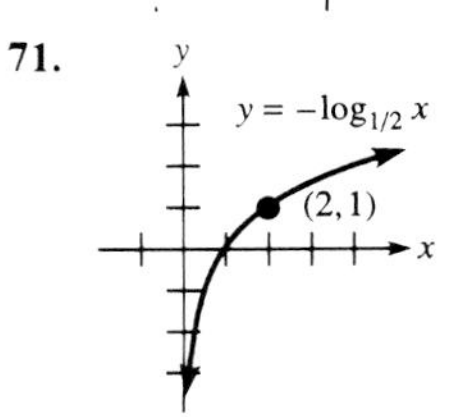

73.

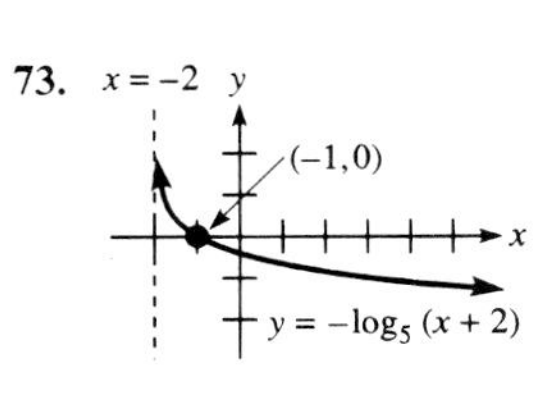

75. 1.3979 **77.** -0.4318 **79.** 0.3729 **81.** 158 **83.** 1.10 **85.** 0.00251 **87.** 0.490 **89.** 4.4; acid **91.** 3.16×10^{-6} **93. a.** 5.27×10^{-6} **b.** 5.28 **95. a.** 2.51×10^{13} joules **b.** 7.94×10^{11} joules **c.** 31.6

Remember This (10.3)

1. b^{m-n} **2.** False **3.** $5^{2/3}$ **4.** 2^{x+7} **5.** $2x^2y^3$ **6.**

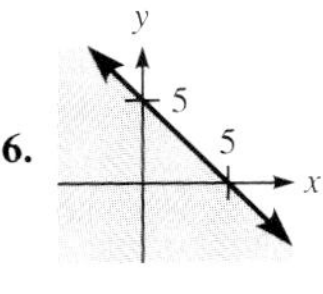

7. $\{3\}$ **8.** $x^3 - 1$ **9.** $A = \dfrac{B}{n-1}$ **10.** $[-10, 6]$ (number line: −10, 0, 6)

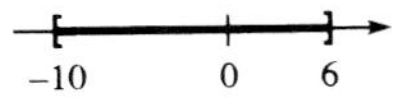

Exercises 10.4

1. $\log_5 7 - \log_5 9$ **3.** $\log_8 6 + \log_8 x$ **5.** $5 \log_3 x$ **7.** $\frac{1}{3} \log_b 5$ **9.** $\log_3 5 + 3 \log_3 x$ **11.** $\frac{1}{2}[\log_b x - (\log_b 2 + \log_b y)]$ **13.** $\frac{1}{2}[\log_b x + \log_b y - \log_b z]$ **15.** $\frac{1}{2}[\log_b(x+5) - \log_b y]$ **17.** 1 **19.** 5 **21.** 0 **23.** -1 **25.** -3 **27.** -3 **29.** 3 **31.** $-\log_b 3$ **33.** $1 - \log_b 7$ **35.** $\log_b 3 - 2$ **37.** $2m + 3n$ **39.** $4m + n$ **41.** $-3n$ **43.** $-5m$ **45.** $3n - 6m$ **47.** $\log_{10} t + k$ **49.** $\log_{10} y + 2k$ **51.** 32 **53.** 49 **55.** $\log_2 8y$ **57.** $\log_b \dfrac{3}{y}$ **59.** $\log_{10} \dfrac{y^3}{x^4}$ **61.** $\log_{10} \sqrt[3]{\dfrac{x+3}{y-1}}$ **63.** $\log_{10} \sqrt{\dfrac{x}{yz}}$ **65.** $\log_{10} \sqrt{\dfrac{y}{xz}}$

67.

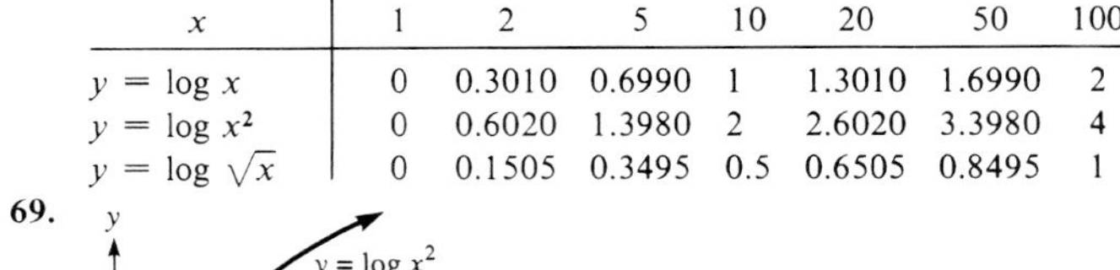

x	1	2	5	10	20	50	100
$y = \log x$	0	0.3010	0.6990	1	1.3010	1.6990	2
$y = \log x^2$	0	0.6020	1.3980	2	2.6020	3.3980	4
$y = \log \sqrt{x}$	0	0.1505	0.3495	0.5	0.6505	0.8495	1

69.

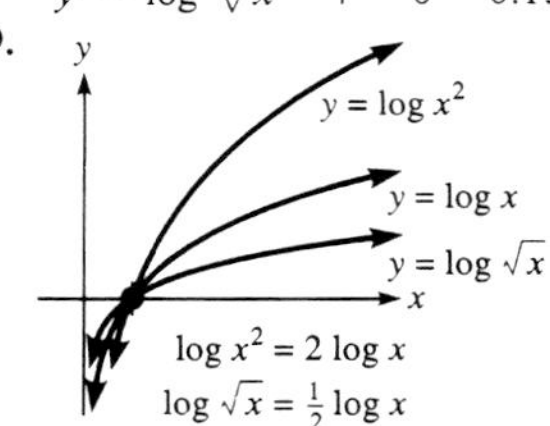

71. a. $x = (\log y - b)/a$ **b.** $x = 3.0103$ **c.** $y = 10^{ax+b}$ **d.** $y = 15.85$ **73. a.** $L = \log(I) + 90 = 10[\log(I) + \log(10^9)] = 10 \log(I \cdot 10^9) = 10 \log \dfrac{I}{10^{-9}}$ **b.** 110 dB **c.** 1,000 ergs/cm²/second

75. Let $x = b^m$ and $y = b^n$. $\dfrac{x}{y} = \dfrac{b^m}{b^n} = b^{m-n}$

$\log_b \dfrac{x}{y} = m - n = \log_b x - \log_b y$

Remember This (10.4)

1. $x = \dfrac{k}{m-n}$ **2.** $\{-2, 4\}$ **3.** b **4.**

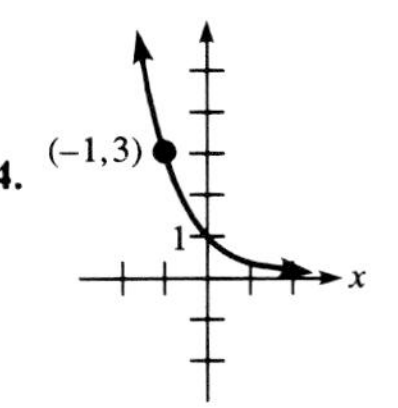

5. $(-\frac{1}{2})^4$ **6.** $f^{-1}(x) = \dfrac{x+2}{3}$ **7.** Distributive **8.** -62 **9.** \$19.50 **10.** $\dfrac{\sqrt{x} - \sqrt{y}}{x - y}$

Exercises 10.5

1. $\{3.585\}$ **3.** $\{1.588\}$ **5.** $\{2.500\}$ **7.** $\{-4.585\}$ **9.** $\{-1.262\}$ **11.** $\{-4.783\}$ **13.** $\{4.593\}$ **15.** $\{10.84\}$ **17.** 2.807 **19.** 1.792 **21.** 2.807 **23.** -1.195 **25.** -4.392 **27.** -1.904 **29.** $\{1\}$ **31.** $\{4\}$ **33.** $\emptyset$ **35.** $\{-40\}$ **37.** $\{42\}$ **39.** $\{-60\}$ **41.** $\{33{,}300\}$ **43.** $\{1\}$ **45.** $\{3\}$ **47.** $\{2\}$ **49.** $\{-1\}$ **51.** 7 years **53. a.** 2009 **b.** 2060 **55.** 14.9 percent **57. a.** 2 **b.** 3

Remember This (10.5)

1. \$8,591 **2.** 2.718 **3.** 0.3124 **4.** 41 **5.** 2

6.

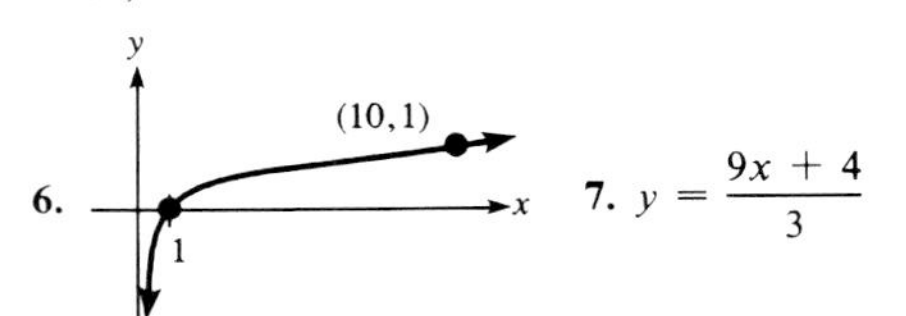

7. $y = \dfrac{9x+4}{3}$ **8.**

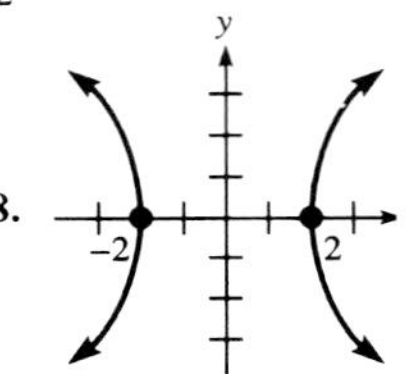

9. $(\frac{8}{9}, \frac{4}{3})$ **10.** 5.03×10^{-3}

Exercises 10.6

1. a. $A = 10{,}000(1.02)^{4t}$ **b.** \$14,859.47 **3. a.** $A = 20{,}000(1.05)^{2t}$ **b.** \$43,657.49 **5. a.** $A = 8{,}000(1.13)^t$ **b.** \$27,156.54 **7. a.** $A = 30{,}000(1.01)^{12t}$ **b.** \$99,011.61 **9. a.** $A = 10{,}000\left(1 + \dfrac{0.1}{365}\right)^{365t}$ **b.** \$20,135.60 **11. a.** $A = 15{,}000e^{0.07t}$ **b.** \$34,745.50 **13. a.** $A = 10{,}000e^{0.10t}$ **b.** \$20,137.53 **15.** $\{7.144\}$ **17.** $\{11.03\}$ **19.** $\{0.2784\}$ **21.** $\{2.744\}$ **23.** $\{2.388\}$ **25.** $\{12.95\}$ **27.** $\{-0.1953\}$ **29.** $\{-1.051\}$ **31. a.** $\{e^{7.4}\}$ **b.** $\{1{,}636\}$ **33. a.** $\{e^{0.21}\}$ **b.** $\{1.234\}$ **35. a.** $\{e^{5.2}\}$ **b.** $\{181.3\}$ **37. a.** $\{e^{0.02}\}$ **b.** $\{1.020\}$ **39. a.** $\{e^{-3.8}\}$ **b.** $\{0.02237\}$ **41. a.** $\{e^{-0.2}\}$ **b.** $\{0.8187\}$ **43. a.** $\{e^6/3\}$ **b.** $\{134.5\}$ **45. a.** $\{e^5/2\}$ **b.** $\{74.21\}$ **47. a.** $\{e/3\}$ **b.** $\{0.9061\}$ **49. a.** $\{e^4/7\}$ **b.** $\{7.800\}$ **51. a.** $\{5e^2\}$ **b.** $\{36.95\}$ **53. a.** $\{3e^5\}$ **b.** $\{445.2\}$ **55. a.** $\{5e^{3/2}\}$ **b.** $\{22.41\}$ **57.** 4.62 percent **59.** 9.90 percent **61.** 10.99 percent **63.** 13.86 percent **65.** \$2 at 5 percent **67.** 13.9 years **69.** 3,939,000; 23,317,000; 92,450,000 **71.** 29.8 percent **73.** 89.7 percent **75.** 72.5 percent

Remember This (10.6)

1. $\frac{1}{8}$ **2.** 0, 3, 8 **3.** 3, 12, 27 **4.** $\frac{63}{32}$ **5.** $\{1.46\}$ **6.** $(1, 5)$ **7.** $2x$ **8.** 0 **9.** $m = \frac{2}{3}$; $(0, -2)$ **10.** $\{0, 1, -1\}$

Chapter 10 Review Exercises

1. 81, $\frac{1}{729}$, 27, 22.36 **3.** $\{\frac{3}{2}\}$ **5.** 12.25 percent **7. a.** Yes **b.** No
9. $f^{-1}(x) = \dfrac{-x+2}{3}$ **11.** 20 percent **13.** 4 **15.** 2.8129
17. $\frac{1}{2}(\log_b 2 + \log_b x - \log_b y)$ **19.** $\log_{10} 2x^5$ **21.** $\{1.7959\}$ **23.** $\emptyset$
25. \$14,555.17 **27.** $\{-4.257\}$ **29.** 5.78 percent **31.** $\frac{1}{8}$, 4, 0.38
33. \$13,974.36 **35.** 17.65 percent **37.** \$15,036.30 **39.** 7.3 **41.** $\frac{5}{2}$
43. $\frac{1}{100}$ **45.** $\log_3 1 = 0$ **47.** $\log_5 125 = x$ **49.** $10^x = 3$
51. $6 \log_{10} x$ **53.** $3[\log_b 8 + \log_b(x+1)]$ **55.** $\log_{10}(x^2 + 2x)$
57. $\log_b 2x^3$ **59.** 32 **61.** $\{0.000631\}$ **63.** $\{12.18\}$ **65.** $\{3.819\}$
67. $\left\{\dfrac{e^4}{2}\right\}$ **69.** $\{1.226\}$ **71.** $\{0.7370\}$ **73.** f^{-1} does not exist.
75. $f^{-1}(x) = \dfrac{x+3}{4}$ **77.** $f^{-1}(x) = 2x + 1$ **79.** f^{-1} does not exist.

81.
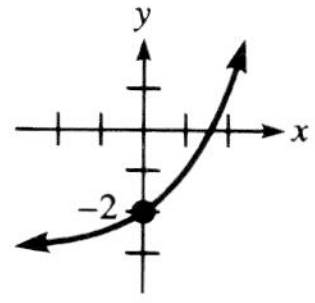

83.
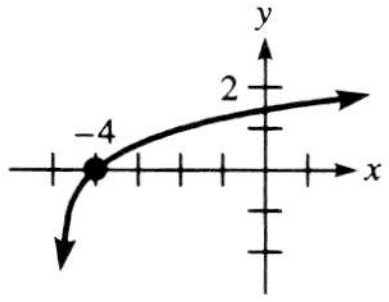

Chapter 10 Test

1. $\log_b 8 = 3$ **2.** $e^{1/2} = x$ **3.** −3 **4.** 36 **5.** $\frac{1}{2}$ **6.** $\log_2 \dfrac{x^4}{3}$
7. $\log_b 8 + \frac{1}{2}\log_b x$ **8.** 1.338 **9.** $f^{-1}(x) = \dfrac{4x+6}{5}$
10. $f[g(x)] = (\sqrt{x+2})^2 - 2 = x;\ g[f(x)] = \sqrt{(x^2-2)+2} = x$
11. \$40,582.49 **12.** 5.49 percent **13.** $\{-0.05414\}$ **14.** 0.251
15. $\{-1.262\}$ **16.** $\{6\}$ **17.** $\{0.8664\}$

18.
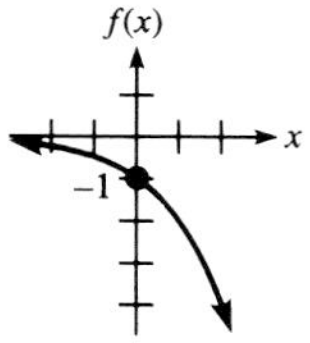

19.
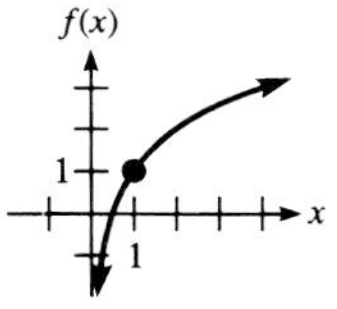

20.
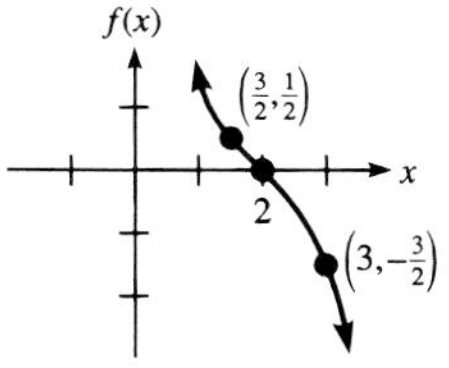

Cumulative Test 10

1. $b = \dfrac{2A}{h}$ **2.** $[1,\infty)$

3. $11x^2 - 3$ **4.** $\dfrac{a^4}{4b^8}$
5. In 2.5 seconds **6.** $\dfrac{x^2 - 3x - 3}{(x-3)(x+3)(2x+1)}$ **7.** $\left\{-3, \dfrac{2}{3}\right\}$
8. $4x + 3y = -17$ **9.**
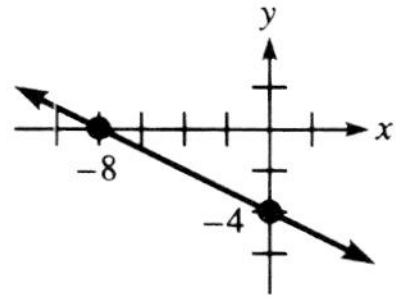

10. $(-\frac{1}{2}, \frac{1}{2})$
11. $(-1,2,-1)$ **12.** $4x^2y\sqrt{3xy}$ **13.** $11i$ **14.** $\dfrac{\sqrt{3}}{3}$
15. $(1,4)$

16. Hyperbola
17. D: $[-1,\infty)$; R: $[0,\infty)$ **18.** 10 **19.** 3.322 **20.** $\{-\frac{3}{2}\}$

Chapter 11
Exercises 11.1

1. −3, −1, 1, 3; 15 **3.** 14, 20, 26, 32; 98 **5.** $\frac{5}{2}$, 3, $\frac{7}{2}$, 4; $\frac{11}{2}$
7. $\frac{1}{5}$, $\frac{4}{5}$, $\frac{7}{5}$, 2; $\frac{73}{5}$ **9.** $4n - 3$; 45 **11.** $6n - 4$; 116 **13.** $7n - 2$; 187
15. $9n - 5$; 355 **17.** $\frac{4}{3}n + \frac{32}{3}$; 32 **19.** $a_n = 2(2)^{n-1}$; 1,024
21. $a_n = 3(5)^{n-1}$; 46,875 **23.** $a_n = 7(3)^{n-1}$; 45,927
25. $a_n = 5(-\frac{1}{5})^{n-1}$; $\frac{1}{3{,}125}$ **27.** $a_n = \frac{1}{2}(-\frac{1}{2})^{n-1}$; $-\frac{1}{1{,}024}$
29. Geometric, $r = 4$; 256, 1,024 **31.** Neither
33. Geometric, $r = 3$; 405, 1,215 **35.** Arithmetic, $d = 6$; 31, 37
37. Neither **39.** \$622.08; geometric
41. 1,080, 1,166.40, 1,259.71, 1,360.49, 1,469.33; geometric
43. a. $\frac{1}{2}$, $\frac{2}{3}$, $\frac{3}{4}$, $\frac{4}{5}$, $\frac{5}{6}$; 1 **b.** 1, $\frac{1}{2}$, $\frac{1}{3}$, $\frac{1}{4}$, $\frac{1}{5}$; 0 **c.** 1, $\frac{1}{4}$, $\frac{1}{9}$, $\frac{1}{22}$, $\frac{1}{49}$; 0
d. 1, $\frac{1}{4}$, $\frac{1}{9}$, $\frac{1}{16}$, $\frac{1}{25}$; 0 **e.** $\frac{5}{3}$, $\frac{17}{9}$, $\frac{53}{27}$, $\frac{161}{81}$, $\frac{485}{243}$; 2
f. 0.4, 0.64, 0.784, 0.8704, 0.92224; 1 **45.** 2, 3, 5, 7, 11; neither
47. 5, 8; neither

Remember This (11.1)

1. $4a_1$ **2.** $2a_1 + (n-1)d$ **3.** $S_n = \dfrac{a_1(1-r^n)}{1-r}$ **4.** 1 **5.** 19,466.40
6. $\log_b 10 = 4$ **7.** $\log_4(5x^3)$ **8.** $\{-0.9589\}$ **9.** $a = \dfrac{b}{1+t}$
10.
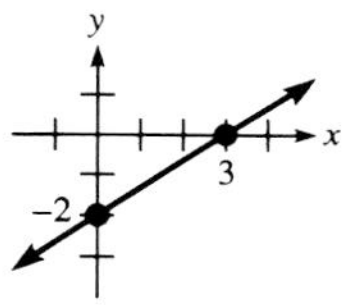

Exercises 11.2

1. 21 **3.** 78 **5.** 171 **7.** 7,875 **9.** 45,150 **11.** 92 **13.** 600
15. 1,639 **17.** 4,056 **19.** 47,541 **21.** 363 **23.** 699,050 **25.** 381
27. 88,573 **29.** 32,767 **31.** $3(1) + 3(2) + 3(3) + 3(4) + 3(5) = 45$
33. $[2(0)+1] + [2(1)+1] + [2(2)+1] + [2(3)+1] + [2(4)+1] = 25$ **35.** $(1^2+2) + (2^2+2) + (3^2+2) + (4^2+2) + (5^2+2) + (6^2+2) = 103$ **37.** $3^2 + 3^3 + 3^4 = 117$
39. $2^{1-2} + 2^{2-2} + 2^{3-2} + 2^{4-2} = 7\frac{1}{2}$
41. $(\frac{1}{3})^1 + (\frac{1}{3})^2 + (\frac{1}{3})^3 + (\frac{1}{3})^4 = \frac{40}{81}$ **43.** $a_i = 6i$; $\sum_{i=1}^{6} 6i$
45. $a_i = 3i - 2$; $\sum_{i=1}^{6}(3i-2)$ **47.** $a_i = (i+1)^2 + 2$; $\sum_{i=1}^{5}[(i+1)^2 + 2]$
49. $a_i = 3^{i-1}$; $\sum_{i=1}^{5} 3^{i-1}$ **51.** $a_i = 3(\frac{1}{4})^{i-1}$; $\sum_{i=1}^{5} 3(\frac{1}{4})^{i-1}$
53. $2 + 3 + 4 + \cdots + 9 = 44$; arithmetic, $d = 1$
55. a. $(6{,}000 - 1{,}500)/900 = 5$ years **b.** Arithmetic series with $d = -72$; $360 + 288 + 216 + 144 + 72 = \$1{,}080$
57. $5 + 4.5 + 4 + 3.5 + 3 + 2.5 = 22.5$; arithmetic with $d = -0.5$
59. $10 + 9 + 8 + 7 + \cdots + 1 = 55$ **61.** 2,304 ft
63. \$20,015.07; \$7,015.07

Remember This (11.2)

1. $\frac{3}{4}$; $\frac{7}{8}$; $\frac{15}{16}$ **2.**
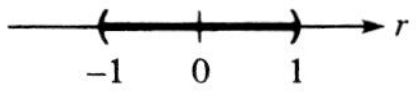

3. $\frac{15}{2}$
4. $\left(-\dfrac{3}{4}\right)^0$ **5.** 0.45, $0.\overline{45}$, 0.46 **6.** $a_n = 3n + 2$ **7.**
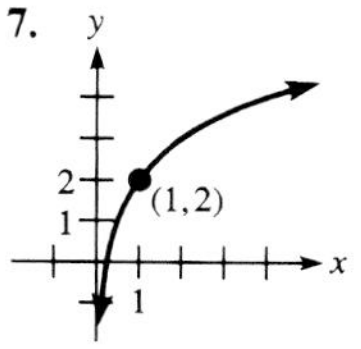

8. $\dfrac{-1 \pm i\sqrt{3}}{2}$ **9.** (10,0) **10.** $\dfrac{a^2}{2b^3}$

Exercises 11.3

1. $\frac{10}{9}$ **3.** $\frac{9}{2}$ **5.** 12 **7.** $\frac{2}{9}$ **9.** $\frac{3}{5}$ **11.** $\frac{10}{3}$ **13.** $\frac{20}{7}$ **15.** 2
17. $-\frac{2}{7}$ **19.** $\frac{4}{3}$ **21.** $\frac{1}{56}$ **23.** $\frac{10}{11}$ **25.** 0.5 **27.** $-\frac{1}{150}$ **29.** $-\frac{25}{24}$
31. $0.2 + 0.02 + 0.002 + \cdots; \frac{2}{9}$ **33.** $0.7 + 0.07 + 0.007 + \cdots; \frac{7}{9}$
35. $0.13 + 0.0013 + 0.000013 + \cdots; \frac{13}{99}$
37. $0.375 + 0.000375 + 0.000000375 + \cdots; \frac{375}{999}$
39. $0.07 + 0.007 + 0.0007 + \cdots; \frac{7}{90}$ **41.** 20 gal **43.** $S = \dfrac{\frac{1}{2}}{1 - \frac{1}{2}} = 1$
45. $S = \frac{4}{3}A$ **47.** 2,000 seconds = $33\frac{1}{3}$ minutes

Remember This (11.3)

1. $a^2 + 2ab + b^2$ **2.** $a^3 - 3a^2b + 3ab^2 - b^3$ **3.** 20 **4.** 6 **5.** 2
6. $1 + 4 + 9 + 16 = 30$ **7.** 2 **8.** $(x - 4)(x + 3)$ **9.** 110
10. $\dfrac{-3}{x + 4}$

Exercises 11.4

1. $x^3 + 3x^2y + 3xy^2 + y^3$ **3.** $x^3 - 3x^2y + 3xy^2 - y^3$
5. $x^5 - 15x^4 + 90x^3 - 270x^2 + 405x - 243$
7. $x^3 + 12x^2 + 48x + 64$
9. $64x^6 - 192x^5n + 240x^4n^2 - 160x^3n^3 + 60x^2n^4 - 12xn^5 + n^6$
11. $64x^6 + 960x^5y + 6{,}000x^4y^2 + 20{,}000x^3y^3 + 37{,}500x^2y^4 + 37{,}500xy^5 + 15{,}625y^6$
13. 1, 7, 21, 35, 35, 21, 7, 1; matches seventh row of Pascal's triangle
15. 10 **17.** 210 **19.** 1 **21.** 1 **23.** $x^4 + 4x^3h + 6x^2h^2 + 4xh^3 + h^4$
25. $x^7 - 7x^6y + 21x^5y^2 - 35x^4y^3 + 35x^3y^4 - 21x^2y^5 + 7xy^6 - y^7$
27. $x^6 + 18x^5 + 135x^4 + 540x^3 + 1{,}215x^2 + 1{,}458x + 729$
29. $27x^3 - 27x^2 + 9x - 1$ **31.** $126x^4y^5$ **33.** $-120x^3y^7$
35. $-13{,}608x^3$ **37.** $280x^3$ **39.** $393{,}216/1{,}953{,}125 \approx 20.13$ percent
41. $\frac{11}{243} = 4.53$ percent **43. a.** 15 **b.** 15

Remember This (11.4)

1. 2 **2.** Because r is greater than 1 **3.** $\frac{345}{999} = \frac{115}{333}$ **4.** c **5.** 670
6. $\emptyset$ **7.** $\pm 2i$ **8.** $\{243\}$ **9.** $\{-16.0944\}$ **10.** $(\frac{11}{2}, 0)$

Chapter 11 Review Exercises

1. 5, 8, 11, 14; 182 **3.** $a_n = 3(-2)^{n-1}$; $-1{,}536$ **5.** $8,589.93
7. 11,184,810 **9.** $a_i = 4i + 3$; $\sum_{i=1}^{6}(4i + 3)$ **11.** 4 **13.** $\frac{7}{11}$
15. $x^3 + 6x^2y + 12xy^2 + 8y^3$ **17.** 28 **19.** $5{,}376x^3y^6$
21. Arithmetic; $d = 3$ **23.** Arithmetic; $d = -4$
25. $a_n = -6n + 1$; -59
27. $8(\frac{1}{2})^1 + 8(\frac{1}{2})^2 + 8(\frac{1}{2})^3 + 8(\frac{1}{2})^4 + 8(\frac{1}{2})^5 + 8(\frac{1}{2})^6 = \frac{63}{8}$
29. $64x^3 + 48x^2y + 12xy^2 + y^3$
31. $x^5 + 5x^4h + 10x^3h^2 + 10x^2h^3 + 5xh^4 + h^5$ **33.** $1{,}120x^4y^4$
35. 39,364 **37.** 20,100 **39.** $a_i = 4i + 2$; $\sum_{i=1}^{10}(4i + 2)$ **41.** $\frac{16}{625}$
43. $1,160.38

Chapter 11 Test

1. Geometric; $\frac{1}{5}, \frac{1}{25}$ **2.** Geometric; 1, 10 **3.** Arithmetic; 17, 21
4. $a_n = n - 4$; 21 **5.** $a_n = -2(-\frac{1}{2})^{n-1}$; $\frac{1}{64}$ **6.** 320 **7.** 22,222,222
8. $a_i = 4i + 1$; $\sum_{i=1}^{10}(4i + 1)$ **9.** -26 **10.** $\frac{121}{27}$ **11.** $\frac{15}{4}$ **12.** 2
13. $\frac{5}{9}$ **14.** 1, 5, 10, 10, 5, 1
15. $x^4 + 12x^3y + 54x^2y^2 + 108xy^3 + 81y^4$ **16.** 84
17. $x^7 + 7x^6y + 21x^5y^2 + 35x^4y^3 + 35x^3y^4 + 21x^2y^2 + 7xy^6 + y^7$
18. $960x^3y^7$ **19.** $14,787.68 **20.** 238 m

Cumulative Test 11 (Final Examination)

1. -5 **2.** $\{-8, \frac{26}{3}\}$ **3.** $y_2 = m(x_2 - x_1) + y_1$ **4.** $\dfrac{2}{a^2b^8}$ **5.** $\{\frac{3}{4}, -\frac{1}{2}\}$
6. $\dfrac{2x - 3}{(x + 2)(x - 2)}$ **7.** $\{-4, -2\}$
8. x-intercept: $(-3,0)$; y-intercept: $(0,2)$ **9.**

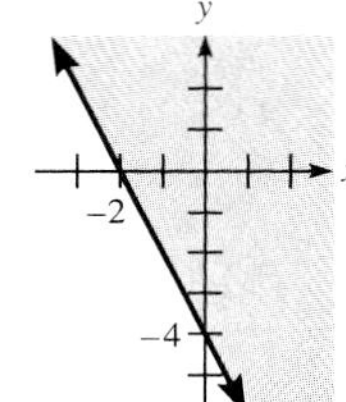

10. (60,600) **11.** $(4,1,-2)$ **12.** $5\sqrt{5}$
13. $-8 - 6i$ **14.** Center: $(6,-3)$; radius: 2
15. $\left\{\dfrac{1 \pm \sqrt{13}}{6}\right\}$ **16.** D: $(-\infty,\infty)$; R: $[-1,4]$
17. 1 **18.** $\{0.4307\}$ **19.** $240x^2$ **20.** 7,400

Index

G

H

I

J

L

M

N

O

P

Q

R

S

M

N

O

P

Q

R

S

ELM Mathematical Skills

The following table lists the California ELM Mathematical Skills and where coverage of these skills can be found in the text. Skills not covered in this text can be found in *Beginning Algebra*. Location of skills are indicated by chapter section or chapter.

Skill	Location in Text
Exponentiation and square roots	1.2
Applications (averages, percents, word problems)	throughout text
Simplifying polynomials by grouping (one and two variables)	3.3
Evaluating polynomials (one and two variables)	3.3
Addition and subtraction of polynomials	3.3
Multiplication of a polynomial by a monomial	3.4
Multiplying two binomials	3.4
Squaring a binomial	3.4
Divide a polynomial by a monomial (no remainder)	4.5
Divide a polynomial by a linear binomial	4.5, 4.6
Factor out the GCF from a polynomial	3.5
Factor a trinomial	3.6
Factor the difference of squares	3.7
Simplify a rational expression by canceling common factors	4.1
Evaluate a rational expression (one or two variables)	1.4
Addition and subtraction of rational expressions	4.3
Multiplication and division of rational expressions	4.2
Simplification of complex fractions	4.4
Positive exponents	1.2
Laws of exponents (positive)	3.1
Simplifying an expression (+ exponents)	3.1
Integer exponents	3.1
Laws of exponents (integers)	3.1
Simplifying an expression (integer exponents)	3.1
Scientific notation	3.2
Radical sign (square roots)	7.1
Simplify products under a radical	7.2
Addition and subtraction of radicals	7.3
Multiplication of radicals	7.4
Solving radical equations	7.5
Solving linear equations, one variable, numerical coefficients	2.1
Solving linear equations, one variable, literal coefficients	2.2
Ratio, proportion and variation	5.5
Solving linear inequalities, one variable, numerical coefficients	2.4, 2.5, 2.6
Solving two equations, two unknowns, numerical coefficients by substitution	6.1
Solving two equations, two unknowns, numerical coefficients by elimination	6.1
Solving quadratic equations from factors	3.9
Solving quadratic equations by factoring	3.9
Graphing points on number lines	1.1
Graphing linear inequalities (one unknown)	2.4, 2.5
Graphing points in the coordinate plane	5.1
Graphing linear equations: $y = mx$, $y = b$, $x = b$	5.1
Reading data from graphs and charts	5.2
Perimeter and area of triangles, squares, rectangles, and parallelograms	1.4
Circumference and area of circles	1.4
Volumes of cubes, cylinders, rectangular solids, and spheres	1.4
Pythagorean theorem and special triangles	3.9
Parallel and perpendicular lines	5.2

TASP Skills

The following table lists the Texas TASP and where coverage of these skills can be found in the text. Skills not covered in this text can be found in *Beginning Algebra*. Location of skills are indicated by chapter section or chapter.

Fundamental Mathematics	
Use number concepts and computation skills	ch. 1
Solve word problems involving integers, fractions, or decimals (including percents, ratios, and proportions)	applications throughout text
Interpret information from a graph, table, or chart	5.2, applications throughout text
Algebra	
Graph numbers or number relationships	1.1, ch. 2, ch. 5
Solve one- and two-variable equations	ch. 2, ch. 3, ch. 4, ch. 6
Solve word problems involving one and two variables	applications throughout text
Understand operations with algebraic expressions	ch. 3, ch. 4
Solve problems involving quadratic equations	ch. 3, ch. 8
Geometry	
Solve problems involving geometric figures	applications throughout text
Apply reasoning skills	applications throughout text

College Level Academic Skills Test (CLAST)

The following table lists the Florida CLAST and where coverage of these skills can be found in the text. Skills not covered in this text can be found in *Beginning Algebra*. Location of skills are indicated by chapter section or chapter.

Skill	Location in Text
Calculate Distances	5.2
Calculate Areas	1.4
Add and Subtract Real Numbers	1.2
Multiply and Divide Real Numbers	1.2
Apply the Order-of-Operations Agreement to Computations Involving Numbers and Variables	1.2, 1.4
Use Scientific Notation in Calculations Involving Very Large or Very Small Measurements	3.2
Calculate Volume	1.4
Solve Linear Equations and Inequalities	ch. 2
Use Given Formulas to Compute Results When Geometric Measurements Are Not Involved	2.2
Find Particular Values of a Function	3.3
Factor a Quadratic Expression	3.5, 3.6, 3.7, 3.8
Find the Roots of a Quadratic Equation	3.9, 8.1, 8.2, 8.3
Identify Information Contained in Bar, Line, and Circle Graphs	5.2
Recognize the Meaning of Exponents	1.2, 3.1
Determine the Order Relation Between Magnitudes	1.1, 1.2, 1.3
Identify Appropriate Types of Measurement of Geometric Objects	applications throughout text
Recognize and Use Properties of Operations	1.3
Determine Whether a Particular Number Is Among the Solutions of a Given Equation	2.1
Recognize Statements and Conditions of Proportionality and Variation	5.5
Recognize Regions of the Coordinate Plane Which Correspond to Specific Conditions	5.1
Identify Applicable Formulas for Computing Measures of Geometric Figures	1.4, applications throughout text
Select Applicable Properties for Solving Equations and Inequalities	ch. 2
Solve Real-World Problems Involving Perimeters, Areas, Volumes of Geometric Figures	applications throughout text
Solve Real-World Problems Involving the Pythagorean Property	3.9
Solve Real-World Problems Involving the Use of Variables	applications throughout text
Solve Problems That Involve the Structure and Logic of Algebra	applications throughout text
Draw Logical Conclusions When the Facts Warrant Them	applications throughout text

To Solve a Word Problem Involving an Equation

1. **Read the problem several times.** The first reading is a preview and is done quickly to obtain a general idea of the problem. The objective of the second reading is to determine exactly what you are asked to find. Write this down. Finally, read the problem carefully and note what information is given. If possible, display the given information in a sketch or chart.
2. **Let a variable represent an unknown quantity** (which is usually the quantity you are asked to find). Write down precisely what the variable represents. If there is more than one unknown, represent these unknowns in terms of the original variable.
3. **Set up an equation** that expresses the relationship between the quantities in the problem.
4. **Solve the equation.**
5. **Answer the question.**
6. **Check the answer** by interpreting the solution in the context of the word problem.

Translating Expressions

OPERATION	VERBAL EXPRESSION	ALGEBRAIC EXPRESSION
Addition	The sum of a number x and 1	$x + 1$ or $1 + x$
	A number y plus 2	$y + 2$ or $2 + y$
	A number w increased by 3	$w + 3$ or $3 + w$
	4 more than a number b	$b + 4$ or $4 + b$
	Add 5 and a number d	$d + 5$ or $5 + d$
Subtraction	The difference of a number x and 6	$x - 6$
	A number y minus 5	$y - 5$
	A number w decreased by 4	$w - 4$
	3 less than a number b	$b - 3$
	Subtract 11 from a number d	$d - 11$
	Subtract a number d from 11	$11 - d$
Multiplication	The product of a number x and 5	$5x$
	4 times a number t	$4t$
	A number y multiplied by 7	$7y$
	Twice a number w	$2w$
	$\frac{1}{3}$ of a number n	$\frac{1}{3}n$
	25 percent of a number p	$0.25p$
Division	The quotient of a number x and 8	$x \div 8$ or $x/8$
	A number y divided by 2	$y \div 2$ or $y/2$
	The ratio of a number a to a number b	$a \div b$ or a/b or $a:b$